HOLT

BIOLOGY

JOHNSON · RAVEN

AUTHORS

George B. Johnson, Ph.D.
Professor of Biology
Washington University
St. Louis, Missouri

Peter H. Raven, Ph.D.
Engelman Professor of Botany
Washington University
St. Louis, Missouri
Director
Missouri Botanical Garden

HOLT, RINEHART AND WINSTON

A Harcourt Education Company

Orlando • **Austin** • New York • San Diego • Toronto • London

About the Authors

George Johnson, Ph.D.
Dr. George Johnson is a professor of biology at Washington University in St. Louis, where he has taught general biology and genetics to undergraduates for 30 years. Dr. Johnson is also a professor of genetics at Washington University School of Medicine. He is the author of more than 50 scientific publications and has published several successful college biology texts.

Peter Raven, Ph.D.
Dr. Peter Raven is the director of the Missouri Botanical Garden and a professor of botany at Washington University in St. Louis. Dr. Raven is a MacArthur Fellow, a member of the National Academy of Sciences, a recipient of the National Medal of Science, and the author of 20 other books and several hundred scientific articles.

Printed in the United States of America

ISBN 0-03-074061-4

4 5 6 7   048   09 08 07 06

Acknowledgments

Contributing Writers

Ann Bekebrede
Science Writer
Sherborn, Massachusetts

Mary Dylewski
Science Writer
Kassel, Germany

Susan Feldkamp
Science Writer
Manchaca, Texas

Frances Jenkins
Science Writer
Sunburg, Ohio

Matt T. Lee, Ph.D.
Science Writer
Coos Bay, Oregon

Annette Ratliff
Science Writer
Austin, Texas

Feature Development

Linda K. Gaul, Ph.D.
Epidemiologist
Texas Department of Health
Austin, Texas

Matt T. Lee, Ph.D.
Science Writer
Coos Bay, Oregon

Inclusion Specialists

Joan A. Solorio
Special Education Director
Austin Independent School
   District
Austin, Texas

John A. Solorio
Multi-Technology Lab Facilitator
Austin Independent School
   District
Austin, Texas

Teacher Edition Development

Linda K. Blumenthal
Science Writer
Columbus, Ohio

Alan Eagy
Biology Teacher
The Dalles High School
The Dalles, Oregon

Linda K. Gaul, Ph.D.
Epidemiologist
Texas Department of Health
Austin, Texas

Erik Hahn
Naturalist
Hartley Nature Center
Duluth, Minnesota

Kevin L. Moore
Science Writer
Denver, Colorado

JoAnne Morgan Mowczko, Ed.D.
Educational Consultant
Gaithersburg, Maryland

Tyson Yager
Science Instructor
Wichita High School East
Wichita, Kansas

Academic Reviewers

David M. Armstrong, Ph.D.
Professor
Environmental, Population, and
 Organismic Biology
University of Colorado
Boulder, Colorado

Nigel Atkinson, Ph.D.
Associate Professor of
 Neurobiology
Institute for Neuroscience
The University of Texas
Austin, Texas

Jerry Baskin, Ph.D.
Professor
School of Biological Sciences
University of Kentucky
Lexington, Kentucky

John A. Brockhaus, Ph.D.
Director of Mapping, Charting,
 and Geodesy Program
Department of Geography and
 Environmental Engineering
United States Military Academy
West Point, New York

John Caprio, Ph.D.
George C. Kent Professor of
 Biological Sciences
Louisiana State University
Baton Rouge, Louisiana

Joe W. Crim, Ph.D.
Professor and Head
Department of Cellular Biology
The University of Georgia
Athens, Georgia

Roger J. Cuffey, Ph.D.
Professor of Paleontology
Department of Geosciences
Pennsylvania State University
University Park, Pennsylvania

James Denbow, Ph.D.
Associate Professor
Department of Anthropology
The University of Texas
Austin, Texas

David Futch, Ph.D.
Department of Biology
San Diego State
 University
San Diego, California

Linda K. Gaul, Ph.D.
Epidemiologist
Texas Department of
 Health
Austin, Texas

**Herbert Grossman,
Ph.D.**
Associate Professor of Botany
 and Biology
Department of
 Environmental Sciences
Pennsylvania State
 University
University Park,
 Pennsylvania

William Guggino, Ph.D.
Professor of Physiology
The Johns Hopkins University
School of Medicine
Baltimore, Maryland

David Haig, Ph.D.
Professor
Department of Organismic and
 Evolutionary Biology
Harvard University
Cambridge, Massachusetts

David R. Hershey, Ph.D.
Education Consultant
Hyattsville, Maryland

David Ho, M.D.
Director and CEO
The Aaron Diamond AIDS
 Research Center
New York, New York

Joan E. N. Hudson, Ph.D.
Associate Professor
Sam Houston State University
Huntsville, Texas

Leland Lim, M.D., Ph.D.
Year II Resident
Department of Neurology and
 Neurological Sciences
Stanford University
School of Medicine
Palo Alto, California

Iris F. Litt, M.D.
Marron and Mary Elizabeth Kendrick
 Professor in Pediatrics
Stanford University
School of Medicine
Palo Alto, California

V. Patteson Lombardi, Ph.D.
Research Assistant Professor
 Human Biology and Medical
 Physiology
Department of Biology
University of Oregon
Eugene, Oregon

C. Riley Nelson, Ph.D.
Associate Professor
Department of Integrative
 Biology
Brigham Young University
Provo, Utah

continued on page 1136

Contents in Brief

UNIT 4

Principles of Ecology 316

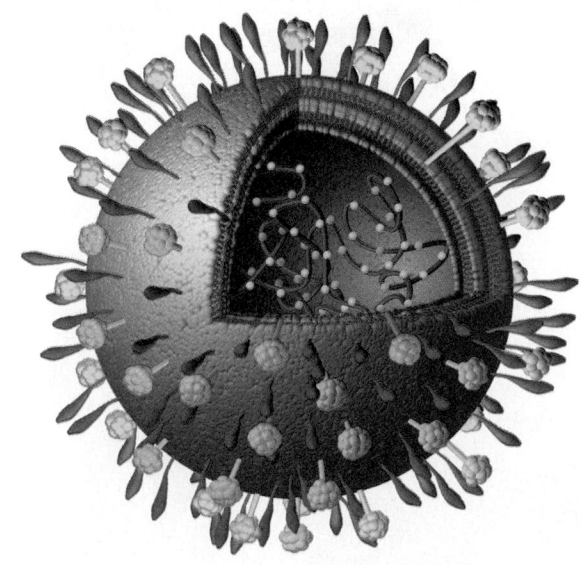

UNIT 6

Exploring Plants *498*

UNIT 7

Exploring Invertebrates 590

Features

Biowatch *features highlight the impact of biology on health and technology issues.*

Forensics Biowatch *features highlight the connection between biology and forensic science.*

Up Close

Up Close features provide detailed looks at important organisms.

Exploring Further

Exploring Further features let you explore key biological topics in greater depth.

SCIENCE • TECHNOLOGY • SOCIETY

Science, Technology, and Society features examine the impact of new technologies on issues in biology.

Lab Program

QUICK LAB

Quick Labs provide hands-on experience yet require few materials.

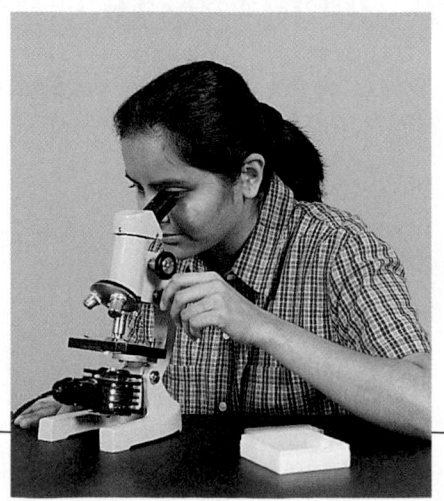

Data Labs allow you to interpret data and draw conclusions.

Math Labs let you practice real-world math skills as you analyze biological problems.

Lab Program *continued*

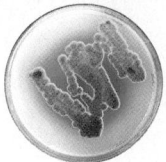

Exploration Labs

allow you to investigate or model a phenomenon and draw conclusions based on your work.

Skills Practice Labs

teach you lab skills used by biologists.

Forensics Labs

*allow you to explore the
techniques used by forensics
scientists in solving crimes and
mysteries of history.*

How to Use Your Textbook

Your Roadmap for Success with *Holt Biology*

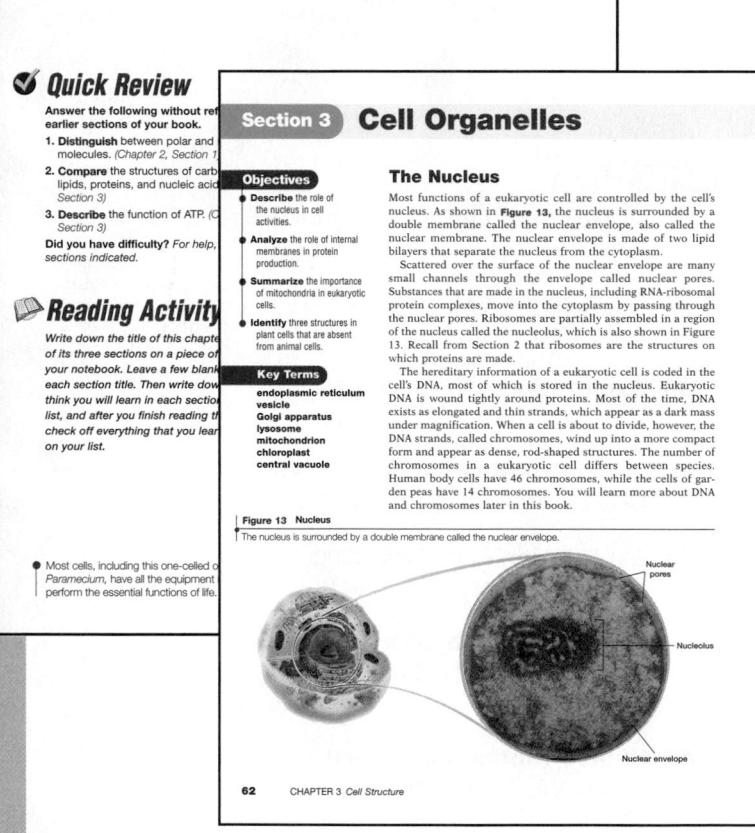

CHAPTER
3 Cell Structure

✔ **Quick Review**

Answer the following without ref earlier sections of your book.

1. **Distinguish** between polar and molecules. *(Chapter 2, Section 1*

2. **Compare** the structures of carb lipids, proteins, and nucleic acid *Section 3)*

3. **Describe** the function of ATP. *(C Section 3)*

Did you have difficulty? *For help, sections indicated.*

📖 **Reading Activity**

Write down the title of this chapte of its three sections on a piece o your notebook. Leave a few blank each section title. Then write dow think you will learn in each sectio list, and after you finish reading th check off everything that you lear on your list.

● Most cells, including this one-celled o *Paramecium*, have all the equipment perform the essential functions of life.

Section 3 **Cell Organelles**

Objectives

● **Describe** the role of the nucleus in cell activities.

● **Analyze** the role of internal membranes in protein production.

● **Summarize** the importance of mitochondria in eukaryotic cells.

● **Identify** three structures in plant cells that are absent from animal cells.

Key Terms

endoplasmic reticulum
vesicle
Golgi apparatus
lysosome
mitochondrion
chloroplast
central vacuole

The Nucleus

Most functions of a eukaryotic cell are controlled by the cell's nucleus. As shown in **Figure 13,** the nucleus is surrounded by a double membrane called the nuclear envelope, also called the nuclear membrane. The nuclear envelope is made of two lipid bilayers that separate the nucleus from the cytoplasm.

Scattered over the surface of the nuclear envelope are many small channels through the envelope called nuclear pores. Substances that are made in the nucleus, including RNA-ribosomal protein complexes, move into the cytoplasm by passing through the nuclear pores. Ribosomes are partially assembled in a region of the nucleus called the nucleolus, which is also shown in Figure 13. Recall from Section 2 that ribosomes are the structures on which proteins are made.

The hereditary information of a eukaryotic cell is coded in the cell's DNA, most of which is stored in the nucleus. Eukaryotic DNA is wound tightly around proteins. Most of the time, DNA exists as elongated and thin strands, which appear as a dark mass under magnification. When a cell is about to divide, however, the DNA strands, called chromosomes, wind up into a more compact form and appear as dense, rod-shaped structures. The number of chromosomes in a eukaryotic cell differs between species. Human body cells have 46 chromosomes, while the cells of garden peas have 14 chromosomes. You will learn more about DNA and chromosomes later in this book.

Figure 13 Nucleus

The nucleus is surrounded by a double membrane called the nuclear envelope.

Nuclear pores

Nucleolus

Nuclear envelope

62 CHAPTER 3 *Cell Structure*

Get Organized

Answer the **Quick Review** questions at the beginning of each chapter to assess your recall of information important to the understanding of new chapter topics. Then complete the **Reading Activity** on the same page to prepare you to read and organize new information.

STUDY TIP Use the **Looking Ahead** outline at the beginning of the chapter to organize your notes on the chapter content in a way that you understand.

Read for Meaning

Read the **Objectives** at the beginning of each section because they will tell you what you'll need to learn. **Key Terms** are also listed for each section. Each key term is highlighted in the text. After reading each chapter, turn to the **Chapter Highlights** page and review the **Key Concepts,** which are brief summaries of the chapter's main ideas. You may want to do this even before you read the chapter.

STUDY TIP If you don't understand a definition, reread the page on which the term is introduced. The surrounding text should help make the definition easier to understand.

↗ Be Resourceful, Use the Web

Internet Connect boxes in your textbook take you to resources that you can use for science projects, reports, and research papers. Go to **scilinks.org** and type in the **SciLinks code** to get information on a topic.

Visit go.hrw.com
Find resources and reference materials that go with your textbook at **go.hrw.com.** Enter the keyword **HX6 Home** to access the home page for your textbook.

Prepare for Tests

Section Reviews and **Chapter Reviews** test your knowledge of the main points of the chapter. Critical Thinking items challenge you to think about the material in different ways and in greater depth. The standardized test prep that is located after each Chapter Review helps you sharpen your test-taking abilities.

STUDY TIP Reread the Objectives and Chapter Highlights when studying for a test to be sure you know the material.

Use the Appendix

Your **Reference and Skills** section near the end of the book contains a variety of resources designed to enhance your learning experience. **Reading and Study Skills** provides helpful study aids; **Math and Problem Solving Skills** sharpens your math skills. **Laboratory Skills** summarizes essential safety information and basic laboratory techniques. **Classification in Kingdoms and Domains** organizes living things according to modern principles of classification.

Structures of Plant Cells

The organelles described in this section are found in both animal cells and plant cells. However, plant cells have three additional structures that are not found in animal cells, shown in **Figure 17.**

Unique Features of Plant Cells

Cell wall The cell membrane of a plant cell is surrounded by a thick cell wall, composed of proteins and carbohydrates, including the polysaccharide cellulose. The cell wall helps support and maintain the shape of the cell, protects the cell from damage, and connects it with adjacent cells.

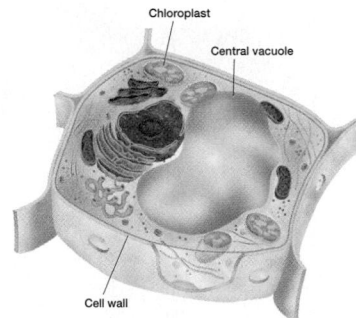

Figure 17   Plant cell. Plant cells have a cell wall, chloroplasts, and a large central vacuole (shown in blue).

Chloroplast

Central vacuole

Cell wall

Chloroplasts Plant cells contain one or more **chloroplasts**. Chloroplasts are organelles that use light energy to make carbohydrates from carbon dioxide and water. Chloroplasts are found not only in plants but also in a wide variety of eukaryotic algae, such as seaweed. Chloroplasts, along with mitochondria, supply much of the energy needed to power the activities of plant cells. Like mitochondria, chloroplasts are surrounded by two membranes, contain their own DNA, and are thought to be the descendents of ancient prokaryotic cells.

Central vacuole As shown in Figure 17, much of a plant cell's volume is taken up by a large, membrane-bound space called the **central vacuole** *(VAK yoo ohl)*. The central vacuole stores water and may contain many substances, including ions, nutrients, and wastes. When the central vacuole is full, it makes the cell rigid. This rigidity enables a plant to stand upright.

Section 3 Review

1. **Describe** the role of the nucleus in cell activities.

2. **Sequence** the course of newly made proteins from the rough ER to the outside of the cell.

3. **Describe** the role of mitochondria in the metabolism of eukaryotic cells.

4. **Explain** how a plant cell's central vacuole and cell wall help make the cell rigid.

5. **Critical Thinking Inferring Relationships** What is the importance of a cell enclosing its digestive enzymes inside lysosomes?

6. **Standardized Test Prep** Which organelle serves as the packaging and distribution center of a eukaryotic cell?
   - **A** nucleus
   - **B** lysosome
   - **C** mitochondrion
   - **D** Golgi apparatus

Visit Holt Online Learning

If your teacher gives you a special password to log onto the **Holt Online Learning** site, you'll find your complete textbook on the Web. In addition, you'll find some great learning tools and practice quizzes. You'll be able to see how well you know the material from your textbook.

How to Use Your Textbook

UNIT 1

Principles of Cell Biology

Chapters

As the runner sprints toward the finish line, 25 trillion red blood cells carry oxygen throughout her body.

Blood: A River of Cells

Yesterday... In the winter of 1667, a French physician named Jean-Baptiste Denis tried a daring new experiment. After examining a man who exhibited fits of rage, Denis transfused the patient with the blood of a gentle calf. His actions reflected the beliefs of his time—that blood carried the characteristics of the creatures in which it flowed.

Illustration from a 1692 medical textbook

Today... Biologists know that red blood cells contain a protein called hemoglobin. **Find out how proteins like hemoglobin form their unique shapes.**

In persons with sickle cell anemia, an incorrect form of hemoglobin is made. When this happens, red blood cells become sickle-shaped and cannot adequately perform their job. The shape of the hemoglobin molecule is the key to its role—carrying oxygen throughout the body. **Discover why the cells of your body need oxygen.**

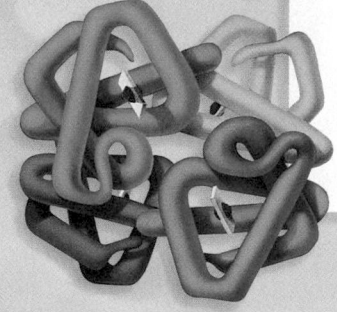

Hemoglobin

Tomorrow... Researchers are discovering new treatments for sickle cell anemia. New drugs that "turn on" the body's production of normal hemoglobin have shown promise in easing the symptoms of sickle cell anemia.

Sickled red blood cell

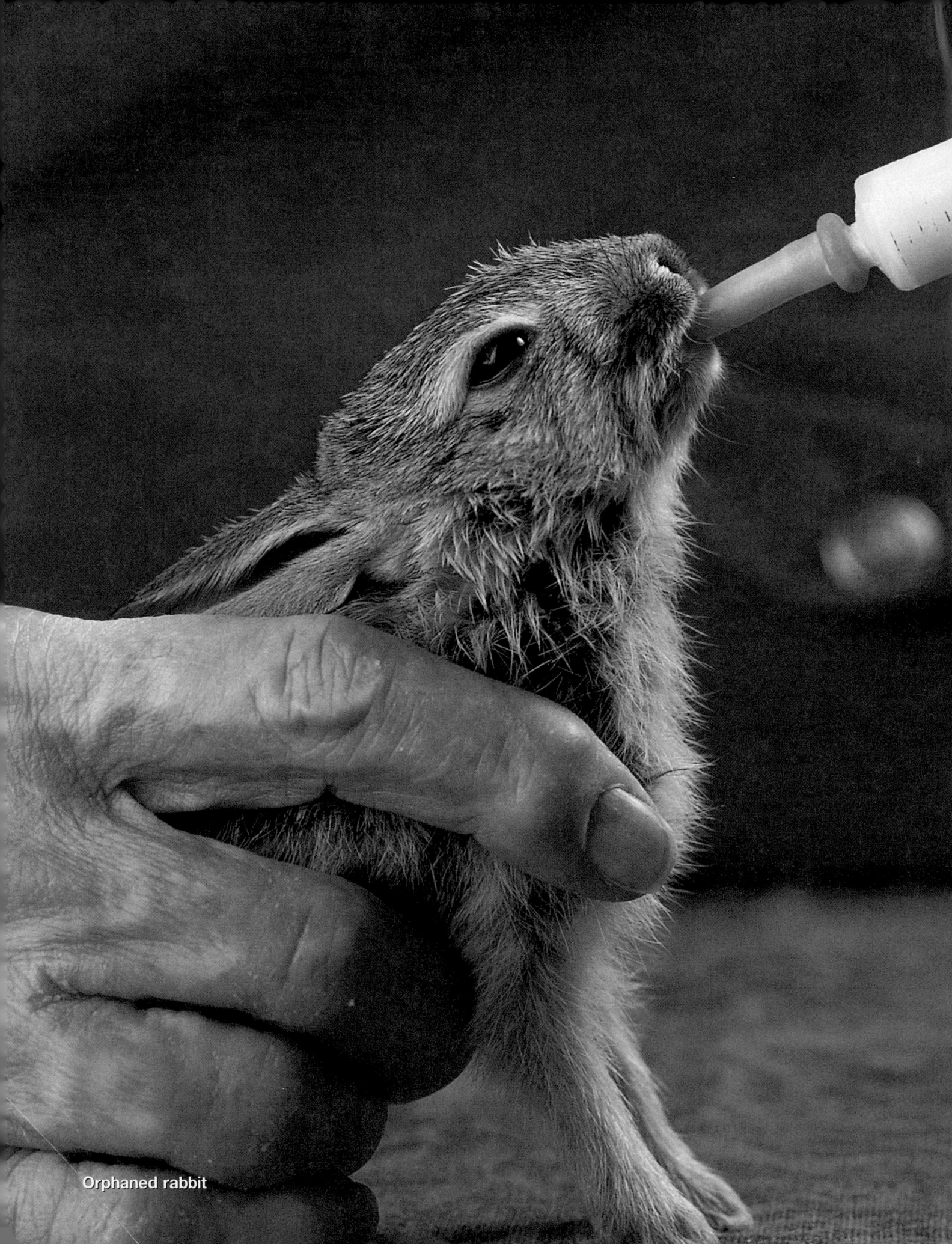

Orphaned rabbit

1 Biology and You

Study *TIP*

The **Reading Activity** that begins each chapter will help you to comprehend more effectively what you read. **Study Tips** in the margins will help you to organize and review information. **Word Origins** notes in the margins will help you understand how scientific terms are constructed from common root words and suffixes. **Real Life** margin notes link the study of biology to real-world phenomena.

Reading Activity

Before you read this chapter, write the word biology *on your paper. Refer to a dictionary, and write out the definition of the prefix* bio– *and the suffix* –logy. *Think of names of other branches of life science that include the suffix* –logy, *and write them down. Use the **Reading and Study Skills** appendix at the back of this book to define the prefixes in the words you thought of.*

Looking Ahead

Section 1
Themes of Biology
Characteristics of Living Organisms
Unifying Themes of Biology

Section 2
Biology in Your World
Solving Real-World Problems
Fighting Disease

Section 3
Scientific Processes
Observation: The Basis of Scientific Research
Stages of Scientific Investigations
Scientific Explanations

internet connect

www.scilinks.org
National Science Teachers Association *sci*LINKS Internet resources are located throughout this chapter.

SC*LINKS*. Maintained by the National Science Teachers Association

Just as this orphaned rabbit needs food and water to live and grow, it also depends on oxygen from plants to survive. Our knowledge of biology helps us understand how all life on Earth is interconnected.

Themes of Biology

Characteristics of Living Organisms

You are surrounded by living things, which a scientist calls *organisms.* Many organisms, such as people, plants, and animals, are obvious. Other living things are so small that you cannot see them without a microscope. How do we know if something is alive? What does it mean to be alive?

While most people are capable of distinguishing between living and nonliving, actually defining life can be quite difficult. Perhaps you consider movement, sensitivity, development, and even death as characteristics of living organisms. While present in all living things, these properties are not enough to describe life.

Clouds, for example, move when stimulated by the wind and develop from moisture that is suspended in the atmosphere. Clouds grow and change shapes. Some might view the breakup of clouds as being similar to death. Disorder, however, is not the same as death. Clouds may break up and vanish, but they do not die.

Biology is the study of life. Biologists recognize that all living organisms, such as the cheetahs shown in **Figure 1,** share certain general properties that separate them from nonliving things. As summarized in Figure 1, every living organism is composed of one or more cells, is able to reproduce, and obtains and uses energy to run the processes of life. Living organisms also maintain a constant internal environment and pass on traits to offspring. Responding and adjusting to the environment as well as growing and developing are other characteristics shared by all living organisms.

As you read further, you will have an opportunity to think more about the properties that help define life. Life is characterized by the presence of *all* of these properties at some stage in an organism's life. Remember this fact as you attempt to determine what is living and what is not.

Figure 1 What does it mean to be alive? Life is characterized by the presence of all seven of these properties at some stage in an organism's life.

Properties of Life

• Cellular organization
• Reproduction
• Metabolism
• Homeostasis
• Heredity
• Responsiveness
• Growth and development

Unifying Themes of Biology

In the study of biology, certain broad themes emerge that both unify living things and help explain biology as a science. The word *science* comes from Latin for "to know." Science is a systematic process of inquiry. As you study the science of biology by reading this textbook, you will repeatedly encounter these themes.

Theme ❶ Cellular Structure and Function

All living things are made of one or more cells. **Cells** are highly organized, tiny structures with thin coverings called membranes. A cell is the smallest unit capable of all life functions. The basic structure of cells is the same in all organisms, although some cells are more complex than others. Some organisms have only a single cell, while others are multicellular (composed of many cells). Your body contains more than 100 trillion cells. **Figure 2** shows a single-celled organism called a paramecium.

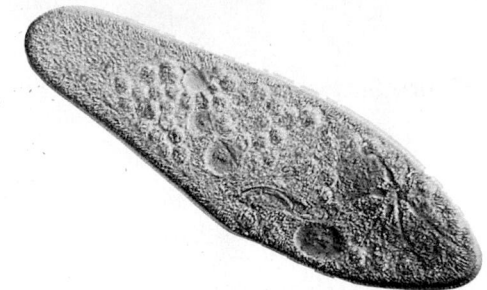

Figure 2
Single-celled paramecium

Theme ❷ Reproduction

All living things can reproduce. **Reproduction** is the process by which organisms make more of their own kind from one generation to the next. Some rapidly growing bacteria divide into offspring cells approximately every 15 minutes, and bristlecone pine trees that are 5,000 years old still produce seedlings. Because no organism lives forever, reproduction, as represented in **Figure 3,** is an essential part of living.

Figure 3   Hatchling snakes

Theme ❸ Metabolism

Living organisms carry out many different chemical reactions in order to obtain and use energy to run the processes of life. All living things use energy to grow, to move, and to process information. Without energy, life soon stops. **Metabolism** is the sum of all of the chemical reactions carried out in an organism.

Almost all the energy used by living organisms is originally captured from sunlight. Plants, algae, and some bacteria capture this solar energy and use it to make complex molecules in a process called photosynthesis. These molecules then serve as the source of energy, or food, for other organisms. For example, paramecia, such as the one shown in Figure 2, eat bacteria. Humans eat plants or animals that, in turn, have eaten plants. Energy flows from the sun to plants, from these plants to plant-eating organisms, and from plant-eating organisms to meat-eating organisms. The teens shown in **Figure 4** are extracting energy from the food they eat.

Figure 4
Extracting energy from food

Figure 5 Harp seal

Theme ❹ Homeostasis

All living organisms must maintain a stable internal environment in order to function properly. Organisms respond to changes in their external environment, and their internal processes adjust accordingly. The maintenance of stable internal conditions in spite of changes in the external environment is called **homeostasis** *(hoh mee oh STAY sihs)*. An organism unable to balance its internal conditions with its environmental conditions could become ill and die. Arctic seals, such as the one shown in **Figure 5,** are able to maintain a constant body temperature in spite of their cold environment because of their body shape and thick layer of body fat.

Theme ❺ Heredity

All living things are able to pass on traits to their offspring through genes that are passed from parent to offspring each generation. A **gene** is the basic unit of heredity. Genes are coded in a molecule called deoxyribonucleic *(dee AHKS ee rie boh nu klay ik)* acid (DNA) and determine an organism's traits. The passing of traits from parent to offspring is called **heredity.** Heredity is the reason children tend to resemble their parents, as shown in **Figure 6.**

Sometimes damage causes genes to change. A change in the DNA of a gene is called a **mutation.** Most mutations are harmful, but sometimes mutations can help an organism survive. For example, in humans a mutation for the blood protein hemoglobin, which carries oxygen to the body's cells, has both a harmful effect and a positive effect. The harmful effect is that the mutated form of the gene results in sickle cell anemia. Sickle cell anemia is a disease in which the defective form of hemoglobin causes many red blood cells to bend into a sickled—that is, a hooked—shape that reduces the oxygen-carrying capability of the cell. The positive effect is that the mutation produces resistance to malaria, a deadly infectious disease.

Mutations that occur in sex cells (egg and sperm) are passed on to other generations. Mutations that occur in body cells are not passed on, but they can disrupt the control of cell reproduction and result in cancer.

Figure 6 Passing on traits

Theme ⑥ Evolution

The great diversity of life on Earth is the result of a long history of change. Change in the inherited characteristics of species over generations is called **evolution.** A **species** is a group of genetically similar organisms that can produce fertile offspring. Individuals in a species are similar, but not identical. Those individuals with genetic traits that better enable them to meet nature's challenges tend to survive and reproduce in greater numbers, causing these favorable traits to become more common. Charles Darwin, the nineteenth-century British naturalist, used the term **natural selection** for the process in which organisms with favorable traits are more likely to survive and reproduce.

Darwin's theory of evolution by natural selection provides a consistent explanation for life's diversity. Most scientists believe that the many different species of animals, plants, and other organisms on Earth today are the result of a long process of evolution. **Figure 7** shows an example of a plant that has flowers modified for attracting insects.

Figure 7 Bee pollinating flower

Figure 8 Owl capturing a rat

Theme ⑦ Interdependence

The organisms in a biological community live and interact with other organisms, as shown in **Figure 8.** A biological community is a group of interacting organisms. **Ecology** is the branch of biology that studies the interactions of organisms with one another and with the nonliving part of their environment. Organisms are dependent on one another and their environment—that is, they are interdependent. Interdependence within biological communities is the result of a long history of evolutionary adjustments. The complex web of interactions in a biological community depends on the proper functioning of all of its members, even those too small to be seen without a microscope.

Section 1 Review

1. **Identify** the seven properties that all living organisms share.

2. **Relate** three of the seven major themes of biology to the life of a harp seal.

3. **Name** the very small, organized structure that is bound by a membrane and that is the basic unit of structure and function in all organisms.

4. **Define** *homeostasis* and *metabolism,* and describe their differences.

5. **Critical Thinking Recognizing Verifiable Facts** If you find an object that looks like an organism, how might you determine if your discovery is indeed alive?

6. **Standardized Test Prep** The mutation that results in sickle cell anemia produces effects that are
   A only harmful.
   B only positive.
   C both harmful and positive.
   D unimportant.

Biology in Your World

Objectives

- **Evaluate** the impact of scientific research on the environment.
- **Evaluate** the impact of scientific research on society with respect to increasing food supplies.
- **Explain** the primary task of the Human Genome Project.
- **Describe** the contributions of scientists in fighting AIDS and cancer.
- **Define** the term *gene therapy.*

Key Terms

genome
HIV
cancer
cystic fibrosis
gene therapy

Solving Real-World Problems

You are unlikely to read a newspaper or magazine today without noticing issues that relate to biology. In this textbook, you will learn about many areas in which biologists are actively working to solve today's problems.

Preserving Our Environment

More than 6 billion people now live on Earth. The increasing human population has had a significant impact on other organisms with which we share this planet. For example, tropical rain forests are home to one-half of the world's species of plants and animals, such as the bird shown in **Figure 9.** The rain forests are being destroyed at the rate of more than one acre every second. At this rate, tropical rain forests—and a million species—may be gone in 30 years. Who knows what potential medicines and foods we are discarding? Like burning a library without reading the books, extinction on this large scale is a tragedy. However, conservation biologists are now exploring ways to achieve a balance between people's growing need for land and the need to preserve the environment.

One of the great achievements of today's biology has been to show the practical benefits of taking better care of our environment. Consider, for example, the fast-food french fry. Because french fries must be formed perfectly to be sold, about one-half of the potatoes used to make french fries were lost as waste. Then a major supplier found a way to use some of the waste: mix it with grain to feed cattle. Leftover potato particles in the potato processing water are also used, serving as a source of methane gas for power plants. Finally, the processing water, rich in nutrients, is used to water and fertilize agricultural crops. The environmental concern that promotes these sensible changes is a major contribution of biology to a better future. Conservation and preservation are now everyday activities of government, industry, and individuals.

Figure 9   Scarlet macaw. The brilliantly colored scarlet macaw, *Ana macao,* lives high in the rain forest canopy in Central America and South America. Its numbers are being reduced by habitat destruction and poaching.

Improving the Food Supply

One of the greatest impacts of modern biology on society in recent years has been the genetic engineering of crop plants. As illustrated in **Figure 10,** biologists have learned how to transfer genes from one kind of plant to another, which changes the hereditary information in its cells. Genetic engineering, shown in Figure 10, has made some crop plants resistant to herbicides (so that weeds can be killed without harming the food crop). Genetic engineering has also made some crop plants poisonous to insect pests, but not humans, and has produced new varieties of crop plants with improved nutritional balance and protein content.

For example, rice, one of the world's most important food crops, lacks iron and vitamin A levels needed for a balanced diet. These deficiencies affect the billions of people worldwide for whom rice is a daily food. The addition of genes from other plants to rice has increased the nutritional value of rice, thus improving the diet of many of the world's people. The long-term safety of genetically engineered food crops is still being studied. These crops may offer great promise, however, of improving the world's food supplies.

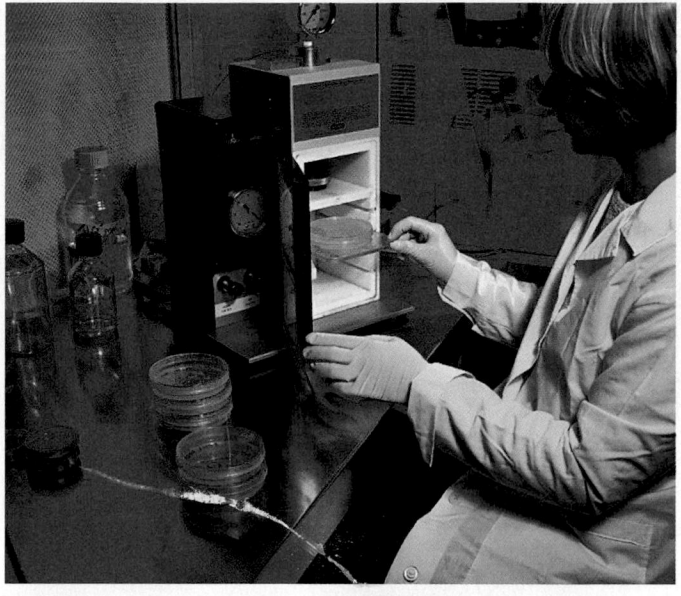

Figure 10 Genetic engineering. Genes for desirable traits can be transferred from one organism to another.

Understanding the Human Genome

Another great achievement of today's biology was completed in April of 2003. Government-funded and private research teams from several countries had been racing to complete the sequencing of the human genome, and on that day they announced joint success. The task had been formidable. A **genome** is the complete genetic material contained in an individual. The human genome contains an astonishing 3 billion individual units! At the height of the research, automated gene-sequencing machines were sequencing DNA fragments at a rate of 1,000 units per second, around the clock.

Biologists are now able to read every human gene, providing them with a detailed road map, as you see in **Figure 11,** of human genes. It will be many years before this information is fully analyzed, but it is already proving to be an invaluable tool in medical research.

internet connect

www.scilinks.org
Topic: Human Genome Project
Keyword: HX4193

SCI**LINKS** Maintained by the National Science Teachers Association

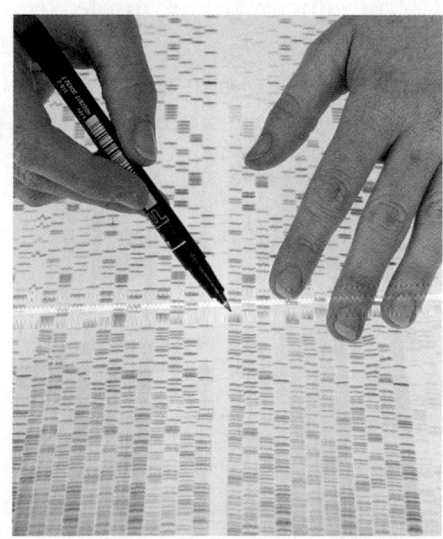

Figure 11 DNA sequencing. The identity of each genetic unit, shown as repeating dark and light bars, has been determined for all human genes found in the human body.

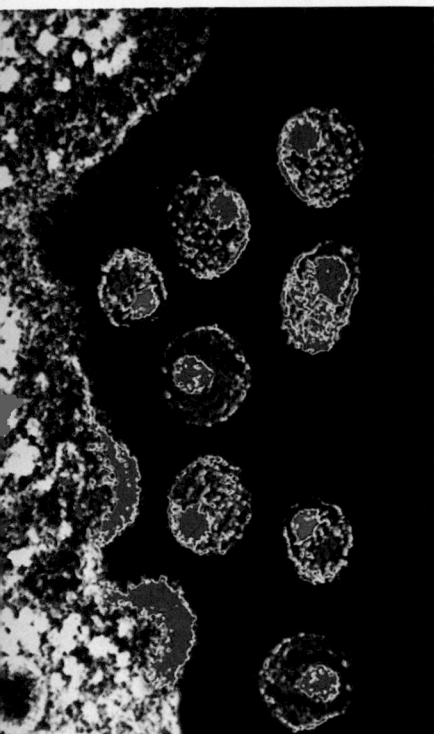

Figure 12 HIV. Individual HIV particles are shown emerging from a white blood cell where they have been assembled.

Fighting Disease

Progress in biology directly affects our lives through medicine, in which scientific advances are curing disease and improving health every day. New technologies have enabled biologists to combat disease in ways scarcely imagined only a few years ago. Among the many diseases that you will study in this text, consider the following.

AIDS

For more than 20 years, biologists have been battling AIDS. AIDS is a fatal disease caused by **HIV,** a virus that attacks and destroys the human immune system. HIV, shown in **Figure 12,** is transmitted by contact with body fluids from an infected person. While biologists have been successful in developing a combination of drugs that slow the progression of AIDS, it has proven very difficult to make a vaccine capable of halting its spread. The problem is that HIV changes as it passes from person to person, altering itself too frequently for any single vaccine to protect many people. This problem soon may be solved. New vaccines now being tested target two or more parts of the virus at the same time. While one part may change, it is very unlikely that two parts will change at the same time in the very same virus particle. For the first time, there is hope of a successful vaccine to control the worldwide outbreak of AIDS.

Cancer

When U.S. President Richard Nixon recruited biologists to join a "War on Cancer" in 1972, we did not know very much about the causes of cancer, although many Americans were dying of it. In the 30 years since then, biologists have learned a lot. **Cancer** is a growth defect in cells, a breakdown of the mechanism that controls cell division.

We now know that many cancers can be largely avoided. To sharply reduce your risk of lung cancer, for example, don't smoke. Many other cancers can be treated successfully when detected early. Colon cancer, for example, develops slowly from intestinal tissue growths called polyps. A simple medical examination enables the detection and removal of the polyps.

Great progress is being made in curing many cancers. More than 25 percent of breast cancers, for example, result from having too many copies of a cell protein that starts cell division. As many as 70 percent of colon and prostate cancers have extra copies of a similar protein. Anticancer drugs that stick to these extra cell proteins, gumming them up so they cannot promote excessive cell division, appear to offer great promise.

Emerging Diseases

The past few years have seen the emergence of new diseases not known in the past and the incidence in the United States of diseases from other parts of the world. West Nile virus is one such disease. West Nile virus was not found in the United States until 1999, when

internet connect

www.scilinks.org
Topic: Cancer Cells
Keyword: HX4030

SC*LINKS.* Maintained by the National Science Teachers Association

birds in the northeastern United States were found to have died from it. The West Nile virus is transmitted by mosquitoes and is known to infect humans as well as birds, horses, and possibly other animals. In humans, the illness that results from infection by the West Nile virus may cause only mild symptoms in some individuals. Other people who have contracted a severe form of the disease, however, have died.

Another disease that has emerged in Europe is commonly known as mad cow disease. Mad cow disease is a fatal disease of cattle caused by eating the body parts of infected animals. Although cattle are grazing animals, their food is sometimes supplemented with protein from parts of other cattle. Humans, too, can be infected by eating meat or other products of infected cattle. No cases of mad cow disease have been reported in the United States, but a similar disorder has been found to affect elk and deer herds throughout the United States.

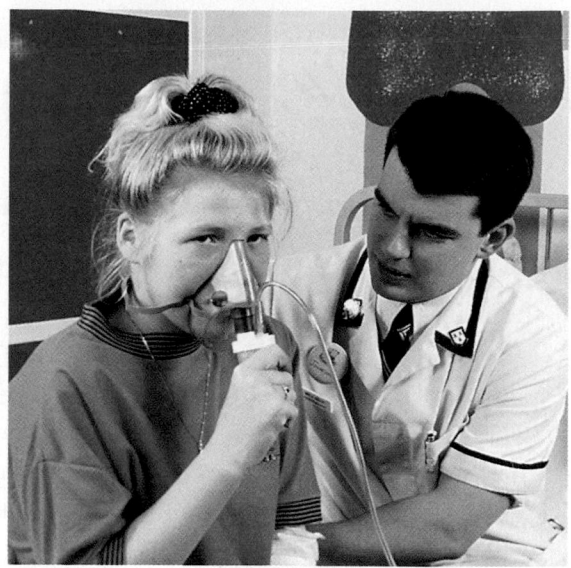

Figure 13 Treatments for cystic fibrosis include respiratory therapy and medication.

Gene Therapy

Biologists have searched for many years for a way to introduce "healthy" genes into people suffering from hereditary disorders. The person shown in **Figure 13** has **cystic fibrosis,** which is caused by an inherited defect in a gene. Cystic fibrosis is a fatal disorder in which a thick, sticky mucus clogs passages in many of the body's organs.

Researchers believe that it is possible to use a virus to transfer a normal copy of the defective gene into the cells of cystic fibrosis patients. Early attempts failed because the virus used to transport the healthy gene posed health problems. New attempts using a different virus are more promising. The replacement of a defective gene with a normal version is called **gene therapy.** Other serious genetic disorders, such as muscular dystrophy and hemophilia, are also good candidates for gene therapy. Though still experimental, the future of gene therapy seems bright.

Section 2 Review

1 **Describe** the relationship between scientific research and the use of environmental resources.

2 **Explain** how genetic engineering can improve food supplies.

3 **Describe** one problem that scientists must overcome to develop a vaccine for HIV.

4 **Explain** how gene therapy might be used to treat cystic fibrosis.

5 **Critical Thinking Evaluating Viewpoints**
Do you agree or disagree that a knowledge of biology is essential in the battle against diseases? Explain your answer.

6 **Standardized Test Prep** One goal of researchers who transplant beneficial plant genes into other plants is to

A accelerate extinction. **C** halt the spread of AIDS.

B cure cancer. **D** reduce pesticide use.

Scientific Processes

Observation: The Basis of Scientific Research

Recognizing the properties of living organisms and knowing why biology is important in your world are good first steps in your exploration of biology. All scientists, including biologists, have a certain way of investigating the world. Studying an actual scientific investigation is an exciting way to learn how science is done. Our story begins many years ago with two biologists, David Bradford and John Harte, but the story continues to develop even today.

Solving Scientific Puzzles

In the summer of 1988, Bradford reflected on the silence that surrounded him. He had spent the summer looking for a species of small frog in the many lakes of Sequoia and Kings Canyon National Parks. The frog species had lived in the parks' lakes for as long as anyone had kept records. In the last count of the frog's populations, the frogs had been everywhere. Now, for some reason, they had disappeared from 98 percent of the lakes.

Observation is the act of noting or perceiving objects or events using the senses. As Bradford reported his observations to other biologists, he found that local populations of amphibians (frogs, toads, and salamanders) elsewhere were also disappearing. Amphibians have been around for 370 million years. The disappearance of amphibians from their natural homes sounded an alarm among biologists that something was altering the environment. Amphibians are particularly sensitive to their environment; their moist skin absorbs chemicals from water.

Between the years 1984 and 1988, John Harte, a professor at the University of California, Berkeley, was also studying amphibians. He was studying the tiger salamander, *Ambystoma tigrinum*, shown in **Figure 14.** Tiger salamanders live in ponds high on the western slopes of the Rocky Mountains of Colorado. Harte had seen their numbers fall by 65 percent as he and his students had collected and analyzed water samples from the ponds in the area over the years.

Harte wanted to discover the facts surrounding the disappearance of the salamanders. Like other scientists, Harte began a scientific investigation that combined knowledge, imagination, and intuition to get a sense of what might be true. Even though scientists might expect certain results, they do not form conclusions until they have enough evidence to support them.

Figure 14 Tiger salamander, *Ambystoma tigrinum*

Stages of Scientific Investigations

Although there is no single "scientific method," scientific investigations tend to have common stages: collecting observations, asking questions, forming hypotheses and making predictions, confirming predictions (with controlled experiments when appropriate), and drawing conclusions. These stages are summarized in **Figure 15.**

Collecting Observations

The core of scientific investigation is careful observation. Harte had studied the Colorado salamander population for years. He had learned what they eat, how they behave, when they reproduce, and what conditions they thrive in. His students had helped him collect water samples from the ponds, as shown in Figure 15. Frequent visits to the ponds helped him realize the salamander population was decreasing in number. Keeping careful records of the lakes' conditions helped him find an explanation.

Asking Questions

Observations of the natural world often raise questions. Harte questioned why the number of salamanders was dropping. He talked to other scientists, carefully observed the organisms and environment in the Rocky Mountains of Colorado, and read scientific reports. He answered many of his questions through his observations, but some key questions remained unanswered.

In the natural world, the moisture that falls as rain and snow is very slightly acidic. In the Rocky Mountains of Colorado, however, the moisture is high in sulfuric acid from power plants that burn high-sulfur coal. This acidic moisture, called acid precipitation, is released into mountain ponds each spring when the snow melts, causing the water in the ponds to become more acidic in late May. Most of the mountains' annual moisture falls as snow. Harte thought acid precipitation was important in the puzzle of the declining salamander population, but he needed evidence.

Figure 15   Testing the acidity of water

Asa Bradman, a student of John Harte, helps in Harte's scientific investigation by collecting water samples from a Colorado pond.

Scientific Processes

• Collecting observations
• Asking questions
• Forming hypotheses and making predictions
• Confirming predictions (with experiments when needed)
• Drawing conclusions

Forming Hypotheses and Making Predictions

A **hypothesis** *(hie PAHTH uh sis)* is an explanation that might be true—a statement that can be tested by additional observations or experimentation. In that respect, a hypothesis (plural form, *hypotheses*) is not just a guess—it is an educated guess based on what is already known. Harte formed two hypotheses that together he believed explained the disappearance of the amphibians:

1. Acids that were formed in the upper atmosphere by pollutants were falling onto the mountains in the winter snows.

2. Melting snow was making the ponds acidic and harming the salamander embryos.

If Harte's hypotheses were correct, he could expect several possible outcomes. A **prediction** is the expected outcome of a test, assuming the hypothesis is correct. For his first hypothesis, Harte predicted he would find acid in the ponds after the snow melted. For his second hypothesis, he predicted that there would be enough acid in the ponds to harm salamander embryos. Using his predictions as a starting point, Harte set out to test his hypotheses.

Confirming Predictions

Harte gathered data from many years of observations, including measurements of the acidity of the ponds before, during, and after snowmelt. Harte and his students had taken water samples at frequent intervals from several ponds. Data for part of one year, after snowmelt, are shown in **Figure 16.**

To describe how acidic a solution is, scientists use a number between 0 and 14 to represent **pH,** which is a relative measure of the hydrogen ion concentration within a solution. Solutions with a low pH (below 7) are acidic, solutions above 7 are basic, and solutions at pH 7 are neutral. Acid rain usually has a pH of between 2 and 6. A solution with a pH of 2 is 10,000 times more acidic than one with a pH of 6.

Figure 16   Pond pH after snowmelt. The pH levels in a pond in the Rocky Mountains are acidic (low pH) at the same time of the year that salamander eggs are developing.

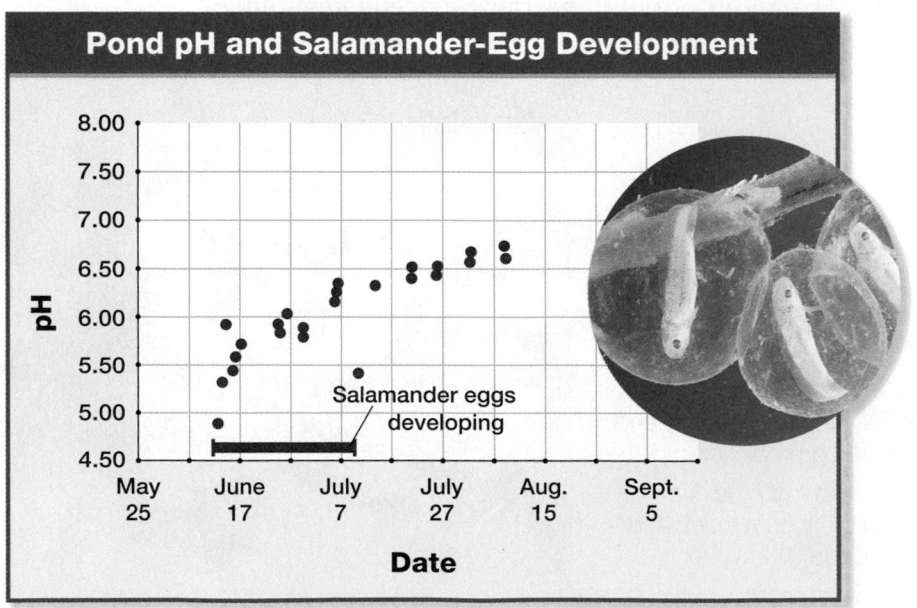

Pond pH and Salamander-Egg Development

Salamander eggs developing

Harte's data indicated that the ponds became more acidic when the snow melted. After a few weeks, the pH rose and then leveled off. The data confirmed Harte's first prediction and supported his first hypothesis—melting snow caused acid to be released into the ponds at snowmelt, as shown in Figure 16. After snowmelt, the acid was neutralized, probably by minerals that dissolved from the rocks in the ponds, and pond pH returned to normal for the rest of the summer.

To confirm his second hypothesis (melting snow was making the ponds acidic and harming the salamander embryos), Harte did an **experiment**—a planned procedure to test a hypothesis. Salamanders lay eggs in the ponds once a year, as soon as pond ice melts. Harte wanted to test whether exposure to the pH levels he had recorded at that time of year would harm the salamanders that hatched from the eggs.

Harte performed a controlled experiment. In a controlled experiment, an experimental group (a group that receives some type of experimental treatment) is compared with a control group. A **control group** is a group in an experiment that receives no experimental treatment. The control and experimental groups are designed to be identical except for one factor, or variable. The factor that is changed in an experiment is called the **independent variable.** In Harte's experiment, the independent variable was the acid (pH) level. The variable that is measured in an experiment is called the **dependent variable.** Harte's dependent variable was the number of salamanders that hatched from the eggs.

Study *TIP*

● **Reviewing Information**
On a separate sheet of paper, make a table with two columns. List the stages common to scientific investigations in the left-hand column. In the right-hand column, describe in your own words what is actually done at that stage and why that is important to scientific inquiry.

Determining the pH of Common Substances

You can use pH indicator paper to determine the pH of various solutions. The pH indicator paper changes color when it is exposed to a solution. The change in color indicates how acidic or basic the solution is.

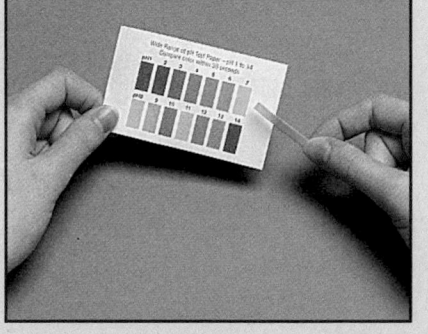

Materials

paper, pencil, wide-range pH paper, three different solutions, beaker or small jar, water

Procedure

1. Make a data table with three columns and the following headings: *Solution, Predicted pH,* and *Measured pH.* Make a row for each solution to be tested.

2. Predict the pH (acid or base) of each solution, and record your predictions in your data table.

3. Test each solution with pH paper, and record the results in the appropriate row in your data table.

Analysis

1. Summarize your findings in two sentences.

2. Determine whether the predictions that you made were correct. Explain any differences between your predictions and your results.

3. Compare your results with those of the rest of the class. Explain any differences.

4. List the steps of scientific methods that you followed in doing this activity.

Figure 17 Deformities in amphibians. Scientists are investigating factors that may play a role in the development of deformities found in amphibians throughout the United States and other parts of the world.

Harte allowed captive salamanders to lay eggs in regular pond water. He collected and then divided the eggs into five groups. One group, the control group, contained eggs placed in pond water with a neutral pH. Each of the other four groups, the experimental groups, contained eggs placed in pond water with different amounts of acid added, similar to the acid levels found in the ponds after snowmelt.

Harte found that acid did indeed affect development. Many of the salamanders never hatched from the eggs placed in acidic water. Some of the salamanders that did hatch were born with developmental abnormalities. Other scientists have found abnormalities in amphibians, as shown in **Figure 17.**

Drawing Conclusions

Once data are collected and analyzed, a conclusion is made as to whether the data support the hypothesis. The hypothesis may be supported or rejected. A hypothesis can be supported but never proven because another experiment with new data and new information may alter the conclusion.

Harte's data supported both of his hypotheses. The pH levels in the ponds before and after snowmelt indicated that the ponds became more acidic after the snow melted. This supported his first hypothesis—acids that were formed in the upper atmosphere by pollutants were falling onto the mountains in the winter snows.

Harte's controlled experiment showed that acidic water reduces the number of salamanders that hatch from eggs. This supported his second hypothesis—melting snow could make the ponds acidic and harm the salamander embryos. Harte concluded that melting snow in the Rocky Mountains of Colorado could cause acid absorbed from atmospheric pollution to be released into the ponds at snowmelt, harming salamander embryos.

Viewing Conclusions in Context

Scientists from many disciplines have been working together to sort out the causes of the global decline in amphibians. Like many important questions, this one does not have a simple answer.

Four factors seem to be contributing in major ways: (1) The animals' habitats are deteriorating and being destroyed. (2) Nonnative species introduced into amphibian habitats out-compete local amphibian populations for resources. (3) Chemical pollutants accumulate in amphibian habitats. Acid rain released into ponds at snowmelt is but one example. (4) Amphibians have a high rate of fatal infections by parasites such as viruses or fungi. In the western United States, infection by ranavirus (a common pathogen in fish) probably has led to declines in populations of mountain salamanders and frogs. A soil fungus called a *chytrid (KI TRID)* also kills amphibians. Amphibian larvae, such as the Pacific tree frog tadpole shown in **Figure 18,** can be infected with the fungus. The fungus dissolves the mouthparts of the larvae, killing them.

Figure 18 Pacific tree frog tadpole. The pacific tree frog's numbers have been reduced by chytrid inflections.

Scientific Explanations

Scientific progress is made the same way a marble statue is, by chipping away the unwanted bits. If a hypothesis does not provide a reasonable explanation for observations, it must be rejected. Harte was able to show that enough acid was being introduced into the ponds to kill the salamander embryos. His hypothesis—that acid from melting snow was killing the salamanders—was therefore supported. The hypothesis that acid rain is contributing to the loss of amphibian populations will require much more evidence before becoming accepted as a broader theory.

It is important in science not to be misled by an isolated observation. Only after many studies like Harte's will scientists be able to assemble a picture that accurately reveals what is harming the amphibians. As you have just read, other environmental factors may play important roles. **Figure 19** summarizes the steps in the development of a theory. A **theory** is a set of related hypotheses that have been tested and confirmed many times by many scientists. A theory unites and explains a broad range of observations.

Sometimes, new theories do not mesh well with mainstream ideas and are criticized by the scientific establishment. However, a theory that clearly explains a broad range of observations, that fits with current theories, and that is useful for predicting new findings may gain acceptance from most scientists over time.

Constructing a Theory

Constructing a theory often involves considering contrasting ideas and conflicting hypotheses. Argument, disagreement, and unresolved questions are a healthy part of scientific research, a true reflection of how science is done. Scientists routinely evaluate one another's work, helping to identify and remove bias. A key requirement of valid scientific research is that it can be replicated—that is, reproduced—by other scientists.

As you study biology, it is important to remember that the word *theory* is used very differently by scientists than by the general public. To scientists, a theory represents that of which they are most certain. In contrast to the general public, *theory* may imply a lack of knowledge, a guess. How often have you heard someone say, "It's only a theory" to imply lack of certainty? As you can imagine, confusion often results. In this textbook, the word *theory* will always be used in its scientific sense—that is, a theory is a well-supported scientific explanation that makes useful predictions.

There is, however, no absolute certainty in a scientific theory. The possibility always remains that future evidence will cause a scientific theory to be revised or rejected. A scientist's acceptance of a theory is always provisional.

Once a scientist completes an investigation, he or she often writes a report for publication in a scientific journal. Before publication, the research report is reviewed

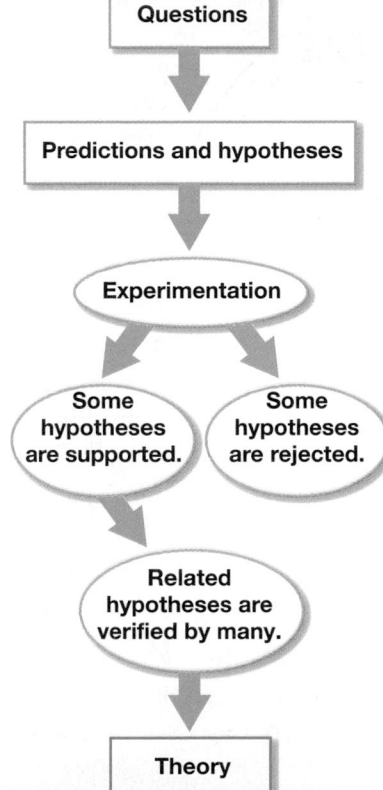

Figure 19 Theories

Scientists build theories from questions, predictions, hypotheses, and the findings of their experiments. When related hypotheses consistently explain scientific events, a theory is formed.

Questions

↓

Predictions and hypotheses

↓

Experimentation

Some hypotheses are supported. / Some hypotheses are rejected.

Related hypotheses are verified by many.

↓

Theory

by other scientists. These reviewers ensure that the investigation was carried out with the appropriate controls, methods, and data analysis. The reviewers also check that the conclusions reached by the author are justified by the data obtained. Publishing an investigation allows other scientists to use the information for hypotheses they are forming. They can also repeat the investigations and confirm the validity of the conclusions.

Analyzing Experimental Design

Background

To study the effects of common substances on the heart rate of a tiny aquatic organism known as *Daphnia*, students placed a *Daphnia* in a drop of water on a glass slide. The students then added 1 or more drops of a test substance dissolved in water to the slide, waited 10 seconds, then counted heart beats for 10 seconds. The students used a clean slide and a new *Daphnia* each time. Their data table is shown below.

Magnification: 16×

Daphnia

| Heart Rate of *Daphnia* in Different Solutions ||
Substance tested	Heart rate (beats per minute)
None	58
Coffee	65
Ethanol	50

Analysis

1. **Identify** the dependent and independent variables in the experiment.

2. **Identify** the experimental groups in the experiment.

3. **Identify** the liquid that should be used for the control group.

4. **Evaluate** how the instructions could be changed to improve the design of the experiment.

5. **Critical Thinking Applying Information** Design an experiment that students can perform to verify the prediction that coffee will increase heart rate in *Daphnia*.

Section 3 Review

1 **Summarize** how scientists use observations, hypotheses, predictions, and experiments in scientific investigations.

2 **Differentiate** independent variables from dependent variables.

3 **Define** the word *theory* in a scientific sense and then in a more general sense.

4 **Critical Thinking Evaluating Results** What hypothesis might another scientist suggest to Harte for the decrease in the number of salamanders?

5 **Standardized Test Prep** A researcher finds that 90 percent of salamanders hatch from eggs in water at pH 7, 80 percent hatch at pH 6, 60 percent at pH 5, and 40 percent at pH 4. What is the approximate percentage that hatch at pH 5.5?

A 55 percent   C 70 percent

B 61 percent   D 85 percent

Study ZONE

CHAPTER HIGHLIGHTS

Key Concepts

1 Themes of Biology

- Living organisms are diverse but share certain characteristics.
- All living organisms are composed of cells, grow and develop, and are able to maintain homeostasis.
- Living organisms reproduce, producing offspring similar to themselves.
- Living organisms obtain and use energy to stay alive, and they respond to their environment.
- Seven themes unify the science of biology: cellular structure and function, reproduction, metabolism, homeostasis, heredity, evolution, and interdependence.

2 Biology in Your World

- Pollution of the atmosphere, extinction of plants and animals, and a growing demand for food are current environmental problems caused by the growing human population.
- Biologists are using genetic engineering to develop crops that require fewer fertilizers and pesticides and to develop new crops.
- Biological research and new technologies will help scientists battle diseases such as AIDS, cancer, and cystic fibrosis.

3 Scientific Processes

- Scientists add to scientific knowledge by sharing observations and posing questions about those observations.
- Although there is no single method, observing, asking questions, and forming and testing hypotheses are important in planning a scientific investigation.
- In a controlled experiment, the independent variable is varied between the experimental and control groups. The measured variable is the dependent variable.
- A collection of hypotheses that have been repeatedly tested and are supported by a great deal of evidence forms a theory.

Key Terms

Section 1

biology (6)
cell (7)
reproduction (7)
metabolism (7)
homeostasis (8)
gene (8)
heredity (8)
mutation (8)
evolution (9)
species (9)
natural selection (9)
ecology (9)

Section 2

genome (11)
HIV (12)
cancer (12)
cystic fibrosis (13)
gene therapy (13)

Section 3

observation (14)
hypothesis (16)
prediction (16)
pH (16)
experiment (17)
control group (17)
independent variable (17)
dependent variable (17)
theory (19)

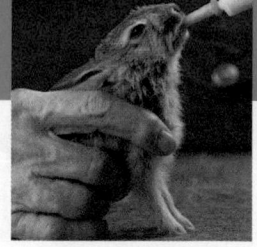

Performance ZONE

CHAPTER REVIEW

Understanding Key Ideas

1. Toads that live in hot, dry regions bury themselves in the soil during the day. What theme of biology does this phenomenon describe?
   - **a.** metabolism
   - **b.** homeostasis
   - **c.** evolution
   - **d.** heredity

2. Which of the following issues is *not* an existing problem that biologists can help solve?
   - **a.** destruction of rain forests
   - **b.** extinction of plants and animals
   - **c.** AIDS
   - **d.** snowmelt in the Rockies

3. The demand for more food is due to
   - **a.** the spread of disease.
   - **b.** the growth of the human population.
   - **c.** uncontrolled scientific experimentation.
   - **d.** extinction.

4. The disorder characterized by cells dividing uncontrollably within the body is called
   - **a.** AIDS.
   - **b.** cystic fibrosis.
   - **c.** cancer.
   - **d.** mad cow disease.

5. The factor that is varied in a controlled experiment is called the
   - **a.** control.
   - **b.** hypothesis.
   - **c.** dependent variable.
   - **d.** independent variable.

6. Which statement is false?
   - **a.** Observations are an important part of the scientific process.
   - **b.** A solution with a low pH is more acidic than a solution with a high pH.
   - **c.** A hypothesis can be proven with a well-designed experiment.
   - **d.** A prediction is the expected outcome of a test.

7. For each pair of terms, write one or more sentences summarizing what you learned in this chapter about those terms.
   - **a.** evolution, natural selection
   - **b.** metabolism, homeostasis
   - **c.** control group, experimental group

8. Is the word *theory* in the newspaper headline shown below used in a scientific sense or in a more general sense? Explain.

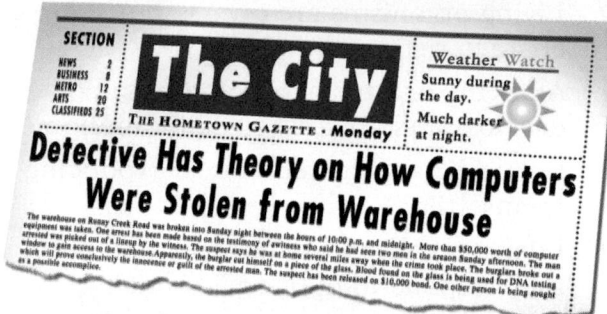

9. Why might it be difficult to determine whether or not microscopic particles are alive?

10. What arguments could you offer for and against the removal of forests to make room for new neighborhoods?

11. **Concept Mapping** Make a concept map that outlines scientific investigations in biology. Try to include the following terms: *biology, observation, communication, hypotheses, predictions, experiments,* and *theories.*

Critical Thinking

12. **Forming Reasoned Opinions** Some people believe that scientists should not tamper with a person's genes. Do you think biologists should use gene therapy to try to cure diseases? Explain your answer.

13. **Applying Information** One of the first branches of biology to be developed was taxonomy, the naming and grouping of organisms. Why is taxonomy important to communication about biology?

Alternative Assessment

14. **Interactive Tutor Unit 5 Heredity** Write a report summarizing how an understanding of heredity allows animal breeders to develop animals that have desirable traits. Find out what kinds of animals are bred for special purposes.

Standardized Test Prep

Understanding Concepts

Directions (1–6): **For *each* question, write on a separate sheet of paper the letter of the correct answer.**

1 What is a statement that can be tested by additional observations or experimentation?
   A. hypothesis   **C.** theory
   B. prediction   **D.** variable

2 What is a group of organisms that can produce fertile offspring?
   F. gene   **H.** reproduction
   G. kingdom   **I.** species

3 What is the basic unit of heredity?
   A. cell   **C.** gene
   B. chromosome   **D.** species

4 What is the sum of all the chemical reactions carried out in an organism?
   F. homeostasis   **H.** reproduction
   G. metabolism   **I.** sensitivity

5 What is a change in inherited characteristics of species over time called?
   A. evolution
   B. homeostasis
   C. reproduction
   D. responsiveness

6 To describe the acidity of a solution, scientists use a number between 0 and 14. What does this number represent?
   F. experimental life
   G. neutrality
   H. pH
   I. solution rate

Directions (7): **For the following question, write a short response.**

7 A scientist on television states that a hypothesis cannot be proven. Assess why this statement is correct.

Test TIP

Carefully read the instructions, the question, and the answer options before choosing an answer.

Reading Skills

Directions (8): **Read the passage below. Then answer the question.**

One of the most important parts of any scientific publication is the section that describes methods and materials used. In this section, the authors describe how they set up the experiment, what instruments they used to collect the data, and how they recorded the data.

8 Why is it important for scientists to include in their scientific publications a section that describes the methods and materials used?
   A. It shows how the data can be applied to other fields of study.
   B. It proves how expensive their experiment is to carry out.
   C. It allows other scientists to reproduce the experiment accurately.
   D. It prevents other scientists from repeating the experiment and claiming it as their own.

Interpreting Graphics

Directions (9): **Base your answer to question 9 on the chart below.**

Municipal Solid Waste by Weight

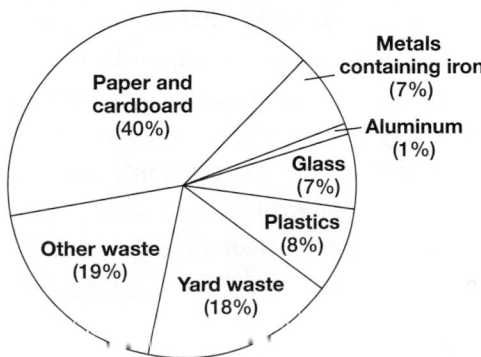

Paper and cardboard (40%)
Metals containing iron (7%)
Aluminum (1%)
Glass (7%)
Plastics (8%)
Yard waste (18%)
Other waste (19%)

9 If each type of solid waste were recycled, which type would have the biggest impact on conserving trees?
   F. aluminum
   G. glass
   H. paper and cardboard
   I. plastics

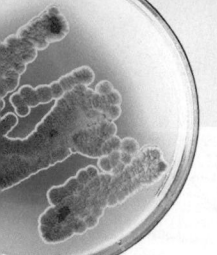

Exploration Lab

Observing the Effects of Acid Rain on Seeds

SKILLS
- Using scientific methods
- Collecting, organizing, and graphing data

OBJECTIVES
- **Use** a scientific method to investigate a problem.
- **Predict** how acid rain affects germination and growth.

MATERIALS
- safety goggles
- protective gloves
- lab apron
- 50 seeds
- 250 mL beakers
- 20 mL mold inhibitor
- distilled water
- paper towels
- solutions of different pH
- wax pencil or marker
- zip-lock plastic bags
- metric ruler
- graph paper

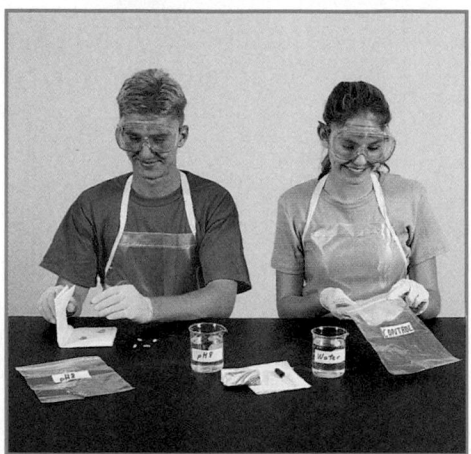

ChemSafety

CAUTION: Always wear safety goggles and a lab apron to protect your eyes and clothing.

CAUTION: Do not touch or taste any chemicals. Know the location of the emergency shower and eyewash station and how to use them. If you get a chemical on your skin or clothing, wash it off at the sink while calling to the teacher. Notify the teacher of a spill. Spills should be cleaned up promptly, according to your teacher's directions.

CAUTION: Glassware is fragile. Notify the teacher of broken glass or cuts. Do not clean up broken glass or spills with broken glass unless the teacher tells you to do so.

Before You Begin

Living things, such as salamander embryos, can be damaged by **acid rain** at certain times during their lives. In this lab, you will investigate the effect of acidic solutions on seeds. One way to investigate a problem is to design and conduct an **experiment.** We begin a scientific investigation by making **observations** and asking questions.

1. Write a definition for each boldface term in the paragraph above and for each of the following terms: pH, hypothesis, prediction, variable, control group.

2. Based on the objectives for this lab, write a question you would like to explore about the effect of acid rain, for example, When is a plant most susceptible to acid rain?

Procedure

PART A: Design an Experiment

1. Work with members of your lab group to explore one of the questions written for step 2 of **Before You Begin.** To explore the question, design an experiment that uses the materials listed for this lab.

You Choose
As you design your experiment, decide the following:
a. what question you will explore
b. what hypothesis you will test
c. how to simulate growing seeds in soil moistened by acid rain
d. how to keep seeds moist during the experiment
e. what your test solutions and control will be
f. how to measure seedling growth
g. what to record in your data table

2. Write a procedure for your experiment. Make a list of all the safety precautions you will take. Have your teacher approve your procedure and safety precautions before you begin the experiment.

PART B: Conduct Your Experiment

3. Put on safety goggles, protective gloves, and a lab apron.

4. Place your seeds in a 250 mL beaker, and slowly add enough mold inhibitor to cover the seeds. **CAUTION: The mold inhibitor contains household bleach, which is a base.** Soak the seeds for 10 minutes, and then pour the mold inhibitor into the proper waste container. Gently rinse the seeds with distilled water, and place them on clean paper towels.

5. Set up your group's experiment. **CAUTION: Solutions with a pH below 7.0 are acids.** Conduct your experiment for 7–10 days. Make observations every 1–2 days, and note any changes. Record each day's observations in a data table, similar to the one below.

DATA TABLE

Solution	Date	Observations

PART C: Cleanup and Disposal

6. Dispose of solutions, broken glass, and seeds in the designated waste containers. Do not pour chemicals down the drain or put lab materials in the trash unless your teacher tells you to do so.

7. Clean up your work area and all lab equipment. Return lab equipment to its proper place. Wash your hands thoroughly before you leave the lab and after you finish all work.

Analyze and Conclude

1. **Summarizing Results** Describe any changes in the look of your seeds during the experiment. Discuss seed type, average seed size, number of germinated seeds, and changes in seedling length.

2. **Analyzing Results** Were there any differences between the solutions? Explain.

3. **Analyzing Methods** What was the control group in your experiment?

4. **Analyzing Data** Make graphs of your group's data. Plot seedling growth (in millimeters) on the y-axis. Plot number of days on the x-axis.

5. **Relating Concepts** What scientific methods did you use to design and conduct your experiment?

6. **Evaluating Methods** How could your experiment be improved?

7. **Inferring Conclusions** How do acidic conditions appear to affect seeds?

8. **Predicting Outcomes** How might acid rain affect the plants in an ecosystem?

9. **Further Inquiry** Write a new question about the effect of acid rain that could be explored with another investigation.

❓ Do You Know?

Do research in the library or media center to answer these questions:

1. Which parts of the United States are most affected by acid rain, and why?

2. How have factories been changed to reduce the amount of acid rain?

Use the following Internet resources to explore your own questions about acid rain.

🔲 **internet** connect

www.scilinks.org
Topic: **Acid Rain**
Keyword: **HX4001**

SC*LINKS*® Maintained by the National Science Teachers Association

Stampeding horses

CHAPTER

2 Chemistry of Life

✔ Quick Review

Answer the following without referring to earlier sections of your book.

1. **Identify** seven properties of life. *(Chapter 1, Section 1)*

2. **List** seven themes of biology. *(Chapter 1, Section 1)*

3. **Distinguish** between metabolism and homeostasis. *(Chapter 1, Section 1)*

Did you have difficulty? *For help, review the sections indicated.*

 Reading Activity

Before you read this chapter, write a short list of all the things you know about the chemistry of organisms. Then write a list of the things that you want to know about the chemistry of organisms. Save your list, and to assess what you have learned, see how many questions you can answer after reading this chapter.

Looking Ahead

Section 1
Nature of Matter
Atoms
Chemical Bonding

Section 2
Water and Solutions
Water in Living Things
Aqueous Solutions

Section 3
Chemistry of Cells
Carbon Compounds

Section 4
Energy and Chemical Reactions
Energy for Life Processes
Enzymes

🔌 internet connect

www.scilinks.org
National Science Teachers Association *sci*LINKS Internet resources are located throughout this chapter.

*sci*LINKS® **Maintained by the National Science Teachers Association**

● What do these horses have in common with the grass under their feet? They and all other organisms are composed of chemical substances that include water, carbohydrates, proteins, and fats.

Nature of Matter

- **Differentiate** between atoms and elements.
- **Analyze** how compounds are formed.
- **Distinguish** between covalent bonds, hydrogen bonds, and ionic bonds.

Key Terms

atom
element
compound
molecule
ion

Figure 1   Atom. The electron cloud is the region of an atom where electrons are most likely to be found. The nucleus of this atom contains six protons and six neutrons.

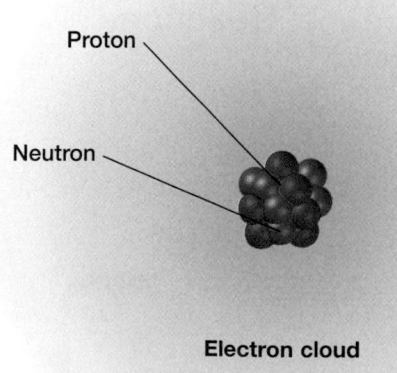

Proton

Neutron

Electron cloud

Atoms

Cooking requires an understanding of how the ingredients in foods interact. A cook's application of chemical principles while preparing recipes affects the flavor and texture of foods. Just as a cook can benefit from a knowledge of basic chemistry, you can better understand principles of biology if you also understand the fundamentals of chemistry. Chemistry will help you learn about biology because organisms, including yourself, are chemical machines.

What does all matter have in common? Matter consists of atoms. An **atom** is the smallest unit of matter that cannot be broken down by chemical means. Atoms are so small and dynamic that their exact structure is difficult to determine. Scientists have developed models, such as the one shown in **Figure 1,** to explain the structure and properties of atoms.

As shown in Figure 1, atoms consist of three kinds of particles: electrons, protons, and neutrons. Protons, shown in red, and neutrons, shown in blue, make up the nucleus, or core, of an atom. The region around the nucleus that electrons may occupy at any time is called the electron cloud, shown as a blue haze around the nucleus. Electrons are negatively charged, so the electron cloud has a negative charge. Protons are positively charged and neutrons have no charge, so the nucleus has a positive charge. Because protons and electrons are oppositely charged, they attract one another. Atoms typically have one electron for each proton, so they have no electrical charge.

Elements

An **element** is a pure substance made of only one kind of atom. There are more than 100 known elements, and each is represented by a one-, two-, or three-letter symbol. For example, the elements hydrogen, oxygen, and carbon are represented by the symbols H, O, and C, respectively. Elements differ in the number of protons their atoms contain. Atoms of the simplest element, hydrogen, each contain one proton and one electron. In contrast, oxygen atoms contain eight protons and eight electrons. The number of neutrons in an atom is often but not always equal to the number of protons in the atom. Atoms of an element that contain different numbers of neutrons are called isotopes. For example, three common isotopes of carbon, C, are carbon-12, carbon-13, and carbon-14. Each contains six protons, however carbon-13 contains seven neutrons, and carbon-14 contains eight neutrons.

Chemical Bonding

Atoms can join with other atoms to form stable substances. A force that joins atoms is called a chemical bond. A **compound** is a substance made of the joined atoms of two or more different elements. For example, when sodium atoms, Na, bond with chlorine atoms, Cl, the compound sodium chloride (table salt) forms. Every compound is represented by a chemical formula that identifies the elements in the compound and their proportions. The formula for sodium chloride, NaCl, shows that there is one sodium atom for every chlorine atom in the compound.

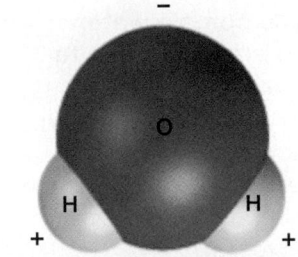

Figure 2　Water molecule. Each water molecule is held together by covalent bonds between two hydrogen atoms and one oxygen atom.

Covalent Bonds

Covalent bonds form when two or more atoms share electrons to form a molecule. A **molecule,** such as the water molecule shown in **Figure 2,** is a group of atoms held together by covalent bonds. Like the rivets and welds that connect steel girders in a skyscraper, covalent bonds join the atoms in molecules. Because the number of protons is equal to the number of electrons in a molecule, the molecule has no net electrical charge. Other examples of molecules include carbon dioxide, CO_2, and oxygen gas, O_2.

The arrangement of their electrons determines how atoms bond together. Electrons are grouped into different levels. The levels closest to the nucleus have less energy than the levels farther from the nucleus. Electron levels can hold a limited number of electrons. The outer electron levels of hydrogen and helium can hold up to two electrons. All other atoms, however, have outer electron levels that can hold up to eight electrons. An atom becomes stable when its outer electron level is full. If the outer electron level is not full, an atom will react readily with atoms that can provide electrons to fill its outer level. As Figure 2 shows, water, H_2O, forms when an oxygen atom, which has six outer electrons, combines with two hydrogen atoms, which have one outer electron each.

Hydrogen Bonds

The electrons in a water molecule are shared by oxygen and hydrogen atoms. However, the shared electrons are attracted more strongly by the oxygen nucleus than by the hydrogen nuclei. The water molecule therefore has partially positive and negative ends, or poles. As shown in **Figure 3,** the partially positive end of one water molecule is attracted to the negative end of another water molecule. Molecules with an unequal distribution of electrical charge, such as water molecules, are called polar molecules. This attraction between two water molecules is an example of a hydrogen bond—a weak chemical attraction between polar molecules.

Figure 3　Hydrogen bonds in water

Water molecules are attracted to each other by hydrogen bonds.

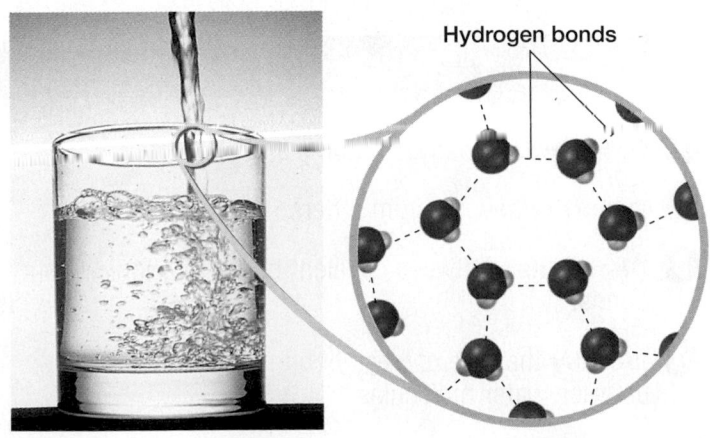

Hydrogen bonds

Ionic Bonds

Sometimes atoms or molecules gain or lose electrons. An atom or molecule that has gained or lost one or more electrons is called an **ion** *(IE ahn)*. Ions have an electrical charge because they contain an unequal number of electrons and protons. An atom that has lost electrons is positively charged, whereas an atom that has gained electrons is negatively charged.

Ions of opposite charge may interact to form an ionic bond. For example, an atom of sodium is unstable because it has only one electron in its outer level. Sodium readily gives up this electron to become a stable, positively charged sodium ion, Na^+. An atom of chlorine is also unstable because it has seven electrons in its outer level. Chlorine readily accepts an electron to become a stable, negatively charged chloride ion, Cl^-. The negative charge of a chloride ion is attracted to the positive charge of a sodium ion. Thus, sodium atoms and chlorine atoms readily form an ionic bond to become sodium chloride, as shown in **Figure 4.**

Figure 4   Ionic bonds in sodium chloride

Ionic bonds in sodium chloride, NaCl, are formed by the interaction between sodium ions, Na^+, and chloride ions, Cl^-.

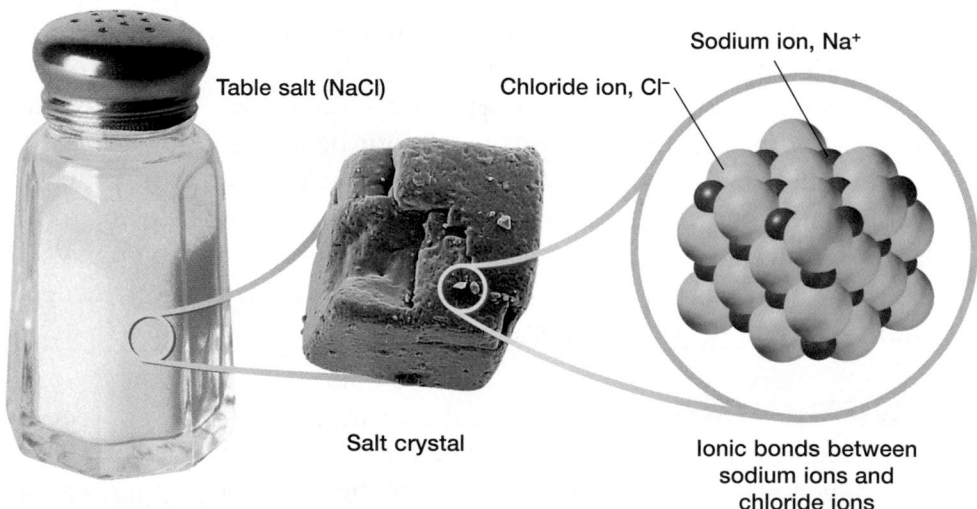

Table salt (NaCl)

Chloride ion, Cl^-

Sodium ion, Na^+

Salt crystal

Ionic bonds between sodium ions and chloride ions

Section 1 Review

1 **Differentiate** between atoms and elements.

2 **Describe** how an atom differs from a molecule.

3 **Distinguish** between covalent bonds and ionic bonds.

4 **Identify** the type of weak bond that forms between water molecules.

5 **Critical Thinking Recognizing Differences** Explain the difference between polar molecules and nonpolar molecules. Give an example of a polar molecule.

6 **Standardized Test Prep** Sodium chloride is an example of

A a compound.   C an isotope.

B a molecule.   D an ion.

Water and Solutions

Water in Living Things

You may not realize it, but nearly 70 percent of your body is made of water. About two-thirds of the molecules in your body are water molecules. Your body's cells are filled with water, and water is the medium in which most cellular events take place. Your cells are also surrounded by water, and water helps move nutrients and other substances into and out of your cells. What are some of the properties of water that make it such an important substance for life?

Storage of Energy

Water absorbs heat more slowly and retains this energy longer than many other substances do. For example, a pot of boiling water removed from a stove takes a long time to cool down. Many organisms release excess heat through water evaporation. For example, humans cool themselves by sweating. The water vapor lost through the evaporation of sweat carries heat away from the body. In organisms, this ability to control temperature enables cells to maintain a constant internal temperature when the external tempera-ture changes drastically. Water thus helps cells maintain homeostasis.

Cohesion and Adhesion

The hydrogen bonds between water molecules cause the cohesion of liquid water. **Cohesion** *(koh HEE zhuhn)* is an attraction between substances of the same kind. Because of cohesion, water and other liquids form thin films and drops, such as those shown in **Figure 5.** Molecules at the surface of water are linked together by hydrogen bonds like a crowd of people linked by holding hands. This attraction between water molecules causes a condition known as *surface tension*. Surface tension prevents the surface of water from stretching or breaking easily.

Water molecules are also attracted to many other similarly polar substances. **Adhesion** *(ad HEE zhuhn)* is an attraction between different substances. Because of adhesion, some substances get wet. Adhesion powers a process, called capillary action, in which water moleculcs move upward through a narrow tube, such as the stem of a plant. The attraction of water to the walls of the tube sucks the water up more strongly than gravity pulls it down. Water moves upward through a plant from roots to leaves through a combination of capillary action, cohesion, and other factors.

Objectives

- **Analyze** the properties of water.
- **Describe** how water dissolves substances.
- **Distinguish** between acids and bases.

Key Terms

cohesion
adhesion
solution
acid
base

Figure 5   Cohesion.
Because of cohesion, water forms drops like those on this plant.

Aqueous Solutions

Many substances dissolve in water. For example, when you add salt to water, the resulting mixture is a saltwater solution. A **solution** is a mixture in which one or more substances are evenly distributed in another substance. Many important substances in the body have been dissolved in blood or other aqueous fluids. Because these substances can dissolve in water, they can more easily move within and between cells. For example, sugar could not be delivered to your cells if it were not dissolved in water.

Polarity

The polarity of water enables many substances to dissolve in water. Ionic compounds and polar molecules dissolve best in water. When ionic compounds are dissolved in water, the ions become surrounded by polar water molecules. As **Figure 6** shows, ions are attracted to the ends of water molecules with the opposite charge. The resulting solution is a mixture of water molecules and ions. A similar attraction results when polar molecules are dissolved in water. In both cases, the ions or molecules become evenly distributed in the water.

Nonpolar molecules do not dissolve well in water. When nonpolar substances, such as oil, are placed in water, the water molecules are more attracted to each other than to the nonpolar molecules. As a result, the nonpolar molecules are shoved together. This explains why oil forms clumps or beads in water. The inability of nonpolar molecules to dissolve in polar molecules is important to organisms. For example, the shape and function of cell membranes depend on the interaction of polar water with nonpolar membrane molecules.

Figure 6 Water dissolves ionic compounds

When sodium chloride, NaCl, is dissolved in water, sodium ions, Na$^+$, and chloride ions, Cl$^-$, become surrounded by water molecules, H$_2$O.

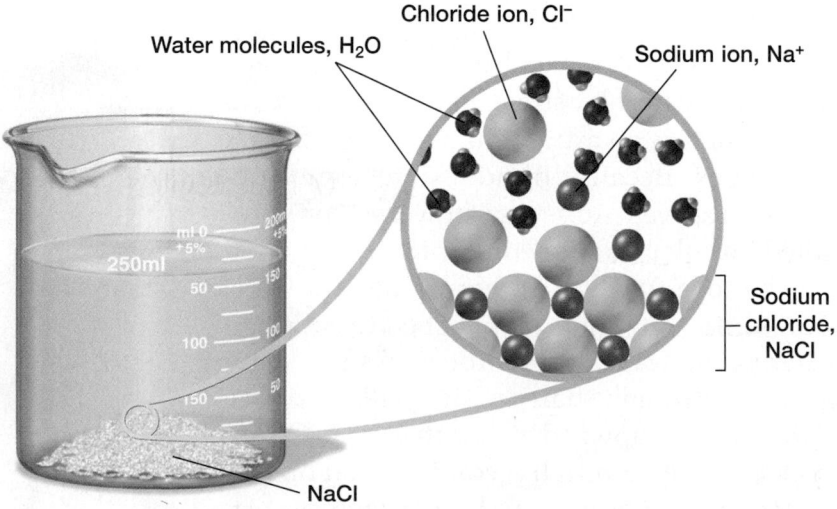

Chloride ion, Cl$^-$

Water molecules, H$_2$O

Sodium ion, Na$^+$

Sodium chloride, NaCl

NaCl

Acids and Bases

While the bonds in water molecules are strong, at any given time a tiny fraction of those bonds might break, forming a hydrogen ion, H^+, and a hydroxide ion, OH^-:

$$H_2O \longrightarrow H^+ + OH^-$$

As a result, pure water always has a low concentration of hydrogen ions and hydroxide ions, which are present in equal numbers. Compounds that form hydrogen ions when dissolved in water are called **acids.** When an acid is added to water, the concentration of hydrogen ions in the solution is increased above that of pure water.

In contrast, compounds that reduce the concentration of hydrogen ions in a solution are called **bases.** Many bases form hydroxide ions when dissolved in water. Such bases lower the concentration of hydrogen ions because hydroxide ions react with hydrogen ions to form water molecules.

The pH scale shown in **Figure 7** is based on the concentration of hydrogen ions in solutions. All solutions have a pH value between 0 and 14. Pure water has a pH value of 7. Acidic solutions have pH values below 7, and basic solutions have pH values above 7. Each whole number represents a factor of 10 on the scale. A solution with a pH value of 5, for example, has 10 times as many hydrogen ions as one with a pH value of 6.

Study TIP

● **Reading Effectively**
As you read, you may encounter the terms *alkaline* or *alkalinity.* Basic solutions—whose pH is above 7—are often called *alkaline solutions.* Solutions with pH values below 7 are usually referred to as *acidic solutions.*

Figure 7   The pH scale.
The pH scale is based on the concentration of hydrogen ions in a solution.

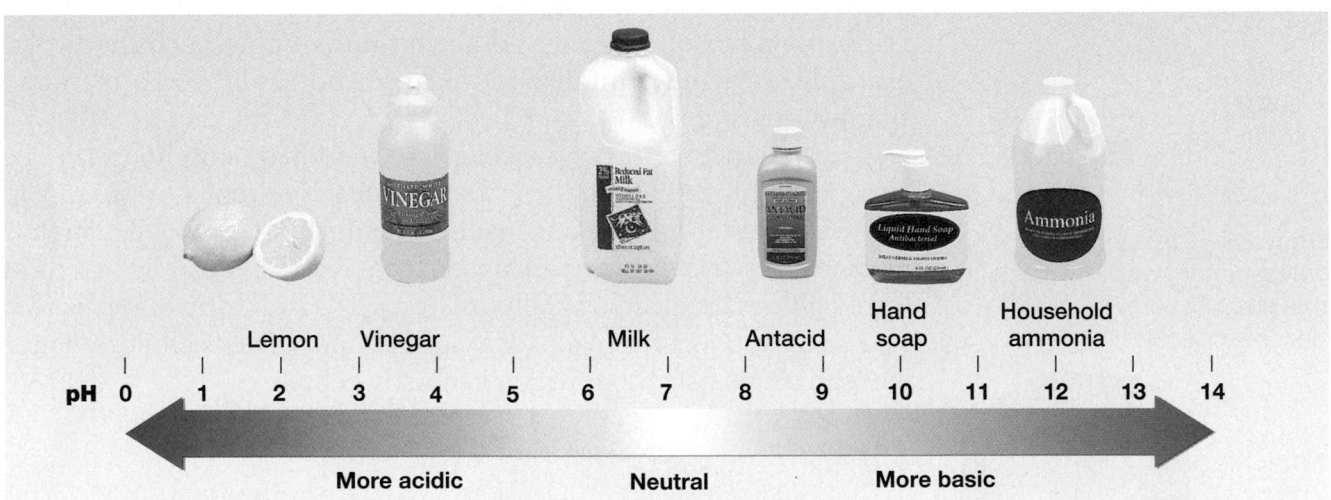

| Lemon | Vinegar | | | Milk | | Antacid | Hand soap | Household ammonia | |
pH 0   1   2   3   4   5   6   7   8   9   10   11   12   13   14

More acidic   **Neutral**   **More basic**

Section 2 Review

❶ **Distinguish** between adhesion and cohesion.

❷ **Identify** a substance that would not dissolve well in water. Explain why.

❸ **Differentiate** between acids and bases.

❹ **Critical Thinking Inferring Relationships**
When salt is added to water, the freezing point of the water decreases. Explain why this occurs.

❺ (**Standardized Test Prep**) The pH of solution *A* is 2. The pH of solution *B* is 4. How does the concentration of hydrogen ions in *A* ($[H^+]_A$) compare with the concentration of hydrogen ions in *B* ($[H^+]_B$)?

A $[H^+]_A = 2 \times [H^+]_B.$ **C** $[H^+]_A = 100 \times [H^+]_B.$
B $[H^+]_B = 2 \times [H^+]_A.$ **D** $[H^+]_B = 100 \times [H^+]_A.$

Chemistry of Cells

Objectives

- **Summarize** the characteristics of organic compounds.
- **Compare** the structures and function of different types of biomolecules.
- **Describe** the components of DNA and RNA.
- **State** the main role of ATP in cells.

Key Terms

carbohydrate
monosaccharide
lipid
protein
amino acid
nucleic acid
nucleotide
DNA
RNA
ATP

Figure 8 Structure of polysaccharides. Starch is a long chain of many linked glucose molecules.

Carbon Compounds

Most matter in your body that is not water is made of organic compounds. Organic compounds contain carbon atoms that are covalently bonded to other elements—typically hydrogen, oxygen, and other carbon atoms. Four principal classes of organic compounds are found in living things: carbohydrates, lipids, proteins, and nucleic acids. Without these compounds, cells could not function.

Carbohydrates

Carbohydrates are organic compounds made of carbon, hydrogen, and oxygen atoms in the proportion of 1:2:1. Carbohydrates are a key source of energy, and they are found in most foods—especially fruits, vegetables, and grains. The building blocks of carbohydrates are single sugars, called **monosaccharides** *(mahn oh SAK uh reyedz)*, such as glucose, $C_6H_{12}O_6$, and fructose. Simple sugars such as glucose are a major source of energy in cells. Disaccharides are double sugars formed when two monosaccharides are joined. For example, sucrose, or common table sugar, consists of both glucose and fructose. Polysaccharides such as starch, shown in **Figure 8,** are chains of three or more monosaccharides. A polysaccharide is an example of a macromolecule, a large molecule made of many smaller molecules.

In organisms, some polysaccharides function as storehouses of the energy contained in sugars. Two polysaccharides that store energy in this way are starch, which is made by plants, and glycogen, which is made by animals. Both starch and glycogen are made of hundreds of linked glucose molecules. Cellulose is a polysaccharide that provides structural support for plants. Humans cannot digest cellulose. Thus, you cannot digest wood, which is mostly cellulose.

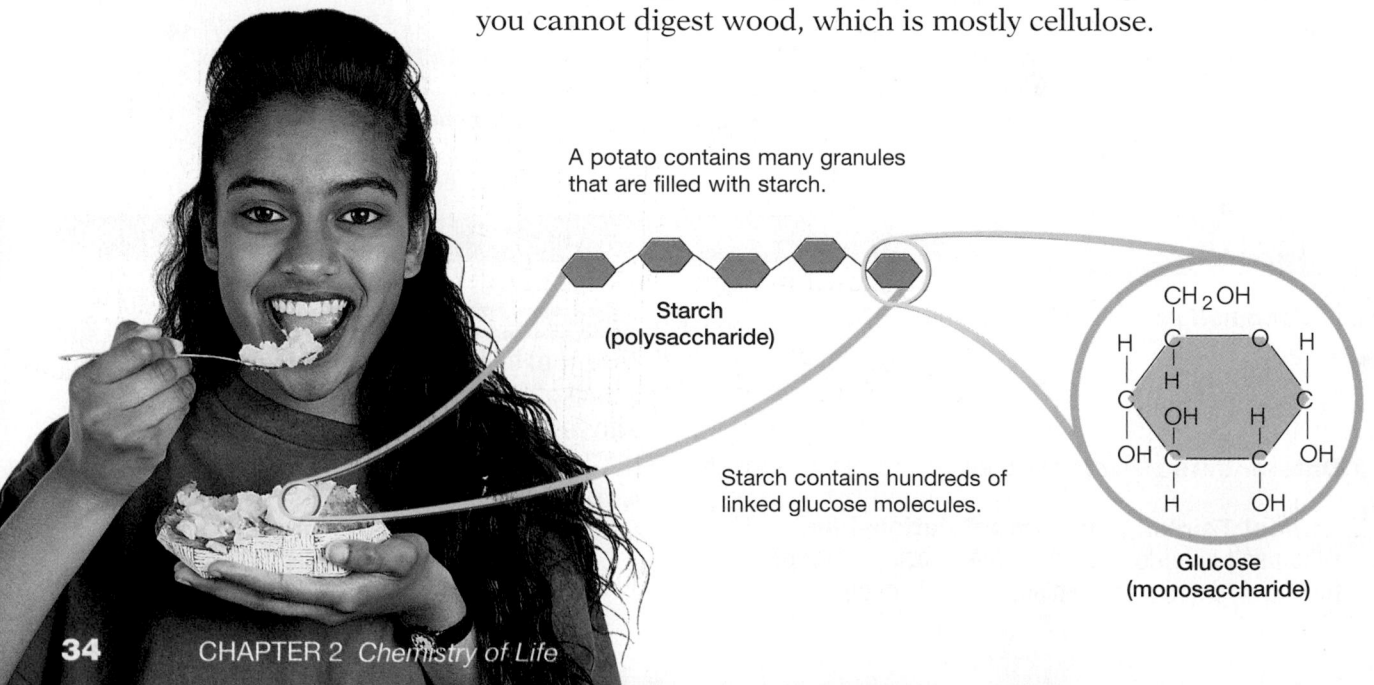

A potato contains many granules that are filled with starch.

Starch
(polysaccharide)

Starch contains hundreds of linked glucose molecules.

Glucose
(monosaccharide)

Figure 9 Structure of fats

Fatty acids can be saturated or unsaturated.

Saturated fatty acid

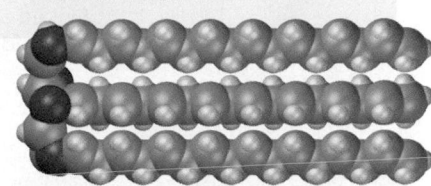

Saturated fats, such as butter, are solid at room temperature.

Unsaturated fatty acid

Unsaturated fats, such as canola oil, are liquid at room temperature.

Lipids

Lipids *(LIHP ihdz)* are nonpolar molecules that are not soluble or mostly insoluble in water. They include fats, phospholipids, steroids, and waxes. Lipids are an important part of the structure and functioning of cell membranes. Phospholipids make up the lipid bilayer of cell membranes. Steroids include cholesterol, which is found in animal cell membranes. Other lipids include some light-absorbing compounds called *pigments,* such as the plant pigment chlorophyll.

Fats are lipids that store energy. As **Figure 9** shows, a typical fat contains three fatty acids bonded to a glycerol molecule backbone. Glycerol is a three-carbon organic molecule. A fatty acid is a long chain of carbon atoms, shown in green, with hydrogen atoms bonded to them. Most carbon atoms in a fatty acid are bonded to either one or two hydrogen atoms, shown in blue. Because bonds between carbon and hydrogen are rich in energy, fats can store a lot of energy.

In a saturated fatty acid, all of the carbon atoms in the chain are bonded to two hydrogen atoms (except the carbon atom on the end, which is bonded to three hydrogen atoms). Most animal fats—such as those in butter, lard, and grease from cooked meats—contain primarily saturated fatty acids. Saturated fatty acids are relatively straight molecules and are generally solid at room temperature.

In an unsaturated fatty acid, some of the carbon atoms are linked by a "double" covalent bond, each with only one hydrogen atom, producing kinks in the molecule, as shown in Figure 9. Most plant oils, such as olive oil, and some fish oils contain mainly unsaturated fatty acids and are generally liquid at room temperature. Hydrogenated vegetable oils contain naturally unsaturated fatty acids that have been saturated artificially by the addition of hydrogen atoms. Thus, hydrogenated vegetable oils, such as those in margarine and vegetable shortening, are generally solid at room temperature.

Real Life

Fat-free potato chips fried in artificial fats contain fewer calories than those fried in natural fats.

Unfortunately, some artificial fats may reduce vitamin absorption and cause indigestion in some people.

Finding Information *Research the benefits and potential shortcomings of artificial fats.*

Proteins

A **protein** *(PROH teen)* is usually a large molecule formed by linked smaller molecules called amino acids. **Amino acids** are the building blocks of proteins. Twenty different amino acids are found in proteins. Some amino acids are polar, and others are nonpolar. Some amino acids are electrically charged, and others are not charged. As **Figure 10** shows, proteins fold into compact shapes, determined in part by how the protein's amino acids interact with water and one another.

Some proteins are enzymes and promote chemical reactions. Other proteins have important structural functions. For example, the protein collagen *(KAHL uh juhn)* is found in skin, ligaments, tendons, and bones. Your hair and muscles contain structural proteins and so do the fibers of a blood clot. Other proteins called antibodies help your body defend against infection. Specialized proteins in muscles enable your muscles to contract. In your blood, a protein called hemoglobin carries oxygen from your lungs to body tissues.

Figure 10　Structure of proteins

Proteins are chains of amino acids which are usually folded into compact shapes.

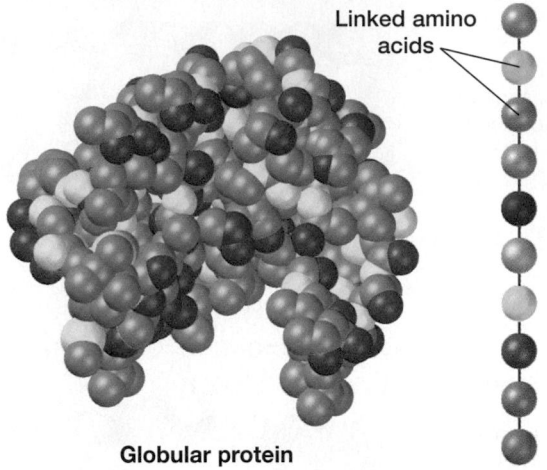

Linked amino acids

Globular protein

BIOWatch

Foods as Fuel

Most foods contain a mixture of carbohydrates, proteins, and fats. The body can use these molecules to build new structures, but it uses them mostly as an energy source. Your body's cells harvest the energy in food molecules for metabolism. The energy value of food molecules is measured in kilocalories (kcal).

The minimal rate of energy use per hour (h), called the basal metabolic rate, is about 70 kcal/h for men and 60 kcal/h for women. Typically, walking uses about 200 kcal/h and jogging uses about 600 kcal/h. If more kilocalories are consumed than are used, the body will store the excess kilocalories as fat, regardless of whether the consumed kilocalories are contained in carbohydrates, proteins, or fats.

Carbohydrates

Most carbohydrates in foods come from plant products, such as fruits, grains, and vegetables. Other sources are milk, which contains the sugar lactose, and various meats, which contain some glycogen. Candy and soft drinks also contain sugars. About 4 kcal of energy are supplied by 1 gram (g) of carbohydrates.

Proteins

Primary sources of dietary protein include legumes, eggs, milk, fish, poultry, and meat. As with carbohydrates, proteins supply about 4 kcal/g. Dietary protein is the source of amino acids. Proteins also provide raw materials for other compounds, such as nucleic acids.

Fats

Fats are found mainly in vegetable oils, such as olive oil; dairy products, such as milk and butter; and meat, such as beef and pork. Fats contain more energy per gram than do carbohydrates and proteins; fats supply about 9 kcal/g of energy.

internet connect

www.scilinks.org
Topic: Foods as Fuel
Keyword: HX4086

SC**LINKS** Maintained by the National Science Teachers Association

Figure 11   Structure of nucleic acids

DNA is made of two strands of multiple nucleotides linked by hydrogen bonds.

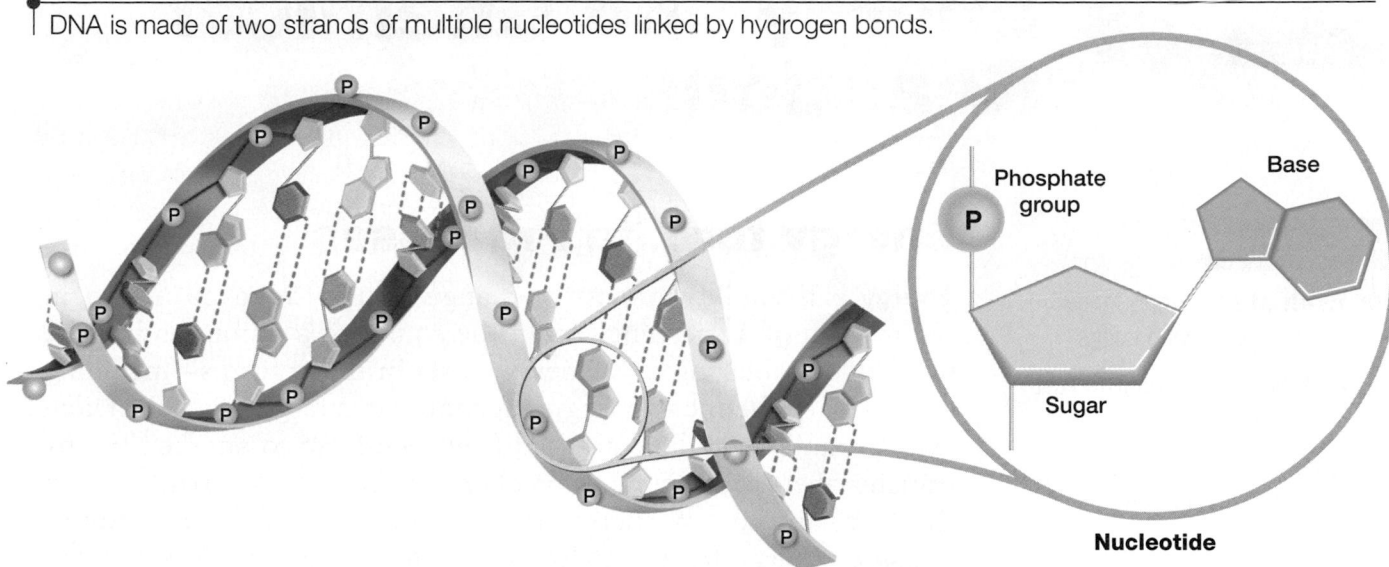

Phosphate group

Base

P

Sugar

Nucleotide

Nucleic Acids

All of your cells contain nucleic acids. A **nucleic acid** is a long chain of smaller molecules called nucleotides. A **nucleotide** has three parts: a sugar, a base, and a phosphate group, which contains phosphorus and oxygen atoms. There are two types of nucleic acids—DNA and RNA—and each type contains four kinds of nucleotides.

DNA, or deoxyribonucleic acid, consists of two strands of nucleotides that spiral around each other, as shown in **Figure 11.** Chromosomes contain long strands of DNA, which stores hereditary information.

RNA, or ribonucleic acid, may consist of a single strand of nucleotides or of based-paired nucleotides. RNA plays many key roles in the manufacture of proteins. RNA can also act as an enzyme, promoting the chemical reactions that link amino acids to form proteins.

ATP

Another important biological molecule is ATP. **ATP,** or adenosine *(uh DEHN uh seen)* triphosphate, is a single nucleotide with two extra energy-storing phosphate groups. When food molecules are broken down inside cells, some of the energy in the molecules is stored temporarily in ATP. Cells need a steady supply of ATP to function.

Section 3 Review

1 Identify what all organic compounds have in common, and list the four principal classes of organic compounds.

2 Compare the structures of saturated and unsaturated lipids.

3 Describe the three parts of a nucleotide and how they are attached to one another.

4 Critical Thinking Inferring Relationships Compare the role of ATP in cells with the roles of RNA.

5 Standardized Test Prep Molecule *X* contains a sugar and a phosphate group. What is molecule *X*?
A a carbohydrate   C a fatty acid
B a nucleotide   D an amino acid

Energy and Chemical Reactions

Objectives

- **Evaluate** the importance of energy to living things.
- **Relate** energy and chemical reactions.
- **Describe** the role of enzymes in chemical reactions.
- **Identify** the effect of enzymes on food molecules.

Key Terms

energy
activation energy
enzyme
substrate
active site

Energy for Life Processes

Energy is the ability to move or change matter. Energy is in food, in the motion of a speeding car, in the sound of a guitar, and in the warmth of a blazing fire. Energy exists in many forms—including light, heat, chemical energy, mechanical energy, and electrical energy—and it can be converted from one form to another. In any transfer of energy or conversion of energy from one form to another, the total amount of energy does not change. The total amount of *usable* energy, however, always decreases. Heat causes a cooking egg to change color and solidify, as shown in **Figure 12.** The energy transferred to the egg by heat rearranges the atoms and molecules in the egg.

Energy can be stored or released by chemical reactions, also shown in Figure 12. A chemical reaction is a process during which chemical bonds between atoms are broken and new ones are formed, producing one or more different substances. At any moment, thousands of chemical reactions are occurring in every cell of your body. The starting materials for chemical reactions are called *reactants*. The newly formed substances are called *products*. Chemical reactions are summarized by chemical equations, which are written in the following form:

$$\text{Reactants} \longrightarrow \text{Products}$$

The arrow is read as "changes to" or "forms." For example, dissolving sodium chloride in water causes the following reaction:

$$\text{NaCl} \longrightarrow \text{Na}^+ + \text{Cl}^-$$

Figure 12   Evidence of chemical reactions

An egg becomes solid when it is heated. A chemical reaction causes the bioluminescent click beetle, *Pyrophorus noctilucus,* to give off light energy.

Figure 13 Energy and chemical reactions

Chemical reactions absorb or release energy.

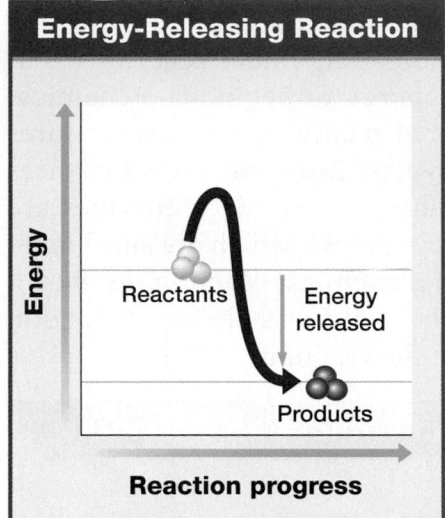

Energy-Releasing Reaction

Energy

Reactants

Energy released

Products

Reaction progress

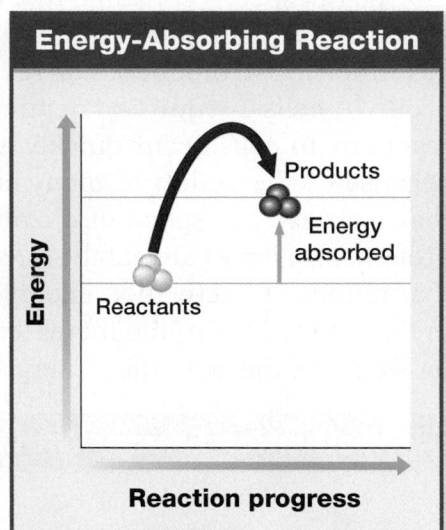

Energy-Absorbing Reaction

Energy

Products

Energy absorbed

Reactants

Reaction progress

Energy in Chemical Reactions

In chemical reactions, energy is absorbed or released when chemical bonds are broken and new ones are formed. The graphs shown in **Figure 13** compare a chemical reaction that releases energy with a chemical reaction that absorbs energy. The freezing and melting of water are physical, not chemical, changes. Freezing and melting, however, are good examples of how energy is released or absorbed. When water freezes, the process that leads to the formation of ice crystals causes heat energy to be released. When you place water in the freezer to make ice, heat is released from the water as the water freezes. When you remove ice cubes from the freezer, the ice begins to melt. When ice melts, it absorbs heat from the environment. When you hold a piece of ice, your hand gets cold and heat is transferred from your hand to the ice as the ice begins to melt.

Metabolism *(muh TAB uh lihz uhm)* is the term used to describe all of the chemical reactions that occur within an organism. Your cells get most of the energy needed for metabolism from the food you eat. As food is digested, chemical reactions convert the chemical energy in food molecules to forms of energy that can be used by cells.

internet connect

www.scilinks.org
Topic: Chemical Reactions
Keyword: HX4040

SCI*LINKS* Maintained by the
National Science
Teachers Association

Activation Energy

The heat from a flame transfers enough energy to ignite the logs in a campfire. The spark from a spark plug causes the gasoline in an automobile engine to ignite. In both cases, energy is needed to start a chemical reaction. The energy needed to start a chemical reaction is called **activation energy.** To better understand activation energy, think of rolling a boulder down a hill. To get the boulder rolling downhill, you must first push it. Activation energy is simply a chemical "push" that starts a chemical reaction. Even in a chemical reaction that releases energy, activation energy must be supplied before the reaction can occur.

Enzymes

Like engines, cells consume fuel because they need energy to function. Just as an engine requires a spark of energy to begin burning gasoline, most biochemical reactions—chemical reactions that occur in cells—require activation energy to begin. The chemical reactions in cells occur quickly and at relatively low temperatures because of the action of many enzymes. **Enzymes** are substances that increase the speed of chemical reactions. Most enzymes are proteins. Enzymes are catalysts *(KAT uh lists)*, which are substances that reduce the activation energy of a chemical reaction. As shown in **Figure 14,** an enzyme increases the speed of a chemical reaction by reducing the activation energy of the reaction.

Figure 14 Enzymes lower activation energy. Enzymes decrease the amount of energy needed to start a chemical reaction. Enzymes do not change the amount of energy contained in either the reactants or the products.

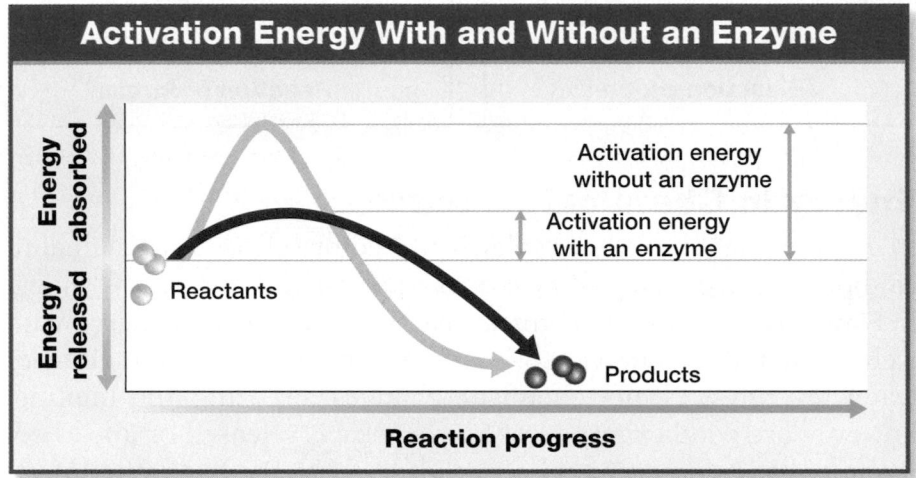

Activation Energy With and Without an Enzyme

Enzymes help organisms maintain homeostasis. Without enzymes, chemical reactions would not occur quickly enough to sustain life. For example, consider a reaction that takes place in your blood. Blood carries carbon dioxide, CO_2, (a waste product made by cells) to your lungs, where it is eliminated as you breathe out. In the lungs, carbon dioxide reacts with water, H_2O, to form carbonic acid, H_2CO_3, as shown by the following equation:

$$CO_2 + H_2O \xrightleftharpoons{\text{carbonic anhydrase}} H_2CO_3$$

The reverse reaction occurs in your lungs, converting carbonic acid back to carbon dioxide and water. Most enzyme-assisted reactions are reversible, meaning they can proceed in the opposite direction.

Without an enzyme, the reaction that produces carbonic acid is very slow; only about 2000 molecules of carbonic acid are produced in an hour. This rate is not fast enough for your blood to carry away the carbon dioxide released by millions of cells. Fortunately, your blood contains the enzyme carbonic anhydrase *(an HIED rays)*. In the presence of carbonic anhydrase, carbon dioxide and water react to form about 600,000 molecules of carbonic acid per second! The enzyme increases the reaction rate about one million times, enabling your body to eliminate carbon dioxide efficiently.

Reading Effectively
As you read, notice that the names of most enzymes, such as amylase and catalase, end with *-ase*. This will help you identify other enzymes you will encounter in this book.

Enzyme Specificity

A substance on which an enzyme acts during a chemical reaction is called a **substrate** *(SUHB strayt)*. Enzymes act only on specific substrates. For example, the enzyme amylase *(AM uh lays)* assists in the breakdown of starch to glucose in the following chemical reaction. In this reaction, starch is amylase's substrate.

$$\text{starch} \xrightleftharpoons{\text{amylase}} \text{glucose}$$

The enzyme catalase *(KAT uh lays)* assists in the breakdown of hydrogen peroxide, H_2O_2, a toxin formed in cells. In this case, hydrogen peroxide is broken down to water, H_2O, and oxygen gas, O_2. In this reaction, hydrogen peroxide is catalase's substrate.

$$2H_2O_2 \xrightleftharpoons{\text{catalase}} 2H_2O + O_2$$

An enzyme's shape determines its activity. Typically, an enzyme is a large protein with one or more deep folds on its surface. These folds form pockets called **active sites.** As shown in **Figure 15,** an enzyme's substrate fits into the active site. An enzyme acts only on a specific substrate because only that substrate fits into its active site.

Step ❶ When an enzyme first attaches to a substrate during a chemical reaction, the enzyme's shape changes slightly so that the substrate fits more tightly in the enzyme's active site.

Step ❷ At an active site, an enzyme and a substrate interact in a way that reduces the activation energy of the reaction, making the substrate more likely to react.

Step ❸ The reaction is complete when products have formed. The enzyme is now free to catalyze further reactions.

Figure 15

BIO graphic — Enzyme Action

Enzymes assist biochemical reactions by bringing key molecules together.

❶ A substrate attaches to an enzyme's active site.

❷ The enzyme reduces the activation energy of the reaction.

❸ The enzyme is not changed by the reaction.

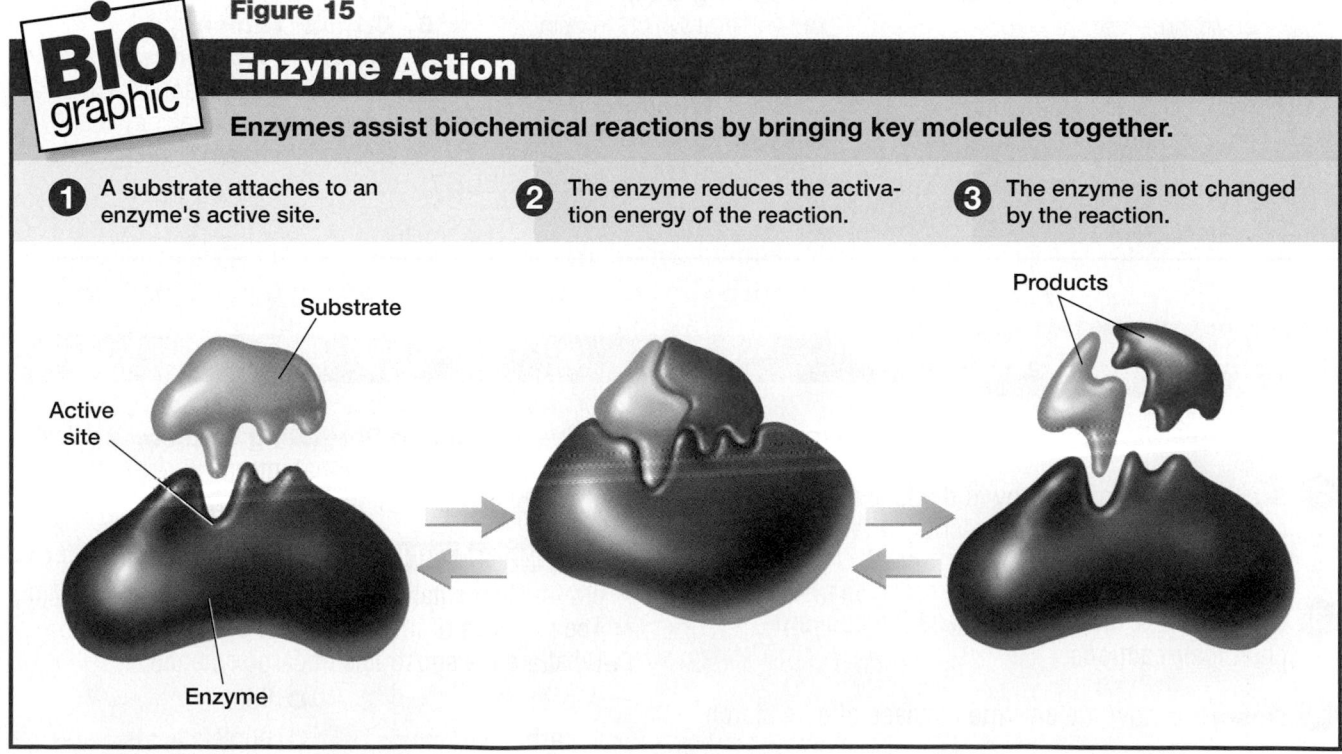

Products

Substrate

Active site

Enzyme

Factors in Enzyme Activity

Any factor that changes the shape of an enzyme can affect the enzyme's activity. For example, enzymes operate most efficiently within a certain range of temperatures. Temperatures outside this range can either break or strengthen some of the enzyme's bonds, changing its shape. Moreover, each enzyme operates best within a certain range of pH values. A pH value outside this range can cause bonds in an enzyme to break, reducing the enzyme's effectiveness.

The enzymes that are active at any one time in a cell determine what happens in that cell. Your body's cells contain many different enzymes, and each enzyme catalyzes a different chemical reaction. Different kinds of cells contain different collections of enzymes. For example, as you read this page, the chemical reactions occurring in nerve cells in your eye are different from the chemical reactions occurring in your red blood cells.

Analyzing the Effect of pH on Enzyme Activity

DATA LAB

Background

The graph at right shows the relationship between pH and the activity of two digestive enzymes, pepsin and trypsin. Pepsin works in the stomach, while trypsin works in the small intestine. Use the graph to answer the following questions.

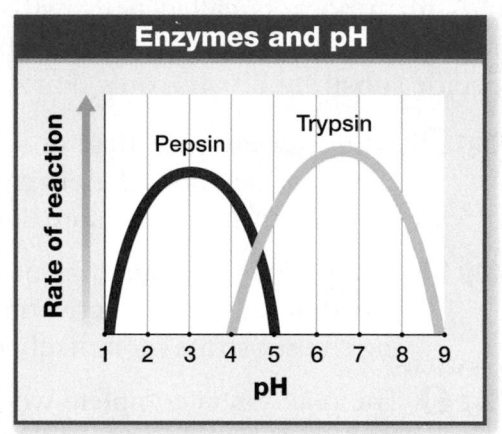

Enzymes and pH

Analysis

1. **Name** the enzyme that works best in highly acidic environments.

2. **Name** the enzyme that works best in less-acidic environments.

3. **Critical Thinking Analyzing Data** Identify the pH value at which trypsin works best.

4. **Critical Thinking Analyzing Data** Identify the pH value at which pepsin works best.

5. **Critical Thinking Inferring Relationships** What does the graph indicate about the relative acidity of the stomach and small intestine?

Section 4 Review

1. **List** three ways that organisms use energy.

2. **Summarize** how energy is made available by chemical reactions, and describe what happens to usable energy.

3. **Explain** how enzymes increase the speed of chemical reactions.

4. **Describe** how the enzyme amylase affects starch.

5. **Critical Thinking Predicting Outcomes** What effect might a molecule that interferes with the action of carbonic anhydrase have on your body?

6. **Standardized Test Prep** Carboxypeptidase is an enzyme that catalyzes reactions in the small intestine. The products of these reactions are amino acids. What are the substrates of carboxypeptidase?
   A proteins   C lipids
   B carbohydrates   D nucleic acids

Key Concepts

1 Nature of Matter

- All matter is made of atoms. Atoms consist of electrons, protons, and neutrons.
- Molecules are groups of atoms linked by covalent bonds.
- Hydrogen bonding occurs between polar molecules.
- An ion is a charged atom or molecule. Ions of opposite charge may form an ionic bond.

2 Water and Solutions

- Water, which is essential for life, stores heat efficiently and binds to itself and other substances.
- Water dissolves polar molecules and ionic compounds.
- Acids increase the hydrogen ion concentration of a solution.
- Bases decrease the hydrogen ion concentration of a solution.
- The pH scale measures the strength of acids and bases.

3 Chemistry of Cells

- Organic compounds are found in living things.
- Carbohydrates, such as glucose, are a source of energy and are used as structural materials in organisms.
- Lipids are nonpolar molecules that store energy and are an important part of cell membranes.
- Proteins are chains of amino acids. The sequence of amino acids determines a protein's shape and specific function.
- Nucleic acids store and transmit hereditary information.
- ATP is the main energy currency of cells.

4 Energy and Chemical Reactions

- Chemical reactions absorb or release energy.
- Starting a chemical reaction requires activation energy.
- Enzymes speed up chemical reactions by decreasing the activation energy of the reactions.
- Enzymes bind only certain substrates.
- Factors such as temperature and pH affect enzyme activity.

Key Terms

Section 1
atom (28)
element (28)
compound (29)
molecule (29)
ion (30)

Section 2
cohesion (31)
adhesion (31)
solution (32)
acid (33)
base (33)

Section 3
carbohydrate (34)
monosaccharide (34)
lipid (35)
protein (36)
amino acid (36)
nucleic acid (37)
nucleotide (37)
DNA (37)
RNA (37)
ATP (37)

Section 4
energy (38)
activation energy (39)
enzyme (40)
substrate (41)
active site (41)

Understanding Key Ideas

1. Most atoms contain one or more
   a. neutrons.
   b. electrons.
   c. protons.
   d. All of the above

2. Water dissolves ionic compounds because water molecules
   a. are nonpolar.
   b. have a pH value of 14 or greater.
   c. have partially charged ends.
   d. do not contain atoms.

3. In cells, ATP temporarily stores
   a. amino acids.
   b. DNA.
   c. energy.
   d. lipids.

4. Energy needed for metabolism does *not* come from
   a. food.
   b. lipids.
   c. carbohydrates.
   d. water.

5. Most enzymes are
   a. lipids.
   b. carbohydrates.
   c. proteins.
   d. nucleic acids.

6. Explain the relationship between an enzyme and the activation energy of the reaction in which the enzyme participates.

7. Look at the water strider in the photograph below. Using what you have learned about the properties of water, explain how the insect can stand on the water's surface.

8. **BIOWatch** Ruby-throated humming-birds migrate 2,000 km every fall. Before migrating, they eat nectar and convert much of the sugar in the nectar to fat. Why is it advantageous for these birds to store energy as fat rather than as glycogen?

9. Describe how molecules such as carbohydrates and lipids are important in homeostasis. (**Hint:** See Chapter 1, Section 1.)

10. **Concept Mapping** Make a concept map that illustrates the structure of matter. Include the following key terms in your map: *atom, element, compound, molecule,* and *ion*.

Critical Thinking

11. **Recognizing Differences** What are two differences between ionic bonds and covalent bonds?

Alternative Assessment

12. **Evaluating Promotional Claims** Analyze the ingredients of various packaged foods. Record the percentage of carbohydrates, fats, and proteins in each food. List any additives that the products contain. Research whether the additives are natural or artificial, and find out why they are added to particular foods. Compare these data to advertising claims about the products you analyzed.

13. **Finding Information** Investigate the laboratory techniques of cell fractionation, centrifugation, and electrophoresis. Find out how each technique enables biologists to experiment with cells and to analyze the substances that cells produce. Prepare an oral report and use graphics to interpret and summarize your findings.

14. **Career Connection Biochemist** Research the field of biochemistry, and write a report on your findings. Your report should include a job description, the training required, names of employers, growth prospects, and an average starting salary.

Standardized Test Prep

Understanding Concepts

Directions (1–4): For *each* question, write on a separate sheet of paper the letter of the correct answer.

1 In what type of bond are electrons shared?
A. covalent
B. hydrogen
C. ionic
D. nuclear

2 What weak bond holds together the two strands of nucleotides in a DNA molecule?
F. covalent
G. hydrogen
H. ionic
I. nuclear

3 When an unknown substance is dissolved in water, it forms hydrogen ions. What can you conclude about the substance?
A. The substance is a base.
B. The substance is an acid.
C. The substance is a carbohydrate.
D. The substance is made up of molecules.

4 Analysis of an unknown substance shows that it has the following characteristics: it contains carbon, hydrogen, and oxygen and is soluble in oil but not in water. What kind of substance is this?
F. carbon dioxide
G. glucose
H. lipid
I. polar amino acid

Directions (5): For the following question, write a short response.

5 In an experiment a student conducted, the rate of an enzyme-catalyzed reaction increased as the student increased the substrate concentration. Deduce why the reaction rate might increase by only a small amount.

Test TIP

Choose your answer to a question based on both what you already know and any information presented in the question.

Reading Skills

Directions (6): Read the passage below. Then answer the question.

Fats are lipids that store energy. A typical fat contains three fatty acids bonded to a glycerol molecule backbone. The fatty acids in fats determine whether they are solids or liquids. Saturated fatty acids do not have any double bonds between carbon atoms and are generally solids at room temperature. Unsaturated fatty acids have one or more double bonds between carbon atoms and are generally liquids at room temperature. Animal fats are usually solid, and plant fats are usually (liquid) oils. However, in many animals of the Arctic and Antarctic, animal fats are mostly oils.

6 What adaptive advantage would the storage of body fat as oil instead of as a solid be to animals that live in freezing climates?
A. Plants can never freeze in the Arctic and Antarctic.
B. Oils are soluble in water while fats are not soluble in water.
C. Oils have lower freezing points and give better protection against freezing.
D. Storing energy in oils as plants do allows the animals to photosynthesize.

Interpreting Graphics

Directions (7): Base your answer to question 7 on the diagram below.

Model of Biological Molecule

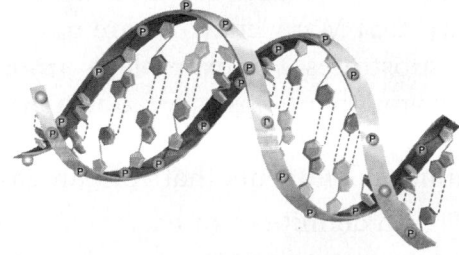

7 What is the main function of this molecule?
F. storing energy
G. making up cell membranes
H. storing hereditary information
I. promoting chemical reactions

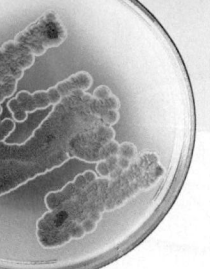

Exploration Lab

Observing Enzyme Detergents

SKILLS
- Using scientific methods
- Measuring volume, mass, and pH

OBJECTIVES
- **Recognize** the function of enzymes in laundry detergents.
- **Relate** temperature and pH to the activity of enzymes.

MATERIALS
- safety goggles and lab apron
- balance
- graduated cylinder
- glass stirring rod

- 150 mL beaker
- 18 g regular instant gelatin or 1.8 g sugar-free instant gelatin
- 0.7 g Na_2CO_3
- tongs or a hot mitt
- 50 mL boiling water
- thermometer
- pH paper
- 6 test tubes
- test-tube rack
- pipet with bulb
- plastic wrap
- tape
- 50 mL beakers (6)

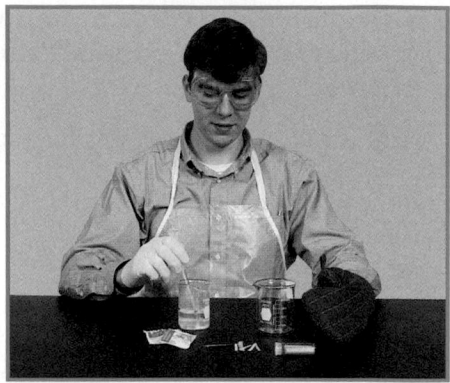

- 50 mL distilled water
- 1 g each of 5 brands of laundry detergent
- wax pencil
- metric ruler

ChemSafety

 CAUTION: Always wear safety goggles and a lab apron to protect your eyes and clothing.

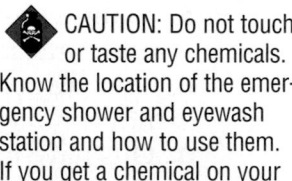 CAUTION: Do not touch or taste any chemicals. Know the location of the emergency shower and eyewash station and how to use them. If you get a chemical on your skin or clothing, wash it off at the sink while calling to the teacher. Notify the teacher of a spill. Spills should be cleaned up promptly, according to your teacher's directions.

CAUTION: Glassware is fragile. Notify the teacher of broken glass or cuts. Do not clean up broken glass or spills with broken glass unless the teacher tells you to do so.

Before You Begin

Enzymes are substances that speed up chemical reactions. Each enzyme operates best at a particular **pH** and temperature. Substances on which enzymes act are called **substrates**. Many enzymes are named for their substrates. For example, a **protease** is an enzyme that helps break down proteins. In this lab, you will investigate the effectiveness of laundry detergents that contain enzymes.

1. Write a definition for each boldface term in the paragraph above.

2. Based on the objectives for this lab, write a question you would like to explore about enzyme detergents.

Procedure

PART A: Make a Protein Substrate

1.  Put on safety goggles and a lab apron.

2.  **CAUTION: Use tongs or a hot mitt to handle heated glassware.** Put 18 g of regular (1.8 g of sugar-free) instant gelatin in a 150 mL beaker. Slowly add 50 mL of boiling water to the beaker, and stir the mixture with a stirring rod. Test and record the pH of this solution.

3. Very slowly add 0.7 g of Na_2CO_3 to the hot gelatin while stirring. Note any reaction. Test and record the pH of this solution.

4. Place 6 test tubes in a test-tube rack. Pour 5 mL of the gelatin-Na_2CO_3 mixture into each tube. Use a pipet to remove any bubbles from the surface of the mixture in each tube. Cover the tubes tightly with plastic wrap and tape. Cool the tubes, and store them at room temperature until you begin **Part C.** Complete step 12.

PART B: Design an Experiment

5. Work with members of your lab group to explore one of the questions written for step 2 of **Before You Begin.** To explore the question, design an experiment that uses the materials listed for this lab.

You Choose

As you design your experiment, decide the following:

a. what question you will explore

b. what hypothesis you will test

c. what detergent samples you will test

d. what your control will be

e. how much of each solution to use for each test

f. how to determine if protein is breaking down

g. what data to record in your data table

---

6. Write a procedure for your experiment. Make a list of all the safety precautions you will take. Have your teacher approve your procedure and safety precautions before you begin the experiment.

PART C: Conduct Your Experiment

7. Put on safety goggles and a lab apron.

8. Make a 10 percent solution of each laundry detergent by dissolving 1 g of detergent in 9 mL of distilled water.

9. Set up your experiment. Repeat step 12.

10. Record your data after 24 hours.

PART D: Cleanup and Disposal

11. Dispose of solutions, broken glass, and gelatin in the designated waste containers. Do not pour chemicals down the drain or put lab materials in the trash unless your teacher tells you to do so.

12. Clean up your work area and all lab equipment. Return lab equipment to its proper place. Wash your hands thoroughly before leaving the lab and after finishing all work.

Analyze and Conclude

1. Analyzing Methods Suggest a reason for adding Na_2CO_3 to the gelatin solution.

2. Analyzing Results Make a bar graph of your data. Plot the amount of gelatin broken down (change in the depth of the gelatin) on the *y*-axis and detergent on the *x*-axis.

3. Inferring Conclusions What conclusions did your group infer from the results? Explain.

4. Further Inquiry Write a new question about enzyme detergents that could be explored with another investigation.

? Do You Know?

Do research in the library or media center to answer these questions:

1. What other household products contain enzymes, and what types of enzymes do they contain?

2. What type of organic compound is broken down by each enzyme that you identified?

Use SciLinks to explore your own questions about products that contain enzymes.

internet connect

www.scilinks.org
Topic: Enzymes
Keyword: HX4072

SCILINKS. Maintained by the National Science Teachers Association

Paramecium (790×)

3 Cell Structure

✔ Quick Review

Answer the following without referring to earlier sections of your book.

1. **Distinguish** between polar and nonpolar molecules. *(Chapter 2, Section 1)*

2. **Compare** the structures of carbohydrates, lipids, proteins, and nucleic acids. *(Chapter 2, Section 3)*

3. **Describe** the function of ATP. *(Chapter 2, Section 3)*

Did you have difficulty? *For help, review the sections indicated.*

📖 Reading Activity

Write down the title of this chapter and the titles of its three sections on a piece of paper or in your notebook. Leave a few blank lines after each section title. Then write down what you think you will learn in each section. Save your list, and after you finish reading this chapter, check off everything that you learned that was on your list.

🖥 internet connect

www.scilinks.org
National Science Teachers Association *sci*LINKS Internet resources are located throughout this chapter.

*sci*LINKS. Maintained by the National Science Teachers Association

● Most cells, including this one-celled organism *Paramecium,* have all the equipment necessary to perform the essential functions of life.

Looking at Cells

- **Describe** how scientists measure the length of objects.
- **Relate** magnification and resolution in the use of microscopes.
- **Analyze** how light microscopes function.
- **Compare** light microscopes with electron microscopes.
- **Describe** the scanning tunneling microscope.

Key Terms

light microscope
electron microscope
magnification
resolution
scanning tunneling
  microscope

Cells Under the Microscope

Most cells are too small to see with the naked eye; a typical human body cell is many times smaller than a grain of sand. Scientists became aware of cells only after microscopes were invented, in the 1600s. When the English scientist Robert Hooke used a crude microscope to observe a thin slice of cork in 1665, he saw "a lot of little boxes." The boxes reminded him of the small rooms in which monks lived, so he called them cells. Hooke later observed cells in the stems and roots of plants. Ten years later, the Dutch scientist Anton van Leeuwenhoek used a microscope to view water from a pond, and he discovered many living creatures. He named them "animalcules," or tiny animals. Today we know that they were not animals but single-celled organisms.

Measuring Cell Structures

Measurements taken by scientists are expressed in metric units. Scientists throughout the world use the metric system. The official name of the metric system is the International System of Measurements, abbreviated as SI. SI is a decimal system, so all relationships between SI units are based on powers of 10. For example, scientists measure the sizes of objects viewed under a microscope using the SI base unit for length, which is the meter. A meter, which is about 3.28 ft (a little more than a yard), equals 100 centimeters (cm), or 1,000 millimeters (mm). A meter also equals 0.001 kilometer (km). Most SI units have a prefix that indicates the relationship of that unit to a base unit. For example, the symbol "μ" stands for the metric prefix micro. A micrometer (μm) is a unit of linear measurement equal to one-millionth of a meter, or one-thousandth of a millimeter. **Table 1** summarizes the SI units used to measure length.

Table 1 Metric Units of Length and Equivalents			
Unit	**Prefix**	**Metric equivalent**	**Real-life equivalent**
Kilometer (km)	*Kilo-*	1,000 m	About two-thirds of a mile
Meter (m)		1 m (SI base unit)	A little more than a yard
Centimeter (cm)	*Centi-*	0.01 m	About half the diameter of a Lincoln penny
Millimeter (mm)	*Milli-*	0.001 m	About the width of a pencil tip
Micrometer (μm)	*Micro-*	0.000001 m	About the length of an average bacterial cell
Nanometer (nm)	*Nano-*	0.000000001 m	About the length of a water molecule

Characteristics of Microscopes

Since Robert Hooke first observed cork cells, microscopes have unveiled the details of cell structure. These powerful instruments provide biologists with insight into how cells work—and ultimately how organisms function. Biologists use different microscopes depending on the organisms they wish to study and the questions they want to answer. Two common kinds of microscopes are light microscopes and electron microscopes. In a **light microscope,** light passes through one or more lenses to produce an enlarged image of a specimen. An **electron microscope** forms an image of a specimen using a beam of electrons rather than light.

An image produced by a microscope, such as the one shown in **Figure 1,** is called a micrograph. Many micrographs are labeled with the kind of microscope that produced the image—such as a light micrograph (LM), a transmission electron micrograph (TEM), or a scanning electron micrograph (SEM). Micrographs often are labeled with the magnification value of the image. **Magnification** is the quality of making an image appear larger than its actual size. For example, a magnification value of 200× indicates that the object in the image appears 200 times larger than the object's actual size. **Resolution** is a measure of the clarity of an image. Both high magnification and good resolution are needed to view the details of extremely small objects clearly. As shown in **Figure 2,** electron microscopes have much higher magnifying and resolving powers than light microscopes.

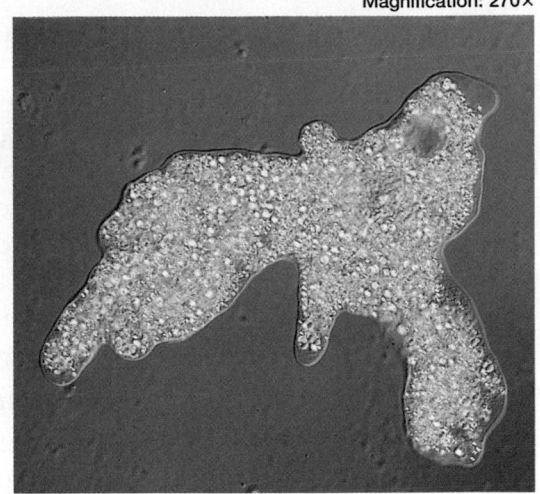

Magnification: 270×

Figure 1 Micrograph. This light micrograph (LM) shows an amoeba.

Figure 2 Magnifying power of microscopes. The scale shows the size range of objects that can be viewed with electron microscopes and light microscopes.

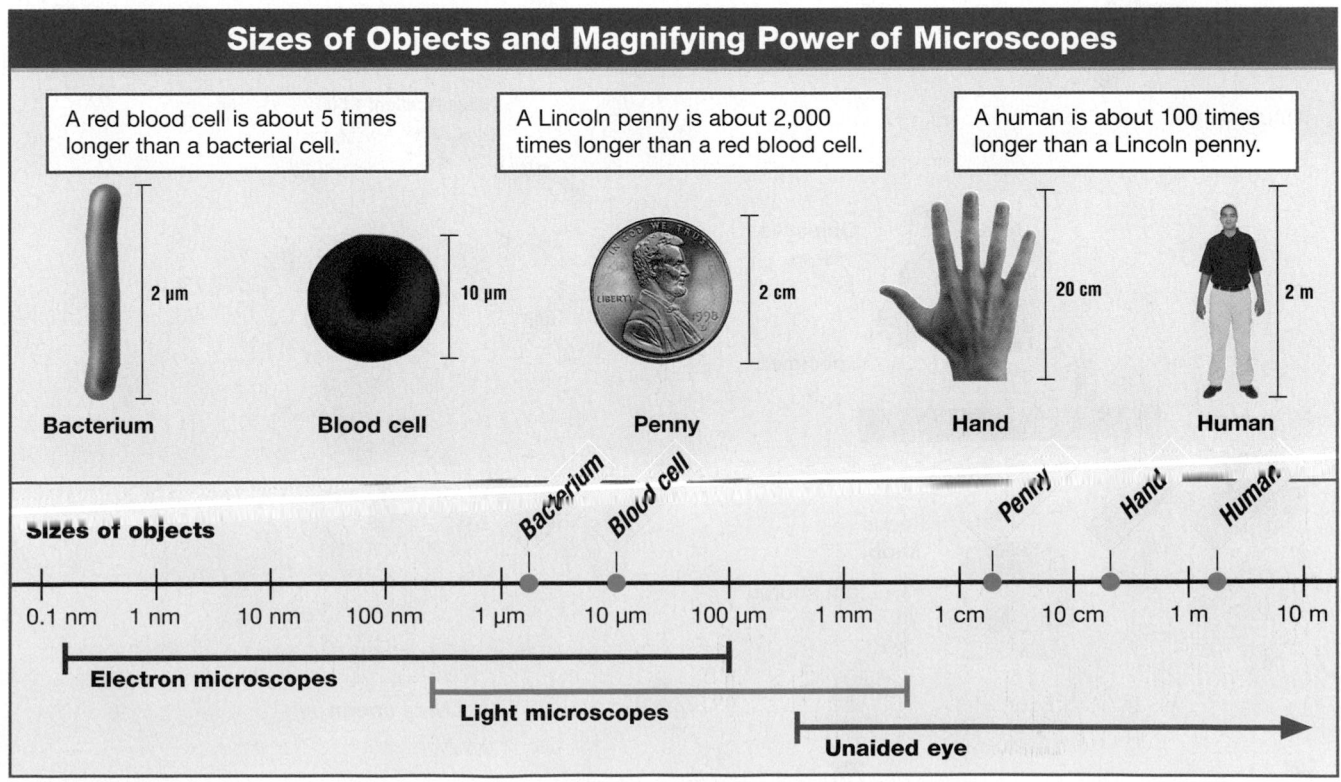

Sizes of Objects and Magnifying Power of Microscopes

A red blood cell is about 5 times longer than a bacterial cell.

A Lincoln penny is about 2,000 times longer than a red blood cell.

A human is about 100 times longer than a Lincoln penny.

2 μm	10 μm	2 cm	20 cm	2 m
Bacterium	Blood cell	Penny	Hand	Human

Sizes of objects

Bacterium · Blood cell · Penny · Hand · Human

0.1 nm | 1 nm | 10 nm | 100 nm | 1 μm | 10 μm | 100 μm | 1 mm | 1 cm | 10 cm | 1 m | 10 m

Electron microscopes

Light microscopes

Unaided eye

Types of Microscopes

Different types of microscopes have different qualities and uses. Microscopes vary in magnification and resolution capabilities, which affect the overall quality of the images they produce. Microscopes also have different limitations. For example, electron microscopes have high magnifying power, but they cannot be used to view living cells. Light microscopes have lower magnifying power, but they can be used to view living cells.

internet connect

www.scilinks.org
Topic: Microscopes
Keyword: HX4122

SCI LINKS. Maintained by the National Science Teachers Association

Compound Light Microscope

Light microscopes that use two lenses are called compound light microscopes. In a typical compound light microscope, such as the one shown in **Figure 3,** a light bulb in the base shines light up through the specimen, which is mounted on a glass slide. The objective lens, closest to the specimen, collects the light, which then travels to the ocular *(AHK yoo luhr)* lens, closest to the viewer's eye. Both lenses magnify the image. Thus, a microscope with a 40× objective lens and a 10× ocular lens produces a total magnification of 400×.

Why not add a third lens and magnify even more? This approach does not work because you cannot distinguish between two objects, or "resolve" them, when they are closer together than a few hundred nm. When the objects are this close, the light beams from the two objects start to overlap!

Figure 3   Compound light microscope

In a compound light microscope, a specimen is mounted on a glass slide and is illuminated with a beam of light from below.

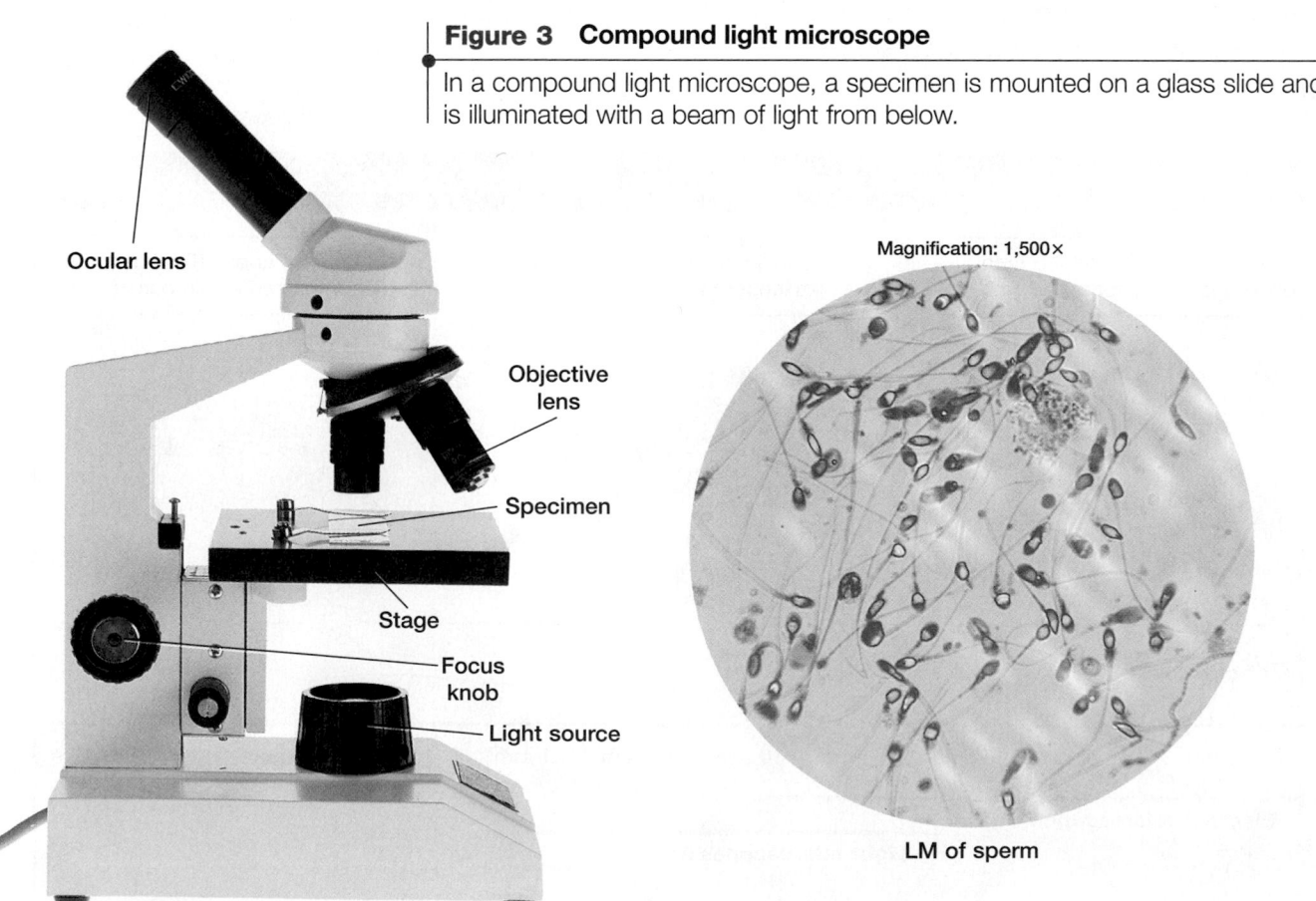

Ocular lens

Objective lens

Specimen

Stage

Focus knob

Light source

Magnification: 1,500×

LM of sperm

The most powerful compound light microscopes have a total magnification of up to 2,000×, which is sufficient for viewing objects as small as 0.5 µm in diameter. For you to see smaller objects, the wavelength of the light beam must be shorter than the wavelength of visible light. Electron beams have a much shorter wavelength than that of visible light, so electron microscopes are much more powerful than light microscopes.

Electron Microscopes

Electron microscopes can magnify an image up to 200,000×, and they can be used to study very small structures inside cells or on cell surfaces. In electron microscopes, both the electron beam and the specimen must be placed in a vacuum chamber so that the electrons in the beam will not bounce off gas molecules in the air. Because living cells cannot survive in a vacuum, they cannot be viewed using electron microscopes.

Transmission electron microscope In a transmission electron microscope, shown in **Figure 4,** the electron beam is directed at a very thin slice of a specimen stained with metal ions. Some structures in the specimen become more heavily stained than others. The heavily stained parts of the specimen absorb electrons, while those that are lightly stained allow electrons to pass through. The electrons that pass through the specimen strike a fluorescent screen, forming an image on the screen. A transmission electron micrograph (TEM), such as the one of sperm cells shown in Figure 4, can reveal a cell's internal structure in fine detail. TEM images are always in black and white. However, with the help of computers, scientists often add artificial colors to make certain structures more visible.

Figure 4 **Transmission electron microscope**

In a transmission electron microscope, electrons pass through a specimen, forming an image of the specimen on a fluorescent screen.

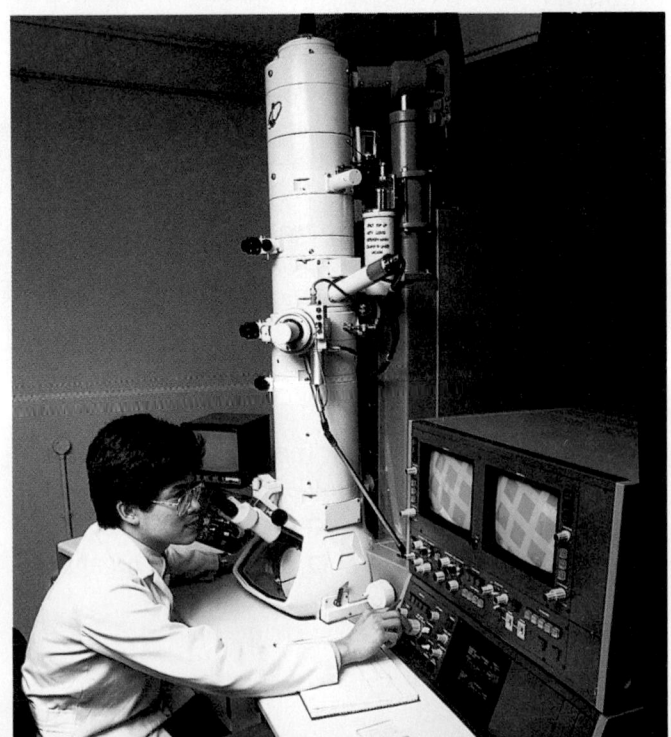

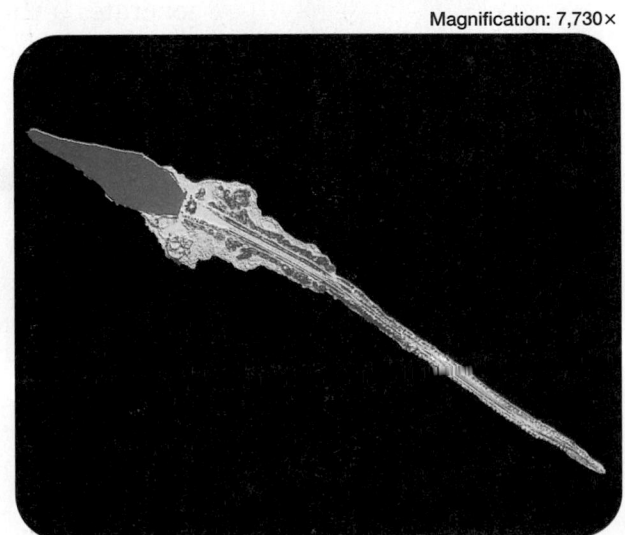

Magnification: 7,730×

TEM of sperm

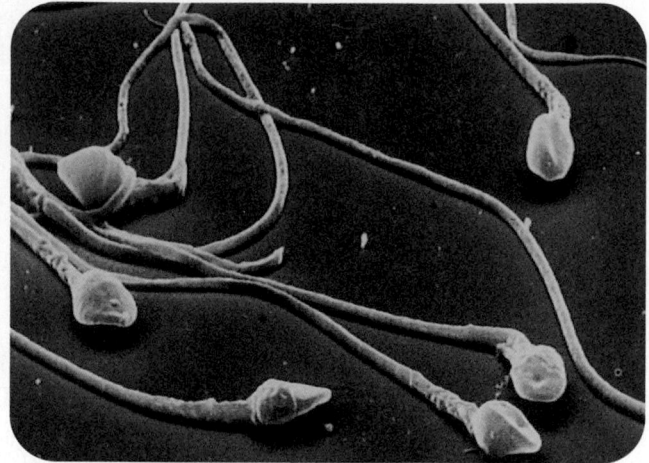

SEM of sperm

Figure 5 Scanning electron microscope

In a scanning electron microscope, electrons bounce off a specimen, forming a three-dimensional image of the specimen on a fluorescent screen.

Scanning electron microscope In a scanning electron microscope, shown in **Figure 5,** the electron beam is focused on a specimen coated with a very thin layer of metal. The electrons that bounce off the specimen form an image on a fluorescent screen. A scanning electron micrograph (SEM) shows three-dimensional images of cell surfaces, such as the image of sperm cells shown in Figure 5. As with the transmission electron microscope, images produced by the scanning electron microscope are black and white, but often they are artificially colored.

Scanning Tunneling Microscope

New video and computer techniques are increasing the resolution and magnification of microscopes. The **scanning tunneling microscope** uses a needle-like probe to measure differences in voltage caused by electrons that leak, or tunnel, from the surface of the object being viewed. A computer tracks the movement of the probe across the object, enabling objects as small as individual atoms to be viewed. The computer generates a three-dimensional image of the specimen's surface. The scanning tunneling microscope can be used to study living organisms.

Section 1 Review

1 **Describe** the relationship between a meter, a millimeter, and a micrometer.

2 **Describe** how magnification and resolution affect the appearance of objects viewed under a microscope.

3 **Compare** the magnifying power of a light microscope with the magnifying power of an electron microscope.

4 **Critical Thinking Recognizing Differences** Explain why electron microscopes cannot be used to view the structure of living cells.

5 **Critical Thinking Comparing Functions** Assume that for the purposes of your investigation, you need detailed images of the internal structure of a bacterium. What type of microscope would you select for that that task? Explain your answer.

6 (**Standardized Test Prep**) The English scientist Robert Hooke used a crude microscope to examine

A electrons

B cork cells

C individual atoms

D single-celled organisms

Cell Features

The Cell Theory

It took scientists more than 150 years to fully appreciate the discoveries of Hooke and Leeuwenhoek. In 1838, the German botanist Mattias Schleiden concluded that cells make up not only the stems and roots but every part of a plant. A year later, the German zoologist Theodor Schwann claimed that animals are also made of cells. In 1858, Rudolph Virchow, a German physician, determined that cells come only from other cells. The observations of Schleiden, Schwann, and Virchow form the **cell theory,** which has three parts:

1. All living things are made of one or more cells.

2. Cells are the basic units of structure and function in organisms.

3. All cells arise from existing cells.

Cell Size

Small cells function more efficiently than large cells. There are about 100 trillion cells in the human body, most ranging from 5 µm to 20 µm in diameter. What is the advantage of having so many tiny cells instead of fewer large ones? All substances that enter or leave a cell must cross that cell's surface. If the cell's surface area–to-volume ratio is too low, substances cannot enter and leave the cell in numbers large enough to meet the cell's needs. Small cells can exchange substances more readily than large cells because small objects have a higher surface area–to-volume ratio than larger objects, as shown in **Table 2.** As a result, substances do not need to travel as far to reach the center of a smaller cell.

Objectives

- **List** the three parts of the cell theory.

- **Determine** why cells must be relatively small.

- **Compare** the structure of prokaryotic cells with that of eukaryotic cells.

- **Describe** the structure of cell membranes.

Key Terms

cell theory
cell membrane
cytoplasm
cytoskeleton
ribosome
prokaryote
cell wall
flagellum
eukaryote
nucleus
organelle
cilium
phospholipid
lipid bilayer

Table 2 Relationship Between Surface Area and Volume

	Side length	Surface area	Volume	Surface area/ volume ratio
1 mm	1 mm	6 mm^2	1 mm^3	6:1
2 mm	2 mm	24 mm^2	8 mm^3	3:1
4 mm	4 mm	96 mm^2	64 mm^3	3:2

Common Features of Cells

Cells share common structural features, including an outer boundary called the **cell membrane.** The cell membrane encloses the cell and separates the cell interior, called the **cytoplasm** (*SITE oh plaz uhm*), from its surroundings. The cell membrane also regulates what enters and leaves a cell—including gases, nutrients, and wastes. Within the cytoplasm are many structures, often suspended in a system of microscopic fibers called the **cytoskeleton.** Most cells have ribosomes. **Ribosomes** (*RIE buh sohmz*) are the cellular structures on which proteins are made. All cells also have DNA, which provides instructions for making proteins, regulates cellular activities, and enables cells to reproduce. Some specialized cells such as red blood cells, however, later lose their DNA.

Calculating Surface Area and Volume

Background

You can improve your understanding of the relationship between a cell's surface area and its volume by practicing with the large cube in Table 2.

1. Find the total surface area of the cube.

- *side length* (l) = 4 mm
- *surface area of one side* = $l \times l = l^2$
- *surface area of one side* (l^2) = 4 mm × 4 mm = 16 mm²
- *total surface area* = $6 \times l^2$ = 6 × 16 mm² = 96 mm²

2. Calculate the volume of the cube.

- *height* $(h) = l$ = 4 mm
- *volume* = $l^2 \times h$ = 16 mm² × 4 mm = 64 mm³

Magnification: 230×

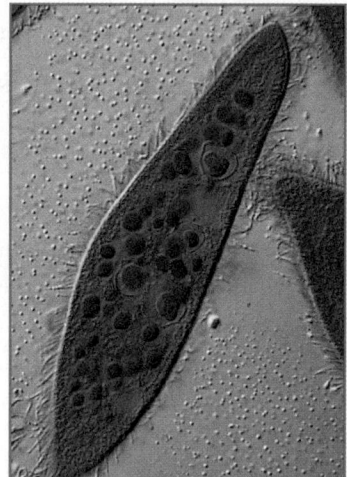

Paramecium (SEM)

3. Determine the surface area–to-volume ratio. A ratio compares two numbers by dividing one number by the other. A ratio can be expressed in three ways:

in words	as a fraction	with a colon
x to y	$\dfrac{x}{y}$	$x{:}y$

For the surface area–to-volume ratio, divide total surface area by volume.

$$\frac{total\ surface\ area}{volume} = \frac{96}{64}$$

Divide both numbers by their greatest common factor:

$$\frac{(96 \div 32)}{(64 \div 32)} = \frac{3}{2}$$

Analysis

1. Calculate the surface area–to-volume ratio of the cube with a side length of 2 mm in Table 2.

2. Calculate the surface area–to-volume ratio of the cube with a side length of 1 mm in Table 2.

3. Critical Thinking Relating Concepts How does the flatness of the single-celled *Paramecium* shown above affect the cell's surface area–to-volume ratio?

Prokaryotes

The smallest and simplest cells are prokaryotes. A **prokaryote** *(proh KAIR ee oht)* is a single-celled organism that lacks a nucleus and other internal compartments. Without separate compartments to isolate materials, prokaryotic cells cannot carry out many specialized functions. Early prokaryotes lived at least 3.5 billion years ago. For nearly 2 billion years, prokaryotes were the only organisms on Earth. They were very simple and small (1–2 μm in diameter). Like their ancestors, modern prokaryotes are also very small (1–15 μm). The familiar prokaryotes that cause infection and cause food to spoil belong to a subset of all prokaryotes that is commonly called *bacteria*.

Characteristics of Prokaryotes

Prokaryotes can exist in a broad range of environmental conditions. Many prokaryotes, including some bacteria that cause infection in humans, grow and divide very rapidly. Some prokaryotes do not need oxygen to survive. Other prokaryotes cannot survive in the presence of oxygen. Some prokaryotes can even make their own food.

The cytoplasm of a prokaryotic cell includes everything inside the cell membrane. As **Figure 6** shows, a prokaryote's enzymes and ribosomes are free to move around in the cytoplasm because there are no internal structures that divide the cell into compartments. In prokaryotes, the genetic material is a single, circular molecule of DNA. This loop of prokaryotic DNA is often located near the center of the cell, suspended within the cytoplasm.

Figure 6   Prokaryotes.
Prokaryotic cells have little internal structure. Many also have a capsule and flagella.

Magnification: 61,850×

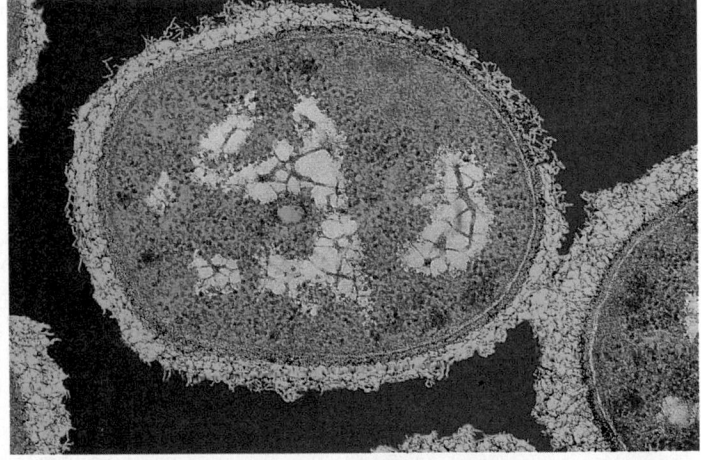

Prokaryotic cells have a **cell wall** surrounding the cell membrane that provides structure and support. The cells of fungi and plants also have cell walls; only animal cells and some protists lack cell walls. Prokaryotes lack an internal supporting skeleton, so they depend on a strong cell wall to give the cell shape. A prokaryotic cell wall is made of strands of polysaccharides connected by short chains of amino acids. Some prokaryotic cell walls are surrounded by a structure called a capsule, which is also composed of polysaccharides. The capsule enables prokaryotes to cling to almost anything, including teeth, skin, and food.

Many prokaryotes have **flagella** *(fluh JEL uh)*, which are long, threadlike structures that protrude from the cell's surface and enable movement. Prokaryotic flagella rotate, propelling the organism through its environment at speeds of up to 20 cell lengths per second. Figure 6 shows a prokaryote with several flagella.

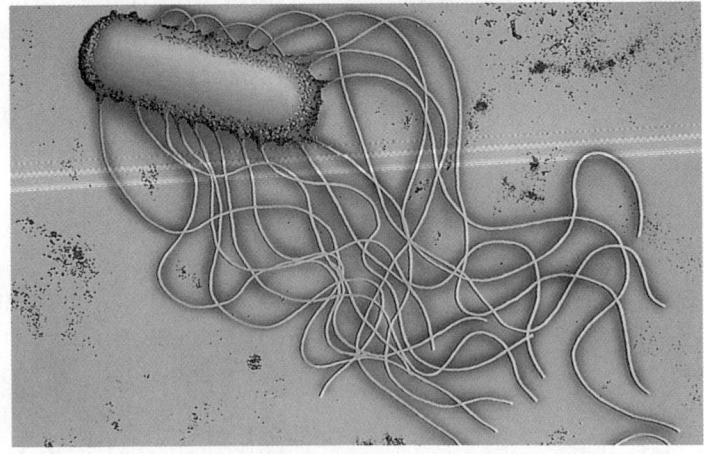

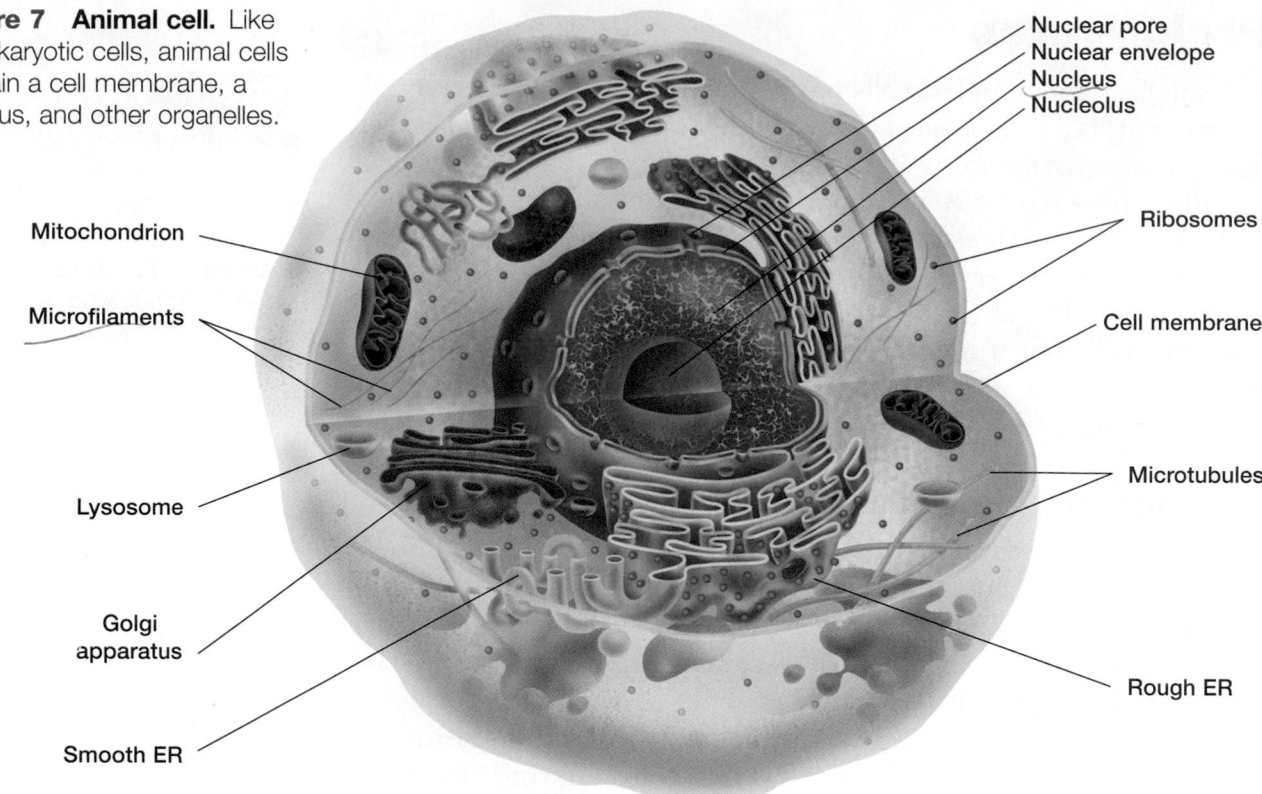

Figure 7 Animal cell. Like all eukaryotic cells, animal cells contain a cell membrane, a nucleus, and other organelles.

Nuclear pore
Nuclear envelope
Nucleus
Nucleolus

Ribosomes

Cell membrane

Microtubules

Rough ER

Mitochondrion

Microfilaments

Lysosome

Golgi apparatus

Smooth ER

Eukaryotic Cells

The first cells with internal compartments were primitive eukaryotic cells, which evolved about 2.5 billion years ago. A **eukaryote** *(yoo KAIR ee oht)* is an organism whose cells have a nucleus. The **nucleus** *(NOO klee uhs)* is an internal compartment that houses the cell's DNA. Other internal compartments, or organelles, enable eukaryotic cells to function in ways different from prokaryotes. An **organelle** is a structure that carries out specific activities in the cell.

The major organelles in an animal cell are shown in **Figure 7.** The cytoplasm includes everything inside the cell membrane but outside the nucleus. A complex system of internal membranes connects some organelles within the cytoplasm. These membranes provide channels that guide the distribution of substances within the cell. The membranes also form envelopes called *vesicles* that move proteins and other molecules from one organelle to another.

Many single-celled eukaryotes use flagella for movement. Short hairlike structures called **cilia** *(SIL ee uh)* protrude from the surface of some eukaryotic cells. Flagella or cilia propel some cells through their environment. In other cells, cilia and flagella move substances across the cell's surface. For example, cilia on cells of the human respiratory system, shown in **Figure 8,** sweep mucus and other debris out of the lungs.

A web of protein fibers, shown in **Figure 9,** makes up the cytoskeleton. The cytoskeleton holds the cell together and keeps the cell's membranes from collapsing. The fluid surrounding the cytoplasm's organelles, internal membranes, and cytoskeleton fibers is called the cytosol.

Figure 8 Cilia. Cilia on cells lining the respiratory system remove debris from air passages.

The Cytoskeleton

The cytoskeleton provides the interior framework of an animal cell, much as your skeleton provides the interior framework of your body. The cytoskeleton is composed of an intricate network of protein fibers anchored to the inside of the plasma membrane. By linking one region to another, they support the shape of the cell, much as steel beams anchor the sides of a building to one another. Other fibers attach the nucleus and other organelles to fixed locations in the cell. Because protein fibers are too small for a light microscope to reveal, biologists visualize the cytoskeleton by attaching fluorescent dyes to antibodies. An antibody is an immune system protein specialized to bind to one particular kind of molecule—in this case—cytoskeleton proteins. When the cell is examined under fluorescent light, the fibers glow because of the fluorescent antibody attached to them.

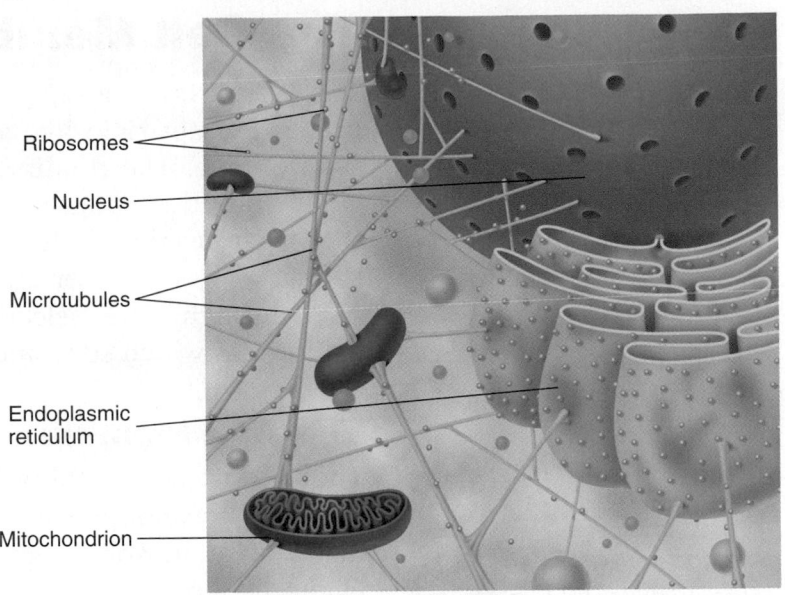

Figure 9 The cytoskeleton. The cytoskeleton's network of protein fibers anchors cells organelles and other components of the cytoplasm.

There are three different kinds of cytoskeleton fibers: (1) long, slender microfilaments made of the protein actin, (2) hollow tubes called *microtubules* made of the protein tubulin, and (3) thick ropes of protein called *intermediate fibers*.

Actin Fibers The actin fibers of the cytoskeleton form a network just beneath the cell surface that is anchored to membrane proteins embedded within the cell membrane. By contracting or expanding, the actin fibers play a major role in determining the shape of animal cells by pulling the plasma membrane in some places and pushing it out in others. If you examine the surface of a protist such as the one shown in **Figure 10,** you will find it alive with motion. Tiny projections extend out from the surface like fingers. Each is a temporary projection of the plasma membrane that shoots out and then retracts.

Microtubules Microtubules within the cytoskeleton act as a highway system for the transportation of information from the nucleus to different parts of the cell. RNA molecules are transported along microtubular "rails" that extend through the interior of the cell like train tracks. The RNA molecules, in complexes with proteins, are attached to so-called motor proteins that chug along microtubules like locomotives on tracks. The motor proteins drag the RNA-protein complexes along with them like freight cars.

Intermediate Fibers The intermediate fibers of the cytoskeleton provide a frame on which ribosomes and enzymes can be confined to particular regions of the cell. The cell can organize complex metabolic activities efficiently by anchoring particular enzymes near one another.

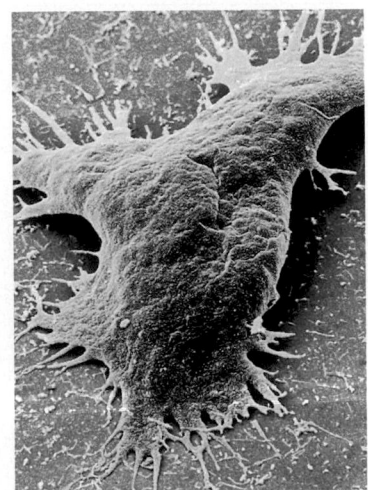

Figure 10 Cytoskeletal projections. The multiple spikes on the surface of this marine amoeba are projections of the cytoskeleton stretching the cell membrane outward.

The Cell Membrane

The cytoplasm of a cell is contained by its membrane. Cell membranes are not rigid like an eggshell. Rather, they are fluid like a soap bubble. The fluidity of cell membranes is caused by lipids, which form the foundation of membranes. The lipids form a barrier that separates the inside of the cell from the outside of the cell. This barrier allows only certain substances in the cell's environment to pass through. This selective permeability of the cell membrane determines which substances enter and leave the cell.

The Cell Membrane as a Barrier

The selective permeability of the cell membrane is caused mainly by the way phospholipids interact with water. A **phospholipid** is a lipid made of a phosphate group and two fatty acids. As shown in **Figure 11,** a phospholipid has both a polar "head" and two nonpolar "tails." You may recall that the polar ends of water molecules will form weak bonds with other polar substances. The head of a phospholipid, which contains a phosphate group, is polar and is attracted to water. In contrast, the two fatty acids, or tails, are nonpolar and therefore are repelled by water.

In a cell membrane, the phospholipids are arranged in a double layer called a **lipid bilayer,** as shown in Figure 11. The nonpolar tails of the phospholipids make up the interior of the lipid bilayer. Because water both inside and outside the cell repels the nonpolar tails, they are forced to the inside of the lipid bilayer. Ions and most polar molecules, including sugars and some proteins, are repelled by the nonpolar interior of the lipid bilayer. The lipid bilayer allows lipids and substances that dissolve in lipids to pass through.

Figure 11   Lipid bilayer

Cell membranes are made of a double layer of phospholipids, called a lipid bilayer.

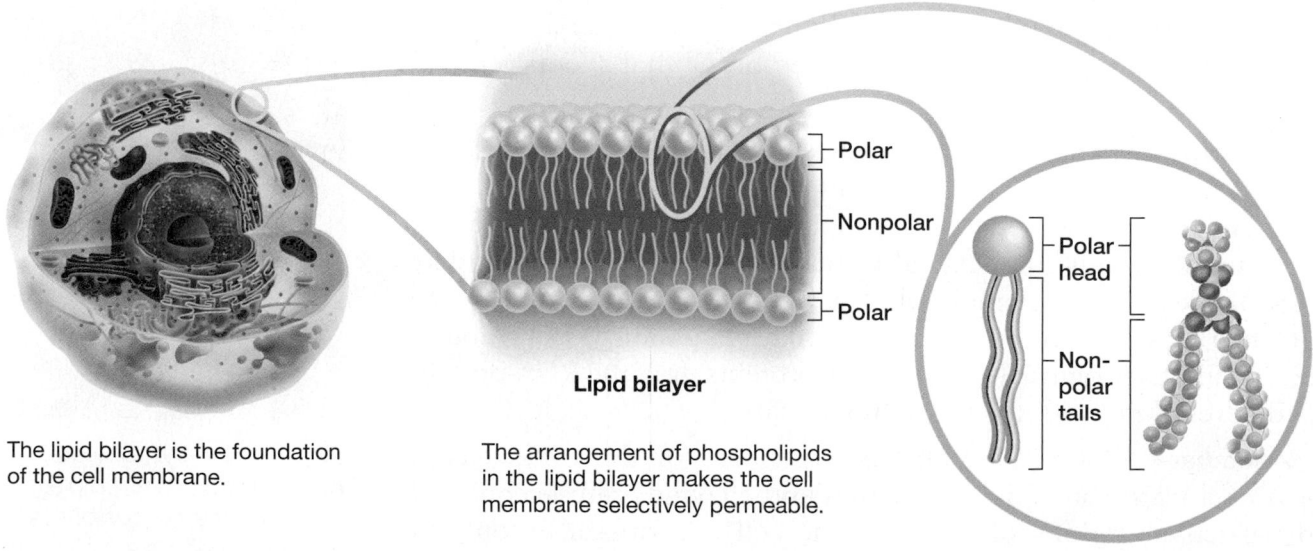

The lipid bilayer is the foundation of the cell membrane.

The arrangement of phospholipids in the lipid bilayer makes the cell membrane selectively permeable.

A phospholipid's "head" is polar, and its two fatty acid "tails" are nonpolar.

Figure 12   Membrane proteins

The cell membrane contains various proteins with specialized functions.

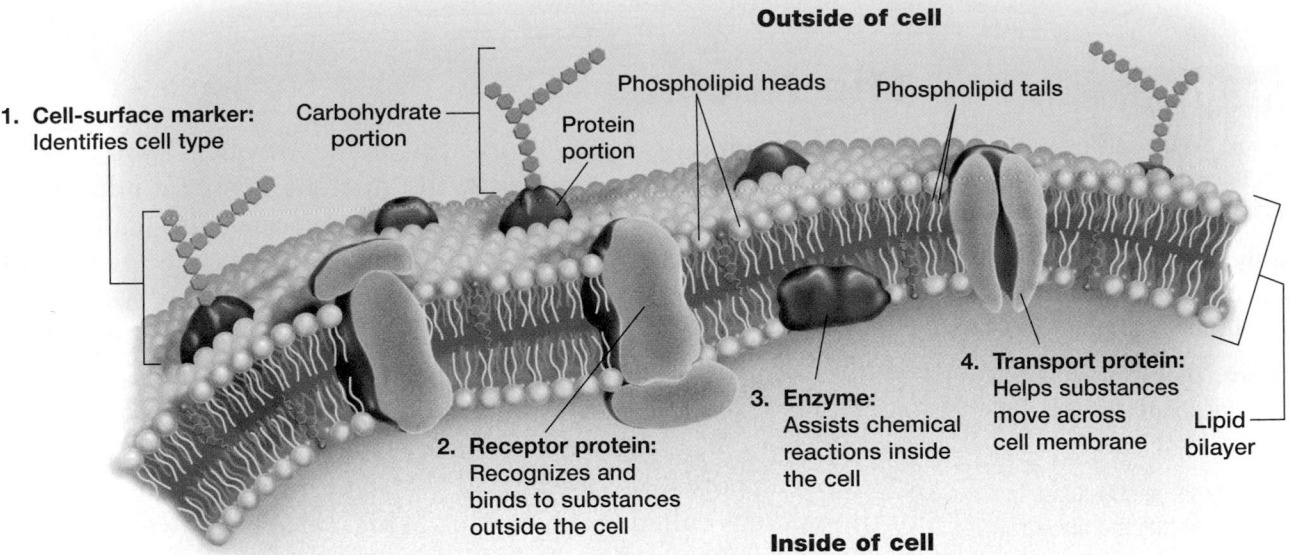

1. **Cell-surface marker:** Identifies cell type
   Carbohydrate portion
   Protein portion

   Outside of cell

   Phospholipid heads
   Phospholipid tails

4. **Transport protein:** Helps substances move across cell membrane
   Lipid bilayer

3. **Enzyme:** Assists chemical reactions inside the cell

2. **Receptor protein:** Recognizes and binds to substances outside the cell

   Inside of cell

Membrane Proteins

Various proteins are located in the lipid bilayer of a cell membrane. What keeps these proteins within the lipid bilayer? You may recall that proteins are made of amino acids and that some amino acids are polar, while others are nonpolar. The nonpolar part of a membrane protein is attracted to the interior of the lipid bilayer but is repelled by the water on either side of the lipid bilayer. In contrast, the polar parts of the protein are attracted to the water on either side of the lipid bilayer. This attraction helps to hold the protein in the lipid bilayer. The motion and fluidity of phospholipids enable some membrane proteins to move around within the lipid bilayer.

As shown in **Figure 12,** cell membranes contain different types of proteins. Marker proteins attached to a carbohydrate on the cell's surface advertise cell type—such as a liver cell or a heart cell. Receptor proteins bind specific substances, such as signal molecules, outside the cell. Enzymes embedded in the cell membrane are involved in important biochemical reactions in the cell. Transport proteins aid the movement of substances into and out of the cell.

Section 2 Review

1. **Describe** the importance of the surface area–to-volume ratio of a cell.

2. **Compare** the structure of a eukaryotic cell with that of a prokaryotic cell.

3. **Critical Thinking Comparing Functions** Describe the functions of two types of cell-membrane proteins.

4. **Analyze** the three parts of the cell theory and describe two observations of early scientists that support it.

5. **Standardized Test Prep** A bacterium that lost its flagella would be unable to
   A move
   B divide
   C make proteins
   D maintain its shape

Cell Organelles

- **Describe** the role of the nucleus in cell activities.

- **Analyze** the role of internal membranes in protein production.

- **Summarize** the importance of mitochondria in eukaryotic cells.

- **Identify** three structures in plant cells that are absent from animal cells.

Key Terms

endoplasmic reticulum
vesicle
Golgi apparatus
lysosome
mitochondrion
chloroplast
central vacuole

The Nucleus

Most functions of a eukaryotic cell are controlled by the cell's nucleus. As shown in **Figure 13,** the nucleus is surrounded by a double membrane called the nuclear envelope, also called the nuclear membrane. The nuclear envelope is made of two lipid bilayers that separate the nucleus from the cytoplasm.

Scattered over the surface of the nuclear envelope are many small channels through the envelope called nuclear pores. Substances that are made in the nucleus, including RNA-ribosomal protein complexes, move into the cytoplasm by passing through the nuclear pores. Ribosomes are partially assembled in a region of the nucleus called the nucleolus, which is also shown in Figure 13. Recall from Section 2 that ribosomes are the structures on which proteins are made.

The hereditary information of a eukaryotic cell is coded in the cell's DNA, most of which is stored in the nucleus. Eukaryotic DNA is wound tightly around proteins. Most of the time, DNA exists as elongated and thin strands, which appear as a dark mass under magnification. When a cell is about to divide, however, the DNA strands, called chromosomes, wind up into a more compact form and appear as dense, rod-shaped structures. The number of chromosomes in a eukaryotic cell differs between species. Human body cells have 46 chromosomes, while the cells of garden peas have 14 chromosomes. You will learn more about DNA and chromosomes later in this book.

Figure 13   Nucleus

The nucleus is surrounded by a double membrane called the nuclear envelope.

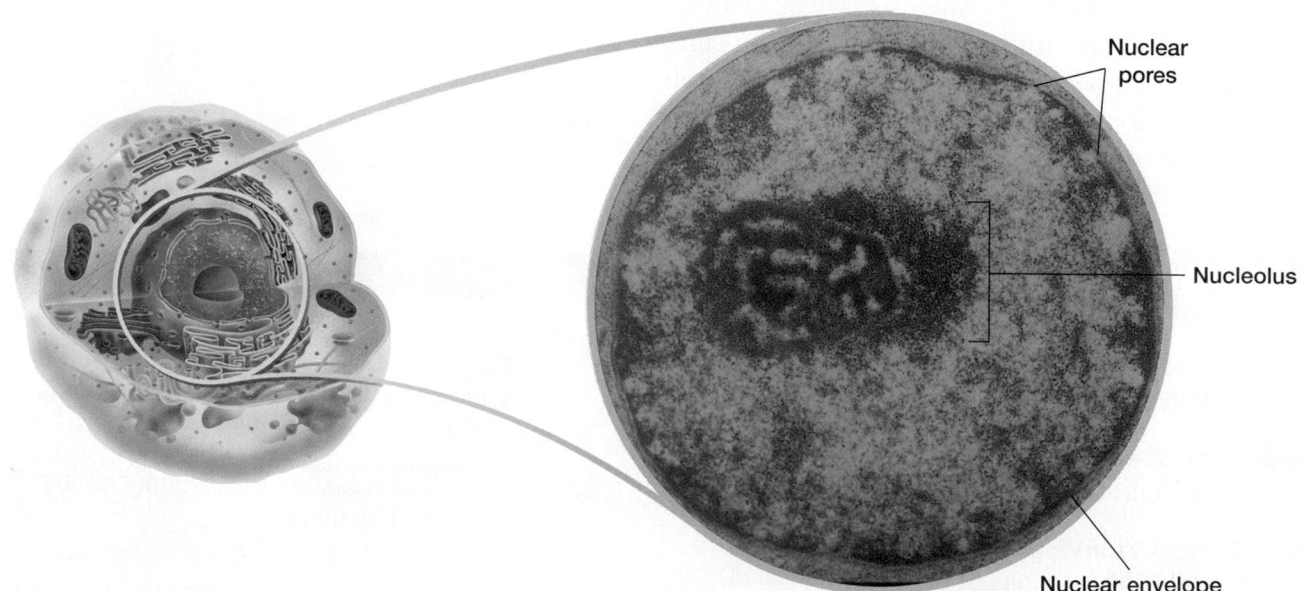

Nuclear pores

Nucleolus

Nuclear envelope

Ribosomes and the Endoplasmic Reticulum

Unlike prokaryotic cells, eukaryotic cells have a system of internal membranes that play an essential role in the processing of proteins. Cells make proteins on ribosomes. Each ribosome is made of dozens of different proteins as well as RNA. Some of the ribosomes in a eukaryotic cell are suspended in the cytosol, as they are in prokaryotic cells. These "free" ribosomes make proteins that remain inside the cell, such as proteins used to build new organelles.

☑ internet connect

www.scilinks.org
Topic: Proteins
Keyword: HX4151

SC*i*LINKS. Maintained by the National Science Teachers Association

Production of Proteins

Proteins that are exported from the cell, such as some signal molecules, are made on the ribosomes that lie on the surface of the endoplasmic reticulum, shown in **Figure 14.** The **endoplasmic reticulum** *(ehn doh PLAZ mihk rih TIHK yuh luhm)*, or ER, is an extensive system of internal membranes that move proteins and other substances through the cell. Like the cell membrane, the membranes of the ER are made of a lipid bilayer with embedded proteins.

The part of the ER with attached ribosomes is called rough ER because it has a rough appearance when viewed in the electron microscope. The rough ER helps transport the proteins that are made by its attached ribosomes. As each protein is made, it crosses the ER membrane and enters the ER. The portion of the ER that contains the completed protein then pinches off to form a vesicle. A **vesicle** is a small, membrane-bound sac that transports substances in cells. Because certain proteins are enclosed inside vesicles, these proteins are kept separate from proteins that are produced by free ribosomes in the cytoplasm.

The rest of the ER is called smooth ER because it lacks ribosomes and thus appears smooth when viewed in the electron microscope. The smooth ER performs various functions, such as making lipids and breaking down toxic substances.

Figure 14 Endoplasmic reticulum

The ER moves proteins and other substances within eukaryotic cells.

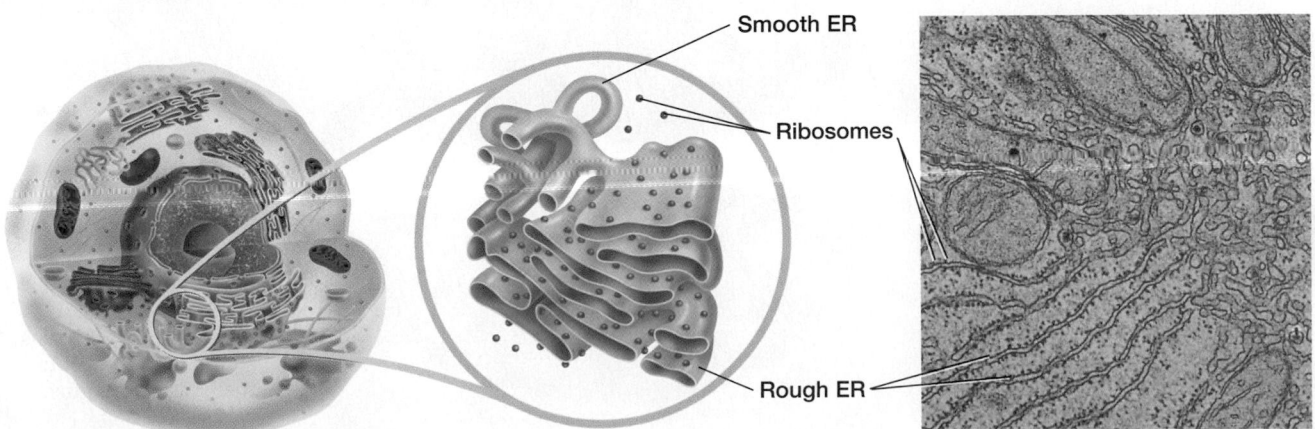

Smooth ER

Ribosomes

Rough ER

Packaging and Distribution of Proteins

Vesicles that contain newly made proteins move through the cytoplasm from the ER to an organelle called the Golgi apparatus. The **Golgi** (*GOHL jee*) **apparatus** is a set of flattened, membrane-bound sacs that serves as the packaging and distribution center of the cell. Enzymes inside the Golgi apparatus modify the proteins that are received in vesicles from the ER. The modified proteins are then enclosed in new vesicles that bud from the surface of the Golgi apparatus. Other vesicles include **lysosomes** (*LIE seh sohms*), which are small, spherical organelles that contain the cell's digestive enzymes. The ER, the Golgi apparatus, and lysosomes work together in the production, packaging, and distribution of proteins, as summarized in **Figure 15.**

Step ❶ Ribosomes make proteins on the rough ER. The proteins are packaged into vesicles.

Step ❷ The vesicles transport the newly made proteins from the rough ER to the Golgi apparatus.

Step ❸ In the Golgi apparatus, proteins are processed and then packaged into new vesicles.

Step ❹ Many of these vesicles move to the cell membrane and release their contents outside the cell.

Step ❺ Other vesicles, including lysosomes, remain within the cytoplasm. Lysosomes digest and recycle the cell's used components by breaking down proteins, nucleic acids, lipids, and carbohydrates.

Study *TIP*

● **Interpreting Graphics**
As you read, use Steps 1–5 in the text, shown in red, to help you follow the same numbered steps shown in Figure 15.

Figure 15

BIOgraphic

Processing of Proteins

Proteins are processed by an internal system of membranes.

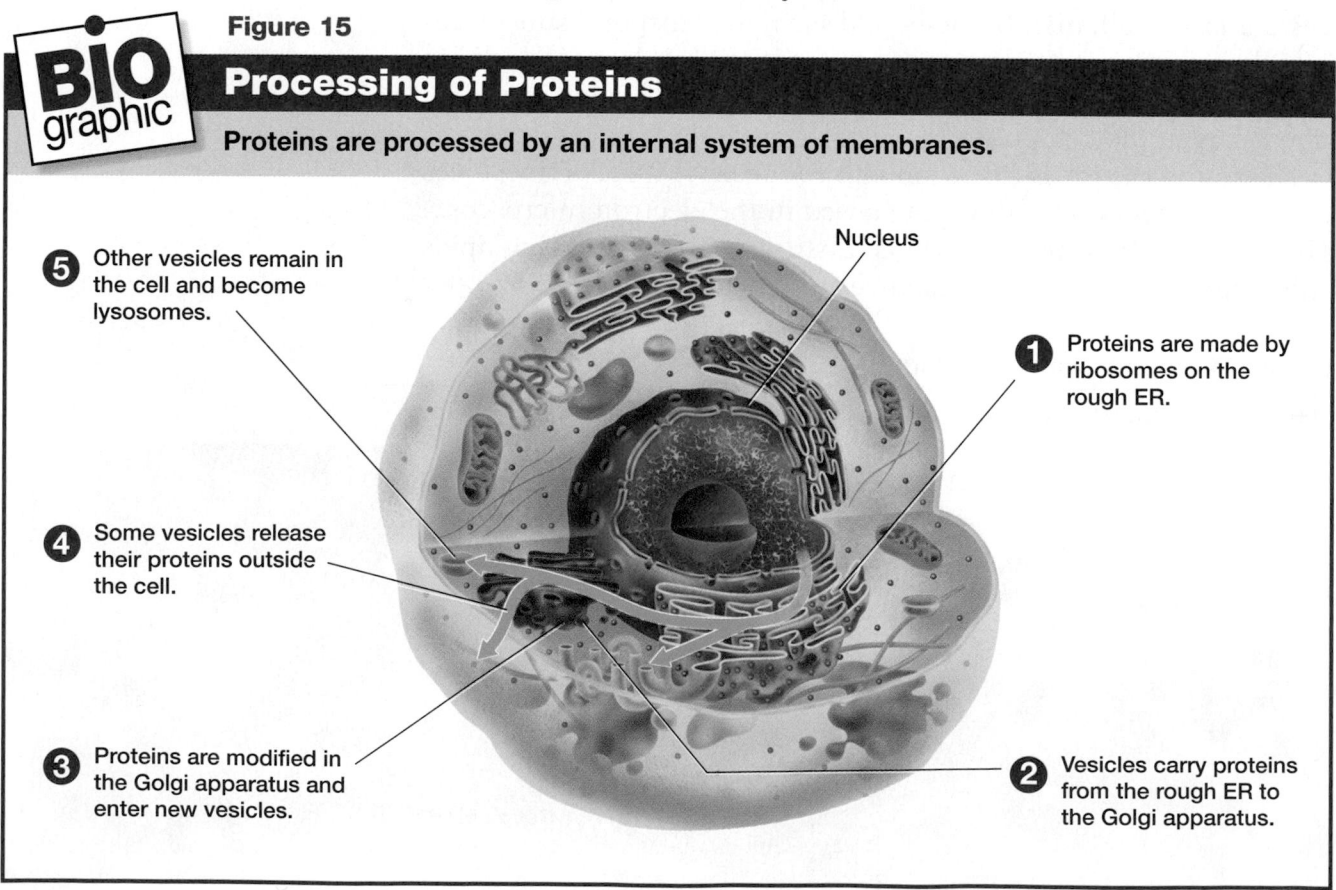

❺ Other vesicles remain in the cell and become lysosomes.

Nucleus

❶ Proteins are made by ribosomes on the rough ER.

❹ Some vesicles release their proteins outside the cell.

❸ Proteins are modified in the Golgi apparatus and enter new vesicles.

❷ Vesicles carry proteins from the rough ER to the Golgi apparatus.

Mitochondria

Nearly all eukaryotic cells contain many **mitochondria** (*miet uh KAHN dree uh*), like the one shown in **Figure 16**. A mitochondrion is an organelle that harvests energy from organic compounds to make ATP, the main energy currency of cells. Although some ATP is made in the cytosol, most of a cell's ATP is made inside mitochondria. Cells that have a high energy requirement, such as a muscle cell, may contain hundreds or thousands of mitochondria. Figure 16 shows that a mitochondrion has two membranes. The outer membrane is smooth. The inner membrane is greatly folded, however, and its surface area is large. The two membranes form two compartments, one inside and one outside the mitochondrion's inner membrane. It is here that the chemical reactions that produce ATP during cell metabolism take place.

Mitochondrial DNA

The nucleus is not the only organelle in the cell that contains nucleic acids. Mitochondria also have DNA and ribosomes, and mitochondria make some of their own proteins. However, most mitochondrial proteins are made by free ribosomes in the cytosol. Mitochondrial DNA is independent of nuclear DNA and similar to the circular DNA of prokaryotic cells. This fact supports the widely accepted theory that primitive prokaryotes are the ancestors of mitochondria. You will learn more about the origin of mitochondrial DNA later in this book.

Figure 16 Mitochondrion

In a eukaryotic cell, mitochondria make most of the ATP.

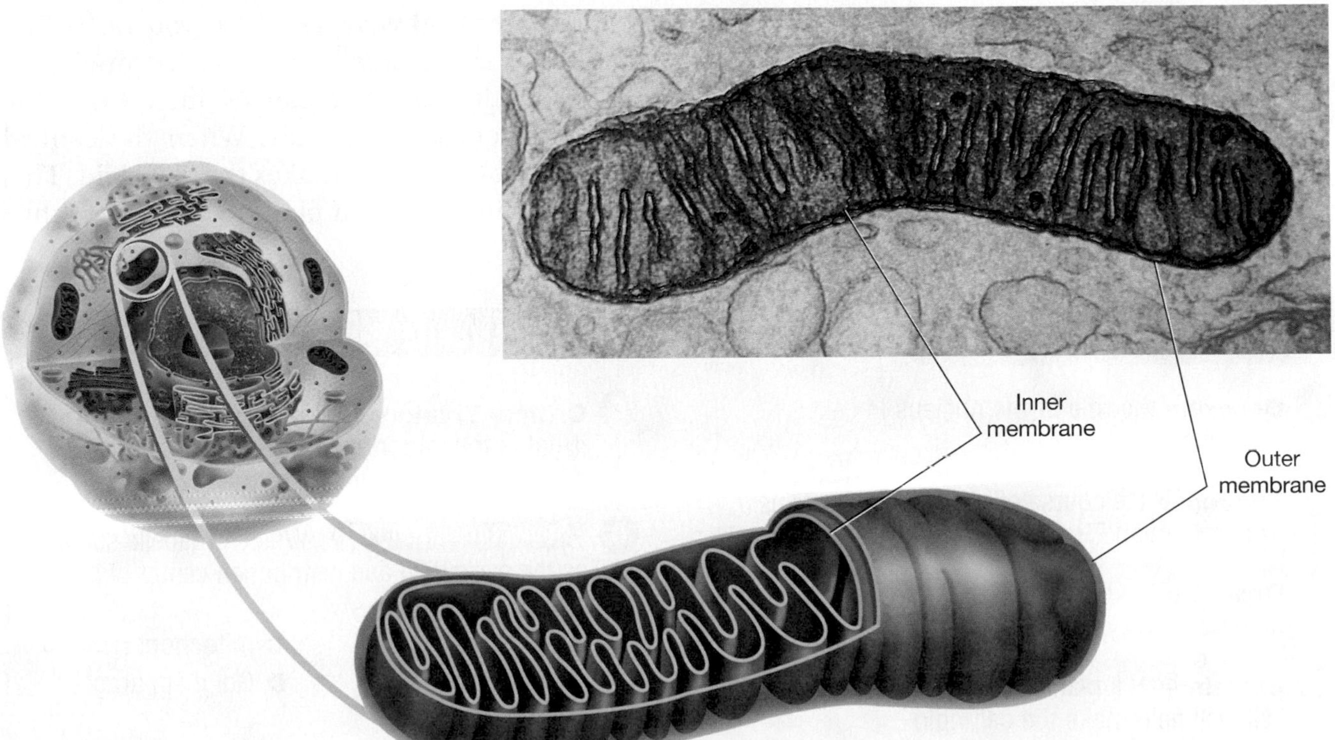

Inner membrane

Outer membrane

Structures of Plant Cells

The organelles described in this section are found in both animal cells and plant cells. However, plant cells have three additional structures that are not found in animal cells, shown in **Figure 17.**

Unique Features of Plant Cells

Cell wall The cell membrane of a plant cell is surrounded by a thick cell wall, composed of proteins and carbohydrates, including the polysaccharide cellulose. The cell wall helps support and maintain the shape of the cell, protects the cell from damage, and connects it with adjacent cells.

Chloroplasts Plant cells contain one or more **chloroplasts.** Chloroplasts are organelles that use light energy to make carbohydrates from carbon dioxide and water. Chloroplasts are found not only in plants but also in a wide variety of eukaryotic algae, such as seaweed. Chloroplasts, along with mitochondria, supply much of the energy needed to power the activities of plant cells. Like mitochondria, chloroplasts are surrounded by two membranes, contain their own DNA, and are thought to be the descendents of ancient prokaryotic cells.

Central vacuole As shown in Figure 17, much of a plant cell's volume is taken up by a large, membrane-bound space called the **central vacuole** *(VAK yoo ohl).* The central vacuole stores water and may contain many substances, including ions, nutrients, and wastes. When the central vacuole is full, it makes the cell rigid. This rigidity enables a plant to stand upright.

Figure 17 Plant cell. Plant cells have a cell wall, chloroplasts, and a large central vacuole (shown in blue).

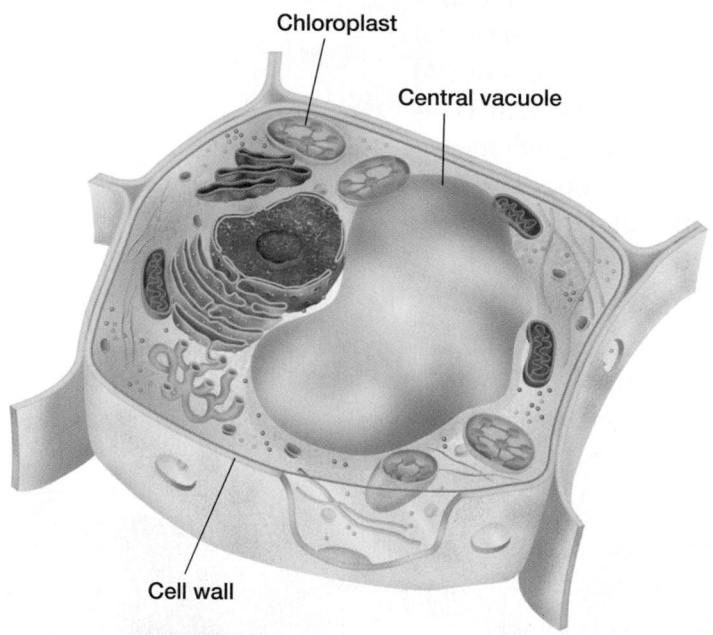

Chloroplast

Central vacuole

Cell wall

Section 3 Review

1. **Describe** the role of the nucleus in cell activities.

2. **Sequence** the course of newly made proteins from the rough ER to the outside of the cell.

3. **Describe** the role of mitochondria in the metabolism of eukaryotic cells.

4. **Explain** how a plant cell's central vacuole and cell wall help make the cell rigid.

5. **Critical Thinking Inferring Relationships** What is the importance of a cell enclosing its digestive enzymes inside lysosomes?

6. **Standardized Test Prep** Which organelle serves as the packaging and distribution center of a eukaryotic cell?
    A nucleus   C mitochondrion
    B lysosome   D Golgi apparatus

 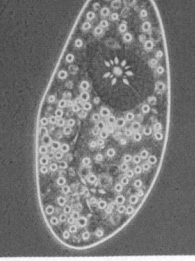

Study ZONE — CHAPTER HIGHLIGHTS

Key Concepts

1 Looking at Cells

- Microscopes enable biologists to examine the details of cell structure and to understand how organisms function.
- Scientists use the metric system to measure the size of objects.
- Light microscopes have a low magnification and can be used to examine living cells.
- Electron microscopes have a high magnification but cannot be used to examine living cells.
- The scanning tunneling microscope uses a computer to generate a three-dimensional image of an object.

2 Cell Features

- The cell theory has three parts.
- Small cells function more efficiently than large cells because small cells have a higher surface-area-to-volume ratio than large cells.
- All cells have a cell membrane, cytoplasm, ribosomes, and DNA.
- Prokaryotic cells lack internal compartments.
- Eukaryotic cells have a nucleus and other organelles, as well as a cytoskeleton of microscopic protein fibers.
- The lipid bilayer of a cell membrane is made of a double layer of phospholipid molecules.
- Proteins in cell membranes include enzymes, receptor proteins, transport proteins, and cell-surface markers.

3 Cell Organelles

- The nucleus of a eukaryotic cell directs the cell's activities and stores DNA.
- In eukaryotic cells, an internal membrane system produces, packages, and distributes proteins.
- Mitochondria harvest energy from organic compounds to ATP.
- Lysosomes digest and recycle a cell's used components.
- Plant cells have three structures that animal cells lack: a cell wall, chloroplasts, and a central vacuole.

Key Terms

Section 1

light microscope (51)
electron microscope (51)
magnification (51)
resolution (51)
scanning tunneling
  microscope (54)

Section 2

cell theory (55)
cell membrane (56)
cytoplasm (56)
cytoskeleton (56)
ribosome (56)
prokaryote (57)
cell wall (57)
flagellum (57)
eukaryote (58)
nucleus (58)
organelle (58)
cilium (58)
phospholipid (60)
lipid bilayer (60)

Section 3

endoplasmic reticulum (63)
vesicle (63)
Golgi apparatus (64)
lysosome (64)
mitochondrion (65)
chloroplast (66)
central vacuole (66)

 **Unit 1**—Use this unit to review the key concepts and terms in this chapter.

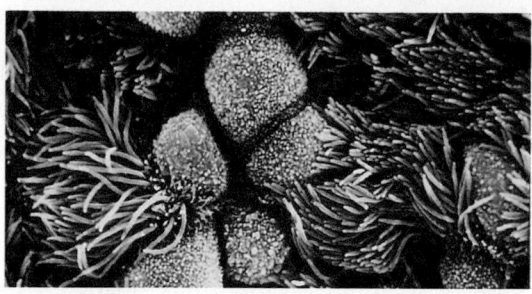

Understanding Key Ideas

1. The main advantage of the transmission electron microscope is that it allows the viewer to see
   a. three-dimensional images of cell surfaces.
   b. the organelles of living cells.
   c. a cell's internal structure in fine detail.
   d. the actual colors of a cell's components.

2. The maximum size of a cell is determined by the ratio between the cell's
   a. surface area and volume.
   b. volume and organelles.
   c. organelles and cytoplasm.
   d. cytoplasm and nucleus.

3. Eukaryotic cells differ from prokaryotic cells in that eukaryotic cells
   a. lack organelles.
   b. have DNA but not ribosomes.
   c. are smaller than prokaryotic cells.
   d. have a nucleus.

4. In the cell membrane, the fatty acids of phospholipid molecules
   a. face the cytoplasm.
   b. face the outside of the cell.
   c. are on both sides of the membrane.
   d. are in the interior of the membrane.

5. One function of the Golgi apparatus is to
   a. store DNA.
   b. make carbohydrates.
   c. modify proteins.
   d. digest and recycle the cell's wastes.

6. Structures present in plant cells but not present in animal cells include
   a. chloroplasts and the central vacuole.
   b. mitochondria and the cell wall.
   c. ribosomes and ER.
   d. lysosomes and the Golgi apparatus.

7. Explain how the cell membrane contributes to a cell's ability to maintain homeostasis.

8. What kind of microscope produced the image of cilia shown below?

9. Transport proteins in the membrane of a lysosome move hydrogen ions into the lysosome. Use this information to predict whether digestive enzymes in a lysosome work best in a neutral, a basic, or an acidic environment. (**Hint:** See Chapter 2, Section 2.)

10. **Concept Mapping** Make a concept map that compares plant cells with animal cells. Include the following terms in your concept map: *cell membrane, cell wall, central vacuole, chloroplasts,* and *mitochondria.*

Critical Thinking

11. **Recognizing Relationships** How does the arrangement of phospholipids influence the permeability of the lipid bilayer?

12. **Inferring Relationships** Muscle cells have more mitochondria than some other kinds of eukaryotic cells. In what way would having many mitochondria be beneficial to muscle cells?

13. **Applying Information** Drugs that rid the body of eukaryotic parasites often have more side effects and are harder on the body than drugs that act on bacterial parasites. Suggest a reason for this difference.

Alternative Assessment

14. **Interactive Tutor Unit 1 Cell Transport and Homeostasis** Write a report summarizing the role of the cell membrane in the preservation of body organs donated for transplant.

Understanding Concepts

Directions (1–4): **For *each* question, write on a separate sheet of paper the letter of the correct answer.**

1 What structure houses a eukaryotic cell's DNA?
   A. cell wall
   B. cytoskeleton
   C. ER
   D. nucleus

2 Which of the following is **not** a protein that might be found in cell membranes?
   F. enzyme
   G. lipid
   H. marker
   I. transporter

3 What structures are found in both eukaryotic cells and prokaryotic cells?
   A. mitochondrion and ER
   B. nucleus and cell membrane
   C. mitochondrion and ribosome
   D. cell membrane and ribosome

4 What property of phospholipids makes them ideal for making up the selectively permeable cell membrane?
   F. They repel small ions.
   G. They react readily with water molecules.
   H. They form triple layers that insulate the cell.
   I. They have a nonpolar region and a polar region.

Directions (5): **For the following question, write a short response.**

5 Analyze why smaller pieces of food cook faster than larger pieces of food, based on the relationship between surface area and volume.

Test TIP

When possible, use the text in the test to answer other questions. For example, use a multiple-choice answer to "jump start" your thinking about another question.

Reading Skills

Directions (6): **Read the passage below. Then answer the question.**

Microbiologists study the growth, structure, development, and many other characteristics of bacteria and other microorganisms. Some microbiologists also study the action of microorganisms on living and dead tissue. This career requires at least a two-year technical training degree from a community college or technical institution. Many microbiologists have a four-year bachelor's degree plus a master's or doctoral degree.

6 What tools would be most useful for a microbiologist?
   A. centrifuges and syringes
   B. nets and specimen cages
   C. electron and light microscopes
   D. balances and graduated cylinders

Interpreting Graphics

Directions (7): **Base your answer to question 7 on the diagram below.**

Animal Cell

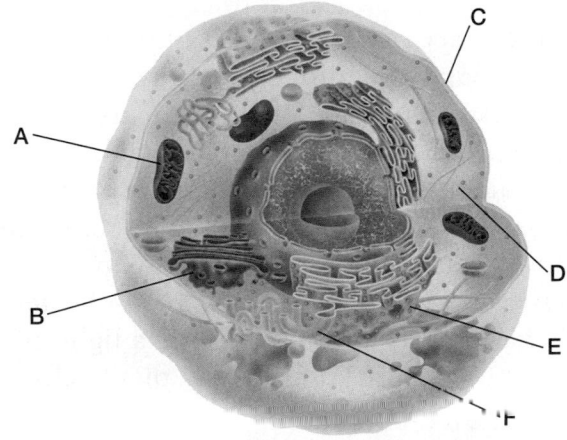

7 What is the function of the structure labeled *A?*
   F. making ATP
   G. making carbohydrates
   H. making proteins
   I. moving proteins through the cell

Skills Practice Lab

Studying Animal Cells and Plant Cells

SKILLS
- Using a compound microscope
- Drawing

OBJECTIVES
- **Identify** the structures you can see in animal cells and plant cells.
- **Compare** and **Contrast** the structure of animal cells and plant cells.

MATERIALS
- compound light microscope
- prepared slide of human epithelial cells
- safety goggles
- lab apron
- polyethylene gloves
- sprig of Elodea
- forceps
- microscope slides and coverslips
- dropper bottle of Lugol's iodine solution

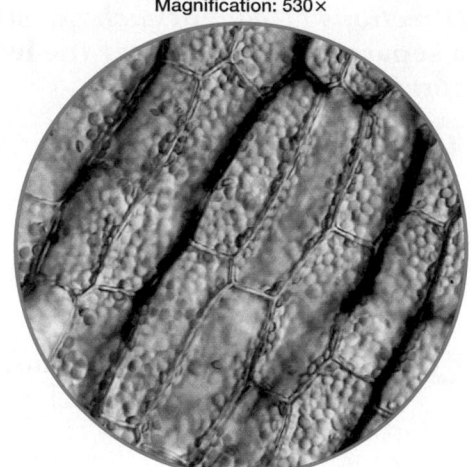

Plant cells

ChemSafety

 CAUTION: Always wear safety goggles and a lab apron to protect your eyes and clothing.

CAUTION: Do not touch or taste any chemicals. Know the location of the emergency shower and eyewash station and how to use them. If you get a chemical on your skin or clothing, wash it off at the sink while calling to the teacher. Notify the teacher of a spill. Spills should be cleaned up promptly, according to your teacher's directions.

CAUTION: Glassware is fragile. Notify the teacher of broken glass or cuts. Do not clean up broken glass or spills with broken glass unless the teacher tells you to do so.

Before You Begin

You can see many cell parts with a **light microscope**. In animal cells, the **cytoplasm, cell membrane, nucleus, nucleolus,** and **vacuoles** can be seen. In plant cells, the **cell wall** and **chloroplasts** can also be seen. Stains add color to cell parts and make them more visible with a light microscope. A stain can even make the **endoplasmic reticulum** visible. In this lab, you will use a light microscope to examine animal and plant cells.

1. Write a definition for each boldface term in the paragraph above.
2. Why might a stain be needed to see cell parts under a microscope?
3. Based on the objectives for this lab, write a question you would like to explore about cell structure.

Procedure

PART A: Animal Cells

1. Examine a prepared slide of human epithelial cells under low power with a compound light microscope. Find cells that are separate from each other, and place them in the center of the field of view. Switch to high power, and adjust the diaphragm until you can see the cells more clearly. Identify as many cell parts as you can. *Note: Remember to use only the fine adjustment to focus at high power.*

2. Draw two or three epithelial cells as they look under high power. Label the cell membrane, the cytoplasm, the nuclear envelope, and the nucleus of at least one of the cells. Make a second drawing of these cells as you imagine they might look in the lining of your mouth.

PART B: Plant Cells

3. Using forceps, carefully remove a small leaf from near the top of an *Elodea* sprig. Place the whole leaf in a drop of water on a slide, and add a cover slip.

4. Observe the leaf under low power. Look for an area of the leaf in which you can see the cells clearly, and move the slide so that this area is in the center of the field of view. Switch to high power, and, if necessary, adjust the diaphragm. Identify as many cell parts as you can.

5. Find an *Elodea* cell in which you can see the chloroplasts clearly. Draw this cell. Label the cell wall, a chloroplast, and any other cell parts that you can see.

6. Notice if the chloroplasts are moving in any of the cells. If you do not see movement, warm the slide in your hand or under a bright lamp for a minute or two. Look for movement of the cell contents again under high power. Such movement is called cytoplasmic streaming.

7. Put on safety goggles, gloves, and a lab apron. Make a wet mount of another *Elodea* leaf, using Lugol's iodine solution instead of water. **CAUTION: Lugol's solution stains skin and clothing. Promptly wash off spills.** Observe these cells under low and high power.

8. Draw a stained *Elodea* cell. Label the cell wall and a chloroplast, as well as the central vacuole, the nucleus, and the cell membrane if they are visible.

PART C: Cleanup and Disposal

9. Dispose of solutions, broken glass, and *Elodea* leaves in the waste containers designated by your teacher. Do not pour chemicals down the drain or put lab materials in the trash unless your teacher tells you to do so.

10. Clean up your work area and all lab equipment. Return lab equipment to its proper place. Wash your hands thoroughly before you leave the lab and after you finish all work.

Analyze and Conclude

1. **Recognizing Patterns** In what observable ways are animal and plant cells similar in structure, and in what observable ways are they different?

2. **Comparing Structures** Compare and contrast the cytoplasm of epithelial cells and *Elodea* cells.

3. **Analyzing Methods** What is the reason for staining *Elodea* cells with iodine?

4. **Inferring Conclusions** Lugol's iodine solution causes the movement of chloroplasts to stop. Explain why.

5. **Inferring Conclusions** If some of the epithelial cells were folded over on themselves but were still transparent, what could you conclude about their thickness?

6. **Further Inquiry** Write a new question about cell structure that could be explored with another investigation.

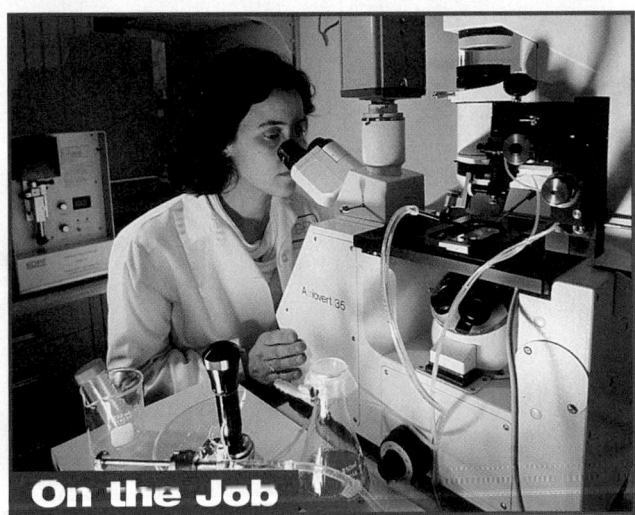

On the Job

Microscopy is an important tool for biologists who study cell structure. Do research to learn more about how biologists use specialized microscopes to study cell structure. For more about careers, visit **go.hrw.com** and type in the keyword **HX4 Careers.**

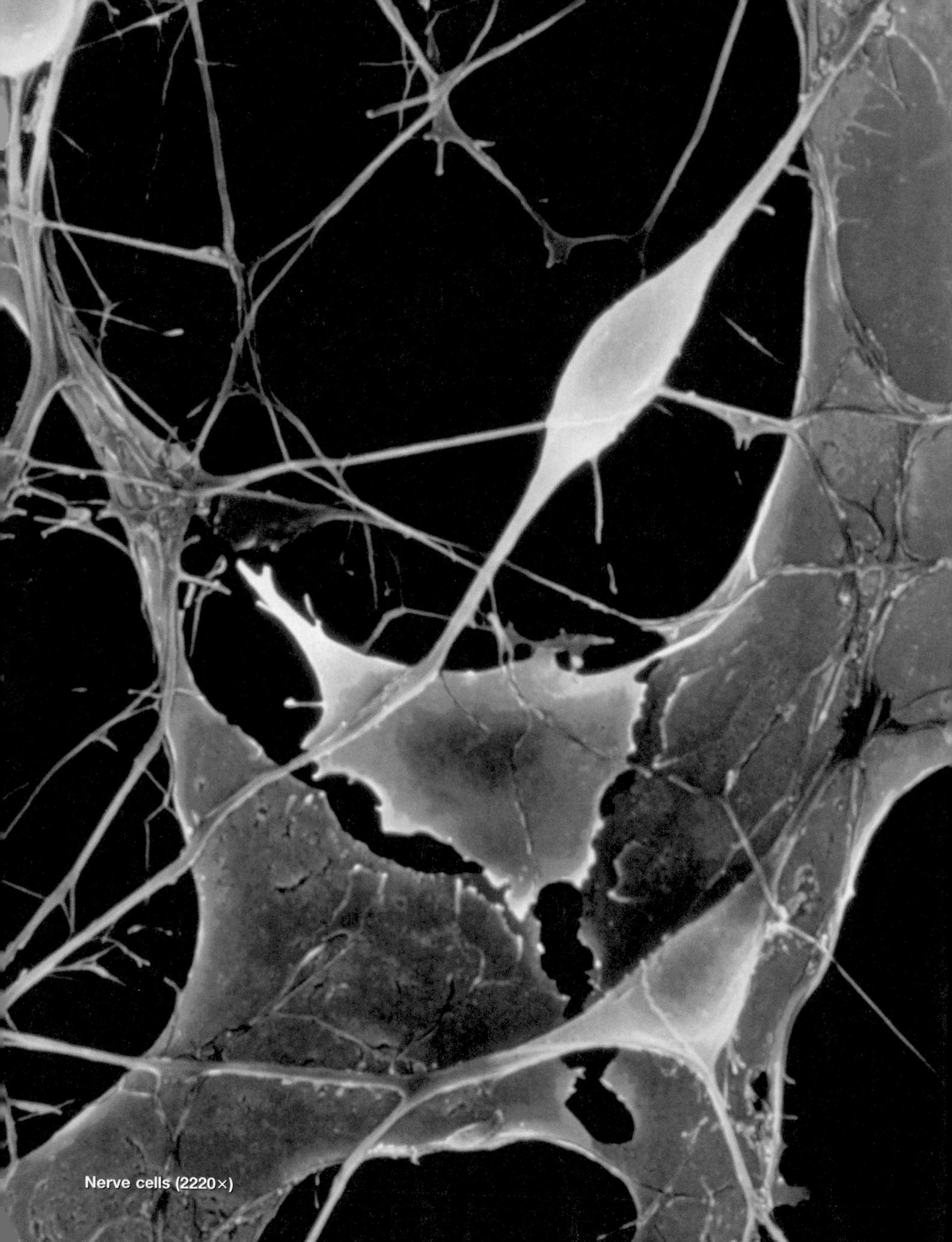

Nerve cells (2220×)

4 Cells and Their Environment

✔ *Quick Review*

Answer the following without referring to earlier sections of your book.

1. **Distinguish** *between polar and nonpolar substances. (Chapter 2, Section 1)*

2. **Describe** *the function of ATP in cells. (Chapter 2, Section 3)*

3. **Identify** *different kinds of proteins that compose the cell membrane. (Chapter 3, Section 2)*

4. **Summarize** *the function of vesicles and the Golgi apparatus. (Chapter 3, Section 3)*

Did you have difficulty? *For help, review the sections indicated.*

Reading Activity

Take a break after reading each section of this chapter, and closely study the figures in the section. Reread the figure captions, and, for each one, write out a question that can be answered by referring to the figure and its caption. Refer to your list of figures and questions as you review the concepts addressed in the chapter.

Looking Ahead

Section 1

Passive Transport
Diffusion
Osmosis
Crossing the Cell Membrane

Section 2

Active Transport
Movement Against a Concentration Gradient
Movement in Vesicles
Membrane Receptor Proteins

☑ internet connect

www.scilinks.org
National Science Teachers Association *sci*LINKS Internet resources are located throughout this chapter.

SC*i*LINKS® Maintained by the National Science Teachers Association

● The transfer of information between these nerve cells requires constant movement of substances across their cell membranes.

Passive Transport

- **Relate** concentration gradients, diffusion, and equilibrium.

- **Predict** the direction of water movement into and out of cells.

- **Describe** the importance of ion channels in passive transport.

- **Identify** the role of carrier proteins in facilitated diffusion.

Key Terms

passive transport
concentration gradient
equilibrium
diffusion
osmosis
hypertonic solution
hypotonic solution
isotonic solution
ion channel
carrier protein
facilitated diffusion

Study TIP

- **Reading Effectively**
As you read this chapter, write the objectives for each section on a sheet of paper. Rewrite each objective as a question, and answer these questions as you read the section.

Diffusion

You constantly interact with your environment, whether you are eating or putting on a raincoat to help keep you dry. Your body also responds to external conditions to maintain a stable internal condition. Just as you must respond to your environment to maintain stability, all other organisms and their cells must respond to external conditions to maintain a constant internal condition. Recall that when organisms adjust internally to changing external conditions, they are maintaining homeostasis. One way cells maintain homeostasis is by controlling the movement of substances across their cell membrane. Cells must use energy to transport some substances across the cell membrane. Other substances move across the cell membrane without any use of energy by the cell.

Random Motion and Concentration

Movement across the cell membrane that does not require energy from the cell is called **passive transport.** To understand passive transport, imagine two rooms of equal size separated by a wall with a closed door, as shown in **Figure 1.** Suppose you release several rubber balls into the first room. The balls move randomly, bouncing off the walls, the floor, the ceiling, and each other. Also suppose the balls can bounce forever without slowing down. The balls become evenly distributed throughout the room. What happens when you open the door between the rooms? Some of the balls in the first room bounce through the doorway and into the second room, as shown in Figure 1. You do not have to use energy to make the balls move into the second room. They enter the second room because of their own random motion. Occasionally, a ball will bounce back into the first room. However, most of the balls that pass through the doorway move from the first room, where their concentration is high, to the second room, where their concentration is low. A difference in the concentration of a substance, such as the balls, across a space is called a **concentration gradient.**

As more balls enter the second room, the concentration of balls in the second room increases, while the concentration of balls in the first room decreases. Eventually the concentration of balls in the two rooms will be equal. The balls will still bounce around the rooms, but they will move from the second room to the first room just as often as they move from the first room to the second room. At this point, the system is said to be in equilibrium, as shown in Figure 1. **Equilibrium** *(ee kwih LIHB ree uhm)* is a condition in which the concentration of a substance is equal throughout a space.

Figure 1   Models of diffusion

Because of diffusion, food coloring (blue) will gradually move through uncolored gelatin (yellow), as shown in the beakers below.

1. Randomly bouncing balls are distributed evenly throughout a closed room.

2. If the door to an adjoining room is opened, the balls begin to enter, or diffuse into, that room.

3. At equilibrium, the concentration of balls inside the two rooms will be equal.

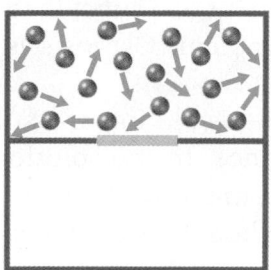
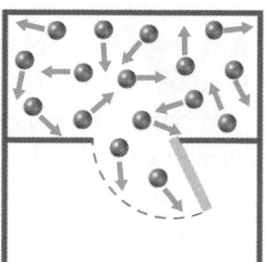

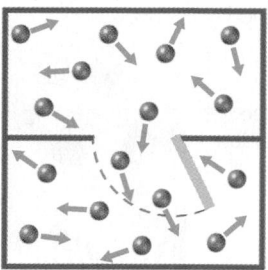

Movement of Substances

Like these imaginary rubber balls, particles of a substance in a solution also move around randomly. If there is a concentration gradient in the solution, the substance will move from an area of high concentration to an area of lower concentration. The movement of a substance from an area of high concentration to an area of lower concentration caused by the random motion of particles of the substance is called **diffusion** *(dih FYOO zhuhn)*. If diffusion is allowed to continue, equilibrium eventually results.

Many substances, such as molecules and ions dissolved in the cytoplasm and in the fluid outside cells, enter or leave cells by diffusing across the cell membrane. Inside the cell, the concentrations of most of these substances are different from their concentrations outside the cell. Thus, for each of these substances a concentration gradient exists across the cell membrane. To diffuse "down" its concentration gradient—from an area of high concentration to an area of lower concentration—a substance must be able to pass through the cell membrane.

The cell membrane is selectively permeable to substances. The nonpolar interior of the lipid bilayer repels ions and most polar molecules. Thus, these substances are prevented from diffusing across the cell membrane. In contrast, molecules that are either very small or nonpolar can diffuse across the cell membrane down their concentration gradient. The diffusion of such molecules across the cell membrane is the simplest type of passive transport.

internet connect

www.scilinks.org
Topic: **Water Movement in Cells**
Keyword: **HX4189**

SC*i*LINKS. Maintained by the National Science Teachers Association

Osmosis

Water molecules can diffuse through channels in the cell membrane, as shown in **Figure 2.** The diffusion of water through a selectively permeable membrane is called **osmosis** *(ahz MOH sihs)*. Like other forms of diffusion, osmosis involves the movement of a substance—water—down its concentration gradient. Osmosis is a type of passive transport.

What causes osmosis? Recall that a solution is a substance dissolved in another substance. In the solutions on either side of the cell membrane, many ions and polar molecules are dissolved in water. When these substances dissolve in water, some water molecules are attracted to them and so are no longer free to move around. If the solutions on either side of the cell membrane have different concentrations of dissolved particles, they will also have different concentrations of "free" water molecules. Then osmosis will occur as free water molecules move into the solution with the lower concentration of free water molecules.

Figure 2   Osmosis

Water diffuses across the cell membrane by osmosis.

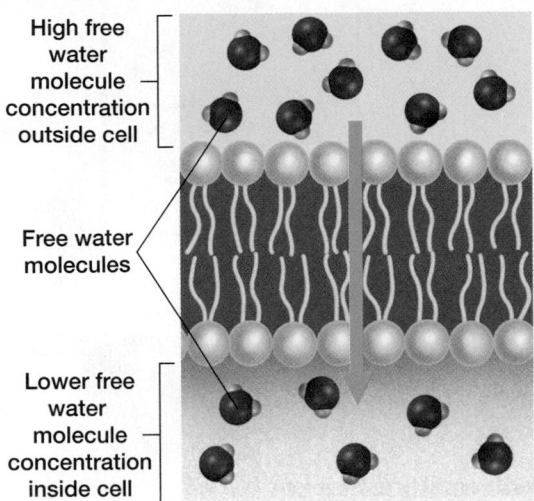

High free water molecule concentration outside cell

Free water molecules

Lower free water molecule concentration inside cell

Observing Osmosis

You can observe the movement of water into or out of a grape under different conditions.

Materials

3 grapes, 3 small jars with lids, saturated sugar solution, grape juice, tap water, marking pen, paper towel, balance

Procedure

1. Make a data table with four columns (Solution, Original mass, Predicted mass, and Actual mass) and a row for each solution (Sugar solution, Grape juice, and Water).

2. Fill one jar with the sugar solution. Fill a second jar with grape juice. (The grape will be more visible inside the jar if you fill the jar with white grape juice, as shown in the middle jar in the photo above right.) Fill the third jar with tap water. Label each jar according to the solution it contains.

3. Using the balance, find the mass of each grape. Place one grape in each jar, and record the mass of each jar in your data table. Put a lid on each jar.

4. Predict whether the mass of each grape will increase or decrease over time. Explain your predictions.

5. After 24 hours, remove each grape from its jar, and dry it gently with a paper towel. Using the balance, find its mass again. Record your results.

6. Clean up your materials before leaving the lab.

Analysis

1. **Identify** the solutions in which osmosis occurred.

2. **Critical Thinking Evaluating Conclusions** How did you determine whether osmosis occurred in each of the three solutions?

3. **Critical Thinking Evaluating Hypotheses** Did the mass of each grape change as you predicted? Why or why not?

Table 1 Hypertonic, Hypotonic, and Isotonic Solutions

If the fluid outside the cell has...	Then outside fluid is...	Water diffuses...		Effect on cell
...lower free water molecule concentration than cytosol	...hypertonic.	...out of cell.	H_2O	Cell shrinks.
...higher free water molecule concentration than cytosol	...hypotonic.	...into cell.	H_2O	Cell swells.
...same free water molecule concentration as cytosol	...isotonic.	...into and out of cell at equal rates.	H_2O	Cell stays same size.

The direction of water movement across the cell membrane depends on the relative concentrations of free water molecules in the cytoplasm and in the fluid outside the cell. There are three possibilities for the direction of water movement:

1. **Water moves out.** When water diffuses out of the cell, the cell shrinks. A solution that causes a cell to shrink because of osmosis is called a **hypertonic** *(hie puhr TAHN ihk)* **solution.** If the fluid outside the cell has a higher concentration of dissolved particles than the cytoplasm has, then the outside fluid also has a lower concentration of free water molecules than the cytoplasm.

2. **Water moves in.** When water diffuses into the cell, the cell swells. A solution that causes a cell to swell because of osmosis is called a **hypotonic** *(hie poh TAHN ihk)* **solution.** If the fluid outside the cell has a lower concentration of dissolved particles than the cytoplasm has, then the outside fluid also has a higher concentration of free water molecules than the cytoplasm.

3. **No net water movement.** If the cytoplasm and the fluid outside the cell have the same concentration of free water molecules, water diffuses into and out of the cell at equal rates. This results in no net movement of water across the cell membrane, and the cell stays the same size—a state of equilibrium. A solution that produces no change in cell volume because of osmosis is called an **isotonic** *(ie soh TAHN ihk)* **solution. Table 1** summarizes the effects of hypertonic, hypotonic, and isotonic solutions on cells.

If left unchecked, the swelling caused by a hypotonic solution could cause a cell to burst. Different kinds of cells have different adaptations that deal with this problem. The cells of plants and fungi have rigid cell walls that keep the cells from expanding too much. Some unicellular eukaryotes have contractile vacuoles *(kuhn TRAK tihl VAK yoo ohlz),* which are organelles that collect excess water inside the cell and force the water out of the cell. Animal cells have neither cell walls nor contractile vacuoles. However, many animal cells can avoid swelling caused by osmosis by removing dissolved particles from the cytoplasm. The removal of dissolved particles from a cell increases the concentration of free water molecules inside the cell.

WORD Origins

The words *hypertonic, hypotonic,* and *isotonic* have the same ending, *–tonic,* which is from the Greek *tonos,* meaning "tension." The prefix *hyper–* is from the Greek *hyper,* meaning "over." The prefix *hypo–* is from the Greek *hypo,* meaning "under." The prefix *iso–* is from the Greek *isos,* meaning "same."

Crossing the Cell Membrane

Recall that most ions and polar molecules cannot pass across the cell membrane because they cannot pass through the nonpolar interior of the lipid bilayer. However, such substances can cross the cell membrane when they are aided by transport proteins. Transport proteins called *channels* provide polar passageways through which ions and polar molecules can move across the cell membrane. Each channel allows only a specific substance to pass through the cell membrane. For example, some channels allow only one type of ion to cross the cell membrane, while others transport a particular kind of sugar or amino acid. This selectivity is one of the most important properties of the cell membrane because it enables a cell to control what enters and leaves.

Diffusion Through Ion Channels

Ions such as sodium, Na^+, potassium, K^+, calcium, Ca^{2+}, and chloride, Cl^-, are involved in many important cell functions. For example, ions are essential to the ability of nerve cells to send electrical signals throughout your body. Muscle cells in your heart could not make your heart beat without the movement of ions between the cells. Although ions cannot diffuse through the nonpolar interior of the lipid bilayer, they can cross the cell membrane by diffusing through ion channels. An **ion channel** is a transport protein with a polar pore through which ions can pass. As **Figure 3** shows, the pore of an ion channel spans the thickness of the cell membrane. Thus, an ion that enters the pore can cross the cell membrane without contacting the nonpolar interior of the lipid bilayer.

The pores of some ion channels are always open. In other ion channels, the pores can be closed by ion channel gates. A model of an ion channel with a gate is shown in Figure 3. Ion channel gates may open or close in response to different kinds of stimuli. These include the stretching of the cell membrane, a change in electrical charge, or the binding of specific molecules to the ion channel. In this way, the stimuli are able to affect the ability of particular ions to cross

Figure 3   Ion channels

Ion channels allow certain ions to pass through the cell membrane.

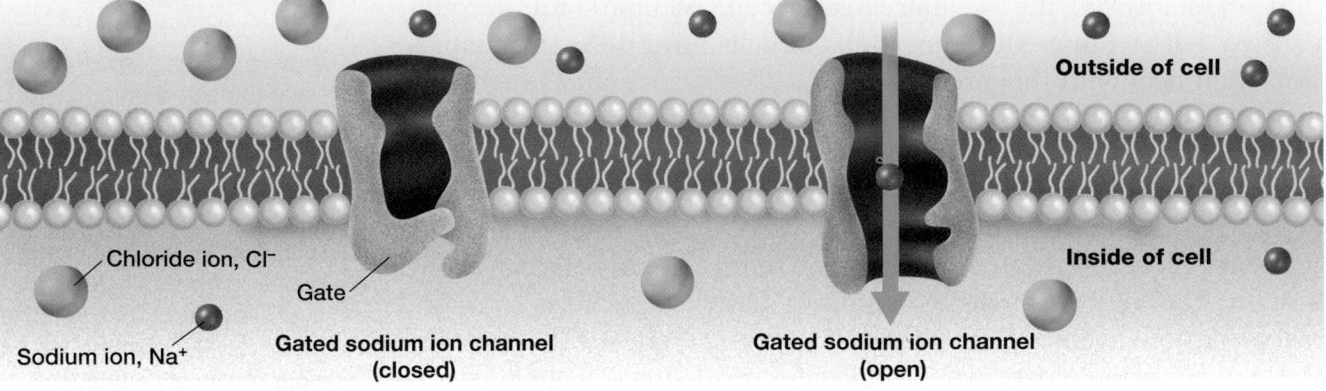

Chloride ion, Cl⁻

Gate

Sodium ion, Na⁺

**Gated sodium ion channel
(closed)**

Outside of cell

Inside of cell

**Gated sodium ion channel
(open)**

the cell membrane. Like the diffusion of small molecules and nonpolar molecules through the lipid bilayer, the diffusion of ions through ion channels is a form of passive transport. No use of energy by the cell is required because the ions move down their concentration gradients.

internet connect

www.scilinks.org
Topic: Ion Channels
Keyword: HX4106

SC*LINKS* Maintained by the National Science Teachers Association

Electrical Charge and Ion Transport

The rate of movement of a substance across the cell membrane is generally determined by the concentration gradient of the substance. The movement of a charged particle, such as an ion, across the cell membrane is also influenced by the particle's positive or negative electrical charge. The inside of a typical cell is negatively charged with respect to the outside of the cell. Opposite charges attract, and like charges repel. Thus, a more positively charged ion located outside the cell is more likely to diffuse into the cell, where the charge is negative. Conversely, a more negatively charged ion located inside the cell is more likely to diffuse out of the cell. The direction of movement caused by an ion's concentration gradient may oppose the direction of movement caused by the ion's electrical charge. Thus, an ion's electrical charge often affects the diffusion of the ion across the cell membrane. This is very important to the functioning of nerve cells in animals.

Analyzing the Effect of Electrical Charge on Ion Transport

Background

The electrical charge of an ion affects the diffusion of the ion across the cell membrane. Some ions are more concentrated inside cells, and some ions are more concentrated outside cells. Use the table below to answer the following questions:

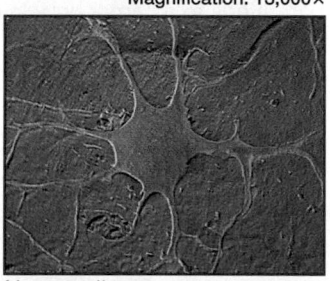

Magnification: 13,000×

Nerve cell

Ion Charges and Concentration Inside and Outside Cell		
Ion	**Charge of ion**	**Concentration of ion outside cell : inside cell**
Sodium (Na^+)	Positive	10:1
Potassium (K^+)	Positive	1:20
Calcium (Ca^{2+})	Positive	10,000:1
Chloride (Cl^-)	Negative	12:1

Analysis

1. **Identify** the ion that is more concentrated inside the cell than outside the cell.

2. **Identify** those ions that are more concentrated outside the cell than inside the cell.

3. **Critical Thinking Recognizing Relationships** Do the positive charges of calcium ions and sodium ions make these ions more likely to move into or out of the cell?

4. **Critical Thinking Inferring Relationships** Which ions' electrical charges oppose the direction of movement that is caused by their concentration gradient?

Facilitated Diffusion

Most cells also have a different kind of transport protein that can bind to a specific substance on one side of the cell membrane, carry the substance across the cell membrane, and release it on the other side. Such proteins are called **carrier proteins.** When carrier proteins are used to transport specific substances—such as amino acids and sugars—down their concentration gradient, that transport is called facilitated diffusion. **Facilitated** *(fah SIHL uh tayt ehd)* **diffusion,** shown in **Figure 4,** is a type of passive transport. It moves substances down their concentration gradient without using the cell's energy.

Step ❶ The carrier protein binds a specific molecule on one side of the cell membrane.

Step ❷ A change in the shape of the carrier protein exposes the molecule to the other side of the cell membrane.

Step ❸ The carrier protein shields the molecule from the interior of the lipid bilayer. The molecule is then released from the carrier protein, which returns to its original shape.

Figure 4

BIOgraphic

Facilitated Diffusion

Carrier proteins transport substances down their concentration gradient.

❶ A molecule outside the cell binds to a carrier protein on the cell membrane.

❷ The carrier protein transports the molecule across the cell membrane.

❸ The molecule is released from the carrier protein inside the cell.

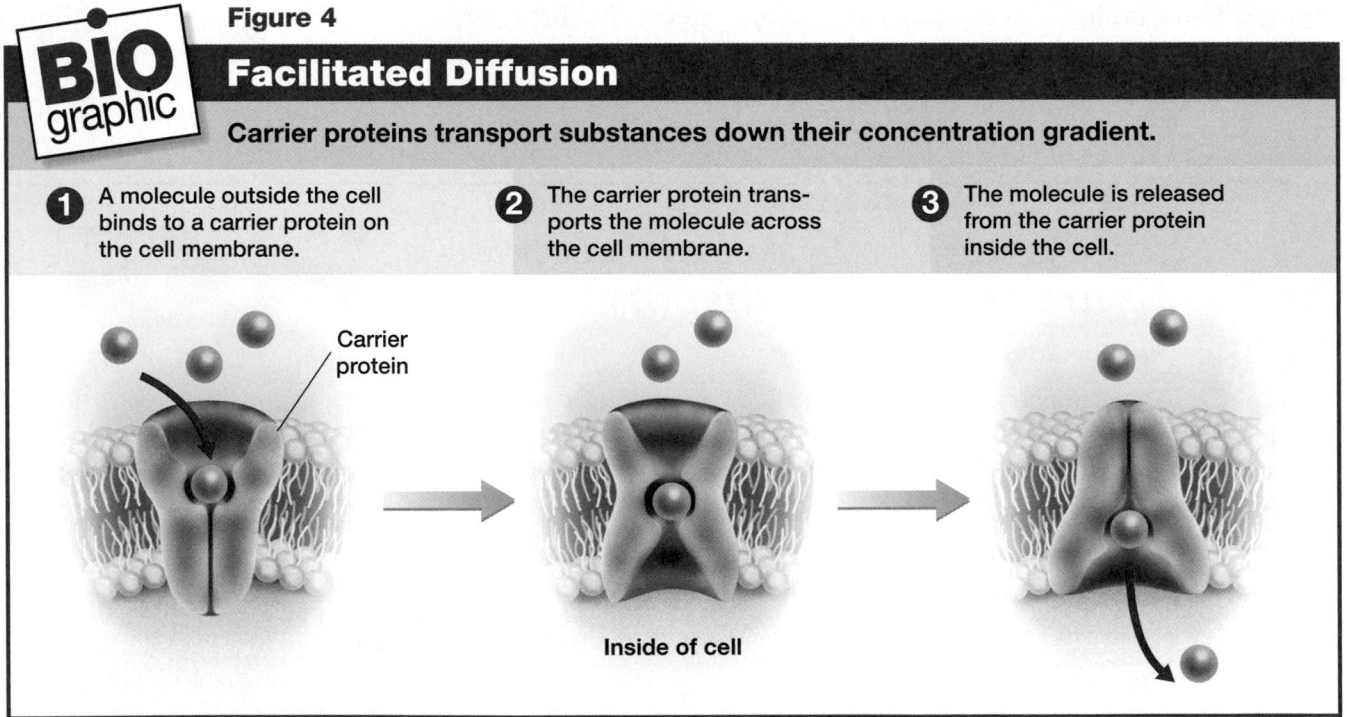

Carrier protein

Inside of cell

Section 1 Review

❶ Distinguish between diffusion and equilibrium.

❷ Describe how the diffusion of ions across a cell membrane differs from the diffusion of nonpolar molecules across the cell membrane.

❸ Explain how some substances cross the cell membrane by facilitated diffusion.

❹ Critical Thinking Predicting Outcomes Predict what would happen to a cell that is placed in a hypertonic solution, and explain why this would occur.

❺ Standardized Test Prep Which substance crosses the cell membrane by facilitated diffusion?

A a sugar   C sodium ion

B water   D chloride ion

Active Transport

Movement Against a Concentration Gradient

Although facilitated diffusion can help move amino acids and sugars across the cell membrane, it can only transport these substances down their concentration gradient. Cells must transport certain amino acids, sugars, and other substances into their cytoplasm from the surrounding fluid. But many of these substances have a low concentration outside cells and a higher concentration inside cells. Their concentration gradients would cause these important substances to move out of the cell rather than into the cell. So, cells also have a way to move some substances against their concentration gradient—from an area of low concentration to an area of higher concentration.

The transport of a substance across the cell membrane against its concentration gradient is called **active transport.** Unlike passive transport, active transport requires the cell to use energy because the substance is being moved against its concentration gradient. Most often, the energy needed for active transport is supplied directly or indirectly by ATP.

Some active-transport processes involve carrier proteins. Like the carrier proteins used in facilitated diffusion, the carrier proteins used in active transport bind to specific substances on one side of the cell membrane and release them on the other side of the cell membrane. But in active transport, the substances bind to carrier proteins where they are low in concentration and are released where they are higher in concentration. Thus, carrier proteins in active transport function as "pumps" that move substances against their concentration gradient. For this reason, these carrier proteins are often called membrane pumps.

Sodium-Potassium Pump

One of the most important membrane pumps in animal cells is a carrier protein called the sodium-potassium pump. In a complete cycle, the **sodium-potassium pump** transports three sodium ions, Na^+, out of a cell and two potassium ions, K^+, into the cell. Sodium ions are usually more concentrated outside the cell than inside the cell, and potassium ions are typically more concentrated inside the cell than outside the cell. Thus, the sodium-potassium pump actively transports both sodium ions and potassium ions against their concentration gradients. The energy needed to power sodium-potassium pumps is supplied by ATP. In some cells, sodium-potassium pumps are so active that they use much of the ATP produced by the cells.

Objectives

- **Compare** active transport with passive transport.
- **Describe** the importance of the sodium-potassium pump.
- **Distinguish** between endocytosis and exocytosis.
- **Identify** three ways that receptor proteins can change the activity of a cell.

Key Terms

active transport
sodium-potassium
  pump
endocytosis
exocytosis
receptor protein
second messenger

Real Life

Why saltwater frogs aren't in a pickle.

Some frogs have urea—a salty product of metabolism that is usually secreted as urine—in their blood. This makes their bodies nearly as salty as seawater, allowing them to live in saltwater environments.

Finding Information Find out the species name of a saltwater frog.

Figure 5

BIO graphic

Sodium-Potassium Pump

The sodium-potassium pump actively transports sodium ions, Na⁺, and potassium ions, K⁺, against their concentration gradient.

❶ Three sodium ions, Na⁺ and a phosphate group (P) from ATP bind to the pump.

❷ The pump changes shape, transporting the three sodium ions across the cell membrane.

❸ Two potassium ions, K⁺, bind to the pump and are transported across the cell membrane.

❹ The phosphate group and the two potassium ions are released inside the cell.

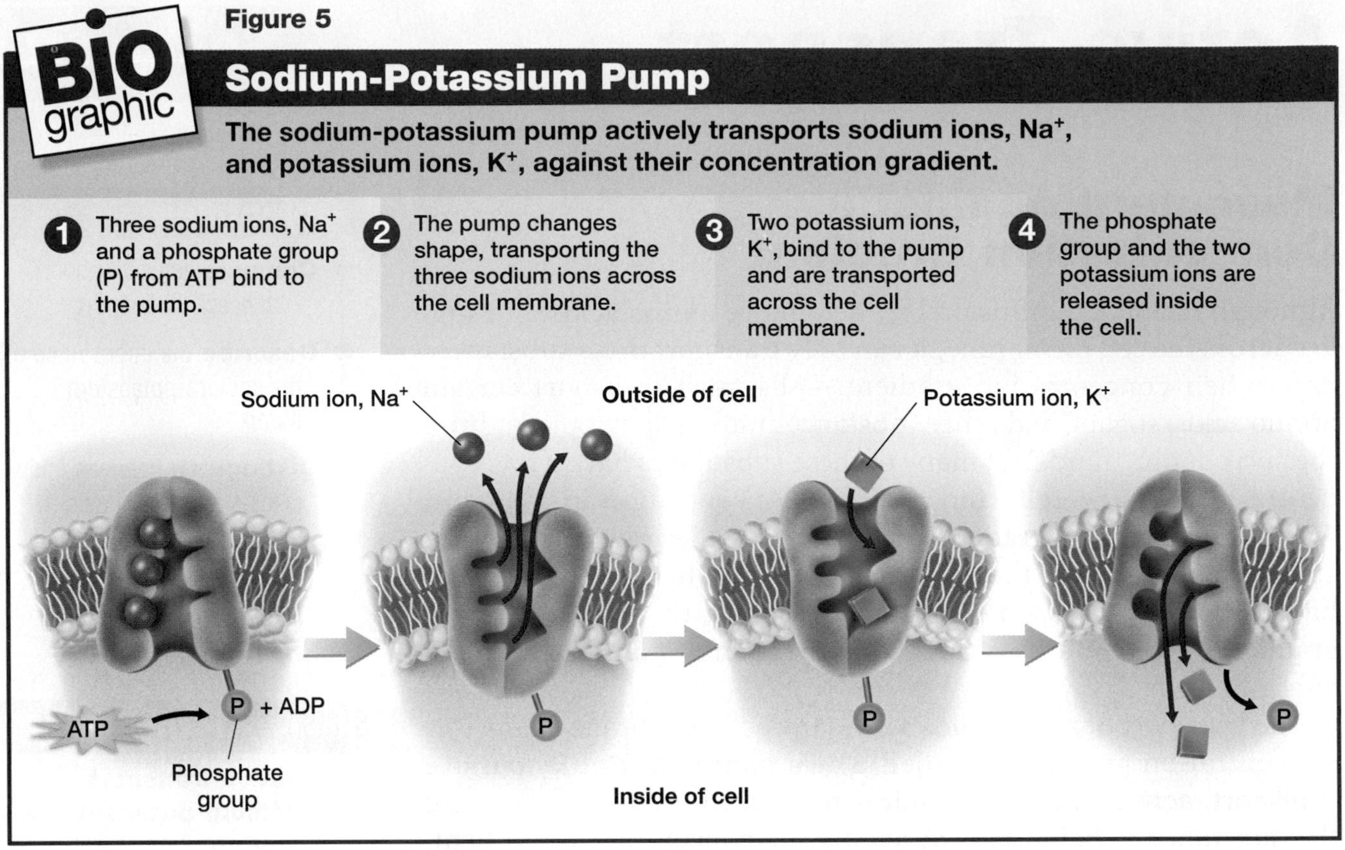

A model of the sodium-potassium pump is shown in **Figure 5.**

Step ❶ Three sodium ions inside the cell bind to the sodium-potassium pump. Because energy is needed to move the sodium ions against their concentration gradient, a phosphate group is removed from ATP and also binds to the pump.

Step ❷ The pump changes shape, transporting the three sodium ions across the cell membrane and releasing them outside the cell.

Step ❸ The pump is now exposed on the surface of the cell. Two potassium ions outside the cell bind to the pump. The phosphate group is released, changing the shape of the pump.

Step ❹ The pump is again exposed to the inside of the cell. The two potassium ions are transported across the cell membrane and are released inside the cell.

The sodium-potassium pump is important for two main reasons. First, the pump prevents sodium ions from accumulating in the cell. Sodium ions continuously diffuse into the cell through ion channels embedded in the lipid bilayer of the cell membrane. The increased concentration of sodium ions would then cause water to enter the cell by osmosis, causing the cell to swell or even burst. Second, the sodium-potassium pump helps maintain the concentration gradients of sodium ions and potassium ions across the cell membrane. Many cells use the sodium-ion concentration gradient to help transport other substances, such as glucose, across the cell membrane.

Movement in Vesicles

Many substances, such as proteins and polysaccharides, are too large to be transported by carrier proteins. These substances are moved across the cell membrane by vesicles. The movement of a substance into a cell by a vesicle is called **endocytosis** *(ehn doh sie TOH sihs)*. During endocytosis, the cell membrane forms a pouch around a substance, as shown in **Figure 6.** The pouch then closes up and pinches off from the membrane to form a vesicle. Vesicles formed by endocytosis may fuse with lysosomes or other organelles.

The movement of a substance by a vesicle to the outside of a cell is called **exocytosis** *(ek soh sie TOH sihs)*, also shown in Figure 6. During exocytosis, vesicles in the cell fuse with the cell membrane, releasing their contents. Cells use exocytosis to export proteins that are modified by the Golgi apparatus. Nerve cells and cells of various glands, for example, release proteins by exocytosis.

Study TIP

● **Interpreting Graphics**
As you look at Figure 6, notice that during endocytosis, the cell membrane pinches off to become the vesicle membrane. Conversely, during exocytosis, the vesicle membrane becomes part of the cell membrane.

Figure 6   Endocytosis and exocytosis

Vesicles transport substances into and out of cells.

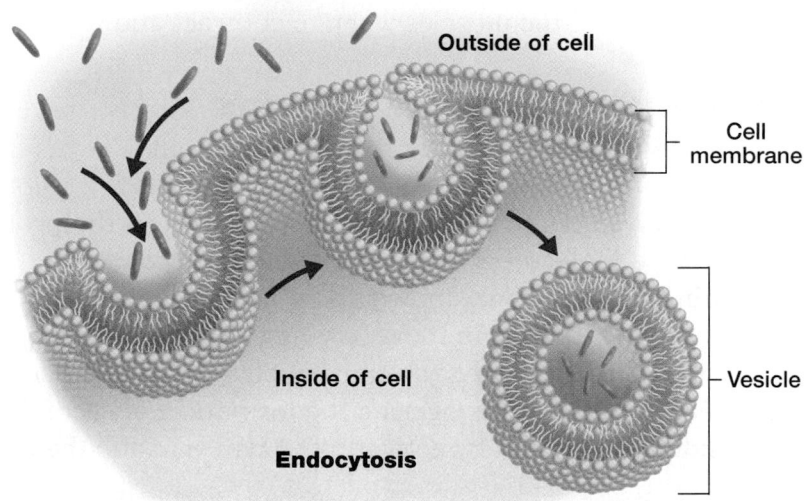

Outside of cell

Cell membrane

Inside of cell

Vesicle

Endocytosis

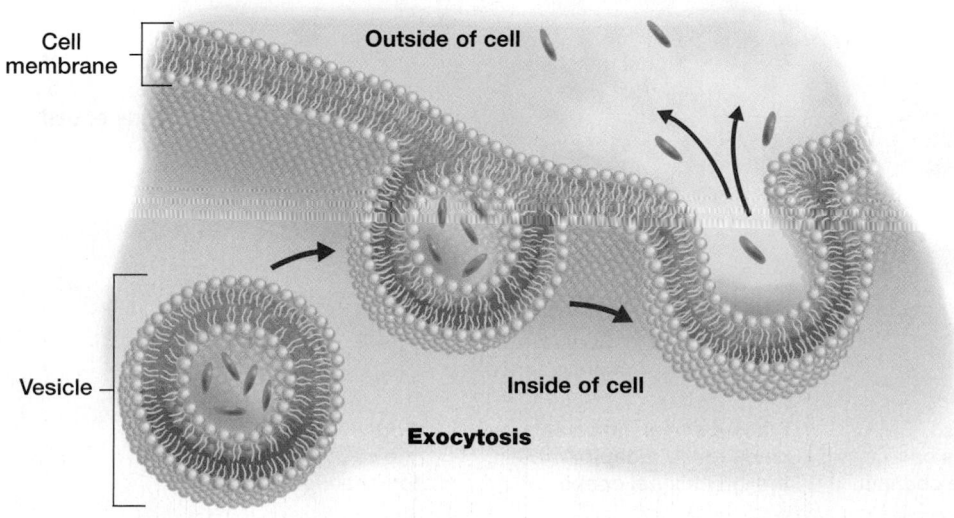

Cell membrane

Outside of cell

Vesicle

Inside of cell

Exocytosis

Membrane Receptor Proteins

We are constantly bombarded with information from other people and through television, the Internet, and many other media. To interpret information, we must be able to communicate and to distinguish between important and unimportant information. Similarly, your body's cells must communicate with each other to coordinate your growth, metabolism, and other activities. Cells that do not lie next to each other cannot communicate directly. Instead, some cells release *signal molecules* that carry information to nearby cells and throughout the body. Hormones are one familiar example of signal molecules. Hormones are made in one part of the body and carried in the bloodstream to other parts, where they have their effects.

Cells must also respond to important information and filter out unimportant information. Cells can receive the messages carried by certain signal molecules because the cell membrane contains specialized proteins that bind these signal molecules. Such proteins are called receptor proteins. A **receptor protein** is a protein that binds to a specific signal molecule, enabling the cell to respond to the signal molecule. For example, the muscles of the person exercising in **Figure 7** could not contract without receptor proteins and signal molecules that tell the muscles when to contract and when to relax.

Figure 7 Action of signal molecules. When you exercise, signal molecules are bound by receptor proteins on your muscle cells, signaling your muscles to contract.

Functions of Receptor Proteins

A signal molecule is bound by a receptor protein that fits that molecule, as shown in **Figure 8.** Most receptor proteins are embedded in the lipid bilayer of the cell membrane. The part of the protein that fits the signal molecule faces the outside of the cell.

The binding of a signal molecule by its complementary receptor protein causes a change in the receiving cell. This change can occur in the following three ways: by causing changes in the permeability of the receiving cell; by triggering the formation of second messengers inside the cell; and by activating enzymes inside the cell.

Figure 8 Changes in permeability

Some receptor proteins are coupled with ion channels.

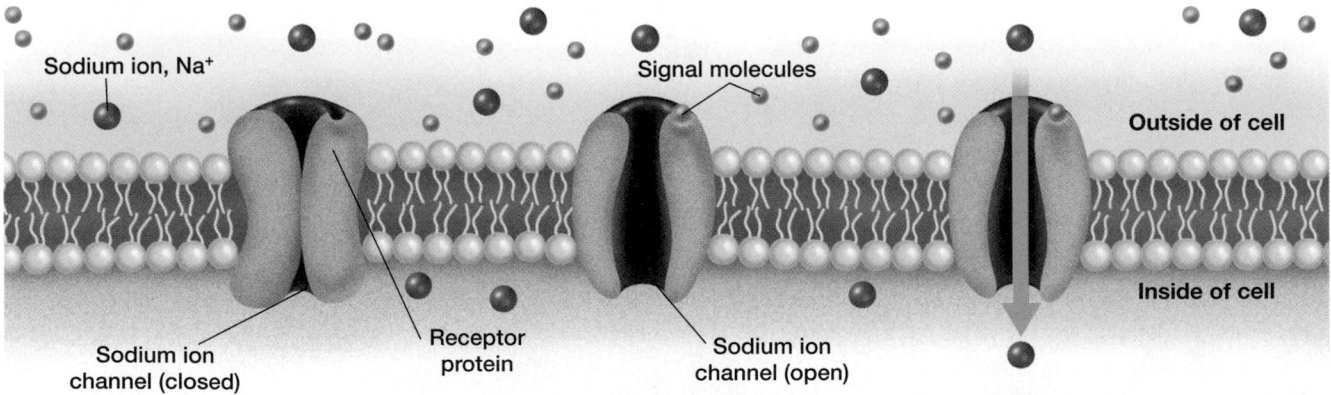

Sodium ion, Na⁺

Signal molecules

Outside of cell

Inside of cell

Sodium ion channel (closed)

Receptor protein

Sodium ion channel (open)

1. The ion channel is closed, so no ions can move through the channel.

2. When a signal molecule binds to the receptor protein, the ion channel opens.

3. Sodium ions diffuse into the cell through the open ion channel.

Figure 9   Second messengers

Some receptor proteins trigger the production of second messengers.

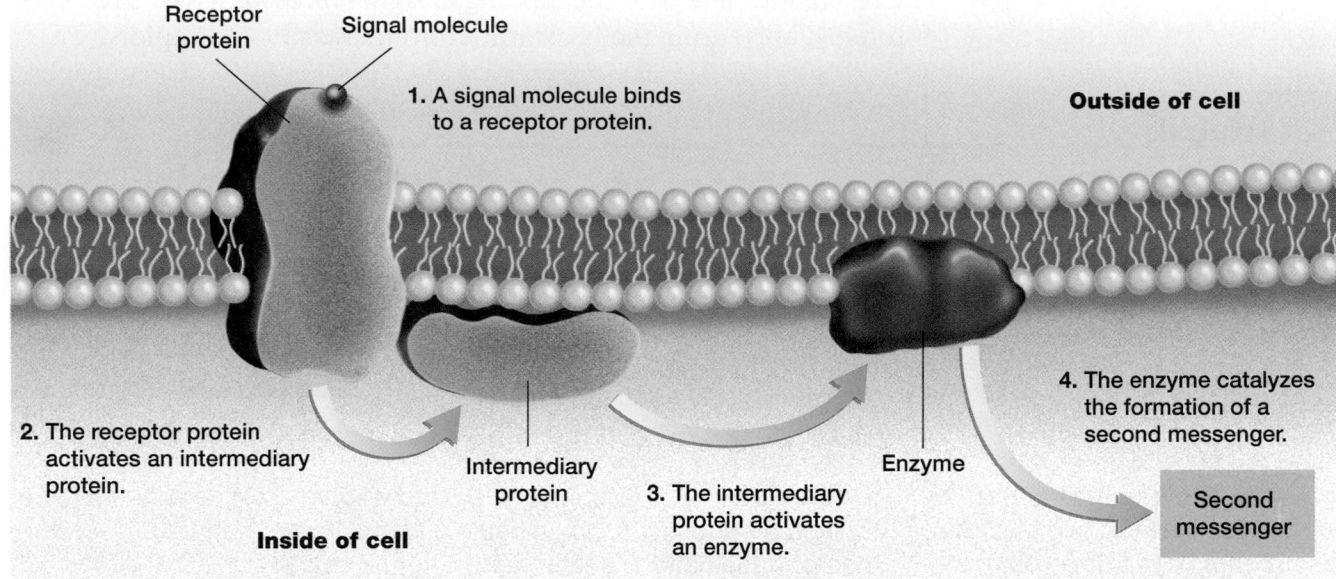

Receptor protein

Signal molecule

1. A signal molecule binds to a receptor protein.

Outside of cell

2. The receptor protein activates an intermediary protein.

Intermediary protein

3. The intermediary protein activates an enzyme.

Enzyme

4. The enzyme catalyzes the formation of a second messenger.

Second messenger

Inside of cell

Changes in Permeability The receptor protein may be coupled with an ion channel, as shown in Figure 8. The binding of a signal molecule to the receptor protein causes the ion channel to open, allowing specific ions to cross the cell membrane. This type of receptor protein is especially important in the nervous system.

Second Messengers The receptor protein may cause the formation of a second messenger inside the cell, as shown in **Figure 9.** When it is activated, a **second messenger** acts as a signal molecule in the cytoplasm. The second messenger amplifies the signal of the first messenger—that is, the original signal molecule. Second messengers can change the functioning of a cell in several ways. For example, some second messengers activate enzymes, triggering a series of biochemical reactions in the cell. Other second messengers change the permeability of the cell by opening ion channels in the cell membrane.

Enzyme Action The receptor protein may act as an enzyme. When a signal molecule binds to the receptor protein, the receptor protein may speed up chemical reactions inside the cell. Receptor proteins may also activate other enzymes located inside the cell or in the cell membrane, triggering chemical reactions in the cell. In this way, the signal molecule can cause many changes in the functioning of the receiving cell.

Many drugs affect the binding of signal molecules to receptor proteins. Some drugs, such as the illegal drug heroin, imitate signal molecules by binding to receptor proteins on a receiving cell, altering the function of the cell. Other drugs block or interfere with receptor proteins, preventing signal molecules from binding to the receptor proteins. For example, signal molecules that bind to receptor proteins on heart-muscle cells stimulate the cells, causing the heart rate to

Real Life

Many medicines are drugs that bind to receptor proteins.
Some of these drugs interfere with the receptor's ability to bind to signal molecules.
Finding Information Research some medicines that bind to receptor proteins.

internet connect

www.scilinks.org
Topic: Receptor Proteins
Keyword: HX4157

SC/LINKS. Maintained by the National Science Teachers Association

increase. Beta blockers, which are drugs prescribed to patients with a rapid heartbeat, bind to some of these receptor proteins. Beta blockers therefore interfere with the binding of signal molecules to the receptor proteins, preventing the heart rate from increasing too rapidly.

BIOWatch

The Shifting Allegiance of HIV

One of the more puzzling aspects of the AIDS epidemic is the slow onset of the disease after infection. In a person infected by HIV, the virus that causes AIDS, it may take 8 to 10 years for full blown AIDS— destruction of the immune system—to develop.

Docking

When HIV is introduced into the human bloodstream, the virus particles circulate throughout the whole body, but they only infect certain cells—large cells called *macrophages.* Why only macrophages? Spikes composed of protein cover the surface of each HIV particle. These spikes come into contact with all cells the virus encounters as it moves through the blood, yet the virus ignores most of the cells. Only when an HIV spike comes into contact with a cell whose surface receptor proteins exactly correspond to the spike's shape does the HIV particle attach to the cell and infect it.

The cell surface receptor protein that matches HIV's spikes is called *CD4,* and it is found on both macrophages and the infection-fighting cells of the immune system called *lymphocytes.* Why then are lymphocytes not infected right away, as macrophages are?

After docking onto the CD4 receptor of a macrophage, the HIV particle requires a second receptor protein to enter the cell. This second receptor, called a co-receptor, pulls the HIV particle across the cell membrane. Macrophages have a co-receptor that HIV recognizes, but lymphocytes lack this specific co-receptor.

Onset of AIDS

During the long period before AIDS develops, HIV is continuously reproduced inside macrophages. While HIV grows in these infected cells, it does not harm them. As the virus reproduces, it accumulates random changes in its genetic material. Eventually and by chance, HIV changes in such a way that its spike proteins now recognize a *new* co-receptor, one present on the surface of lymphocytes. When the body's lymphocytes become infected with HIV, the consequences are deadly—HIV eventually destroys most of the body's supply of lymphocytes. This shift in the allegiance of HIV from one type of co-receptor to another leads directly to the onset of AIDS.

Section 2 Review

1. **Distinguish** between passive transport and active transport.

2. **Describe** how the sodium-potassium pump helps prevent animal cells from bursting.

3. **Compare** two ways that the binding of a signal molecule to a receptor protein causes a change in the activity of the receiving cell.

4. **Identify** the terms *endocytosis* and *exocytosis* and distinguish between them.

5. **Critical Thinking Applying Information** During exercise, potassium ions accumulate in the fluid that surrounds muscle cells. Which cell membrane protein helps muscle cells counteract this tendency? Explain your answer.

6. **Standardized Test Prep** The concentration of molecule X is greater inside a cell than outside. If the cell acquires X from its surroundings, X must cross the cell membrane by means of
   A exocytosis.
   B active transport.
   C receptor proteins.
   D second messengers.

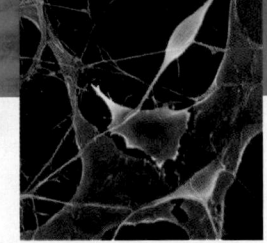

Key Concepts

1 Passive Transport

- Passive transport is the movement of substances across the cell membrane without the use of energy by the cell.

- Diffusion is the movement of a substance from an area of high concentration to an area of lower concentration.

- Osmosis is the diffusion of free water molecules across a selectively permeable membrane.

- Ion channels are proteins that have a pore through which ions can cross the cell membrane.

- In facilitated diffusion, a carrier protein transports a substance across the cell membrane down the concentration gradient of the substance.

2 Active Transport

- Active transport is the movement of a substance against the concentration gradient of the substance. Active transport requires cells to use energy.

- In animal cells, the sodium-potassium pump uses energy supplied by ATP to transport sodium ions out of the cell and potassium ions into the cell.

- During endocytosis, substances are moved into a cell by a vesicle that pinches off from the cell membrane.

- During exocytosis, substances inside a vesicle are released from a cell as the vesicle fuses with the cell membrane.

- Communication between cells often involves signal molecules that are bound by receptor proteins on cells.

- A signal molecule that is bound by a receptor protein on a cell can change the activity of the cell in three ways: by enabling specific ions to cross the cell membrane, by causing the formation of a second messenger, or by speeding up chemical reactions inside the cell.

Key Terms

Section 1

passive transport (74)
concentration gradient (74)
equilibrium (74)
diffusion (75)
osmosis (76)
hypertonic solution (77)
hypotonic solution (77)
isotonic solution (77)
ion channel (78)
carrier protein (80)
facilitated diffusion (80)

Section 2

active transport (81)
sodium-potassium pump (81)
endocytosis (83)
exocytosis (83)
receptor protein (84)
second messenger (85)

Unit 1—*Cell Transport and Homeostasis*
Use Topics 1–6 in this unit to review the key concepts and terms in this chapter.

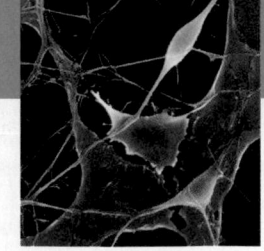

Performance ZONE

CHAPTER REVIEW

Understanding Key Ideas

1. In diffusion, a substance moves
   a. against its concentration gradient.
   b. from an area of lower concentration to an area of higher concentration.
   c. independent of its concentration.
   d. down its concentration gradient.

2. Substances enter or leave a cell through the
   a. cytoplasm.   c. nucleus.
   b. Golgi apparatus. d. cell membrane.

3. Facilitated diffusion
   a. is driven by energy from ATP.
   b. is a type of active transport.
   c. employs receptor proteins.
   d. employs carrier proteins.

4. The sodium-potassium pump moves
   a. sodium ions into the cell and potassium ions out of the cell.
   b. sodium ions out of the cell and potassium ions into the cell.
   c. sodium and potassium into the cell.
   d. sodium and potassium out of the cell.

5. The binding of a signal molecule by a receptor protein can
   a. activate a second messenger inside the receiving cell.
   b. trigger enzyme activity in the cell.
   c. change the permeability of the cell.
   d. All of the above

6. The drawing below shows a plant cell that has become shriveled after having been placed in a solution. Is the solution most likely hypertonic, hypotonic, or isotonic? Explain your reasoning.

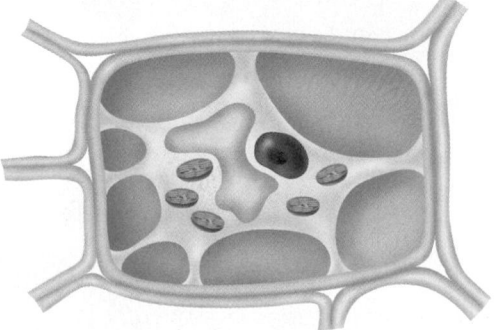

7. Define the term *homeostasis,* and explain how the sodium-potassium pump contributes to homeostasis in an animal.

8. When a cell takes in a food particle by endocytosis, the vesicle that is formed may fuse with a lysosome. How would fusion with a lysosome help the cell digest the food particle?

9. **BIOWatch** Which types of cells does HIV infect, and which of these does it destroy?

10. **Concept Mapping** Make a concept map that shows how cells maintain homeostasis. Include the following terms in your map: *concentration gradient, diffusion, osmosis,* and *carrier protein.*

Critical Thinking

11. **Applying Information** Identify the property of a cell membrane that allows particles to pass through it. Predict how a cell membrane that allows some types of ions and molecules through but not others would affect equilibrium between the cell and the fluid around it.

Alternative Assessment

12. **Finding Information** Use library or Internet resources to investigate problems in transport of molecules across the cell membrane. Several human diseases, including cystic fibrosis, hypercholesterolemia, and myasthenia gravis affect the function of specific transport proteins. Describe the symptoms of and treatments for the diseases. Summarize your findings in a written report.

13. **Interactive Tutor Unit 1 Cell Transport and Homeostasis** Write a report summarizing the roles of osmosis and diffusion in the preservation and maintenance of body organs donated for transplants. Why must the organs be preserved in special solutions prior to a transplant? Find out what kinds of substances these solutions contain.

Standardized Test Prep

Understanding Concepts

Directions (1–3): For *each* question, write on a separate sheet of paper the letter of the correct answer.

1 A cell begins to swell when placed in an unknown solution. What can you conclude about the solution?
  A. The solution is hypertonic.
  B. The solution is hypotonic.
  C. The solution is isotonic.
  D. The solution is saturated.

2 Which of the following processes allows the cell to dispose of wastes?
  F. endocytosis
  G. exocytosis
  H. facilitated diffusion
  I. sodium-potassium pumping

3 A gelatin block is prepared with a chemical indicator that turns pink in the presence of a base. The block is enclosed in a membrane and placed in a beaker of basic solution. After half an hour, the block begins to turn pink. Why does the gelatin turn pink?
  A. The membrane is impermeable to acid, so the gelatin becomes an acid.
  B. The membrane is impermeable to base, so the gelatin becomes a base.
  C. The base diffuses down its concentration gradient through the membrane and into the gelatin.
  D. The acid diffuses down its concentration gradient through the membrane and into the gelatin.

Directions (4): For the following question, write a short response.

4 How do the processes of osmosis and diffusion explain why cooking dried pasta in boiling water makes the pasta soft?

Test TIP

Pay close attention to words such as *not, only, rather,* and *some* that appear in questions.

Reading Skills

Directions (5): Read the passage below. Then answer the question.

The diffusion of water through a selectively permeable membrane is called osmosis. Osmosis is a form of diffusion that involves the movement of water down its concentration gradient. When solutions on either side of a membrane have different concentrations of dissolved particles, they also have different concentrations of water molecules that are not interacting with the dissolved particles. The water molecules move from the side with a lower concentration of dissolved particles to the side with a higher concentration.

5 A student wants to model osmosis by placing a mesh bag in a solution of salt water. The bag fills up with a saltwater solution that is the same concentration as the solution outside of the bag. Why is this **not** a good model for osmosis?
  F. Salt water is not a solution.
  G. The bag is impermeable to salt.
  H. The bag is not selectively permeable.
  I. The solution is not concentrated enough.

Interpreting Graphics

Directions (6): Base your answer to question 6 on the diagram below.

Sodium-Potassium Pump

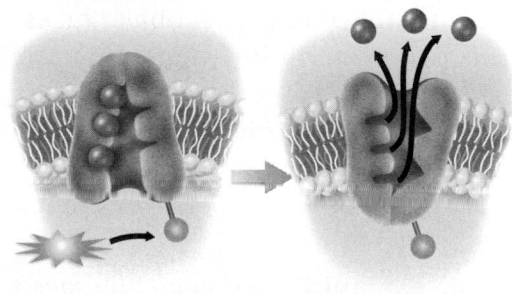

A   *B*

6 What substance is released from the pump in step *B*?
  A. ATP   **C.** potassium ion
  B. phosphate group   **D.** sodium ion

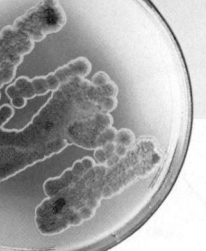

Exploration Lab

Analyzing the Effect of Cell Size on Diffusion

SKILLS
- Using scientific methods
- Collecting, organizing, and graphing data

OBJECTIVES
- **Relate** the size of a cell to its surface area–to-volume ratio.
- **Predict** how the surface area–to-volume ratio of a cell will affect the diffusion of substances into the cell.

MATERIALS
- safety goggles
- lab apron
- disposable gloves
- block of phenolphthalein agar (3 × 3 × 6 cm)
- plastic knife
- metric ruler
- 250 mL beaker
- 150 mL of vinegar
- plastic spoon
- paper towel

CAUTION: Always wear safety goggles and a lab apron to protect your eyes and clothing.

CAUTION: Do not touch or taste any chemicals. Know the location of the emergency shower and eyewash station and how to use them. If you get a chemical on your skin or clothing, wash it off at the sink while calling to the teacher. Notify the teacher of a spill. Spills should be cleaned up promptly, according to your teacher's directions.

CAUTION: Glassware is fragile. Notify the teacher of broken glass or cuts. Do not clean up broken glass or spills with broken glass unless the teacher tells you to do so.

Before You Begin

Substances enter and leave a cell in several ways, including by **diffusion.** How efficiently a cell can exchange substances depends on the **surface area–to-volume ratio** (surface area ÷ volume) of the cell. **Surface area** is the size of the outside of an object. **Volume** is the amount of space an object takes up. In this lab, you will investigate how cell size affects the diffusion of substances into a cell. To do this, you will make cell models using agar that contains an indicator. This indicator will change color when an acidic solution diffuses into it.

1. Write a definition for each boldface term in the paragraph above.

2. Based on the objectives for this lab, write a question you would like to explore about cell size and diffusion.

Procedure

PART A: Design an Experiment

1. Work with members of your lab group to explore one of the questions written for step 2 of **Before You Begin.** To explore the question, design an experiment that uses the materials listed for this lab.

You Choose

As you design your experiment, decide the following:

a. what question you will explore

b. what hypothesis you will test

c. how many "cells" (agar cubes) you will have and what sizes they will be

d. how long to leave the "cells" in the vinegar

e. how to determine how far the vinegar diffused into a "cell"

f. how to prevent contamination of agar cubes as you handle them

g. what data to record in your data table

2. Write a procedure for your experiment. Make a list of all the safety precautions you will take. Have your teacher approve your procedure and safety precautions before you begin the experiment.

PART B: Conduct Your Experiment

3. Put on safety goggles, a lab apron, and disposable gloves.

4. Carry out the experiment you designed. Record your observations in your data table.

PART C: Cleanup and Disposal

5. Dispose of solutions, broken glass, and agar in the designated waste containers. Do not pour chemicals down the drain or put lab materials in the trash unless your teacher tells you to do so.

6. Clean up your work area and all lab equipment. Return lab equipment to its proper place. Wash your hands thoroughly before you leave the lab and after you finish all work.

Analyze and Conclude

1. **Summarizing Results** Describe any changes in the appearance of the cubes.

2. **Summarizing Results** Make a graph using your group's data. Plot "Diffusion Distance (mm)" on the vertical axis. Plot "Surface Area–to-Volume Ratio" on the horizontal axis.

3. **Analyzing Results** Using the graph you made in item 2, make a statement about the relationship between the surface area–to-volume ratio and the distance a substance diffuses.

4. **Summarizing Results** Make a graph using your group's data. Plot "Rate of Diffusion (mm/min)" (distance vinegar moved ÷ time) on the vertical axis. Plot "Surface Area–to-Volume Ratio" on the horizontal axis.

5. **Analyzing Results** Using the graph you made in item 4, make a statement about the relationship between the surface area–to-volume ratio and the rate of diffusion of a substance.

6. **Evaluating Methods** In what ways do your agar models simplify or fail to simulate the features of real cells?

7. **Calculating** Calculate the surface area and volume of a cube with a side length of 5 cm. Calculate the surface area and volume of a cube with a side length of 10 cm. Determine the surface area–to-volume ratio of each of these cubes. Which cube has the greater surface area–to-volume ratio?

8. **Evaluating Conclusions** How does the size of a cell affect the diffusion of substances into the cell?

9. **Further Inquiry** Write a new question about cell size and diffusion that could be explored with another investigation.

? Do You Know?

Do research in the library or media center to answer these questions:

1. How does cell transport in prokaryotic cells differ from cell transport in eukaryotic cells?

2. Which of the following molecules can diffuse across the cell membrane without the help of a transport protein: water, carbohydrates, lipids, proteins?

Use the following Internet resources to explore your own questions about cell size and cell transport.

internet connect

www.scilinks.org
Topic: Cell Membrane
Keyword: HX4035

SCiLINKS. Maintained by the National Science Teachers Association

Deer eating leaves

CHAPTER

5 Photosynthesis and Cellular Respiration

✓ Quick Review

Answer the following without referring to earlier sections of your book.

1. **Describe** different kinds of chemical bonds. *(Chapter 2, Section 1)*

2. **List** the properties of organic compounds. *(Chapter 2, Section 3)*

3. **Distinguish** between mitochondria and chloroplasts. *(Chapter 3, Section 3)*

4. **Differentiate** between passive transport and active transport. *(Chapter 4, Section 2)*

Did you have difficulty? *For help, review the sections indicated.*

Reading Activity

Before you read this chapter, write a short list of all of the things you know about photosynthesis and cellular respiration. Then write a list of the things that you want to know about photosynthesis and cellular respiration. Save your list, and to assess what you have learned, see how many of your own questions you can answer after reading this chapter.

internet connect

www.scilinks.org
National Science Teachers Association *sci*LINKS Internet resources are located throughout this chapter.

*sci*LINKS® Maintained by the National Science Teachers Association

This deer, like many other organisms, depends upon the energy stored in plants for food. Food supplies the energy needed for cellular activities.

Energy and Living Things

Objectives

- **Analyze** the flow of energy through living systems.
- **Compare** the metabolism of autotrophs with that of heterotrophs.
- **Describe** the role of ATP in metabolism.
- **Describe** how energy is released from ATP.

Key Terms

photosynthesis
autotroph
heterotroph
cellular respiration

Energy in Living Systems

You get energy from the food you eat. Where does the energy in food come from? Directly or indirectly, almost all of the energy in living systems needed for metabolism comes from the sun. **Figure 1** shows how energy flows through living systems. Energy from the sun enters living systems when plants, algae, and certain prokaryotes absorb sunlight. Some of the energy in sunlight is captured and used to make organic compounds. These organic compounds store chemical energy and can serve as food for organisms.

Building Molecules That Store Energy

Metabolism involves either using energy to build molecules or breaking down molecules in which energy is stored. **Photosynthesis** is the process by which light energy is converted to chemical energy. Organisms that use energy from sunlight or from chemical bonds in inorganic substances to make organic compounds are called **autotrophs** *(AWT oh trohfs)*. Most autotrophs, especially plants, are photosynthetic organisms. Some autotrophs, including certain prokaryotes, use chemical energy from inorganic substances to make organic compounds. Prokaryotes found near deep-sea volcanic vents live in perpetual darkness. Sunlight does not reach the bottom of the ocean. These prokaryotes get energy, however, from chemicals flowing out of the vents.

Figure 1   Flow of energy

Energy flows from sunlight or inorganic substances to autotrophs, such as grasses, and then to heterotrophs, such as rabbits and foxes.

Light energy

1. Plants convert light energy to chemical energy.

2. Rabbits get energy by eating plants.

3. Foxes get energy by eating rabbits.

Breaking Down Food for Energy

The chemical energy in organic compounds can be transferred to other organic compounds or to organisms that consume food. Organisms that must get energy from food instead of directly from sunlight or inorganic substances are called **heterotrophs** (*HEHT uhr oh trohfs*). Heterotrophs, including humans, get energy from food through the process of cellular respiration. **Cellular respiration** is a metabolic process similar to burning fuel. While burning converts almost all of the energy in a fuel to heat, cellular respiration releases much of the energy in food to make ATP. This ATP provides cells with the energy they need to carry out the activities of life.

WORD Origins

● The words *autotroph* and *heterotroph* have the same suffix, *-troph*, which is from the Greek word *trophikos*, meaning "to feed." The prefix *auto-* is from the Greek word *autos*, meaning "self," and the prefix *hetero-* is from the Greek word *heteros*, meaning "other."

Transfer of Energy to ATP

The word *burn* is often used to describe how cells get energy from food. Although the overall processes are similar, the "burning" of food in living cells clearly differs from the burning of a log in a campfire. When a log burns, the energy stored in wood is released quickly as heat and light. But in cells, chemical energy stored in food molecules is released gradually in a series of enzyme-assisted chemical reactions. As shown in **Figure 2,** the product of one chemical reaction becomes a reactant in the next reaction. In the breakdown of starch, for example, each reaction releases energy.

When cells break down food molecules, some of the energy in the molecules is released as heat. Much of the remaining energy is stored temporarily in molecules of ATP. Like money, ATP is a portable form of energy "currency" inside cells. ATP delivers energy wherever energy is needed in a cell. The energy released from ATP can be used to power other chemical reactions, such as those that build molecules. In cells, most chemical reactions require less energy than is released from ATP. Therefore, enough energy is released from ATP to drive most of a cell's activities.

Figure 2 Breakdown of starch

Energy is released from starch in a series of enzyme-assisted chemical reactions.

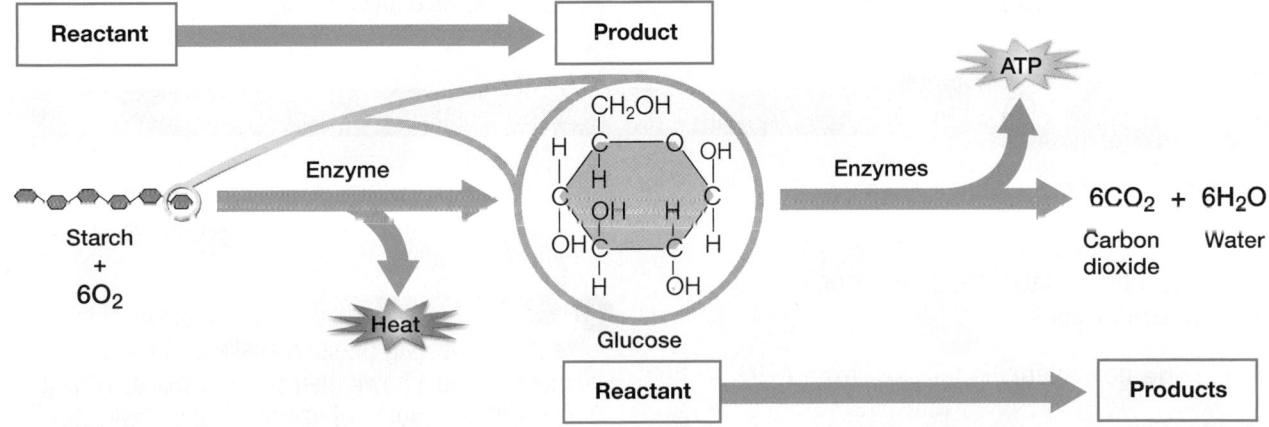

ATP

Recall that ATP (adenosine triphosphate) is a nucleotide with two extra energy-storing phosphate groups. As shown in **Figure 3,** the three phosphate groups in ATP form a chain that branches from a five-carbon sugar called ribose *(RIE bohs)*. This phosphate "tail" is unstable because the phosphate groups are negatively charged and therefore repel each other. The phosphate groups store energy like a compressed spring does. This energy is released when the bonds that hold the phosphate groups together are broken.

Breaking the outer phosphate bond requires an input of energy. Much more energy is released, however, than is consumed by the reaction. As shown in Figure 3, the removal of a phosphate group from ATP produces adenosine diphosphate, or ADP. This reaction releases energy in a way that enables cells to use the energy. The following equation summarizes this reaction:

$$H_2O + ATP \rightarrow ADP + P + energy$$

Cells use the energy released by this reaction to power metabolism. In some chemical reactions, two phosphate groups are removed from ATP instead of just one. This tends to make the reaction irreversible because the pair of phosphate groups that is removed is not available for the reverse reaction. Rather, the pair is quickly split into two single phosphate groups.

Figure 3   ATP releases energy

When the outer phosphate group detaches from ATP, energy is released.

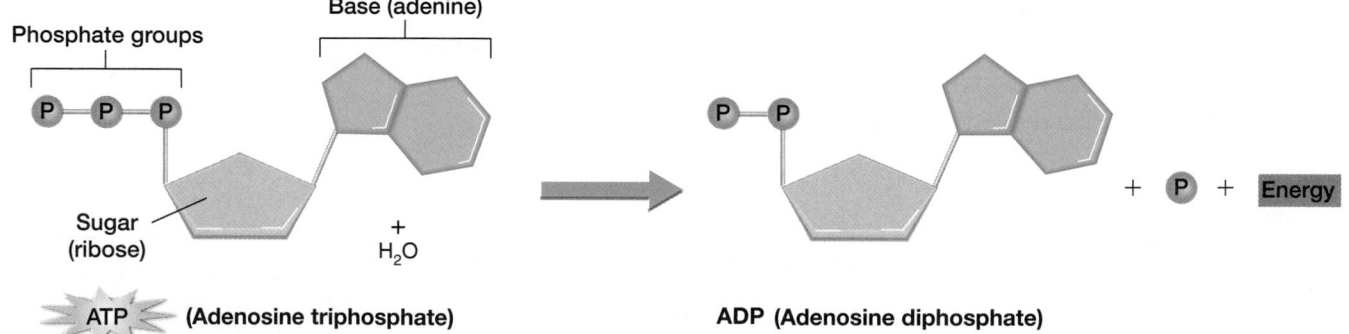

Section 1 Review

1. **Identify** the primary source of energy that flows through most living systems.

2. **Compare** the metabolism of autotrophs with that of heterotrophs.

3. **Describe** how energy is released from ATP.

4. **Critical Thinking Inferring Relationships** How can the energy in the food that a fox eats be traced back to the sun?

5. **Critical Thinking Analyzing Patterns** Explain how life involves a continuous flow of energy.

6. **Standardized Test Prep** A grasshopper obtains energy by eating grass. A snake eats the grasshopper, and a hawk then eats the snake. What is the original source of energy for the hawk?
   - **A** the snake
   - **B** the grasshopper
   - **C** the grass
   - **D** the sun

Photosynthesis

Using the Energy in Sunlight

When you eat a hamburger, you get energy from the sun indirectly. Plants, such as grass, capture the energy in sunlight. The beef in a hamburger comes from a cow that ate grass. The bun, lettuce, and tomato come from plants. With few exceptions, you end up with plants whenever you trace your food back to its origin. Plants, algae, and some bacteria capture about 1 percent of the energy in the sunlight that reaches Earth and convert it to chemical energy through the process of photosynthesis.

The Stages of Photosynthesis

Photosynthesis is the process that provides energy for almost all life. As **Figure 4** shows, photosynthesis has three stages:

Stage 1 Energy is captured from sunlight.

Stage 2 Light energy is converted to chemical energy, which is temporarily stored in ATP and the energy carrier molecule NADPH.

Stage 3 The chemical energy stored in ATP and NADPH powers the formation of organic compounds, using carbon dioxide, CO_2.

Photosynthesis occurs in the chloroplasts of plant cells and algae and in the cell membrane of certain prokaryotes. Photosynthesis can be summarized by the following equation:

$$6CO_2 + 6H_2O \xrightarrow{\text{light}} C_6H_{12}O_6 + 6O_2$$

carbon   water   sugars   oxygen
dioxide   gas

This equation, however, does not show how photosynthesis occurs. It merely says that six carbon dioxide molecules, six water molecules, and light are needed to form one six-carbon organic compound and six molecules of oxygen. Plants use the organic compounds they make during photosynthesis to carry out their life processes. For example, some of these sugars are used to form starch, which can be stored in stems or roots. The plant may later break down the starch to make ATP used to power metabolism. All of the proteins, nucleic acids, and other molecules of the cell are assembled from fragments of these sugars.

Figure 4 Photosynthesis

The process of photosynthesis occurs in three stages.

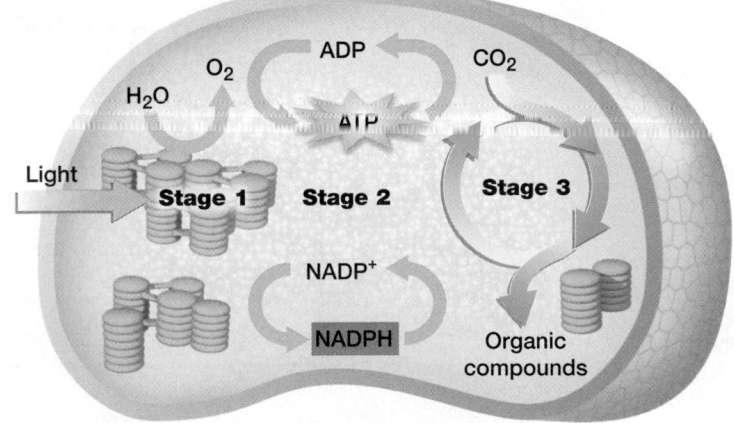

Stage One: Absorption of Light Energy

The chemical reactions that occur in the first and second stages of photosynthesis are sometimes called "light reactions," or light-dependent reactions. Without the absorption of light, these reactions could not occur. Light energy is used to make energy-storing compounds. Light is a form of radiation—energy in the form of waves that travel through space. Different types of radiation, such as light and heat, have different wavelengths (the distance between two consecutive waves). When the sun shines on you, your body is bombarded by many kinds of radiation from the sun. However, you can see only radiation known as visible light. You see wavelengths of visible light as different colors. As shown in **Figure 5,** sunlight contains all the wavelengths of visible light, red through violet.

Pigments

How does a human eye or a leaf absorb light? These structures contain light-absorbing substances called **pigments.** Pigments absorb only certain wavelengths and reflect all the others. **Chlorophyll** *(KLOR uh fihl),* the primary pigment involved in photosynthesis, absorbs mostly blue and red light and reflects green and yellow light. This reflection of green and yellow light makes many plants, especially their leaves, look green. Plants contain two types of chlorophyll, chlorophyll *a* and chlorophyll *b.* Both types of chlorophyll play an important role in plant photosynthesis.

The pigments that produce yellow and orange fall leaf colors, as well as the colors of many fruits, vegetables, and flowers, are called **carotenoids** *(kuh RAH tuh noydz).* Carotenoids absorb wavelengths of light different from those absorbed by chlorophyll, so having both pigments enables plants to absorb more light energy during photosynthesis. The graph in **Figure 6** shows the wavelengths of light absorbed by chlorophyll *a,* chlorophyll *b,* and carotenoids.

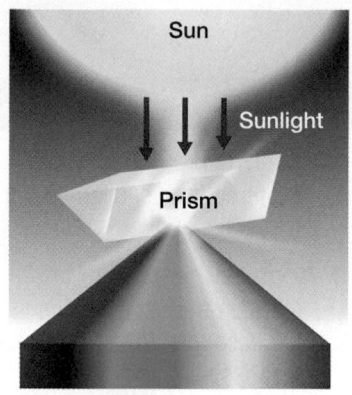

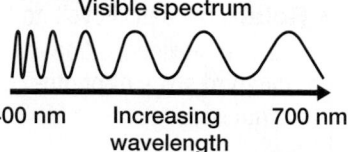

Figure 5   Visible spectrum. Sunlight contains a mixture of all the wavelengths (colors) of visible light. When sunlight passes through a prism, the prism separates the light into different colors.

Figure 6   Light absorption during photosynthesis. Chlorophylls absorb mostly violet, blue, and red light, while carotenoids absorb mostly blue and green light.

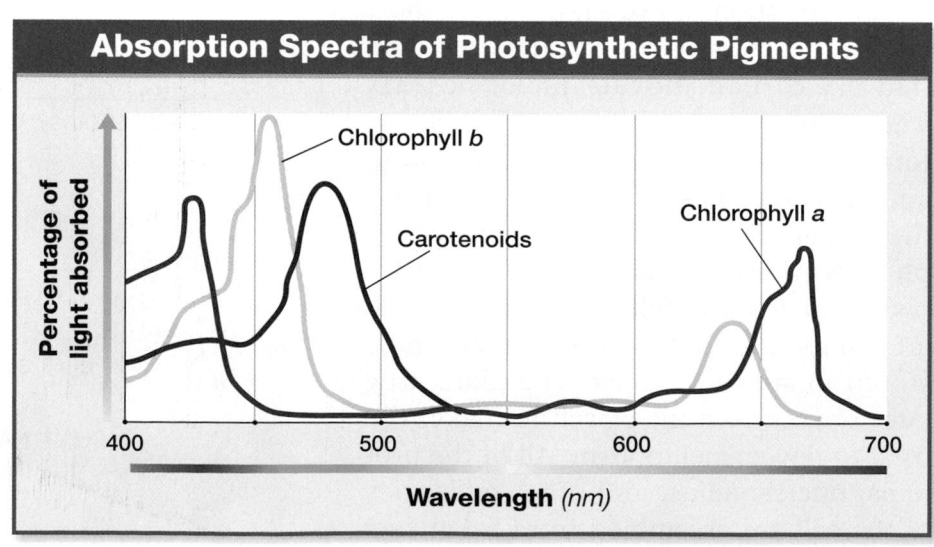

Production of Oxygen

As shown in **Figure 7,** pigments involved in plant photosynthesis are located in the chloroplasts of leaf cells. Clusters of pigments are embedded in the membranes of disk-shaped structures called **thylakoids** *(THIE luh koydz).* When light strikes a thylakoid in a chloroplast, energy is transferred to electrons in chlorophyll. This energy transfer causes the electrons to jump to a higher energy level. Electrons with extra energy are said to be "excited." This is how plants first capture energy from sunlight.

Excited electrons jump from chlorophyll molecules to other nearby molecules in the thylakoid membrane, where the electrons are used to power the second stage of photosynthesis. The excited electrons that leave chlorophyll molecules must be replaced by other electrons. Plants get these replacement electrons from water molecules, H_2O. Water molecules are split by an enzyme inside the thylakoid. When water molecules are split, chlorophyll molecules take the electrons from the hydrogen atoms, H, leaving hydrogen ions, H^+. The remaining oxygen atoms, O, from the disassembled water molecules combine to form oxygen gas, O_2.

internet connect

www.scilinks.org
Topic: Light Absorption
Keyword: HX4116

SCLINKS Maintained by the National Science Teachers Association

Figure 7 Chloroplast

Pigment molecules are embedded in thylakoid membranes, as are other molecules that participate in photosynthesis.

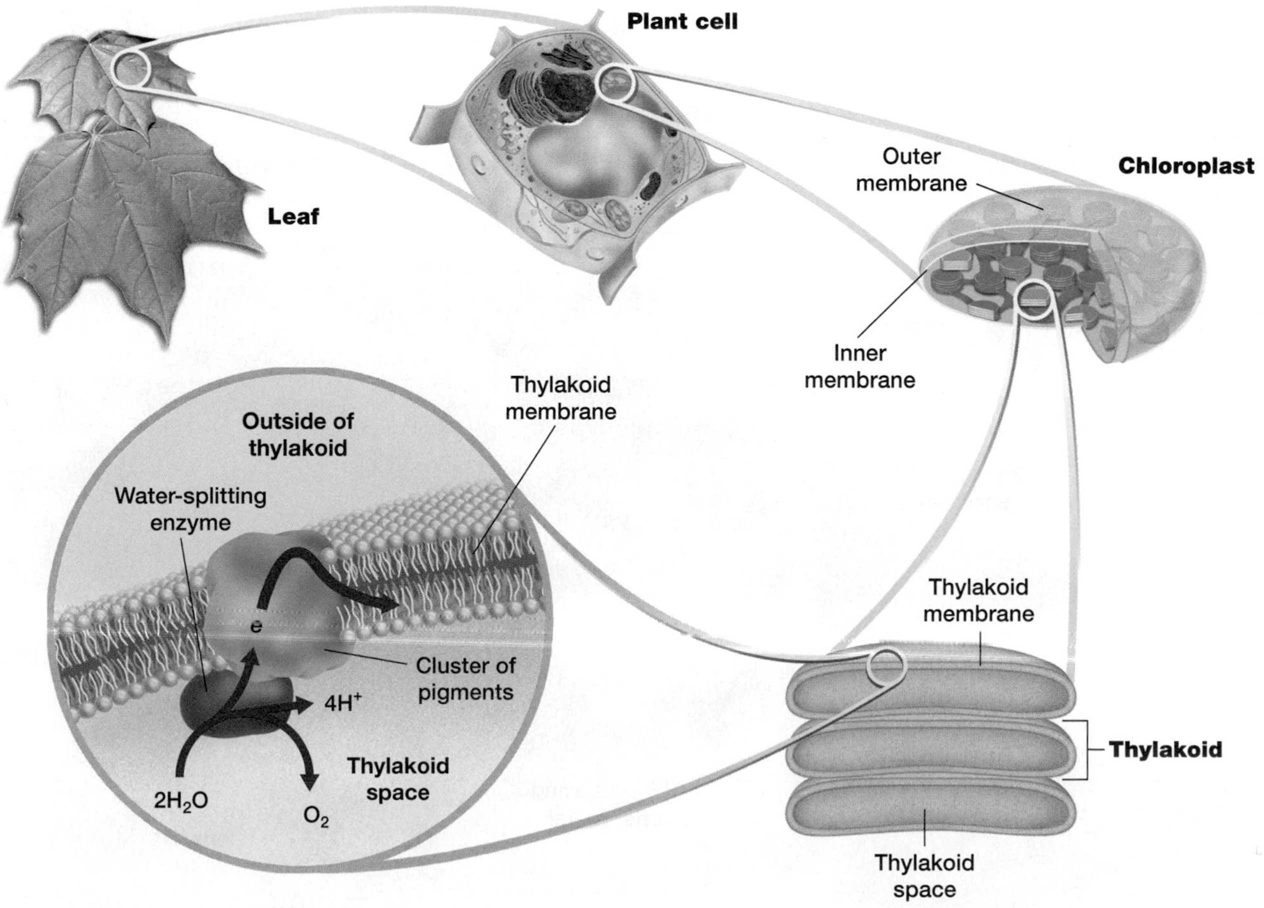

Plant cell

Leaf

Chloroplast

Outer membrane

Inner membrane

Thylakoid membrane

Outside of thylakoid

Water-splitting enzyme

e

2H₂O

4H⁺

O₂

Cluster of pigments

Thylakoid space

Thylakoid membrane

Thylakoid

Thylakoid space

Stage Two: Conversion of Light Energy

Excited electrons that leave chlorophyll molecules are used to produce new molecules, including ATP, that temporarily store chemical energy. First an excited electron jumps to a nearby molecule in the thylakoid membrane. Then the electron is passed through a series of molecules along the thylakoid membrane like a ball being passed down a line of people. The series of molecules through which excited electrons are passed along a thylakoid membrane are called **electron transport chains.** Trace the path taken by excited electrons in the electron transport chains shown in **Figure 8.**

Electron Transport Chains

How are electron transport chains used to make molecules that temporarily store energy in the cell? The first electron transport chain shown in Figure 8 lies between the two large green clusters of pigment molecules. This type of electron transport chain contains a protein (the large purple molecule) that acts as a membrane pump. Excited electrons lose some of their energy as they each pass through this protein. The energy lost by the electrons is used to pump hydrogen ions, H^+, into the thylakoid. Recall that hydrogen ions are also produced when water molecules are split inside the thylakoid.

As the process continues, hydrogen ions become more concentrated inside the thylakoid than outside, producing a concentration gradient across the thylakoid membrane. As a result, hydrogen ions have a tendency to diffuse back out of the thylakoid down their

Study *TIP*

● **Interpreting Graphics**
Look closely at Figure 8. Electrons are represented by the symbol e^-. The red arrows show the path of excited electrons. Hydrogen ions are represented by the symbol H^+. The blue arrows show the path of hydrogen ions that cross the thylakoid membrane.

Figure 8　Electron transport chains of photosynthesis

Electron transport chains (represented by the red lines) convert light energy to chemical energy.

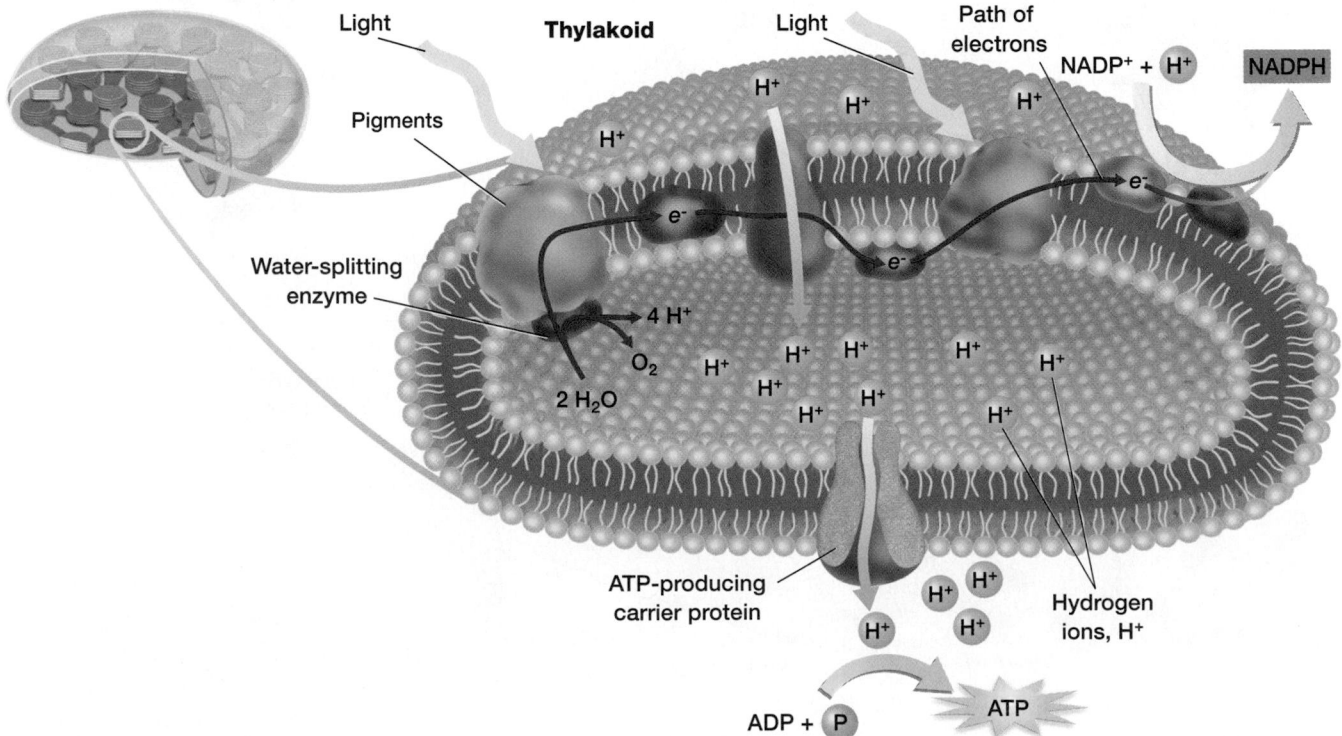

concentration gradient through specialized carrier proteins (illustrated on the lower surface of the thylakoid). These carrier proteins are unusual because they function both as an ion channel and as an enzyme. As hydrogen ions pass through the channel portion of the protein, the protein catalyzes a reaction in which a phosphate group is added to a molecule of ADP, making ATP. Thus, the movement of hydrogen ions across the thylakoid membrane through these proteins provides the energy needed to make ATP, which is used to power the third stage of photosynthesis.

While one electron transport chain provides energy used to make ATP, a second electron transport chain provides energy used to make NADPH. **NADPH** is an electron carrier that provides the high-energy electrons needed to make carbon-hydrogen bonds in the third stage of photosynthesis. The second electron transport chain shown in Figure 8 lies to the right of the second green pigment molecule. In this second chain, excited electrons combine with hydrogen ions as well as an electron acceptor called NADP⁺, forming NADPH.

The light-dependent reactions of photosynthesis can be summarized as follows. Pigment molecules in the thylakoids of chloroplasts absorb light energy. Electrons in the pigments are excited by light and move through electron transport chains in thylakoid membranes. These electrons are replaced by electrons from water molecules, which are split by an enzyme. Oxygen atoms from water molecules combine to form oxygen gas. Hydrogen ions accumulate inside thylakoids, setting up a concentration gradient that provides the energy to make ATP.

Identifying a Product of Photosynthesis

You can use the following procedure to identify the gas given off by a photosynthetic organism.

Materials

MBL or CBL system with appropriate software, test tube or small glass jar, sprig of *Elodea,* distilled water, cool light source, dissolved oxygen (DO) probe

Procedure

1. Set up an MBL/CBL system to collect and graph data from a dissolved oxygen probe at 30-second intervals for 60 data points. Calibrate the DO probe.

2. Place a sprig of *Elodea* in a test tube or glass jar, and fill the test tube or jar with distilled water.

3. Place the test tube or glass jar under a cool light source, and lower a DO probe into the water. Collect data for 30 minutes.

4. When data collection is complete, view the graph of your data. If possible, print the graph. Otherwise, sketch the graph on paper.

Analysis

1. **Infer** the cause of any change you observed.

2. **Propose** a control for this experiment.

3. **Critical Thinking Evaluating Hypotheses** Explain how your data support or do not support the hypothesis that photosynthetic organisms give off oxygen.

Stage Three: Storage of Energy

In the first and second stages of photosynthesis, light energy is used to make ATP and NADPH, which temporarily store chemical energy. These stages are therefore considered light-dependent. In the third (final) stage of photosynthesis, however, carbon atoms from carbon dioxide in the atmosphere are used to make organic compounds in which chemical energy is stored. The transfer of carbon dioxide to organic compounds is called **carbon dioxide fixation.** The reactions that "fix" carbon dioxide are sometimes called "dark reactions," or light-independent reactions. Among photosynthetic organisms, there are several ways in which carbon dioxide is fixed.

Calvin Cycle

The most common method of carbon dioxide fixation is the Calvin cycle. The **Calvin cycle** is a series of enzyme-assisted chemical reactions that produces a three-carbon sugar. The Calvin cycle is summarized in **Figure 9.**

Step ❶ In carbon dioxide fixation, each molecule of carbon dioxide, CO_2, is added to a five-carbon compound by an enzyme.

Figure 9

Calvin Cycle

The Calvin cycle is a common method of carbon dioxide fixation.

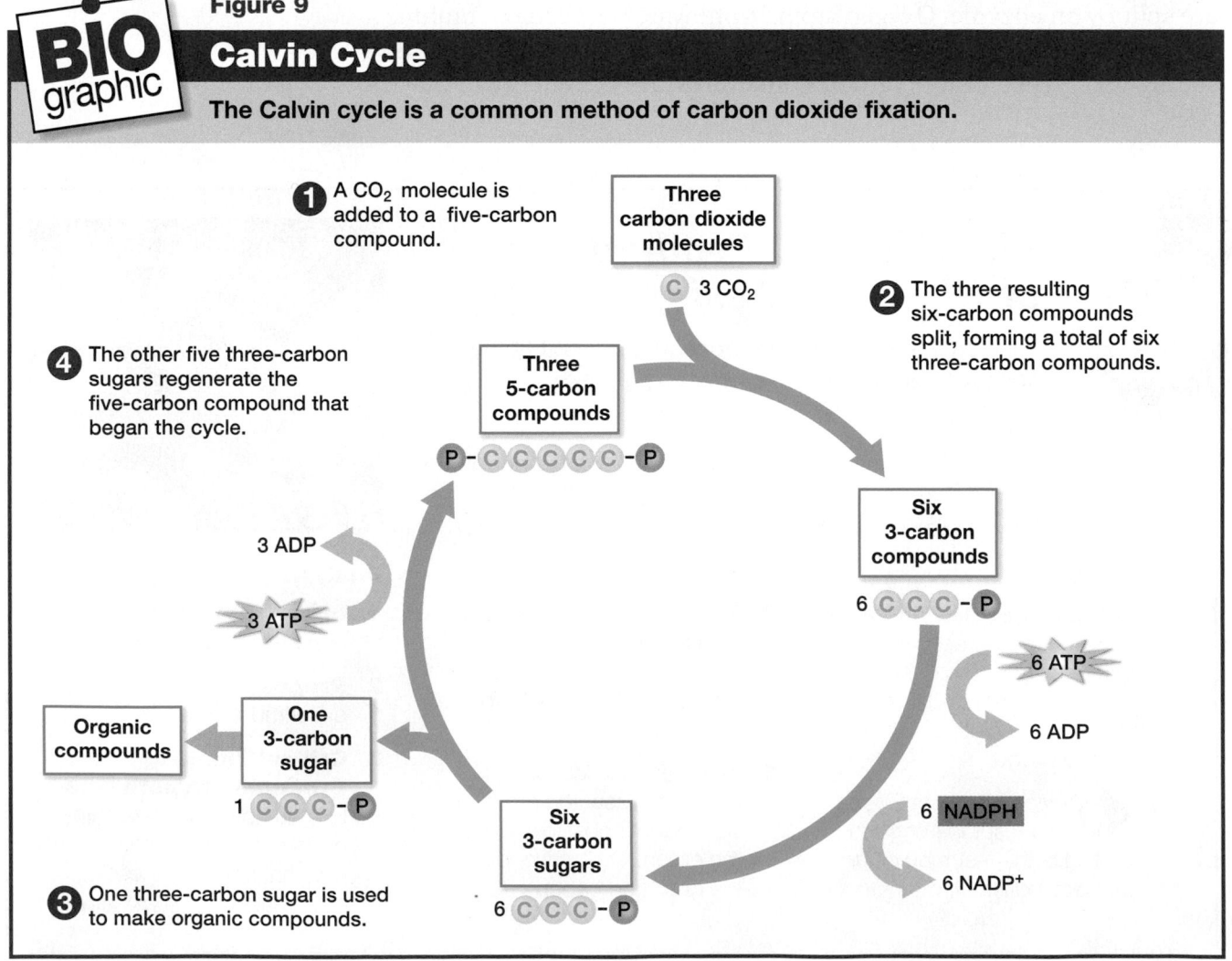

❶ A CO_2 molecule is added to a five-carbon compound.

Three carbon dioxide molecules

ⓒ 3 CO_2

❷ The three resulting six-carbon compounds split, forming a total of six three-carbon compounds.

❹ The other five three-carbon sugars regenerate the five-carbon compound that began the cycle.

Three 5-carbon compounds

P-ⒸⒸⒸⒸⒸ-P

3 ADP

3 ATP

Six 3-carbon compounds

6 ⒸⒸⒸ-P

6 ATP

6 ADP

Organic compounds

One 3-carbon sugar

1 ⒸⒸⒸ-P

6 NADPH

6 NADP+

❸ One three-carbon sugar is used to make organic compounds.

Six 3-carbon sugars

6 ⒸⒸⒸ-P

Step ② The resulting six-carbon compound splits into two three-carbon compounds. Phosphate groups from ATP and electrons from NADPH are added to the three-carbon compounds, forming three-carbon sugars.

Step ③ One of the resulting three-carbon sugars is used to make organic compounds—including starch and sucrose—in which energy is stored for later use by the organism.

Step ④ The other three-carbon sugars are used to regenerate the initial five-carbon compound, thereby completing the cycle.

The Calvin cycle is named for Melvin Calvin, the American biochemist who worked out the chemical reactions in the cycle. The reactions are cyclic—they recycle the five-carbon compound needed to begin the cycle again. A total of three carbon dioxide molecules must enter the Calvin cycle to produce each three-carbon sugar that will be used to make other organic compounds. These organic compounds provide the organism with energy for growth and metabolism. The energy used in the Calvin cycle is supplied by ATP and NADPH made during the second stage of photosynthesis.

Factors that Affect Photosynthesis

Photosynthesis is directly affected by various environmental factors. The most obvious of these factors is light. In general, the rate of photosynthesis increases as light intensity increases until all the pigments are being used. At this saturation point, the reactions of the Calvin cycle cannot proceed any faster. The overall rate of photosynthesis is thus limited by the slowest step, which occurs in the Calvin cycle. The carbon dioxide concentration affects the rate of photosynthesis in a similar manner. Once a certain concentration of carbon dioxide is present, photosynthesis cannot proceed any faster.

Photosynthesis is most efficient within a certain range of temperatures. Like all metabolic processes, photosynthesis involves many enzyme-assisted chemical reactions. Recall that unfavorable temperatures may inactivate certain enzymes.

Section 2 Review

① **Summarize** how photosynthetic organisms capture the energy in sunlight.

② **Compare** the roles of water molecules and hydrogen ions in electron transport chains.

③ **Describe** the role of the Calvin cycle in the third stage of photosynthesis.

④ **Critical Thinking Organizing Information** Make a table in which you identify the role of each of the following in photosynthesis: light, water, pigments, ATP, NADPH, and carbon dioxide.

⑤ **Critical Thinking Inferring Relationships** What combination of environmental factors affects the rate of photosynthesis?

⑥ **Standardized Test Prep** During photosynthesis, plants store energy in
A ADP.
B carbon dioxide.
C 3-carbon sugars.
D water.

Cellular Respiration

Objectives

- **Summarize** how glucose is broken down in the first stage of cellular respiration.

- **Describe** how ATP is made in the second stage of cellular respiration.

- **Identify** the role of fermentation in the second stage of cellular respiration.

- **Evaluate** the importance of oxygen in aerobic respiration.

Key Terms

aerobic
anaerobic
glycolysis
NADH
Krebs cycle
FADH₂
fermentation

Cellular Energy

Most of the foods we eat contain usable energy. Much of the energy in a hamburger, for example, is stored in proteins, carbohydrates, and fats. But before you can use that energy, it is transferred to ATP. Like in most organisms, your cells transfer the energy in organic compounds, especially glucose, to ATP through a process called cellular respiration. Oxygen in the air you breathe makes the production of ATP more efficient, although some ATP is made without oxygen. Metabolic processes that require oxygen are called **aerobic** *(ehr OH bihk)*. Metabolic processes that do not require oxygen are called **anaerobic** *(AN ehr oh bihk)*, meaning "without air."

The Stages of Cellular Respiration

Cellular respiration is the process cells use to harvest the energy in organic compounds, particularly glucose. The breakdown of glucose during cellular respiration can be summarized by the following equation:

$$\underset{\text{glucose}}{C_6H_{12}O_6} + \underset{\substack{\text{oxygen}\\\text{gas}}}{6O_2} \xrightarrow{\text{enzymes}} \underset{\substack{\text{carbon}\\\text{dioxide}}}{6CO_2} + \underset{\text{water}}{6H_2O} + \underset{\text{ATP}}{\text{energy}}$$

As **Figure 10** shows, cellular respiration occurs in two stages:

Stage 1 Glucose is converted to pyruvate *(PIE roo vayt)*, producing a small amount of ATP and NADH.

Stage 2 When oxygen is present, pyruvate and NADH are used to make a large amount of ATP. This process is called aerobic respiration. Aerobic respiration occurs in the mitochondria of eukaryotic cells and in the cell membrane of prokaryotic cells. When oxygen is not present, pyruvate is converted to either lactate *(LAK tayt)* or ethanol (ethyl alcohol) and carbon dioxide.

The equation above does not show *how* cellular respiration occurs. It simply shows that the complete enzyme-assisted breakdown of a glucose molecule uses six oxygen molecules and forms six carbon dioxide molecules, six water molecules, and ATP. Aerobic respiration produces most of the ATP made by cells. Intermediate products of aerobic respiration form the organic compounds that help build and maintain cells.

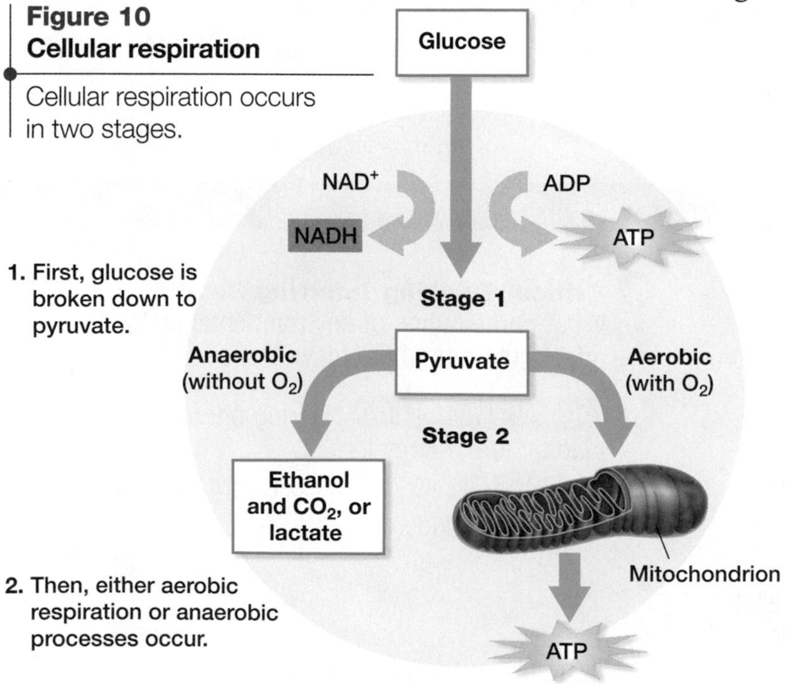

Figure 10
Cellular respiration

Cellular respiration occurs in two stages.

1. First, glucose is broken down to pyruvate.

2. Then, either aerobic respiration or anaerobic processes occur.

Stage One: Breakdown of Glucose

The primary fuel for cellular respiration is glucose, which is formed when carbohydrates such as starch and sucrose are broken down. If too few carbohydrates are available to meet an organism's glucose needs, other molecules, such as fats, can be broken down to make ATP. In fact, one gram of fat contains more energy than two grams of carbohydrates. Proteins and nucleic acids can also be used to make ATP, but they are usually used for building important cell parts.

Glycolysis

In the first stage of cellular respiration, glucose is broken down in the cytoplasm during a process called **glycolysis** *(glie KAHL uh sihs)*. Glycolysis is an enzyme-assisted anaerobic process that breaks down one six-carbon molecule of glucose to two three-carbon pyruvate ions. Recall that a molecule that has lost or gained one or more electrons is called an ion. Pyruvate is the ion of a three-carbon organic acid called pyruvic acid. The pyruvate produced during glycolysis still contains some of the energy that was stored in the glucose molecule.

As glucose is broken down, some of its hydrogen atoms are transferred to an electron acceptor called NAD$^+$. This forms an electron carrier called **NADH**. For cellular respiration to continue, the electrons carried by NADH are eventually donated to other organic compounds. This recycles NAD$^+$, making it available to accept more electrons. Glycolysis is summarized in **Figure 11.**

Step ❶ In a series of three reactions, phosphate groups from two ATP molecules are transferred to a glucose molecule.

Step ❷ In two reactions, the resulting six-carbon compound is broken down to two three-carbon compounds, each with a phosphate group.

Step ❸ Two NADH molecules are produced, and one more phosphate group is transferred to each three-carbon compound.

Step ❹ In a series of four reactions, each three-carbon compound is converted to a three-carbon pyruvate, producing four ATP molecules in the process.

Glycolysis uses two ATP molecules but produces four ATP molecules, yielding a net gain of two ATP molecules. Glycolysis is followed by another set of reactions that use the energy temporarily stored in NADH to make more ATP.

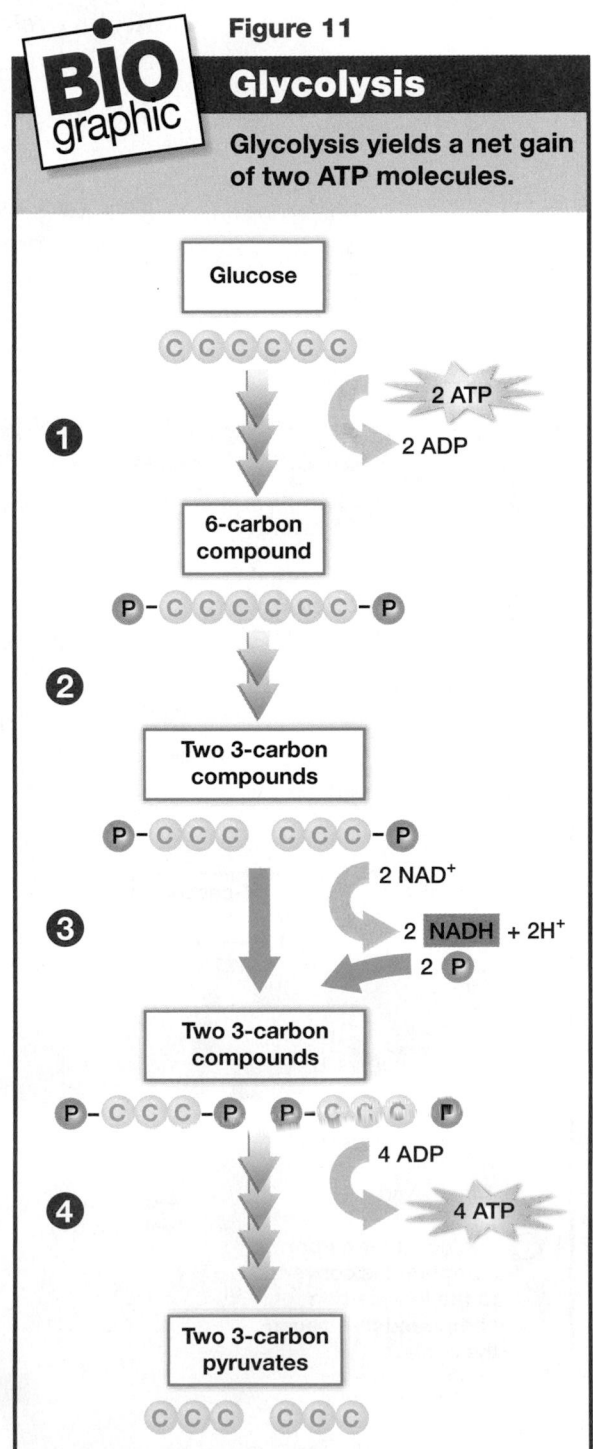

Figure 11

Glycolysis

Glycolysis yields a net gain of two ATP molecules.

Stage Two: Production of ATP

When oxygen is present, pyruvate produced during glycolysis enters a mitochondrion and is converted to a two-carbon compound. This reaction produces one carbon dioxide molecule, one NADH molecule, and one two-carbon acetyl *(uh SEET uhl)* group. The acetyl group is attached to a molecule called coenzyme A (CoA), forming a compound called acetyl-CoA *(uh SEET uhl-koh ay)*.

Krebs Cycle

Acetyl-CoA enters a series of enzyme-assisted reactions called the **Krebs cycle,** summarized in **Figure 12.** The cycle is named for the biochemist Hans Krebs, who first described the cycle in 1937.

Step ❶ Acetyl-CoA combines with a four-carbon compound, forming a six-carbon compound and releasing coenzyme A.

Step ❷ Carbon dioxide, CO_2, is released from the six-carbon compound, forming a five-carbon compound. Electrons are transferred to NAD^+, making a molecule of NADH.

Figure 12

BIOgraphic

Krebs Cycle

The Krebs cycle produces electron carriers that temporarily store chemical energy.

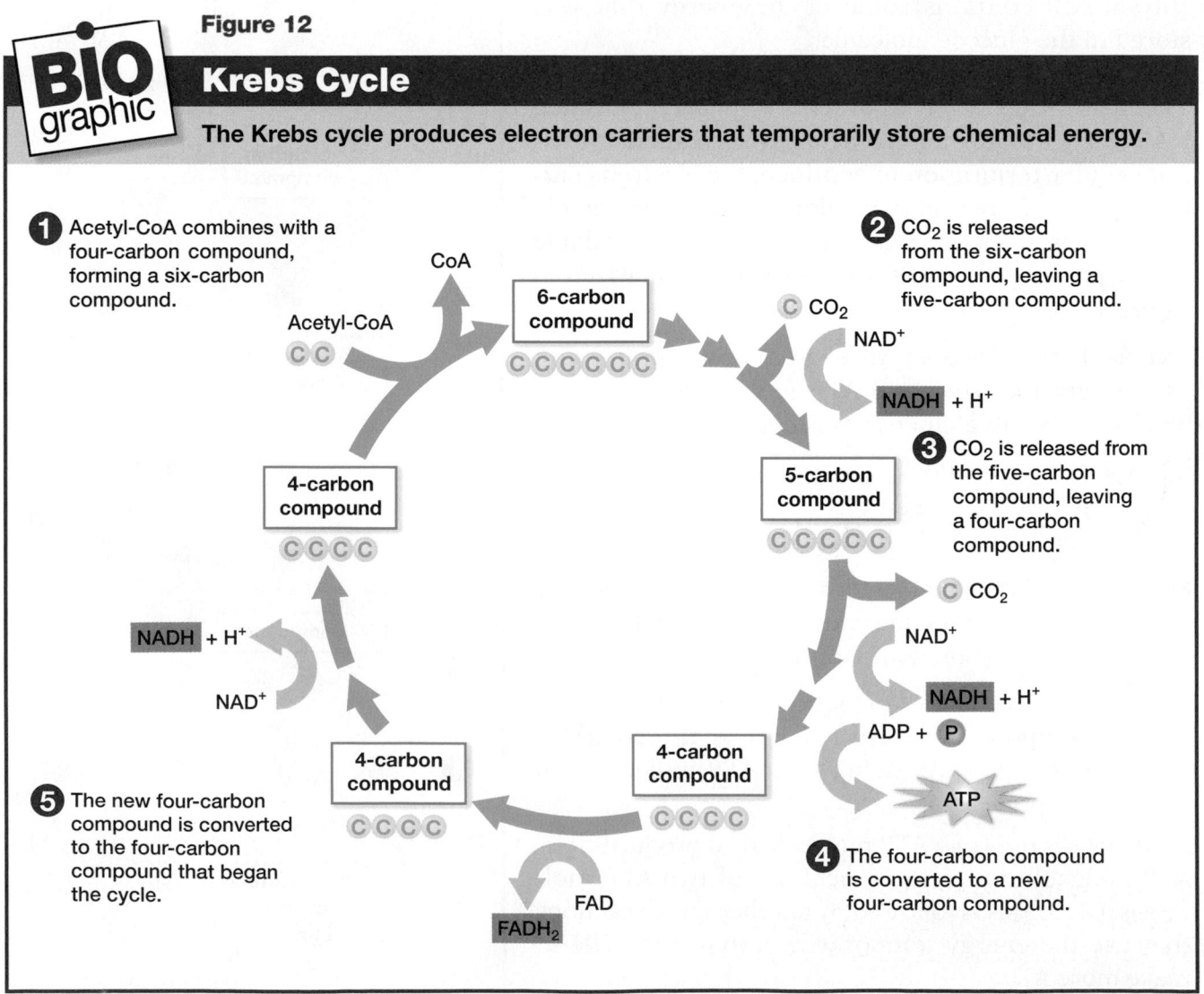

❶ Acetyl-CoA combines with a four-carbon compound, forming a six-carbon compound.

❷ CO_2 is released from the six-carbon compound, leaving a five-carbon compound.

❸ CO_2 is released from the five-carbon compound, leaving a four-carbon compound.

❹ The four-carbon compound is converted to a new four-carbon compound.

❺ The new four-carbon compound is converted to the four-carbon compound that began the cycle.

Step ❸ Carbon dioxide is released from the five-carbon compound, resulting in a four-carbon compound. A molecule of ATP is made, and a molecule of NADH is also produced.

Step ❹ The existing four-carbon compound is converted to a new four-carbon compound. Electrons are transferred to an electron acceptor called FAD, making a molecule of $FADH_2$. **$FADH_2$** is another type of electron carrier.

Step ❺ The new four-carbon compound is then converted to the four-carbon compound that began the cycle. Another molecule of NADH is produced.

After the Krebs cycle, NADH and $FADH_2$ now contain much of the energy that was previously stored in glucose and pyruvate. When the Krebs cycle is completed, the four-carbon compound that began the cycle has been recycled, and acetyl-CoA can enter the cycle again.

🔲 **internet** connect
www.scilinks.org
Topic: Aerobic Respiration
Keyword: HX4004
SC*LINKS* Maintained by the National Science Teachers Association

Electron Transport Chain

In aerobic respiration, electrons donated by NADH and $FADH_2$ pass through an electron transport chain, as shown in **Figure 13.** In eukaryotic cells, the electron transport chain is located in the inner membranes of mitochondria. The energy of these electrons is used to pump hydrogen ions out of the inner mitochondrial compartment. Hydrogen ions accumulate in the outer compartment, producing a concentration gradient across the inner membrane. Hydrogen ions diffuse back into the inner compartment through a carrier protein that adds a phosphate group to ADP, making ATP. At the end of the electron transport chain, hydrogen ions and spent electrons combine with oxygen molecules, O_2, forming water molecules, H_2O.

Figure 13   Electron transport chain of aerobic respiration

In the inner membranes of mitochondria, electron transport chains (represented by the red lines) make ATP.

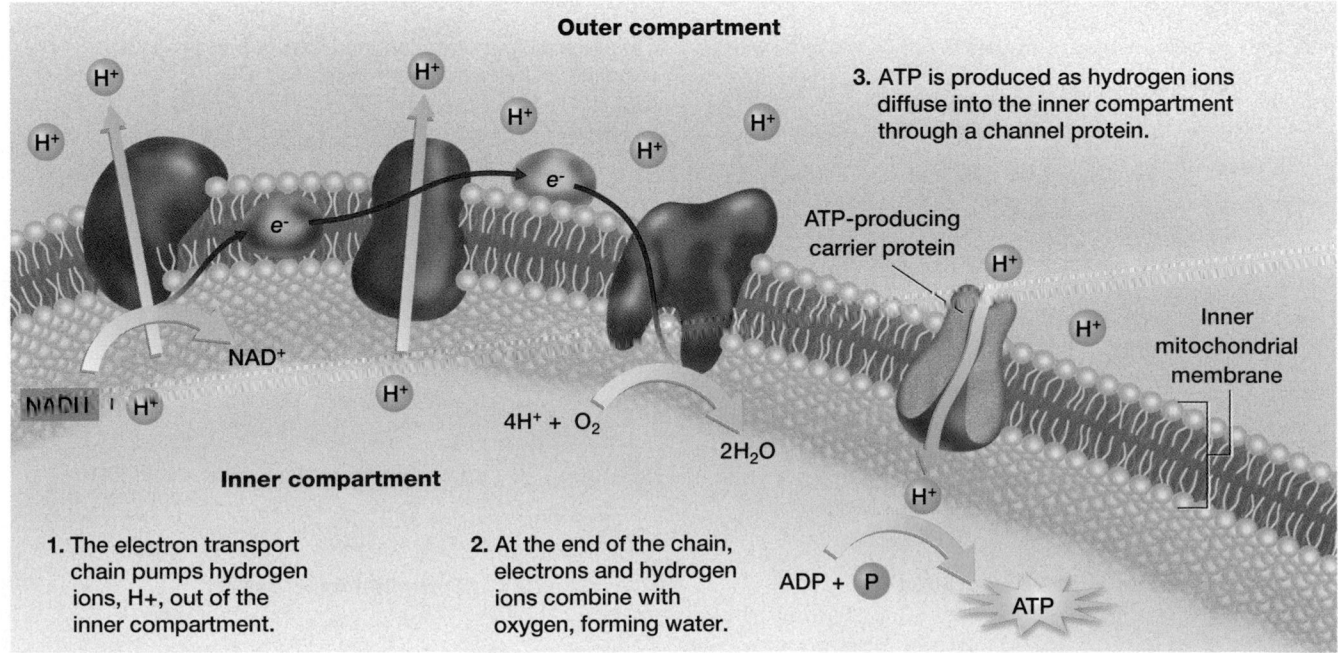

Respiration in the Absence of Oxygen

What happens when there is not enough oxygen for aerobic respiration to occur? The electron transport chain does not function because oxygen is not available to serve as the final electron acceptor. Electrons are not transferred from NADH, and NAD^+ therefore cannot be recycled. When oxygen is not present, NAD^+ is recycled in another way. Under anaerobic conditions, electrons carried by NADH are transferred to pyruvate produced during glycolysis. This process recycles NAD^+ needed to continue making ATP through glycolysis. The recycling of NAD^+ using an organic hydrogen acceptor is called **fermentation.** Prokaryotes carry out more than a dozen kinds of fermentation, all using some form of organic hydrogen acceptor to recycle NAD^+. Two important types of fermentation are lactic acid fermentation and alcoholic fermentation. Lactic acid fermentation by some prokaryotes and fungi is used in the production of foods such as yogurt and some cheeses, as shown in **Figure 14.**

Figure 14 Fermentation.
In cheese making, fungi or prokaryotes added to milk carry out lactic acid fermentation on some of the sugar in the milk.

Lactic Acid Fermentation

In some organisms, a three-carbon pyruvate is converted to a three-carbon lactate through lactic acid fermentation, as shown in **Figure 15.** Lactate is the ion of an organic acid called lactic acid. For example, during vigorous exercise pyruvate in muscles is converted to lactate when muscle cells must operate without enough oxygen. Fermentation enables glycolysis to continue producing ATP in muscles as long as the glucose supply lasts. Blood removes excess lactate from muscles. Lactate can build up in muscle cells if it is not removed quickly enough, sometimes causing muscle soreness.

Figure 15 Two types of fermentation

When oxygen is not present, cells recycle NAD+ through fermentation.

In lactic acid fermentation, pyruvate is converted to lactate.

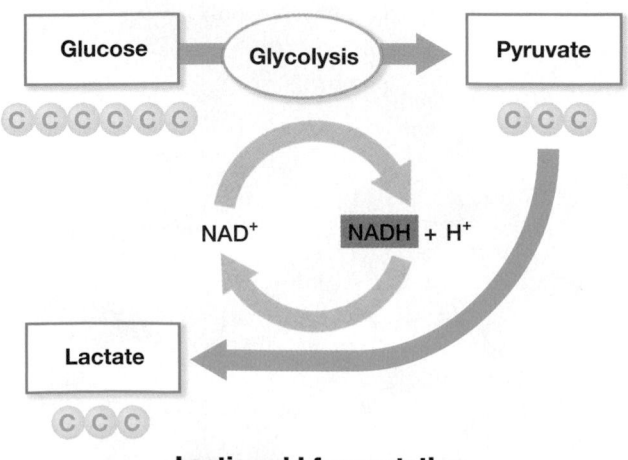

Lactic acid fermentation

In alcoholic fermentation, pyruvate is broken down to ethanol, releasing carbon dioxide, CO_2.

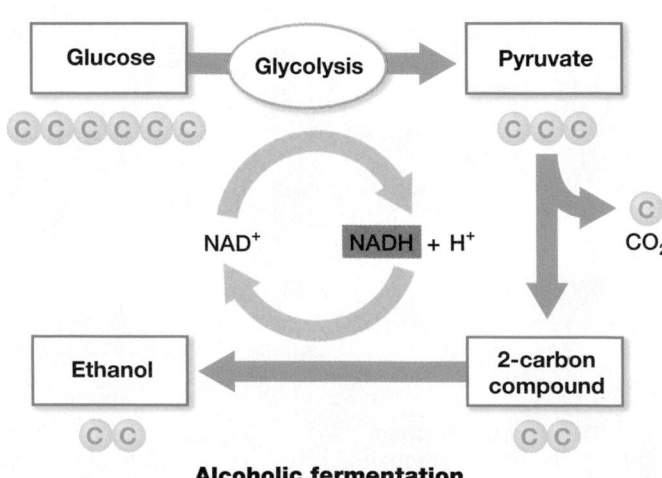

Alcoholic fermentation

Alcoholic Fermentation

In other organisms, the three-carbon pyruvate is broken down to ethanol (ethyl alcohol), a two-carbon compound, through alcoholic fermentation. Carbon dioxide is released during the process. As shown in Figure 15, alcoholic fermentation is a two-step process. First, pyruvate is converted to a two-carbon compound, releasing carbon dioxide. Second, electrons are transferred from a molecule of NADH to the two-carbon compound, producing ethanol. As in lactic acid fermentation, NAD^+ is recycled, and glycolysis can continue to produce ATP.

Alcoholic fermentation by yeast, a fungus, has been used in the preparation of many foods and beverages. Wine and beer contain ethanol made during alcoholic fermentation by yeast. Carbon dioxide released by the yeast causes the rising of bread dough and the carbonation of some alcoholic beverages, such as beer. Ethanol is actually toxic to yeast. At a concentration of about 12 percent ethanol kills yeast. Thus, naturally fermented wine contains about 12 percent ethanol.

BIO Watch

Muscle Fatigue and Endurance Training

Anyone who runs or exercises for a long period of time soon learns about muscle fatigue. As you continue vigorous exercise, the muscles you are using become fatigued—that is, tired and less able to generate force. The reasons for muscle fatigue are not fully understood, but in most cases the fatigue increases when the production of lactic acid by the exercising muscle increases.

Anaerobic Threshold

Why does an exercising muscle produce lactic acid? A resting muscle obtains most of its energy from aerobic respiration. A continuously exercising muscle, however, soon depletes its available oxygen. At this point, called the *anaerobic threshold*, the exercising muscle begins to obtain the ATP needed anaerobically. In the absence of oxygen, glycolysis extracts the required ATP from glycogen in the muscle. *Glycogen* is a storable form of glucose that acts as an energy reserve. Glycolysis converts the muscle glycogen to pyruvate, which is then fermented to lactic acid.

The ability to perform continuous exercise is limited by the body's stored glycogen. So, physical endurance can increase if glycogen stored in muscles is spared during exercise. Trained athletes such as cyclist Lance Armstrong, shown at right, get a relatively large portion of their energy from aerobic respiration. Thus, their muscle glycogen reserve is depleted more slowly than that in untrained individuals. In fact, the greater the level of physical training, the higher the proportion of energy the body derives from aerobic respiration.

Athletic Endurance

Endurance-trained athletes generally have more muscle mass than untrained people. But it is

Lance Armstrong

endurance-trained athletes' high aerobic capacity—rather than their greater muscle mass—that allows these athletes to exercise more before lactic acid production and glycogen depletion cause muscle fatigue.

Figure 16 Effect of oxygen on ATP production

Most ATP is produced during aerobic respiration.

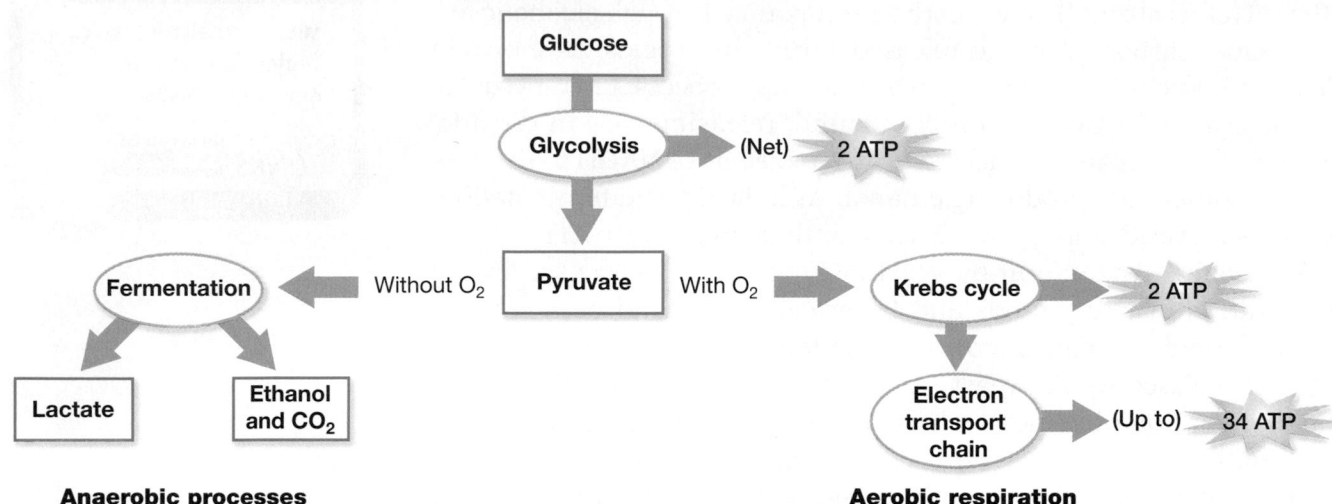

Anaerobic processes   Aerobic respiration

Production of ATP

The total amount of ATP that a cell is able to harvest from each glucose molecule that enters glycolysis depends on the presence or absence of oxygen. As shown in **Figure 16,** cells use energy most efficiently when oxygen is present. In the first stage of cellular respiration, glucose is broken down to pyruvate during glycolysis. Glycolysis is an anaerobic process, and it results in a net gain of two ATP molecules. In the second stage of cellular respiration, the pyruvate passes through either aerobic respiration or (anaerobic) fermentation. When oxygen is present, aerobic respiration occurs. When oxygen is not present, fermentation occurs instead. The NAD$^+$ that gets recycled during fermentation allows glycolysis to continue producing ATP. Thus, a small amount of ATP is produced even during fermentation. Most of a cell's ATP is made, however, during aerobic respiration. For each molecule of glucose that is broken down, as many as two ATP molecules are made directly during the Krebs cycle, and up to 34 ATP molecules are produced later by the electron transport chain.

Section 3 Review

1 **List** the products of glycolysis. What is the role of each of these products in cellular respiration?

2 **Summarize** the roles of the Krebs cycle and the electron transport chain during aerobic respiration.

3 **Describe** the role of fermentation in the second stage of cellular respiration.

4 **Critical Thinking Comparing Functions** Explain why cellular respiration is more efficient when oxygen is present in cells.

5 **Critical Thinking Inferring Conclusions** Excess glucose in your blood is stored in your liver as glycogen. How might your body sense when to convert glucose to glycogen and glycogen back to glucose?

6 **Standardized Test Prep** When oxygen is present, most of the ATP made in cellular respiration is produced by

A aerobic respiration.   C alcoholic fermentation.

B glycolysis.   D lactic acid fermentation.

Study ZONE

CHAPTER HIGHLIGHTS

Key Concepts

1 Energy and Living Things

- Energy from sunlight flows through living systems, from autotrophs to heterotrophs.

- Photosynthesis and cellular respiration form a cycle because one process uses the products of the other.

- ATP supplies cells with energy needed for metabolism.

2 Photosynthesis

- Photosynthesis has three stages. First, energy is captured from sunlight. Second, energy is temporarily stored in ATP and NADPH. Third, organic compounds are made using ATP, NADPH, and carbon dioxide.

- Pigments absorb light energy during photosynthesis.

- Electrons excited by light travel through electron transport chains, in which ATP and NADPH are produced.

- Through carbon dioxide fixation, often by the Calvin cycle, carbon dioxide in the atmosphere is used to make organic compounds, which store energy.

- Photosynthesis is directly affected by environmental factors such as the intensity of light, the concentration of carbon dioxide, and temperature.

3 Cellular Respiration

- Cellular respiration has two stages. First, glucose is broken down to pyruvate during glycolysis, making some ATP. Second, a large amount of ATP is made during aerobic respiration. When oxygen is not present, NAD^+ is recycled during the anaerobic process of fermentation.

- The Krebs cycle is a series of reactions that produce energy-storing molecules during aerobic respiration.

- During aerobic respiration, large amounts of ATP are made in an electron transport chain.

- When oxygen is not present, fermentation follows glycolysis, regenerating NAD^+ needed for glycolysis to continue.

Key Terms

Section 1

photosynthesis (94)
autotroph (94)
heterotroph (95)
cellular respiration (95)

Section 2

pigment (98)
chlorophyll (98)
carotenoid (98)
thylakoid (99)
electron transport chain (100)
NADPH (101)
carbon dioxide fixation (102)
Calvin cycle (102)

Section 3

aerobic (104)
anaerobic (104)
glycolysis (105)
NADH (105)
Krebs cycle (106)
$FADH_2$ (107)
fermentation (108)

Unit 2, Unit 3—Use Topics 1–6 in these units to review the key concepts and terms in this chapter.

Understanding Key Ideas

1. Energy flows through living systems from?
   a. the sun, to heterotrophs, and then to autotrophs.
   b. autotrophs, to the environment, and then to heterotrophs.
   c. the sun, to autotrophs, and then to heterotrophs.
   d. the environment, to heterotrophs, and then to autotrophs.

2. The products of photosynthesis that begin cellular respiration are
   a. organic compounds and oxygen.
   b. carbon dioxide and water.
   c. $NADP^+$ and hydrogen.
   d. ATP and water.

3. The thylakoid membranes of a chloroplast are the sites where
   a. electron transport chains operate.
   b. NADPH and ATP are produced.
   c. pigments are located.
   d. all of the above

4. The oxygen produced during photosynthesis comes directly from the
   a. splitting of carbon dioxide molecules.
   b. splitting of water molecules.
   c. mitochondrial membranes.
   d. absorption of light.

5. Study the micrograph of a chloroplast shown below, and identify the structures labeled *X*. During photosynthesis, are hydrogen ions more concentrated in these structures or in the spaces around them?

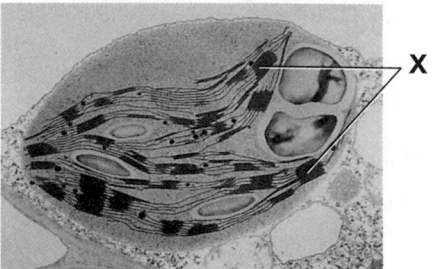

6. Which of the following is the correct pairing of a process and its requirement for oxygen?
   a. glycolysis: no oxygen required
   b. fermentation: oxygen required
   c. Krebs cycle: no oxygen required
   d. none of the above

7. Analyze the flow of energy that enables you to get energy from the food you eat.

8. **BIOWatch** What change occurs in muscles at the anaerobic threshold?

9. How is starch broken down to glucose prior to glycolysis? (**Hint:** See Chapter 2, Section 4.)

10. **Concept Mapping** Make a concept map that shows how photosynthesis and cellular respiration are related. Try to include the following terms in your map: *glycolysis, Krebs cycle, electron transport chain, Calvin cycle, fermentation,* and *NADH.*

Critical Thinking

11. Distinguishing Relevant Information The enzyme that aids in the conversion of pyruvate to acetyl-CoA requires vitamin B1, also called *thiamine.* Thiamine is not made in the human body and must be taken in in foods. How would a deficiency of thiamine in cells affect cellular respiration?

12. Inferring Relationships How might the folding of the inner membrane of mitochondria affect the rate of aerobic respiration? Explain your answer.

Alternative Assessment

13. Analyzing Methods Research several ways that fermentation is used in food preparation. Find out what kinds of microorganisms are used in cultured dairy products, such as yogurt, sour cream, and some cheeses. Research the role of alcoholic fermentation by yeast in bread making. Prepare an oral report to summarize your findings.

Understanding Concepts

Directions (1–4): For *each* question, write on a separate sheet of paper the letter of the correct answer.

1 What pigment causes a plant to look green?
 A. carotenoid **C.** NADH
 B. chlorophyll **D.** NAPH

2 A scientist makes the following statement: "If Earth's early atmosphere had been rich in oxygen, photosynthetic organisms would not have been able to evolve." Which of the following statements can be used to argue **against** this hypothesis?
 F. Earth's present atmosphere is rich in oxygen.
 G. The main component of the atmosphere is nitrogen gas.
 H. Nonphotosynthetic organisms require atmospheric or dissolved oxygen to survive.
 I. Photosynthetic organisms today carry out photosynthesis in the presence of oxygen.

3 Which of the following is **not** involved in the aerobic part of cellular respiration?
 A. ATP   **C.** the Krebs cycle
 B. glycolysis **D.** mitochondria

4 What is the product of the electron transport chain of photosynthesis?
 F. ATP and NADPH
 G. glucose
 H. pyruvate
 I. water

Directions (5): For the following question, write a short response.

5 Differentiate between heterotrophs and autotrophs.

Test TIP

If you come upon a word you do not know, try to identify its prefix, suffix, or root. Sometimes knowing even one part of the word will help you answer the question

Reading Skills

Directions (6): Read the passage below. Then answer the question.

Exercise physiologists study the best ways to regulate diet and training to maximize the performance and health of athletes. The diet of an athlete depends on the energy requirements of the athlete's sport. Some sports, such as weight lifting, involve mainly anaerobic metabolism. Others, such as jogging and swimming, involve more aerobic respiration.

6 The heart pumps oxygen-rich blood to the body's cells. Why do aerobic sports generally condition heart muscle faster than anaerobic sports?
 A. You do not use your heart during anaerobic respiration.
 B. You have to breathe more when you are doing an anaerobic sport.
 C. Some muscle cells do not use aerobic respiration to generate ATP.
 D. Muscle cells need more oxygen when they are undergoing aerobic respiration.

Interpreting Graphics

Directions (7): Base your answer to question 7 on the chart below.

Effect of Temperature on Photosynthesis

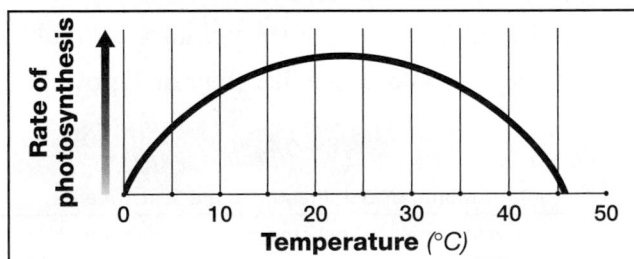

7 Which statement is supported by the data?
 F. Photosynthesis does not occur at 0°C.
 G. The rate of photosynthesis at 40°C is greater than the rate at 20°C.
 H. The optimum temperature for photosynthesis is approximately 46°C.
 I. The rate of photosynthesis increases as temperature increases from 25°C to 30°C.

Skills Practice Lab

Observing Oxygen Production from Photosynthesis

SKILLS
- Measuring
- Collecting Data
- Graphing

OBJECTIVE
- **Measure** amount of oxygen produced by an *Elodea* sprig.

MATERIALS
- 500 mL of 5 percent baking-soda-and-water solution
- 600 mL beaker
- 20 cm long *Elodea* sprigs (2–3)
- glass funnel
- test tube
- metric ruler
- protective gloves

Elodea

ChemSafety

CAUTION: Always wear safety goggles and a lab apron to protect your eyes and clothing.

CAUTION: Glassware is fragile. Notify the teacher of broken glass or cuts. Do not clean up broken glass or spills with broken glass unless the teacher tells you to do so.

CAUTION: Wear disposable polyethlene gloves when handling any plant. Do not eat any part of a plant or plant seed used in the lab. Wash hands thoroughly after handling any part of a plant.

Before You Begin

Plants use **photosynthesis** to produce food. One product of photosynthesis is oxygen. In this activity, you will observe the process of photosynthesis and determine the rate of photosynthesis for *Elodea*.

1. Write a definition for the boldface term above.

2. Create a Data Table like the one below.

DATA TABLE

Amount of Gas Present in the Test Tube		
Days of exposure to light	Total amount of gas present (mm)	Amount of gas produced per day (mm)
0		
1		
2		
3		
4		
5		

Procedure

1. Add 450 mL of baking-soda-and-water solution to a beaker.

2. Put two or three sprigs of *Elodea* in the beaker. The baking soda will provide the *Elodea* with the carbon dioxide it needs for photosynthesis.

3. Place the wide end of the funnel over the *Elodea*. The end of the funnel with the small opening should be pointing up. The *Elodea* and the funnel should be completely under the solution, as shown on the facing page.

4. Fill a test tube with the remaining baking-soda-and-water solution. Place your thumb over the end of the test tube. Turn the test tube upside-down, taking care that no air enters. Hold the opening of the test tube under the solution and place the test tube over the small end of the funnel. Try not to let any solution leak out of the test tube as you do this.

5. Place the beaker setup in a well-lit area near a lamp or in direct sunlight.

6. Record that there was 0mm gas in the test tube on day 0. (If you were unable to place the test tube without getting air in the tube, measure the height of the column of air in the test tube in millimeters. Record this value for day 0.) In this lab, change in gas volume is indicated by a linear measurement expressed in millimeters.

7. For days 1 through 5, measure the amount of gas in the test tube. Record the measurements in your data table under the heading, "Total amount of gas present (mm)."

8. Calculate the amount of gas produced each day by subtracting the amount of gas present on the previous day from the amount of gas present today. Record these amounts under the heading, "Amount of gas produced per day (mm)."

9. Plot the data from your table on a graph.

Analyze and Conclude

1. **Summarizing Results** Using information from your graph, describe what happened to the amount of gas in the test tube.

2. **Analyzing Data** How much gas was produced in the test tube after day 5?

3. **Drawing Conclusions** Write the equation for photosynthesis. Explain each part of the equation. For example, what ingredients are necessary for photosynthesis to take place? What substances are produced by photosynthesis? What gas is produced that we need in order to live?

4. **Predicting Patterns** What may happen to the oxygen level if an animal, such as a snail, were put in the beaker with the *Elodea* sprig while the *Elodea* sprig was making oxygen?

5. **Further Inquiry** Write a new question about photosynthesis that could be explored with another investigation.

? Do You Know?

Do research in the library or media center to answer these questions:

1. What is hydroponic farming?

2. How do coral reefs depend on photosynthesis?

Use the following Internet resources to explore your own questions about photosynthesis.

internet connect

www.scilinks.org
Topic: Photosynthesis
Keyword: HX4136

SCiLINKS. Maintained by the National Science Teachers Association

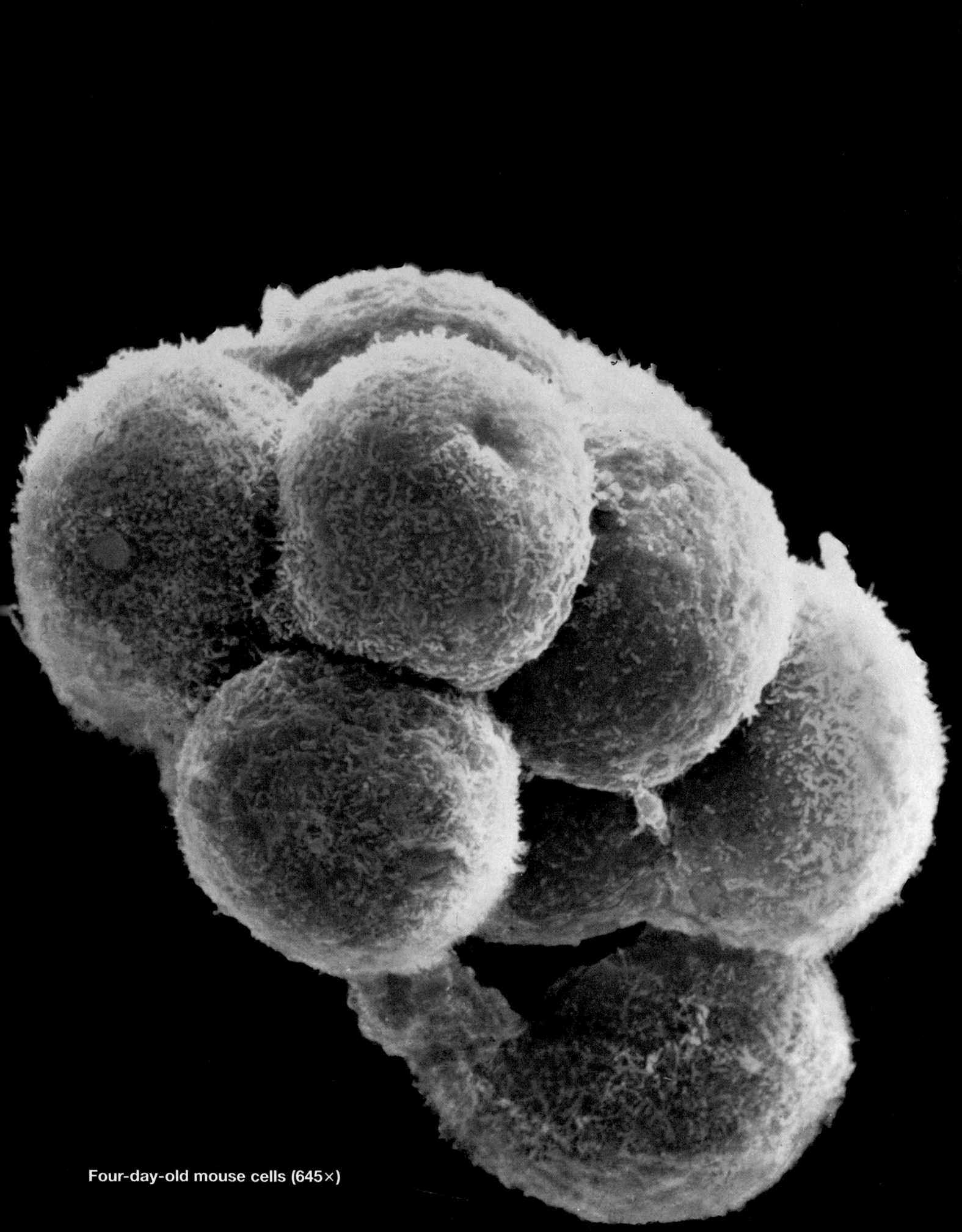

Four-day-old mouse cells (645×)

CHAPTER

6 Chromosomes and Cell Reproduction

✔️ Quick Review

Answer the following without referring to earlier sections of your book.

1. **Define** the term *mutation*. *(Chapter 1, Section 1)*

2. **Describe** the structure of proteins and of DNA. *(Chapter 2, Section 3)*

3. **Summarize** the function and structure of the nucleus and of microtubules. *(Chapter 3, Section 2)*

Did you have difficulty? *For help, review the sections indicated.*

Reading Activity

Copy the following statements in your notebook:

• *Chromosomes from females determine the sex of humans.*

• *Every human cell contains 46 chromosomes.*

• *Healthy cells cannot become cancerous cells.*

Before you read the chapter, write down if you agree with each statement. After you have finished reading the chapter, decide whether you still agree with your first response.

This cluster of cells is smaller than the head of a pin, but over the next 17 days, they will divide repeatedly to form a new mouse. Chromosomes inside each cell carry the instructions for growth and development of an individual.

Looking Ahead

Section 1
Chromosomes
Formation of New Cells by Cell Division
How Chromosome Number and Structure Affect Development

Section 2
The Cell Cycle
The Life of a Eukaryotic Cell
Control of the Cell Cycle

Section 3
Mitosis and Cytokinesis
Chromatid Separation in Mitosis
Mitosis and Cytokinesis

Chromosomes

Objectives

- **Identify** four examples of cell division in eukaryotes and one example in prokaryotes.

- **Differentiate** between a gene, a DNA molecule, a chromosome, and a chromatid.

- **Differentiate** between homologous chromosomes, autosomes, and sex chromosomes.

- **Compare** haploid and diploid cells.

- **Predict** how changes in chromosome number or structure can affect development.

Key Terms

gamete
binary fission
gene
chromosome
chromatid
centromere
homologous
  chromosome
diploid
haploid
zygote
autosome
sex chromosome
karyotype

Formation of New Cells by Cell Division

About 2 trillion cells are produced by an adult human body every day. This is about 25 million new cells per second! These new cells are formed when older cells divide. Cell division, also called cell reproduction, occurs in humans and other organisms at different times in their life. In **Figure 1,** the cells of the fawn that is growing and developing and the cells in the wound that is healing are undergoing cell division. The type of cell division differs depending on the organism and why the cell is dividing. For example, bacterial cells undergoing reproduction divide by one type of cell division. Eukaryotic organisms undergoing growth, development, repair, or asexual reproduction divide by a different type of cell division. And the formation of gametes involves yet a third type of cell division. **Gametes** are an organism's reproductive cells, such as sperm or egg cells.

Regardless of the type of cell division that occurs, all of the information stored in the molecule DNA (deoxyribonucleic acid) must be present in each of the resulting cells. Recall from Chapter 3 that DNA stores the information that tells cells which proteins to make and when to make them. This information directs a cell's activities and determines its characteristics. Thus, when a cell divides, the DNA is first copied and then distributed. Each cell ends up with a complete set (copy) of the DNA.

Figure 1 Cell division

The cells of these organisms are undergoing some type of cell division.

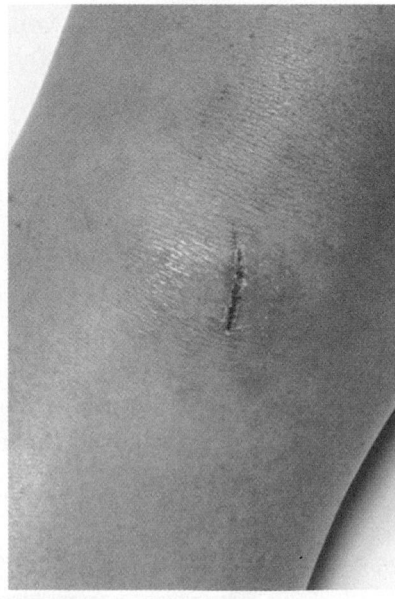

Repair   Growth and development

Prokaryotic Cell Reproduction

A prokaryote's single DNA molecule is circular and is attached to the inner cell membrane. Prokaryotes reproduce by a type of cell division called binary fission. **Binary fission** is a form of asexual reproduction that produces identical offspring. In asexual reproduction, a single parent passes exact copies of all of its DNA to its offspring.

Binary fission occurs in two stages: first, the DNA is copied (so that each new cell will have a copy of the genetic information), and then the cell divides. The prokaryote divides by adding a new cell membrane to a point on the membrane between the two DNA copies. As new material is added, the growing cell membrane pushes inward and the cell is constricted in the middle, like a long balloon being squeezed near the center. A new cell wall forms around the new membrane. Eventually the dividing prokaryote is pinched into two independent cells. Each cell contains one of the circles of DNA and is a complete functioning prokaryote.

Eukaryotic Cell Reproduction

The vast amount of information encoded in DNA is organized into units called genes. A **gene** is a segment of DNA that codes for a protein or RNA molecule. A single molecule of DNA has thousands of genes lined up like train cars. Genes play an important role in determining how a person's body develops and functions. When genes are being used, the DNA is stretched out so that the information it contains can be used to direct the synthesis of proteins.

As a eukaryotic cell prepares to divide, the **chromosomes**—the DNA and the proteins associated with the DNA—become visible, as shown in **Figure 2.** Before the DNA coils up, however, the DNA is copied. The two exact copies of DNA that make up each chromosome are called **chromatids** *(KROH muh tihdz)*. The two chromatids of a chromosome are attached at a point called a **centromere.** The chromatids become separated during cell division and placed into each new cell, ensuring that each new cell will have the same genetic information as the original cell.

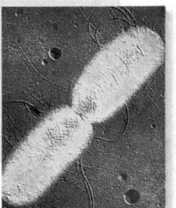

Figure 2 Chromosome structure. A chromosome consists of DNA tightly coiled around proteins. The chromosomes condense as a cell prepares to divide.

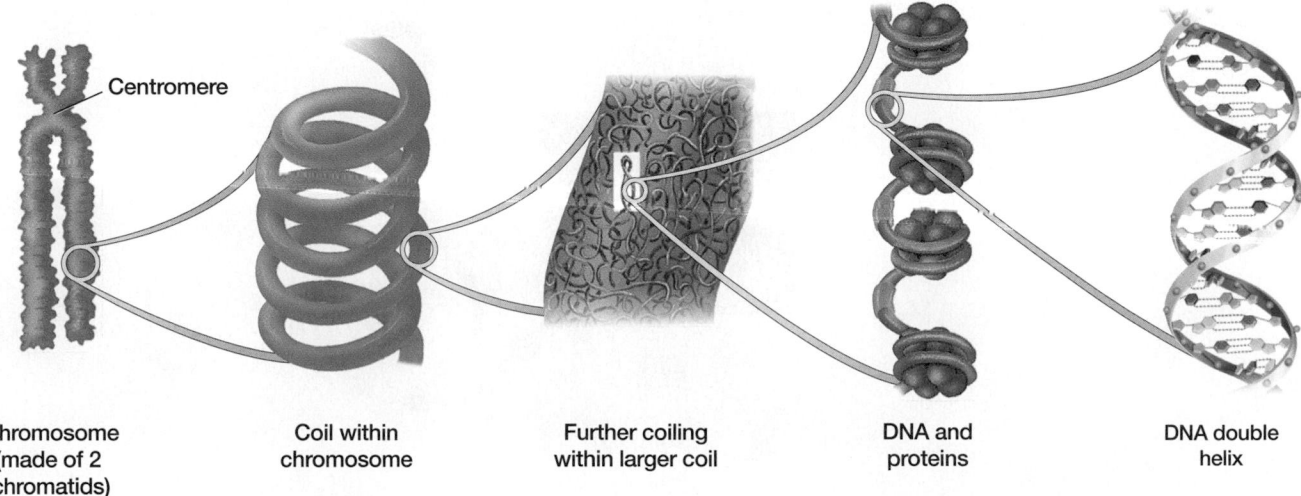

Chromosome (made of 2 chromatids)   Coil within chromosome   Further coiling within larger coil   DNA and proteins   DNA double helix

Centromere

How Chromosome Number and Structure Affect Development

Each human somatic cell (any cell other than a sperm or egg cell) normally has two copies of 23 different chromosomes, for a total of 46 chromosomes. The 23 chromosomes differ in size, shape, and set of genes. Each chromosome contains thousands of genes that play important roles in determining how a person's body develops and functions. For this reason, a complete set of all chromosomes is essential to survival.

Sets of Chromosomes

Each of the 23 pairs of chromosomes consists of two homologous *(hoh MAHL uh gus)* chromosomes, or homologues *(HOH muh logs)*. **Homologous chromosomes** are chromosomes that are similar in size, shape, and genetic content. Each homologue in a pair of homologous chromosomes comes from one of the two parents, as shown in **Figure 3.** Thus, the 46 chromosomes in human somatic cells are actually two sets of 23 chromosomes. One set comes from the mother, and one set comes from the father. A human chromosome is shown in **Figure 4.**

Figure 3 Fertilization

When haploid gametes fuse, they produce a diploid zygote.

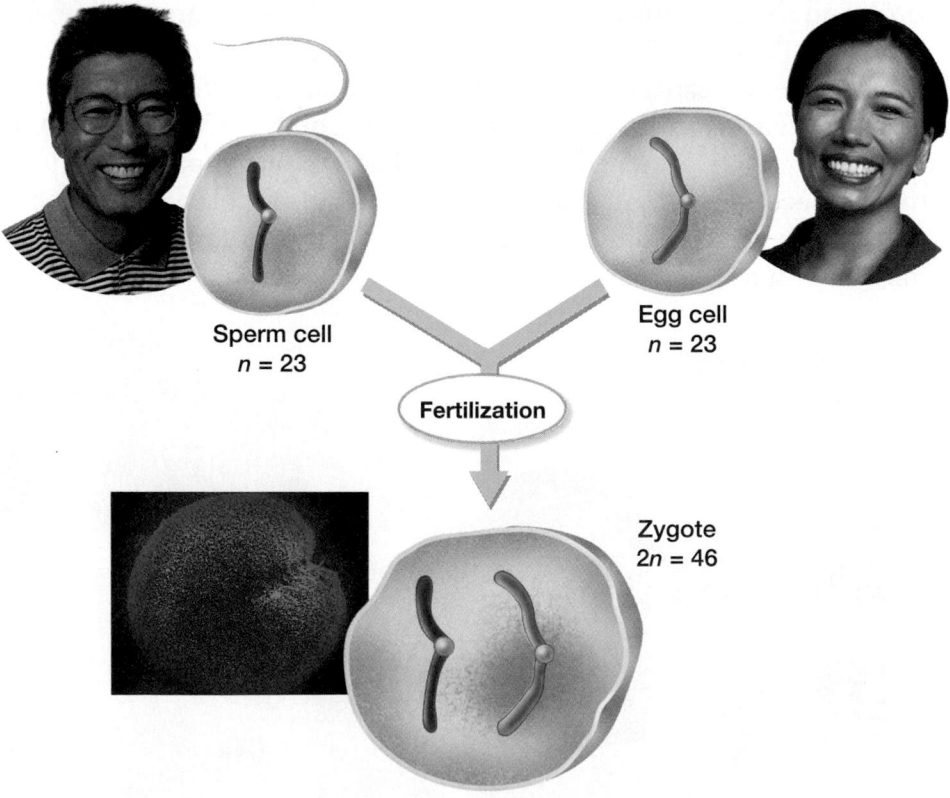

Sperm cell
n = 23

Egg cell
n = 23

Fertilization

Zygote
2*n* = 46

All of the cells in the body, other than gametes, are somatic cells. When a cell, such as a somatic cell, contains two sets of chromosomes, it is said to be **diploid** *(DIHP loyd)*. Unlike somatic cells, human gametes contain only one set of chromosomes (23 total). When a cell, such as a gamete, contains one set of chromosomes, it is said to be **haploid** *(HAP loyd)*. Biologists use the symbol n to represent one set of chromosomes. The haploid number in a human gamete can be written as $n = 23$. The diploid number in a somatic cell can be written as $2n = 46$. The fusion of two haploid gametes—a process called fertilization—forms a diploid zygote, as shown in Figure 3. A **zygote** *(ZY goht)* is a fertilized egg cell, the first cell of a new individual.

As seen in **Table 1,** each organism has a characteristic number of chromosomes. The number of chromosomes in cells is constant within a species. Fruit flies, for example, have only eight chromosomes in each cell. Although most species have different numbers of chromosomes, some species by chance have the same number. For example, potatoes, plums, and chimpanzees all have 48 chromosomes in each cell. Many plants have far more chromosomes. Some ferns have more than 500. A few kinds of organisms—such as the Australian ant *Myrmecia,* the plant *Haplopappus* (a desert relative of the sunflower), and the fungus *Penicillium* (from which the antibiotic penicillin is obtained)—have only one pair of chromosomes.

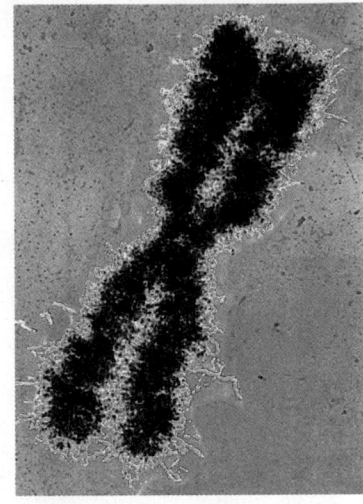

Magnification: 12,542×

Figure 4 Human chromosome. As many as 500 chromosomes lined up end to end would fit in a 0.2 cm space—about the thickness of a nickel. The chromosome above has replicated and consists of two identical chromatids.

Table 1 Chromosome Number of Various Organisms	
Organism	**Number (2*n*) of chromosomes**
Penicillium	1–4
Saccharomyces (yeast)	16
Mosquito	6
Housefly	12
Garden pea	14
Corn	20
Adder's tongue fern	480–1,020
Frog	26
Human	46
Orangutan	48
Dog	78

Sex Chromosomes

Of the 23 pairs of chromosomes in human somatic cells, 22 pairs are called autosomes. **Autosomes** are chromosomes that are not directly involved in determining the sex (gender) of an individual. The **sex chromosomes,** one of the 23 pairs of chromosomes in humans, contain genes that will determine the sex of the individual.

In humans and many other organisms, the two sex chromosomes are referred to as the X and Y chromosomes. The genes that cause a fertilized egg to develop into a male are located on the Y chromosome. Thus, any individual with a Y chromosome is male, and any individual without a Y chromosome is female. For example, in human males, the sex chromosomes are made up of one X chromosome and one Y chromosome (XY). The sex chromosomes in human females consist of two X chromosomes (XX). Because a female can donate only an X chromosome to her offspring, the sex of an offspring is determined by the male, who can donate either an X or a Y.

The structure and number of sex chromosomes vary in different organisms. In some insects, such as grasshoppers, there is no Y chromosome—the females are characterized as XX and the males are characterized as XO (the O indicates the absence of a chromosome). In birds, moths, and butterflies, the male has two X chromosomes and the female has only one.

Change in Chromosome Number

Each of an individual's 46 chromosomes has thousands of genes. Because genes play an important role in determining how a person's body develops and functions, the presence of all 46 chromosomes is essential for normal development and function. A person must have the characteristic number of chromosomes in his or her cells. Humans with more than two copies of a chromosome, a condition called trisomy *(TRY soh mee)*, will not develop properly. Abnormalities in chromosome number can be detected by analyzing a **karyotype** *(KAR ee uh tiep)*, a photo of the chromosomes in a dividing cell that shows the chromosomes arranged by size. **Figure 5** shows a typical karyotype. A portion of a karyotype from an individual with an extra copy of chromosome 21 is also shown in Figure 5. This condition is called Down syndrome, or trisomy 21. Short stature, a round face with

Figure 5   A human karyotype

Karyotypes are used to examine an individual's chromosomes.

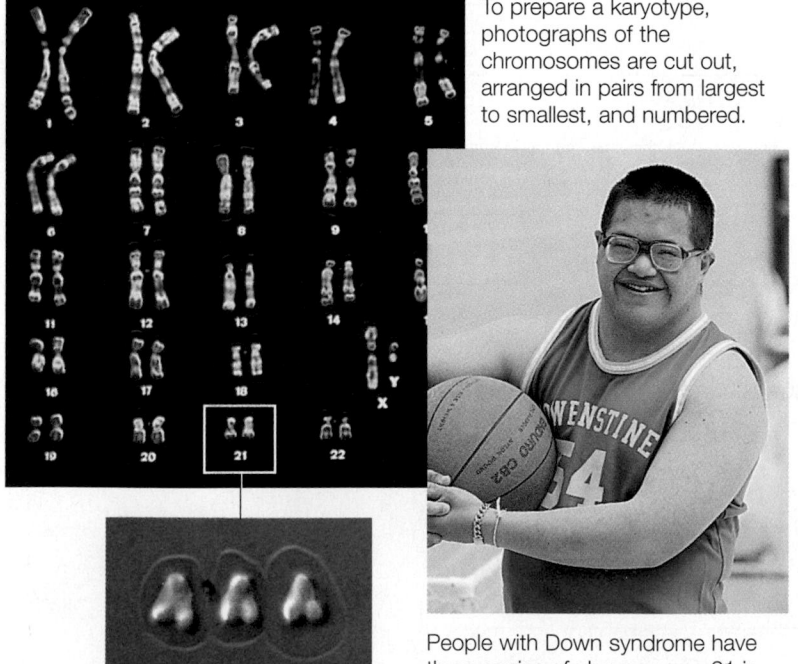

To prepare a karyotype, photographs of the chromosomes are cut out, arranged in pairs from largest to smallest, and numbered.

People with Down syndrome have three copies of chromosome 21 in their karyotype.

upper eyelids that cover the inner corners of the eyes, and varying degrees of mental retardation are characteristics of people with Down syndrome.

In mothers younger than 30, Down syndrome occurs in about 1 in 1,500 births. In mothers 37 years old, the incidence doubles to 1 in 290 births. In mothers over 45, the risk is as high as 1 in 46 births. Older mothers are more likely to have a baby with Down syndrome because all the eggs a female will ever produce are present in her ovaries when she is born, unlike males who produce new sperm throughout adult life. As a female ages, her eggs can accumulate an increasing amount of damage. Because of this risk, a pregnant woman over the age of 35 may be advised to undergo prenatal testing that includes fetal karyotyping.

What events can cause an individual to have an extra copy of a chromosome? When sperm and egg cells form, each chromosome and its homologue separate, an event called disjunction *(dihs JUHNK shuhn)*. If one or more chromosomes fail to separate properly—an event called nondisjunction—one new gamete ends up receiving both chromosomes and the other gamete receives none. Trisomy occurs when the gamete with both chromosomes fuses with a normal gamete during fertilization, resulting in offspring with three copies of that chromosome instead of two. In Down syndrome, nondisjunction involves chromosome 21.

BIOWatch

Prenatal Testing

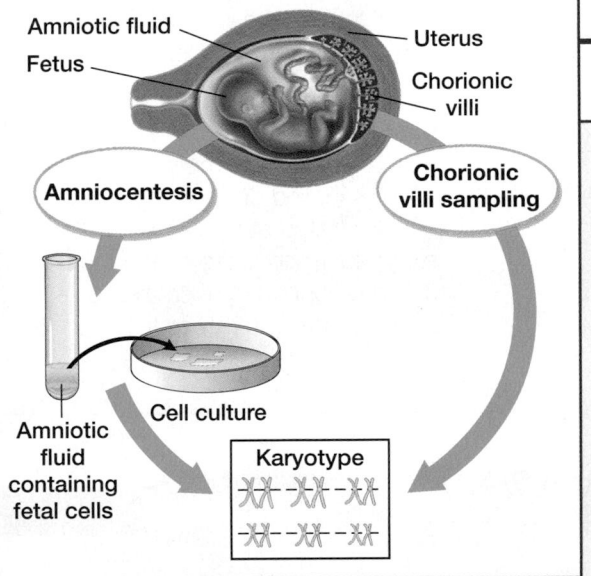

Will our baby be normal? For many expectant parents, prenatal testing can help answer this question. In prenatal testing, the cells of a fetus are tested for normal chromosome number and cell structure by a procedure called fetal karyotyping. Fetal karyotyping allows parents and doctors to view the chromosomes found in the cells of the fetus. The doctor can then check for any abnormalities, such as Down syndrome. There are two ways to obtain fetal cells.

Amniocentesis

In amniocentesis *(am nee oh sehn TEE sihs)*, a needle and syringe are used to remove a small amount of the amniotic fluid that surrounds the fetus. The fluid contains fetal cells. The fetal cells are grown in a laboratory for 1–4 weeks to obtain enough actively dividing fetal cells to make a karyotype, which is then analyzed.

Chorionic Villi Sampling (CVS)

In chorionic villi *(kawr ee AHN ihk VIHL ie)* sampling, a tissue sample is collected from the chorionic villi, fingerlike extensions of the placenta that grow into the mother's uterus. Enough actively dividing cells are obtained to produce a karyotype without having to culture cells. Since the villi have the same genetic makeup as the fetus, the doctor is able to detect abnormalities in the fetal chromosome number.

Change in Chromosome Structure

Changes in an organism's chromosome structure are called *mutations*. Breakage of a chromosome can lead to four types of mutations. In a deletion mutation, a piece of a chromosome breaks off completely. After cell division, the new cell will lack a certain set of genes. In many cases this proves fatal to the zygote. In a duplication mutation, a chromosome fragment attaches to its homologous chromosome, which will then carry two copies of a certain set of genes. A third type of mutation is an inversion mutation, in which the chromosome piece reattaches to the original chromosome but in a reverse orientation. If the piece reattaches to a nonhomologous chromosome, a translocation mutation results.

QUICK LAB

Modeling Chromosomal Mutations

You can use paper and a pencil to model the ways in which chromosome structure can change.

Materials

14 note-card pieces, pencils, tape

Procedure

1. Write the numbers 1–8 on note-card pieces (one number per piece). Tape the pieces together in numerical order to model a chromosome with eight genes.

2. Use the "chromosome" you made to model the four alterations in chromosome structure discussed on this page and illustrated at right. For example, remove the number 3 and reconnect the remaining chromosome pieces to represent a deletion.

3. Reconstruct the original chromosome before modeling a duplication, an inversion, and a translocation. Use the extra note-card pieces to make the additional numbers you need.

Analysis

Describe how a cell might be affected by each mutation if the cell were to receive a chromosome with that mutation.

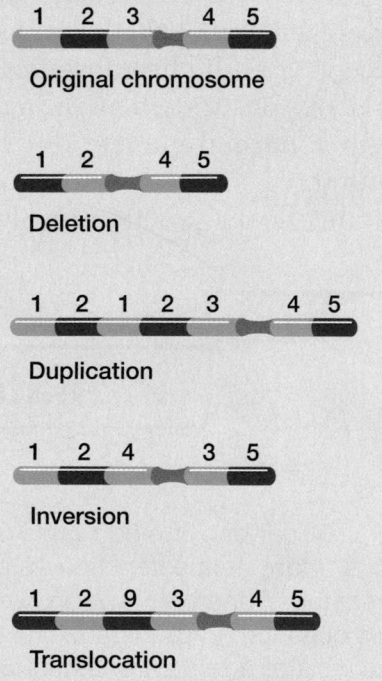

1 2 3   4 5
Original chromosome

1 2   4 5
Deletion

1 2 1 2 3   4 5
Duplication

1 2 4   3 5
Inversion

1 2 9 3   4 5
Translocation

Section 1 Review

1 **Summarize** how prokaryotic cells divide by binary fission.

2 **Identify** the point in a eukaryotic cell cycle at which DNA condenses to form visible chromosomes.

3 **Summarize** the difference between a haploid cell and a diploid cell.

4 **Critical Thinking Evaluating Conclusions** Do you agree or disagree that homologous chromosomes are found in gametes. Explain.

5 **Standardized Test Prep** How does the karyotype of a person with Down syndrome differ from a normal karyotype?

A It lacks a chromosome.

B It has two sex chromosomes.

C It occurs in XO individuals.

D It has an extra copy of a chromosome.

The Cell Cycle

The Life of a Eukaryotic Cell

Cell division in eukaryotic cells is more complex than cell division in bacteria because it involves dividing both the cytoplasm and the chromosomes inside the nucleus. Many internal organelles must be correctly rearranged before the eukaryotic cell can properly divide and form two fully functioning cells.

The Cell Cycle

The life of a eukaryotic cell is traditionally shown as a cycle, as illustrated in **Figure 6.** The **cell cycle** is a repeating sequence of cellular growth and division during the life of an organism. A cell spends 90 percent of its time in the first three phases of the cycle, which are collectively called **interphase.** A cell will enter the last two phases of the cell cycle only if it is about to divide. The five phases of the cell cycle are summarized below:

1. **First growth (G_1) phase.** During the G_1 phase, a cell grows rapidly and carries out its routine functions. For most organisms, this phase occupies the major portion of the cell's life. Cells that are not dividing remain in the G_1 phase. Some somatic cells, such as most muscle and nerve cells, never divide. Therefore, if these cells die, the body cannot replace them.

2. **Synthesis (S) phase.** A cell's DNA is copied during this phase. At the end of this phase, each chromosome consists of two chromatids attached at the centromere.

3. **Second growth (G_2) phase.** In the G_2 phase, preparations are made for the nucleus to divide. Hollow protein fibers called microtubules are rearranged during G_2 in preparation for mitosis.

4. **Mitosis.** The process during cell division in which the nucleus of a cell is divided into two nuclei is called **mitosis** *(mie TOH sihs)*. Each nucleus ends up with the same number and kinds of chromosomes as the original cell.

5. **Cytokinesis.** The process during cell division in which the cytoplasm divides is called **cytokinesis** *(SIET oh kih nee sihs)*.

Mitosis and cytokinesis produce new cells that are identical to the original cells and allow organisms to grow, replace damaged tissues, and, in some organisms, reproduce asexually.

Objectives

- **Identify** the major events that characterize each of the five phases of the cell cycle.

- **Describe** how the cell cycle is controlled in eukaryotic cells.

- **Relate** the role of the cell cycle to the onset of cancer.

Key Terms

cell cycle
interphase
mitosis
cytokinesis
cancer

Figure 6   The eukaryotic cell cycle. The cell cycle consists of phases of growth, DNA replication, preparation for cell division, and division of the nucleus and cytoplasm.

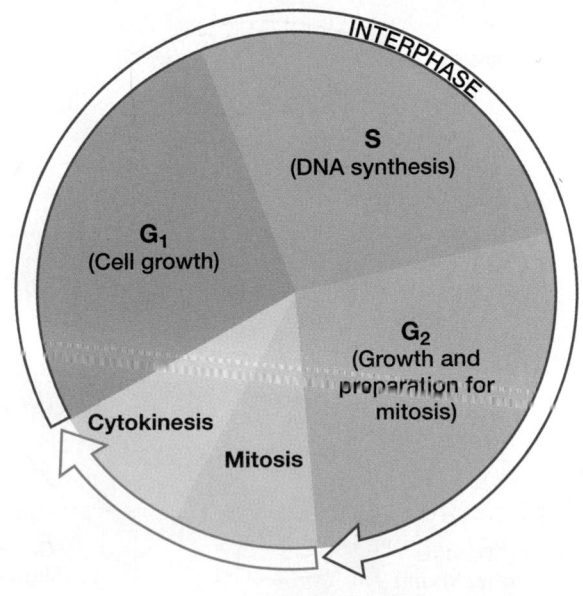

INTERPHASE

S
(DNA synthesis)

G_1
(Cell growth)

G_2
(Growth and preparation for mitosis)

Cytokinesis

Mitosis

Control of the Cell Cycle

Study TIP

● **Reviewing Information**

Learn the stages of interphase by reviewing the steps numbered 1–5 on the previous page. You can see in Figures 6 and 7 that the cell cycle is a repeating series of three steps followed by mitosis and cytokinesis.

If a cell spends 90 percent of its time in interphase, how do cells "know" when to divide? How is the cycle controlled? Just as traffic lights control the flow of traffic, cells have a system that controls the phases of the cell cycle. Cells have a set of "red light–green light" switches that are regulated by feedback information from the cell. The cell cycle has key checkpoints (inspection points) at which feedback signals from the cell can trigger the next phase of the cell cycle (green light). Other feedback signals can delay the next phase to allow for completion of the current phase (yellow or red light).

The cell cycle in eukaryotes is controlled by many proteins. Control occurs at three principal checkpoints, as shown in **Figure 7.**

1. **Cell growth (G$_1$) checkpoint.** This checkpoint makes the decision of whether the cell will divide. If conditions are favorable for division and the cell is healthy and large enough, certain proteins will stimulate the cell to begin the synthesis (S) phase. During the S phase, the cell will copy its DNA. If conditions are not favorable, cells can typically stop the cell cycle at this checkpoint. The cell cycle will also stop at this checkpoint if the cell needs to pass into a resting period. Certain cells, such as some nerve and muscle cells, remain in this resting period permanently and never divide.

2. **DNA synthesis (G$_2$) checkpoint.** DNA replication is checked at this point by DNA repair enzymes. If this checkpoint is passed, proteins help to trigger mitosis. The cell begins the many molecular processes that are needed to proceed into mitosis.

3. **Mitosis checkpoint.** This checkpoint triggers the exit from mitosis. It signals the beginning of the G$_1$ phase, the major growth period of the cell cycle.

Figure 7   Control of the cell cycle. The cell cycle in eukaryotes is controlled at three inspection points, or checkpoints. Many proteins are involved in the control of the cell cycle.

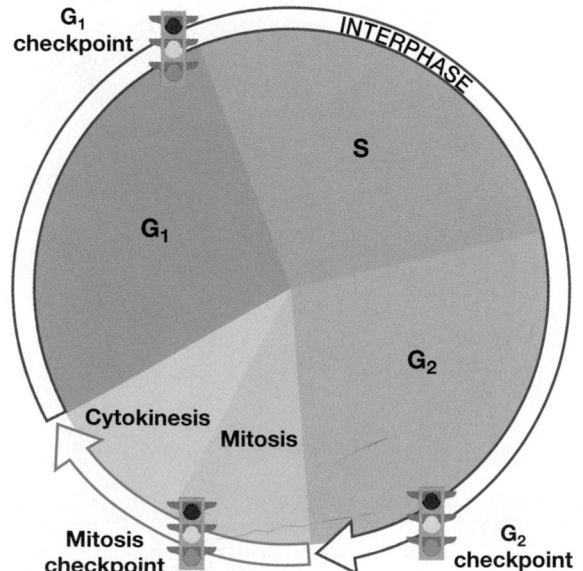

When Control Is Lost: Cancer

Certain genes contain the information necessary to make the proteins that regulate cell growth and division. If one of these genes is mutated, the protein may not function, and regulation of cell growth and division can be disrupted. **Cancer,** the uncontrolled growth of cells, may result. Cancer is essentially a disorder of cell division. Cancer cells do not respond normally to the body's control mechanisms.

Some mutations cause cancer by overproducing growth-promoting molecules, thus speeding up the cell cycle. Others cause cancer by inactivating the control proteins that normally act to slow or stop the cell cycle.

Cancer

Although all cancers are not curable, great progress has been made in cancer research over the last 30 years. We now know that cancer results from damage to a small set of genes that, in normal cells, limits the ability of cells to divide.

What causes this damage? Certain environmental factors appear to be associated with cancer. For example, the incidence of cancer per thousand people is not uniform throughout the United States. Rather, it is higher in cities and in the Mississippi delta, suggesting that pollution and pesticide runoff may contribute to cancer. When pollutants, radiation, and other environmental factors associated with cancer are analyzed, a clear pattern emerges. Most cancer-causing agents are powerful mutagens—that is, they readily damage DNA. The conclusion that cancer is caused by mutation of a cell's DNA is now supported by a very large body of evidence.

How many mutations are required to produce cancer? Research in the last several years indicates that mutation of only a few genes can transform normal cells into cancerous ones. All of these cancer-causing genes are involved with regulating how fast cells grow and divide. How is cell division regulated? As a crude analogy, imagine a car parked on the side of a road. To get it going, you must step on the accelerator and release the brake.

Stepping on the Accelerator

A cell divides when it receives a signal to do so. A "divide" signal is usually in the form of a chemical substance released by another cell. The substance is bound by a protein on the surface of the receiving cell. This binding activates a second protein inside the cell—relaying the signal from the outside of the cell to the inside. Here, a family of proteins then relay the signal inward to the

nucleus. One protein molecule passes the signal to the next like a baton in a relay race. The genes for these signal-carrying proteins are called *onco-genes* (*onco* is Greek for "mass" or "tumor."). If oncogenes are changed by mutation to become more active, cancer can result. Like stepping on the accelerator of a car, an increase in the activity of these proteins amplifies the "divide" signal. This causes the cell to divide more often.

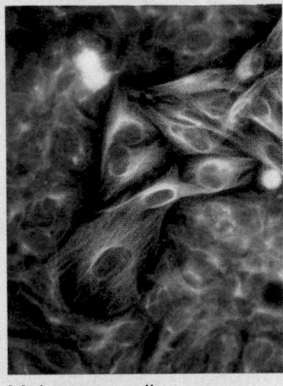

Melanoma cells

Releasing the Brakes

At the nucleus, the divide signal overrides a set of genes that act as "brakes." These braking genes—called *tumor suppressor genes*—prevent cell division from occurring too often. In cancer, these tumor suppressor genes are damaged. Like removing pressure from the brakes of a car increases a car's speed, decreasing the activity of tumor suppressors speeds up cell division.

Cells have three kinds of tumor suppressors, all of which must be disabled before cancer can occur. First, cells have proteins that inhibit DNA replication for limited periods. In cancer cells they are permanently inactivated. Second, cells have error-correcting proteins that detect damage to genes. In most cancers this error-detection has been disabled. Third, cancer cells rebuild the tips of their chromosomes. A little is lost from the ends of chromosomes at each replication, limiting the number of times a normal cell can divide. Adding the deleted material back to the tips removes this limit to a cell's life span.

internet connect

www.scilinks.org
Topic: Cancer Cells
Keyword: HX4030

SC*LINKS* Maintained by the National Science Teachers Association

Section 2 Review

1. **Differentiate** between the G_1, G_2, and S phases of the eukaryotic cell cycle.

2. **Relate** what occurs at each of the three principal checkpoints in the cell cycle.

3. **Critical Thinking** **Evaluating Information**

Why are individual chromosomes more difficult to see during interphase than during mitosis?

4. **Standardized Test Prep** In the cell cycle of typical cancer cells, mutations have caused

   A slower growth. **C** uncontrolled growth.

   B a failure in mitosis. **D** a halt in cell division.

Mitosis and Cytokinesis

- **Describe** the structure and function of the spindle during mitosis.

- **Summarize** the events of the four stages of mitosis.

- **Differentiate** cytokinesis in animal and plant cells.

Key Terms

spindle

Chromatid Separation in Mitosis

Every second about 2 million new red blood cells are produced in your body by cell divisions occurring in the bone marrow. These cells have received the signal to divide. The cells continue past the G_2 phase and enter into the last two phases of the cell cycle—mitosis and cytokinesis. During mitosis the nucleus divides to form two nuclei, each containing a complete set of the cell's chromosomes. During cytokinesis the cytoplasm is divided between the two resulting cells.

During mitosis, the chromatids on each chromosome are physically moved to opposite sides of the dividing cell with the help of the spindle, shown in **Figure 8. Spindles** are cell structures made up of both centrioles and individual microtubule fibers that are involved in moving chromosomes during cell division.

Forming the Spindle

At each of the cell's poles lies a centrosome. The centrosome is an organelle that organizes the assembly of the spindle. In animal cells, a pair of centrioles is found inside each centrosome. As you can see in Figure 8, centrioles are conspicuous. They are not necessary, however, for spindle formation.

Centrioles and spindle fibers are both made of hollow tubes of protein called microtubules. Each spindle fiber is made of an individual microtubule. Each centriole, however, is made of nine triplets of

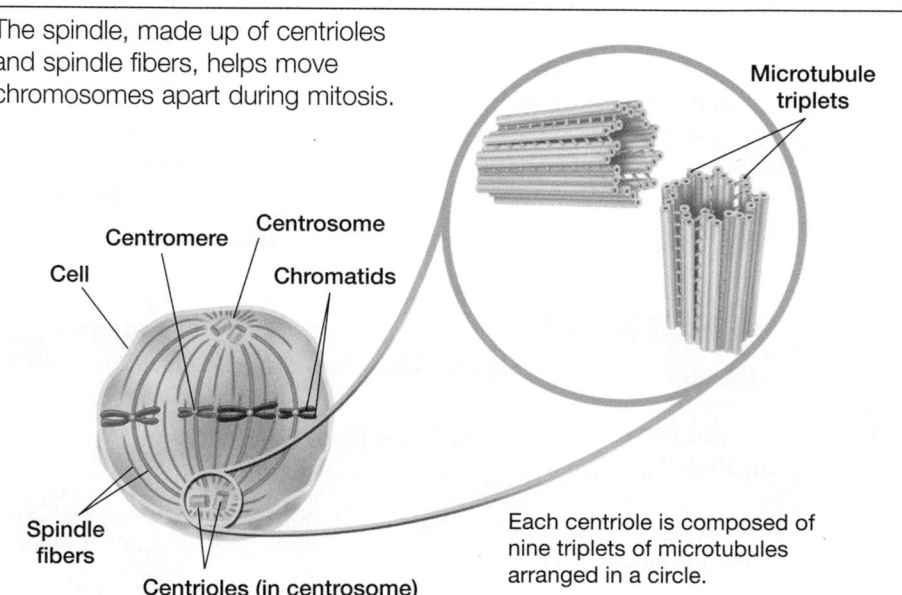

Figure 8 The spindle

The spindle, made up of centrioles and spindle fibers, helps move chromosomes apart during mitosis.

Microtubule triplets

Centromere

Centrosome

Cell

Chromatids

Spindle fibers

Centrioles (in centrosome)

Each centriole is composed of nine triplets of microtubules arranged in a circle.

microtubules arranged in a circle. Unlike animal cells, plant cells do not have centrioles, but they form a spindle that is almost identical to that of an animal cell.

Separation of Chromatids by Attaching Spindle Fibers

Some of the microtubules in the spindle interact with each other. Others attach to a protein structure found on each side of the centromere. The two sets of microtubules extend out toward opposite poles of the cell. Once the microtubules attach to the centromeres and poles, the two chromatids in each chromosome can be separated.

The paired chromatids separate. One of the pair of chromatids will move to one pole of the cell. The second member of the pair will move to the other pole. Once separated, the chromatids move along paths described by microtubules to which they are attached. The chromatids draw closer to the poles of the cell as these microtubules are broken down bit by bit and become shorter.

As soon as the chromatids separate from each other they are called chromosomes. When the chromosomes finally arrive, each pole has one complete set of chromosomes.

MATH LAB

Calculating the Number of Cells Resulting from Mitosis

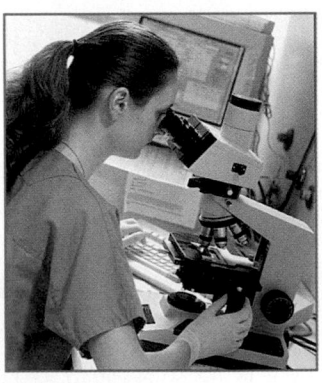

Background

Scientists investigating cancer might need to know the number of cells produced in a certain amount of time. In the human body the rate of mitosis is about 25 million (2.5×10^7) cells produced every second! You can calculate the number of cells produced by mitosis in a given amount of time.

1. **Calculate the number of cells produced by mitosis in the given time.** For example, to find the number of cells produced in 3 minutes, determine how many seconds there are in 3 minutes (since the rate is given in seconds).

$$\frac{60 \text{ seconds}}{1 \text{ minute}} \times 3 \text{ minutes} = 180 \text{ seconds}$$

2. **Multiply the rate of mitosis by the time (in seconds) asked for in the problem (180 seconds).**

$$\frac{2.5 \times 10^7 \text{ cells}}{\text{second}} \times 180 \text{ seconds} = 4.5 \times 10^9 \text{ cells} (1,500,000,000 \text{ cells})$$

Analysis

1. **Calculate** the number of cells that would be produced in 1 hour.

2. **Calculate** the number of cells that would be produced in 1 day.

3. **Critical Thinking Predicting Patterns** Identify factors that might increase or decrease the rate of mitosis.

Mitosis and Cytokinesis

Although mitosis is a continuous process, biologists traditionally divide it into four stages, as shown in **Figure 9.**

Mitosis

Step ❶ Prophase Chromosomes coil up and become visible during prophase. The nuclear envelope dissolves and a spindle forms.

Step ❷ Metaphase During metaphase the chromosomes move to the center of the cell and line up along the equator. Spindle fibers link the chromatids of each chromosome to opposite poles.

Step ❸ Anaphase Centromeres divide during anaphase. The two chromatids (now called chromosomes) move toward opposite poles as the spindle fibers attached to them shorten.

Step ❹ Telophase A nuclear envelope forms around the chromosomes at each pole. Chromosomes, now at opposite poles,

Figure 9

BIOgraphic

Stages of Mitosis

The chromosome copies in the nucleus of a dividing cell are separated into two nuclei.

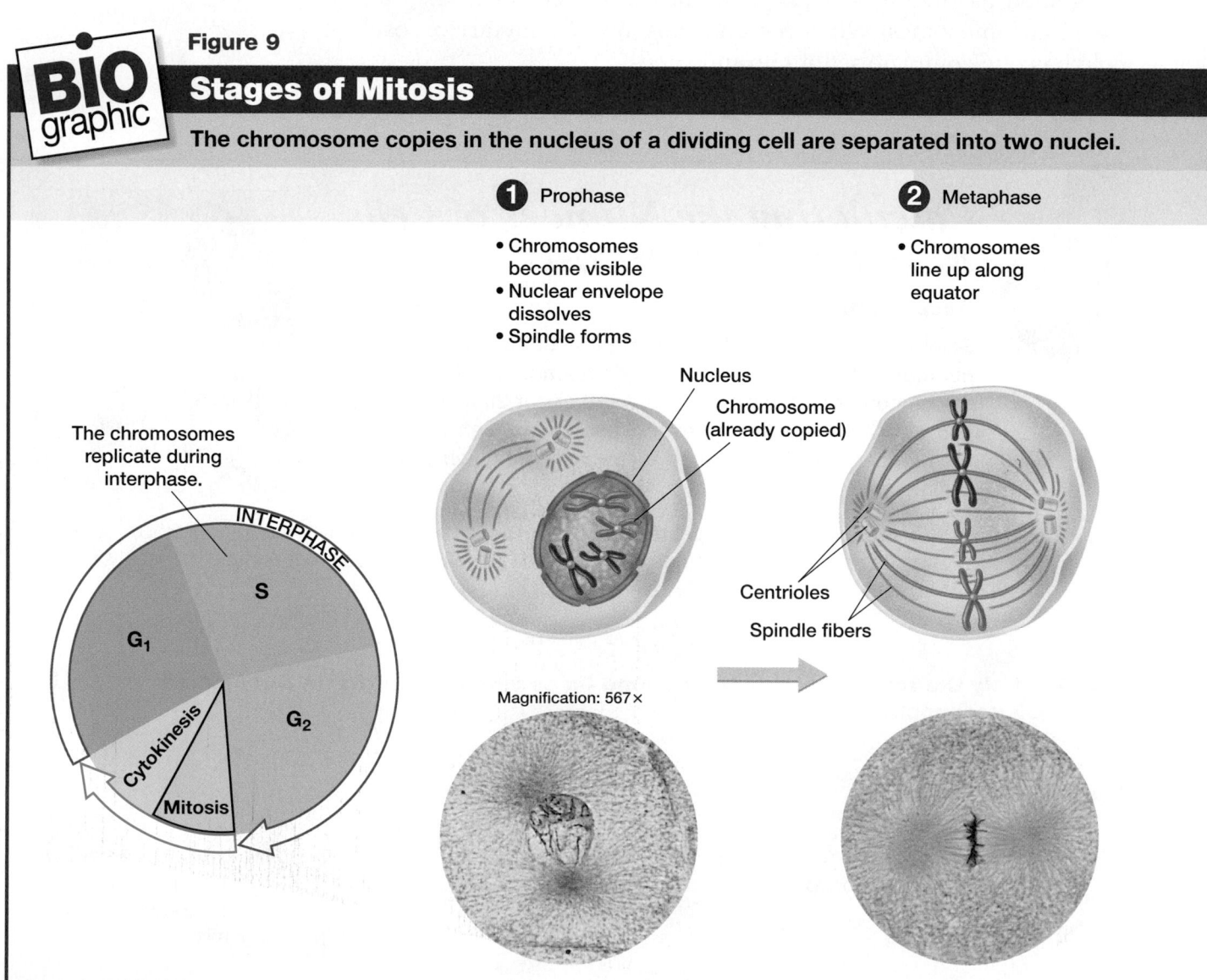

❶ Prophase
- Chromosomes become visible
- Nuclear envelope dissolves
- Spindle forms

❷ Metaphase
- Chromosomes line up along equator

The chromosomes replicate during interphase.

INTERPHASE

S

G₁

Cytokinesis

G₂

Mitosis

Nucleus

Chromosome (already copied)

Centrioles

Spindle fibers

Magnification: 567×

uncoil and the spindle dissolves. The spindle fibers break down and disappear. Mitosis is complete.

Cytokinesis

As mitosis ends, cytokinesis begins. During cytokinesis, the cytoplasm of the cell is divided in half, and the cell membrane grows to enclose each cell, forming two separate cells as a result. The end result of mitosis and cytokinesis is two genetically identical cells where only one cell existed before.

During cytokinesis in animal cells and other cells that lack cell walls, the cell is pinched in half by a belt of protein threads, as shown in **Figure 10**.

Belt of protein threads

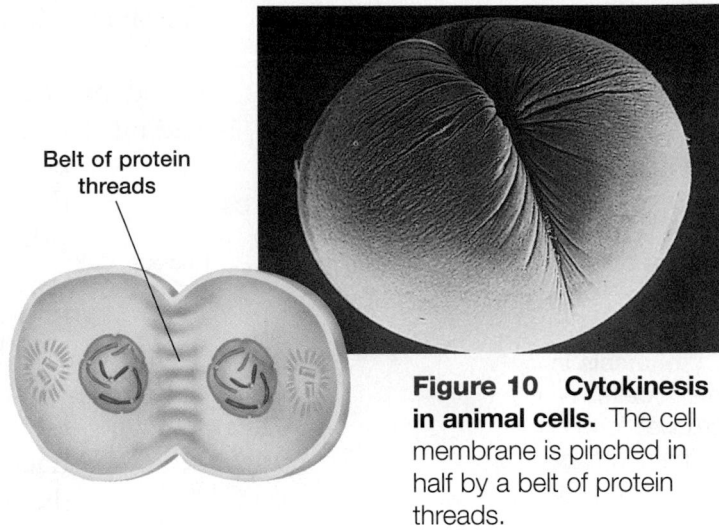

Figure 10 Cytokinesis in animal cells. The cell membrane is pinched in half by a belt of protein threads.

3 Anaphase

- Centromeres divide
- Chromatids (now called chromosomes) move toward opposite poles

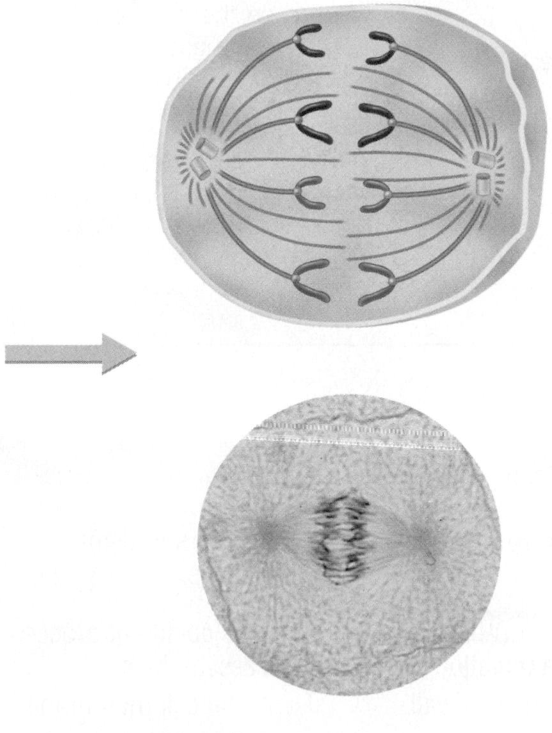

4 Telophase

- Nuclear envelope forms at each pole
- Chromosomes uncoil
- Spindle dissolves
- Cytokinesis begins

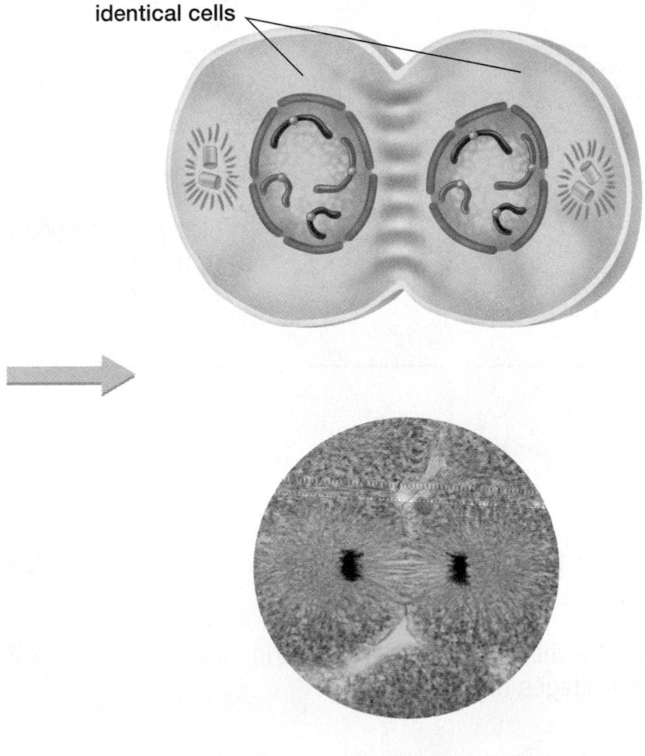

Two genetically identical cells

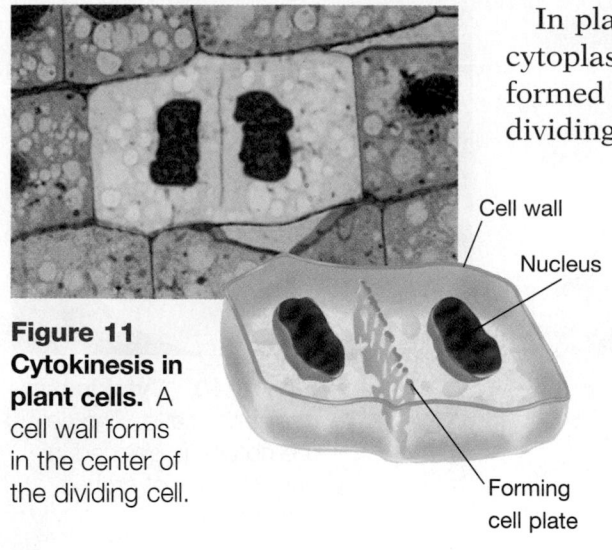

**Figure 11
Cytokinesis in
plant cells.** A
cell wall forms
in the center of
the dividing cell.

Cell wall

Nucleus

Forming
cell plate

In plant cells and other cells that have rigid cell walls, the cytoplasm is divided in a different way. In plant cells, vesicles formed by the Golgi apparatus fuse at the midline of the dividing cell and form a cell plate. A cell plate is a membrane-bound cell wall that forms across the middle of the plant cell. A new cell wall then forms on both sides of the cell plate, as shown in **Figure 11.** When complete, the cell plate separates the plant cell into two new plant cells.

In both animal and plant cells, offspring cells are about equal in size. Each offspring cell receives an identical copy of the original cell's chromosomes. Each offspring cell also receives about one-half of the original cell's cytoplasm and organelles.

Section 3 Review

1 **Describe** the function of the microtubules during anaphase.

2 **Describe** the events that occur during each of the four stages of mitosis.

3 **Compare** how cytokinesis occurs in plant cells with how it occurs in animal cells.

4 **Standardized Test Prep** Mitosis could not proceed if a mutation interrupted the assembly of

  A the cell wall.   **C** the cell membrane.

  B spindle fibers.   **D** the nuclear envelope.

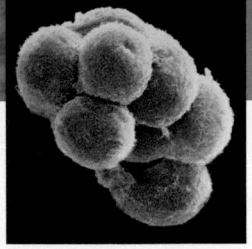

Key Concepts

1 **Chromosomes**

- Cell division allows organisms to reproduce asexually, grow, replace worn-out or damaged tissues, and form gametes.

- Bacteria reproduce by binary fission.

- Before cell division, DNA coils around proteins and the chromosomes condense. At cell division, each chromosome consists of two chromatids attached at the centromere.

- Each organism has a characteristic number of chromosomes.

- Human somatic cells are diploid, with 23 pairs of homologous chromosomes. Human gametes are haploid, with 23 chromosomes.

- Sex chromosomes carry information that determines an organism's sex.

- Changes in chromosome number or structure can cause abnormal development. Karyotypes are used to examine an individual's chromosomes.

2 **The Cell Cycle**

- The life of a eukaryotic cell—the cell cycle—includes interphase, mitosis, and cytokinesis.

- Interphase consists of 3 phases: growth, DNA synthesis (replication), and preparation for cell division. A cell about to divide enters the mitosis and cytokinesis phases of the cell cycle.

- The cell cycle is carefully controlled; failure of cellular control can result in cancer.

3 **Mitosis and Cytokinesis**

- During mitosis, spindle fibers drag the chromatids to opposite poles of the cell. A nuclear envelope forms. Each resulting nucleus contains a copy of the original cell's chromosomes.

- Cytokinesis in animal cells occurs when a belt of protein threads pinches the cell membrane in half. Cytokinesis in plant cells occurs when vesicles from the Golgi apparatus fuse to form a cell plate.

Key Terms

Section 1

gamete (118)
binary fission (119)
gene (119)
chromosome (119)
chromatid (119)
centromere (119)
homologous chromosome (120)
diploid (121)
haploid (121)
zygote (121)
autosome (122)
sex chromosome (122)
karyotype (122)

Section 2

cell cycle (125)
interphase (125)
mitosis (125)
cytokinesis (125)
cancer (126)

Section 3

spindle (128)

 **Unit 4—*Cell Reproduction*** Use Topics 1–4 in this unit to review the key concepts and terms in this chapter.

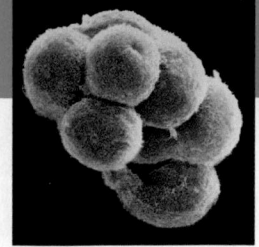

Performance ZONE

Understanding Key Ideas

1. In humans, females have _____ sex chromosomes.
  a. XY   **c.** YY
  b. XX   **d.** XO

2. The diagram below represents a(n) _____ mutation.
  a. deletion
  b. translocation
  c. inversion
  d. duplication

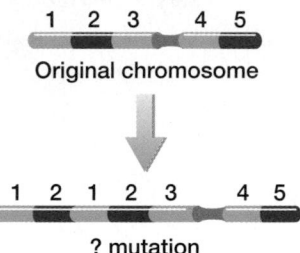

1 2 3   4 5
Original chromosome

1 2 1 2 3   4 5
? mutation

3. When the cell cycle is not controlled, _____ may result.
  a. Down syndrome
  b. binary fission
  c. cancer
  d. a spindle

4. As a result of mitosis, each resulting cell
  a. receives an exact copy of all of the chromosomes present in the original cell.
  b. receives most of the chromosomes from the original cell.
  c. donates a chromosome to the original cell.
  d. receives exactly half the chromosomes from the original cell.

5. During the metaphase stage of mitosis,
  a. the cell membrane folds inward.
  b. chromosomes line up at the cell's equator.
  c. spindle fibers shorten, pulling chromosomes to the poles of the cell.
  d. chromosomes are at opposite ends of the cell.

6. List five organelles that must divide or fragment before the cytoplasm divides. (**Hint:** See Chapter 3, Section 2.)

7. How does cell division differ between animal and plant cells?
  a. Plant cells do not have centrioles.
  b. Animal cells form a cell plate.
  c. Plant cells are always haploid.
  d. Animal cells do not have centrioles.

8. Summarize how normal cells can become cancer cells.

9. ⊕ **BIOWatch** What information, besides chromosome number, can amniocentesis and chorionic villi sampling reveal?

10. ⊞ **Concept Mapping** Make a concept map that shows the events in the cell cycle. Try to include the following words in your map: *cell cycle, interphase, synthesis phase, chromosomes, cytokinesis, mitosis, second growth phase,* and *first growth phase.*

Critical Thinking

11. Inferring Relationships Explain the relationships between mitosis in eukaryotic cells and binary fission in prokaryotes.

12. Evaluating Conclusions Damage to the brain or the spinal cord is usually permanent. Use your knowledge of the cell cycle to explain why damaged cells in the brain or spinal cord are not replaced.

Alternative Assessment

13. Finding and Communicating Information Scientists have determined that telomeres (the tips of chromosomes) are shaved down slightly every time a cell divides. When the telomeres reach a certain length, the cell may lose its ability to divide. Find out what scientists have recently uncovered about telomeres and their association with cell division and cancer. Prepare a brief written report to share with your class.

Standardized Test Prep

Understanding Concepts

Directions (1–4): **For *each* question, write on a separate sheet of paper the letter of the correct answer.**

1 What term describes the asexual reproduction of prokaryotes?
  A. binary fission
  B. cytokinesis
  C. disjunction
  D. mitosis

2 What might happen if cytokinesis were omitted from the cell cycle?
  F. The cell would lose its mitochondria.
  G. The cell would become a cancer cell.
  H. The cell would not divide into two offspring cells.
  I. The offspring cells would not have enough DNA.

3 In what stage of the cell cycle is a cell's DNA copied?
  A. cytokinesis
  B. G_1
  C. mitosis
  D. S

4 What are chromatids?
  F. prokaryotic chromosomes
  G. dense patches of protein within the nucleus
  H. structures that move chromosomes during mitosis
  I. two exact copies of DNA that make up each chromosome

Directions (5): **For the following question, write a short response.**

5 A newspaper article describes the concern that more infants with Down syndrome will be born in the United States as more women delay having children. What information might account for this statistic?

Test TIP

Test questions may not be arranged in order of increasing difficulty. If you are unable to answer a question, mark it and move on to another question.

Reading Skills

Directions (6): **Read the passage below. Then answer the question.**

There are many environmental factors that appear to be associated with cancer, such as pollution and radiation. Cancer occurs when cell division does not respond to the normal signals that regulate the cell cycle. Some cancer-fighting drugs kill cancer cells by interrupting the cell cycle. Examples of such drugs include vincristine and taxol, which prevent the mitosis spindle microtubules from functioning.

6 How is cancer a disorder of cell division?
  A. It is mitosis without cytokinesis.
  B. It is nondisjunction of gamete cells.
  C. It is the uncontrolled growth of cells.
  D. It is cell division without replication of DNA.

Interpreting Graphics

Directions (7): **Base your answer to question 7 on the chart below.**

The Cell Cycle

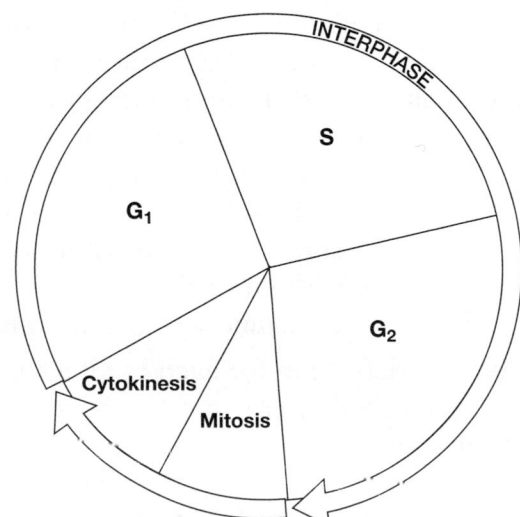

7 What are the phases of the cell cycle starting immediately after cell division?
  F. cytokinesis, G_1, S, G_2, mitosis
  G. G_1, cytokinesis, mitosis, G_2
  H. G_1, S, G_2, mitosis, cytokinesis
  I. S, G_2, mitosis, cytokinesis, G_1

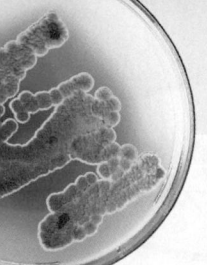

Exploration Lab

Modeling Mitosis

SKILLS
- Modeling
- Using scientific methods

OBJECTIVES
- **Describe** the events that occur in each stage of mitosis.
- **Relate** mitosis to genetic continuity.

MATERIALS
- pipe cleaners of at least two different colors
- yarn
- wooden beads
- white labels
- scissors

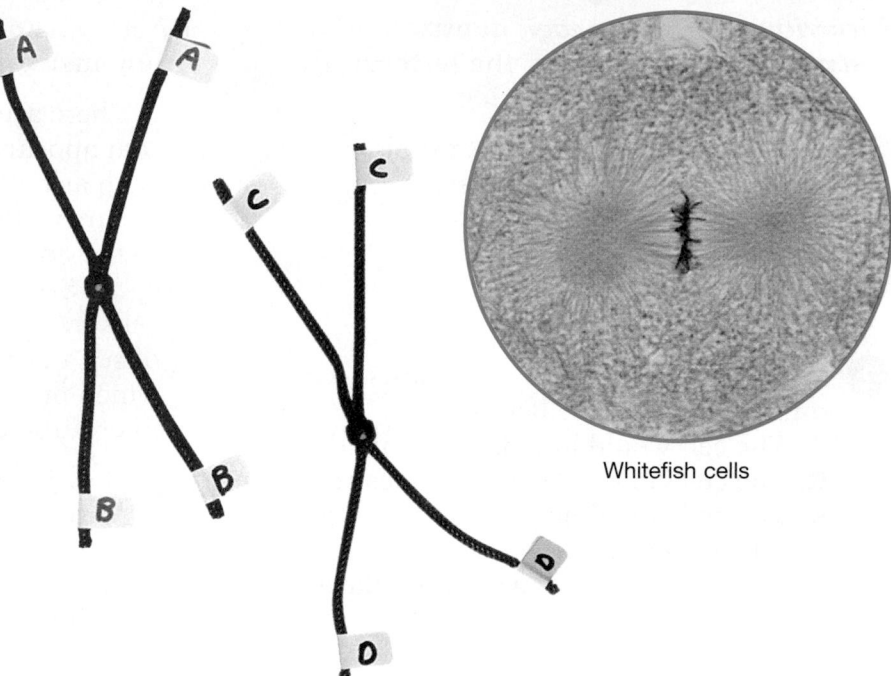

Magnification: 670×

Whitefish cells

Before You Begin

The cell cycle includes all of the phases in the life of a cell. The **cell cycle** is a repeating sequence of cellular growth and division during the life of an organism. Mitosis is one of the phases in the cell cycle. **Mitosis** is the process by which the material in a cell's nucleus is divided during cell reproduction. In this lab, you will build a model that will help you understand the events of mitosis. You can also use the model to demonstrate the effects of **nondisjunction** and **mutations.**

1. Write a definition for each boldface term in the paragraph above and for the following terms: chromatid, centromere, spindle fiber, cytokinesis.

2. Where in the human body do cells undergo mitosis?

3. How does a cell prepare to divide during interphase of the cell cycle?

4. Based on the objectives for this lab, write a question you would like to explore about mitosis.

Procedure

PART A: Design a Model

1. Work with the members of your lab group to design a model of a cell that uses the materials listed for this lab. Be sure your model cell has at least two pairs of chromosomes and is about to undergo mitosis.

> ### You Choose
> **As you design your model, decide the following:**
> **a.** what question you will explore
> **b.** how to construct a cell membrane
> **c.** how to show that your cell is diploid
> **d.** how to show the locations of at least two genes on each chromosome
> **e.** how to show that chromosomes are duplicated before mitosis begins

2. Write out the plan for building your model. Have your teacher approve the plan before you begin building the model.

3. Build the cell model your group designed. **CAUTION: Sharp or pointed objects can cause injury. Handle scissors carefully. Promptly notify your teacher of any injuries.** Use your model to demonstrate the phases of mitosis. Draw and label each phase you model.

4. Use your model to explore one of the questions written for step 4 of **Before You Begin.** Describe the steps you took to explore the question.

PART B: Test Hypotheses

Answer each of the following questions by writing a hypothesis. Use your model to test each hypothesis, and describe your results.

5. Cytokinesis follows mitosis. How will the size of each new cell that is formed following cytokinesis compare with that of the original cell?

6. Sometimes two chromatids fail to separate during mitosis. How might this failure affect the chromosome number of the two new cells?

7. A mutation is a permanent change in a gene or chromosome. What effect might a mutation in a parent cell have on future generations of cells that result from the parent cell?

PART C: Cleanup and Disposal

8. Dispose of paper and yarn scraps in the designated waste container.

9. Clean up your work area and all lab equipment. Return lab equipment to its proper place. Wash your hands thoroughly before you leave the lab and after you finish all work.

Analyze and Conclude

1. Analyzing Results How do the nuclei you made by modeling mitosis compare with the nucleus of the model cell you started with? Explain your result.

2. Evaluating Methods How could you modify your model to better illustrate the process of mitosis?

3. Recognizing Patterns How does the genetic makeup of the cells that result from mitosis compare with the genetic makeup of the original cell?

4. Inferring Conclusions How is mitosis important?

5. Further Inquiry Write a new question about mitosis or the cell cycle that could be explored with your model.

? Do You Know?

Do research in the library or media center to answer these questions:

1. How often do different cells of the human body undergo mitosis?

2. What are some common chemicals that disrupt the cell cycle?

Use the following Internet resources to explore your own questions about mitosis or the cell cycle.

internet connect

www.scilinks.org
Topic: Cell Cycle
Keyword: HX4033

SCiLINKS. Maintained by the National Science Teachers Association

Do you know what cancer is
and what may cause it?

Understanding Cancer

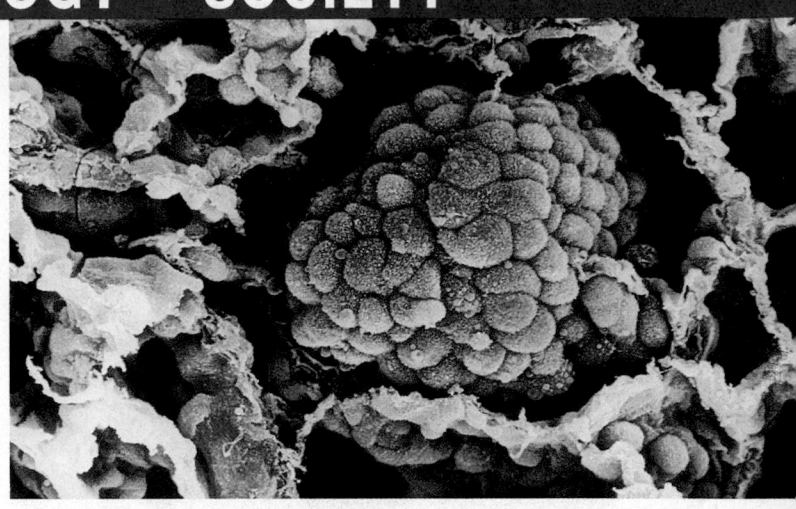

Most healthy cells of the body grow, divide a certain number of times, age, and die. Sometimes, this orderly process is disrupted when cells lose the ability to limit and direct their growth. Cells may divide too often and produce an excess of cells, called a *tumor*. A tumor can be benign or malignant. Benign tumors do not spread to other parts of the body, and can usually be surgically removed. A malignant tumor—cancer—invades and destroys nearby healthy tissues and organs. Cancerous tumors can metastasize, that is, can spread to other parts of the body and form new tumors. There are probably at least 100 different kinds of cancer, with each one affecting different kinds of cells and having different characteristics.

Cancer cells do not respond normally to the chemical signals that regulate the cell cycle. In some cancer cells, the way cell division signals are transmitted to the nucleus is abnormal. In other cancer cells, the entire cell cycle control system may be abnormal. If cancer cells stop dividing, they do so at random points in the cell cycle rather than at the normal checkpoints. When cultured in the laboratory with adequate nutrients, cancer cells can divide indefinitely and are referred to as "immortal." Most mammalian cells grown in culture divide only 20 to 50 times before they stop dividing, age, and die.

How Cancer Begins

Changes in the genes that influence the cell cycle can cause the transformation of a normal cell into a cancer cell. There are two types of these genes. The first type codes for proteins that stimulate cell division. Genes of this type are normally turned off in cells that are not dividing. This type of gene can be converted to an oncogene, that is, a "cancer gene," by mutation. One common oncogene, a gene called *ras*, is present in mutated form in about 30 percent of human cancers and in some forms of leukemia.

The second type of gene associated with cancer is a tumor suppressor gene. Tumor suppressor genes code for proteins that normally restrain cell division. In many cancers, tumor suppressor genes have been inactivated by mutation. An inherited mutation in one copy of a tumor suppressor gene results in higher risk of cancer. However, cancer does not occur unless and until the remaining, healthy copy of the gene is also inactivated by mutation. If a person is born with two normal copies of a tumor suppressor gene, both must be inactivated before cancer can develop.

Almost 50 percent of human cancers are associated with a mutation in the tumor suppressor gene *p53*. These cancers include many breast, colon, lung, prostate, and skin cancers. The protein produced by the *p53* gene normally acts as an emergency brake in the cell cycle. The p53 protein also induces the death of damaged cells. To perform these functions, p53 protein must

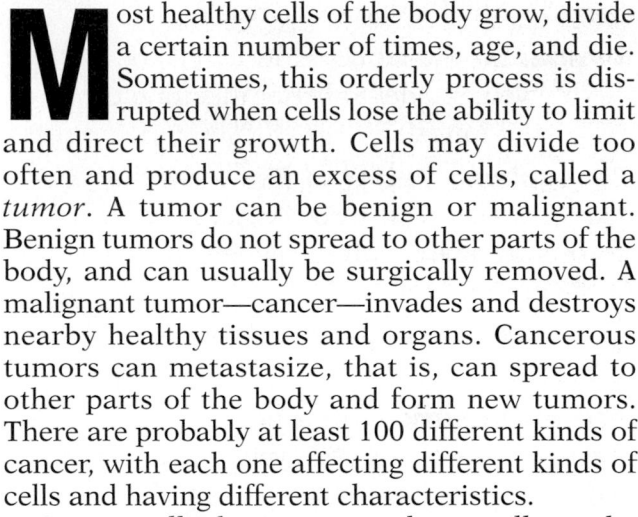

Bronchial tumor
Tumor cells have replaced the
normal, ciliated tissue in the bronchi.

Lung tumor
A malignant tumor differs from the tissue that surrounds it.

bind to DNA. Mutant p53 protein cannot bind to DNA, thus cell division occurs unchecked.

Mutations that result in cancer, whether they involve oncogenes or tumor suppressor genes, can occur spontaneously. Mutations can also be induced by factors in the environment, such as X-rays and ultraviolet radiation, cigarette smoke, asbestos, and even by the human papilloma virus and the hepatitis B virus.

Surviving Cancer

In the United States, about one of every five deaths—more than half a million each year—are caused by cancer. Only heart disease kills more people. The number of new cancer cases and the number of cancer deaths for every 100,000 persons have decreased over the past 10 years. The death rates for children and adults under the age of 50 have decreased dramatically.

Some kinds of cancer kill more people than others. Lung cancer is the number one cancer killer. Colorectal cancer, breast cancer, and prostate cancer are the next most common. Cancer survival is measured by the percentage of cancer patients who survive a specific number of years. Five-year cancer survival rates for common cancers are shown in the table at the right.

Cancer survival is influenced by the type of cancer and by the stage at which a cancer is diagnosed. For example, the 5-year survival rate for colorectal cancer diagnosed in its first stage is 96 percent. When diagnosed in its fourth and most advanced stage, however, the 5-year survival rate is only 5 percent. Cancer screenings and early detection of cancers, such as those of the breast and colon, greatly increase a person's chances of surviving cancer. ■

Cancer site	5-year survival (percent)
Prostate (males only)	98
Breast (females only)	88
Colorectal	63
Leukemia	47
Brain	32
Lung	15
Pancreas	4
All cancers	64

Analyzing STS Issues

Science and Society

1 Use library resources or the Internet to research factors that increase the risk of cancer. List some of the risk factors and what types of cancer they could lead to. Why are factors in lifestyle or the environment difficult to identify? How can people protect themselves from exposure to known risk factors?

Technology

2 The type, size, stage, and location of a person's cancer determines which cancer treatment is most appropriate for that person. A patient should take an active part in researching and developing an informed treatment plan. Using library resources or the Internet, research and identify some therapies used to treat cancer. What are some advantages and disadvantages to these treatments?

UNIT 2

Principles of Genetics

A DNA sequencing gel, shown here being examined by a technician, reveals an individual's unique chemical fingerprint.

in perspective

Hemophilia

Yesterday... Queen Victoria of England carried the gene for hemophilia. "Our poor family," she wrote in her diary, "seems persecuted by this awful disease, the worst I know." One son died of hemophilia and two daughters inherited the gene for the disease. **Why are carriers of hemophilia always female?**

Queen Victoria

Today... The blood of normal individuals contains a protein called factor VIII that enables the blood to clot after an injury. But in people with hemophilia, the gene that produces factor VIII is defective and bleeding continues uncontrolled. **Discover how a gene controls the production of a protein.** Injections of factor VIII at the first sign of bleeding allow many hemophiliacs to control bleeding episodes.

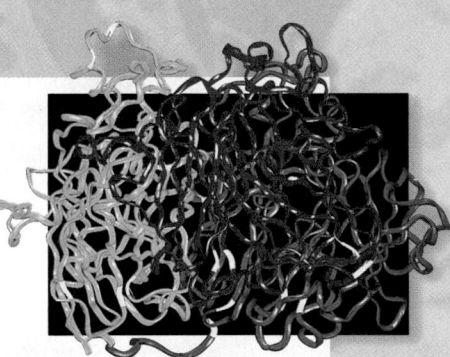

Factor VIII protein

Tomorrow... Biologists have recently learned to make a genetically engineered form of factor VIII. New research in genetic engineering has yielded female pigs that produce factor VIII in their milk. **What are the safety advantages of genetically-engineered factor VIII?**

☑ **internet** connect

www.scilinks.org
Topic: Hemophilia
Keyword: HX4097

SCI**LINKS**® Maintained by the National Science Teachers Association

141

Sperm on the surface of an egg (2890×)

7 Meiosis and Sexual Reproduction

 ## Quick Review

Answer the following without referring to earlier sections of your book.

1. **Define** the terms *evolution* and *natural selection.* (Chapter 1, Section 1)

2. **Define** the term *homologous chromosomes,* and identify chromatids. (Chapter 6, Section 1)

3. **Differentiate** between haploid cells and diploid cells. (Chapter 6, Section 1)

4. **Describe** the structure and function of the spindle. (Chapter 6, Section 3)

5. **Summarize** the steps of mitosis. (Chapter 6, Section 3)

Did you have difficulty? *For help, review the sections indicated.*

Reading Activity

Take a few moments to study the first two pages in Section 1, including Figure 1. Then on a sheet of paper or in your notebook, answer the following questions:

- *What is the topic of Section 1?*

- *How are meiosis I and meiosis II similar?*

- *How are meiosis I and meiosis II different?*

A special form of cell reproduction produces the egg and sperm cells shown here. When an egg joins with a single sperm cell, genetic instructions from a male and female are combined, and a new individual is formed.

Looking Ahead

Section 1

Meiosis
Formation of Haploid Cells
Meiosis and Genetic Variation
Meiosis and Gamete Formation

Section 2

Sexual Reproduction
Sexual and Asexual Reproduction
Sexual Life Cycles in Eukaryotes

⧉ internet connect

www.scilinks.org
National Science Teachers Association *sci*LINKS Internet resources are located throughout this chapter.

SCiLINKS. **Maintained by the National Science Teachers Association**

Objectives

- **Summarize** the events that occur during meiosis.

- **Relate** crossing-over, independent assortment, and random fertilization to genetic variation.

- **Compare** spermatogenesis and oogenesis.

Key Terms

meiosis
crossing-over
independent
  assortment
spermatogenesis
sperm
oogenesis
ovum

Formation of Haploid Cells

Some organisms reproduce by joining gametes to form the first cell of a new individual. The gametes are haploid—they contain one set of chromosomes. Imagine how the chromosome number would increase with each generation if chromosome reduction did not occur!

Meiosis *(meye OH sihs)* is a form of cell division that halves the number of chromosomes when forming specialized reproductive cells, such as gametes or spores. Meiosis involves two divisions of the nucleus—meiosis I and meiosis II.

Before meiosis begins, the DNA in the original cell is replicated. Thus, meiosis starts with homologous chromosomes. Recall that homologous chromosomes are similar in size, shape, and genetic content. The stages of meiosis are summarized in **Figure 1.**

Step ❶ Prophase I The chromosomes condense, and the nuclear envelope breaks down. Homologous chromosomes pair along their length. **Crossing-over** occurs when portions of a chromatid on one homologous chromosome are broken and exchanged with the corresponding chromatid portions of the other homologous chromosome.

Figure 1

Stages of Meiosis

Four cells are produced, each with half as much genetic material as the original cell.

❶ Prophase I	❷ Metaphase I	❸ Anaphase I	❹ Telophase I and cytokinesis
Chromosomes become visible. The nuclear envelope breaks down. Crossing-over occurs.	Pairs of homologous chromosomes move to the equator of the cell.	Homologous chromosomes move to opposite poles of the cell.	Chromosomes gather at the poles of the cell. The cytoplasm divides.

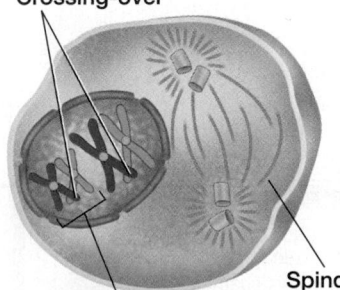

Crossing-over

Homologous
chromosomes

Spindle

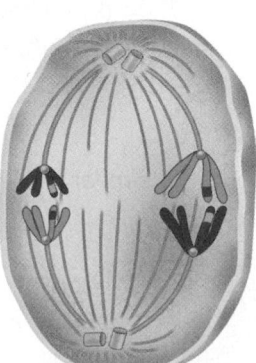

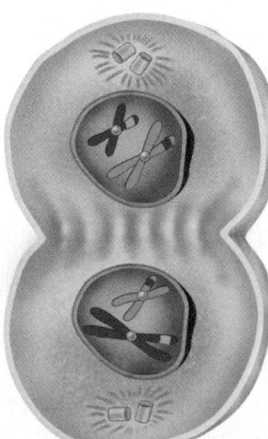

Step ② Metaphase I The pairs of homologous chromosomes are moved by the spindle to the equator of the cell. The homologous chromosomes remain together.

Step ③ Anaphase I The homologous chromosomes separate. As in mitosis, the chromosomes of each pair are pulled to opposite poles of the cell by the spindle fibers. *But the chromatids do not separate at their centromeres—each chromosome is still composed of two chromatids. The genetic material, however, has recombined.*

Step ④ Telophase I Individual chromosomes gather at each of the poles. In most organisms, the cytoplasm divides (cytokinesis), forming two new cells. Both cells or poles contain one chromosome from each pair of homologous chromosomes. *Chromosomes do not replicate between meiosis I and meiosis II.*

Step ⑤ Prophase II A new spindle forms around the chromosomes.

Step ⑥ Metaphase II The chromosomes line up along the equator and are attached at their centromeres to spindle fibers.

Step ⑦ Anaphase II The centromeres divide, and the chromatids (now called chromosomes) move to opposite poles of the cell.

Step ⑧ Telophase II A nuclear envelope forms around each set of chromosomes. The spindle breaks down, and the cell undergoes cytokinesis. The result of meiosis is four haploid cells.

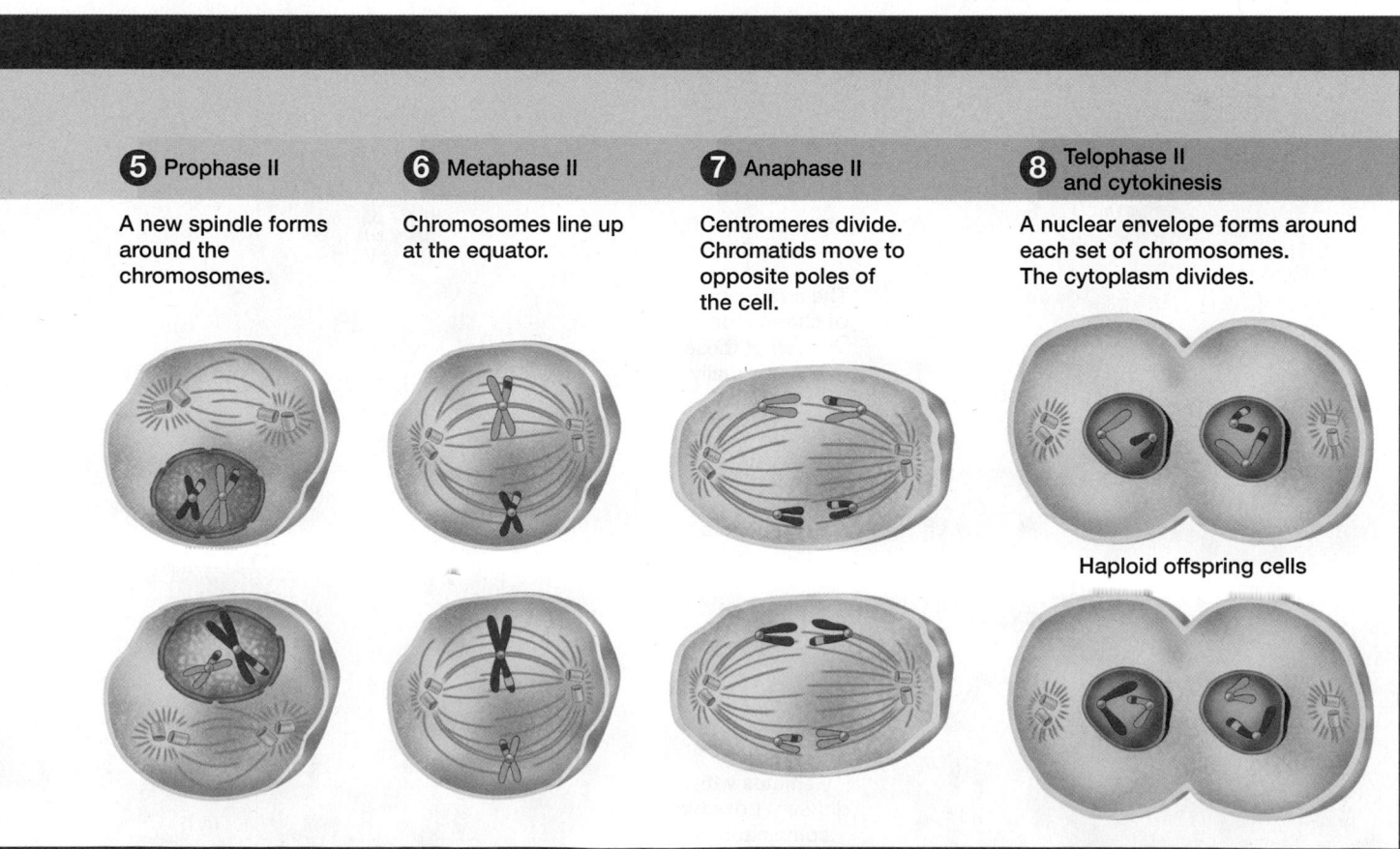

⑤ Prophase II

A new spindle forms around the chromosomes.

⑥ Metaphase II

Chromosomes line up at the equator.

⑦ Anaphase II

Centromeres divide. Chromatids move to opposite poles of the cell.

⑧ Telophase II and cytokinesis

A nuclear envelope forms around each set of chromosomes. The cytoplasm divides.

Haploid offspring cells

Meiosis and Genetic Variation

Meiosis is an important process that allows for the rapid generation of new genetic combinations. Three mechanisms make key contributions to this genetic variation: independent assortment, crossing-over, and random fertilization.

Independent Assortment

Most organisms have more than one chromosome. In humans, for example, each gamete receives one chromosome from each of 23 pairs of homologous chromosomes. But, which of the two chromosomes that an offspring receives from each of the 23 pairs is a matter of chance. This random distribution of homologous chromosomes during meiosis is called **independent assortment**. Independent assortment is summarized in **Figure 2.** Each of the 23 pairs of chromosomes segregates (separates) independently. Thus, 2^{23} (about 8 million) gametes with different gene combinations can be produced from one original cell by this mechanism.

Crossing-Over and Random Fertilization

The DNA exchange that occurs during crossing-over adds even more recombination to the independent assortment of chromosomes that occurs later in meiosis. Thus, the number of genetic combinations that can occur among gametes is practically unlimited.

Figure 2 Independent assortment

The same cell is shown twice. Because each pair of homologous chromosomes separates independently, four different gametes can result in each case.

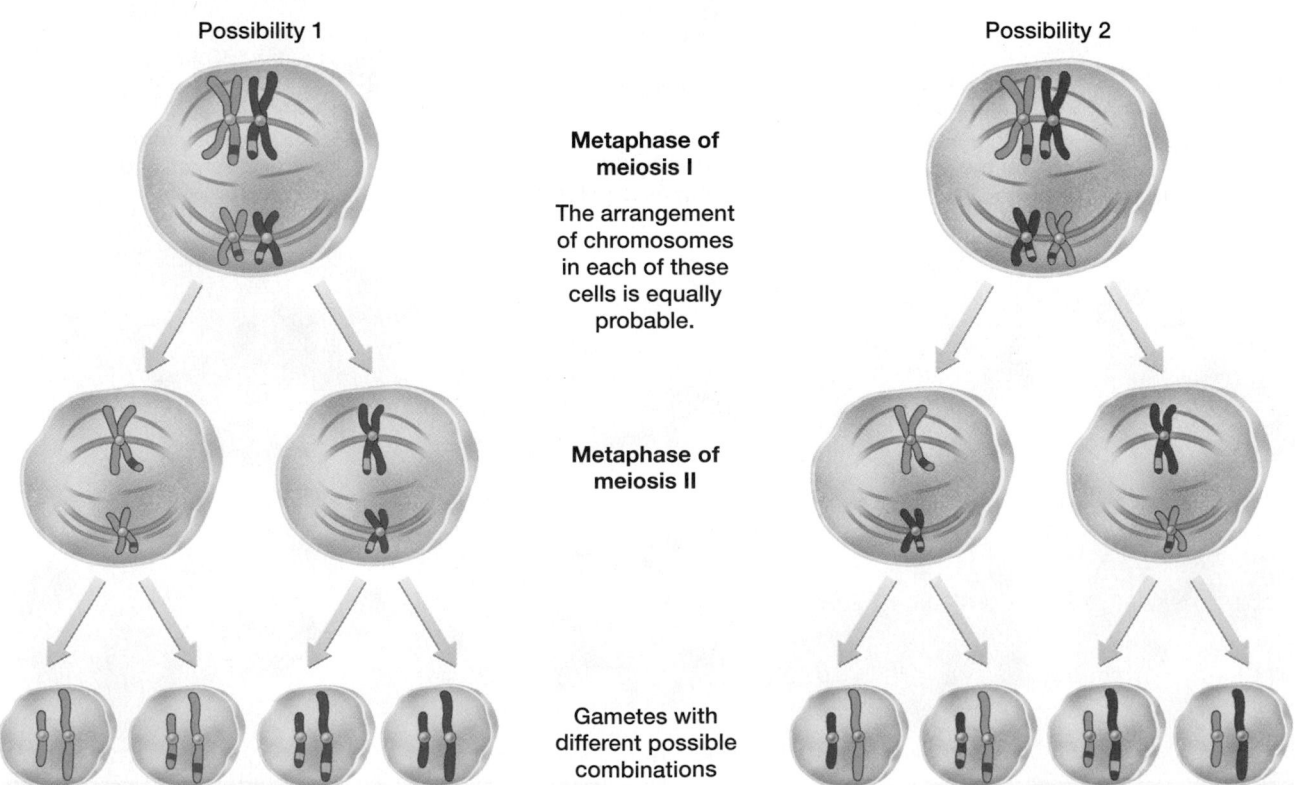

Possibility 1   Possibility 2

Metaphase of meiosis I

The arrangement of chromosomes in each of these cells is equally probable.

Metaphase of meiosis II

Gametes with different possible combinations

Modeling Crossing-Over

You can use paper strips and pencils to model the process of crossing-over.

Homologous chromosomes

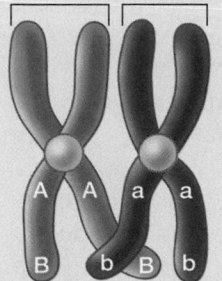

Materials

4 paper strips, pens or pencils (two colors), scissors, tape

Procedure

1. Using one color, write the letters *A* and *B* on two paper strips. These two strips will represent one of the two homologous chromosomes shown above.

2. Using a second color, write the letters *a* and *b* on two paper strips. These two strips will represent the second homologous chromosome shown above.

3. Use your chromosome models, scissors, and tape to demonstrate crossing-over between the chromatids of two homologous chromosomes.

Analysis

1. Determine what the letters *A, B, a,* and *b* represent.

2. Infer why the chromosomes you made are homologous.

3. Compare the number of different types of chromatids (combinations of *A, B, a,* and *b*) before crossing-over with the number after crossing-over.

4. Critical Thinking Applying Information How does crossing-over relate to genetic recombination?

Furthermore, the zygote that forms a new individual is created by the random joining of two gametes (each gamete produced independently). Because fertilization of an egg by a sperm is random, the number of possible outcomes is *squared* ($2^{23} \times 2^{23} = 64$ trillion).

Importance of Genetic Variation

Meiosis and the joining of gametes are essential to evolution. No genetic process generates variation more quickly. In many cases, the pace of evolution appears to increase as the level of genetic variation increases. For example, when domesticated animals such as cattle and sheep are bred for large size, many large animals are produced at first. But as the existing genetic combinations become used up, the ability to obtain larger and larger animals slows down. Further progress must then wait for the formation of new gene combinations.

Racehorse breeding provides another example. Thoroughbred racehorses are all descendants of a small number of individuals, and selection for speed has accomplished all it can with this limited amount of genetic variation. The winning times in major races stopped dramatically improving decades ago.

The pace of evolution is sped up by genetic recombination. The combination of genes from two organisms results in a third type, not identical to either parent. But bear in mind that natural selection does not always favor genetic change. Indeed, many modern organisms are little changed from their ancestors of the distant past. Natural selection may favor existing combinations of genes, slowing the pace of evolution.

WORD Origins

● The word *meiosis* is from the Greek word *meioun,* meaning "to make smaller." Knowing this makes it easier to remember that during meiosis, the chromosome number is reduced by half to form haploid gametes.

Meiosis and Gamete Formation

The fundamental events of meiosis occur in all sexually reproducing organisms. However, organisms vary in timing and structures associated with gamete formation. Meiosis is the primary event in the formation of gametes—gametogenesis.

Meiosis in Males

The process by which sperm are produced in male animals is called **spermatogenesis** *(spur mat uh JEHN uh sihs)*. Spermatogenesis occurs in the testes (male reproductive organs). As illustrated in **Figure 3,** a diploid cell first increases in size and becomes a large immature cell (germ cell). The large cell then undergoes meiosis I. Two cells are produced, each of which undergoes meiosis II to form a total of four haploid cells. The four cells change in form and develop a tail to become male gametes called **sperm.**

Meiosis in Females

The process by which gametes are produced in female animals is called **oogenesis** *(oh oh JEHN uh sihs)*. Oogenesis, summarized in Figure 3, occurs in the ovaries (female reproductive organs). Notice that during cytokinesis following meiosis I, the cytoplasm divides unequally. One of the resulting cells gets nearly all of the cytoplasm. It is this cell that will ultimately give rise to an egg cell. The other cell is very small and is called a polar body. The polar body may divide again, but its offspring cells will not survive.

Figure 3 Meiosis in male and female animals

Meiosis of a male diploid cell results in four haploid sperm, while meiosis of a female diploid cell results in only one functional haploid egg cell.

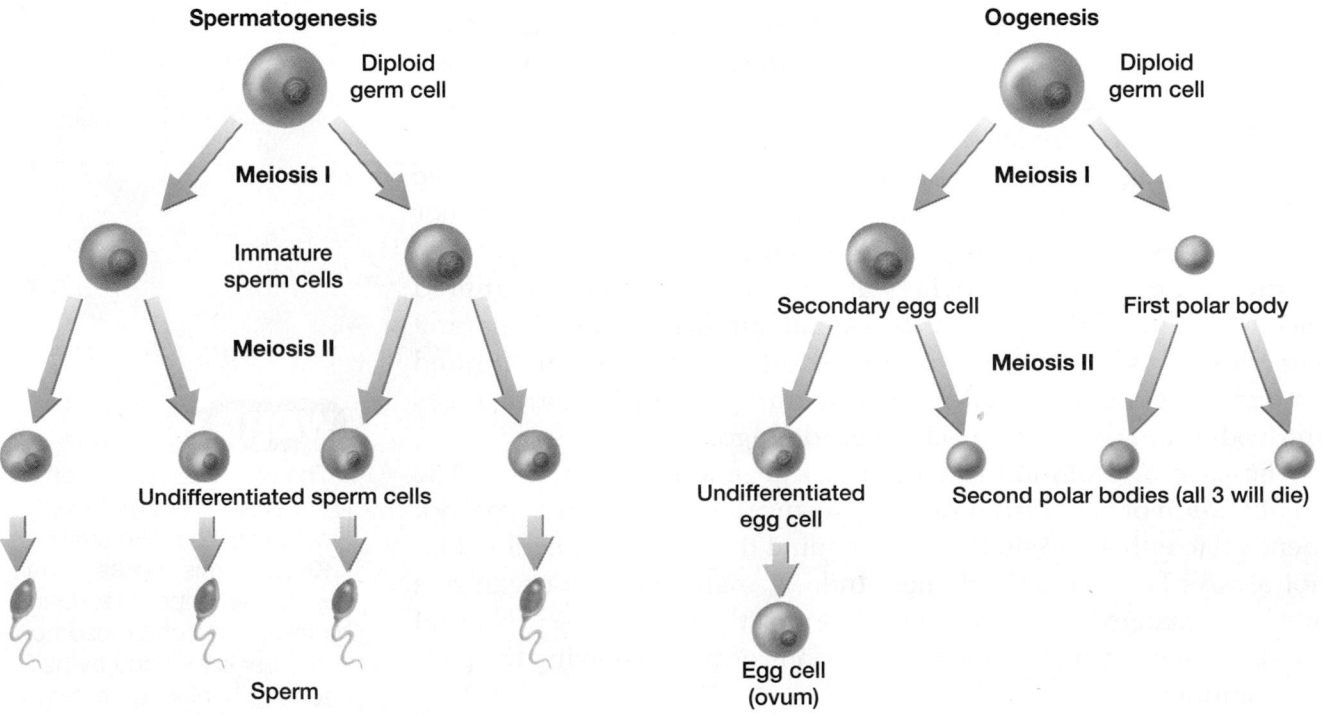

Spermatogenesis

Diploid germ cell

Meiosis I

Immature sperm cells

Meiosis II

Undifferentiated sperm cells

Sperm

Oogenesis

Diploid germ cell

Meiosis I

Secondary egg cell

First polar body

Meiosis II

Undifferentiated egg cell

Second polar bodies (all 3 will die)

Egg cell (ovum)

The larger cell undergoes meiosis II, and the division of the egg cell during cytokinesis is again unequal. The larger cell develops into a gamete called an **ovum** (plural, ova) or, more commonly, egg. The smaller cell, the second polar body, dies. Because of its larger share of cytoplasm, the mature ovum has a rich storehouse of nutrients. These nutrients nourish the young organism that develops if the ovum is fertilized.

BIOWatch

The Making of an Egg

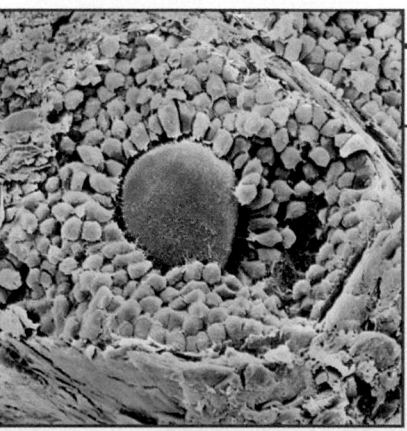

Human ovarian follicle

An intricate series of steps controls the formation of an egg. As oogenesis proceeds, eggs become very different from other cells in the organism. One obvious difference is the size of the egg cells. In many animals, eggs swell to gigantic proportions as they accumulate cellular components. How is this transformation accomplished?

Oogenesis in *Drosophila*
Researchers have examined oogenesis in great detail in the fruit fly, *Drosophila melanogaster.* As in other animals, eggs in *Drosophila* are produced when germ cells divide. Four rapid cell divisions produce 16 cells, which form a cluster known as a germ cell cyst. Bridges called *ring canals* interconnect all cells in the

cyst. However, only one of the cells develops into an egg. The other 15 cells become nurse cells, which donate organelles—including mitochondria and parts of the endoplasmic reticulum—to the growing egg. The organelles move through the ring canals by traveling along a network of microtubules. Some scientists believe that this movement reflects an organized sorting process, in which functional organelles collect in the egg and damaged organelles collect in the nurse cells. The nurse cells die as the egg completes its development.

Oogenesis in Other Organisms
A comparable process of cyst formation takes place during oogenesis in the mouse. As in the

fruit fly, many cellular components are redistributed among the cells in the cyst. These findings suggest that the early steps in egg formation may be very similar in a wide range of organisms.

internet connect

www.scilinks.org
Topic: Oogenesis
Keyword: HX4194

SC*I*LINKS. Maintained by the National Science Teachers Association

Section 1 Review

1 **Explain** the significance of meiosis in sexual reproduction.

2 **Name** the stage of meiosis during which chromatids are separated to opposite poles of the cell.

3 **Compare** the processes of crossing-over and independent assortment.

4 **Differentiate** gamete formation in male animals from gamete formation in female animals.

5 **Critical Thinking Evaluating Information** If one cell in a dog ($2n = 78$) undergoes meiosis and another cell undergoes mitosis, how many chromosomes will each resulting cell contain?

6 **Standardized Test Prep** If a cell begins meiosis with two pairs of homologous chromosomes, how many chromatids will be in each cell that is produced at the end of meiosis I?

A 1   **C** 4

B 2   **D** 8

Objectives

- **Differentiate** between asexual and sexual reproduction.

- **Identify** three types of asexual reproduction.

- **Evaluate** the relative genetic and evolutionary advantages and disadvantages of asexual and sexual reproduction.

- **Differentiate** between the three major sexual life cycles found in eukaryotes.

Key Terms

asexual reproduction
clone
sexual reproduction
life cycle
fertilization
sporophyte
spore
gametophyte

Sexual and Asexual Reproduction

Some organisms look exactly like their parents and siblings. Others share traits with family members but are not identical to them. Some organisms have two parents, while others have one. The type of reproduction that produces an organism determines how similar the organism is to its parents and siblings. Reproduction, the process of producing offspring, can be asexual or sexual.

In **asexual reproduction** a single parent passes copies of all of its genes to each of its offspring; there is no fusion of haploid cells such as gametes. An individual produced by asexual reproduction is a **clone,** an organism that is genetically identical to its parent. As you have read, prokaryotes reproduce by a type of asexual reproduction called binary fission. Many eukaryotes, as shown in **Figure 4,** also reproduce asexually.

In contrast, in **sexual reproduction** two parents each form reproductive cells that have one-half the number of chromosomes. A diploid mother and father would give rise to haploid gametes, which join to form diploid offspring. Because both parents contribute genetic material, the offspring have traits of both parents but are not exactly like either parent. As shown in **Figure 5,** sexual reproduction, with the formation of haploid cells, occurs in eukaryotic organisms, including humans.

Types of Asexual Reproduction

There are many different types of asexual reproduction. For example, amoebas reproduce by fission, the separation of a parent into two or more individuals of about equal size. Some multicellular eukaryotes undergo fragmentation, a type of reproduction in which the body breaks into several pieces. Some or all of these fragments later develop into complete adults when missing parts are regrown. Other organisms, like the hydra shown in Figure 4, undergo budding, in which new individuals split off from existing ones. The bud may break from the parent and become an independent organism, or it may remain attached to the parent. An attached bud can eventually give rise to a group of many individuals.

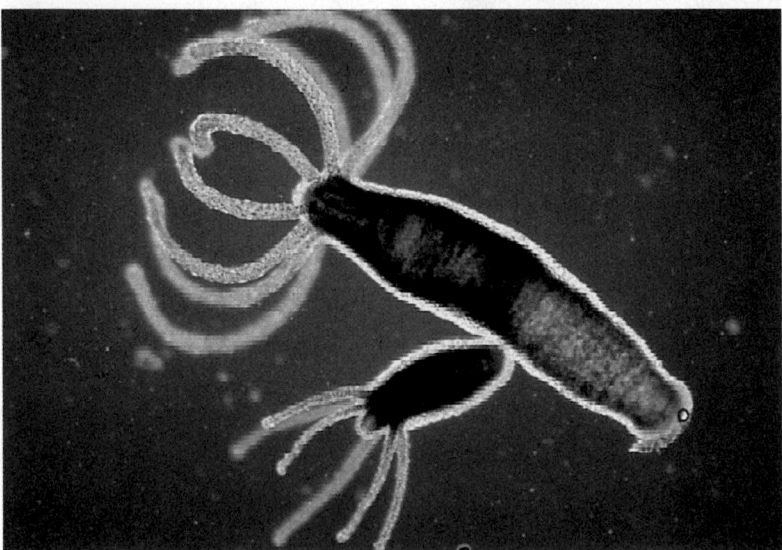

Figure 4 Asexual reproduction. Asexual reproduction creates clones. This hydra is in the process of reproducing asexually. The smaller hydra budding from the parent is genetically identical to the parent.

Genetic Diversity

Asexual reproduction is the simplest and most primitive method of reproduction. In a stable environment, asexual reproduction allows organisms to produce many offspring in a short period of time, without using energy to produce gametes or to find a mate. However, the DNA of these organisms varies little between individuals. This may be a disadvantage in a changing environment because a population of organisms may not be able to adapt to a new environment.

On the other hand, sexual reproduction provides a powerful means of quickly making different combinations of genes among individuals. Such genetic diversity is the raw material for evolution.

Evolution of Sexual Reproduction

The evolution of sexual reproduction may have allowed early protists to repair their own DNA. Only diploid cells can repair certain kinds of chromosome damage, such as breaks in both strands of DNA. Many modern protists are haploid most of the time, and they reproduce asexually. (They form a diploid cell only in response to stress in the environment.) Thus the process of meiosis and the pairing of homologous chromosomes may have allowed early protistan cells to repair damaged DNA. This hypothesis is further supported by the fact that many enzymes that repair DNA damage are involved in meiosis.

Figure 5   Sexual reproduction

Sexual reproduction creates genetic diversity. Human gametes contain 23 chromosomes. (Only one chromosome is shown in each gamete below.) After fertilization the resulting zygote has 23 pairs of chromosomes.

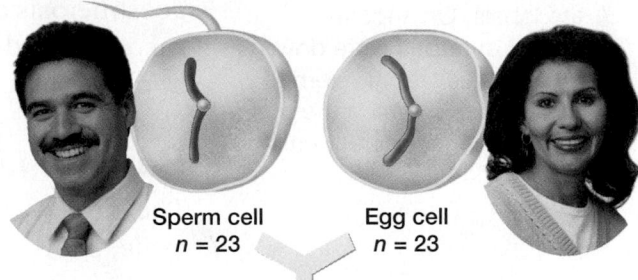

Sperm cell
$n = 23$

Egg cell
$n = 23$

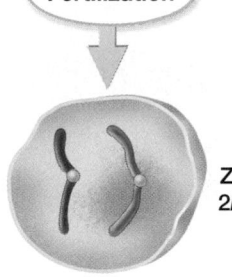

Fertilization

Zygote
$2n = 46$

QUICK LAB

Observing Reproduction in Yeast

Yeast are unicellular organisms that live in liquid or moist environments. You can examine a culture of yeast to observe one of the types of reproduction that yeast can undergo.

Materials

microscope, microscope slides, dropper, culture of yeast

Procedure

1. Make a wet mount of a drop of yeast culture.

2. Observe the yeast with a compound microscope under low power.

3. Look for yeast that appear to be in "pairs."

4. Observe the pairs under high power, and then make drawings of your observations.

Analysis

1. **Infer** the type of reproduction you observed when the yeast appeared to be in pairs.

2. **Identify** the reason for your answer.

3. **Determine,** by referring to your textbook, the name of the type of reproduction you observed.

Sexual Life Cycles in Eukaryotes

Study TIP

● **Interpreting Graphics**

After reading this chapter, trace or make a sketch of Figures 6, 7, and 8 without the labels. On separate pieces of paper, write down the labels. Without referring to your book, match the labels with the correct part of your sketch.

The entire span in the life of an organism from one generation to the next is called a **life cycle.** The life cycles of all sexually reproducing organisms follow a basic pattern of alternation between the diploid and haploid chromosome numbers. The type of sexual life cycle that a eukaryotic organism has depends on the type of cell that undergoes meiosis and on when meiosis occurs. Eukaryotes that undergo sexual reproduction can have one of three types of sexual life cycles: haploid, diploid, or alternation of generations.

Haploid Life Cycle

The haploid life cycle is the simplest of sexual life cycles. In this life cycle, shown in **Figure 6,** haploid cells occupy the major portion of the life cycle. The zygote is the only diploid cell, and it undergoes meiosis immediately after it is formed, creating new haploid cells. The haploid cells give rise to haploid multicellular individuals that produce gametes by mitosis (not meiosis). In a process called fusion, the gametes fuse to produce a diploid zygote, and the cycle continues.

When the diploid zygote undergoes meiosis it provides an opportunity for the cell to correct any genetic damage, as discussed earlier. The damage is repaired during meiosis, when the two homologous chromosomes are lined up side-by-side in preparation for crossing over. Special repair enzymes remove any damaged sections of double stranded DNA, and fill in any gaps. This type of life cycle is found in many protists, as well as in some fungi and algae, such as the unicellular *Chlamydomonas (KLUH mih duh moh nuhs)*, shown in Figure 6.

Figure 6   Haploid life cycle. Some organisms, such as *Chlamydomonas,* have haploid cells as a major portion of their life cycle.

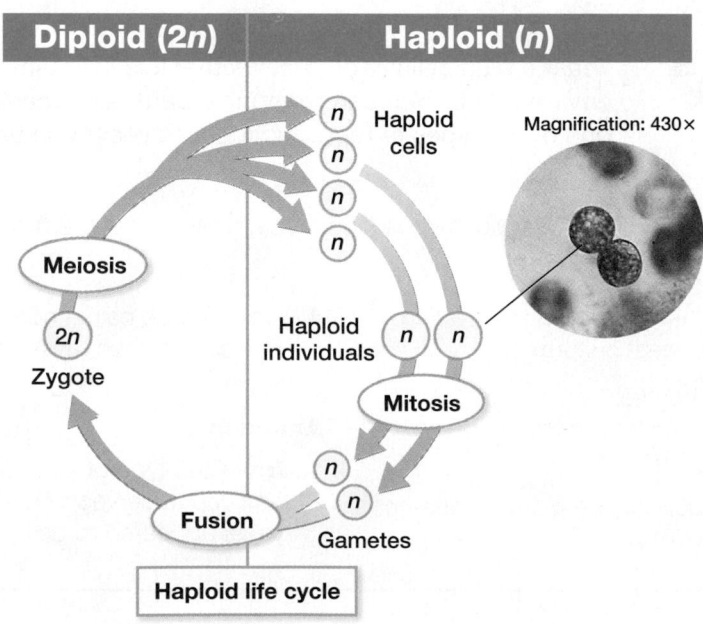

Diploid Life Cycle

The outstanding characteristic of the diploid life cycle is that adult individuals are diploid, each individual inheriting chromosomes from two parents. In most animals, including humans, a diploid reproductive cell undergoes meiosis to produce gametes.

As shown in **Figure 7,** the gametes (sperm and egg cells) join in a process called **fertilization,** which results in a diploid zygote. After fertilization, the resulting zygote begins to divide by mitosis. This single diploid cell eventually gives rise to all of the cells of the adult. The cells of the adult are also diploid since they are produced by mitosis.

The diploid individual that develops from the zygote occupies the major portion of the diploid life cycle. The gametes are the only haploid cells in the diploid life cycle; all of the other cells are diploid.

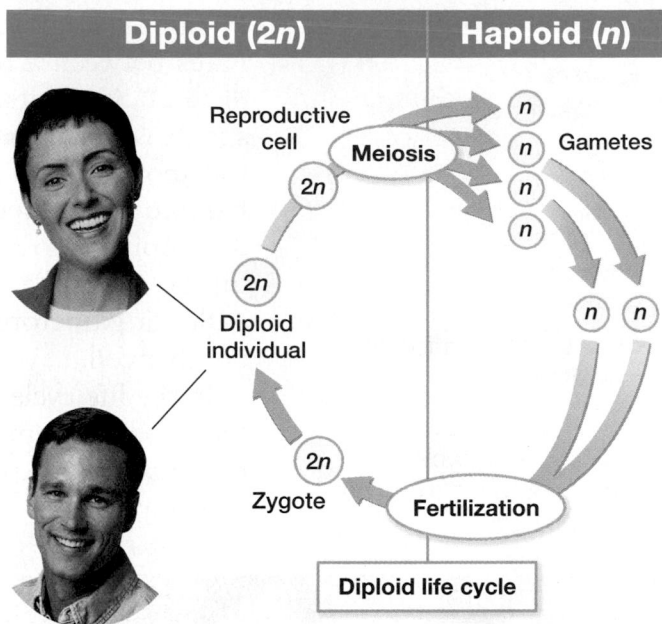

Figure 7 Diploid life cycle. Humans and other organisms have a life cycle dominated by a diploid individual.

Exploring Further

Cloning by Parthenogenesis

A snake is born to a mother that did not have a mate. Although this may sound impossible, or like some headline in a tabloid magazine, this can actually occur in nature. Parthenogenesis *(pahr thuh noh JEHN uh sihs)* is a type of reproduction in which a new individual develops from an unfertilized egg. Since there is no male that contributes genetic material, the offspring is a clone (genetically identical) of the mother. Clones are usually produced in nature by asexual reproduction. Parthenogenesis, however, is a special form of cloning.

How Does Parthenogenesis Occur?

Parthenogenesis in snakes has usually occurred in older females that have lived many years without male companionship, such as those in a zoo. It is hypothesized that in the mother snake, her own chromosomes are copied in place of the missing father's chromosomes, thereby self-fertilizing her egg. Other scientists think that after a long absence of males, some unknown signal (such as a hormone) triggers the egg to start dividing.

Whiptail lizard

Organisms That Undergo Parthenogenesis

Organisms capable of reproducing by parthenogenesis include dandelions, hawkweeds, and some fishes, lizards, and frogs. Whiptail lizards are all females that lay eggs that hatch without any male contributions. Honeybees also produce male drones by parthenogenesis.

Parthenogenesis is not thought to be possible in mammals. Embryos of mammals that do not have genes from both a female and a male parent do not develop normally. The only natural mammalian clones known are identical twins, which develop when a fertilized egg splits and two individuals develop.

Alternation of Generations

Plants, algae, and some protists have a life cycle that regularly alternates between a haploid phase and a diploid phase. As shown in **Figure 8,** in plants, the diploid phase in the life cycle that produces spores is called a **sporophyte** (*SPOH ruh fiet*). Spore-forming cells in the sporophyte undergo meiosis to produce spores. A **spore** is a haploid reproductive cell produced by meiosis that is capable of developing into an adult without fusing with another cell. Thus, unlike a gamete, a spore gives rise to a multicellular individual called a gametophyte (*guh MEET uh fiet*) without joining with another cell.

In the life cycle of a plant, the **gametophyte** is the haploid phase that produces gametes by mitosis. The gametophyte produces gametes that fuse and give rise to the diploid phase. Thus, the sporophyte and gametophyte generations take turns, or alternate, in the life cycle.

In moss, for example, haploid spores develop in a capsule at the tip of the sporophyte "stalk." When the lid of the capsule pops off, the spores scatter. The spores germinate by mitosis and eventually form sexually mature gametophytes. The male gametophytes release sperm which swim through a film of moisture to the eggs in the female gametophyte. The diploid zygote develops as a sporophyte within the gametophyte and the life cycle continues.

It is important not to lose sight of the basic similarity of all three types of sexual life cycles. All three involve an alternation of haploid and diploid phases. The three types of sexual life cycles differ from each other only in which phases become multicellular.

Figure 8 Alternation of generations

Some organisms, such as roses, have a life cycle that alternates between diploid and haploid phases.

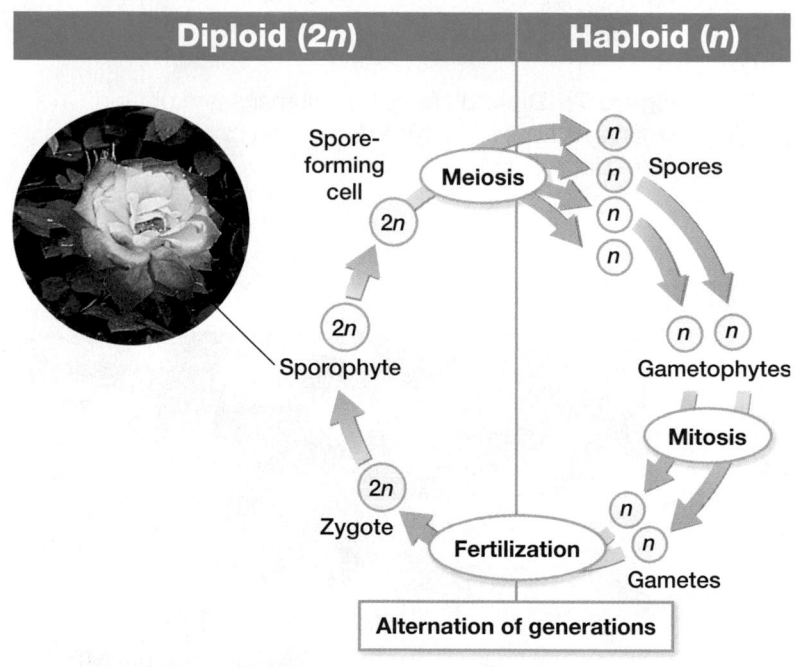

Diploid (2*n*) — Haploid (*n*)

Spore-forming cell · Meiosis · 2*n* · Spores · *n n n n*

2*n* · Sporophyte · Gametophytes · *n n* · Mitosis

2*n* · Zygote · Fertilization · Gametes · *n n*

Alternation of generations

Section 2 Review

1 Identify the type of reproduction that results in offspring that are genetically identical to their parent.

2 Describe two different types of eukaryotic asexual reproduction.

3 Compare the haploid life cycle found in *Chlamydomonas* with a diploid life cycle.

4 Summarize the process of alternation of generations.

5 Critical Thinking Evaluating Information Evaluate the significance of mutations and repair of mutations to the evolution of sexual reproduction.

6 Standardized Test Prep The amount of genetic variation in offspring is greatest in organisms that reproduce

A sexually through meiosis.

B sexually through fission.

C asexually through mitosis.

D asexually through budding.

Study ZONE

CHAPTER HIGHLIGHTS

Key Concepts

1 Meiosis

- Meiosis reduces the number of chromosomes by half to form reproductive cells. When the reproductive cells unite in fertilization, the normal diploid number is restored.

- During meiosis I, homologous chromosomes separate. Crossing-over during prophase I results in the exchange of genetic material between homologous chromosomes.

- During meiosis II, the two chromatids of each chromosome separate. As a result of meiosis, four haploid cells are produced from one diploid cell.

- Independent assortment, crossing-over, and random fertilization contribute to produce genetic variation in sexually reproducing organisms.

- In sexually reproducing eukaryotic organisms, gametes form through the process of spermatogenesis in males and oogenesis in females.

2 Sexual Reproduction

- Asexual reproduction is the formation of offspring from one parent. The offspring are genetically identical to the parent.

- Sexual reproduction is the formation of offspring through the union of gametes. The offspring are genetically different from their parents.

- A disadvantage to asexual reproduction in a changing environment is the lack of genetic diversity among the offspring.

- Sexual reproduction increases variation in the population by making possible genetic recombination.

- Sexual reproduction may have begun as a mechanism to repair damaged DNA.

- Eukaryotic organisms can have one of three kinds of sexual life cycles, depending on the type of cell that undergoes meiosis and on when meiosis occurs.

Key Terms

Section 1

meiosis (144)
crossing-over (144)
independent assortment (146)
spermatogenesis (148)
sperm (148)
oogenesis (148)
ovum (149)

Section 2

asexual reproduction (150)
clone (150)
sexual reproduction (150)
life cycle (152)
fertilization (153)
sporophyte (154)
spore (154)
gametophyte (154)

Unit 4—Cell Reproduction
Use Topics 5–6 in this unit to review the key concepts and terms in this chapter.

Understanding Key Ideas

1. Which of the following events occurs during prophase I of meiosis?
   a. crossing-over
   b. duplication of chromatids
   c. reduction in chromosome number
   d. separation of chromatids to opposite poles

2. Homologous pairs of chromosomes move to opposite poles during
   a. prophase I.   c. metaphase II.
   b. anaphase I.   d. anaphase II.

3. Spermatogenesis produces
   a. four haploid cells.
   b. four diploid cells.
   c. four polar bodies.
   d. two haploid cells.

4. Sexual reproduction may have originated as a way for cells to
   a. shuffle genetic material.
   b. repair damaged DNA.
   c. produce diploid individuals.
   d. increase their population growth at a maximum rate.

5. In plants, the sporophyte generation produces _____ spores through meiosis.
   a. haploid
   b. triploid
   c. diploid
   d. mutated

6. After crossing-over as shown below, what would the sequence of genes be for each of the chromatids?

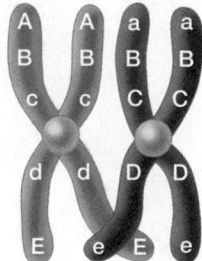

7. Compare and contrast the processes of mitosis and meiosis. (**Hint:** See Chapter 6, Section 4.)

8. **🌐 BIOWatch** What is the function of a nurse cell in egg formation?

9. **Exploring Further** Review two hypotheses that have been proposed to explain parthenogenesis according to the strengths and weaknesses of those hypotheses. Which hypothesis do you think is more supportable, and why?

10. **Concept Mapping** Make a concept map that shows the three sexual life cycles in eukaryotic organisms. Include the following words in your map: *meiosis, fusion, gametes, spores, fertilization, zygote, gametophyte, sporophyte, haploid,* and *diploid*.

Critical Thinking

11. **Evaluating Results** Occasionally homologous chromosomes fail to separate during meiosis I. Using the hypothetical example of an adult organism that has two pairs of chromosomes, describe the chromosomal makeup of the eggs that would result from this error in meiosis.

12. **Evaluating Results** If normal human sperm fertilized the eggs described above, what would the chromosomal makeup of the resulting zygote be?

13. **Applying Information** How do independent assortment, crossing-over, and random fertilization affect the rate of evolution?

14. **Critiquing Hypotheses** A student states that organisms that reproduce asexually are at a disadvantage in a stable environment. If you agree with this hypotheses, name one or more of its strengths. If you disagree, name one or more of its weaknesses.

Alternative Assessment

15. **Interactive Tutor Unit 4** Cell Reproduction Write a report summarizing the effects of various treatments for infertility. Find out how the production of gametes may be affected in some people who are infertile.

Understanding Concepts

Directions (1–4): **For *each* question, write on a separate sheet of paper the letter of the correct answer.**

1 How do most multicellular eukaryotes form specialized reproductive cells?
A. binary fission
B. fragmentation
C. meiosis
D. mitosis

2 How does meiosis differ from mitosis?
F. Mitosis includes cytokinesis, while meiosis does not.
G. The DNA is replicated before mitosis but not before meiosis.
H. Meiosis produces haploid cells, while mitosis produces diploid cells.
I. Mitosis produces gametes, while meiosis produces offspring cells.

3 What is the process that contributes to the formation of an embryo from the zygote?
A. fission
B. meiosis
C. mitosis
D. oogenesis

4 During what process are genes exchanged between homologous chromosomes?
F. crossing-over   **H.** meiosis
G. fertilization   **I.** telophase II

Directions (5): **For the following question, write a short response.**

5 Plants experience alternation of generations with a sporophyte phase and a gametophyte phase. How are sporophytes and gametophytes different in terms of the number of chromosomes they have?

Test TIP

If you are unsure of the answer to a particular question, put a question mark beside it and go on to the next question. If you have time, go back and reconsider any question that you skipped. (Do not write in this book.)

Reading Skills

Directions (6): **Read the passage below. Then answer the question.**

Many zoos have captive breeding programs for endangered species. These programs attempt to increase genetic diversity through selective breeding. However, this can be difficult if the number of individuals in a breeding program is low. Researchers can document the genetic makeup of each individual in the breeding program in order to mate individuals that are genetically varied.

6 Why might the genetic variation of captive-bred animals be different from the genetic variation of wild animals?
A. Wild animals usually produce offspring that cannot reproduce.
B. Wild animals are more likely to live longer than captive-bred animals.
C. Captive-bred animals are more likely to have mutations that lead to diversity.
D. Captive-bred animals can mate only with the individuals in their enclosed habitat.

Interpreting Graphics

Directions (7): **Base your answer to question 7 on the diagram below.**

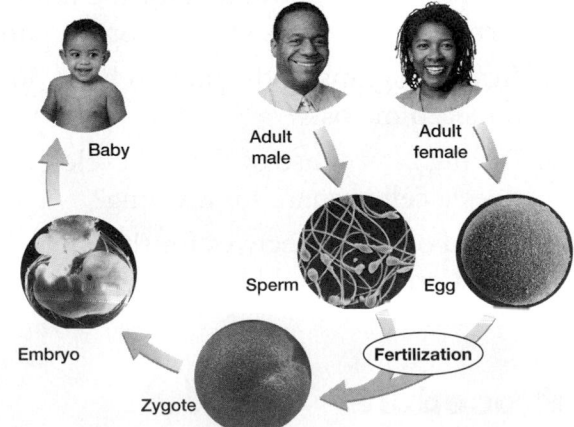

Human Reproduction

7 Which type of eukaryotic life cycle does this diagram represent?
F. asexual   **H.** diploid
G. cloning   **I.** haploid

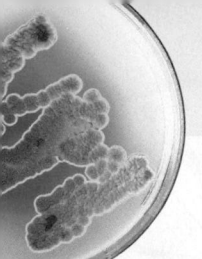

Exploration Lab

Modeling Meiosis

SKILLS
- Modeling
- Using scientific methods

OBJECTIVES
- **Describe** the events that occur in each stage of the process of meiosis.
- **Relate** the process of meiosis to genetic variation.

MATERIALS
- pipe cleaners of at least two different colors
- yarn
- wooden beads
- white labels
- scissors

Glass frog with eggs

Before You Begin

Meiosis is the process that results in the production of cells with half the normal number of chromosomes. It occurs in all organisms that undergo **sexual reproduction.** In this lab, you will build a model that will help you understand the events of meiosis. You can also use the model to demonstrate the effects of events such as **crossing-over** to explain results such as **genetic recombination.**

1. Write a definition for each boldface term in the paragraph above and for the following terms: homologous chromosomes, gamete.

2. In what organs in the human body do cells undergo meiosis?

3. During interphase of the cell cycle, how does a cell prepare for dividing?

4. Based on the objectives for this lab, write a question you would like to explore about meiosis.

Procedure

PART A: Design a Model

1. Work with the members of your lab group to design a model of a cell using the materials listed for this lab. Be sure that your model cell has at least two pairs of chromosomes.

2. Write out the plan for building your model. Have your teacher approve the plan before you begin building the model.

You Choose

As you design your experiment, decide the following:

a. what question you will explore

b. how to construct a cell membrane

c. how to show that your cell is diploid

d. how to show the locations of at least two genes on each chromosome

e. how to show that chromosomes are duplicated before meiosis begins

---

3. ◆ Build the cell model your group designed. **CAUTION: Sharp or pointed objects can cause injury. Handle scissors carefully.** Use your model to demonstrate the phases of meiosis. Draw and label each phase you model.

4. Use your model to explore one of the questions written by your group for step 4 of **Before You Begin.** Describe the steps you took to explore your question.

PART B: Test Hypotheses

Answer each of the following questions by writing a hypothesis. Use your model to test each hypothesis, and describe your results.

5. In humans, gametes (eggs and sperm) result from meiosis. Will all gametes produced by one parent be identical?

6. When an egg and a sperm fuse during sexual reproduction, the resulting cell (the first cell of a new organism) is called a zygote. How many copies of each chromosome and each gene will be found in a zygote?

7. Crossing-over frequently occurs between the chromatids of homologous chromosomes during meiosis. Under what circumstances does crossing-over result in new combinations of genes in gametes?

8. Synapsis (the pairing of homologous chromosomes) must occur before crossing-over can take place. How would the outcome of meiosis be different if synapsis did not occur?

PART C: Cleanup and Disposal

9. Dispose of paper and yarn scraps in the designated waste container.

10. Clean up your work area and all lab equipment. Return lab equipment to its proper place. Wash your hands thoroughly before you leave the lab and after finishing all work.

Analyze and Conclude

1. **Analyzing Results** How do the nuclei you made by modeling meiosis compare with the nucleus of the cell you started with? Explain your result.

2. **Recognizing Relationships** How are homologous chromosomes different from chromatids?

3. **Forming Reasoned Opinions** How is synapsis important to the outcome of meiosis? Explain.

4. **Evaluating Methods** How could you modify your model to better illustrate the process of meiosis?

5. **Drawing Conclusions** How are the processes of meiosis similar to those of mitosis? How are they different?

6. **Predicting Outcomes** What would happen to the chromosome number of an organism's offspring if the gametes for sexual reproduction were made by mitosis instead of by meiosis?

7. **Further Inquiry** Write a new question about meiosis or sexual reproduction that could be explored with your model.

? **Do You Know?**

Do research in the library or media center to answer these questions:

1. What types of human abnormalities arise when chromosomes do not separate properly during meiosis?

2. How do chemicals such as nicotine affect meiosis?

Use the following Internet resources to explore your own questions about meiosis and gamete formation.

internet connect

www.scilinks.org
Topic: Meiosis
Keyword: HX4120

SCI**LINKS**® Maintained by the National Science Teachers Association

Show jumper

CHAPTER
8 Mendel and Heredity

✔ Quick Review

Answer the following without referring to earlier sections of your book.

1. **Define** the term *gamete*. *(Chapter 6, Section 1)*
2. **Summarize** the relationship between chromosomes and genes. *(Chapter 6, Section 1)*
3. **Differentiate** between autosomes and sex chromosomes. *(Chapter 6, Section 1)*
4. **Describe** how independent assortment during meiosis contributes to genetic variation. *(Chapter 7, Section 1)*

Did you have difficulty? *For help, review the chapters indicated.*

📖 Reading Activity

Before you read this chapter, write a short list of all the things you know about inheritance. Then write a list of the things that you want to know about inheritance. Save your list, and to assess what you have learned, see how many questions you can answer after reading this chapter.

🖳 internet connect

www.scilinks.org
National Science Teachers Association *sci*LINKS Internet resources are located throughout this chapter.

 **sci*LINKS.** Maintained by the National Science Teachers Association

● Horses like this show jumper are bred for certain characteristics, such as speed and agility. Most characteristics are inherited—passed down from parents to offspring.

The Origins of Genetics

- **Identify** the investigator whose studies formed the basis of modern genetics.

- **List** characteristics that make the garden pea a good subject for genetic study.

- **Summarize** the three major steps of Gregor Mendel's garden pea experiments.

- **Relate** the ratios that Mendel observed in his crosses to his data.

Key Terms

heredity
genetics
monohybrid cross
true-breeding
P generation
F_1 generation
F_2 generation

Mendel's Studies of Characters

Many of your characteristics—or *characters*—including the color and shape of your eyes and the texture of your hair resemble those of your parents. The passing of characters from parents to offspring is called **heredity.** From the beginning of recorded history, humans have attempted to alter crop plants and domestic animals to give them traits that are more useful to us. Before DNA and chromosomes were discovered, heredity was one of the greatest mysteries of science.

Mendel's Breeding Experiments

The scientific study of heredity began more than a century ago with the work of an Austrian monk named Gregor Johann Mendel, shown in **Figure 1.** Mendel carried out experiments in which he bred different varieties of the garden pea *Pisum sativum,* shown in **Figure 2** and in **Table 1.** British farmers had performed similar breeding experiments more than 200 years earlier. But Mendel was the first to develop rules that accurately predict patterns of heredity. The patterns that Mendel discovered form the basis of **genetics,** the branch of biology that focuses on heredity.

Mendel's parents were peasants, so he learned much about agriculture. This knowledge became invaluable later in his life. As a young man, Mendel studied theology and was ordained as a priest. Three years after being ordained, he went to the University of Vienna to study science and mathematics. There he learned how to study science through experimentation and how to use mathematics to explain natural phenomena.

Figure 1 Gregor Mendel. Mendel's experiments with garden peas led to our modern understanding of heredity.

Mendel later repeated the experiments of a British farmer, T. A. Knight. Knight had crossed a variety of the garden pea that had purple flowers with a variety that had white flowers. (The term *cross* refers to the mating or breeding of two individuals.) All of the offspring of Knight's crosses had purple flowers. However, when two of the purple-flowered offspring were crossed, their offspring showed both white and purple flowers. The white trait had reappeared in the second generation!

Mendel's experiments differed from Knight's because Mendel counted the number of each kind of offspring and analyzed the data.

Figure 2 Pollen transfer in Mendel's experiments

To cross-pollinate flowers of different colors, Mendel first removed the stamens—the pollen-producing structures—from one flower.

Mendel transferred pollen from a second flower to the pistil of the original flower.

Useful Features in Peas

The garden pea is a good subject for studying heredity for several reasons. **Table 1** shows the seven characters that Mendel chose to study.

1. Several characters of the garden pea exist in two clearly different forms. For example, the flower color is either purple or white—there are no intermediate forms. Note that the term *character* is used to mean inherited characteristic, such as flower color. *Trait* refers to a single form of a character—having purple flowers is a trait.

2. The male and female reproductive parts of garden peas are enclosed within the same flower. You can control mating by allowing a flower to fertilize itself (self-fertilization), or you can transfer the pollen to another flower on a different plant (cross-pollination). To cross-pollinate two pea plants, Mendel removed the stamens (the male reproductive organs that produce pollen) from the flower of one plant. As shown in Figure 2, he then dusted the pistil (the female reproductive organ that produces eggs) of that plant with pollen from a different pea plant.

3. The garden pea is small, grows easily, matures quickly, and produces many offspring. Thus, results can be obtained quickly, and there are plenty of subjects to count.

Table 1 The Seven Characters Mendel Studied and Their Contrasting Traits						
Flower color	Seed color	Seed shape	Pod color	Pod shape	Flower position	Plant height

Traits Expressed as Simple Ratios

Mendel's initial experiments were monohybrid crosses. A **monohybrid cross** is a cross that involves *one* pair of contrasting traits. For example, crossing a plant with purple flowers and a plant with white flowers is a monohybrid cross. Mendel carried out his experiments in three steps, as summarized in **Figure 3.**

Step ❶ Mendel allowed each variety of garden pea to self-pollinate for several generations. This ensured that each variety was **true-breeding** for a particular character; that is, all the off-spring would display only one form of the character. For example, a true-breeding purple-flowering plant should produce only plants with purple flowers in subsequent generations.

These true-breeding plants served as the parental generation in Mendel's experiments. The parental generation, or **P generation,** are the first two individuals that are crossed in a breeding experiment.

Step ❷ Mendel then cross-pollinated two P generation plants that had contrasting traits, such as purple flowers and white flowers. Mendel called the offspring of the P generation the first *filial* generation, or **F_1 generation.** He then examined each F_1 plant and recorded the number of F_1 plants expressing each trait.

Step ❸ Finally, Mendel allowed the F_1 generation to self-pollinate. He called the offspring of the F_1 generation plants the second filial generation, or **F_2 generation.** Again, each F_2 plant was characterized and counted.

WORD *Origins*

● The word *filial* is from the Latin *filialis*, meaning "of a son or daughter." Thus F (filial) generations are all those generations that follow a P (parental) generation.

Figure 3

BIOgraphic

Three Steps of Mendel's Experiments

Mendel studied traits in three generations of plants.

❶ Producing a true-breeding P generation   ❷ Producing an F_1 generation   ❸ Producing an F_2 generation

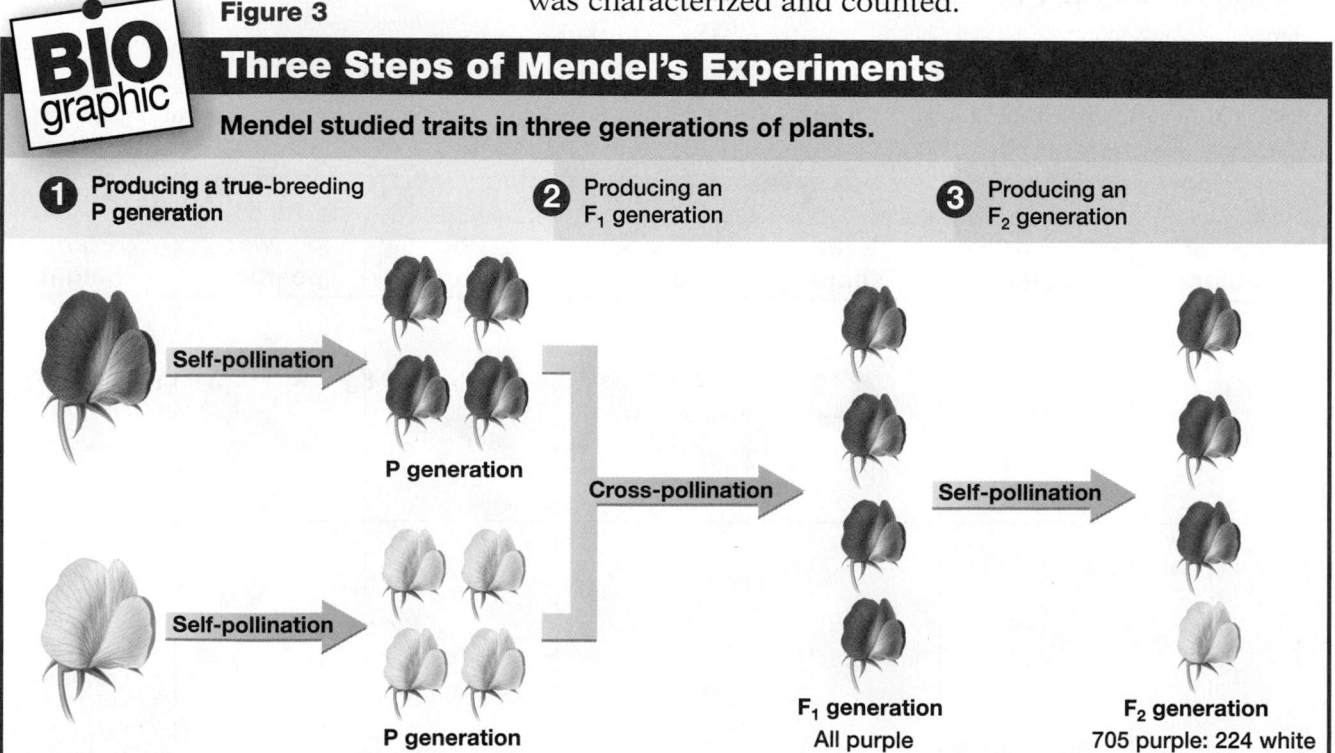

Self-pollination — P generation

Self-pollination — P generation

Cross-pollination

F_1 generation
All purple

Self-pollination

F_2 generation
705 purple: 224 white

Mendel's Results

Each of Mendel's F_1 plants showed only one form of the character. The contrasting trait had disappeared! But when the F_1 generation was allowed to self-pollinate, the missing trait *reappeared* in some of the plants in the F_2 generation. When Mendel crossed purple flowers with white flowers, all of the offspring in the F_1 generation had purple flowers. In the F_2 generation, 705 plants had purple flowers and 224 plants had white flowers—a ratio of 705 to 224.

A ratio is a comparison of two numbers and can be written as a fraction ($\frac{705}{224}$) or with a colon (705:224). You can see patterns more easily in data if you reduce a ratio to its simplest form. To do this, divide each term by the smaller of the two terms. This reduces 705:224 to 3.15, which is then rounded to 3:1.

$$\frac{705}{224} \div \frac{224}{224} = 3.15 \text{ (or about 3)}$$

For each of the seven characters Mendel studied, he found the same 3:1 ratio of plants expressing the contrasting traits in the F_2 generation.

Calculating Mendel's Ratios

Background

You can calculate the ratios Mendel obtained in the F_2 generation for the characters he studied. First copy the partially completed table below on a separate piece of paper.

Analysis

1. **Calculate** the ratio for each contrasting trait. Use colon form.

2. **State** the ratio for each pair of contrasting traits in words and as a fraction.

3. **Critical Thinking Interpreting Results** Do the data confirm a 3:1 ratio in the F_2 generation for each of the characters he studied?

Characters	F_2 generation results		Ratio
Flower color	705 purple	224 white	3.15:1
Seed color	6,022 yellow	2,001 green	
Seed shape	5,474 round	1,850 wrinkled	
Pod color	428 green	152 yellow	
Pod shape	882 round	299 constricted	
Flower position	651 axial	207 top	
Plant height	787 tall	277 dwarf	

Section 1 Review

1. **Describe** the contribution of Mendel to the foundation of modern genetics.

2. **Describe** why garden-pea plants are good subjects for genetic experiments.

3. **Summarize** the design of Mendel's pea-plant studies.

4. **State** the ratio Mendel obtained in each F_2 generation for each of the characters he studied.

5. **Critical Thinking Evaluating Outcomes** What differences would be expected in experiments with squash plants, which usually do not self-pollinate?

6. **(Standardized Test Prep)** When two true-breeding pea plants that show contrasting traits are crossed, all of the offspring show

   A both forms of the character.

   B one form of the character.

   C one-fourth of each trait.

   D a different trait.

Mendel's Theory

Objectives

- **Describe** the four major hypotheses Mendel developed.
- **Define** the terms *homozygous, heterozygous, genotype,* and *phenotype.*
- **Compare** Mendel's two laws of heredity.

Key Terms

allele
dominant
recessive
homozygous
heterozygous
genotype
phenotype
law of segregation
law of independent
  assortment

A Theory of Heredity

Before Mendel's experiments, many people thought offspring were a *blend* of the traits of their parents. For example, if a tall plant were crossed with a short plant, the offspring would be medium in height. Mendel's results did not support the blending hypothesis. Mendel correctly concluded that each pea has two separate "heritable factors" for each character—one from each parent. As shown in **Figure 4,** when gametes (sperm and egg cells) form, each receives only one of the organism's two factors for each character. When gametes fuse during fertilization, the offspring has two factors for each character, one from each parent.

Mendel's Hypotheses

The four hypotheses Mendel developed were based directly on the results of his experiments. These four hypotheses now make up the Mendelian theory of heredity—the foundation of genetics.

1. *For each inherited character, an individual has two copies of the gene—one from each parent.*

2. *There are alternative versions of genes.* For example, the gene for flower color in peas can exist in a "purple" version

---

Figure 4   Mendel's factors

Each parent has two separate "factors," or genes, for a particular trait.

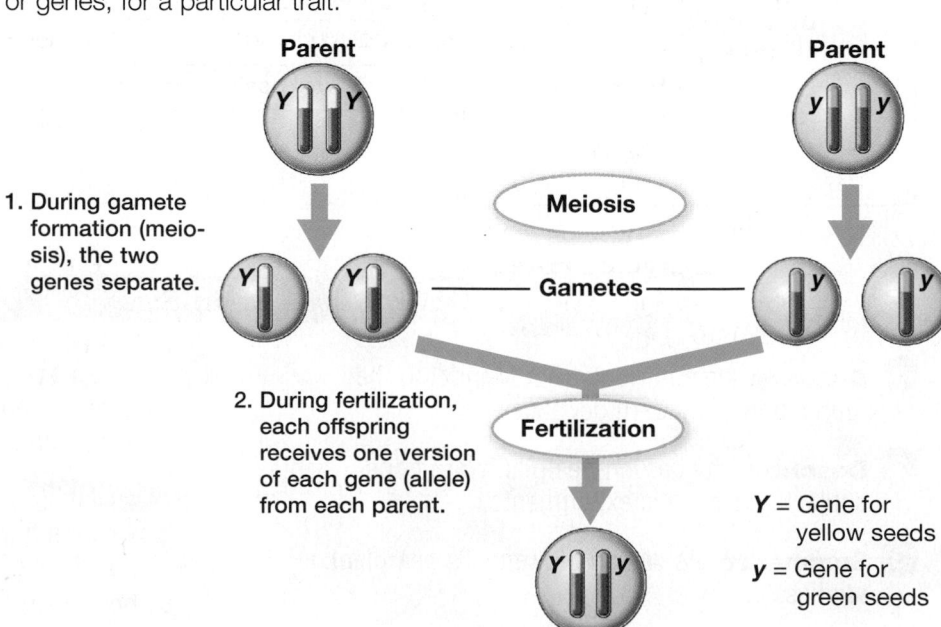

Parent

Parent

1. During gamete formation (meiosis), the two genes separate.

Meiosis

Gametes

2. During fertilization, each offspring receives one version of each gene (allele) from each parent.

Fertilization

Y = Gene for yellow seeds

y = Gene for green seeds

---

or a "white" version. Today the different versions of a gene are called its **alleles.** As shown in Figure 4, an individual receives one allele from each parent. Each allele can be passed on when the individual reproduces.

3. *When two different alleles occur together, one of them may be completely expressed, while the other may have no observable effect on the organism's appearance.* Mendel described the expressed form of the character as **dominant.** The trait that was not expressed when the dominant form of the character was present was described as **recessive.** For every pair of contrasting traits that Mendel studied, the allele for one form of the character was always dominant and the allele for the other form of the character was always recessive. For example, if a plant has both purple and white alleles for flower color but blooms purple flowers, then purple is the dominant form of the character; white is the recessive form. This is shown in **Figure 5.**

4. *When gametes are formed, the alleles for each gene in an individual separate independently of one another. Thus, gametes carry only one allele for each inherited character. When gametes unite during fertilization, each gamete contributes one allele.* As shown in Figure 4, each parent can contribute only one of the alleles because of the way gametes are produced during the process of meiosis.

Mendel's Findings in Modern Terms

Geneticists have developed specific terms and ways of representing an individual's genetic makeup. For example, letters are often used to represent alleles. Dominant alleles are indicated by writing the first letter of the character as a capital letter. For instance, in pea plants, purple flower color is a dominant trait and is written as *P.* Recessive alleles are also indicated by writing the first letter of the dominant trait, but the letter is lowercase. For example, white flower color is recessive and is written as *p.*

If the two alleles of a particular gene present in an individual are the same, the individual is said to be **homozygous** *(hoh moh ZIE guhs)* for that character. For example, a plant with two white flower alleles is homozygous for flower color, as shown in Figure 5. The allele for yellow peas, *Y,* is dominant to the allele for green peas, *y.* A plant with two yellow-pea alleles, *YY,* is homozygous for seed color.

If the alleles of a particular gene present in an individual are different, the individual is **heterozygous** *(heht uhr oh ZIE guhs)* for that character. As shown in Figure 5, a plant with one "purple flower" allele and one "white flower" allele is heterozygous for flower color. A plant with one "yellow pea" allele and one "green pea" allele is heterozygous for seed color.

PP
Purple flowers, homozygous dominant
Pp
Purple flowers, heterozygous

pp
White flowers, homozygous recessive

Figure 5   Recessive alleles. Alleles can be present but not expressed. The allele for purple flowers, *P*, is dominant to the recessive allele, *p*.

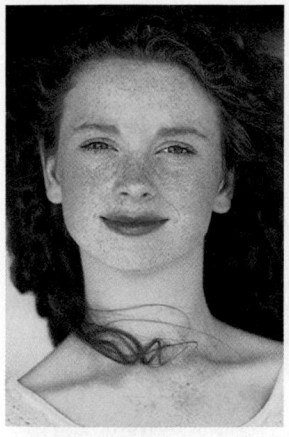

Figure 6 Dominent alleles. In heterozygous individuals, freckles, *F,* is the dominant allele. Similarly, the allele for a cleft chin is dominant to the allele for a chin without a cleft.

In heterozygous individuals, only the dominant allele is expressed; the recessive allele is present but unexpressed. An example of a human trait that is expressed in a heterozygous individual is freckles. Freckles *F,* is a dominant allele. The recessive allele is *f,* no freckles. The recessive allele may be present but not expressed. As shown in **Figure 6,** people who are heterozygous for freckles (*Ff*) will have freckles even though they also have the allele for no freckles, *f.*

The set of alleles that an individual has for a character is called its **genotype** *(JEE noh tiep).* The physical appearance of a character is called its **phenotype** *(FEE noh tiep).* Phenotype is determined by which alleles are present. For example, if *Pp* is the genotype of a pea plant, its phenotype is purple flowers. If *pp* is the genotype of a pea plant, its phenotype is white flowers. When considering seed color, if *Yy* is the genotype of a pea plant, its phenotype is yellow seeds. If *yy* is the genotype of a pea plant, its phenotype is green seeds. Note that by convention, the dominant form of the character is written first, followed by the lowercase letter for the recessive form of the character.

Identifying Dominant or Recessive Traits

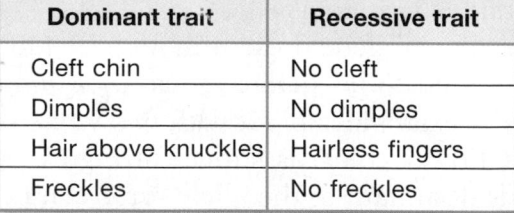

You can determine some of the genotypes and all of the phenotypes for human characters that are inherited as simple dominant or recessive traits.

Materials

pencil, paper

Procedure

1. **Make** a table like the one at right. For each character, circle the phenotype that best matches your own phenotype.

2. **Determine** how many students in your class share your phenotype by recording your results in a table on the chalkboard.

Dominant trait	Recessive trait
Cleft chin	No cleft
Dimples	No dimples
Hair above knuckles	Hairless fingers
Freckles	No freckles

Analysis

1. **Summarize** the class results for each character.

2. **Calculate** the class dominant:recessive ratio for each character.

3. **Critical Thinking Applying Information** For which phenotypes in the table can you determine a person's genotype without ever having seen his or her parents? Explain.

The Laws of Heredity

Mendel's hypotheses brilliantly predicted the results of his crosses and also accounted for the ratios he observed. Similar patterns of heredity have since been observed in countless other organisms. Because of their importance, Mendel's ideas are often referred to as the laws of heredity.

The Law of Segregation

The first law of heredity describes the behavior of chromosomes during meiosis. At this time, homologous chromosomes and then chromatids are separated. The first law, the **law of segregation,** states that the two alleles for a character segregate (separate) when gametes are formed (as shown in Figure 4).

The Law of Independent Assortment

Mendel went on to study whether the inheritance of one character (such as plant height) influenced the inheritance of a different character (such as flower color). To study how different pairs of genes are inherited, Mendel conducted dihybrid crosses. A dihybrid cross is a cross that considers two pairs of contrasting characters. For example, a cross that considers both plant height and flower color is a dihybrid cross.

Mendel found that for the characters he studied, the inheritance of one character did not influence the inheritance of any other character. The **law of independent assortment** states that the alleles of different genes separate independently of one another during gamete formation. For example, the alleles for the height of the plant shown in **Figure 7** separate independently of the alleles for its flower color. We now know that this law applies only to genes that are located on different chromosomes or that are far apart on the same chromosome.

The search for the physical nature of Mendel's "factors" dominated biology for more than half a century after Mendel's work was rediscovered in 1900. We now know that the units of heredity are portions of DNA called *genes,* which are found on the chromosomes that an individual inherits from its parents.

Figure 7   The law of independent assortment. Mendel found that the inheritance of one character, such as plant height, did not influence the inheritance of another character, such as flower color.

Section 2 Review

1. **Differentiate** between alleles and genes.

2. **Apply** the terms *homozygous, heterozygous, dominant,* or *recessive* to describe plants with the genotypes *PP* and *Pp.*

3. **Identify** the phenotypes of rabbits with the genotypes *Bb* and *bb,* where *B* = black coat and *b* = brown coat.

4. **Determine** whether the rabbits in item 3 are heterozygous or homozygous.

5. **Critical Thinking Critiquing Explanations** Review Mendel's two laws according to their strengths and weaknesses in terms of our modern understanding of meiosis.

6. **Standardized Test Prep**   If a pea plant is heterozygous for a particular character, how can the alleles that control the character be described?
   A two recessive   **C** one dominant, one recessive
   B two dominant   **D** three dominant, one recessive

Studying Heredity

- **Predict** the results of monohybrid genetic crosses by using Punnett squares.

- **Apply** a test cross to determine the genotype of an organism with a dominant phenotype.

- **Predict** the results of monohybrid genetic crosses by using probabilities.

- **Analyze** a simple pedigree.

Key Terms

Punnett square
test cross
probability
pedigree
sex-linked gene

Punnett Squares

Animal breeders try to breed animals with very specific characteristics. Thus, breeders must be able to predict how often a trait will appear when two animals are crossed (bred). Likewise, horticulturists (plant breeders) need to produce plants with very specific characteristics. One simple way of predicting the expected results (not necessarily the actual results) of the genotypes or phenotypes in a cross is to use a Punnett square.

A **Punnett square** is a diagram that predicts the outcome of a genetic cross by considering all possible combinations of gametes in the cross. Named for its inventor, Reginald Punnett, the simplest Punnett square consists of four boxes inside a square. As shown in **Figure 8,** the possible gametes that one parent can produce are written along the top of the square. The possible gametes that the other parent can produce are written along the left side of the square. Each box inside the square is filled in with two letters obtained by combining the allele along the top of the box with the allele along the side of the box. The letters in the boxes indicate the possible genotypes of the offspring.

One Pair of Contrasting Traits

Punnett squares can be used to predict the outcome of a monohybrid cross (a cross that considers one pair of contrasting traits between two individuals). For example, a Punnett square can be used to predict the outcome of a cross between a pea plant that is homozygous for yellow seed color (*YY*) and a pea plant that is homozygous for green seed color (*yy*). Figure 8 shows that 100 percent of the offspring in this type of cross are expected to be heterozygous (*Yy*), expressing the dominant trait of yellow seed color.

Figure 8   Monohybrid cross: homozygous plants

A cross between a pea plant that is homozygous for yellow seeds (*YY*) and a pea plant that is homozygous for green seeds (*yy*) will produce only yellow heterozygous offspring (*Yy*).

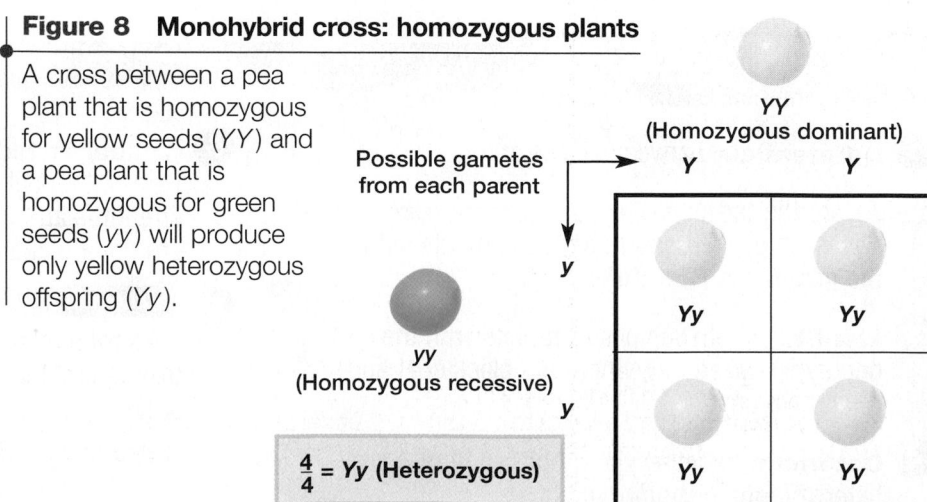

Possible gametes from each parent

YY
(Homozygous dominant)

yy
(Homozygous recessive)

$\frac{4}{4}$ = *Yy* (Heterozygous)

Figure 9 shows a Punnett square that predicts the results of a monohybrid cross between two pea plants that are both heterozygous (*Yy*) for seed color. One-fourth of the offspring would be expected to have the genotype *YY*, two-fourths (or one-half) would be expected to have the genotype *Yy*, and one-fourth would be expected to have the genotype *yy*. Another way to express this is to say that the genotypic ratio is 1 *YY* : 2 *Yy* : 1 *yy*. Because the *Y* allele is dominant over the *y* allele, three-fourths of the offspring would be yellow, and one-fourth would be green. The phenotypic ratio is 3 yellow : 1 green.

Punnett squares allow direct and simple predictions to be made about the outcomes of genetic crosses. Although animal breeders and horticulturists are not always certain what characteristics will turn up in the offspring, they can use the predictions from Punnett squares to cross individuals that they know will be most likely to produce offspring with the desired phenotypes.

Figure 9   Monohybrid cross: heterozygous plants

Crossing two pea plants that are heterozygous for seed color (*Yy*) will produce offspring in the ratio shown in the Punnett square.

	Yy (Heterozygous)	
	Y	**y**
Y	YY	Yy
Yy (Heterozygous) **y**	Yy	yy

$\frac{1}{4}$ = *YY* (Homozygous dominant)

$\frac{2}{4}$ = *Yy* (Heterozygous)

$\frac{1}{4}$ = *yy* (Homozygous recessive)

Exploring Further

Crosses That Involve Two Characters

Suppose a horticulturist has two characters that she wants to consider when crossing two plants. A cross that involves two pairs of contrasting traits is called a dihybrid cross. For example, she may want to predict the results of a cross between two pea plants that are heterozygous for seed shape (*R* = round, *r* = wrinkled) and seed color (*Y* = yellow, *y* = green).

Determine possible gametes
To use a Punnett square to predict the results of this cross, first consider how the four alleles from either parent (*RrYy*) can combine to form gametes that are either *RY*, *Ry*, *rY*, or *ry* (Figure A).

Then write the genotypes of these gametes on the top and left sides of a Punnett square (Figure B).

Complete the Punnett square
On a separate sheet of paper, make a copy of the Punnett square in Figure B, which has been partially filled in with the predicted genotypes. Fill in the remaining genotypes, then do the following:

- **List** all of the possible genotypes that can result.
- **Calculate** the genotypic ratio for this cross.
- **List** all of the possible phenotypes that can result.
- **Calculate** the phenotypic ratio for this cross.

Figure A   Gametes

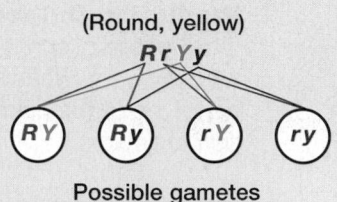

Parent

(Round, yellow)
RrYy

Possible gametes

Figure B   Punnett square

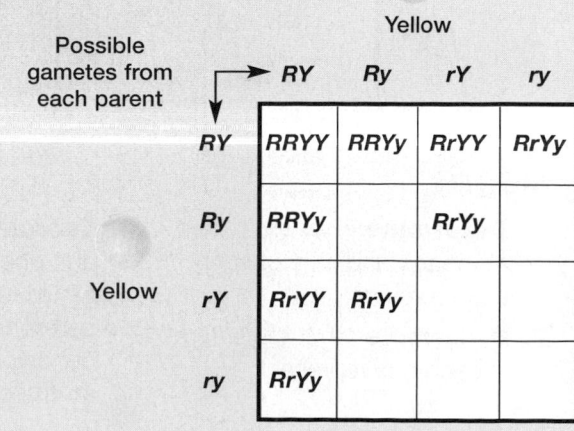

Yellow

Possible gametes from each parent	**RY**	**Ry**	**rY**	**ry**
RY	RRYY	RRYy	RrYY	RrYy
Ry	RRYy		RrYy	
rY	RrYY	RrYy		
ry	RrYy			

Determining Unknown Genotypes

Animal breeders, horticulturists, and others involved in breeding organisms often need to know whether an organism with a dominant phenotype is heterozygous or homozygous for a character. How do they determine this? For example, how might a horticulturist determine whether a pea plant with a dominant phenotype, such as yellow seeds, is homozygous (*YY*) or heterozygous (*Yy*)? The horticulturist could perform a test cross. In a **test cross**, an individual whose phenotype is dominant, but whose genotype is not known, is crossed with a homozygous recessive individual.

For example, a plant with yellow seeds but of unknown genotype (*Y?*) is test-crossed with a plant with green seeds (*yy*). If all of the offspring produce yellow seeds, the offspring must be *Yy*. Thus, the genotype of the "unknown" plant must be *YY*. If half of the offspring produce yellow seeds and half produce green seeds, the genotype of the unknown plant must be *Yy*. In reality, if the cross produces even one plant that produces green seeds, the genotype of the unknown parent plant is likely to be heterozygous. After performing a test cross, the horticulturist can continue breeding the original plant with more certainty of its genotype.

Analyzing a Test Cross

Background

You can use a test cross to determine whether a plant with purple flowers is heterozygous (*Pp*) or homozygous dominant (*PP*). On a separate sheet of paper, copy the two Punnett squares shown below, and fill in the boxes in each square.

Is this purple flowering pea plant *Pp* or *PP*?

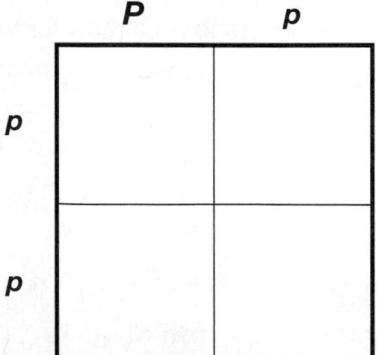

Figure A Heterozygous (*Pp*) plant

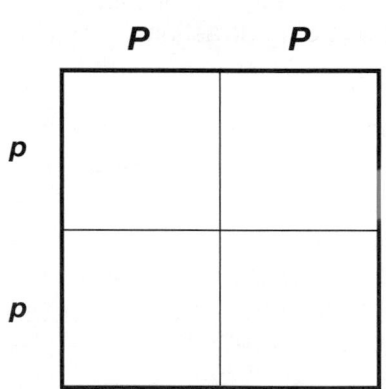

Figure B Homozygous (*PP*) plant

Analysis

1. **Determine** what the letters at the top and side of each box represent.

2. **Determine** what the letters in each box represent.

3. **Calculate** the genotypic and phenotypic ratios that would be predicted if the parent of the unknown genotype were homozygous for the character (Figure B).

4. **Critical Thinking Predicting Outcomes**
   If half of the offspring have white flowers, what is the genotype of the plant with purple flowers?

Outcomes of Crosses

Like Punnett squares, probability calculations can be used to predict the results of genetic crosses. **Probability** is the likelihood that a specific event will occur. Probabilities can be expressed in words, as decimals, as percentages, or as fractions. For example, if an event definitely will occur, its probability can be expressed as either 1 out of 1 (in words), 1 (as a decimal numeral), 100 percent (as a percentage), or $\frac{1}{1}$ (as a fraction). If an event definitely will not occur, its probability can be expressed as either 0 out of 0, 0, 0 percent, or $\frac{0}{0}$.

In order to simplify our discussion of probability, we will express probabilities as fractions. Probability can be determined by the following formula:

$$\text{Probability} = \frac{\text{number of one kind of possible outcome}}{\text{total number of all possible outcomes}}$$

Consider the possibility that a coin tossed into the air will land on heads (one possible outcome). The total number of all possible outcomes is two—heads or tails. Thus, the probability that a coin will land on heads is $\frac{1}{2}$, as shown in **Figure 10.**

Probability of a Specific Allele in a Gamete

The same formula can be used to predict the probability of an allele being present in a gamete. If a pea plant has two alleles for seed color, the plant can contribute either allele (yellow or green) to the gamete it produces (the law of independent assortment). For a plant with two alleles for seed color, the total number of possible outcomes is two—green or yellow. The probability that a gamete will carry the allele for green seed color is $\frac{1}{2}$. The probability that a gamete from this plant will carry the allele for yellow seed color is also $\frac{1}{2}$.

Probability of the Outcome of a Cross

Because two parents are involved in a genetic cross, both parents must be considered when calculating the probability of the outcome of a genetic cross. Consider the analogy of two coins being tossed at the same time. The probability of a penny landing on heads is $\frac{1}{2}$, and the probability of a nickel landing on heads is $\frac{1}{2}$. The way one coin falls does not depend on how the other coin falls. Similarly, the allele carried by the gamete from the first parent does not depend on the allele carried by the gamete from the second parent. The outcomes are independent of each other.

To find the probability that a *combination* of two independent events will occur, multiply the separate probabilities of the two events. Thus, the probability that a nickel *and* a penny will both land on heads is

$$\frac{1}{2} \times \frac{1}{2} = \frac{1}{4}$$

Study *TIP*

● **Reviewing Information**
Because probability is a ratio of a subset of all possible outcomes to all possible outcomes, the value for probability is never greater than 1. When it is less than one, it can be expressed as a fraction or as a percentage of the whole.

Figure 10 Probability of heads or tails. The probability that a tossed coin will land on heads is $\frac{1}{2}$. The probability that a tossed coin will land on tails is $\frac{1}{2}$.

Figure 11 Probability with two coins

The probability of the results of flipping two coins is easy to compute.

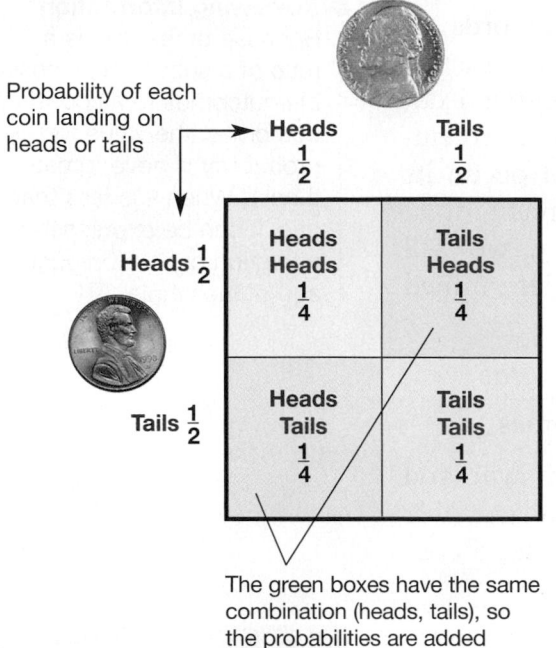

Probability of each coin landing on heads or tails

	Heads $\frac{1}{2}$	Tails $\frac{1}{2}$
Heads $\frac{1}{2}$	Heads Heads $\frac{1}{4}$	Tails Heads $\frac{1}{4}$
Tails $\frac{1}{2}$	Heads Tails $\frac{1}{4}$	Tails Tails $\frac{1}{4}$

The green boxes have the same combination (heads, tails), so the probabilities are added together.

$$\frac{1}{4} + \frac{1}{4} = \frac{1}{2}$$

The possible results of tossing a nickel and a penny at the same time and the probability of each outcome are shown in **Figure 11.** Since the combination of heads and tails can occur in two possible ways, those two probabilities are added together.

$$\frac{1}{4} + \frac{1}{4} = \frac{2}{4} \text{ or } \frac{1}{2}$$

Consider the possible results that can occur in a cross between two pea plants that are heterozygous for seed shape (*Rr*). The *R* allele for round seed shape is dominant over the *r* allele for wrinkled seed shape. The probability of each parent carrying gametes with *R* or *r* alleles is $\frac{1}{2}$. The probability of offspring with *RR* alleles is

$$\frac{1}{2} \times \frac{1}{2} = \frac{1}{4}$$

Similarly, the probability of offspring with *rr* alleles is

$$\frac{1}{2} \times \frac{1}{2} = \frac{1}{4}$$

The combination of *Rr* alleles can occur in two possible ways. One parent can contribute the *R* allele, and the second parent the *r* allele, or vice versa. Thus, the probability of offspring with *Rr* alleles is

$$\frac{1}{4} + \frac{1}{4} = \frac{1}{2}$$

Predicting the Results of Crosses Using Probabilities

Background

In rabbits, the allele *B* for black hair is dominant over the allele *b* for brown hair. You can practice using probabilities to predict the outcome of genetic crosses by completing the genetic problems below. Draw Punnett squares for each problem.

Analysis

1. **Calculate** the probability of homozygous dominant (*BB*) offspring resulting from a cross between two heterozygous (*Bb*) parents.

2. **Calculate** the probability of heterozygous offspring resulting from a cross between a heterozygous parent and a homozygous recessive (*bb*) parent.

3. **Calculate** the probability of heterozygous offspring resulting from a cross between a homozygous dominant parent and a homozygous recessive parent.

4. **Calculate** the probability of homozygous dominant offspring resulting from a cross between a heterozygous parent and a homozygous recessive parent.

Inheritance of Traits

Imagine that you want to learn about an inherited trait present in your family. How would you find out the chances of passing the trait to your children? Geneticists often prepare a **pedigree,** a family history that shows how a trait is inherited over several generations. Pedigrees are particularly helpful if the trait causes a genetic disorder and the family members want to know if they are carriers or if their children might get the disorder. Carriers are individuals who are heterozygous for an inherited disorder but do not show symptoms of the disorder. Carriers can pass the allele for the disorder to their offspring.

Figure 12 shows an example of a pedigree for a family with albinism. In the genetic disorder albinism, the body is unable to produce an enzyme necessary for the production of melanin. Melanin is a pigment that gives dark color to hair, skin, scales, eyes, and feathers. Without melanin, an organism's surface coloration may be milky white and its eyes may be pink, as shown in Figure 12.

Scientists can determine several pieces of genetic information from a pedigree:

Autosomal or Sex-Linked? If a gene is autosomal, it will appear in both sexes equally. Recall that an autosome is a chromosome other than an X or Y sex chromosome. If a trait is sex-linked, its effects are usually seen only in males. A **sex-linked gene's** allele is located only on the X or Y chromosome. Most sex-linked genes are carried on the X chromosome and are recessive. Because males have only one X chromosome, a male who carries a recessive allele on the X chromosome will exhibit the sex-linked condition.

A female who carries a recessive allele on one X chromosome will not exhibit the condition if there is a dominant allele on her other X chromosome. She will express the recessive condition only if she inherits two recessive alleles. Thus, her chances of inheriting and exhibiting a sex-linked condition are significantly less.

Figure 12   Albinism pedigree

Albinism is a genetic disorder transmitted by a recessive allele.

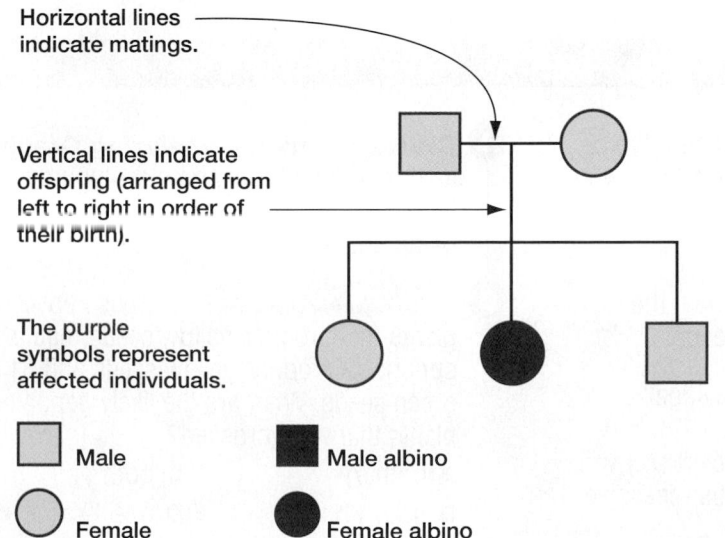

Horizontal lines indicate matings.

Vertical lines indicate offspring (arranged from left to right in order of their birth).

The purple symbols represent affected individuals.

☐ Male   ■ Male albino

◯ Female   ● Female albino

In the wild, albino animals have little chance of survival. They lack the pigments that provide protection from the sun's ultraviolet rays.

Dominant or Recessive? If the gene is autosomal dominant, every individual with the condition will have a parent with the condition. If the condition is recessive, an individual with the condition can have one, two, or neither parent exhibit the condition.

Heterozygous or Homozygous? If individuals with autosomal traits are homozygous dominant or heterozygous, their phenotype will show the dominant allele. If individuals are homozygous recessive, their phenotype will show the recessive allele. Two people who are heterozygous carriers of a recessive mutation will not show the mutation, but they can produce children who are homozygous for the recessive allele.

Evaluating a Pedigree

Background

The photo shows a family with an *albino* member. Pedigrees, such as the one below, can be used to track different genetic traits, including albinism. Use the pedigree below to practice interpreting a pedigree.

Analysis

1. **Interpret** the pedigree to determine whether the trait is sex-linked or *autosomal* and whether the trait is inherited in a dominant or recessive manner.

2. **Determine** whether Female A is homozygous or heterozygous.

3. **Critical Thinking Applying Information** If Female B has children with a homozygous individual, what is the probability that the children will be heterozygous?

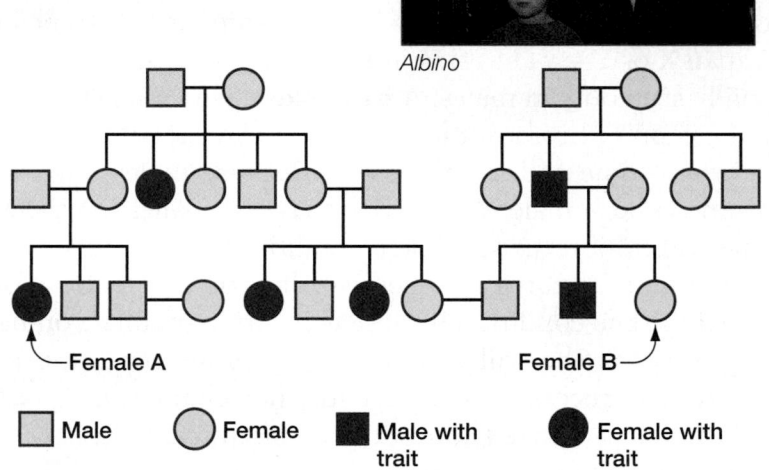

Albino

Female A

Female B

☐ Male   ⬤ Female   ■ Male with trait   ⬤ Female with trait

Section 3 Review

1. **Predict** the expected phenotypic and genotypic ratios among the offspring of two individuals who are heterozygous for freckles (*Ff*) by using a Punnett square.

2. **Summarize** how a test cross can reveal the genotype of a pea plant with round seeds.

3. **Calculate** the probability that an individual heterozygous for a cleft chin (*Cc*) and an individual homozygous for a chin without a cleft (*cc*) will produce offspring that are homozygous recessive for a chin without a cleft. (*cc*)

4. **Critical Thinking Analyzing Graphics** When analyzing a pedigree, how can you determine if an individual is a carrier (heterozygous) for the trait being studied?

5. **Standardized Test Prep** A cross between two pea plants that produce yellow seeds results in 124 offspring: 93 produce yellow seeds and 31 produce green seeds. What are the likely genotypes of the plants that were crossed?

   A both *Yy*

   B both *YY*

   C both *yy*

   D one *YY*, one *Yy*

Complex Patterns of Heredity

Complex Control of Characters

A horse with red hair mates with a horse with white hair, and their offspring has both red and white hair. How can this be? If characters are controlled by single genes with simple dominant and recessive alleles, the colt's hair should be one color or the other. Not always! Most of the time, characters, such as hair color in horses, display more-complex patterns of heredity than the simple dominant-recessive patterns discussed so far.

Characters Influenced by Several Genes

When several genes influence a character, this is an example of **polygenic inheritance.** The genes for a polygenic character may be scattered along the same chromosome or located on different chromosomes. Determining the effect of any one of these genes is difficult. Due to independent assortment and crossing-over during meiosis, many different combinations appear in offspring. Familiar examples of polygenic characters in humans include eye color, height, weight, and hair and skin color. All of these characters have degrees of intermediate conditions between one extreme and the other, as shown in **Figure 13.**

Intermediate Characters

Recall that in Mendel's pea-plant crosses, one allele was completely dominant over another. In some organisms, however, an individual displays a phenotype that is intermediate between the two parents, a condition known as **incomplete dominance.** For example, when a snapdragon with red flowers is crossed with a snapdragon with white flowers, a snapdragon with pink flowers is produced. Neither the red nor the white allele is completely dominant over the other allele. The flowers appear pink because they have less red pigment than the red flowers. In Caucasians, the child of a straight-haired parent and a curly-haired parent will have wavy hair. Straight and curly hair are homozygous dominant traits. Wavy hair is heterozygous and is intermediate between straight and curly hair.

Objectives

- **Identify** five factors that influence patterns of heredity.
- **Describe** how mutations can cause genetic disorders.
- **List** two genetic disorders, and describe their causes and symptoms.
- **Evaluate** the benefits of genetic counseling.

Key Terms

polygenic inheritance
incomplete dominance
multiple alleles
codominance

Figure 13 Polygenic inheritance. Many characters—height, weight, hair color, and skin color—are influenced by many genes.

Characters Controlled by Genes with Three or More Alleles

Genes with three or more alleles are said to have **multiple alleles.** For example, in the human population, the ABO blood groups (blood types) are determined by three alleles, I^A, I^B, and i. The letters A and B refer to two carbohydrates on the surface of red blood cells. In the i allele, neither carbohydrate is present. The I^A and I^B alleles are both dominant over i. But neither I^A nor I^B is dominant over the other. When I^A and I^B are both present they are codominant. Even for characters controlled by genes with multiple alleles, an individual can have only two of the possible alleles for that gene. **Figure 14** shows how combinations of the three different alleles can produce four different blood types—A, B, AB, and O. Notice that a person who inherits two i alleles has type O blood.

Characters with Two Forms Displayed at the Same Time

For some characters, two dominant alleles are expressed at the same time. In this case, both forms of the character are displayed, a phenomenon called **codominance.** Codominance is different from incomplete dominance because both traits are displayed.

The situation of human ABO blood groups, as discussed above, is an example of co-dominance. The genotype of a person who has blood type AB is $I^A I^B$, and neither allele is dominant over the other. Type AB blood cells carry both A- and B-types of carbohydrate molecules on their surfaces.

Figure 14   Multiple alleles control the ABO blood groups

Different combinations of the three alleles I^A, I^B, and i result in four different blood phenotypes, A, AB, B, and O. For example, a person with the alleles I^A and i would have blood type A.

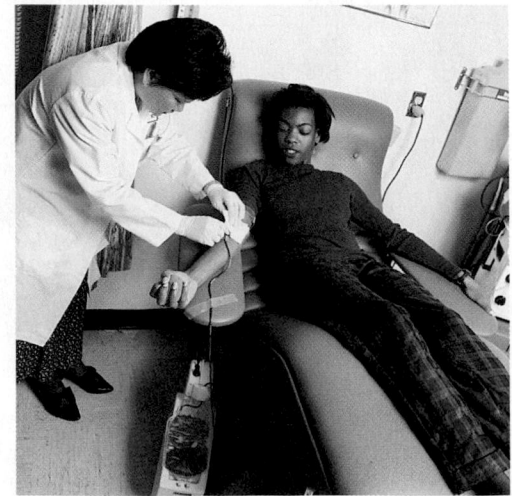

Possible alleles	I^A	I^B	i
I^A	$I^A I^A$	$I^A I^B$	$I^A i$
I^B	$I^A I^B$	$I^B I^B$	$I^B i$
i	$I^A i$	$I^B i$	ii

Blood types   A   AB   B   O

Characters Influenced by the Environment

An individual's phenotype often depends on conditions in the environment. In plants, hydrangea *(hie DRAYN juh)* flowers of the same genetic variety range in color from blue to pink, as shown in **Figure 15.** Hydrangea plants in acidic soil bloom blue flowers, while those in neutral to basic soil will bloom pink flowers.

The color of the arctic fox is affected by temperature. During summer, the fox produces enzymes that make pigments. These pigments darken the fox's coat to a reddish brown, as shown in **Figure 16,** enabling the fox to blend in with the summer landscape. During the winter, the pigment-producing genes of the arctic fox do not function because of the cold temperature. As a result, the coat of the fox is white, and the animal blends in with the snowy background.

Fur color in Siamese cats is also influenced by temperature. In a Siamese cat, the fur on its ears, nose, paws, and tail is darker than on the rest of its body. The Siamese cat has a genotype that results in dark fur at locations on its body that are cooler than the normal body temperature. Thus, the darkened parts have a lower body temperature than the light parts.

In humans many characters, such as height, are influenced by the environment. For example, height is influenced by nutrition, an internal environmental condition. Exposure to the sun, an external environmental condition, alters the color of the skin. Many aspects of human personality, such as aggressive behavior, are strongly influenced by the environment, although genes appear to play an important role. Because identical twins have identical genes, they are often used to study environmental influences. Because identical twins are genetically identical, any differences between them are attributed to environmental influences.

Figure 15 Environmental influences on flower color.
Hydrangea with the same genotype for flower color express different phenotypes depending on the acidity of the soil.

Figure 16 Environmental influences on fur color

Can the same species of fox look so different? Many arctic mammals, such as the arctic fox, develop white fur during the winter and dark fur during the summer.

Genetic Disorders

In order for a person to develop and function normally, the proteins encoded by his or her genes must function precisely. Unfortunately, sometimes genes are damaged or are copied incorrectly, resulting in faulty proteins. Changes in genetic material are called mutations. Mutations are rare because cells have efficient systems for correcting errors. But mutations sometimes occur, and they may have harmful effects.

The harmful effects produced by inherited mutations are called genetic disorders. Many mutations are carried by recessive alleles in heterozygous individuals. This means that two phenotypically normal people who are heterozygous carriers of a recessive mutation can produce children who are homozygous for the recessive allele. In such cases, the effects of the mutated allele cannot be avoided. Several human genetic disorders are summarized in **Table 2.**

Sickle Cell Anemia

An example of a recessive genetic disorder is sickle cell anemia, a condition caused by a mutated allele that produces a defective form of the protein hemoglobin. Hemoglobin is found within red blood cells, where it binds oxygen and transports it through the body. In sickle cell anemia, the defective form of hemoglobin causes many red blood cells to bend into a sickle shape, as seen in **Figure 17.** The sickle-shaped cells rupture easily, resulting in less oxygen being carried by the blood. Sickle-shaped cells also tend to get stuck in blood vessels; this can cut off blood supply to an organ.

The recessive allele that causes sickle-shaped red blood cells also helps protect the cells of heterozygous individuals from the effects of malaria. Malaria is a disease caused by a parasitic protozoan that invades red blood cells. The sickled red blood cells of heterozygous individuals cause the death of the parasite. But the individual's normal red blood cells can still transport enough oxygen. Therefore, these people are protected from the effects of malaria that threaten individuals who are homozygous dominant for the hemoglobin gene.

Cystic Fibrosis (CF)

Cystic fibrosis is the most common fatal, hereditary, recessive disorder among Caucasians. One in 25 Caucasian individuals has at least one copy of a defective gene that makes a protein necessary to move chloride into and out of cells. About 1 in 2,500 Caucasian infants in the United States is homozygous for the *cf* allele. The airways of the lungs become clogged with thick mucus, and the ducts of the liver and pancreas become blocked. While treatments can relieve some of the symptoms, there is no known cure.

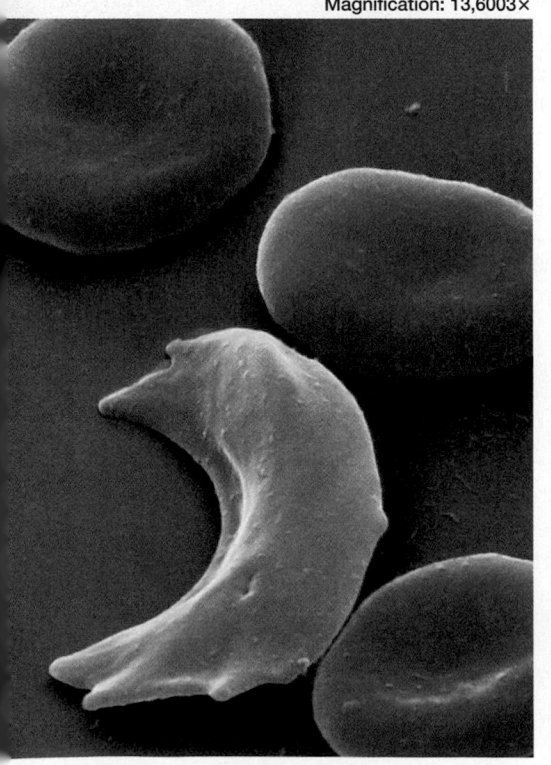

Figure 17 Sickle cell. One out of 500 African Americans has sickle cell anemia, which is caused by a gene mutation that produces a defective form of hemoglobin.

Magnification: 13,6003×

Hemophilia

Another recessive genetic disorder is hemophilia *(hee moh FIHL ee uh)*, a condition that impairs the blood's ability to clot. Hemophilia is a sex-linked trait. More than a dozen genes code for the proteins involved in blood clotting. A mutation on one of these genes on the X chromosome causes the form of hemophilia called hemophilia A. If the mutation appears on the X chromosome, which a male receives from his mother, he does not have a normal gene on the Y chromosome to compensate. Therefore, he will develop hemophilia.

internet connect

www.scilinks.org
Topic: Genetic Disorders
Keyword: HX4091

SCI *LINKS.* Maintained by the National Science Teachers Association

Huntington's Disease (HD)

Huntington's disease is a genetic disorder caused by a dominant allele located on an autosome. The first symptoms of HD—mild forgetfulness and irritability—appear in victims in their thirties or forties. In time, HD causes loss of muscle control, uncontrollable physical spasms, severe mental illness, and eventually death. Unfortunately, most people who have the HD allele do not know they have the disease until after they have had children. Thus, the disease is unknowingly passed on from one generation to the next.

Table 2 Some Human Genetic Disorders

Disorder	Dominant or Recessive	Symptom	Defect	Frequency Among Human Births
Sickle Cell Anemia	Recessive	Poor blood circulation	Abnormal hemoglobin molecules	1 in 500 (African Americans)
Hypercholesterolemia	Dominant	Excessive cholesterol levels in blood, leading to heart disease	Abnormal form of cell surface receptor for cholesterol	1 in 500
Tay-Sachs Disease	Recessive in early childhood	Deterioration of central nervous system; death	Defective form of a brain enzyme	1 in 3,500 (Ashkenazi Jews)
Cystic Fibrosis	Recessive	Mucus clogs organs including the lungs, liver, and pancreas; affected individuals usually do not survive to old age	Defective chloride-ion transport protein	1 in 2,500 (Caucasians)
Hemophilia A (Classical)	Sex-linked recessive	Failure of blood to clot	Defective form of a blood-clotting factor	1 in 10,000 (males)
Huntington's Disease	Dominant	Gradual deterioration of brain tissue in middle age; shortened life expectancy	Inhibitor of brain-cell metabolism is made	1 in 10,000

Treating Genetic Disorders

Most genetic disorders cannot be cured, although progress is being made. A person with a family history of genetic disorders may wish to undergo genetic counseling before becoming a parent. Genetic counseling is a form of medical guidance that informs people about genetic problems that could affect them or their offspring.

In some cases, a genetic disorder can be treated if it is diagnosed early enough. For example, an individual with the genetic disorder phenylketonuria (PKU) lacks an enzyme that converts the amino acid phenylalanine into the amino acid tyrosine. As a result, phenylalanine builds up in the body and causes severe mental retardation. If PKU is diagnosed soon after birth, however, the newborn can be placed on a low-phenylalanine diet. Because this disorder can be easily diagnosed by inexpensive laboratory tests, many states require PKU testing of all newborns.

Gene Therapy

Gene technology may soon allow scientists to correct certain recessive genetic disorders by replacing defective genes with copies of healthy ones, an approach called *gene therapy*. The essential first step in gene therapy is to isolate a copy of the gene. The defective *cf* gene was isolated in 1989. In 1990, a working *cf* gene was successfully transferred into human lung cells growing in tissue culture by attaching the *cf* gene to the DNA of a cold virus. The cold virus—carrying the normal *cf* gene piggyback—easily infects lung cells. The *cf* gene enters the lung cells and begins producing functional CF protein. Thus, the defective cells are "cured" and are able to transport chloride ions across their plasma membranes.

Similar attempts in humans, however, were not successful. Most people have had colds and, as a consequence, have built up a natural immunity to the cold virus. Their lungs therefore reject the cold virus and its *cf* passenger. In the last few years, similar attempts using a different virus to transport the *cf* gene into lung cells have been initiated. This virus, called *AAV*, produces almost no immune response and so seems a much more suitable vehicle for introducing *cf* into cells. Clinical trials are underway, and the outlook is promising.

Section 4 Review

1. **Differentiate** between incomplete dominance and codominance.

2. **Identify** two examples of characters that are influenced by environmental conditions.

3. **Summarize** how a genetic disorder can result from a mutation.

4. **Describe** how males inherit hemophilia.

5. **Critical Thinking Justifying Conclusions** A nurse states that a person cannot have the blood type ABO. Do you agree or disagree? Explain.

6. **Standardized Test Prep** The mutated allele that causes Huntington's disease is
   A sex-linked and recessive.
   B sex-linked and dominant.
   C autosomal and recessive.
   D autosomal and dominant.

Key Concepts

1 The Origins of Genetics

- Gregor Mendel bred varieties of the garden pea in an attempt to understand heredity. Mendel observed that contrasting traits appear in offspring according to simple ratios.

- In Mendel's experiments, only one of the two contrasting forms of a character was expressed in the F_1 generation. The other form reappeared in the F_2 generation in a 3:1 ratio.

2 Mendel's Theory

- Different versions of a gene are called alleles. An individual usually has two alleles for a gene, each inherited from a different parent.

- Individuals with the same two alleles for a gene are homozygous; those with two different alleles for a gene are heterozygous.

- The law of segregation states that the two alleles for a gene separate when gametes are formed. The law of independent assortment states that two or more pairs of alleles separate independently of one another during gamete formation.

3 Studying Heredity

- The results of genetic crosses can be predicted with the use of Punnett squares and probabilities.

- A test cross can be used to determine whether an individual expressing a dominant trait is heterozygous or homozygous.

- A trait's pattern of inheritance within a family can be determined by analyzing a pedigree.

4 Complex Patterns of Heredity

- Characters usually display complex patterns of heredity, such as incomplete dominance, codominance, and multiple alleles.

- Mutations can cause genetic disorders, such as sickle cell anemia, hemophilia, and Huntington's disease.

- Genetic counseling can help patients concerned about a genetic disorder.

Key Terms

Section 1

heredity (162)
genetics (162)
monohybrid cross (164)
true-breeding (164)
P generation (164)
F_1 generation (164)
F_2 generation (164)

Section 2

allele (167)
dominant (167)
recessive (167)
homozygous (167)
heterozygous (167)
genotype (168)
phenotype (168)
law of segregation (169)
law of independent assortment (169)

Section 3

Punnett square (170)
test cross (172)
probability (173)
pedigree (175)
sex-linked gene (175)

Section 4

polygenic inheritance (177)
incomplete dominance (177)
multiple alleles (178)
codominance (178)

Unit 5—Heredity
Use this unit to review the key concepts and terms in this chapter.

Performance ZONE

Understanding Key Ideas

1. The scientist whose studies formed the basis of modern genetics is
   a. T. A. Knight.
   b. Gregor Mendel.
   c. Louis Pasteur.
   d. Robert Hooke.

2. Which of the following is *not* a good reason why *Pisum sativum* makes an excellent subject for genetic study?
   a. Many varieties exist.
   b. They require cross-pollination.
   c. They grow quickly.
   d. They demonstrate complete dominance.

3. If smooth peas are dominant over wrinkled peas, the allele for smooth peas should be represented as
   a. *W*.   b. *S*.   c. *w*.   d. *s*.

4. The law of segregation states that pairs of alleles
   a. separate when gametes form.
   b. separate independently of one another during gamete formation.
   c. are always the same.
   d. are always different.

5. The trait shown below is

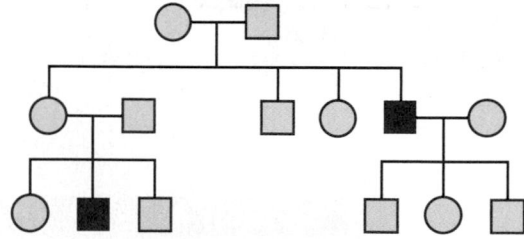

   a. sex-linked and dominant.
   b. autosomal and dominant.
   c. sex-linked and recessive.
   d. autosomal and recessive.

6. *D*, dimples, is the dominant allele to the recessive allele, *d*, no dimples. The probability of parents with *Dd* and *dd* genotypes having a child with no dimples (*dd*) is
   a. $\frac{1}{8}$.
   b. $\frac{1}{4}$.
   c. $\frac{1}{2}$.
   d. 1.

7. The unknown genotype of an individual with a dominant phenotype can be determined using
   a. a ratio.
   b. a dihybrid cross.
   c. probability.
   d. a test cross.

8. Explain how working genes have been inserted into defective cells during gene therapy.

9. Relate the events of meiosis to the law of segregation. (**Hint:** See Chapter 7, Section 1.)

10. **Exploring Further** State the genotypic and phenotypic ratios that would result from a cross between two *YyRR* pea plants.

11. **Concept Mapping** Make a concept map about Mendel's experiments. Try to include the following words in your map: *Pisum sativum*, *P generation*, *F_1 generation*, *F_2 generation*, *dominant trait*, and *recessive trait*.

Critical Thinking

12. **Evaluating Results** Mendel based his conclusion about inheritance patterns on experiments involving large numbers of plants. Why do you think the use of large numbers of individuals is advantageous when studying patterns of inheritance?

13. **Justifying Conclusions** A 20-year-old man who has cystic fibrosis has a sister who is planning to have a child. The man encourages his sister to see a genetic counselor. What do you think the man's reasons are for giving such advice?

Alternative Assessment

14. **Technology and Learning** Find out how new technologies have changed plant-breeding methods since Mendel's time. Prepare an oral report to summarize your findings. Or create a display that compares the methods and equipment Mendel might have used with those used by plant breeders today.

Standardized Test Prep

Understanding Concepts

Directions (1–4): For *each* question, write on a separate sheet of paper the letter of the correct answer.

1 What are the offspring of true-breeding parents called?
A. F_1 generation
B. F_2 generation
C. dominant offspring
D. recessive offspring

2 What term describes a gene with two dominant alleles that are expressed at the same time?
F. codominant
G. incompletely dominant
H. mutational
I. polygenic

3 What characteristic is described in the statement: The dog's coat is brown?
A. dominance **C.** pedigree
B. genotype **D.** phenotype

4 What does the law of segregation state?
F. The two alleles for a gene separate when gametes are formed.
G. A species can have a variety of different alleles that code for a single characteristic.
H. The alleles of different genes separate independently from one another during gamete formation.
I. Populations of a single species divided geographically will change over time to form two separate species.

Directions (5): For the following question, write a short response.

5 Albinism is rare among wild animals, but common among some domesticated species. What factor might account for this difference?

Test **TIP**

Before answering word problems that involve genetics, write the problem down by using letters to symbolize genotypes.

Reading Skills

Directions (6): Read the passage below. Then answer the question.

Genetic counselors use various types of information, including pedigrees, laboratory tests, and karyotypes, to determine the odds of a person or a couple's child having a genetic disorder. Genetic counselors also outline the options for dealing with those risks and offer emotional support.

6 A man and a woman who do not have hemophilia visit a genetic counselor. What tool might the counselor use to determine the risk of the couple having a child with hemophilia?
A. blood test
B. DNA fingerprint
C. karyotype
D. pedigree

Interpreting Graphics

Directions (7): Base your answer to question 7 on the diagram below.

Pea Plant Cross

	?	?
?	*Tt*	*tt*
?	*Tt*	*tt*

7 The diagram above shows the expected results of a cross between two pea plants. *T* and *t* represent the alleles for tall and dwarf traits, respectively. What genotypic ratio is expected in the offspring of this cross?
F. 1 *Tt* : 1 *tt*
G. 3 *Tt* : 1 *tt*
H. 1 *Tt* : 3 *tt*
I. 1 *TT* : 1 *tt*

Skills Practice Lab

Modeling Monohybrid Crosses

SKILLS
- Predicting outcomes
- Calculating data
- Organizing data
- Analyzing data

OBJECTIVES
- **Predict** the genotypic and phenotypic ratios of offspring resulting from the random pairing of gametes.
- **Calculate** the genotypic ratio and phenotypic ratio among the offspring of a monohybrid cross.

MATERIALS
- lentils
- green peas
- 2 Petri dishes

Before You Begin

A **monohybrid cross** is a cross that involves one pair of contrasting traits. Different versions of a gene are called **alleles.** When two different alleles are present and one is expressed completely and the other is not, the expressed allele is **dominant** and the unexpressed allele is **recessive.**

1. Write a definition for each boldface term in the paragraph above.

2. Based on the objectives for this lab, write a question you would like to explore about heredity.

Procedure

PART A: Simulating a Monohybrid Cross

1. You will model the random pairing of alleles by choosing lentils and peas from Petri dishes. These dried seeds will represent the alleles for seed color. A green pea will represent *G*, the dominant allele for green seeds, and a lentil will represent *g*, the recessive allele for yellow seeds.

2. Each Petri dish will represent a parent. Label one Petri dish "female gametes" and the other Petri dish "male gametes." Place one green pea and one lentil in the Petri dish labeled "female gametes" and place one green pea and one lentil in the Petri dish labeled "male gametes."

3. Each parent contributes one allele to each offspring. Model a cross between these two parents by choosing a random pairing of the dried seeds from the two Petri dishes. Do this by simultaneously picking one seed from each Petri dish *without looking*. Place the pair of seeds together on the lab table. The pair of seeds represents the genotype of one offspring.

4. Record the genotype of the first offspring in your lab report in a table like Table A shown below.

DATA TABLE A

Gamete Pairings		
Trial	Offspring genotype	Offspring phenotype
1		
2		
3		
4		
5		
6		
7		
8		
9		
10		

5. Return the seeds to their original dishes and repeat step 3 nine more times. Record the genotype of each offspring in your data table.

6. Based on each offspring's genotype, determine and record each offspring's phenotype.

PART B: Calculating Genotypic and Phenotypic Ratios

7. In your lab report, prepare a data table similar to Table B shown below.

DATA TABLE B

Offspring Ratios		
Genotypes	Total	Genotypic ratios
Homozygous dominant (GG)		
Heterozygous (Gg)		___ : ___ : ___
Homozygous recessive (gg)		
Phenotypes		Phenotypic ratios
Green seeds		
Yellow seeds		_____ : _____

8. Determine the genotypic and phenotypic ratios among the offspring. First count and record the number of homozygous dominant, heterozygous, and homozygous recessive individuals you recorded in Table A. Then record the number of offspring that produce green seeds and the number that produce yellow seeds under "Phenotypes" in your data table.

9. Calculate the genotypic ratio for each genotype using the following equation:
phenotypic ratio =
$$\frac{\text{number of offspring with a given genotype}}{\text{total number of offspring}}$$

10. Calculate the phenotypic ratio for each phenotype using the following equation:
phenotypic ratio =
$$\frac{\text{number of offspring with a given phenotype}}{\text{total number of offspring}}$$

11. Now pool the data for the whole class, and record the data in your lab report in a second table like Table B.

12. Compare the class's sample with your small sample of 10. Calculate the genotypic and phenotypic ratios for the class data, and record them in your data table.

13. Construct a Punnett square showing the parents and their offspring in your lab report.

14. Clean up your materials before leaving the lab.

Analyze and Conclude

1. Summarizing Results What character is being studied in this investigation?

2. Analyzing Data What are the genotypes of the parents? Describe the genotypes of both parents using the terms homozygous or heterozygous, or both. Did Table B reflect a classic monohybrid-cross phenotypic ratio of 3:1?

3. Drawing Conclusions If a genotypic ratio of 1:2:1 is observed, what must the genotypes of both parents be?

4. Predicting Patterns Show what the genotypes of the parents would be if 50 percent of the offspring were green and 50 percent of the offspring were yellow.

5. Further Inquiry Construct a Punnett square for the cross of a heterozygous black guinea pig and an unknown guinea pig whose offspring include a recessive white-furred individual. What are the possible genotypes of the unknown parent?

? Do You Know?

Do research in the library or media center to answer these questions:

1. How are hybrids of plants such as orchids produced?

2. What is *hybrid vigor*?

Use the following Internet resources to explore your own questions about genetics.

internet connect

www.scilinks.org
Topic: Genetic Code
Keyword: HX4089

SCLINKS. Maintained by the National Science Teachers Association

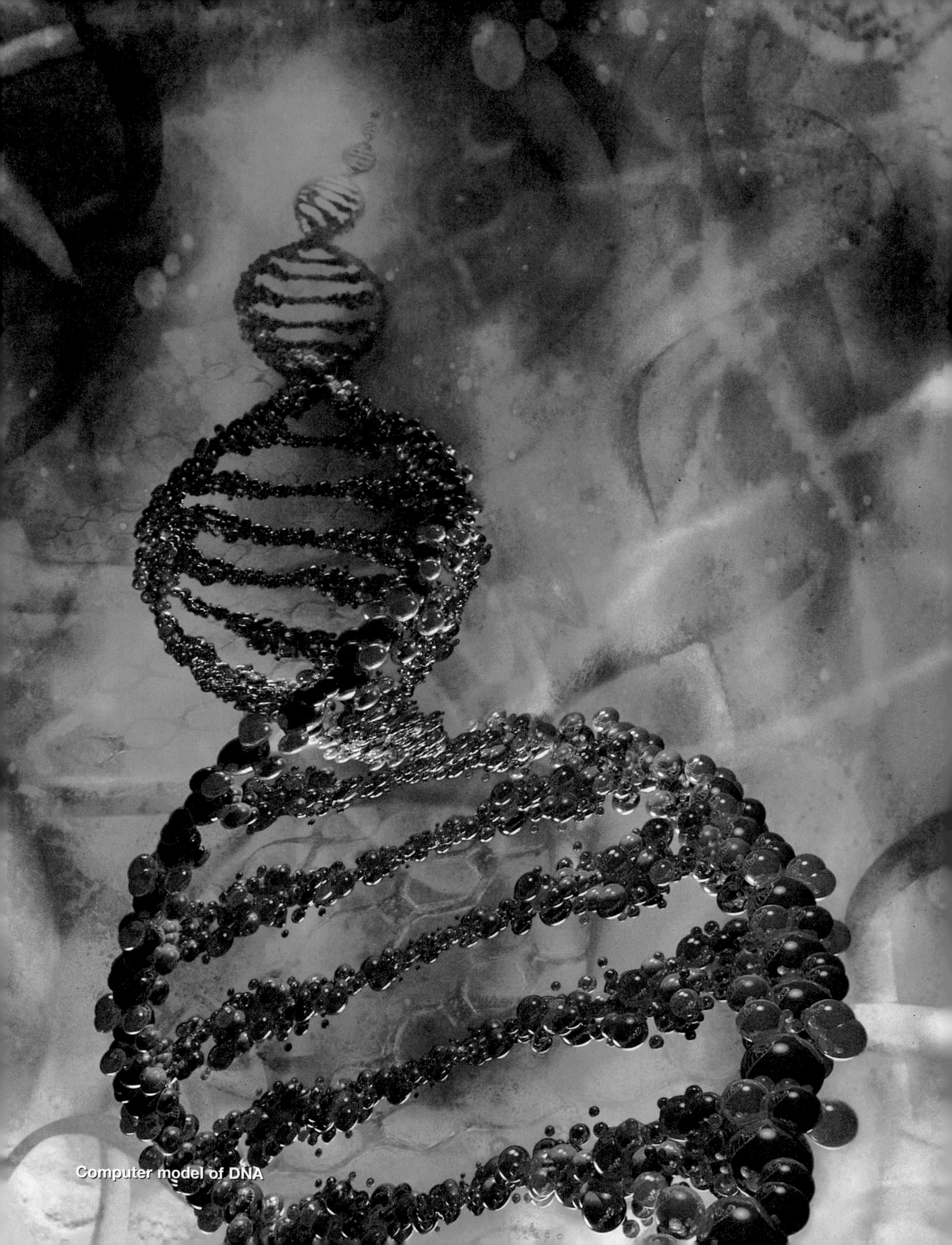

Computer model of DNA

9 DNA: The Genetic Material

Quick Review

Answer the following without referring to earlier sections of your book.

1. **Describe** the structure of chromosomes. *(Chapter 6, Section 1)*

2. **Define** the term *gene*. *(Chapter 6, Section 1)*

3. **Identify** the stage in a cell cycle in which DNA is copied. *(Chapter 6, Section 2)*

4. **Identify** changes in DNA as mutations. *(Chapter 6, Section 3)*

5. **Summarize** Mendel's theory of heredity. *(Chapter 8, Section 2)*

Did you have difficulty? *For help, review the sections indicated.*

Reading Activity

Before you read this chapter, write a short list of all the things you know about DNA. Then write a list of the things that you want to know about DNA. Save your list, and to assess what you have learned, see how many of your own questions you can answer after reading this chapter.

internet connect

www.scilinks.org
National Science Teachers Association *sci*LINKS Internet resources are located throughout this chapter.

*sci*LINKS. Maintained by the National Science Teachers Association

● Each human cell contains about 6 billion pairs of nucleotides—one pair for each "step" in the winding staircase of the DNA model shown on the facing page.

Identifying the Genetic Material

- **Relate** Griffith's conclusions to the observations he made during the transformation experiments.

- **Summarize** the steps involved in Avery's transformation experiments, and state the results.

- **Evaluate** the results of the Hershey and Chase experiment.

Key Terms

vaccine
virulent
transformation
bacteriophage

Transformation

Mendel's experiments and results answered the question of why you resemble your parents. You resemble your parents because you have copies of their chromosomes, which contain sets of instructions called genes. But Mendel's work created more questions, such as, What are genes made of? Scientists believed that if they could answer this question they would understand how chromosomes function in heredity.

Griffith's Experiments

In 1928, an experiment completely unrelated to the field of genetics led to an astounding discovery about DNA. Frederick Griffith, a bacteriologist, was trying to prepare a vaccine *(vahk SEEN)* against pneumonia. *Streptococcus pneumoniae* (abbreviated *S. pneumoniae*), is shown in **Figure 1.** *S. pneumoniae* is a prokaryote (of the type commonly called a bacterium) that causes pneumonia. A **vaccine** is a substance that is prepared from killed or weakened disease-causing agents, including certain bacteria. The vaccine is introduced into the body to protect the body against future infections by the disease-causing agent.

Griffith worked with two types, or strains, of *S. pneumoniae,* as shown in **Figure 2.** The first strain is enclosed in a capsule composed of polysaccharides. The capsule protects the bacterium from the body's defense systems. This helps make the microorganism **virulent** *(VIHR yoo luhnt),* or able to cause disease. Because of the capsule, this strain of *S. pneumoniae* grows as smooth-edged (*S*) colonies when grown in a Petri dish. The second strain of *S. pneumoniae* lacks the polysaccharide capsule and does not cause disease. When grown in a Petri dish, the second strain forms rough-edged *(R)* colonies.

Griffith knew that mice infected with the *S* bacteria grew sick and died, while mice infected with the *R* bacteria were not harmed, as shown in Figure 2. To determine whether the capsule on the *S* bacteria was causing the mice to die, Griffith injected the mice with dead *S* bacteria. The mice remained healthy. Griffith then prepared a vaccine of weakened *S* bacteria by raising their temperature to a point at which the bacteria were "heat-killed," meaning that they could no longer reproduce. (The capsule remained on the bacteria).

Figure 1 *Streptococcus pneumoniae*

Certain types of *S. pneumoniae* bacteria can cause the lung disease pneumonia.

Magnification: 17,250×

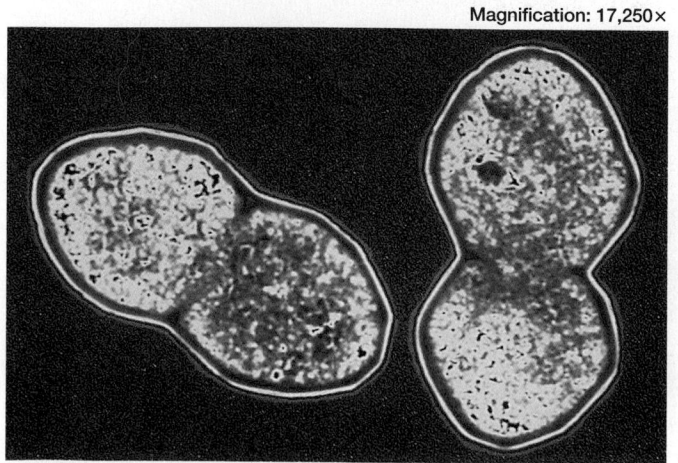

Figure 2 Griffith's discovery of transformation

Griffith discovered that harmless bacteria could turn virulent when mixed with bacteria that cause disease.

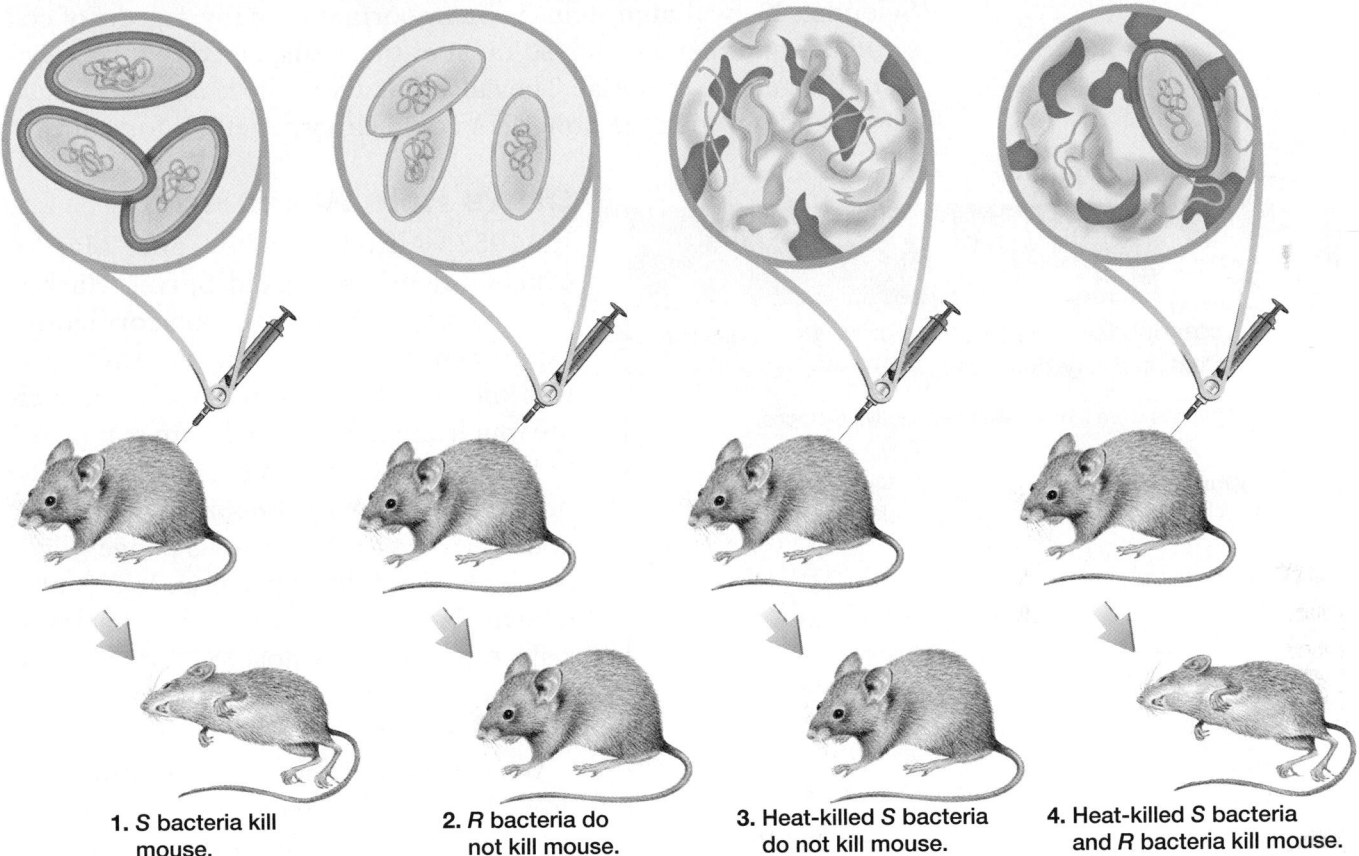

1. *S* bacteria kill mouse.

2. *R* bacteria do not kill mouse.

3. Heat-killed *S* bacteria do not kill mouse.

4. Heat-killed *S* bacteria and *R* bacteria kill mouse.

When Griffith injected mice with heat-killed *S* bacteria, the mice still lived. Thus, Griffith knew it was not the capsule on the *S* bacteria that killed the mice. He then mixed the harmless live *R* bacteria with the harmless heat-killed *S* bacteria. Mice injected with this mixture of previously harmless preparations died. When Griffith examined the blood of the dead mice, he found that the live *R* bacteria had acquired capsules. Somehow, the harmless *R* bacteria had changed and became virulent *S* bacteria. Griffith had discovered what is now called transformation. **Transformation** is a change in genotype caused when cells take up foreign genetic material. But the cause of the transformation was not known at the time.

Avery's Experiments

The search for the substance responsible for transformation continued until 1944. Then, an elegant series of experiments showed that the activity of the material responsible for transformation is not affected by protein-destroying enzymes. The activity is stopped, however, by a DNA-destroying enzyme. Thus, almost 100 years after Mendel's experiments, Oswald Avery and his co-workers at the Rockefeller Institute, in New York City, demonstrated that DNA is the material responsible for transformation. DNA contains the instructions for the making of the capsule in the *S* strain of *S. pneumoniae*.

WORD *Origins*

● The word *virulent* is from the Latin *virulentus*, which means "full of poison." Knowing this makes it easier to remember that a microorganism's virulence is its ability to cause disease.

Viral Genes and DNA

Even though Avery's experiments clearly indicated that the genetic material is composed of DNA, many scientists remained skeptical. Scientists knew that proteins were important to many aspects of cell structure and metabolism, so most of them suspected that proteins were the genetic material. They also knew very little about DNA, so they could not imagine how DNA could carry genetic information.

DNA's Role Revealed

In 1952, Alfred Hershey and Martha Chase, scientists at Cold Spring Harbor Laboratory, in New York, performed an experiment that settled the controversy. It was known at that time that viruses, which are much simpler than cells, are composed of DNA or RNA surrounded by a protective protein coat. A **bacteriophage** *(bak TIHR ee uh fayj)*, also referred to as phage *(fayj)*, is a virus that infects bacteria. It was also known that when phages infect bacterial cells, the phages are able to produce more viruses, which are released when the bacterial cells rupture.

What was not known at the time was how the bacteriophage reprograms the bacterial cell to make viruses. Does the phage DNA, the protein, or both issue instructions to the bacteria?

Hershey and Chase used the bacteriophage T2, shown in **Figure 3,** to answer this question. Hershey and Chase knew that the only molecule in the phage that contains phosphorus is its DNA. Likewise, the only phage molecules that contain sulfur are the proteins in its coat. Hershey and Chase used these differences to carry out the experiment shown in Figure 3.

Step ❶ Hershey and Chase first grew T2 with *Escherichia coli* (abbreviated *E. coli*) bacteria in a nutrient medium that contained radioactive sulfur (^{35}S). The protein coat of the virus would incorporate the ^{35}S. They grew a second batch of phages with *E. coli* bacteria in a nutrient medium that contained radioactive phosphorus (^{32}P). The radioactive phosphorus would become part of the phages' DNA.

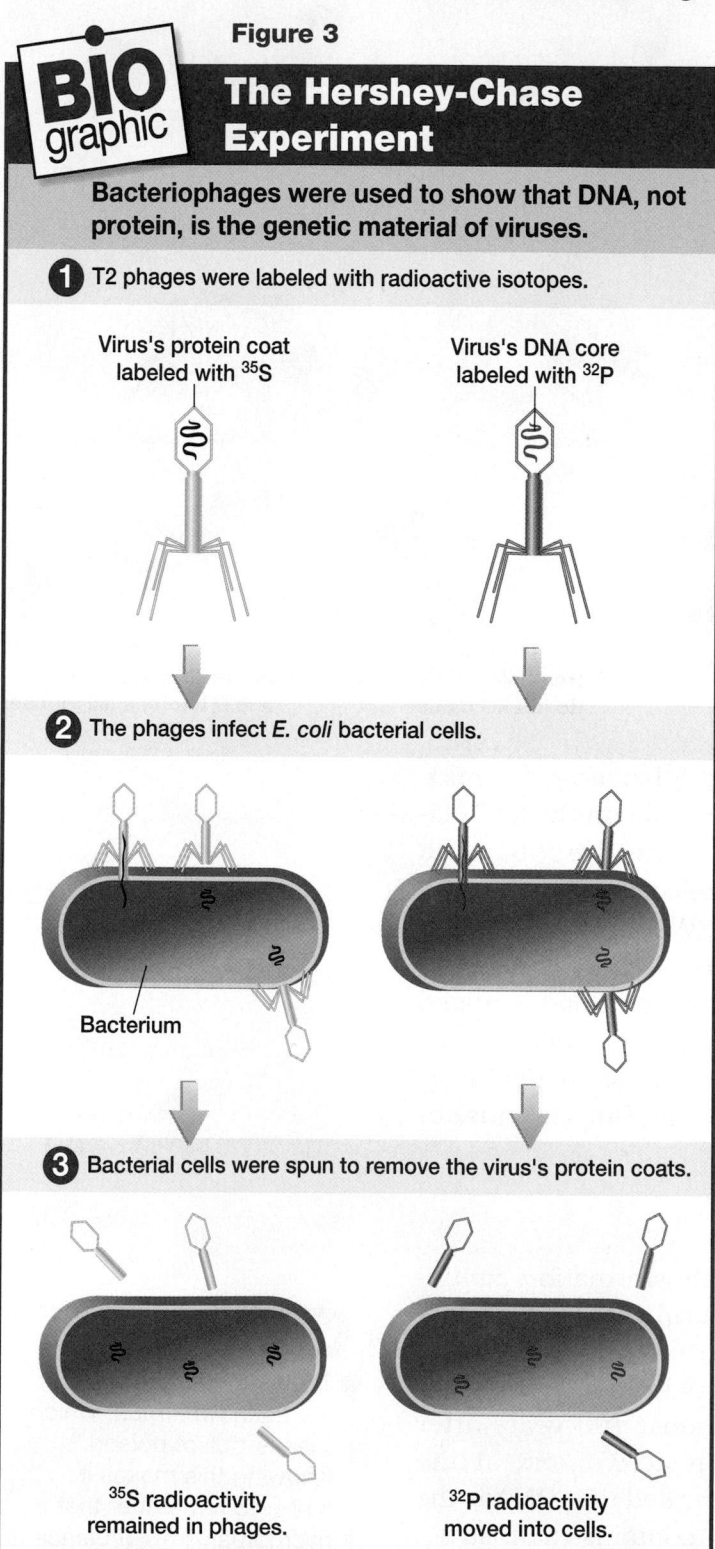

Figure 3

BIOgraphic

The Hershey-Chase Experiment

Bacteriophages were used to show that DNA, not protein, is the genetic material of viruses.

❶ T2 phages were labeled with radioactive isotopes.

Virus's protein coat labeled with ^{35}S

Virus's DNA core labeled with ^{32}P

❷ The phages infect *E. coli* bacterial cells.

Bacterium

❸ Bacterial cells were spun to remove the virus's protein coats.

^{35}S radioactivity remained in phages.

^{32}P radioactivity moved into cells.

Step ➋ The ^{35}S-labeled and ^{32}P-labeled phages were used to infect two separate batches of *E. coli* bacteria. Because radioactive elements release particles that can be detected with machines, they can be followed, or traced, in a biological process. Scientists could determine whether it was the DNA, the protein, or both that were being transferred into the bacterial cells to reprogram the bacteria.

Step ➌ After a few minutes, the scientists tore the ^{35}S-labeled phages off the surfaces of the bacteria (with the help of a blender). The bacteria infected with the ^{32}P-labeled phage were likewise mixed in a blender. The investigators used a centrifuge to separate the bacteria and phages. The heavier, bacterial cells formed a solid layer at the bottom of the centrifuge tubes. The lighter, viral parts remained in the upper, liquid layer.

Hershey and Chase examined the layers from the ^{35}S-infected bacteria. The scientists found that most of the ^{35}S label was still part of the phage (the upper layer), meaning the protein was not injected into the bacteria. When they examined the layers from the ^{32}P-infected bacteria, the scientists found the ^{32}P label mostly in the layer containing the bacterial cells (the lower layer). The DNA had been injected into the hosts. Moreover, the new generation of phages that was produced by these bacteria also contained radioactive DNA.

Hershey and Chase concluded that the DNA of viruses is injected into the bacterial cells, while most of the viral proteins remain outside. The injected DNA molecules causes the bacterial cells to produce more viral DNA and proteins. This meant that the DNA, rather than proteins, is the hereditary material, at least in viruses.

These important experiments, and many others since, have shown that DNA is the molecule that stores genetic information in living cells. As you will see in the next section, the structure of DNA makes DNA particularly well suited to this function.

Section 1 Review

➊ Summarize Griffith's transformation experiments.

➋ Describe how Avery's experiment supplied evidence that DNA, and not protein, is the genetic material.

➌ Describe the contributions of Hershey and Chase to the understanding that DNA is the genetic material.

➍ Critical Thinking Evaluating Methods Why did heat kill Griffith's *S* bacteria?

➎ Critical Thinking Applying Information What might Hershey and Chase have concluded if they had found ^{32}P and ^{35}S in the bacterial cells?

➏ (Standardized Test Prep) The first experiments that correctly identified the molecule that carries genetic information were performed by
   A Oswald Avery.   **C** Frederick Griffith.
   B Alfred Hershey.   **D** Martha Chase.

The Structure of DNA

- **Describe** the three components of a nucleotide.
- **Develop** a model of the structure of a DNA molecule.
- **Evaluate** the contributions of Chargaff, Franklin, and Wilkins in helping Watson and Crick determine the double-helical structure of DNA.
- **Relate** the role of the base-pairing rules to the structure of DNA.

Key Terms

double helix
nucleotide
deoxyribose
base-pairing rules
complementary base pair

A Winding Staircase

By the early 1950s, most scientists were convinced that genes were made of DNA. They hoped that the mystery of heredity could be solved by understanding the structure of DNA. The research of many scientists led two young researchers at Cambridge University, James Watson and Francis Crick, to piece together a model of the structure of DNA. The discovery of DNA's structure was important because it clarified *how* DNA could serve as the genetic material.

Watson and Crick determined that a DNA molecule is a **double helix**—two strands twisted around each other, like a winding staircase. As shown in **Figure 4,** each strand is made of linked nucleotides (*NOO klee oh tiedz*). **Nucleotides** are the subunits that make up DNA. Each nucleotide is made of three parts: a phosphate group, a five-carbon sugar molecule, and a nitrogen-containing base. Figure 4 shows how these three parts are arranged to form a nucleotide. The five-carbon sugar in DNA nucleotides is called **deoxyribose** (*dee ahk see RIE bohs*), from which DNA gets its full name, deoxyribonucleic acid.

Figure 4 DNA double helix

Watson and Crick's model of DNA is a double helix composed of two nucleotide chains that are twisted around a central axis and held together by hydrogen bonds.

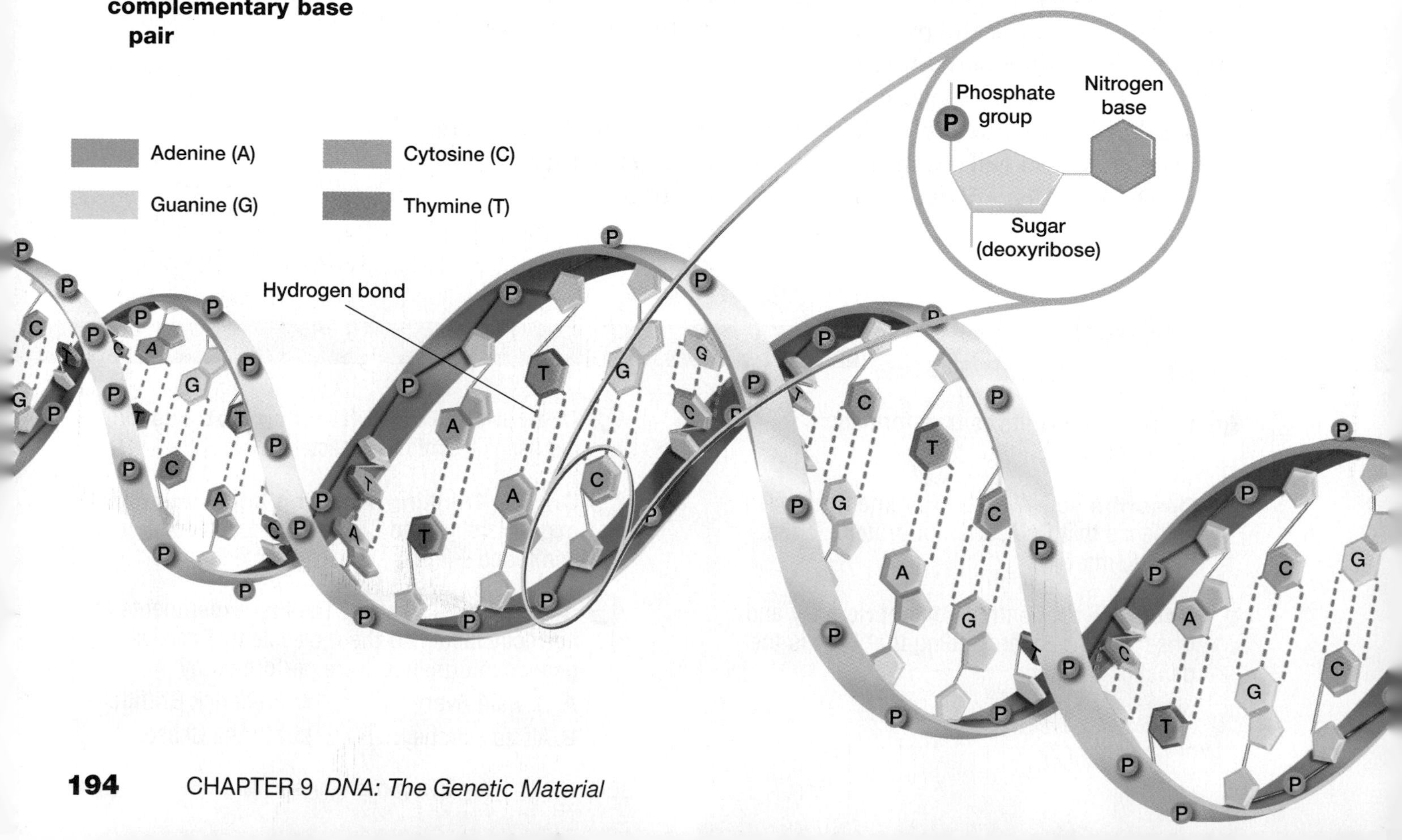

Adenine (A)

Cytosine (C)

Guanine (G)

Thymine (T)

Hydrogen bond

Nucleotide

Phosphate group

Nitrogen base

Sugar (deoxyribose)

While the sugar molecule and the phosphate group are the same for each nucleotide in a molecule of DNA, the nitrogen base may be any one of four different kinds. **Figure 5** illustrates the four different nitrogen bases in DNA: adenine *(AD uh neen)*, guanine *(GWAH neen)*, thymine *(THIE meen)*, and cytosine *(SIET oh seen)*. Adenine (A) and guanine (G) are classified as purines *(PYUR eenz)*, nitrogen bases made of two rings of carbon and nitrogen atoms. Thymine (T) and cytosine (C) are classified as pyrimidines *(pih RIHM uh deenz)*, nitrogen bases made of a single ring of carbon and nitrogen atoms.

Note how the DNA shown in Figure 4 resembles a ladder twisted like a spiral staircase. The sugar-phosphate backbones (the blue "ribbons") are similar to the side rails of a ladder. The paired nitrogen bases are similar to the rungs of the ladder. The nitrogen bases face each other. The double helix is held together by weak hydrogen bonds between the pairs of bases.

Figure 5   Purines and pyrimidines

The nitrogen base in a nucleotide can be either a bulky, double-ring purine, or a smaller, single-ring pyrimidine.

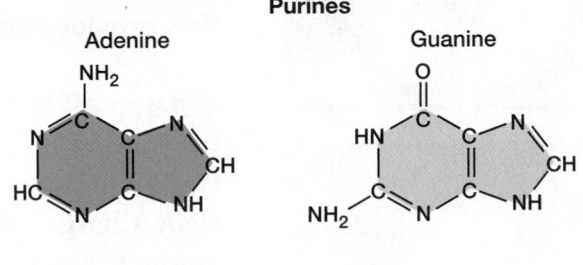

Purines

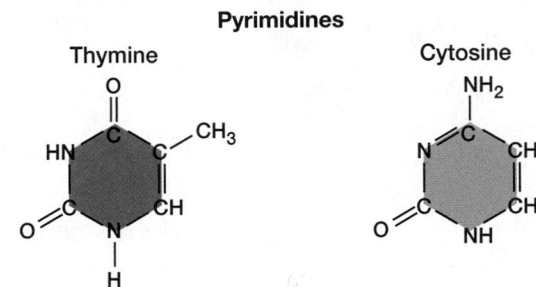

Pyrimidines

Observing Properties of DNA

QUICK LAB

You can extract DNA from onion cells using ethanol and a stirring rod.

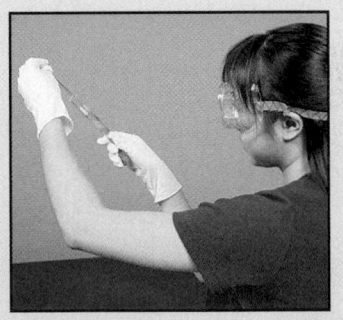

Materials

safety goggles and plastic gloves, 5 mL of onion extract, test tube, 5 mL of ice-cold ethanol, plastic pipet, glass stirring rod, test tube rack

Procedure

1. Place 5 mL of onion extract in a test tube.

2. CAUTION: *Ethanol is flammable. Do not use it near a flame.* Hold the test tube at a 45° angle. Use a pipet to add 5 mL of ice-cold ethanol to the tube one drop at a time. *NOTE: Allow the ethanol to run slowly down the side of the tube so that it forms a distinct layer.*

3. Let the test tube stand for 2–3 minutes.

4. Insert a glass stirring rod into the boundary between the onion extract and ethanol. Gently twirl the stirring rod by rolling the handle between your thumb and finger.

5. Remove the stirring rod from the liquids, and examine any material that has stuck to it. Touch the material to the lip of the test tube, and observe how the material acts as you try to remove it.

6. Clean up your materials and wash your hands before leaving the lab.

Analysis

1. **Describe** any material that stuck to the stirring rod.

2. **Relate** the characteristics of your sample to the structural characteristics of DNA.

3. **Propose** a way to determine if the material on the stirring rod is DNA.

Discovering DNA's Structure

How were Watson and Crick able to determine the double helical structure of DNA? As with most discoveries in science, other scientists provided crucial pieces that helped them solve this puzzle.

Chargaff's Observations

In 1949, Erwin Chargaff, a biochemist working at Columbia University, in New York City, made an interesting observation about DNA. Chargaff's data showed that for each organism he studied, the amount of adenine always equaled the amount of thymine (A=T). Likewise, the amount of guanine always equaled the amount of cytosine (G=C). However, the amount of adenine and thymine and of guanine and cytosine varied between different organisms.

Wilkins and Franklin's Photographs

The significance of Chargaff's data became clear in the 1950s when scientists began using X-ray diffraction to study the structures of molecules. In X-ray diffraction, a beam of X rays is directed at an object. The X rays bounce off the object and are scattered in a pattern onto a piece of film. By analyzing the complex patterns on the film, scientists can determine the structure of the molecule (much like shining a light on an object and then analyzing its shadow).

In the winter of 1952, Maurice Wilkins and Rosalind Franklin, two scientists working at King's College in London, developed high-quality X-ray diffraction photographs of strands of DNA. These photographs, such as the one in **Figure 6,** suggested that the DNA molecule resembled a tightly coiled helix and was composed of two or three chains of nucleotides.

Figure 6 Franklin and her X-ray diffraction photo. The photographs revealed the X pattern characteristic of a helix. Franklin died of cancer when she was 37 years old.

Watson and Crick's DNA Model

The three-dimensional structure of the DNA molecule, however, was yet to be discovered. Any model had to take into account both Chargaff's findings and Franklin and Wilkins's X-ray diffraction data. In 1953, Watson and Crick used this information, along with their knowledge of chemical bonding, to come up with a solution. With tin-and-wire models of molecules, they built a model of DNA with the configuration of a double helix, a "spiral staircase" of two strands of nucleotides twisting around a central axis. **Figure 7** shows Watson (left) and Crick next to their tin-and-wire model of DNA.

Figure 7 Watson and Crick's model. The double-helical model of DNA takes into account Chargaff's observations and the patterns on Franklin's X-ray diffraction photographs.

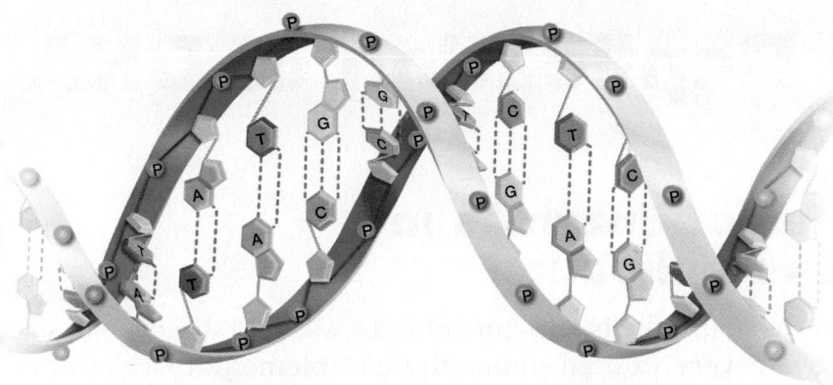

Figure 8 Base-pairing in DNA

The diagram of DNA below the helix makes it easier to visualize the base-pairing that occurs between DNA strands.

T	A	T	G	G	A	G	A	G	T	C
A	T	A	C	C	T	C	T	C	A	G

Pairing Between Bases

Watson and Crick determined that a purine on one strand of DNA is always paired with a pyrimidine on the opposite strand, as you can see in **Figure 8.** More specifically, an adenine on one strand always pairs with a thymine on the opposite strand, and a guanine on one strand always pairs with a cytosine on the opposite strand. The structure and size of the nitrogen bases allows for only these two paired combinations. These **base-pairing rules** are supported by Chargaff's observations. One easy way to visualize base-pairing is by simplifying the way in which DNA structure is represented, as shown in Figure 8.

Adenine forms two hydrogen bonds with thymine, and cytosine forms three hydrogen bonds with guanine. The hydrogen bonds between the nitrogen bases keep the two strands of DNA together. The strictness of base-pairing results in two strands that contain **complementary base pairs.**   That is, the sequence of bases on one strand determines the sequence of bases on the other strand. For example, if the sequence of nitrogen bases on one strand of a DNA molecule is TCGAACT, the sequence of nitrogen bases on the other strand must be AGCTTGA.

Study TIP

● **Organizing Information**
Create a timeline that summarizes the people and events that led to the discovery that DNA is the molecule where genetic information is stored. Start with 1928, and end with 1953.

Section 2 Review

1 **Describe** the three parts of a DNA nucleotide.

2 **Relate** the base-pairing rules to the structure of DNA.

3 **Describe** the two pieces of information from other scientists that enabled James Watson and Francis Crick to discover the double-helical structure of DNA.

4 **Explain** why the two strands of the double helix are described as complementary.

5 **Critical Thinking Applying Information** Suppose a strand of DNA has the nucleotide sequence CCAGATTG. What is the nucleotide sequence of the complementary strand?

6 **Standardized Test Prep** Which pattern shows how bases pair in complementary strands of DNA?
A A-C and T-G
B A-T and C-G
C A-G and T-C
D A-A and C-C

The Replication of DNA

- **Summarize** the process of DNA replication.

- **Describe** how errors are corrected during DNA replication.

- **Compare** the number of replication forks in prokaryotic and eukaryotic DNA.

Key Terms

DNA replication
DNA helicase
replication fork
DNA polymerase

Roles of Enzymes in DNA Replication

When the double helix structure of DNA was first discovered, scientists were very excited about the complementary relationship between the sequences of nucleotides. They predicted that the complementary structure was used as a basis to make exact copies of the DNA each time a cell divided. Watson and Crick proposed that one DNA strand serves as a template, or pattern, on which the other strand is built. Within five years of the discovery of DNA's structure, scientists had firm evidence that the complementary strands of the double helix do indeed serve as templates for building new DNA.

The process of making a copy of DNA is called **DNA replication.** DNA replication is summarized in **Figure 9.** Recall from your reading of earlier chapters that DNA replication occurs during the synthesis (S) phase of the cell cycle, before a cell divides.

Step ❶ Before DNA replication can begin, the double helix unwinds. This is accomplished by enzymes called DNA helicases. **DNA helicases** open the double helix by breaking the hydrogen bonds that link the complementary nitrogen bases between the two strands.

Figure 9

BIO graphic

DNA Replication

DNA replication results in two identical DNA strands.

❶ The two original DNA strands separate.

❷ DNA polymerases add complementary nucleotides to each strand.

❸ Two DNA molecules form that are identical to the original DNA molecule.

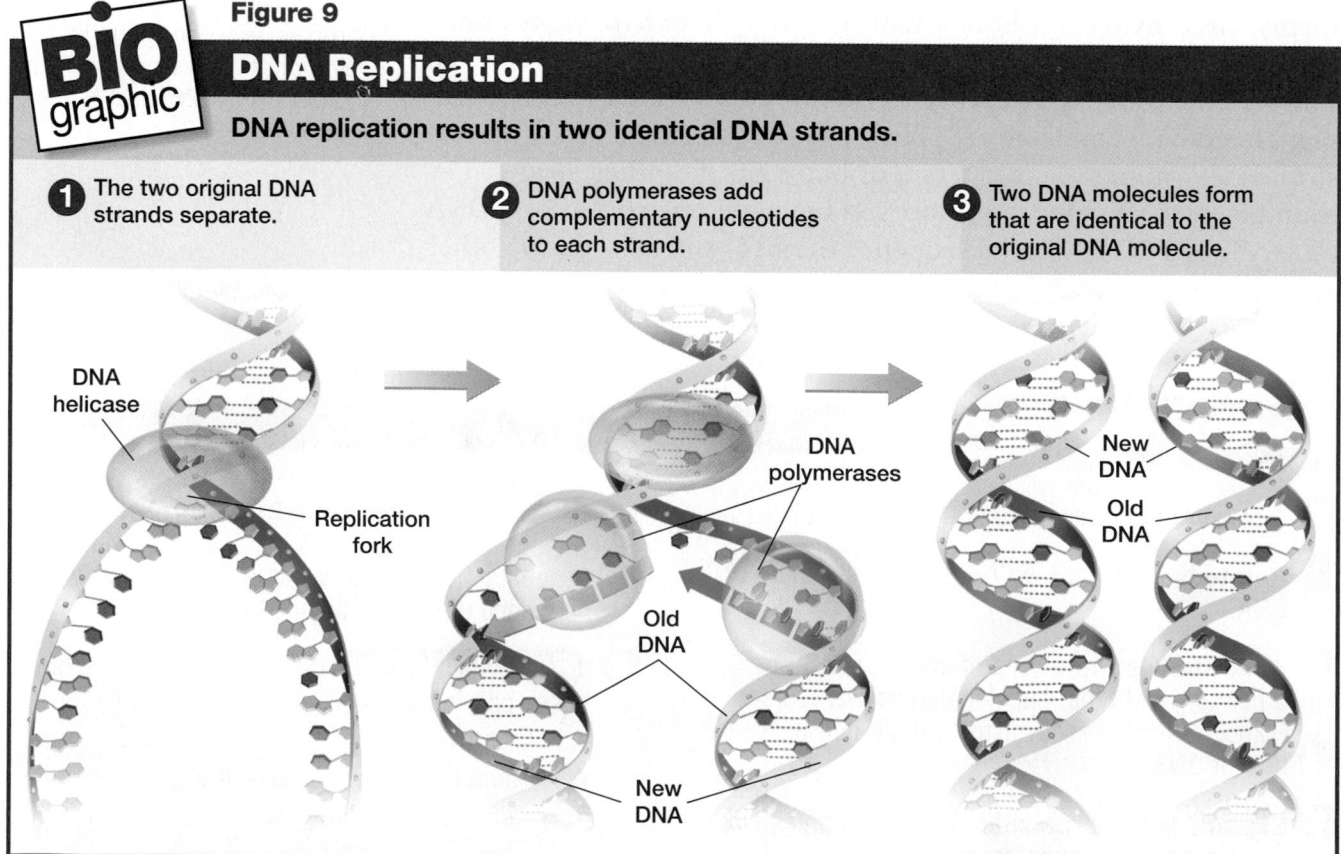

DNA helicase

Replication fork

DNA polymerases

Old DNA

New DNA

New DNA

Old DNA

Once the two strands are separated, additional proteins attach to each strand, holding them apart and preventing them from assuming their double-helical shape. The areas where the double helix separates are called **replication forks** because of their Y shape, as shown in Figure 9.

Step ❷ At the replication fork, enzymes known as **DNA polymerases** move along each of the DNA strands. DNA polymerases add nucleotides to the exposed nitrogen bases, according to the base-pairing rules. As the DNA polymerases move along, two new double helixes are formed.

Step ❸ Once DNA polymerases have begun adding nucleotides to a growing double helix, the process continues until all of the DNA has been copied and the polymerases are signaled to detach. This process produces two DNA molecules, each composed of a new and an original strand. The nucleotide sequences in both of these DNA molecules are identical to each other and to the original DNA molecule.

Checking for Errors

In the course of DNA replication, errors sometimes occur and the wrong nucleotide is added to the new strand. An important feature of DNA replication is that DNA polymerases have a "proofreading" role. They can add nucleotides to a growing strand only if the previous nucleotide is correctly paired to its complementary base. In the event of a mismatched nucleotide, the DNA polymerase can backtrack. The DNA polymerase removes the incorrect nucleotide and replaces it with the correct one. This proofreading reduces errors in DNA replication to about one error per 1 billion nucleotides.

Analyzing the Rate of DNA Replication

Background

Cancer is a disease caused by cells that divide uncontrollably. Scientists studying drugs that prevent cancer often measure the effectiveness of a drug by its effect on DNA replication. During normal DNA replication, nucleotides are added at a rate of about 50 nucleotides per second in mammals and 500 nucleotides per second in bacteria.

Magnification: 83,640×

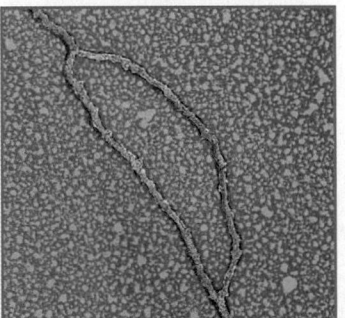

DNA replication forks

Analysis

1. **Calculate** the time it would take a bacterium to add 4,000 nucleotides to one DNA strand undergoing replication.

2. **Calculate** the time it would take a mammalian cell to add 4,000 nucleotides to one DNA strand undergoing replication.

3. **Critical Thinking**
   Predicting Outcomes How would the total time needed to add the 4,000 nucleotides be affected if a drug that inhibits DNA polymerases were present?

The Rate of Replication

Replication does not begin at one end of the DNA molecule and end at the other. The circular DNA molecules found in prokaryotes usually have two replication forks that begin at a single point. The replication forks move away from each other until they meet on the opposite side of the DNA circle, as shown in **Figure 10.**

In eukaryotic cells, each chromosome contains a single, long strand of DNA. The length presents a challenge: The replication of a typical human chromosome with one pair of replication forks spreading from a single point, as occurs in prokaryotes, would take 33 days! To understand how eukaryotes meet this challenge, imagine that your class has to carry 25 boxes to another building. Carrying one box over, returning, carrying the second box, and so on, would be very slow. It would be much faster if everyone in the class picked up a box so that all of the boxes could be carried in one trip. That is similar to replication in eukaryotic cells, as shown in Figure 10. Each human chromosome is replicated in about 100 sections that are 100,000 nucleotides long, each section with its own starting point. With multiple replication forks working in concert, an entire human chromosome can be replicated in about 8 hours.

Figure 10   Replication forks

Prokaryotic and eukaryotic DNA have a different number of replication forks.

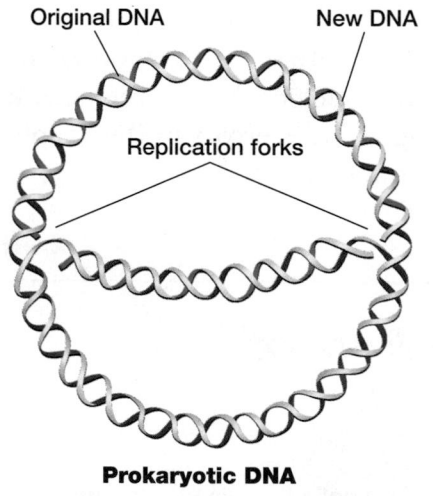

Original DNA   New DNA

Replication forks

Prokaryotic DNA

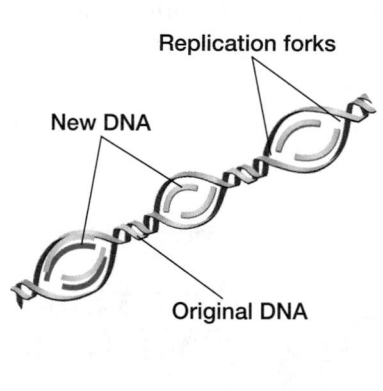

Replication forks

New DNA

Original DNA

Eukaryotic DNA

Section 3 Review

1 **Explain** the two roles that enzymes play in DNA replication as is illustrated in Figure 9 in this section.

2 **Explain** the relationship between DNA polymerases and mutations.

3 **State** the effect of multiple replication forks on the speed of replication in eukaryotes.

4 **Critical Thinking Evaluating Information** If a mutation occurs during the formation of an egg cell or sperm cell, is that mutation more significant or less significant than a mutation that occurs in a body cell? Explain your answer.

5 **Standardized Test Prep** How many DNA strands exist after one molecule of DNA has been replicated?

A 1   C 4

B 2   D 8

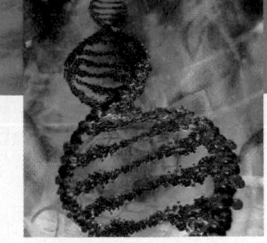

Key Concepts

1 Identifying the Genetic Material

- The experiments of Griffith and of Avery yielded results that suggested DNA was the genetic material.

- Hershey and Chase used the bacteriophage T2 and radioactive labels to show that viral genes are made of DNA, not protein.

- DNA stores the information that tells cells which proteins to make and when to make them.

2 The Structure of DNA

- DNA is made of two strands of nucleotides twisted into the form of a double helix.

- Each nucleotide in DNA is made of the sugar deoxyribose, a phosphate group, and one of four nitrogen bases. The four nitrogen bases found in DNA nucleotides are adenine (A), thymine (T), guanine (G), and cytosine (C).

- The two strands of DNA are complementary—each A on one strand pairs with a T on the opposite strand, and each G on one strand pairs with a C on the opposite strand.

- Watson and Crick announced their discovery of the structure of DNA in 1953 with the help of data gathered by Wilkins, Franklin, and Chargaff.

3 The Replication of DNA

- Before a cell divides, it copies its DNA by a process called DNA replication.

- In DNA replication, enzymes work to unwind and separate the double helix and add complementary nucleotides to the exposed strands.

- The result of DNA replication is two exact copies of the cell's original DNA. Each new double helix is composed of one original DNA strand and one new DNA strand.

- DNA polymerase proofreads DNA during its replication so that very few errors occur.

Key Terms

Section 1
vaccine (190)
virulent (190)
transformation (191)
bacteriophage (192)

Section 2
double helix (194)
nucleotide (194)
deoxyribose (194)
base-pairing rules (197)
complementary base pair (197)

Section 3
DNA replication (198)
DNA helicase (198)
replication fork (199)
DNA polymerase (199)

BIOLOGY INTERACTIVE TUTOR

Unit 6—*Gene Expression*
Use Topics 1–2 in this unit to review the key concepts and terms in this chapter.

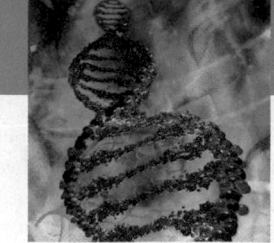

Performance ZONE

Understanding Key Ideas

1. In his experiments on *Streptococcus pneumoniae*, Griffith found that
   a. harmless *R* bacteria changed and became virulent.
   b. the *S* bacteria were transformed.
   c. the capsule did not protect the bacterium.
   d. mice injected with the *R* bacteria died.

2. Hershey and Chase showed that
   a. bacteriophages can infect human cells.
   b. DNA controls heredity.
   c. bacteria undergo transformation.
   d. a vaccine for pneumonia could be produced.

3. James Watson and Francis Crick
   a. built a structural model of DNA.
   b. discovered DNA replication.
   c. used X-ray diffraction.
   d. discovered DNA polymerases.

4. Multiple replication forks along the DNA
   a. correct replication errors.
   b. reduce DNA replication time.
   c. ensure that the new and old DNA strands are complementary.
   d. signal DNA polymerase to stop.

5. The table below summarizes the percentage of each nitrogen base found in an organism's DNA.

Percentage of Each Nitrogen Base			
	A	**T**	**G**	**C**
Human	30.4	30.1	19.6	19.9
Wheat	27.3	27.1	22.7	22.8
E. coli	24.7	23.6	26.0	25.7

   a. What is the ratio of purines to pyrimidines?
   b. Within each organism, which nucleotides are found in similar percentages?
   c. Do the ratio and percentages in (a) and (b) follow Chargaff's rule?

6. If the sequence of nucleotides on one strand of a DNA molecule is GCCATTG, the sequence on the complementary strand is
   a. GGGTAAG.
   b. CCCTAAC.
   c. CGGTAAC.
   d. GCCATTC.

7. Does DNA replication occur immediately before asexual reproduction, before sexual reproduction, or before both?

8. What are two functions of DNA polymerases during DNA replication?

9. Differentiate between DNA, genes, chromatids, and chromosomes. (**Hint:** See Chapter 6, Section 1.)

10. ⌂ **Concept Mapping** Make a concept map that shows the structure of DNA and how it is copied. Try to include the following words in your concept map: *nucleotides, phosphate group, five-carbon sugar, nitrogen base, purine, pyrimidine, double helix, replication, DNA polymerases,* and *gene.*

Critical Thinking

11. **Evaluating Models** Explain why you do or do not think Watson and Crick's model of DNA illustrated in **Figure 7** in this chapter is a good representation of the structure of DNA. What existing information about DNA did Watson and Crick's model have to take into account?

12. **Predicting Results** Identify the process by which new molecules of DNA are synthesized, and predict the effect on this process of reducing available DNA helicases.

Alternative Assessment

13. **Selecting Technology** Research two methods used to sequence the nucleotides in a gene. Compare and contrast the two methods. Give examples of how this technology might be used in a clinical setting. Prepare a poster to summarize the nucleotide-sequencing methods you researched.

Standardized Test Prep

Understanding Concepts

Directions (1–4): For *each* question, write on a separate sheet of paper the letter of the correct answer.

1 What is the name of the process that was involved in changing Griffith's *R* bacteria to *S* bacteria?
A. crossing over
B. DNA replication
C. polymerization
D. transformation

2 A scientist extracted 4.6 picograms (or 4.6×10^{-12} grams) of DNA from mouse muscle cells. How much DNA could be extracted from the same number of mouse sperm?
F. 2.3 picograms
G. 4.6 picograms
H. 9.2 picograms
I. 10^{-12} picograms

3 Which of the following is **not** a component of a DNA nucleotide?
A. double helix
B. five-carbon sugar
C. nitrogen base
D. phosphate group

4 What molecule did Hershey and Chase's work show was the genetic material of the T2 bacteria?
F. DNA
G. a protein
H. DNA helicase
I. DNA polymerase

Directions (5): For the following question, write a short response.

5 X-rays damage DNA in organisms. Rosalind Franklin died of cancer at an early age. Analyze how her work with X-ray diffraction might have led to her death.

Test TIP

You can sometimes figure out an answer to a question before you look at the answer choices. After you answer the question in your mind, compare your answer with each answer choice. Choose the answer that most closely matches your own answer.

Reading Skills

Directions (6): **Read the passage below. Then answer the question.**

The polymerase chain reaction (PCR) is a technique used in genetic engineering and applied in procedures such as DNA fingerprinting and genetic screening. This technique allows a small sample of DNA to be copied many times. Genetic analysis, which determines the presence of restriction length polymorphisms (DNA fingerprinting), or mutations on specific genes, can be performed with *PCR-amplified DNA* samples.

6 What could be inferred as the meaning of the term "PCR-amplified DNA" above?
A. a kind of PCR that is used on large samples of DNA
B. DNA fingerprinting that results in amplification of a PCR technique
C. a sample of DNA that has been replicated many times using PCR
D. a sequence of DNA that codes for an auditory trait such as sensitive hearing

Interpreting Graphics

Directions (7): **Base your answer to question 7 on the diagram below.**

DNA Replication

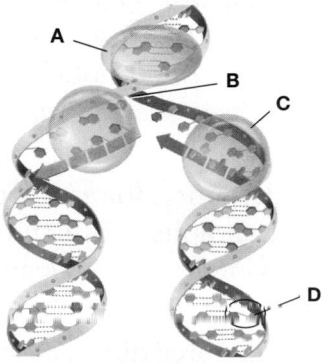

7 What is the function of the structure labeled *A*?
F. separating DNA strands
G. reconnecting DNA strands
H. checking the new DNA strands for errors
I. adding nucleotides to make new DNA strands

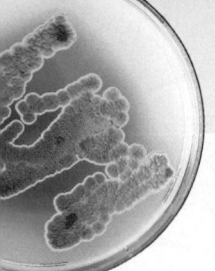

Exploration Lab

Modeling DNA Structure

SKILLS
- Modeling
- Using scientific methods

OBJECTIVES
- **Design** and analyze a model of DNA.
- **Describe** how replication occurs.
- **Predict** the effect of errors during replication.

MATERIALS
- plastic soda straws, 3 cm sections
- metric ruler
- pushpins (red, blue, yellow, and green)
- paper clips

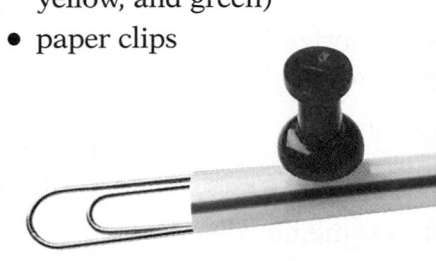

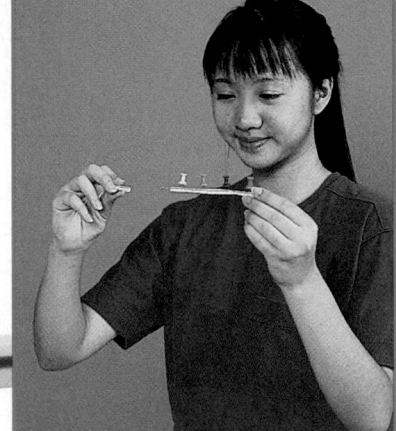

Before You Begin

DNA contains the instructions that cells need in order to make every **protein** required to carry out their activities and to survive. DNA is made of two strands of **nucleotides** twisted around each other in a **double helix.** The two strands are **complementary,** that is, the sequence of bases on one strand determines the sequence of bases on the other strand. The two strands are held together by hydrogen bonds.

In this lab, you will build a model to help you understand the structure of DNA. You can also use the DNA model to illustrate and explore processes such as **replication** and **mutation.**

1. Write a definition for each boldface term in the paragraphs above and for each of the following terms: replication fork, base-pairing rules.
2. Identify the three different components of a nucleotide.
3. Identify the four different nitrogen bases that can be found in DNA nucleotides.
4. Based on the objectives for this lab, write a question you would like to explore about DNA structure.

Procedure

PART A: Design a Model

1. Work with the members of your lab group to design a model of DNA that uses the materials listed for this lab. Be sure that your model has at least 12 nucleotides on each strand.

> ### You Choose
> **As you design your model, decide the following:**
> **a.** what question you will explore
> **b.** how to use the straws, pushpins, and paper clips to represent the three components of a nucleotide
> **c.** how to link (bond) the nucleotides together
> **d.** in what order you will place the nucleotides on each strand

2. Write out the plan for building your model. Have your teacher approve the plan before you begin building the model.
3. Build the DNA model your group designed. **CAUTION: Sharp or pointed objects may cause injury. Handle pushpins carefully.** Sketch and label the parts of your DNA model.
4. Use your model to explore one of the questions written for step 4 of **Before You Begin.**

PART B: DNA Replication

5. Discuss with your lab group how the model you built for Part A may be used to illustrate the process of replication.

6. Write a question you would like to explore about replication. Use your model to explore the question you wrote. Sketch and label the steps of replication.

PART C: Test Hypothesis

Answer each of the following questions by writing a hypothesis. Use your model to test each hypothesis, and describe your results.

7. Mitosis follows replication. How might the cells produced by mitosis be affected if nucleotides on one DNA strand were incorrectly paired during replication?

8. What would happen if only one strand in a DNA molecule were copied during replication?

PART D: Cleanup and Disposal

9. Dispose of damaged pushpins in the designated waste container.

10. Clean up your work area and all lab equipment. Return lab equipment to its proper place. Wash your hands thoroughly before you leave the lab and after you finish all work.

Analyze and Conclude

1. Analyzing Results In your original DNA model, were the two strands identical to each other?

2. Relating Concepts How does DNA structure ensure that the two DNA molecules made by replication are the same as the original DNA molecule?

3. Drawing Conclusions Did the two DNA molecules you made in step 6 have the same nitrogen-base sequence as your original model DNA molecule?

4. Inferring Relationships The order of nitrogen bases on a DNA strand is a code for making proteins. What does this mean has happened to the "code" in one of the DNA molecules you made in step 7?

5. Predicting Outcomes What would happen if the DNA in a cell that is about to divide were not replicated?

6. Inferring Information What are the advantages of having DNA remain in the nucleus of a cell?

7. Further Inquiry Write a new question about DNA that could be explored with your model.

? Do You Know?

Do research in the library or media center to answer these questions:

1. Are there any pollutants in the environment that disrupt replication when an organism is exposed to the pollutant?

2. How do DNA molecules differ among various species of animals and plants? How are they similar?

Use the following Internet resources to explore your own questions about DNA.

internet connect

www.scilinks.org
Topic: DNA
Keyword: HX4058

*SCI*LINKS. Maintained by the National Science Teachers Association

Firefly (19×)

10 How Proteins Are Made

✔ *Quick Review*

Answer the following without referring to earlier sections of your book.

1. **Summarize** the structure and function of proteins. *(Chapter 2, Section 3)*
2. **Describe** the function of ribosomes. *(Chapter 3, Section 2)*
3. **Differentiate** between DNA and genes. *(Chapter 6, Section 1)*
4. **Describe** the structure and function of DNA. *(Chapter 9, Section 2)*
5. **State** the base-pairing rules. *(Chapter 9, Section 2)*

Did you have difficulty? *For help, review the sections indicated.*

📖 *Reading Activity*

Before you read this chapter, write a short list of all the things you know about how proteins are made. Then, write a list of the things that you want to know about how proteins are made. Save your list, and to assess what you have learned, see how many of your own questions you can answer after reading this chapter.

🔲 internet connect

www.scilinks.org
National Science Teachers Association *sci*LINKS Internet resources are located throughout this chapter.

SC*LINKS.* **Maintained by the National Science Teachers Association**

● The firefly shown here gives off light because of a chemical reaction activated by an enzyme (which is a type of protein) made by its cells.

From Genes to Proteins

Objectives

- **Compare** the structure of RNA with that of DNA.
- **Summarize** the process of transcription.
- **Relate** the role of codons to the sequence of amino acids that results after translation.
- **Outline** the major steps of translation.
- **Discuss** the evolutionary significance of the genetic code.

Key Terms

ribonucleic acid (RNA)
uracil
transcription
translation
gene expression
RNA polymerase
messenger RNA
codon
genetic code
transfer RNA
anticodon
ribosomal RNA

Decoding the Information in DNA

Traits, such as eye color, are determined by proteins that are built according to instructions coded in DNA. Recall that proteins have many functions, including acting as enzymes and cell membrane channels. Proteins, however, are not built directly from DNA. Ribonucleic *(rie boh noo KLAY ihk)* acid is also involved.

Like DNA, **ribonucleic acid (RNA)** is a nucleic acid—a molecule made of nucleotides linked together. RNA differs from DNA in three ways. First, RNA consists of a single strand of nucleotides instead of the two strands found in DNA, as shown in **Figure 1.** Second, RNA nucleotides contain the five-carbon sugar ribose *(RIE bohs)* rather than the sugar deoxyribose, which is found in DNA nucleotides. Ribose contains one more oxygen atom than deoxyribose contains. And third, in addition to the A, G, and C nitrogen bases found in DNA, RNA nucleotides can have a nitrogen base called **uracil** *(YUR uh sihl)*—abbreviated as U. No thymine (T) bases are found in RNA. Like thymine, uracil is complementary to adenine whenever RNA base-pairs with another nucleic acid.

A gene's instructions for making a protein are coded in the sequence of nucleotides in the gene. The instructions for making a protein are transferred from a gene to an RNA molecule in a process called **transcription.** Cells then use two different types of RNA to read the instructions on the RNA molecule and put together the amino acids that make up the protein in a process called **translation.** The entire process by which proteins are made based on the information encoded in DNA is called **gene expression,** or protein synthesis. This process is summarized in Figure 1.

Figure 1   Gene expression

The instructions for building a protein are found in a gene and are "rewritten" to a molecule of RNA during transcription. The RNA is then "deciphered" during translation.

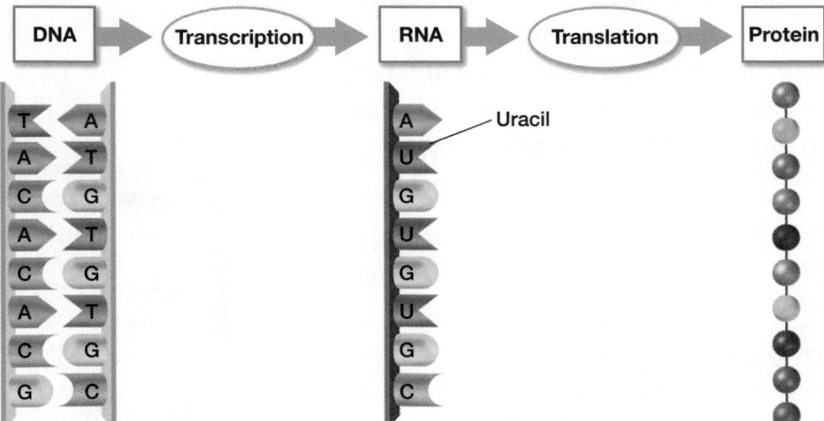

Transfer of Information from DNA to RNA

The first step in the making of a protein, transcription, takes the information found in a gene in the DNA and transfers it to a molecule of RNA. **RNA polymerase,** an enzyme that adds and links complementary RNA nucleotides during transcription, is required. **Figure 2** summarizes the steps of transcription.

Step ❶ Transcription begins when RNA polymerase binds to the gene's promoter—a specific sequence of DNA that acts as a "start" signal for transcription.

Step ❷ RNA polymerase then unwinds and separates the two strands of the double helix, exposing the DNA nucleotides on each strand.

Step ❸ RNA polymerase adds and then links complementary RNA nucleotides as it "reads" the gene. RNA polymerase moves along the nucleotides of the DNA strand that has the gene, much like a train moves along on a track. Transcription follows the base-pairing rules for DNA replication except that in RNA, uracil, rather than thymine, pairs with adenine.

As transcription proceeds, the RNA polymerase eventually reaches a "stop" signal in the DNA. This "stop" signal is a sequence of bases that marks the end of each gene in eukaryotes, or the end of a set of genes in prokaryotes.

Figure 2

BIOgraphic

Transcription: Making RNA

RNA polymerase adds complementary RNA nucleotides as it reads the gene.

❶ RNA polymerase binds to the gene's promoter.

❷ The two DNA strands unwind and separate.

❸ Complementary RNA nucleotides are added.

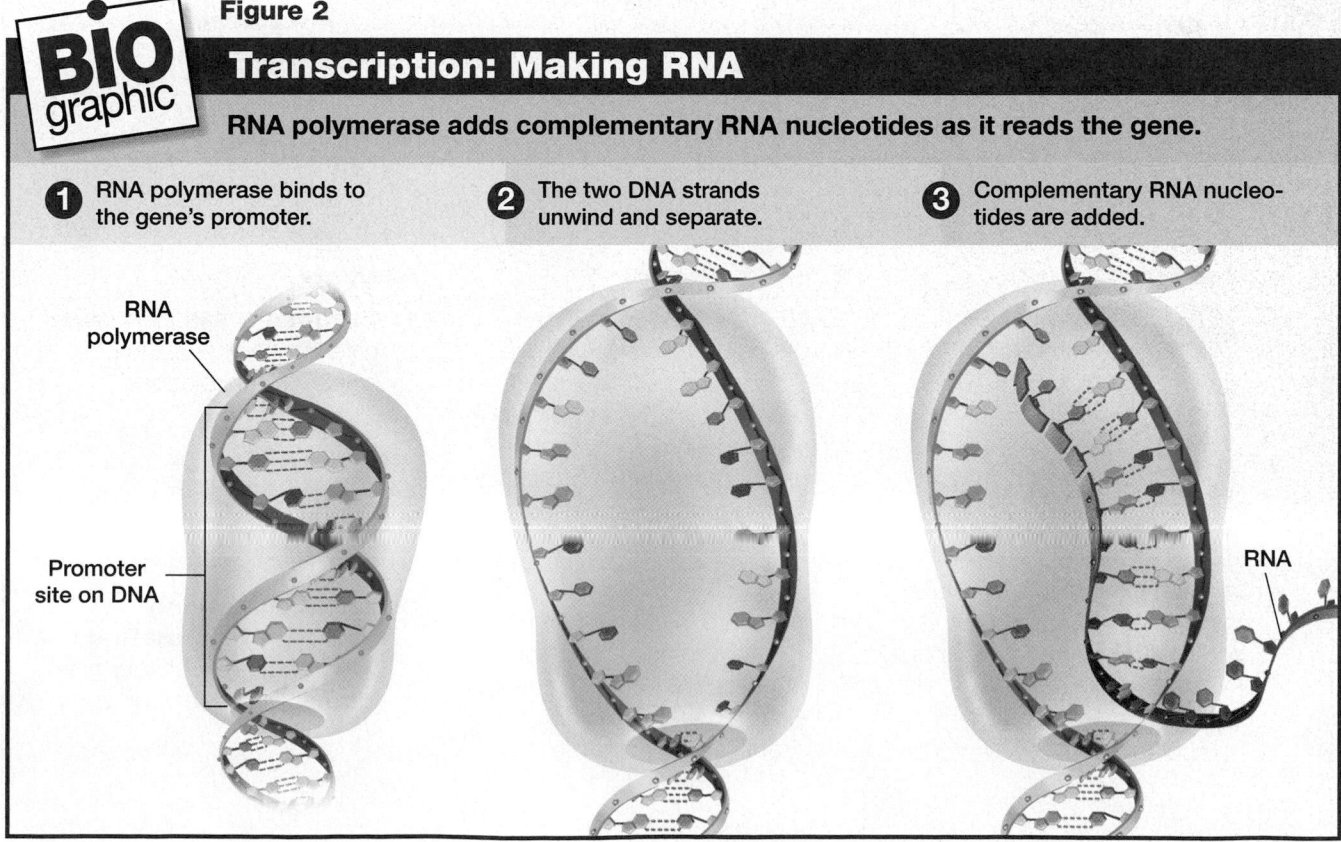

RNA polymerase

Promoter site on DNA

RNA

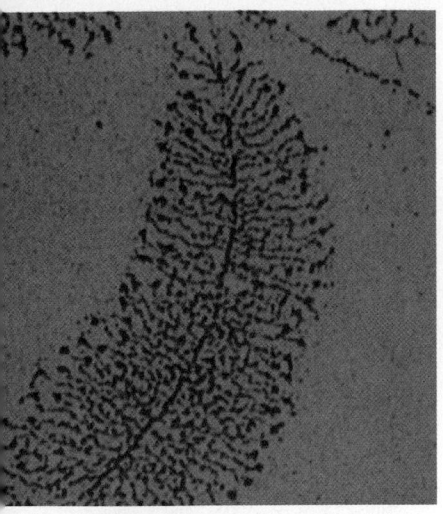

Figure 3   Multiple copies of RNA. In eukaryotes, RNA polymerase adds about 60 nucleotides per second. There are typically about 100 RNA polymerase molecules per gene.

When the RNA nucleotides are added during transcription, they are linked together with covalent bonds. As RNA polymerase moves down the strand, a single strand of RNA grows. Behind RNA polymerase, the two strands of DNA close up by forming hydrogen bonds between complementary bases, re-forming the DNA double helix.

Like DNA replication, transcription uses DNA nucleotides as a template for making a new molecule. However, in DNA replication, the new molecule made is DNA. In transcription, the new molecule made is RNA. In addition, in DNA replication, both strands of DNA serve as templates, whereas in transcription, only part of one of the two strands of DNA (a gene) serves as a template.

Transcription in prokaryotic cells occurs in the cytoplasm (because prokaryotic cells have no nucleus); transcription in eukaryotic cells occurs in the nucleus, where the DNA is located. During transcription, many identical RNA molecules are made simultaneously from a single gene, as shown in **Figure 3.** The RNA being made fans out from the gene to give a "feathery" appearance. The long line along the length of the "feather" is the DNA being transcribed. The circles along the length are the RNA polymerase molecules. The "hairs" on the feather are the RNA chains being made.

Modeling Transcription

You can use paper and pens to model the process of transcription.

Materials

paper, scissors, pens or pencils (two colors), tape

Procedure

1. Cut a sheet of paper into 36 squares, each about 2.5 × 2.5 cm (1 × 1 in.) in size.

2. To make one side of your DNA model, line up 12 squares in a column. Using one color, randomly label each square with one of the following letters: A, C, G, or T. Each square represents a DNA nucleotide. Use tape to keep the squares in a column.

3. To make the second side of your DNA model, line up 12 squares next to the first column. Use the same color you used in step 2 to label each square with the complementary DNA nucleotide. Tape the squares together in a column.

4. Separate the two columns. The remaining 12 squares represent RNA nucleotides. Use a different color to "transcribe" one of the DNA strands.

Analysis

1. **Propose** a reason for using different colors for the DNA and RNA "nucleotides."

2. **Predict** how a change in the sequence of nucleotides in a DNA molecule would affect the mRNA transcribed from the DNA molecule.

3. **Critical Thinking Applying Information** Use your model to test your prediction. Describe your results.

The Genetic Code: Three-Nucleotide "Words"

Different types of RNA are made during transcription, depending on the gene being expressed. When a cell needs a particular protein, it is messenger RNA that is made. **Messenger RNA** (mRNA) is a form of RNA that carries the instructions for making a protein from a gene and delivers it to the site of translation. The information is translated from the language of RNA—nucleotides—to the language of proteins—amino acids. The RNA instructions are written as a series of three-nucleotide sequences on the mRNA called **codons** *(KOH dahnz)*. Each codon along the mRNA strand corresponds to an amino acid or signifies a start or stop signal for translation.

In 1961, Marshall Nirenberg, an American biochemist, deciphered the first codon by making artificial mRNA that contained only the base uracil (U). The mRNA was translated into a protein made up entirely of phenylalanine amino-acid subunits. Nirenberg concluded that the codon UUU is the instruction for the amino acid phenylalanine. Later, scientists deciphered the other codons. **Figure 4** shows the **genetic code**—the amino acids and "start" and "stop" signals that are coded for by each of the possible 64 mRNA codons.

Figure 4 Interpreting the genetic code

The amino acid coded for by a specific mRNA codon can be determined by following the three steps below.

1. Find the first base of the mRNA codon along the left side of the table.

2. Follow that row to the right until you are beneath the second base of the codon.

3. Move up or down in that section until you are even, on the right side of the chart, with the third base of the codon.

Codons in mRNA					
First base	**Second base**				**Third base**
	U	**C**	**A**	**G**	
U	UUU ⎤ UUC ⎦ Phenylalanine UUA ⎤ UUG ⎦ Leucine	UCU ⎤ UCC ⎥ UCA ⎥ UCG ⎦ Serine	UAU ⎤ UAC ⎦ Tyrosine UAA ⎤ UAG ⎦ Stop	UGU ⎤ UGC ⎦ Cysteine UGA − Stop UGG − Tryptophan	U C A G
C	CUU ⎤ CUC ⎥ CUA ⎥ CUG ⎦ Leucine	CCU ⎤ CCC ⎥ CCA ⎥ CCG ⎦ Proline	CAU ⎤ CAC ⎦ Histidine CAA ⎤ CAG ⎦ Glutamine	CGU ⎤ CGC ⎥ CGA ⎥ CGG ⎦ Arginine	U C A G
A	AUU ⎤ AUC ⎥ Isoleucine AUA ⎦ AUG − Start	ACU ⎤ ACC ⎥ ACA ⎥ ACG ⎦ Threonine	AAU ⎤ AAC ⎦ Asparagine AAA ⎤ AAG ⎦ Lysine	AGU ⎤ AGC ⎦ Serine AGA ⎤ AGG ⎦ Arginine	U C A G
G	GUU ⎤ GUC ⎥ GUA ⎥ GUG ⎦ Valine	GCU ⎤ GCC ⎥ GCA ⎥ GCG ⎦ Alanine	GAU ⎤ GAC ⎦ Aspartic Acid GAA ⎤ GAG ⎦ Glutamic Acid	GGU ⎤ GGC ⎥ GGA ⎥ GGG ⎦ Glycine	U C A G

RNA's Roles in Translation

Translation takes place in the cytoplasm. Here transfer RNA molecules and ribosomes help in the synthesis of proteins. **Transfer RNA** (tRNA) molecules are single strands of RNA that temporarily carry a specific amino acid on one end. Each tRNA is folded into a compact shape and has an anticodon *(an tee KOH dahn)*. An **anticodon** is a three-nucleotide sequence on a tRNA that is complementary to an mRNA codon. As shown in **Figure 5,** the amino acid that a tRNA molecule carries corresponds to a particular mRNA codon.

Ribosomes, shown in Figure 5, are composed of both proteins and ribosomal RNA (rRNA). **Ribosomal RNA** molecules are RNA molecules that are part of the structure of ribosomes. A cell's cytoplasm contains thousands of ribosomes. Each ribosome temporarily holds one mRNA and two tRNA molecules. Figure 5 summarizes the process of translation:

Step ❶ Translation begins when the mRNA leaves the nucleus and enters the cytoplasm. The mRNA, the two ribosomal subunits, and a tRNA carrying the amino acid methionine *(muh THIE uh neen)* together form a functional ribosome. The mRNA "start" codon AUG, which signals the beginning of a protein chain, is oriented in a region of the ribosome called the P site, where the tRNA molecule carrying methionine can bind to the start codon.

Figure 5

BIOgraphic

Translation: Assembling Proteins

Amino acids are assembled from information encoded in mRNA.

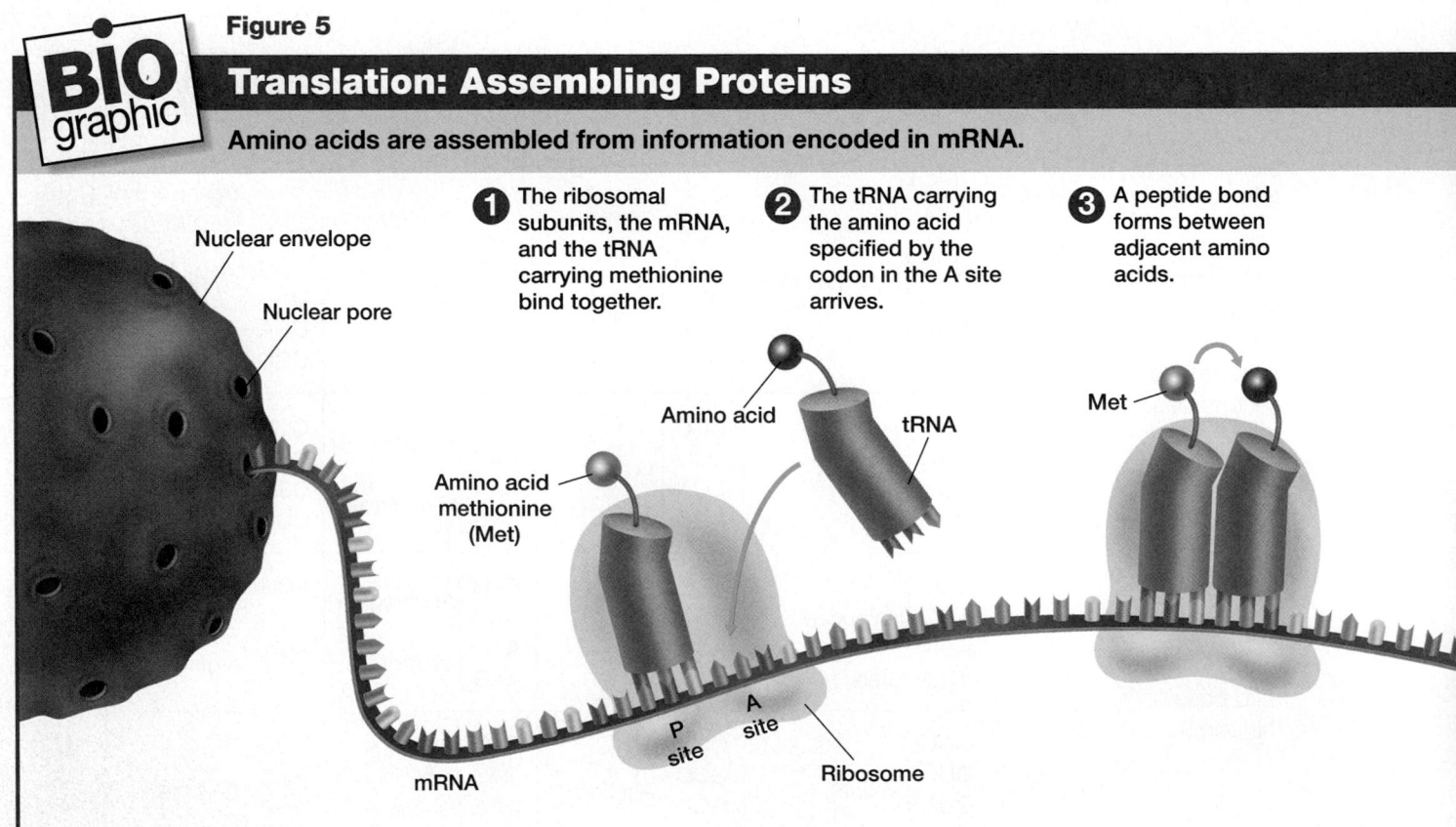

Nuclear envelope

Nuclear pore

❶ The ribosomal subunits, the mRNA, and the tRNA carrying methionine bind together.

❷ The tRNA carrying the amino acid specified by the codon in the A site arrives.

❸ A peptide bond forms between adjacent amino acids.

Amino acid

tRNA

Met

Amino acid methionine (Met)

P site

A site

mRNA

Ribosome

Step ② The codon in the area of the ribosome called the A site is ready to receive the next tRNA. A tRNA molecule with the complementary anticodon arrives and binds to the codon. The tRNA is carrying its specific amino acid.

Step ③ Now both the A site and the P site are holding tRNA molecules, each carrying a specific amino acid. Enzymes then help form a peptide bond between the adjacent amino acids.

Step ④ Afterward, the tRNA in the P site detaches, leaves behind its amino acid, and moves away from the ribosome.

Step ⑤ The tRNA (with its protein chain) in the A site moves over to fill the empty P site. Because the anticodon remains attached to the codon, the tRNA molecule and mRNA molecule move as a unit. As a result, a new codon is present in the A site, ready to receive the next tRNA and its amino acid. An amino acid is carried to the A site by a tRNA and then bonded to the growing protein chain.

Step ⑥ The tRNA in the P site detaches and leaves its amino acid.

Step ⑦ Steps 2 through 6 are repeated until a stop codon is reached. A stop codon is one of three codons (UAG, UAA, or UGA) for which there is no tRNA molecule with a complementary anticodon. Because there is no tRNA to fit into the empty A site in the ribosome, protein synthesis stops. The newly made protein is released into the cell.

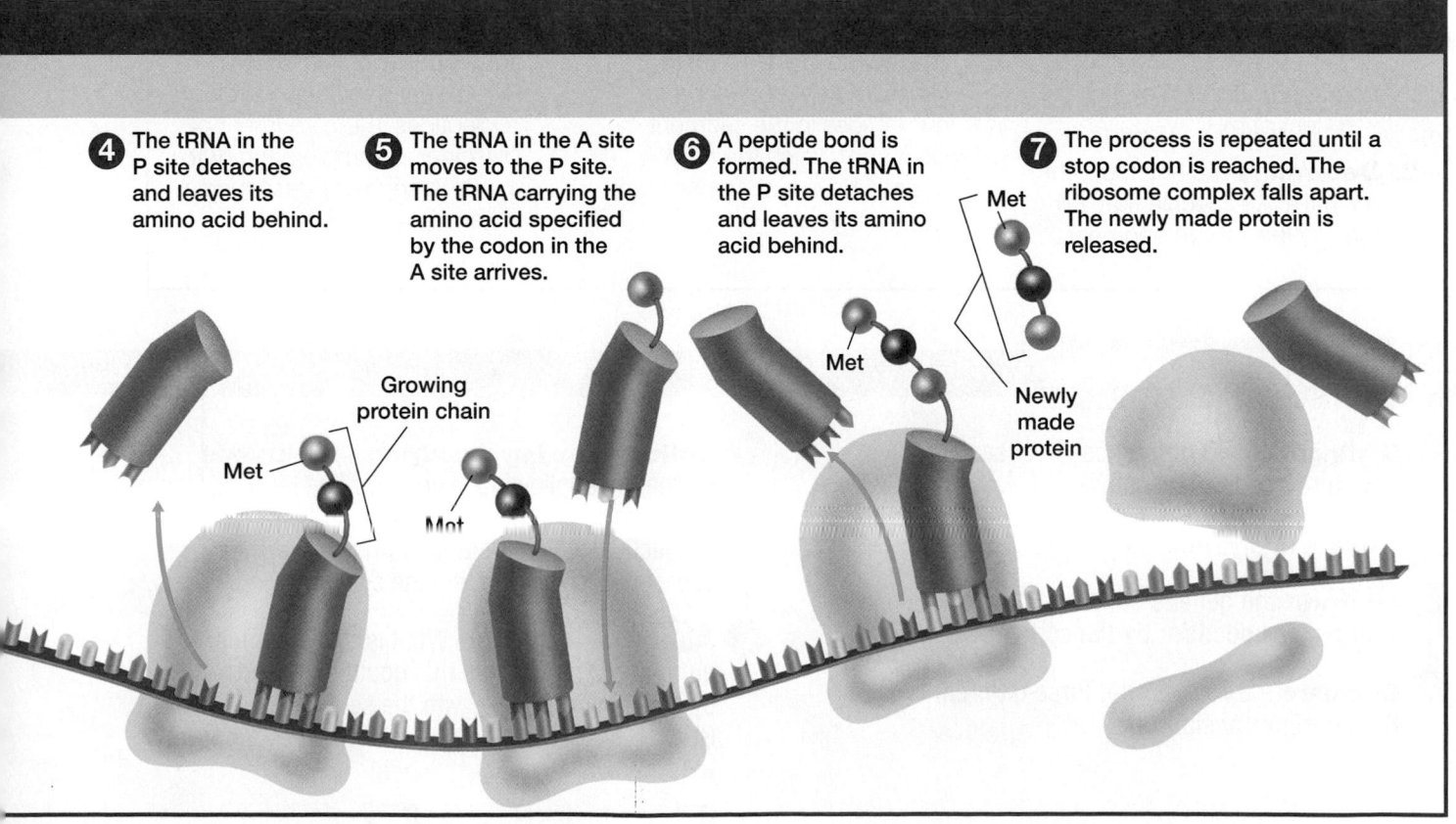

④ The tRNA in the P site detaches and leaves its amino acid behind.

⑤ The tRNA in the A site moves to the P site. The tRNA carrying the amino acid specified by the codon in the A site arrives.

⑥ A peptide bond is formed. The tRNA in the P site detaches and leaves its amino acid behind.

⑦ The process is repeated until a stop codon is reached. The ribosome complex falls apart. The newly made protein is released.

Growing protein chain

Met

Met

Met

Met

Newly made protein

As the mRNA moves across the ribosome, another ribosome can find the AUG codon on the same mRNA and begin making a second copy of the same protein. In this way many copies of the same protein are made from a single mRNA molecule.

With few exceptions, the genetic code is the same in all organisms. For example, the codon GUC codes for the amino acid valine in bacteria, in eagles, in plants, and in your own cells. For this reason, the genetic code is often described as being nearly universal. It appears that all life-forms have a common evolutionary ancestor with a single genetic code. Some exceptions include the ways cell organelles that contain DNA (such as mitochondria and chloroplasts) and a few microscopic protists read "stop" codons.

Decoding the Genetic Code

Background

Keratin is one of the proteins in hair. The gene for keratin is transcribed and translated by certain skin cells. The series of letters below represents the sequence of nucleotides in a portion of an mRNA molecule transcribed from the gene for keratin. This mRNA strand and the genetic code in Figure 4 can be used to determine some of the amino acids in keratin.

U C U C G U G A A U U U U C C

Analysis

1. **Determine** the sequence of amino acids that will result from the translation of the segment of mRNA above.

2. **Determine** the anticodon of each tRNA molecule that will bind to this mRNA segment.

3. **Critical Thinking Recognizing Patterns** Determine the sequence of nucleotides in the segment of DNA from which the mRNA strand above was transcribed.

4. **Critical Thinking Recognizing Patterns** Determine the sequence of nucleotides in the segment of DNA that is complementary to the DNA segment described in item 3.

Section 1 Review

1. **Distinguish** two differences between RNA structure and DNA structure.

2. **Explain** how RNA is made during transcription.

3. **Interpret** the genetic code to determine the amino acid coded for by the codon CCU.

4. **Compare** the roles of the three different types of RNA during translation.

5. **Critical Thinking Justifying Conclusions** Evaluate the following statement: The term *transcription* is appropriate for describing the production of RNA, and the term *translation* is appropriate for describing the synthesis of proteins.

6. (**Standardized Test Prep**) What is the maximum number of amino acids that could be coded for by a section of mRNA with the sequence GUUCAGAACUGU?

A 3   C 6

B 4   D 12

Gene Regulation and Structure

Protein Synthesis in Prokaryotes

Although prokaryotic organisms, such as bacteria, might seem simple because of their small size, prokaryotic cells typically have about 2,000 genes. The human genome, which is the largest genome sequenced to date, has about 30,000 genes. Not all of the genes, however, are transcribed and translated all of the time; this would waste the cell's energy and materials. Both prokaryotic and eukaryotic cells are able to regulate which genes are expressed and which are not, depending on the cell's needs.

An example of gene regulation that is well understood in prokaryotes is found in the bacterium *Escherichia coli*. When you eat or drink a dairy product, the disaccharide lactose ("milk sugar") reaches the intestinal tract and becomes available to the *E. coli* living there. The bacteria can absorb the lactose and break it down for energy or for making other compounds. In *E. coli*, recognizing, consuming, and breaking down lactose into its two components, glucose and galactose, requires three different enzymes, each of which is coded for by a different gene.

As shown in **Figure 6,** the three lactose-metabolizing genes are located next to each other and are controlled by the same promoter site. There is an on-off switch that "turns on" (transcribes and then translates) the three genes when lactose is available and "turns off" the genes when lactose is not available.

Objectives

- **Describe** how the *lac* operon is turned on or off.
- **Summarize** the role of transcription factors in regulating eukaryotic gene expression.
- **Describe** how eukaryotic genes are organized.
- **Evaluate** three ways that point mutations can alter genetic material.

Key Terms

operator
operon
lac operon
repressor
intron
exon
point mutation

Figure 6 Turning prokaryotic genes on and off

The *lac* operon allows a bacterium to build the proteins needed for lactose metabolism only when lactose is present.

Lactose absent—the *lac* operon is *off*.

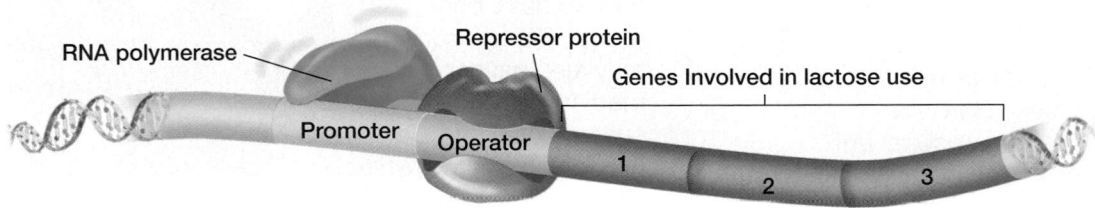

Lactose present—the *lac* operon is *on*.

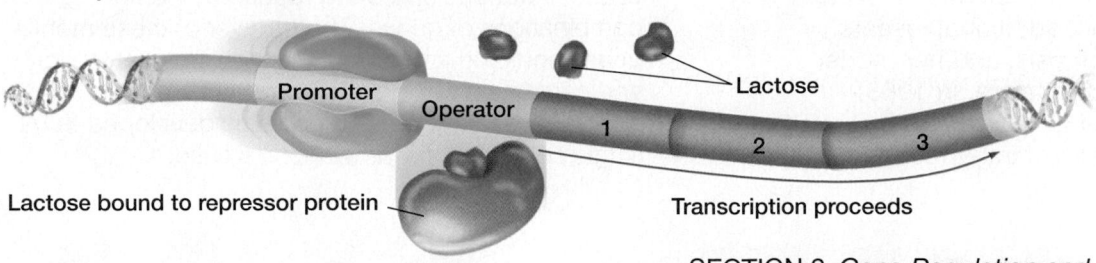

The piece of DNA that overlaps the promoter site and serves as the on-off switch is called an **operator.** Because of its position, the operator is able to control RNA polymerase's access to the three lactose-metabolizing genes.

In bacteria, a group of genes that code for enzymes involved in the same function, their promoter site, and the operator that controls them all function together as an **operon** *(AHP uhr ahn).* The operon that controls the metabolism of lactose is called the **lac operon** and is shown in Figure 6.

What determines whether the *lac* operon is in the "on" or "off" mode? When there is no lactose in the bacterial cell, a repressor turns the operon off. A **repressor** is a protein that binds to an operator and physically blocks RNA polymerase from binding to a promoter site. The blocking of RNA polymerase consequently stops the transcription of the genes in the operon, as shown in Figure 6.

When lactose is present, the lactose binds to the repressor and changes the shape of the repressor. The change in shape causes the repressor to fall off of the operator, as shown in Figure 6. Now the bacterial cell can begin transcribing the genes that code for the lactose-metabolizing enzymes. By producing the enzymes only when the nutrient is available, the bacterium saves energy.

Exploring Further

Jumping Genes

The spotted and streaked patterns seen in Indian corn result from genes that have moved from one chromosomal location to another. Such genes are called transposons *(trans POH zahns).* When a transposon jumps to a new location, it often inactivates a gene or causes mutations. In Indian corn, some pigment genes are not expressed in some cells because they have been disrupted by jumping genes.

The Discovery of Transposons
In the 1950s, the geneticist Barbara McClintock discovered transposons while studying corn. Most scientists rejected her ideas for more than 20 years. The idea that genes could change locations on the chromosome contradicted the prevailing view that genes and chromosomes are stable parts of the cell. Over time, additional research supported her hypothesis, and her model gradually gained acceptance. In 1983, McClintock received a Nobel Prize for her discoveries involving transposons.

Importance of Transposons
All organisms, including humans, appear to have transposons. Transposons probably play a role in spreading genes for antibiotic resistance among bacteria. Transposons that affect flower color in morning glory flowers have been found. Transposons may also have medical applications, such as helping scientists discover how white blood cells make antibodies and what causes cancer.

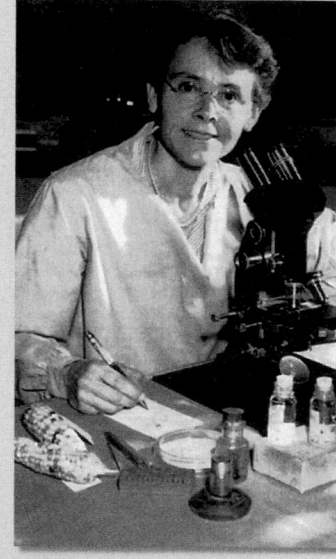
Barbara McClintock

Although the movement of transposons is very rare, transposons are important because they can cause mutations and bring together different combinations of genes. The transfer of these mobile genes could be a powerful mechanism in evolution and could help solve certain mysteries about evolution, such as how larger organisms developed from single cells and how new species arise.

Protein Synthesis in Eukaryotes

Eukaryotic cells contain much more DNA than prokaryotic cells do. Like prokaryotic cells, eukaryotic cells must continually turn certain genes on and off in response to signals from their environment. Operons have not been found often in eukaryotic cells. Instead, genes with related functions are often scattered on different chromosomes.

Because a nuclear envelope physically separates transcription from translation in a eukaryotic cell, more opportunities exist for regulating gene expression. For example, gene regulation can occur before, during, and after transcription. Gene regulation can also occur after mRNA leaves the nucleus or after translation, when the protein is functional.

Controlling the Onset of Transcription

Most gene regulation in eukaryotes controls the onset of transcription—when RNA polymerase binds to a gene. Like prokaryotes, eukaryotic cells use regulatory proteins. But many more proteins are involved in eukaryotes, and the interactions are more complex. These regulatory proteins in eukaryotes are called transcription factors.

As shown in **Figure 7,** transcription factors help arrange RNA polymerases in the correct position on the promoter. A gene can be influenced by many different transcription factors.

An enhancer is a sequence of DNA that can be bound by a transcription factor. Enhancers typically are located thousands of nucleotide bases away from the promoter. A loop in the DNA may bring the enhancer and its attached transcription factor (called an activator) into contact with the transcription factors and RNA polymerase at the promoter. As shown in Figure 7, transcription factors bound to enhancers can activate transcription factors bound to promoters.

Study TIP

● **Organizing Information**
Make a table to organize information about the regulation of protein synthesis. Across the top write the headings *Prokaryotes* and *Eukaryotes.* Along the sides write *Protein(s) that regulate(s) the genes* and *Details of regulation.* Add information to the table as you read Section 2.

Figure 7   Controlling transcription in eukaryotes

Transcription factors bind to the enhancer and to the RNA polymerase. The binding activates transcription factors bound to the promoter.

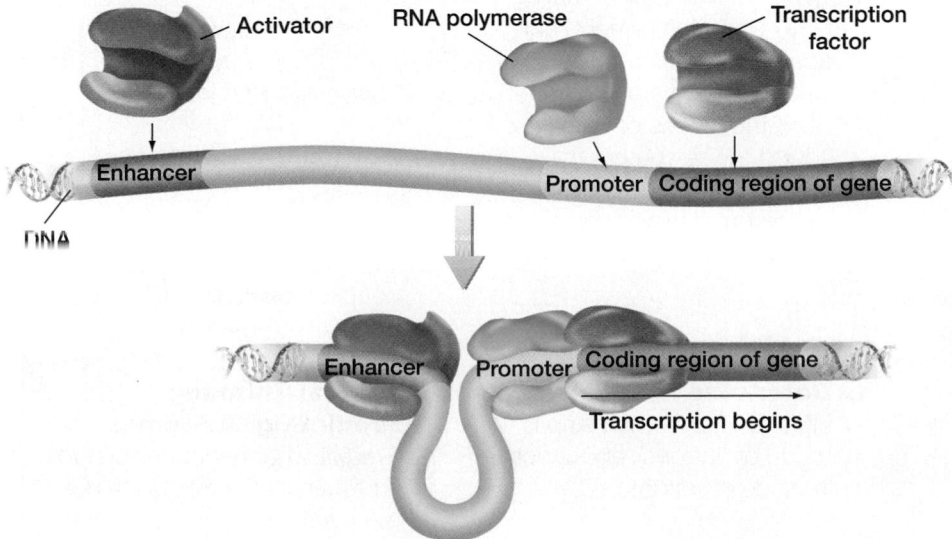

Intervening DNA in Eukaryotic Genes

WORD *Origins*

● The "int" in the word *intron* comes from the "int" in the word *intervening*. The "ex" in the word *exon* comes from the "ex" in the word *expressed*.

While it is tempting to think of a gene as an unbroken stretch of nucleotides that code for a protein, this simple arrangement is usually found only in prokaryotes. In eukaryotes, many genes are interrupted by **introns** *(IN trahnz)*—long segments of nucleotides that have no coding information. **Exons** *(EK sahnz)* are the portions of a gene that are translated (expressed) into proteins. After a eukaryotic gene is transcribed, the introns in the resulting mRNA are cut out by complex assemblies of RNA and protein called *spliceosomes*. The exons that remain are "stitched" back together by the spliceosome to form a smaller mRNA molecule that is then translated.

Many biologists think this organization of genes adds evolutionary flexibility. Each exon encodes a different part of a protein. By having introns and exons, cells can occasionally shuffle exons between genes and make new genes. The thousands of proteins that occur in human cells appear to have arisen as combinations of only a few thousand exons. Some genes in your cells exist in multiple copies, in clusters of as few as three or as many as several hundred. For example, your cells each contain 12 different hemoglobin genes, all of which arose as duplicates of one ancestral hemoglobin gene.

Modeling Introns and Exons

QUICK LAB

You can use masking tape to represent introns and exons.

Materials

masking tape, pens or pencils (two colors), metric ruler, scissors

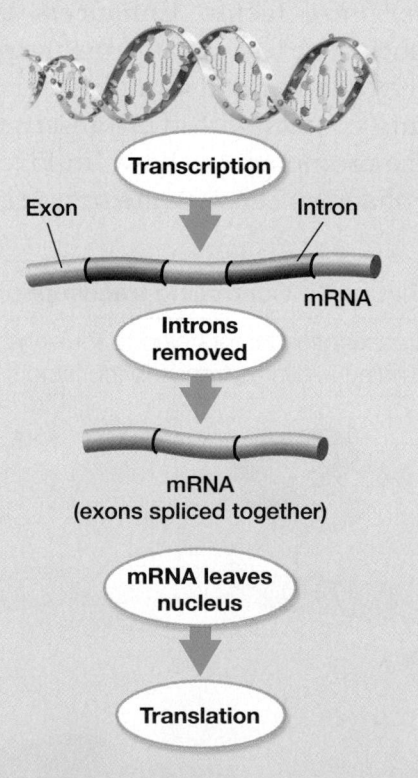

Transcription

Exon | Intron

mRNA

Introns removed

mRNA (exons spliced together)

mRNA leaves nucleus

Translation

Procedure

1. Place a 15–20 cm strip of masking tape on your desk. The tape represents a gene.

2. Use two colors to write the words *appropriately joined* on the tape exactly as shown in the diagram below. Space the letters so that they take up the entire length of the strip of tape. The segments in one color represent introns; those in the other color represent exons.

appropriately joined

3. ◆ Lift the tape. Working from left to right, cut apart the groups of letters written in the same color. Stick the pieces of tape to your desk as you cut them, making two strips according to color and joining the pieces in their original order.

Analysis

1. Determine from the resulting two strips which strip is made of "introns" and which is made of "exons."

2. Critical Thinking Predicting Outcomes Predict what might happen to a protein if an intron were not removed.

Mutations

Although changes in an organism's hereditary information are relatively rare, they can occur. As you learned in Chapter 6, a change in the DNA of a gene is called a mutation. Mutations in gametes can be passed on to offspring of the affected individual, but mutations in body cells affect only the individual in which they occur.

Mutations that move an entire gene to a new location are called *gene rearrangements*. Changes in a gene's position often disrupt the gene's function because the gene is exposed to new regulatory controls in its new location—like what would happen if you moved to France and couldn't speak French. Two types of gene rearrangements are shown in **Figure 8.** Genes sometimes move as part of a transposon. That is, the genes are carried by the moving transposon like fleas on a dog. Other times, the portion of the chromosome containing a gene may be rearranged during meiosis.

Mutations that change a gene are called *gene alterations*. Gene alterations such as those shown in Figure 8 usually result in the placement of the wrong amino acid during protein assembly. This error will usually disrupt a protein's function. In a **point mutation,** a single nucleotide changes. In an *insertion* mutation, a sizable length of DNA is inserted into a gene. Insertions often result when mobile segments of DNA, called transposons, move randomly from one position to another on chromosomes. Transposons make up 45 percent of the human genome. In a *deletion* mutation, segments of a gene are lost, often during meiosis.

Figure 8   Major types of mutations

The substitution, addition, or removal of one or more nucleotides is called a gene alteration. If the mutation changes the original position of a gene of the chromosome, the gene may not function normally.

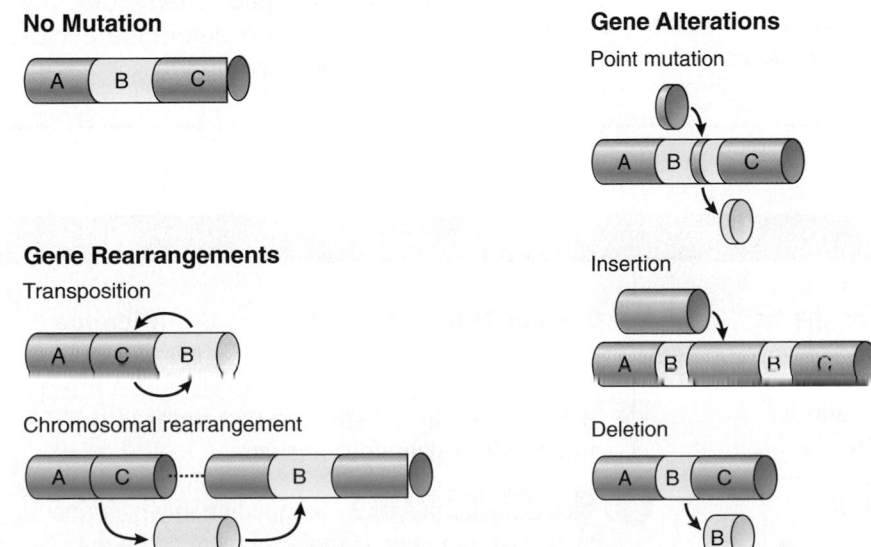

No Mutation

Gene Alterations
Point mutation

Gene Rearrangements
Transposition

Chromosomal rearrangement

Insertion

Deletion

Because the genetic message is read as a series of triplet nucleotides, insertions and deletions of one or two nucleotides can upset the triplet groupings. Imagine deleting the letter C from the sentence "THE CAT ATE." Keeping the triplet groupings, the message would read "THE ATA TE," which is meaningless. A mutation that causes a gene to be read in the wrong three-nucleotide sequence is called a *frameshift mutation*.

BIOWatch

Gene Sequencing

Many genetic disorders, such as sickle cell anemia, are caused by single nucleotide mutations. Today, certain genetic disorders can be detected by comparing the sequence of nucleotides in the genes involved to the sequence in corresponding healthy genes.

Special Nucleotides

One technique to find the sequence of nucleotides in a gene uses nucleotides that each have a different colored fluorescent dye "tagged" on. The tagged nucleotides are added to a test tube containing single strands of the gene of interest, "untagged" nucleotides, enzymes needed to make DNA, and small single-stranded pieces of DNA called primers.

The primers base-pair with the single strands of the gene of interest. The tagged and untagged nucleotides compete to make the primer longer by matching the nucleotides on the gene of interest, according to the base-pairing rules. The tagged nucleotides are altered so that once a tagged nucleotide is added on to the primer, the synthesis reaction stops on that primer strand.

Base-Pairing Rules Help

The researcher separates the different-sized strands using a method called gel electrophoresis. The different fluorescent dyes help the researcher determine the sequence of the nucleotides on the gene.

Sequence Information is Important

Today, the sequence of nucleotides in genes from many different

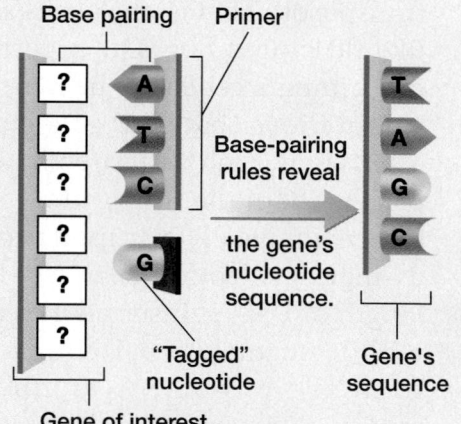

organisms are kept on databases on the Internet. Scientists use this information to look for similarities to the gene they are studying. Comparing the sequences can help them find genes with similar functions, help them classify organisms, and determine evolutionary relationships.

Section 2 Review

1 **Describe** the effect a repressor has on the *lac* operon when lactose is present.

2 **Explain** the role of transcription factors and enhancers in eukaryotic gene expression.

3 **Differentiate** between exons and introns.

4 **Critical Thinking Evaluating Significance** Which type of mutation would have a greater effect on the sequence of amino acids in a protein, a base-pair substitution or a frameshift mutation? Explain your answer.

5 **Standardized Test Prep** A mutation in which one nucleotide in a gene is replaced with a different nucleotide is called

A a deletion.　**C** a substitution.

B an insertion.　**D** a frameshift mutation.

Study ZONE   CHAPTER HIGHLIGHTS

Key Concepts

1 From Genes to Proteins

- The instructions needed to make proteins are coded in the nucleotides that make up a gene. The instructions are transferred to an mRNA molecule during transcription. The RNA is complementary to the gene, and the RNA nucleotides are put together with the help of RNA polymerase.

- During translation, the mRNA molecule binds to a ribosome, and tRNAs carry amino acids to the ribosome according to the codons on the mRNA. Each codon specifies an amino acid. The amino acids are joined to form a protein.

- The genetic code (codons) used by most organisms to translate mRNA is nearly universal.

2 Gene Regulation and Structure

- Prokaryotic and eukaryotic cells are able to control which genes are expressed and which are not, depending on the cell's needs.

- In prokaryotes, gene expression is regulated by operons. Gene expression is switched off when repressor proteins block RNA polymerase from transcribing a gene.

- In eukaryotes, an enhancer must be activated for a eukaryotic gene to be expressed. Transcription factors initiate transcription by binding to enhancers and to RNA polymerases.

- Many eukaryotic genes are interrupted by segments of DNA that do not code for proteins; these segments are called introns. The segments of DNA that are expressed are called exons. After transcription, the introns are cut out, and the exons are joined. The exons are then translated.

- Mutations are changes in DNA. Gene alterations are mutations that change a gene. These mutations can involve a change in a single nucleotide or an entire gene.

Key Terms

Section 1

ribonucleic acid (RNA) (208)
uracil (208)
transcription (208)
translation (208)
gene expression (208)
RNA polymerase (209)
messenger RNA (211)
codon (211)
genetic code (211)
transfer RNA (212)
anticodon (212)
ribosomal RNA (212)

Section 2

operator (216)
operon (216)
lac operon (216)
repressor (216)
intron (218)
exon (218)
point mutation (219)

Unit 6—*Gene Expression*
Use Topics 3–6 in this unit to review the key concepts and terms in this chapter.

Performance ZONE

Understanding Key Ideas

1. Anticodons are found on _____ molecules.
   a. mRNA
   b. DNA
   c. rRNA
   d. tRNA

2. Unlike DNA, RNA contains
   a. the sugar deoxyribose.
   b. the nitrogen base uracil.
   c. a phosphate group.
   d. nucleotides.

3. A short chain of DNA has the nucleotide sequence ATA CCG. Its complementary mRNA nucleotide sequence is
   a. TAT GCC.
   b. UAU GCC.
   c. TUT GCC.
   d. UAU GGC.

4. The *lac* operon allows a bacterium to build the proteins needed for lactose metabolism when
   a. RNA polymerase is not bound to the promoter.
   b. lactose is absent.
   c. lactose is present.
   d. the repressor is bound to the operator.

5. Transcription of lactose-metabolizing genes is blocked when the _____ is bound to the operator.
   a. repressor
   b. operon
   c. inducer
   d. enhancer

6. In eukaryotes, gene expression can be regulated by
   a. mutations.
   b. transcription factors.
   c. repressors.
   d. operons.

7. **Exploring Further** Compare the way transposons and exons affect genes.

8. Does the drawing below represent a strand of RNA or a strand of DNA? Explain your answer.

U C A U C G U C G A A C U C

9. **BIOWatch** A researcher trying to determine the sequence of nucleotides on a particular gene obtained the following sequence with the primer and tagged nucleotide: TCCGGAAG. What was the sequence of nucleotides on the gene?

10. **Concept Mapping** Make a concept map that shows the role of RNA in gene expression. Try to include the following words in your map: *transcription, translation, mRNA, tRNA, rRNA, gene, promoter, codons, anticodons, proteins, amino acids, ribosome,* and *cytoplasm.*

Critical Thinking

11. **Evaluating Results** A molecular biologist isolates mRNA from the brain and from the liver of a mouse and finds that the mRNA molecules are different from each other. Can these results be correct or has the biologist made an error? Explain your answer.

12. **Evaluating an Argument** A classmate states that damage to exons is very likely to affect the synthesis of a protein, while damage to introns is not. Evaluate that statement.

13. **Evaluating Significance** Compare and contrast chromosomal mutations with point mutations, and evaluate the significance of each.

Alternative Assessment

14. **Interactive Tutor Unit 6 Gene Expression** Write a report summarizing how antibiotics inhibit protein synthesis in bacteria. How do some antibiotics interfere with translation?

Standardized Test Prep

Understanding Concepts

Directions (1–5): For *each* question, write on a separate sheet of paper the letter of the correct answer.

1 Which of the following shows the correct order of events in producing a protein from a DNA sequence?
A. exon splicing, transcription, translation
B. exon splicing, translation, transcription
C. transcription, exon splicing, translation
D. translation, transcription, exon splicing

2 What process involves making proteins from the information carried by mRNA?
F. DNA replication
G. gene regulation
H. transcription
I. translation

3 What is a change in the genetic material of an organism called?
A. codon   C. operator
B. mutation   D. operon

4 What term describes mutations that change one nucleotide in a gene?
F. codon mutation
G. operon mutation
H. point mutation
I. repressor protein

5 What process involves making RNA based on the sequence of nucleotides in DNA?
A. DNA replication
B. gene regulation
C. transcription
D. translation

Directions (6): For the following question, write a short response.

6 How does gene replication of the *lac* operon promote homeostasis in intestinal *E. coli* bacteria?

Test TIP

Slow, deep breathing may help you relax. If you suffer from test anxiety, focus on your breathing in order to calm down.

Reading Skills

Directions (7): Read the passage below. Then answer the question.

Many antibiotics fight bacterial infections by interfering with bacterial protein synthesis. Some antibiotics combine with ribosomal proteins. Erythromycin and chloramphenicol combine with the 50S ribosomal subunit. The tetracyclines, streptomycin, gentamicin, kanamycin, and the nitrofurans combine with the 30S ribosomal subunit. Mupirocin and puromycin inhibit protein synthesis at the tRNA level.

7 How do the antibiotics mupirocin and erythromycin differ in the types of biological molecules they act on?
F. Mupirocin acts on nucleic acids while erythromycin acts on proteins.
G. Mupirocin inhibits protein synthesis while erythromycin inhibits DNA replication.
H. Mupirocin acts on the 30S subunit while erythromycin acts on the 50S subunit.
I. Mupirocin inhibits bacterial protein synthesis while erythromycin inhibits human protein synthesis.

Interpreting Graphics

Directions (8): Base your answer to question 8 on the diagram below.

Translation Model

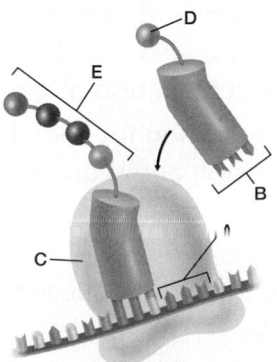

8 Which part of the model represents a codon?
A. *A*   C. *C*
B. *B*   D. *E*

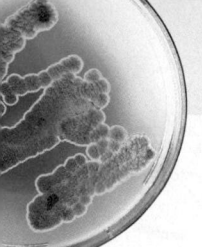

Exploration Lab

Modeling Protein Synthesis

SKILLS
- Modeling
- Using scientific methods

OBJECTIVES
- **Compare** and **Contrast** the structure and function of DNA and RNA.
- **Model** protein synthesis.
- **Demonstrate** how a mutation can affect a protein.

MATERIALS
- masking tape
- plastic soda-straw pieces of one color
- plastic soda-straw pieces of a different color
- paper clips
- pushpins of five different colors
- marking pens of the same colors as the pushpins
- 3 × 5 in. note cards
- oval-shaped card
- transparent tape

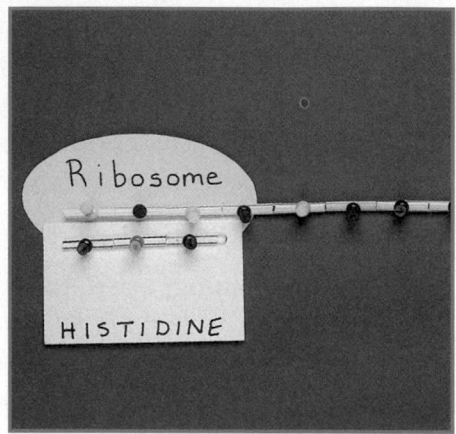

Before You Begin

The nature of a **protein** is determined by the sequence of amino acids in its structure. During **protein synthesis,** the sequence of nitrogen bases in an **mRNA** molecule is used to assemble **amino acids** into a protein chain.

A **mutation** is a change in the nitrogen-base sequence of DNA. Many mutations lead to altered or defective proteins. For example, the genetic blood disorder **sickle cell anemia** is caused by a mutation in the gene for **hemoglobin.**

In this lab, you will build models that will help you understand how protein synthesis occurs. You can also use the models to explore how a mutation affects a protein.

1. Write a definition for each boldface term in the paragraph above and for each of the following terms: transcription, translation, tRNA, ribosome, codon, anticodon.

2. Describe three differences between DNA and RNA.

3. Based on the objectives for this lab, write a question you would like to explore about protein synthesis.

Procedure

PART A: Design a Model

1. Work with the members of your lab group to design models of DNA, RNA, and a cell. Use the materials listed for this lab.

> ### You Choose
> **As you design your models, decide the following:**
> a. what question you will explore
> b. how to represent DNA nucleotides
> c. how to represent RNA nucleotides
> d. how to represent five different nitrogen bases
> e. how to link (bond) nucleotides together
> f. how to represent tRNA molecules with amino acids
> g. how to represent the locations of DNA and ribosomes

2. Write out the plan for building your models. Have your teacher approve the plan before you begin building the models.

3. Build the models your group designed. **CAUTION: Sharp or pointed objects may cause injury. Handle pushpins carefully.** Start your model of DNA with a strand of nucleotides that has the following sequence of nitrogen bases: TTTGGTCTCCTC.

PART B: Model Protein Synthesis

4. Use your models and Figure 10-5 on pp. 212–213 to demonstrate how transcription and translation occur. Draw and label the steps of each process.

5. Use your models to explore one of the questions written for step 3 of **Before You Begin.**

PART C: Test Hypothesis

Answer each of the following questions by writing a hypothesis. Use your models to test each hypothesis, and describe your results.

6. The DNA model you built for step 3 represents a portion of a gene for hemoglobin. Sickle cell anemia results from the substitution of an A for the T in the third codon of the nitrogen-base sequence given in step 3. How will this substitution affect a hemoglobin molecule?

7. The addition of a nucleotide to a strand of DNA is a type of mutation called an *insertion*. What happens when an insertion occurs in the first codon in a DNA strand, before the DNA strand is transcribed?

PART D: Cleanup and Disposal

8. Dispose of damaged pushpins in the designated waste container.

9. Clean up your work area and all lab equipment. Return lab equipment to its proper place. Wash your hands thoroughly before you leave the lab and after you finish all work.

Analyze and Conclude

1. Comparing Structures How did the nitrogen-base sequence of the mRNA you made compare with that of the DNA it was transcribed from?

2. Recognizing Relationships How is the nitrogen-base sequence of a gene related to the structure of a protein?

3. Recognizing Patterns What is the relationship between the anticodon of a tRNA and the amino acid the tRNA carries?

4. Drawing Conclusions How does a mutation in the gene for a protein affect the protein?

5. Further Inquiry Write a new question about protein synthesis that could be explored with your model.

? Do You Know?

Do research in the library or media center to answer these questions:

1. What are mutagens, and how do they affect DNA?

2. What are two other genetic disorders that result from mutations?

Use the following Internet resources to explore your own questions about DNA.

internet connect

www.scilinks.org
Topic: Genetic Disorders
Keyword: HX4091

SCI*LINKS*. Maintained by the National Science Teachers Association

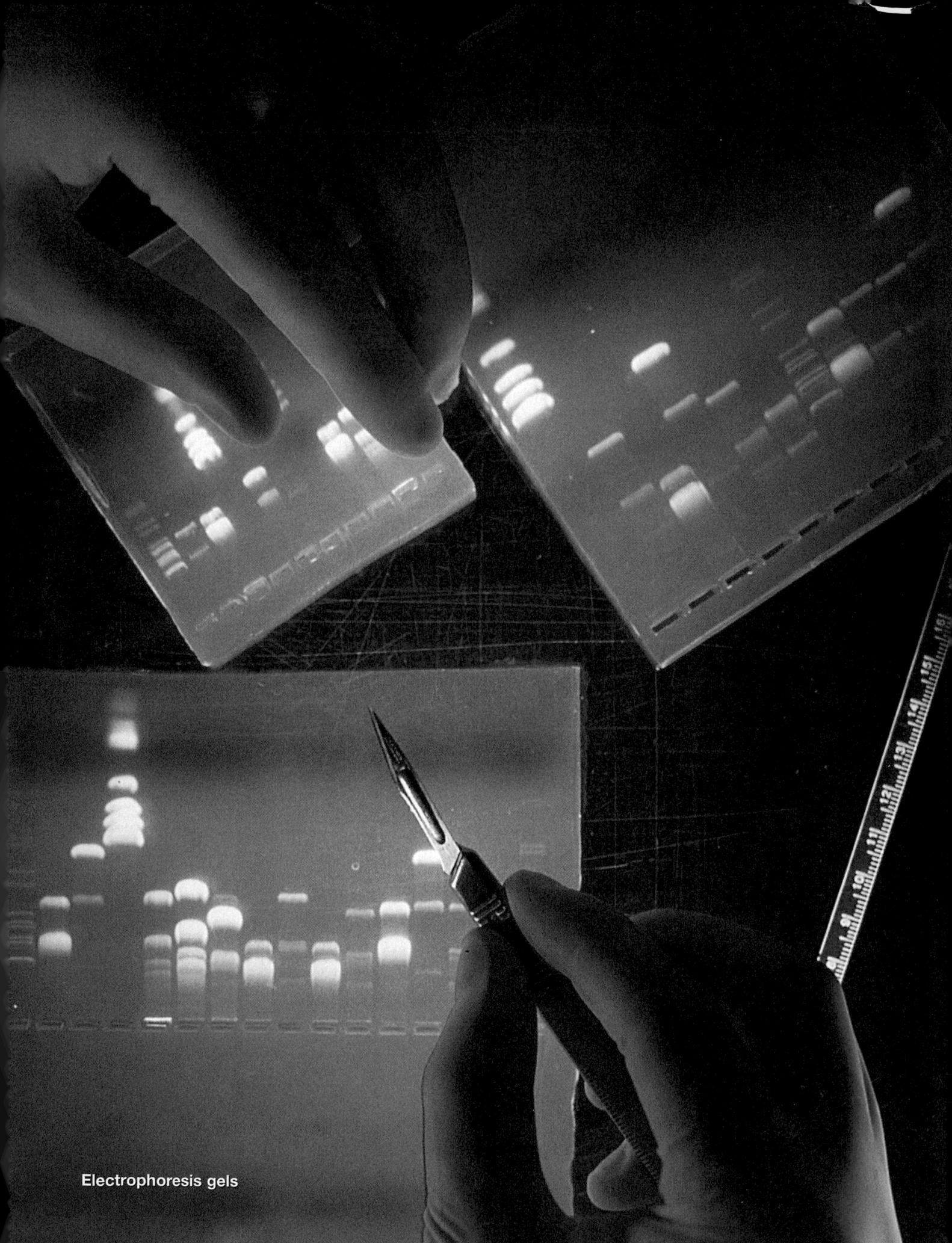

Electrophoresis gels

CHAPTER
11 Gene Technology

✔ Quick Review

Answer the following without referring to earlier sections of your book.

1. **Define** the term *gene*. *(Chapter 6, Section 1)*
2. **Describe** the structure of DNA. *(Chapter 9, Section 2)*
3. **State** the base-pairing rules that determine the structure of DNA. *(Chapter 9, Section 1)*
4. **Explain** why the genetic code is said to be universal. *(Chapter 10, Section 1)*

Did you have difficulty? *For help, review the sections indicated.*

📖 Reading Activity

Before you read this chapter, write a short list of all the things you know about gene technology. Then, write a list of the things that you want to know about gene technology. Save your list, and to assess what you have learned, see how many of your own questions you can answer after reading this chapter.

internet connect

www.scilinks.org
National Science Teachers Association *sci*LINKS Internet resources are located throughout this chapter.

*SCI*LINKS. Maintained by the National Science Teachers Association

● Electrophoresis is a technique used in a laboratory that results in the separation of charged particles. DNA is a negatively charged molecule, and is moved by electric current through an electrophoresis gel.

Genetic Engineering

Objectives

- **Describe** four basic steps commonly used in genetic engineering experiments.

- **Evaluate** how restriction enzymes and the antibiotic tetracycline are used in genetic engineering.

- **Relate** the role of electrophoresis and probes in identifying a specific gene.

Key Terms

genetic engineering
recombinant DNA
restriction enzyme
vector
plasmid
gene cloning
electrophoresis
probe

Basic Steps of Genetic Engineering

Not too long ago, using bacteria to produce human insulin and inserting genes into tomatoes and human cells were ideas that existed only in science fiction books and movies. But now, the techniques required to carry out these ideas have been developed and are used daily.

In 1973, Stanley Cohen and Herbert Boyer conducted an experiment that revolutionized genetic studies in biology. They isolated the gene that codes for ribosomal RNA from the DNA of an African clawed frog and then inserted it into the DNA of *Escherichia coli* bacteria, as summarized in **Figure 1.** During transcription, the bacteria produced frog rRNA, thereby becoming the first genetically altered organisms. The process of manipulating genes for practical purposes is called **genetic engineering.** Genetic engineering may involve building **recombinant DNA**—DNA made from two or more different organisms.

The basic steps in genetic engineering can be explored by examining how the human gene for insulin is transferred into bacteria. Insulin is a protein hormone that controls sugar metabolism. Diabetics cannot produce enough insulin, so they must take doses of insulin regularly. Before genetic engineering, insulin was extracted from the pancreases of slaughtered cows and pigs and then purified. Today, the human insulin gene is transferred to bacteria through genetic engineering. Because the genetic code is universal, bacteria can transcribe and translate a human insulin gene using the same code a human cell uses in order to produce human insulin.

Figure 1   Genetic alteration of an organism

Cohen and Boyer produced the first genetically engineered organisms.

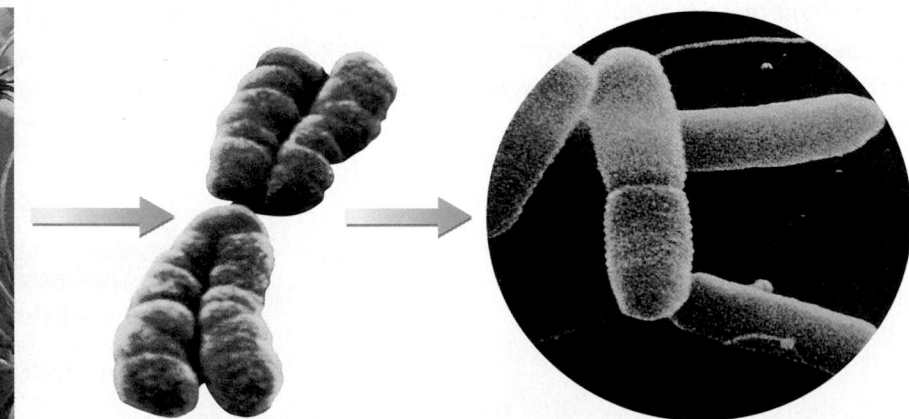

1. Cohen and Boyer used an African clawed frog as their experimental organism.

2. They isolated an rRNA gene from one of its chromosomes.

3. They inserted the gene into bacteria. The bacteria produced frog rRNA.

Steps in a Genetic Engineering Experiment

Genetic engineering experiments use different approaches, but most share four basic steps, as illustrated in **Figure 2.**

Step ❶ Cutting DNA The DNA from the organism containing the gene of interest (in our example, the insulin gene) is cut by restriction enzymes. **Restriction enzymes** are bacterial enzymes that recognize and bind to specific short sequences of DNA, and then cut the DNA between specific nucleotides within the sequences. The DNA from a vector also is cut. A **vector** is an agent that is used to carry the gene of interest into another cell. Commonly used vectors include viruses, yeast, and plasmids. **Plasmids,** shown in Figure 2, are circular DNA molecules that can replicate independently of the main chromosomes of bacteria.

Step ❷ Making recombinant DNA The DNA fragments from the organism containing the gene of interest are combined with the DNA fragments from the vector. An enzyme called DNA ligase is added to help bond the ends of DNA fragments together. In our example, human DNA fragments are combined with plasmid DNA fragments. The host cells then take up the recombinant DNA.

Step ❸ Cloning In a process called **gene cloning,** many copies of the gene of interest are made each time the host cell reproduces. Recall from your reading that bacteria reproduce by binary fission, producing identical offspring. When a bacterial cell replicates its DNA, its plasmid DNA also replicates.

Step ❹ Screening Cells that have received the particular gene of interest are distinguished, or separated, from the cells that did not take up the vector with the gene of interest. The cells can transcribe and translate the gene of interest to make the protein coded for in the gene.

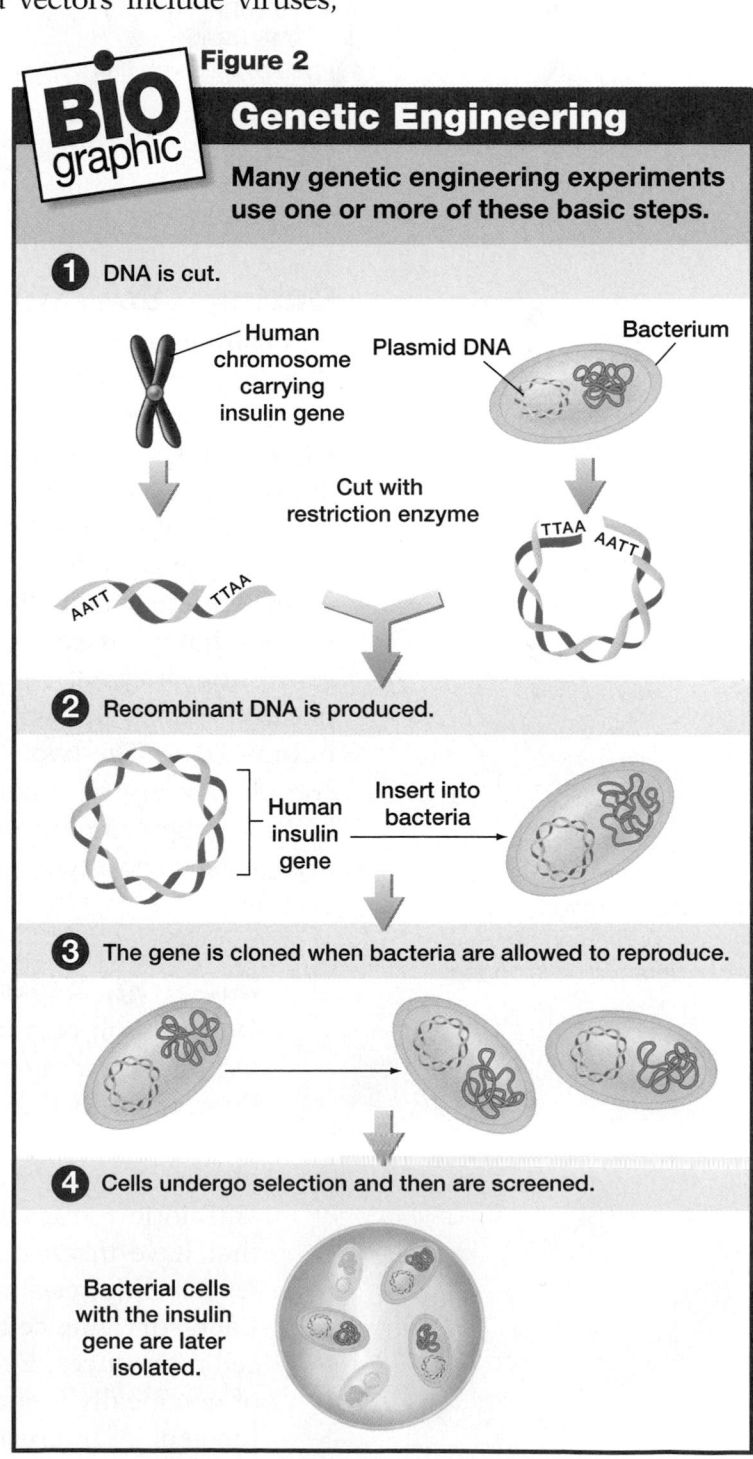

Figure 2

BIO graphic

Genetic Engineering

Many genetic engineering experiments use one or more of these basic steps.

❶ DNA is cut.

Human chromosome carrying insulin gene   Plasmid DNA   Bacterium

Cut with restriction enzyme

AATT   TTAA   TTAA   AATT

❷ Recombinant DNA is produced.

Human insulin gene   Insert into bacteria

❸ The gene is cloned when bacteria are allowed to reproduce.

❹ Cells undergo selection and then are screened.

Bacterial cells with the insulin gene are later isolated.

Figure 3   Restriction enzymes cut DNA

The restriction enzyme *Eco*RI recognizes the nucleotide sequence
GAATTC and makes its cut between the G and the A.

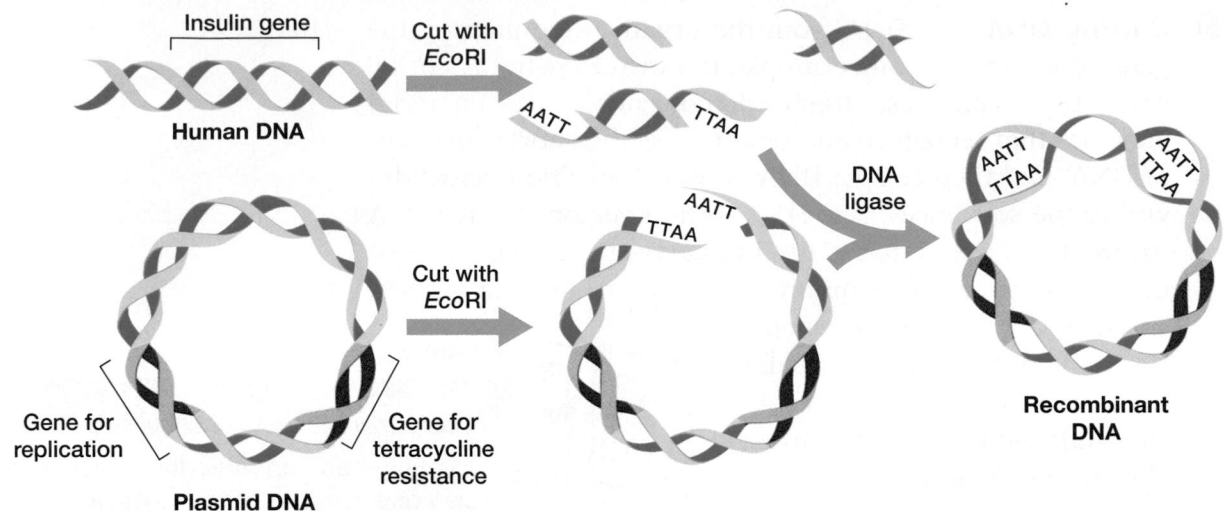

Cutting DNA and Making Recombinant DNA

An example of how restriction enzymes work is shown in **Figure 3.**
The enzyme recognizes a specific sequence of DNA. The sequence
the enzyme recognizes and the sequence on the complementary
DNA strand are palindromes—they read the same backward as they
do forward (such as the word *noon*).

The cuts of most restriction enzymes produce pieces of DNA with
short single strands on each end that are complementary to each
other. The ends are called *sticky ends*. As illustrated in Figure 3, the
vectors that are used contain only one nucleotide sequence that the
restriction enzyme recognizes. Thus, vectors such as the circular
plasmids "open up" with the same sticky ends as those of the cut
human DNA. The two DNA molecules bond together by means of
complementary base pairing at the sticky ends. The plasmid DNA
has both the gene for plasmid DNA replication and the gene that
makes the cell carrying the plasmid resistant to the antibiotic tetracycline.

Figure 4   Screening.
Only the cells that take up
the vectors are resistant
to tetracycline and survive
when tetracycline is added.

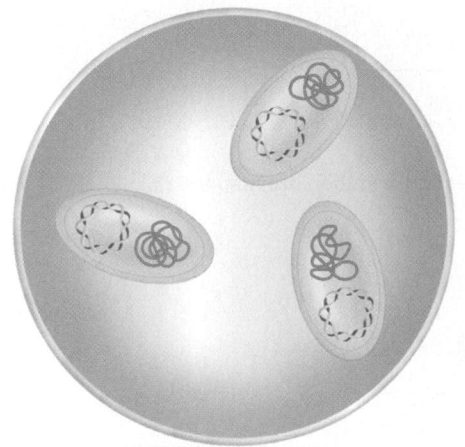

Cloning, Selecting, and Screening Cells

One difficult part in a genetic engineering experiment is finding and isolating the cells that contain the gene of interest.
First, the cells that have taken up the plasmid must be identified. The bacterial cells that have taken up the plasmid are
identified by growing the bacteria on plates that contain the
antibiotic tetracycline. As shown in **Figure 4,** only the cells
that have taken up the vectors (which contain the gene for
tetracycline resistance) survive when exposed to tetracycline.
Each surviving cell makes a copy of the vector every time the
cell reproduces. Eventually, each surviving cell forms a colony
of genetically identical cells, or clones. Some vectors contain
the gene of interest, and some do not.

Confirmation of a Cloned Gene

The surviving bacterial colonies are tested for the presence of the gene of interest. One method used to identify a specific gene is a technique called a Southern blot, as summarized in **Figure 5**.

Step 1 In a Southern blot, the DNA from each bacterial clone colony is isolated and cut into fragments by restriction enzymes.

Step 2 The DNA fragments are separated by gel **electrophoresis** *(ee LEK troh fuh REE sis)*, a technique that uses an electric field within a gel to separate molecules by their size. The gel is a rectangular slab of gelatin with a line of little rectangular wells near the top edge. The DNA sample is placed in the pits. Because DNA is negatively charged, it migrates toward the positive pole when the electric field is applied. The DNA fragments move through the gel, with the smallest DNA fragments moving fastest. A pattern of bands is formed. The gel is soaked in a chemical solution that separates the double strands in each DNA fragment into single-stranded DNA fragments.

Step 3 The DNA bands are then transferred (blotted) directly onto a piece of filter paper. The filter paper is moistened with a probe solution. **Probes** are radioactive- or fluorescent-labeled RNA or single-stranded DNA pieces that are complementary to the gene of interest.

Step 4 Only the DNA fragments complementary to the probe will bind with the probe and form visible bands.

WORD Origins

● The word *electrophoresis* is from the Latin *electrocus*, meaning "electricity," and the Greek *phoresis*, meaning "to carry." Knowing this makes it easier to remember that electrophoresis uses electricity to separate DNA fragments.

Figure 5

BIOgraphic

Southern Blot: Identifying a Gene of Interest

A DNA or RNA probe can be used to identify a cloned gene.

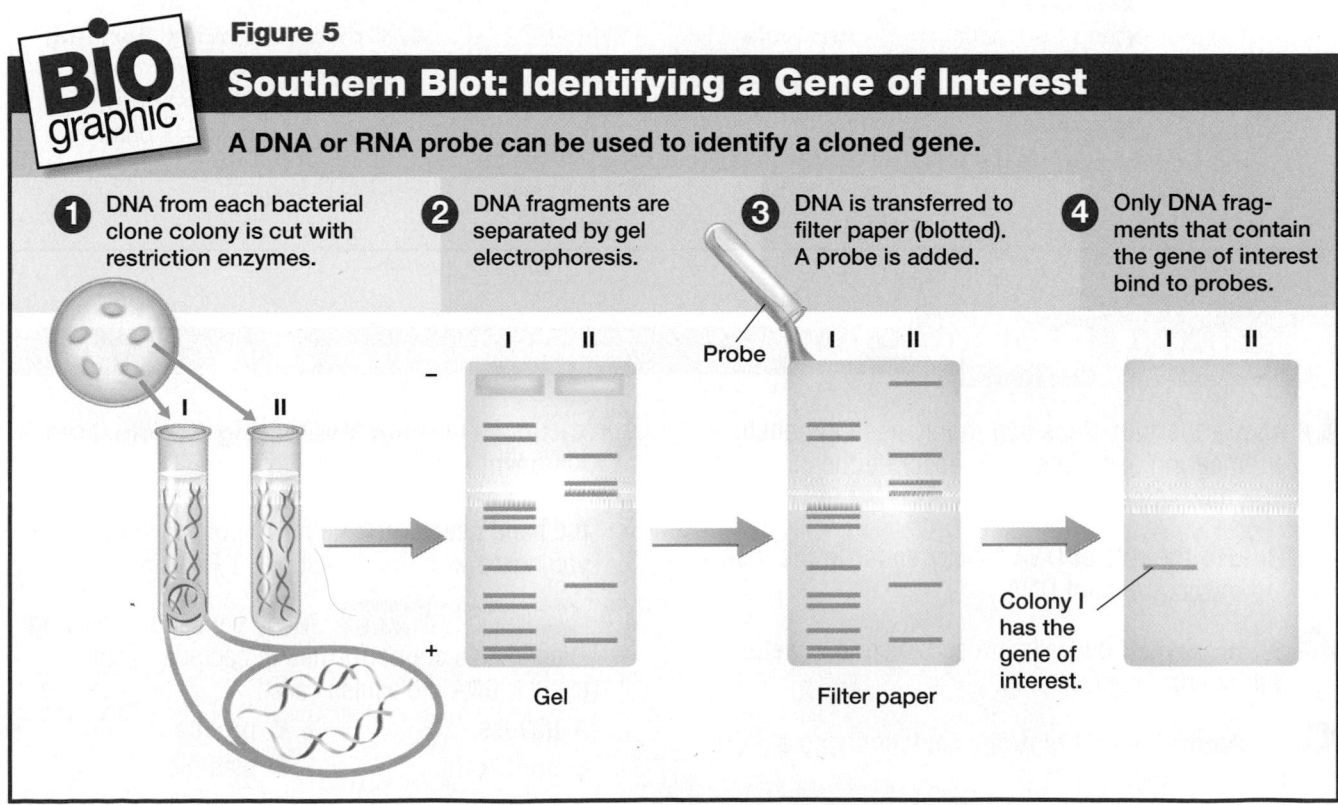

1 DNA from each bacterial clone colony is cut with restriction enzymes.

2 DNA fragments are separated by gel electrophoresis.

3 DNA is transferred to filter paper (blotted). A probe is added.

4 Only DNA fragments that contain the gene of interest bind to probes.

Probe

Gel

Filter paper

Colony I has the gene of interest.

Once the bacterial colonies containing the gene of interest are identified, the researcher can manipulate the genetically engineered bacteria in many different ways. For example, the gene of interest can be isolated so that the researcher has pure DNA to use in genetic studies. The researcher can then study how the gene is controlled. Pure DNA allows the researcher to determine the sequence of nucleotides that make up the gene. By comparing the nucleotide sequence of several different organisms, researchers can study the evolution of a particular gene.

The gene of interest can also be isolated and then transferred to other organisms. The bacterial colonies can be used to produce large quantities of the protein coded for by the gene so that the protein can be studied further or used to make drugs, such as insulin.

QUICK LAB

Modeling Gel Electrophoresis

You can use beads to model how DNA fragments are separated in a gel during electrophoresis.

Materials

500 mL beaker, large jar, 3 sets of beads—each set a different size and different color

Procedure

1. Fill a large jar with the largest beads. The filled jar represents a gel.

2. Mix the two smaller beads in the beaker and then pour them slowly on top of the "gel." The two smaller size beads represent DNA fragments of different sizes.

3. Observe the flow of the beads through the "gel." Lightly agitate the jar if the beads do not flow easily.

Analysis

1. **Identify** which beads flowed through the "gel" the fastest.

2. **Relate** the sizes of the beads to the sizes of DNA fragments.

3. **Determine** whether the top or the bottom of the jar represents the side of the gel with the positively charged pole.

4. **Critical Thinking Forming Conclusions** Why do the beads you identified in Analysis question 1 pass through the "gel" more quickly?

Section 1 Review

1. **Apply** the four steps commonly used in genetic engineering experiments to describe the cloning of a human gene.

2. **Relate** the role of DNA "sticky ends" in the making of recombinant DNA.

3. **Summarize** how cells are screened in genetic engineering experiments.

4. **Evaluate** the role of probes in identifying a specific gene.

5. **Critical Thinking Evaluating Conclusions** A student performing electrophoresis on a DNA sample believes that her smallest DNA fragment is the band nearest the negative pole of the gel. Do you agree with her conclusion? Explain.

6. **Standardized Test Prep** Many genetic engineering experiments are performed in bacteria using circular DNA molecules called

   A phages.   C probes.
   B promoters.   D plasmids.

Human Applications of Genetic Engineering

The Human Genome Project

In February of 2001, scientists working on the Human Genome Project published a working draft of the human genome sequence. The sequence of an organism's genome is the identification of all base pairs that compose the DNA of the organism. The **Human Genome Project** is a research project that has linked over 20 scientific laboratories in six countries. Teams of scientists, such as those shown in **Figure 6,** cooperated to identify all 3.2 *billion* base pairs of the DNA that makes up the human genome. Scientists were surprised by some of the discoveries they made.

The Geography of the Genome

One of the most surprising things about the human genome is the large amount of DNA that does *not* encode proteins. In fact, only 1 to 1.5 percent of the human genome is DNA that codes for proteins. Each human cell contains about six feet of DNA, but less than 1 inch of that is devoted to exons. Recall that exons are sequences of nucleotides that are transcribed and then translated. Exons are scattered about the human genome in clumps that are not spread evenly among chromosomes. For example, chromosome number 19 is small and is packed with transcribed genes. The much larger chromosomes 4 and 8, by contrast, have few transcribed genes. On most human chromosomes, great stretches of untranscribed DNA fill the chromosomes between scattered clusters of transcribed genes.

The Number of Human Genes

When they examine the complete sequence of the human genome, scientists were surprised at how few genes there actually are. Human cells contain only about 20,000 to 25,000 genes. This is only about double the number of genes in a fruit fly. And it is only about one fifth of the 120,000 genes that scientists had expected to find. How had scientists made this prediction of the number of human genes, and why was it wrong? When scientists had counted unique human messenger RNA (mRNA) molecules, they had found over 120,000. Each of these different forms of mRNA molecules can, in turn, be translated into a unique protein. So the scientists expected to find as many genes as there are types of mRNA molecules.

Summarize two major goals of the Human Genome Project.

Describe how drugs produced by genetic engineering are being used.

Summarize the steps involved in making a genetically engineered vaccine.

Identify two different uses for DNA fingerprints.

Key Terms

Human Genome Project
vaccine
DNA fingerprint

Figure 6 Genetic Research. Hundreds of scientists around the world worked to identify the human genome sequence.

Genetically Engineered Drugs and Vaccines

Much of the excitement about genetic engineering has focused on its potential uses in our society. The possibilities for the applications of these techniques in medicine and research are endless. Many applications are already commonplace, such as the production of genetically engineered proteins used to treat illnesses and the creation of new vaccines used to combat infections.

Drugs

Many genetic disorders and other human illnesses occur when the body fails to make critical proteins. Juvenile diabetes is such an illness. The body is unable to control levels of sugar in the blood because a critical protein, insulin, cannot be made. These failures can be overcome if the body can be supplied with the protein it lacks. The proteins that regulate the body's functions are typically present in the body in very low amounts. Today hundreds of pharmaceutical companies around the world produce medically important proteins in bacteria using genetic engineering techniques as summarized in **Figure 7.**

Factor VIII, a protein that promotes blood clotting, is an example of a GM medicine (**g**enetically **m**odified; a drug manufactured by genetic engineering). A deficiency in factor VIII leads to one type of hemophilia, an inherited disorder characterized by prolonged bleeding. For a long time, hemophiliacs received blood factors that had been isolated from donated blood. Unfortunately, some of the donated blood was infected with viruses such as HIV and hepatitis B. The viruses were sometimes unknowingly transmitted to people who received blood transfusions. Today, the use of genetically engineered factor VIII eliminates these risks.

Figure 7 Use of genetically engineered medicines. Many medicines, such as medicines used to treat burns, are produced by genetic engineering techniques.

Genetically Engineered Medicines

Product:	Used for treatment of:
• Erythropoetin	Anemia
• Growth factors	Burns, ulcers
• Human growth hormone	Growth defects
• Insulin	Diabetes
• Interferons	Viral infections and cancer
• Taxol	Ovarian cancer

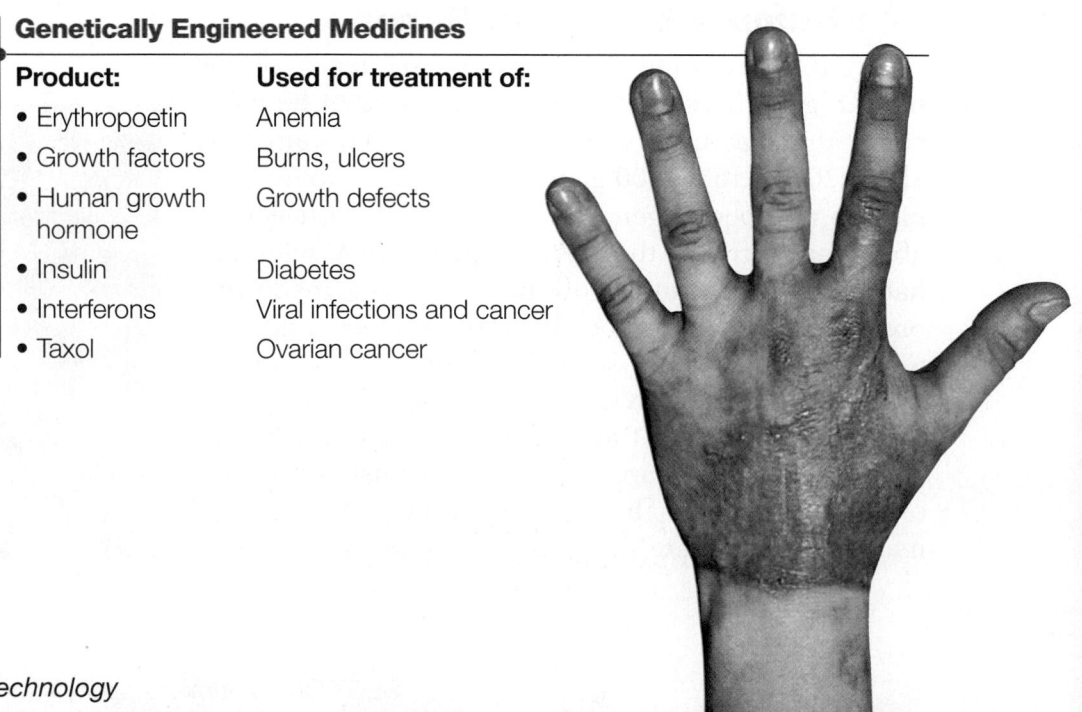

Vaccines

Many viral diseases, such as smallpox and polio, cannot be treated effectively by existing drugs. Instead they are combated by prevention—using vaccines. A **vaccine** is a solution containing all or part of a harmless version of a pathogen (disease-causing microorganism). When a vaccine is injected, the immune system recognizes the pathogen's surface proteins and responds by making defensive proteins called antibodies. In the future, if the same pathogen enters the body, the antibodies are there to combat the pathogen and stop its growth before it can cause disease.

Traditionally, vaccines have been prepared either by killing a specific pathogenic microbe or by making the microbe unable to grow. This ensures that the vaccine itself will not cause the disease. The problem with this approach is that there is a small but real danger that a failure in the process to kill or weaken a pathogen will result in the transmission of the disease to the very patients seeking protection. This danger is one of the reasons why, for example, rabies vaccines are administered only when a person has actually been bitten by an animal suspected of carrying rabies.

Vaccines made by genetic engineering techniques avoid this danger. As illustrated in **Figure 8,** the genes that encode the pathogen's surface proteins can be inserted into the DNA of harmless viruses such as cowpox (Vaccinia). The modified but harmless cowpox virus becomes an effective and safe vaccine, as illustrated in Figure 8. The surfaces of the modified virus display herpes surface proteins in addition to the virus's own surface proteins. When the modified virus is injected into a human body, the body's immune system quickly responds to this challenge. The immune system makes antibodies that attack any virus displaying the herpes surface protein. As a result, the body is thereafter protected against infection by the herpes virus.

Real Life

You might get a vaccine in a banana.

Genetic engineers are putting genes from disease-causing microbes into fruits and vegetables to create vaccines that are inexpensive and easy to take. Clinical trials using different foods, including potatoes, are underway.

Finding Information
What are the most common ways vaccines are now administered?

Figure 8 Making a genetically engineered vaccine

A person vaccinated with a genetically engineered vaccine, such as the genital herpes vaccine, will make antibodies against the virus.

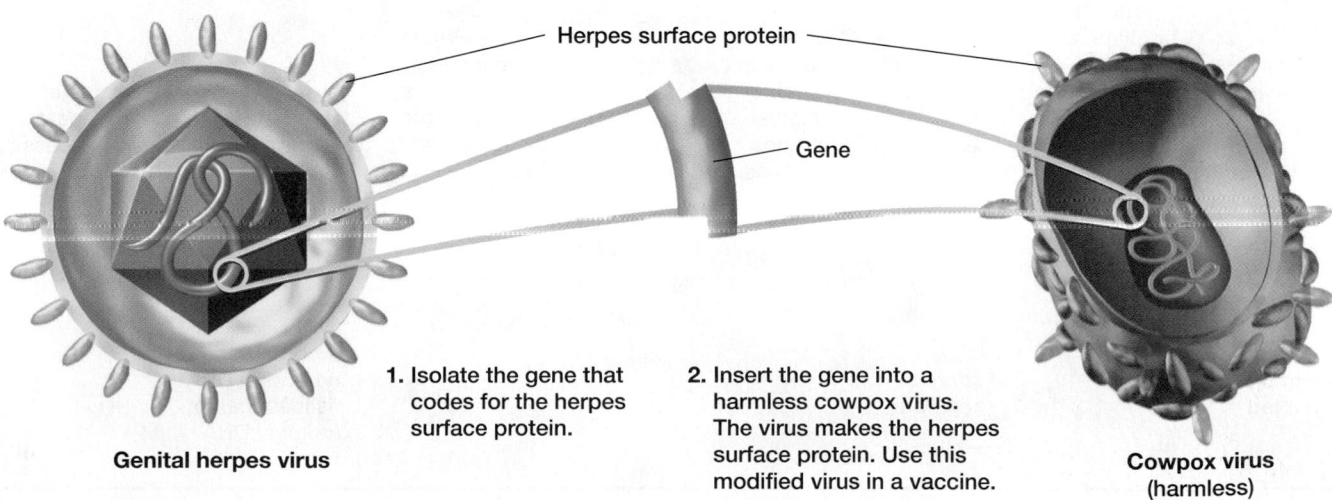

Herpes surface protein

Gene

1. Isolate the gene that codes for the herpes surface protein.

2. Insert the gene into a harmless cowpox virus. The virus makes the herpes surface protein. Use this modified virus in a vaccine.

Genital herpes virus

Cowpox virus (harmless)

Vaccines for the herpes II virus and for the hepatitis B virus are now being made through genetic engineering. The herpes II virus produces small blisters on the genitals (the external sex organs). The hepatitis B virus causes an inflammation of the liver that can be fatal. A major effort is underway to produce a vaccine that will protect people against malaria, a protozoan-caused disease for which there is currently no effective protection.

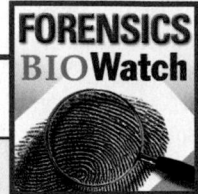

FORENSICS BIOWatch

Polymerase Chain Reaction (PCR)

A detective finds a single hair as the only evidence left behind at a crime scene. Will this hair provide enough DNA to analyze? For DNA fingerprinting and many of the genetic engineering uses discussed throughout this chapter, a certain amount of DNA is needed. Sometimes, however, only a very tiny amount of DNA is available.

Today scientists use a technique called the polymerase chain reaction (PCR) to quickly make many copies of selected segments of the available DNA. With PCR, a scientist can produce a billionfold increase in DNA material within a few hours!

Heating and Replication Cycles

In PCR the double-stranded DNA sample to be copied is heated, which separates the strands. The mixture is cooled, and short pieces of artificially made DNA called primers are added. The primers bind to places on the DNA where the copying can begin.

DNA polymerase and free nucleotides are added to the mixture. The DNA polymerase extends the DNA by attaching complementary free nucleotides to the primer. The result is two strands of DNA that are identical to each other and to the original strand. The heating and replica-

tion process is repeated over and over again. Every 5 minutes, the sample of DNA doubles again, resulting in many copies of the sample in a short amount of time. Today, scientists use PCR machines, which automatically cycle the reaction temperature.

PCR's Many Uses

PCR can duplicate DNA from as few as 50 white blood cells, which might be found in a nearly invisible speck of blood. PCR is important for diagnosing genetic disorders and for solving crimes. PCR is also used in different types of research and for studying ancient fragments of DNA found in fossils or in preserved material.

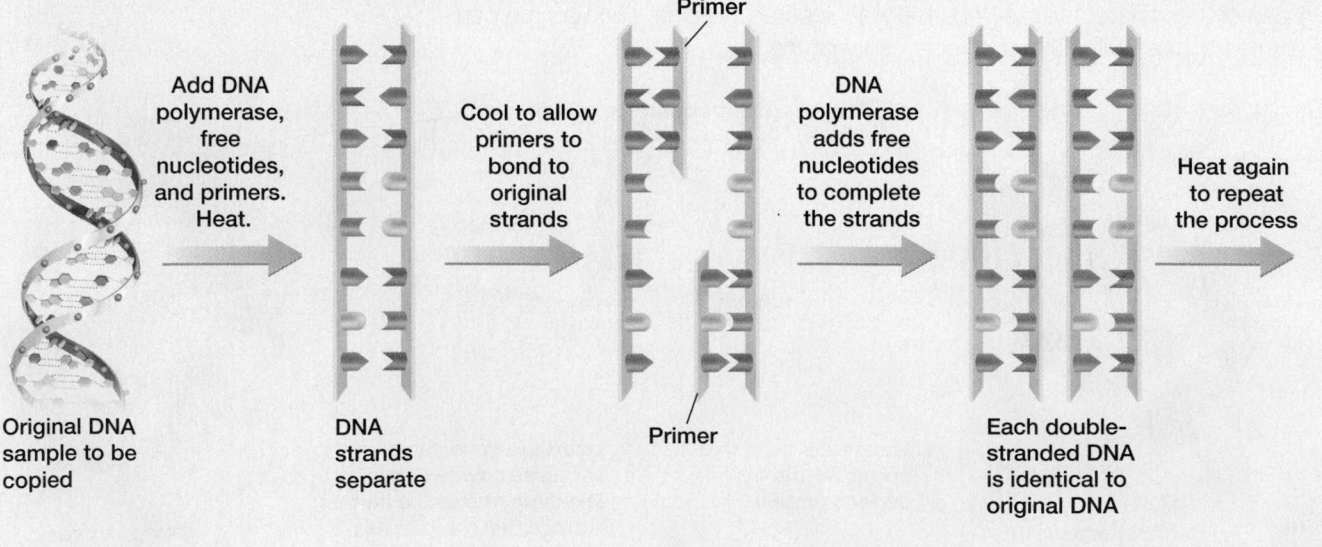

Original DNA sample to be copied

Add DNA polymerase, free nucleotides, and primers. Heat.

DNA strands separate

Cool to allow primers to bond to original strands

Primer

Primer

DNA polymerase adds free nucleotides to complete the strands

Each double-stranded DNA is identical to original DNA

Heat again to repeat the process

DNA Fingerprinting

Other than identical twins, no two individuals have the same genetic material. Scientists use DNA sequencing gel technology to determine a DNA fragment's nucleotide sequence, as shown in **Figure 9.** Because the places a restriction enzyme can cut depend on the DNA sequence, the lengths of DNA restriction fragments will differ between two individuals. Such DNA fragments of different lengths (polymorphisms) are called restriction fragment length polymorphisms, or RFLPs.

RFLPs can be used to identify individuals and to determine how closely related members of a population are to one another. The Southern blot technique, shown in Figure 5, is used to show an individual's RFLP profile. The result is called a *DNA fingerprint*. A **DNA fingerprint** is a pattern of dark bands on photographic film that is made when an individual's DNA restriction fragments are separated by gel electrophoresis, probed, and then exposed to an X-ray film. Because restriction enzymes cut the DNA from different individuals into DNA fragments of different lengths (RFLPs), each individual (other than identical twins) has a unique pattern of banding, or DNA fingerprint.

The banding patterns from two individuals can be compared to establish whether they are related, such as in a paternity case. Because it can be performed on a sample of DNA found in blood, semen, bone, or hair, DNA fingerprinting is useful in forensics. Forensics is the scientific investigation of the causes of injury and death when criminal activity is suspected. DNA fingerprints are also valuable for identifying the genes that cause genetic disorders, such as Huntington's disease and sickle cell anemia.

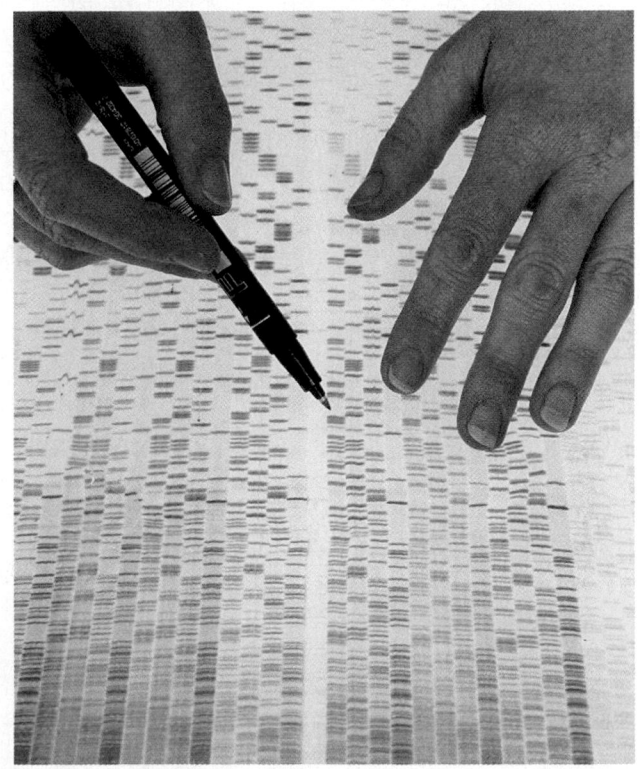

Figure 9   DNA sequence.
The nucleotide sequence of DNA fragments can be determined using DNA sequencing gel technology.

Section 2 Review

1 **Relate** the use of genetic engineering to the treatment of human illnesses such as hemophilia.

2 **Relate** genetic engineering techniques to the making of vaccines.

3 **List** two ways in which DNA fingerprinting has been useful to society.

4 **Critical Thinking Distinguishing Relevant Information** A student states that genetic engineering is "perfectly safe and sound." What safety and ethical issues do you think might arise over the use of genetic engineering?

5 **Standardized Test Prep** One medicine made in bacteria using genetic engineering techniques is insulin, which is used to treat

A heart attacks   C diabetes

B smallpox   D cystic fibrosis

Genetic Engineering in Agriculture

Objectives

- **Describe** three ways in which genetic engineering has been used to improve plants.

- **Summarize** two ways in which genetic engineering techniques have been used to modify farm animals.

- **Summarize** the cloning of sheep through the use of differentiated cells.

Key Terms

transgenic animal

Improving Crops

Farmers began primitive genetic breeding by selecting seeds from their best plants, replanting them, and gradually improving the quality of successive generations. In the twentieth century, plant breeders started using the principles of genetics to select plants. Today, genetic engineers can add favorable characteristics to a plant by manipulating the plant's genes, as shown in **Figure 10.**

Genetic engineers can change plants in many ways, including making crop plants more tolerant to drought conditions and creating plants that can adapt to different soils, climates, and environmental stresses.

Genetic engineers have developed crop plants that are resistant to a biodegradable weedkiller called *glyphosate.* This has enabled farmers to apply glyphosate to kill weeds without killing their crops. Because the field does not need to be tilled to control weeds, less topsoil is lost to erosion. Half of the 72 million acres of soybeans planted in the United States in 2000 were genetically modified to be glyphosate resistant.

Scientists have also developed crops that are resistant to insects by inserting a certain gene isolated from soil bacteria into crop plants. This gene makes a protein that injures the gut of chewing insects. Crops that are resistant to insects do not need to be sprayed with pesticides, many of which can harm the environment.

More Nutritious Crops

Genetic engineers have been able, in many instances, to improve the nutritional value of crop plants. For example, in Asia many people use rice as a major source of food, yet rice has low levels of iron and beta carotene, which your body uses to make vitamin A (necessary for vision). As a result, millions suffer from iron deficiency and poor vision. Genetic engineers have added genes to rice from other plants, as shown in **Figure 11,** to overcome this deficiency.

Figure 10 Genetically engineered plants. At least 50 plants have been genetically engineered, including potatoes, soybeans, and corn. The researcher Athanasios Theologis genetically engineered tomatoes to ripen without becoming soft.

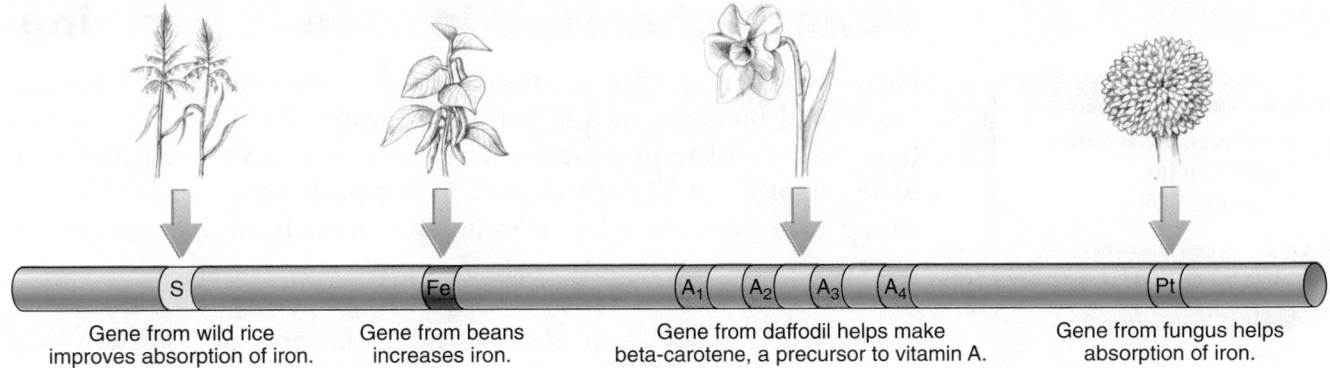

Gene from wild rice improves absorption of iron.

Gene from beans increases iron.

Gene from daffodil helps make beta-carotene, a precursor to vitamin A.

Gene from fungus helps absorption of iron.

Risks of Genetically Modified Crops

Many people, including influential scientists, have expressed concern that genetically modified crops (GM crops) might turn out to be dangerous. What kind of unforeseen negative effects might "improved" GM crops have?

Potential Problems

Some food crops, such as corn and soybeans, have been genetically rendered resistant to glyphosate, a weed killer that is harmless to humans. Glyphosate, when used on a food crop, will kill the weeds but will not harm the GM crop, thus increasing food crop yields. Some scientists are concerned that the use of GM crops and the subsequent use of glyphosate will eventually lead to glyphosate-resistant weeds. This will leave farmers with few weed-control alternatives.

Some GM crops have genes added to improve nutritional character, as was done in rice. It is important to check that consumers are not allergic to the product of the introduced gene. For this reason, screening of GM crops for causes of allergy problems is now routine.

Are GM Crops Harmful to the Environment?

Will introduced genes pass from GM crops to their wild or weedy relatives? This sort of gene flow happens naturally all the time, so this concern is legitimate. For most crops, no closely related wild plant is around to receive the gene. The GM gene cannot pass to a nonrelative, because crop plants cannot successfully reproduce with unrelated species, any more than a cat can breed with a giraffe. There are wild relatives of corn in Mexico and Guatamala, which frequently exchange genes with corn crops. Scientists are divided about whether it makes any difference if one of the genes is a GM gene.

Might pests become resistant to GM toxins? Pests are becoming resistant to GM toxins just as they have become resistant to the chemical pesticides that are sprayed on crops.

Scientists, the public, and regulatory agencies must work together to evaluate the risks and benefits of GM products.

Figure 11 Rice enriched with iron and vitamin A. Genetically modified "golden" rice offers the promise of improving the diets of people in rice-consuming countries, where iron and vitamin A deficiencies are a serious problem.

Gene Technology in Animal Farming

Farmers have long tried to improve farm animals and crops through traditional breeding and selection programs. In the past, the cow that produced the most milk on a farm may have been mated to male offspring of high producers in hopes that the cow's offspring would also produce a lot of milk. But these traditional processes were slow and inefficient.

Now, many farmers use genetic-engineering techniques to improve or modify farm animals. Some farmers add growth hormone to the diet of cows to increase milk production. Previously, the growth hormone was extracted from the brains of dead cows. But now the cow growth hormone gene is introduced into bacteria. The bacteria produce the hormone so cheaply that it is practical to add it as a supplement to the cows' diet.

By altering the gene responsible for GH production, scientists have stimulated natural GH in pigs, increasing their weight. Though these procedures are still new, they may lead to the creation of new breeds of very large and fast-growing cattle and hogs.

Making Medically Useful Proteins

Another way in which gene technology is used in animal farming is in the addition of human genes to the genes of farm animals in order to get the farm animals to produce human proteins in their milk. This is used especially for complex human proteins that cannot be made by bacteria through gene technology. The human proteins are

internet connect

www.scilinks.org
Topic: Cloning
Keyword: HX4047

SCILINKS. Maintained by the National Science Teachers Association

Figure 12   Cloning a sheep from mammary cells

In 1997 scientists announced the first successful cloning using differentiated cells—a lamb named Dolly.

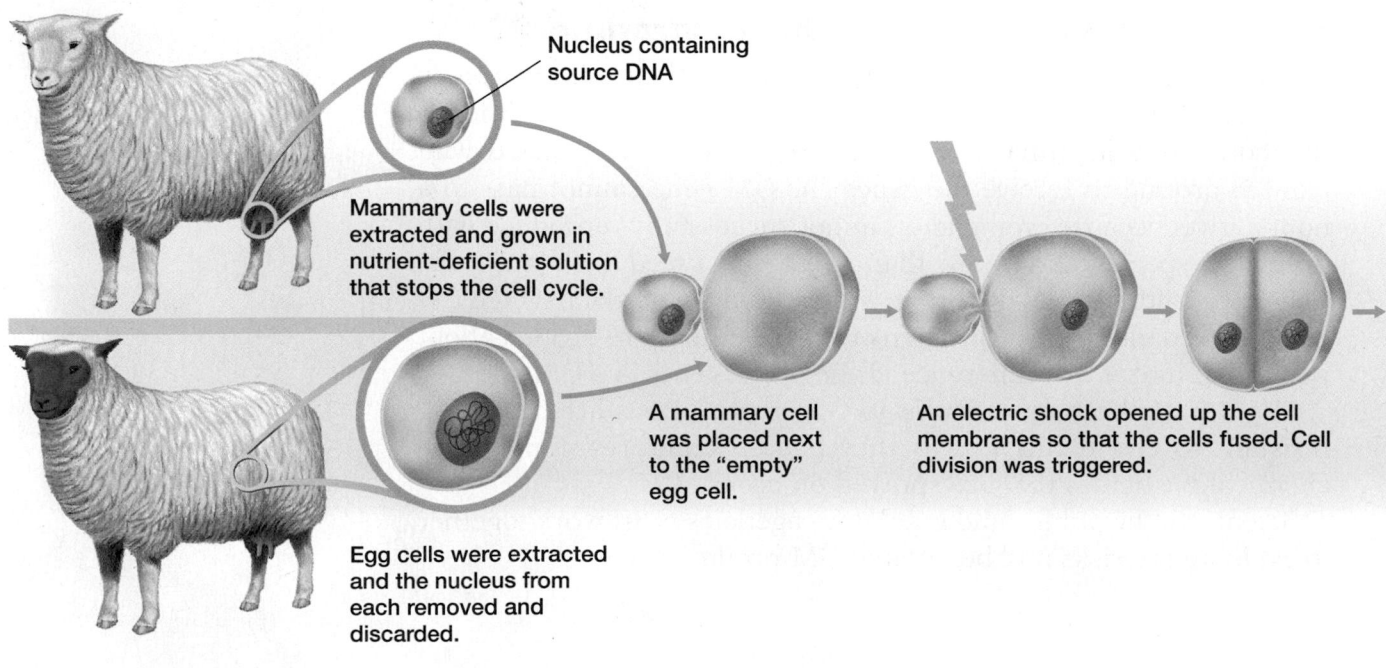

Nucleus containing source DNA

Mammary cells were extracted and grown in nutrient-deficient solution that stops the cell cycle.

A mammary cell was placed next to the "empty" egg cell.

An electric shock opened up the cell membranes so that the cells fused. Cell division was triggered.

Egg cells were extracted and the nucleus from each removed and discarded.

extracted from the animals' milk and sold for pharmaceutical purposes. The animals are called **transgenic animals** because they have foreign DNA in their cells.

Most recently, scientists have turned to cloning animals as a way of creating herds of identical animals that can make medically useful proteins. The intact nucleus of an embryonic or fetal cell (whose DNA has been recombined with a human gene) is placed into an egg whose nucleus has been removed. The egg with the new nucleus is then placed into the uterus of a surrogate, or substitute, mother and is allowed to develop.

Cloning From Adult Animals

In 1997, a scientist named Ian Wilmut captured worldwide attention when he announced the first successful cloning using differentiated cells from an adult animal. A differentiated cell is a cell that has become specialized to become a specific type of cell (such as a liver or udder cell). As summarized in **Figure 12,** a lamb was cloned from the nucleus of a mammary cell taken from an adult sheep. Previously, scientists thought that cloning was possible only using embryonic or fetal cells that have not yet differentiated. Scientists thought that differentiated cells could not give rise to an entire organism. Wilmut's experiment proved otherwise.

An electric shock was used to fuse mammary cells from one sheep with egg cells without nuclei from a different sheep. The fused cells divided to form embryos, which were implanted into surrogate mothers. Only one embryo survived the cloning process. Dolly, born on July 5, 1996, was genetically identical to the sheep that provided the mammary cell.

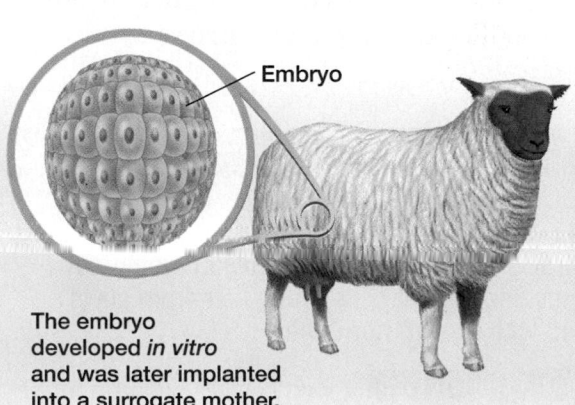

Embryo

The embryo developed *in vitro* and was later implanted into a surrogate mother.

After a 5-month pregnancy, a lamb was born that was genetically identical to the sheep from which the mammary cell was extracted.

Problems With Cloning

Since Dolly's birth in 1996, scientists have successfully cloned animals. Only a few of the cloned offspring survived for long, however. Many become fatally oversized. Others encounter problems in development. For example, three cloned calves were born healthy in March, 2001, only to die a month later of immune system failure.

The Importance of Genomic Imprinting

Technical problems in reproductive cloning lie within a developmental process that conditions eggs and sperm so that the right combination of genes are turned "on" or "off" during early development. When cloned offspring become adults, a different combination of genes is activated. The process of conditioning the DNA during an early stage of development is called genomic imprinting.

In genomic imprinting, chemical changes made to DNA prevent a gene's expression without altering its sequence. Usually, a gene is locked into the "off" position by adding methyl ($-CH_3$) groups to its cytosine nucleotides, as shown in **Figure 13.** The bulky methyl groups prevent polymerase enzymes from reading the gene, so the gene cannot be transcribed. Later in development, the methyl groups are removed and the gene is reactivated.

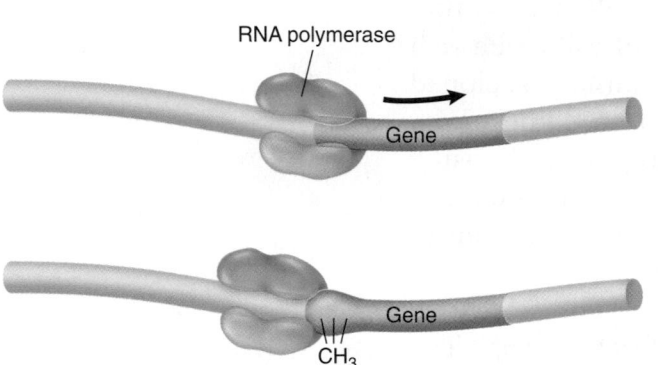

Figure 13 Methylated gene. In one model of genomic imprinting, methyl groups attached to a gene prevent the gene from being expressed.

Why Cloning Fails

Normal vertebrate development depends on precise genomic imprinting. This process, which takes place in adult reproductive tissue, takes months for sperm and years for eggs. Reproductive cloning fails because the reconstituted egg begins to divide within minutes. There is simply not enough time in these few minutes for the reprogramming to process properly. Key genes fail to become properly methylated, and this leads to critical errors in development.

Because of these technical problems, and because of ethical problems, efforts to clone humans are illegal in most countries.

Section 3 Review

1. **List** three ways in which food crops have been improved through genetic engineering.

2. **Compare** the cloning of sheep through the use of differentiated cells with the cloning of sheep through the use of embryonic cells.

3. **Critical Thinking Analyzing Methods** In the movie *Jurassic Park*, scientists used DNA to bring back extinct species. How is that different from the creation of cloned sheep using differentiated cells?

4. **Critical Thinking Forming Reasoned Opinions** List reasons you would or would not be concerned about consuming milk from cows treated with growth hormone.

5. **Standardized Test Prep** Using genetic engineering to produce rice with high levels of beta-carotene should help people who suffer from a deficiency in
   A vitamin A.
   B growth hormone.
   C glyphosate.
   D complex proteins.

CHAPTER HIGHLIGHTS

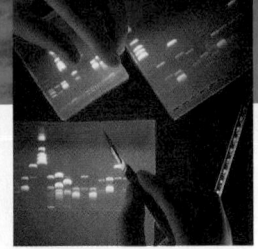

Key Concepts

1 Genetic Engineering

- Genetic engineers manipulate DNA for practical purposes.

- Restriction enzymes cleave DNA into fragments that have short sticky ends. Sticky ends allow DNA fragments from different organisms to join together to form recombinant DNA.

- Recombinant DNA is inserted into host cells. The cells are screened to identify cells that have the recombinant DNA. Each time the cells reproduce, the gene of interest is cloned.

- Electrophoresis uses an electric field within a gel to separate DNA fragments by their size.

- Specific genes can be identified with the Southern blot technique.

2 Genetic Engineering in Medicine and Society

- Genetic engineering is used to manufacture human proteins for use as drugs and to make safer and more effective vaccines.

- Some human genetic disorders are being treated with gene therapy.

- DNA fingerprinting is used to identify individuals and determine relationships between individuals.

- The Human Genome Project is an effort to determine the nucleotide sequence of and map the location of every gene on each human chromosome by the year 2003. The sequence of the genomes of many organisms has already been determined.

3 Genetic Engineering in Agriculture

- Crop plants can be genetically engineered to have favorable characteristics, including improved yields and resistance to herbicides and destructive pests.

- Genetically engineered growth hormone increases milk production in dairy cows and weight gain in cattle and hogs.

- Success in cloning animals using differentiated cells was announced in 1997. In addition, transgenic animals can be cloned and used to make proteins that are useful in medicine.

Key Terms

Section 1

genetic engineering (228)
recombinant DNA (228)
restriction enzyme (229)
vector (229)
plasmid (229)
gene cloning (229)
electrophoresis (231)
probe (231)

Section 2

Human Genome Project (233)
vaccine (235)
DNA fingerprint (237)

Section 3

transgenic animal (241)

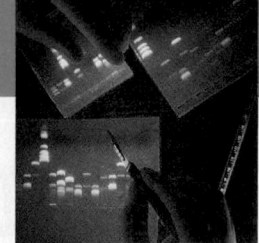

Understanding Key Ideas

1. Gel electrophoresis is used to _____ DNA fragments.
   a. separate   **c.** cut
   b. join   **d.** copy

2. Which of the following human illnesses can be treated using a product of genetic engineering?
   a. malaria   **c.** flu
   b. hemophilia   **d.** a sinus cold

3. Injecting a healthy copy of a gene into a person who has a defective gene is called
   a. probing.
   b. gene therapy.
   c. PCR.
   d. DNA cloning.

4. The major effort to map and sequence all human genes is called
   a. the RFLP Project.
   b. the PCR Project.
   c. the Human Genome Project.
   d. DNA fingerprinting.

5. A transgenic organism is produced as a result of
   a. hybridization.
   b. recombinant DNA.
   c. mutation.
   d. RFLPs.

6. The process of making recombinant DNA is *least* related to
   a. clones.
   b. DNA fragments.
   c. restriction enzymes.
   d. sticky ends.

7. Genetic engineers can make plants
   a. resistant to insects.
   b. more tolerant to droughts.
   c. that are adapted to different soils.
   d. All of the above

8. Describe how molecule *A* was produced.

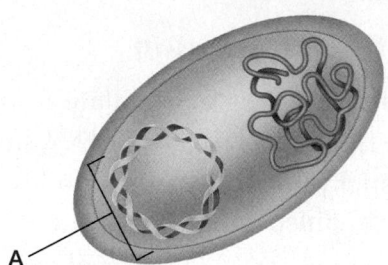

9. **BIOWatch** You have discovered a fossilized bone. How can you use PCR to obtain sufficient DNA for DNA analysis?

10. **Concept Mapping** Make a concept map about genetic engineering. Try to include the following words in your map: *DNA of interest, vectors, recombinant DNA, plasmids, restriction enzymes, sticky ends,* and *research.*

Critical Thinking

11. **Forming Reasoned Opinions** In the United States, government regulations require researchers to contain experimental genetically engineered organisms inside a laboratory and to ensure that the organisms could not survive outside the laboratory. Why do you think these strict regulations are necessary?

12. **Distinguishing Fact from Opinion** A judge presiding over a highly publicized murder trial dismissed the prosecution's request to admit DNA fingerprints as evidence, calling it "unproven." Do you agree with the judge? Explain your answer.

13. **Distinguishing Relevant Information** Organize and videotape a class debate about the safety questions raised by the potential release of genetically engineered plants, bacteria, and animals into the environment. Use library references and on-line databases to back up your arguments.

Standardized Test Prep

Understanding Concepts

Directions (1–4): **For *each* question, write on a separate sheet of paper the letter of the correct answer.**

1 What term describes a molecule containing DNA from two different organisms?
   A. plasmid
   B. probe
   C. recombinant DNA
   D. RFLP DNA

2 Which of the following is an extra ring of DNA in bacteria?
   F. clone
   G. plasmid
   H. probe
   I. restriction enzyme

3 What agent allows genetic engineers to cut DNA at specific sites?
   A. DNA ligase
   B. DNA polymerase
   C. plasmid DNA
   D. restriction enzyme

4 What technique is used to identify individuals in paternity cases and criminal cases?
   F. DNA fingerprinting
   G. gene therapy
   H. genomic imprinting
   I. vaccination

Directions (5–6): **For *each* question, write a short response.**

5 Examine how natural selection could be affected by genetic engineering.

6 Analyze the difference in the meanings of the terms recombinant DNA and restriction enzyme.

Reading Skills

Directions (7): **Read the passage below. Then answer the question.**

The question of awarding patents on genetically engineered organisms arose when a microbiologist named Ananda Chakrabarty filed for a patent on a bacterium capable of digesting the components of crude oil. Chakrabarty identified enzymes that degrade different components of crude oil and added the enzymes to *Pseudomonas* bacteria. His patent request was brought before the U. S. Supreme Court, which ruled in 1980 that human-engineered organisms are patentable under federal law.

7 What type of genetic engineering did Chakrabarty use to add enzymes to *Pseudomonas* bacteria?
   A. DNA fingerprinting
   B. gel electrophoresis
   C. human cloning
   D. recombinant DNA

Interpreting Graphics

Directions (8): **Base your answer to question 8 on the diagram below.**

DNA Cut with a Restriction Enzyme

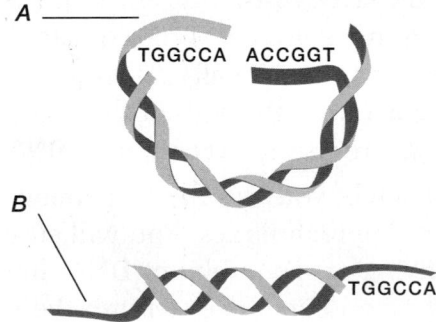

8 The diagram above shows two pieces of DNA that have been cut with the same restriction enzyme. What nucleotide sequence must the sticky end labeled *B* have if it is to bond with the sticky end labeled *A*?
   F. ACCGGT   **H.** TCCGGA
   G. CTTAAG   **I.** UGGCCU

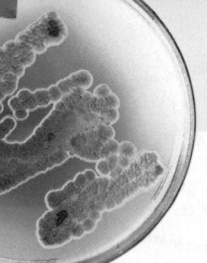

Exploration Lab

Modeling Recombinant DNA

SKILLS
- Modeling
- Comparing

OBJECTIVES
- **Construct** a model that can be used to explore the process of genetic engineering.
- **Describe** how recombinant DNA is made.

MATERIALS
- paper clips (56)
- plastic soda straw pieces (56)
- pushpins (15 red, 15 green, 13 blue, and 13 yellow)

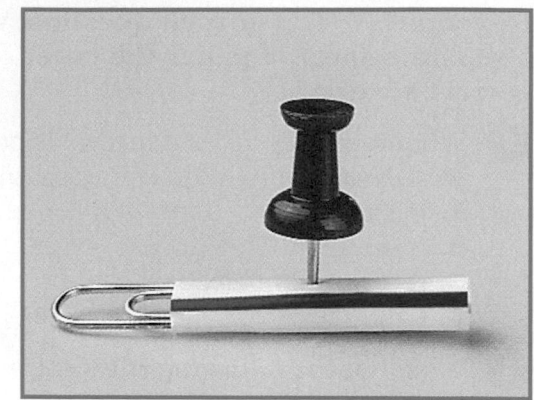

Before You Begin

Genetic engineering is the process of taking a gene from one organism and inserting it into the DNA of another organism. The gene is delivered by a **vector,** such as a virus, or a bacterial **plasmid.**

First, a fragment of a chromosome that contains the gene is isolated by using a **restriction enzyme,** which cuts DNA at a specific nucleotide-base sequence. Some restriction enzymes cut DNA unevenly, producing single-stranded **sticky ends.** The DNA of the vector is cut by the same restriction enzyme. Next, the chromosome fragment is mixed with the cut DNA of the vector. Finally, an enzyme called **DNA ligase** joins the ends of the two types of cut DNA, producing **recombinant DNA.**

In this lab, you will model genetic engineering techniques. You will simulate the making of recombinant DNA that has a human gene inserted into the DNA of a plasmid.

1. Write a definition for each boldface term in the paragraph above and for the term *base-pairing rules.*

2. Based on the objectives for this lab, write a question you would like to explore about the process of genetic engineering.

Procedure

PART A: Model Genetic Engineering

1. Make 56 model nucleotides. To make a nucleotide, insert a pushpin midway along the length of a 3 cm piece of a soda straw. **CAUTION: Handle pushpins carefully. Pointed objects can cause injury.** Push a paper clip into one end of the soda-straw piece until it touches the pushpin.

2. Begin a model of a bacterial plasmid by arranging nucleotides for one DNA strand in the following order: blue, red, green, yellow, red, red, blue, blue, green, red, blue, green, red, blue, blue, green, yellow, and red. Join two adjacent nucleotides by inserting the paper clip end of one into the open end of the other.

3. Using your first DNA strand and the base-pairing rules, build the complementary strand of plasmid DNA. **Note:** *Yellow is complementary to blue, and green is complementary to red.*

4. Complete your model of a circular plasmid by joining the opposite ends of each DNA strand. Make a sketch showing the sequence of bases in your model plasmid. Use the abbreviations B, Y, G, and R for the pushpin colors. Your sketch should be similar to the one at the top of the next page.

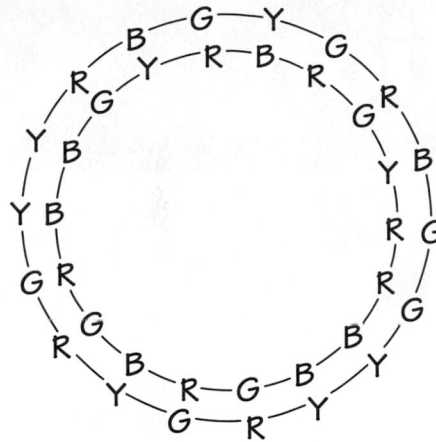

5. Begin a model of a human chromosome fragment made by a restriction enzyme. Place nucleotides for one DNA strand in the following order: BBRRYGGBRY. Build the second DNA strand by arranging the remaining nucleotides in the following order: BRRYGBYYGG.

6. Match the complementary portions of the two strands of DNA you made in step 5. Pair as many base pairs in a row as you can. Make a sketch showing the sequence of bases in your model of a human chromosome fragment.

7. Imagine that the restriction enzyme that cut the human chromosome fragment you made in steps 5 and 6 is moving around your model plasmid until it finds the sequence YRRBBG and its complementary sequence, BGGYYR. This restriction enzyme cuts each sequence between a B and a G. Find such a section in your sketch of your model plasmid's DNA.

8. Simulate the action of the restriction enzyme on the section you identified in step 7. Open both strands of your model plasmid's DNA by pulling apart the adjacent green and blue nucleotides in each strand. Make a sketch of the split plasmid DNA molecule.

9. Move your model human DNA fragment into the break in your model plasmid's DNA molecule. Imagine that a ligase joins the ends of the human and plasmid DNA. Make a sketch of your final model DNA molecule.

PART B: Cleanup and Disposal

10. Dispose of damaged pushpins in the designated waste container.

11. Clean up your work area and all lab equipment. Return lab equipment to its proper place. Wash your hands thoroughly before you leave the lab and after you finish all work.

Analyze and Conclude

1. **Comparing Structures** Compare your models of plasmid DNA and human DNA.

2. **Relating Concepts** What do the sections of four unpaired nucleotides in your model human DNA fragment represent?

3. **Comparing Structures** How did your original model plasmid DNA molecule differ from your final model DNA molecule?

4. **Drawing Conclusions** What does the molecule you made in step 9 represent?

5. **Further Inquiry** Write a new question that could be explored with another investigation.

On the Job

Genetic engineering is used to produce many products that are useful to humans. Do research to discover how genetic engineering is used to better our lives. For more about careers, visit **go.hrw.com** and type in the keyword **HX4 Careers.**

Tropical rain forests contain more than one-half of all the world's animal and plant species, such as this red-eyed tree frog from Central America.

in perspective

Tropical Rain Forests

Yesterday... During their travels, early European visitors were fascinated by the variety and number of tropical rain-forest plants and animals so different from those they knew at home. **Discover how early scientists named and classified living things.**

Exploring the Brazilian jungle in 1820

Today... Today's explorers are faced with the challenge of classifying living things in lesser-known areas, such as tropical rain forests, that have not yet been studied by scientists. **What organisms are found in tropical rain forests, and what is the impact of human activity on these species?**

Rain-forest researcher at Madre de Dios, Peru

Tomorrow... To preserve species, new ways of protecting tropical habitats are being explored. For example, modern coffee-growing techniques eliminate native species as fields are planted solely with coffee, a major source of revenue for many tropical countries. But shade coffee plantations grow coffee in the midst of the forest canopy, thus preserving habitat for tropical species.

Coffee beans

249

Lightning at sea

12 History of Life on Earth

✔ Quick Review

Answer the following without referring to earlier sections of your book.

1. **Compare** the structure of proteins, lipids, and nucleic acids. *(Chapter 2, Section 3)*

2. **Describe** the role of enzymes in catalyzing chemical reactions. *(Chapter 2, Section 4)*

3. **Contrast** prokaryotes and eukaryotes. *(Chapter 3, Section 2)*

4. **Identify** the structure and function of chloroplasts and mitochondria. *(Chapter 3, Section 3)*

5. **Summarize** the role of DNA in heredity. *(Chapter 9, Section 1)*

Did you have difficulty? *For help, review the sections indicated.*

📖 Reading Activity

Before you begin to read this chapter, write down all of the key words for each section of the chapter. Then, write a definition next to each word that you have heard of. As you read the chapter, write definitions next to the words that you did not previously know, and modify as needed your original definitions of words familiar to you.

- Billions of years ago, the combination of simple molecules and energy from sources such as lightning may have given rise to the complex organic molecules necessary for life.

Looking Ahead

Section 1

How Did Life Begin?
The Age of Earth
Formation of the Basic Chemicals of Life
Precursors of the First Cells

Section 2

The Evolution of Cellular Life
The Evolution of Prokaryotes
The Evolution of Eukaryotes
Multicellularity
Mass Extinctions

Section 3

Life Invaded the Land
The Ozone Layer
Plants and Fungi on Land
Arthropods
Vertebrates

🔲 **internet** connect

www.scilinks.org
National Science Teachers Association *sci*LINKS Internet resources are located throughout this chapter.

*sci*LINKS. **Maintained by the National Science Teachers Association**

How Did Life Begin?

Objectives

- **Summarize** how radioisotopes can be used in determining Earth's age.

- **Compare** two models that describe how the chemicals of life originated.

- **Describe** how cellular organization might have begun.

- **Recognize** the importance that a mechanism for heredity has to the development of life.

Key Terms

radiometric dating
radioisotope
half-life
microsphere

The Age of Earth

When Earth formed, about 4.5 billion years ago, it was a fiery ball of molten rock. Eventually, the planet's surface cooled and formed a rocky crust. Water vapor in the atmosphere condensed to form vast oceans. Most scientists think life first evolved in these oceans and that the evolution of life occurred over hundreds of millions of years. Evidence that Earth has existed long enough for this evolution to have taken place can be found by measuring the age of rocks found on Earth.

Measuring Earth's Age

Scientists have estimated the age of Earth using a technique called radiometric dating. **Radiometric dating** is the estimation of the age of an object by measuring its content of certain radioactive isotopes (*IE soh tohps*). An isotope is a form of an element whose atomic mass (the mass of each individual atom) differs from that of other atoms of the same element. Radioactive isotopes, or **radioisotopes,** are unstable isotopes that break down and give off energy in the form of charged particles (radiation). This breakdown, called radioactive decay, results in other isotopes that are smaller and more stable.

For example, certain rocks contain traces of potassium-40, an isotope of the element potassium. As **Figure 1** shows, the decay of potassium-40 produces two other isotopes, argon-40 and calcium-40. The time it takes for one-half of a given amount of a radioisotope to decay is called the radioisotope's **half-life.** By measuring the proportions of certain radioisotopes and their products of decay, scientists can compute how many half-lives have passed since a rock was formed.

Figure 1 Rate of decay for potassium-40. This graph shows the rate of decay for the radioisotope potassium-40. After one *half-life* has passed, half of the original amount of the radioisotope remains.

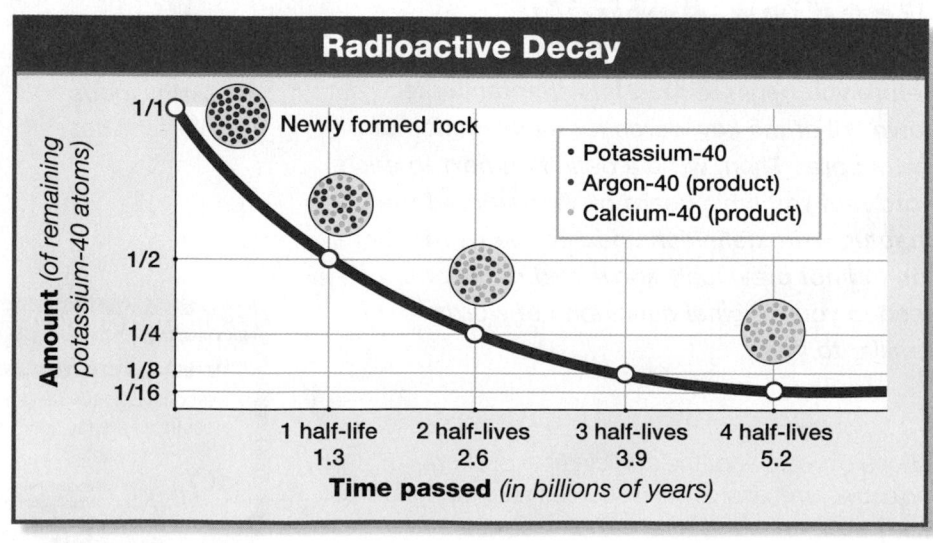

Radioactive Decay

Amount *(of remaining potassium-40 atoms)*

1/1 — Newly formed rock

- Potassium-40
- Argon-40 (product)
- Calcium-40 (product)

1/2
1/4
1/8
1/16

| 1 half-life | 2 half-lives | 3 half-lives | 4 half-lives |
| 1.3 | 2.6 | 3.9 | 5.2 |

Time passed *(in billions of years)*

Formation of the Basic Chemicals of Life

Most scientists think that life on Earth developed through natural chemical and physical processes. It is thought that the path to the development of living things began when molecules of nonliving matter reacted chemically during the first billion years of Earth's history. These chemical reactions produced many different simple, organic molecules. Energized by the sun and volcanic heat, these simple, organic molecules formed more-complex molecules that eventually became the building blocks of the first cells. The hypothesis that many of the organic molecules necessary for life can be made from molecules of nonliving matter has been tested and supported by results of laboratory experiments.

internet connect

www.scilinks.org
Topic: Radioactive Decay
Keyword: HX4154

SCLINKS Maintained by the National Science Teachers Association

QUICK LAB

Modeling Radioactive Decay

You can use some dried corn, a box, and a watch to make a model of radioactive decay that will show you how scientists measure the age of objects.

Materials

approximately 100 dry corn kernels per group, cardboard box, clock or watch with a second hand

Procedure

1. On a separate sheet of paper, make a data table like the one below.

2. Assign one member of your team to keep time.

3. Place 100 dry corn kernels into a box.

4. Shake the box gently from side to side for 10 seconds.

5. Keep the box still and remove and count the kernels that "point" to the left side of the box, as shown below. Record in your data table the number of kernels you removed.

6. Repeat steps 4 and 5 until all kernels have been counted and removed.

7. Calculate the number of kernels remaining for each time interval.

8. Make a graph using your group's data. Plot "Total shake time (seconds)" on the x-axis. Plot "Number of kernels remaining" on the y-axis.

Analysis

1. **Identify** what the removed kernels represent in each step.

2. **Calculate** the half-life of your sample, in seconds, that is represented in this activity.

3. **Calculate** the age of your sample, in years, if each 10-second interval represents 5,700 years.

4. **Evaluate** the ability of this model to demonstrate radioactive decay.

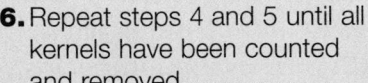

DATA TABLE		
Total shake time (seconds)	Number of kernels removed	Number of kernels remaining
10		
20		
30		

The "Primordial Soup" Model

In the 1920s, the Russian scientist A. I. Oparin and the British scientist J.B.S. Haldane both suggested that the early Earth's oceans contained large amounts of organic molecules. This hypothesis became known as the primordial *(prie MAWR dee uhl)* soup model. Earth's vast oceans were thought to be filled with many different organic molecules, like a soup that is filled with many different vegetables and meats. Oparin and Haldane hypothesized that these molecules formed spontaneously in chemical reactions activated by energy from solar radiation, volcanic eruptions, and lightning.

Oparin, together with the American scientist Harold Urey, and other scientists also proposed that Earth's early atmosphere lacked oxygen. They hypothesized that the early atmosphere was instead rich in nitrogen gas, N_2; hydrogen gas, H_2; and hydrogen-containing gases such as water vapor, H_2O; ammonia, NH_3; and methane, CH_4. They reasoned that electrons in these gases would have been frequently pushed to higher energy levels by light particles from the sun or by electrical energy in lightning. Today, high-energy electrons are quickly soaked up by the oxygen in Earth's atmosphere because oxygen atoms have a great "thirst" for such electrons. But without oxygen, high-energy electrons would have been free to react with hydrogen-rich molecules, forming a variety of organic compounds.

In 1953, the primordial soup model was tested by Stanley Miller, who was then working with Urey. Miller placed the gases that he and Urey proposed had existed on early Earth into a device like the one seen in **Figure 2.** To simulate lightning, he provided electrical sparks. After a few days, Miller found a complex collection of organic molecules in his apparatus. These chemicals included some of life's basic building blocks: amino acids, fatty acids, and other hydrocarbons (molecules made of carbon and hydrogen). These results support the hypothesis that some basic chemicals of life could have formed spontaneously under conditions like those in the experiment.

Reevaluating the Miller-Urey Model

Recent discoveries have caused scientists to reevaluate the Miller-Urey experiment. We now know that the reductant molecules used in Miller's experiment could not have existed in abundance on the early Earth. Four billion years ago, Earth did not have a protective layer of ozone gas, O_3. Today ozone protects Earth's surface from most of the sun's damaging ultraviolet radiation. Without ozone, ultraviolet radiation would have destroyed any ammonia and methane present in the atmosphere. When these gases are absent from the Miller-Urey experiment, key biological molecules are not made. This raises a very important question: If the chemicals needed to form life were not in the atmosphere, where did they come from? Some scientists argue that the chemicals were produced within ocean bubbles. Others say that the chemicals arose in deep sea vents. The correct answer has not been determined yet.

Figure 2   Miller-Urey experiment. Miller simulated an atmospheric composition that Oparin and other scientists incorrectly hypothesized existed on early Earth. His experiment produced several different organic compounds.

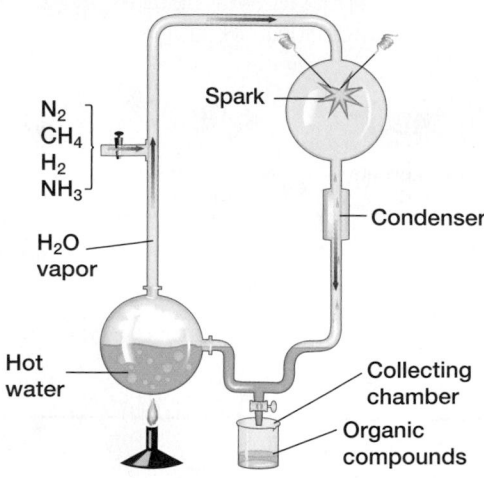

Spark

N₂
CH₄
H₂
NH₃

H₂O vapor

Condenser

Hot water

Collecting chamber

Organic compounds

The Bubble Model

In 1986, the geophysicist Louis Lerman suggested that the key processes that formed the chemicals needed for life took place within bubbles on the ocean's surface. Lerman's hypothesis, also known as the bubble model, is summarized in **Figure 3**.

Step ❶ Ammonia, methane, and other gases resulting from the numerous eruptions of undersea volcanoes were trapped in underwater bubbles.

Step ❷ Inside the bubbles, the methane and ammonia needed to make amino acids might have been protected from damaging ultraviolet radiation. Chemical reactions would take place much faster in bubbles (where reactants would be concentrated) than in the primordial soup proposed by Oparin and Haldane.

Step ❸ Bubbles rose to the surface and burst, releasing simple organic molecules into the air.

Step ❹ Carried upward by winds, the simple organic molecules were exposed to ultraviolet radiation and lightning, which provided energy for further reactions.

Step ❺ More complex organic molecules that formed by further reactions fell into the ocean with rain, starting another cycle.

Thus, the molecules of life could have appeared more quickly than is accounted for by the primordial soup model alone.

Study TIP

● *Reading Effectively*
Before reading this chapter, write the Objectives for each section on a sheet of paper. Rewrite each Objective as a question, and answer these questions as you read the section.

Figure 3

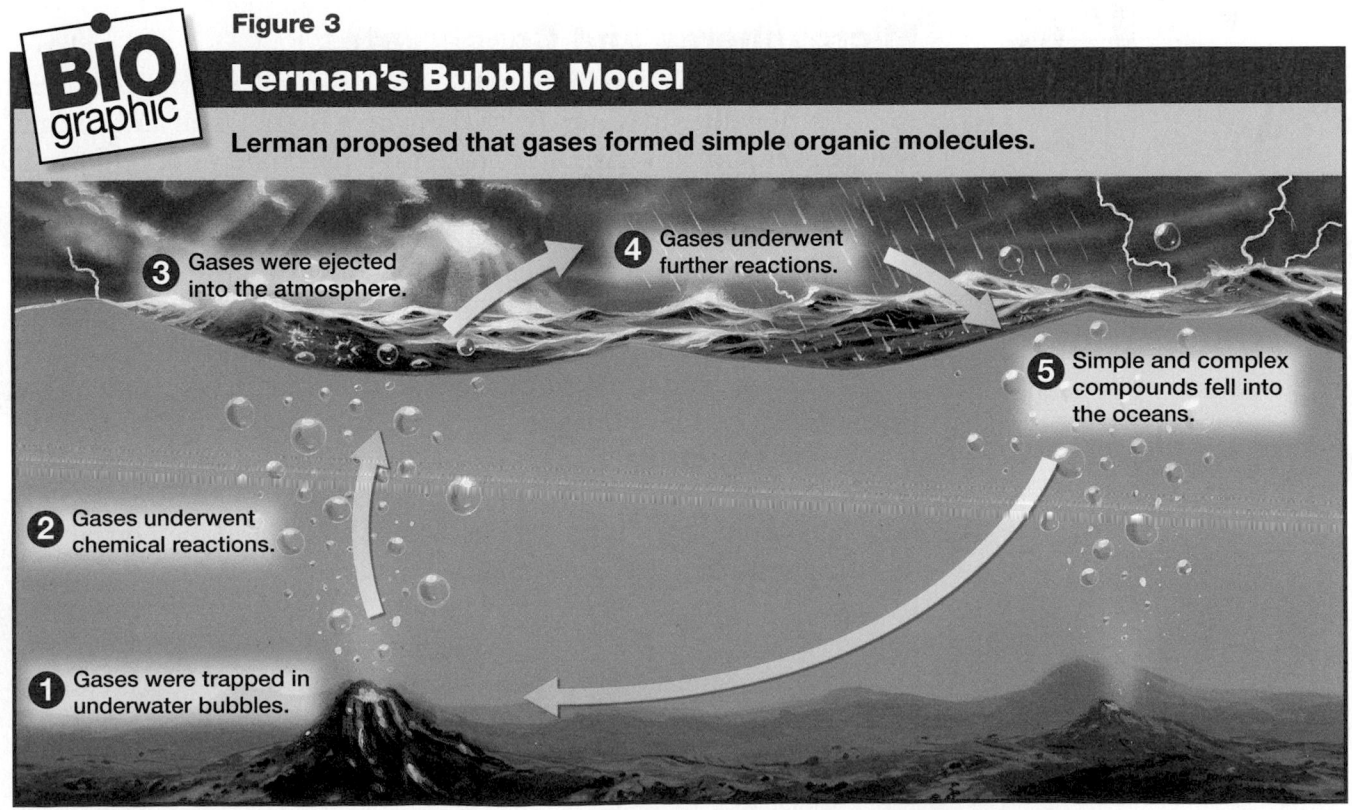

BIOgraphic

Lerman's Bubble Model

Lerman proposed that gases formed simple organic molecules.

❸ Gases were ejected into the atmosphere.

❹ Gases underwent further reactions.

❺ Simple and complex compounds fell into the oceans.

❷ Gases underwent chemical reactions.

❶ Gases were trapped in underwater bubbles.

Precursors of the First Cells

Scientists disagree about the details of the process that led to the origin of life. Most scientists, however, accept that under certain conditions, the basic molecules of life could have formed spontaneously through simple chemistry. But there are enormous differences between simple organic molecules and large organic molecules found in living cells. How did amino acids link to form proteins? How did nucleotides form the long chains of DNA that store the instructions for making proteins? In the laboratory, scientists have not been able to make either proteins or DNA form spontaneously in water. However, short chains of RNA, the nucleic acid that helps carry out DNA's instructions, have been made to form on their own in water.

A Possible Role As Catalysts

In the 1980s, American scientists Thomas Cech of the University of Colorado and Sidney Altman of Yale University found that certain RNA molecules can act like enzymes. RNA's three-dimensional structure provides a surface on which chemical reactions can be catalyzed. Messenger RNA acts as an information-storing molecule. As a result of Cech's and Altman's work and other experiments showing that RNA molecules can form spontaneously in water, a simple hypothesis was formed: RNA was the first self-replicating information-storage molecule and it catalyzed the assembly of the first proteins. More important, such a molecule would have been capable of changing from one generation to the next. This hypothesis is illustrated in **Figure 4.**

Microspheres and Coacervates

Observations show that lipids, which make up cell membranes, tend to gather together in water. By shaking up a bottle of oil and vinegar, you can see something similar happen—small spheres of oil form in the vinegar. Certain lipids, when combined with other molecules, can form a tiny droplet whose surface resembles a cell membrane. Similarly, laboratory experiments have shown that, in water, short chains of amino acids can gather into tiny droplets called **microspheres.** Another type of droplet, called a *coacervate (koh AS suhr VAYT)*, is composed of molecules of different types, including linked amino acids and sugars.

Scientists think that formation of microspheres might have been the first step toward cellular organization. According to this hypothesis, microspheres formed, persisted for a while, and then dispersed. Over millions of years, those microspheres that could persist longer by incorporating molecules and energy would have become more common than shorter lasting microspheres were. Microspheres could not be considered true cells, however, unless they had the characteristics of living things, including heredity.

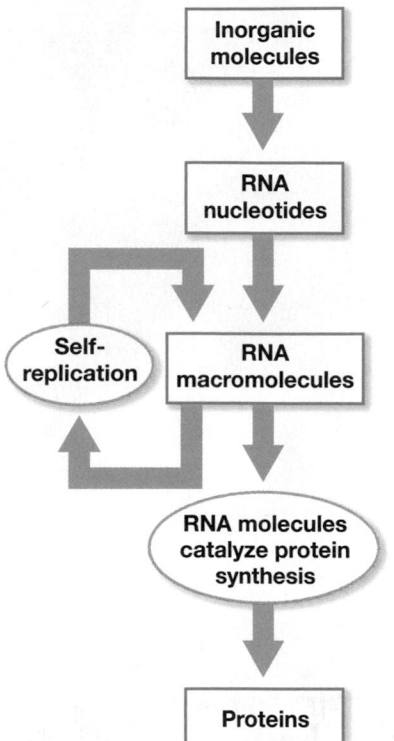

Figure 4 Proposed stages leading to RNA self-replication and protein synthesis. Chemical reactions between inorganic molecules formed RNA nucleotides. The nucleotides assembled into RNA macromolecules. These molecules were able to self-replicate and to catalyze the formation of proteins.

Origin of Heredity

Although scientists disagree about the details of the origin of heredity, many agree that double-stranded DNA evolved after RNA and that RNA "enzymes" catalyzed the assembly of the earliest proteins. Many scientists also tentatively accept the hypothesis that some microspheres or similar structures that contained RNA developed a means of transferring their characteristics to offspring. But researchers do not yet understand how DNA, RNA and hereditary mechanisms first developed. Therefore, the subject of how life might have originated naturally and spontaneously remains a subject of intense interest, research, and discussion.

QUICK LAB

Modeling Coacervates

By using simple chemistry, you will see that some properties of coacervates resemble the properties of cells.

Materials

safety goggles, protective gloves and lab apron, graduated cylinder, 1 percent gelatin solution, 1 percent gum arabic solution, test tube, 0.1 M HCl, pipet, microscope slide and coverslip, microscope

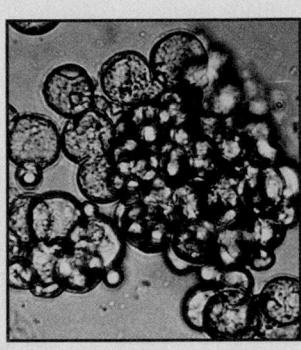

Procedure

1. **CAUTION: Hydrochloric acid is corrosive. Put on safety goggles, gloves, and apron. Avoid contact with skin and eyes. Avoid breathing vapors. If any of this solution should spill on you, immediately flush the area with water, and notify your teacher.**

2. Mix 5 mL of a 1 percent gelatin solution with 3 mL of a 1 percent gum arabic solution in a test tube.

3. Add 0.1 M HCl to the gelatin–gum arabic solution one drop at a time until the solution turns cloudy.

4. Prepare a wet mount of the cloudy solution, and examine it under a microscope at high power.

5. Prepare a drawing of the structures that you see. They should resemble the structures in the micrograph above.

Analysis

1. **Describe** what happened to the solutions after the acid was added.

2. **Compare** the appearance of coacervates with that of cells.

3. **Predict** what would happen to the coacervates if a base were added to the solution.

4. **Critical Thinking Evaluating Hypotheses** Based on the evidence you obtained, defend the hypothesis that coacervates could have been the basis of life on Earth.

Section 1 Review

1. **Explain** how radioisotopes are used to determine the age of a rock.

2. **Critique** two scientific models that explain the origin of life.

3. **Describe** the first step that may have led toward cellular organization.

4. **Explain** how heredity may have arisen.

5. **Standardized Test Prep** Miller and Urey's model is inconsistent with the finding that Earth's early atmosphere lacked
   A nitrogen. C water.
   B hydrogen. D ozone.

Objectives

- **Distinguish** between the two groups of prokaryotes.

- **Describe** the evolution of eukaryotes.

- **Recognize** an evolutionary advance first seen in protists.

- **Summarize** how mass extinctions have affected the evolution of life on Earth.

Key Terms

fossil
cyanobacteria
eubacteria
archaebacteria
endosymbiosis
protist
extinction
mass extinction

The Evolution of Prokaryotes

When did the first organisms form? To find out, scientists study the best evidence of early life that we have, fossils. A **fossil** is the preserved or mineralized remains (bone, tooth, or shell) or imprint of an organism that lived long ago. The oldest known fossils, which are microscopic fossils of prokaryotes, come from rock that is 2.5 billion years old.

Recall that prokaryotes are single-celled organisms that lack internal membrane-bound organelles. Among the first prokaryotes to appear were marine cyanobacteria. **Cyanobacteria** *(SIE an oh bak TIR ee ah)* are photosynthetic prokaryotes. Before cyanobacteria appeared, oxygen gas was scarce on Earth. But as ancient cyanobacteria carried out photosynthesis, they released oxygen gas into Earth's oceans. After hundreds of millions of years, the oxygen produced by cyanobacteria began to escape into the air, as shown in **Figure 5.** Over time, more oxygen was added to the air. Today oxygen gas makes up 21 percent of the Earth's atmosphere.

Two Groups of Prokaryotes

Early in the history of life, two different groups of prokaryotes evolved—eubacteria (which are commonly called *bacteria*) and archaebacteria. Living examples include *Escherichia coli,* a species of eubacteria, and *Sulfolobus,* a group of archaebacteria. **Eubacteria** are prokaryotes that contain a chemical called peptidoglycan *(PEP tih doh GLIE kan)* in their cell walls. Eubacteria include many bacteria that cause disease and decay.

Archaebacteria are prokaryotes that lack peptidoglycan in their cell walls and have unique lipids in their cell membranes. Modern archaebacteria are thought to closely resemble early archaebacteria. Chemical evidence indicates that archaebacteria and eubacteria diverged very early.

Figure 5 Evolutionary timeline. This timeline shows some of the major events that occurred during the evolution of life on Earth.

Age (in Millions of Years Ago)

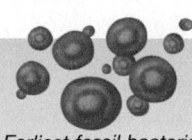

Earliest fossil bacteria   *Origin of O₂ by photosynthesis*

| 3,500 | 2,500 |

PRECAMBRIAN ERA

The Evolution of Eukaryotes

About 1.5 billion years ago, the first eukaryotes appeared. A eukaryotic cell is much larger than a prokaryote is. Eukaryotic cells have a complex system of internal membranes. Eukaryotic DNA is enclosed within a nucleus. Almost all eukaryotes have mitochondria. Chloroplasts, which carry out photosynthesis, are found only in protists and plants. Mitochondria and chloroplasts are the size of prokaryotes, and they contain their own DNA.

The Origins of Mitochondria and Chloroplasts

Most biologists think that mitochondria and chloroplasts originated as described by the theory of **endosymbiosis** that was proposed in 1966 by the American biologist Lynn Margulis. This theory proposes that mitochondria are the descendants of symbiotic, aerobic (oxygen-requiring) eubacteria and chloroplasts are the descendants of symbiotic, photosynthetic eubacteria.

Analyzing Signs of Endosymbiosis

Magnification: 6930x

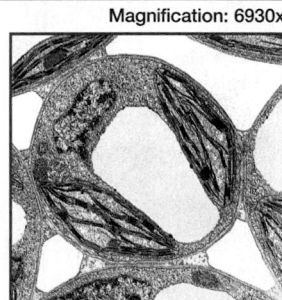

Background

You may recall that mitochondria have their own DNA and produce their own proteins. The data below were collected by scientists studying the proteins produced by mitochondrial DNA. The scientists found that the three-nucleotide sequences (codons) in the nucleus of an organism's cells can code for different amino acids than those coded for in the cell's mitochondria. Examine the data below, and answer the questions that follow.

Analysis

1. **Defend** the theory of endosymbiosis using these data.

2. **Infer** what these data indicate about the evolution of plant cells.

3. **Describe** how these data can be used to support the idea that more than one type of cell evolved early in the history of life.

Amino Acids Made in the Nucleus and Mitochondria			
	Amino acids or other instructions coded for in the nucleus	Amino acids or other instructions coded for in mitochondria	
Codon	Plants and mammals	Plants	Mammals
UGA	Stop	Stop	Tryptophan
AGA	Arginine	Arginine	Stop
AUA	Isoleucine	Isoleucine	Methionine
AUU	Isoleucine	Isoleucine	Methionine
CUA	Leucine	Leucine	Leucine

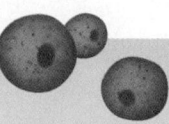

First eukaryotes

1,500

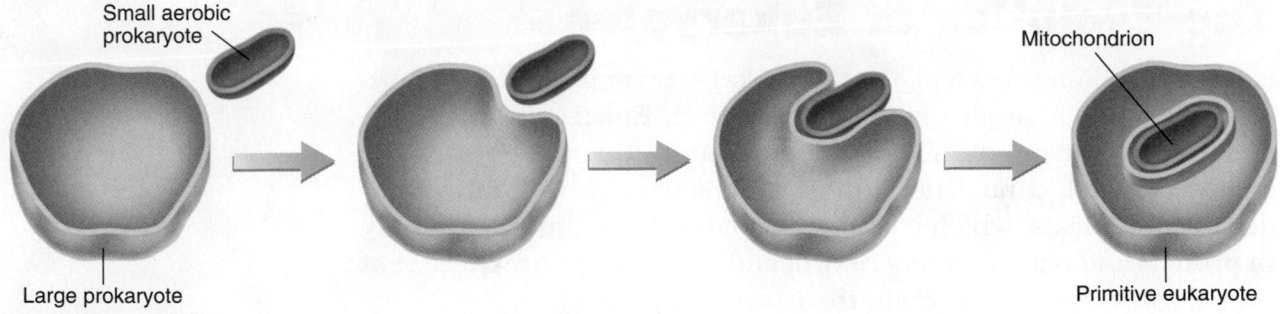

Figure 6 Endosymbiosis

Mitochondria are thought to have evolved from small, aerobic prokaryotes that began to live inside larger prokaryotes.

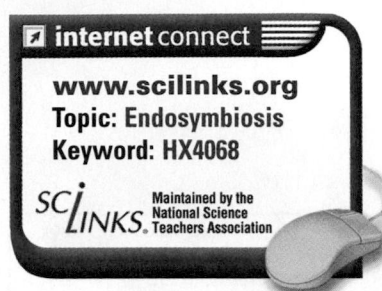

According to Lynn Margulis's theory of endosymbiosis, bacteria entered large cells either as parasites or as undigested prey as illustrated in **Figure 6.** Instead of being digested, the bacteria began to live inside the host cell, where they performed either cellular respiration (mitochondria) or photosynthesis (chloroplasts). The invading bacteria that became chloroplasts were probably closely related to cyanobacteria. Both mitochondria and chloroplasts have characteristics that are similar to those of bacteria. The following observations support the idea that mitochondria and chloroplasts descended from bacteria:

1. **Size and structure.** Mitochondria are about the same size as most eubacteria, and chloroplasts are the same size as some cyanobacteria. Both mitochondria and chloroplasts are surrounded by two membranes. The smooth outer membrane of mitochondria is thought to be derived from the endoplasmic reticulum of the larger host cell. The inner membrane of mitochondria is folded into many layers, so it looks like the cell membranes of aerobic eubacteria. Inside this membrane are proteins that carry out cellular respiration. Both chloroplasts and cyanobacteria contain thylakoids, structures in which photosynthesis takes place.

2. **Genetic material.** Mitochondria and chloroplasts have circular DNA similar to the chromosomes found in bacteria. Both chloroplasts and mitochondria contain genes that are different from those found in the nucleus of the host cell.

3. **Ribosomes.** Mitochondrial and chloroplast ribosomes have a size and structure similar to the size and structure of bacterial ribosomes.

4. **Reproduction.** Like bacteria, chloroplasts and mitochondria reproduce by simple fission. This replication takes place independently of the cell cycle of the host cell.

Age (in Millions of Years Ago)

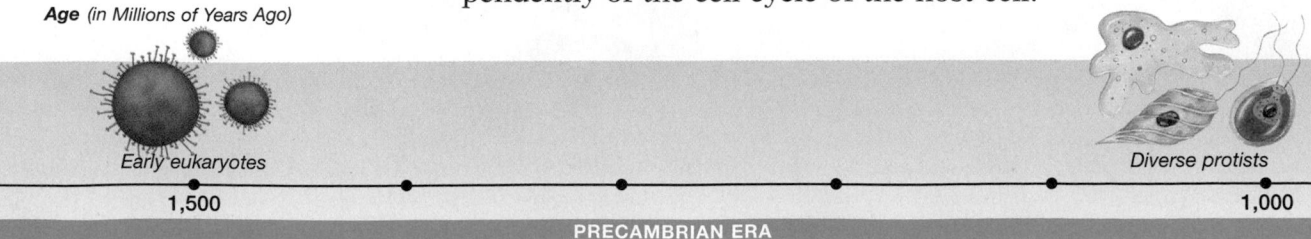

Early eukaryotes

1,500

Diverse protists

1,000

PRECAMBRIAN ERA

Multicellularity

Many biologists group all living things into six broad categories called kingdoms. The two oldest kingdoms, Eubacteria and Archaebacteria, are made up of single-celled prokaryotes. The first eukaryotic kingdom was the kingdom Protista. **Protists** make up a large, varied group that includes both multicellular and unicellular organisms. The other three kingdoms (fungi, plants, and animals) evolved later and also consist of eukaryotes.

The unicellular body plan has been tremendously successful, with unicellular organisms today constituting about half the biomass (the total weight of all living things) on Earth. But a single cell must carry out all of the activities of the organism. Distinct types of cells in one body can have specialized functions. For example, some organisms may have specific cells that help the organism protect itself from predators or disease. Other cells may help the organism resist drying out. Other examples of specialized cells include cells that help a multicellular organism move about in order to find a mate or food. With all these advantages, it is not surprising that multicellularity has arisen independently many times.

Almost every organism large enough to see with the naked eye is multicellular. Most protists, such as those shown in **Figure 7,** are single celled, but there are many multicellular forms. The development of multicellular organisms of the kingdom Protista marked an important step in the evolution of life on Earth. The oldest known fossils of multicellular organisms were found in 700 million year-old rocks.

Some of the multicellular lines that resulted did not produce diverse groups of organisms. Among those groups of organisms that survive today are plantlike red, green, and brown algae, shown in **Figure 8.** You may know these algae as seaweed. Three of the multicellular groups that evolved from the protists were very successful, producing three separate kingdoms—Fungi, Plantae, and Animalia. Each of these three kingdoms evolved from a protistan ancestor.

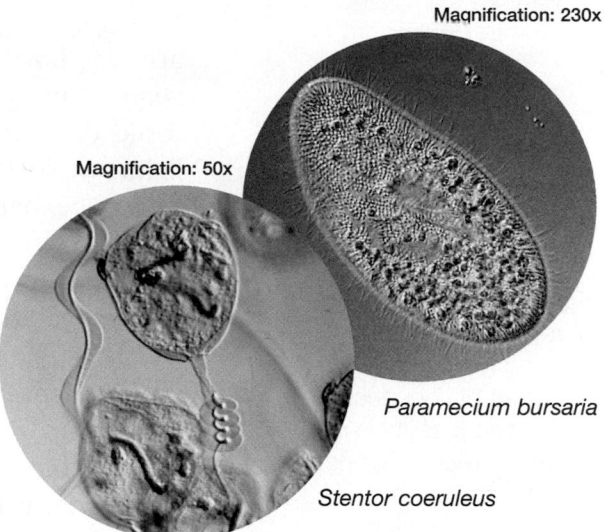

Magnification: 230x

Magnification: 50x

Paramecium bursaria

Stentor coeruleus

Figure 7 Single-celled protists. Single-celled protists occur in many shapes and can live in many different types of environments, including water and land.

Figure 8 Brown algae. Brown algae, called kelps, are multicellular protists that form vast underwater "forests" in some coastal waters.

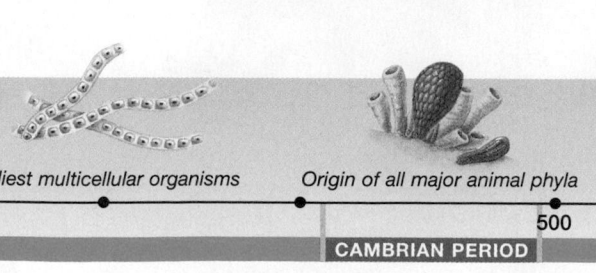

Earliest multicellular organisms

Origin of all major animal phyla

500

PRECAMBRIAN ERA

CAMBRIAN PERIOD

ORDOVICIAN PERIOD

Origins of Modern Organisms

Most animal phyla that exist today probably originated during a relatively short time (most estimates range from 10 to 100 million years) during the late Precambrian and early Cambrian periods. This rapid diversification of animals is sometimes known as the "Cambrian explosion." The Cambrian period was a time of great evolutionary expansion, as shown in **Figure 9.** Many unusual marine animals also appeared at this time, animals for which there are no close living relatives. A very rich collection of Cambrian fossils was uncovered in 1909 in a geological formation in Canada called the Burgess Shale. The fossils in the Burgess Shale include those of strange animals that are not like anything alive today.

The Ordovician period, which followed the Cambrian period, lasted from about 505 million to 438 million years ago. During this time, many different animals continued to abound in the seas. Among them were trilobites, marine arthropods that became extinct about 250 million years ago, shown in Figure 9.

Figure 9 Appearance of a Cambrian sea. By studying fossils from the Cambrian period, such as the trilobite fossil below, artists re-create a scene from the shallow seas of the Cambrian period.

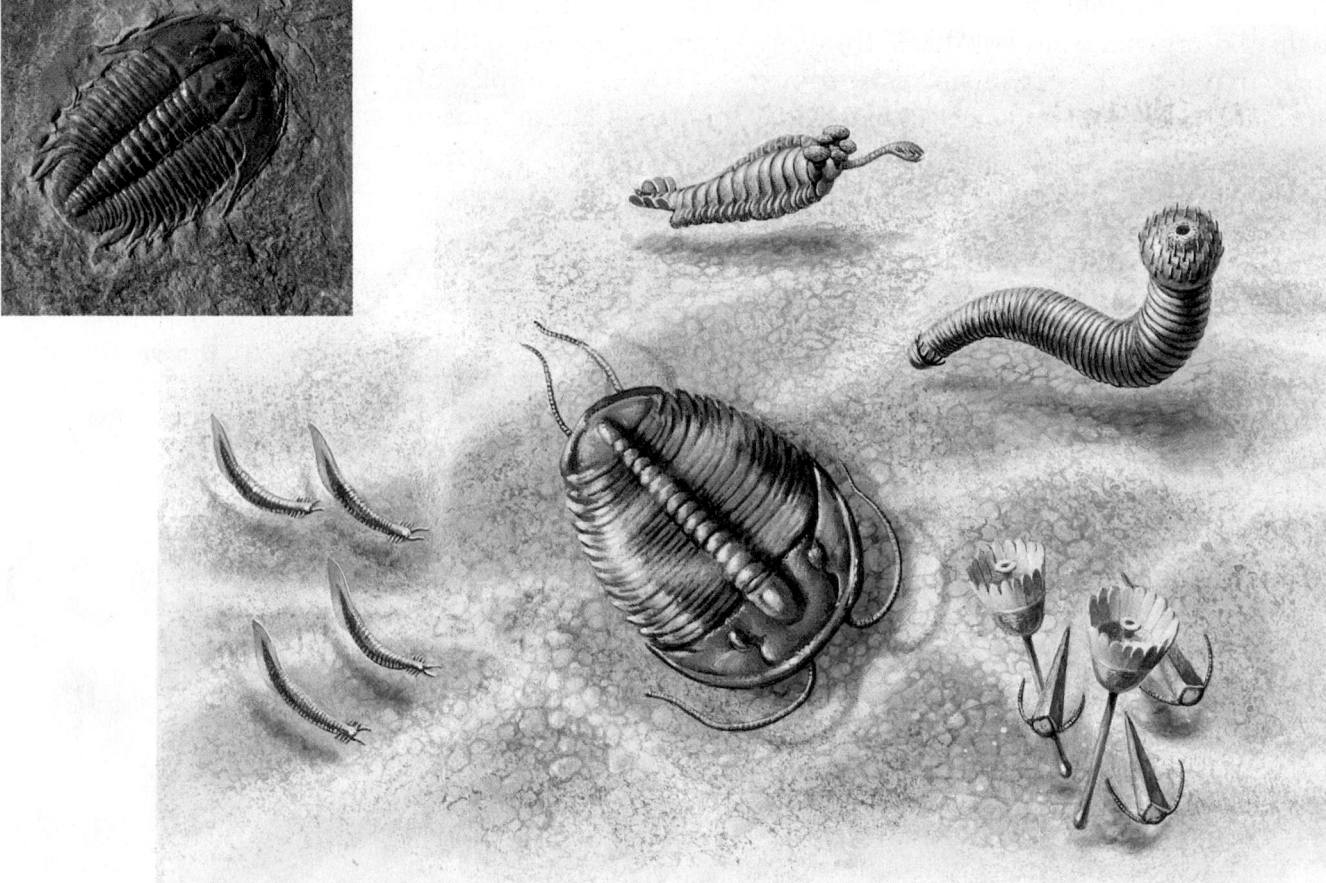

Age (in Millions of Years Ago)

Origin of all major animal phyla

560

510

PRECAMBRIAN ERA

CAMBRIAN PERIOD

Mass Extinctions

The fossil record indicates that a sudden change occurred at the end of the Ordovician period. About 440 million years ago, a large percentage of the organisms on Earth suddenly became extinct. **Extinction** is the death of all members of a species. This was the first of five major mass extinctions that have occurred on Earth. A **mass extinction** is an episode during which large numbers of species become extinct.

Another mass extinction of about the same size happened about 360 million years ago. The third and most devastating of all mass extinctions occurred at the end of the Permian period, about 245 million years ago. About 96 percent of all species of animals living at the time became extinct. About 35 million years later, a fourth, less devastating mass extinction occurred. Although the specific causes of these extinctions are unknown, evidence indicates that worldwide geological and weather changes were likely factors. The fifth mass extinction will be discussed in more detail in a later chapter. It occurred 65 million years ago and brought about the extinction of about two-thirds of all land species, including most of the dinosaurs.

Some scientists think that another mass extinction is occurring today. These scientists reason that this new extinction is taking place because the Earth's ecosystems, especially tropical rain forests, are being destroyed by human activity, as shown in **Figure 10.** The world has already lost half its tropical rain forests. If the current rate of destruction continues, from 22 percent to 47 percent of Earth's plant species will be lost, along with 2,000 of the world's 9,000 species of birds and countless insect species. This would be an astonishing loss of biodiversity.

internet connect

www.scilinks.org
Topic: Extinction
Keyword: HX4078

SC*LINKS* Maintained by the National Science Teachers Association

Figure 10 Rain forests are being destroyed at an alarming rate. Although tropical rain forests cover only 7 percent of the Earth's land surface, they contain more than one-half of all the world's animal and plant species.

Section 2 Review

1 Contrast the two major groups of prokaryotes.

2 Analyze Margulis's theory of endosymbiosis, citing its strengths and weaknesses.

3 Compare bacteria with eukaryotes.

4 Summarize how multicellularity advanced the evolution of protists.

5 Critical Thinking Justify the argument that today's organisms would not exist if mass extinctions had not occurred.

6 Standardized Test Prep The kingdom that includes both multicellular and unicellular eukaryotes is called
A Plantae.   **C** Eubacteria.
B Protista.   **D** Archaebacteria.

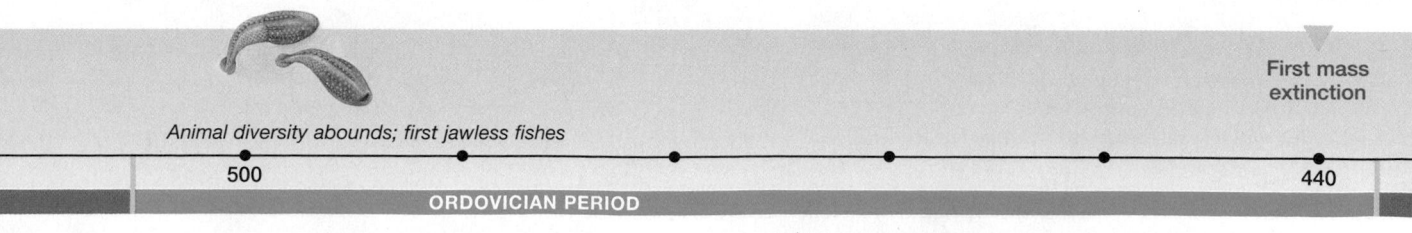

Animal diversity abounds; first jawless fishes

500

First mass extinction

440

ORDOVICIAN PERIOD

SILURIAN

Life Invaded the Land

Objectives

- **Relate** the development of ozone to the adaptation of life to the land.
- **Identify** the first multicellular organisms to live on land.
- **Name** the first animals to live on land.
- **Explain** the relationship between plants and fossil fuels.
- **List** the first vertebrates to leave the oceans.

Key Terms

mycorrhizae
mutualism
arthropod
vertebrate
continental drift

The Ozone Layer

The sun provides both life-giving light and dangerous ultraviolet radiation. Early in Earth's history, life formed in the seas, where early organisms were protected from ultraviolet radiation. These organisms could not leave the water because ultraviolet radiation made life on dry ground unsafe. What enabled life-forms to leave the protection of the seas and live on the land?

Formation of the Ozone Layer

During the Cambrian period and for millions of years afterward, organisms did not live on the dry, rocky surface of Earth. However, a slow change was taking place. About 2.5 billion years ago, photosynthesis by cyanobacteria began adding oxygen to Earth's atmosphere. As oxygen began to reach the upper atmosphere, the sun's rays caused some of the molecules of oxygen, O_2, to chemically react and form molecules of ozone, O_3. In the upper atmosphere, ozone blocks the ultraviolet radiation of the sun, as shown in **Figure 11**. After millions of years, enough ozone had accumulated to make the Earth's land a safe place to live.

Figure 11 Ozone shields the Earth

As ancient cyanobacteria added oxygen to the atmosphere, ozone began to form.

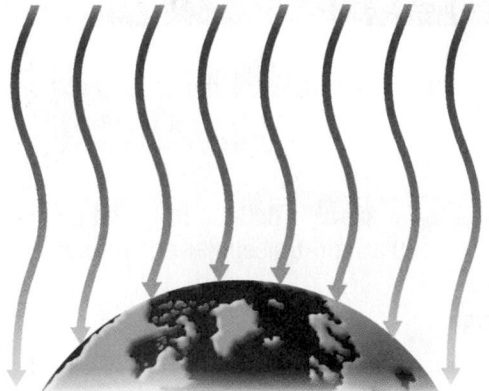
Ultraviolet radiation

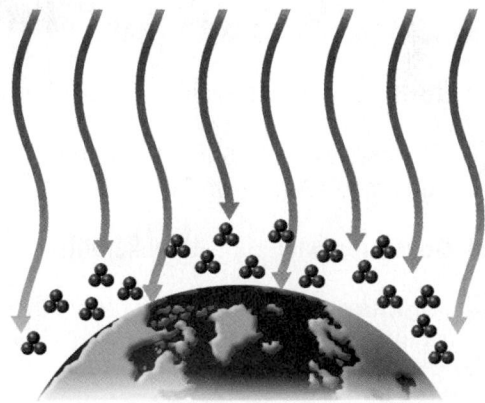
Ultraviolet radiation

Age (in Millions of Years Ago)

Plants and arthropods invade land; jawed fishes first appear

Bony fishes become abundant

440	430	410	400	390
	SILURIAN PERIOD		**DEVONIAN PERIOD**	

Plants and Fungi on Land

The first multicellular organisms to live on land may have been fungi living together with plants or algae. Such paired organisms were able to live on land because each group possessed a quality needed by the other.

Plants, which likely evolved from photosynthetic protists, could carry out photosynthesis. In photosynthesis, plants use the energy from sunlight to make carbohydrates. Plants cannot, however, harvest needed minerals from bare rock. In contrast, fungi cannot make nutrients from sunlight but can absorb minerals—even from bare rock.

Early plants and fungi formed biological partnerships called mycorrhizae *(MIE koh RIE zee)*, which enabled them to live on the harsh habitat of bare rock. **Mycorrhizae,** which exist today, are symbiotic associations between fungi and the roots of plants, as shown in **Figure 12.** The fungus provides minerals to the plant, and the plant provides nutrients to the fungus. This kind of partnership is called mutualism. **Mutualism** is a relationship between two species in which both species benefit. Plants and fungi began living together on the surface of the land about 430 million years ago.

Fossilized *Cooksonia*

Example of living mycorrhizae

Magnification: 15x

Root

Fungus

Figure 12 Mycorrhizae formed on the roots of the first plants. This fossil is of *Cooksonia*, the first known vascular plant, which lived 410 million year ago. *Cooksonia* was only a few centimeters tall. *Cooksonia*'s roots formed mycorrhizae similar to the living mycorrhizae shown in color to the left.

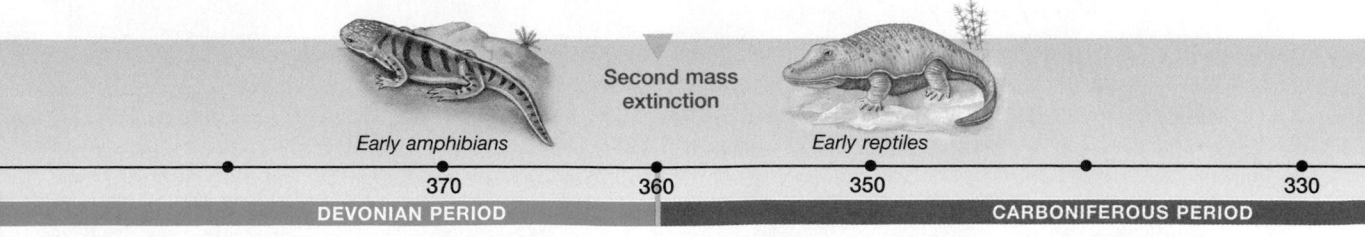

Second mass extinction

Early amphibians

Early reptiles

370	360	350		330
DEVONIAN PERIOD		CARBONIFEROUS PERIOD		

Arthropods

By 100 million years after their first union with fungi, plants had covered the surface of the Earth, forming large forests. These land plants provided a food source for land-dwelling animals. The first animals to successfully invade land from the sea were arthropods. An **arthropod** is a kind of animal with a hard outer skeleton, a segmented body, and paired, jointed limbs. Examples of arthropods include lobsters, crabs, insects, and spiders, like the one in **Figure 13.** Biologists think a type of scorpion was the first arthropod to live on land.

A unique kind of terrestrial arthropod—the insect—evolved from the first land dwellers. Insects have since become the most plentiful and diverse group of animals in Earth's history. The success of the insects is probably connected to their ability to fly. Insects were the first animals to have wings. Flying allowed insects, like the dragonfly shown in **Figure 14,** to efficiently search for food, mates, and nesting sites. It also led to partnerships between insects and flowering plants. In the great coal swamps of the Carboniferous period, organic materials such as the plants also shown in Figure 14 were subjected to pressure from overlying earth. Over millions of years this produced fossil fuels—beds of coal and reservoirs of oil. Humans now burn both oil and coal to release stored energy, in the process, also releasing carbon dioxide.

Figure 13   An arthropod. This marbled spider is a member of the phylum Arthropoda, which includes about 1 billion billion (10^{18}) individuals in about 1.5 million described species.

Figure 14   Swamp 320 million years ago. Forested swamps were dominated by tall, seedless canopy trees and shorter tree ferns. Dragonflies had wingspans of more than 1 m (about 3.25 ft).

Age (in Millions of Years Ago)

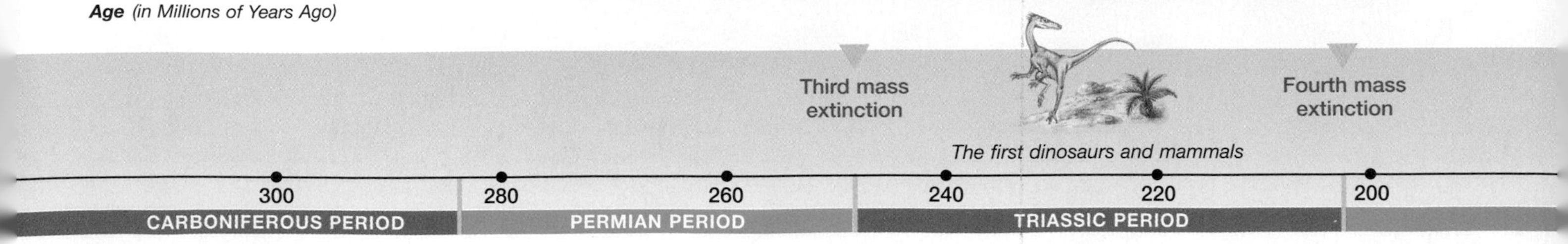

Third mass extinction

Fourth mass extinction

The first dinosaurs and mammals

300	280	260	240	220	200
CARBONIFEROUS PERIOD	PERMIAN PERIOD		TRIASSIC PERIOD		

Vertebrates

A **vertebrate** is an animal with a backbone—vertebrates are the animals most familiar to us. Humans are vertebrates, and almost all other land animals bigger than our fist are vertebrates as well.

Fishes

According to the fossil record, the first vertebrates were small, jawless fishes that evolved in the oceans about 530 million years ago. Jawed fishes first appeared about 430 million years ago. Jaws enabled fishes to bite and chew their food instead of sucking up their food. As a result, jawed fishes were efficient predators. A fossilized example of a jawed fish is shown in **Figure 15.** Fishes soon came to be among the most abundant animals in the seas, and for hundreds of millions of years the sea is where vertebrates stayed. Fishes are the most successful living vertebrates—they make up more than half of all modern vertebrate species. After nearly 200 million years of living in the sea, fishes have become uniquely adapted for success in water. Major changes had to occur in fish body organization, however, before some descendants of fishes became capable of living on land.

Figure 15   Fossilized fish skeleton. This fish skeleton clearly shows the backbone, the structure that is characteristic of all vertebrate animals.

Amphibians

The first vertebrates to inhabit the land did not come out of the sea until 370 million years ago. Those first land vertebrates were early amphibians. Amphibians are smooth-skinned, four-legged animals that today include frogs, toads, and salamanders.

Several structural changes in the bodies of amphibians occurred as they adapted to life on land. Amphibians had moist breathing sacs—lungs—which allowed the animals to absorb oxygen from air. The limbs of amphibians are thought to have derived from the bones of fish fins. The evolution of a strong support system of bones in the region just behind the head made walking possible. This system of bones provided a rigid base for the limbs to work against. Because of their strong, flexible internal skeleton, the bodies of vertebrates can be much larger than those of insects. While amphibians were well adapted to their environment, a new group of animals more suited to a drier environment evolved from them.

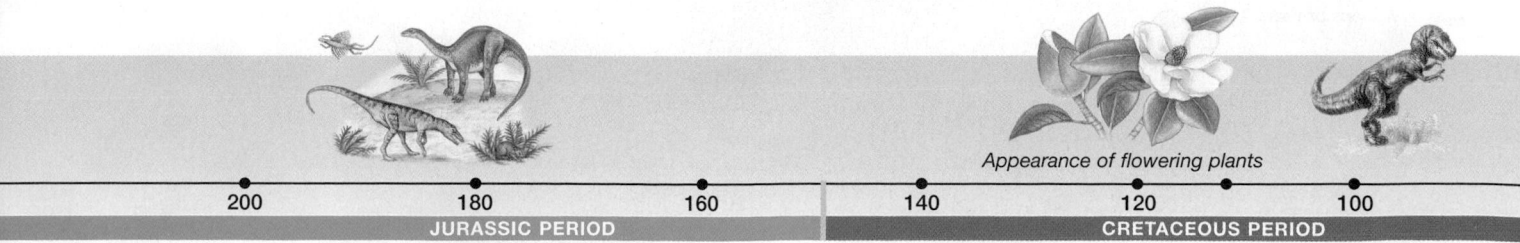

Appearance of flowering plants

200	180	160	140	120	100
JURASSIC PERIOD			**CRETACEOUS PERIOD**		

Reptiles

Reptiles evolved from amphibian ancestors about 340 million years ago. Modern reptiles include snakes, lizards, turtles, and crocodiles. Reptiles are better suited to dry land than amphibians because reptiles' watertight skin slows the loss of moisture. Reptiles also have a watertight egg, such as the one shown in **Figure 16.** Unlike amphibians, reptiles can lay their eggs on dry land. Amphibians must lay their eggs in water or in very moist soil because their eggs are unable to retain enough water to remain alive.

Figure 16 Reptiles. Reptiles, such as this crocodile, were the largest group of land-dwelling organisms until the end of the Cretaceous period.

Mammals and Birds

Birds apparently evolved from feathered dinosaurs during or after the Jurassic period. Therapsids, reptiles with complex teeth and legs positioned beneath their body, gave rise to mammals about the same time dinosaurs evolved, during the Triassic period. Sixty-five million years ago, during the fifth mass extinction, most species disappeared forever. All of the dinosaurs except for the ancestors of birds became extinct. The smaller reptiles, mammals, and birds survived. Although many resources were available to the surviving animals, the world's climate was no longer largely dry. Thus the reptiles' advantages in dry climates were not so important. Birds and mammals then became the dominant vertebrates on land.

Both extinctions and continental drift played important roles in evolution. **Continental drift** is the movement of Earth's land masses over Earth's surface through geologic time. Continental drift resulted in the present-day position of the continents. The movement of continents helps explain why there are a large number of marsupial (pouched) mammal species in both Australia and South America, continents that were once connected.

Section 3 Review

1. **Summarize** why ozone was important in enabling organisms to live on land.

2. **Name** the first multicellular organisms that colonized land.

3. **Identify** the first kinds of animals to live on land.

4. **Describe** the first kinds of vertebrates that inhabited land.

5. **Critical Thinking Defend** the argument that fossil fuels are *not* a renewable resource.

6. **Standardized Test Prep** Mycorrhizae are mutualistic relationships between the roots of plants and
   A amphibians.   C cyanobacteria.
   B insects.   D fungi.

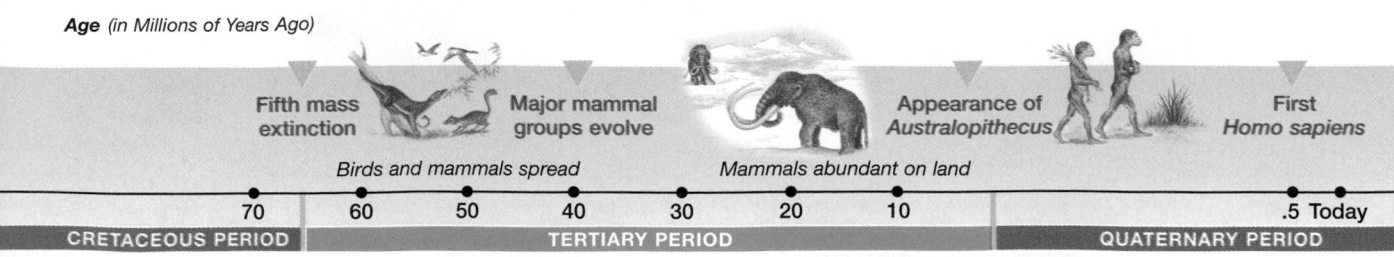

Age (in Millions of Years Ago)

Fifth mass extinction — Major mammal groups evolve — Appearance of *Australopithecus* — First *Homo sapiens*

Birds and mammals spread   Mammals abundant on land

| 70 | 60 | 50 | 40 | 30 | 20 | 10 | .5 Today |
| CRETACEOUS PERIOD | | TERTIARY PERIOD | | | | | QUATERNARY PERIOD |

Study ZONE

Key Concepts

1 How Did Life Begin?

- The Earth formed about 4.5 billion years ago according to evidence obtained by radiometric dating.

- The primordial soup model and the bubble model propose explanations of the origin of the chemicals of life.

- Scientists think RNA formed before DNA or proteins formed.

- Scientists think that the first cells may have developed from microspheres.

- The development of heredity made it possible for organisms to pass traits to subsequent generations.

2 Complex Organisms Developed

- Prokaryotes are the oldest organisms and are divided into two groups, archaebacteria and eubacteria.

- Prokaryotes likely gave rise to eukaryotes through the process of endosymbiosis.

- Mitochondria and chloroplasts are thought to have evolved through endosymbiosis.

- Multicellularity arose many times and resulted in many different groups of multicellular organisms.

- Extinctions influenced the evolution of the species alive today.

3 Life Invaded the Land

- Ancient cyanobacteria produced oxygen, some of which became ozone. Ozone enabled organisms to live on land.

- Plants and fungi formed mycorrhizae and were the first multicellular organisms to live on land.

- Arthropods were the first animals to leave the ocean.

- The first vertebrates to invade dry land werc amphibians.

- The extinction of many reptile species enabled birds and mammals to become the dominant vertebrates on land.

- The movement of the continents on the surface of the Earth has contributed to the geographic distribution of some species.

Key Terms

Section 1

radiometric dating (252)
radioisotope (252)
half-life (252)
microsphere (256)

Section 2

fossil (258)
cyanobacteria (258)
eubacteria (258)
archaebacteria (258)
endosymbiosis (259)
protist (261)
extinction (263)
mass extinction (263)

Section 3

mycorrhizae (265)
mutualism (265)
arthropod (266)
vertebrate (267)
continental drift (268)

Performance ZONE

Understanding Key Ideas

1. After three half-lives of a radioisotope have passed, how much of the original radioisotope has decayed?
   a. 1/8
   b. 1/2
   c. 3/4
   d. 7/8

2. Unlike the primordial soup model, Lerman's bubble model takes _____ into account.
   a. ozone
   b. ultraviolet radiation
   c. lightning
   d. volcanoes

3. Cells are different from microspheres because cells
   a. contain amino acids.
   b. have a two-layer outer boundary.
   c. grow by taking in molecules from their surroundings.
   d. transfer information through heredity.

4. Cell specialization came about as a result of
   a. endosymbiosis.
   b. archaebacteria.
   c. the Cambrian period.
   d. multicellularity.

5. The first multicellular organisms to invade the land were
   a. reptiles.
   b. amphibians.
   c. fungi and plants.
   d. mammals.

6. Describe the evidence that supports the theory of endosymbiosis.

7. The half-life of carbon-14 is 5,700 years. If a sample originally had 26 g of carbon-14, how much would it contain after 22,800 years?

8. What is the half-life of the radioisotope represented in the graph?

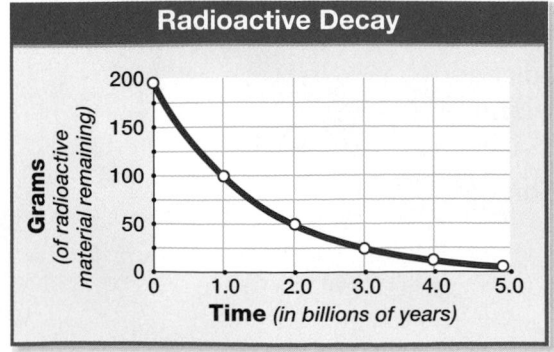

Radioactive Decay

(Graph: Grams (of radioactive material remaining) vs. Time (in billions of years))

9. Relate the order in which different types of organisms invaded the land to the flow of energy. (**Hint:** See Chapter 5, Section 1.)

10. **Concept Mapping** Construct a concept map that shows how life might have originated by natural forces. Include the following items in your map: *spontaneous origin, primordial soup model, bubble model, RNA, proteins, microspheres,* and *radioisotopes.*

Critical Thinking

11. **Evaluating Viewpoints** Several scientists have said that if a large asteroid struck the Earth, the impact could result in a mass extinction. If an asteroid impact did not kill all organisms, would evolution continue or stop? Explain.

12. **Recognizing Relationships** Propose a hypothesis for the appearance of all animal phyla on Earth within a relatively short period during the late Precambrian and early Cambrian periods.

Alternative Assessment

13. **Being a Team Member** Work together in groups to design a poster to illustrate the different models that describe how life's chemicals may have originated. Show how the compounds on early Earth would have participated in each of these models.

14. **Finding and Communicating Information** Thomas Cech and Sidney Altman shared a Nobel prize in 1989 for their work on RNA. Research their work and the rewards associated with winning a Nobel prize. Relate your findings in an oral report.

15. **Finding and Communicating Information** Use the media center or Internet resources to study scientific hypotheses for the origin of life that are alternatives to the hypotheses proposed by Oparin and Lerman. Analyze either Oparin's or Lerman's hypotheses as presented in your textbook along with one alternative scientific hypotheses that you discover in your research.

Understanding Concepts

Directions (1–3): For *each* question, write on a separate sheet of paper the letter of the correct answer.

1 What did the development of heredity allow organisms to store and pass on to their offspring?
   A. energy
   B. information
   C. radioisotopes
   D. UV radiation

2 A cell's surface-area-to-volume ratio limits the size that unicellular organisms can achieve. How can multicellularity solve this problem?
   F. As a cell grows in size, its surface-area-to-volume ratio increases, which improves the efficiency of the cell.
   G. Multicellular organisms tend to have larger cells than unicellular organisms, which makes them stronger.
   H. Communication between the nucleus and other parts of the cell is faster in larger cells than in smaller cells.
   I. An organism with many small cells has a greater surface-area-to-volume ratio than an organism with one large cell.

3 Photosynthesis in what organisms originally formed the oxygen that became ozone in Earth's atmosphere?
   A. archaebacteria
   B. arthropods
   C. cyanobacteria
   D. mycorrhizae

Directions (4): For the following question, write a short response.

4 Identify and describe the event that resulted in the origin of eukaryotes.

Test TIP

If time permits, take short mental breaks to improve your concentration during a test.

Reading Skills

Directions (5): Read the passage below. Then answer the question.

Originally, there were no organisms living on dry land. One factor that contributed to this occurrence was that outside of an aquatic environment, organisms could not get all of the nutrients and minerals they needed to survive. Then, a symbiotic relationship developed between plants and fungi, which provided nutrients for the fungi and minerals for the plants. This allowed them to colonize the bare land. Land plants served as food for insects, which then served as food for amphibians and other organisms.

5 What type of relationship did the plants and fungi have?
   F. aquatic   **H.** mutualistic
   G. microscopic   **I.** predatory

Interpreting Graphics

Directions (6): Base your answer to question 6 on the diagram below.

Experimental Setup

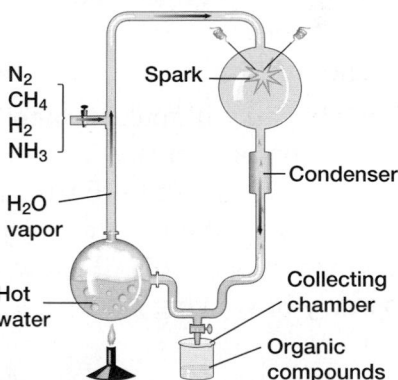

6 What conclusion could be drawn from the results of the experiment shown?
   A. Earth's early atmosphere lacked N_2.
   B. N_2 and H_2 can be converted into NH_3 when heated.
   C. Water can be changed into organic compounds if it is heated vigorously.
   D. Organic compounds can form under conditions such as those in the experiment.

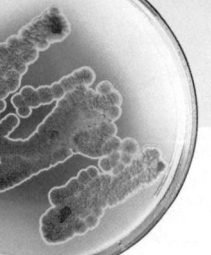

Exploration Lab

Making a Timeline of Life on Earth

SKILLS
- Observing
- Inferring relationships
- Organizing data

OBJECTIVES
- **Compare** and **contrast** the distinguishing characteristics of representative organisms of the six kingdoms.
- **Organize** the appearance of life on Earth in a timeline.

MATERIALS
- 5 m roll of adding-machine tape
- meterstick
- colored pens or pencils
- photographs or drawings of organisms from ancient Earth to present day

Before You Begin

About 4.5 billion years ago, Earth was a ball of molten rock. As the surface cooled, a rocky crust formed and water vapor in the atmosphere condensed to form rain. By 3.9 billion years ago, oceans covered much of the Earth's surface. Rocks formed in these oceans contain **fossils** of bacterial cells that lived about 3.5 billion years ago. The **fossil record** shows a progression of life-forms and contains evidence of many changes in Earth's surface and atmosphere.

In this lab, you will make a **timeline** showing the major events in Earth's history and in the history of life on Earth, such as the evolution of new groups of organisms and the mass extinctions. This timeline can be used to study how living things have changed over time.

1. Write a definition for each boldface term in the paragraphs above.
2. Make a data table similar to the one at right.
3. Based on the objectives for this lab, write a question you would like to explore about the history of life on Earth.

Procedure

PART A: Making a Timeline

1. Make a mark every 20 cm along a 5 m length of adding-machine tape. Label one end of the tape "5 billion years ago" and the other end "Today." Write "20 cm = 200 million years" near the beginning of your timeline.

2. Locate and label a point representing the origin of Earth on your timeline. Use your textbook as a reference. See the timeline at the bottom of Section 2 and Section 3 of this chapter. Also locate and label the 11 periods of the geologic time scale beginning with the Cambrian period.

3. Using your textbook as a reference, mark the following events on your timeline: the first cyanobacteria appear; oxygen enters the atmosphere; the five mass extinctions; the first eukaryotes appear; the first multicellular organisms appear; the first vertebrates appear; the first plants,

DATA TABLE		
Organism	Kingdom	Characteristics/adaptation for life on Earth

fungi, and land animals appear; the first dinosaurs and mammals appear; the first flowering plants appear; the first humans appear.

4. Look at the photographs of organisms provided by your teacher. Identify the major characteristics of each organism. Record your observations in your data table.

5. Lay out your timeline on the floor in your classroom. Place photographs (or drawings) of the organisms you examined on your timeline to show when they appeared on Earth.

6. Fold the timeline at the mark representing 4.8 billion years ago. This leaves 24 segments, each representing 200 million years, in your timeline. Now you can think of each segment as 1 hour in a 24-hour day.

7. When you are finished, walk slowly along your timeline. Note the sequence of events in the history of life on Earth and the relative amount of time between each event.

PART B: Cleanup and Disposal

8. Dispose of paper scraps in the designated waste container.

9. Clean up your work area and all lab equipment. Return lab equipment to its proper place.

Analyze and Conclude

1. **Analyzing Information** Think of each segment of your timeline as 1 hour in a 24-hour day as you answer each of the following questions:
   a. How long has life existed on Earth?
   b. For what part of the day did only unicellular life-forms exist?
   c. At what time of day did the first plants appear on Earth?
   d. At what time of day did mammals appear on Earth?

2. **Summarizing Information** Identify the major developments in life-forms that have occurred over the last 3.5 billion years.

3. **Inferring Relationships** How do mass extinctions appear to be related to the appearance of new major groups of organisms?

4. **Justifying Conclusions** Cyanobacteria are thought to be responsible for adding oxygen to Earth's atmosphere. Use your timeline to justify this conclusion.

5. **Calculating** Determine the amount of time, as a percentage of the time that life has existed on Earth, that humans (*Homo sapiens*) have existed.

6. **Further Inquiry** Write a new question about the history of life on Earth that could be explored in another investigation.

On the Job

Timelines are used to organize events in chronological order. Do research to discover how other scientists use timelines in their work. For more about careers, visit **go.hrw.com** and type in the keyword **HX4 Careers**.

Phyllium pulchrifolium
(beautiful moving leaf)

13 The Theory of Evolution

✔ *Quick Review*

Answer the following without referring to earlier sections of your book.

1. **Describe** the structure of proteins. *(Chapter 2, Section 3)*

2. **Relate** the sequence of nucleotides in DNA to the amino acid sequence in proteins. *(Chapter 10, Section 2)*

3. **Define** genetic mutations. *(Chapter 10, Section 2)*

4. **Describe** gene sequencing. *(Chapter 11, Section 1)*

5. **Summarize** the concept of radiometric dating. *(Chapter 12, Section 1)*

Did you have difficulty? *For help, review the sections indicated.*

📖 *Reading Activity*

Create a Reader Response Log to record your personal responses to the concepts presented in this chapter. Divide your paper in half. On the left side of the paper, copy a word, phrase, or passage from the text. On the right side, write your reactions, thoughts, or questions about your entries from the text.

Looking Ahead

Section 1

The Theory of Evolution by Natural Selection
Darwin Proposed a Mechanism for Evolution
Evolution by Natural Selection
Darwin's Ideas Updated

Section 2

Evidence of Evolution
The Fossil Record
Anatomy and Development
Biological Molecules

Section 3

Examples of Evolution
Natural Selection at Work
Formation of New Species

■ internet connect

www.scilinks.org
National Science Teachers Association *sci*LINKS Internet resources are located throughout this chapter.

SC*I*LINKS. Maintained by the National Science Teachers Association

● The body of the beautiful moving leaf insect closely resembles the leaves on which it lives. Camouflage such as this helps protect animals from predators.

The Theory of Evolution by Natural Selection

Objectives

- **Identify** several observations that led Darwin to conclude that species evolve.
- **Relate** the process of natural selection to its outcome.
- **Summarize** the main points of Darwin's theory of evolution by natural selection as it is stated today.
- **Contrast** the gradualism and punctuated equilibrium models of evolution.

Key Terms

population
natural selection
adaptation
reproductive isolation
gradualism
punctuated
  equilibrium

Darwin Proposed a Mechanism for Evolution

The idea that life evolves may have been first proposed by Lucretius, a Roman philosopher who lived about 2,000 years ago before the modern theory of evolution was proposed. Then, in 1859, the English naturalist Charles Darwin, shown in **Figure 1,** published convincing evidence that species evolve, and he proposed a reasonable mechanism explaining how evolution occurs.

Like all scientific theories, the theory of evolution has developed through decades of scientific observation and experimentation. The modern theory of evolution began to take shape as a result of Darwin's work. Today almost all scientists accept that evolution is the basis for the diversity of life on Earth.

As a youth, Darwin struggled in school. His father was a wealthy doctor who wanted him to become either a doctor or a minister. Not interested in the subjects his father urged him to study, Darwin frequently spent more time outdoors than in class. At the age of 16, Darwin was sent to Edinburgh, Scotland, to study medicine. Repelled by surgery, which at the time was done without anesthetics, Darwin repeatedly skipped lectures to collect biological specimens. In 1827, Darwin's father sent him to Cambridge University, in England, to become a minister. Although he completed a degree in theology, Darwin spent much of his time with friends who were also interested in natural science.

Figure 1 Charles Darwin. Darwin was born in England in 1809 and died in 1882.

In 1831, one of Darwin's professors at Cambridge recommended him for a position as a naturalist on a voyage of HMS *Beagle*. Although the ship had an official naturalist, the *Beagle*'s captain preferred to have someone aboard who was of his own social class. At the age of 22, Darwin set off on a journey that would both change his life and forever change how we think of ourselves. The ship and its route are shown in **Figure 2.**

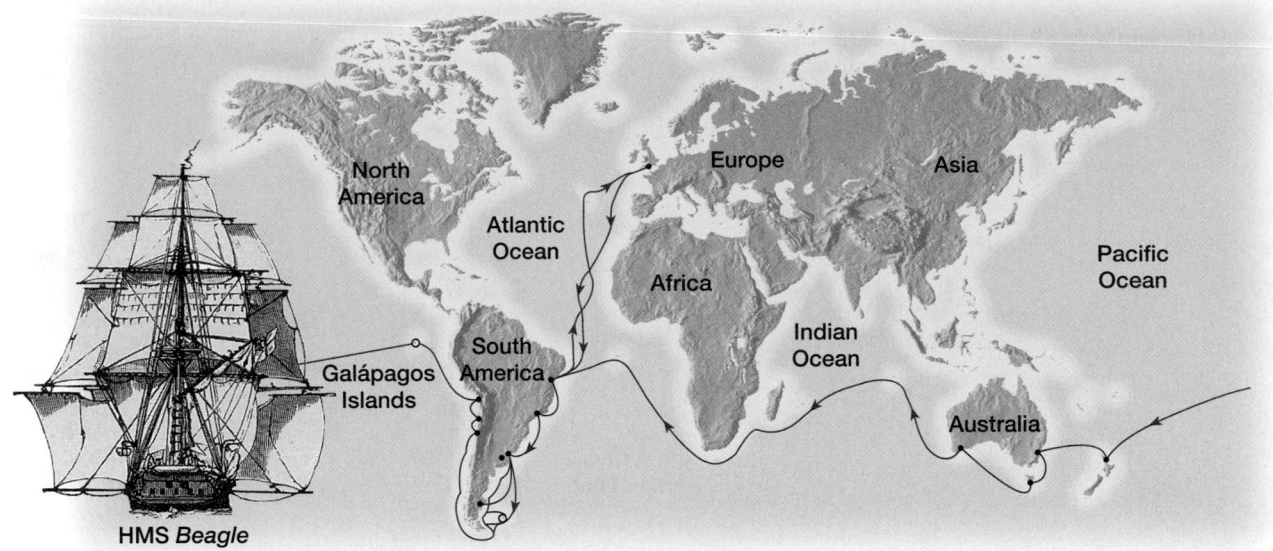

North America

Europe

Asia

Atlantic Ocean

Africa

Pacific Ocean

Galápagos Islands

South America

Indian Ocean

Australia

HMS *Beagle*

Figure 2   The route of HMS *Beagle*. HMS *Beagle* sailed around the world along the route shown on this map. The purpose of the ship's 5-year voyage was to survey the coast of South America.

Science Before Darwin's Voyage

In Darwin's time, most people—including scientists—held the view that each species is a divine creation that exists, unchanging, as it was originally created. But scientists had begun to seek to explain the origins of fossils. Some scientists tried to explain their observations by altering traditional explanations of creation. Others (including Darwin's own grandfather) proposed various mechanisms to explain how living things change over time.

In 1809, the French scientist Jean Baptiste Lamarck (1744–1829) proposed a hypothesis for how organisms change over generations. Lamarck believed that over the lifetime of an individual, physical features increase in size because of use or reduce in size because of disuse. Further, according to Lamarck, these changes are then passed on to offspring. This part of Lamarck's hypothesis is now known to be incorrect. However, Lamarck correctly pointed out that change in species is linked to the "physical conditions of life," referring to an organism's environmental conditions.

Darwin's Observations

During his voyage on the *Beagle*, Darwin found evidence that challenged the traditional belief that species are unchanging. During the voyage, Darwin read Charles Lyell's book *Principles of Geology*. Lyell proposed that the surface of Earth changed slowly over many years. As Darwin visited different places, he also saw things that he thought could be explained only by a process of gradual change. For example, in South America, Darwin found fossils of extinct armadillos. These fossilized animals closely resembled, but were not identical to, the armadillos living in the area.

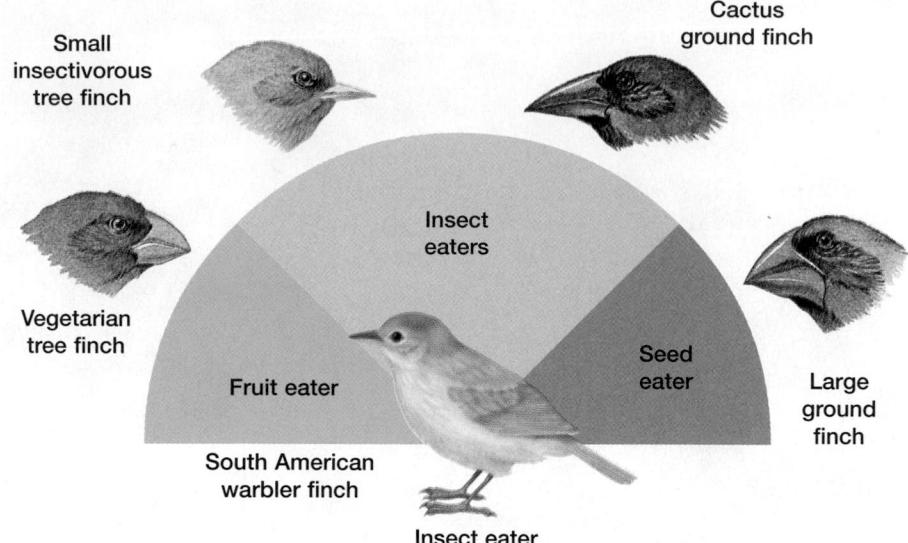

Figure 3 Darwin's finches.
Darwin discovered that these finches closely resembled South American finches.

Cactus ground finch

Small insectivorous tree finch

Insect eaters

Vegetarian tree finch

Fruit eater

Seed eater

Large ground finch

South American warbler finch

Insect eater

Darwin visited the Galápagos Islands, located about 1,000 km (620 mi) off the coast of Ecuador. Darwin was struck by the fact that many of the plants and animals of the Galápagos Islands resembled those of the nearby coast of South America. Darwin later suggested that the simplest explanation for this was that the ancestors of Galápagos species such as those shown in **Figure 3,** migrated to the islands from South America long ago and changed after they arrived. Darwin later called such a change "descent with modification"—evolution.

When Darwin returned from his voyage at the age of 27, he continued his lifelong study of plants, animals, and geology. However, he did not report his ideas about evolution until many years later. During those years, Darwin studied the data from his voyage. As Darwin studied his data, his confidence that organisms had evolved grew ever stronger. But he was still deeply puzzled about how evolution occurs.

Growth of Populations

The key that unlocked Darwin's thinking about how evolution takes place was an essay written in 1798 by the English economist Thomas Malthus. Malthus wrote that human populations are able to increase faster than the food supply can. Malthus pointed out that unchecked populations grow by geometric progression, as shown in **Figure 4.** Food supplies, however, increase by an arithmetic progression at best, also shown in Figure 4. He suggested that human populations do not grow unchecked because death caused by disease, war, and famine slows population growth.

The term *population*, as it is used in biology, does not only refer to the human population. In the study of biology, a **population** consists of all the individuals of a species that live in a specific geographical area and that can interbreed.

Figure 4 Geometric and arithmetic progressions.
The blue graph line shows uncontrolled population growth, in which the numbers increase by a multiplied constant. The red graph line shows increased food supply, in which the numbers increase by an added constant.

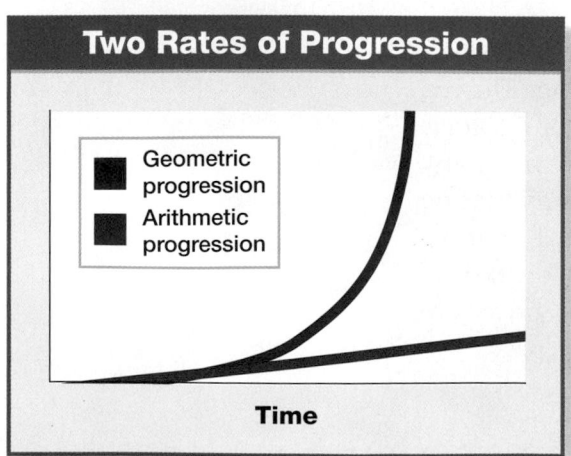

Two Rates of Progression

■ Geometric progression
■ Arithmetic progression

Time

Evolution by Natural Selection

Darwin realized that Malthus's hypotheses about human populations apply to all species. Every organism has the potential to produce many offspring during its lifetime. In most cases, however, only a limited number of those offspring survive to reproduce. Considering Malthus's view and his own observations and experience in breeding domestic animals, Darwin made a key association. *Individuals that have physical or behavioral traits that better suit their environment are more likely to survive and will reproduce more successfully than those that do not have such traits.* Darwin called this differential rate of reproduction **natural selection.** In time, the number of individuals that carry favorable characteristics that are also inherited will increase in a population. And thus the nature of the population will change—a process called *evolution*.

Darwin further suggested that organisms differ from place to place because their habitats present different challenges to, and opportunities for, survival and reproduction. Each species has evolved and has accumulated adaptations in response to its particular environment. An **adaptation** is an inherited trait that has become common in a population because the trait provides a selective advantage.

☑ **internet** connect

www.scilinks.org
Topic: Natural Selection
Keyword: HX4128

SC*LINKS* Maintained by the National Science Teachers Association

Publication of Darwin's Work

In 1844, Darwin finally wrote down his ideas about evolution and natural selection in an early outline that he showed to only a few scientists he knew and trusted. At about this time, both a newly published book that claimed that evolution occurred, and Lamarck's hypotheses about evolution were harshly criticized. Shrinking from such controversy, Darwin put aside his manuscript.

Darwin decided to publish after he received a letter and essay in June 1858 from the young English naturalist Alfred Russel Wallace (1823–1913), who was in Malaysia at the time. Wallace's essay described a hypothesis of evolution by natural selection! In his letter, he asked if Darwin would help him get the essay published. Darwin's friends arranged for a summary of Darwin's manuscript to be presented with Wallace's paper at a public scientific meeting.

Figure 5   Political cartoon of Charles Darwin. This 1874 cartoon of Darwin with a monkeylike "ancestor" is an example of how some people ridiculed Darwin because of his work.

Darwin's Theory

Darwin's book *On the Origin of Species by Means of Natural Selection* appeared in November of 1859. Many people were deeply disturbed by Darwin's theory, including the suggestion, made in a later work, that humans are related to apes, as **Figure 5** on the previous page suggests. But Darwin's arguments and evidence that evolution occurs slowly convinced biologists around the world. Darwin's theory of evolution by natural selection is supported by four major points:

1 Inherited variation exists within the genes of every population or species (the result of random mutation and translation errors).

2 In a particular environment, some individuals of a population or species are better suited to survive (as a result of variation) and have more offspring (natural selection).

3 Over time, the traits that make certain individuals of a population able to survive and reproduce tend to spread in that population.

4 There is overwhelming evidence from fossils and many other sources that living species evolved from organisms that are extinct.

QUICK LAB

Modeling Natural Selection

By making a simple model of natural selection you can begin to understand how natural selection changes a population.

Materials

paper, pencil, watch or stopwatch

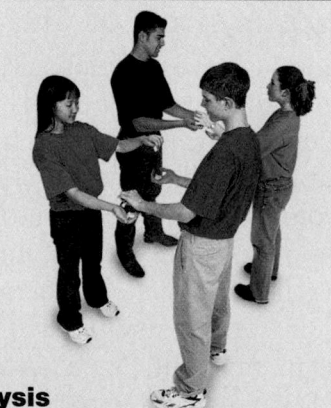

Procedure

1. On a chalkboard or overhead projector, make a data table like the one shown below.

2. Write each of the following words on separate pieces of paper: live, die, reproduce, mutate. Fold each piece of paper in half twice so that you cannot see the words. Shuffle your folded pieces of paper.

3. Exchange two of your pieces of paper with those of a classmate. Make as many exchanges with additional classmates as you can in 30 seconds. Mix your pieces of paper between each exchange you make.

4. Look at your pieces of paper. If you have two pieces that say "die" or two pieces that say "mutate," then sit down. If you do not, then you are a "survivor." Record your results in your class table.

5. If you are a "survivor," record the words you are holding in the data table. Then refold your pieces of paper and repeat steps 2 and 3 two more times with other "survivors."

Analysis

1. Identify what the four slips of paper represent.

2. Describe what happens to most mutations in this model.

3. Identify what factor(s) determined who "survived." Explain.

4. Evaluate the shortcomings of this model of natural selection.

DATA TABLE			
Student name	Trial 1	Trial 2	Trial 3

Darwin's Ideas Updated

Since the time Darwin's work was published, his hypothesis—that natural selection explains how evolution happens—has been carefully examined by biologists. New discoveries, especially in the area of genetics, have given scientists new insight into how natural selection brings about the evolution of species.

Change Within Populations

Darwin's key inference was based on the idea that in any population, individuals that are best suited to survive and do well in their environment will produce the most offspring. So, the traits of those individuals will become more common in each new generation.

Scientists now know that genes are responsible for inherited traits. Therefore, certain forms of a trait become more common in a population because more individuals in the population carry the alleles for those forms. In other words, natural selection causes the *frequency* of certain alleles in a population to increase or decrease over time. Mutations and the recombination of alleles that occurs during sexual reproduction provide endless sources of new variations for natural selection to act upon.

Species Formation

The environment differs from place to place. Thus, populations of the same species living in different locations tend to evolve in different directions. **Reproductive isolation** is the condition in which two populations of the same species do not breed with one another because of geographic separation, a difference in mating periods, or other barrier to reproduction. As two isolated populations of the same species become more different over time, they may eventually become unable to breed with one another. Generally, when the individuals of two related populations can no longer breed with one another, the two populations are different species. As shown in **Figure 6,** the Kaibab squirrel, which lives on the North Rim of the Grand Canyon in Arizona, has a black belly and other characteristics that distinguish it from the Abert squirrel. The Abert squirrel, which has a white belly, lives on the South Rim of the Grand Canyon. Because they have been so isolated from one another, they have become different enough that some biologists consider them separate species.

Figure 6 Reproductive isolation in action

These two squirrel populations became isolated from each other about 10,000 years ago, thus preventing their interbreeding.

Kaibab squirrel

Abert squirrel

The Tempo of Evolution

For decades, most biologists have understood evolution as a gradual process that occurs continuously. The model of evolution in which gradual change over a long period of time leads to species formation is called **gradualism.** But American biologists Stephen Jay Gould and Niles Eldredge have suggested that successful species may stay unchanged for long periods of time. Gould and Eldredge have hypothesized that major environmental changes in the past have caused evolution to occur in spurts. This model of evolution, in which periods of rapid change in species are separated by periods of little or no change, is called **punctuated equilibrium.**

Exploring Further

Punctuated Equilibrium

How could major environmental changes lead to spurts in evolution? The fossil record shows that drastic environmental changes have occurred very infrequently, separated by periods of time that often last tens of millions of years. Events such as volcanic eruptions, asteroid impacts, and ice ages have been linked to sudden and drastic changes in climate. Such changes have also been linked to the extinction of many groups of organisms. As a result, environments that were once inhabited became empty. This provided opportunities for colonization by species that could quickly adapt to the new conditions through natural selection.

What Fossils Reveal

Despite large gaps, due most likely to poor conditions for fossilization, there is some evidence of both gradualism and punctuated equilibrium in the fossil record. Many groups of organisms appear suddenly in the fossil record. Some of these groups remain virtually unchanged for millions of years, while other groups disappear as suddenly as they appear. Still other groups of organisms appear to

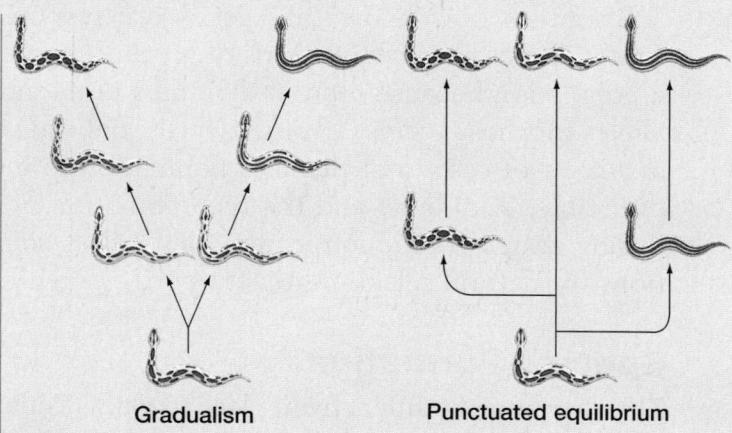

Gradualism   Punctuated equilibrium

change slowly through time, as predicted by the gradualism model of evolution. More study of the fossil record may reveal additional examples of one or both types of evolution.

Section 1 Review

1 **List** two observations made by Charles Darwin during his 5-year voyage that led him to conclude that living species evolved from extinct species.

2 **Describe** how natural selection occurs.

3 **Summarize** the modern theory of evolution by natural selection.

4 **Compare** the punctuated equilibrium model of evolution with the gradualism model.

5 **Standardized Test Prep** Speciation can result when two populations have become

A extinct.

B reproductively isolated.

C interbred.

D one population.

Evidence of Evolution

The Fossil Record

Have you ever looked at a series of maps that show how a city has grown? Buildings and streets are added, changed, or destroyed as the years pass by. In the same way, fossils of animals show a pattern of development from early ancestors to modern descendants. Fossils offer the most direct evidence that evolution takes place. Recall that a fossil is the preserved or mineralized remains or imprint of an organism that lived long ago. Fossils, therefore, provide an actual record of Earth's past life-forms. Change over time (evolution) can be seen in the fossil record. Fossilized species found in older rocks are different from those found in newer rocks, as you can see in **Figure 7.**

After observing such differences, Darwin predicted that intermediate forms between the great groups of organisms would eventually be found. Since Darwin's time, some of these intermediates have been found, while others have not. For example, fossil intermediaries have been found between fishes and amphibians, between reptiles and birds, and between reptiles and mammals, adding valuable evidence about the fossil history of the vertebrates.

Today, Darwin's theory is almost universally accepted by scientists as the best available explanation for the biological diversity on Earth. Based on a large body of supporting evidence, most scientists agree on the following three major points:

1. Earth is about 4.5 billion years old.

2. Organisms have inhabited Earth for most of its history.

3. All organisms living today share common ancestry with earlier, simpler life-forms.

Objectives

- **Describe** how the fossil record supports evolution.

- **Summarize** how biological molecules such as proteins and DNA are used as evidence of evolution.

- **Infer** how comparing the anatomy and development of living species provides evidence of evolution.

Key Terms

paleontologist
vestigial structure
homologous
   structure

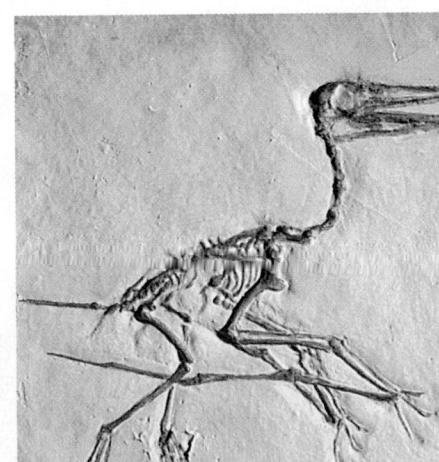

Crinoid

Pterodactyl

Figure 7 Fossils. Fossils of early multicellular life-forms, such as the crinoid, occur in 800-million-year-old rocks found in Indiana. Fossils of the pterodactyl, an extinct reptile, occur in 140- to 210-million-year-old rocks.

Formation of Fossils

Study TIP

● **Reading Effectively**

Read the heading "Formation of Fossils," and ask one or more Who, What, Where, When, Why, or How questions. For example, How are fossils formed? As you read, answer your questions.

The fossil record, and thus the record of the evolution of life, is not complete. Many species have lived in environments where fossils do not form. Most fossils form when organisms and traces of organisms are rapidly buried in fine sediments deposited by water, wind, or volcanic eruptions. The environments that are most likely to cause fossil formation are wet lowlands, slow-moving streams, lakes, shallow seas, and areas near volcanoes that spew out volcanic ash. The chances that organisms living in upland forests, mountains, grasslands, or deserts will die in just the right place to be buried in sediments and fossilized are very low. Even if an organism lives in an environment where fossils can form, the chances are slim that its dead body will be buried in sediment before it decays. For example, it is very likely to be eaten and scattered by scavengers.

Figure 8   Evidence of whale evolution

Whales are thought to have evolved from an ancestral line of four-legged mammals, which are represented here by their fossils and artistic reconstructions showing what scientists think they may have looked like.

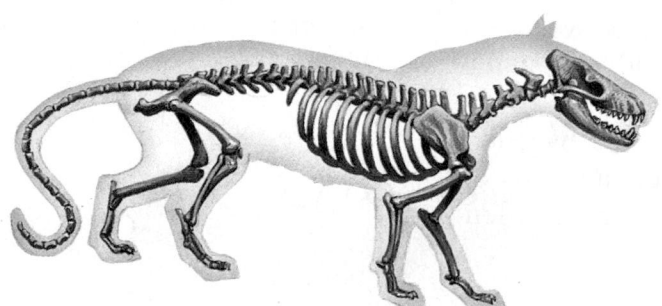

Mesonychids are one hypothesized link between modern whales and certain hoofed mammals. They were about 2 m (6 ft) long. They are thought to have lived about 60 million years ago. Some scientists favor an alternative hypothesis linking whales to other ancestral hooved mammals. These hooved mammals are also ancestral to hippopotamuses or pigs.

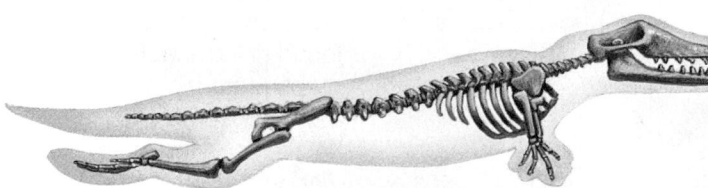

Ambulocetus natans apparently walked on land like modern sea lions and swam by flexing its backbone and paddling with its hind limbs (as do modern otters). They were about 3 m (10 ft) long. They existed about 50 million years ago.

Furthermore, the bodies of some organisms decay faster than others do. For example, an animal with a hard exoskeleton (such as a crab) would have a better chance of becoming fossilized than would a soft-bodied organism, such as an earthworm.

Although the fossil record will never be complete, it presents strong evidence that evolution has taken place. When a fossil is discovered, **paleontologists** (scientists who study fossils) analyze the sediments around it. By radiometric dating certain types of rocks and minerals in those sediments, paleontologists can arrange the fossils in order from oldest to youngest. When this is done, orderly patterns of evolution can be seen. Based on existing fossils, **Figure 8** shows an artist's idea of the appearance of three extinct species that might have been ancestral to modern whales. They are arranged in the order that they evolved, based on their fossil's age as determined by radiometric dating.

☐ internet connect
www.scilinks.org
Topic: Paleontology
Keyword: HX4134
SCLINKS. Maintained by the National Science Teachers Association

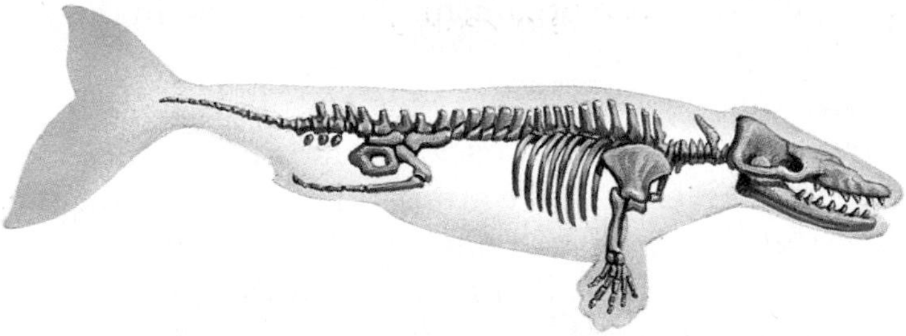

Rodhocetus kasrani, a more recent ancestor of modern whales, probably spent little time on land. Its reduced hind limbs could not have aided in walking or swimming. It is thought to have existed about 40 million years ago.

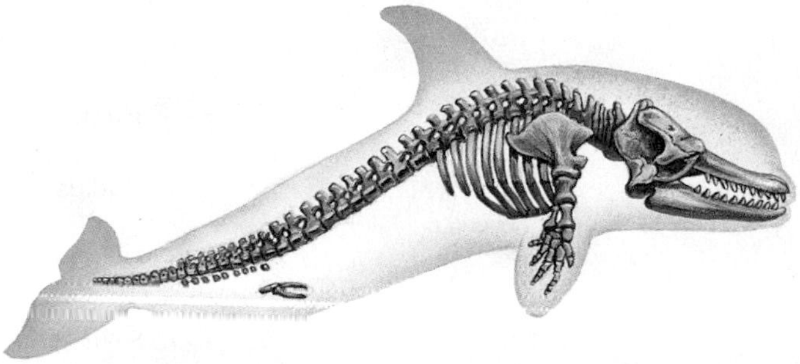

Modern whales have forelimbs that are flippers and hind limbs that have been reduced to only a few internal functionless hind-limb bones.

Anatomy and Development

WORD *Origins*

● The word *vestigial* comes from the Latin word *vestigium,* meaning "footprint." *Homologous* is from the Greek word *homologos,* meaning "agreeing."

Comparisons of the anatomy of different types of organisms often reveal basic similarities in body structures even though the structure's functions may differ between organisms. For example, sometimes bones are present in an organism but are reduced in size and either have no use or have a less important function than they do in other, related organisms. Such structures, which are considered to be evidence of an organism's evolutionary past, are called **vestigial** *(vehs TIJ ee uhl)* **structures.** For example, the hind limbs of whales are vestigial structures.

As different groups of vertebrates evolved, their bodies evolved differently. But similarities in bone structure can still be seen, suggesting that all vertebrates share a relatively recent common ancestor. As you can see in **Figure 9,** the forelimbs of the vertebrates shown are composed of the same basic groups of bones. Such structures are referred to as homologous *(hoh MAHL uh guhs).* **Homologous structures** are structures that share a common ancestry. That is, a similar structure in two organisms can be found in the common ancestor of the organisms.

Most scientists believe that the evolutionary history of organisms is also seen in the development of embryos. At some time in their development, all vertebrate embryos have a tail, buds that become limbs, and pharyngeal *(fuh RIN jee uhl)* pouches. The tail remains in most adult vertebrates. Only adult fish and immature amphibians retain pharyngeal pouches (which contain their gills). In humans, the tail disappears during fetal development, and pharyngeal pouches develop into structures in the throat.

Figure 9 Homologous structures

The forelimbs of vertebrates contain the same kinds of bones, which form in the same way during embryological development.

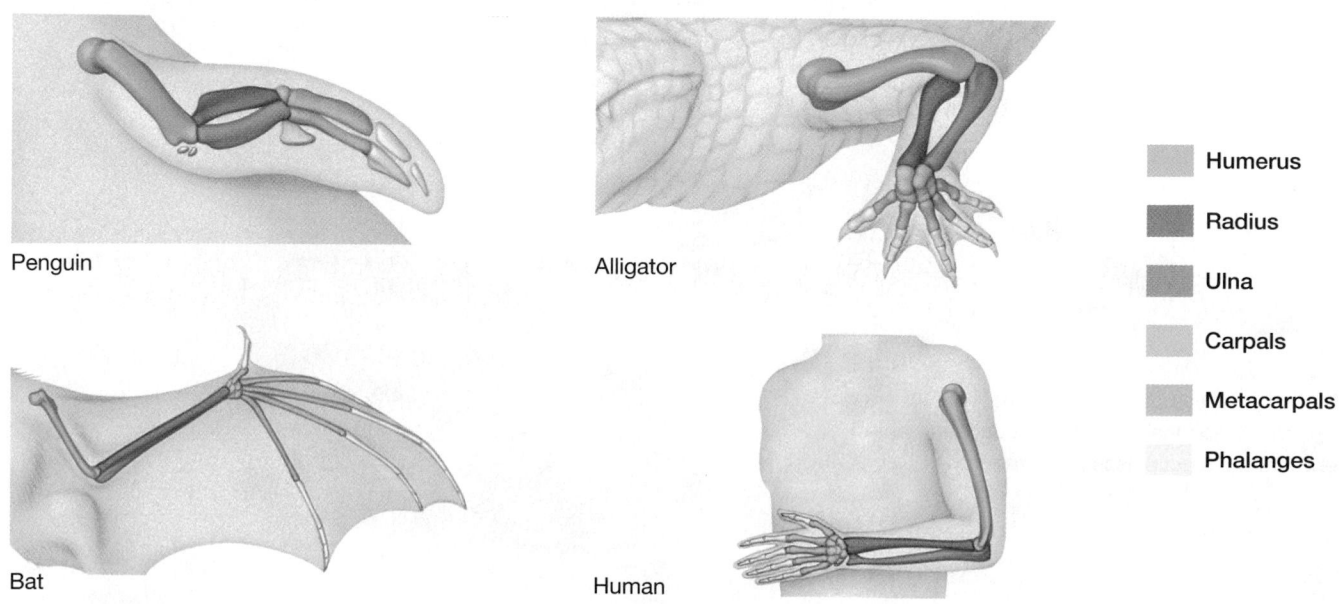

Penguin

Alligator

Bat

Human

Humerus

Radius

Ulna

Carpals

Metacarpals

Phalanges

Biological Molecules

The picture of successive change seen in the fossil record allows scientists to make a prediction that can be tested. If species have changed over time as the fossil record indicates, then the genes that determine the species' characteristics should also have changed by mutation and selection. As species evolved, one change after another should have become part of their genetic instructions. Therefore, changes in a gene's nucleotide sequence should build up over time.

Proteins

This prediction was first tested by analyzing the amino acid sequences of similar proteins found in several species. If evolution has taken place, then, in general, species descended from a recent common ancestor should have fewer amino acid differences between their proteins than do species that share a common ancestor in the more distant past.

Comparing the same hemoglobin protein in several species reveals the pattern shown in **Figure 10**. Species that are thought to have shared a common ancestor more recently (for example, humans and gorillas) have few amino acid sequence differences. However, those species that are thought to have shared a common ancestor in the more distant past (such as humans and mice) have many amino acid sequence differences.

DNA Sequences

This pattern, however, does not hold true for all proteins. A certain protein may evolve more rapidly in some groups than others. Comparisons of proteins, therefore, may not reflect evolutionary relationships supported by the fossil record and other evidence. Evolutionary histories, however, are generally not inferred from any single protein's amino acid sequences. More accurate hypotheses about evolutionary histories are based on large numbers of gene sequences. These evolutionary histories based on DNA sequences tend be very similar to evolutionary histories inferred by biologists based comparative anatomy and evidence from the fossil record.

Hemoglobin Comparison	
Species	Amino Acid Differences from Human Hemoglobin Protein
Gorilla	1
Rhesus monkey	8
Mouse	27
Chicken	45
Frog	67
Lamprey	125

Figure 10   Hemoglobin differences. The more similar organisms' hemoglobin proteins are, the more recent the organisms' common ancestor is likely to have been.

Section 2 Review

1. **Relate** how the fossil record provides evidence that evolution has occurred.

2. **State** how comparing the amino acid sequence of a protein can provide evidence that evolution has taken place.

3. **Describe** how comparing the anatomy of living species provides evidence of evolution.

4. (**Standardized Test Prep**) Which two organisms would likely have the least-similar nucleotide sequences in a given gene?
   A chimpanzee and gorilla
   B gorilla and dog
   C dog and shark
   D shark and butterfly

Examples of Evolution

Natural Selection at Work

How does evolution occur? The heart of Darwin's theory of evolution is that natural selection is the mechanism that drives evolution. Darwin wrote: "Can we doubt . . . that individuals having any advantage, however slight, over others, would have the best chance of surviving and of procreating their kind? On the other hand, we may feel sure that any variation in the least degree injurious would be rigidly destroyed. This preservation of favorable variations, I call Natural Selection." In his writings, Darwin offered examples of how natural selection has shaped life on Earth. There are now many well-known examples of natural selection in action.

The key lesson scientists have learned about evolution by natural selection is that the *environment* dictates the direction and amount of change. If the environment changes in the future, the set of characteristics that most help an individual reproduce successfully may change. For example, the polar bear's white fur, shown in **Figure 11,** enables it to hunt successfully in its snowy environment. In a warmer environment, having white fur would no longer be an advantage.

Factors in Natural Selection

The process of natural selection is driven by four important points that are true for all real populations:

❶ All populations have genetic variation. That is, in any population there is an array of individuals that differ slightly from each other in genetic makeup. While this may be obvious in humans, it is also true in species whose members may appear identical, such as a species of bacteria.

❷ The environment presents challenges to successful reproduction. Naturally, an organism that does not survive to reproduce or whose offspring die before the offspring can reproduce does not pass its genes on to future generations.

❸ Individuals tend to produce more offspring than the environment can support. Thus individuals of a population often compete with one another to survive.

❹ Individuals that are better able to cope with the challenges presented by their environment tend to leave more offspring than those individuals less suited to the environment do.

Figure 11 Polar bear.
Camouflage benefits predators and prey alike.

Example of Natural Selection

The lung disease tuberculosis (TB) is usually caused by the bacterium *Mycobacterium tuberculosis*, shown in **Figure 12,** and it kills more adults than any other infectious disease in the world. In the 1950s, two effective antibiotics, isoniazid and rifampin, became available, and they have saved millions of lives. In the late 1980s, however, new strains of *M. tuberculosis* that are largely or completely resistant to isoniazid and rifampin appeared. Rates of TB infection began to skyrocket in many countries, and in 1993 the World Health Organization declared a global TB health emergency.

How did antibiotic-resistant strains of *M. tuberculosis* evolve? A detailed look at a single typical case reveals how: through natural selection. This case is of a 35-year-old man living in Baltimore who was treated with rifampin for an active TB infection. After 10 months, the antibiotics cleared up the infection. Two months later, however, the man was readmitted to the hospital with a severe TB infection, and despite rifampin treatment, he died 10 days later. The strain of *M. tuberculosis* isolated from his body was totally resistant to rifampin.

How had TB bacteria within his body become resistant to rifampin? Doctors compared DNA of the rifampin-resistant bacteria to DNA from samples of normal, rifampin-sensitive *M. tuberculosis*. There seemed to be only one difference: a single base change from cytosine to thymine in a gene called rpoB.

Figure 12 Tuberculosis.
TB may be diagnosed from an X-ray of the lungs. TB is caused by *Mycobacterium tuberculosis*.

Evolution of Antibiotic Resistance

Rifampin acts by binding to *M. tuberculosis* RNA polymerase, preventing transcription and so killing the bacterial cell. The mutation in the polymerase's rpoB gene prevents rifampin from binding to the polymerase. The mutation, however, does not destroy the polymerase's ability to transcribe mRNA. The mutation likely occurred in a single *M. tuberculosis* bacterial cell sometime during the first infection. Because its polymerase function was no longer normal, the mutant bacterium could not divide as rapidly as normal bacteria can, but it still could divide. The antibiotic caused the normal bacterial cells to eventually die. The mutant bacteria continued to grow and reproduce in the antibiotic-containing environment.

Because the total number of *M. tuberculosis* bacteria was reduced drastically by the first antibiotic treatment, the patient's infection had seemed to clear. However, mutant, antibiotic-resistant bacteria survived and continued to grow in his body. The mutant bacteria could reproduce more effectively in the presence of the antibiotic than the normal bacteria could. Therefore, the mutant bacteria became more common in the bacterial population, and they eventually became the predominant type. When the patient became acutely ill again with TB, the *M. tuberculosis* bacterial cells in his lungs were the rifampin-resistant cells. In this way, natural selection led to the evolution of rifampin resistance in *M. tuberculosis*.

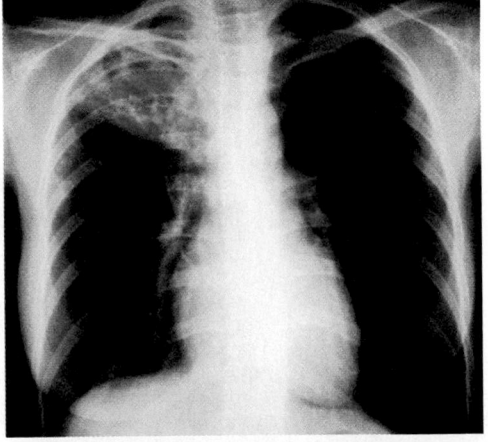

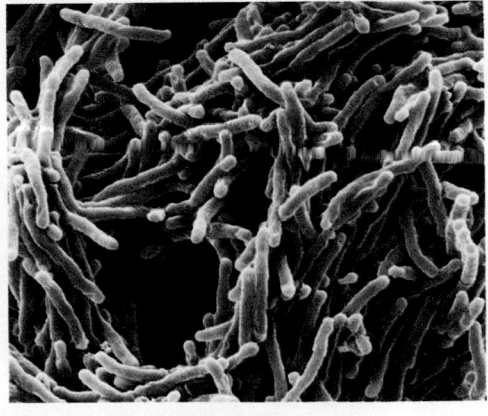

Mycobacterium tuberculosis

Evolution in Darwin's Finches

Darwin collected 31 specimens of finches from three islands when he visited the Galápagos Islands. In all, he collected 9 distinct species, all very similar to one another except for their bills. Two ground finches with large bills feed on seeds that they crush in their beaks, while two with narrower bills eat insects. One finch is a fruit eater, one picks insects out of cactuses, and yet another creeps up on sea birds and uses its sharp beak to drink their blood.

Darwin suggested that the nine species of Galápagos finches evolved from an original ancestral species. Changes occurred as different populations accumulated adaptations to different food sources. This idea was first tested in 1938 by the naturalist David Lack. He watched the birds closely for five months and found little evidence to support Darwin's hypothesis. Stout-beaked finches and slender-beaked finches were feeding on the same sorts of seeds. A second, far more thorough study was carried out over 25 years beginning in 1973 by Peter and Rosemary Grant of Princeton University. The Grants' study presents a much clearer picture that supports Darwin's interpretation.

It was Lack's misfortune to study the birds during a wet year, when food was plentiful. The size of the beak of the finch is of little importance in such times. Slender and stout beaks both work well to gather the small, soft seeds which were plentiful.

During dry years, however, plants produce few seeds, large or small. During these leaner years, few small, tender seeds were available. The difference between survival and starvation is the ability to eat the larger, tougher seeds that most birds usually pass by. The Grants measured the beaks of many birds every year. They found that after several dry years, the birds that had longer, more-massive beaks had better feeding success and produced more offspring.

When wet seasons returned, birds tended to have smaller beaks again, as shown in **Figure 13.** The numbers of birds with different beak shapes are changed by natural selection in response to the available food supply, just as Darwin had suggested.

Figure 13   Natural selection in finches

By relating the environment to beak size, the Grants showed that natural selection influences evolution.

Beak size measured

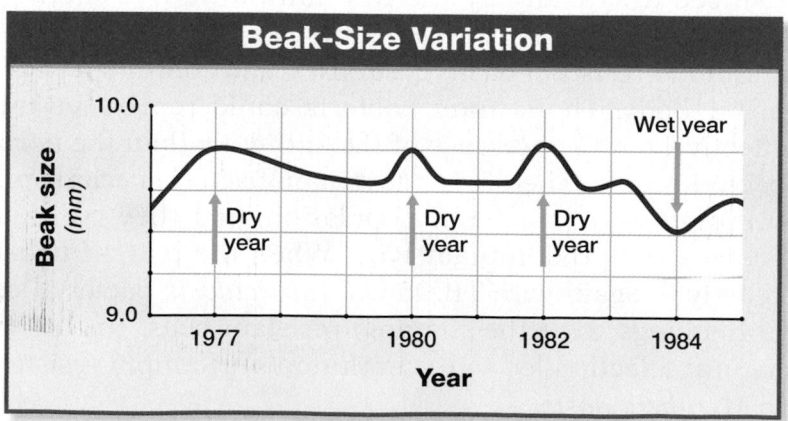

Formation of New Species

Species formation occurs in stages. Recall that natural selection favors changes that increase reproductive success. Therefore, a species molded by natural selection has an improved "fit" to its environment. The accumulation of differences between groups is called **divergence** *(die VUHR jehns)*. Divergence leads to the formation of new species. Biologists call the process by which new species form **speciation** *(spee see AY shun)*.

Forming Subspecies

Separate populations of a single species often live in several different kinds of environments. In each environment, natural selection acts on the population. Natural selection results in the evolution of offspring that are better adapted to that environment. If their environments differ enough, separate populations of the same species can become very dissimilar. Over time, populations of the same species that differ genetically because of adaptations to different living conditions become what biologists call **subspecies.** The members of newly formed subspecies have taken the first step toward speciation. Eventually, the subspecies may become so different that they can no longer interbreed successfully. Biologists then consider them separate species.

Maintaining New Species

What keeps new species separate? Why are even closely related species usually unable to interbreed? Once subspecies become different enough, a barrier to reproduction, like the one shown in **Figure 14,** usually prevents different groups from breeding with each other.

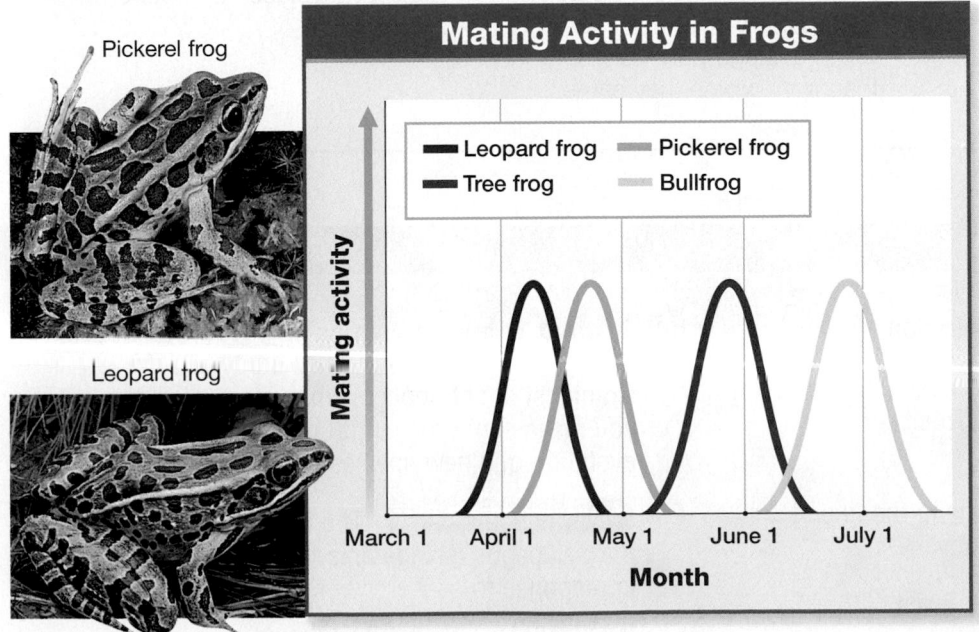

Pickerel frog

Leopard frog

Mating Activity in Frogs

- ▬ Leopard frog
- ▬ Tree frog
- ▬ Pickerel frog
- ▬ Bullfrog

Mating activity

March 1 | April 1 | May 1 | June 1 | July 1

Month

Figure 14 Mating activity in various frogs. Though they appear to be similar, pickerel frogs (*Rana palustris*) and leopard frogs (*Rana pipiens*) are different species. The graph shows that the time of peak mating activity varies between four species of frogs.

There are several types of barriers that may isolate two or more closely related groups. For example, groups may be geographically isolated or may reproduce at different times. Physical differences may also prevent mating, or they may not be attracted to one another for mating. The hybrid offspring may not be fertile or suited to the environment of either parent.

Biologists have seen the stages of speciation in many different organisms. Thus, the way that natural selection leads to the formation of new species has been thoroughly documented. As changes continue to build up over time, living species may become very different from their ancestors and from other species that evolved from the same recent common ancestor, leading to the appearance of new species.

MATH LAB

Analyzing Change in Lizard Populations

Background

In 1991, Jonathan Losos, an American scientist, measured hind-limb length of lizards from several islands and the average perch diameter of the island plants. The lizards were descended from a common population 20 years earlier, and the islands had different kinds of plants on which the lizards perched. Examine the graph at right and answer the following questions:

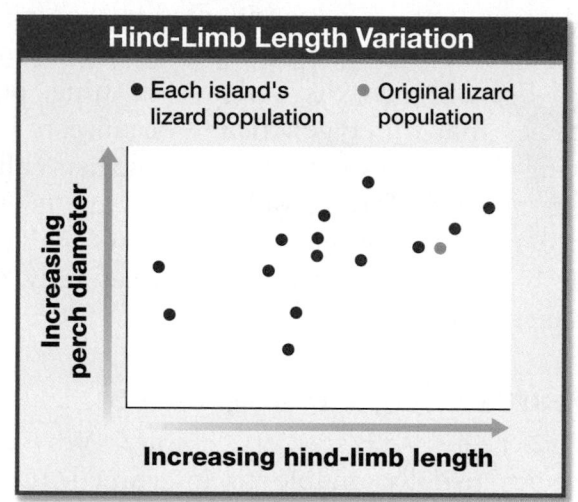

Hind-Limb Length Variation

● Each island's lizard population ● Original lizard population

Increasing perch diameter

Increasing hind-limb length

Analysis

1. **Interpreting Graphics** How did the average hind-limb length of each island's lizard population change from that of the original population?

2. **Predict** what would happen to a population of lizards with short hind limbs if they were placed on an island with a larger average perch diameter than from where they came.

3. **Justify** the argument that this experiment supports the theory of evolution by natural selection.

Section 3 Review

1 **List** four elements of natural selection.

2 **Describe** the mechanism that causes population changes in antibiotic-resistant bacteria.

3 **Identify** what caused the change in the finch's beaks as seen in the Grants' study.

4 **Describe** how speciation takes place.

5 **Critical Thinking Evaluating Results** Based on the results of David Lack's study and the Grants' study of finches, what conclusion can you make about the length of time required for evolution of a new species to take place?

6 **Standardized Test Prep** The beaks of finches on the Galápagos Islands enlarged over generations in response to

A isolation.   **C** rain.

B pollution.   **D** limited food supply.

Key Concepts

1 The Theory of Evolution by Natural Selection

- Charles Darwin concluded that animals on the coast of South America that resembled those on the nearby islands evolved differences after separating from a common ancestor.

- Darwin was influenced by Thomas Malthus, who wrote that populations tend to grow as much as the environment allows.

- Darwin proposed that natural selection favors individuals that are best able to survive and reproduce.

- Under certain conditions, change within a species can lead to new species.

- Gradualism is a process of evolution in which speciation occurs gradually, and punctuated equilibrium is a process in which speciation occurs rapidly between periods of little or no change.

2 Evidence of Evolution

- Evidence of orderly change can be seen when fossils are arranged according to their age.

- Differences in amino acid sequences and DNA sequences are greater between species that are more distantly related than between species that are more closely related.

- Similarities of structures in different vertebrates provide evidence that all vertebrates share a common ancestor.

3 Examples of Evolution

- Individuals that have traits that enable them to survive in a given environment can reproduce and pass those traits to their offspring.

- Experiments show that evolution through natural selection has occurred within populations of antibiotic-resistant bacteria and in Darwin's finches.

- Speciation begins as a population adapts to its environment.

- Reproductive isolation keeps newly forming species from breeding with one another.

Key Terms

Section 1

population (278)
natural selection (279)
adaptation (279)
reproductive isolation (281)
gradualism (282)
punctuated equilibrium (282)

Section 2

paleontologist (285)
vestigial structure (286)
homologous structure (286)

Section 3

divergence (291)
speciation (291)
subspecies (291)

Performance ZONE CHAPTER REVIEW

Understanding Key Ideas

1. According to the modern theory of evolution,
   a. Lamarck was completely wrong.
   b. random gene mutation is a part of evolution.
   c. punctuated equilibrium has replaced natural selection.
   d. the diversity of life-forms resulted from the inheritance of acquired characteristics.

2. With respect to the problem of antibiotic-resistant tuberculosis, which entity evolves?
   a. the patient
   b. the bacterium
   c. the antibiotic
   d. None of the above.

3. What is true about gradualism with respect to punctuated equilibrium?
   a. Each is a model of evolution.
   b. Neither is a model of evolution.
   c. Only gradualism portrays true evolution.
   d. Only punctuated equilibrium portrays true evolution.

4. The process by which isolated populations of the same species become new species is called
   a. speciation.
   b. reproductive isolation.
   c. genetic variation.
   d. natural seclection.

5. For each pair of terms, explain the differences in their meanings.
   a. adaptation, natural selection
   b. extinction, reproductive isolation
   c. population, subspecies
   d. homologous, vestigial
   e. divergence, speciation

6. Adult lobsters and barnacles look very different. The larvae of barnacles and lobsters, however, are practically identical. What does this indicate about the evolutionary history of these organisms?

7. Could a population of identical organisms undergo natural selection? Why or why not?

8. Explain the relationship between the number of nucleotide differences between two species and the time since the species shared a common ancestor.

9. What is a subspecies, and how is formation of a subspecies related to the process of speciation?

10. How is meiosis beneficial to the evolution of a species by natural selection? (**Hint:** See Chapter 7, Section 1.)

11. **Exploring Further** Other than punctuated equilibrium, what naturally occurring phenomena might explain large gaps in the fossil record?

12. **Concept Mapping** Make a concept map that shows how natural selection leads to speciation. Try to include the following terms in your map: *evolution, natural selection, genetic variation, environment, speciation,* and *divergence.*

Critical Thinking

13. **Applying Information** If a favorable trait increases the life span of an organism without affecting reproductive success, does evolution occur?

14. **Evaluating** Analyze Darwin's theory of evolution by natural selection and describe one strength and one weakness.

15. **Justifying Conclusions** About 40 years after the publication of *On the Origin of Species,* genetics was recognized as a science. Explain how information about genetics might support Darwin's theory of evolution.

Alternative Assessment

16. **Career Connection** **Paleontologist** Research the field of paleontology, and write a report on your findings. Your report should include a job description, training required, kinds of employers, growth prospects, and starting salary.

Standardized Test Prep

Understanding Concepts

Directions (1–4): **For *each* question, write on a separate sheet of paper the letter of the correct answer.**

1 What term describes the process by which a species becomes better suited to its environment?
A. adaptation
C. gradualism
B. equilibrium
D. natural selection

2 What are anatomical structures that share a common ancestry?
F. analogous structures
G. evolutionary structures
H. homologous structures
I. vestigial structures

3 In the Grants' study, the effect of climate on the size of the finch's beak provides an example of which of the following?
A. fossilization
B. natural selection
C. speciation
D. reproductive isolation

4 The woodpecker finch and the warbler finch are different species. Which of the following can you conclude about these two birds?
F. They cannot interbreed.
G. The lack a common ancestor.
H. They lack homologous structures.
I. They have very different embryos.

Directions (5): **For the following question, write a short response.**

5 Generation time is the time from the beginning of an organism's life to the point of reproduction. What effect would the time from the beginning of an organism's life to the point of reproduction have on the rate of evolution of a species?

Test TIP

For a question about a structure or phenomenon that has a complex name, write down the name and review its meaning before answering the question.

Reading Skills

Directions (6): **Read the passage below. Then answer the question.**

Alfred Russel Wallace was a biologist who collected insects on an 1848 expedition to the Amazon. He also made observations in the Malay Archipelago between 1854 and 1862. Wallace discovered that animals on the western islands of the Malay Archipelago differed sharply from those on the eastern islands.

6 What condition might have caused these animals to evolve into different species?
A. fossilization
B. population growth
C. punctuated equilibrium
D. reproductive isolation

Interpreting Graphics

Directions (7): **Base your answer to question 7 on the diagram below.**

Vertebrate Evolution

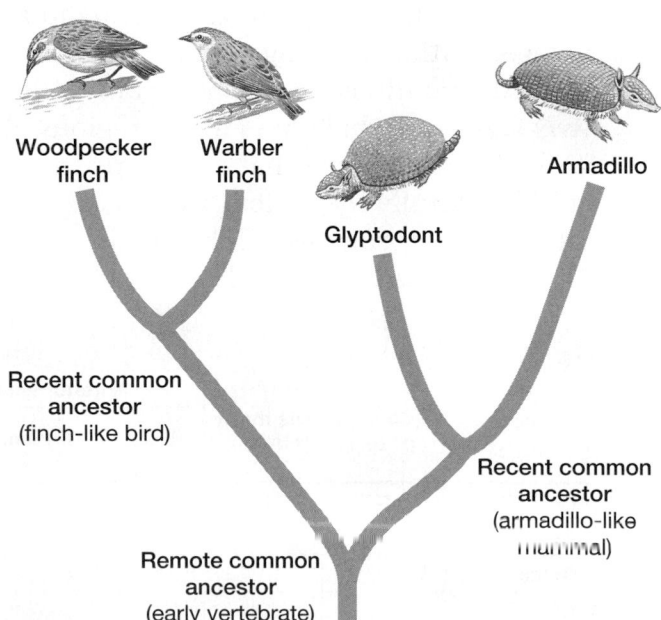

7 Which organism has DNA that is probably most similar to the glyptodont's DNA?
F. armadillo
H. warbler finch
G. finch-like bird
I. woodpecker finch

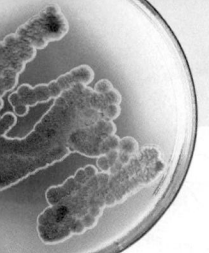

Exploration Lab

Modeling Natural Selection

SKILLS
- Modeling a process
- Inferring relationships

OBJECTIVES
- **Model** the process of selection.
- **Relate** favorable mutations to selection and evolution.

MATERIALS
- scissors
- construction paper
- cellophane tape
- soda straws
- felt-tip marker
- meterstick or tape measure
- penny or other coin
- six-sided die

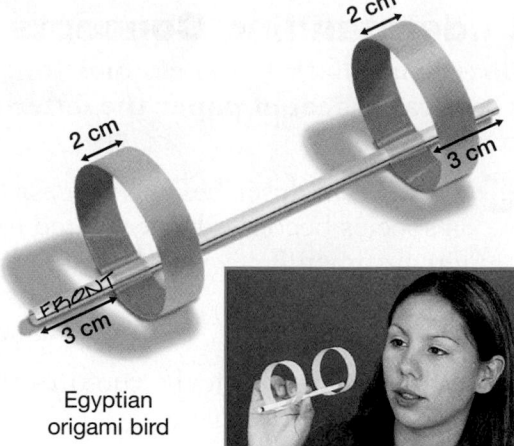

Egyptian origami bird

Before You Begin

Natural selection occurs when organisms that have certain **traits** survive to reproduce more than organisms that lack those traits do. A population evolves when individuals with different **genotypes** survive or reproduce at different rates. In this lab, you will model the selection of favorable traits in a new generation by using a paper model of a bird—the fictitious Egyptian origami bird (*Avis papyrus*), which lives in dry regions of North Africa. Assume that only birds that can successfully fly the long distances between water sources will live long enough to breed successfully.

1. Write a definition for each boldface term in the preceding paragraph.
2. Make a data table similar to the one shown below.
3. Based on the objectives for this lab, write a question you would like to explore about the process of selection.

Procedure

PART A: Parental Generation

1. Cut two strips of paper, 2 × 20 cm each. Make a loop with one strip of paper, letting the paper overlap by 1 cm, and tape the loop closed. Repeat for the other strip.

Bird	Coin flip (H or T)	Die throw (1–6)	Anterior wing (cm)			Posterior wing (cm)			Average distance flown (m)
			Width	Circum.	Distance from front	Width	Circum.	Distance from back	
Parent	NA	NA	2	19	3	2	19	3	
Generation 1									
Chick 1									
Chick 2									
Chick 3									
Generation 2									
Chick 1									
Chick 2									
Chick 3									

DATA TABLE

2. Tape one loop 3 cm from each end of the straw, as shown above. Mark the front end of the bird with a felt-tip marker. This bird represents the parental generation.

3. Test how far your parent bird can fly by releasing it with a gentle overhand pitch. Test the bird twice. Record the bird's average flight distance in your data table.

PART B: First (F_1) Generation

4. Each origami bird lays a clutch of three eggs. Assume that one of the chicks is a clone of the parent. Use the parent to represent this chick in step 6.

5. Make two more chicks. Assume that these chicks have mutations. Follow Steps A–C below for each chick to determine the effects of its mutation.

Step A Flip a coin to determine which end is affected by a mutation.

Heads = anterior (front)

Tails = posterior (back)

Step B Throw a die to determine how the mutation affects the wing.

- = Wing position moves 1 cm toward the end of the straw.
- = Wing circumference decreases by 2 cm.
- = Wing position moves 1 cm toward the middle of the straw.
- = Wing width increases by 1 cm.
- = Wing circumference increases by 2 cm.
- = Wing width decreases by 1 cm.

Step C A mutation is lethal if it causes a wing to fall off the straw or a wing with a circumference smaller than that of the straw. If you get a lethal mutation, disregard it and produce another chick.

6. Record the mutations and the wing dimensions of each offspring.

7. Test each bird twice by releasing it with a gentle overhand pitch. Release the birds as uniformly as possible. Record the distance each bird flies. The most successful bird is the one that flies the farthest.

PART C: Subsequent Generations

8. Assume that the most successful bird in the previous generation is the sole parent of the next generation. Repeat steps 4–7 using this bird.

9. Continue to breed, test, and record data for eight more generations.

PART D: Cleanup and Disposal

10. Dispose of paper scraps in the designated waste container.

11. Clean up your work area and all lab equipment. Return lab equipment to its proper place. Wash your hands thoroughly before you leave the lab and after you finish all work.

Analyze and Conclude

1. Analyzing Results Did the birds you made by modeling natural selection fly farther than the first bird you made?

2. Inferring Conclusions How might this lab help explain the variety of species of Galápagos finches?

3. Further Inquiry Write another question about natural selection that could be explored with another investigation.

On the Job

Population biology is the study of populations. Do research to discover how population biologists study evolution. For more about careers, visit **go.hrw.com** and type in the keyword **HX4 Careers.**

Desert organisms

14 Classification of Organisms

✔ *Quick Review*

Answer the following without referring to earlier sections of your book.

1. **Define** the term *species*. *(Chapter 13, Section 2)*

2. **Relate** macromolecules to evolutionary history. *(Chapter 13, Section 2)*

3. **Relate** homologous structures to evolutionary relationships. *(Chapter 13, Section 2)*

4. **Summarize** speciation. *(Chapter 13, Section 3)*

Did you have difficulty? *For help, review the sections indicated.*

📖 *Reading Activity*

Take a break after reading each section of this chapter, and closely study the figures in the section. Reread the figure captions, and, for each one, write out a question that can be answered by referring to the figure and its caption. Refer to your list of figures and questions as you review the concepts addressed in the chapter before you complete the Performance Zone chapter review.

Looking Ahead

Section 1

Categories of Biological Classification
Taxonomy
Classifying Organisms

Section 2

How Biologists Classify Organisms
What Is a Species?
Evolutionary History

📶 internet connect

www.scilinks.org
National Science Teachers Association *sci*LINKS Internet resources are located throughout this chapter.

SC*L*INKS Maintained by the
National Science Teachers Association

● As illustrated by this variety of desert plants, most of Earth's surface—including its oceans—is populated by a great diversity of organisms. Scientists have developed systems of naming and classifying organisms.

Categories of Biological Classification

Objectives

- **Describe** Linnaeus's role in developing the modern system of naming organisms.

- **Summarize** the scientific system for naming a species.

- **List** the seven levels of biological classification.

Key Terms

taxonomy
binomial nomenclature
genus
family
order
class
phylum
kingdom
domain

Taxonomy

Just as it is impossible for postal workers to sort mail bearing only the addressee's first name, it is impossible for biologists to memorize every name of the estimated 10–30 million organisms on Earth. To make sorting mail easier, postal workers sort first by zip code, then by street name and house number. In the same way, biologists group organisms into large categories that in turn are assigned to smaller and more specific categories.

More than 2,000 years ago, the Greek philosopher and naturalist Aristotle grouped plants and animals according to their structural similarities. Later Greeks and Romans grouped plants and animals into basic categories such as oaks, dogs, and horses. Eventually each unit of classification came to be called a *genus (JEE nuhs)* (plural, *genera*), the Latin word for "group." Starting in the Middle Ages, genera were named in Latin. The science of naming and classifying organisms is called **taxonomy** *(tak SAH nuh mee)*.

Until the mid-1700s, biologists named a particular type of organism by adding descriptive phrases to the name of the genus. These phrases sometimes consisted of 12 or more Latin words. They were called polynomials (from *poly*, meaning "many," and *nomen*, meaning "name"). As you can see in **Figure 1,** the polynomial for the European honeybee became very large and awkward. Polynomials were often changed by biologists, so organisms were rarely known to everyone by the same name.

Figure 1   European honeybee. The European honeybee once had a 12-part scientific name.

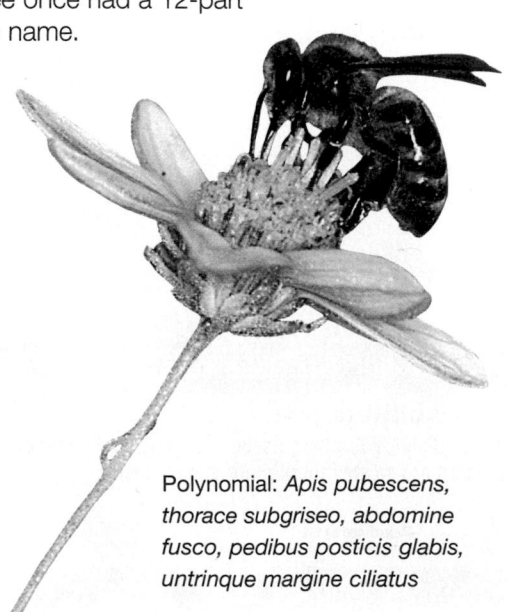

Polynomial: *Apis pubescens, thorace subgriseo, abdomine fusco, pedibus posticis glabis, untrinque margine ciliatus*

A Simpler System

A simpler system for naming organisms was developed by the Swedish biologist Carl Linnaeus. His ambition was to catalog all the known kinds of organisms. In the 1750s, he wrote several books that used the polynomial system. But Linnaeus also included a two-word Latin name for each species. Linnaeus's two-word system for naming organisms is called **binomial** *(bie NOH mee uhl)* **nomenclature** (from *bi*, meaning "two"). His two-part name for the European honeybee was *Apis mellifera*, the genus name followed by a single descriptive word. Over the past 250 years since Linnaeus first used two-part binomial species names, his approach has been universally adopted. Most of the species he described in 1753 still have the two-part names he gave them.

Scientific Names

The unique two-part name for a species is now referred to as its *scientific name*. The first word is the genus to which the organism belongs. A **genus** is a taxonomic category containing similar species. Organisms in a genus share important characteristics. For example, the genus *Quercus* is composed of oak trees. The second word in a scientific name identifies one particular kind of organism within the genus, called a species. A species is the basic biological unit in the Linnaean system of classification. **Table 1** lists and describes two species of oaks in the genus *Quercus*.

The first letter of the genus name is always capitalized, and the first letter of the second word is always lowercase. Scientific names are italicized or underlined. After the first use of the full scientific name, the genus name can be abbreviated as a single letter. For example, *Quercus rubra* can be abbreviated *Q. rubra*.

The scientific name of an organism gives biologists a common way of communicating, regardless of their native language. One species may have many common names, and one common name may be used for more than one species. For example, the bird called a robin in Great Britain is *Erithacus rubicula*. The very different bird called a robin in North America is *Turdus migratorius*.

The name given to a species must conform to the rules established by an international commission of scientists. All scientific names must have two Latin words or terms created according to the rules of Latin grammar. Two different types of organisms cannot have the same scientific name. Since all the members of a genus share the genus name, the second word in the name of each member of that genus must be different. For example, only one species of the genus *Homo* can be given the name *sapiens*.

WORD *Origins*

● The word *Quercus* is Latin for "oak." The word *rubra* is Latin for "red," and the word *phellos* is Greek for "cork." (Cork is part of the bark of a tree.)

🔲 internet connect

www.scilinks.org
Topic: Naming Species
Keyword: HX4127

SC*i*LINKS. Maintained by the National Science Teachers Association

Table 1 Two Species of Oak			
Common name	**Genus**	**Scientific name**	**Traits**
Red oak	*Quercus*	*Quercus rubra*	Lobed leaves; produces acorns approximately 25 mm (1 in.) long
Willow oak	*Quercus*	*Quercus phellos*	Unlobed leaves; produces acorns approximately 15 mm (0.6 in.) long

Classifying Organisms

Linnaeus worked out a broad system of classification for plants and animals in which an organism's form and structure are the basis for arranging specimens in a collection. The genera and species that he described were later organized into a ranked system of groups that increase in inclusiveness. The different groups into which organisms are classified have expanded since Linnaeus's time and now consist of eight levels, as shown in **Figure 2.**

Similar genera are grouped into a **family.** Similar families are combined into an **order.** Orders with common properties are united in a **class.** Classes with similar characteristics are assigned to a **phylum** *(FIE luhm)*. Similar phyla are collected into a **kingdom.** Similar kingdoms are grouped into **domains.** All living things are grouped into one of three domains. Two domains, Archaea and Bacteria, are each composed of a single kingdom of prokaryotes. The third domain, Eukarya, contains all four kingdoms of eukaryotes.

In order to remember the eight categories of classification in their proper order, it may prove useful to memorize a phrase, such as **Do Kindly Pay Cash Or Furnish Good Security**, to remember **Domain Kingdom Phylum Class Order Family Genus Species.**

Figure 2 System of classification. Each living thing is assigned to a series of groups, beginning with domain (most inclusive) and ending with species (least inclusive).

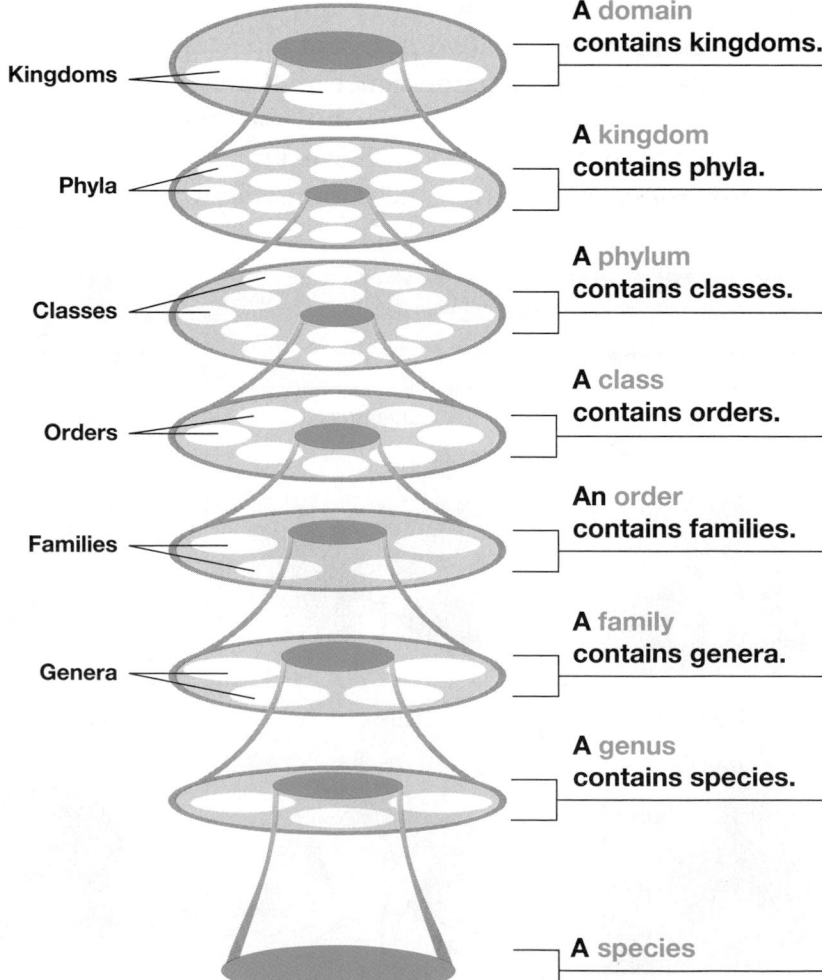

Kingdoms

A domain
contains kingdoms.

Phyla

A kingdom
contains phyla.

Classes

A phylum
contains classes.

Orders

A class
contains orders.

Families

An order
contains families.

Genera

A family
contains genera.

A genus
contains species.

A species

Classification of the Honeybee

Scientific names are particularly powerful because they tell you so much about the organism. Each level of classification is based on characteristics shared by all the organisms it contains. For example, consider the classification of the honeybee, shown in **Figure 3.**

The honeybee's scientific name, *Apis mellifera*, indicates that it belongs to the genus *Apis*, which is classified in the family Apidae. Knowing the honeybee's family is Apidae tells you a great deal about the honeybee. All members of the family Apidae are bees that live either alone or in hives, as does *Apis mellifera*.

Knowing the order bees belong in tells you even more. The order to which the honeybee belongs, Hymenoptera, includes ants, bees, and wasps, which usually have two pairs of wings and are likely to be able to sting. At each higher level, the information becomes more general.

At the next higher level of classification, *A. mellifera* belongs to the class Insecta, meaning it is an insect with three major body parts and three pairs of legs. Its phylum, Arthropoda, indicates that *A. mellifera* is an arthropod, an organism with a coelom, segmented body, jointed appendages, and a hard outer skin made of a complex

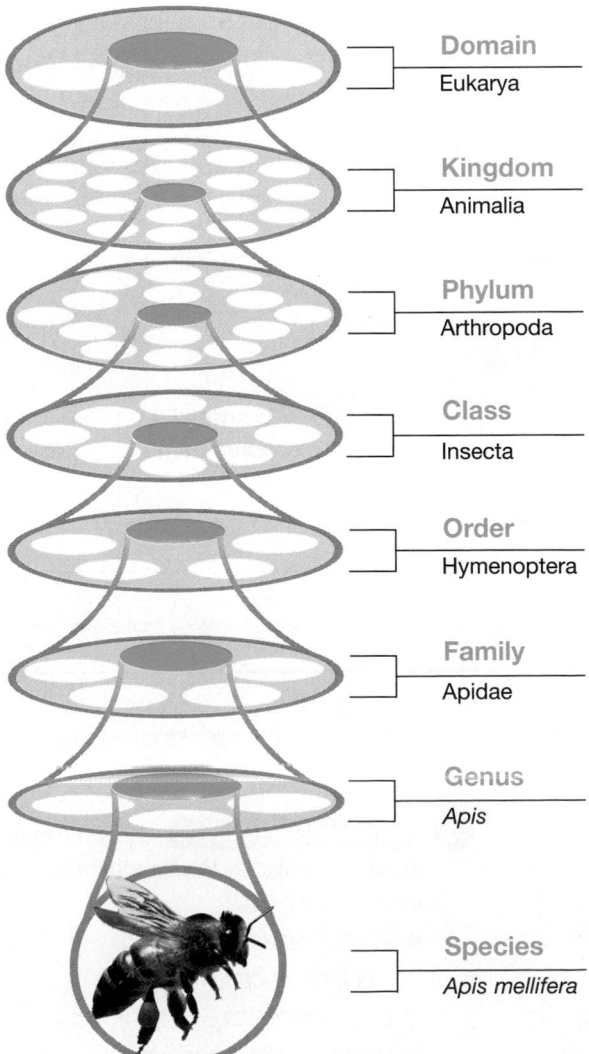

Domain
Eukarya

Kingdom
Animalia

Phylum
Arthropoda

Class
Insecta

Order
Hymenoptera

Family
Apidae

Genus
Apis

Species
Apis mellifera

Figure 3 Classification of a bee. The European honeybee is classified in eight different levels of biological classification.

carbohydrate called chitin. Arthropods have been the most successful of all animals. Two-thirds of all the named species on Earth are arthropods.

Its kingdom, Animalia, tells you that *A. mellifera* is a multicellular heterotroph whose cells lack walls. And finally, *A. mellifera* is classified in the domain Eukarya, which is composed of all eukaryotes. Recall that eukaryotes are composed of one or more cells that have membrane-bound organelles and whose DNA is enclosed within a nucleus.

Using a Field Guide

You can use a standard pictoral field guide or a dichotomous key to help you identify species of plants, animals, or other organisms.

Materials

paper and pencil, a plant or animal field guide

Procedure

1. CAUTION: Wear protective gloves when handling any wild plant. Keep your hands away from your face. Using a dichotomous key or other field guide, identify several species of plants that share the same phylum and class. Collect specimens only if your teacher tells you to do so.

2. Try to identify two plants of the same genus.

3. In a table like the one on the right, record the scientific name of each specimen.

4. Read the description of each species in the field guide. Determine the set of charateristics that fit each specimen. Write these characteristics in a table like the one below.

Analysis

1. **List** the characteristics shared by two specimens that are in the same genus but are different species.

2. **Describe** how the binomial name of these two species show that they are members of the same genus.

3. **Identify** the key characteristics your field guide uses to tell these two species apart.

4. **Critical Thinking Analyzing Data** Based on your observations, are two species from the same genus more similar or less similar than two species from different genera?

Specimen	Genus name	Binomial species name	Identifying characteristics
A			
B			
C			

Section 1 Review

1. **Explain** how Linnaeus helped develop the modern system of taxonomy.

2. **List** rules that scientists use to name organisms.

3. **Sequence** the eight levels of biological classification, beginning with the most specific level.

4. **Standardized Test Prep** Which organism is most similar to the sandhill crane, *Grus canadensis*?

A *Branta canadensis*

B *Falcipennis canadensis*

C *Grus americana*

D *Recurvirostra americana*

How Biologists Classify Organisms

What Is a Species?

Have you ever wondered how scientists tell one species from another? For example, how can you tell a mushroom that can be eaten from a similar looking mushroom that is poisonous? Scientists traditionally have used differences in appearance and structure to group organisms. Sometimes, however, these differences cannot be used to determine if two organisms are of the same species. For example, all single-celled protists classified as *Paramecium syngens* were once thought to be a single species. These organisms actually represent several protistan species that look very similar but have other characteristics that differ.

In 1942, the biologist Ernst Mayr of Harvard University proposed a biologically based definition of species, which is called the biological species concept. Mayr defined a **biological species** as a group of natural populations that are interbreeding or that could interbreed, and that are reproductively isolated from other such groups.

Reproductive isolation occurs when a barrier separates two or more groups of organisms and prevents them from interbreeding. In nature, however, reproductive barriers between sexually reproducing species are not always complete. Sometimes individuals of different species interbreed and produce offspring called *hybrids*. For example, wolves and dogs are members of separate species in the genus *Canis*. But interbreeding between wolves and dogs produces fertile offspring, such as the hybrid shown in **Figure 4.** Another example of a fertile hybrid is triticale, a hybrid of wheat and rye. It is important to remember that when reproductive barriers between two species are not complete, the two species are closely related.

Objectives

- **List** the characteristics that biologists use to classify organisms.
- **Summarize** the biological species concept.
- **Relate** analogous structures to convergent evolution.
- **Describe** how biologists use cladograms to determine evolutionary histories.

Key Terms

biological species
phylogeny
convergent evolution
analogous character
cladistics
ancestral character
derived character
cladogram
evolutionary
  systematics
phylogenetic tree

Figure 4   Dog-wolf hybrid. Wolves and dogs can produce fertile offspring.

Dog (*Canis familiaris*)

Wolf (*Canis lupus*)

Dog-wolf hybrid

Figure 5 New species of tree kangaroo. *Dendrolagus mbaiso* is a black-and-white tree kangaroo that lives in the mountains of New Guinea.

Evaluating the Biological Species Concept

The biological species concept works well for most members of the kingdom Animalia, in which strong barriers to hybridization usually exist. For example, Asian elephants and African elephants do not interbreed in nature. But the biological species concept fails to describe species that reproduce asexually, such as all species of bacteria and some species of protists, fungi, plants, and even some animals.

Within many groups of organisms, there are no barriers to interbreeding between the species. Many species of plants, some mammals, and many fishes are able to form fertile hybrids with one or a few closely related species. In practice, modern biologists recognize species by studying an organism's features.

Number of Species

The number of species in the world is much greater than the number described. Large numbers of species, such as the kangaroo shown in **Figure 5,** are still being discovered. Only about 1.5 million species have been described to date. Scientists estimate that 5 million to 10 million more species may live in the tropics alone. Since no more than 500,000 tropical species have been named, it is clear that our knowledge of Earth's diversity of species is limited.

Analyzing Taxonomy of Mythical Organisms

Background

Classification of organisms often requires grouping organisms based on their characteristics. Use the following list of mythological organisms and their characteristics to complete the analysis.

- **Pegasus** stands 6 ft tall, has a horse's body, a horse's head, four legs, and two wings.
- **Centaur** stands 6 ft tall, has a horse's body with a human torso, a male human head, and four legs.
- **Griffin** stands 4–6 ft tall, has a lion's body, an eagle's head, four legs, two wings, fur on its body, and feathers on its head and wings.
- **Dragon** can grow to several hundred feet, has a snake-like

body, from 1–3 reptile-like heads, four legs, scales, and breathes fire.
- **Chimera** stands 6 ft tall, has a goat's body, snake's tail, four legs, a lion's head, fur on its body and head, scales on its tail, and breathes fire.
- **Hydra** is several hundred feet long, has a long body with four legs and a spiked tail, 100 snake heads, scales, and is poisonous.

Analysis

1. **Identify** the characteristics that you think are the most useful for grouping the organisms into separate groups.

2. **Classify** the organisms into at least three groups based on the characteristics you think are most important.

3. **Evaluate** the use of the biological species concept to classify these mythical organisms.

Evolutionary History

Linnaeus's classification system was based on his observation that organisms have different degrees of similarity. For instance, a tiger resembles a gorilla more closely than either resembles a fish. According to Darwin's views, organisms that are more similar to one another than they are to other organisms have descended from a more recent common ancestor. Therefore, classification based on similarities should reflect an organism's **phylogeny,** that is, its evolutionary history. Inferring evolutionary connections from similarities, however, can be misleading. Not all features—or *characters*—are inherited from a common ancestor. Consider the wings of a bird and the wings of an insect. Both enable flight, but the structures of the two kinds of wings differ. Moreover, fossil evidence indicates that they evolved independently of one another. Through the process called **convergent evolution,** similarities evolve in organisms not closely related to one another, often because the organisms live in similar habitats. Similarities that arise through convergent evolution are called **analogous** *(ah NAHYL uh guhs)* **characters. Figure 6** shows an example of convergent evolution.

Cladistics

Most biologists today analyze evolutionary relationships using cladistics *(kluh DIHS tihks)*. **Cladistics** is a method of analysis that reconstructs phylogenies by inferring relationships based on shared characters. Cladistics can be used to hypothesize the sequence in which different groups of organisms evolved. To do this, cladistics focuses on the nature of the characters in different groups of organisms. With respect to two different groups, a character is defined as an **ancestral character** if it evolved in a common ancestor of both groups. Thus when considering the relationship between birds and mammals, a backbone is an ancestral character. Having feathers, however, is a derived character. A **derived character** evolved in an ancestor of one group but not of the other. Feathers evolved in an ancestor of birds that was not also ancestral to mammals.

internet connect

www.scilinks.org
Topic: Taxonomy
Keyword: HX4174

SC*LINKS*. Maintained by the National Science Teachers Association

---

Figure 6 Similar structures in the cactus and spurge families

Although they evolved in different parts of the world, cactuses and spurges look similar.

Organ pipe cactus, *Sternocereus thurber*

Gifboom, *Euphorbia virosa*

Cladistics is based on the principle that shared *derived* characters provide evidence that two groups are relatively closely related. Shared *ancestral* characters, however, do not. For example, lizards and dogs have limbs, but a whale has no limbs. Having limbs does not provide evidence that dogs and lizards are more closely related than are dogs and whales. Recall that whales are descended from an ancestor that had limbs. Therefore, the presence of limbs is a shared ancestral character of all three groups. However, both dogs and whales have mammary glands—a shared derived character not found in lizards or lizard ancestors. This provides evidence that dogs and whales share a more recent common ancestor than either shares with lizards.

Cladograms

How many different ways can you organize your possessions? For example, should all your clothes be grouped according to their type or according to color? Biologists sometimes disagree about how to organize groups of organisms.

Why Study Cladograms?
Some biologists use cladograms to study the evolutionary relationships among certain groups of organisms, such as species within a genus or genera within a family. Cladograms show how closely two or more groups are related, based on important characteristics. Cladograms convey comparative information about relationships. Organisms that are grouped more closely on a cladogram share a more recent common ancestor than those farther apart. Because the analysis is comparative, cladistic analysis deliberately includes an organism that is only distantly related to the other organisms. This distantly related organism is called an *out-group.* The out-group serves as a base line for comparisons with the other organisms being evaluated, the in-group.

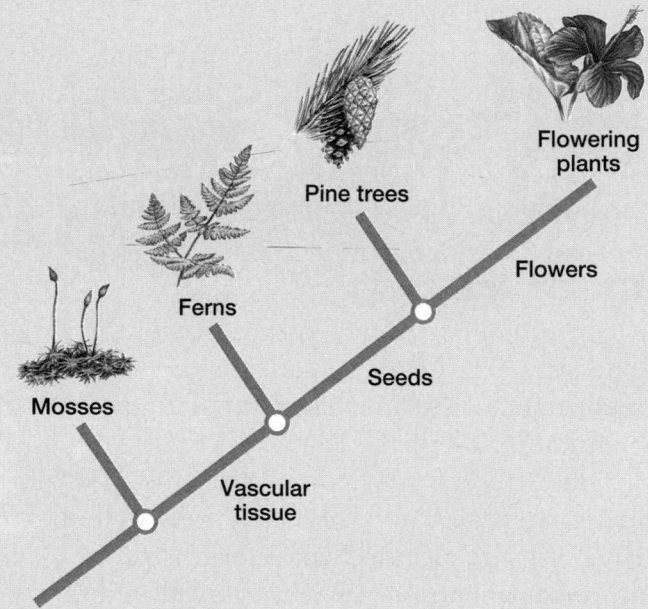

Constructing a Cladogram
This example shows the evolutionary relationships among plants.

1. In the table at the left, the characters in the row for the out-group are marked with a 0. A character not found in the out-group is considered a derived character and is marked with a 1.

2. Starting with a diagonal line, as shown above, the out-group (mosses) is placed on the first branch. Just past this first branch, the most common derived character is listed—vascular tissue. Vascular tissue is a series of tubes and vessels within a plant.

3. Next the second most common derived character—seeds—is listed. Ferns lack seeds and so are placed on the second branch of the cladogram.

4. The third most common derived character—flowers—is then listed. Conifers do not have flowers and so are placed on one branch, and the flowering plants are placed on the other.

Organisms	Characters		
	Vascular tissue	Seeds	Flowers
Mosses (out-group)	0	0	0
Pine trees	1	1	0
Flowering plants	1	1	1
Ferns	1	0	0

A biologist using cladistics constructs a branching diagram called a **cladogram,** which shows the evolutionary relationships among groups of organisms. Organisms that share derived characters are grouped together on the cladogram. As groups evolve, new derived characters appear on the cladogram that were not present in earlier organisms.

Considering Characters

The great strength of cladistics is objectivity. If a computer is fed the same set of data repeatedly, it will make exactly the same cladogram every time. The disadvantage of cladistics is that the *degree* of difference between organisms is not considered. Cladistic analysis simply indicates that a character does or does not exist.

Cladistic analysis does not take into account variations in the "strength" of a character, such as the size or location of a fin or the effectiveness of a lung. Each character is treated equally. Thus in a cladogram of vertebrate evolution, birds are grouped with reptiles. This accurately reflects their true ancestry—birds evolved from dinosaurs. The immense evolutionary impact of a derived character like feathers, however, is ignored.

Evolutionary success often depends on high-impact events, such as the evolution of feathers. Some modern cladistic studies therefore attempt to weigh the evolutionary significance of the characters being studied.

Study TIP

● **Organizing Information**
The numbered list in the Exploring Further feature tells you how to make a cladogram. Use this list to prepare a cladogram of the organisms listed in the Data Lab.

DATA LAB

Making a Cladogram

Background

A cladogram is a model that represents a hypothesis about the order in which organisms evolved from a common ancestor. Scientists construct a cladogram by first analyzing characters in a data table. The absence of a vascular system and the absence of seeds is ancestral. Use the data below to construct a cladogram on a separate sheet of paper.

Horsetail

Liverwort

Analysis

1. **Identify** the out-group.

2. **Name** the least common derived trait.

3. **List** the order in which the plants in the table would be placed on a cladogram.

Plants	Characters		
	Seeds	Vascular system
Horsetails	No	Yes
Liverworts	No	No
Pine trees	Yes	Yes

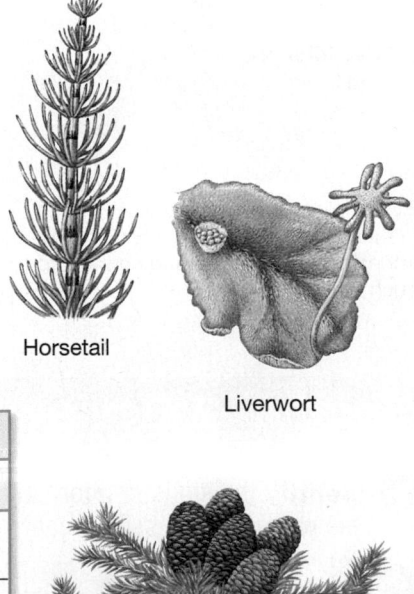

Pine tree

Evolutionary Systematics

In **evolutionary systematics,** taxonomists give varying degrees of importance to characters and thus produce a subjective analysis of evolutionary relationships. In this type of analysis, evolutionary relationships are displayed in a branching diagram called a **phylogenetic tree.** As you can see in **Figure 7,** evolutionary systematics places birds in an entirely separate class from reptiles, giving more importance to characters like feathers that made powered flight possible. Evolutionary systematics involves the full observational power of the biologist, along with any biases he or she may have. A phylogenetic tree and a cladogram are similar in that each represents a hypothesis of evolutionary history, which must be inferred because it was not observed.

Figure 7   Evolutionary systematics and cladistic taxonomy

Biologists differ in the ways that they classify organisms.

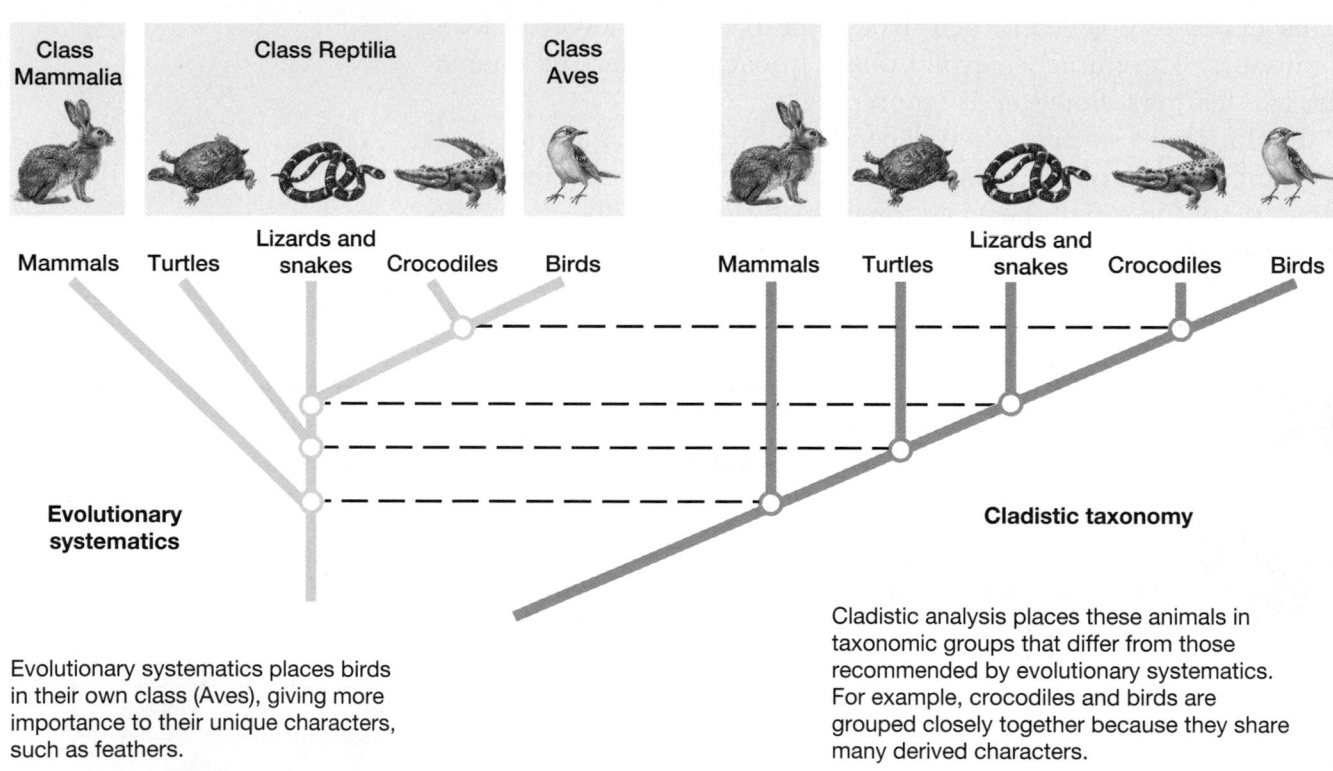

Evolutionary systematics places birds in their own class (Aves), giving more importance to their unique characters, such as feathers.

Cladistic analysis places these animals in taxonomic groups that differ from those recommended by evolutionary systematics. For example, crocodiles and birds are grouped closely together because they share many derived characters.

Section 2 Review

1 Identify the kinds of information that scientists use when they classify organisms.

2 Summarize what scientists mean by the biological species concept.

3 Define *cladistics*, and list the kind of information cladistic analysis reveals about evolutionary histories.

4 Critical Thinking Inferring Relationships Explain the relationship between convergent evolution and analogous characters.

5 Standardized Test Prep Different species are placed in the same group on a cladogram if they have the same

  A appearance.   **C** derived characters.

  B scientific name.   **D** analogous characters.

CHAPTER HIGHLIGHTS

Key Concepts

1 Categories of Biological Classification

- Swedish biologist Carl Linnaeus developed binomial nomenclature, the two-word system of naming organisms.
- Scientific names are written in Latin and give biologists a common way of communicating, regardless of the language they speak.
- The scientific name of an organism consists of its genus name followed by a second name, which identifies its species.
- Each category of classification is based on characteristics that are shared by all the organisms in the category.
- Scientists use an eight-level system to classify organisms.
- The modern system of classification includes the groups domain, kingdom, phylum, class, order, family, genus, and species.

2 How Biologists Classify Organisms

- Biologists usually define species according to their appearance and structure.
- The biological species concept defines species according to their sexual reproductive potential.
- The biological species concept cannot be used to classify asexually reproducing species.
- Similar organisms may have analogous structures that arose through convergent evolution.
- Cladistics focuses on sets of unique characteristics found in a particular group of organisms to reconstruct an evolutionary history.
- Evolutionary systematics is a more subjective method of classification than is cladistics, but evolutionary systematics allows greater evolutionary importance to be placed on certain characters.

Key Terms

Section 1

taxonomy (300)
binomial nomenclature (300)
genus (301)
family (302)
order (302)
class (302)
phylum (302)
kingdom (302)
domain (302)

Section 2

biological species (305)
phylogeny (307)
convergent evolution (307)
analogous character (307)
cladistics (307)
ancestral character (307)
derived character (307)
cladogram (309)
evolutionary systematics (310)
phylogenetic tree (310)

Understanding Key Ideas

1. The system of binomial nomenclature was developed by
   a. Linneaus.   **c.** Mayr.
   b. Aristotle.   **d.** Darwin.

2. The scientific name for humans is correctly written as
   a. Homo sapiens.   **c.** *Homo sapiens.*
   b. Homo Sapiens.   **d.** *Homo Sapiens.*

3. A difference between the scientific name of an organism and the classification of that organism is that
   a. the scientific name includes the family and class of the organism.
   b. the scientific name always contains three words (trinomial nomenclature).
   c. the classification includes more categories than the scientific name.
   d. classification can vary from place to place.

4. Explain the relationship of domains to kingdoms.

5. The offspring of a donkey *(Equus asinus)* and a horse *(Equus caballus)* is a mule, shown below. The mule is sterile. Is the classification of donkeys and horses into different species justified according to the biological species concept? Explain.

6. Biologists classify organisms based on
   a. their appearance.
   b. their structure.
   c. their ability to interbreed.
   d. All of the above

7. What features common to both bats and birds suggest that convergent evolution has occurred?

8. **Exploring Further** Explain how constructing a cladogram might help scientists understand evolutionary relationships.

9. Contrast analogous characters with homologous characters. (**Hint:** See Chapter 13, Section 2.)

10. ⌂ **Concept Mapping** Construct a concept map that shows how biologists determine the classification of a new species. Use the following terms: *genus, species, binomial nomenclature, kingdom, biological species concept, derived characters, cladogram,* and *evolutionary systematics.*

Critical Thinking

11. **Evaluating Hypotheses** Scientists infer that groups of organisms that have homologous traits must be related, so they use cladograms to study relationships. Explain the reasoning that supports this inference.

12. **Evaluating Viewpoints** Explain this statement: Diversity is the result of evolution; classification systems are the inventions of humans.

13. **Forming Reasoned Opinions** In the laboratory, a scientist studied two identical-looking daisies that belong to the genus *Aster*. The two plants produce fertile hybrids in the laboratory, but they never interbreed in nature because one plant flowers only in the spring and the other only in autumn. Do the plants belong to the same species? Explain.

Standardized Test Prep

Understanding Concepts

Directions (1–4): **For *each* question, write on a separate sheet of paper the letter of the correct answer.**

1 For grasshoppers and locusts to be in the same family, what other category must they also share?
  A. genus
  B. group
  C. order
  D. species

2 In which category are two members most closely related?
  F. class
  G. family
  H. genus
  I. order

3 What kind of features can convergent evolution lead to?
  A. analogous characters
  B. ancestral characters
  C. derived characters
  D. homologous structures

4 What kind of relationship does a cladogram reveal?
  F. analogous
  G. binomial
  H. convergent
  I. evolutionary

Directions (5–6): **For *each* question, write a short response.**

5 What can be inferred about the relationship between *Escherichia coli* and *Entamoeba coli* from their names?

6 A derived character evolves in an ancestor of one group but not another. Analyze what shared derived characters in two different groups provide evidence for.

Test TIP

After you finish writing your answer to a short-response question, proofread it for errors in spelling, grammar, and punctuation.

Reading Skills

Directions (7): **Read the passage below. Then answer the question.**

Monotremes, such as the duckbill platypus and echidna, have hair and produce milk. Unlike other mammals, they reproduce by laying eggs. Placental mammals (such as horses and humans) and marsupials (such as kangaroos and koalas) have live birth. Marsupials differ from placental mammals in that marsupial young are born before they are completely developed. They develop outside of the mother's body for several months in a pouch located on their mother's abdomen. A cladogram shows that in the evolution of mammals, hair and milk production developed first. Then, live birth and finally the placenta developed.

7 In what order did the ancestors of each of these species first diverge from the mammal ancestral line?
  A. echidna, koala, horse
  B. horse, echidna, koala
  C. koala, echidna, horse
  D. koala, horse, echidna

Interpreting Graphics

Directions (8): **Base your answer to question 8 on the diagram below.**

Seven Levels of Biological Classification

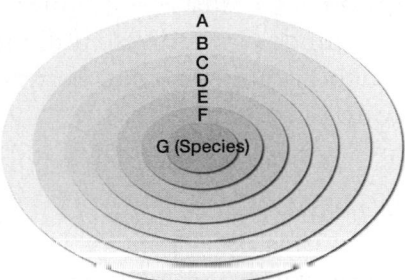

8 What can be inferred about different species that belong to the same group at level *D*?
  F. They belong to the same group at *C*.
  G. They belong to different groups at *C*.
  H. They belong to the same group at *E*.
  I. They belong to different groups at *E*.

Skills Practice Lab

Making a Dichotomous Key

SKILLS
- Identifying and comparing
- Organizing data

OBJECTIVES
- **Identify** objects using dichotomous keys.
- **Design** a dichotomous key for a group of objects.

MATERIALS
- 6 to 10 objects found in the classroom (e.g., shoes, books, writing instruments)
- stick-on labels
- pencil

A B C

D E F

Before You Begin

One way to identify an unknown organism is to use an **identification key,** which contains the major characteristics of groups of organisms. A **dichotomous key** is an identification key that contains pairs of contrasting descriptions. After each description, a key either directs the user to another pair of descriptions or identifies an object. In this lab, you will design and use a dichotomous key. A dichotomous key can be written for any group of objects.

1. Write a definition for each boldface term in the paragraph above.

2. Based on the objectives for this lab, write a question you would like to explore about making or using a dichotomous key.

Procedure

PART A: Using a Dichotomous Key

1. Use the **Key to Forest Trees** to identify the tree that produced each of the leaves

shown here. Identify one leaf at a time. Always start with the first pair of statements (**1***a* and **1***b*). Follow the direction beside the statement that describes the leaf. Proceed through the key until you get to the name of a tree.

Key to Forest Trees

1*a*	Leaf edge has no teeth, waves, or lobes	go to **2**
1*b*	Leaf edge has teeth, waves, or lobes	go to **3**
2*a*	Leaf has a bristle at its tip	**shingle oak**
2*b*	Leaf has no bristle at its tip	go to **4**
3*a*	Leaf edge is toothed	**Lombardy poplar**
3*b*	Leaf edge has waves or lobes	go to **5**
4*a*	Leaf is heart-shaped	**red bud**
4*b*	Leaf is not heart-shaped	**live oak**
5*a*	Leaf edge has lobes	**English oak**
5*b*	Leaf edge has waves	**chestnut oak**

PART B: Design a Dichotomous Key

2. Work with the members of your lab group to design a dichotomous key using the materials listed for this lab.

3. Before you begin writing your key, have your teacher approve the objects your group has decided to work with.

4. Using the **Key to Forest Trees** as a guide, write a key for the objects your group selected. Remember, a dichotomous key includes pairs of contrasting descriptions.

5. Use your key to explore one of the questions written for step 2 of **Before You Begin.**

6. After each group has completed step 5, exchange keys and the objects they identify with another group. Use the key you receive to identify the objects. If the key does not work, return it to the group so corrections can be made.

PART C: Cleanup

7. Clean up your work area and all lab equipment. Return lab equipment to its proper place. Wash your hands thoroughly before you leave the lab and after you finish all work.

Analyze and Conclude

1. **Drawing Conclusions** What tree produced each of the leaves shown in this lab?

2. **Forming Hypotheses** What other characteristics might be used to identify leaves with a dichotomous key?

3. **Analyzing Methods** How was the key your group designed dichotomous?

4. **Evaluating Results** Were you able to use another group's key to identify the objects for which it was written? If not, describe the problems you encountered.

5. **Analyzing Methods** Does a dichotomous key begin with general descriptions and then proceed to more specific descriptions or vice versa? Explain your answer, giving an example from your key.

6. **Further Inquiry** Write a new question about making or using keys that could be explored with another investigation.

? Do You Know?

Do research in the library or media center to answer these questions:

1. Are the identification keys used in biology always dichotomous?

2. What other types of identification keys do biologists use?

Use the following Internet resources to explore your own questions about the diversity of life on Earth.

internet connect

www.scilinks.org
Topic: Classification
Keyword: HX4044

SCI LINKS Maintained by the National Science Teachers Association

Fewer than 4,000 gray wolves exist in the lower 48 states.

in perspective

The Gray Wolf

Yesterday...

The gray wolf once roamed North America from coast to coast. In an effort to protect their livestock, early settlers hunted and killed wolves nearly to the point of extinction. By the mid-1950s, however, few wolves existed in the lower 48 states. **How does the relationship between predator and prey help to maintain balance in an ecosystem?**

Steel trap

Today...

Gray wolves play a vital role in ensuring the diversity and health of ecosystems in which they live. Because they tend to prey upon the old, the sick, and the injured, wolves help to produce stronger populations of deer, elk, and moose. **Discover how competition between species affects a biological community.** In each of the lower 48 states except Minnesota, the gray wolf is listed as endangered.

Biologist David Mech

Tomorrow...

Because the numbers and range of the gray wolf have been increasing, the U.S. Fish and Wildlife Service is reviewing potential changes to the Endangered Species Act protection for gray wolves. Researchers must continue to study wolf populations before a long-term plan to conserve gray wolves can be implemented. **Read to learn why species diversity is important.**

Wolf researcher Diane Boyd

internet connect

www.scilinks.org
Topic: Wolves
Keyword: HX4191

SC*LINKS* Maintained by the National Science Teachers Association

Atlantic puffins

15 Populations

✔ Quick Review

Answer the following without referring to earlier sections of your book.

1. **Describe** Mendel's laws of inheritance. *(Chapter 8, Section 2)*

2. **Define** *phenotype* and *genotype*. *(Chapter 8, Section 2)*

3. **Define** *probability*. *(Chapter 8, Section 3)*

4. **Evaluate** the significance of mutations. *(Chapter 10, Section 2)*

5. **Define** *natural selection*. *(Chapter 13, Section 1)*

Did you have difficulty? *For help, review the sections indicated.*

📖 Reading Activity

Copy the following statements on a piece of paper or in your notebook, leaving a few blank lines after each statement.

1. *Very small populations are more likely to become extinct than larger populations.*

2. *A single bacterium that divides every 30 minutes will become a population of more than a million in only 10 hours.*

3. *Natural selection acts only on genes themselves, not on phenotypes.*

Before you read the chapter, write down whether you agree or disagree with each statement. Save your responses, and after you have finished reading the chapter, decide whether or not you still agree with your first response.

• Individuals of a species that live together form a population. This group of puffins is part of the population of puffins that live on the far northern Atlantic coast of North America.

Looking Ahead

Section 1

How Populations Grow
What Is a Population?
Modeling Population Growth
Growth Patterns in Real Populations

Section 2

How Populations Evolve
The Change of Population Allele Frequencies
Action of Natural Selection on Phenotypes
Natural Selection and the Distribution of Traits

⏸ internet connect

www.scilinks.org
National Science Teachers Association *sci*LINKS Internet resources are located throughout this chapter.

SCi_LINKS® Maintained by the National Science Teachers Association

How Populations Grow

Objectives

- **Distinguish** among the three patterns of dispersion in a population.
- **Contrast** exponential growth and logistic growth.
- **Differentiate** *r*-strategists from *K*-strategists.

Key Terms

population
population size
population density
dispersion
population model
exponential growth
  curve
carrying capacity
density-dependent
  factor
logistic model
density-independent
  factor
r-strategist
K-strategist

What Is a Population?

The people on a city sidewalk shown in **Figure 1** are members of a population. Since 1930, the world's human population has nearly tripled. What causes populations to grow? What determines how fast they grow? What factors can slow their growth?

A **population** consists of all the individuals of a species that live together in one place at one time. This definition allows scientists to use similar terms when speaking of the world's human population, the population of *Escherichia coli* bacteria that live in your intestine, or the population of Devil's Hole pupfish that swim in the tiny pool shown in **Figure 1.**

Every population tends to grow because individuals tend to have multiple offspring over their lifetime. But eventually, limited resources in an environment limit the growth of a population. The statistical study of all populations is called demography *(dih MAH gruh fee).* Demographers study the composition of a population and try to predict how the size of the population will change.

Figure 1   Populations

A population can be can be widely distributed, as is Earth's human population. Or a population can be confined to a small area, such as the population of Devil's Hole pupfish, which lives in this small pool and nowhere else.

Three Key Features of Populations

Every population has features that help determine its future. One of the most important features of any population is its size. The number of individuals in a population, or **population size,** can affect the population's ability to survive. Studies have shown that very small populations are among those most likely to become extinct. Random events or natural disturbances, such as a fire or flood, endanger small populations more than they endanger larger populations. Small populations also tend to experience more inbreeding (breeding with relatives) because only relatives are available as mates. Inbreeding produces a more genetically uniform population and is therefore likely to reduce the population's fitness—more individuals will be homozygous for harmful recessive traits. For example, the worldwide cheetah population is very small, and the individuals are almost genetically identical. Many biologists think that a disaster, such as a new disease, could cause their extinction.

A second important feature of a population is its density. **Population density** is the number of individuals that live in a given area. If the individuals of a population are few and are spaced widely apart, they may seldom encounter one another, making reproduction rare.

A third feature of a population is the way the individuals of the population are arranged in space. This feature is called **dispersion.** Three main patterns of dispersion are possible within a population, and each is shown in **Figure 2.** If individuals are randomly spaced, the location of each individual is self-determined or determined by chance. If individuals are evenly spaced, they are located at regular intervals. In a clumped distribution, individuals are bunched together in clusters. Each of these patterns reflects the interactions between the population and its environment.

■ internet connect

www.scilinks.org
Topic: Population Characteristics
Keyword: HX4143

SCiLINKS. Maintained by the National Science Teachers Association

Figure 2 Patterns of dispersion

These are the three possible patterns of dispersion in a population.

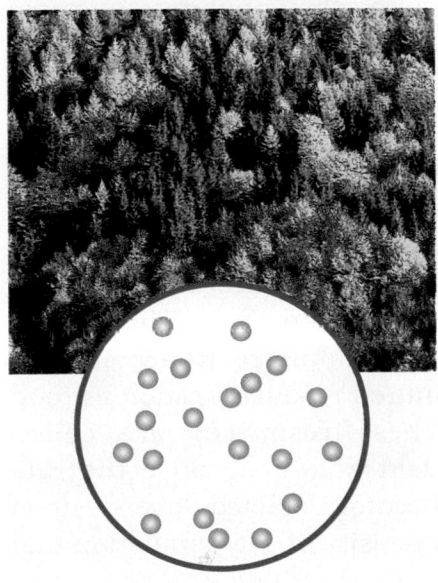

Pine trees in a *random distribution*

Birds in an *even distribution*

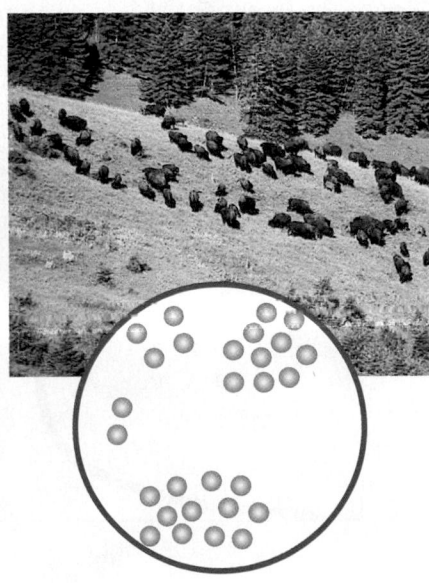

Buffalo in a *clumped distribution*

Modeling Population Growth

When demographers try to predict how a population will grow, they make a model of the population. A **population model** is a hypothetical population that attempts to exhibit the key characteristics of a real population. By making a change in the model and observing the outcome, demographers can predict what might occur in a real population. To learn how demographers study a population, consider a simple model of population growth in three stages of complexity.

Growth Rate

A population grows when more individuals are born than die in a given period. So a simple population model describes the rate of population growth as the difference between the birthrate and the death rate. For human populations, birth and death rates are usually expressed as the number of births and deaths per thousand people per year.

Growth Rate and Population Size

When population size is plotted against time on a graph, the population growth curve resembles a J-shaped curve and is called an exponential *(ehks poh NEHN shuhl)* growth curve. An **exponential growth curve** is a curve in which the rate of population growth stays the same, as a result the population size increases steadily.

Figure 3 shows an exponential growth curve. For example, a single bacterial cell that divides every 30 minutes will produce more than 1 million bacteria after only 10 hours. To calculate the number of individuals that will be added to the population as it grows, multiply the size of the current population (N) by the rate of growth (r).

However, populations do not usually grow unchecked. Their growth is limited by predators, disease, and the availability of resources. Eventually, growth slows, and the population may stabilize. The population size that an environment can sustain is called the **carrying capacity** (K).

Real Life

Uncle Sam wants to count you.

The United States census, conducted every 10 years, collects detailed information on the country's population.

Finding Information
Explore Internet resources to find out more about the United States census. Why should every household complete a census form? What steps has the government taken in the past few years to improve the accuracy of the census?

Figure 3 Exponential growth. This J-shaped curve is characteristic of exponential growth.

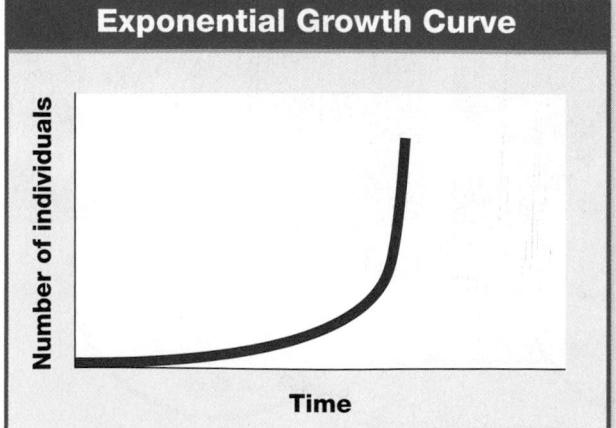

Exponential Growth Curve

Number of individuals

Time

Resources and Population Size

As a population grows, limited resources (that is, resources in short supply) eventually become depleted. When this happens, the growth of the population slows. The population model can be adjusted to account for the effect of limited resources, such as food and water. These resources are called **density-dependent factors** because the rate at which they become depleted depends upon the population density of the population that uses them.

The population model that takes into account the declining resources available to populations is called the logistic model of population growth, after the mathematical form of the equation. The **logistic model** is a population model in which exponential growth is limited by a density-dependent factor. The everyday meaning of the word *logistics* refers to the ability to obtain, maintain, and transport materials. In other words, logistics is about solving the day-to-day problems of living. Unlike the simple model, the logistic model assumes that birth and death rates vary with population size. When a population is below carrying capacity, the growth rate is rapid. However, as the population approaches the carrying capacity, death rates begin to rise and birthrates begin to decline. As a result, the rate of growth slows. The population eventually stops growing when the death rate equals the birthrate. In real situations, the population may, for a short time, actually exceed the carrying capacity of its environment. If this happens, deaths will increase and outnumber births until the population falls down to the carrying capacity. Many scientists are concerned that the Earth's human population, which passed 6 billion in 1999, may have exceeded its carrying capacity. A curve that shows logistic growth is illustrated in **Figure 4.**

The logistic model of population growth, though simple, provides excellent estimates of how populations grow in nature. Competition for food, shelter, mates, and limited resources tends to increase as a population approaches its carrying capacity. The accumulation of wastes also increases. Demographers try to make logistic models based on current population sizes and predict how much a population will increase. **Figure 5** summarizes the three stages of a population model.

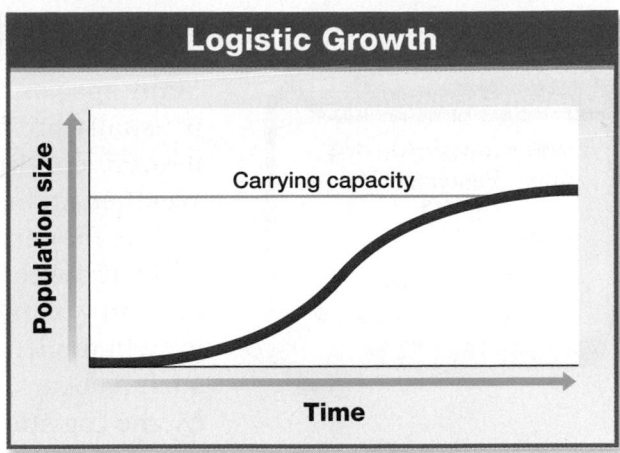

Logistic Growth

Population size / Carrying capacity / Time

Figure 4 Logistic growth. The curve of logistic growth looks like a stretched-out letter *S*.

Figure 5   Population model. A population model can be described using three stages.

Population Growth Models

- **Simple model (part one): calculating the population growth rate**

    r (rate of growth) = birthrate − death rate

    The rate of population growth equals the rate of births minus the rate of deaths.

- **Simple model (part two): exponential growth curve**

    Δ*N* (change in population, read as "delta N") = *rN*

    Once *r* has been determined for a population (part one), the number of individuals that will be added to a population as it grows is equal to the rate of growth multiplied by the number of individuals in the current population (*N*).

- **More realistic model: logistic model**

    $$\Delta N = rN\frac{(K - N)}{K}$$

    Population size calculations often need to be adjusted by the number of members of the population at carrying capacity (*K*).

Growth Patterns in Real Populations

Many species of plants and insects reproduce rapidly. Their growth is usually limited not by density-dependent factors but by environmental conditions, also known as **density-independent factors.** Weather and climate are the most important density-independent factors. For example, mosquito populations increase in the summer, while the weather is warm, but decrease in the winter. The growth of many plants and insects is often described by an exponential growth model. The population growth of slower growing organisms, such as bears, elephants, and humans however, is better described by the logistic growth model. Most species have a strategy somewhere between the two models; other species change from one strategy to the other as their environment changes. (Note that the use of the word *strategy* here means "pattern of living." An organism does not consciously plan its strategy.)

Rapidly Growing Populations

Many species, including bacteria, some plants, and many insects like cockroaches and mosquitos, are found in rapidly changing environments. Such species, called **r-strategists,** grow exponentially when environmental conditions allow them to reproduce. This strategy results in temporarily large populations. When environmental conditions worsen, the population size drops quickly. In general, r-strategists have a short life span. In addition they reproduce early in life and have many offspring each time they reproduce. Their offspring are small, and they mature rapidly with little or no parental care. The cockroaches shown in **Figure 6** are r-strategists.

Figure 6   Different species have different growth patterns

Cockroaches are r-strategists, while whales are K-strategists.

Cockroaches

Humpback whale

Slowly Growing Populations

Organisms that grow slowly, such as whales, often have small population sizes. These species are called **K-strategists** because their population density is usually near the carrying capacity (K) of their environment. K-strategists are characterized by a long life span, few young, a slow maturing process, and reproduction late in life. K-strategists often provide extensive care of their young and tend to live in stable environments. Many endangered species, such as tigers, gorillas, and the whale shown in **Figure 6,** are K-strategists.

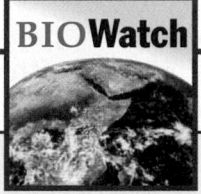

BIOWatch

Population Pyramids

A picture is worth a thousand words, according to an old proverb. Some kinds of pictures are worth more. For example, one way of representing the structure of a large human population is a graph in which age groups are plotted on the y-axis and the numbers of individuals are plotted on the x-axis. The younger age groups appear at the bottom, and the older groups appear at the top. The resulting graphic often resembles a pyramid and thus is called a population pyramid.

Predicting Future Health Needs

The construction of a population pyramid has many applications.

From the late 1940s until 1960, for example, population pyramids for the United States were bottom-heavy with "baby boom" children, who were born up to 15 years after World War II. During this period, there was an increased demand for child-care products and pediatric care.

By 1997, the baby-boom segment of the population had moved up to the 30–54 age-group levels. Baby boomers were competing for opportunities to work, marry, and buy houses. Demands for goods and services by this age group showed increases over previous years. As the baby-boom generation ages, the need for geriatric medical care will increase.

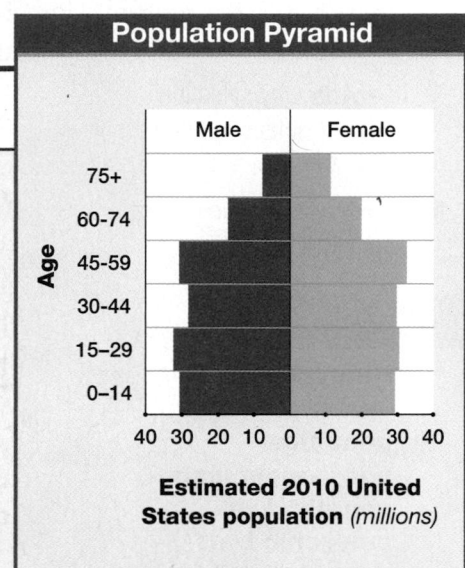

Population Pyramid

Estimated 2010 United States population (millions)

☑ internet connect

www.scilinks.org
Topic: Population Pyramids
Keyword: HX4146

SCLINKS® Maintained by the National Science Teachers Association

Section 1 Review

1. **Identify** the pattern of dispersion of fans attending a basketball game as random, even, or clumped. Explain your answer.

2. **Differentiate** a logistic growth pattern from an exponential growth pattern.

3. **Describe** why an r-strategist might be better suited for an unpredictable environment than a K-strategist is.

4. **Critical Thinking Applying Information** If healthcare improvements lead to a decreased death rate, how must the birth rate change in order to achieve a stable population size? Explain your reasoning.

5. (**Standardized Test Prep**) Which model best describes population growth that is limited by density-independent factors?
   A logistic model
   B exponential growth model
   C stage I model
   D stage III model

How Populations Evolve

Objectives

- **Summarize** the Hardy-Weinberg principle.
- **Describe** the five forces that cause genetic change in a population.
- **Identify** why selection against unfavorable recessive alleles is slow.
- **Contrast** directional and stabilizing selection.

Key Terms

Hardy-Weinberg principle
gene flow
nonrandom mating
genetic drift
polygenic trait
normal distribution
directional selection
stabilizing selection

Figure 7   Fruit fly mutation. *Drosophila melanogaster* experiences about one mutation in every 840,000,000,000 base pair replications.

The Change of Population Allele Frequencies

In the more than a century that has passed since Darwin's death, the science of genetics has blossomed. This has allowed biologists to better understand how natural selection changes the proportions of alleles within populations. But before you can understand how populations change in response to evolutionary forces, you need to learn how populations behave in the absence of these forces.

Allele Frequencies

When Mendel's work was rediscovered in 1900, biologists began to study how frequencies of alleles change in a population. Specifically, they wondered if dominant alleles, which are usually more common than recessive alleles, would spontaneously replace recessive alleles within populations.

In 1908, the English mathematician G. H. Hardy and the German physician Wilhelm Weinberg independently demonstrated that dominant alleles do not automatically replace recessive alleles. Using algebra and a simple application of the theories of probability, they showed that the frequency of alleles in a population does not change. Moreover, the ratio of heterozygous individuals to homozygous individuals does not change from generation to generation unless the population is acted on by processes that favor particular alleles. If a dominant allele is lethal, for example, it will not become more common just because it is dominant. Their discovery, called the **Hardy-Weinberg principle,** states that the frequencies of alleles in a population do not change unless evolutionary forces act on the population.

The Hardy-Weinberg Principle

The Hardy-Weinberg principle holds true for any population as long as the population is large enough that its members are not likely to mate with relatives and as long as evolutionary forces are not acting. There are five principle evolutionary forces: *mutation,* illustrated in **Figure 7,** *gene flow, nonrandom mating, genetic drift,* and *natural selection.* These evolutionary forces can cause the ratios of genotypes in a population to differ significantly from those predicted by the Hardy-Weinberg principle. The Hardy-Weinberg principle can be expressed as an equation that can be used to predict genotype frequencies in a population.

Mutation

Although mutation from one allele to another can eventually change allele frequencies, mutation rates in nature are very slow. Most genes mutate only about 1 to 10 times per 100,000 cell divisions, so mutation does not significantly change allele frequencies, except over very long periods of time. Furthermore, not all mutations result in phenotypic changes. Recall that more than one codon can code for the same amino acid. Therefore, some mutations may result in no change in the amino acid coded for in a protein, and other changes in an amino acid that do occur may not affect how the protein works. Mutation is, however, the source of variation and thus makes evolution possible.

internet connect

www.scilinks.org
Topic: Hardy-Weinberg Equation
Keyword: HX4095

SCiLINKS. Maintained by the National Science Teachers Association

Exploring Further

Using the Hardy-Weinberg Equation

You can use the Hardy-Weinberg principle to predict genotype frequencies. The Hardy-Weinberg principle is usually stated as an equation.

$$p^2 \quad + \quad 2pq \quad + \quad q^2 = 1$$

frequency of individuals that are homozygous for allele A frequency of heterozygous individuals with alleles A and a frequency of individuals that are homozygous for allele a

By convention, the frequency of the more common of the two alleles is referred to as p, and the frequency of the rarer allele is referred to as q. A frequency is the proportion of a group that is of one type. The frequency of allele A is the proportion of all alleles that are A for this gene in the population. Similarly, the frequency of allele a is the proportion

of alleles that are a. The sum of the allele frequencies must always equal 1.

Individuals that are homozygous for allele A occur at a frequency of p times p, or p^2. Individuals that are homozygous for allele a occur at the frequency of q times q, or q^2. Heterozygotes have one copy of A and one copy of a, but heterozygotes can occur in two ways—A from the father and a from the mother or a from the father and A from the mother. Therefore, the frequency of heterozygotes is $2pq$.

Calculating the frequency of cystic fibrosis
How do you calculate the number of people in a crowd, like the one below, who are likely to be carriers of the cystic fibrosis gene?

1. **Calculate the frequency of the recessive allele.** Recall from Chapter 9 that cystic fibrosis is caused by the recessive allele c. If q^2, the frequency of recessive homozygotes, is 0.00048, then q is $\sqrt{0.00048}$, or 0.022.

2. **Calculate the frequency of the dominant allele C.**

   Because
   $$p + q = 1, p = 1 - q.$$
   So
   $$p = 1 - 0.022,$$
   or 0.978.

3. **Determine the frequency of heterozygotes.**

   $$2pq = 2 \times 0.978 \times 0.022 = 0.043$$

This means that 43 of every 1,000 Caucasian North Americans are predicted to carry the cystic fibrosis allele unexpressed (without disease).

Hardy-Weinberg proportions seldom, if ever, occur in nature because at least one of the five causes of evolution is always affecting populations.

Figure 8 Nonrandom mating. Female widowbirds prefer to mate with males, such as the one shown, that have long tails over males that have short tails. This increases the proportion of alleles for long tails in the population.

Gene Flow

The movement of individuals from one population to another can cause genetic change. The movement of individuals to or from a population, called migration, creates **gene flow,** the movement of alleles into or out of a population. Gene flow occurs because new individuals (immigrants) add alleles to the population and departing individuals (emigrants) take alleles away.

Nonrandom Mating

Sometimes individuals prefer to mate with others that live nearby or are of their own phenotype, a situation called **nonrandom mating.** Mating with relatives (inbreeding) is a type of nonrandom mating that causes a lower frequency of heterozygotes than would be predicted by the Hardy-Weinberg principle. Inbreeding does not change the frequencies of alleles, but it does increase the proportion of homozygotes in a population. For example, populations of self-fertilizing plants consist mostly of homozygous individuals. Nonrandom mating also results when organisms choose their mates based on certain traits. In animals, females often select males based on their size, color, ability to gather food, or other characteristics, as shown in **Figure 8.**

Genetic Drift

In small populations the frequency of an allele can be greatly changed by a chance event. For example, a fire or landslide can reduce a large population to a few survivors. When an allele is found in only a few individuals, the loss of even one individual from the population can have major effects on the allele's frequency. Because this sort of change in allele frequency appears to occur randomly, as if the frequency were drifting, it is called **genetic drift.** Small populations that are isolated from one another can differ greatly as a result of genetic drift.

The cheetah, shown in **Figure 9,** is a species whose evolution has been seriously affected by genetic drift. Cheetahs have undergone drastic population declines over the last 5,000 years. As a result, the

Figure 9 Cheetahs are endangered. Cheetahs have gone through at least two drastic declines in population size.

cheetahs alive today are descendants of only a few individuals, and each cheetah is almost genetically uniform with other members of the population. One consequence of this genetic uniformity is reduced disease resistance—cheetah cubs are more likely to die from disease than are the cubs of lions or leopards. This reduction in genetic diversity of cheetahs may hasten their extinction.

Natural Selection

Natural selection causes deviations from the Hardy-Weinberg proportions by directly changing the frequencies of alleles. The frequency of an allele will increase or decrease, depending on the allele's effects on survival and reproduction. For example, the allele for sickle cell anemia is slowly declining in frequency in the United States because individuals who are homozygous for this allele rarely have children. Heterozygotes are resistant to malaria, a significant health problem in many parts of the world. Heterozygotes, however, do not have an advantage over normal homozygotes as they would have in a malaria area. As a result, homozygotes are selected *against* in the United States, and the frequency of the sickle cell allele decreases. Natural selection is one of the most powerful agents of genetic change.

Study TIP

● **Compare and Contrast**
To compare and contrast the five forces that cause evolution, list each force and describe how it causes populations to evolve. Write the ways in which the five forces are alike and the ways in which they are different.

QUICK LAB

Demonstrating the Hardy-Weinberg Principle

You can model the allele frequencies in a population with this simple exercise.

Materials
equal numbers of cards marked *A* or *a* to represent the dominant and recessive alleles for a trait, paper bag

Procedure

1. Make a data table like the one below.

2. Work in a group, which will represent a population. Count the individuals in your group, and obtain *that* number of both *A* and *a* cards.

3. Place the cards in a paper bag, and mix them. Have each individual draw two cards, which represent a genotype. Record the genotype and phenotype in your data table.

4. Randomly exchange one "allele" with another individual in your group. Record the resulting genotypes.

5. Repeat Step 4 four more times.

Analysis

1. **Determine** the genotype and phenotype ratios in your group for each trial. Do the ratios vary among the trials?

2. **Hypothesize** what could cause a change in the "genetic makeup" of your group. Test one of your hypotheses.

DATA TABLE					
	Trial 1	**Trial 2**	**Trial 3**	**Trial 4**	**Trial 5**
Genotype					
Phenotype					

Action of Natural Selection on Phenotypes

Natural selection constantly changes populations through actions on individuals within the population. However, natural selection does not act directly on genes. It enables individuals who express favorable traits to reproduce and pass those traits on to their offspring. This means that natural selection acts on phenotypes, not genotypes.

How Selection Acts

Think carefully about how natural selection might operate on a mutant allele. Only characteristics that are expressed can be targets of natural selection. Therefore, selection cannot operate against rare recessive alleles, even if they are unfavorable. Only when the allele becomes common enough that heterozygous individuals come together and produce homozygous offspring does natural selection have an opportunity to act. For example, Alexei Nikolayevich, the only son of the last Tsar of Russia, shown in **Figure 10,** suffered from *hemophilia,* a disease caused by a recessive gene. Without modern medical care, a cut can lead to uncontrollable bleeding and death. This kind of selection would remove a homozygous person from the gene pool. However, such an act of natural selection does not affect heterozygotes, who do not express hemophilia. Therefore, the gene is not eliminated from the population.

Why Genes Persist

To better understand this limitation on natural selection, consider this example. If a recessive allele (*a*) is homozygous in only 1 out of 100 individuals, then 18 out of 100 individuals will be heterozygous (*Aa*) and will therefore carry the allele unexpressed. So natural selection can act on only 1 out of every 19 individuals that carry the allele. As a result, this leaves 18 individuals that maintain the allele in the population. Many human diseases caused by recessive alleles have frequencies similar to this. For example, cystic fibrosis, the most common fatal genetic disorder among Caucasians, produces a thick mucus that clogs the lungs and other organs. About 1 in 25 Caucasians has a copy of the defective gene but shows no symptoms. Homozygous recessive individuals, which include about 1 in 2,500, die from the disease. Genetic conditions are not eliminated by natural selection because very few of the individuals bearing the alleles express the recessive phenotype.

Figure 10   Hemophilia in a family. The last Tsar of Russia's only son, Alexei Nikolayevich (yellow circle), had hemophilia, a blood-clotting disorder that affects males who have a single copy of a recessive gene.

Natural Selection and the Distribution of Traits

Natural selection shapes populations affected by phenotypes that are controlled by one or by a large number of genes. A trait that is influenced by several genes is called a **polygenic** (*pah lee JEHN ihk*) **trait.** Human height and human skin color, for example, are influenced by dozens of genes. Natural selection can change the allele frequencies of many different genes governing a single trait, influencing most strongly those genes that make the greatest contribution to the phenotype. Like following one duck in a flock, it is difficult to keep track of a particular gene. Biologists measure changes in a polygenic trait by measuring each individual in the population. These measurements are then used to calculate the average value of the trait for the population as a whole.

Because genes can have many alleles, polygenic traits tend to exhibit a range of phenotypes clustered around an average value. If you were to plot the height of everyone in your class on a graph, the values would probably form a hill-shaped curve called a **normal distribution,** as illustrated in **Figure 11.**

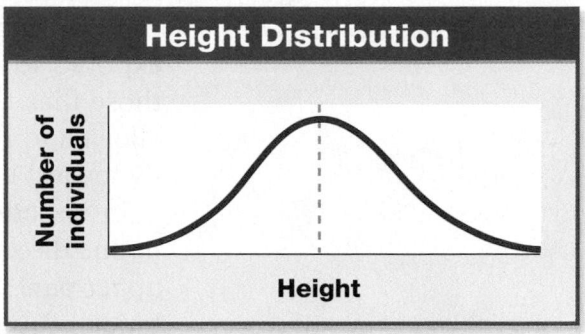

Height Distribution

Figure 11 Normal distribution. This hill-shaped curve represents a normal distribution. The blue, dashed line represents the average height for this population.

Building a Normal Distribution Curve

Background

You can help your class build a normal distribution curve by measuring the length of your shoes and plotting the data.

Materials

paper, pencil, measuring tape, graph paper

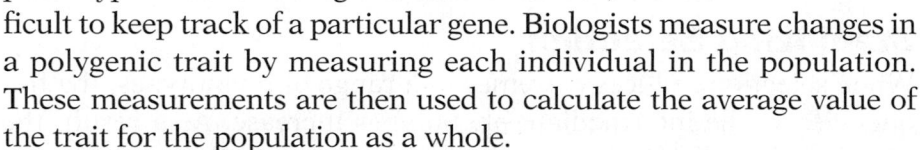

DATA TABLE

Shoe length (centimeters)	Gender

Procedure

1. Prepare a table like the one above.

2. Measure and record the length of one of your shoes to the nearest centimeter. Record your measurement and your gender.

3. Formulate a hypothesis about whether female shoes as a group are longer, shorter, or the same as shoes from males.

4. Determine the number of shoes of each length represented in the class.

5. Make a graph showing the distribution of shoe length in your class. Show the number of students on the y-axis and shoe length on the x-axis. Your graph should resemble the graph in Figure 11.

6. Make a second graph using data only from females.

7. Make a third graph using data only from males.

Analysis

1. **Describe** the shape of the curve that resulted from the graph you made in step 5.

2. **Distinguish** how the distribution curve for shoe length of females differs from the curve for the shoe length of males.

3. **Predict** how the distribution curve that you made in step 5 would change if the data for males were deleted.

Directional Selection

When selection eliminates one extreme from a range of phenotypes, the alleles promoting this extreme become less common in the population. In one experiment, when fruit flies raised in the dark were exposed to light, some flew toward light and some did not. Only those flies that had the strongest tendency to fly toward light were allowed to reproduce. After 20 generations, the average tendency to fly toward light increased.

In **directional selection,** the frequency of a particular trait moves in one direction in a range. Directional selection is illustrated in the upper panel of **Figure 12.** This type of selection has a role in the evolution of single-gene traits, such as pesticide resistance in insects.

Stabilizing Selection

When selection reduces extremes in a range of phenotypes, the frequencies of the intermediate phenotypes increase. As a result, the population contains fewer individuals that have alleles promoting extreme types.

As you can see in the lower panel of Figure 12, in **stabilizing selection,** the distribution becomes narrower, tending to "stabilize" the average by increasing the proportion of similar individuals. Stabilizing selection is very common in nature.

Figure 12 Two kinds of selection on polygenic traits. Directional selection is the change of the average value of a population. Stabilizing selection is the increase of the number of average individuals in a population.

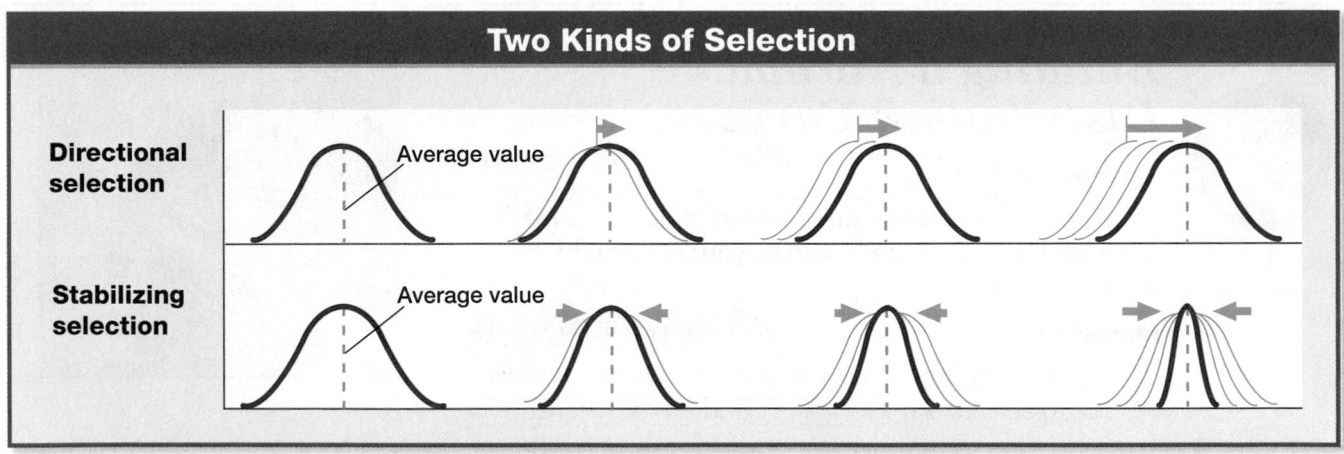

Two Kinds of Selection

Directional selection — Average value

Stabilizing selection — Average value

Section 2 Review

1. **Describe** the Hardy-Weinberg principle.

2. **List** the five forces that cause genetic change in a population.

3. **Describe** why natural selection against an unfavorable recessive allele is a slow process.

4. **Critical Thinking Comparing Concepts** Explain how directional selection and stabilizing selection differ, and whether they tend to increase or decrease diversity.

5. **Critical Thinking Justifying Conclusions** Individuals who are heterozygous for the allele for sickle cell anemia are resistant to malaria. Explain the effects of natural selection on the frequency of the sickle cell allele in an area where malaria occurs.

6. **Standardized Test Prep** Which evolutionary force decreases the genetic diversity of a population by increasing the proportion of similar individuals?
   - **A** mutation
   - **B** gene flow
   - **C** directional selection
   - **D** stabilizing selection

Study ZONE

CHAPTER HIGHLIGHTS

Key Concepts

1 How Populations Grow

- A population consists of all the individuals of a species that live together in one place at one time.

- A population's future survival is determined by its size, density, and dispersion.

- Though a population's growth is limited by factors such as predation and availability of resources, a population can grow rapidly and may eventually stabilize at a size that the environment can sustain.

- Some populations grow quickly in response to density-independent factors, and other populations grow more slowly and their size is controlled by density-dependent factors.

2 How Populations Evolve

- The Hardy-Weinberg principle states that the frequencies of alleles and genotypes remain constant in populations in which evolutionary forces are absent.

- Allele frequencies in a population can change if evolutionary forces, such as mutation, migration, nonrandom mating, genetic drift, and natural selection, act on the population.

- Natural selection acts only on phenotype, not on genotype.

- Natural selection reduces the frequency of a harmful recessive allele slowly; very few individuals are homozygous recessive, so very few express the allele.

- The range of phenotypes that are controlled by polygenic traits result in a normal distribution when plotted on a graph.

- Directional selection results in the range of phenotypes shifting toward one extreme.

- Stabilizing selection results in the range of phenotypes narrowing.

Key Terms

Section 1

population (320)
population size (321)
population density (321)
dispersion (321)
population model (322)
exponential growth curve (322)
carrying capacity (322)
density-dependent factor (322)
logistic model (323)
density-independent factor (324)
r-strategist (324)
K-strategist (325)

Section 2

Hardy-Weinberg principle (326)
gene flow (328)
nonrandom mating (328)
genetic drift (328)
polygenic trait (331)
normal distribution (331)
directional selection (332)
stabilizing selection (332)

Unit 7—*Ecosystem Dynamics*
Use Topic 2 in this unit to review the key concepts and terms in this chapter.

Understanding Key Ideas

1. A colony of bacteria that has a limited food supply likely will undergo _____ growth.
   a. exponential
   b. logistic
   c. natural
   d. random

2. According to the Hardy-Weinberg principle, allele frequencies in randomly mating populations without selection
   a. change when birth rate exceeds death rate.
   b. increase and then decrease.
   c. decrease and then increase.
   d. do not change.

3. Which of the following is *not* a cause of genetic change?
   a. genetic drift
   b. random mating
   c. natural selection
   d. mutation

4. Why is it unlikely that natural selection will quickly reduce the frequency of hemophilia?
   a. Natural selection acts only on recessive homozygotes.
   b. Hemophilia is not a genetic disorder.
   c. The frequency of recessive homozygotes is too great.
   d. Dominant homozygotes can have affected children.

5. Biologists introduced pheasants onto an island in Washington State in the 1930s. Using the data shown below, determine the island's carrying capacity.

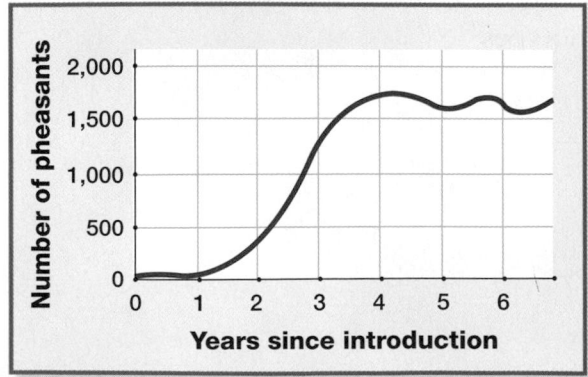

6. Can evolution *not* be associated with genetic change?

7. Relate the concept of directional selection to extreme traits.

8. **BIOWatch** By the year 2030, where will the "baby boom" age group be represented on the population pyramid?

9. **Exploring Further** The frequency of homozygous recessive albino rats in a population is 0.01. Calculate the expected frequency of the dominant allele in this population.

10. **Concept Mapping** Draw a concept map that shows how the forces of genetic change cause evolution. Try to include the following terms in your map: *Hardy-Weinberg principle, genetic drift, nonrandom mating, natural selection, mutation,* and *gene flow.*

Critical Thinking

11. **Applying Information** After a forest fire, certain plants quickly recolonize the burned area. Are these plants more likely to be *r*-strategists or *K*-strategists? Explain your answer.

12. **Evaluating Models** Is a population growth model based on exponential growth more or less realistic than a logistic population model? Explain your answer.

13. **Applying Information** Why might purebred dogs and cats be subject to more inherited disorders than are mixed breeds?

Alternative Assessment

14. **Identifying Variables** Formulate a hypothesis about human population growth. Then use library or Internet resources to find estimates of the current rate of human population growth and forecasts for future growth. Predict trends from the data and communicate your conclusions in the form of a report to your class.

Standardized Test Prep

Understanding Concepts

Directions (1–4): **For *each* question, write on a separate sheet of paper the letter of the correct answer.**

1 What is an organism that tends to grow exponentially when allowed to reproduce?
  A. dispersed organism
  B. *K*-strategist
  C. polygenic organism
  D. *r*-strategist

2 In a population model, what is a limited resource known as?
  F. density-dependent factor
  G. density-independent factor
  H. logistic model
  I. polygenic trait

3 If spacing of individuals in a population is self-determined, what kind of dispersion is occurring?
  A. clumped
  B. exponential
  C. random
  D. regular

4 Which of the following results in a population whose individuals have extreme traits?
  F. directional selection
  G. exponential growth
  H. random mating
  I. stabilizing selection

Directions (5–6): **For *each* question, write a short response.**

5 A scientist hypothesizes that natural selection causes human populations to evolve. Justify this hypothesis using the bubonic plague in the Middle Ages.

6 Differentiate between the terms population size and population density.

Test TIP

For short-response questions, be sure to answer the prompt as fully as possible. Include supporting details in your response.

Reading Skills

Directions (7): **Read the passage below. Then answer the question.**

The availability of resources on a ranch helps ranchers determine the carrying capacity of the ranch. Ranchers consider abundance and distribution of natural food sources and water, particularly if raising livestock that roam freely and graze. Ranchers monitor these assets to be sure that the population of livestock does not exceed the carrying capacity of the ranch.

7 What could happen if a rancher maintains a population of livestock that exceeds the carrying capacity of the ranch?
  A. The population of livestock would evolve to need fewer resources.
  B. There could be food and water shortages or rapid spread of diseases.
  C. The population would increase causing the depletion of even more resources.
  D. The natural resources on the ranch would increase naturally to compensate.

Interpreting Graphics

Directions (8): **Base your answer to question 8 on the chart below.**

Distribution of Body Colors in Bark Beetles

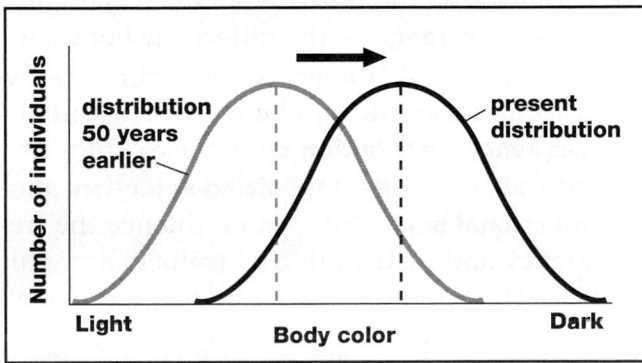

8 Which evolutionary force is represented by the chart?
  F. gene flow
  G. directional selection
  H. mutation
  I. stabilizing selection

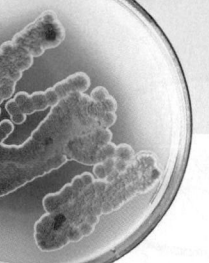

Exploration Lab

Observing How Natural Selection Affects a Population

SKILLS
- Using scientific methods
- Collecting, graphing, and analyzing data

OBJECTIVES
- **Measure** and collect data for a trait in a population.
- **Graph** a frequency distribution curve of your data.
- **Analyze** your data by determining its mean, median, mode, and range.

- **Predict** how natural selection can affect the variation in a population.

MATERIALS
- metric ruler
- graph paper (optional)
- green beans or snow peas
- calculator
- balance

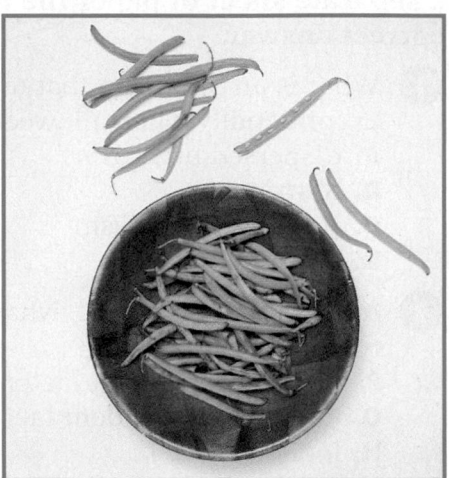

Before You Begin

Natural selection can occur when there is **variation** in a **population.** You can analyze the variation in certain traits of a population by determining the mean, median, mode, and range of the data collected on several individuals. The **mean** is the sum of all data values divided by the number of values. The **median** is the midpoint in a series of values. The **mode** is the most frequently occurring value. The **range** is the difference between the largest and smallest values. The variation in a characteristic can be visualized with a **frequency distribution curve.** Two kinds of natural selection—**stabilizing selection** and **directional selection**—can influence the frequency and distribution of traits in a population. This changes the shape of a frequency distribution curve. In this lab, you will investigate variation in fruits and seeds.

1. Write a definition for each boldface term in the paragraph above.

2. Based on the objectives for this lab, write a question you would like to explore about variation in green beans or snow peas.

Procedure

PART A: Design an Experiment

1. Work with the members of your lab group to explore one of the questions written for step 2 of **Before You Begin.** To explore the question, design an experiment that uses the materials listed for this lab.

> ### You Choose
> **As you design your experiment, decide the following:**
> **a.** what question you will explore
> **b.** what hypothesis you will test
> **c.** which trait (length, color, weight, etc.) you will measure
> **d.** how you will measure the trait
> **e.** how many members of the population you will measure (keep in mind that the more data you gather, the more revealing your frequency distribution curve will be)
> **f.** what data you will record in your data table

2. Write a procedure for your experiment. Make a list of all the safety precautions you will take. Have your teacher approve your procedure and safety precautions before you begin the experiment.

3. Conduct your experiment.

PART B: Cleanup and Disposal

4. Dispose of seeds in the designated waste containers. Do not put lab materials in the trash unless your teacher tells you to do so.

5. Clean up your work area and all lab equipment. Return lab equipment to its proper place. Wash your hands thoroughly before you leave the lab and after you finish all work.

Analyze and Conclude

1. **Summarizing Results** Make a frequency distribution curve of your data. Plot the trait you measured on the x-axis (horizontal axis) and the number of times that trait occurred in your population on the y-axis (vertical axis).

2. **Calculating** Determine the mean, median, mode, and range of the data for the trait you studied.

3. **Analyzing Results** How does the mean differ from the mode in your population?

4. **Drawing Conclusions** What type of selection appears to have produced the type of variation observed in your experiment?

5. **Evaluating Data** The graph below shows the distribution of wing length in a population of birds on an island. Notice that the mean and the mode are quite different. Is the mean always useful in describing traits in a population? Explain.

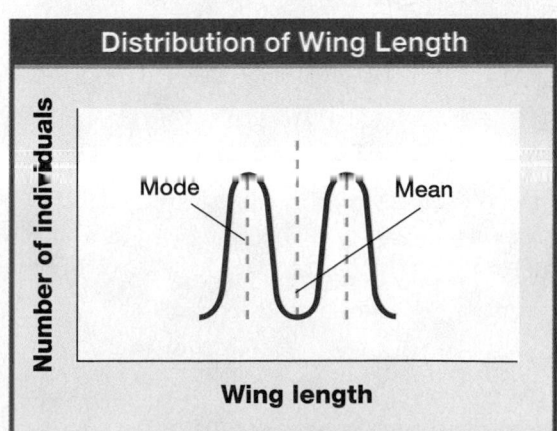

6. **Forming Hypotheses** What type of selection (stabilizing or directional) would be indicated if the mean of a trait you measured shifted, over time, to the right of a frequency distribution graph?

7. **Further Inquiry** Write a new question about variation in populations that could be explored in another investigation.

? Do You Know?

Do research in the library or media center to answer these questions:

1. What factors have contributed to the rise in bacterial resistance to antibiotic drugs?

2. How do farmers use directional selection to improve domestic plants and animals?

Use the following Internet resources to explore your own questions about how populations change.

🔲 **internet** connect

www.scilinks.org
Topic: **Populations**
Keyword: **HX4147**

SC**LINKS**® Maintained by the National Science Teachers Association

Coral-reef ecosystem

16 Ecosystems

✓ *Quick Review*

Answer the following without referring to earlier sections of your book.

1. **Contrast** autotrophs with heterotrophs. *(Chapter 5, Section 1)*

2. **Summarize** the process of photosynthesis. *(Chapter 5, Section 2)*

3. **Describe** the process of cellular respiration. *(Chapter 5, Section 3)*

4. **Compare** the energy flow in photosynthesis with the energy flow in cellular respiration. *(Chapter 5, Sections 2 and 3)*

Did you have difficulty? *For help, review the sections indicated.*

📖 *Reading Activity*

Copy the following statements on a piece of paper or in your notebook, leaving a few blank lines after each.

1. *In an ecosystem, more energy is stored in plants than in consumers.*

2. *The extinction of one species in an ecosystem can have an impact on all other species.*

Before you read the chapter, write down whether you agree or disagree with each statement. After you have finished reading the chapter, decide whether or not you still agree with your first response.

• Materials and energy cycle continuously through the components of this coral reef. The complex relationship of organisms and their physical environment makes up an ecological system, or *ecosystem*.

Looking Ahead

Section 1

What Is an Ecosystem?
Interactions of Organisms and Their Environment
Diverse Communities in Ecosystems
Change of Ecosystems over Time

Section 2

Energy Flow in Ecosystems
Movement of Energy Through Ecosystems
Loss of Energy in a Food Chain

Section 3

Cycling of Materials in Ecosystems
Biogeochemical Cycles
The Water Cycle
The Carbon Cycle
The Phosphorus and Nitrogen Cycles

🖥 internet connect

www.scilinks.org
National Science Teachers Association *sci*LINKS Internet resources are located throughout this chapter.

*SCi*LINKS. **Maintained by the National Science Teachers Association**

What Is an Ecosystem?

Objectives

- **Distinguish** an ecosystem from a community.

- **Describe** the diversity of a representative ecosystem.

- **Sequence** the process of succession.

Key Terms

ecology
habitat
community
ecosystem
abiotic factor
biotic factor
biodiversity
pioneer species
succession
primary succession
secondary succession

Interactions of Organisms and Their Environment

It is easy to think of the environment as being around but not part of us—something we always use, sometimes enjoy, and sometimes damage. But in fact, we are part of the environment along with all of Earth's other organisms. All of Earth's inhabitants are interwoven in a complex web of relationships, such as the one illustrated in **Figure 1.** To understand how the interactions of the parts can affect a whole system, think about how a computer operates. Removing one circuit from a computer can change or limit the interactions of the computer's many components in ways that influence the computer's overall operation. In a similar way, removing one species from our environment can have many consequences, not all of them easily predictable.

In 1866, the German biologist Ernst Haeckel gave a name to the study of how organisms fit into their environment. He called this study *ecology,* which comes from the Greek words *oikos,* meaning "house," or "place where one lives," and *logos,* meaning "study of." **Ecology** is the study of the interactions of living organisms with one another and with their physical environment (soil, water, climate, and so on). The place where a particular population of a species lives is its **habitat.** The many different species that live together in a habitat are called a **community.** An **ecosystem,** or ecological system, consists of a community and all the physical aspects of its habitat, such as the soil, water, and weather. The physical aspects of a habitat are called **abiotic** *(ay bie AHT ihk)* **factors,** and the organisms in a habitat are called **biotic factors.**

Figure 1 Organisms interact within an ecosystem. Organisms within an ecosystem continually change and adjust. This plant species is dependent on the bat for its reproduction, and the bat uses part of the flower for food.

Diverse Communities in Ecosystems

The variety of organisms, their genetic differences, and the communities and ecosystems in which they occur is termed **biodiversity.** Consider a pine forest in the southeastern United States, such as the one shown in **Figure 2.** If you could fence in a square kilometer (0.4 mi²) of this forest and then collect every organism, what would you expect to get? Which of the six kingdoms of organisms would be represented in your collection?

Figure 2   Pine forest.
Pine forests like this one are common in the southeastern United States.

Ecosystem Inhabitants

Large animals in the forest might include a bear or a white-tailed deer. The woods also contain smaller mammals—raccoons, foxes, squirrels, rabbits, and chipmunks. Snakes and toads often remain hidden among the leaves. Many birds can be found, including hawks, warblers, and sparrows. If the square kilometer included a lake, you might find catfish, bass, perch, a variety of turtles, and perhaps an alligator.

There are pine trees, a variety of smaller trees, and shrubs. Beneath the trees, grasses and many kinds of flowers grow on the forest floor.

The soil contains an immense number of worms. Hidden under the bark of trees and beneath the leaves covering the ground are many different species of insects and spiders, such as those shown in **Figure 3.**

Many of the life-forms in the soil and water of a pine forest are too small to be seen without a microscope. Protists, which include algae and related microscopic eukaryotes, thrive in water. There may be billions of bacteria in a handful of soil.

internet connect

www.scilinks.org
Topic: Biodiversity
Keyword: HX4020

SC*LINKS* Maintained by the National Science Teachers Association

Figure 3   Forest spider and insect

The jumping spider is found in sunny, dry parts of the forest. The larvae of the stag beetle live in and eat decaying wood and bark.

Jumping spider

Male stag beetle

Figure 4   Forest fungi

These fungi digest plants and other materials they find in the forest.

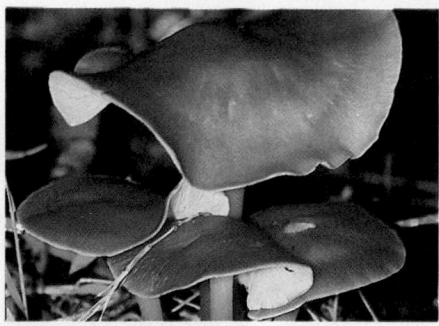

Mushrooms are often found on moist forest floors.

Shelf fungi grow on and digest trees.

You might find many kinds of fungi growing on fallen trees and spreading as fine threads through the decaying material on the forest floor, as illustrated in **Figure 4.** Other fungi are found on the surface of trees or rocks as lichens. Lichens are associations between fungi and algae or cyanobacteria.

If you were to remove every organism from your square kilometer, the nonliving surroundings that remain make up the abiotic factor. This would include the minerals, organic compounds, water, wind that blows over the Earth, rain, and sunlight.

Ecosystem Boundaries

The physical boundaries of an ecosystem are not always obvious, and they depend on how the ecosystem is being studied. For example, a scientist might consider a single rotting log on the forest floor to be an ecosystem if he or she is interested only in the fungi and insects living in the log. Often individual fields, forests, or lakes are studied as isolated ecosystems. Of course, no location is ever totally isolated. Even oceanic islands get occasional migrant visitors, such as birds blown off course.

QUICK LAB

Evaluating Biodiversity

By making simple observations, you can draw some conclusions about biodiversity in an ecosystem.

Materials

note pad, pencil

Procedure

1. **CAUTION: Do not approach or touch any wild animals. Do not disturb plants.** Prepare a list of biotic and abiotic factors that you observe around your home or in a nearby park.

Analysis

1. **Identify** the habitat and community that you observed.

2. **Calculate** the number of different species as a percentage of the total number of organisms that you saw.

3. **Rank** the importance of biotic factors within the ecosystem you observed.

4. **Infer** what the relationships are between biotic factors and abiotic factors in the observed ecosystem.

Change of Ecosystems over Time

When a volcano forms a new island, a glacier recedes and exposes bare rock, or a fire burns all of the vegetation in an area, a new habitat is created. This change sets off a process of colonization and ecosystem development. The first organisms to live in a new habitat where soil is present tend to be small, fast-growing plants, called **pioneer species.** They may make the ground more hospitable for other species. Later waves of plant immigrants may then outcompete and replace the pioneer species.

Succession

A somewhat regular progression of species replacement is called **succession.** Succession that occurs where life has not existed before is called **primary succession.** Succession that occurs in areas where there has been previous growth, such as in abandoned fields or forest clearings, is called **secondary succession.** It was once thought that the stages of succession were predictable and that succession always led to the same final community of organisms within any particular ecosystem. Ecologists now recognize that initial conditions and chance play roles in the process of succession. For example, if two species are in competition, a sudden change in the climate may favor the success of one species over the other. For this reason, no two successions are alike.

Glacier Bay: an Example of Succession

A good example of primary succession is a receding glacier because land is continually being exposed as the face of the glacier moves back. The glacier that composes much of the head of Glacier Bay, Alaska, has receded some 100 km (62 mi) over the last 200 years. **Figure 5** shows the kinds of changes that have taken place as time passed.

The most recently exposed areas are piles of rock and gravel that lack the usable nitrogen essential to plant and animal life. The seeds and spores of pioneer species are carried in by the wind. These include lichens, mosses, fireweed, willows, cottonwood, and *Dryas,* a sturdy plant with clumps about 30 cm (1 ft) across. At first all of these plants grow close to the ground, severely stunted by mineral deficiency, but *Dryas* eventually crowds out the other plants.

After about 10 years, alder seeds blown in from distant sites take root. Alder roots have nitrogen-fixing nodules, so they are able to grow more rapidly than *Dryas.* Dead leaves and fallen branches from the alder trees add more usable nitrogen to the soil. The added nitrogen allows willows and cottonwoods to invade and grow with vigor. After about 30 years, dense thickets of alder, willow, and cottonwood shade and eventually kill the *Dryas.*

Figure 5   Glacier Bay

A receding glacier makes primary succession possible.

Recently exposed land has few nutrients.

Alders, grasses, and shrubs later take over from pioneer plants.

As the amount of soil increases, spruce and hemlock trees become plentiful.

About 80 years after the glacier first exposes the land, Sitka spruce invades the thickets. Spruce trees use the nitrogen released by the alders and eventually form a dense forest. The spruce blocks the sunlight from the alders, and the alders then die, just as the *Dryas* did before them. After the spruce forest is established, hemlock trees begin to grow. Hemlocks are very shade tolerant and have a root system that competes well against spruce for soil nitrogen. Hemlock trees soon become dominant in the forest. This community of spruce and hemlock proves to be a very stable ecosystem from the perspective of human time scales, but it is not permanent. As local climates change, this forest ecosystem may change too.

QUICK LAB

Modeling Succession

You can create a small ecosystem and measure how organisms modify their environment.

Materials

1 qt glass jar with a lid, one-half quart of pasteurized milk, pH strips

Procedure

1. Prepare a table like the one below.

2. Half fill a quart jar with pasteurized milk, and cover the jar loosely with a lid. Measure and record the pH. Place the jar in a 37°C incubator.

3. Check and record the pH of the milk with pH strips every day for seven days. As milk spoils, its pH changes. Different populations of microorganisms become established, alter substances in the milk, and then die off when conditions no longer favor their survival.

4. Record any visible changes in the milk each day.

Analysis

1. **Identify** what happened to the pH of the milk as time passed.

2. **Infer** what the change in pH means about the populations of microorganisms in the milk.

3. **Critical Thinking Evaluating Results** How does this model confirm the model of succession in Glacier Bay?

DATA TABLE

Day	pH	Appearance
1		
2		
3		

Section 1 Review

1. **Identify** what components of an ecosystem are *not* part of a community.

2. **Relate** how gardening or agriculture affects succession.

3. **Differentiate** primary succession from secondary succession.

4. **Critical Thinking Applying Information** Why do some ecosystems remain stable for centuries, while others undergo succession?

5. (Standardized Test Prep) In the succession that occurs as a glacier recedes, alders can grow relatively rapidly because alders have
   A nitrogen-fixing nodules.   C no roots.
   B no need for minerals.   D shade tolerance.

Energy Flow in Ecosystems

Movement of Energy Through Ecosystems

Everything that organisms do in ecosystems—running, breathing, burrowing, growing—requires energy. The flow of energy is the most important factor that controls what kinds of organisms live in an ecosystem and how many organisms the ecosystem can support. In this section you will learn where organisms get their energy.

Primary Energy Source

Most life on Earth depends on photosynthetic organisms, which capture some of the sun's light energy and store it as chemical energy in organic molecules. These organic compounds are what we call food. The rate at which organic material is produced by photosynthetic organisms in an ecosystem is called **primary productivity.** Primary productivity determines the amount of energy available in an ecosystem. Most organisms in an ecosystem can be thought of as chemical machines driven by the energy captured in photosynthesis. Organisms that first capture energy, the **producers,** include plants, some kinds of bacteria, and algae. Producers make energy-storing molecules. All other organisms in an ecosystem are consumers. **Consumers** are those organisms that consume plants or other organisms to obtain the energy necessary to build their molecules.

Trophic Levels

Ecologists study how energy moves through an ecosystem by assigning organisms in that ecosystem to a specific level, called a **trophic** *(TROHF ihk)* **level,** in a graphic organizer based on the organism's source of energy. Energy moves from one trophic level to another, as illustrated in **Figure 6.**

Objectives

- **Distinguish** between producers and consumers.
- **Compare** food webs with food chains.
- **Describe** why food chains are rarely longer than three or four links.

Key Terms

primary productivity
producer
consumer
trophic level
food chain
herbivore
carnivore
omnivore
detritivore
decomposer
food web
energy pyramid
biomass

Figure 6   Trophic levels

The sun is the ultimate source of energy for producers and all consumers.

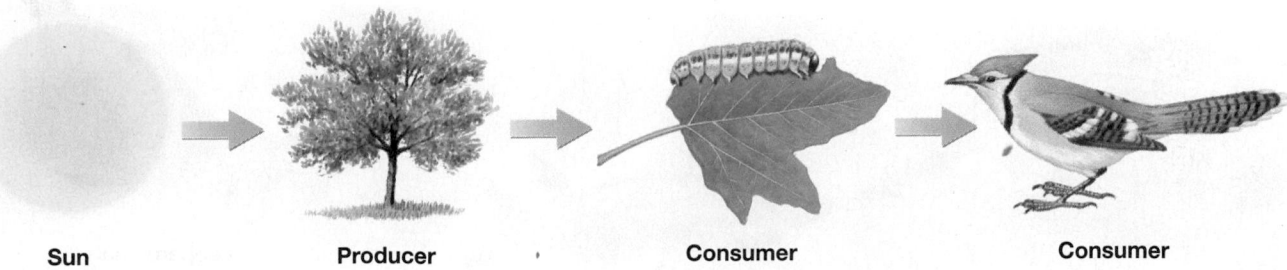

| Sun | Producer | Consumer | Consumer |

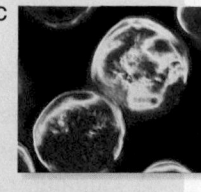
First Level The path of energy through the trophic levels of an ecosystem is called a **food chain.** An example is shown in **Figure 7.** The lowest trophic level of any ecosystem is occupied by the producers, such as plants, algae, and bacteria. Producers use the energy of the sun to build energy-rich carbohydrates. Many producers also absorb nitrogen gas and other key substances from the environment and incorporate them into their biological molecules.

Second Level At the second trophic level are **herbivores** *(HUHR beh vohrz),* animals that eat plants or other primary producers. They are the primary consumers. Cows and horses are herbivores, as are caterpillars and some ducks. A herbivore must be able to break down a plant's molecules into usable compounds. However, the ability to digest cellulose is a chemical feat that only a few organisms have evolved. As you will recall, cellulose is a complex carbohydrate found in plants. Most herbivores rely on microorganisms, such as bacteria and protists, in their gut to help digest cellulose. Humans cannot digest cellulose because we lack these particular microorganisms.

Third Level At the third trophic level are secondary consumers, animals that eat other animals. These animals are called **carnivores.** Tigers, wolves, and snakes are carnivores. Some animals, such as bears, are both herbivores and carnivores; they are called **omnivores** *(AHM nih vohrz).* They use the simple sugars and starches stored in plants as food, but they cannot digest cellulose.

In every ecosystem there is a special class of consumers called detritivores, which include worms and fungal and bacterial decomposers. **Detritivores** *(deh TRIH tih vohrz)* are organisms that obtain their energy from the organic wastes and dead bodies that are produced at

Figure 7 Aquatic food chain

This food chain shows one path of energy flow in an Antarctic ecosystem.

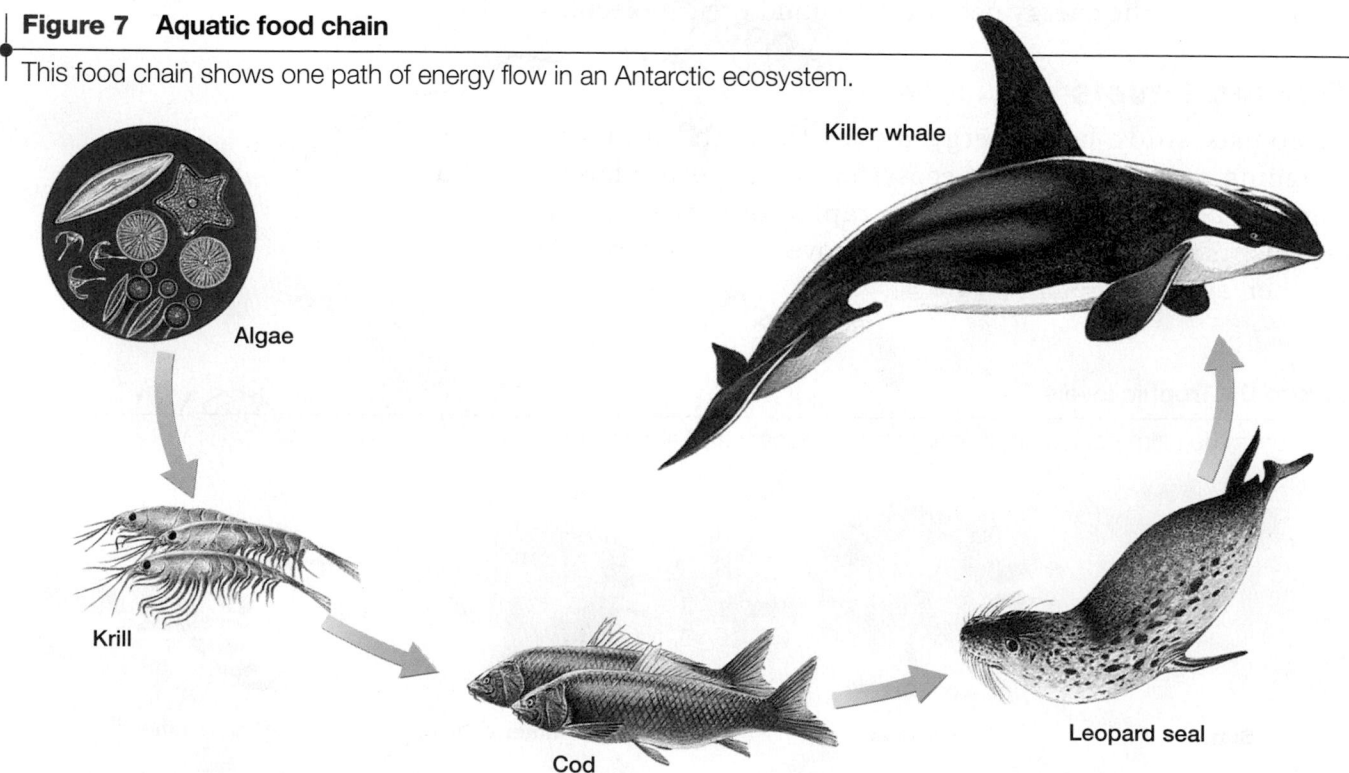

Killer whale

Algae

Krill

Cod

Leopard seal

all trophic levels. Bacteria and fungi are known as **decomposers** because they cause decay. Decomposition of bodies and wastes releases nutrients back into the environment to be recycled by other organisms.

Many ecosystems contain a fourth trophic level composed of those carnivores that consume other carnivores. They are called tertiary consumers, or top carnivores. A hawk that eats a snake is a tertiary consumer. Very rarely do ecosystems contain more than four trophic levels.

In most ecosystems, energy does not follow simple straight paths because individual animals often feed at several trophic levels. This creates a complicated, interconnected group of food chains called a **food web,** as illustrated in **Figure 8.**

■ internet connect

www.scilinks.org
Topic: Food Chains
and Webs
Keyword: HX4085

SCLINKS. Maintained by the National Science Teachers Association

Figure 8   Aquatic food web

This food web shows a more complete picture of the feeding relationships in an Antarctic ecosystem.

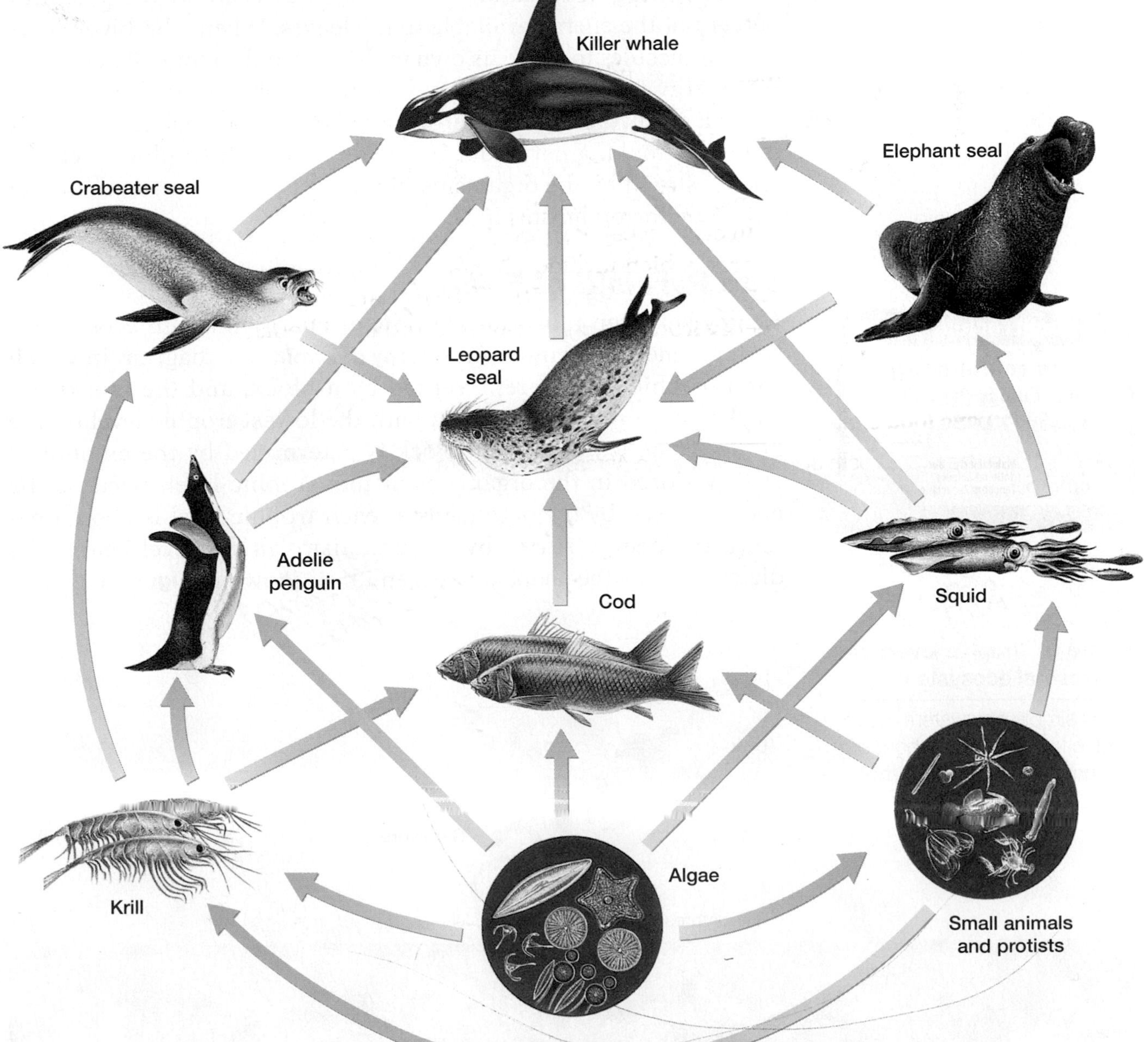

Killer whale

Crabeater seal

Elephant seal

Leopard seal

Adelie penguin

Cod

Squid

Krill

Algae

Small animals and protists

Loss of Energy in a Food Chain

A deer browsing on leaves is acquiring energy. Potential energy is stored in the chemical bonds within the molecules of the leaves. Some of this energy is transformed to other forms of potential energy, such as fat. Some of it aids the deer in running and breathing, and in fueling cellular processes. But much of the energy is dispersed into the environment as heat.

Energy Transfer

During every transfer of energy within an ecosystem, energy is lost as heat. Although heat can be used to do work (as in a steam engine), it is generally not a useful source of energy in biological systems. Thus, the amount of useful energy available to do work decreases as energy passes through an ecosystem. The loss of useful energy limits the number of trophic levels an ecosystem can support. When a plant harvests energy from sunlight, photosynthesis captures only about 1 percent of the energy available to the leaves. When a herbivore uses plant molecules to make its own molecules, only about 10 percent of the energy in the plant ends up in the herbivore's molecules. And when a carnivore eats the herbivore, about 90 percent of the energy is lost in making carnivore molecules. At each trophic level, the energy stored by the organisms in a level is about one-tenth of that stored by the organisms in the level below.

The Pyramid of Energy

Ecologists often illustrate the flow of energy through ecosystems with an energy pyramid. An **energy pyramid** is a diagram in which each trophic level is represented by a block, and the blocks are stacked on top of one another, with the lowest trophic level on the bottom. The width of each block is determined by the amount of energy stored in the organisms at that trophic level. Because the energy stored by the organisms at each trophic level is about one-tenth the energy stored by the organisms in the level below, the diagram takes the shape of a pyramid, as shown in **Figure 9.**

WORD Origins

- The word *ecosystem* is from the Greek words *oikos*, meaning "house," and *systematos*, meaning "to place together." Knowing this information makes it easier to remember that an ecosystem includes a community of living things as well as all physical aspects of its environment.

Figure 9 Trophic levels of a terrestrial ecosystem

In this simple ecosystem, each trophic level contains about 90 percent less energy than the level below it.

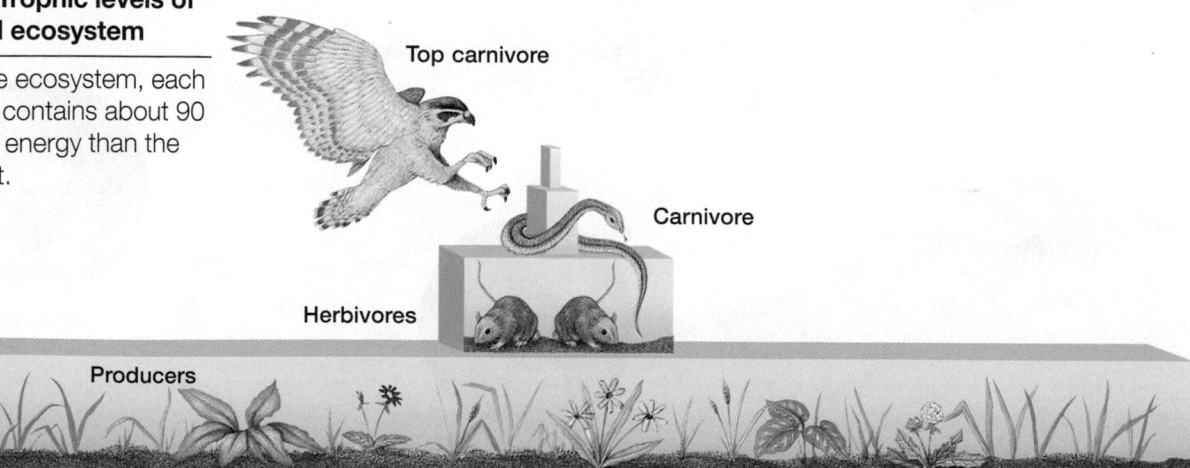

Top carnivore

Carnivore

Herbivores

Producers

Limitations of Trophic Levels

Most terrestrial ecosystems involve only three or, on rare instances, four levels. Too much energy is lost at each level to allow more levels. For example, a large human population could not survive by eating lions captured on the Serengeti Plain of Africa because there are too few lions to make this possible. The amount of grass in that ecosystem cannot support enough zebras to maintain a large enough population of lions to feed lion-eating humans. In other words, the number of trophic levels that can be maintained in a community is limited by the dispersal of potential energy.

Humans are omnivores, and unlike lions, we can choose to eat either meat or plants. As illustrated in **Figure 10,** about 10 kg (22 lb) of grain are needed to build about 1 kg (2.2 lb) of human tissue if the grain is directly ingested by a human. If a cow eats the grain and a human eats the cow, then about 100 kg (220 lb) of grain are needed to build about 1 kg (2.2 lb) of human tissue.

Also, the number of individuals in a trophic level may not be an accurate indicator of the amount of energy in that level. Some organisms are much bigger than others and therefore use more energy. Because of this, the number of organisms often does not form a pyramid when one compares different trophic levels. For instance, caterpillars and other insect herbivores greatly outnumber the trees they feed on. To better determine the amount of energy present in trophic levels, ecologists measure biomass. **Biomass** is the dry weight of tissue and other organic matter found in a specific ecosystem. Each higher level on the pyramid contains only 10 percent of the biomass found in the trophic level below it.

Figure 10   Energy efficiency in food consumption

Adding a trophic level to a food chain increases the energy demand of consumers by a factor of about 10.

It takes a certain amount of grain

to produce enough bread

to provide one person with a certain amount of energy.

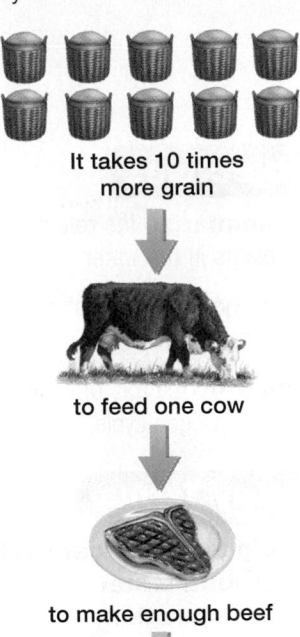

It takes 10 times more grain

to feed one cow

to make enough beef

to provide one person with the same amount of energy.

Section 2 Review

1 **Explain** how producers differ from consumers.

2 **Analyze** the flow of energy through a food chain that contains four tropic levels, one of which is a carnivore.

3 **Construct** a food web, and explain the interactions of the organisms that compose it.

4 **List** the reasons why food chains do *not* tend to exceed four links.

5 **Critical Thinking Justifying an Argument** Explain why scientists believe that most animals would become extinct if all plants died.

6 (**Standardized Test Prep**) Which series shows a correct path of energy flow in a marine food chain?
   A krill → cod → algae
   B cod → leopard seal → krill
   C leopard seal → algae → krill
   D algae → krill → cod

Cycling of Materials in Ecosystems

- **Summarize** the role of plants in the water cycle.

- **Analyze** the flow of energy through the carbon cycle.

- **Identify** the role of bacteria in the nitrogen cycle.

biogeochemical cycle
ground water
transpiration
nitrogen fixation

Biogeochemical Cycles

Humans throw away tons of garbage every year as unwanted, unneeded, and unusable. Nature, however, does not throw anything away. Most energy flows through the Earth's ecosystems from the sun to producers to consumers. The physical parts of the ecosystems, however, cycle constantly. Carbon atoms, for example, are passed from one organism to another in a great circle of use. Producers are eaten by herbivores, herbivores are eaten by carnivores, and carnivores are eaten by top carnivores. Eventually the top carnivores die and decay; their carbon atoms then become part of the soil to feed the producers in a long and complex cycle that reuses this important element. Carbon is not the only element that is constantly recycled in this way. Other recycled elements include many of the inorganic (noncarbon) substances that make up the soil, water, and air, such as nitrogen, sulfur, calcium, and phosphorus.

All materials that cycle through living organisms are important in maintaining the health of ecosystems, but four substances are particularly important: water, carbon, nitrogen, and phosphorus. All organisms require carbon, hydrogen, oxygen, nitrogen, phosphorus, and sulfur in relatively large quantities. They require other elements, such as magnesium, sodium, calcium, and iron, in smaller amounts. Some elements, such as cobalt and manganese, are required in trace amounts.

The paths of water, carbon, nitrogen, and phosphorus pass from the nonliving environment to living organisms, such as the trees in **Figure 11,** and then back to the nonliving environment. These paths form closed circles, or cycles, called biogeochemical (*bie oh jee oh KEHM ih kuhl*) cycles. In each **biogeochemical cycle,** a pathway forms when a substance enters living organisms such as trees from the atmosphere, water, or soil; stays for a time in the living organism; then returns to the nonliving environment. Ecologists refer to such substances as cycling within an ecosystem between a living reservoir (an organism that lives in the ecosystem) and a nonliving reservoir. In almost all biogeochemical cycles, there is much less of the substance in the living reservoir than in the nonliving reservoir.

Figure 11 Trees and the carbon cycle. Approximately 500 million tons of carbon were taken up as a result of forest regrowth in the Northern Hemisphere between 1980 and 1989.

The Water Cycle

Of all the nonliving components of an ecosystem, water has the greatest influence on the ecosystem's inhabitants. In the nonliving portion of the water cycle, water vapor in the atmosphere condenses and falls to the Earth's surface as rain or snow. Some of this water seeps into the soil and becomes part of the **ground water,** which is water retained beneath the surface of the Earth. Most of the remaining water that falls to the Earth does not remain at the surface. Instead, heated by the sun, it reenters the atmosphere by evaporation. The path of water within an ecosystem is shown in **Figure 12.**

In the living portion of the water cycle, much water is taken up by the roots of plants. After passing through a plant, the water moves into the atmosphere by evaporating from the leaves, a process called **transpiration.** Transpiration is also a sun-driven process. The sun heats the Earth's atmosphere, creating wind currents that draw moisture from the tiny openings in the leaves of plants.

In aquatic ecosystems (lakes, rivers, and oceans), the nonliving portion of the water cycle is the most important. In terrestrial ecosystems, the nonliving and living parts of the water cycle both play important roles. In thickly vegetated ecosystems, such as tropical rain forests, more than 90 percent of the moisture in the ecosystem passes through plants and is transpired from their leaves. In a very real sense, plants in rain forests create their own rain. Moisture travels from plants to the atmosphere and falls back to the Earth as rain.

internet connect

www.scilinks.org
Topic: Water Cycle
Keyword: HX4188

SC*LINKS*. Maintained by the National Science Teachers Association

Figure 12 **Water cycle**

This diagram shows the major steps in the water cycle.

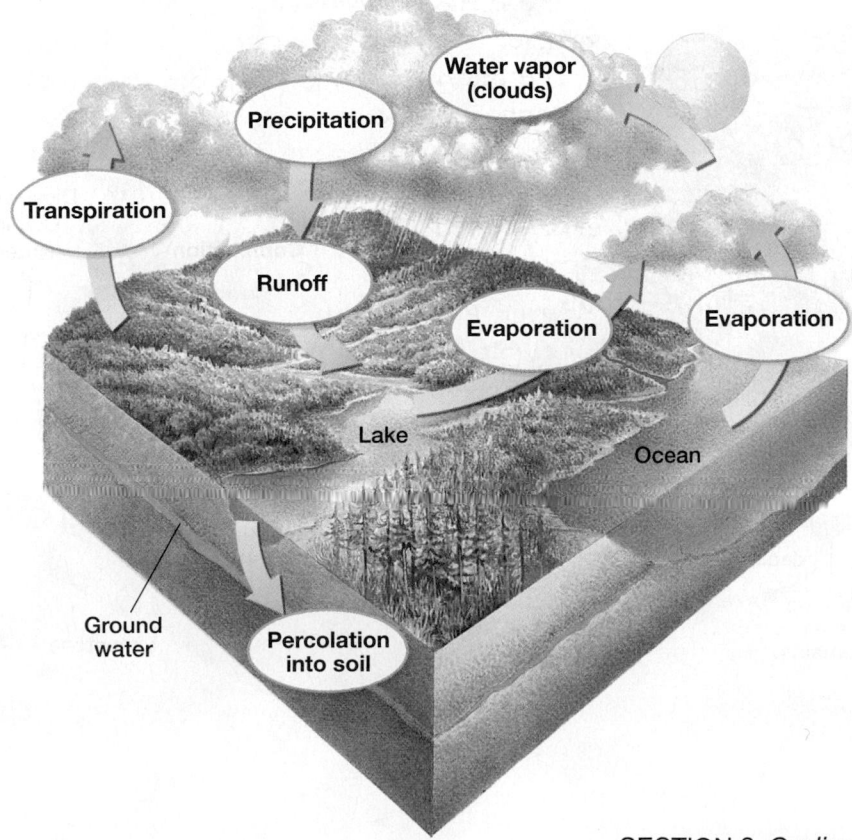

The Carbon Cycle

Carbon also cycles between the nonliving environment and living organisms. You can follow the carbon cycle in **Figure 13.** Carbon dioxide in the air or dissolved in water is used by photosynthesizing plants, algae, and bacteria as a raw material to build organic molecules. Carbon atoms may return to the pool of carbon dioxide in the air and water in three ways.

1. **Respiration.** Nearly all living organisms, including plants, engage in cellular respiration. They use oxygen to oxidize organic molecules during cellular respiration, and carbon dioxide is a byproduct of this reaction.

2. **Combustion.** Carbon also returns to the atmosphere through combustion, or burning. The carbon contained in wood may stay there for many years, returning to the atmosphere only when the wood is burned. Sometimes carbon can be locked away beneath the Earth for thousands or even millions of years. The remains of organisms that become buried in sediments may be gradually transformed by heat and pressure into fossil fuels—coal, oil, and natural gas. The carbon is released when the fossil fuels are burned.

3. **Erosion.** Marine organisms use carbon dioxide dissolved in sea water to make calcium carbonate shells. Over millions of years, the shells of the dead organisms form sediments, which form limestone. As the limestone becomes exposed and erodes, the carbon becomes available to other organisms.

Figure 13 Carbon cycle

This diagram shows the major steps of the carbon cycle.

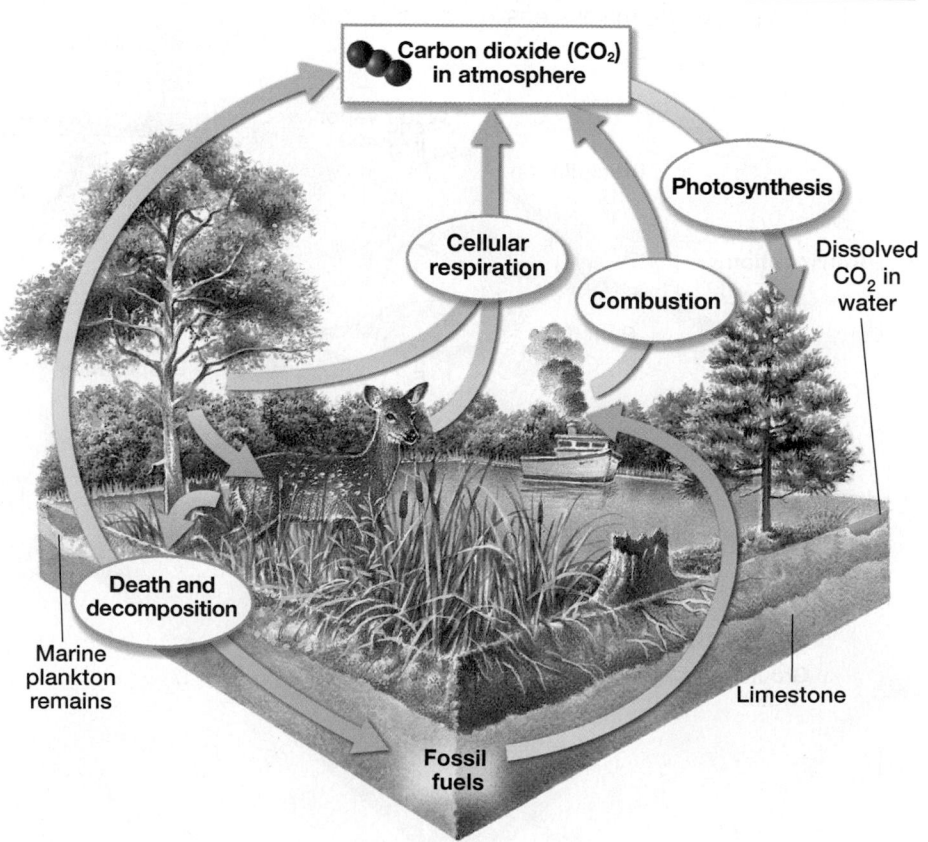

The Phosphorus and Nitrogen Cycles

Organisms need nitrogen and phosphorus to build proteins and nucleic acids. Phosphorus is an essential part of both ATP and DNA. Phosphorus is usually present in soil and rock as calcium phosphate, which dissolves in water to form phosphate ions, PO_4^{3-}. This phosphate is absorbed by the roots of plants and used to build organic molecules. Animals that eat the plants reuse the organic phosphorus.

The atmosphere is about 78 percent nitrogen gas, N_2. However, most organisms are unable to use it in this form. The two nitrogen atoms in a molecule of nitrogen gas are connected by a strong triple covalent bond that is very difficult to break. However, a few bacteria have enzymes that can break it, and they bind nitrogen atoms to hydrogen to form ammonia, NH_3. The process of combining nitrogen with hydrogen to form ammonia is called **nitrogen fixation.** Nitrogen-fixing bacteria live in the soil and are also found within swellings, or nodules, on the roots of beans, alder trees, and a few other kinds of plants.

The nitrogen cycle, diagramed in **Figure 14,** is a complex process with four important stages.

1. **Assimilation** is the absorption and incorporation of nitrogen into organic compounds by plants.

2. **Ammonification** is the production of ammonia by bacteria during the decay of organic matter.

3. **Nitrification** is the production of nitrate from ammonia.

4. **Denitrification** is the conversion of nitrate to nitrogen gas.

Study *TIP*

● **Reviewing Information**
Using your own words, write four sentences, each one describing one of the four biogeochemical cycles.

Figure 14 Nitrogen cycle

Bacteria carry out many of the important steps in the nitrogen cycle, including the conversion of atmospheric nitrogen into a usable form, ammonia.

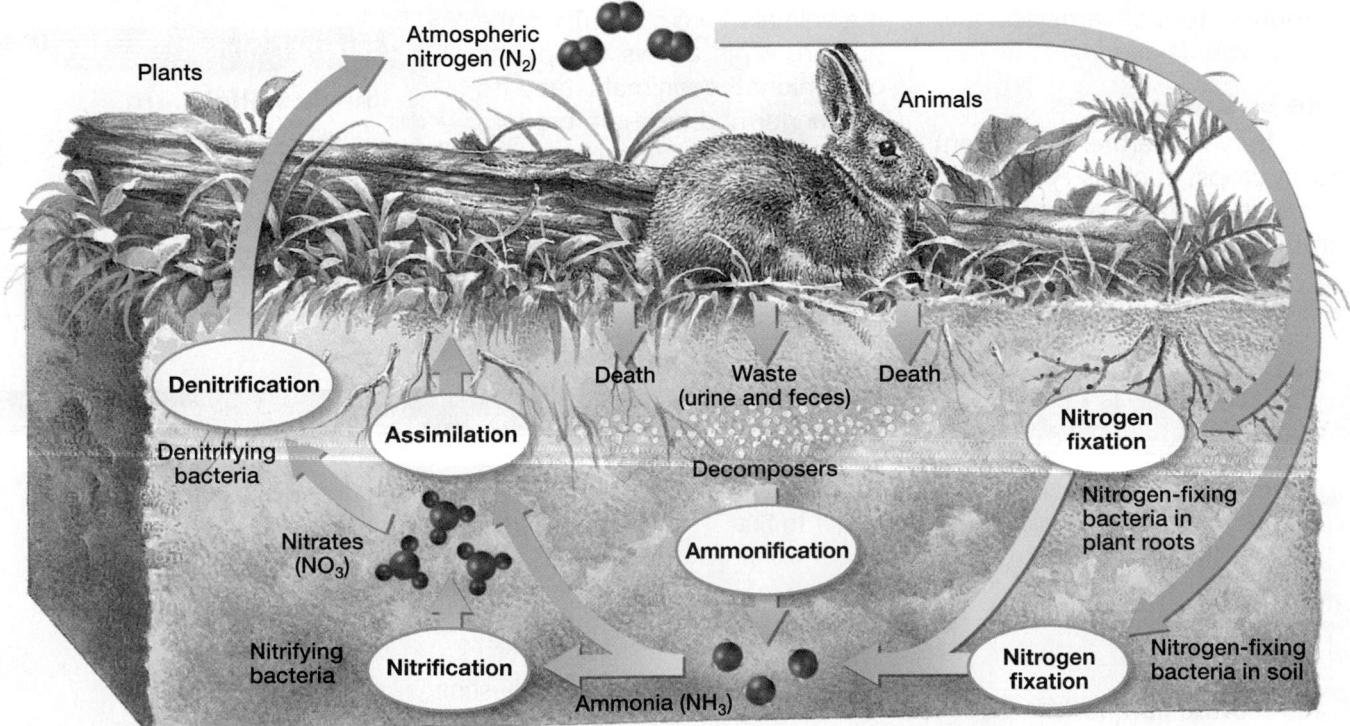

The growth of plants in ecosystems is often limited by the availability of nitrate and ammonia in the soil. Today most of the ammonia and nitrate that farmers add to soil is produced chemically in factories, rather than by bacterial nitrogen fixation. Genetic engineers are trying to place nitrogen-fixing genes from bacteria into the chromosomes of crop plants. If these attempts are successful, the plants themselves will be able to fix nitrogen, thus eliminating the need for nitrogen-supplying fertilizers. Some farmers adjust their farming methods to increase natural recycling of nitrogen.

BIOWatch

Sustainable Agriculture

In an ecosystem, decomposers return mineral nutrients to the soil. However, when the plants are harvested and shipped away, there is a net loss of nutrients from the soil where the plants were growing. The amount of organic matter in the soil also decreases, making the soil less able to hold water and more likely to erode.

What is Sustainable Agriculture?

Sustainable agriculture refers to farming that remains productive and profitable through practices that help replenish the soil's nutrients, reduce erosion, and control weeds and insect pests.

Use of Cover Crops

After harvest, farmers can plant cover crops, such as rye, clover, or vetch, instead of letting the ground lie bare. Cover crops keep the soil from compacting and washing away, and they help the soil absorb water. They also provide a habitat for beneficial insects, slow the growth of weeds, and keep the ground from overheating. When cover crops are plowed under, as illustrated in the figure at right, they return nutrients to the soil.

Rotational Grazing

Farmers who raise cattle and sheep can divide their pastures into several grazing areas. By rotating their livestock from one area to another, they can prevent the animals from overgrazing the pasture. This allows the plants on which the animals feed to live longer and be more productive. Water quality improves as the pasture vegetation becomes denser. Animals distribute manure more evenly with rotational grazing than they do in feed lots or unmanaged pastures.

There are many other methods used in sustainable agriculture. Farmers must determine which methods work best for their crops, soil conditions, and climate.

internet connect

www.scilinks.org
Topic: Sustainable Agriculture
Keyword: HX4170

SC*LINKS*® Maintained by the National Science Teachers Association

Section 3 Review

1 **Identify** the role of energy in the part of the water cycle in which plants transfer water to the atmosphere.

2 **Analyze** the carbon cycle's relationship to the flow of energy.

3 **Describe** how bacteria participate in the nitrogen cycle.

4 **Critical Thinking Defend** the argument that nutrients can cycle but energy cannot.

5 **Standardized Test Prep** Which component of the carbon cycle removes carbon dioxide from the atmosphere?

A combustion   **C** erosion

B cellular respiration   **D** photosynthesis

Study ZONE

Key Concepts

1 What Is an Ecosystem?

- Ecology is the study of how organisms interact with each other and with their environment.
- A community of organisms and their nonliving environment constitute an ecosystem.
- Ecosystems contain diverse organisms.
- Ecosystems change through the process of succession.
- Succession on a newly formed habitat is primary succession.
- Secondary succession occurs on a habitat that has previously supported growth.

2 Energy Flow in Ecosystems

- Energy moves through communities in food chains, passing from photosynthesizers (producers) to herbivores (consumers) to carnivores (consumers), creating a food web.
- Energy transfers between trophic levels transfer only 10 percent of the energy in a trophic level to the next level.
- Most terrestrial communities have only three or four trophic levels because energy transfers between trophic levels are inefficient.

3 Cycling of Materials in Ecosystems

- Minerals and other materials cycle within ecosystems among organisms and between organisms and the physical environment.
- In the water cycle, water falls as precipitation and either evaporates from bodies of water, is stored in ground water, or cycles through plants and then evaporates.
- Carbon enters the living portion of the carbon cycle through photosynthesis. Organisms release carbon through cellular respiration. Carbon trapped in rocks and fossil fuels is released by erosion and burning.
- Bacteria fix atmospheric nitrogen, thus making ammonia available to other organisms.

Key Terms

Section 1

ecology (340)
habitat (340)
community (340)
ecosystem (340)
abiotic factor (340)
biotic factor (340)
biodiversity (341)
pioneer species (343)
succession (343)
primary succession (343)
secondary succession (343)

Section 2

primary productivity (345)
producer (345)
consumer (345)
trophic level (345)
food chain (346)
herbivore (346)
carnivore (346)
omnivore (346)
detritivore (346)
decomposer (347)
food web (347)
energy pyramid (348)
biomass (349)

Section 3

biogeochemical cycle (350)
ground water (351)
transpiration (351)
nitrogen fixation (353)

BIOLOGY

Unit 7—*Ecosystem Dynamics*
Use Topics 1, 3–6 in this unit to review the key concepts and terms in this chapter.

Understanding Key Ideas

1. Ecosystems differ from communities in that ecosystems usually contain
a. several climates.
b. several communities.
c. only one habitat.
d. only one food web.

2. What critical role is played by fungi and bacteria in any ecosystem?
a. primary production
b. decomposition
c. boundary setting
d. physical weathering

3. Which sequence shows the correct order of succession at Glacier Bay, Alaska?
a. alder, *Dryas*, hemlock
b. *Dryas*, hemlock, alder
c. *Dryas*, alder, Sitka spruce
d. mosses, hemlock, Sitka spruce

4. Which role is *not* performed by bacteria in the nitrogen cycle?
a. fixing nitrogen
b. changing urea to ammonia
c. turning nitrates into nitrogen gas
d. changing nitrates to ammonia

5. How would the food web below be affected if the plants were eliminated?

a. Herbivores would become carnivores.
b. The food web would collapse.
c. The herbivores would change trophic levels.
d. Nothing would happen.

6. How much energy is available at the third trophic level of an energy pyramid if 1,000 kcal is available in the first level?
a. 1,000 kcal   c. 10 kcal
b. 100 kcal   d. 1 kcal

7. Humans, raccoons, and bears are omnivores. What adaptive advantage might this feeding strategy provide?

8. 🌎 **BIOWatch** After harvesting, a farmer could either plow the remaining cornstalks into the field or burn them. Which option is best for sustainable agriculture? Explain your answer.

9. Relate photosynthesis to the nitrogen cycle. (**Hint:** See Chapter 5, Section 2.)

10. 🔲 **Concept Mapping** Make a concept map that describes the flow of energy through an ecosystem. Try to include the following terms in your map: *trophic level, food web, food chain, producer, consumer, carnivore, detritivore,* and *herbivore.*

Critical Thinking

11. Inferring Relationships Analyze the flow of energy between an ecosystem and one of its top carnivores, such as a hawk.

12. Applying Information Is nitrogen cycling or carbon cycling more important to a pioneer species during primary succession? Explain your answer.

13. Predicting Results Describe the probable effects on an ecosystem if all decomposers were to die.

Alternative Assessment

14. Identifying Functions Obtain photocopies of nature paintings by American painters such as John James Audubon or Edward Hicks. Choose three animals, and write a report that compares the animals, the ecosystems in which they live, their roles in biogeochemical cycles, and the trophic level they occupy.

Standardized Test Prep

Understanding Concepts

Directions (1–5): **For *each* question, write on a separate sheet of paper the letter of the correct answer.**

1 Which of the following situations describes a carnivore and an herbivore?
  A. A horse eats an apple.
  B. A rabbit eats a dandelion.
  C. A mountain lion eats a rabbit.
  D. A fungus breaks down a dead oak tree.

2 What term applies to most humans?
  F. carnivore   **H.** herbivore
  G. detrivore   **I.** omnivore

3 What is an organism that obtains energy from organic wastes and dead bodies called?
  A. carnivore   **C.** herbivore
  B. detrivore   **D.** omnivore

4 What is the process by which materials pass between the nonliving environment and living organisms?
  F. biogeochemical cycle
  G. energy pyramid
  H. food web
  I. primary succession

5 Through what process do plants return water to the atmosphere?
  A. assimilation
  B. nitrification
  C. succession
  D. transpiration

Directions (6): **For the following question, write a short response.**

6 Ecologists once referred to stable ecosystems as a final or climax community. Now most ecologists say that no ecosystem can truly have a final end point. Analyze why ecologists have changed their viewpoint.

Test TIP

For multiple-choice questions, try to eliminate any answer choices that are obviously incorrect, and then consider the remaining answer choices.

Reading Skills

Directions (7): **Read the passage below. Then answer the question.**

Artificial ecosystems used in the treatment of waste water and pollutants can demonstrate succession. Artificial wastewater-treatment ecosystems tend to undergo eutrophication, just as natural wetlands do. However, the high nutrient levels in waste water promote rapid algae growth. If the systems are not manipulated, they will eventually fill with algae and decaying organic matter, providing nutrients for other species. The system can then form a marsh and eventually a meadow.

7 Why don't meadow grasses populate the new ecosystem before the marsh plants and algae begin to grow there?
  F. The presence of algae is harmful to meadow grasses.
  G. The presence of decaying organic matter is harmful to meadow grasses.
  H. Meadow grasses require that pioneer species first make nutrient-rich soil.
  I. Meadow grasses cannot compete with marsh plants in established ecosystems.

Interpreting Graphics

Directions (8): **Base your answer to question 8 on the graph below.**

Atmospheric Carbon Dioxide Variation

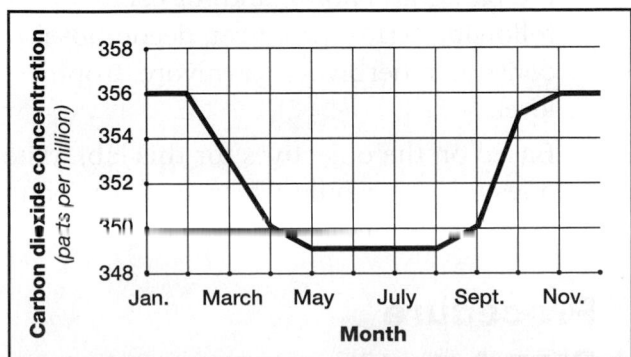

8 During which of the following months is the rate of photosynthesis greatest?
  A. January   **C.** May
  B. March   **D.** September

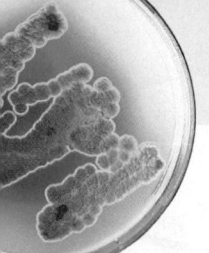

Exploration Lab

Modeling Ecosystem Change over Time

SKILLS
- Using scientific methods
- Modeling
- Observing

OBJECTIVES
- **Construct** a model ecosystem.
- **Observe** the interactions of organisms in a model ecosystem.

- **Predict** how the number of each species in a model ecosystem will change over time.
- **Compare** a model ecosystem with a natural ecosystem.

MATERIALS
- coarse sand or pea gravel
- large glass jar with a lid or terrarium
- soil
- pinch of grass seeds

- pinch of clover seeds
- rolled oats
- mung bean seeds
- earthworms
- isopods (pill bugs)
- mealworms (beetle larva)
- crickets

Before You Begin

Organisms in an **ecosystem** interact with each other and with their environment. One of the interactions that occurs among the organisms in an ecosystem is feeding. A **food web** describes the feeding relationships among the organisms in an ecosystem. In this lab, you will model a natural ecosystem by building a **closed ecosystem** in a bottle or a jar. You will then observe the interactions of the organisms in the ecosystem and note any changes that occur over time.

1. Write a definition for each boldface term in the paragraph above and for each of the following terms: producer, decomposer, consumer, herbivore, carnivore, trophic level.

2. Based on the objectives for this lab, write a question you would like to explore about ecosystems.

Procedure

PART A: Building an Ecosystem in a Jar

1. Place 2 in. of sand or pea gravel in the bottom of a large, clean glass jar with a lid. **CAUTION: Glassware is fragile.**

Notify your teacher promptly of any broken glass or cuts. Do not clean up broken glass or spills with broken glass unless your teacher tells you to do so. Cover the gravel with 2 in. of soil.

2. Sprinkle the seeds of two or three types of small plants, such as grasses and clovers, on the surface of the soil. Put a lid on the jar, and place it in indirect sunlight. Let the jar remain undisturbed for a week.

3. After one week, place a handful of rolled oats into the jar. Place the mealworms in the oats, and then place the other animals into the jar and replace the lid. Place the lid on the jar loosely to enable air entry.

You Choose

As you design your experiment, decide the following:
a. what question you will explore
b. what hypothesis you will test
c. how you will plant the seeds
d. where you will place the ecosystem for one week so that it remains undisturbed and in indirect sunlight
e. how often you will add water to the ecosystem after the first week
f. how many of each organism you will use
g. what data you will record in your data table

PART B: Design an Experiment

4. Work with the members of your lab group to explore one of the questions written for step 2 of **Before You Begin.** To explore the question, design an experiment that uses the materials listed for this lab.

5. Write a procedure for your experiment. Make a list of all the safety precautions you will take. Have your teacher approve your procedure and safety precautions before you begin the experiment.

6. Set up your group's experiment. Conduct your experiment for at least 14 days.

PART C: Cleanup and Disposal

7. Dispose of solutions, broken glass, and other materials in the designated waste containers. Do not put lab materials in the trash unless your teacher tells you to do so.

8. Clean up your work area and all lab equipment. Return lab equipment to its proper place. Wash your hands thoroughly before you leave the lab and after you finish all work.

Analyze and Conclude

1. **Summarizing Results** Make graphs showing how the number of individuals of each species in your ecosystem changed over time. Plot time on the *x*-axis and the number of organisms on the *y*-axis.

2. **Analyzing Results** How did your results compare with your hypothesis? Explain any differences.

3. **Inferring Conclusions** Construct a food web for the ecosystem you observed.

4. **Recognizing Relationships** Does your model ecosystem resemble a natural ecosystem? Explain.

5. **Analyzing Methods** How might you have built your model ecosystem differently to better represent a natural ecosystem?

6. **Evaluating Methods** Was your model ecosystem truly a "closed ecosystem"? List your model's strengths and weaknesses as a closed ecosystem.

7. **Further Inquiry** Write a new question about ecosystems that you could explore with another investigation.

? Do You Know?

Do research in the library or media center to answer these questions:

1. What is Biosphere 2?

2. What problems were encountered by the Biosphere 2 crew during the 1991–1993 project?

Use the following Internet resources to explore your own questions about ecosystems.

internet connect

www.scilinks.org
Topic: Ecosystems
Keyword: HX4066

SCI**LINKS** Maintained by the
National Science Teachers Association

Desert biome in winter

✔️ *Quick Review*

Answer the following without referring to earlier sections of your book.

1. **Compare** the energy flow in photosynthesis to the energy flow in cellular respiration. *(Chapter 5, Section 1)*

2. **Differentiate** between the terms *habitat, community,* and *ecosystem. (Chapter 16, Section 1)*

3. **Analyze** the relationship between primary productivity and energy flow in ecosystems. *(Chapter 16, Section 2)*

Did you have difficulty? *For help, review the sections indicated.*

📖 *Reading Activity*

Before you read this chapter, create a list of all the ways that two species in an ecosystem can interact. Then develop a list of all the different types of communities or ecosystems you can think of. Can any of the ecosystems be grouped into larger systems (biomes)? Read the chapter to see how scientists have defined interactions and biomes.

🔲 **internet** connect

www.scilinks.org
National Science Teachers Association *sci*LINKS Internet resources are located throughout this chapter.

*sci*LINKS. Maintained by the National Science Teachers Association

● Harsh and unforgiving, the desert is home to plants and animals equipped to thrive in the face of environmental challenges. No other terrestrial biome displays a wider range of extreme conditions.

How Organisms Interact in Communities

Evolution in Communities

What are the most important members of an ecosystem? When you try to answer this question, you soon realize that you cannot view an ecosystem's inhabitants as single organisms, but only as members of a web of interactions.

Interactions Among Species

Some interactions among species are the result of a long evolutionary history in which many of the participants adjust to one another over time. Thus, adaptations appeared in flowering plants that promoted efficient dispersal of their pollen by insects and other animals. In turn, adaptations appeared in pollinators that enabled them to obtain food or other resources from the flowers they pollinate. Natural selection has often led to a close match between the characteristics of the flowers of a plant species and its pollinators, as you can see in **Figure 1.** Back-and-forth evolutionary adjustments between interacting members of a community are called **coevolution.**

Predators and Prey Coevolve

Predation is the act of one organism killing another for food. Familiar examples of predation include lions eating zebras and snakes eating mice. Less familiar, but no less important, examples occur among arthropods. Spiders are exclusively predators, as are centipedes.

Figure 1   Coevolution. With its long beak and tongue, the hummingbird is able to reach the nectar deep within this flower.

In **parasitism,** one organism feeds on and usually lives on or in another, typically larger, organism. Parasites do not usually kill their prey (known as the "host"). Rather, they depend on the host for food and a place to live. The host often serves to transmit the parasite's offspring to new hosts. Many parasites (such as lice) feed on the host's outside surface. Among the external parasites that may have fed on you at some time are ticks, mosquitoes, and fleas. More highly specialized parasites like hookworms live entirely within the body of their host.

Plant Defenses Against Herbivores

As you might expect, animal prey species have ways to escape, avoid, or fight off predators. But predation is also a problem for plants, which live rooted in the ground. The most obvious way that plants protect themselves from herbivores is with thorns, spines, and prickles. But it is even more common for a plant to contain chemical compounds that discourage herbivores. Virtually all plants contain defensive chemicals called **secondary compounds.** For some plants, secondary compounds are the primary means of defense.

As a rule, each group of plants makes its own special kind of defensive chemical. For example, the mustard plant family produces a characteristic group of chemicals known as mustard oils. These oils give pungent aromas and tastes to such plants as mustard, cabbage, radish, and horseradish. The same tastes that we enjoy signal the presence of chemicals that are toxic to many groups of insects.

How Herbivores Overcome Plant Defenses

Surprisingly, certain herbivores are able to feed on plants that are protected by particular defensive chemicals. For example, the larvae of cabbage butterflies feed almost exclusively on plants of the mustard and caper families. Yet these plants produce mustard oils that are toxic to many groups of insects. How do the butterfly larvae manage to avoid the chemical defenses of the plants? Cabbage butterflies have the ability to break down mustard oils and thus feed on mustards and capers without harm.

Real Life

Leaflets three, let it be.
Members of the genus *Toxicodendron*, which includes poison ivy, produce a defensive chemical called urushiol *(OO roo shee awl),* which  causes a severe, itchy rash in some people.
Finding Information
Do research to discover effective treatments for the rash caused by poison ivy.

Predicting How Predation Would Affect a Plant Species

Background

Grazing is the predation of plants by animals. Some plant species, such as *Gilia*, respond to grazing by growing new stems. Consider a field in which a large number of these plants are growing and being eaten by herbivores.

Analysis

1. **Identify** the plant that is likely to produce more seeds?

2. **Explain** how grazing affects this plant species.

3. **Evaluate** the significance to its environment of the plant's regrowth pattern.

4. **Hypothesize** how this plant species might be affected if individual plants did not produce new stems in response to grazing.

Ungrazed plant

Grazed plant

Regrowth after grazing

Symbiotic Species

In **symbiosis** *(sim bie OH sis)*, two or more species live together in a close, long-term association. Symbiotic relationships can be beneficial to both organisms or benefit one organism and leave the other harmed or unaffected. Parasitism, mentioned earlier, is one type of symbiotic relationship that is detrimental to the host organism. While it is relatively easy to determine that an organism in a symbiotic relationship is being helped, it can be difficult to determine that an organism is neither harmed nor helped.

Mutualism

Mutualism is a symbiotic relationship in which both participating species benefit. A well-known instance of mutualism involves ants and aphids, as shown in **Figure 2.** Aphids are small insects that use their piercing mouthparts to suck fluids from the sugar-conducting vessels of plants. They extract a certain amount of the sucrose and other nutrients from this fluid. However, much of the fluid—so-called honeydew—runs out in an altered form through their anus. Certain ants have taken advantage of this fact and "milk" the aphids for the honeydew, which they use as food. The ants, in turn, protect the aphids against insect predators. Thus, both species benefit from the relationship.

Commensalism

A third form of symbiosis is **commensalism,** a symbiotic relationship in which one species benefits and the other is neither harmed nor helped. Among the best-known examples of commensalism are the relationships between certain small tropical fishes and sea anemones, marine animals that have stinging tentacles. These fishes, such as the clown fish shown in **Figure 3,** have the ability to live among and be protected by the tentacles of the sea anemones, even though these tentacles would quickly paralyze other fishes.

Figure 2   Mutualism. The small green insects on this plant stem are aphids. They are protected by their ant guards.

Figure 3   Commensalism. The clown fish can survive the stings of the sea anemone, which protects it from predators.

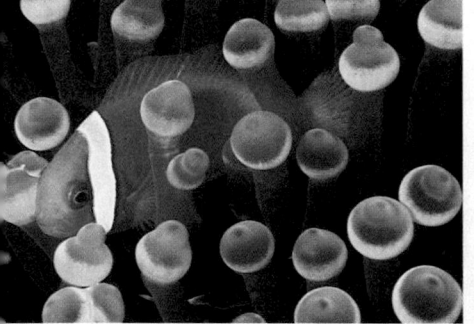

Section 1 Review

1 **Explain** why predator-prey coevolution can be described as an "arms race."

2 **Critical Thinking Applying Information** Is the relationship between a plant and its pollinator mutualistic? Why or why not.

3 **Critical Thinking Interpreting Interactions** In a relationship that is an example of commensalism, would the species that is neither helped nor harmed evolve in response to the other species? Defend your answer.

4 **Critical Thinking Illustrating Principles** In Japan, native honeybees have an effective defense strategy against giant Japanese hornets. Imported European honeybees, however, are unable to defend themselves. Use this example to illustrate the results of natural selection in adaptation.

5 **Standardized Test Prep** Which pair of organisms has a mutualistic relationship?

A clown fish and sea anemone

B aphid and ant

C lion and zebra

D flea and dog

How Competition Shapes Communities

Common Use of Scarce Resources and Competition

When two species use the same resource, they participate in a biological interaction called **competition.** Resources for which species compete include food, nesting sites, living space, light, mineral nutrients, and water. Competition occurs for resources in short supply. In Africa, for example, lions and hyenas compete for prey. Fierce rivalry between these species can lead to battles that cause injuries to both sides. But most competitive interactions do not involve fighting. In fact, some competing species never encounter one another. They interact only by means of their effects on the abundance of resources.

To understand how competition influences the makeup of communities, you must focus on the day-to-day events within the community. What do organisms eat? Where do they live? The functional role of a particular species in an ecosystem is called its **niche** *(NICH).* A niche is how an organism lives—the "job" it performs within the ecosystem.

A niche may be described in terms of space utilization, food consumption, temperature range, requirements for moisture or mating, and other factors. A niche is not to be confused with a habitat, the place where an organism lives. A habitat is a location; a niche is a pattern of living. **Figure 4** summarizes some aspects of the jaguar's niche in the Central American rain forest.

A niche is often described in terms of how the organism affects energy flow within the ecosystem in which it lives. For example, the niche of a deer that eats a shrub is that of a herbivore. The niches of some organisms overlap. If the resources that these organisms share are in short supply, it is likely that there will be competition between the organisms.

Objectives

- **Describe** the role of competition in shaping the nature of communities.

- **Distinguish** between fundamental and realized niches.

- **Describe** how competition affects an ecosystem.

- **Summarize** the importance of biodiversity.

Key Terms

competition
niche
fundamental niche
realized niche
competitive exclusion
biodiversity

A Jaguar's Niche

- **Diet** Jaguars feed on mammals, fish, and turtles.

- **Reproduction** Jaguars give birth from June to August, during the rainy season.

- **Time of activity** Jaguars hunt by day and by night.

Figure 4 Each organism has its own niche. All of the ways that this jaguar interacts with its environment make up its niche.

Size of a Species' Niche

To gain a better understanding of what a niche is, you can look more closely at a particular species. Imagine a Cape May warbler (a small, insect-eating songbird) flying in a forest and landing to search for dinner in a spruce tree. The niche of this bird is influenced by several variables. These variables include the temperature it prefers, the time of year it nests, what it likes to eat, and where on the tree it finds its food. (The Cape May warbler spends its summers almost exclusively in the northeastern United States and Canada. It nests in midsummer, eats small insects, and searches for food high on spruce trees at the tips of the branches.) The entire range of resource opportunities an organism is potentially able to occupy within an ecosystem is its **fundamental niche.**

Dividing Resources Among Species

Now reconsider what the Cape May warbler is doing. It feeds mainly at the very top of the spruce tree even though insects that the warbler could eat are located all over the tree. In other words, Cape May warblers occupy only a portion of their fundamental niche. Why?

Closer study reveals that this surprising behavior is part of a larger pattern of niche restriction. In the late 1950s, the ecologist Robert MacArthur, while a graduate student at Yale University, carried out a classic investigation of niche usage, summarized in **Figure 5.** He studied the feeding habits of five warbler species—the Cape May warbler and four of its potential competitors. MacArthur found that all five species fed on insects in the same spruce trees at the same time. As Figure 5 shows, however, each species concentrated on a different part of the tree. Although all five species of warbler had very similar fundamental niches, they did not use the same resources. In effect,

Figure 5   Niche restriction

Each of these five warbler species feeds on insects in a different portion of the same tree, as indicated by the five colors shown below.

■ Cape May warbler   ■ Blackburnian warbler   ■ Black-throated warbler   ■ Bay-breasted warbler   ■ Myrtle warbler

they divided the range of resources among them, each taking a different portion. A different color is used to represent the feeding areas of each of the five warbler species shown in Figure 5.

The part of its fundamental niche that a species occupies is called its **realized niche.** Stated in these terms, the realized niche of the Cape May warbler is only a small portion of its fundamental niche. How does this species of warbler benefit from hunting for food in only a portion of the tree? MacArthur suggested that this feeding pattern reduces competition among the five species of warblers. Because each of the five warbler species uses a different set of resources by occupying a different realized niche, the species are not in competition with one another. MacArthur concluded that natural selection has favored a range of preferences and behaviors among the five species that "carve up" the available resources. Most ecologists agree with this conclusion.

Study TIP

● **Reading Effectively**
To better understand the relationship between fundamental and realized niches, draw two circles, one within the other. Label the larger circle "Fundamental niche, entire tree." Label the smaller circle "Realized niche."

Predicting Changes in a Realized Niche

DATA LAB

Background

Two features of a niche that can be readily measured are the location where the species feeds and the size of its preferred prey. The darkest shade in the center of the graph below indicates the prey size and feeding location most frequently selected by one bird species (called Species A).

Analysis

1. **State** the range of lengths of Species A's preferred prey.

2. **Identify** the maximum height at which Species A feeds.

3. **Critical Thinking Predicting Outcomes**
Species B is introduced into Species A's feeding range. Species B has exactly the same feeding preferences, but it hunts at a slightly different time of day. How might this affect Species A?

4. **Interpreting Graphics**
Species C is now introduced into Species A's feeding range. Species C feeds at the same time of day as Species A, but it prefers prey that are between 10 and 13 mm long. How might this affect Species A?

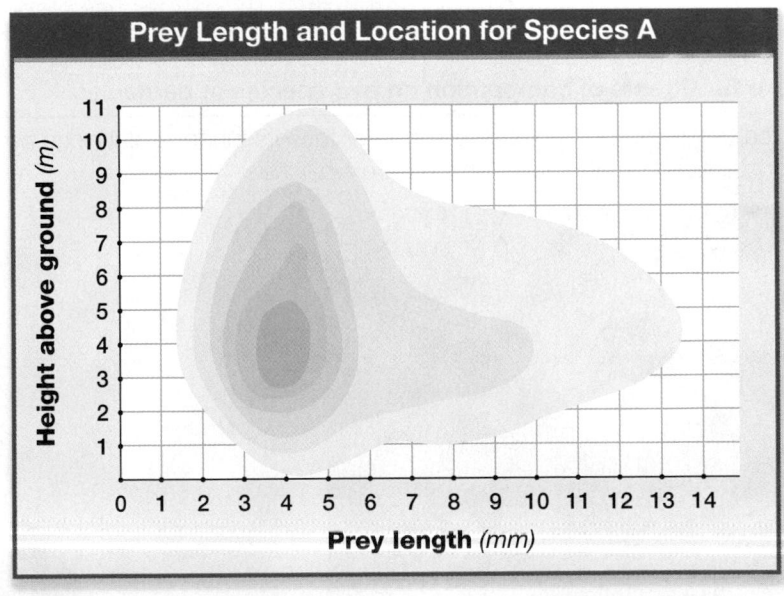

5. **Critical Thinking Predicting Outcomes**
How would the introduction of a species with exactly the same feeding habits as Species A affect the graph?

6. **Interpreting Graphics** What does the lightest shade at the edge of the contour lines represent?

Competition and Limitations of Resource Use

WORD Origins

● The word *niche* is from the Latin word *nidus*, meaning "nest." The place an organism occupies in its environment is part of its niche—its overall functional role.

A very clear case of competition was shown by experiments carried out in the early 1960s by Joseph Connell of the University of California. Connell worked with two species of barnacles that grow on the same rocks along the coast of Scotland. Barnacles are marine animals that are related to crabs, lobsters, and shrimp. Young barnacles attach themselves to rocks and remain attached there for the rest of their lives. As you can see in **Figure 6,** one species, *Chthamalus stellatus,* lives in shallow water, where it is often exposed to air by receding tides. A second species, *Semibalanus balanoides,* lives lower down on the rocks, where it is rarely exposed to the atmosphere.

When Connell removed *Semibalanus* from the deeper zone, *Chthamalus* was easily able to occupy the vacant surfaces. This indicates that it was not intolerance of the deeper environment that prevented *Chthamalus* from becoming established there. *Chthamalus's* fundamental niche clearly includes the deeper zone. However, when *Semibalanus* was reintroduced, it could always outcompete *Chthamalus* by crowding it off the rocks. In contrast, *Semibalanus* could not survive when placed in the shallow-water habitats where *Chthamalus* normally occurs. *Semibalanus* apparently lacks the adaptations that permit *Chthamalus* to survive long periods of exposure to air. Connell's experiments show that *Chthamalus* occupies only a small portion of its fundamental niche. The rest is unavailable because of competition with *Semibalanus.* As MacArthur suggested, competition can limit how species use resources.

Figure 6 Effects of competition on two species of barnacles

The realized niche of *Chthamalus* is smaller than its fundamental niche because of competition from the faster-growing *Semibalanus*.

■ *Chthamalus stellatus*

■ *Semibalanus balanoides*

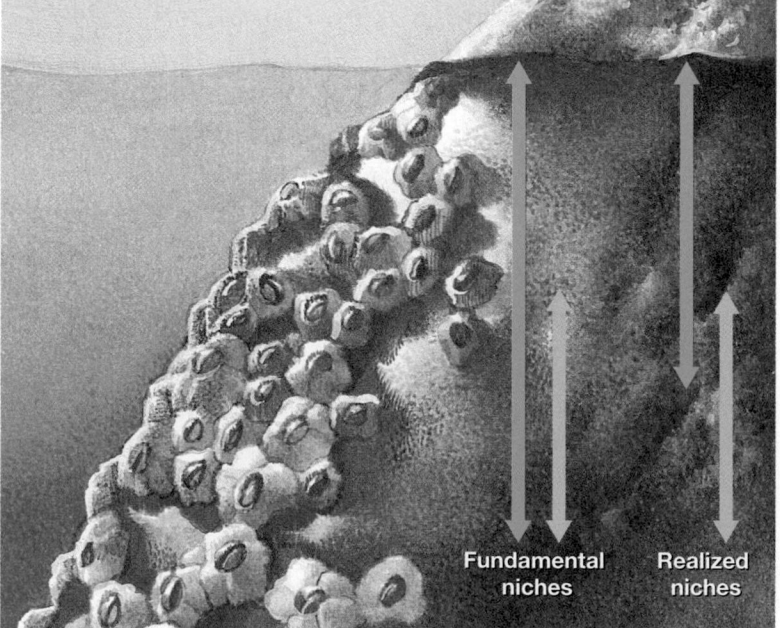

Fundamental niches   Realized niches

Competition Without Division of Resources

In nature, shortage is the rule, and species that use the same resources are almost sure to compete with each other. Darwin noted that competition should be most acute between very similar kinds of organisms because they tend to use the same resources in the same way. Can we assume, then, that when very similar species compete, one species will always become extinct locally? In a series of carefully controlled laboratory experiments performed in the 1930s, the Russian biologist G. F. Gause looked into this question.

In his experiments, Gause grew two species of *Paramecium* in the same culture tubes, where they had to compete for the same food (bacteria). Invariably, the smaller of the two species, which was more resistant to bacterial waste products, drove the larger one to extinction, as shown in the first graph in **Figure 7.** Gause hypothesized that if two species are competing, the species that uses the resource more efficiently will eventually eliminate the other. This elimination of a competing species is referred to as **competitive exclusion.**

When Can Competitors Coexist?

Is competitive exclusion the inevitable outcome of competition for limited resources, as Gause suggests? No. When it is possible for two species to avoid competing, they may coexist.

In a revealing experiment, Gause challenged *Paramecium caudatum*—the defeated species in his earlier experiments—with a third species, *P. bursaria*. These two species were also expected to compete for the limited bacterial food supply. Gause thought one species would win out, as had happened in his previous experiments.

But that's not what happened. As shown in the second graph in Figure 7, both species survived in the culture tubes. Like MacArthur's warblers, the two species of *Paramecium* divided the food resources. How did they do it? In the upper part of the culture tubes, where oxygen concentration and bacterial density were high, *P. caudatum* was dominant. It was better able to feed on bacteria than was *P. bursaria*. But in the lower part of the tubes, the lower oxygen concentration favored the growth of a different potential food—yeast. *Paramecium bursaria* was better able to eat the yeast, so it used this resource more efficiently. The fundamental niche of each species was the whole culture tube, but the realized niche of each species was only a portion of the tube. Because the niches of the two species did not overlap too much, both species were able to survive.

Figure 7 Gause's experiments. The outcome of competition depends on the degree of similarity between the fundamental niches of the competing species.

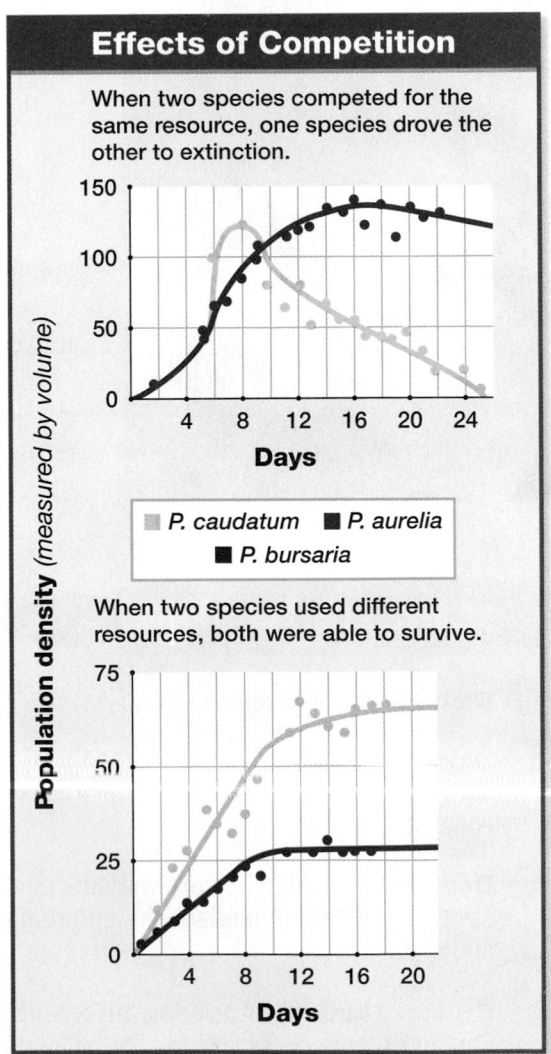

Effects of Competition

When two species competed for the same resource, one species drove the other to extinction.

Population density (measured by volume)

Days

■ *P. caudatum*   ■ *P. aurelia*
■ *P. bursaria*

When two species used different resources, both were able to survive.

Days

Predation and Competition

Many studies of natural ecosystems have demonstrated that predation reduces the effects of competition. A very clear example is provided by the studies of Robert Paine of the University of Washington. Paine examined how sea stars affect the numbers and types of species within marine intertidal communities. Sea stars are fierce predators of marine animals such as clams and mussels. When sea stars were kept out of experimental plots, the number of their prey species fell from 15 to 8. The 7 eliminated species were crowded out by the sea stars' chief prey, mussels, shown in **Figure 8.** Mussels can outcompete other species for space on the rocks. By preying on mussels, sea stars keep the mussel populations too low to drive out other species.

Because predation can reduce competition, it can also promote **biodiversity,** the variety of living organisms present in a community. Biodiversity is a measure of both the number of different species in a community (species richness) and the relative numbers of each of the species (species diversity).

Figure 8   Effect of removing sea stars. When the sea star *Pisaster* was removed from an ecosystem, the diversity of its prey species decreased. Mussels, the superior competitor, crowded seven other prey species out of the ecosystem.

Biodiversity and Productivity

A key investigation carried out in the early 1990s by David Tilman of the University of Minnesota illustrates the relationship between biodiversity and productivity. Tilman and some co-workers and students tended 207 experimental plots in a Minnesota prairie. Each plot contained a mix of up to 24 native prairie plant species. The biologists monitored the plots, measuring how much growth was occurring. Tilman found that the greater the number of species a plot had, the greater the amount of plant material produced in that plot. Tilman's experiments clearly demonstrated that increased species richness leads to greater productivity.

In addition to increased productivity, Tilman also found that the plots with greater numbers of species recovered more fully from a major drought. Thus, the biologically diverse plots were also more stable than the plots with fewer species.

Section 2 Review

1 Distinguish between *niche* and *habitat*.

2 Describe the conclusions reached by Connell and Paine about how competition affects ecosystems.

3 Describe how Tilman's experiments demonstrate the effects of biodiversity on productivity and stability.

4 Critical Thinking Applying Information
Can an organism's realized niche be larger than its fundamental niche? Justify your answer.

5 Critical Thinking Evaluating Conclusions
A scientist finds no evidence that species in a community are competing and concludes that competition never played a role in the development of this community. Is this conclusion valid? Justify your answer.

6 Standardized Test Prep When two species use the same resource, one species may drive the other to extinction. This phenomenon is called

A space utilization.   **C** niche restriction.

B competitive exclusion.   **D** resource division.

Major Biological Communities

Climate's Effect on Where Species Live

If you traveled across the country by car you would notice dramatic changes in the plants and animals outside your window. For example, the drought-tolerant cactuses in the deserts of Arizona do not live in the wetlands of Florida. Why is this? The climate of any physical environment determines what organisms live there. **Climate** refers to the prevailing weather conditions in any given area.

Temperature and Moisture

The two most important elements of climate are temperature and moisture. **Figure 9** illustrates the different types of ecosystems that occur under particular temperature and moisture conditions.

Temperature Most organisms are adapted to live within a particular range of temperatures and will not thrive if temperatures are colder or warmer. The growing season of plants, for example, is primarily influenced by temperature.

Moisture All organisms require water. On land, water is sometimes scarce, so patterns of rainfall often determine an area's life-forms. The moisture-holding ability of air increases when it is warmed and decreases when it is cooled.

Objectives

- **Recognize** the role of climate in determining the nature of a biological community.

- **Describe** how elevation and latitude affect the distribution of biomes.

- **Summarize** the key features of the Earth's major biomes.

- **Compare** features of plants and animals found in different biomes.

- **Compare** and contrast the major freshwater and marine habitats.

Key Terms

climate
biome
littoral zone
limnetic zone
profundal zone
plankton

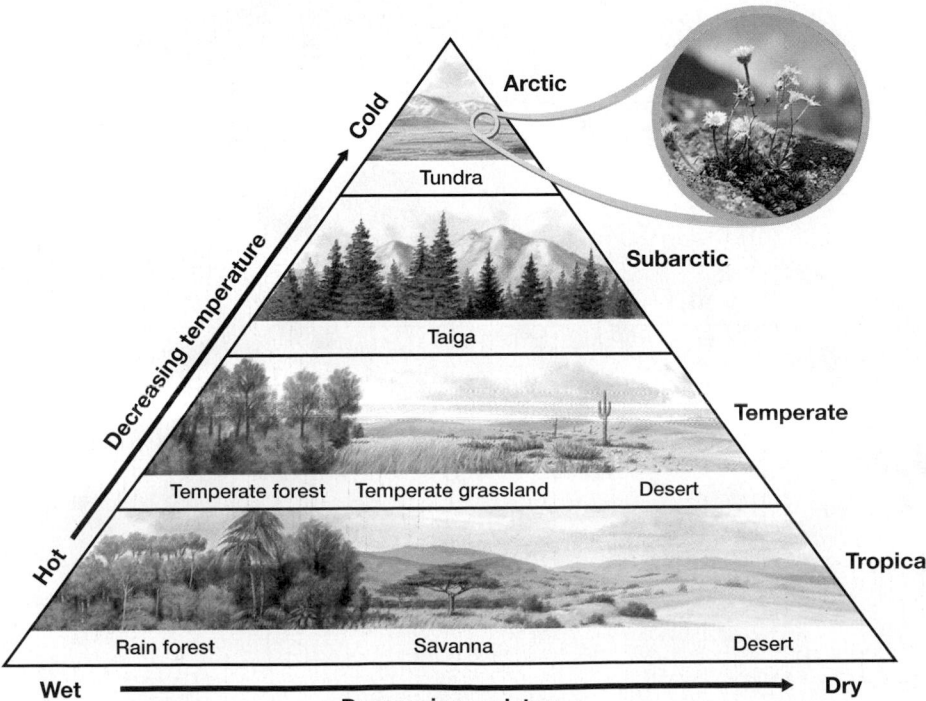

Figure 9 Elements of climate. Temperature and moisture help determine ecosystem distribution. For example, the asters and the saxifrage shown are able to produce flowers and seeds in the cold temperatures of the tundra.

Major Biological Communities

If you were to tour the world and look at biological communities on land and in the oceans, you would soon learn a general rule of ecology: very similar communities occur in many different places that have similar climates and geographies. A major biological community that occurs over a large area of land is called a **biome.**

A biome's structure and appearance are similar throughout its geographic distribution. While there are different ways of classifying biomes, the classification system used here recognizes seven of the most widely occurring biomes: (1) tropical rain forest, (2) savanna, (3) taiga, (4) tundra, (5) desert, (6) temperate grassland, and (7) temperate forest (deciduous and evergreen). These biomes differ greatly from one another because they have developed in regions with very different climates. The global distribution of these biomes is shown in **Figure 10.**

Many factors such as soil type and wind play important roles in determining where biomes occur. Two key factors are particularly important: temperature and precipitation. **Figure 11** is based on the work of ecologist Robert Whittaker. The graph shows the relationship between temperature and humidity and the biological communities that exist under different conditions. In general, temperature and available moisture decrease as latitude (distance from the equator) increases. They also decrease as elevation (height above sea level) increases. As a result, mountains often show the same sequence of change in ecosystems that is found as one goes north or south from the equator.

Figure 10 Earth's biomes.
Seven major biomes cover most of the Earth's land surface. Because mountainous areas do not belong to any one biome, they are given their own designation.

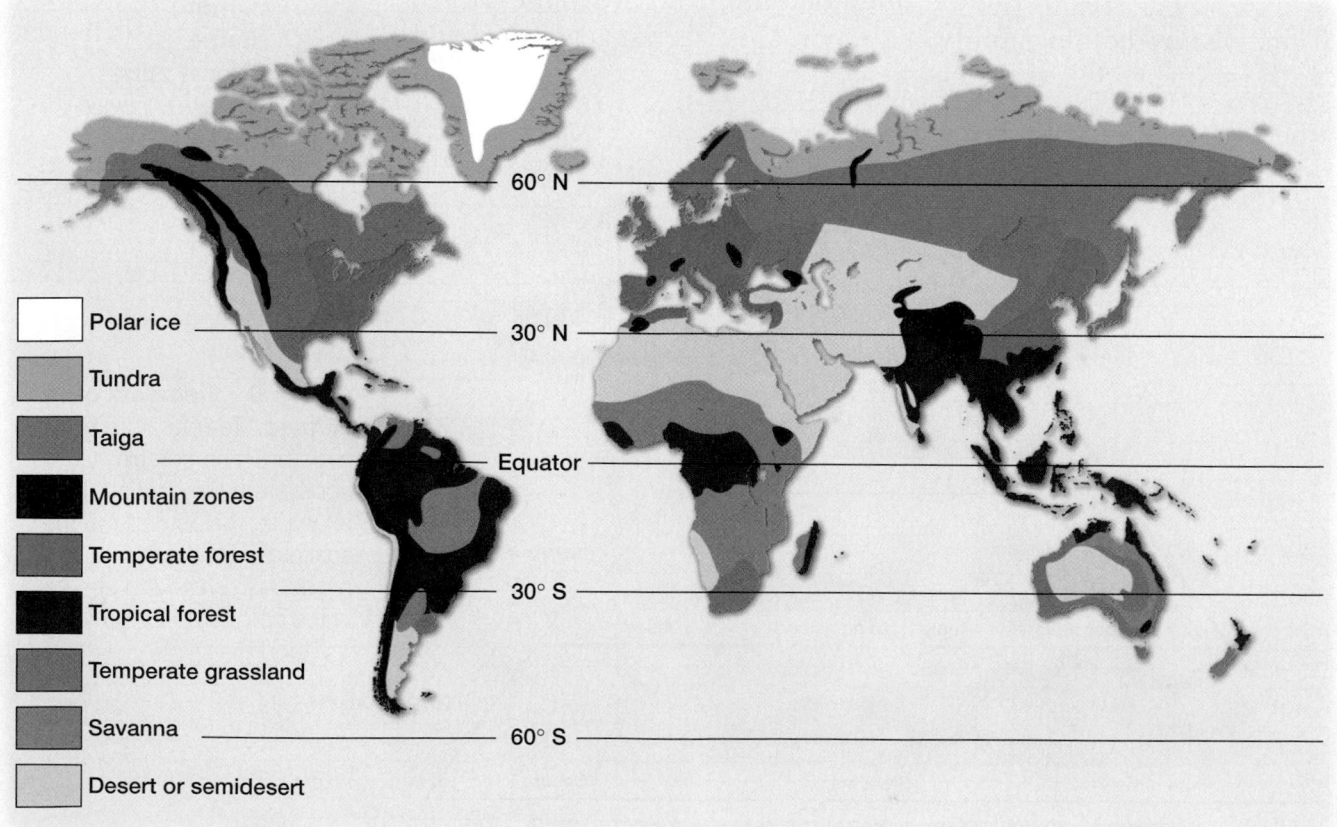

- Polar ice
- Tundra
- Taiga
- Mountain zones
- Temperate forest
- Tropical forest
- Temperate grassland
- Savanna
- Desert or semidesert

60° N
30° N
Equator
30° S
60° S

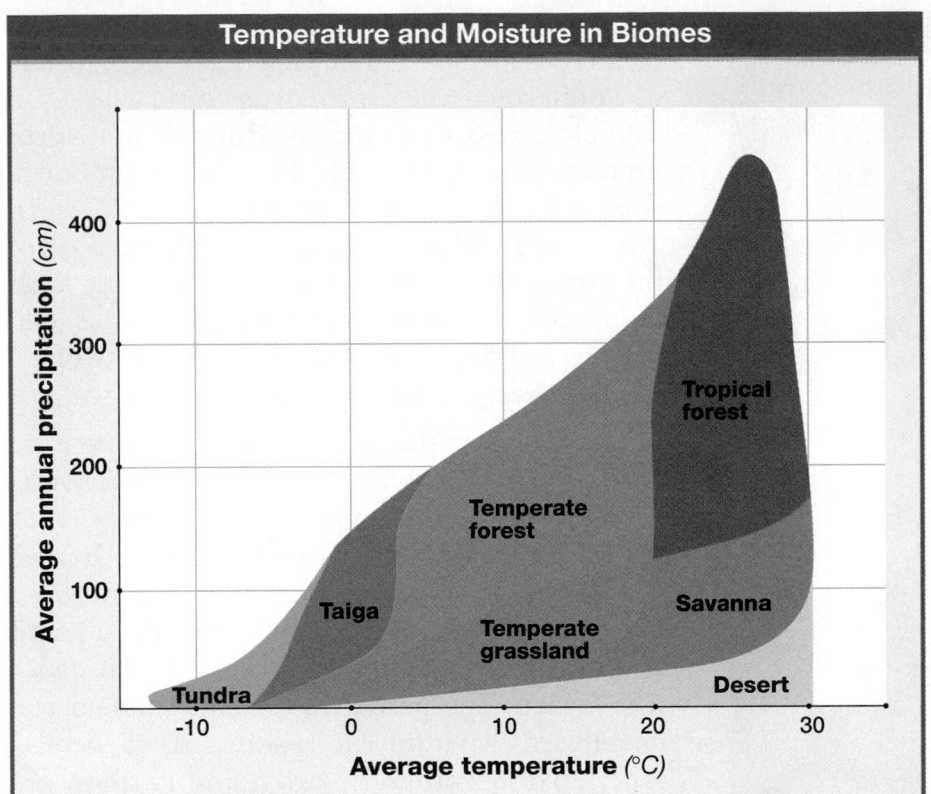

Temperature and Moisture in Biomes

Average annual precipitation (cm)

400
300
200
100

Tropical forest

Temperate forest

Taiga

Temperate grassland

Savanna

Tundra

Desert

-10   0   10   20   30

Average temperature (°C)

Figure 11 Conditions in biomes. Different biomes have characteristics of temperature and humidity.

Terrestrial Biomes

Tropical Rain Forests Tropical rain forests receive an average of as much as 450 cm (180 in.) of rain per year, with little difference in distribution from season to season. The richest biome in terms of number of species is the tropical rainforest. Tropical rainforests may contain at least half of the Earth's species of terrestrial organisms—more than 2 million species. Tropical rain forests have a high primary productivity even though they exist mainly on quite infertile soils. Most of the nutrients are held within the plants; the soil itself contains few nutrients.

Savannas The world's great dry grasslands, called savannas, are found in tropical areas that have relatively low annual precipitation or prolonged annual dry seasons. Annual rainfall is generally 90 to 150 cm (35 to 60 in.) in savannas. There is a wider fluctuation in temperature during the year than in the tropical rain forests, and there is seasonal drought. These factors have led to an open landscape with widely spaced trees. Many of the animals are active only during the rainy season. Huge herds of grazing mammals are found on the savannas of East Africa.

Tropical rain forest in Puerto Rico

Savanna in East Africa

Taiga in Manitoba, Canada

Taiga Cold, wet climates promote the growth of coniferous forests. A great ring of northern forests of coniferous trees, primarily spruce and fir, extends across vast areas of Eurasia and North America. This biome, one of the largest on Earth, is called by its Russian name, taiga *(TIE guh)*. Winters in the taiga are long and cold, and most of the precipitation falls in the summer. Many large mammals, including herbivores such as elk, moose, and deer and carnivores such as wolves, bears, lynxes, and wolverines live in the taiga.

Tundra Between the taiga and the permanent ice surrounding the North Pole is the open, sometimes boggy biome known as the tundra. This enormous biome covers one-fifth of the Earth's land surface. Annual precipitation in the tundra is very low, usually less than 25 cm (10 in.), and water is unavailable for most of the year because it is frozen. The permafrost, or permanent ice, usually exists within 1 m (about 3 ft) of the surface. Foxes, lemmings, owls, and caribou are among the vertebrate inhabitants.

Tundra in Denali National Park, Alaska

QUICK LAB

Investigating Factors That Influence the Cooling of Earth's Surface

You can discover how the amount of water in an environment affects the rate at which that environment cools.

Materials

MBL or CBL system with appropriate software, temperature probes, test tubes, beaker, hot plate, one-holed stoppers, water, sand, test-tube tongs, test-tube rack

Procedure

1. Set up an MBL/CBL system to collect and graph data from each temperature probe at 5-second intervals for 240 data points. Calibrate the probe using stored data.

2. Fill one test tube with water. Fill another test tube halfway with sand.

3. Place a temperature probe in the sand, and suspend another temperature probe at the same depth in the water,

using one-holed stoppers to hold each temperature probe in place.

4. Place both test tubes in a beaker of hot water. Heat them to a temperature of about 70°C. **Caution: Hot water can burn skin.**

5. Using test-tube tongs, remove the test tubes and place them in the test-tube rack. Record the drop in temperature for 20 minutes.

Analysis

1. **Critical Thinking Analyzing Results** Did the two test tubes cool at the same rate? Offer an explanation for your observations.

2. **Critical Thinking Predicting Outcomes** In which biome—tropical rain forest or desert—would you expect the air temperature to drop most rapidly? Explain your answer.

Terrestrial Biomes

Deserts Typically, less than 25 cm (10 in.) of precipitation falls annually in the world's desert areas. The scarcity of water is the overriding factor influencing most biological processes in the desert. In desert regions, the vegetation is characteristically sparse. Deserts are most extensive in the interiors of continents. Less than 5 percent of North America is open desert. The amount of water that actually falls on a particular place in a desert can vary greatly, both during a given year and between years.

Temperate Grasslands Moderate climates halfway between the equator and the poles promote the growth of rich temperate grasslands called prairies. Temperate grasslands once covered much of the interior of North America. Such grasslands are often highly productive when converted to agriculture. The roots of grasses characteristically penetrate far into the soil, which tends to be deep and fertile. Herds of grazing animals often populate temperate grasslands. In North America, huge herds of bison once inhabited the prairies.

Temperate Deciduous Forests Relatively mild climates and plentiful rain promote the growth of forests. Temperate deciduous forests (deciduous trees shed their leaves in the fall) grow in areas with relatively warm summers, cold winters, and annual precipitation that generally ranges from 75 to 250 cm (30 to 100 in.). Temperate deciduous forests cover much of the eastern United States and are home to deer, bears, beavers, raccoons, and other familiar animals. The trees are hardwoods (oak, hickory, and beech).

Temperate Evergreen Forests In other temperate areas, drier weather and different soil conditions favor the growth of evergreens. Large portions of the southeastern and western United States have temperate evergreen forests—extensive areas where pine forests predominate over deciduous forests. Where conditions are even drier, temperate forests give way to areas of dry shrubs, such as in the chaparral areas of coastal California and in the Mediterranean.

Desert in Texas

Temperate grasslands in Kansas

Temperate deciduous forest in Pennsylvania

Temperate evergreen forest in Washington

Aquatic Communities

At a glance, you might at first think that freshwater and marine communities are separate from terrestrial biomes. Yet large amounts of organic and inorganic material continuously enter both bodies of fresh water and ocean habitats from communities on the land.

Freshwater Communities

Freshwater habitats—lakes, ponds, streams, and rivers—are very limited in area. Lakes cover only about 1.8 percent of the Earth's surface, and rivers and streams cover about 0.3 percent. All freshwater habitats are strongly connected to terrestrial ones, with freshwater marshes and wetlands constituting intermediate habitats. Many kinds of organisms are restricted to freshwater habitats, including plants, fish, and a variety of arthropods, mollusks, and other invertebrates too small to be seen without a microscope.

Ponds and lakes have three zones in which organisms live, as illustrated in **Figure 12.** The **littoral zone** is a shallow zone near the shore. Here, aquatic plants live along with various predatory insects, amphibians, and small fish. The **limnetic zone** refers to the area that is farther away from the shore but close to the surface. It is inhabited by floating algae, zooplankton, and fish. The **profundal zone** is a deep-water zone that is below the limits of effective light penetration. Numerous bacteria and wormlike organisms that eat debris on the lake's bottom live in this zone. The breakdown of this debris releases large amounts of nutrients. Not all freshwater systems are deep enough to include a profundal zone.

Study *TIP*

● **Organizing Information**
Make a concept map that describes the zones of a pond or lake as described at right and in Figure 12. For each zone, include the plant and animal life found there.

Figure 12   Three lake zones. Each region, or zone, of a lake contains characteristic organisms.

Limnetic zone

Littoral zone

Profundal zone

Wetlands

Swamps, such as the one shown in **Figure 13,** as well as marshes, bogs, and other communities that are covered with a layer of water are called *wetlands*. Wetlands typically are covered with a variety of water-tolerant plants, called *hydrophytes* ("water plants"). Marsh grasses and cattails are hydrophytes. Wetlands are dynamic communities that support a diverse array of invertebrates, birds, and other animals. Wetlands are among the most productive ecosystems on Earth, exceeded only by coral reefs in diversity and concentration of species. They also play a key ecological role by providing water storage basins that moderate flooding. Many wetlands are being disrupted by human development of what is sometimes perceived as otherwise useless land, but government efforts are now underway to protect the remaining wetlands.

Figure 13 Forested wetlands.
This swampy terrain is typical of the forested wetlands found in the southeastern United States.

Estuaries

If you've ever eaten seafood caught in a saltwater marsh, you've experienced one of the benefits of estuaries. Estuaries are unique transition zones between marine and freshwater environments. Nutrients washed from nearby land stimulate the growth of plants and algae. As a result, estuaries are among the most productive ecosystems on Earth. One hundred acres of healthy estuary can produce 4 to 10 times as much organic matter as a cultivated cornfield of the same size! The estuary's plants, invertebrates, fishes, birds, mammals, and other animals are part of a complex food web.

Where the River Meets the Sea

In addition to serving as wildlife habitats, estuaries filter sediment and nutrients, purifying the water that drains off the land. Porous salt-marsh soils absorb floodwaters and protect coastal communities from erosion. Estuaries also provide jobs for people in the seafood and recreation industries.

Estuaries in Peril

Sadly, the public has long regarded estuaries as wastelands. People have drained estuaries to provide land for housing and agriculture. Pollutants and improperly treated sewage have poisoned some estuaries' habitats. As a result, estuaries in the United States have greatly decreased in size over the past century. However, a variety of environmental organizations are now working to restore estuaries by cleaning up pollutants and replanting native vegetation. For example, restoration efforts in Tampa Bay, Florida, improved sewage treatment facilities, enabling sea grass meadows to return. When the sea grasses returned, so did the fishes and other animals that depend on them.

internet connect

www.scilinks.org
Topic: Estuaries
Keyword: HX4073

SciLINKS Maintained by the National Science Teachers Association

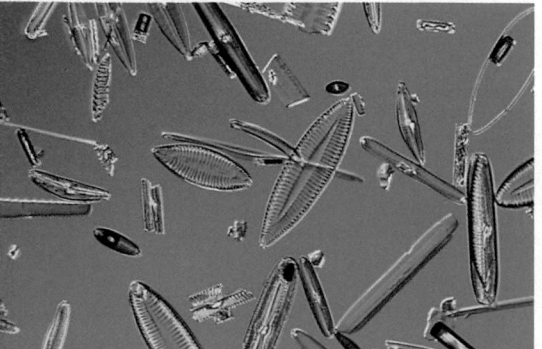
Blue stripe snapper from shallow ocean waters

Diatoms from surface of the open sea

Anglerfish from ocean depths

Marine Communities

Nearly three-fourths of the Earth's surface is covered by ocean, which consists of three major kinds of marine communities.

Shallow Ocean Waters The zone of shallow water is small in area, but compared with other parts of the ocean, it is inhabited by large numbers of species. The seashore between high and low tide, called the *intertidal zone,* is home to many species of marine invertebrates. Coral reef communities, the world's most diverse, occur in shallow tropical waters. The world's great fisheries are located in the coastal zones of cooler waters, where nutrients washed out from land support huge numbers of fishes.

Surface of the Open Sea Drifting freely in the upper waters of the ocean is a diverse community of **plankton,** composed of bacteria, algae, fish larvae, and many small invertebrate animals. Fishes, whales, and invertebrates such as jellyfishes feed on plankton. And larger fishes and birds, in turn, feed on some of these animals. Photosynthetic plankton (algae such as diatoms and some bacteria) that form the base of this food chain account for about 40 percent of all the photosynthesis that takes place on Earth. Because light penetrates water only to the depth of about 100 m (328 ft), this rich community is confined to the ocean's surface.

Ocean Depths In the deepest waters of the sea, the marine community lives in total darkness, in deep cold, and under great pressure. Despite what seem like hostile conditions, the deep ocean supports a diverse community of bizarre invertebrates and fishes. This includes great squids and angler fishes that attract prey with projections from their head that emit light. On the ocean floor, at an average depth of more than 3 km (1.9 mi), researchers have found an unexpected abundance of species, a diversity that rivals the tropical rain forest.

Section 3 Review

1 **Describe** the relationship between climate and location of species.

2 **Compare** the tolerance to lack of water needed by plants and animals in savannas and tropical rain forests.

3 **Critical Thinking Analyzing Information** Why can't photosynthesis occur in the deepest parts of the ocean or in a deep lake?

4 **Critical Thinking Forming Reasoned Opinions** The equator passes across the country of Ecuador. But the climate there can range from hot and humid to cool and dry. What might explain this?

5 **Standardized Test Prep** In which biome would you most likely find plants that are adapted to infertile soils and fairly constant, plentiful precipitation?
A tropical rain forest   C temperate grassland
B tundra   D savanna

Key Concepts

1 How Organisms Interact in Communities

- Species within communities coevolve, making many adjustments to living together.

- In a predator-prey interaction, prey often evolve ways to escape being eaten. Predators evolve ways to overcome the defenses of the prey.

- In mutualism and commensalism, species evolve in ways that benefit one or both parties.

2 How Competition Shapes Communities

- Interactions among species help shape communities.

- Competition occurs when two species use the same limited resource.

- An organism's niche is its way of life. An organism may occupy only a part of its fundamental niche, which is called its realized niche.

- Competition can limit how species use resources.

- Biodiversity tends to promote stability and productivity.

3 Major Biological Communities

- Climate and genes largely determines where species live.

- Temperature and moisture are key factors in determining where biomes occur.

- The seven major biomes are tropical rain forest, desert, savanna, temperate deciduous forest, temperate grassland, taiga, and tundra.

- Freshwater communities have three zones of life—littoral, limnetic, and profundal.

- The three major marine communities are shallow ocean waters, open sea surface, and deep-sea waters.

Key Terms

Section 1

coevolution (362)
predation (362)
parasitism (362)
secondary compound (363)
symbiosis (364)
mutualism (364)
commensalism (364)

Section 2

competition (365)
niche (365)
fundamental niche (366)
realized niche (367)
competitive exclusion (369)
biodiversity (370)

Section 3

climate (371)
biome (372)
littoral zone (376)
limnetic zone (376)
profundal zone (376)
plankton (378)

Understanding Key Ideas

1. In predator-prey coevolution, if the prey gains a defense to stop predation, then the predator may evolve
  a. in a way that enables it to overcome the prey's defense.
  b. so that it can parasitize the prey.
  c. secondary compounds.
  d. into a prey species.

2. The principle of competitive exclusion indicates that
  a. a niche can be shared by two species if their niches are very similar.
  b. niche subdivision may occur.
  c. one species will eliminate a competing species if their niches are very similar.
  d. competition ends in worldwide elimination of a species.

3. Describe the niches of a lion, a zebra, and the grass that grows on the African plain in terms of how each species affects energy flow in the ecosystem.

4. Which abiotic factor is likely *not* a reason for the desert biome's low primary productivity?
  a. extreme temperatures
  b. frequent flooding
  c. high predation
  d. availability of sunlight

5. When populations of similar species occupy the same area at the same time, these populations often
  a. share all their resources equally.
  b. divide their range of resources.
  c. compete for resources to the death.
  d. look in other areas for different resources.

6. Which of the words sets below describes a vulture eating a dead rabbit?
  a. heterotroph, scavenger
  b. parasite, predator
  c. herbivore, mutualism
  d. competitor, commensalisms

7. Describe how elevation and latitude affect the distribution of biomes.

8. **BIOWatch** Why might digging a deep channel in an estuary change the types of living things that thrive there?

9. How does the flow of energy through living systems help determine the components of a biological community? (**Hint:** See Chapter 5, Section 1.)

10. **Concept Mapping** Make a concept map that shows how the biomes can be classified based on precipitation, temperature, and geographical location. Try to include the following terms in your map: *tropical rain forest, savanna, desert, temperate deciduous forest, temperate grassland, taiga,* and *tundra*.

Critical Thinking

11. Justifying Conclusions In Gause's experiments, *Paramecium caudatum* could coexist with *P. bursaria* but not with *P. aurelia*. Predict what would happen if *P. aurelia* and *P. bursaria* were grown together, and justify your conclusions.

12. Justifying Conclusions Newly introduced predators often prove devastating to native animals. Explain why prey are often more vulnerable to introduced predators than to native predators.

13. Analyzing Data Using the data presented in this chapter, explain why many ecologists refer to the tundra as a frozen desert.

Alternative Assessment

14. Summarizing Information Work with a small group of students to develop a map that shows the most prominent terrestrial and aquatic communities within your state, and for coastal states, those immediately offshore as well. Be certain to include any large swamps or wetlands that connect terrestrial and aquatic communities.

Standardized Test Prep

Understanding Concepts

Directions (1–4): **For *each* question, write on a separate sheet of paper the letter of the correct answer.**

1 Both a spruce tree and a hemlock tree require nitrogen from the soil. What is the interaction between these two species?
   A. competition   **C.** mutualism
   B. commensalism   **D.** succession

2 What term describes the ways in which an organism interacts with its environment?
   F. ecosystem   **H.** niche
   G. habitat   **I.** space

3 Which of the following is a transition zone between tropical rain forest and desert?
   A. savanna
   B. taiga
   C. temperate deciduous forest
   D. tundra

4 Why is the open ocean biome considered only slightly more productive than the desert biome?
   F. Neither of these biomes receive very much rain, which restricts productivity.
   G. Light penetrates only the top 100 meters of water in the open ocean, which restricts productivity.
   H. The open ocean receives the same amount of light, yet has significantly more water than the desert.
   I. Both lack large trees that block sunlight, but the open ocean lacks the soil needed for plants to grow.

Directions (5): **For the following question, write a short response.**

5 Compare and contrast mutualism and commensalism.

Test **TIP**

If you are unsure of the correct answer to a multiple-choice question, start by crossing out answers that you know are wrong. Reducing your choices in this way may help you choose the correct answer.

Reading Skills

Directions (6): **Read the passage below. Then answer the question.**

A keystone predator is one who regulates the populations of various competitors in an ecosystem. The *Pisaster* sea star is one example of a keystone predator. The sea star preys on mussels who might otherwise out compete other species that live in a marine intertidal community. Keystone predators reduce the occurrence of competitive exclusion of weak competitors.

6 How does a keystone predator promote biodiversity?
   A. It reduces competition by decreasing the populations of superior competitors.
   B. It reduces competition by decreasing the populations of inferior competitors.
   C. It increases competition by increasing the populations of superior competitors.
   D. It increases competition by decreasing the populations of inferior competitors.

Interpreting Graphics

Directions (7): **Base your answer to question 7 on the graph below.**

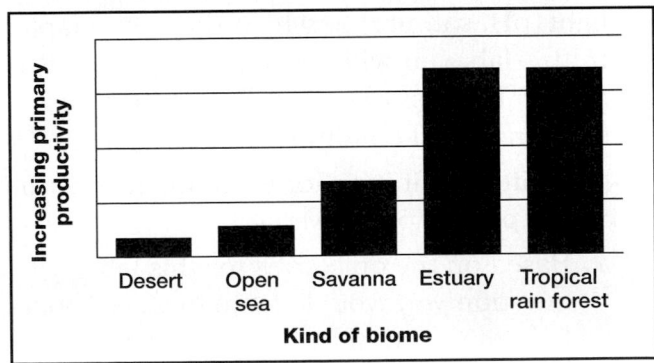

Comparative Productivity of Ecosystems

7 Where would you place a bar representing the primary productivity of the temperate grassland biome, if it could be added to this graph?
   F. between desert and open seas
   G. between open seas and savanna
   H. between savanna and estuary
   I. between estuary and tropical rain forest

Skills Practice Lab

Observing How Brine Shrimp Select a Habitat

SKILLS
- Using scientific methods
- Collecting, organizing, and graphing data

OBJECTIVES
- **Observe** the behavior of brine shrimp.
- **Assess** the effect of environmental variables on habitat selection by brine shrimp.

MATERIALS
- clear, flexible plastic tubing
- metric ruler
- marking pen
- corks to fit tubing
- brine shrimp culture
- screw clamps
- test tubes with stoppers and test-tube rack
- pipet
- Petri dish
- Detain™ or methyl cellulose
- aluminum foil
- calculator
- fluorescent lamp or grow light
- funnel
- graduated cylinder or beaker
- hot-water bag
- ice bag
- pieces of screen
- tape

Magnification 20×

Brine shrimp

Before You Begin

Different organisms are adapted for life in different **habitats.** For example, **brine shrimp** are small crustaceans that live in salt lakes. Given a choice, organisms select habitats that provide the conditions (e.g., temperature, light, pH, salinity) to which they are adapted. In this lab, you will investigate habitat selection by brine shrimp and determine which environmental conditions they prefer.

1. Write a definition for each boldface term in the paragraph above.

2. Based on the objectives for this lab, write a question you would like to explore about habitat selection by brine shrimp.

Procedure

PART A: Making and Sampling a Test Chamber

1. Divide a piece of plastic tubing into 4 sections by making a mark at 12 cm, 22 cm, and 32 cm from one end. Label the sections *1, 2, 3,* and *4.*

2. Place a cork in one end of the tubing. Then transfer 50 mL of brine shrimp culture to the tubing. Place a cork in the open end of the tubing.

3. When you are ready to count shrimp, divide the tubing into four sections by placing a screw clamp at each mark on the tubing. *While someone holds the corks firmly in place,* first tighten the middle clamp and then the outer clamps.

4. Starting at one end, pour the contents of each section into a test tube labeled with the same number. After you empty a section, loosen the adjacent clamp and fill the next test tube.

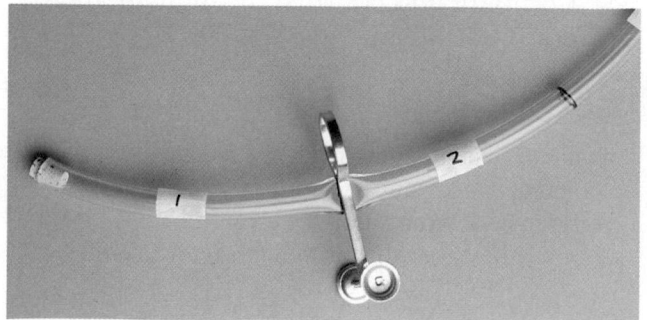

5. Stopper one test tube, and invert it gently to distribute the shrimp. Use a pipet to transfer a 1 mL sample of shrimp culture to a Petri dish. Add a few drops of Detain™ to the sample. Count and record the number of live shrimp.

6. Repeat step 5 three more times for the same test tube. Record the average number of shrimp for this test tube.

7. Repeat steps 5 and 6 for each of the remaining test tubes.

PART B: Design an Experiment

8. Work with the members of your lab group to explore one of the questions written for step 2 of **Before You Begin.** To explore the question, design an experiment that uses the materials listed for this lab.

You Choose

As you design your experiment, decide the following:

a. what question you will explore

b. what hypothesis you will test

c. how to set up your control

d. how to expose the brine shrimp to the conditions you chose

e. how long to expose the brine shrimp to the environmental conditions

f. how you will set up your data table

9. Write a procedure for your group's experiment. Make a list of all the safety precautions you will take. Have your teacher approve your procedure and safety precautions before you begin the experiment.

10. Set up and conduct your group's experiment. Do *not* use water over 70°C, which can burn you. **CAUTION: If you are working with the hot water bag, handle it carefully. If you are working with a lamp, do not touch the bulb. Light bulbs get very hot and can burn your skin.**

PART C: Cleanup and Disposal

11. Dispose of broken glass in the designated waste container. Put brine shrimp in the designated container. Do not pour chemicals down the drain or put lab materials in the trash unless your teacher tells you to do so.

12. Clean up your work area and all lab equipment. Return lab equipment to its proper place. Wash your hands thoroughly before you leave the lab and after you finish all work.

Analyze and Conclude

1. Summarizing Results Make a bar graph of your data. Plot the environmental variable on the *x*-axis and the number of shrimp on the *y*-axis.

2. Analyzing Results How did the shrimp react to changes in the environment?

3. Analyzing Methods Why was a control necessary?

4. Analyzing Methods Why was it necessary to take many counts in each test tube (step 6 of Part A)?

5. Further Inquiry Write a new question about brine shrimp that could be explored with another investigation.

? Do You Know?

Do research in the library or media center to answer these questions:

1. What are some predators of brine shrimp?

2. What is the ideal habitat for one of your favorite animals?

Use the following Internet resources to explore your own questions about habitat selection.

☑ internet connect

www.scilinks.org
Topic: Adaptation
Keyword: HX4002

*SCi*LINKS. Maintained by the National Science Teachers Association

Bald eagle

Quick Review

Answer the following without referring to earlier sections of your book.

1. **Define** *pH* and describe acid rain. *(Chapter 1, Section 3)*

2. **Identify** the role of the ozone layer in Earth's atmosphere. *(Chapter 12, Section 1)*

3. **Describe** the relationship between the long-term survival of species and the resources on which they depend. *(Chapter 15, Section 1)*

4. **Summarize** the events of the water cycle and the carbon cycle. *(Chapter 16, Section 3)*

Did you have difficulty? *For help, review the sections indicated.*

Reading Activity

Before you read this chapter, write a short list of environmental issues and efforts that you are familiar with. Then, write a list of questions about the environment and environmental issues. Save your list, and to assess what you have learned, see how many of your own questions you can answer after reading this chapter.

Looking Ahead

The bald eagle was a familiar sight in colonial America. By 1963, however, fewer than 1,000 bald eagles remained in the lower 48 states—their numbers diminished by habitat loss, sport hunting, and, in the mid-twentieth century, the pesticide DDT. After efforts to protect and manage our national bird, the bald eagle has made a remarkable comeback.

internet connect

www.scilinks.org
National Science Teachers Association *sci*LINKS Internet resources are located throughout this chapter.

*sci*LINKS. Maintained by the National Science Teachers Association

Global Change

Objectives

- **Recognize** the causes and effects of acid rain.

- **Evaluate** the long-term consequences of atmospheric ozone depletion.

- **Explain** how the burning of fossil fuels has changed the atmosphere.

- **Analyze** the proposed relationship between the greenhouse effect and global warming.

Key Terms

acid rain
chlorofluorocarbon
global warming
greenhouse effect

The Atmosphere and Ecosystems

You may be surprised to learn that some kinds of human activity can ultimately influence every ecosystem on Earth. Human-induced environmental changes are affecting ecosystems worldwide and may lead to global change.

Acid Rain

Coal-burning power plants send smoke high into the atmosphere through tall smokestacks, often more than 300 m (984 ft) tall. This smoke contains high concentrations of sulfur because the coal that the plant burns is rich in sulfur. The intent of those who designed the power plant was to release the sulfur-rich smoke high into the atmosphere, where winds would disperse and dilute it. Tall smokestacks, first introduced in the mid-1950s, are common in the United States and in Europe.

Scientists have since discovered that the sulfur introduced into the atmosphere by smokestacks can combine with water vapor to produce sulfuric acid. Rain and snow carry the sulfuric acid back to Earth's surface. This acidified precipitation is called **acid rain.** In North America, acid rain is most severe in the northeastern United States and in southeastern Canada, areas that are downwind from coal-burning plants in the Midwest. Recall that the pH of pure water is 7.0. In the northeastern United States, rain and snow have an average pH of about 4.0–4.5. This is over 10 times as acidic as the typical pH values for precipitation in the rest of the United States.

What is the impact of acid rain? Rainwater and some soils are naturally slightly acidic. Research suggests, however, that the acidity added by human activity is having a dramatic effect. In the United States and Canada, thousands of lakes are "dying" as their pH levels fall below 5.0. Forests in the eastern United States and southern Canada are being damaged. The acid pH may be harming symbiotic fungi in tree roots. The trees shown in **Figure 1** may have been affected by acid rain along with other factors.

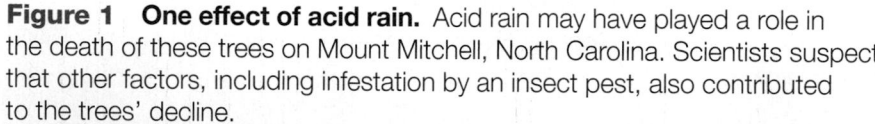
Figure 1 One effect of acid rain. Acid rain may have played a role in the death of these trees on Mount Mitchell, North Carolina. Scientists suspect that other factors, including infestation by an insect pest, also contributed to the trees' decline.

The Ozone Layer

You may recall from your reading that organisms were able to leave the oceans and colonize the land only after a protective shield of ozone, O_3, had developed in the upper atmosphere. Imagine what would happen if that shield were destroyed. Alarmingly, it appears that this is just what is happening. The ozone layer is being reduced, and human activity may play a large role in its reduction.

The Ozone Hole

In 1985, a researcher in Antarctica noticed that ozone levels in the atmosphere seemed to be as much as 35 percent lower than the average values during the 1960s. Satellite images taken over the South Pole revealed that the ozone concentration was unexpectedly lower over Antarctica than elsewhere in the Earth's atmosphere, as shown in **Figure 2.** It was as if an "ozone eater" were causing a mysterious zone of below-normal concentration, an area that researchers called the ozone hole. Alarmed, scientists examined satellite images taken in previous years. They found that the disintegration of the Earth's ozone shield was evident as far back as 1978. Every year since then, more ozone has disappeared, and the ozone hole has grown larger. Moreover, a smaller hole has appeared over the Arctic.

Because the decrease in ozone allows more ultraviolet radiation to reach the Earth's surface, scientists expect an increased incidence of diseases caused by exposure to ultraviolet radiation. These diseases include skin cancer, cataracts (a disorder in which the lens of the eye becomes cloudy), and cancer of the retina, the light-sensitive part of the eye. In fact, in the United States, the number of cases of malignant melanoma, a potentially lethal form of skin cancer, has almost doubled since 1980.

What Is Destroying Ozone?

The major cause of ozone destruction is a class of chemicals called **chlorofluorocarbons** (CFCs). Invented in the 1920s, CFCs were considered extremely stable, supposedly harmless, and a nearly ideal heat exchanger. Throughout the world, CFCs were commonly used as coolants in refrigerators and air conditioners, as aerosol propellants in spray cans, and as foaming agents in the production of plastic-foam cups and containers. Though CFCs were escaping into the atmosphere, at first no one worried.

By 1985, the scientific community had learned that CFCs are the primary cause of the ozone hole. High in the atmosphere, ultraviolet radiation from the sun is able to break the usually stable bonds in CFCs. The resulting free chlorine atoms then enter into a series of reactions that destroy ozone. As a result of this discovery, CFCs have been banned as aerosol propellants in spray cans in the United States. Today many countries limit or ban the use of CFCs.

Real Life

Ozone can be harmful.
In many cities and towns, summer weather reports include warnings of high ozone levels. Ozone at ground level irritates people's nasal passages, throats and lungs and may harm the immune system.
Finding Information
Find out why ground-level ozone is increasing even though the ozone layer in the upper atmosphere is decreasing.

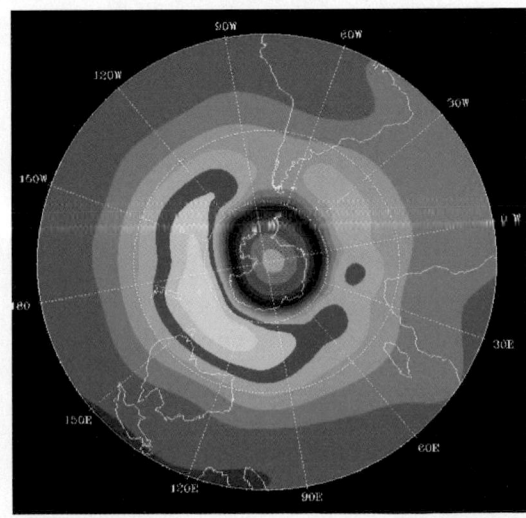

Figure 2 Ozone "hole" over Antarctica. In this satellite view of the South Pole, the pink area indicates the region with the least amount of ozone.

Global Temperatures

The average global temperature has been steadily increasing for more than a century, particularly since the 1950s. In Earth's long history there have been many such periods of **global warming,** often followed by centuries of cold. Scientists hypothesize that sunspot cycles may contribute to these cyclical changes in global temperature. Many scientists suspect, however, that human activity may be significantly contributing to global warming in modern times.

The Greenhouse Effect

Our planet would be as cold as the moon except for the insulating effects of certain gases—called greenhouse gases—such as water vapor, carbon dioxide (CO_2), methane, and nitrous oxide in the atmosphere. The chemical bonds in carbon dioxide molecules absorb solar energy as heat radiates from Earth. As shown in **Figure 3,** this process, called the **greenhouse effect,** traps heat within the atmosphere in the same way glass traps heat within a greenhouse. There has been a large increase in carbon dioxide in the Earth's atmosphere in recent times. This increase seems to be related to the burning of fossil fuels that has accompanied clearing of forests and urban industrialization.

Is Global Warming Occurring?

Figure 4 shows the average change in global temperature since 1960, and, on the same graph, the average concentration of atmospheric carbon dioxide. The correlation of increasing temperatures with increasing carbon dioxide levels is very close. Therefore many scientists are convinced temperature and carbon dioxide levels are related.

Figure 3   Earth's atmosphere traps heat.
Just as the glass panes of a greenhouse retain heat, CO_2 and other greenhouse gases in the atmosphere capture heat radiated from Earth's surface.

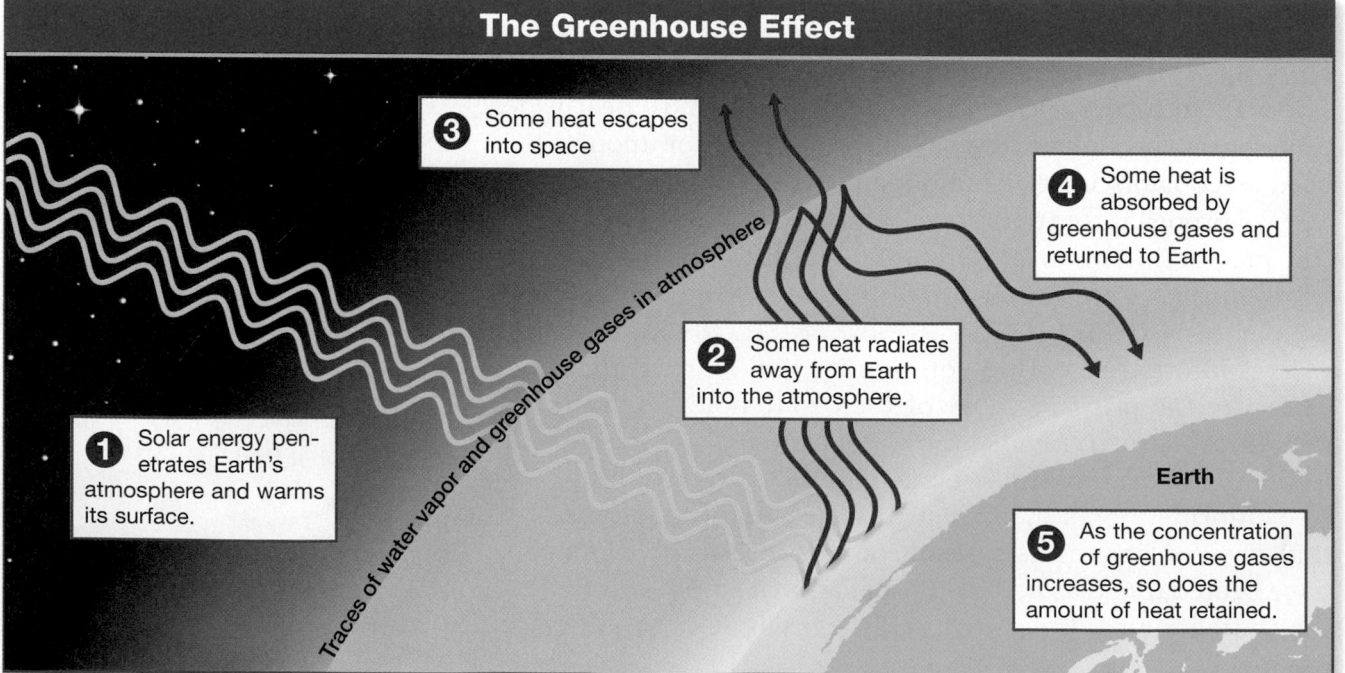

The Greenhouse Effect

❸ Some heat escapes into space

❹ Some heat is absorbed by greenhouse gases and returned to Earth.

❷ Some heat radiates away from Earth into the atmosphere.

❶ Solar energy penetrates Earth's atmosphere and warms its surface.

Traces of water vapor and greenhouse gases in atmosphere

Earth

❺ As the concentration of greenhouse gases increases, so does the amount of heat retained.

In science, however, correlation does not prove cause and effect. Both global temperature and levels of greenhouse gases may be changing because of other variables that have not been recognized yet. Some countries take seriously the possibility that increasing greenhouse gases play a role in global warming. These countries are attempting to formulate international treaties that place limits on greenhouse-gas emissions. The matter remains controversial, however, and the role of greenhouse gases in global warming is hotly contested.

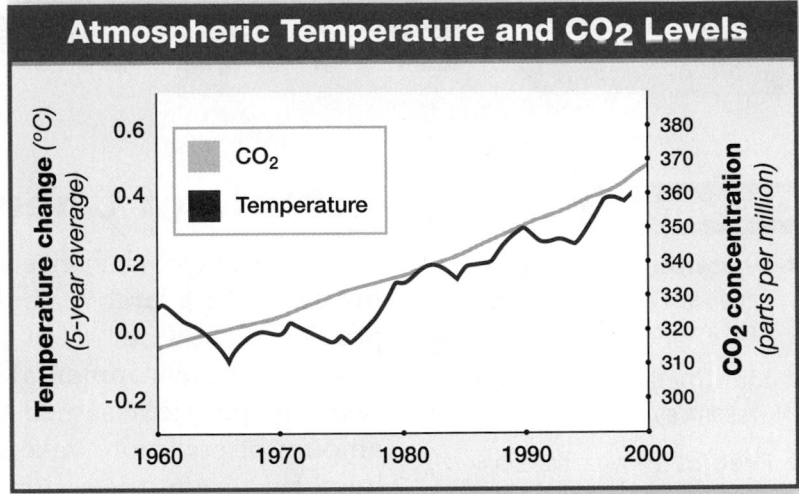

Figure 4 Global changes. Atmospheric CO_2 concentration and the average change in global temperature have risen together over the past several years.

Modeling the Greenhouse Effect

You can use a quart jar to explore the greenhouse effect.

Materials

MBL or CBL system with appropriate software, 2 temperature probes, 1qt jar, lid with a 0.5 cm hole in the center, tape, heat source

Procedure

1. Set up an MBL/CBL system to collect data from two temperature probes at 6 second intervals for 150 data points.

2. Insert the end of one probe into the hole in the lid of a quart jar, and tape the probe in place. Place the other probe about 4 in. from the jar and at the same height as the first probe.

3. Place the jar about 30 cm from a heat-radiating source, and begin collecting data.

4. After 5 minutes, turn off (or remove) the heat source. Collect data for another 10 minutes.

Analysis

1. **Propose** an explanation for any differences between the two probes.

2. **Critical Thinking Comparing Functions** How does carbon dioxide gas in the atmosphere function in a way similar to the glass jar?

3. **Critical Thinking Predicting Outcomes** How would the temperature on Earth be different if there were no carbon dioxide in the atmosphere?

Section 1 Review

1. **Summarize** two harmful effects of acid rain.

2. **Differentiate** the greenhouse effect from global warming.

3. **Critical Thinking Predicting Outcomes** How might the depletion of the ozone layer affect human health?

4. **Critical Thinking Evaluating Models** Is the greenhouse-gases model adequate to explain all warming of Earth over its existence? Why or why not?

5. **Standardized Test Prep** Which activity might reduce the severity of the greenhouse effect?
   A cutting trees   C promoting decomposition
   B burning vegetation D planting more trees

Effects on Ecosystems

Objectives

- **Describe** the effects of chemical pollutants on the environment.

- **Identify** three nonrenewable resources.

- **Predict** the potential consequences of uncontrolled population growth.

- **Contrast** population growth in developing countries with that in industrialized countries.

Key Terms

biological
  magnification
aquifer

Effects of Chemical Pollution

Examples of global change such as ozone depletion or global warming occur on a large scale. But what about serious environmental problems that occur in our own backyard? For example, one important urban environmental problem is chemical pollution. Until recently, people assumed that the environment can absorb any amount of pollution. Lake Erie and other large lakes became polluted because of the assumption that they could absorb unlimited amounts of industrial chemicals. Because of this incorrect assumption, pollution has often risen to a serious level.

In a highly publicized example of pollution, a very large oil tanker ran aground off the coast of Alaska in 1989. Oil from the tanker heavily polluted 1,600 km (1,000 mi) of coastline and injured or killed thousands of marine animals, such as the one shown in **Figure 5.** Despite costly and heroic cleanup efforts, damage to local wildlife was extensive. This dramatic example is not an isolated occurrence. Smaller oil spills and leaks that receive little or no publicity account for more than 90 percent of all pollution from oil seepage.

Many of the most disastrous incidents of pollution involve industrial chemicals that are toxic or carcinogenic (cancer-causing). Until recently, there has been relatively little regulation of the manufacture, transportation, storage, and destruction of such chemicals. A particularly clear example of this problem occurred in Basel, Switzerland, in 1986. Firefighters putting out a warehouse fire accidentally washed 27,000 kg (30 tons) of mercury and pesticides that were stored in the warehouse into the Rhine River. These poisons flowed down the Rhine, through Germany and the Netherlands, and into the North Sea, killing fish and other aquatic animals and plants. Today the river is recovering, but its species diversity remains far lower than it was before the disaster.

Agricultural Chemicals

In many countries, modern agriculture introduces large amounts of chemicals into the global ecosystem. These chemicals include pesticides, herbicides, and fertilizers. Industrialized countries, like the United States, now attempt to carefully monitor side effects of these chemicals. Unfortunately, large quantities of many toxic chemicals that are no longer manufactured still circulate in the ecosystem.

Figure 5 Oil spill victim. This common scoter was one of thousands of animals injured or killed when an accident released a large volume of oil off the Alaskan coast in 1989.

Pesticides For example, molecules of chlorinated hydrocarbons—a class of compounds that includes the pesticides DDT, chlordane, lindane, and dieldrin—break down slowly in the environment. They also accumulate in the fatty tissue of animals. As these molecules pass up through the trophic levels of the food chain, they become increasingly concentrated. This process is called **biological magnification,** illustrated in **Figure 6.** The presence of DDT in birds causes thin, fragile eggshells, which can break during incubation. Because of DDT use, many predatory birds in the United States and elsewhere failed to reproduce, and their numbers dwindled. In 1972, the use of DDT was severely restricted in the United States, and threatened bird populations slowly began to increase. However, chlorinated hydrocarbons are still manufactured in the United States and exported to other countries, where their use continues.

In order for us to meet the needs of an increasingly crowded world, the use of chemicals is necessary. We must learn to use them as intelligently as possible. Doing so will enable us to protect the productive capacity of the Earth. Failure is not a rational option.

Figure 6   Biological magnification of DDT

Because DDT accumulates in fatty tissue, DDT concentrations (in parts per million, ppm) increase as this chemical moves up the food chain.

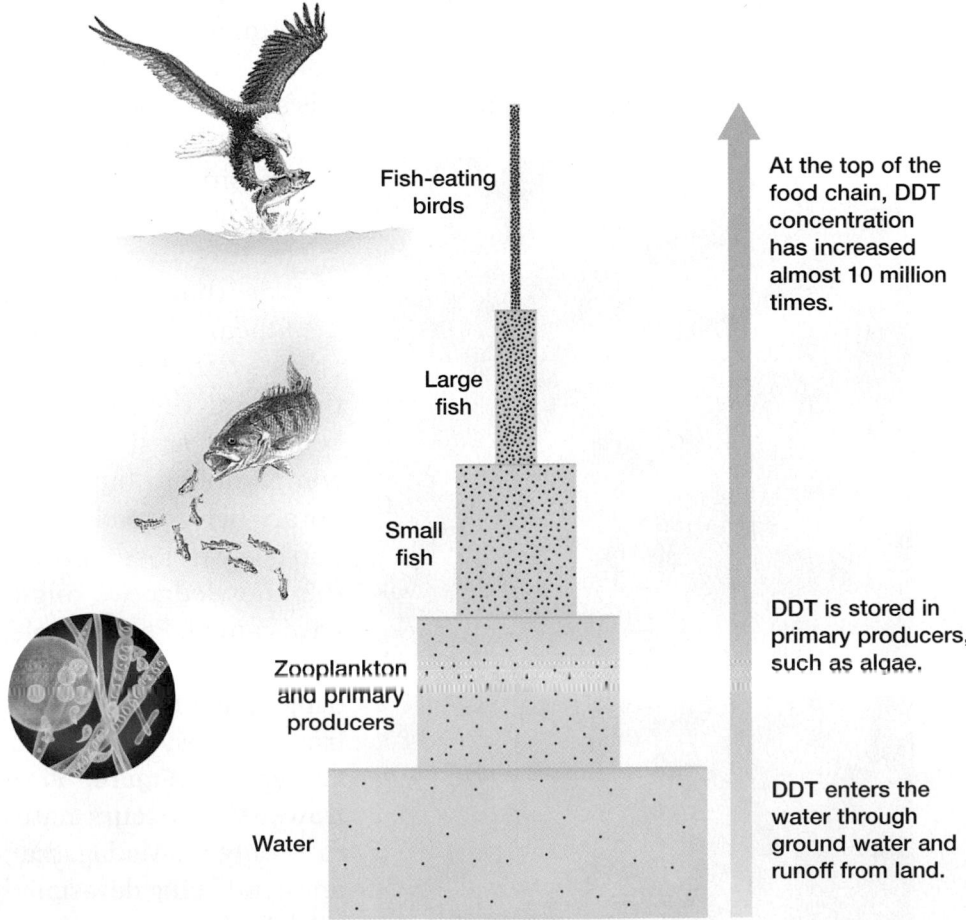

Fish-eating birds

At the top of the food chain, DDT concentration has increased almost 10 million times.

Large fish

Small fish

DDT is stored in primary producers, such as algae.

Zooplankton and primary producers

Water

DDT enters the water through ground water and runoff from land.

∴ Red dots represent DDT.

Loss of Resources

Among the many ways that ecosystems are being damaged, one problem stands out as potentially more serious than the rest—the consumption or destruction of resources that we cannot replace. Though a polluted stream can be cleaned up, no one can restore an extinct species. Three kinds of nonrenewable resources are being consumed or destroyed at alarming rates: species of living things, topsoil, and ground water.

Extinction of Species

Over the last 50 years, about half of the world's tropical rain forests have been burned to make pasture and farmland or have been cut for timber. Many thousands of square miles more will be destroyed this year. The people responsible, often poor farmers, view the forest lands as a resource to be developed, much as Americans viewed North American forests a century ago.

The problem is that as the rain forests disappear, so do their inhabitants. No one knows how many species are being lost. To find out, scientists carefully catalogue all of the residents of one small segment of forest and then extrapolate their data. That is, scientists use what they know to predict what they don't know. The resulting estimates vary widely, but it is clear Earth is losing many species. Some 10 percent of well-known species teeter on the brink of extinction. Worst-case estimates are that we will lose up to one-fifth of the world's species of plants and animals—about 1 million species—during the next 50 years. An extinction of this size has not occurred in at least 65 million years, since the end of the age of dinosaurs.

The tragedy of extinction is that as species disappear, so do our chances to learn about them and their possible benefits. This situation is comparable to burning a library before reading the books—we lose forever the knowledge we might have gained. For example, two potent anticancer drugs have been isolated from the rosy periwinkle, shown in **Figure 7,** a flower that occurs naturally only on Madagascar, an island being devastated by deforestation.

Study TIP

● **Organizing Information**

Make a table to organize information about *loss of resources.* Across the top, write the headings *Extinction of species, Loss of topsoil,* and *Ground-water pollution and depletion.* Along the sides, write *Causes, Effects,* and *Possible solutions.* Add information to the table as you read this chapter.

Figure 7 Beneficial species. Two potent anti-cancer drugs have been isolated from the leaves of the rosy periwinkle, a native of Madagascar.

Loss of Topsoil

The United States is one of the most productive agricultural countries on Earth, largely because of its fertile soils. These soils have accumulated over tens of thousands of years. The Midwestern farm belt sits astride what was once a great prairie. The topsoil of that ecosystem accumulated slowly as the remains of countless animals and plants decayed. By the time humans came to plow the prairie, the topsoil was more than a meter thick.

This rich topsoil cannot be replaced, and it is being lost at a rate of several centimeters each decade. Turning over the soil to eliminate weeds, allowing animals to overgraze ranges and pastures, and practicing poor land management all permit wind and rain to remove more and more of the topsoil. Since 1950, the world has lost one-third of its topsoil, primarily because of human activity.

Ground-Water Pollution and Depletion

A third resource that we cannot replace is ground water. Much ground water is stored within porous rock reservoirs called **aquifers** *(AHK wuh furz)*, as shown in **Figure 8.** Water seeps into aquifers too slowly to replace the large amount of water now being withdrawn. In most areas of the United States, there is relatively little effort to conserve ground water. Consequently, a very large portion of it is wasted on watering lawns, on washing cars, and through leaky and inefficient faucets and toilets. A great deal more ground water is being polluted by irresponsible disposal of chemical wastes. Once pollution enters the ground water, there is no effective way to remove it.

internet connect

www.scilinks.org
Topic: Aquifers
Keyword: HX4013

SC*LINKS* Maintained by the National Science Teachers Association

Figure 8 Aquifer

Large amounts of ground water are being removed from many aquifers far faster than natural processes can replenish it.

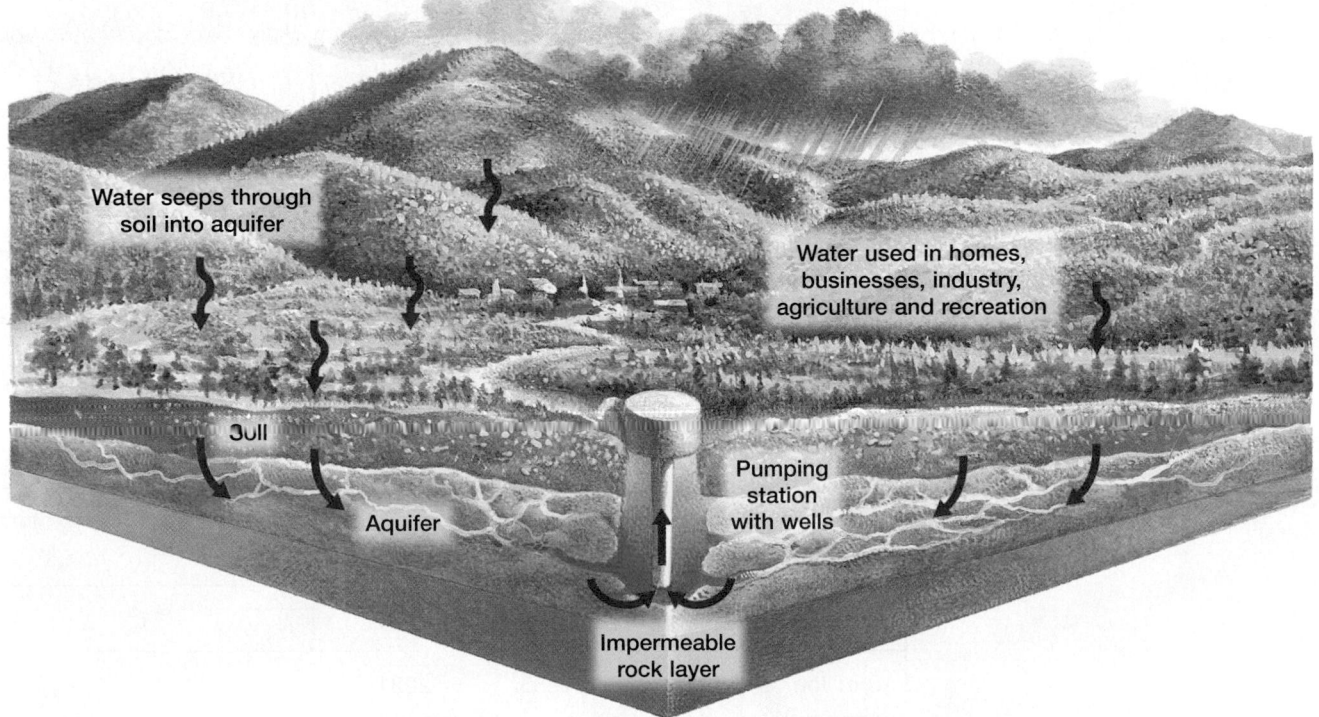

Water seeps through soil into aquifer

Water used in homes, businesses, industry, agriculture and recreation

Soil

Aquifer

Pumping station with wells

Impermeable rock layer

Growth of the Human Population

If all of the problems associated with pollution mentioned earlier in this chapter were solved, would the future of our environment be free from danger? Not necessarily. Scientists recognize that we would still need to address a more fundamental problem: the rapid growth of the human population, as shown in **Table 1.**

Humans first reached North America at least 12,000 years ago, by crossing the land bridge between Siberia and Alaska. Humans then spread throughout North America and South America. Ten thousand years ago the continental ice sheets that covered northern Europe and North America receded, and agriculture soon developed. There were only about 5 million people on Earth then. As agriculture produced more dependable sources of food, the human population began to grow. By about 2,000 years ago, there were an estimated 130 million people on Earth. By 1650, the world's population had reached 500 million.

Since then, the average global birthrate has remained near 30 births per 1,000 people per year. However, with the development of technology to ensure better sanitation and improved medical care, the death rate has fallen steadily. In 2002, the estimated death rate was about 9 deaths per 1,000 people per year. The difference between the annual birthrate in 2002 (now estimated to be 21 births per 1,000 people) and death rate results in an annual worldwide increase in the human population of approximately 1.3 percent. This number may seem small, but don't be deceived. The world's population would double in just over 60 years if it continued to grow at this rate!

Figure 9 Population growth. The world's population has topped 6 billion.

www.scilinks.org
Topic: Population Growth
Keyword: HX4144

Table 1 Number of Years to Add 1 Billion People		
Human population	Year	Years to add
1 billion	1800	All of human history
2 billion	1930	130
3 billion	1960	30
4 billion	1974	14
5 billion	1987	13
6 billion	1999	12
Projected		
7 billion	2009	11
8 billion	2021	11

Worldwide Rates of Growth

The world's population exceeded 6 billion in October 1999, and the annual increase is now about 94 million people. About 260,000 people are added to the world population each day, or about 180 every minute. Population growth is fastest in the developing countries of Asia, Africa, and Latin America. It is slowest in the industrialized countries of North America, Europe, Japan, and in New Zealand, and Australia. In industrialized countries like the United States, about one fifth of the population is under 15 years of age. In developing countries like Nigeria, the percentage is typically twice as high, which may lead to explosive population growth in coming decades.

Figure 10 shows the population growth rates of developing and developed countries. The population growth rate in the United States is only 0.8 percent. Most European countries are growing even more slowly, and the populations of Germany and Russia are actually declining. By contrast, as of 1996, Nigeria's population was increasing by about 3.05 percent per year.

The global rate of population growth has been declining. The United Nations projects that the world's population will stabilize at 9.7 billion by the year 2050. As you have read, however, population growth rates are uneven across Earth. Population growth tends to be the highest in countries that can least afford it. Already limited resources are strained further, and natural resources—ground water, land for farming, forests—are ever more quickly depleted or polluted.

No one knows whether the Earth can support six billion people indefinitely, much less the far larger population that lies in our future. Building a sustainable world is the most important task facing humanity's future. The quality of life available to your children in the new century will depend to a large extent on our success.

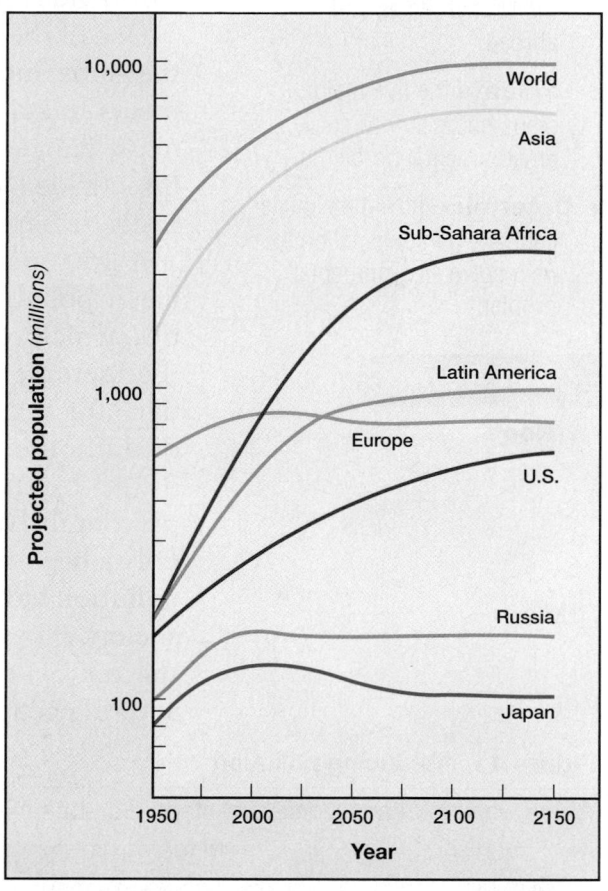

Figure 10 World population growth patterns. Most of the world population increase since 1950 has been in developing countries.

Section 2 Review

1 **Evaluate** how research showing that chlorinated hydrocarbons are an environmental threat may impact people in the United States.

2 **Summarize** why supplies of soil and ground water are dwindling even though these resources are replenished by natural processes.

3 **Critical Thinking Inferring Relationships** Describe two instances in which technology has caused the growth rate of the human population to increase.

4 **Critical Thinking Evaluating Viewpoints** A classmate claims that the growth of the human population will not affect populations of other species. Evaluate this viewpoint.

5 **Standardized Test Prep** Which organism would have the highest concentration of a pollutant that enters the water and is absorbed by aquatic primary producers?

A large fish   C fish-eating bird
B small fish   D zooplankton

Solving Environmental Problems

Objectives

- **Describe** two effective approaches that have been taken to reduce pollution in the United States and abroad.

- **Evaluate** the five major steps necessary to solve environmental problems.

- **Determine** how individuals can take personal action to help solve environmental problems.

Key Terms

None

A Worldwide Effort

As you have seen, environmental problems affect all inhabitants of an ecosystem without regard to state or national boundaries. As human activities continue to place severe stresses on ecosystems, worldwide attention must be focused on solving these problems.

One of the most encouraging developments of the early 1990s was the global increase in efforts to reduce pollution. International agreements to stop CFC production are one example. And the release of many dangerous industrial and agricultural chemicals—notably the insecticide DDT and the carcinogens asbestos and dioxin—has been restricted in the United States.

A great deal of progress has also been made in reducing air and water pollution. The number of secondary sewage treatment facilities, which remove chemicals as well as bacteria from sewage, is on the increase. **Figure 11** shows a "scrubber," a device that reduces harmful sulfur emissions from industrial smokestacks. Emissions of sulfur dioxide, carbon monoxide, and soot—three pollutants produced by the burning of coal—have been cut by more than 30 percent in 10 years. However encouraging, this progress represents only a beginning. Serious attempts to address the overall problem of pollution have also brought about more fundamental changes in our society. For example, a reduction of the number of automobiles on the road is encouraged by providing dedicated lanes to cars with several occupants, as shown in **Figure 12.**

Figure 11   Reducing pollution

Waste water is cleaned in several steps. Scrubbers remove many pollutants from factory emissions.

Sewage treatment plant

Smokestacks with scrubbers

In the United States

Two effective approaches have been taken to reduce pollution in the United States. The first approach has been to pass laws forbidding it. In the last 30 years, laws have begun to significantly slow the spread of pollution. These laws impose strict standards for what can be released into the environment. For example, all cars are required to have catalytic converters to reduce emissions. Similarly, the Clean Air Act of 1990 requires scrubbers on the smokestacks of power plants. Converters and scrubbers make cars and energy more expensive. The effect is that the consumer pays more to avoid polluting the environment.

A second effective approach to reducing pollution has been to make it more expensive by placing a tax on it. The gasoline tax is a good example of such a tax. To be fully effective, however, a tax must be high enough to reflect the actual cost of the pollution. By adjusting the tax, the government attempts to balance the conflicting demands of environmental safety and economic growth. Such taxes, often imposed on industry in the form of "pollution permits," are becoming increasingly common.

Figure 12 Carpooling.
Some cities reserve certain lanes for carpool use during peak travel times.

 **BIOWatch**

The Real Costs of Pollution

To find economic solutions to environmental problems, it is first necessary to understand that the economy of much of the industrialized world is based on a system of supply and demand. As something gets scarce, its price increases. This increased profit on an item in short supply acts as an incentive for the production of more of the item. If too much of the item is available, the price falls. Because it is less profitable to produce the item, less of it is made.

A Price Too Low?
This system works very well and is responsible for the economic strength of our nation. But if demand is set by price, then it is very important that *all* of the production costs be included in the price of an item. If a person selling an item were able to pass off part of its production cost to a third person, the seller would be able to

set a lower price and sell more of the item. Stimulated by the lower price, the buyer would purchase more of the item.

The True Cost of Pollution
The true costs of energy and manufactured goods are composed of direct and indirect production costs. Direct costs include materials and wages. Indirect costs include pollution and other damage to the environment. For example, the true costs of fossil fuels include the indirect costs of reduced harvests of fish and shellfish due to oil spills. But the indirect costs are *not* included in the price that the consumer pays for fossil fuels. As a result, far more is consumed than if these costs were included. The indirect costs do not disappear because we ignore them. They are passed on to future generations, who

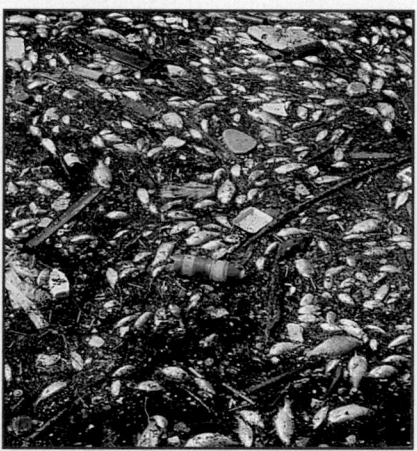

must pay the bill in terms of damage to the ecosystems on which they depend.

internet connect

www.scilinks.org
Topic: Solving Environmental Problems
Keyword: HX4166

SC*LINKS* Maintained by the National Science Teachers Association

Solving Environmental Problems

It is easy to get discouraged when considering the world's many serious environmental problems. But do not lose track of the conclusion that emerges from our examination of these environmental problems—each of the world's many problems is solvable. If one looks at how environmental problems have been overcome, a clear pattern emerges.

Five Steps to Success

Viewed simply, there are five components to successfully solving any environmental problem.

1. **Assessment.** The first stage is scientific analysis of the problem, the gathering of information about what is happening. To construct a scientific model of an ecosystem, data must be collected and analyzed. A model makes it possible to describe the current state of the ecosystem. A model would also allow scientists to make predictions about the future of the ecosystem.

2. **Risk analysis.** Using the information obtained by scientific analysis, scientists predict the consequences of different types of environmental intervention. It is also essential to evaluate any negative effects associated with a plan of action.

3. **Public education.** When it is possible to describe alternative courses of action, the public must be informed. This involves explaining the problem in understandable terms, such as at the public meeting shown in **Figure 13,** presenting the alternative actions available, and explaining the probable costs and results of the different choices.

4. **Political action.** The public, through its elected officials, selects and implements a course of action. Individuals can be influential at this stage by exercising their right to vote and by contacting their elected officials.

5. **Follow-through.** The results of any action should be carefully monitored to see if the environmental problem is being solved.

Real Life

Better the second time around?
Although many communities have recycling programs, some researchers argue that recycling doesn't pay off.

Evaluating Viewpoints
Find out the main arguments for and against recycling, and propose the kinds of waste-management programs that would work in your community.

Figure 13 Public participation. Public discussion of environmental problems helps citizens to evaluate alternative courses of action.

Two Success Stories

The development of appropriate solutions to the world's environmental problems often rests partly on the shoulders of politicians, economists, bankers, scientists, and engineers. However, it is important not to lose sight of the key role often played by informed individuals. Two examples serve to illustrate the point.

The Nashua River Running through the heart of New England, the Nashua River was severely polluted by mills established in Massachusetts in the early 1900s. When Marion Stoddart, shown in **Figure 14,** moved to a town along the river in 1962, she was appalled. Stoddart organized the Nashua River Cleanup Committee. The committee presented bottles of dirty river water to politicians, spoke at town meetings, recruited business people to help finance a waste treatment plant, and began to clean garbage from the Nashua River's banks. This citizen's campaign contributed to the passage of the Massachusetts Clean Water Act of 1966. Industrial dumping into the river is now banned, and the river has largely recovered.

Lake Washington Following World War II, this very large lake east of Seattle became surrounded by a ring of 10 suburbs, each with its own municipal sewage treatment plant. Between 1940 and 1953, these 10 municipal sewage plants discharged their treated outflow into the lake. Safe enough to drink, the outflow was believed to be harmless. Starting in the early 1940s, the combined daily discharge in the lake was 80 million liters (20 million gallons).

In 1954, an ecology professor at the University of Washington in Seattle, W. T. Edmondson, noted that his research students were reporting blue-green algae growing in the lake. Such algae require an abundance of the nutrients nitrogen and phosphorus to grow. Because deep freshwater lakes like Lake Washington usually lack these nutrients, the presence of the algae was surprising. The researchers found that phosphates and nitrates in the sewage had been fertilizing the lake! Edmondson was alarmed and began a campaign in 1956 to educate public officials about the danger: Bacteria decomposing the dead algae would soon deplete the lake's oxygen. This would kill all life in the lake, and it would never recover. After five years, as a direct result of his efforts, joint municipal taxes financed the cleanup of Lake Washington with a massive trunk sewer that rings the lake and carries treated discharge far out into Puget Sound. Today, through the efforts of many people, the lake is healthy, its waters clean and blue, as shown in **Figure 15.**

Figure 14 Marion Stoddart. The recovery of the Nashua River shows that polluted environments can be restored when committed individuals, like Marion Stoddart, work to bring about a change.

Figure 15 Lake Washington. Once choked with algae that were nourished by the outflow from sewage treatment plants, Lake Washington is a healthy lake today.

Figure 16 Conserving energy. By riding a bicycle instead of traveling in a car, bus, or train, this commuter helps save energy.

Your Contribution

You cannot hope to preserve what you do not understand. Humans rely on the Earth's ecosystems for food and all of the other materials our civilization depends on. It has been said that we do not inherit the Earth from our parents but borrow it from our children. Therefore, we must preserve for them a world in which they can live.

Although solving the world's environmental problems will take the efforts of many people, including politicians, economists, and engineers, the issues are largely biological. When all is said and done, your knowledge of ecology is the essential tool that you can contribute to the effort. **Figures 16** and **17** show some simple ways you can participate in solving the problems described in this chapter by conserving energy and reducing pollution and waste. You can save energy by walking, riding a bicycle, or taking public transportation to work or school. Newspapers, aluminum products, glass containers, and many plastic containers can be recycled. A resource- and energy-use inventory of your home can identify additional ways to help the environment. For example, installing inexpensive, low-flow shower heads can reduce shower water use by up to 50 percent.

Figure 17 Your lifestyle affects the environment. Choices that you make in your day-to-day activities can benefit the environment.

You Can Help

- Conserve energy by walking, riding a bicycle, or taking public transportation
- Do a resource- and energy-use inventory of your home
- Learn about the environment
- Recycle newspapers, aluminum products, glass, and plastic
- Create rich soil by making your own compost heap from leaves, grass, and fruit peelings

Section 3 Review

1 **Describe** how a tax can reduce pollution.

2 **Critical Thinking** **Justifying Conclusions**
Of the five steps listed in this chapter for solving environmental problems, which step might be the most difficult to implement?

3 **Critical Thinking** **Analyzing Information**
At which step in the solution of an environmental problem could you have the greatest influence? Explain your answer.

4 How does the release of certain nutrients into a lake harm the lake?

5 **Standardized Test Prep** One good way to conserve natural resources is to
A use high-flow shower heads.
B use public transportation.
C throw away cans and jars.
D drive instead of walk.

Key Concepts

1 Global Change

- Acid rain, which is caused by airborne pollutants that lower the pH of rain, has damaged many forests and lakes, especially in the Northeast.

- Destruction of the ozone layer is caused by chlorofluorocarbons (CFCs) and several other manufactured chemicals.

- The greenhouse effect occurs when greenhouse gases, such as carbon dioxide, trap heat within Earth's atmosphere.

- Many scientists think that increased concentrations of CO_2 and other greenhouse gases in the atmosphere have led to global warming.

2 Effects on Ecosystems

- The release of toxic chemicals into the environment can have serious effects, particularly when their concentration is magnified by food chains.

- Three nonreplaceable resources—species of living things, topsoil, and ground water—are being consumed or destroyed at a rapid rate.

- Rapid growth of the human population places serious stress on the Earth's ecosystems.

3 Solving Environmental Problems

- Worldwide efforts to reduce pollution are being made, but they are only part of the solution to the overall pollution problem.

- Taxing products or services that create pollution and creating laws requiring pollution-control devices are two methods that have been used to reduce pollution.

- Each of the world's many environmental problems can be solved if seriously addressed. A combination of scientific investigation and public action can solve many environmental problems.

Key Terms

Section 1

acid rain (386)
chlorofluorocarbon (387)
global warming (388)
greenhouse effect (388)

Section 2

biological magnification (391)
aquifer (393)

Section 3

None

Unit 7—*Ecosystem Dynamics*
Use this unit to review the key concepts and terms in this chapter.

Understanding Key Ideas

1. Which of the following causes acid rain?
a. releasing chlorofluorocarbons
b. burning high-sulfur coal
c. polluting ground water
d. scrubbing smokestack emissions

2. The burning of fossil fuels has changed the atmosphere by
a. increasing the global concentration of ozone in the upper atmosphere.
b. reducing the amount of CFCs.
c. producing an ozone hole.
d. increasing the concentration of carbon dioxide.

3. Which of the following statements about the extinction of species that is now occurring is *not* true?
a. It is the largest extinction event since the dinosaurs disappeared.
b. One of its causes is the destruction of tropical rain forests.
c. It is confined to tropical countries.
d. Potentially useful species are becoming extinct.

4. Which of these countries has the largest population growth rate?
a. Germany
b. United States
c. Nigeria
d. Russia

5. Of the five major steps to solving environmental problems, which involves determining the potential outcomes of an environmental plan before it is tried?
a. assessment
b. risk analysis
c. follow-through
d. political action

6. Explain why the loss of topsoil and the extinction of living species are a threat to increased food production.

7. Analyze the relationship between biological magnification and the reduction of the bald eagle population by the pesticide DDT.

8. **BIOWatch** Name some of the indirect costs associated with recovering and using fossil fuels such as oil.

9. Describe how the global decrease in available ground water and the loss of topsoil could affect the carrying capacity of the Earth. (**Hint:** See Chapter 16, Section 1.)

10. **Concept Mapping** Make a concept map that shows how human activities are disrupting the atmosphere and that describes the effects of these disruptions. Try to use the following terms in your map: *greenhouse effect, carbon dioxide, greenhouse gases, global warming, CFCs, ozone layer, acid rain,* and *high-sulfur coal.*

Critical Thinking

11. Evaluating an Argument Evaluate the statement that rising levels of carbon dioxide in the atmosphere might lead to increased food production.

12. Predicting Outcomes How would stopping all pesticide use likely affect rates of food production and of diseases, such as malaria, that are spread by insects?

Alternative Assessment

13. Summarizing Information Write a report that identifies several alternatives to CFCs and describes the requirements established for CFC replacements.

14. Career Connection **Environmental Scientist** Use library or Internet resources to research the educational background necessary to become an environmental scientist. Describe the degrees or training that is recommended for this career. Summarize the employment outlook for this field.

15. Interactive Tutor Unit 7 **Ecosystem Dynamics** Write a report summarizing ways that humans can work to reduce the depletion and pollution of ground water. How would more-efficient use and recycling of ground water benefit ecosystems?

Standardized Test Prep

Understanding Concepts

Directions (1–4): **For *each* question, write on a separate sheet of paper the letter of the correct answer.**

1 What is liquid precipitation with a low pH that results from sulfur emissions reacting with water usually called?
   A. acid rain
   B. greenhouse gas
   C. sulfuric acid
   D. thermal pollution

2 What process involves the warming of Earth by radiated energy trapped by gases such as carbon dioxide, methane, and nitrous oxide?
   F. biological magnification
   G. greenhouse effect
   H. ozone destruction
   I. thermal pollution

3 Which of the following is a class of stable chemicals that break down and release free chlorine atoms high in the atmosphere?
   A. CFC
   B. DDT
   C. greenhouse gas
   D. hydrocarbon

4 What term describes the increasing concentration of substances in animal tissue toward the top of the food chain?
   F. biological magnification
   G. chlorofluorocarbon toxicity
   H. DDT concentration
   I. hydrocarbon pollution

Directions (5): **For the following question, write a short response.**

5 Tall smokestacks are part of many coal-burning power plants. Define the purpose of these tall smokestacks and evaluate their effectiveness.

Test TIP

When using a graph to answer a question, make sure you know what variables are represented on the *x*- and *y*-axes before answering the question.

Reading Skills

Directions (6): **Read the passage below. Then answer the question.**

Species diversity has declined drastically several times during Earth's history. Each time, new species evolved. Many unique species alive today are near extinction. However, some people hypothesize that losing these organisms to extinction is not a great loss, because extinction ultimately increases biodiversity.

6 What is a weakness of this hypothesis?
   A. Evolution takes millions of years.
   B. Extinction does not affect ecosystems.
   C. Organisms that are near extinction always play a minor role in ecosystems.
   D. Damage to ecosystems can be easily repaired by introducing new species.

Interpreting Graphics

Directions (7): **Base your answer to question 7 on the graph below.**

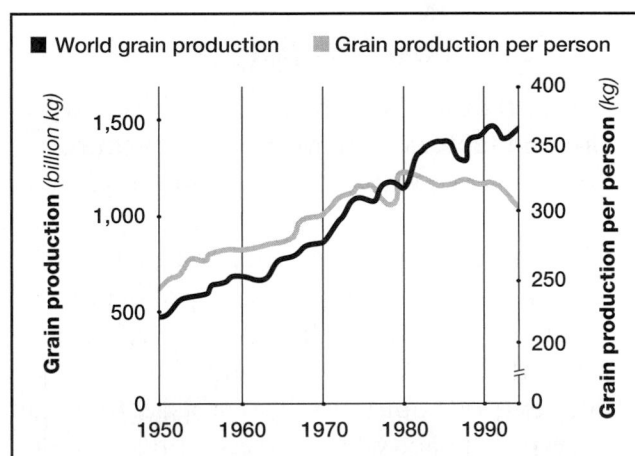

Grain Production

7 How have world grain production and the human population changed since 1980?
   F. Both have decreased at the same rate.
   G. World grain production has increased; the human population has decreased.
   H. Both have decreased, but the grain production has decreased more rapidly.
   I. Human population has increased more rapidly than grain production.

Skills Practice Lab

Studying Population Growth

SKILLS
- Using a microscope
- Collecting, graphing, and analyzing data
- Calculating

OBJECTIVES
- **Observe** the growth and decline of a population of yeast cells.
- **Determine** the carrying capacity of a yeast culture.

MATERIALS
- safety goggles
- lab apron
- yeast culture
- (2) 1 mL pipets
- 2 test tubes
- 1% methylene blue solution
- ruled microscope slide (2 × 2 mm)
- coverslip
- compound microscope

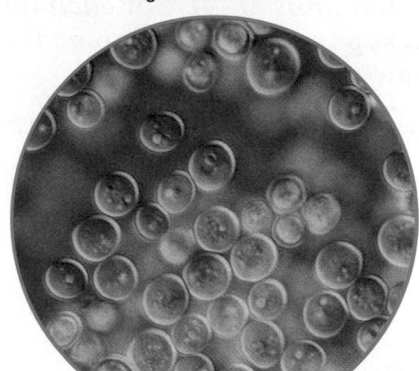

Yeast cells

ChemSafety

CAUTION: Always wear safety goggles and a lab apron to protect your eyes and clothing.

CAUTION: Do not touch or taste any chemicals. Know the location of the emergency shower and eyewash station and how to use them. If you get a chemical on your skin or clothing, wash it off at the sink while calling to the teacher. Notify the teacher of a spill. Spills should be cleaned up promptly, according to your teacher's directions.

CAUTION: Glassware is fragile. Notify the teacher of broken glass or cuts. Do not clean up broken glass or spills with broken glass unless the teacher tells you to do so.

Before You Begin

Recall that population size is controlled by **limiting factors**—environmental resources such as food, water, oxygen, light, and living space. **Population growth** occurs when a population's **birthrate** is greater than its **death rate.** A decline in population size occurs when a population's death rate surpasses its birthrate. In this lab, you will study the concepts of population growth, decline, and carrying capacity by growing and observing yeast.

1. Write a definition for each boldface term in the previous paragraph.
2. Make a data table similar to the one below at left.
3. Based on the objectives for this lab, write a question about population growth that you would like to explore.

Procedure

PART A: Counting Yeast Cells

1.  Put on safety goggles and a lab apron.
2. Transfer 1 mL of a yeast culture to a test tube. Add 2 drops of methylene blue to the tube. **Caution: Methylene blue will stain your skin and clothing.** The methylene blue will remain blue in dead cells but will turn colorless in living cells.

DATA TABLE

Time	Number of cells per square		Population size
(hours)	Squares 1–6	Average	(cells/0.1 mL)
0			
24			
48			
72			
96			

3. Make a wet mount by placing 0.1 mL (one drop) of the yeast and methylene blue mixture on a ruled microscope slide. Cover the slide with a coverslip.

4. Observe the wet mount under the low power of a compound microscope. Notice the squares on the slide. Then switch to the high power. *Note: Adjust the light so that you can clearly see both stained and unstained cells.* Move the slide so that the top left-hand corner of one square is in the center of your field of view. This will be area 1, as shown in the diagram below.

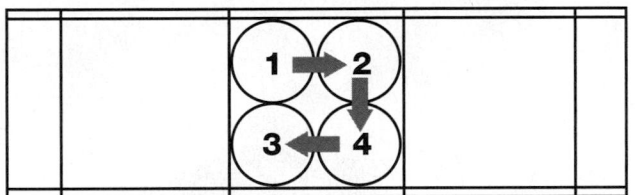

5. Count the live (unstained) cells and the dead (stained) cells in the four corners of a square using the pattern shown in the diagram above. In your data table, record the numbers of live cells and dead cells in the square.

6. Repeat step 5 until you have counted 6 squares on the slide. Complete Part B.

7. Find the total number of live cells in the 6 squares. Divide this total by 6 to find the average number of live cells per square. Record this number in your data table. Repeat this procedure for dead cells.

8. Estimate the population of live yeast cells in 1 mL (the amount in the test tube) by multiplying the average number of cells per square by 2,500. Record this number in your data table. Repeat this procedure for dead cells.

9. Repeat steps 1 through 8 each day for 4 more days.

PART B: Cleanup and Disposal

10. Dispose of solutions and broken glass in the designated waste containers. Do not pour chemicals down the drain or put lab materials in the trash unless your teacher tells you to do so.

11. Clean up your work area and all lab equipment. Return lab equipment to its proper place. Wash your hands thoroughly before you leave the lab and after you finish all work.

Analyze and Conclude

1. **Analyzing Methods** Why were several areas and squares counted and then averaged each day?

2. **Summarizing Results** Graph the changes in the numbers of live yeast cells and dead yeast cells over time. Plot the number of cells in 1 mL of yeast culture on the y-axis and the time (in hours) on the x-axis.

3. **Inferring Conclusions** What limiting factors probably caused the yeast population to decline?

4. **Further Inquiry** Write a new question about population growth that could be explored in another investigation.

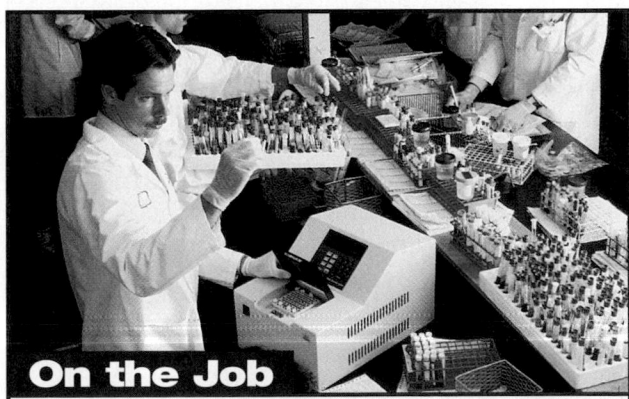

On the Job

Biologists must sometimes estimate the number of cells in a given volume. Do research to discover how a **hemacytometer** is used in medical tests. For more about careers, visit **go.hrw.com** and type in the keyword **HX4 Careers.**

Biologists race against time to save species threatened with extinction.

Saving Biodiversity

How many species of organisms live on Earth? That seems like a simple question, but no one knows the answer. The uncertainty over this basic question illustrates how little scientists actually know about biodiversity.

One thing is certain about biodiversity, though: it is disappearing fast. Biologists need good information about biodiversity in order to choose the actions that will save the most species. To focus conservation efforts, biologists agree that the first priority must be to take an inventory of global biodiversity, determining which species live where.

A New Approach

Beginning in the late 1980s, a group of biologists and conservationists decided a shortcut was needed. They launched the Rapid Assessment Program, or RAP, to speed the study of biodiversity. RAP sends small teams of experts to tropical habitats to quickly determine what kinds of organisms live there. The program focuses on "hot spots," threatened areas that are high in

biodiversity and contain large numbers of unique species. The aim is to survey as many of these hot spots as possible and identify areas that need to be protected.

To conduct these surveys, Conservation International, the nonprofit organization that sponsors the program, brings together several of the world's leading tropical biologists. Each expedition also includes several scientists from the country where the team is working. (One of the goals of RAP is to train biologists in tropical countries.) The RAP team uses satellite photos, aerial observation, and discussions with scientists and government officials in tropical countries to choose the locations it will visit.

Instead of taking an exhaustive, time-consuming inventory of all species, the team may focus on a few familiar groups. An inventory of mammals, birds, fishes, flowering plants, butterflies, and reptiles gives a good indication of an area's total biodiversity. A habitat with many species of plants and birds, for example, also probably has many species of bacteria, insects, and other less conspicuous organisms.

Rare treasures
Loss of habitat threatens many tropical rain forest species.

Tropical rain forest
Tropical rain forests are home to many species.

The Rewards of RAP

RAP scientists describe their work as exhausting but exhilarating. "It's what we live for," says Tom Schulenberg, an ornithologist (a biologist who studies birds) and RAP team leader. "We're always scheming to get back to the field." While exploring new places, Schulenberg feels "an incredible sense of excitement, knowing no other biologist has been there and everything you see is being seen for the first time."

How RAP Helps Conserve Biodiversity

Though less exciting than exploring a rain forest, the next stage of RAP is just as important. The scientists return to the United States to analyze their data. They then present a report containing their recommendations to the host country. RAP scientists stress that their role is to provide scientifically sound advice, not to tell tropical countries how to manage their natural resources. They leave all decisions to the individual governments. Using the information from RAP reports, tropical countries can guide their land-use decisions to help preserve biodiversity. ■

Biologist

RAP biologists

Profile

When biologists recognized that traditional field research was gathering data too slowly, the Rapid Assessment Program was born. Biologists help conduct short, intensive surveys to quickly fill some of the gaps in the knowledge of biodiversity.

Job Description

Biologists study all aspects of the biology of living things—anatomy, physiology, behavior, ecology, and evolutionary relationships. Their research may involve lab work, field studies, or a combination of both. Many biologists work for universities, museums, or government agencies.

Job Duties

Biologists who are members of a RAP team apply their knowledge of basic biology in parts of the world that are most important for saving species for future generations. Working in remote areas far from medical care has its dangers. Team members have been laid low by bubonic plague, malaria, and hepatitis. Besides disease, they must also watch out for poisonous snakes, biting insects and spiders, and falling trees.

Science/Math Career Preparation

Biology	Mathematics
Zoology	Genetics
Botany	Biochemistry
Evolutionary biology	

Analyzing STS Issues

Science and Society

1 What are the benefits of preserving biodiversity? About 25 percent of medicines are derived from chemicals made by plants. Research a medicine derived from a tropical plant. What plant was it isolated from? Where does the plant live? What disease or diseases is the medicine used to treat?

2 What are other countries doing to preserve biodiversity? The government of many countries throughout the world are participating in the comprehensive protection of the Earth's biological resources. Using library resources or the Internet, research the various measures these governments have taken to catalog and preserve biodiversity and write a report summarizing your findings.

Technology: PCR

3 Who should get the benefits of biodiversity? The enzyme that copies DNA during the polymerase chain reaction (PCR) was isolated from an archaebacterium that lives in Yellowstone National Park. Although PCR generates more than $200 million in income each year, the federal government receives no royalties from the use of the enzyme. Research this issue and then write an essay supporting or opposing this statement: Companies that profit from PCR should be required to compensate the government for using a species discovered on federal land.

Exploring Diversity

Surrounded by protective gear, a worker in a bio-technology clean room is isolated from contaminants in the surrounding environment.

in perspective

Microbes: Unseen Agents of Disease

 **Yesterday...** Surgeons worked without masks, gowns, and gloves? Although this early operation looks strange to us today, Dr. Joseph Lister was actually ahead of his time. In the mid-1860s, Dr. Lister became the first physician to treat patients with an antiseptic during surgery. Lister recognized that spraying an airborne mist of carbolic acid over a patient reduced the likelihood of infections. **Read to learn how bacteria can cause disease.**

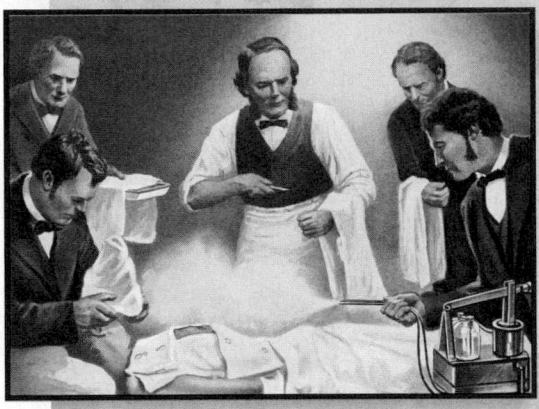

Dr. Joseph Lister and patient in 1867

Today... Physicians fight bacteria with a far more sophisticated array of weapons than those used by Dr. Lister and his colleagues. Antibiotic drugs target infections throughout the body. **What is the Gram reaction?**

Antibiotics

Tomorrow... Some kinds of bacteria have become resistant to certain antibiotics. When an antibiotic fails to stop a bacterial infection, a physician will usually prescribe a different antibiotic—and hope that it works. As antibiotics are used with increasing frequency, strains of antibiotic-resistant bacteria (such as those that cause drug-resistant tuberculosis) continue to become more widespread. **Read to discover how antibiotics kill bacteria.**

Tuberculosis bacterium

internet connect

www.scilinks.org
Topic: Bacteria
Keyword: HX4018

SCiLINKS
Maintained by the
National Science
Teachers Association

A meadow

19 Introduction to the Kingdoms of Life

✓ *Quick Review*

Answer the following without referring to earlier sections of your book.

1. **Calculate** the surface area–to–volume ratio of a cell. *(Chapter 3, Section 2)*

2. **Summarize** the characteristics of prokaryotes and eukaryotes. *(Chapter 3, Section 2)*

3. **Differentiate** introns from exons. *(Chapter 10, Section 2)*

4. **Identify** the terms *bacteria* and *archaebacteria* *(Chapter 12, Section 2)* and *kingdom*. *(Chapter 14, Section 2)*

5. **Summarize** the system of classification of organisms. *(Chapter 14, Section 1)*

Did you have difficulty? *For help, review the sections indicated.*

📖 *Reading Activity*

When you look at a living thing, how do know if it is an animal or some other type of organism? Animals have a particular set of characteristics that distinguish them from other organisms. Develop a list of the characteristics you look for when determining whether a living thing is an animal.

🖥 **internet** connect

www.scilinks.org
National Science Teachers Association *sci*LINKS Internet resources are located throughout this chapter.

sciLINKS Maintained by the
National Science Teachers Association

● Members of the plant kingdom, such as the trees and herbs shown, have many similar features, including body organization and mode of nutrition. Scientists group organisms according to similarities.

Introduction to Kingdoms and Domains

- **Identify** the characteristics used to classify kingdoms.

- **Differentiate** bacteria from archaebacteria.

Key Terms

None

Figure 1   Six kingdoms. Living organisms are divided into six kingdoms and are grouped according to their cell type, complexity, and method for obtaining nutrition.

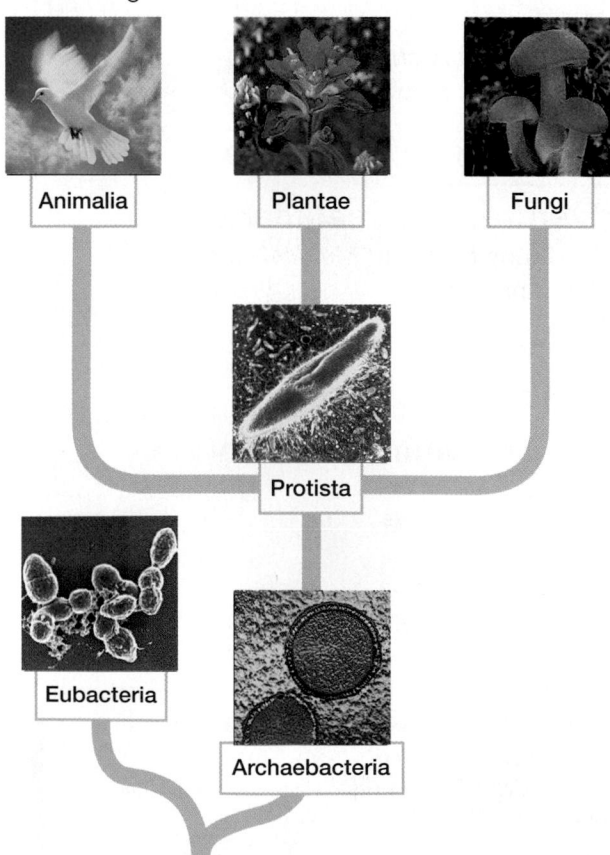

Animalia

Plantae

Fungi

Protista

Eubacteria

Archaebacteria

The Six Kingdoms of Life

Biologists have long organized living things into large groups called kingdoms. For example, a hummingbird, an earthworm, an elephant, and a butterfly are all members of the animal kingdom. But what *is* an animal? Given the diversity of living things, it is easy to forget that all living things have much in common. Focusing on a few fundamental characteristics makes it easier to see the relatedness within kingdoms.

Cell Type Organisms are either prokaryotes, which have prokaryotic cells, or eukaryotes, which have eukaryotic cells. Scientists generally recognize two kingdoms of prokaryotes and four kingdoms of eukaryotes.

Cell Walls In four of the six kingdoms organisms have cell walls. In one of the six kingdoms, organisms lack cell walls. In the remaining kingdom, some organisms have cell walls and some do not.

Body Type Organisms are either unicellular or multicellular. Two kingdoms consist only of unicellular organisms. Two other kingdoms have both unicellular and multicellular organisms. The two remaining kingdoms consist entirely of multicellular organisms, many of which have tissues and organs.

Nutrition Many organisms are autotrophs. Autotrophs make nutrients from inorganic materials. Many other organisms are heterotrophs. Heterotrophs get nutrients by consuming other organisms. Three kingdoms consist of both autotrophic and heterotrophic organisms. One kingdom consists of autotrophs. The other two kingdoms have only heterotrophic organisms.

Today, biologists group organisms into six kingdoms, based on their similarities: Eubacteria, Archaebacteria, Protista, Fungi, Plantae, and Animalia. Eubacteria and the Archaebacteria were once grouped in the kingdom Monera, which contained all the prokaryotes. However, data from RNA and DNA sequencing led biologists to divide the monerans into two distinct kingdoms. **Figure 1** shows how biologists think the six kingdoms are related to one another.

The Three Domains of Life

For many decades, scientists recognized two basic forms of life, prokaryotes and eukaryotes. Then in 1977, the American scientist Carl Woese and his colleagues proposed that some prokaryotes are so fundamentally different from others that they merit their own broad division.

Woese and his team based their proposal on comparisons of ribosomal RNA sequences. The scientists showed that the group of prokaryotes that make up the kingdom Archaebacteria are more closely related to eukaryotes than they are to the other kingdom of prokaryotes, Eubacteria.

In 1996, scientists made the first comparison between complete DNA sequences of an archaebacterium and a bacterium. In recognition of the vast differences between the two groups of prokaryotes, biologists have adopted a classification system that divides all organisms into three superkingdoms, or domains: Bacteria, Archaea, and Eukarya. **Figure 2** shows how the three domains are related.

The domain thought to be the oldest is Bacteria, which is composed of the organisms in the kingdom Eubacteria. Archaea is the second prokaryotic domain and is also composed of a single kingdom, Archaebacteria. A third domain, Eukarya, contains all four of the eukaryotic kingdoms: Animalia (animals), Fungi (fungi), Plantae (plants), and Protista (protists). As you can see in Figure 2, the rRNA sequences of protists indicate that they are a very diverse group.

internet connect

www.scilinks.org
Topic: **Domains of Life**
Keyword: **HX4060**

SC*i*NKS. Maintained by the National Science Teachers Association

---

Figure 2   Three domains

This phylogenetic tree, based on rRNA sequences, demonstrates the division of all living things into three broad domains.

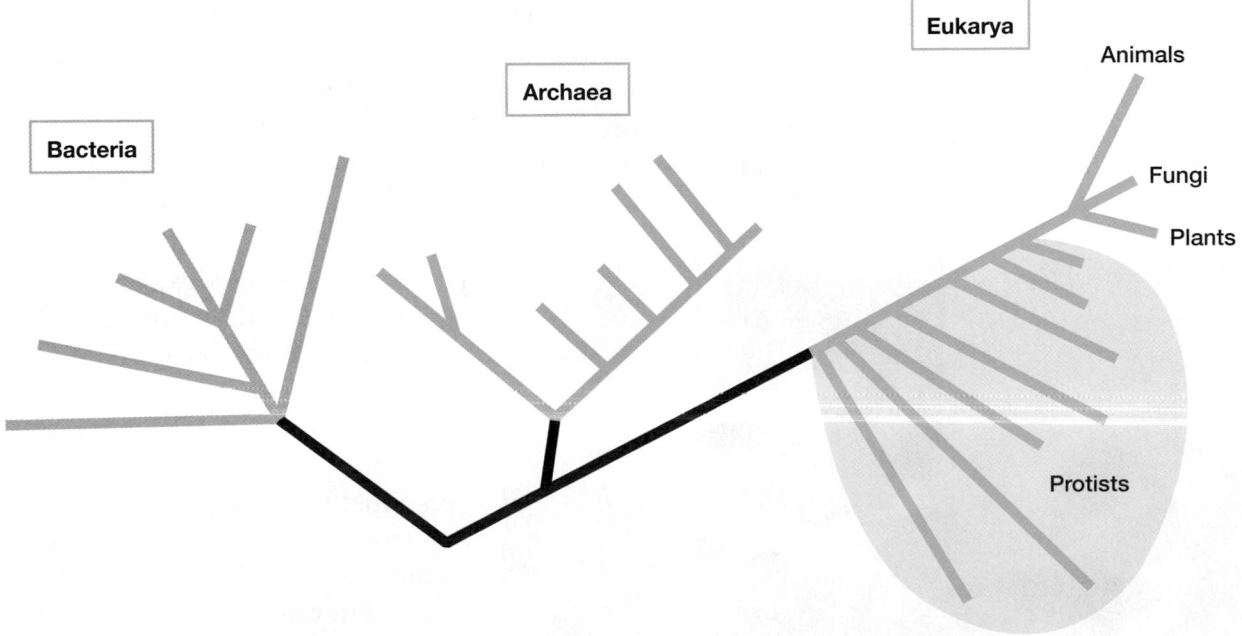

The Domain Bacteria

The domain Bacteria contains a single kingdom, the kingdom Eubacteria. Some scientists call this kingdom Bacteria. The common name for its members is *bacteria*. Bacteria are prokaryotes that have the same kind of lipid in their cell membranes as do eukaryotes. Most bacterial cells vary in size from 0.1 to 15 µm. Bacteria are found in practically every environment on Earth, and they affect humans every day. Bacteria are similar in physical structure, with no internal compartments, and they obtain nutrients in one of many different ways. There are several key characteristics common to all bacteria.

Characteristics of Bacteria

Cell Wall Bacteria have strong exterior cell walls made of peptidoglycan *(pep tih doh GLY kan)*, a weblike molecule complex made of carbohydrate strands cross-linked by short peptide bridges.

Gene Translation Apparatus Scientists infer evolutionary relationships between organisms based in part on the amino acid sequences in their proteins. The amino acid sequences of the ribosome proteins and RNA polymerases found in bacteria differ from those found in eukaryotes or in archaebacteria. This information, together with data from ribosomal RNA, is used to infer evolutionary relationships of groups within the kingdom Eubacteria.

Kinds of Bacteria

Bacteria are the most abundant organisms on Earth. There are more bacteria living in your mouth than there are mammals living on Earth. Some bacteria cause disease. Other bacteria are used by humans to process foods, such as the bacterium shown in **Figure 3.** Bacteria are used to control agricultural pests, to produce various chemicals, and to perform genetic engineering. Traditionally, bacteria have been classified according to their shape, the nature of their cell wall, and their type of metabolism. Traditional groupings of bacteria may change, however, as we get more information about their DNA and RNA.

Some bacteria obtain energy from inorganic compounds such as hydrogen sulfide, ammonia, and methane. Some bacteria are photosynthetic and are found in ocean and freshwater ecosystems, where they are primary producers. Other bacteria are heterotrophs. Some heterotrophic bacteria are capable of living in the absence of oxygen, while others must have oxygen to live. Heterotrophic bacteria are important decomposers in many ecosystems. They are responsible for the recycling of carbon, nitrogen, and phosphorus.

Study TIP
● **Organizing Information**
Draw two overlapping circles. Label one circle "Bacteria" and the other circle "Archaea." In the circle labeled "Bacteria," write down the characteristics of bacteria. Do the same for the circle labeled "Archaea." Finally, in the area where the circles overlap, write down the characteristics that the two kingdoms share.

Figure 3   Useful bacterium.
Bacteria, such as *Lactobacillus bulgaricus,* which turns milk into yogurt, can be useful to humans.

Magnification: 3,150×

The Domain Archaea

The domain Archaea also contains a single kingdom, the kingdom Archaebacteria. Archaebacteria are prokaryotes that seem to have diverged very early from the bacteria. They are more closely related to eukaryotes than to bacteria. Although they are a diverse group, all archaebacteria have certain features in common.

Characteristics of Archaebacteria

Cell Wall and Membrane The cell walls of archaebacteria do not contain peptidoglycan, as the cell walls of bacteria do. Archaebacteria contain lipids very different from those of bacteria or eukaryotes.

Gene Structure and Translation As with the genes of eukaryotes, the genes of archaebacteria are interrupted by introns. The ribosomal proteins of archaebacteria are very similar to those of eukaryotes and different from those of bacteria.

Kinds of Archaebacteria

The archaebacteria first identified by scientists live in extreme environments, such as hot springs and salty lakes. This led scientists who studied archaebacteria to think that all of these unusual organisms live in very harsh environments. In recent years, however, scientists have learned that archaebacteria are much more common than was first thought. Several "signature sequences" of DNA are common to all archaebacteria. Using these sequences as probes, scientists were surprised to find archaebacteria in ordinary soil and even in seawater. There are three basic kinds of archaebacteria.

Methanogens These archaebacteria obtain energy by combining hydrogen gas, H_2, and carbon dioxide, CO_2, to form methane gas, CH_4. Methanogens live deep in the mud of swamps and are poisoned by even traces of oxygen.

Extremophiles A group of extremophiles called *thermophiles* lives in very hot places—up to 106°C. **Figure 4** shows a thermophilic species. *Halophiles* inhabit very salty lakes that can be three times as salty as seawater. Still other extremophiles live in very acidic places with a pH below 1 or under enormous pressure—up to 800 atmospheres.

Nonextreme Archaebacteria Nonextreme archaebacteria grow in all the same environments that bacteria do.

Figure 4 Thermophile. This species of *Acidianus brierleyi* is a thermophilic bacterium and is found near volcanic vents.

The Domain Eukarya

The third domain of life, Eukarya, is made up of four kingdoms: Protista, Fungi, Plantae, and Animalia. Members of the domain Eukarya are eukaryotes, organisms composed of eukaryotic cells. A complex internal structure enabled eukaryotic cells to become larger and, eventually, led to the evolution of multicellular life. While different in many fundamental respects, eukaryotes share several key features.

Characteristics of Eukarya

Highly Organized Cell Interior All eukaryotes have cells with a nucleus and other internal compartments. This allows specialization of functions within a single cell.

Multicellularity True multicellularity, in which the activities of individual cells are coordinated and the cells themselves are in contact, occurs only in eukaryotes.

Sexual Reproduction Although exchange of genes occurs in bacteria, genetic exchange in eukaryotes is a more regular process. Eukaryotes have a life cycle that involves sexual reproduction. In this type of reproduction, meiotic cell division forms haploid gametes, and two gametes unite to form a diploid cell in fertilization. Genetic recombination during meiosis and fertilization causes the offspring of eukaryotes to vary widely, thus providing raw material for evolution.

Kinds of Eukaryotes

A wide variety of eukaryotes are unicellular. Most unicellular eukaryotes are grouped in the kingdom Protista. Protista contains both unicellular and multicellular organisms, many of which are aquatic. The protists are grouped together primarily because they do not fit in any other kingdom of eukaryotes.

Fungi are a group of heterotrophs that are mostly multicellular. Fungi are composed of cells with cell walls of chitin. One group of fungi, the yeasts, is unicellular. Many fungi live on and decompose dead organisms, and many other fungi are parasitic.

Figure 5 Multicellular eukaryotes. This jaguar and the vegetation it lives among are examples of complex multicellular eukaryotes.

Plants and animals, such as those in **Figure 5,** are all multicellular organisms. Almost all plants are autotrophs and have cells with cell walls composed of cellulose. All animals are heterotrophs composed of cells that do not have cell walls. Most plants and animals have tissues and organs.

Table 1 below summarizes the major characteristics of the organisms in the six kingdoms and three domains. You will learn more about the structure and diversity of living things as you continue to study biology.

Table 1 Kingdom and Domain Characteristics

Domain	Kingdom	Characteristics				
		Cell type	Cell structure	Body type	Nutrition	Example
Bacteria	Eubacteria	Prokaryotic	Cell wall, peptidoglycan	Unicellular	Autotrophic and heterotrophic	Enterobacteria Spirochetes
Archaea	Archae-bacteria	Prokaryotic	Cell wall, no peptidoglycan	Unicellular	Autotrophic and heterotrophic	Methanogens
Eukarya	Protista	Eukaryotic	Mixed	Unicellular and multicellular	Autotrophic and heterotrophic	Amoebas Euglenas Kelps
Eukarya	Fungi	Eukaryotic	Cell wall, chitin	Unicellular and multicellular	Heterotrophic	Yeasts Mushrooms
Eukarya	Plantae	Eukaryotic	Cell wall, cellulose	Multicellular	Autotrophic	Ferns Pine trees
Eukarya	Animalia	Eukaryotic	No cell wall	Multicellular	Heterotrophic	Birds Earthworms

Section 1 Review

1 **Analyze** the relationship between kingdoms and domains.

2 **Identify** the characteristics that distinguish the six kingdoms.

3 **Describe** how kingdom Eubacteria differs from kingdom Archaebacteria.

4 **Critical Thinking Evaluating Conclusions** Justify the division of prokaryotes into two kingdoms.

5 **Standardized Test Prep** Organism *X* is a multicellular, heterotrophic eukaryote whose cells lack cell walls. To which kingdom does organism *X* belong?

A Animalia   **C** Fungi

B Plantae   **D** Archaebacteria

Advent of Multicellularity

Objectives

- **Contrast** the terms *colony* and *aggregate*.
- **List** the characteristics of protists.
- **List** the characteristics of fungi.

Key Terms

colonial organism
aggregation
multicellular
differentiation
tissue
organ
organ system
protist
hypha

The Many Forms of Multicellularity

More than half of the biomass on Earth is composed of unicellular organisms—prokaryotes and some eukaryotes. For these organisms, unicellularity has been tremendously successful. However, many other organisms have found success not as individual cells but as members of a coordinated group of cells. Groups of cells that live together can have different levels of cooperation, as shown in **Figure 6.**

Colonies

Occasionally, the cell walls of bacteria adhere to one another. Some bacteria, such as cyanobacteria, form filaments, sheets, or three-dimensional formations of cells. These formations are not considered multicellular if the cells do not communicate and coordinate their activities. Such bacteria may properly be considered colonial. A **colonial organism** is a group of cells that are permanently associated but that do not communicate with one another. A colonial in the kingdom Protista is shown in Figure 6.

Aggregations

An **aggregation** *(a gruh GAY shuhn)* is a temporary collection of cells that come together for a period of time and then separate. For example, a plasmodial slime mold, such as the one shown in Figure 6, is a unicellular organism (a member of the kingdom Protista) that spends most of its life moving about and feeding as single-celled amoebas. When starved, however, these cells aggregate into a large group. This weblike mass produces spores, which are then dispersed to distant locations where there may be more food.

Figure 6 Multicellularity. *Volvox* is a colonial organism. Each is a hollow ball of hundreds or thousands of flagellated cells embedded in a jellylike layer. A plasmodial slime mold is an aggregate organism. Its cells form a large mass temporarily.

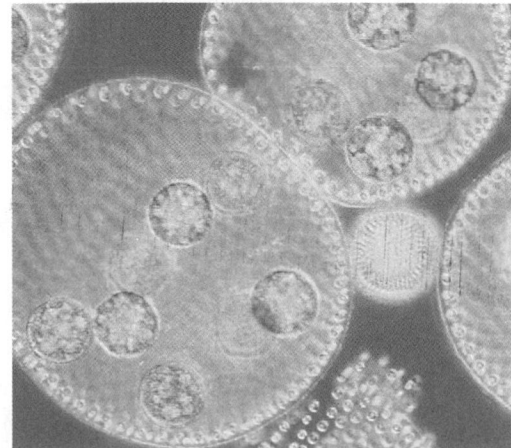

Volvox sp.

Plasmodial slime mold

True Multicellularity

A **multicellular** organism is an organism composed of many cells that are permanently associated with one another, such as the green alga shown in **Figure 7.** Multicellularity occurs only in eukaryotes. While single cells cannot grow larger than a certain size, multicellular organisms can be large. True multicellularity occurs when individual cells are in contact with each other and when their activities are coordinated.

Multicellularity enables cells to specialize in different functions. With this division of labor, a multicellular organism can have cells that protect it. Other cells help the organism move about, and still others play roles in reproduction and feeding. Cell specialization begins as a new organism develops. For example, as a chicken develops from an egg, new cells form by cell division. These cells grow and undergo **differentiation,** the process by which cells develop a specialized form and function.

Figure 7   Cell specialization.
This green alga has specialized cells that hold it to the bottom of ocean tidal pools. It has many other specialized cells.

Complex Multicellularity

Plants and animals have complex multicellularity. The specialized cells of most plants and animals are organized into structures called tissues and organs. A **tissue** is a distinct group of cells with similar structure and function. Muscle, for example, is a tissue composed of many muscle cells that work together. Different tissues may be organized into an **organ,** which is a specialized structure with a specific function. An example of an organ is the heart, which is composed of muscle, nerve, and other tissues that work together as a pump. Various organs that carry out a major body function make up an **organ system.** The circulatory system, which is composed of the heart, the blood vessels, and the blood within them, is an example of an organ system. The relationships between tissues, organs, and organ systems are shown in **Figure 8.**

Figure 8   Complex multicellularity

Specialized cells form tissue that makes up an organ called the lung. The lungs and other organs constitute an organ system.

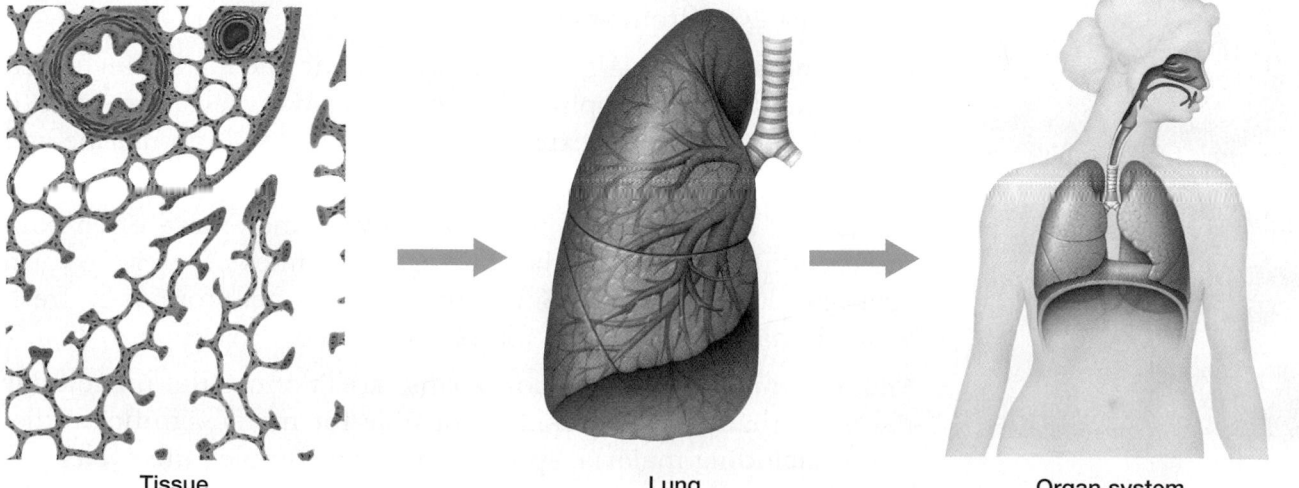

Tissue   Lung   Organ system

Kingdom Protista

Of the six kingdoms of organisms, the kingdom Protista is the most diverse. Members of the kingdom Protista, **protists,** are defined on the basis of a single characteristic: they are eukaryotes that are not fungi, plants, or animals. Many are unicellular; in fact, all single-celled eukaryotes (except yeasts) are protists. Some protists, such as some kinds of algae, have cell specialization. Protists can vary widely in size, as shown in **Figure 9.** Most are microscopic, but some are as large as trees.

Protists are very diverse in many other ways. While all protists have a cell membrane, some—including algae and slime molds—have strong cell walls. Others—including diatoms and forams—produce shells of glassy silica. Some protists are photosynthetic autotrophs. Other protists are heterotrophs. Many protists move about, using structures such as flagella, cilia, or pseudopods. Protists normally reproduce asexually by means of mitotic cell division. When they are under stress, many protists reproduce sexually. The most ecologically important protists are probably the algae that live in the ocean and form the base of many ocean food chains.

Kinds of Protists

Biologists recognize several general groups of protists. The six groups listed below are based on physical or nutritional characteristics.

Protists That Use Pseudopodia Amoebas are protists that have flexible surfaces with no cell walls or flagella; they move by using extensions of cytoplasm called pseudopodia *(soo doh POH dee uh)*. Forams, by contrast, have porous shells through which long, thin projections of cytoplasm can be extended.

Protists That Use Flagella Many protists, including autotrophs and heterotrophs, move by using flagella. The ciliates, which have large numbers of cilia, are so different from other protists that some biologists place them in a separate kingdom.

Protists with Double Shells Diatoms are photosynthetic protists with unique double shells made of silica, like boxes with lids. Diatoms are part of the plankton and may be found in fresh water or in marine environments.

Photosynthetic Algae Algae are photosynthetic and are distinguished by the kinds of chlorophyll they contain. Many algae are multicellular and reproduce sexually. Algae may be found in marine and freshwater environments.

Funguslike Protists Slime molds and water molds are often confused with fungi because they aggregate in times of stress to form spore-producing bodies. Slime molds are often found in fresh water, in damp soil, and on forest floors.

Spore-Forming Protists Sporozoans are nonmotile unicellular parasites that form spores. Responsible for many significant diseases, including malaria, sporozoans have complex life cycles.

Figure 9 Protists range in size. Protists can be so small that a microscope is required to see them. Some multicellular algae, such as the giant kelp can be as tall as 100 m.

Magnification: 1,500×

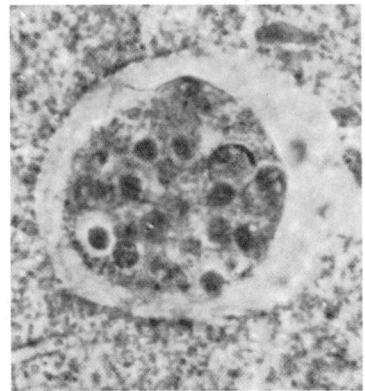

Entamoeba histolytica

Giant kelp

Kingdom Fungi

Fungi are a very unusual and successful group of organisms. Although most fungi are multicellular, one group, the yeasts, are unicellular. The cell walls of all fungal cells contain chitin, the same tough material found in a crab's shell. The bodies of fungi consist of long strands of cells that are connected end to end and that share cytoplasm. The slender strands of fungi, like those in **Figure 10,** are called **hyphae** *(HIE fee)* (singular, *hypha*). Often, hyphae are packed together to form complex reproductive structures, such as mushrooms. Fungi reproduce by a variety of asexual and sexual methods.

Figure 10   Fungus. The hyphae of fungi look like threads or filaments when they are viewed up close.

In the past, fungi were classified in the plant kingdom. Like plants, fungi do not move from place to place. In addition, the general appearance of many fungi is similar to that of plants. But, fungi lack the green pigment chlorophyll and the ability to conduct photosynthesis. Like animals, fungi are heterotrophs. Unlike most animals, however, fungi do not ingest their food. Instead, fungi obtain food by secreting digestive enzymes onto whatever they grow on. Many fungi are saprophytes that live on dead organisms. Many other fungi are parasites that live on living organisms and cause diseases that affect plants and animals.

Kinds of Fungi

There are three phyla of fungi, and they are distinguished by their type of reproductive structures.

Zygomycetes Zygomycetes *(zie goh MIE seets)* form structures for sexual reproduction called zygosporangia. Zygomycetes include species such as *Rhizopus stolonifer,* common bread mold.

Basidiomycetes Basidiomycetes *(buh sih dee oh MIE seets)* include fungi that make mushrooms. Mushrooms are the sexual reproductive structures produced by basidiomycetes. Basidiomycetes almost always reproduce sexually.

Ascomycetes Ascomycetes *(as koh MIE seets)* form sexual spores in special saclike structures called asci. The sexual reproductive structure formed by ascomycetes often resembles a cuplike structure called an ascocarp.

Section 2 Review

1. **Differentiate** a colony from an aggregation.

2. **Describe** the characteristics of the protists.

3. **Describe** the characteristics of the fungi.

4. **Evaluate** the argument that insects and fungi are closely related because both have chitin covering their bodies.

5. **Standardized Test Prep** The process by which cells become specialized in form and function during development is called
   A association.
   B aggregation.
   C differentiation.
   D coordination.

Complex Multicellularity

- **List** the levels of cellular organization that occur in plants and animals.
- **Name** the characteristics of plants.
- **Identify** the characteristics of animals.
- **Differentiate** plants from animals.

Key Terms

vascular tissue
invertebrate
vertebrate

Kingdom Plantae

Plants are complex multicellular autotrophs; they have specialized cells and tissues. Most plants have several different types of cells that are organized into many specialized tissues. For example, **vascular** *(VAS kyoo lur)* **tissue** is made up of specialized cells that play a role in transporting water and dissolved nutrients. Plant cells are different from all other cells in that their cell walls are composed of cellulose, a complex carbohydrate.

Unlike many other organisms, plants cannot move from one place to another. A few groups have motile sperm, but most plants are rooted in the ground. Portable reproductive structures, such as spores and seeds, enable the dispersal of plants.

As autotrophs, plants are the primary producers in most terrestrial food webs. Thus, they provide the nutritional foundation for most terrestrial ecosystems. Plants also release oxygen gas to the atmosphere. They are also very important in the cycling of phosphorus, water, nitrogen, and carbon.

Like the fungi, plants evolved on land and are the dominant organisms on the surface of Earth. Plants cover every part of the terrestrial landscape, except for the extreme polar regions and the highest mountaintops.

Plants are sources of food for humans and other animals. They are also a source of medicines, dyes, cloth, paper, and many other products. As shown in **Figure 11,** plants vary in size from the 1 mm tall duckweed *(Wolffia microscopica)* to the giant sequoia redwood *(Sequoia sempervirens),* which can grow to over 90 m (296 ft) tall.

Figure 11 Range of plant size. Plants can be as small as duckweed or as large as a redwood.

Duckweed

Redwood tree

Mosses

Ferns

Figure 12   Seedless plants. Mosses and ferns are two groups of plants that do not produce seeds.

Kinds of Plants

There are four basic kinds of plants, as shown in **Figure 12** and **Figure 13.** They differ from one another according to the type of vascular tissue and reproductive structures that they have.

Figure 13   Seed plants. Plants that make seeds are either nonflowering, such as the pine trees, or flowering, such as these bluebonnets.

Nonvascular Plants Plants without a well-developed system of vascular tissues are called nonvascular plants. These plants are all relatively small. They lack the tissue to transport water and dissolved nutrients. They also lack true roots, stems, and leaves. Mosses, such as the one shown in Figure 12, are the most familiar example of nonvascular plants.

Plants with a well-developed system of vascular tissues are called vascular plants. Their larger, more-complex bodies are organized into roots, stems, and leaves. Most plants are vascular plants. One group of vascular plants, called seedless vascular plants, does not produce seeds.

Seedless Vascular Plants Ferns are the most common and familiar seedless vascular plants. They have roots, stems, and leaves, and their surfaces are coated with a waxy covering that reduces water loss. They reproduce with spores that are resistant to drying. Both haploid and diploid phases occupy significant parts of the life cycle.

Most vascular plants produce seeds. Vascular plants that produce seeds are called seed plants. There are two general types.

Pine trees

Nonflowering Seed Plants Gymnosperms *(JIHM noh spuhrmz)* are vascular plants that reproduce using seeds but do not produce flowers. Gymnosperms include plants that produce seeds in cones, such as pines and spruces. Seeds enable plants to scatter offspring and to survive long periods of harsh environmental conditions, such as drought and extreme temperatures.

Flowering Seed Plants Most plants that produce seeds also produce flowers. Flowering plants are called *angiosperms (AN jee oh spuhrmz)*. Angiosperms, such as roses, grasses, and oaks, produce seeds in fruits. Fruits are structures that enable the dispersal of seeds.

Bluebonnets

Figure 14 A running animal. This impala is demonstrating the ability to avoid predators. Running is the result of several organ systems working well together.

Kingdom Animalia

Animals are complex multicellular heterotrophs. Their cells are mostly diploid, lack a cell wall, and are organized as tissues. In addition, their zygotes develop through several stages. These adaptations have enabled animals to be successful in different habitats. The specialized tissue called *muscle* enables animals to move about readily. As illustrated in **Figure 14,** the ability of animals to move more rapidly and in more complex ways than members of other kingdoms is one of their most interesting characteristics. A remarkable form of movement unique to animals is flight, an ability that is well developed among both insects and vertebrates. Movement enables animals to avoid predators and to look for food and mates.

Most animals reproduce sexually. In animals, cells formed in meiosis function directly as gametes. The haploid cells do not divide by mitosis first, as they do in plants and fungi, but rather fuse directly with one another to form the zygote. The zygote then gradually develops into an adult, going through several developmental stages.

Almost all animals (99 percent) are **invertebrates;** that is, they lack a backbone. Of the more than 1 million living species, only about 42,500 have a backbone; they are referred to as **vertebrates.** The animal kingdom includes about 35 phyla, most of which live in the sea. Far fewer phyla live in fresh water, and fewer still live on land.

Kinds of Animals

Figure 15 Variety of animals. Organisms as small as the mite share basic characteristics with organisms as large as the whale.

Animals are very diverse in form, as shown in **Figure 15.** They can range in size from 0.5 mm (0.02 in.) microscopic mites *(Demodex follicularum)* that live on your skin to enormous whales, which are vertebrates, and giant squids, which are invertebrates. Blue whales can reach a length of 30 m (100 ft) and weigh up to 220 tons. The many kinds of animals can be grouped by phylum.

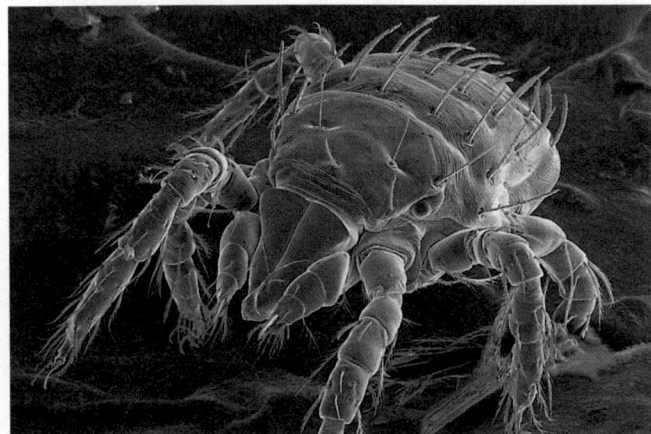

Demodex follicularum

Blue whale

Sponges Sponges, like the one shown in **Figure 16,** are the only animals that do not have tissues, but they do have specialized cells.

Cnidarians Cnidarians *(nih DAYR ee uhnz)* are mostly marine animals; they include jellyfish, sea anemones, and corals.

Flatworms Flatworms have flat, ribbonlike bodies. Some are parasitic and live inside the bodies of animals. Others are freeliving and may live in soil or water.

Roundworms Nematodes, or roundworms, are small worms that have long, very slender bodies. Some roundworms are freeliving in soil or water, while others are parasites of animals and plants.

Segmented Worms Annelids, or segmented worms, live in both water and soil and include the familiar earthworm. Bristled marine worms are segmented worms, as are leeches, which can be blood-sucking parasites.

Mollusks Mollusks have a saclike cavity called a coelom that encloses internal organs. Mollusks are very diverse aquatic and terrestrial animals. They include snails, oysters, clams, octopuses, and squids. Most mollusks have a hard external skeleton (a shell).

Arthropods By far the most diverse of all animals, arthropods have an external skeleton. They also have jointed appendages, such as antennae and jaws. These structures enable arthropods to sense their environment and obtain food. Two-thirds of all named species of animals are arthropods, most of them insects. The high rate of reproduction of insects has contributed to their success.

Echinoderms This group of invertebrates includes sea stars, sea urchins, and sand dollars. Many echinoderms *(ee KIE noh duhrmz)* are able to regenerate a lost limb. In fact, some echinoderms can lose a limb to a predator in order to escape that predator.

Invertebrate Chordates Invertebrate chordates are aquatic animals that have much in common with vertebrates, though they do not have a backbone. Some invertebrate chordate are swimmers that resemble fish, while others live attached to a rock or other object.

Vertebrates have an internal skeleton made of bone, a vertebral column (backbone) that surrounds and protects the spinal cord, and a head with a brain contained in a bony skull. Vertebrates include mammals, fish, birds, reptiles, and amphibians.

Figure 16 Sponge.
A sponge has specialized cells in its body that enable it to eat and reproduce sexually.

Ecological Roles

Animals fulfill various roles in an ecosystem. Some animals are detritivores, animals that feed on waste and dead tissue. The microscopic mites that live on your body and eat dead skin cells are detritivores. Other animals, such as buffalo that eat grass, are primary consumers. Many animals, such as humans, bears, and lions, are secondary consumers that eat primary consumers. Finally, some animals, such as intestinal worms, act as parasites and may cause disease in other animals.

Modeling True Multicellularity

In order to understand the advantage that true multicellular organisms have over colonial organisms, you will model multicellular and colonial life.

Materials

two 15 ft lengths of rope, several objects in the classroom

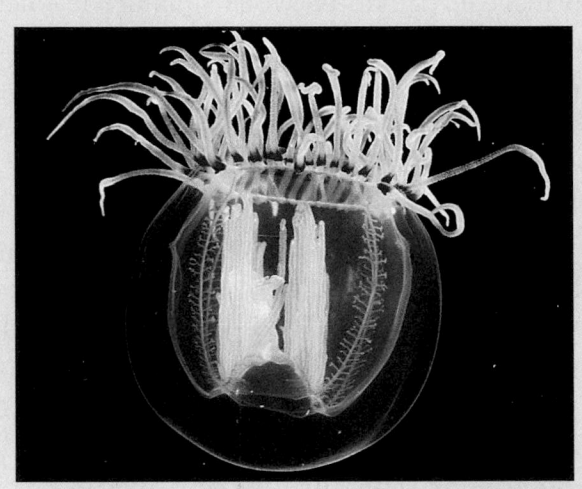

Procedure

1. Working as a class, divide into two groups. One group will model a colonial organism, and the other will model a true multicellular organism. Your teacher will loosely tie a rope around each group.

2. One student in each group will receive a set of instructions for collecting objects from around the classroom.

3. As each group carries out its instructions, students modeling the true multicellular organism may talk with one another, but students modeling the colonial organism must remain silent.

Analysis

1. **Identify** which group finished the assigned task first.

2. **Infer** why the first group to finish was able to accomplish its task so quickly.

3. **Choose** which type of organism is more advanced. Explain.

4. **Predict** how the more advanced organism could become more efficient.

Section 3 Review

1. **Identify** the type of cellular organization found only in plants and animals.

2. **Describe** the identifying characteristics of plants.

3. **Relate** the characteristics of animals to a dog or a cat.

4. **Compare** the characteristics of plants and animals.

5. **Standardized Test Prep** Tissues are composed of distinct types of

   A cells.
   B organs.
   C organ systems.
   D organisms.

Key Concepts

1 Introduction to Kingdoms and Domains

- Members of the six kingdoms are grouped according to their cell type, cell complexity, body type, and nutrition.

- Bacteria are heterotrophic and autotrophic prokaryotes that have peptidoglycan in their cell wall.

- Bacteria are classified according to their nutrition, their cell shape, and the nature of their cell wall.

- Archaebacteria are prokaryotes that have unusual lipids in their cell membrane, have no peptidoglycan in their cell wall, and have introns in their genes.

2 Advent of Multicellularity

- A colonial organism is a group of cells that live together permanently but do not coordinate most cell activity. Aggregations are collections of cells that come together for a limited period of time.

- Only eukaryotes exhibit true multicellularity, which occurs when the activities of the individual cells are coordinated and the cells are in contact with one another.

- Protists include multicellular and unicellular eukaryotes and can be heterotrophs or autotrophs.

- Fungi are eukaryotic, principally multicellular heterotrophs that exist mainly as slender hyphae.

3 Complex Multicellularity

- Specialized cells are organized into structures called tissues, organs, and organ systems. These cells have special functions and coordinate their activities with one another.

- Plants are photosynthetic eukaryotes with tissues. Their cells have cell walls.

- Plants are the primary producers in most terrestrial food webs. They release oxygen gas and aid in resource cycling.

- Animals are multicellular heterotrophs with cells that lack a cell wall, that are organized as tissues, and that are mostly diploid. They reproduce sexually, and their zygotes develop through several stages.

Key Terms

Section 1

None

Section 2

colonial organism (418)
aggregation (418)
multicellular (419)
differentiation (419)
tissue (419)
organ (419)
organ system (419)
protist (420)
hypha (421)

Section 3

vascular tissue (422)
invertebrate (424)
vertebrate (424)

Understanding Key Ideas

1. Which of the following characteristics is *not* used to classify organisms into kingdoms?
  a. cell type
  b. photosynthetic pigment
  c. body type
  d. nutrition

2. Eubacteria exhibit all of the following except
  a. nuclei.
  b. strong exterior walls made of peptidoglycan.
  c. a lack of internal compartments.
  d. size less than 15 μ.

3. Archaebacteria differ from eubacteria because archaebacteria
  a. have cell walls made of peptidoglycan.
  b. have cell membranes made of phospholipids.
  c. contain genes interrupted by introns.
  d. cannot live in harsh environments.

4. Tissues occur only in
  a. animals and plants.
  b. animals.
  c. protists and plants.
  d. animals and protists.

5. Animals differ from plants in that animals
  a. have chlorophyll.
  b. have no cell walls.
  c. have eukaryotic cells.
  d. reproduce sexually.

6. What information can you find in the photograph below that indicates that the bird is a complex multicellular organism?

7. How do cell walls differ among bacteria, fungi, and plants?

8. What evidence led Carl Woese to propose that archaebacteria should be classified separately from other prokaryotes?

9. What criteria are used to classify organisms as members of the kingdom Protista?

10. **Concept Mapping** Construct a concept map that shows the characteristics of protists and fungi. Try to include the following terms in your map: *protists, fungi, multicellularity, tissues, organs, organ systems,* and *hyphae.*

Critical Thinking

11. **Applying Information** How did the advent of multicellularity enable organisms to grow in complexity?

Alternative Assessment

12. **Finding Information** Use the media center or Internet resources to find out what kind of cell types and cell organization occurs in different kinds of organisms. What kind of animals are scientists using now to research basic questions about embryo development? What questions are scientists asking, and what information are they finding? Prepare a poster to summarize your findings.

13. **Comparing Structures** Visit the zoo, and list the scientific names of all the animals you see, or use your library to research 10 organisms. Record the scientific and common names of these organisms. For each organism, identify a trait that led taxonomists to classify the organism in its particular genus or family.

14. **Career Connection** Zoologist Research the field of zoology, and write a report on your findings. Your report should include a job description, training required, kinds of employers, growth prospects, and a starting salary.

Standardized Test Prep

Understanding Concepts

Directions (1–4): **For *each* question, write on a separate sheet of paper the letter of the correct answer.**

1 In which kind of organization does multicellularity occur?
  A. aggregates
  B. archaebacteria
  C. eukaryotes
  D. prokaryotes

2 Through what process do cells become specialized?
  F. aggregation
  G. colonialism
  H. differentiation
  I. vascularization

3 What term is used to describe an organism with cells that are permanently associated but do not communicate with one another?
  A. aggregate
  B. colonial
  C. complex
  D. heterotrophic

4 How are the cells of multicellular fungi arranged?
  F. in sheets
  G. in tissues
  H. in colonies
  I. in filaments

Directions (5): **For the following question, write a short response.**

5 Paleontologists think that many different organisms evolved in the history of Earth, but not one kind of cell nor level of complexity completely dominated Earth at any one time. What mechanism could explain the diversity of life?

Test TIP

When using an illustration that has labels to answer a question, read the labels carefully, and then check that the answer you choose matches your interpretation of the labels.

Reading Skills

Directions (6): **Read the passage below. Then answer the question.**

Developmental biologists often study the tissues and cells of simple animals in order to understand more complex animals. For example, they might study embryonic development in fruit flies and echinoderms in order to understand embryonic development of more-complex animals.

6 Why are the mechanisms of the processes in simple and complex organisms often similar?
  A. There is a correspondence between cell structures and their functions.
  B. Processes can have only one kind of mechanism that functions.
  C. Simple and complex organisms generally belong to the same kingdom.
  D. Simple organisms are usually smaller copies of more-complex organisms.

Interpreting Graphics

Directions (7): **Base your answer to question 7 on the diagram below.**

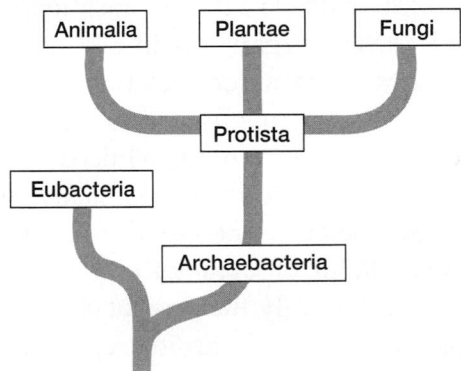

Phylogenetic Tree of Kingdoms

7 Archaebacteria and Protista are placed on the same branch in the phylogenetic tree because organisms of both kingdoms have what in common?
  F. similar ribosomal proteins
  G. similar lipids in their cell membranes
  H. similar nuclear structures
  I. similar forms of multicellular organization

Skills Practice Lab

Surveying Kingdom Diversity

SKILLS
- Using a microscope
- Comparing

OBJECTIVES
- **Observe** representatives of each of the six kingdoms.
- **Compare** and **contrast** the organisms within a kingdom.
- **Analyze** the similarities and differences among the six kingdoms.

MATERIALS
- specimens from each of the six kingdoms
- compound microscopes
- hand lenses or stereo-microscopes

Before You Begin

Many biologists classify living things into six **kingdoms.** The organisms in a kingdom have fundamental characteristics in common. For example, the organisms of two kingdoms are made of **prokaryotic cells,** while the organisms in the other four kingdoms are made of **eukaryotic cells.** Some kingdoms contain only **unicellular** or **colonial** organisms, while others contain only **multicellular** organisms. In this lab, you will examine representatives of six kingdoms of organisms. You will see that each kingdom is distinct from the others.

1. Write a definition for each boldface term in the previous paragraph and for each of the following terms: tissue, organ, organ system, autotroph, heterotroph.

2. Make a data table similar to the one below.

3. Based on the objectives for this lab, write a question you would like to explore about the kingdoms of organisms.

DATA TABLE				
Kingdom name	Type of cells	Level of organization	Other characteristics	Examples

Procedure

PART A: Conducting a Survey

1. Put on safety goggles and a lab apron.

2. Visit the station for each kingdom listed below, and examine the specimens there. Answer the questions, and record observations in your data table.

3. **Archaebacteria** Examine the prepared slides.

   a. What does a microscope reveal about the structure of archaebacteria?

   b. How do these organisms get energy for life processes?

4. **Eubacteria** Examine the prepared slides.

   a. What does a microscope reveal about the structure of eubacteria?

   b. Would you consider *Anabaena* to be unicellular or multicellular? Explain.

   c. How does *Anabaena* appear to obtain energy for life processes? Explain.

5. **Protists** Examine the prepared slides.

   a. What does a microscope reveal about the structure of protozoans?

   b. How do protozoans appear to obtain energy for life processes? Explain.

   c. Are the algae unicellular or multicellular? Explain.

   d. How do algae differ from protozoans?

6. **Fungi** Examine the specimens.

   a. Are fungi unicellular or multicellular? Explain.

   b. What does a microscope reveal about the structure of fungi?

   c. How do the fungi appear to obtain energy for life processes? Explain.

7. **Plants** Examine the specimens.

   a. What is the most striking characteristic shared by these plants?

   b. What does a microscope reveal about the structure of plants?

8. **Animals** Examine the specimens.

   a. What is the most striking characteristic shared by these animals?

   b. What is the most striking difference among these animals?

PART B: Cleanup and Disposal

9. Dispose of broken glass and solutions in the designated waste containers. Do not pour chemicals down the drain or put lab materials in the trash unless your teacher tells you to do so.

10. Wash your hands thoroughly before you leave the lab and after you finish all work.

Analyze and Conclude

1. **Summarizing Data** What are the main differences observed among the six kingdoms?

2. **Recognizing Patterns** How does the size of bacterial cells compare with the cell size in the other kingdoms?

3. **Analyzing Methods** How did you determine the cell type for each kingdom?

4. **Inferring Conclusions** Which kingdom exhibits the most diversity?

5. **Further Inquiry** Write a new question about the kingdoms of life that could be explored with another investigation.

On the Job

A **survey** is a detailed study that is conducted through observation and analysis. Do research to learn about the surveys conducted by a famous biologist, such as Charles Darwin, Carl Linnaeus, or John James Audubon. For more about careers, visit **go.hrw.com** and type in the keyword **HX4 Careers**.

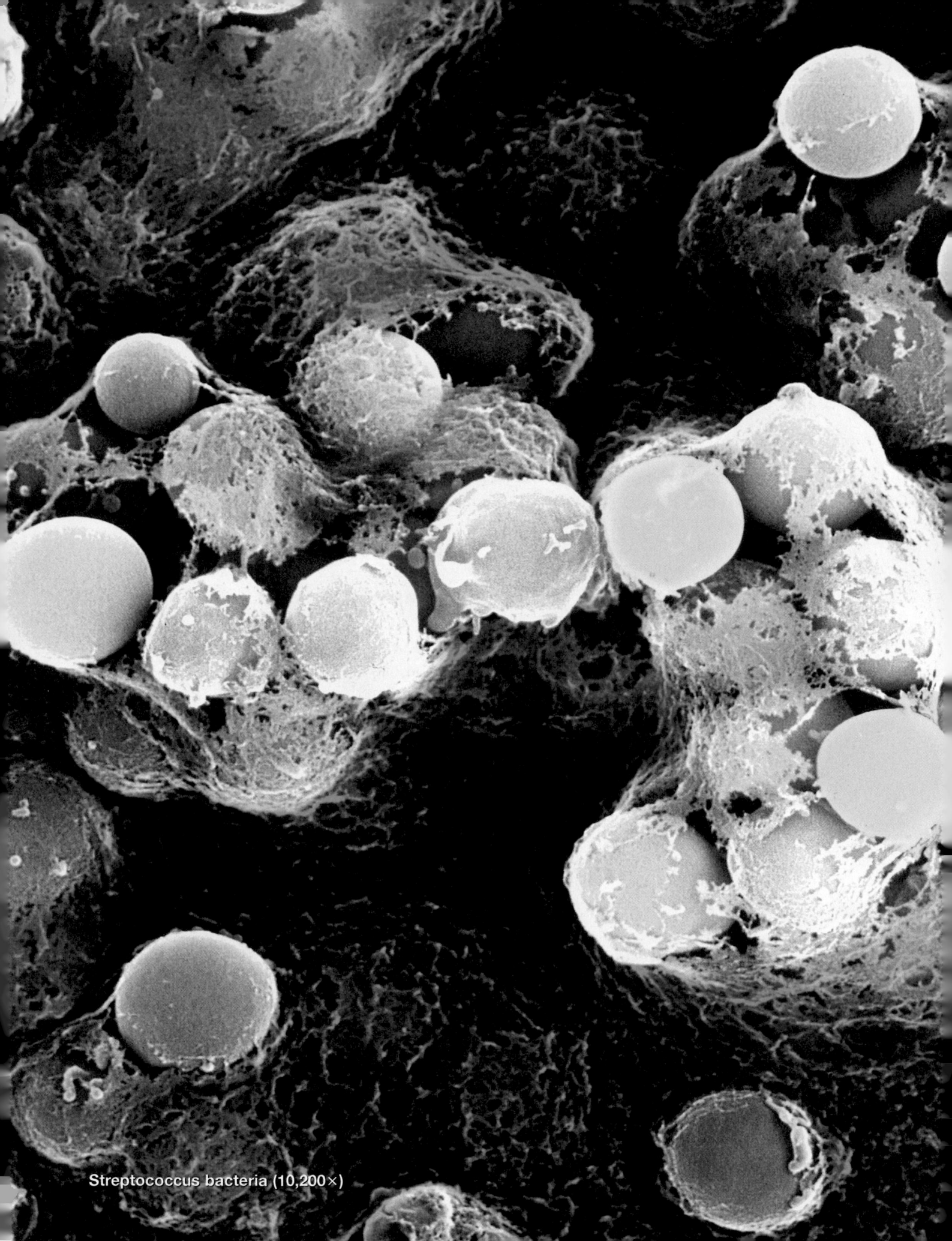

Streptococcus bacteria (10,200×)

CHAPTER

20 Viruses and Bacteria

✔ *Quick Review*

Answer the following without referring to earlier sections of your book.

1. **List** the properties of life. *(Chapter 1, Section 1)*

2. **Define** *prokaryote*. *(Chapter 3, Section 2)*

3. **Describe** a bacteriophage. *(Chapter 9, Section 1)*

4. **Differentiate** DNA from RNA. *(Chapter 10, Section 1)*

Did you have difficulty? *For help, review the sections indicated.*

📖 *Reading Activity*

Take a break after reading each section of this chapter, and closely study the figures in the section. Reread the figure captions, and, for each one, write out a question that can be answered by referring to the figure and its caption. Refer to your list of figures and questions as you review the concepts addressed in the chapter before you complete the Performance Zone chapter review.

internet connect

www.scilinks.org
National Science Teachers Association *sci*LINKS Internet resources are located throughout this chapter.

*sci*LINKS. Maintained by the National Science Teachers Association

• *Streptococcus* bacteria include a number of strains that can produce a wide range of illnesses. Some, like "strep throat," are easily treated. Others are rare and require immediate medical attention.

Is a Virus Alive?

Throughout the book, you have learned about the properties of life. All living things are made of cells, are able to grow and reproduce, and are guided by information stored in their DNA. The smallest organisms that have these properties are prokaryotes. **Viruses** are segments of nucleic acids contained in a protein coat. Viruses are not cells. Viruses are smaller than prokaryotes and range in size from about 20 nm to 250 nm (0.02–0.25 μm) in diameter. (One nanometer is equal to 0.001 μm or 0.00000004 in.) Most viruses, such as the Ebola virus shown in **Figure 1,** can be seen only with an electron microscope. Viruses are **pathogens**—agents that cause disease. Viruses replicate by infecting cells and using the cell to make more viruses. Because viruses do not have all the properties of life, biologists do not consider them to be living. Viruses do not grow, do not have homeostasis, and do not metabolize. Because they cause diseases in many organisms, viruses have a major impact on the living world.

Discovery of Viruses

Near the end of the nineteenth century, scientists were trying to find the cause of tobacco mosaic disease, which stunts the growth of tobacco plants. Scientists filtered bacteria from the sap of infected plants. They were surprised to find that the filtered sap could still cause uninfected plants to become infected. The scientists concluded that the pathogen is smaller than a bacterium. The pathogen was called a *virus*, a Latin word meaning "poison."

For many years after this discovery, viruses were thought to be tiny cells. In 1935, biologist Wendell Stanley of the Rockefeller Institute purified tobacco mosaic virus (TMV). He crystallized the purified virus. Stanley concluded that TMV is a chemical rather than an organism.

Each particle of TMV is composed of RNA and protein. Scientists were able to separate the RNA from the protein and reassemble the virus so that it could infect plants.

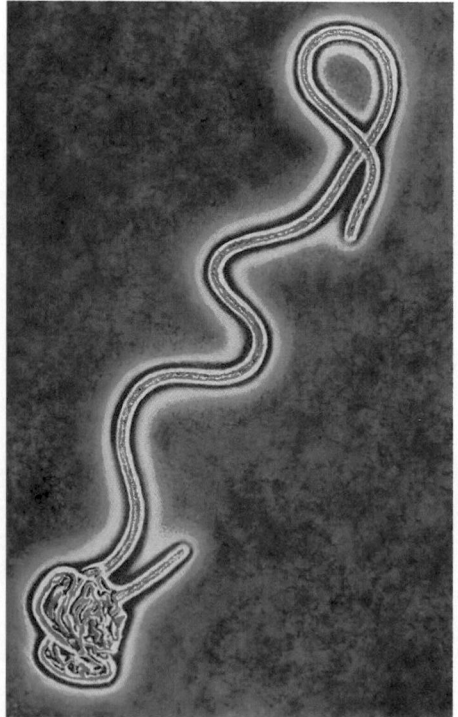

Figure 1 Ebola virus. This virus causes an often-fatal disease and has been recognized only since 1976.

Viral Structure

The virus protein coat, or **capsid,** may contain either RNA or DNA, but not both. RNA viruses include the human immunodeficiency virus (HIV), which causes AIDS, influenza viruses, and rabies virus. DNA viruses include those viruses that cause warts, chickenpox, and mononucleosis. Many viruses, such as the influenza virus shown in **Figure 2,** have a membrane, or **envelope,** surrounding the capsid. The envelope helps the virus enter cells. It consists of proteins, lipids, and **glycoproteins** *(glie koh PROH teenz),* which are proteins with attached carbohydrate molecules. Some viruses also contain specific enzymes.

Viruses exist in a variety of shapes. Some viruses, such as the Ebola virus, shown in Figure 1, are long rods that form filaments. Spherical viruses, such as the influenza virus shown in Figure 2, typically are studded with receptors. These receptors help the virus enter cells. A helical virus, like the tobacco mosaic virus shown in Figure 2, is rodlike in appearance, with capsid proteins winding around the core in a spiral. Polyhedral viruses have many sides and are roughly spherical. The capsid of most polyhedral viruses has 20 triangular faces and 12 corners. This odd shape is an efficient one for containing a viral genome. Figure 2 shows the polyhedral shape of an adenovirus, which can cause several different kinds of infections in humans.

Viruses that infect bacteria, called **bacteriophages,** have a complicated structure. A T4 bacteriophage, for example, has a polyhedron capsid attached to a helical tail. A long DNA molecule is coiled within the polyhedron.

Study *TIP*

● **Compare and Contrast**
To compare and contrast the properties of life as defined in Chapter 1 and the properties of viruses, make a two-column list. In one column, write the properties of life. In the other column, write the properties of life that viruses have.

Figure 2 Viral structures

Viruses can have characteristic shapes.

Magnification: 202,500×

Magnification: 1,250,000×

Magnification: 135,000×

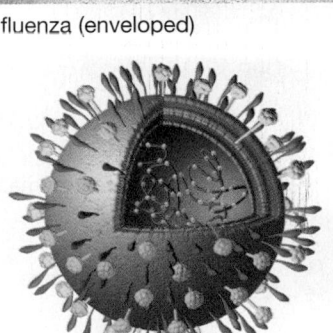

Influenza (enveloped)

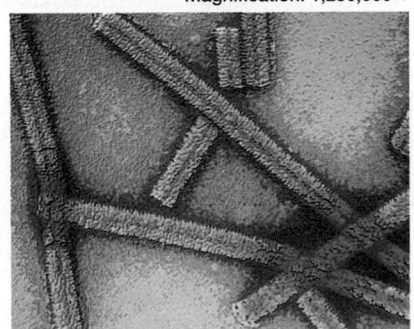
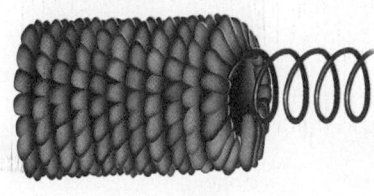

Tobacco mosaic virus (helical)

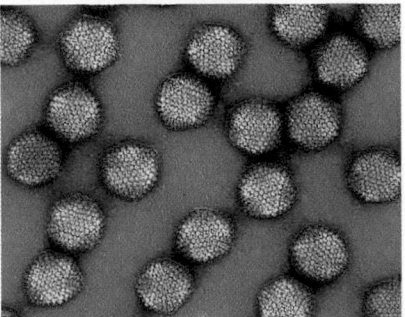

Adenovirus (polyhedral)

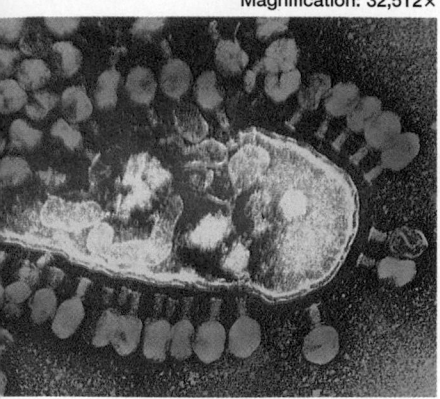

Magnification: 32,512×

Figure 3 Bacteriophage infecting a bacterium. Bacteriophages (pink) first attach to a bacterial cell (blue) and then push their DNA into it. The cell then produces more viruses.

Viral Reproduction

Viruses lack the enzymes necessary for metabolism and have no structures to make protein. Therefore, viruses must rely on living cells (host cells) for replication, as shown in **Figure 3.** Before a virus can replicate, it must first infect a living cell. A plant virus, like TMV, enters a plant cell through tiny tears in the cell wall at points of injury. An animal virus enters its host cell by endocytosis. A bacterial virus, or bacteriophage, punches a hole in the bacterial cell wall and injects its DNA into the cell.

Lytic Cycle

The reproduction of bacterial viruses has been well studied. Inside a cell, some viruses, such as T4—a virus that infects *E. coli*— will set out on one of two paths: the lytic cycle or the lysogenic cycle.

In bacterial viruses, the cycle of viral infection, replication, and cell destruction is called the **lytic** cycle. After the viral genes have entered the cell, they use the host cell to replicate viral genes and to make viral proteins, such as capsid proteins. The proteins are then assembled with the replicated viral genomes to form complete viruses. The host cell is broken open and releases newly made viruses. Though reproduction in a single bacterial virus is illustrated here, these stages are common to infections by some other viruses as well. The lytic cycle is shown in **Figure 4.**

Figure 4 Viral replication in bacteria. The bacterial virus, T4, provides a model by which viruses replicate through the lytic cycle or lysogenic cycle.

Lysogenic Cycle

During an infection, some viruses stay inside the cells but do not make new viruses. Instead of producing virus particles, the viral gene is inserted into the host chromosome and is called a **provirus.**

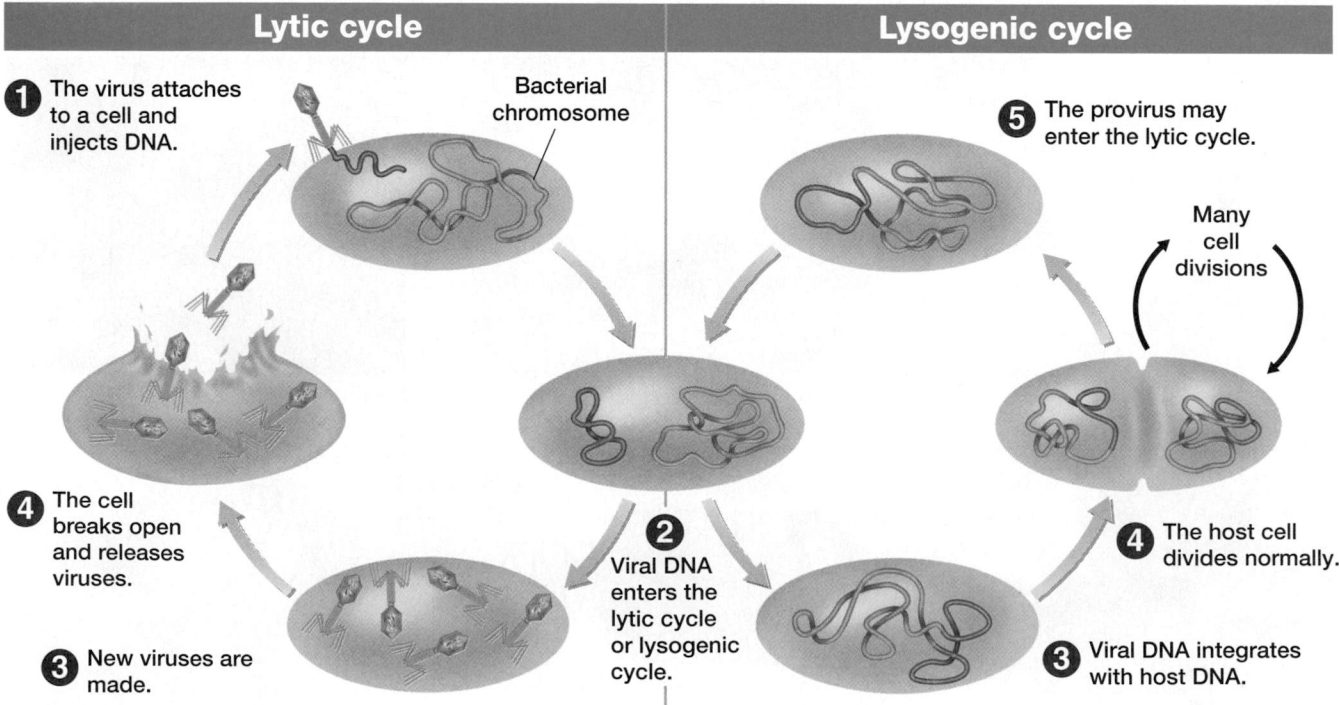

Lytic cycle

1 The virus attaches to a cell and injects DNA.

Bacterial chromosome

4 The cell breaks open and releases viruses.

3 New viruses are made.

2 Viral DNA enters the lytic cycle or lysogenic cycle.

Lysogenic cycle

5 The provirus may enter the lytic cycle.

Many cell divisions

4 The host cell divides normally.

3 Viral DNA integrates with host DNA.

Whenever the cell divides, the provirus also divides, resulting in two infected host cells. In this cycle, called the **lysogenic** *(lie soh JEHN ihk)* cycle, the viral genome replicates without destroying the host cell. This cycle is shown in Figure 4. In some lysogenic viruses, a change in the environment can cause the provirus to begin the lytic cycle. This results in the destruction of the host cell.

In animal cells, viruses can replicate slowly so that the host cell is not destroyed by the virus. For example, the virus that causes cold sores in humans hides deep in the nerves of the face. When the conditions in the body become favorable for the virus, such as when a person is under stress, the virus then begins to cause tissue damage that is seen as a cold sore or fever blister.

Host Cell Specificity

Viruses are often restricted to certain kinds of cells. For example, TMV infects tobacco and related plants, but does not infect animals. Scientists hypothesize that this specificity may be due to the viruses' origin. Viruses may have originated when fragments of host genes escaped or were expelled from cells. The hypothesis that viruses originated from a variety of host cells may explain why there are so many different kinds of viruses. Biologists think there are at least as many kinds of viruses as there are kinds of organisms.

Structure of HIV—an Enveloped Virus

Many viruses that infect only animals, such as the influenza virus shown in Figure 2, have an exterior viral envelope. **Figure 5** shows human immunodeficiency virus (HIV), the virus that causes acquired immune deficiency syndrome (AIDS). Figure 5 illustrates the envelope and other features common to several animal viruses. In many cases, the viral envelope is composed of a lipid bilayer derived from the membrane of the host cell. On the surface of the virus, glycoproteins are embedded within the envelope.

Within the envelope lies the capsid, which in turn encloses the virus's genetic material. In the case of HIV, the genetic material is composed of two molecules of single-stranded RNA. The approximately 9,000 nucleotides of HIV make up nine genes. Three of these genes are common to many different viruses.

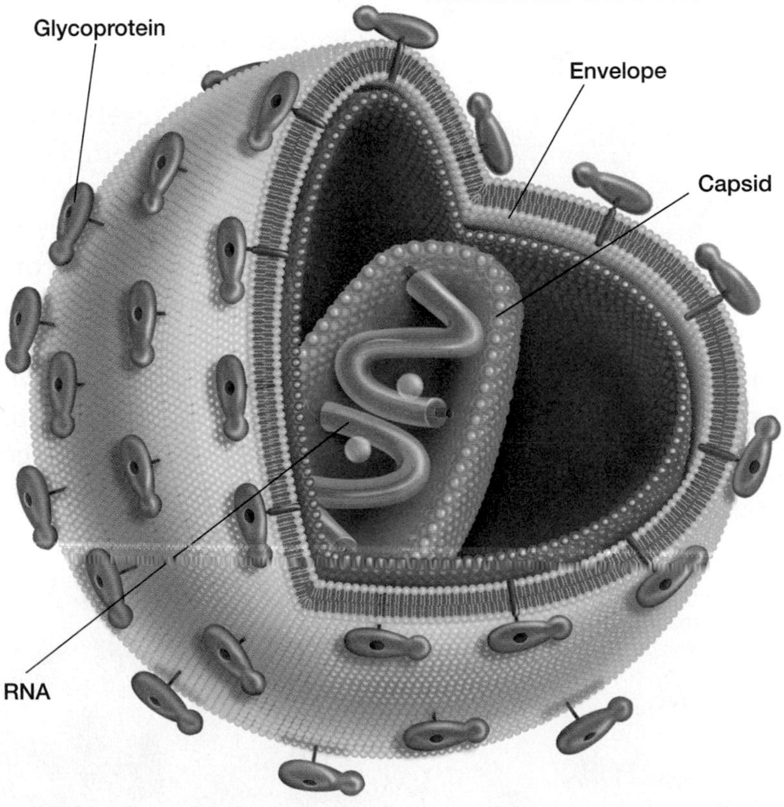

Figure 5 HIV. HIV infects human white blood cells.

Glycoprotein

Envelope

Capsid

RNA

How HIV Infects Cells

HIV, shown in **Figure 6**, provides a good example of how animal viruses enter cells. HIV entry is a two-step process. First, the virus attaches to the cell at specific sites called receptors. Second, the viral envelope fuses with the cell membrane, opens, and releases the capsid into the host cell.

Attachment

Studding the surface of each HIV are spikes composed of a glycoprotein. This particular glycoprotein precisely fits a human cell surface receptor called CD4. Thus the HIV glycoprotein can bind to any cell whose membrane has CD4 receptors. In humans, immune system cells called lymphocytes and macrophages, as well as certain cells in the brain, possess CD4 receptors.

Entry into Macrophages

HIV cannot enter a cell merely by docking onto a CD4 receptor. Rather, the HIV glycoprotein must also bind to a co-receptor called CCR5. This binding to CCR5 allows the viral capsid to enter the cell. **Figure 7** shows the process of HIV infecting a cell. Because human macrophages possess both CD4 and CCR5 receptors, HIV can enter macrophages. Lymphocytes, which are critical to immune system function, do not have CCR5 receptors. HIV therefore does not enter lymphocytes until later in the course of HIV infection.

Replication

Once inside a cell, the HIV capsid comes apart and releases its contents, which include the viral RNA. Accompanying the RNA is an enzyme called reverse transcriptase. Reverse transcriptase uses the viral RNA as a template for making a DNA version of the viral genome. This process is mistake prone, so it creates many new mutations in the viral genome. The viral DNA molecule then enters the cell nucleus and becomes a part of, or integrates into, the host's DNA. The integrated viral DNA uses the host cell's machinery to direct the production of many copies of the virus. HIV doesn't kill the macrophage by rupturing it; instead, the new viruses are released from the cell by budding. The new virus particle's envelope is thus derived from the cell membrane.

Figure 6   HIV. The spherical structure of HIV is visible in this transmission electron micrograph of individual virus particles.

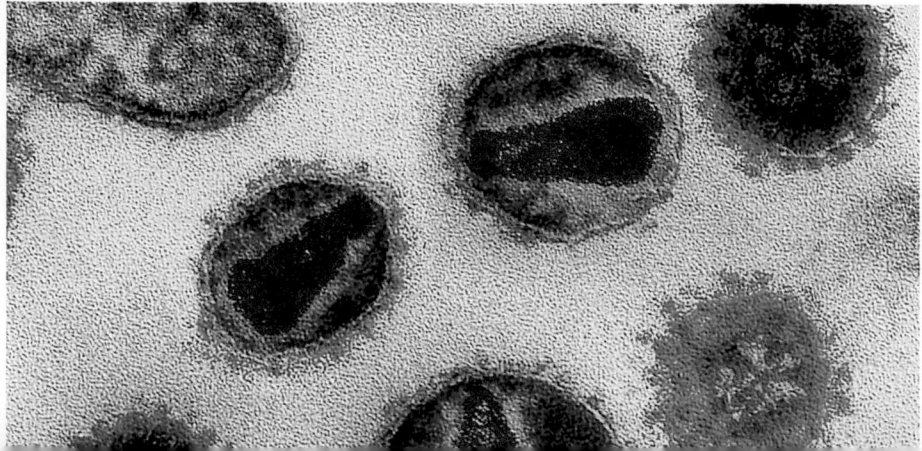

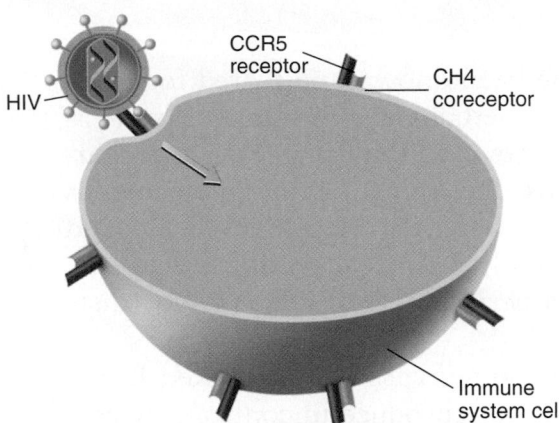

❶ The glycoprotein on HIV's surface docks at a CD4 receptor. A co-receptor, CCR5, helps HIV enter the cell.

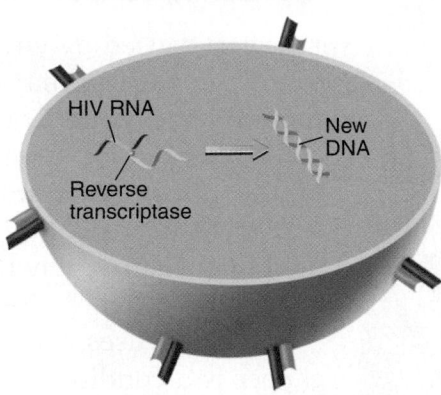

❷ The viral envelope fuses with the cell membrane. The capsid enters the cell. A DNA copy of the HIV RNA is made by viral reverse transcriptase.

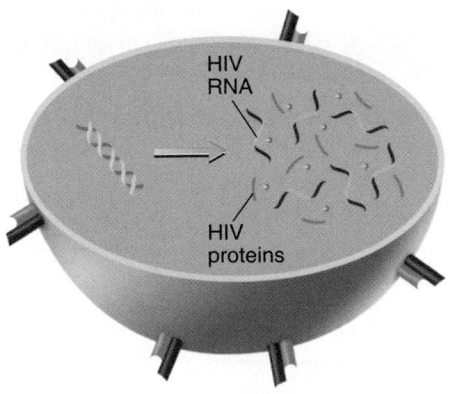

❸ The viral DNA integrates into the host genome and directs synthesis of HIV proteins and HIV RNA.

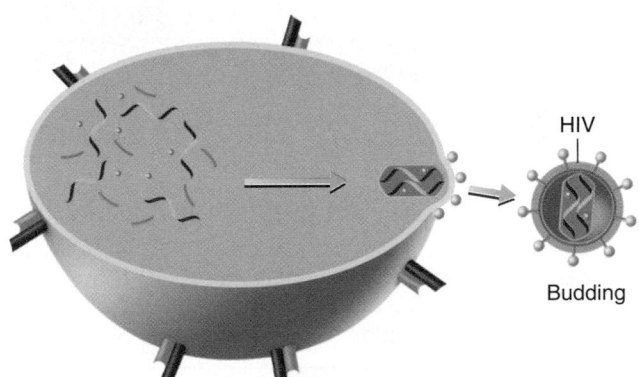

❹ New HIV particles exit macrophages by budding. HIV particles exit T cells by budding or bursting through the membrane.

AIDS

For years after the initial infection, HIV continues to replicate (and mutate). Eventually and by chance, HIV's surface glycoproteins change such that they now recognize a new co-receptor. This co-receptor is found on the subset of lymphocytes called T cells.

Unlike its activity in macrophages, HIV reproduces in T cells and then destroys them. This increases the number of virus particles in the blood, which then infect other T cells, widening the circle of cell death. It is this destruction of the body's T cells that blocks the body's immune response and signals the onset of AIDS. AIDS is a disease in which a person's immune system is unable to defend against infections that do not normally occur in healthy individuals.

Usually, HIV-infected people do not develop AIDS symptoms until years after infection. As a result, an HIV-infected individual can feel healthy and still spread the virus to others. HIV is not passed from an infected person to a healthy one through casual contact. It is transmitted in body fluids (such as semen or vaginal fluid) through sexual contact and in blood through the sharing of nonsterile needles. It is also transmitted to infants during pregnancy or through breast milk.

Figure 7   HIV infection.
HIV docks at specific receptors on cell membranes. The virus is reproduced by the infected cell.

Viral Diseases

Diseases caused by viruses have been known and feared for thousands of years. Perhaps the most lethal virus in human history is the influenza virus. Commonly known as the flu, influenza is characterized by chills, fever, and muscular aches. The virus infects cells of the upper respiratory tract. There the viruses replicate and spread to new cells. About 22 million Americans and Europeans died of flu during an 18-month period in 1918–1919. **Table 1** lists some familiar viral diseases.

Certain viruses can also cause some types of cancer. Recall that cancer is a condition in which cells reproduce uncontrollably as a result of the failure of mechanisms that control cell growth and division. Viruses associated with human cancers include hepatitis B (liver cancer), Epstein-Barr virus (Burkitt's lymphoma), and human papilloma virus (cervical cancer).

Table 1 Important Viral Diseases		
Disease	**Description of illness**	**How the disease is transmitted**
AIDS	Immune system failure	Sexual contact, contaminated blood, or contaminated needles
Common cold	Sinus congestion, muscle aches, cough, fever	Inhalation, direct contact
Ebola	High fever, uncontrollable bleeding	Body fluids
Hepatitis A	Flulike symptoms, swollen liver, yellow skin, painful joints	Contaminated blood, food, or water
Hepatitis B	Flulike symptoms, swollen liver, yellow skin, painful joints; can cause liver cancer	Sexual contact, contaminated blood, or contaminated needles
Influenza (flu)	Fever, chills, fatigue, cough, sore throat, muscle aches, weakness, headache	Inhalation
Mumps	Painful swelling in salivary glands	Inhalation
Polio	Fever, headache, stiff neck, possible paralysis	Contaminated food or water
Rabies	Mental depression, fever, restlessness, difficulty swallowing, paralysis, convulsions; fatal	Bite of infected animal
SARS (Severe acute respiratory syndrome)	High fever, headache, dry cough; can be fatal	Inhalation, direct contact
Smallpox	Blisters, lesions, fever, malaise, blindness, disfiguring scars; often fatal	Inhalation
Yellow fever	Fever, weakness, yellow skin; often fatal	Bite of infected mosquito

Emerging Viruses

Newly recognized viruses or viruses that have reappeared or spread to new areas are called emerging viruses. These new pathogens are dangerous to public health. People become infected when they have contact with the normal hosts of these viruses.

In 1999, a mosquito-borne virus called West Nile virus began to spread across the United States. West Nile virus probably was brought from overseas to America by an infected bird. While it is an emerging virus in North America, West Nile virus is common in Africa, eastern Europe, and western Asia. People who are infected with the virus from mosquito bites typically experience mild flulike symptoms. In some people, particularly the elderly, inflammation of the brain may occur, which can be fatal.

First detected in the southwestern United States, hantavirus is spread in rodent droppings and can cause a lethal illness in humans. At least 38 percent of its human victims die.

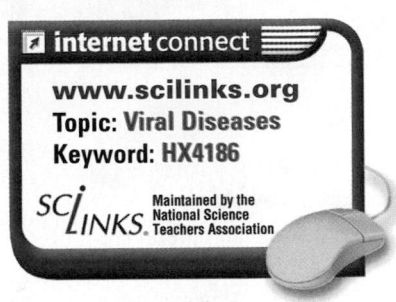

internet connect

www.scilinks.org
Topic: **Viral Diseases**
Keyword: **HX4186**

SCLINKS. Maintained by the National Science Teachers Association

Prions and Viroids

In 1982, the American scientist Stanley Prusiner, of Stanford University, described a new class of pathogens that he called **prions** *(PREE awnz)*. Prions are composed of proteins but have no nucleic acid. A disease-causing prion is folded into a shape that does not allow the prion to function. Contact with a prion will cause a normal version of the protein to misfold, too. In this way the misfolding spreads like a chain reaction.

Prions were first linked to a brain disease in sheep called scrapie. Later, brain diseases such as mad cow disease, displayed by the cow shown in **Figure 8,** and Creutzfeldt-Jakob disease were also associated with prions. Eating meat containing the disease-causing prion can cause infection.

A **viroid** *(VEER oid)* is a single strand of RNA that has no capsid. Viroids are important infectious disease agents in plants. Viroids have affected economically important plants such as cucumbers, potatoes, avocados, and oranges.

Figure 8 Infected cow.
This cow, which is unable to stand and walk, is showing signs of mad cow disease.

Section 1 Review

1. **Compare** the properties of viruses with the properties of cells.

2. **Describe** Stanley's experiment with the tobacco mosaic virus.

3. **Name** the parts of a virus.

4. **List** the steps by which viruses replicate.

5. **Describe** how HIV causes AIDS.

6. **Critical Thinking Evaluate** the argument that emerging viruses are new viruses.

7. **Standardized Test Prep** Viruses differ from cells because viruses
   A can grow.
   B do not metabolize.
   C have homeostasis.
   D lack nucleic acids.

Objectives

- **List** seven differences between bacteria and eukaryotic cells.

- **Describe** three different ways bacteria can obtain energy.

- **Describe** the external and internal structure of *Escherichia coli.*

- **Distinguish** two ways that bacteria cause disease.

- **Identify** three ways that bacteria benefit humans.

Key Terms

pilus
bacillus
coccus
spirillum
capsule
antibiotic
endospore
conjugation
anaerobic
aerobic
toxin

Bacterial Structure

The prokaryotes referred to in this chapter as bacteria include the organisms that compose the kingdom Eubacteria (Domain Bacteria) and the organisms that compose the kingdom Archaebacteria (Domain Archaea). Bacteria differ from eukaryotes in at least seven ways.

1. **Internal compartmentalization.** Bacteria are prokaryotes. Unlike eukaryotes, prokaryotes lack a cell nucleus. Many species of bacteria have no internal compartments or membrane systems.

2. **Cell size.** Most bacterial cells range in size between 1μ and 5μ, while eukaryotic cells tend to range between 10μ and 100μ. There are, however, some very large bacteria—up to 750μ—and some very small eukaryotic cells.

3. **Multicellularity.** All bacteria are single cells. Some bacteria may stick together or may form strands. These formations, however, are not truly multicellular, and the activities of individual cells are not specialized.

4. **Chromosomes.** Bacterial chromosomes consist of a single circular piece of DNA. Eukaryotic chromosomes are linear pieces of DNA that are associated with proteins.

5. **Reproduction.** Bacteria reproduce by binary fission, a process in which one cell pinches into two cells. In eukaryotes, however, microtubules pull chromosomes to opposite poles of the cell during mitosis. Afterward, the cytoplasm of the eukaryotic cell divides in half, forming two cells.

6. **Flagella.** Bacterial flagella are simple structures composed of a single fiber of protein that spins like a corkscrew to move the cell. Eukaryotic flagella are more-complex structures made of microtubules that whip back and forth rather than spin. Some bacteria also have shorter, thicker outgrowths called **pili** *(PIHL ee)* (singular, pilus), shown in **Figure 9.** Pili enable bacteria to attach to surfaces or to other cells.

7. **Metabolic diversity.** Bacteria have many metabolic abilities that eukaryotes lack. For example, bacteria perform several different kinds of anaerobic and aerobic processes, while eukaryotes are mostly aerobic organisms.

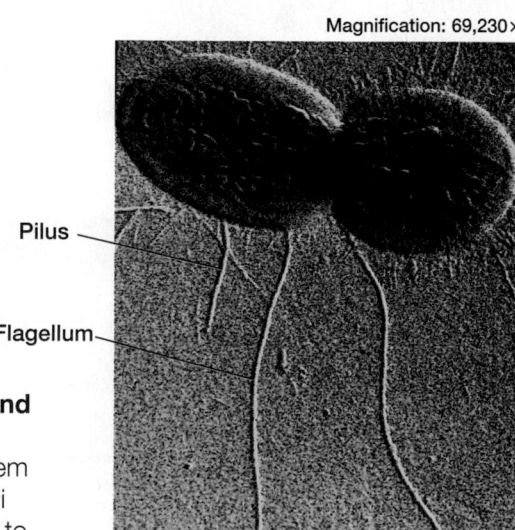

Magnification: 69,230×

Pilus

Flagellum

Figure 9 Flagella and pili. Bacteria have flagella that provide them with movement and pili that enable adherence to surfaces.

Proteus mirabilis

Bacterial Cell Shapes

A bacterial cell is usually one of three basic shapes, as shown in **Figure 10: bacillus** *(buh SIHL uhs)*, a rod-shaped cell; **coccus** *(KAHK us)*, a round-shaped cell; or **spirillum** *(spy RIHL uhm)*, a spiral cell. A few kinds of bacteria aggregate into strands. Species that form filaments are indicated by the prefix *strepto-*, while species that form clusters are indicated by the prefix *staphylo-*.

Members of the kingdom Eubacteria have a cell wall made of peptidoglycan, a network of polysaccharide molecules linked together with chains of amino acids. Outside the cell wall and membrane, many bacteria have a gel-like layer called a **capsule.** Members of the kingdom Archaebacteria often lack cell walls.

Cell walls Eubacteria can have two types of cell walls, distinguished by a dye staining technique called the Gram stain. One group is called Gram-negative, and the other Gram-positive.

Gram staining is important in medicine because the two groups of eubacteria differ in their susceptibility to different antibiotics. **Antibiotics** are chemicals that interfere with life processes in bacteria. Thus, Gram staining can help determine which antibiotic would be most useful in fighting an infection.

Endospores Some bacteria form thick-walled **endospores** *(EHN doh spohrz)* around their chromosomes and a small bit of cytoplasm when they are exposed to harsh conditions. These conditions can be the depletion of nutrients, a drought, or high temperatures. Endospores can survive environmental stress and may germinate years after they were formed, releasing new, active bacteria.

Pili Pili enable bacteria to adhere to the surface of sources of nutrition, such as your skin. Some kinds of pili enable bacteria to exchange genetic material through a process called conjugation. **Conjugation** *(kahn juh GAY shuhn)* is a process in which two organisms exchange genetic material. In prokaryotes, pili from one bacterium adhere to a second bacterium, and genetic material is transferred from the first bacterium to the second bacterium. Conjugation enables bacteria to spread genes within a population.

Study *TIP*

● **Reviewing Information**
Prepare flash cards for each of the **Key Terms** in this chapter. On each card, write the term on one side and its definition on the other side. Use the cards to review meanings of the Key Terms.

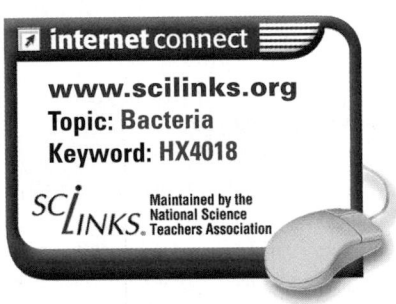

▣ **internet** connect

www.scilinks.org
Topic: Bacteria
Keyword: HX4018

SC*LINKS.* Maintained by the National Science Teachers Association

Figure 10 Bacterial shapes. Bacteria are usually one of three shapes.

Magnification: 117,300×

Magnification: 2,295×

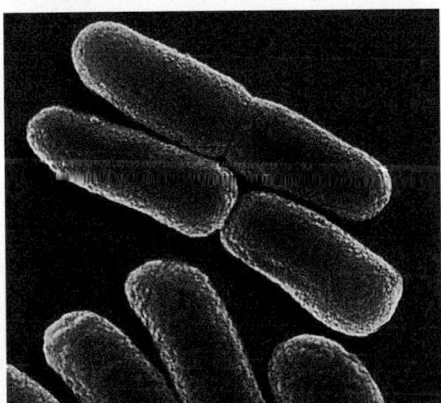

Bacillus (rod-shaped)
E. coli

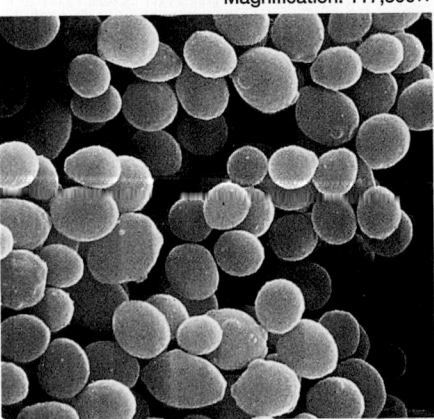

Coccus (round-shaped)
Micrococcus luteus

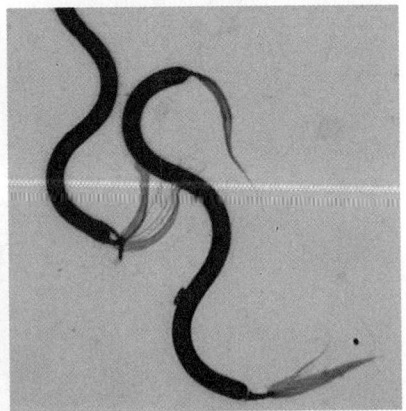

Spirillum (spiral-shaped)
Spirillum volutans

Obtaining Energy

Over 4,000 species of bacteria have been named, and probably many more haven't yet been discovered. Bacteria occur in the widest possible range of habitats and play key ecological roles in nearly all of them. As you may recall from an earlier chapter, bacteria thrive in hot springs, frigid arctic seas, and groundwater. They are even found at high pressures in the deep sea and inside solid rock.

Bacteria can be classified in several different ways. Classifying bacteria by the different ways in which they obtain energy, for example, gives a good general sense of the great diversity among bacteria.

Bacteria can also be classified according to their phylogenetic relationships. By comparing the sequence of their ribosomal RNA, scientists have determined that there are at least 12 phyla of eubacteria and four phyla of archaebacteria.

Photosynthesizers

A significant fraction of the world's photosynthesis is carried out by bacteria. Photosynthetic bacteria can be classified into four major groups based on the photosynthetic pigments they contain: purple nonsulfur bacteria, green sulfur bacteria, purple sulfur bacteria, and cyanobacteria. Green sulfur bacteria and purple sulfur bacteria grow in **anaerobic** (oxygen-free) environments. They cannot use water as a source of electrons for photosynthesis and instead use sulfur compounds, such as hydrogen sulfide, H_2S. Purple nonsulfur bacteria use organic compounds, such as acids and carbohydrates, as a source of electrons for photosynthesis.

Of particular importance are the cyanobacteria, which often clump together in large mats of filaments. Recall that cyanobacteria are thought to have made the Earth's oxygen atmosphere. Each filament is a chain of cells encased in a continuous jellylike capsule. Many cyanobacteria, such as species of *Anabaena*, shown in **Figure 11,** are capable of fixing nitrogen.

Chemoautotrophs

Bacteria called chemoautotrophs *(KEE moh AW toh trohfs)* obtain energy by removing electrons from inorganic molecules such as ammonia, NH_3, and hydrogen sulfide, H_2S, or from organic molecules such as methane, CH_4. In the presence of one of these hydrogen-rich chemicals, chemoautotrophic bacteria can manufacture all their own amino acids and proteins. Chemoautotrophic bacteria that live in the soil, such as *Nitrosomonas* and *Nitrobacter,* are of great importance to the environment and to agriculture. They have an important role in the nitrogen cycle called nitrification. Nitrification, as you may recall from an earlier chapter, is the process in which bacteria oxidize ammonia into nitrate. Nitrate is the form of nitrogen most commonly used by plants.

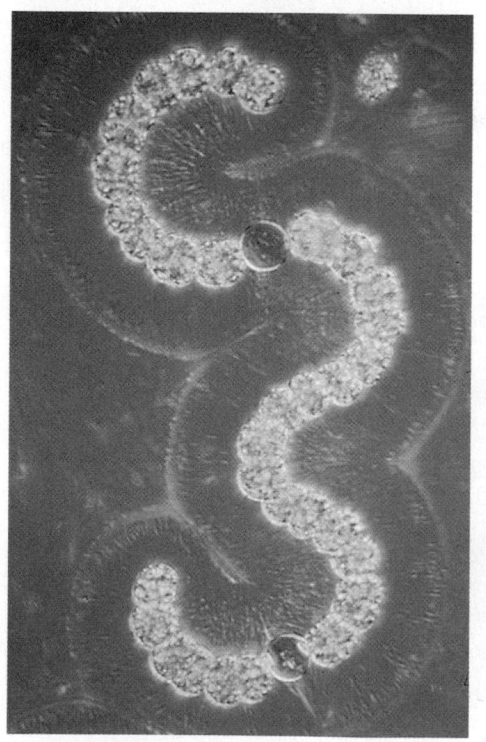

Figure 11 Photosynthetic bacterium. *Anabaena* is a photosynthetic cyanobacterium in which individual cells adhere in filaments. The two large orange-colored cells are encased in a structure where nitrogen fixation occurs.

Heterotrophs

Most bacteria are heterotrophs. Together with fungi, heterotrophic bacteria are the principal decomposers of the living world; they break down the bodies of dead organisms and make the nutrients available to other organisms. Many are **aerobic,** that is, they live in the presence of oxygen. Some other bacteria can live without oxygen.

Other activities of heterotrophic bacteria may be helpful or harmful to humans. For example, more than half of our antibiotics are produced by several species of *Streptomyces,* a filamentous bacterium found in soil. On the other hand, one species of *Staphylococcus* can secrete a poison into food. This poison causes nausea, diarrhea, and vomiting in people who eat the *Staphylococcus*-contaminated food.

Species of the symbiotic bacteria *Rhizobium* are by far the most important of all nitrogen-fixing organisms. *Rhizobium* species are heterotrophic bacteria that usually live within lumps on the roots of legumes (plants such as soybeans, beans, peas, peanuts, alfalfa, and clover), as shown in **Figure 12.** Farmers take advantage of *Rhizobium*'s nitrogen-fixing abilities when they "rotate" their crops every few years and grow legumes, which replenish the soil with nitrogen-containing compounds.

Magnification: 1,440×

Figure 12 Nitrogen-fixing bacteria. The bacteria inside the lumps on these soybean roots contain *Rhizobium,* a nitrogen-fixing bacteria.

FORENSICS BIOWatch

Bioterror Detectives

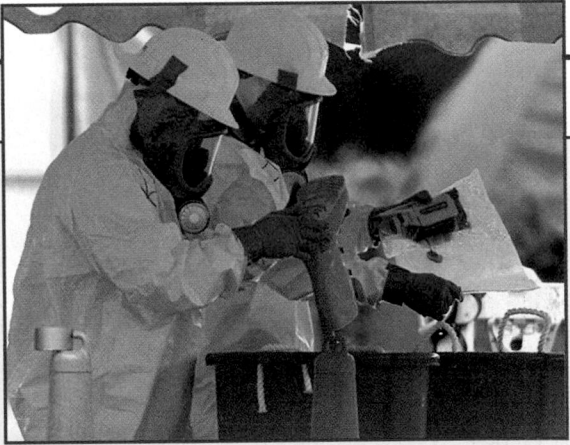

When a criminal commits a crime, investigators look for fingerprints at the crime scene. But what happens when the criminal is a bioterrorist and the weapon is a deadly bacterium?

The new field of forensic genomics offers investigators a new way to track bacteria to their source. Forensic genomics uses genome sequencing to identify a specific bacterial type—a subset of a species—called a *strain.* Just as a fingerprint can belong to only one individual, the genomic sequence of each different strain is unique.

Anthrax is a lethal bacterium that produces tough spores. When anthrax spores were sent through the mail in the fall of 2001, five people died. Investigators wanted to know the source of the bacterium. They thought it might have been stolen from a research laboratory. Because labs work with specific strains, a bacterium could possibly be linked to one—and only one—lab.

Scientists sequenced the genome of the anthrax strain from the first victim. Then they compared it to the genome sequence of a known strain. In this case, researchers identified more than 60 genetic markers that differed between the two strains. With time, genome sequencing of other anthrax strains will create a "library" of genome markers, increasing the speed and success rate of future identification attempts.

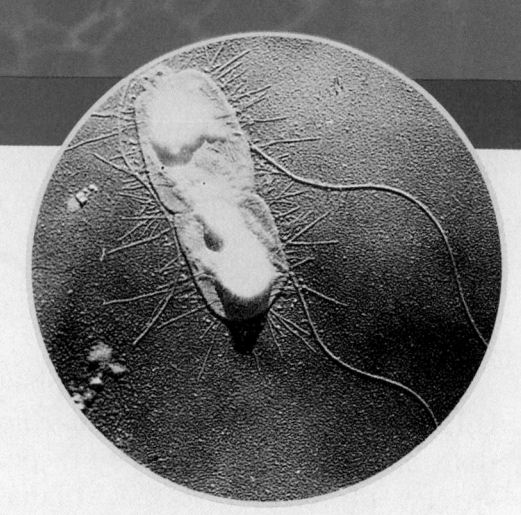

Escherichia coli

- **Scientific name:** *Escherichia coli*
- **Size:** Up to 1 μm
- **Habitat:** Inhabits the intestines of many mammals
- **Mode of nutrition:** Heterotrophic

Characteristics

Cell structure *E. coli* is a Gram-negative eubacterium. It has a rigid cell wall composed of peptidoglycan. Outside the cell wall lies the outer membrane, which is composed of lipids and polysaccharides.

Genetic material Like all bacteria, *E. coli* has a single DNA molecule in the form of a loop. *E. coli* has approximately 5,000 genes.

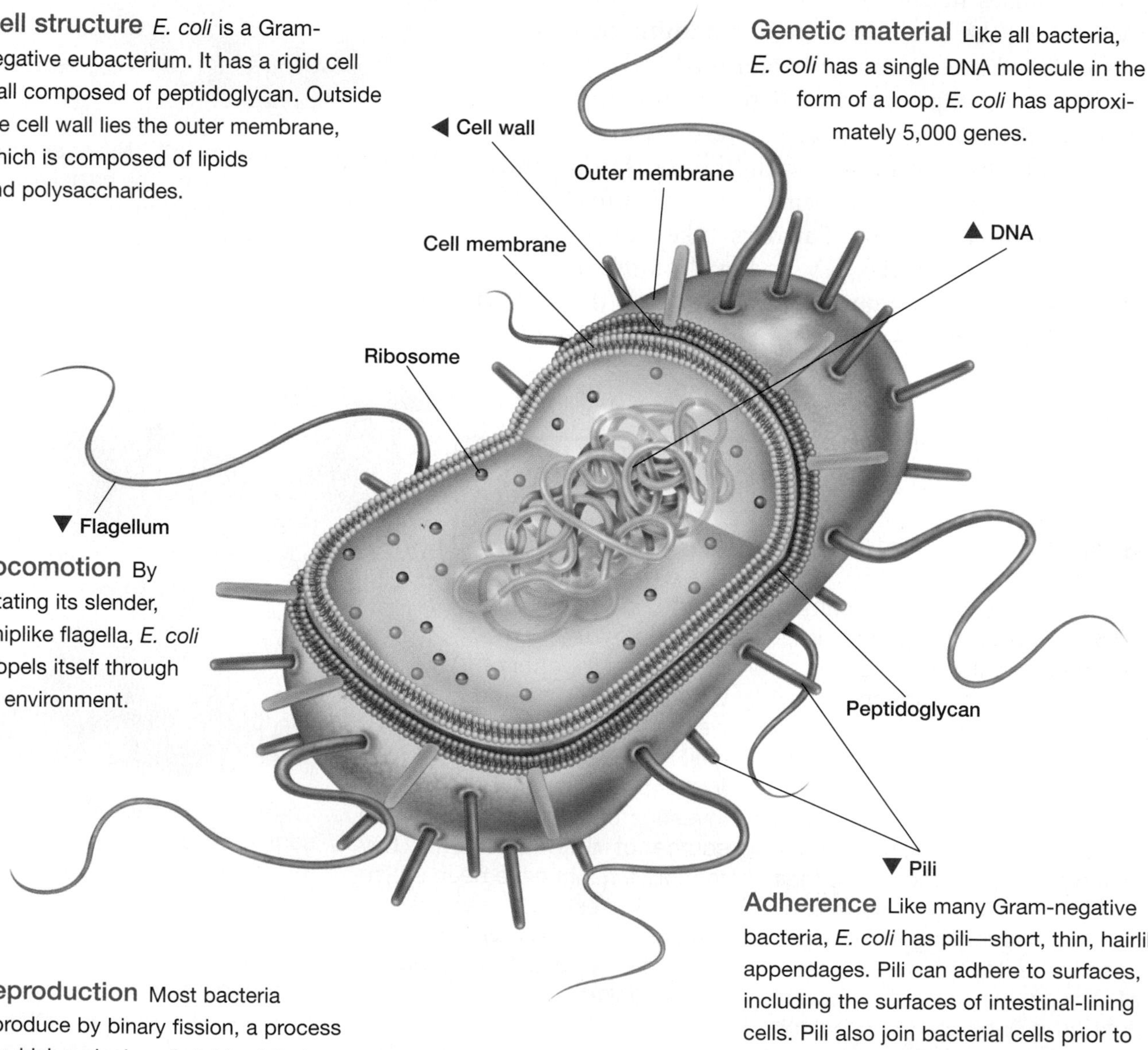

◀ Cell wall

Outer membrane

Cell membrane

▲ DNA

Ribosome

▼ Flagellum

Locomotion By rotating its slender, whiplike flagella, *E. coli* propels itself through its environment.

Peptidoglycan

▼ Pili

Adherence Like many Gram-negative bacteria, *E. coli* has pili—short, thin, hairlike appendages. Pili can adhere to surfaces, including the surfaces of intestinal-lining cells. Pili also join bacterial cells prior to conjugation.

Reproduction Most bacteria reproduce by binary fission, a process by which a single cell divides into two identical new cells. *E. coli* can divide as often as every 20 minutes.

Pathogenic Bacteria

In order to understand infectious diseases, think of your body as a treasure chest full of resources. Your body has protein, minerals, fats, carbohydrates, and vitamins. You may want to keep and use these resources, but so do many other organisms, including the bacteria on and in your body. Bacteria have evolved various means of obtaining these resources from you. In some cases, the competition for the resources in your body can result in your becoming ill.

Bacteria Can Metabolize Their Host

Heterotrophic bacteria obtain nutrients by secreting enzymes that break down complex organic structures in their environment and then absorbing them. If that environment is your throat or lungs, this can cause serious problems.

For example, tuberculosis, a disease of the lungs, is caused by *Mycobacterium tuberculosis,* shown in **Figure 13.** Tuberculosis was once one of the most common causes of death. In most cases, infection occurs when tiny droplets of moisture that contain the bacteria are inhaled. Some bacteria settle in the lungs, where they grow using human tissue as their nutrients. The bacteria may also spread to other parts of the body. Symptoms may include coughing up sputum and blood, chest pain, fever, fatigue, weight loss, and loss of appetite. If left untreated, death may occur as quickly as within 18 months but more commonly within 5 years. Other important bacterial diseases are described in **Table 2.**

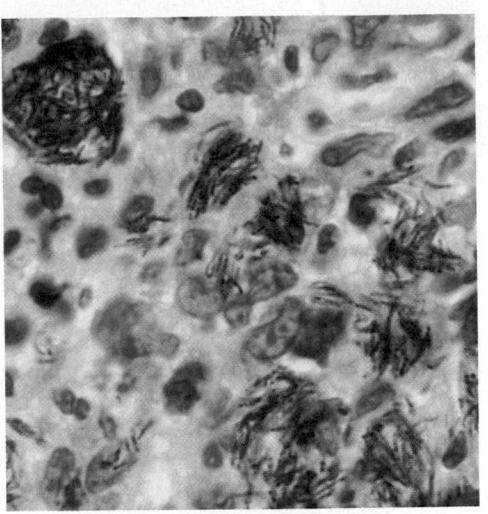

Figure 13 Tuberculosis in a lung. The red-stained structures in this light micrograph are *Mycobacterium tuberculosis,* which cause tuberculosis.

Table 2	Important Bacterial Diseases		
Disease	**Description of illness**	**Bacterium**	**How the disease is transmitted or caused**
Anthrax (respiratory)	Fever, severe difficulty breathing	*Bacillus anthracis*	Inhalation of spores
Bubonic plague	Fever, bleeding, lymph nodes that form swellings called buboes; often fatal	*Yersinia pestis*	Bite of an infected flea
Cholera	Severe diarrhea and vomiting; often fatal if not treated	*Vibrio cholerae*	Drinking contaminated water
Dental cavities	Destruction of minerals in tooth	*Streptococcus mutans*	Dense collections of bacteria in mouth
Lyme disease	Rash, pain, swelling in joints	*Borrelia burgdorferi*	Bite of an infected tick
Tuberculosis	Fever, cough, difficulty breathing	*Mycobacterium tuberculosis*	Inhalation
Typhus	Headache, high fever	*Rickettsia prowazekii*	Bite of infected flea or louse

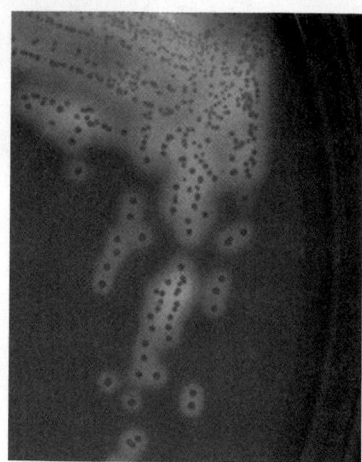

Figure 14 The effect of bacterial toxins. This species of *Streptococcus* secretes a toxin that destroys red blood cells. The agar contains red blood cells and clearly shows a zone around the bacteria where the toxin has destroyed the red blood cells.

Not all bacteria are lethal. For example, some bacteria cause everyday health problems, such as acne. Acne occurs in about 85 percent of teenagers. Bacteria, such as *Propionibacterium acnes*, normally grow in an oil gland of the skin. They metabolize a certain kind of oil produced by those glands. During puberty the oil glands increase the amount of oil produced, and the bacterial population on the skin increases greatly. The bacteria grow in the pores where the oil normally flows, forming pimples and blackheads.

Bacterial Toxins

The second way bacteria cause disease is by secreting chemical compounds into their environment. These chemicals, called **toxins,** are poisonous to eukaryotic cells, as shown in **Figure 14.** Toxins can be secreted into the body of an infected person or into a food in which bacteria are growing.

When bacteria grow in food and produce toxins, the toxins can cause illness in humans who eat those contaminated foods. This kind of illness is called an intoxication. For example, *Staphylococcus aureus* causes the most common type of food poisoning. The symptoms include nausea, vomiting, and diarrhea. This type of poisoning is painful but is seldom fatal.

Another type of intoxication that is fatal occurs when food is not canned properly. Sometimes canned food is not heated enough to kill endospore-forming bacteria, such as *Clostridium botulinum*. The bacteria can then grow and produce a deadly toxin that affects the nervous system. A person who eats food that contains this toxin then becomes ill with a disease called botulism, whose symptoms include double vision and paralysis. People with botulism may die because they are unable to breathe.

Some bacteria are responsible for other diseases reported in the news, such as *E. coli* O157:H7, the cause of several outbreaks of food poisoning in the United States. *E. coli* normally lives in our intestines. However, if it acquires DNA that codes for the toxin through conjugation, it can produce the toxin. *E. coli* poisoning is associated with raw or improperly cooked ground beef.

Most bacteria can be killed by boiling water or various chemicals. Using hot, soapy water to prevent contamination of our food utensils and food supply is one way of preventing disease. Many commercial antibacterial products can also be used to prevent bacterial contamination in the kitchen and in industrial food factories.

Biowarfare

Biowarfare is the deliberate exposure of people to biological toxins or pathogens such as bacteria or viruses. The United States government is justifiably concerned about the use of *bioweapons*—biological toxins or pathogens suitable for mass infection—against military personnel overseas and against civilians within the United States. Biologists are working on new approaches to recognize the onset of an attack with a bioweapon, to treat infected people, and to slow the spread of any outbreak of disease.

Antibiotics

In 1928, the British bacteriologist Alexander Fleming noticed a fungus of the genus *Penicillium* growing on a culture of *S. aureus*. He saw that bacteria did not grow near the fungus. He concluded that the fungus was secreting a substance that killed the bacteria, as shown in **Figure 15.** Fleming isolated the substance and named it penicillin. In the early 1940s, scientists found that penicillin was effective in treating many bacterial diseases, such as pneumonia.

Different antibiotics interfere with different cellular processes. Because these processes do not occur in viruses, antibiotics are not effective against them. Other antibiotics, such as tetracycline and ampicillin, have been discovered in nature or imitated chemically.

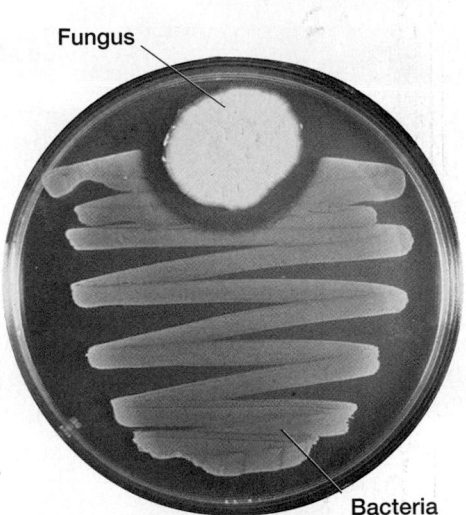

Figure 15 Antibiotics are naturally produced. Alexander Fleming saw a plate of agar very similar to this one. Notice how the bacteria do not grow next to this fungus.

Antibiotic-Resistant Bacteria

In recent years, some bacteria have become resistant to antibiotics. Susceptible bacteria are eliminated from the population, and resistant bacteria survive and reproduce, thus passing on their resistance traits. Mutations for antibiotic resistance arise spontaneously in bacterial populations as errors in DNA replication. There are many individuals in a bacterial population, and bacteria multiply very rapidly (doubling their numbers in as few as 20 minutes). Therefore, a mutation that gives the bacteria a selective advantage can quickly spread throughout a population.

Antibiotic Misuse Mutations that confer resistance to antibiotics are strongly favored in bacterial populations being treated with an antibiotic. Usually, if the full course of the antibiotic is administered, all the targeted bacteria are killed and there is no chance for a resistant strain to develop. If antibiotic treatment ends prematurely, some of the bacteria may survive. Which ones? The ones most resistant to the antibiotic. A patient who does not take the full course of a prescribed antibiotic is setting the stage for the development of antibiotic-resistant bacteria.

Multiple-antibiotic Resistance A related problem can arise in a patient being treated with two or more antibiotics at the same time. This practice selects for bacteria that have acquired several antibiotic-resistance genes. A number of strains of *Staphylococcus aureus* associated with severe infections of hospital patients (so-called hospital staph) have appeared in recent years. These strains are resistant to penicillin and a wide variety of other antibiotics, so infections caused by these strains are very difficult to treat.

Recently, concern has arisen over the common use of antibacterial soaps. Antibacterial soaps are marketed as a means of protecting people from harmful bacteria. Their routine use, however, may favor bacteria resistant to the antibacterial agents in the soap. Ultimately, routine use of antibacterial soaps could reduce our ability to treat common bacterial infections.

Importance of Bacteria

Despite the misery that some bacteria cause humans in the form of disease and food spoilage, much of what bacteria do is extremely important to our health and economic well-being.

Food and Chemical Production

Many of the foods that we eat are processed by specific kinds of bacteria. For example, many fermented foods are produced with the assistance of bacteria, as shown in **Figure 16.** These foods include pickles, buttermilk, cheese, sauerkraut, olives, vinegar, sourdough bread, and even some kinds of sausages.

Humans are able to use different bacteria to produce different kinds of chemicals for industrial uses, as shown in **Figure 17.** For example, different kinds of *Clostridium* species can make either acetone or butanol. These chemicals can be used to produce a large variety of other useful chemicals.

Genetic engineering companies use genetically engineered bacteria to produce their many products, such as drugs for medicine and complex chemicals for research.

Mining and Environmental Uses of Bacteria

Mining companies can use bacteria to concentrate desired elements from low-grade ore. Low-grade ore has a low percentage of the desired mineral, but it also has sulfur compounds. Chemoautotrophic bacteria can convert the sulfur into a soluble compound, leaving the desired mineral behind. The sulfur compound can be washed away with water, leaving only the desired mineral. This technique can be used to harvest copper or uranium.

Bacteria metabolize different organic chemicals and are therefore used to help clean up environmental disasters such as petroleum and chemical spills. Powders containing petroleum-metabolizing bacteria are used to help clean oil spills.

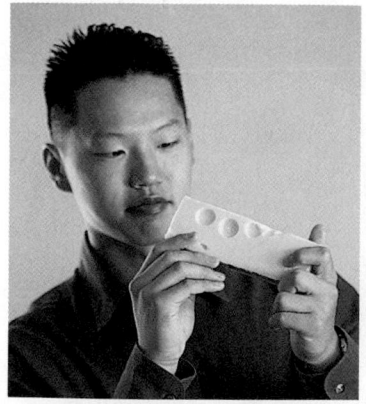

Figure 16 Swiss cheese. In making Swiss cheese, bacteria grow in the cheese and produce gas. As the cheese hardens these pockets of gas remain, giving the cheese its characteristic holes.

Figure 17 Industrial fermenter. Bacteria can be used to produce useful chemicals such as in this fermenter.

Section 2 Review

1 Construct a table that lists the seven ways bacteria differ from eukaryotic cells.

2 List the structures found in *E. coli.*

3 Identify the relationship between photosynthesis, heterotrophic metabolism, and chemoautotrophic metabolism.

4 Describe the relationship between metabolism, toxins, bacteria, and disease.

5 List three ways bacteria are helpful.

6 Critical Thinking Defending a Theory How does the growth of antibiotic resistance in bacteria support the theory of evolution by natural selection?

7 Standardized Test Prep Which disease is caused by inhaling a bacterium?
A cholera
B botulism
C *E. coli* food poisoning
D tuberculosis

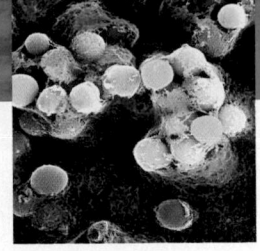

Study ZONE

CHAPTER HIGHLIGHTS

Key Concepts

1 Viruses

- Viruses consist of segments of a nucleic acid contained in a protein coat, and some have an envelope.
- Viruses do not have all of the characteristics of life and are therefore not considered to be alive.
- Viruses replicate inside living cells. They enter a cell by injecting their genetic material into the cell, by slipping through tears in the plant cell wall, or by binding to molecules on the cell surface.
- Some viruses replicate through a lytic cycle or a lysogenic cycle.
- HIV replicates inside immune system cells, eventually destroying them, leaving the host without adequate defense against disease.
- Emerging viruses are newly recognized viruses or viruses that have reappeared or spread to new areas.
- Viroids are infectious RNA molecules that cause disease in plants, and prions are infectious proteins that cause disease in certain animals.

2 Bacteria

- Bacteria differ from eukaryotes in their cellular organization, cell structures, and metabolic diversity.
- Bacteria can be classified into two groups according to their cell wall structure. Gram staining can be used to distinguish these two groups.
- Bacteria can transfer genes to one another by conjugation.
- Bacteria are grouped according to their ribosomal RNA sequences and the way they obtain energy.
- Bacteria cause disease by metabolizing nutrients in their host or by releasing toxins, which damage their host.
- Bacterial disease can usually be fought with soap, chemicals, and antibiotics.
- Bacteria are used to make foods, antibiotics, and chemicals; to fix nitrogen; to clean the environment; and to cycle important chemicals in the environment.

Key Terms

Section 1

virus (434)
pathogen (434)
capsid (435)
envelope (435)
glycoprotein (435)
bacteriophage (435)
lytic (436)
provirus (436)
lysogenic (437)
prion (441)
viroid (441)

Section 2

pilus (442)
bacillus (443)
coccus (443)
spirillum (443)
capsule (443)
antibiotic (443)
endospore (443)
conjugation (443)
anaerobic (444)
aerobic (445)
toxin (448)

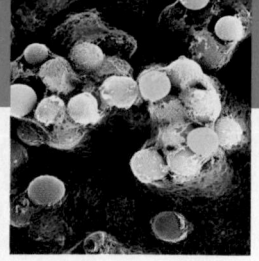

Understanding Key Ideas

1. Unlike cells, viruses do not
 a. grow.
 b. have homeostasis.
 c. metabolize.
 d. All of the above

2. What evidence led Stanley to conclude that TMV is not a living organism?
 a. The extract of TMV crystallized.
 b. TMV is made of RNA and protein.
 c. TMV reproduces only in cells.
 d. The virus poisons tobacco plants.

3. HIV infects and destroys
 a. skin cells.   **c.** immune cells.
 b. red blood cells. **d.** bacterial cells.

4. Bacteria
 a. always have flagella.
 b. are smaller than viruses.
 c. have aerobic or anaerobic metabolism.
 d. have a nucleus.

5. Bacteria that do not require sunlight and obtain energy by removing electrons from hydrogen-rich chemicals are called
 a. heterotrophs.
 b. photosynthetic bacteria.
 c. cyanobacteria.
 d. chemoautotrophs.

6. Environmental spills of petroleum are sometimes cleaned up using
 a. viroids.   **c.** bacteria.
 b. prions.   **d.** bacteriophages.

7. Identify the pilus in the photo below.

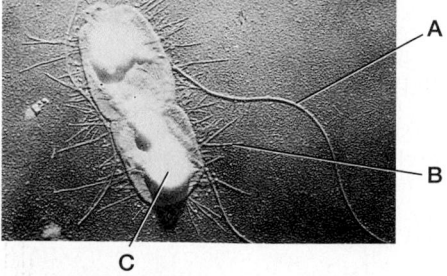

8. If cold viruses invade your body, your body's immune system may destroy most but not all of these viruses. How does your body's immune system affect the evolution of the cold viruses?

9. **BIOWatch** How can a "library" of anthrax genome markers help scientists identify unknown anthrax strains?

10. **Concept Mapping** Make a concept map describing the relationships of bacteria and viruses to diseases. Try to include the following terms in your map: *bacteria, viruses, pathogen, emerging viruses, antibiotics,* and *toxin*.

Critical Thinking

11. **Applying Information** How does the increase of resistance to antiviral drugs in HIV relate to the theory of evolution by natural selection?

12. **Evaluating Results** In the 1520s, the Spanish explorer Cortés and his armies introduced smallpox to the Americas. The death rate among the Native American people ranged from 50 to 90 percent compared with a death rate of about 10 percent among people in Europe. What accounts for the difference in death rates?

13. **Career Connection** Virologist Research the field of virology. Write a report on your findings that includes job description, training required, kinds of employers, growth prospects, and starting salary.

Alternative Assessment

14. **Finding Information** Research and write a report on a preventable viral disease, such as polio or smallpox. In your report, discuss the process scientists followed in identifying the cause of the disease, isolating the virus, formulating a vaccine, and testing the vaccine.

Standardized Test Prep

Understanding Concepts

Directions (1–5): For *each* question, write on a separate sheet of paper the letter of the correct answer.

1 Which of the following is a type of virus that infects bacteria?
A. bacteriophage   **C.** glycoprotein
B. emerging virus   **D.** viroid

2 The basic components of all viruses are a nucleic acid and what other structure?
F. endospore   **H.** icosahedron
G. glycoprotein   **I.** protein coat

3 What structure allows *E. coli* to move?
A. flagellum   **C.** peptidoglycan
B. nucleus   **D.** pili

4 When does a bacteriophage kill its host cell?
F. while conjugating
G. during the lytic cycle
H. during the lysogenic cycle
I. while assembling the capsid

5 A person is infected by HIV, and the virus does not mutate while in the person's body. Why would that person not likely experience immune system failure?
A. HIV must mutate before entering the human body.
B. Only mutant HIV can be picked up by the body's macrophages.
C. Only after HIV mutates does it infect T cells and destroy the immune system.
D. The immune system fails only when the person is infected with a cold virus.

Directions (6): For the following item, write a short response.

6 Compare and contrast *prions, viroids,* and *viruses.*

Test *TIP*

When using a diagram to answer a question, look in the image for evidence that supports your potential answer.

Reading Skills

Directions (7): Read the passage below. Then answer the question.

Many microbiologists oppose the use of antibiotics in patients with viral infections. They claim that antibiotics will not help the patient, because a virus does not carry out any of the life processes that antibiotics interrupt. Such overuse of antibiotics encourages the proliferation of resistant strains of bacteria.

7 How does using an antibiotic promote the proliferation of resistant strains of bacteria?
F. A virus is more effective at reproducing when there is an antibiotic.
G. Antibiotics cause some bacteria to change into more dangerous bacteria.
H. Mutated bacteria that are resistant to the antibiotic will be naturally selected.
I. Antibiotics weaken the immune system, which helps bacterial reproduction.

Interpreting Graphics

Directions (8): Base your answer to question 8 on the diagram below.

Viral Replication in Bacteria

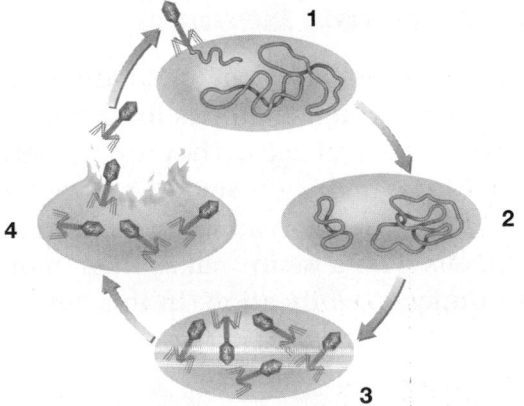

8 What is the virus doing to the bacteria in step 1?
A. injecting its DNA
B. injecting its capsid
C. withdrawing DNA
D. withdrawing proteins

Skills Practice Lab

Staining and Observing Bacteria

SKILLS
- Using aseptic techniques
- Using a microscope

OBJECTIVES
- **Prepare** and stain wet mounts of bacteria.
- **Identify** different types of bacteria by their shape.

MATERIALS
- wax pencil
- 3 microscope slides
- safety goggles
- lab apron
- disposable gloves
- rubbing alcohol
- paper towels
- 3 culture tubes of bacteria (A, B, and C)
- test-tube rack
- sterile cotton swabs
- Bunsen burner with striker
- microscope slide
- forceps or wooden alligator-type clothespin
- 150 mL beaker
- methylene blue stain in dropper bottle
- compound microscope

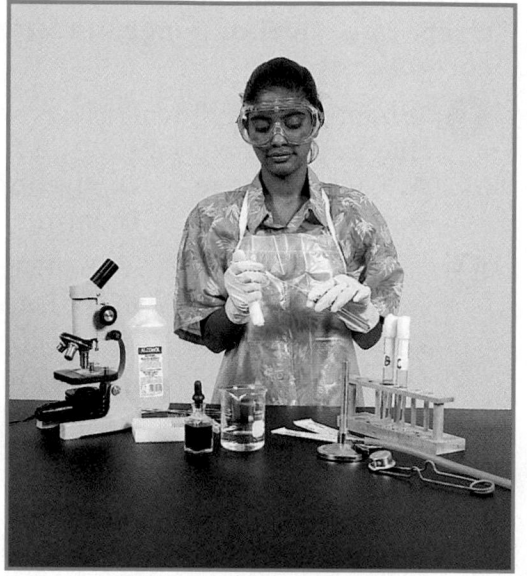

ChemSafety

CAUTION: Always wear safety goggles and a lab apron to protect your eyes and clothing.

CAUTION: Do not touch or taste any chemicals. Know the location of the emergency shower and eyewash station and how to use them. If you get a chemical on your skin or clothing, wash it off at the sink while calling to the teacher. Notify the teacher of a spill. Spills should be cleaned up promptly, according to your teacher's directions.

CAUTION: Glassware is fragile. Notify the teacher of broken glass or cuts. Do not clean up broken glass or spills with broken glass unless the teacher tells you to do so.

Before You Begin

Like all **prokaryotes,** bacteria are unicellular organisms that sometimes form filaments or loose clusters of cells. They are prepared for viewing by making a **smear,** a slide on which cells have been spread and dried. Treating the cells with a **stain** makes them more visible under magnification. In this lab, you will stain, identify, and compare and contrast different types of bacteria.

1. Write a definition for each boldface term in the paragraph above and for each of the following terms: strepto, staphylo, coccus, bacillus, spirillum.

2. Based on the objectives for this lab, write a question you would like to explore about different kinds of bacteria.

Procedure

PART A: Observing Live Bacteria

1.  Put on safety goggles, a lab apron, and disposable gloves.

2. Use a wax pencil to label three microscope slides *A, B,* and *C.*

3. Use rubbing alcohol and paper towels to clean the surface of your lab table and gloves. Allow the table to air dry. **CAUTION: Alcohol is flammable. Do not use alcohol near an open flame.**

4. Light a Bunsen burner with a striker. **CAUTION: Keep combustibles away from flames. Do not light a Bunsen burner when others in the room are using alcohol.**

5. Beginning with culture A, make a smear of three different bacteria as follows. Remove the cap from a culture tube. *Note: Do not place the cap on the table.* Pass the opening of the tube through the flame of a Bunsen burner. Lightly touch the tip of a sterile swab to the bacterial culture. Pass the opening of the tube through the flame again, and replace the cap. Transfer a small amount of the culture to the appropriately labeled microscope slide by rubbing the swab on the slide, and spread out to cover about half of the total slide area. Dispose of the swab in a proper container. Repeat for cultures B and C.

6. Allow your smears to air dry. Using microscope slide forceps, pick up each slide one at a time and pass it over the flame several times. Let each slide cool.

7. Using microscope slide forceps, place one of your slides across the mouth of a 150 mL beaker half-filled with water. Place 2–3 drops of methylene blue stain on the dried bacteria. *Note: Do not allow the stain to spill into the beaker.* **CAUTION: Methylene blue will stain your skin and clothing.** Let the stain stay on the slide for 2 minutes. Then dip the slide into the water in the beaker several times to rinse it. Blot the slide dry with a paper towel. *Note: Do not rub the slide.*

8. Repeat step 7 for your other two slides.

9. Allow each slide to air dry, and then observe them with a microscope. Make a sketch of a few cells on each slide. Compare your sketches to Figure 10, and identify the type of bacteria on each slide.

PART B: Cleanup and Disposal

10. Dispose of slides, used swabs, solutions, and broken glass in the designated waste containers. Do not pour chemicals down the drain or put lab materials in the trash unless your teacher tells you to do so.

11. Clean up your work area and all lab equipment. *Clean the surface of your lab table with rubbing alcohol.* Return lab equipment to its proper place. Wash your hands thoroughly before you leave the lab and after you finish all work.

Analyze and Conclude

1. **Summarizing Results** Describe the shape and grouping of the cells of each type of bacteria you observed.

2. **Analyzing Methods** Why should the test tube caps from the culture tubes (in Part B) not be placed on the table?

3. **Evaluating Viewpoints** Evaluate the following advice: Always use caution when handling bacteria, even if the bacteria is known to be harmless.

4. **Drawing Conclusions** How did you classify the bacteria in cultures A, B, and C, as a coccus, a bacillus, or a spirillum?

5. **Further Inquiry** Write a new question about bacteria that could be explored with another investigation.

On the Job

Microbiologists are scientists who study organisms too small to be seen by the naked eye. Do research to discover how microbiologists better our lives. For more about careers, visit **go.hrw.com** and type in the keyword **HX4 Careers.**

A thorough knowledge of biology—coupled with insight and hard work—helped scientists track down a killer.

Disease Detectives

In May of 1993, a killer was on the loose in the Four Corners area—New Mexico, Arizona, Utah, and Colorado—in the southwestern United States. Over the previous month, three young, previously healthy people had died in a disturbingly similar way in northwestern New Mexico. All of the victims had come down with what seemed to be a mild case of the flu—with fever, muscle aches, weakness, and lung congestion. All three had been rushed to the hospital gasping for breath, but even emergency treatment could not save them.

Autopsies revealed that all three victims had drowned. Their lungs had filled with plasma—the liquid part of the blood—that had leaked from the tiny blood vessels within the lungs. What could cause such severe damage? Laboratory tests ruled out diseases with similar symptoms, including influenza, bubonic plague, and anthrax.

As doctors and researchers began to investigate further, additional deaths came to light. Fearing that these cases might signal the beginning of a disease outbreak, officials decided to ask for help from the federal Centers for Disease Control and Prevention (CDC) in Atlanta, Georgia.

Catching a Killer Virus

The CDC sent a team of disease investigators to the Four Corners area while researchers in Atlanta tested blood and tissue samples from the victims. One of the tests revealed that the victims had been exposed to a kind of virus known as a hantavirus.

This result was surprising. All known hantaviruses from North America were harmless to humans. Furthermore, the hantaviruses that cause disease in Asia and Europe attack the kidneys, not the lungs. But further testing confirmed the result. The killer was a hantavirus.

What is the virus like? Scientists used one of biology's most powerful tools to find out—the polymerase chain reaction, or PCR. PCR is a method of making many copies of a particular

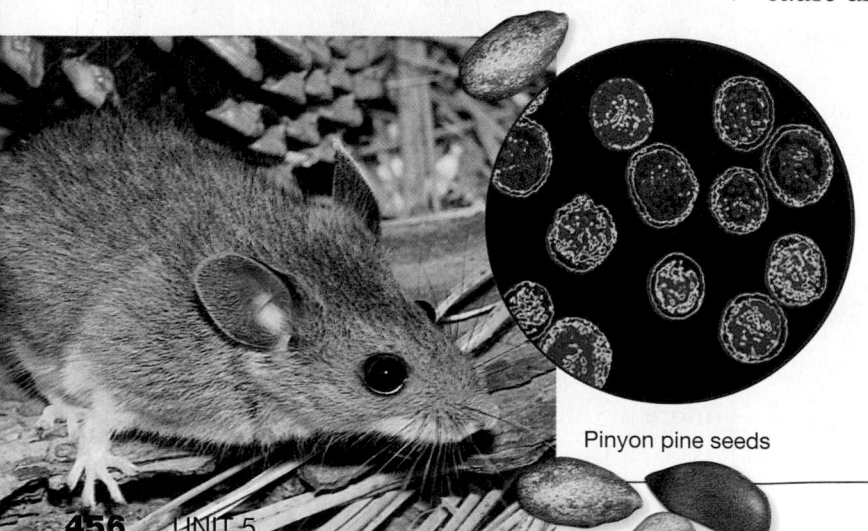

Pinyon pine seeds

Source of the virus
Deer mice, which thrive on the seeds of pinyon pine, were found to carry hantavirus, a dangerous pathogen.

Four Corners area
Researchers were surprised to find a deadly hantavirus here.

DNA strand. The copying process requires a primer, a short piece of DNA only a few nucleotides long that becomes the beginning of the new strand. To investigate the characteristics of the new hantavirus, CDC researchers made several primers that shared nucleotide sequences with known hantaviruses. By adding different primers to tissue samples from the victims, the scientists could "fish out" pieces of the virus's genes and then determine their nucleotide sequence.

Determining the Cause of the Outbreak

Within a month of being called in, CDC scientists had caught the killer and sequenced much of its genetic material. They called this new disease *hantavirus pulmonary syndrome,* or HPS.

But scientists still needed to learn how HPS is transmitted and why the outbreak occurred when it did. Elsewhere, researchers knew that hantaviruses are carried by rodents—rats, mice, and their relatives. To find out whether rodents also carried the new hantavirus, researchers trapped thousands of rodents in and around the homes of the victims. About 30 percent of the deer mice they caught tested positive for the virus. These small mice—about 15 cm long—live throughout most of the United States.

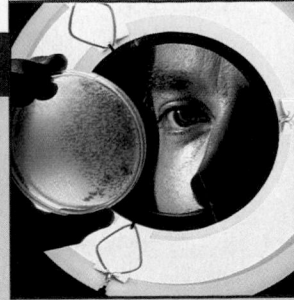
Why had the disease not appeared before in the Four Corners area? An unusually wet winter the year before led to a bumper crop of pinyon pine seeds, the deer mouse's favorite food. In turn, the deer mouse population boomed. The larger population probably meant that mice entered homes more often and people were more likely to contact them outdoors. ■

Analyzing STS Issues

Science and Society

1 **What does the CDC do?** Researchers from the CDC played a key role in solving the hantavirus mystery. Use library resources or the Internet to find out more about the CDC. What is the CDC's mission? What kinds of people work there? Where does the CDC get its funding?

2 **How is the hantavirus controlled?** While researching the CDC, investigate further for information on the hantavirus. Hantavirus infections come from activities that involve humans and infected rodents coming into contact with one another. What measures can be taken to limit exposure to rodents?

Technology: PCR

3 **What are some other newly discovered diseases?** HPS is just one of many emerging, or newly discovered, diseases. Use library resources or the Internet to research three other emerging diseases that have come to light in the last 30 years. Find out where each disease was discovered, its cause, and its symptoms. How is PCR used in the study of emerging diseases?

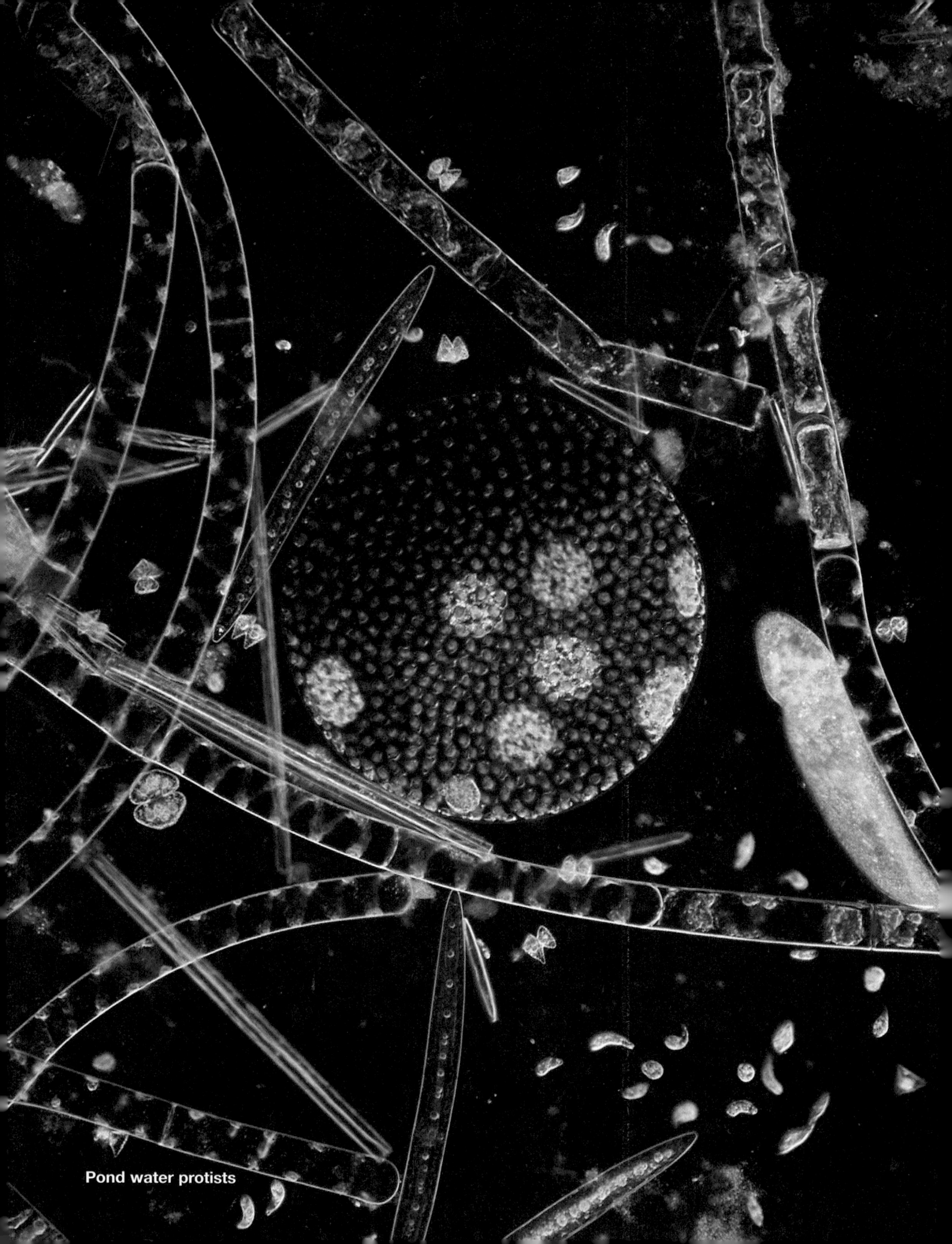

Pond water protists

CHAPTER
21 Protists

Quick Review

Answer the following without referring to earlier sections of your book.

1. **Label** the structures of a eukaryotic cell. *(Chapter 3, Section 3)*

2. **Describe** the different kinds of photosynthetic pigments that make photsynthesis possible. *(Chapter 5, Section 2)*

3. **Summarize** the importance of sexual reproduction. *(Chapter 7, Section 2)*

4. **Define** the term *plankton*. *(Chapter 16, Section 3)*

5. **Summarize** the advantage of multicellularity. *(Chapter 18, Section 2)*

Did you have difficulty? *For help, review the sections indicated.*

Reading Activity

List on a separate piece of paper all the words, phrases, and ideas that you associate with the word protist. *Share one or two of your ideas with your classmates. Inquire of each other or the teacher to clarify each of the ideas listed. After all discussion is completed, write notes on your paper about what you remember. Look over your notes to see what you know about protists based on your experience and discussion.*

internet connect

www.scilinks.org
National Science Teachers Association *sci*LINKS Internet resources are located throughout this chapter.

*sci*LINKS® Maintained by the National Science Teachers Association

● An amazing world exists in a drop of pond water. There you will find a whole landscape of single cells that are independent organisms—the protists—the most diverse of all the kingdoms.

Characteristics of Protists

Objectives

- **List** the characteristics of protists.
- **List** three environments where protists can be found.
- **Identify** the unifying features of protists.
- **Distinguish** asexual and sexual reproduction of *Chlamydomonas*.
- **Differentiate** two ways multicellular protists reproduce sexually.

Key Terms

protozoan
alga
zygospore
alternation of
   generations
sporangium

Diversity

The most diverse of all organisms, protists are mostly unicellular, microscopic organisms, such as the green scum you might find growing on a rock in a pond. But a few protists are as complex and multicellular as the massive house-high seaweed called kelp, which is found in the oceans.

Characteristics

The kingdom Protista *(proh TEES tuh)* consists of an unusually diverse assortment of eukaryotes that exhibit a broad array of characteristics, as summarized in **Figure 1.** For example, some protists are photosynthetic (like plants), some ingest food (like animals), and some absorb their food (like fungi).

Some protists have flagella or cilia, which they can use for locomotion or to collect nutrients. Other protists use other means of locomotion. Protists are found almost everywhere there is water. Many live in lakes and oceans, floating as plankton or anchored to rocks. They are also common inhabitants of damp soil and sand, and they thrive in moist environments, such as leaf litter. Some species are parasites.

Many protists have mechanisms for monitoring and responding to stimuli in their environment. For example, some protists have eyespots, small organelles containing light-sensitive pigments that detect changes in the quality and intensity of light.

The First Eukaryotes

Protists—the first eukaryotes—are thought to have evolved about 1.5 billion years ago through the process of endosymbiosis. The kingdom Protista contains life-forms similar to those that gave rise to the three kingdoms of multicellular organisms—fungi, plants, and animals.

Figure 1 Protist characteristics. These characteristics, which are found in many eukaryotes, first evolved in protists.

Protist Characteristic

- Sexual reproduction
- Multicellularity
- Mitosis and meiosis
- Complex flagella and cilia

Magnification: 1,174×

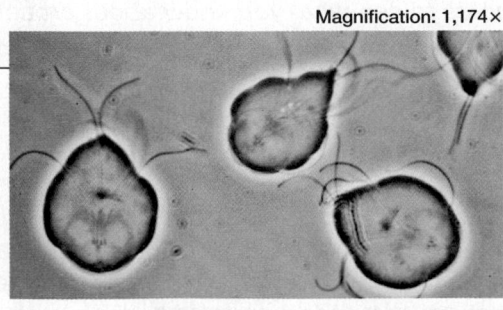

Two important eukaryotic features that evolved among the protists are sexual reproduction and multicellularity. Many protists reproduce only asexually, by mitosis; some use meiosis and sexual reproduction only in times of environmental stress; and others reproduce sexually most of the time.

Multicellularity involves a significant amount of coordination among specialized cells. This trait evolved independently in different groups of protists at different times. Early during the evolution of protists, complex flagella and cilia also appeared.

What Unites Protists?

The kingdom Protista contains all eukaryotes that cannot be classified as animals, plants, or fungi. Protists lack the specialized features that characterize the three other multicellular kingdoms. Unlike plants and animals, protists do not reproduce by forming embryos. Protists do not develop complex multicellular reproductive structures. The major phyla of protists are very different from one another and, with a few exceptions, are only distantly related.

Historically, scientists have referred to heterotrophic protists as **protozoa** and to photosynthetic protists as **algae**. These commonly used terms are not formal classification categories. The major phyla of protists are classified according to some of their distinguishing features in **Table 1.** This classification, however, does not reflect the evolutionary relationships of protists.

Table 1 Some Important Protistan Phyla

Phylum	Distinguishing features	Mode of nutrition
Rhizopoda (amoebas) Foraminifera (forams)	Move using pseudopodia	Heterotrophic
Chlorophyta (green algae) Rhodophyta (red algae) Phaeophyta (brown algae)	Typically multicellular	Photosynthetic
Bacillariophyta (diatoms) Dinoflagellata Euglenophyta	Typically unicellular	Typically photosynthetic
Kinetoplastida Ciliophora (ciliates)	Move using flagella/cilia	Heterotrophic
Acrasiomycota (cellular slime molds) Myxomycota (plasmodial slime molds) Oomycota	Funguslike	Heterotrophic
Apicomplexa (sporozoans)	Form spores	Heterotrophic

Reproduction

Reproduction in the unicellular green alga *Chlamydomonas* is typical of unicellular protists. Chlamydomonas is unusual in that some species can form colonies. Colonial *Chlamydomonas* lose their flagella and secrete a sticky fluid that binds each cell to its neighbors. *Chlamydomonas* species reproduce sexually and asexually.

As a mature organism, the single-celled protist is haploid. When it reproduces asexually, *Chlamydomonas* first absorbs its tail and divides by mitosis, producing two to eight haploid cells called zoospores, which remain within the wall of the parent cell. Mature zoospores break out of the parent cell and grow to become mature haploid cells.

Sexual Reproduction in Unicellular Protists

During environmental stress, such as a shortage of nutrients, *Chlamydomonas* species reproduce sexually. The haploid cell divides first by mitosis to produce haploid gametes. After they are released, a pair of gametes from different *Chlamydomonas* individuals fuse to form a pair. This pair of gametes then shed their cell walls and fuse into a diploid zygote with a thick protective wall called a **zygospore** *(ZIE goh spohr)*.

A zygospore can withstand unfavorable environmental conditions for long periods of time. When environmental conditions become favorable again, meiosis within the zygospore produces haploid cells, which break out of the zygospore wall. These haploid cells grow into mature cells, completing the sexual life cycle, as shown in **Figure 2**.

Figure 2 *Chlamydomonas* **reproduction.** The unicellular green algae in the genus *Chlamydomonas* reproduce sexually and asexually.

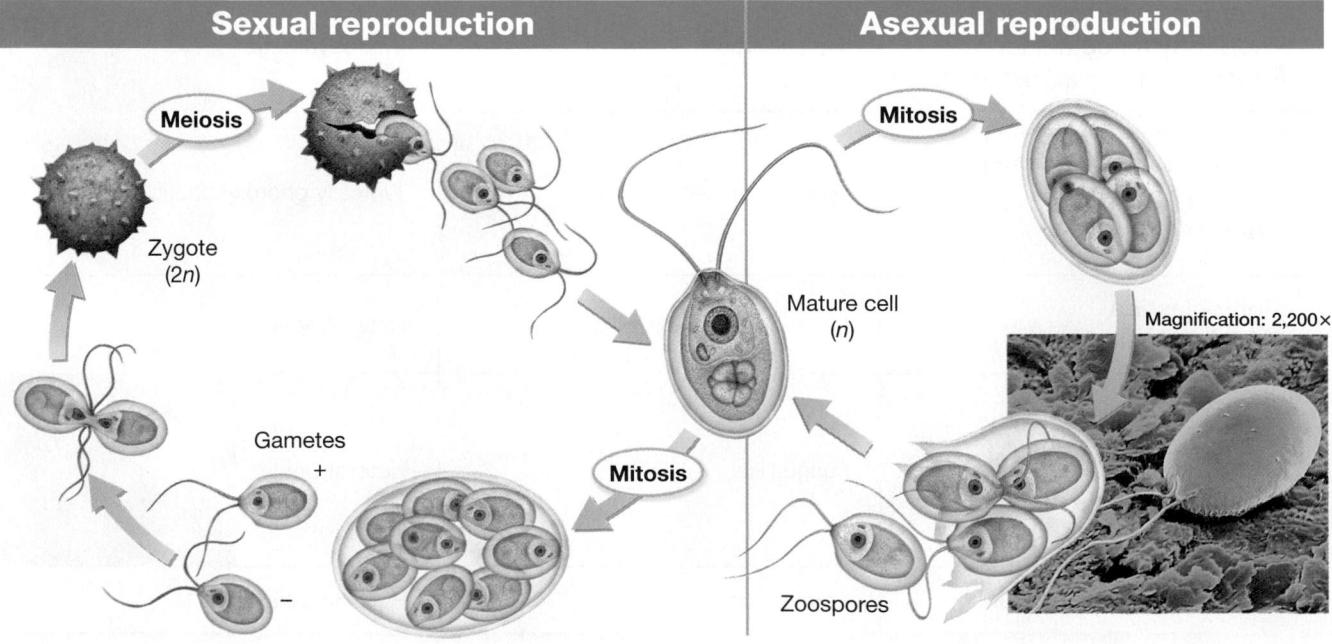

Sexual reproduction

Meiosis

Zygote (2n)

Gametes +

−

Asexual reproduction

Mitosis

Magnification: 2,200×

Mature cell (n)

Mitosis

Zoospores

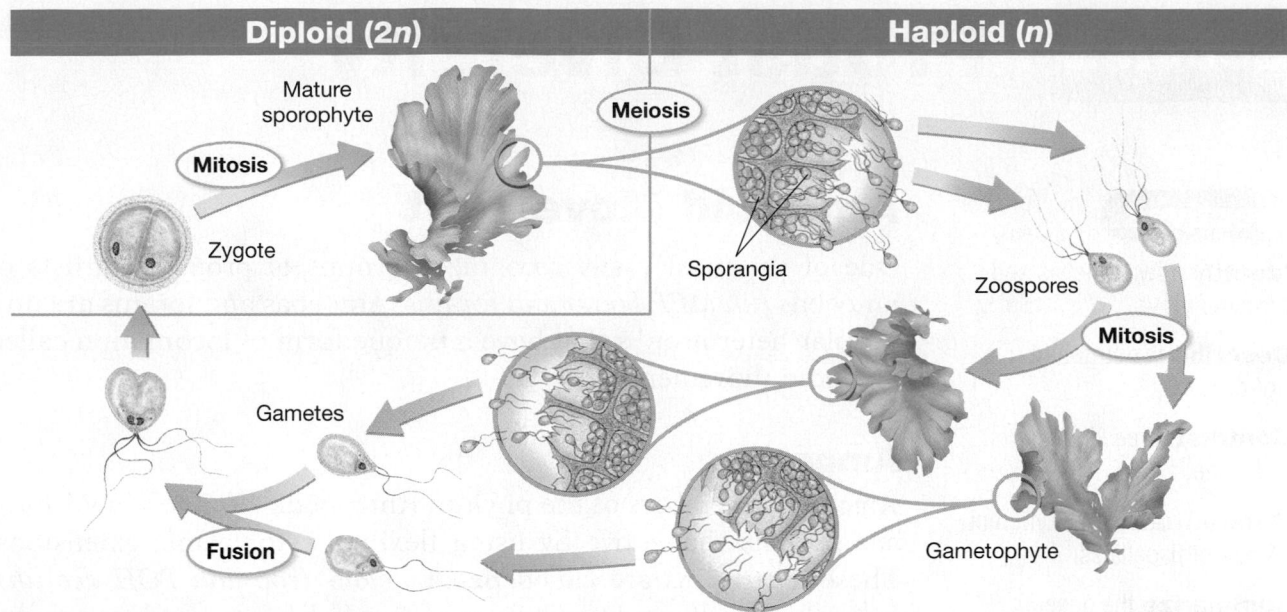

Diploid (2n)	Haploid (n)

Mature sporophyte

Mitosis

Meiosis

Zygote

Sporangia

Zoospores

Mitosis

Gametes

Fusion

Gametophyte

Figure 3 Life cycle of
Ulva. The multicellular green alga *Ulva*, or sea lettuce, has a life cycle in which haploid and diploid individuals alternate.

Sexual Reproduction in Multicellular Protists

Sexual reproduction among multicellular protists occurs in many different ways. Two of these are alternation of generations and conjugation.

Alternation of generations *Ulva* is a common genus of marine green alga. **Figure 3** shows that the reproductive cycle of *Ulva*, called **alternation of generations,** is characterized by two distinct multi-cellular phases. The diploid, spore-producing phase is called the sporophyte generation. The haploid, gamete-producing phase is called the gametophyte generation. The adult sporophyte alga has reproductive cells called **sporangia** *(spoh RAN jee uh),* which pro-duce haploid spores by meiosis. The spores grow into multicellular haploid gametophytes. The mature gametophyte produces haploid gametes that fuse and complete the life cycle by dividing through mitosis to form a new diploid sporophyte.

Conjugation *Spirogyra,* a filamentous green alga species, reproduces sexually by conjugation. Conjugation is the temporary union of two protists to exchange nuclear material. The process begins when two filaments align side by side. Part of the cell walls between adjacent algae form a bridge between the cells. The nucleus of one cell then passes through the tube into the adjacent cell. The two nuclei even-tually form a resting spore, which produces a new haploid filament.

Section 1 Review

1 Critical Thinking Relating Concepts
Describe two ways sexual reproduction can occur in multicellular protists.

2 Summarize why protists are not classified with the other three eukaryotic kingdoms.

3 List three characteristics of protists.

4 Standardized Test Prep All organisms in the kingdom Protista are
A unicellular.
B parasitic.
C eukaryotic.
D photosynthetic.

Protist Diversity

Objectives

- **Identify** how amoebas and forams move.
- **Describe** the structure of diatoms.
- **Contrast** three kinds of algae.
- **Differentiate** three different kinds of flagellates.
- **Summarize** the general characteristics of a *Paramecium*.

Key Terms

amoeba
pseudopodium
diatom
euglenoid
kinetoplastid
cilium
plasmodium

Figure 4   Pseudopodia. This amoeba is using its pseudopodia to engulf a smaller organism (in yellow circle).

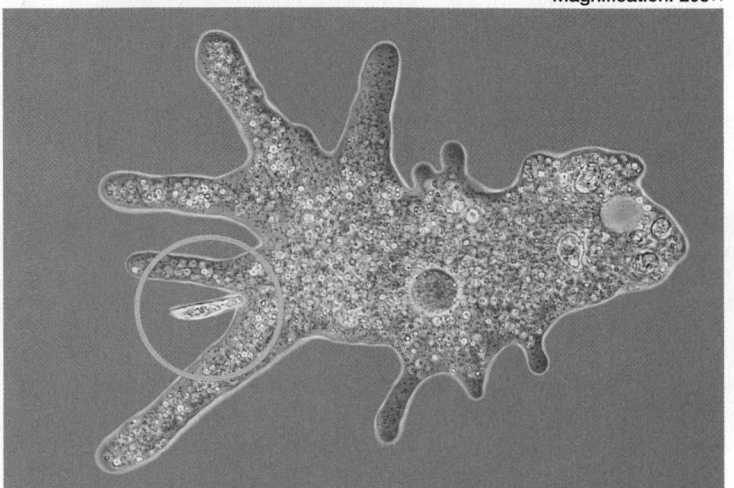

Magnification: 205×

Ameboid Movement

One of the most easily recognized groups of protists consists of amoebas *(uh MEE buhs)* and forams. Amoebas and forams are unicellular heterotrophs that have a unique form of locomotion called ameboid movement.

Amoebas

Amoebas, members of the phylum Rhizopoda *(RIE zoh POH duh)*, are protists that move by using flexible, cytoplasmic extensions. These extensions are called **pseudopodia** *(soo doh POH dee uh)*, from the Greek words *pseudo*, meaning "false," and *podium*, meaning "foot." Because an amoeba has no cell walls or flagella, it is extremely flexible. During ameboid movement, a pseudopodium bulges from the cell surface, stretches outward, and anchors itself to a nearby surface. The cytoplasm from the rest of the amoeba then flows into the pseudopodium. Pseudopodia can surround and engulf food particles, as shown in **Figure 4.** Amoebas live in both fresh water and salt water and are especially abundant in soil. Meiosis and sexual reproduction do not occur in most amoebas. They reproduce by fission, dividing into two new cells. The majority of amoebas are free-living, but some species are parasites, such as *Entamoeba histolytica*, the protist that causes amebic dysentery in humans. These organisms are transmitted in contaminated food or water.

Foraminifera

Members of the phylum Foraminifera *(foh ram ih NIHF ur uh)*, or forams, are marine protists that typically live in sand or attach themselves to other organisms or rocks. Forams are characterized by their porous shells, called tests. Tests usually have many chambers arranged in a spiral shape. They resemble a tiny snail and consist of organic material that contains grains of calcium carbonate. Long, thin projections of cytoplasm extend through the pores in the tests to aid in swimming and in catching prey. Some forams also obtain nourishment from algae that live symbiotically under their tests.

The tests of dead forams have accumulated on ocean floors over millions of years. Their calcium carbonate shells helped form the limestone deposits that are important components of many land formations.

Algae

Algae are protists that are strict phototrophs. Some are unicellular, others are multicellular. Algae are distinguished by the type of photosynthetic pigment they contain and by their cell or body shape.

Green Algae

Most green algae (phylum Chlorophyta) are freshwater unicellular organisms, but some are large, multicellular marine organisms, as shown in **Figure 5.** They also exist as a major part of microscopic marine plankton, inhabit damp soil, or even thrive within the cells of other organisms as symbionts. Green algae contain the same pigments found in the chloroplasts of plants. Most green algae have sexual and asexual reproductive stages.

Red Algae

Red algae (phylum Rhodophyta) are mostly multicellular organisms found in warm ocean waters. Their red pigments are efficient at absorbing the light that penetrates deep waters. Some red algae, as shown in Figure 5, have calcium carbonate in their cell walls. Others are used to make agar and carrageenan *(kayr uh GEE nuhn)*. Red algae have a complex life cycle, usually involving alternation of generations.

Brown Algae

Brown algae (phylum Phaeophyta) are multicellular and are found mostly in marine environments. The larger brown algae known as kelp grow along coasts and provide food and shelter for many different kinds of organisms. They are among the largest organisms on Earth. Most brown algae reproduce by alternation of generations.

Figure 5   Three phyla of algae. *Ulva* is a species of green algae composed of a sheet of cells that is only two cells thick. Coralline alga is a red alga that contributes to the great coral reefs. Some species of the brown algae *Macrocystis* can grow to a length of more than 60 m.

Ulva species

Macrocystis species

Coralline alga

Diatoms

WORD Origins

● The word *amoeba* is from the Greek *amoibe,* meaning "change." Knowing this makes it easier to remember that amoebas are able to change their shape.

Diatoms *(DIE uh tahmz),* members of the phylum Bacillariophyta, are photosynthetic, unicellular protists with unique double shells. The shells are made of silica and often have unique markings. Their shells are like small boxes with lids, one half fitting inside the other. Abundant in oceans and lakes, diatoms are important producers in the food chain. Diatoms can have one of two types of symmetry: radial (like a wheel) or bilateral (two-sided). The empty shells of diatoms form thick deposits that are mined commercially as diatomaceous earth. Diatomaceous earth is used as an abrasive or to add the sparkling quality to paint used on roads. Diatomaceous earth is also sold as a natural control for pests, such as slugs, and some insects, such as fleas. The sharp edges of the diatom shells cut into the body of the pest, leading to its death.

Diatoms secrete chemicals through holes in their shells, enabling them to move by gliding. Individuals are diploid and usually reproduce asexually. The two halves of the shell separate, and each half regenerates another matching half. As a consequence of this model of reproduction, diatoms tend to get smaller and smaller with each generation. When an individual gets too small because of repeated division, it slips out of its shell, grows to full size, and regenerates a new shell. Sexual reproduction in diatoms is rare.

Observing Characteristics of Diatoms

Try this activity to find out why diatomaceous earth is used to make abrasives, fine filters, and reflective paints.

Magnification: 240×

Materials

pipet, water, microscope slide, toothpick, diatomaceous earth, coverslip, compound microscope

Procedure

1. Using a pipet, place a drop of water in the center of a clean microscope slide.

2. Use a toothpick to scoop up a small amount of diatomaceous earth and mix it with the water drop. Add a coverslip.

3. Observe your wet mount under both low and high power of a compound microscope.

4. Draw some of the diatoms you see.

5. Observe the wetmount under low power as your partner shines a flashlight (at a 45° angle) on the slide. Turn off the microscope's light source so that only the flashlight is lighting the slide. Record your observations.

Analysis

1. **Label** your drawings as radial or bilateral. Find out the meanings of these terms and how they apply to your diatom drawings.

2. **Select** some characteristics you observed that are useful in classifying particular species of diatoms.

3. **Interpret** what you observed when the flashlight was shone on the slide.

Flagellates

Flagellates are protists that move using flagella. The three major phyla of flagellates are the dinoflagellates, the zoomastigotes, and the euglenoids.

Dinoflagellates

Dinoflagellates *(DIE noh FLAJ uh layts)*, members of the phylum Dinoflagellata are unicellular and most have two flagella. A few kinds of dinoflagellates are found in fresh water, but most are marine and make up part of the plankton. Most dinoflagellates have a protective coat made of cellulose that is often encrusted with silica, giving them unusual shapes, as shown in **Figure 6.** Their flagella beat in two grooves—one encircling the body like a belt, the other perpendicular to it. As a result, dinoflagellates spin through the water like a top. A few dinoflagellates produce powerful toxins. The poisonous "red tides" that occur frequently in coastal areas are often associated with population explosions of dinoflagellates. Dinoflagellates usually reproduce asexually by mitosis, and they can be photosynthetic, heterotrophic, or both.

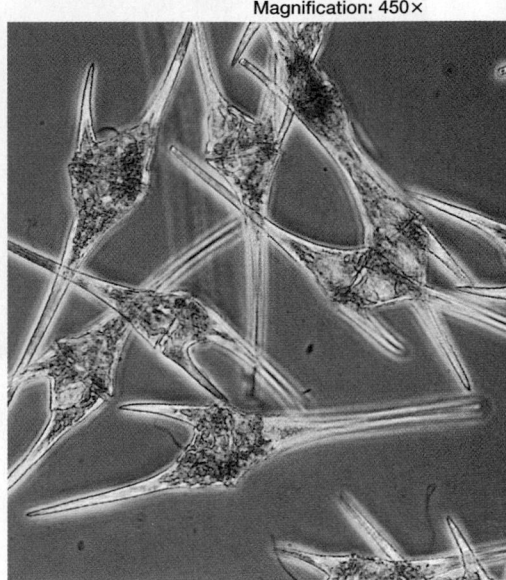

Magnification: 450×

Figure 6   Dinoflagellates.
This dinoflagellate, *Gonyaulax tamarensis*, moves using flagella.

Euglenoids

Euglenoids *(yoo GLEE noyds)*, members of the phylum Euglenophyta, are freshwater protists with two flagella. This group clearly shows the difficulty of classifying protists as animals or plants. About one-third of the 1,000 known species of euglenoids have chloroplasts and are photosynthetic; other species lack chloroplasts, ingest their food, and are heterotrophic. A member of *Euglena*, shown in **Figure 7,** has a protein scaffold called a pellicle *(PEHL ih kuhl)* inside the cell membrane. Since the pellicle is flexible, the euglenoid can change shape. A light-sensitive organ called the eyespot helps orient the movements of these organisms toward light. Reproduction in this phylum occurs by mitosis.

Figure 7   *Euglena.*
Euglena is a versatile protist. It contains chloroplasts and is photosynthetic, but it is also heterotrophic and can live without light.

Kinetoplastids

Kinetoplastids, members of the phylum Kinetoplastida, are unicellular heterotrophs that have at least one flagellum, and some species have thousands. While most reproduce only asexually, some are known to produce gametes and reproduce sexually. Kinetoplastids are clearly related to euglenoids, and many taxonomists merge the two phyla together. Some kinetoplastids, such as the trypanosomes, cause diseases such as African sleeping sickness in humans and domestic animals.

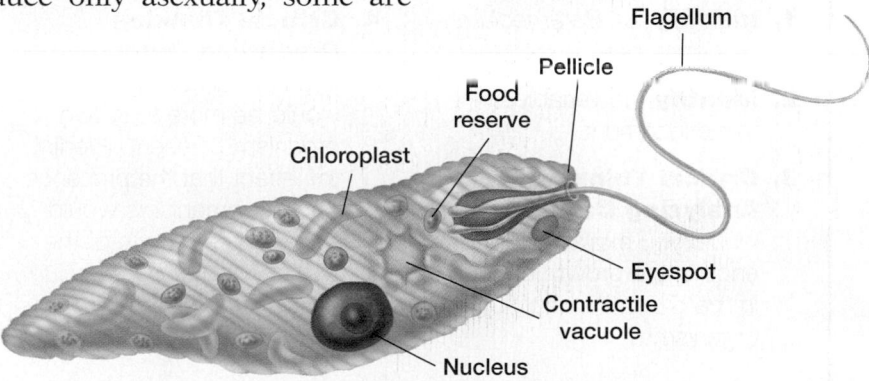

Flagellum

Pellicle

Food reserve

Chloroplast

Eyespot

Contractile vacuole

Nucleus

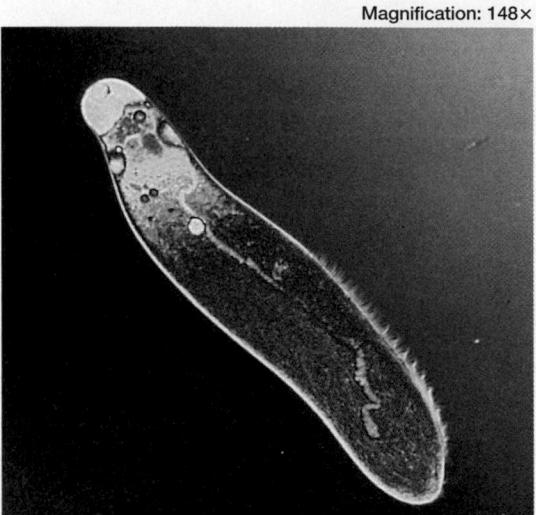

Magnification: 148×

Figure 8 Ciliate. The cilia that surround this ciliate enable it to move and feed.

Ciliates

Ciliates are the most complex and unusual of the protists. All members of the phylum Ciliophora *(sih lee AWF oh ruh)* have large numbers of **cilia,** tightly packed rows of short flagella used for movement, as shown in **Figure 8.** Ciliates are complex unicellular heterotrophs. The body wall of ciliates is a tough but flexible outer pellicle that enables the organism to squeeze through or move around many obstacles. The pellicle consists of an outer membrane with numerous fluid-filled cavities beneath it. Ciliates, such as *Paramecium,* shown in *Up Close: Paramecium,* on the following page, form vacuoles for ingesting nutrients and regulating their water balance.

Most ciliates have two types of nuclei within their cells: micronuclei and larger macronuclei. The micronuclei contain normal chromosomes that divide by mitosis. Macronuclei contain small pieces of DNA derived from the micronuclei.

Reproduction in ciliates is usually by mitosis, with the body splitting in half. Cells divide asexually for about 700 generations and then die if sexual reproduction has not occurred. Most ciliates can engage in sexual reproduction through conjugation, a process in which two cells unite and exchange genetic material.

DATA LAB

Interpreting Competition Among Protists

Background

Protists, like all organisms, must compete with one another for nutrients. To examine the effects of competition between two species of *Paramecium,* equal numbers of the paramecia were grown together (dashed lines) and separately (solid lines). Study the graph at right, and answer the following questions.

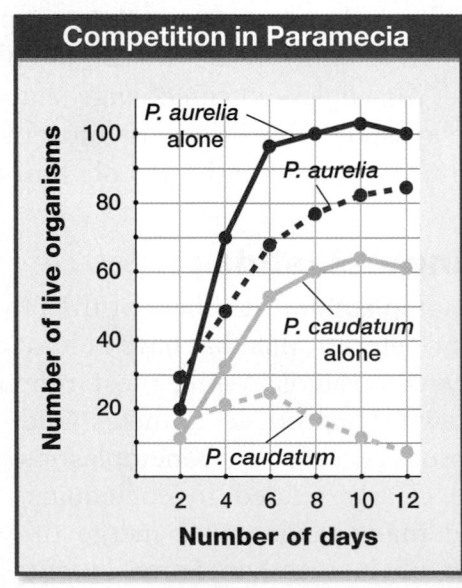

Analysis

1. **Identify** the *Paramecium* that grew best alone.

2. **Identify** the variables in this experiment.

3. **Critical Thinking Analyzing Data** How would you explain the difference in the growth curves in the group that had both organisms?

4. **Critical Thinking Predicting Outcomes** In a natural setting there would be more than two organisms present. Predict the effect that the presence of other organisms would have on the growth of the two species of *Paramecium.*

Up Close

Paramecium

- **Scientific name:** *Paramecium caudatum*
- **Size:** Microscopic; up to 1 mm long
- **Habitat:** Lives in freshwater streams and ponds
- **Diet:** Bacteria, small protists, organic debris

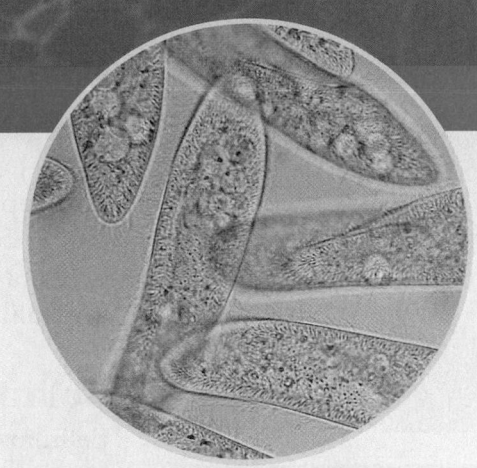

Magnification: 1,200×

Characteristics

Nuclei Members of the genus *Paramecium* have two nuclei. The macronucleus contains fragmented chromosomes used in routine cellular functions, and it divides by pinching in two. The micronucleus contains the cell's chromosomes and divides by mitosis.

▲ **Macronucleus**

Maintaining water concentration
Like other freshwater protists, *P. caudatum* constantly absorbs water by osmosis. *P. caudatum* gets rid of excess water using its contractile vacuoles, saclike organelles that expand, collecting excess water, and then contract, squeezing water out of the cell.

▲ **Contractile vacuole**

Food vacuole

Surface *P. caudatum,* a ciliate, is covered with thousands of cilia arranged in rows along the cell. Cilia beat in waves that move diagonally across the cell, causing the protist to spin through the water. *P. caudatum* is surrounded by a rigid protein covering called a pellicle.

▲ **Cilia**

Micronucleus

▼ **Oral groove**

Nutrition Cilia lining the oral groove create a "whirlpool" that helps capture small bits of food. Food moves down the funnel-shaped groove and is engulfed in a food vacuole combining with digestive enzymes as it moves through the cell. Undigested food is released from the cell by exocytosis.

Genetic variation *P. caudatum* generally reproduces asexually by binary fission. Genes are shuffled during a sexual process called conjugation.

Protistan Molds

Protistan molds are heterotrophs with some mobility. They were once thought to be related to fungi because of a similar appearance and reproductive structures. However, the cell walls of protistan molds contain different carbohydrates than fungal cell walls. Also, protists carry out normal mitosis, whereas mitosis in fungi is unusual, as you will learn later in this book.

Cellular Slime Molds

Cellular slime molds, members of the phylum Acrasiomycota *(uh KRAZ ee oh mie koh tuh)*, resemble amoebas but have distinct features. The individual organisms behave as separate amoebas, moving through the soil and ingesting bacteria. During environmental stress, the individual amoebas gather together and move toward a fixed center. There they form multicellular colonies called slugs. Each slug develops a base, a stalk, and a swollen tip that develops spores, as shown in **Figure 9.** Each of these spores, when released, becomes a new amoeba, which then begins to feed and repeat the life cycle.

Figure 9   Cellular slime mold. The individual amoebas of *Dictyostelium discoideum* aggregate, move in a mass called a slug, and form a stalked structure that contains spores.

Plasmodial Slime Molds

Plasmodial slime molds, members of the phylum Myxomycota *(MIHKS oh MIE koh tuh)*, are a group of organisms that stream along as a **plasmodium,** a mass of cytoplasm that looks like an oozing slime. As they move, they engulf bacteria and other organic material, as shown in **Figure 10.** A plasmodial slime mold contains many nuclei, but they are not separated by cell walls. If the plasmodium begins to dry out or starve, it divides into many small mounds. Each mound produces a stalk tipped with a capsule in which haploid spores develop. The spores are highly resistant to hostile environmental conditions. When conditions are favorable, the spores germinate and become haploid cells that are either amoeboid or flagellated. These haploid cells are able to fuse into diploid zygotes, which undergo mitosis and form a new plasmodium.

Figure 10   Plasmodial slime mold. A plasmodial slime mold is a mass of cytoplasm containing many nuclei.

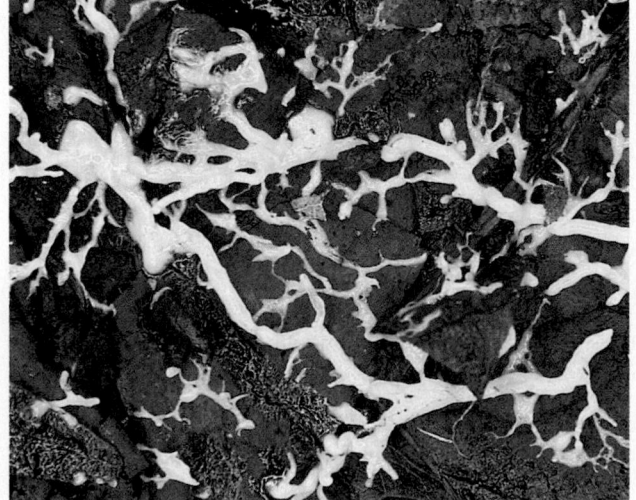

Other Molds

Oomycetes *(oh oh MIE seets),* members of the phylum Oomycota, are the water molds, white rusts, and downy mildews that often grow on dead algae and dead animals in fresh water. All members of the group either are parasites or feed on dead organic matter. Oomycetes are unusual in that their spores have two flagella: one pointed forward, the other pointed backward. Many oomycetes are plant pathogens, including *Phytophthora infestans,* which causes late blight in potatoes. This protist caused the failure of potato crops in Ireland between 1845–1850, which led to a famine that killed about 1 million people.

Sporozoans

Parasitic protists that form spores during their reproduction cycle are called sporozoans. They are members of the phylum Apicomplexa and are nonmotile, unicellular parasites. All sporozoans are parasitic and cause many serious diseases. Malaria, a sporozoan disease, kills more people than any other infectious disease. Sporozoans infect animals and are transmitted from host to host. There are about 4,500 named species in the phylum Apicomplexa.

Sporozoans have complex life cycles that involve both asexual and sexual reproduction. Sexual reproduction involves the fertilization of a large female gamete by a small, flagellated male gamete. The resulting zygote forms a thick-walled structure that makes the zygote resistant to drought and other unfavorable environmental conditions. A few sporozoans are listed in **Table 2.**

Many sporozoans are transmitted from one host to another by blood-feeding insects, such as mosquitoes, black flies, and midges. Other sporozoans are transmitted in the feces of an infected animal, such as the one shown in **Figure 11.** The parasites infect a new individual when that individual eats or drinks food or water that has been contaminated with infected feces. Some sporozoans are even transmitted through predation, such as when a cat becomes infected by eating an infected mouse.

Figure 11 *Cryptosporidium.*
Cryptosporidia (stained purple) are found in the feces of infected animals. In the 1980s and 1990s, water supplies in several cities in the United States became infected with these sporozoans.

Table 2 Some Sporozoans and Their Hosts

Sporozoans	Disease	Hosts
Plasmodium sp.	Malaria	Humans and other vertebrates
Toxoplasma sp.	Toxoplasmosis	Humans and cats
Babesia sp.	Cattle tick fever	Cattle, mice, humans, deer, dogs
Cryptosporidium sp.	Cryptosporidiosis (diarrhea)	Cattle, humans, birds, deer, dogs, cats

Section 2 Review

1 Describe a pseudopodium.

2 Identify the role of diatoms in the aquatic food chain.

3 Construct a table that compares three kinds of algae.

4 Explain how protistan molds differ from fungi.

5 Critical Thinking Evaluating Viewpoints "*Euglena* is a protozoan, not an alga." Is this an accurate statement? Why or why not?

6 Standardized Test Prep The disease *cryptosporidiosis* is caused by a parasitic protist that forms resistant stages called
A spores.
B slime molds.
C kinetoplastids.
D pseudopodia.

Protists and Health

Objectives

- **Identify** two ways that protists affect human health.

- **Name** three human diseases, other than malaria, caused by protists.

- **Summarize** how malaria is transmitted.

- **Evaluate** the methods used to control malaria.

Key Terms

sporozoite
merozoite

Protists and Humans

One of the greatest effects protists have on humans is that they cause disease. This effect can be measured in pain, death, and the medical costs of preventing and treating diseases. Some diseases caused by protists are listed in **Table 3.**

Protists also affect humans through the diseases they cause in livestock. The added cost of treating diseased livestock is passed on to consumers in the form of higher meat prices. This increased cost of living hinders progress in the developing world.

Beneficial Protists

There are many commensal protists that live in the digestive tracts of humans and in the digestive tracts of the animals that humans eat. Cattle could not digest the cellulose in the hay and grass they eat without the aid of commensal protists in their digestive tract.

Protists, which make up much of the plankton in the ocean, help to support food chains. Protists are also the single largest group of photosynthesizers on the planet. Because we all breathe oxygen, we all benefit from the gas that protists produce. Many protists are also detritivores, so they help recycle important chemicals, such as nitrogen, carbon, and phosphorus, in the environment.

Table 3 Diseases Caused by Protists

Disease	Description of illness	Protist	How the disease is transmitted
African sleeping sickness	Fever, weakness, lethargy	*Trypanosoma gambiense,* *Trypanosoma rhodesiense*	Bite from infected tsetse fly
Amebic dysentery	Bloody diarrhea, vomiting, extremely strong stomach cramps, fever	*Entamoeba histolytica*	Contaminated food or water
Giardiasis	Cramps, nausea, diarrhea, vomiting	*Giardia lamblia*	Contaminated food or water
Malaria	Fever, chills, sweats	*Plasmodium* sp.	Bite from infected mosquito
Toxoplasmosis	Primary danger is fetal infection; can cause convulsions, brain damage, blindness, and death in fetuses	*Toxoplasma gondii*	Contact with infected cats or improperly cooked meat

Malaria

Malaria is one of the most deadly diseases in humans. Over 100 million people have malaria at any one time, and up to 3 million people, mostly children, die from it every year.

The symptoms include severe chills, fever, sweating, confusion, and great thirst. Victims die of anemia, kidney failure, or brain damage unless the disease is treated.

Malaria Life Cycle

Malaria is caused by several species of *Plasmodium* and is spread by the bite of certain mosquitoes. There are three stages in the *Plasmodium* life cycle, as shown in **Figure 12.** When an infected mosquito bites a human to obtain blood, it injects saliva that contains a chemical that prevents the blood from clotting. If the mosquito is infected with *Plasmodium*, it will also inject about 1,000 protists with its saliva. This infective stage of *Plasmodium* is called the **sporozoite.** Sporozoites infect the liver, where they rapidly divide and produce millions of cells of the second stage of the life cycle, called the **merozoite.** Merozoites infect red blood cells and divide rapidly. In about 48 hours, the blood cells rupture, releasing more merozoites and toxic substances. This begins a cycle of fever and chills that characterizes malaria. The cycle repeats every 48–72 hours (depending on which species is causing the infection) as more blood cells are infected and destroyed.

In the third stage, some of the merozoites in the blood develop into gametes. After these gametes are eaten by a mosquito, male and female gametes fuse to form zygotes. The zygotes divide many times to form many sporozoites, which migrate to the salivary glands of the mosquito. Unlike humans, the mosquito cannot regurgitate food or mix saliva and food in its mouth; therefore, the malaria parasite must mature in the mosquito before it can infect another human.

Figure 12 Life cycle of Plasmodium. *Plasmodium* has a complex life cycle that involves a mosquito, human blood, and liver cells.

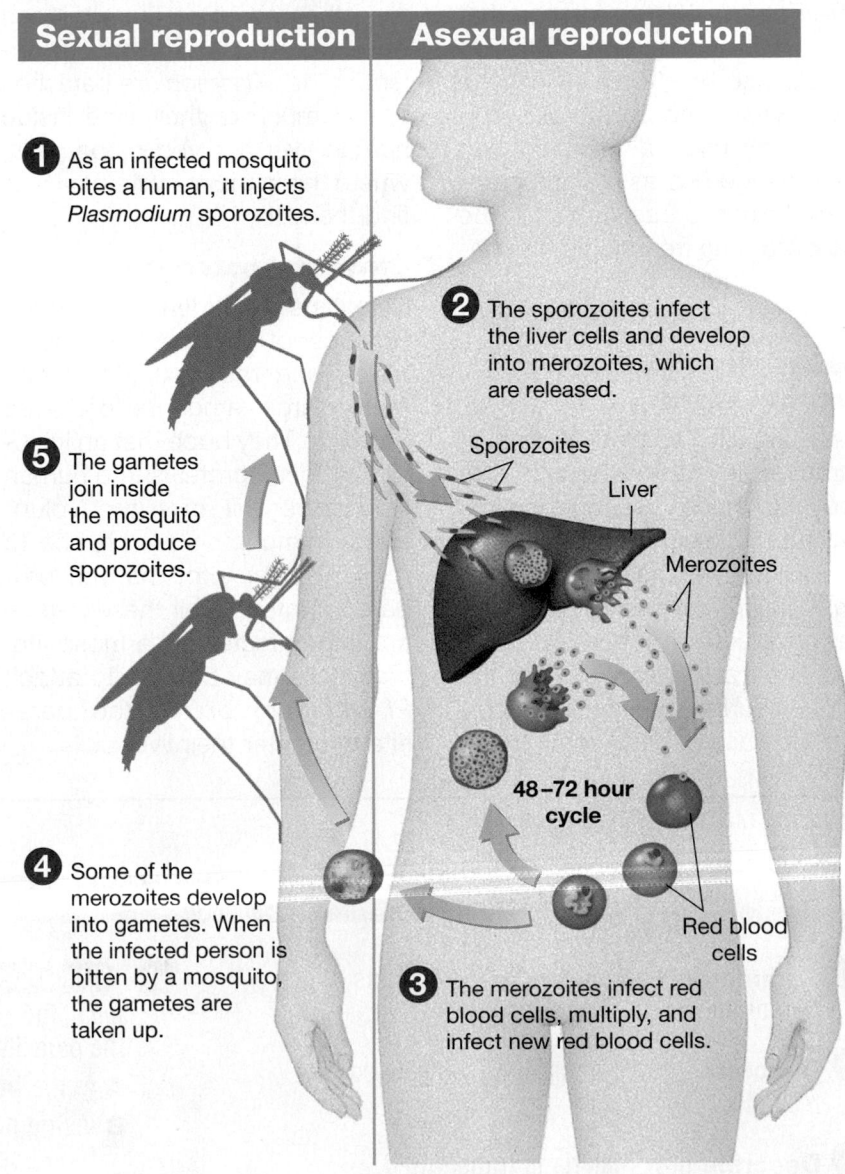

Sexual reproduction | **Asexual reproduction**

1. As an infected mosquito bites a human, it injects *Plasmodium* sporozoites.

2. The sporozoites infect the liver cells and develop into merozoites, which are released.

Sporozoites

Liver

Merozoites

5. The gametes join inside the mosquito and produce sporozoites.

48–72 hour cycle

4. Some of the merozoites develop into gametes. When the infected person is bitten by a mosquito, the gametes are taken up.

Red blood cells

3. The merozoites infect red blood cells, multiply, and infect new red blood cells.

Treating and Preventing Malaria

In the middle of the seventeenth century, quinine *(KWIE nien)*, a chemical derived from the bark of the cinchona tree *(Cinchona officialis,* found in South America) was discovered to be a remedy for malaria. Derivatives of quinine, such as chloroquine and primaquine, are now used to treat malaria.

Malaria can also be controlled by reducing mosquito populations. This is done by spraying insecticides, reducing mosquito breeding places, and introducing animals that will eat mosquito larvae, such as mosquito fish.

BIOWatch

Making a Malaria Vaccine

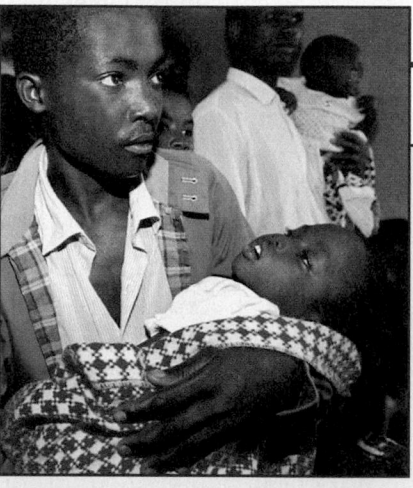

With their bodies racked by fever as high as 40°C (104°F) and chills so intense that their teeth chatter, up to 3 million people needlessly die of malaria every year. The worldwide epidemic of malaria is becoming more dangerous as malaria parasites become resistant to the standard drug treatments. To control the epidemic, scientists are trying to make a malaria vaccine.

Making the Malaria Vaccine Work

For a vaccine to work against a parasite, it must stimulate the body's immune cells to recognize and attack specific molecules on the surface of the parasite. But *Plasmodium* goes through several stages inside the human body, and each stage can have many different surface molecules. Therefore, a vaccine may cause the immune system to attack only one stage. If a few parasites survive, they could produce millions of new parasites. A second problem is that *Plasmodium* parasites spend much of their time inside human liver and red blood cells, where the immune system cannot find them.

Current Approaches

Scientists have identified surface proteins of *Plasmodium* sporozoites and merozoites and can make large amounts of these proteins. They hope that an injection of these proteins into human volunteers will give the volunteers' immune cells a chance to recognize the proteins and prepare for infection. If these people are later infected by a mosquito, their cells may be able to attack *Plasmodium* before the parasite can enter their liver cells.

World-wide efforts to develop an effective malaria vaccine have produced many failures and a few partial successes. Scientists are still working on ways to make the immune system's response to the vaccines stronger and longer lasting.

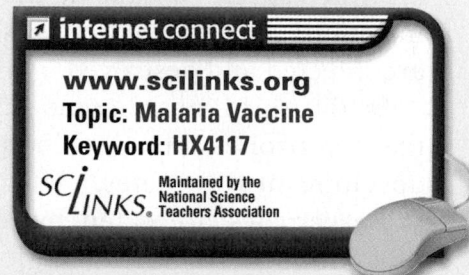

internet connect

www.scilinks.org
Topic: Malaria Vaccine
Keyword: HX4117

SC*LINKS*. Maintained by the National Science Teachers Association

---

Section 3 Review

1 **Summarize** two different ways protists affect human health.

2 **Describe** three human diseases caused by protists.

3 **Describe** how malaria is transmitted.

4 **Standardized Test Prep** In the life cycle of *Plasmodium*, the parasite that causes malaria, where do the parasite's gametes combine to form a zygote?

A in the human liver

B inside human blood cells

C inside a tsetse fly

D inside a mosquito

Study ZONE

CHAPTER HIGHLIGHTS

Key Concepts

1 Characteristics of Protists

- The kingdom Protista contains the most diverse groups of eukaryotic organisms of any kingdom.
- Protists live in moist environments and can be either free-living or parasitic.
- Some protists are able to reproduce sexually in times of stressful environmental conditions.

2 Protist Diversity

- Amoebas and forams are protists that move using cytoplasmic extensions called pseudopodia.
- Diatoms are unicellular protists with glasslike double shells.
- Algae are strictly photosynthetic protists that can be multicellular or unicellular. Algae are classified according to the type of photosynthetic pigment they contain.
- Flagellates move with the use of flagella.
- Ciliates are unicellular protists that use cilia to move.
- Protistan molds resemble fungi, but they are considered protists.
- Cellular slime molds normally live as individual cells and aggregate during times of stress.
- Plasmodial slime molds live as colonial organisms and form sexual reproductive structures that form and spread spores.
- Oomycetes are molds that live as saprophytes or parasites.
- All sporozoans are parasitic and have complex life cycles.

3 Protists and Health

- Protists negatively affect human health by causing diseases in humans and their food sources.
- Protists positively affect human health through their participation in food webs, through commensal relationships with humans and their food sources, and by recycling vital resources.
- Protists cause diseases such as dysentery, giardiasis, malaria, and toxoplasmosis.
- Malaria is the most serious protist disease that affects humans. Drugs and mosquito control can be used to control malaria.

Key Terms

Section 1

protozoan (461)
alga (461)
zygospore (462)
alternation of generations (463)
sporangium (463)

Section 2

amoeba (464)
pseudopodium (464)
diatom (466)
euglenoid (467)
kinetoplastid (467)
cilium (468)
plasmodium (470)

Section 3

sporozoite (473)
merozoite (473)

Understanding Key Ideas

1. Two important eukaryotic features that evolved among protists are
  a. photosynthesis and silica shells.
  b. forams and pseudopodia.
  c. sexual reproduction and multicellularity.
  d. spores and microtubules.

2. Which habitat is *least* likely to harbor any species of protist?
  a. ocean waters
  b. the human liver
  c. a desert
  d. leaf litter

3. Which pair is *not* a correct match between a protist and its manner of reproduction?
  a. Spirogyra: conjugation
  b. Ulva: alternation of generations
  c. Paramecium: alternation of generations
  d. Chlamydomonas: gametes

4. Pseudopodia are used by members of the phylum
  a. Bacillariophyta.
  b. Euglenophyta.
  c. Rhizopoda.
  d. Chlorophyta.

5. Red algae differ from green in that red algae
  a. always inhabit marine environments.
  b. lack pigment.
  c. are always multicellular.
  d. display alternation of generations.

6. A photosynthetic single-celled protist that moves using flagella would likely be classified as a member of which phylum?
  a. Apicomplexa
  b. Oomycota
  c. Bacillariophyta
  d. Euglenophyta

7. How is amebic dysentery spread?
  a. by drinking contaminated water
  b. from the bite of a mosquito
  c. from the bite of the tsetse fly
  d. by eating overcooked meat

8. 🌐 **BIOWatch** Why has a successful malaria vaccine been so difficult to make?

9. ⌗ **Concept Mapping** Make a concept map describing the protists. Try to include the following terms in your map: *red algae, malaria, protozoa, slime mold, autotrophs, plants, diseases, heterotrophs,* and *animals.* Include additional terms as needed.

Critical Thinking

10. Evaluating Methods People in a small village were being infected with a parasitic protist from water in a nearby river. Scientists decided to dump a large amount of sulfur into the river to kill the protists. What possible errors are there in using this method to prevent sickness?

11. Justifying Conclusions A scientist found two euglenas, below, and concluded that specimen A is heterotrophic, but specimen B is not Justify the scientist's conclusions.

Magnification: 206×

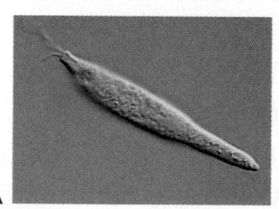 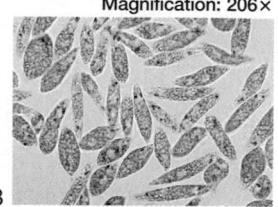

A   B

12. Applying Information If while drawing blood from a patient with malaria, some of the blood splattered on your bare arm, could you catch this disease?

Alternative Assessment

13. Finding Information Use the media center or Internet resources to learn more about the history of plant diseases caused by protistan molds. Present an oral report that describes how scientists controlled these diseases in the past and how the information gained through research has created new strategies for disease prevention. Use photos, illustrations, and graphics to enhance your presentation.

Standardized Test Prep

Understanding Concepts

Directions (1–5): For *each* question, write on a separate sheet of paper the letter of the correct answer.

1 What are photosynthetic protists with box-like double shells?
A. diatoms
C. kinetoplastids
B. euglenoids
D. plankton

2 What do slime molds form in order to reproduce?
F. plasmodia
H. slugs
G. pseudopodia
I. spores

3 Which of the following is **not** a characteristic of paramecia?
A. chloroplasts
C. micronucleus
B. cilia
D. pellicle

4 In what stage is *Plasmodium* infective?
F. larva
H. mosquito
G. merozoite
I. sporozoite

5 Malaria-transmitting mosquitoes prefer a warm climate. What might happen to the incidence of malaria in the world if global warming continues?
A. The geographical range of the disease might expand, possibly increasing the incidence of malaria.
B. Fewer mosquitoes would carry the disease, so the incidence of malaria would possibly decrease.
C. Fewer people would be infected by the disease, because the increased temperatures would kill the protist.
D. Malaria-transmitting mosquitoes would not be able to breed, possibly decreasing the incidence of malaria.

Directions (6): For the following question, write a short response.

6 How are kinetoplastids and dinoflagellates alike and different?

Reading Skills

Directions (7): **Read the passage below. Then answer the question.**

The Irish potato famine of 1845–1850 resulted in many deaths and an increase in Irish immigration to the United States. The famine was caused by a protist called *Phytophthora infestans*, which caused potatoes to turn black and mushy in a disease called late blight. *Phytophthora infestans* is part of the phylum Oomycota, which are water molds. Like fungi, water molds have thread-like structures that produce spores. Oomycetes are unusual in that their spores have two flagella that point in opposite directions.

7 Why were members of the phylum Oomycota once thought to be related to fungi?
F. The carbohydrates in their cell walls are similar to that of fungi.
G. Like fungi and unlike most protists, they carry out normal mitosis.
H. Their appearance and reproductive structures are similar to fungi.
I. Like most fungi, their spores have two flagella pointing in opposite directions.

Interpreting Graphics

Directions (8): **Base your answer to question 8 on the graph below.**

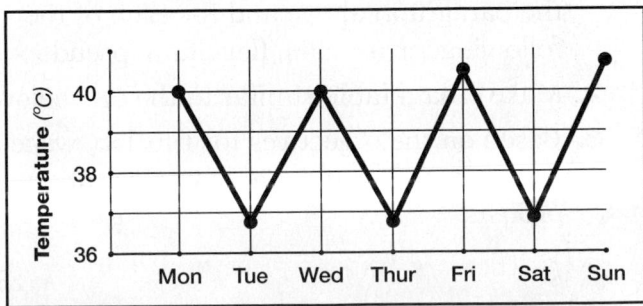

Daily Temperature of a Malaria Patient

8 At this rate, what will be the approximate temperature of the patient on Monday?
A. 37°C
C. 41°C
B. 39°C
D. 45°C

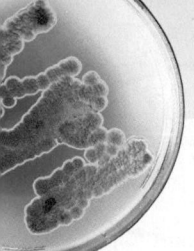

Exploration Lab

Observing Protistan Responses to Light

SKILLS
- Using scientific methods
- Using a microscope

OBJECTIVES
- **Identify** several different types of protists.
- **Compare** the structures, methods of locomotion and feeding, and behaviors of several different protists.
- **Relate** a protist's response to light to its method of feeding.

MATERIALS
- Detain™ (protist-slowing agent)
- microscope slides
- plastic pipets with bulbs
- mixed culture of protists
- toothpicks
- coverslips
- compound microscope
- protist references
- black construction paper
- scissors
- paper punch
- white paper
- sunlit window sill or lamp
- forceps

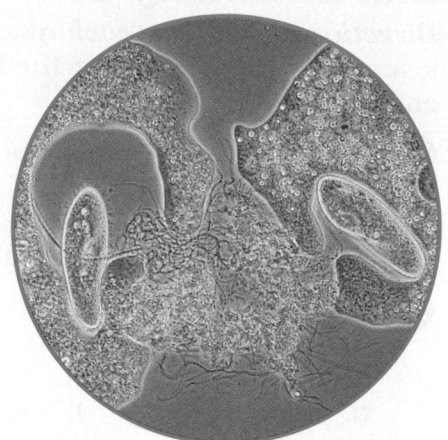

Before You Begin

Protists belong to the kingdom **Protista**, which is a diverse group of **eukaryotes** that cannot be classified as animals, plants, or fungi. Many protists are unicellular. Among the protists, there are **producers, consumers,** and **decomposers.** In this lab, you will observe live protists and compare their structures, methods of locomotion and feeding, and behaviors.

1. Write a definition for each boldface term in the paragraph above and for each of the following terms: cilia, flagellum, pseudopod.
2. Make a data table similar to the one below.
3. Based on the objectives for this lab, write a question you would like to explore about protists.

Procedure

PART A: Make Observations

1. **Caution: Do not touch your face while handling microorganisms.** Place a drop of Detain™ on a microscope slide. Add a drop of liquid from the bottom of a mixed culture of protists. Mix the drops with a toothpick. Add a coverslip. View the slide under low power of a microscope. Switch to high power.
2. Use references to identify the protists. Record data for each type of protist.
3. Repeat step 1 *without* using Detain™.
4. Punch a hole in a 40 × 20 mm piece of black construction paper that has a slight curl, as shown in the photo.

DATA TABLE				
Protist	Color	Method of locomotion	Method of feeding	Other observations

5. Place a wet mount of protists on a piece of white paper. Then put the paper and slide on a sunlit window sill or under a table lamp. Position the sun shade on top of the slide so that the hole is in the center of the coverslip.

6. To examine a slide, first view the area in the center of the hole under low power. *Note: Do not disturb the sun shade. Do not switch to high power.* Then have a partner carefully remove the sun shade with forceps while you observe the slide.

PART B: Design an Experiment

7. Work with members of your lab group to explore one of the questions written for step 3 of **Before You Begin.** To explore the question, design an experiment that uses the materials listed for this lab.

You Choose

As you design your experiment, decide the following:

a. what question you will explore

b. what hypothesis you will test

c. how long you will expose protists to light

d. how many times you will repeat your experiment

e. what your control will be

f. what data you will record and how you will make your data table

8. Write a procedure for your experiment. Make a list of all the safety precautions you will take. Have your teacher approve your procedure and safety precautions before you begin the experiment.

9. Set up and carry out your experiment.

PART C: Cleanup and Disposal

10. Dispose of lab materials and broken glass in the designated waste containers. Put protists in the designated containers. Do not put lab materials in the trash unless your teacher tells you to do so.

11. Clean up your work area and all lab equipment. Return lab equipment to its proper place. Wash your hands thoroughly before you leave the lab and after you finish all work.

Analyze and Conclude

1. **Summarizing Results** Describe the different types of locomotion you observed in protists, and give examples of each.

2. **Analyzing Results** Identify which protists were affected by light, and describe how they were affected.

3. **Inferring Conclusions** What is the relationship between a protist's response to light and its method of feeding?

4. **Further Inquiry** Write a new question about protists that could be explored with another investigation.

(?) Do You Know?

Do research in the library or media center to answer these questions:

1. What livestock diseases caused by parasitic protists are most commonly found in the United States?

2. How do backpackers avoid getting diseases caused by protists that are transmitted in water?

Use the following Internet resources to explore your own questions about protists.

internet connect

www.scilinks.org
Topic: Protists
Keyword: HX4153

SCI**LINKS**® Maintained by the National Science Teachers Association

Scarlet waxy
cap mushrooms

22 Fungi

✓ Quick Review

Answer the following without referring to earlier sections of your book.

1. **Distinguish** meiosis from mitosis. *(Chapter 7, Section 1)*

2. **Summarize** the importance of mycorrhizae to plants. *(Chapter 12, Section 3)*

3. **Describe** the process of succession. *(Chapter 16, Section 1)*

4. **Describe** the meaning of symbiosis. *(Chapter 17, Section 1)*

5. **Define** the term *mutualism*. *(Chapter 17, Section 1)*

Did you have difficulty? *For help, review the sections indicated.*

📖 Reading Activity

Before you begin this chapter, write the following statements on a separate piece of paper:

1. Fungi are very similar to plants.

2. Fungi are economically valuable.

3. Many fungi are beneficial to other organisms.

Take a minute or two to consider the statements and decide whether you agree or disagree with each. Share your opinions with the class. Read the chapter, and then reconsider the statements to see if your opinions are confirmed or changed.

Looking Ahead

Section 1

Characteristics of Fungi
Kingdom Fungi
Structures and Nutrients
Reproduction

Section 2

Fungal Diversity
Reproductive Structures
Zygomycetes
Ascomycetes
Basidiomycetes

Section 3

Fungal Partnerships
Symbiotic Relationships

🔲 internet connect

www.scilinks.org
National Science Teachers Association *sci*LINKS Internet resources are located throughout this chapter.

SCI*LINKS.* Maintained by the National Science Teachers Association

● Many fungi are edible, but some people are made sick by this mushroom. Never eat mushrooms growing in the wild. Some are fatally poisonous.

Characteristics of Fungi

Objectives

- **List** the characteristics of the kingdom Fungi.

- **Describe** the structure of a typical fungus body.

- **Identify** how fungi obtain nutrients.

- **Relate** the way fungi obtain nutrients to their role in ecosystems.

- **Distinguish** the ways that fungi reproduce.

Key Terms

chitin
hypha
mycelium

Study TIP

Organizing Information

The numbered list on this page gives four important characteristics of fungi. Use this information to start a concept map that summarizes the characteristics of fungi.

Kingdom Fungi

Some of the most unusual organisms that exist today are members of the kingdom Fungi. Mushrooms and molds are common fungi that grow so rapidly they sometimes appear overnight.

The first fungi were probably unicellular eukaryotes. Fungi first appeared on Earth about 430 million years ago. In the past, biologists grouped fungi with plants because fungi are immobile, have a cell wall, and appear "rooted" in the soil, as the mushrooms do in **Figure 1.** However, the unique features of fungi indicate that they should be classified as a separate kingdom.

1. **Fungi are heterotrophic.** The stalk and cap of the mushroom are not green like the stem and leaves of a plant. Plants appear green because they contain chlorophyll, the pigment that aids in photosynthesis. Fungi do not contain chlorophyll. Rather, they obtain energy by breaking down organic material that they absorb from their environment.

2. **Fungi have filamentous bodies.** Plants are made of many cell and tissue types, but fungi are made of long, slender filaments. These filaments weave tightly together to form the fungus body and reproductive structures, such as the mushroom.

3. **Fungal cells contain chitin.** The cells of the mushroom, like the cells of all fungi, have walls made of **chitin** *(KIE tihn)*, the tough polysaccharide found in the hard outer covering of insects. Plant cells have walls made of cellulose, a different polysaccharide.

4. **Fungi exhibit nuclear mitosis.** Mitosis in fungi is different from mitosis in plants and in most other eukaryotes. In most eukaryotes, the nuclear envelope disintegrates in prophase and re-forms in telophase. In dividing mushroom cells, however, the nuclear envelope remains intact from prophase to anaphase. As a result, spindle fibers form within the nucleus. The spindle fibers then drag chromosomes to opposite poles of the nucleus, rather than to opposite poles of the cell. Mitosis is complete when the nuclear envelope pinches in two.

Figure 1   Mushrooms

These mushrooms are actually the reproductive structures of a large network of filaments that makes up the body of a fungus.

Structures and Nutrients

Structures

In **Figure 2,** the fungus *Penicillium* is shown growing on an orange. The green and white fuzz you recognize as mold is actually the reproductive structures of the fungus. The body of the fungus lies within the tissues of the orange. All fungi except yeasts have bodies composed of slender filaments called **hyphae** *(HIE fee).*

When hyphae grow, they branch and form a tangled mass called a **mycelium** *(mie SEE lee uhm),* shown in Figure 2. A mycelium can be made of many meters of individual hyphae. This body organization provides a high surface-area-to-volume ratio, which makes a fungus well suited for absorbing nutrients from its environment.

Each hypha is a long string of cells divided by partial walls. Some species do not have walls between cells. Cytoplasm flows freely throughout the hypha.

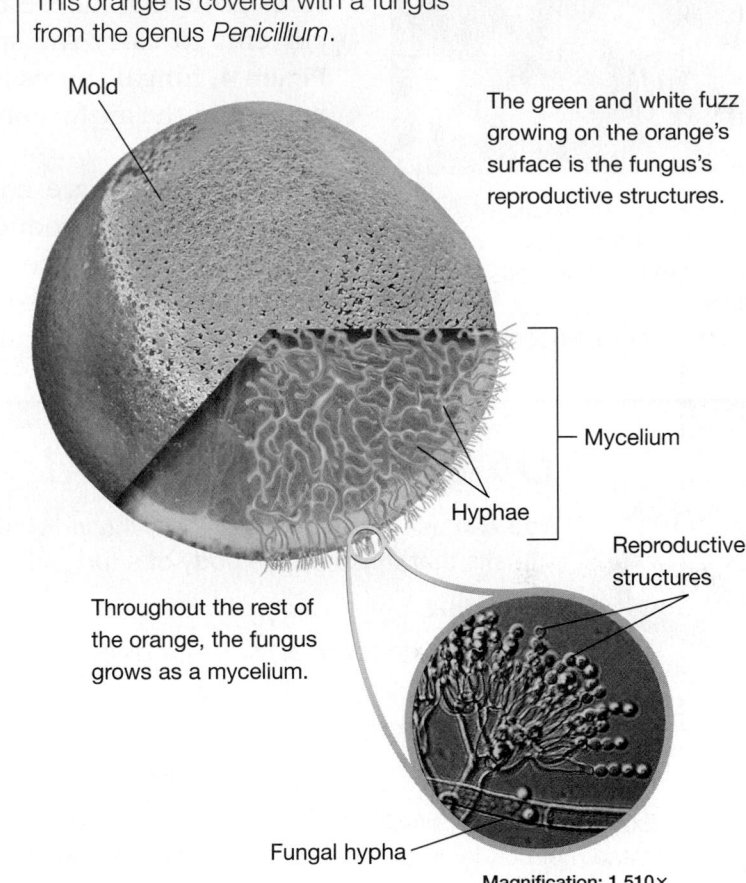

Figure 2 *Penicillium* mold

This orange is covered with a fungus from the genus *Penicillium.*

Mold

The green and white fuzz growing on the orange's surface is the fungus's reproductive structures.

Mycelium

Hyphae

Reproductive structures

Throughout the rest of the orange, the fungus grows as a mycelium.

Fungal hypha

Magnification: 1,510×

Nutrients

All fungi obtain nutrients by secreting digestive enzymes that break down organic matter in their environment. Fungi then absorb the decomposed molecules. Many fungi decompose nonliving organic matter, such as leaves, branches, dead animals, and waste. So, fungi are resource recyclers. Other fungi, such as the fungus that causes ringworm, are parasites that absorb nutrients from living hosts.

Parasitic fungi compete for nutrients with their hosts. In humans, they sometimes cause infectious diseases, such as athlete's foot and yeast infections. **Figure 3** warns of the fungus *Histoplasma capsulatum,* which invades the body's organs and causes life-threatening infections.

Fungi often grow on human foods, such as bread and fruit, making the food undesirable. Fungi are also known to attack nonfood materials, such as paper, cardboard, cloth, paint, and leather.

Some fungi are commercially valuable. Unicellular fungi called yeasts are useful in baking, brewing, and wine-making. Other fungi provide the flavor and aroma of certain cheeses. Many kinds of antibiotics, such as penicillin, are produced by fungi.

RECOMMENDATIONS BEFORE ENTERING THE CAVERNS

STOP

You enter at your own risk

- WEAR A FACE MASK
- DO NOT BRING ANY FOOD OR DRINK
- DO NOT CARRY A BACKPACK OR ANY TYPE OF EQUIPMENT
- DO NOT COLLECT ANY TYPE OF MATERIAL
- TAKE A SHOWER AFTER LEAVING OF
- WASH YOUR CLOTHES   THE CAVERNS

Figure 3 Dangerous fungi. *Histoplasma capsulatum* grows in the feces of bats and birds and infects humans when the dried spores are inhaled. Symptoms of infection include fever, cough, chills, chest pain, and muscle aches.

Reproduction

Fungi reproduce by releasing spores formed sexually or asexually in reproductive structures at the tips of hyphae. Reproductive structures grow high above the food source. This adaptation allows air currents to carry the spores to a new habitat. As you can see in **Figure 4,** fungal spores are so small and light that they remain suspended in the air for long periods of time; the wind can carry them great distances.

Fungal spores are haploid. Most spores are formed by mitosis during asexual reproduction. In sexual reproduction, hyphae from two mating types fuse. The fused hyphae form a sexual reproductive structure. On this structure, the fungus forms spores through fusion of the two genetically different nuclei.

Figure 4 Puffball.
Lycoperdon perlatum, a puffball, releases hundreds of thousands of spores through a small opening.

QUICK LAB

Observing Bread Mold

You can use a microscope to see the individual threads of cells that make up the body of a fungus.

Materials

prepared slide of *Rhizopus*—black bread mold, compound microscope

Procedure

1. Examine a slide of black bread mold under low power of a microscope.

2. Move the slide to an area where you can clearly see threadlike structures.

3. Draw what you see in your lab notebook. Be sure to use at least one third of the page.

4. Move the slide to examine an area where you can clearly see the round bulblike structures.

5. Draw what you see in your notebook.

Analysis

1. Label the drawings you made, using **Figure 2** as a guide.

2. Explain where you would find each structure on the loaf of bread shown above.

3. Critical Thinking Recognizing Relationships Relate the structures you drew to their functions described in the text.

Section 1 Review

1 Distinguish the characteristics of fungi from those of plants.

2 Compare the characteristics of the mycelium with those of the reproductive structures.

3 Summarize the way fungi obtain nutrients.

4 Describe the role fungi play in the environment.

5 Summarize the different ways that fungi reproduce.

6 Critical Thinking Evaluating Conclusions Two of your fellow students insist that yeasts should be classified as protists because they are eukaryotic unicellular organisms. Evaluate their claim.

7 Standardized Test Prep An organism cannot be a fungus if it

A is unicellular.

B releases spores.

C is photosynthetic.

D has cell walls.

Fungal Diversity

Reproductive Structures

You can see how diverse fungi are if you examine the types of reproductive structures that they have. Based on the types of structures produced during sexual reproduction, fungi can be classified in three phyla. **Table 1** below lists some of these characteristics.

Asexual Reproduction

A fourth group, the deuteromycetes, is composed of fungi in which no sexual stage has been seen. Traditionally, this group has been called a phylum. Through the use of molecular techniques, scientists have reclassified most of these asexually reproducing organisms into the phylum Ascomycota.

There are about 17,000 species without a sexual stage. Many of these fungi are economically important. For example, some species of *Penicillium* produce the antibiotic penicillin. Other species produce the unique flavors of some cheeses. Species of *Aspergillus* are used for fermenting soy sauce and producing citric acid. Most of the fungi that cause skin diseases, such as athlete's foot and ringworm, are also deuteromycetes.

Objectives

- **Describe** the characteristics used to classify fungi.
- **List** two commercial uses for fungi.
- **Describe** three phyla of fungi.
- **Distinguish** between the life cycles of zygomycetes, ascomycetes, and basidiomycetes.
- **Describe** the mushroom *Amanita muscaria*.

Key Terms

zygosporangium
stolon
rhizoid
ascus
yeast
budding
basidium

Table 1 Three Sexually Reproducing Phyla of Fungi

Phylum	Distinctive characteristics	Examples
Zygomycota	Sexual spores are formed in zygosporangia; hyphae have no walls	black bread molds
Ascomycota	Sexual spores are formed in asci; hyphae are divided by walls	morels, truffles, yeasts, cup fungi
Basidiomycota	Sexual spores are formed in basidia; hyphae are divided by walls	mushrooms, puffballs, rusts, smuts

internet connect

www.scilinks.org
Topic: Characteristics of Fungi
Keyword: HX4037

SCiLINKS. Maintained by the National Science Teachers Association

Zygomycetes

Magnification: 10×

If you place an uncovered loaf of bread near a windowsill, after a few days a cottony mold, shown in **Figure 5,** will cover its surface. Common black bread mold, *Rhizopus stolonifer,* is a member of the phylum Zygomycota *(zie goh my COHT uh)*. Members of the phylum Zygomycota are named for the thick-walled sexual structures called **zygosporangia** *(zie goh spohr AN jee uh)* that characterize these members.

Species of *Rhizopus* and other zygomycetes usually live in the soil and feed on decaying plant and animal matter. The mycelia that grow along the surface of the bread are called **stolons** *(STOH lahnz)*. The hyphae that anchor the fungus in the bread are called **rhizoids** *(RIE zoydz)*. The hyphae of zygomycetes usually do not have walls.

Asexual reproduction in zygomycetes is much more common than sexual reproduction. During asexual reproduction, haploid spores are produced in the tips of specialized hyphae. It is the spores in fungi that cause fungal allergies in people. When they mature, these spores are shed and carried by the wind to new locations, where they germinate and start new mycelia. Reproduction in *Rhizopus* is shown in **Figure 6.**

Figure 5   Bread mold. *Rhizopus stolonifer* is often found growing on bread.

Figure 6   Life cycle of zygomycetes. Zygomycetes may reproduce sexually or asexually.

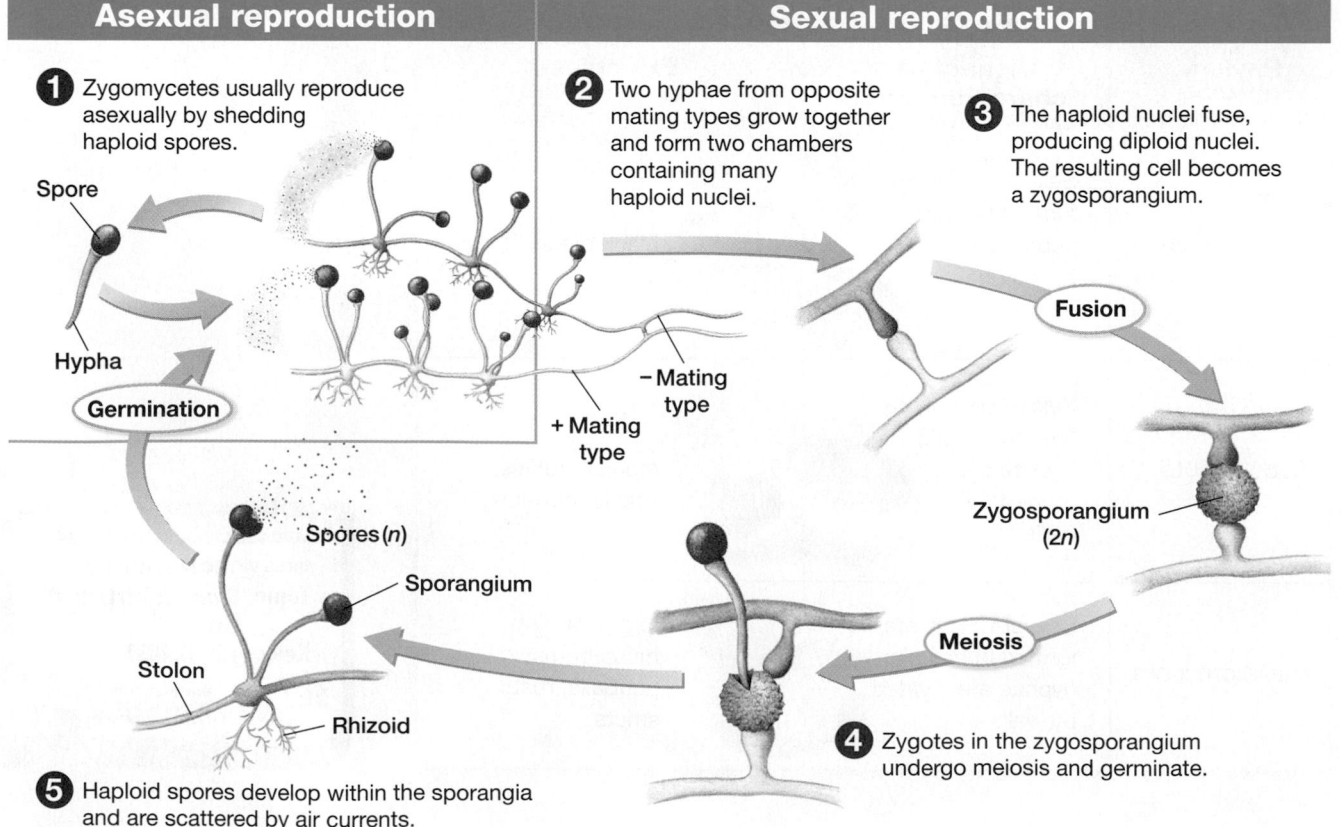

Asexual reproduction

1 Zygomycetes usually reproduce asexually by shedding haploid spores.

Spore

Hypha

Germination

Spores (n)

Sporangium

Stolon

Rhizoid

5 Haploid spores develop within the sporangia and are scattered by air currents.

Sexual reproduction

2 Two hyphae from opposite mating types grow together and form two chambers containing many haploid nuclei.

3 The haploid nuclei fuse, producing diploid nuclei. The resulting cell becomes a zygosporangium.

− Mating type

+ Mating type

Fusion

Zygosporangium (2n)

Meiosis

4 Zygotes in the zygosporangium undergo meiosis and germinate.

Ascomycetes

The chestnut tree, *Castanea dentata,* was once a common tree in the eastern United States. Around 1890, a disease called chestnut blight wiped out virtually all the chestnut trees within a few years. Chestnut blight is caused by *Endothia parasitica,* a member of the phylum Ascomycota *(AS koh mie koh tuh).* Other ascomycetes include flavorful morels and truffles prized by gourmet chefs.

The ascomycetes are named for their characteristic sexual reproductive structure. The microscopic **ascus** *(AS kuhs)* is a saclike structure in which haploid spores are formed. Asci usually form within the interwoven hyphae of a cup-shaped fruiting body. Reproduction in a typical ascomycete is shown in **Figure 7.**

Ascomycetes usually reproduce asexually. Asexual spores form at the tips of the hyphae. The spores are not contained in any sac or structure. When the spores are released, air currents carry them to other places, where they may germinate and form new mycelia.

Yeast is the common name given to unicellular ascomycetes. There are about 350 species of yeasts. *Saccharomyces cerevisiae,* or baker's yeast, has been used for thousands of years to make bread and alcoholic beverages, such as beer. Other yeasts, such as *Candida albicans,* cause human disease. *C. albicans* causes thrush, a disease in which milk-white lesions form on the mouth, lips, and throat.

Most yeasts reproduce asexually by fission or budding. In **budding,** a small cell forms from a large cell and pinches itself off from the large cell.

Figure 7   Life cycle of ascomycetes. Ascomycetes can reproduce sexually or asexually.

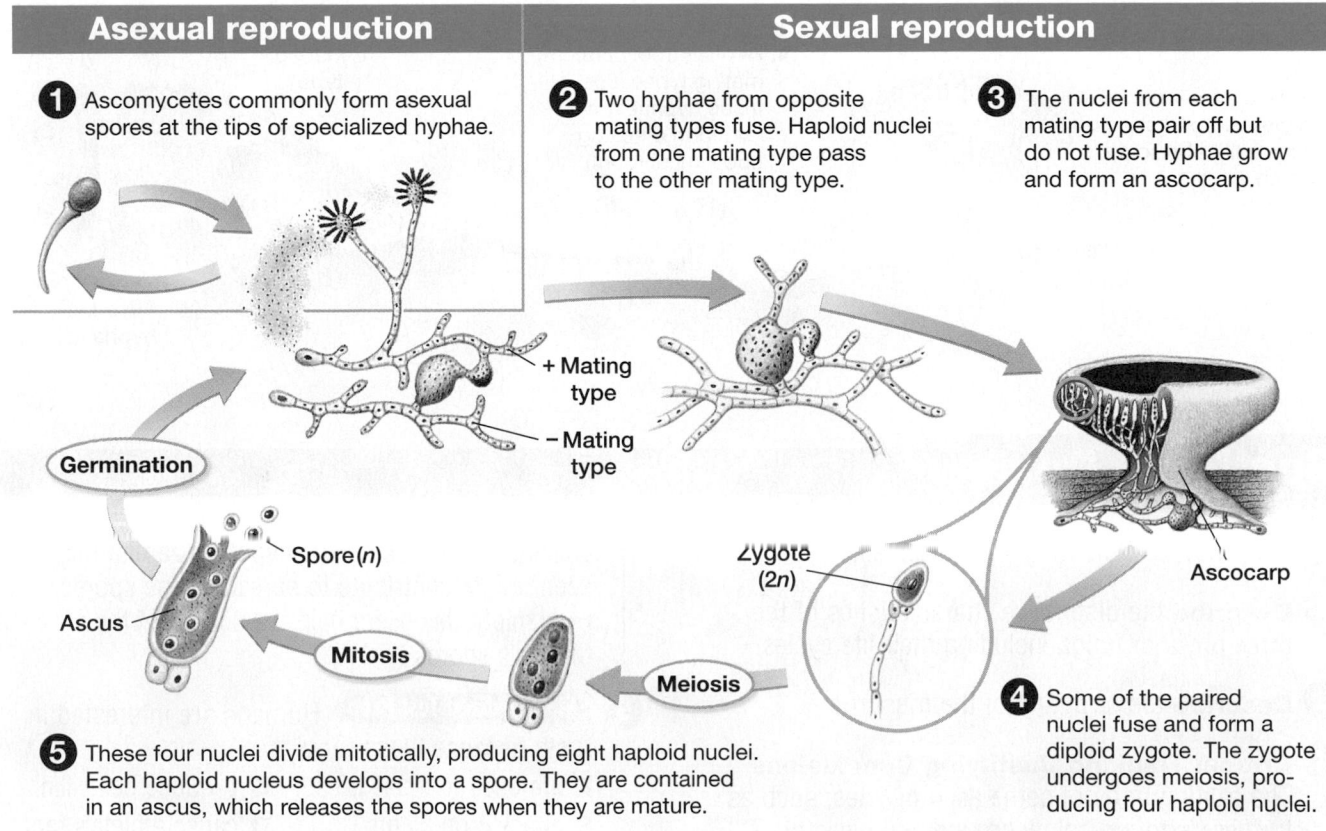

Asexual reproduction

① Ascomycetes commonly form asexual spores at the tips of specialized hyphae.

Germination

Spore(*n*)

Ascus

Mitosis

⑤ These four nuclei divide mitotically, producing eight haploid nuclei. Each haploid nucleus develops into a spore. They are contained in an ascus, which releases the spores when they are mature.

Sexual reproduction

② Two hyphae from opposite mating types fuse. Haploid nuclei from one mating type pass to the other mating type.

③ The nuclei from each mating type pair off but do not fuse. Hyphae grow and form an ascocarp.

+ Mating type

− Mating type

Zygote (2*n*)

Meiosis

Ascocarp

④ Some of the paired nuclei fuse and form a diploid zygote. The zygote undergoes meiosis, producing four haploid nuclei.

Basidiomycetes

The kind of fungi with which you are probably most familiar—mushrooms—are members of the phylum Basidiomycota *(buh SIHD ee oh mie koh tuh)*. Other basidiomycetes include toadstools, puffballs, jelly fungi, and shelf fungi. The **basidium** *(buh SIHD ee uhm)* is the club-shaped sexual reproductive structure for which the basidiomycetes are named. Spores are produced on this structure. You can see these spores in the *Up Close: Mushroom* feature later in this chapter. Asexual reproduction is rare among the basidiomycetes, except in some rusts and smuts. These two important groups of plant pathogens affect many crop plants, as shown in **Figure 8.** Sexual reproduction of a typical basidiomycete is illustrated in **Figure 9.** Many mushrooms are harmless, but many are also deadly, such as the *Amanita muscaria* (fly agaric), shown in the *Up Close* feature. Other *Amanita* species have names, such as death angel and destroying angel, that reflect the danger of their toxins for humans.

Figure 8　Rust on wheat.
Rust is a basidiomycete that attacks cereal crops, making them unfit for humans to eat.

Figure 9　Life cycle of basidiomycetes

Basidiomycetes usually reproduce sexually by means of a fruiting body, also called a mushroom.

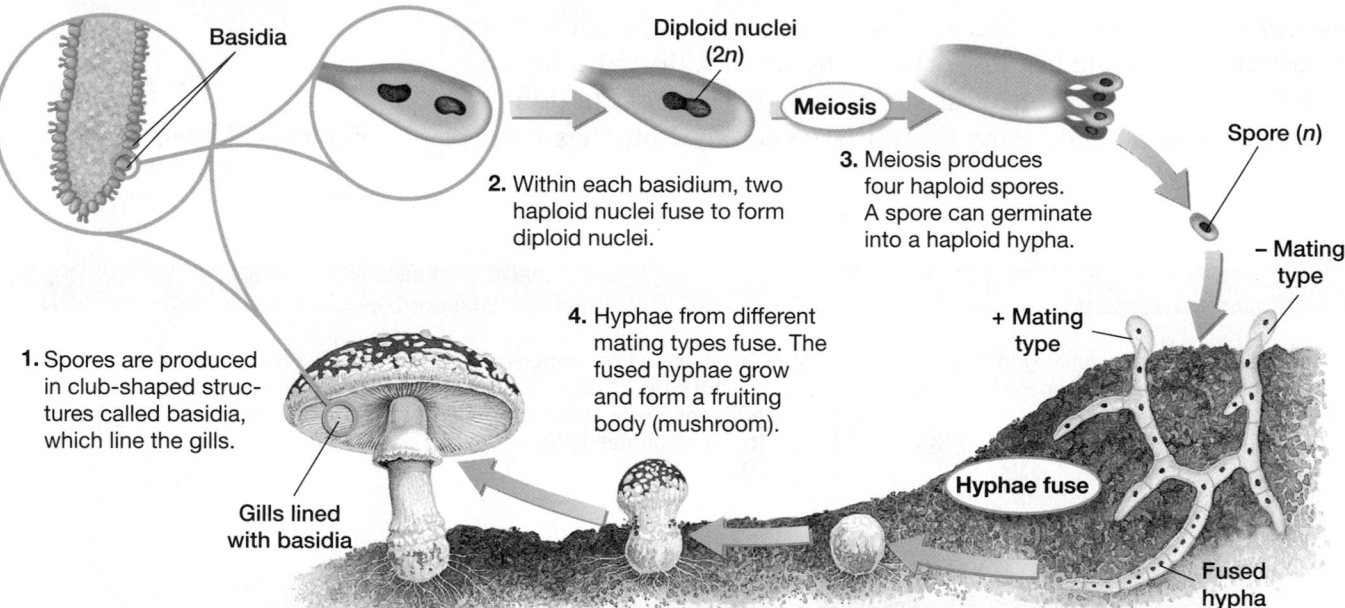

Basidia

Diploid nuclei (2n)

Meiosis

Spore (n)

2. Within each basidium, two haploid nuclei fuse to form diploid nuclei.

3. Meiosis produces four haploid spores. A spore can germinate into a haploid hypha.

– Mating type

+ Mating type

4. Hyphae from different mating types fuse. The fused hyphae grow and form a fruiting body (mushroom).

1. Spores are produced in club-shaped structures called basidia, which line the gills.

Gills lined with basidia

Hyphae fuse

Fused hypha

Section 2 Review

1. **Summarize** how fungi are classified.

2. **Describe** the distinctive characteristics of the three phyla of fungi, including their life cycles.

3. **Describe** the structure of the mushroom.

4. **Critical Thinking Justifying Conclusions**
The fruiting body of some ascomycetes, such as truffles, is found below ground and gives off a delicious scent. Scientists hypothesize that this scent might contribute to spreading the spores. How might this scent help the spread of the fungus's spores?

5. **Standardized Test Prep** Humans are interested in rusts because these fungi
  A are used to make bread.　C produce penicillin.
  B attack crop plants.　D cause athlete's foot.

Up Close

Mushroom

- **Scientific name:** *Amanita muscaria*
- **Size:** 10–15 cm
- **Habitat:** Moist organic soils
- **Nutrition:** Heterotrophic

Characteristics

Cell structure *A. muscaria* and other fungi have cell walls made of chitin, a complex polysaccharide also found in the external skeleton of insects. In some fungi, hyphae are not divided into separate cells but have many nuclei in the same cytoplasm. In other fungi, hyphae are divided into cells by perforated walls called septa.

Reproduction Under proper conditions, underground hyphae grow upward and weave together to produce a mushroom. Mushrooms are the reproductive structures of fungi such as *A. muscaria*. A mushroom has a flattened cap attached to a stem called a stalk. The underside of the cap is lined with rows of gills. Thousands of club-shaped reproductive cells called basidia form on the gills. Through fusion and meiosis, each basidium produces spores that are released and form new hyphae.

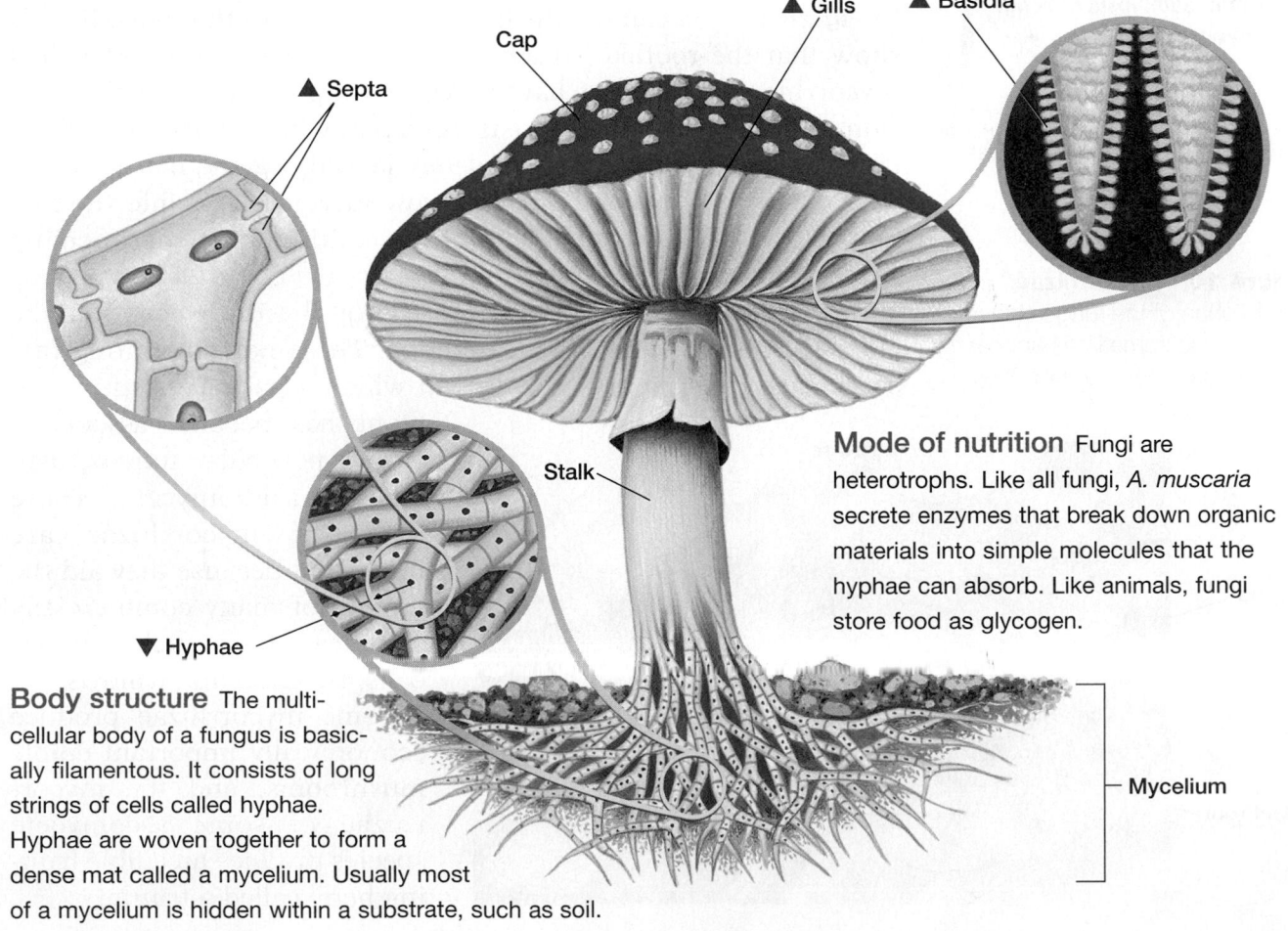

▲ Septa

▲ Gills   ▲ Basidia

Cap

Mode of nutrition Fungi are heterotrophs. Like all fungi, *A. muscaria* secrete enzymes that break down organic materials into simple molecules that the hyphae can absorb. Like animals, fungi store food as glycogen.

Stalk

▼ Hyphae

Mycelium

Body structure The multicellular body of a fungus is basically filamentous. It consists of long strings of cells called hyphae. Hyphae are woven together to form a dense mat called a mycelium. Usually most of a mycelium is hidden within a substrate, such as soil.

Fungal Partnerships

Objectives

- **Distinguish** two symbiotic relationships that involve fungi.

- **Summarize** the ecological importance of mycorrhizae.

- **Describe** lichens.

Key Terms

mycorrhiza
lichen

internet connect

www.scilinks.org
Topic: Symbiosis of Fungi
Keyword: HX4172

SCI LINKS. Maintained by the National Science Teachers Association

Figure 10 Mycorrhizae.
The hyphae of the fungus in mycorrhizae sometimes appear as a tangled mass around the root of the plant.

Hypha

Plant root

Symbiotic Relationships

Fungi are involved in many kinds of symbiotic associations with algae and plants. These mutualistic relationships play important roles in ecology. You may recall that mutualism is a type of symbiosis in which each partner benefits. In these symbiotic associations, the fungus (a heterotroph) provides minerals and other nutrients that it absorbs from the environment. The algae or plant (a photosynthesizer) provides the ability to use sunlight to power the building of carbohydrates.

Mycorrhizae

A **mycorrhiza** *(MIE koh RIE zuh)* is a type of mutualistic relationship formed between fungi and vascular plant roots. The hyphae help transfer phosphorus and other minerals from the soil to the roots of the plant, while the plant supplies carbohydrates to the fungus.

In the mycorrhizae of most species of plants, the hyphae penetrate the outer cells of the root. The fungus is usually a zygomycete. In **Figure 10,** you can see the hyphae that grow in the roots. Fossils show that the rootlike structures of the earliest plants often had mycorrhizae, which may have played an important role in the invasion of land by plants. Scientists think that when plants invaded the land the soil of that time completely lacked organic matter. However, plants with mycorrhizae can grow successfully in infertile soil. Some vascular plants survive today by continuing this partnership as mycorrhizae.

In many plants, the mycorrhizae do not physically penetrate the plant root but instead wrap around it. These nonpenetrating mycorrhizae represent relationships in which a particular species of plant has become associated with a particular fungus, usually a basidiomycete. These kinds of mycorrhizae are important because they aid the growth of many commercially significant trees, such as pines, oaks, beeches, and willows.

Some mycorrhizae produce economically important edible mushrooms, and the mycorrhizae of some ascomycete species produce an edible fruiting body called a truffle.

Lichens

A **lichen** *(LIE kuhn)* is a symbiosis between a fungus and a photosynthetic partner such as a green alga, a cyanobacterium, or both. The photosynthetic partner provides carbohydrates. It is protected from the environment by the fungal partner, which helps it absorb mineral nutrients. In most lichens, the fungus is an ascomycete. When you look at a lichen, such as the ones in **Figure 11,** you are seeing the fungus. The photosynthetic partner is hidden between the layers of hyphae. Sunlight penetrates the layers of hyphae and enables the photosynthetic partner to carry out photosynthesis.

Magnification: 100×

Figure 11   Lichens. The algae, shown as green cells in the micrograph, are the photosynthetic partners of the fungus growing in this British soldier lichen.

The tough construction of the fungus combined with the photosynthetic abilities of the alga, or cyanobacterium, has enabled lichens to colonize harsh habitats. Lichens have been found in arid desert regions and in the Arctic; they grow on bare soil, on tree trunks, and on sunbaked rocks. Recall that during succession, lichens are often the first colonists. They break down rocks and prepare the environment for other organisms. Lichens are a key component of primary succession because they are able to grow on rock and help break the rock down into soil. Lichens containing cyanobacteria carry out nitrogen fixation and introduce useful forms of nitrogen into the soil.

Analyzing the Effect of Mycorrhizae

Background

Two groups of plants were planted in similar soils under similar conditions, but group A was grown in sterilized soil and group B was grown in nonsterilized soil. After 18 weeks of growth, a photograph was taken of the plants. Examine the photographs, and answer the following questions:

Analysis

1. **Compare** the growth of the two groups. Which grew faster?

2. **Explain** why one group grew better than the other group.

3. **Critical Thinking Inferring Relationships** Suggest a possible cause of slower growth in the smaller plants.

4. **Recommend** a course of action to restore growth in the stunted plants.

Lichens are able to survive drought and freezing by becoming dormant. When moisture and warmth return, lichens resume normal activities. In harsh environments, lichens may grow slowly. Some lichens that grow in the mountains appear to be thousands of years old and cover an area no larger than a fist. These lichens are among the oldest living organisms on Earth. Although lichens are known to survive extremes of temperatures, they are susceptible to chemical changes in their environment and so have become a living indicator of the amount of pollution in the environment in which they live.

Exploring Further

Lichens as Environmental Watchdogs

Since the 1950s scientists have found that most lichens require clean air to thrive. In the Los Angeles Basin, for example, rising smog levels have been linked to the disappearance of lichens. In the Pacific Northwest, lichens are most abundant in old-growth forests with good air quality. For these reasons, scientists have been using lichens to monitor air pollution.

Lichens as Indicators of Air Pollution

Lichens have no roots, so the nutrients they take up must come from the air. Rain, fog, and dew wet the surface of a lichen. When they are wet, lichens absorb nutrients and any pollutants that are in the air.

Lichens can live for centuries, making them well suited for studies of air-pollution changes that occur over a long time. Many lichen species also have large geographical ranges. Thus, a single species can indicate air quality at different distances from a source of pollution, such as a factory or power plant.

How Are Lichens Used?

To monitor air quality with lichens, scientists often map the distribution of lichens in an area. They count the number of lichen species, note how often each species occurs, and measure the total area covered by each species. Mapping studies done over many years can reveal long-term changes in lichen survival.

Scientists can obtain more detailed data on the effects of air pollution by determining the concentration of metals and other pollutants in lichen samples. They also assess the health of a lichen by measuring its chlorophyll content and its rate of photosynthesis.

To test air quality in places where lichens do not exist, scientists sometimes transplant healthy lichens from areas where they occur naturally. They then analyze the transplanted lichens for pollutants and look for any changes in the health of the lichens that may be caused by the move.

Pollutants from the stove pipe keep lichens from growing on part of this roof.

Section 3 Review

1. Describe two types of symbioses that involve fungi.

2. Explain how mycorrhizae are thought to have helped plants to colonize land.

3. Identify the organisms found in lichens.

4. Summarize how lichens promote the process of biological succession.

5. Standardized Test Prep The relationship between a fungus and an alga in a lichen is an example of

A mutualism.   C parasitism.

B commensalism.   D predation.

CHAPTER HIGHLIGHTS

Key Concepts

1 Characteristics of Fungi

- Fungi are eukaryotic heterotrophs. Their bodies are made up of slender woven filaments. Fungal cells contain chitin and go through nuclear mitosis.

- Fungi obtain nutrients by secreting digestive enzymes and absorbing the decomposed nutrients from their environment.

- Fungi decompose dead organic matter; they are an important resource recycler.

- Most fungi reproduce by releasing spores that are produced asexually and sexually.

2 Fungal Diversity

- Fungi are classified by their sexual reproductive structures.

- Fungi in which sexual reproduction has not been observed are referred to as deuteromycetes.

- Fungi in the phylum Zygomycota produce spores in thick-walled sexual structures called zygosporangia.

- Fungi in the phylum Ascomycota produce spores in a saclike structure called an ascus.

- Yeasts are unicellular ascomycetes that reproduce by budding.

- Fungi in the phylum Basidiomycota produce spores in a club-shaped structure called a basidium.

3 Fungal Partnerships

- Fungi can be involved in two types of symbioses, mycorrhizae or lichens.

- Mycorrhizae are symbiotic associations in which a fungus transfers minerals to a plant's roots, which in turn supply carbohydrates to the fungus.

- The fungal partner in a lichen protects the photosynthetic partner and provides the lichen with minerals. The photosynthetic partner provides the fungus with carbohydrates.

Key Terms

Section 1
chitin (482)
hypha (483)
mycelium (483)

Section 2
zygosporangium (486)
stolon (486)
rhizoid (486)
ascus (487)
yeast (487)
budding (487)
basidium (488)

Section 3
mycorrhiza (490)
lichen (491)

Understanding Key Ideas

1. Fungi differ from plants in that fungi
   a. are multicellular.
   b. are immobile.
   c. have cell walls.
   d. are heterotrophic.

2. Which of the following characteristics is shared by all fungi and helps them obtain their nutrients?
   a. external digestion
   b. phagocytosis
   c. feeding on nonliving matter
   d. catching prey

3. Deuteromycetes are more difficult to classify than fungi of other phyla because
   a. they develop from zygosporangia.
   b. they are parasitic.
   c. they undergo meiosis.
   d. they do not reproduce sexually.

4. Some fungal associations no larger than a fist appear to be thousands of years old. These have been found
   a. in temperate forests.
   b. on well-irrigated alluvial plains.
   c. in fields of corn.
   d. in harsh environments, high in the mountains.

5. One might expect that plants without mycorrhizae are
   a. more likely to get fungal diseases.
   b. unsuccessful in the transfer of minerals from the soil to the roots.
   c. best suited to poor soil conditions.
   d. primitive and might soon become extinct.

6. In basidiomycetes, most of the mycelium grows
   a. above the ground as a mushroom.
   b. above the ground as an ascocarp.
   c. as a network of hyphae in the soil.
   d. as a network of hyphae within bread.

7. In many fungi, both asexual and sexual reproduction involve
   a. the fusion of opposite mating types.
   b. the shedding of haploid spores.
   c. the formation of basidia.
   d. budding.

8. Fungi differ from most eukaryotes in that they
   a. are heterotrophic.
   b. have a cell wall.
   c. reproduce sexually.
   d. exhibit nuclear mitosis.

9. **Exploring Further** Explain how scientists use lichens to monitor air quality and why lichens were chosen by scientists to monitor air quality.

10. **Concept Mapping** Construct a concept map that describes the structure and reproductive methods of different fungi. Use the following terms in your map: *chitin, hyphae, zygosporangia, stolon, rhizoid, ascus, yeast,* and *budding*. Use additional terms as necessary.

Critical Thinking

11. **Predicting Outcomes** If all fungi suddenly disappeared from Earth, what types of changes would you notice immediately? after a period of time?

12. **Forming Reasoned Opinions** When purchasing garden plants, is it better to buy bare root plants or plants that are established in a pot of soil? Explain your reasoning from an ecological point of view.

Alternative Assessment

13. **Summarizing Information** Use the media center or Internet resources to learn about the economic impact of fungi. Research fungi used as foods and the importance of certain fungi in the preparation or manufacture of food. Create a display to summarize your findings.

Understanding Concepts

Directions (1–4): For *each* question, write on a separate sheet of paper the letter of the correct answer.

1 What carbohydrate makes up the cell walls of fungi?
A. cellulose
B. chitin
C. fructose
D. glucose

2 What kind of structure has a symbiotic relationship with fungi in mycorrhizae?
F. algae
G. chloroplasts
H. lichens
I. roots

3 How does the zygosporangium of *Rhizopus stolonifer* function to ensure the survival of the species?
A. It anchors the fungus to the food source.
B. It causes allergies in humans who want to eat it.
C. It dissolves organic matter for the fungus to absorb.
D. It remains dormant during unsuitable conditions.

4 What part of a fungus grows within the food source?
F. ascus
G. mycelium
H. reproductive structure
I. zygosporangium

Directions (5): For the following question, write a short response.

5 A scientist concludes that several cultures of cup-shaped ascocarps are a species of ascomycetes that only reproduce asexually. Why is the scientist's conclusion incorrect?

Test TIP

When using a graph to answer a question, study the data plotted on the graph to identify any trends in the data before you answer the question.

Reading Skills

Directions (6): Read the passage below. Then answer the question.

Downy mildew caused by the fungus *Plasmopara viticola* devastated French grape crops during the nineteenth century. Pierre Millardet developed the Bordeaux mixture, which contained lime and copper sulfate, to prevent people from stealing and eating the grapes. He noticed that the mixture prevented the fungus from growing. In fact, it killed the fungal spores.

6 Which of the following is true of fungal spores?
A. They contain cellulose.
B. They are usually haploid.
C. They are not reproductive cells.
D. They usually spread underground.

Interpreting Graphics

Directions (7): Base your answer to question 7 on the graph below.

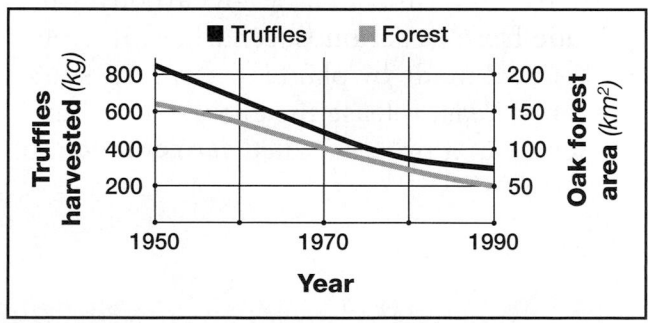

Changes in the Truffles Harvested and Oak Forest Area over 40 Years

7 Which of the statements is supported by the data in the chart?
F. The oak forest covered 125 km² in 1970.
G. The truffle harvest decreased constantly between 1950 and 1975.
H. The truffle harvest decreased most rapidly between 1980 and 1990.
I. The oak forest area decreased by 50 percent between 1955 and 1970.

Skills Practice Lab

Observing Yeast and Fermentation

SKILLS
- Observing
- Measuring
- Collecting data
- Analyzing data

OBJECTIVE
- **Observe** the release of energy by yeast during fermentation.

MATERIALS
- 500 mL vacuum bottle
- 10 cm glass tubing
- 2-hole rubber stopper
- 250 mL beaker
- 75 g sucrose
- one package dry yeast
- thermometer
- 50 cm rubber tubing
- 150 mL limewater

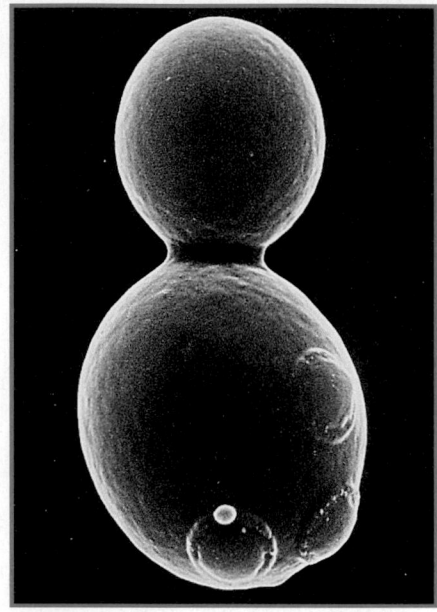

Yeast cells

| ⚠️ CAUTION: Always wear safety goggles and a lab apron to protect your eyes and clothing. | ⚠️ CAUTION: Glassware is fragile. Notify the teacher of broken glass or cuts. Do not clean up broken glass or spills with broken glass unless the teacher tells you to do so. |

Before You Begin

Sucrose is a disaccharide—a carbohydrate made from two monosaccharides. It is one chemical made by plants to store the sun's energy. Yeast release the energy stored in sucrose in a process called **fermentation**. In this investigation you will have a chance to observe and measure the products of fermentation.

1. Write a definition for the boldface term in the paragraph above.

2. Make a data table like the one below.

DATA TABLE								
Fermentation by Yeast								
Time	Date	Temperature	Time	Date	Temperature	Time	Date	Temperature
1.			8.			15.		
2.			9.			16.		
3.			10.			17.		
4.			11.			18.		
5.			12.			19.		
6.			13.			20.		
7.			14.			21.		

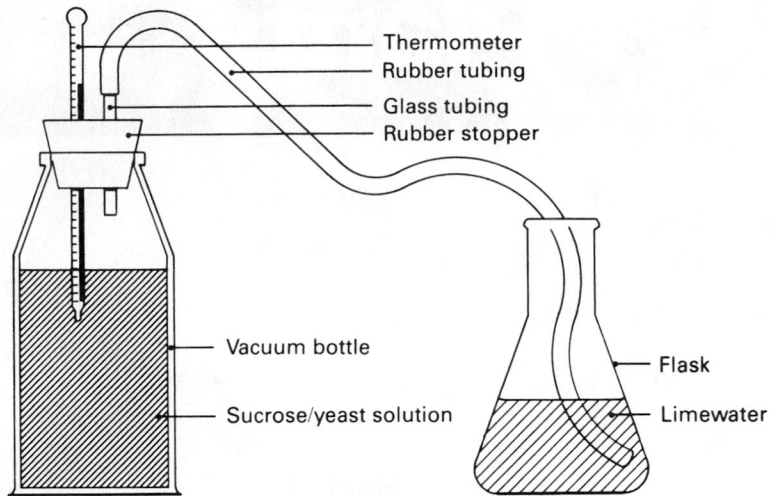

Labels on diagram:
Thermometer
Rubber tubing
Glass tubing
Rubber stopper
Vacuum bottle
Sucrose/yeast solution
Flask
Limewater

Procedure

1. Set up your vacuum bottle according to the diagram above.

2. Mix 75g of sucrose in 400 mL of water.

3. When the sucrose has dissolved, add one-half package of fresh yeast and stir.

4. Pour the sucrose-yeast solution into a vacuum bottle until it is approximately three-quarters full.

5. Adjust the thermometer so that it extends down into the sugar-yeast solution.

6. Record the temperature of the solution on the observation chart as soon as possible. Continue to record the temperature as often as possible during the next two days.

Analyze and Conclude

1. **Summarizing Results** Prepare a graph of your data, illustrating the temperature over time. Complete the graph by drawing a curve through the plotted points.

2. **Analyzing Data** What does the curved line plotted on the graph indicate?

3. **Drawing Conclusions** What can you conclude about the energy contained in sucrose?

4. **Predicting Patterns** What do you think would happen if there were only one hole in the stopper for the thermometer?

5. **Further Inquiry** If you know that fermentation liberates energy and gives off carbon dioxide and alcohol as waste products, how would you prove that fermentation is really taking place in the sugar-yeast solution?

 **Do You Know?**

Do research in the library or media center to answer these questions:

1. How do mammals get energy from glucose when they cannot use respiration?

2. How do "naturally carbonated" sodas generate bubbles?

Use the following Internet resources to explore your own questions about fermentation.

internet connect

www.scilinks.org
Topic: Fermentation
Keyword: HX4080

SCI LINKS. Maintained by the National Science Teachers Association

UNIT 6 Exploring Plants

More than 20 billion bushels of corn are harvested worldwide every year.

in perspective

Corn: America's Crop

Yesterday... Used for food by Native Americans beginning 10,000 years ago, corn was first cultivated in what is today called Mexico in 6000 to 5000 B.C. The Aztec Indians, the Maya, and the Inca all grew corn and featured it prominently in their traditions and ceremonies. Corn is an important food crop. **Where is the Corn Belt?**

Incas harvesting the corn crop

Today... Cooked, steamed, roasted, ground into cornmeal, or popped, corn has long been a mainstay of the American diet. Corn is used to make cornstarch, corn syrup, breakfast cereals, cornmeal, salad dressing, corn oil, margarine, coloring agents, stabilizing agents, and countless other products. **Read to find out what corn and wheat have in common.**

Agricultural scientist

Tomorrow... Scientists continue to develop new uses for corn and its products. For example, ethanol, or ethyl alcohol distilled from corn, can be used as a lead-free octane booster and as a replacement for regular leaded gasoline. **Discover how corn is able to conduct photosynthesis efficiently in intense heat.**

Automobile that runs on ethanol made from corn

internet connect

www.scilinks.org
Topic: Corn
Keyword: HX4052

SC LINKS. Maintained by the National Science Teachers Association

499

Texas wildflowers

✔ *Quick Review*

Answer the following without referring to earlier sections of your book.

1. **Describe** the process of mitosis. *(Chapter 6, Section 3)*

2. **Describe** the process of meiosis. *(Chapter 7, Section 1)*

3. **Identify** life cycles that have a gametophyte and life cycles that have a sporophyte. *(Chapter 7, Section 2)*

4. **Describe** the role of mycorrhizae. *(Chapter 12, Section 3 and Chapter 22, Section 3.)*

5. **List** the characteristics of the kingdom Plantae. *(Chapter 19, Section 3)*

Did you have difficulty? *For help, review the sections indicated.*

Reading Activity

Before you begin to read this chapter, survey each section and identify any subtitles, headings, and captions that signal the topic of discussion. As you read, locate other words in the body of the text that signal the sequential pattern.

🖸 **internet** connect

www.scilinks.org
National Science Teachers Association *sci*LINKS Internet resources are located throughout this chapter.

*SCI*LINKS. Maintained by the National Science Teachers Association

● Plants provide us with the food and oxygen that make life possible. They enrich our lives with beauty and sweet scents. Plants also provide buildings, paper, furniture, clothing, and medicines.

Adaptations of Plants

Objectives

- **Summarize** how plants are adapted to living on land.

- **Distinguish** nonvascular plants from vascular plants.

- **Relate** the success of plants on land to seeds and flowers.

- **Describe** the basic structure of a vascular plant sporophyte.

Key Terms

cuticle
stoma
guard cell
vascular system
nonvascular plant
vascular plant
seed
embryo
seed plant
flower
phloem
xylem
shoot
root
meristem

Establishment of Plants on Land

Plants are the dominant group of organisms on land, based on weight. The kingdom Plantae is a very diverse group. Individuals range from less than 2 mm across to more than 100 m tall. Most plants are photosynthetic; they produce organic materials from inorganic materials by photosynthesis. A few plant species, like the one shown in **Figure 1,** live as parasites. Many parasitic plants cannot photosynthesize.

Plants probably evolved from multicellular aquatic green algae that could not survive on land. Multicellularity enabled plants to develop features that helped them live more successfully on land. Before plants could thrive on land, they had to be able to do three things: absorb nutrients from their surroundings, prevent their bodies from drying out, and reproduce without water to transmit sperm.

Absorbing Nutrients

Aquatic algae and plants take nutrients from the water around them. On land, most plants take nutrients from the soil with their roots. Although the first plants had no roots, fossils show that fungi lived on or within the underground parts of many early plants. So botanists think that fungi may have helped early land plants to get nutrients from Earth's rocky surface. Symbiotic relationships between fungi and the roots of plants are called *mycorrhizae.* Today, about 80 percent of all plant species form mycorrhizae.

Preventing Water Loss

The first plants lived at the edges of bodies of water, where drying out was not a problem. A watertight covering, which reduces water loss, made it possible for plants to live in drier habitats. This covering, called the **cuticle,** is a waxy layer that covers the nonwoody aboveground parts of most plants. But like the wax on a shiny car, the cuticle does not let oxygen or carbon dioxide pass through it. Pores called **stomata** *(STOH muh tuh)* (singular, stoma) permit plants to exchange oxygen and carbon dioxide. Stomata, which extend through the cuticle and the outer layer of cells, are found on at least some parts of most

Figure 1   Rafflesia. The flowers of *Rafflesia keithii,* of Malaysia, measure almost 1 m across and weigh up to 11 kg (24 lb). The plant, which has no stems or leaves, is parasitic on the roots of grape vines.

Figure 2 Stomata and guard cells

The surface of a leaf has numerous stomata, each of which is surrounded by a pair of guard cells.

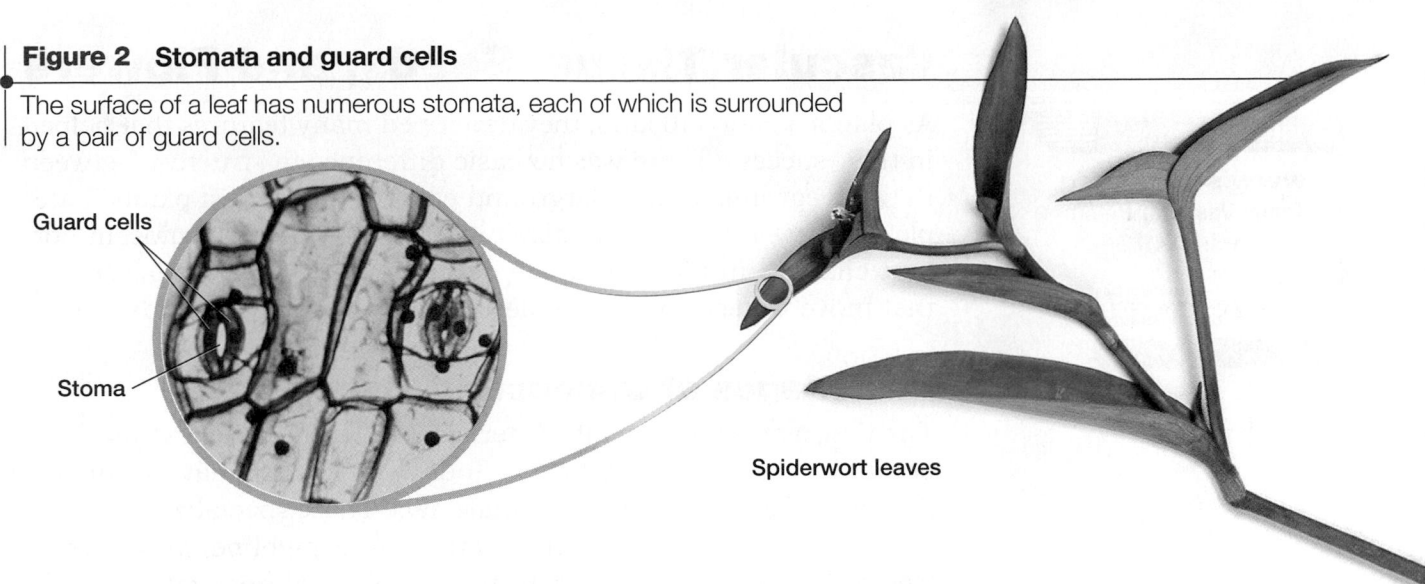

Guard cells

Stoma

Spiderwort leaves

plants. A pair of specialized cells called **guard cells** border each stoma, as seen in **Figure 2.** Stomata open and close as the guard cells change shape.

Reproducing on Land

Aquatic algae reproduce sexually when sperm swim through the water and fertilize eggs. The sperm of most plants, however, must be able to move without water. In most plants, sperm are enclosed in a structure that keeps them from drying out. The structures that contain sperm make up pollen. Pollen permits the sperm of most plants to be carried by wind or animals rather than by water.

QUICK LAB

Observing the Behavior of Stomata

You can use nail polish to see that a leaf has many stomata.

Materials

clear nail polish, plant kept in light, plant kept in darkness, two 4–5 cm strips of clear tape, 2 microscope slides, compound microscope

Procedure

1. Paint a thin layer of clear nail polish on a 1 × 1 cm area of a leaf on a plant kept in light. Do the same using a plant kept in darkness. Let the nail polish dry for 5 minutes.

2. Place a 4–5 cm strip of clear tape over the nail polish on each leaf. Press the tape firmly to the nail polish.

3. Carefully pull the tape off each leaf. Stick each piece of tape to a microscope slide. Label it appropriately.

4. View each slide with a microscope, first under low power and then under high power.

5. Draw and label what you see on each slide.

Analysis

1. **Describe** any differences in the stomata of the two plants.

2. **Critical Thinking Drawing Conclusions** Which plant will lose water more quickly? Explain.

Vascular Tissue, Seeds, and Flowers

As plants adapted to land, they developed many features that helped in their success. There was no basic difference in structure between the aboveground and underground parts of the earliest plants. Later plants, however, had roots, stems, and leaves. One of the most important changes in plants was the development of *conducting tissues* that move water and other materials through the plant body.

Advantages of Conducting Tissue

The first plants were small. Materials were transported within their bodies by osmosis and diffusion. Today, most plants have strands of specialized cells that transport materials. These specialized cells are connected end to end like the sections in a pipeline, as shown in **Figure 3.** Some strands carry water and mineral nutrients from the roots to the leaves. Other strands carry organic nutrients from the leaves to wherever they are needed.

Specialized cells that transport water and other materials within a plant are found in vascular tissues. The existence of vascular tissue allowed for larger and more-complex plants. The larger, more-complex plants have a **vascular system,** a system of well-developed vascular tissues that distribute materials more efficiently. Three groups of plants alive today lack a vascular system. These relatively small plants that have no vascular system are called **nonvascular plants.** Plants that have a vascular system are called **vascular plants.**

Advantages of Seeds

After vascular tissue, the seed was the next important adaptation to appear in plants. A **seed** is a structure that contains the embryo of a plant. An **embryo** is an early stage in the development of plants and animals. Most plants living today are **seed plants**—vascular plants that produce seeds. The first seed plants appeared about 380 million years ago. Seeds offer a plant's offspring several survival advantages, which are summarized in **Figure 4.**

Figure 3   Vascular tissue. Thick-walled, tubular cells like these carry water from the tips of roots to the tips of leaves. Stacked end to end, these cells form tiny pipes called vessels.

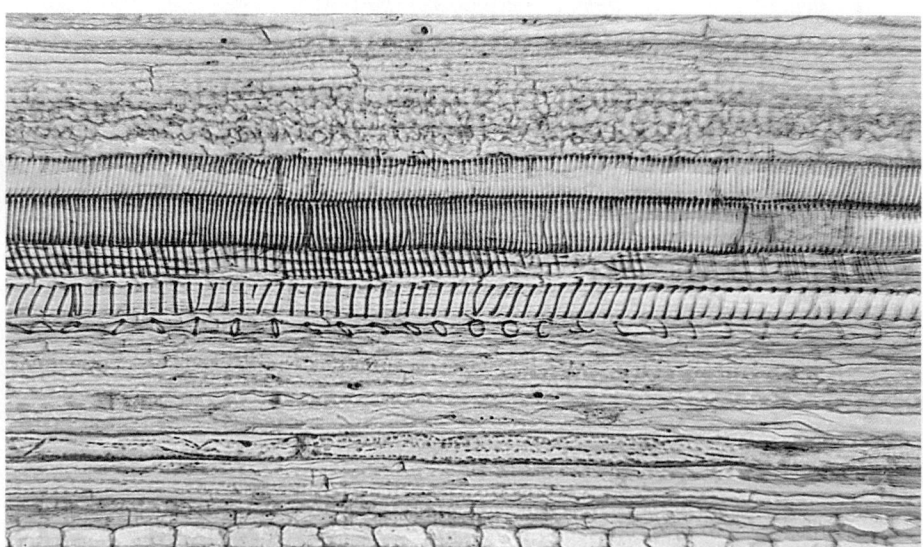

Figure 4   Structure and function of seeds

The structure of a seed helps it to perform its functions.

Wing

Pine seed

Stored food

Seed coat

Embryo

Pine cone

Seeds

Pine seedling

The seed coat of a pine seed covers and protects the embryo.

The stored food supply will nourish the embryo as it starts to grow.

A wing helps pine seeds disperse.

Pine seeds may not germinate for several years.

1. **Protection.** Seeds are surrounded by a protective cover called the seed coat. The seed coat protects the embryo from drying out and from mechanical injury and disease.

2. **Nourishment.** Most kinds of seeds have a supply of nutrients stored in them. These nutrients are a ready source of nourishment for a plant embryo as it starts to grow.

3. **Plant dispersal.** Seeds disperse (spread) the offspring of seed plants. Many seeds have structures that help wind, water, or animals carry them away from their parent plant. Dispersal prevents competition for water, nutrients, light, and living space between parents and offspring.

4. **Delayed growth.** The embryo in a seed is in a state of suspended animation. Most seeds will not sprout until conditions are favorable, such as when moisture is present and the weather is warm. Thus, seeds make it possible for plant embryos to survive through unfavorable periods such as droughts or cold winters.

internet connect

www.scilinks.org
Topic: Structure and Function of Seeds
Keyword: HX4169

SC*LINKS*. Maintained by the National Science Teachers Association

Advantages of Flowers

The last important adaptation to appear as plants evolved was the **flower,** a reproductive structure that produces pollen and seeds. Flowers make plant reproduction more efficient. The pollen of the first seed plants was carried by wind. Large amounts of pollen are needed to ensure cross-pollination by wind—an inefficient system. Most plants living today are flowering plants—seed plants that produce flowers. The first flowering plants appeared more than 130 million years ago. Many flowers attract animals, such as insects, bats, and birds. As **Figure 5** shows, tiny pollen grains stick to animals, which carry pollen directly from one flower to another. Flowering plants that are pollinated by animals produce less pollen, and cross-pollination can occur between individuals that live far apart.

Figure 5   Pollination. This honeybee is covered with pollen grains containing the sperm of the plant it has just visited. The bee will transfer some of the pollen to the next flower it visits.

Plant Life Cycles

In many algae, the zygote is the only diploid ($2n$) cell. It undergoes meiosis right after fertilization. So the bodies of these algae consist of haploid cells. In the ancestors of plants, however, meiosis was delayed. The zygote divided by mitosis and grew into a multicelled sporophyte that was diploid and produced haploid (n) spores by meiosis. The spores grew into multicelled gametophytes that were haploid and produced gametes by mitosis. As a result, plants have life cycles in which haploid plants that make gametes (gametophytes) alternate with diploid plants that make spores (sporophytes). A life cycle in which a gametophyte alternates with a sporophyte is called alternation of generations. The basic plant life cycle is shown in **Figure 6**.

Unlike the green algae with alternation of generations, plants have gametophytes and sporophytes that look very different. In addition, the relative sizes of gametophytes and sporophytes changed as plants evolved, as **Figure 7** shows. In nonvascular plants, such as mosses, the gametophyte generation is dominant (most noticeable). In vascular plants, such as the flowering plants, the sporophyte generation is dominant. Like the presence of a vascular system, the relative sizes of gametophytes and sporophytes is a fundamental difference between the nonvascular plants and the vascular plants.

Figure 6 Alternation of generations. In the life cycle of a plant, a diploid sporophyte generation alternates with a haploid gametophyte generation.

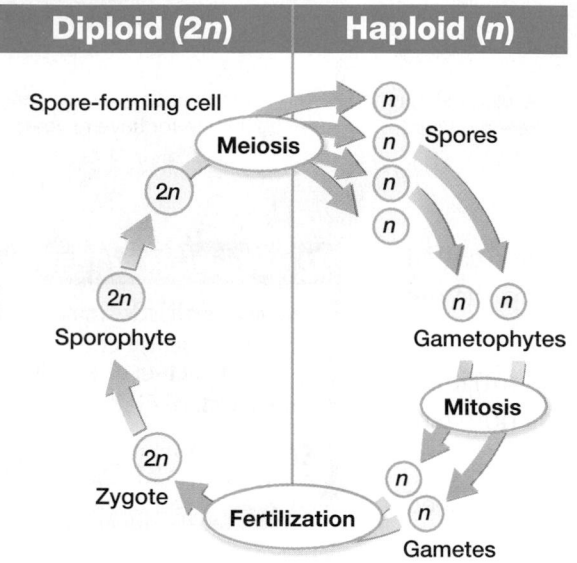

Figure 7 Nonvascular plants versus vascular plants

Gametophytes and sporophytes vary greatly in size.

Sphagnum moss

A sporophyte grows atop a gametophyte.

The gametophytes form inside a flower.

Tomato sporophytes

The Vascular-Plant Sporophyte

As the size of plant sporophytes increased, so did the complexity of their structure. An increase in size enables cell specialization and, therefore, the development of complex tissues and specialized structures. The following features characterize the sporophytes of most vascular plants.

Vascular System Larger bodies require an efficient vascular system for transporting materials internally. The sporophytes of vascular plants have a vascular system with two types of vascular tissue. Each type of vascular tissue contains strands of long, tubelike cells that are lined up end to end like sections of pipe. These strands of cells transport water and nutrients within a plant's body. Relatively soft-walled cells transport organic nutrients in a kind of tissue called **phloem** (*FLOH uhm*). Hard-walled cells transport water and mineral nutrients in a kind of tissue called **xylem** (*ZIE luhm*). The walls of the water-conducting cells in xylem are thickened, which helps support the plant body. This makes it possible for vascular plants to grow to great heights.

Distinctive Body Form Nearly all plants have a body that consists of a vertical shaft from which specialized structures branch, as shown in **Figure 8**. The part of a plant's body that grows mostly upward is called the **shoot.** In most plants, the part of the body that grows downward is called the **root.** Zones of actively dividing plant cells, called **meristems** (*MEHR uh stehmz*), produce plant growth. The vertical body form results as new cells are made at the tips of the plant body. As vascular plants became better adapted to life on land, most developed the familiar plant structures—roots, stems, and leaves—which are complex structures made of several different types of specialized tissues.

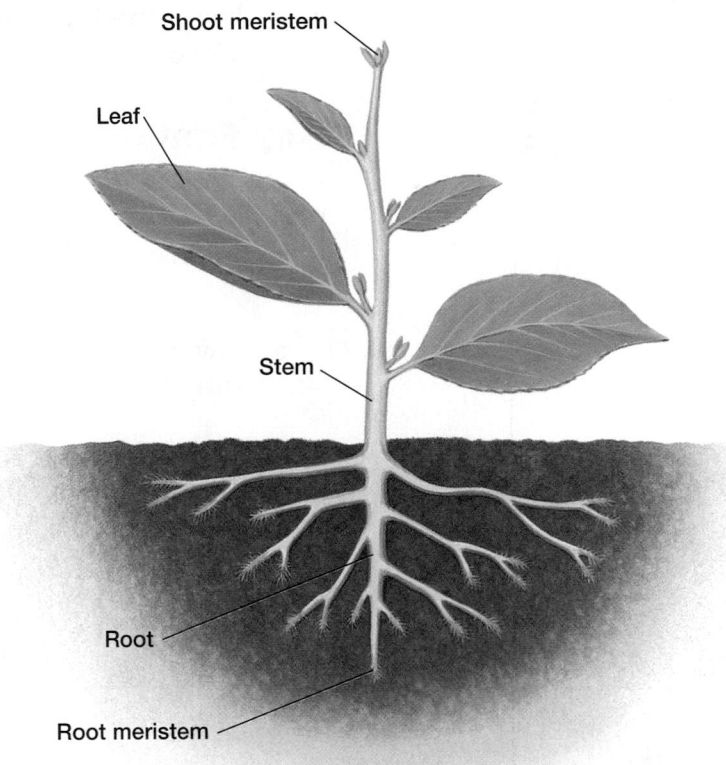

Figure 8   Vascular plant sporophyte. The sporophytes of the vast majority of vascular plants have an aboveground shoot with stems and leaves and an underground root. Growth occurs in regions called *meristems*.

Shoot meristem

Leaf

Stem

Root

Root meristem

Section 1 Review

1 **Summarize** how plants are adapted to living successfully on land.

2 **Describe** two basic differences between nonvascular plants and vascular plants.

3 **Critical Thinking Relating Concepts** How have seeds and flowers made plants more successful on land?

4 **Critical Thinking Evaluating Conclusions** Why do you think vascular plants are more successful as land plants than are nonvascular plants?

5 **Standardized Test Prep** Water loss from most plants is reduced by a waxy layer called the
A xylem.   C stomata.
B phloem.   D cuticle.

Kinds of Plants

- **Describe** the key features of the four major groups of plants.

- **Classify** plants into one of the 12 phyla of living plants.

Key Terms

rhizoid
rhizome
frond
cone
gymnosperm
angiosperm
fruit
endosperm
monocot
dicot

Nonvascular Plants

The brilliant green carpet of mosses you see in **Figure 9** is made up of thousands of individual plants. Living carpets of mosses are often found near streams, coastlines, and other moist places. But these tough little plants also live in some surprising places, such as cracks in city sidewalks and rocky mountaintops—any place where a little moisture can collect. The moisture makes it possible for these nonvascular plants to survive.

Nonvascular plants do not have a vascular system for transporting water and other nutrients within their bodies. This means that all nonvascular plants lack true roots, stems, and leaves, although most have structures that resemble them. True roots, stems, and leaves are complex structures that contain vascular tissues.

Key Features of Nonvascular Plants

In addition to the lack of true roots, stems, and leaves, nonvascular plants share several other features. These features are key adaptations that have enabled them to survive on land.

Small Size All nonvascular plants are small and relatively simple. Water and other nutrients are transported within their bodies mostly by osmosis and diffusion, which move materials short distances. This greatly limits the size of a nonvascular plant's body.

Larger Gametophyte The gametophytes of nonvascular plants are larger and more noticeable than the sporophytes. Hairlike projections called **rhizoids** anchor the gametophytes to the surfaces on which they grow. The smaller, usually nongreen sporophytes grow on the gametophytes and depend on them for nutrients.

Require Water for Sexual Reproduction Nonvascular plants must be covered by a film of water in order for fertilization to occur. Eggs and sperm form in separate structures, which are often on separate plants. The gametophytes grow in mats of tightly packed individuals. When these mats are covered by a film of water, the sperm can easily swim to neighboring individuals and fertilize their eggs.

Figure 9 A carpet of mosses. Mosses grow in tightly packed mats that may contain dozens of plants per square inch.

Kinds of Nonvascular Plants

The nonvascular plants include the mosses and the two simplest groups of plants—liverworts and hornworts. Examples of these plants are shown in **Figure 10**.

Mosses The mosses (phylum Bryophyta) are the most familiar nonvascular plants. The "leafy" green plants that you recognize as mosses are gametophytes. Moss sporophytes, which are not green, grow from the tip of a gametophyte. Each sporophyte consists of a bare stalk topped by a spore capsule. Most mosses have a cuticle, stomata, and some simple conducting cells. The walls of the water-conducting cells in mosses are not thickened, as they are in a vascular plant. Mosses never get very large because their water-conducting cells carry water only short distances.

Liverworts Like the mosses, liverworts (phylum Hepatophyta) grow in mats of many individuals. Liverworts have no conducting cells, no cuticle, and no stomata. Their gametophytes are green. In some species, such as the common liverwort shown in Figure 10, the gametophytes of liverworts are flattened and have lobes. Structures that resemble stems and leaves make up the gametophytes of most liverworts, like those of the mosses. The sporophytes of liverworts are very small and consist of a short stalk topped by a spore capsule.

Hornworts The hornworts (phylum Anthocerophyta) are a small group of nonvascular plants that, like the liverworts, completely lack conducting cells. The sporophyte of a hornwort has both stomata and a cuticle. The gametophyte of a hornwort is green and flattened. Green hornlike sporophytes grow upward from the gametophytes.

WORD *Origins*

- The word *liverwort* combines the familiar word *liver* and the Old English word *wort*, meaning "herb." The word *liverwort* dates back to the Middle Ages, when it was thought that plants resembling certain body parts could cure diseases of those body parts.

Figure 10 Nonvascular plants

There are three phyla of nonvascular plants.

Polytrichum, a moss
(Phylum Bryophyta)

Marchantia, a liverwort
(Phylum Hepatophyta)

Anthoceros, a hornwort
(Phylum Anthocerophyta)

Seedless Vascular Plants

Vascular plants that do not produce seeds are called seedless vascular plants. The earliest known seedless vascular plant, *Cooksonia,* is illustrated in **Figure 11.** The sporophytes of these ancient plants had branched, leafless stems that were only a few centimeters long. Spore-forming sporangia were located at the tips of the stems. *Rhynia*, another early seedless vascular plant, also had horizontal underground stems, or **rhizomes.**

Key Features of Seedless Vascular Plants

Seedless vascular plants are much larger and more complex than the nonvascular plants. Other key features enabled them to spread and adapt to drier habitats on land.

Vascular System Seedless vascular plants have a vascular system with both xylem and phloem. The water-conducting cells in the xylem are reinforced with *lignin,* a major part of wood. Because of their vascular system, seedless vascular plants grow much larger than nonvascular plants and also develop true roots, stems, and leaves.

Larger Sporophyte The sporophytes of seedless vascular plants are larger than the gametophytes. Their larger size makes it easier for the wind to carry away spores, which makes dispersal more efficient. The much smaller gametophytes of most seedless vascular plants develop on or below the surface of soil. As in the nonvascular plants, water is needed for fertilization. When there is enough water on or in the soil, the sperm swim to eggs and fertilize them.

Drought-Resistant Spores The spores of the seedless vascular plants have thickened walls that are resistant to drying. Such spores make it possible for a plant to live in drier habitats.

Real Life

The spores of a common club moss, *Lycopodium,* **form a powder that has several uses.**

Herbalists use the spores to make a powder for treating skin disorders. The spores are also used to make photographic flash powder.

Finding Information
Find out how **Lycopodium** *powder is used by the pharmaceutical industry.*

Figure 11   The earliest known vascular plant

Cooksonia, the oldest known vascular plant, lived about 410 million years ago.

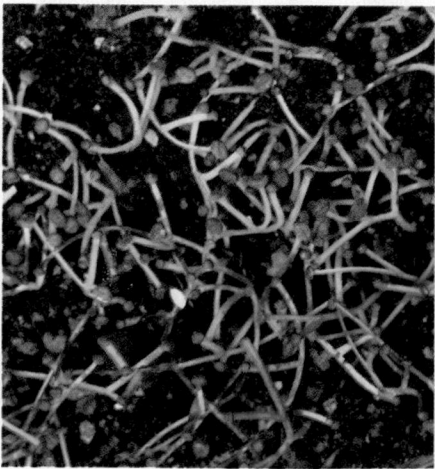

Model of *Cooksonia*

Artist's rendition of *Cooksonia*

Kinds of Seedless Vascular Plants

The seedless vascular plants include ferns and three other groups of plants known as fern allies—whisk ferns, club mosses, and horsetails. Like the ferns, the fern allies usually grow in moist places.

Ferns The ferns (phylum Pterophyta) are the most common and most familiar seedless vascular plants. Ferns grow throughout the world, but they are most abundant in the tropics. The plants you recognize as ferns are sporophytes. Most fern sporophytes have a rhizome that is anchored by roots and leaves called **fronds.** The coiled young leaves of a fern, shown in **Figure 12,** are called fiddleheads. Spores are produced in sporangia that grow in clumps on the lower side of fronds. The gametophytes of ferns are flattened, heart-shaped green plants that are usually less than 1 cm (0.4 in.) across.

Club Mosses Unlike true mosses, the club mosses (phylum Lycophyta), have roots, stems, and leaves. Their leafy green stems branch from an underground rhizome. Spores develop in sporangia that form on specialized leaves. In some species, such as the one seen in **Figure 13,** clusters of nongreen spore-bearing leaves form a structure called a **cone.**

Horsetails The horsetails (phylum Sphenophyta) also have roots, stems, and leaves. The vertical stems of horsetails, which grow from a rhizome, are hollow and have joints. Whorls of scalelike leaves grow at the joints. Spores form in cones located at the tips of stems.

Whisk Ferns The whisk ferns (phylum Psilotophyta) probably most closely resemble the earliest vascular plants. Whisk ferns have highly branched stems and no leaves or roots. They produce spores in sporangia that form at the tips of short branches.

Figure 12 A fern. This sword fern sporophyte has many fronds and fiddleheads. The inset shows a gametophyte at twice its actual size.

Figure 13 Fern allies

In addition to ferns, there are three other living phyla of seedless vascular plants that are known as fern allies.

Lycopodium, a club moss
(Phylum Lycophyta)

Equisetum, a horsetail
(Phylum Sphenophyta)

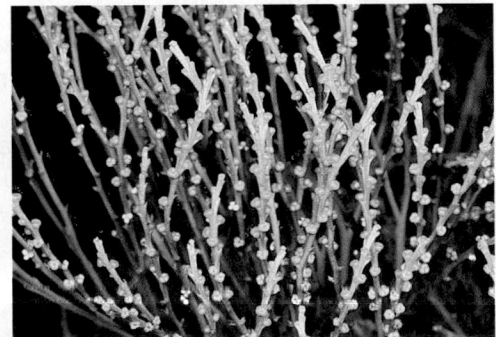

Psilotum, a whisk fern
(Phylum Psilotophyta)

Gymnosperms

Gymnosperms *(JIHM noh spurmz)* are seed plants whose seeds do not develop within a sealed container (a fruit). The word *gymnosperm* comes from the Greek words *gymnos,* meaning "naked," and *sperma,* meaning "seed."

Key Features of Gymnosperms

Gymnosperms are among the most successful groups of plants. The following key features have made them successful on land.

Seeds All gymnosperms produce seeds. Seeds protect plant embryos, provide them with nutrients, and permit them to survive long periods of unfavorable conditions. In some plants, seeds also disperse new plants far from their parents.

Greatly Reduced Gametophytes All seed plants produce very tiny gametophytes of two types—male and female. The gametophytes form within the tissues of the sporophytes. Grains of pollen are male gametophytes. Female gametophytes form within structures that become seeds. In all but one species of gymnosperm, male and female gametophytes develop in male and female cones, respectively.

Wind Pollination The sperm of gymnosperms do not swim through water to reach and fertilize eggs. Instead, the sperm are carried to the structures that contain eggs by pollen, which can drift on the wind, as seen in **Figure 14.** Wind pollination makes sexual reproduction possible even when conditions are very dry.

Figure 14 Juniper pollen. This juniper, a type of gymnosperm, releases clouds of pollen in late fall or early winter.

Analyzing the Effect of Climate on Plants

Background

The map at right shows the taiga of North America. The taiga is a vast forest of conifers, a type of gymnosperm. The graph shows average annual temperature and precipitation data for Anchorage, Alaska, which is located at the western edge of the taiga. Use the map and graph to answer the following questions.

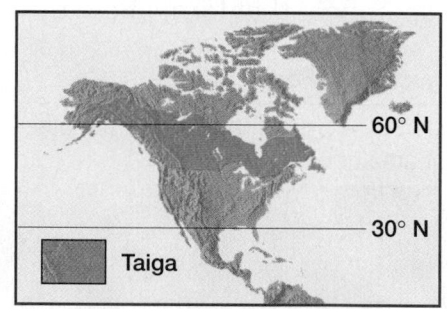

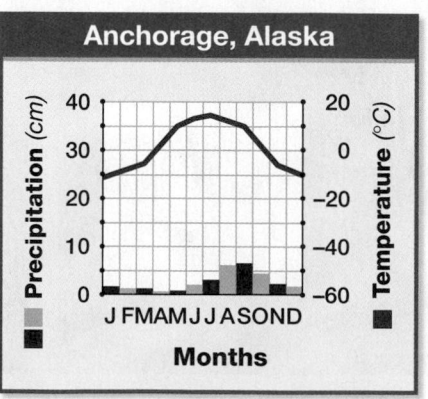

Analysis

1. **Describe** the climate of Anchorage, Alaska.

2. **Critical Thinking Predicting Patterns** What type of climate would you expect to find in other parts of the taiga?

3. **Critical Thinking Drawing Conclusions** Does climate appear to be an important factor in where the conifers of the taiga grow? Explain.

Kinds of Gymnosperms

Four groups of living seed plants are referred to as gymnosperms—conifers, cycads, ginkgo, and gnetophytes. Examples of each of these four groups are shown in **Figure 15** and **Figure 16**.

Conifers The conifers (phylum Coniferophyta) are the most familiar, and most successful, gymnosperms. Conifers have leaves that are either needle-like or reduced to tiny scales, as Figure 15 shows. These leaves are an adaptation for limiting water loss. Some of the tallest living plants, the redwoods of coastal California and Oregon, are conifers. The oldest trees in the world are thought to be bristlecone pines, another species of conifer that grows in the Rocky Mountains and Great Basin. Some bristlecone pines are about 5,000 years old. Vast forests of conifers grow in cool, dry regions of the world.

Cycads The cycads (phylum Cycadophyta) have short stems and palmlike leaves. Cones that produce pollen and those that produce seeds develop on different plants. Cycads are widespread throughout the tropics.

Ginkgo The only living species of ginkgo (phylum Ginkgophyta), or maidenhair tree, has fan-shaped leaves that resemble the leaves of the maidenhair fern. The male and female gametophytes of ginkgo develop on separate trees. Ginkgo seeds do not develop within a cone.

Gnetophytes The gnetophytes (phylum Gnetophyta) are a diverse group of trees, shrubs, and vines that produce pollen and seeds in cones that resemble flowers. One type of gnetophyte, *Ephedra*, is common in the western United States.

Figure 15 Juniper leaves and cones. The tiny scalelike leaves of junipers—a type of conifer—are an adaptation that limits water loss. The blue, berrylike structures are the female cones of this juniper.

Figure 16 Other gymnosperms

In addition to conifers, there are three other living phyla of gymnosperms.

Encephalartos, a cycad
(Phylum Cycadophyta)

Leaves and seeds of *Ginkgo*
(Phylum Ginkgophyta)

Ephedra (Mormon tea), a gnetophyte
(Phylum Gnetophyta)

Angiosperms

Most seed plants are flowering plants, or **angiosperms** *(AN jee oh spurmz)*. Angiosperms produce seeds that develop enclosed within a specialized structure called a **fruit,** as seen in **Figure 17.** The word *angiosperm* comes from the Greek words *angeion,* meaning "case," and *sperma,* meaning "seed."

Key Features of Angiosperms

Angiosperms are the most recent group of plants to evolve. The following key features made them the most successful group of plants.

Flowers The male and female gametophytes of angiosperms develop within flowers, which promote pollination and fertilization more efficiently than do cones. Some flowers, such as roses, are brightly colored or have strong scents. This attracts insects and other animals that carry pollen and increases the likelihood of cross-pollination. Other flowers, such as garden peas, are adapted for self-pollination, which often occurs before the flowers open. The flowers of many angiosperms, such as oaks and grasses, have small greenish flowers that are adapted for wind pollination. The female reproductive part of a flower also provides a pathway that enables sperm to reach and fertilize eggs without swimming through water.

Fruits Although fruits provide some protection for developing seeds, their primary function is to promote seed dispersal. The angiosperms produce many different types of fruits, which develop from parts of flowers. Many fruits are eaten by animals. The seeds are dispersed as they pass undigested from the animals' bodies. Other fruits have structures that help them float on wind or water. Some fruits even forcefully eject their seeds, flinging them away from the parent plant.

Endosperm The seeds of angiosperms have a supply of stored food called **endosperm** at some time during their development. In many angiosperms, the endosperm is absorbed by the embryo before the seeds mature.

Figure 17 Seeds in a fruit.
These melons, which contain seeds, are the fruits of an angiosperm.

Kinds of Angiosperms

Botanists divide the angiosperms into two subgroups—monocots and dicots as **Figure 18** shows. The **monocots** are flowering plants that produce seeds with one seed leaf (cotyledon). Most monocots also produce flowers with parts that are in multiples of three and have long, narrow leaves with parallel veins. The **dicots** are flowering plants that produce seeds with two seed leaves. Most dicots also produce flowers with parts in multiples of two, four, or five and have leaves with branching veins. **Table 1** lists examples of some of the most familiar families of angiosperms.

Figure 18   A monocot and a dicot

Monocots and dicots differ in several ways.

Daylilies are monocots.

Roses are dicots.

Table 1 Familiar Families of Angiosperms

Subgroup	Family	Examples
Monocots (class Monocotyledonae)	Iridaceae (iris)	Irises, gladiolus, crocus, blue-eyed grass
	Liliaceae (lily)	Daylilies, tulips, asparagus, aloe vera
	Poaceae (grass)	Wheat, corn, rice, lawn grasses
Dicots (class Dicotyledonae)	Asteraceae (aster)	Daisies, sunflowers, lettuce, ragweed
	Brassicaceae (mustard)	Broccoli, cauliflower, turnips, cabbage
	Fabaceae (legume)	Beans, clovers, peas, peanuts, soybeans
	Rosaceae (rose)	Roses, apples, peaches, pears, plums
	Solanaceae (nightshade)	Potatoes, tomatoes, peppers, petunias

Section 2 Review

1 **Identify** three key features of each of the four major groups of plants.

2 **Classify** each of the following plants as one of the four major groups of plants: pine trees, carnations, sphagnum moss, and wood fern.

3 **Critical Thinking Evaluating Conclusions** Why are angiosperms said to be the most successful group of plants?

4 **Critical Thinking Recognizing Patterns** How are spores and pollen grains adapted for their functions, and how do their numbers impact their environment?

5 **Standardized Test Prep** Ferns reproduce by producing

A spores.   **C** flowers.

B cones.   **D** seeds.

Plants in Our Lives

Plants as Food

Humans depend on plants in many ways. For one thing, plants store the extra nutrients they make or absorb in their bodies. Thus, plant parts contain organic nutrients (carbohydrates, fats, and proteins) and minerals (calcium, magnesium, and iron). All types of plant parts—roots, stems, leaves, flowers, fruits, and seeds—are eaten as food.

Fruits and Vegetables

The United States government identifies each of the foods that comes from a plant as an agricultural commodity. Each type of food is classified by a term—such as *fruit* or *vegetable*—that is registered in Washington, D.C. But, these terms have different meanings in botany. For example, to a botanist, a fruit is the part of a plant that contains seeds, and a **vegetative part** is any nonreproductive part of a plant. The foods that you think of as fruits—such as apples, bananas, and melons—are also fruits in the botanical sense. Vegetables, on the other hand, may be any botanical part of a plant, as you can see in **Figure 19.** Fruits and vegetables provide dietary fiber and are important sources of essential vitamins and minerals.

Figure 19   Plant parts eaten as food

The foods you eat come from different parts of plants.

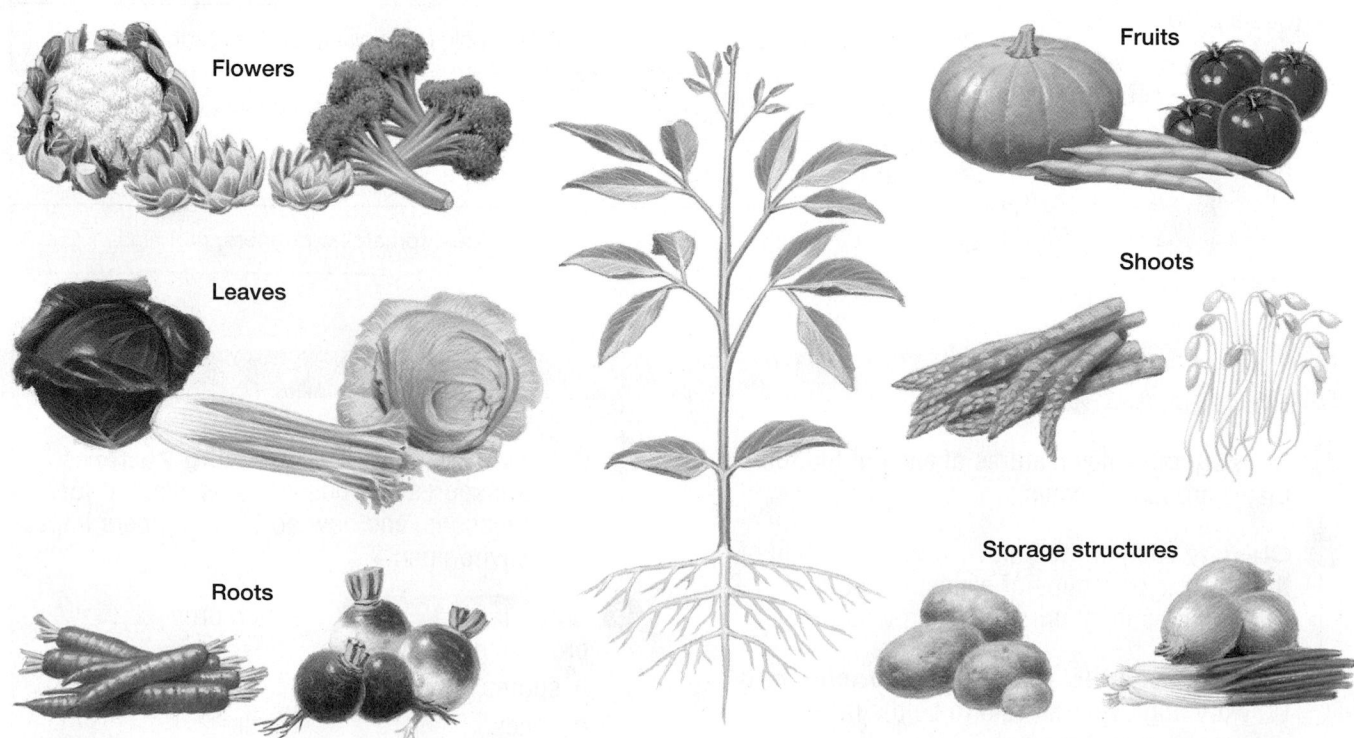

Flowers

Fruits

Leaves

Shoots

Roots

Storage structures

Root Crops

Potatoes are an important food staple in many parts of the world. Rich in calories and easy to grow, potatoes are an ideal crop for a small farm. Potatoes are classified as a root crop because they grow underground. But, potatoes are actually tubers, modified underground stems that store starch. Yams, an essential food crop in many tropical parts of the world, are roots. Sweet potatoes, carrots, radishes, turnips, beets, and cassava *(kuh SAH vuh)* are also important root crops. These vegetables are enlarged roots that store starch. Cassava, seen in **Figure 20,** is the staple food of more than 500 million people. Also known as manioc *(MAN ee awk),* cassava supplies more than one-third of the calories eaten in Africa.

Legumes

Many members of the pea family, which are called *legumes,* produce protein-rich seeds in long pods. For example, about 45 percent of a soybean, the most important legume grown for food, is protein. Soybeans are often cooked and pressed into cakes called tofu *(TOH foo).* Peas, peanuts, and the many different types of beans are the seeds of legumes. Alfalfa, which is fed to livestock, is another important legume. Like many legumes, alfalfa has nitrogen-fixing bacteria, which add nitrogen compounds to the soil, in its roots. Therefore, alfalfa is also grown to enrich the soil.

Figure 20   Cassava.
Cassava develops thick starch-filled roots up to 120 cm (4 ft) long. The roots are eaten like potatoes. Tapioca, shown above in the bowl, is made from cassava.

QUICK LAB

Distinguishing Between Fruits and Vegetables

You can find out if a plant product is a fruit by cutting it open and examining its internal structure.

Materials

apple, banana, green bean, potato, squash, tomato, plastic knife

Procedure

1. Look at several familiar fruits and vegetables. Classify each one as either a fruit or a vegetable in the familiar sense.

2. ⚠ **CAUTION: Sharp objects can cause injury. Handle knives carefully.** Use a plastic knife to cut open each fruit and vegetable.

3. Look at the fruits and vegetables again. Classify each by its botanical function—either a fruit or a vegetative part.

Analysis

1. **Compare** the familiar and botanical classifications you gave each fruit and vegetable.

2. **Critical Thinking Analyzing Data** Which fruits and vegetables did you classify differently?

3. **Critical Thinking Analyzing Results** Defend the classifications you made for item 2.

4. **Critical Thinking Drawing Conclusions** Based on your data, when is a vegetable a fruit?

Cereals

Most of the foods that people eat come directly or indirectly from the fruits of cereals. **Cereals** are grasses that are grown as food for humans and livestock. Cereal grasses produce large numbers of a type of edible, dry fruit called a **grain.** A grain contains a single seed with a large supply of endosperm. Each grain develops from a flower. The flowers of cereal grasses form in tightly packed clusters of many individual flowers. A grain is covered by a dry, papery husk called the *bran,* which includes the wall of the ovary and the seed coat. Cereal grains are rich in carbohydrates and also contain protein, vitamins, and dietary fiber. More than 70 percent of the world's cultivated farmland is used for growing cereal grains. In fact, more than half of the calories that humans consume come from just three cereal grasses: wheat, corn, and rice.

Wheat

For more than one-third of the world's population, wheat, seen in **Figure 21,** is the primary source of food. The endosperm of wheat grains, which is high in carbohydrates, is commonly ground into white flour and used to make breads and pasta. Vitamin-rich wheat germ consists of the embryos of wheat grains. Whole-wheat flour consists of the endosperm plus the germ and bran layers. Wheat grains are not always ground into flour. In the Middle East, wheat grains are often boiled or soaked, dried, and then pounded until they crack. The cracked grains, called *bulgur (BUL guhr),* are used in dishes such as tabbouleh *(tuh BOO lee)* and pilaf *(pih LAHF).* Most wheat is grown in temperate regions that have fertile soil and moderate rainfall. One of the world's best wheat-growing areas is the Great Plains region of the United States and Canada—a temperate grassland biome.

Corn

Corn, seen in **Figure 22,** is the most widely cultivated crop in the United States. American colonists of the 1600s and 1700s first learned how to grow corn from Native Americans. In the southeastern United States, corn was more widely grown than wheat, which does not grow as well in hot climates. Thus, foods that are made from corn—corn bread, corn pone, hominy, and grits—are a traditional component of the southeastern American diet. Corn is also one of the world's chief foods for farm animals. About 70 percent of the corn crop harvested in the United States is consumed by livestock. Other uses for corn include the production of corn syrup, margarine, corn oil, cornstarch, and fuel-grade ethanol. Most of the corn grown in the United States today comes from a region known as the Corn Belt, which includes Iowa, Nebraska, Minnesota, Illinois, and Indiana.

Figure 21 Wheat. Modern bread wheat is a hybrid of three wild species. The ripenend heads of bread wheat turn golden brown.

Figure 22 Corn. Each ear of corn developed from a flower spike that consisted of more than 500 flowers.

Rice

For more than half of the people in the world, rice is the main part of every meal. Although it is low in protein, rice is an excellent source of energy-rich carbohydrates. While brown rice still has its vitamin-rich bran layers, white rice has been processed to remove the bran layers. This processing helps to prevent spoilage in stored rice. In societies where people eat mainly rice, vitamin-rich sauces such as soy sauce are often added to white rice to make meals more nutritious. The white rice you buy at a grocery store is enriched with added vitamins. Rice is often added to processed foods such as breakfast cereal, soup, baby food, and flour. In the United States, rice is grown in central California, in the Southeast, and along the Gulf Coast in fields such as the one shown in **Figure 23.**

Figure 23   Gulf coast rice field. Rice plants are grown in standing water.

Vegetarian Diets

Many people are vegetarian; they eat only foods from plants. Like any diet, vegetarian diets must satisfy the body's nutritional needs to be healthy. Two important considerations in eating a healthy vegetarian diet are the essential amino acids and vitamins B-12 and D.

Getting the Essential Amino Acids

Essential amino acids is the term for those amino acids that the human body cannot make. Most plant proteins contain all of the essential amino acids, but in different relative amounts. For example, cereal proteins tend to be low in lysine and high in methionine; the opposite is true for the proteins in beans. Traditionally, cereals and beans are eaten together to obtain enough of the essential amino acids. However, vegetarians can get the essential amino acids they need even if they do not eat a variety of plant foods, as long as they eat enough protein. For example, if you eat enough rice to satisfy your daily protein requirement, you will get more than twice your daily requirement for lysine. The recommended daily requirement for protein is 44 g for a 55 kg (121 lb) woman and 56 g for a 70 kg (154 lb) man. Legumes, grains, nuts, broccoli, and potatoes are good sources of protein.

Vitamins B-12 and D

A vegetarian diet generally can provide enough of all but two vitamins, vitamin B-12 and vitamin D. To get enough vitamin B-12,

vegetarians can eat fortified foods, add eggs or dairy products to their diet, or take vitamin B-12 supplements. Vitamin D is made in the skin when it is exposed to sunlight. Vegetarians may need to take vitamin D supplements if their exposure to sunlight is limited.

internet connect

www.scilinks.org
Topic: Vegetarian Diets
Keyword: HX4184

SCiLINKS Maintained by the National Science Teachers Association

Figure 24 Latex. The milky sap of certain plants is called latex. This man is collecting latex from a rubber tree in Java, Indonesia.

Nonfood Uses of Plants

Plants are used by people for many purposes other than food. For example, rubber was first made from latex, the milky white sap of tropical trees of the genus *Hevea*. Latex is extracted from rubber trees by the method seen in **Figure 24.** Guayule *(gwah YOO lee),* a member of the sunflower family that is native to the southwestern United States, is another source of natural rubber. (Most of today's rubber, however, is synthesized from petroleum, a nonrenewable resource.) The most important nonfood products obtained from plants are wood and fibers.

Wood

After food, wood is the single most valuable resource obtained from plants. Countless products, such as those shown in **Figure 25,** are made from wood. The wood from trees that have been cut down and sawed into boards is called lumber. Nearly 75 percent of the lumber cut in the United States is used for building construction. The rest is used to make products that contain wood, or it is ground and moistened to make wood pulp. Wood pulp is made into paper, rayon, and many other products. Finally, for more than a quarter of the world's people, wood is still the main source of fuel for heating and cooking.

Figure 25 Items made with wood

Furniture, buildings, boats, cabinets, and violins are made from wood.

Figure 26   Sources of medicines

These two common garden plants are the sources of important medicines.

Rosy periwinkle, the original source of two cancer-fighting drugs

Foxglove, the source of a drug used to treat cardiac disorders

Medicines

People have always used substances obtained from plants to treat a variety of ailments. By studying the plants traditionally used to treat human ailments, researchers have developed many "modern" medicines. For example, solutions made by soaking the bark of willow trees, *Salix,* were a traditional cure for aches and pains. The pain-relieving chemical found in willows is called salicin *(SAL uh sihn).* Acetylsalicylic *(uh SEET l sal uh SIHL ihk)* acid, a derivative of salicin, was first sold in 1899 under the name "aspirin." Today, aspirin is the most widely used pain-relieving drug in the world.

Two familiar garden plants, seen in **Figure 26,** are important sources of life-saving medicines. The extremely poisonous leaves of the foxglove, *Digitalis purpurea,* yield digitalis *(dihj ih TAL ihs),* a drug that is used to stabilize irregular heartbeats and to treat cardiac disorders. The rosy periwinkle, *Catharanthus roseus,* is the source of two cancer-treatment drugs—vinblastine *(vihn BLAS teen)* and vincristine *(vihn KRIHS teen).* Vinblastine is often used to treat Hodgkin's disease, a type of cancer that affects the lymph nodes. Vincristine is used to treat childhood leukemia and other types of cancer. **Table 2** contains other examples of medicines that originally derived from plants.

Table 2 Some Drugs Originally Derived from Plants		
Name	**Source**	**Action**
Caffeine	Tea leaves	Acts as a stimulant
Codeine	Poppy fruits	Relieves pain
Cortisone	Yam tubers	Relieves symptoms of allergies
Ephedrine	Ephedra stems	Acts as a decongestant
Taxol	Yew tree bark	Reduces the size of cancerous tumors

internet connect

www.scilinks.org
Topic: Medicines from
Plants
Keyword: HX4119

SCI*LINKS* Maintained by the
National Science
Teachers Association

Figure 27 Cotton

Cotton is the plant fiber that is most widely used to make cloth.

Cotton bolls that have split open, revealing cotton fibers

Indian woman spinning cotton into thread that will be woven into cloth

Fibers

If you were to look at this sheet of paper very closely through a magnifying glass, you would see that it is made of many interlocking fibers. These fibers are strands of cellulose, which is a component of the cell walls of plants. In plants, fibers help provide support for the plant body. The strength and flexibility of plant fibers make them ideal materials for making paper, cloth, and rope. Most of the fibers used to make paper come from wood. Paper-making fibers are also obtained from many other plants, including cotton, flax, rice, bamboo, and papyrus *(puh PIE ruhs)*.

For centuries, people have made clothing with cloth made of cotton, the world's most important plant fiber. As **Figure 27** shows, white fibers fill up the inside of a cotton boll *(bohl)*, the fruit of the cotton plant. Cotton thread is spun from the fine white fibers, which grow on cotton seeds. The stems of flax yield softer, more durable fibers that are used to make linen. More than 30 percent of the world's clothing is now made of synthetic fibers, but natural plant fibers are still prized for their durability and comfort. Sturdy fibers of hemp and sisal *(SIE suhl)* plants are used to make rope.

Section 3 Review

1 **Describe** several ways in which wood is used.

2 **List** five medicines that are derived from plants, and state how each is used.

3 **Name** two types of plants that provide fiber used in clothing.

4 **Critical Thinking Predicting Results** Name the three most important cereal grains and predict the results if one of them ceased to exist.

5 **Critical Thinking Evaluating Viewpoints** Justify the viewpoint that wood is the most important nonfood plant product.

6 **Standardized Test Prep** Which plant has nitrogen-fixing bacteria in its roots?

A potato

B wheat

C alfalfa

D corn

Study ZONE   *CHAPTER HIGHLIGHTS*

Key Concepts

1 Adaptations of Plants

- To survive on land, plants must absorb mineral nutrients, prevent their bodies from drying out, and reproduce without water to transmit male gametes.

- Vascular plants have a system of well-developed tissues that transport water within a plant. The nonvascular plants lack a vascular system.

- Seeds protect and nourish a plant's embryo, disperse the offspring, and delay the growth of the embryo until conditions are favorable. Flowers make reproduction more efficient by promoting pollination.

- The sporophytes of vascular plants have a vascular system. Their bodies consist of an aboveground shoot and an underground root.

2 Kinds of Plants

- Nonvascular plants are small and lack vascular tissue. Mosses, liverworts, and hornworts are nonvascular plants.

- Seedless vascular plants produce spores with thickened walls that prevent them from drying out. Ferns, club mosses, horsetails, and whisk ferns are seedless vascular plants.

- Gymnosperms are seed plants that produce cones. Conifers, cycads, ginkgoes, and gnetophytes are gymnosperms.

- Angiosperms are seed plants that produce flowers and fruits. The angiosperms are classified as either monocots or dicots.

3 Plants In Our Lives

- All types of plant parts—roots, stems, leaves, flowers, fruits, and seeds—provide food for humans. Rice, corn, and wheat are cereal grasses and are our most important sources of food.

- Wood is a source of wood pulp used for making paper, lumber used for building materials, and fuel.

- Many important medicines are currently made from plants or were originally derived from plants.

- Plant fibers are used to make paper, cloth, and rope. The most important sources of plant fibers are wood and cotton.

Key Terms

Section 1

cuticle (502)
stoma (502)
guard cell (503)
vascular system (504)
nonvascular plant (504)
vascular plant (504)
seed (504)
embryo (504)
seed plant (504)
flower (505)
phloem (507)
xylem (507)
shoot (507)
root (507)
meristem (507)

Section 2

rhizoid (508)
rhizome (510)
frond (511)
cone (511)
gymnosperm (512)
angiosperm (514)
fruit (514)
endosperm (514)
monocot (515)
dicot (515)

Section 3

vegetative part (516)
cereal (518)
grain (518)

Understanding Key Ideas

1. Seeds helped plants adapt to life on land by
  a. providing nourishment for embryos.
  b. protecting embryos from air pollution.
  c. sprouting during unfavorable weather.
  d. limiting the dispersal of plant offspring.

2. Which of the following is *not* a characteristic of vascular plants?
  a. xylem and phloem
  b. stems and leaves
  c. a dominant gametophyte
  d. a diploid sporophyte

3. Unlike angiosperms, gymnosperms
  a. are pollinated by wind.
  b. do not have seeds.
  c. have a diploid sporophyte generation.
  d. do not bear fruit.

4. Which of the following are *not* dicots?
  a. grass family
  b. rose family
  c. mustard family
  d. legume family

5. The most important sources of food are
  a. legumes.
  b. root crops.
  c. cereal grains.
  d. vegetables.

6. Drugs derived from the rosy periwinkle are used in the treatment of
  a. heart disease.
  b. leukemia.
  c. allergies.
  d. headaches.

7. Which of the following is *not* a source of fibers for both paper and cloth?
  a. flax   **c.** wood
  b. cotton   **d.** sisal

8. 🌐**BIOWatch** A friend is concerned that your vegetarian diet is not healthy. Make a list of the measures you would take to ensure that your diet will provide you with all the nutrients you need.

9. How does meiosis result in the production of haploid spores? (**Hint:** See Chapter 7, Section 1.)

10. Look at the cocklebur in the photograph below. It is a fruit that contains the seeds of a cocklebur plant. Suggest how this plant's seeds might be dispersed.

11. 🔲 **Concept Mapping** Make a concept map that shows how plants are classified. Include the following terms in your map: *vascular plants, nonvascular plants, ferns, angiosperms, gymnosperms, mosses, cones, vascular tissue, seeds,* and *flowers*.

Critical Thinking

12. Inferring Causes Most plants have a vascular system and a sporophyte that is much larger that the gametophyte. How have these features contributed to the success of plants as they evolved on land?

13. Forming Reasoned Opinions Why is the loss of tropical rain forests and other types of forests of concern to medical science?

Alternative Assessment

14. Selecting Technology You are asked to compare the length and thickness of the fibers in several types of paper. Make a list of the laboratory equipment you would need for this task, and explain why you would need each piece.

Standardized Test Prep

Understanding Concepts

Directions (1–4): **For *each* question, write on a separate sheet of paper the letter of the correct answer.**

1 What structure made it possible for plants to prevent water loss and to spread onto land?
  A. cuticle
  B. mycorrhizae
  C. pollen
  D. seed

2 Which of the following is a diploid individual in plant life cycles?
  F. epiphyte
  G. gametophyte
  H. sporophyte
  I. zygospore

3 Why might it be a reproductive advantage for a woody plant to be tall?
  A. Seeds on tall plants are less likely to be eaten by insects.
  B. Seeds on tall plants are more likely to be dispersed by grazing animals.
  C. Seeds can disperse farther in the wind when located farther from the ground.
  D. Seeds grown up high are more likely to get the sunlight they need to germinate.

4 Which of the following are **not** seedless vascular plants?
  F. club mosses
  G. ferns
  H. gymnosperms
  I. horsetails

Directions (5): **For the following question, write a short response.**

5 Many people consider corn to be a cereal crop instead of a vegetable. Why is corn considered a cereal crop agriculturally and a fruit botanically?

Test TIP

Carefully read the instructions, the question, and the answer options before choosing an answer.

Reading Skills

Directions (6): **Read the passage below. Then answer the question.**

People use many paper products every day. These include items such as notebook paper, paper towels, newspaper, and packaging. Some paper products, such as notebook paper and newspaper, are recyclable. When recycling paper products isn't possible, people can reduce their impact on the environment by using reusable instead of disposable products, buying products made with recycled paper, and buying products that have minimal paper packaging.

6 What biological molecule gives paper its strength?
  A. cellulose
  B. chitin
  C. chlorophyll
  D. glucose

Interpreting Graphics

Directions (7): **Base your answer to question 7 on the chart below.**

Some Families of Angiosperms

Subgroup	Family	Examples
Monocots	Liliaceae	garlic, asparagus, onion
	Bromeliaceae	pineapple, Spanish moss
	Palmae	raffia, date, palmetto
Dicots	Umbelliferae	carrot, celery, parsley
	Malvaceae	cotton, hibiscus, okra
	Labiatae	rosemary, sage, thyme

7 In which pair of families do all of the plants produce seeds with two seed leaves?
  F. Liliaceae and Bromeliaceae
  G. Liliaceae and Labiatae
  H. Palmae and Malvaceae
  I. Umbelliferae and Labiatae

Skills Practice Lab

Surveying Plant Diversity

SKILLS
- Observing
- Comparing

OBJECTIVES
- **Identify** similarities and differences among four phyla of living plants.
- **Relate** structural adaptations of plants to their success on land.

MATERIALS
- live or preserved specimens of mosses, ferns, conifers, and flowering plants
- stereomicroscope or hand lens
- compound microscope
- prepared slides of fern gametophytes

Before You Begin

Most plants are complex photosynthetic organisms that live on land. The ancestors of plants lived in water. As plants evolved on land, however, they developed adaptations that made it possible for them to be successful in dry conditions. All plant life cycles are characterized by **alternation of generations,** in which a haploid **gametophyte** stage alternates with a diploid **sporophyte** stage. Distinct differences in the relative sizes and structures of gametophytes and sporophytes are seen among the 12 phyla of living plants. In this lab, you will examine representatives of the four most familiar plant phyla.

1. Write a definition for each boldface term in the paragraph above and for the following terms: sporangium, spore, frond, cone, flower, fruit.

2. Make a data table similar to the one below.

3. Based on the objectives for this lab, write a question you would like to explore about the characteristics of plants.

Procedure
PART A: Conducting a Survey

1. Visit the station for each of the plants listed below, and examine the specimens there. Answer the questions, and record observations in your data table.

2. **Mosses** Examine a clump of moss with a stereomicroscope or hand lens. Make a sketch of what you see.

3. **Mosses** Examine a moss gametophyte with a sporophyte attached to it. Draw what you see, and label the parts you recognize. Label each part as haploid or diploid.
   a. Which stage of a moss has rootlike structures?
   b. Where are the spores of a moss produced?

DATA TABLE			
Phylum name	**Dominant generation**	**Major characteristics**	**Examples**

4. **Ferns** Examine the sporophyte of a fern, and look for evidence of reproductive structures on the underside of the fronds. Draw what you see. Label a leaf (frond), stem, root, and reproductive structure.

   a. How does water travel through a fern? List observations supporting your answer.

   b. What kind of reproductive cells are produced by fern fronds?

5. **Ferns** Examine a slide of a fern gametophyte with a compound microscope. Draw what you see, and label any structures you recognize.

6. **Conifers** Draw a part of a branch of one of the conifers at this station. Label a leaf, stem, and cone (if present).

   a. Is a branch of a pine tree part of a gametophyte or part of a sporophyte?

   b. In what part of a conifer would you look to find its reproductive structures?

7. **Conifers** Examine a prepared slide of pine pollen. Draw a few of the grains.

   a. What reproductive structure is found within a pollen grain?

   b. How does the structure of pine pollen aid in its dispersal by wind?

8. **Angiosperms** Draw one of the representative angiosperms at this station. Label a leaf, stem, root, and flower (if present). Indicate the sporophyte and location of gametophytes.

   a. Where do angiosperms produce sperm and eggs?

   b. How do the seeds of angiosperms differ from those of gymnosperms?

9. **Angiosperms** Examine several fruits. Draw and label the parts of one fruit.

PART B: Cleanup and Disposal

10. ⚠ Dispose of broken glass in the designated waste containers. Do not put lab materials in the trash unless your teacher tells you to do so.

11. ⚠ Wash your hands thoroughly before you leave the lab and after you finish all work.

Analyze and Conclude

1. **Analyzing Information** How are bryophytes different from the other major groups of plants?

2. **Recognizing Patterns** How do the gametophytes of gymnosperms and angiosperms differ from the gametophytes of bryophytes and ferns?

3. **Drawing Conclusions** What structures are present in both gymnosperms and angiosperms but absent in both bryophytes and ferns?

4. **Evaluating Hypotheses** Dispersal is the main function of fruits in angiosperms. Defend or refute this hypothesis. List observations you made during this lab to support your position.

5. **Inferring Conclusions** Based on their characteristics, which phylum of plants appears to be the most successful? Justify your conclusion.

6. **Further Inquiry** Write a new question about plant diversity that could be explored with another investigation.

On the Job

Drawing accurate diagrams of organisms is an important part of a plant taxonomist's research. Do research to discover the names of some famous plant taxonomists and scientific artists and to learn about some of the techniques used in their work. For more about careers, visit **go.hrw.com** and type in the keyword **HX4 Careers**.

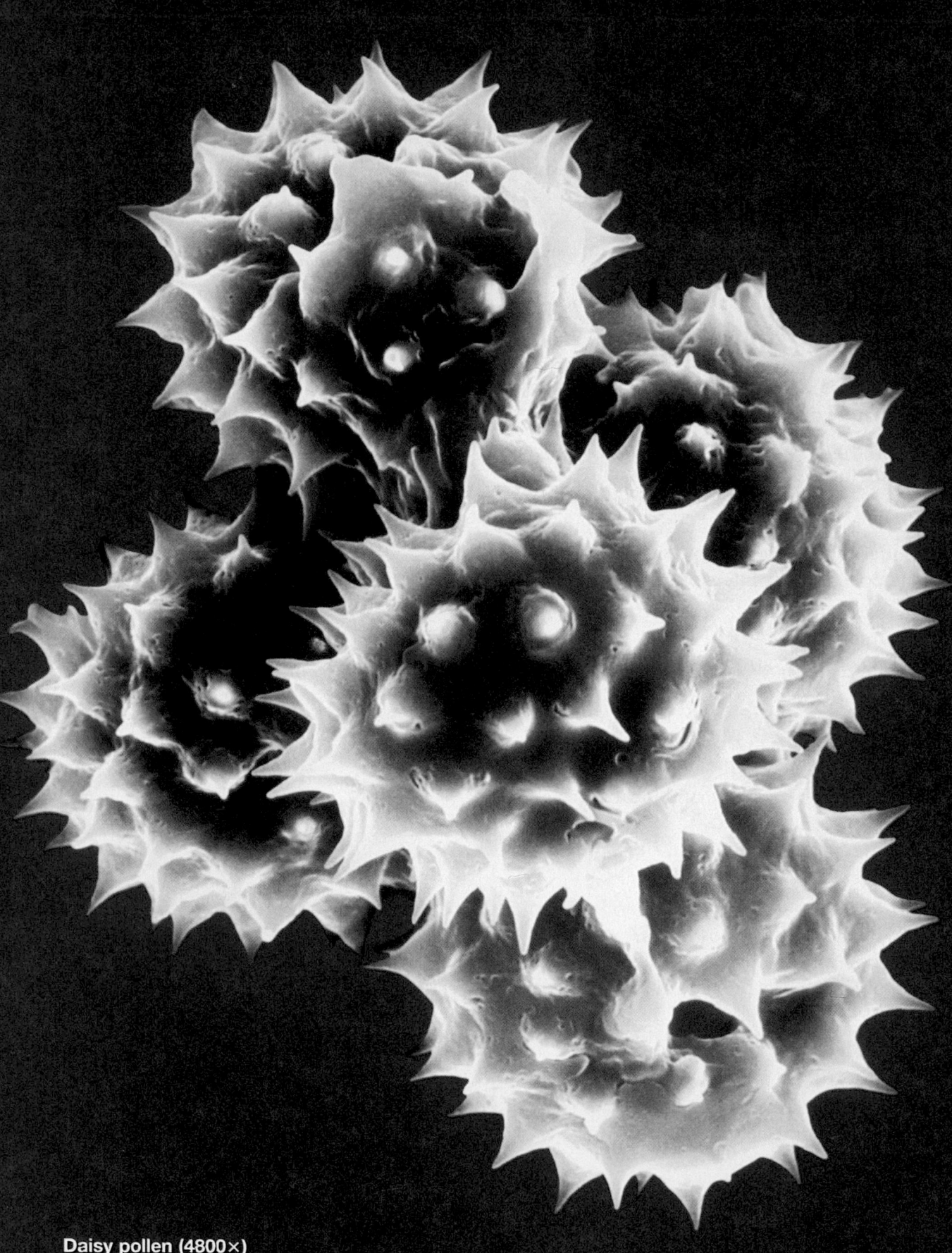

Daisy pollen (4800×)

24 Plant Reproduction

✔️ Quick Review

Answer the following without referring to earlier sections of your book.

1. **Compare** the processes of mitosis and meiosis. *(Chapter 6, Section 3 and Chapter 7, Section 1)*
2. **Distinguish** sexual reproduction from asexual reproduction. *(Chapter 7, Section 2)*
3. **Differentiate** between plant gametophytes and sporophytes. *(Chapter 23, Section 1.)*
4. **Define** the terms *cone* and *fruit*. *(Chapter 23, Section 2)*

Did you have difficulty? *For help, review the sections indicated.*

Reading Activity

Before you begin to read this chapter, write down all of the key words for each of the three sections of the chapter. Then write a definition next to each word that you have heard of. As you read the chapter, write definitions next to the words that you did not previously know, and modify as needed any definitions you have written.

Looking Ahead

Section 1

Sexual Reproduction in Seedless Plants
Reproduction in Nonvascular Plants
Reproduction in Seedless Vascular Plants

Section 2

Sexual Reproduction in Seed Plants
Reproductive Structures of Seed Plants
Seeds
Cones
Flowers

Section 3

Asexual Reproduction
Vegetative Reproduction
Plant Propagation

🔗 internet connect

www.scilinks.org
National Science Teachers Association *sci*LINKS Internet resources are located throughout this chapter.

*sci*LINKS. **Maintained by the National Science Teachers Association**

● The characteristic shape of pollen grains differs from one plant species to the next. The spiky, outer layer is composed chiefly of a polymer of carotinoids—the pigments that give fall leaves their characteristic colors.

Sexual Reproduction in Seedless Plants

Objectives

- **Summarize** the life cycle of a moss.
- **Summarize** the life cycle of a fern.
- **Compare** and **Contrast** the life cycle of a moss with the life cycle of a fern.

Key Terms

archegonium
antheridium
sorus

Reproduction in Nonvascular Plants

The carpet of green you often see near streams and in moist, shady places is usually made up of mosses or liverworts. As you learned in the previous chapter, these small, relatively simple plants are non-vascular plants. They do not have a vascular system for distributing water and nutrients. Mosses and liverworts do not usually thrive outside moist places because they must be covered by a film of water to reproduce sexually.

Like all plants, nonvascular plants have a life cycle called alternation of generations. In this type of life cycle, a gamete-producing stage, or gametophyte, alternates with a spore-producing stage, or sporophyte. Gametophytes produce gametes (eggs and sperm) in separate multicellular structures. The structure that produces eggs is called an **archegonium** *(ark uh GOHN ee uhm)*. The structure that produces sperm is called an **antheridium** *(an thuhr IHD ee uhm)*. Sporophytes produce spores in a sporangium. The gametophytes of nonvascular plants are larger and more noticeable than are the sporophytes. This difference in size is very pronounced in the liver-worts, as you can see in **Figure 1.**

Figure 1   Reproductive structures of a liverwort

The gametophytes of Marchantia, a common liverwort, produce male and female gametes on separate stalks.

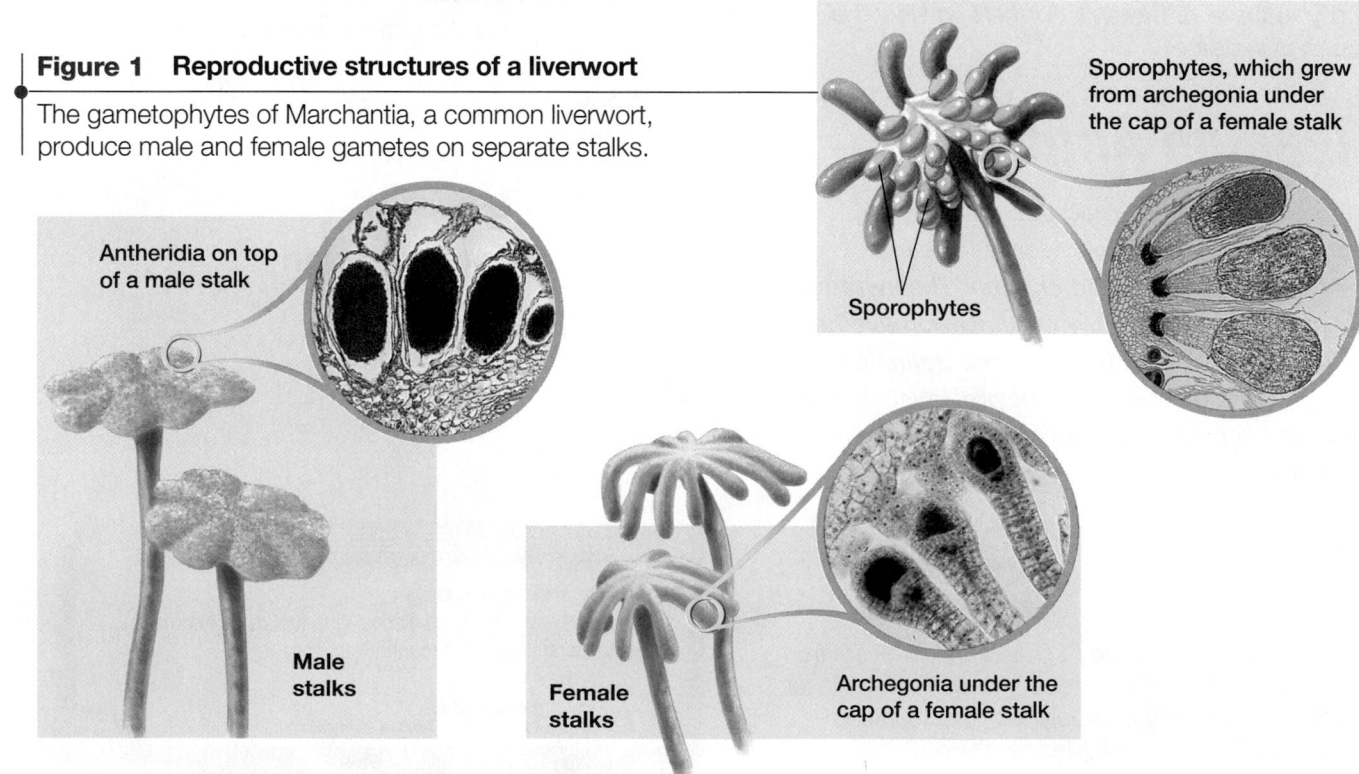

Sporophytes, which grew from archegonia under the cap of a female stalk

Antheridia on top of a male stalk

Sporophytes

Male stalks

Female stalks

Archegonia under the cap of a female stalk

Life Cycle of a Moss

The life cycle of a moss is summarized in **Figure 2.** Sexual reproduction results in a fertilized egg, or zygote. The diploid zygote grows into a new diploid sporophyte. As you can see, a moss sporophyte grows from a gametophyte and remains attached to it. The sporophyte consists of a bare stalk with a spore capsule (sporangium) at its tip. Spores form by meiosis inside the spore capsule. Therefore, as in all plants, the spores are haploid. The spore capsule opens when the spores are mature, and the spores are carried away by wind or water. When a moss spore settles to the ground, it germinates and grows into a "leafy" green gametophyte. Archegonia and antheridia form at the tips of the haploid gametophytes. Eggs and sperm form by mitosis inside the archegonia and antheridia. Remember, moss gametophytes grow in tightly packed clumps of many individuals. When water covers a clump of mosses, sperm can swim to nearby archegonia and fertilize the eggs inside them.

WORD Origins

● The word *archegonium* comes from the Greek words *archegonos,* meaning "first of a race." Knowing this makes it easier to remember that a new and genetically different individual grows from an archegonium when its egg is fertilized.

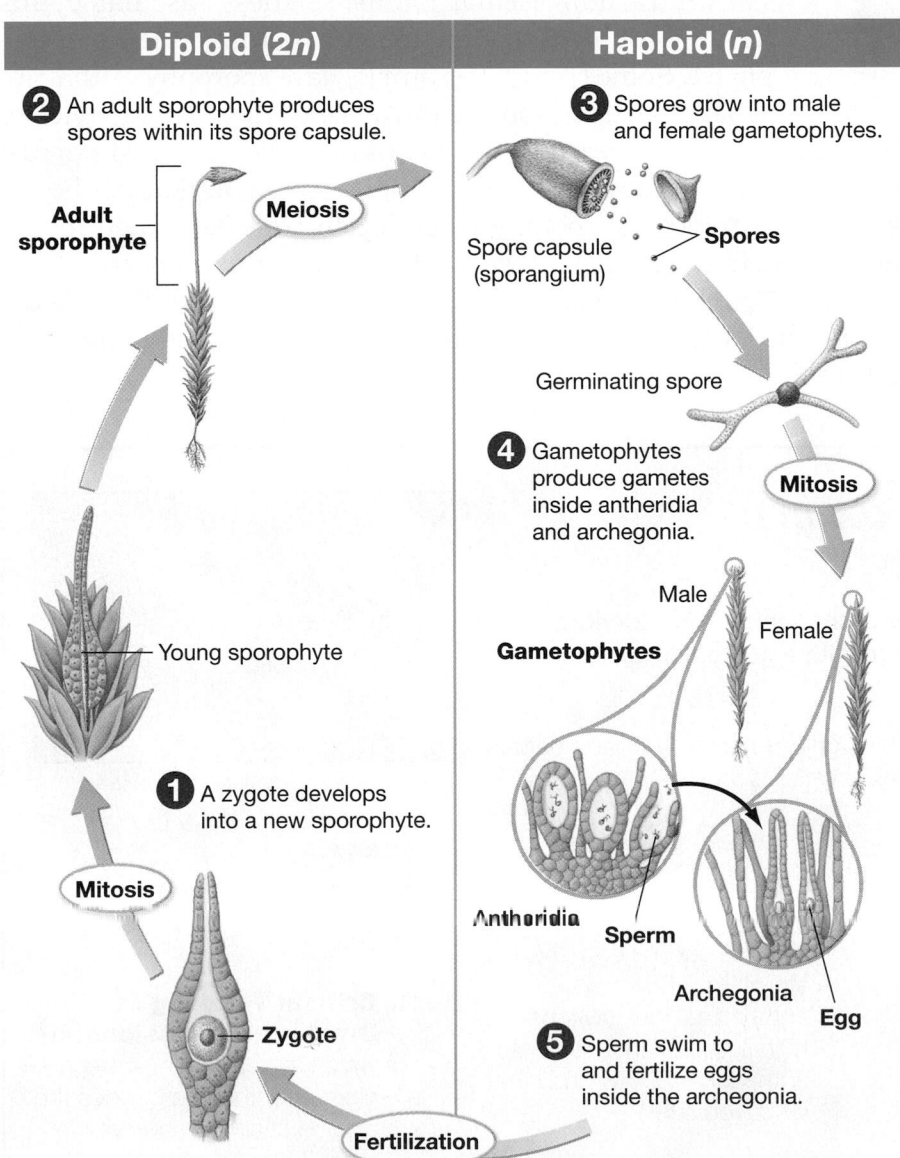

Figure 2 Moss life cycle. In mosses, a sporophyte that consists of a spore capsule on a bare stalk alternates with a "leafy," green gametophyte.

Diploid (2*n*)	Haploid (*n*)

❷ An adult sporophyte produces spores within its spore capsule.

Adult sporophyte

Meiosis

❸ Spores grow into male and female gametophytes.

Spore capsule (sporangium)

Spores

Germinating spore

❹ Gametophytes produce gametes inside antheridia and archegonia.

Mitosis

Young sporophyte

Male

Female

Gametophytes

❶ A zygote develops into a new sporophyte.

Mitosis

Antheridia

Sperm

Archegonia

Egg

Zygote

❺ Sperm swim to and fertilize eggs inside the archegonia.

Fertilization

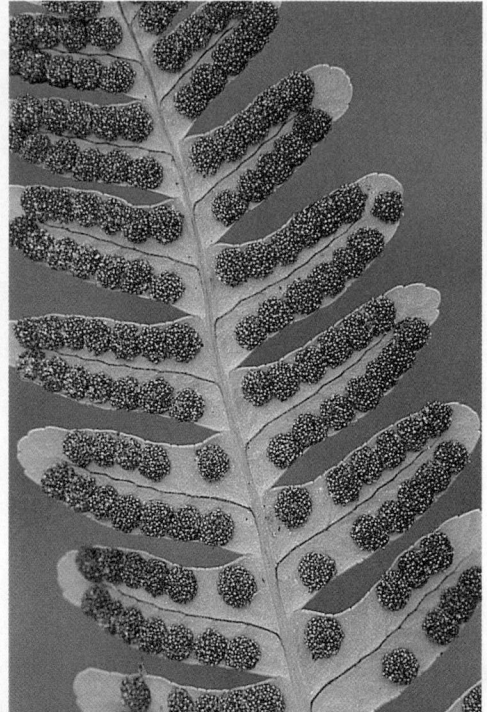

Figure 3   Sori on a fern frond. Many sori are visible on this portion of a frond from a polypody fern. Each sorus consists of about 20–30 sporangia.

Reproduction in Seedless Vascular Plants

You may recall that the seedless vascular plants include the whisk ferns, horsetails, club mosses, and ferns. The seedless vascular plants differ from the nonvascular plants because they have efficient water- and food-conducting systems of vascular tissue. Like the nonvascular plants, the seedless vascular plants thrive in moist, shady places. They can reproduce sexually only when a film of water covers the gametophyte. Eggs form in archegonia, and sperm form in antheridia. The archegonia and antheridia develop on the lower surfaces of the gametophytes. In many species of seedless vascular plants, both eggs and sperm are produced by the same individual. In some species, however, eggs and sperm are produced by separate gametophytes.

Unlike nonvascular plants, seedless vascular plants have sporophytes that are much larger than their gametophytes. Some ferns, for example, have sporophytes that are as large as trees. On the other hand, the gametophytes of ferns are thin, green, heart-shaped plants that are less than 1 cm (0.4 in.) across. The sporophytes produce spores in sporangia. In horsetails and club mosses, sporangia develop in conelike structures. In ferns, clusters of sporangia form on the lower surfaces of fronds, as shown in **Figure 3**. A cluster of sporangia on a fern frond is called a **sorus.** The word *sorus* comes from the Greek word *soros,* meaning "a heap."

Observing a Fern Gametophyte

You can observe the archegonia and antheridia of a fern gametophyte with a microscope.

Materials

prepared slide of a fern gametophyte with archegonia and antheridia, compound microscope

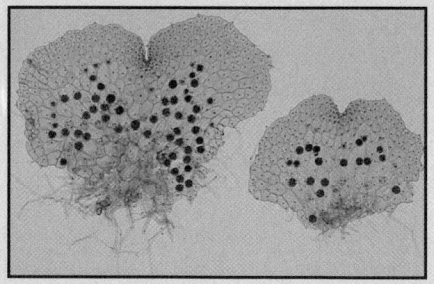

Fern Gametophytes (56×)

Procedure

1. Examine a slide of a fern gametophyte under low power of a microscope. Move the slide until you can see a cluster of archegonia. Now, switch to high power, and focus on one archegonium. Draw and label what you see.

2. Switch back to low power, and move the slide until you can see several egg-shaped structures. These are antheridia. Now, switch to high power, and focus on one antheridium. Draw and label what you see.

Analysis

1. **Describe** the appearance of an archegonium and an antheridium.

2. **Critical Thinking Drawing Conclusions** In which structure, an archegonium or antheridium, does the growth of a new sporophyte begin? Explain.

Life Cycle of a Fern

The life cycle of a fern is summarized in **Figure 4.** A fertilized egg, or zygote, grows into a new sporophyte. The diploid sporophyte produces spores by meiosis. The haploid spores fall to the ground and grow into haploid gametophytes. Fern gametophytes produce gametes by mitosis—eggs in archegonia and sperm in antheridia. Sperm swim to archegonia and fertilize the eggs inside them.

Figure 4 **Fern life cycle.** In ferns, a large sporophyte with leaves called fronds alternates with a small, green, heart-shaped gametophyte.

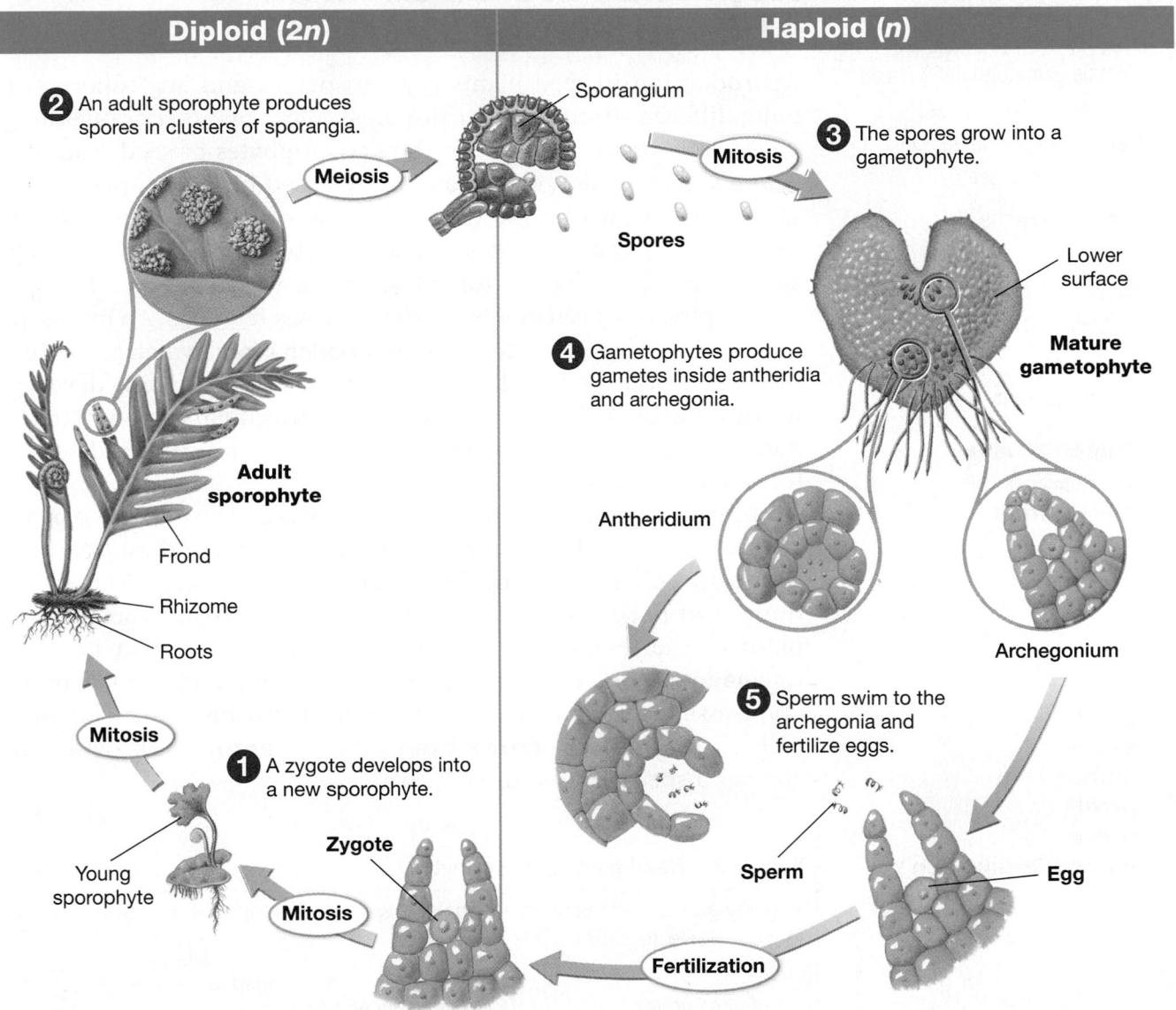

Diploid (2n) | **Haploid (n)**

2 An adult sporophyte produces spores in clusters of sporangia.

Sporangium

Meiosis

3 The spores grow into a gametophyte.

Mitosis

Spores

Lower surface

4 Gametophytes produce gametes inside antheridia and archegonia.

Mature gametophyte

Adult sporophyte

Antheridium

Frond

Rhizome

Roots

Archegonium

5 Sperm swim to the archegonia and fertilize eggs.

Mitosis

1 A zygote develops into a new sporophyte.

Young sporophyte

Zygote

Sperm

Egg

Mitosis

Fertilization

Section 1 Review

1 **List** five major steps in the life cycle of a moss.

2 **List** five major steps in the life cycle of a fern.

3 **Critical Thinking Forming Reasoned Opinions** Which reproductive structures, gametes or spores, are responsible for the dispersal (spread) of seedless plants? Justify your answer.

4 **Critical Thinking Analyzing Information** What are the major differences between the moss life cycle and the fern life cycle?

5 **Standardized Test Prep** What is the function of an archegonium?

A to produce sperm
B to produce eggs
C to produce spores
D to conduct water

Sexual Reproduction in Seed Plants

- **Distinguish** the male and female gametophytes of seed plants.
- **Describe** the function of each part of a seed.
- **Summarize** the life cycle of a conifer.
- **Relate** the parts of a flower to their functions.
- **Summarize** the life cycle of an angiosperm.

Key Terms

pollen grain
ovule
pollination
pollen tube
seed coat
cotyledon
sepal
petal
stamen
anther
pistil
ovary
double fertilization

Reproductive Structures of Seed Plants

Reproduction in seed plants (gymnosperms and angiosperms) is quite different from reproduction in seedless plants. For one thing, you need a microscope to see the gametophytes of seed plants, as **Figure 5** shows. Also, spores are not released from seed plants. The spores remain within the tissue of a sporophyte and develop into two kinds of gametophytes—male gametophytes, which produce sperm, and female gametophytes, which produce eggs. The tiny gametophytes of seed plants consist of only a few cells. An immature male gametophyte of a seed plant is a **pollen grain,** which has a thick protective wall. A female gametophyte of a seed plant develops inside an **ovule** *(AHV yool),* which is a multicellular structure that is part of the sporophyte. Following fertilization, the ovule and its contents develop into a seed.

Because the gametophytes of seed plants are very small, seed plants are able to reproduce sexually without water. Wind and animals transport pollen grains to the structures that contain ovules. The transfer of pollen grains from the male reproductive structures of a plant to the female reproductive structures of a plant is called **pollination.** When a pollen grain reaches a compatible female reproductive structure, a tube emerges from the pollen grain. This tube, called a **pollen tube,** grows from a pollen grain to an ovule and enables a sperm to pass directly to an egg.

Figure 5   Seed plant gametophytes

The tiny gametophytes of angiosperms develop within specialized structures that form in the reproductive parts of a flower.

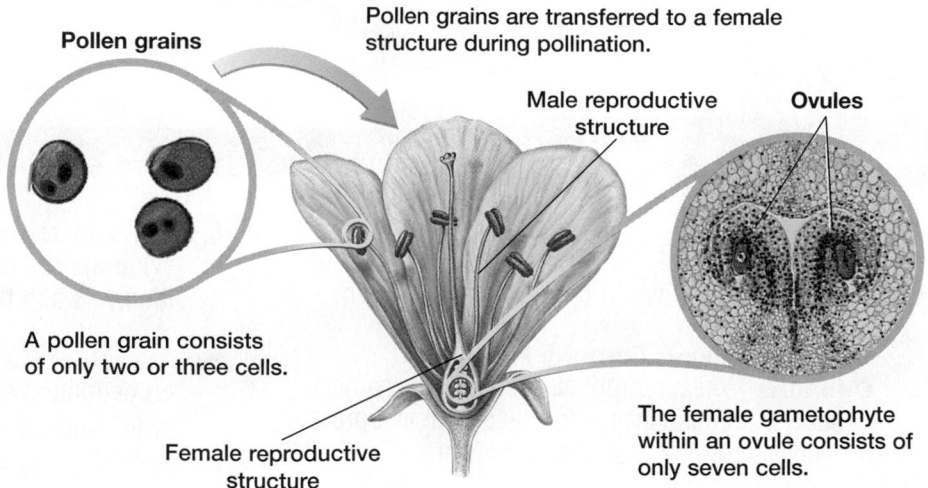

Pollen grains

Pollen grains are transferred to a female structure during pollination.

Male reproductive structure

Ovules

A pollen grain consists of only two or three cells.

Female reproductive structure

The female gametophyte within an ovule consists of only seven cells.

Seeds

As you learned in the previous chapter, seeds contain the embryos of seed plants. A plant embryo is a new sporophyte. A seed forms from an ovule after the egg within it has been fertilized. The outer cell layers of an ovule harden to form the **seed coat** as a seed matures. The tough seed coat protects the embryo in a seed from mechanical injury and from a harsh environment. Different kinds of seeds germinate under different conditions. Some seeds must dry out before they germinate. Other seeds must be exposed to light. In still other seeds, chemicals that inhibit germination must be washed away by rain. Often, a seed must be exposed to cold temperatures, or the seed coat must be damaged, before the seed can germinate. Thus, seeds enable the embryos of seed plants to survive conditions that are unfavorable for plant growth for long periods of time.

Seeds also contain tissue that provides nutrients to plant embryos. In gymnosperms, this nutritious tissue is part of the female gametophyte. The seeds of angiosperms, however, develop a nutritious tissue called *endosperm*. Endosperm originates at the same time an egg is fertilized. In some angiosperms, such as corn and wheat, endosperm is still present in mature seeds. In other angiosperms, such as beans and peas, the nutrients in the endosperm have already been transferred to the embryo by the time a seed is mature.

Leaflike structures called **cotyledons** *(kah tuh LEE duhnz)*, or seed leaves, are a part of a plant embryo. Cotyledons function in the transfer of nutrients to the embryo. The embryos of gymnosperms have two or more cotyledons. For example, pine embryos have eight cotyledons. In the flowering plants, the embryos of monocots have one cotyledon, and the embryos of dicots have two cotyledons. The structure of three types of seeds is shown in **Figure 6.**

Study TIP

● **Interpreting Graphics**
After reading the chapter, trace or make a sketch of Figure 6 without the labels. On separate pieces of paper, write down the labels. Without referring to your book, match the labels with the correct parts of your sketch.

Figure 6   Seed structure

Seeds have many similarities and differences in structure.

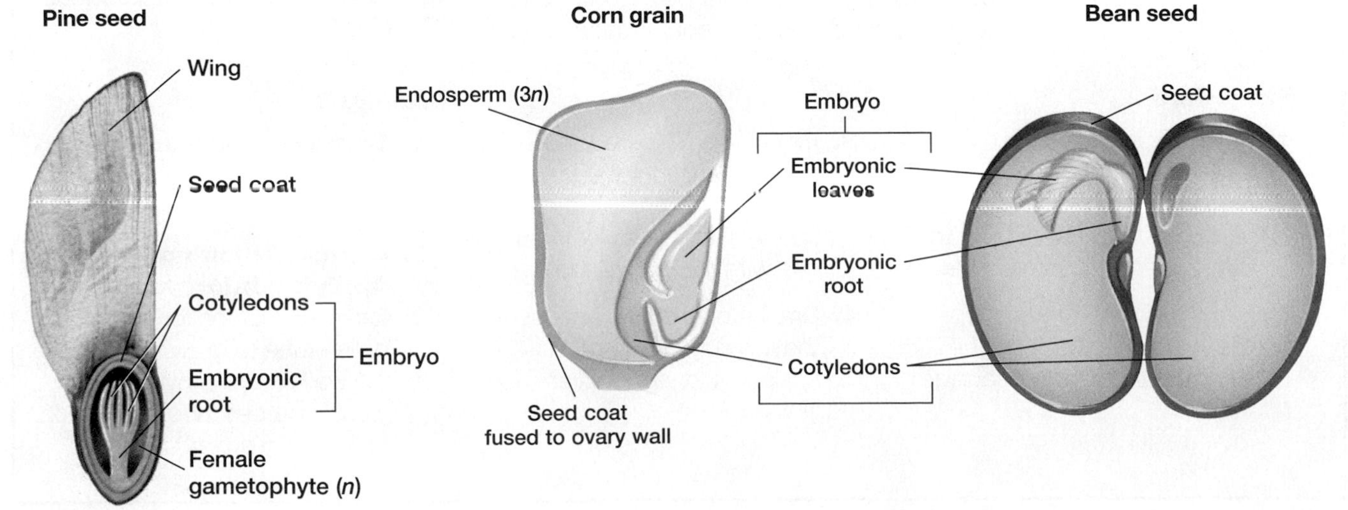

Pine seed — Wing, Seed coat, Cotyledons, Embryonic root, Female gametophyte (n), Embryo

Corn grain — Endosperm (3n), Seed coat fused to ovary wall, Embryo, Embryonic leaves, Embryonic root, Cotyledons

Bean seed — Seed coat, Cotyledons

Figure 7 Male and female pine cones. This branch of an Austrian pine has an immature seed cone and many pollen cones.

Cones

Seed plants are the most successful of all plants. The success of the seed plants is due in part to the specialized structures in which seeds develop. In angiosperms, the ovules (immature seeds) are completely enclosed by sporophyte tissue at the time of pollination. In gymnosperms, the ovules are not completely enclosed by sporophyte tissue until after pollination.

The gametophytes of gymnosperms develop in cones, which consist of whorls (circles) of modified leaves called scales. Gymnosperms produce two types of cones. Male cones, or pollen cones, produce pollen grains within sacs that develop on the surface of their scales. Female cones, or seed cones, produce ovules on the surface of their scales. Many gymnosperms produce both male and female cones on the same plant. As shown in **Figure 7,** the numerous small pollen cones lie to the left of the large seed cone. In some gymnosperms, male and female cones form on separate plants.

Pollen cones produce large quantities of pollen grains that are carried by wind to female cones. At the time of pollination, the scales of a female cone are open, exposing the ovules. When a pollen grain lands near an ovule, a slender pollen tube grows out of the pollen grain and into the ovule. The sperm moves through the pollen tube and enters the ovule. Thus, the pollen tube delivers a sperm to the egg inside the ovule. Seed cones close up after pollination and remain closed until the seeds within them are mature. This process can take up to two years.

Observing the Gametophytes of Pines

You can observe the gametophytes of a pine with a microscope.

Materials

prepared slides of the following: male pine cone, female pine cone, pine ovule; hand lens; compound microscope

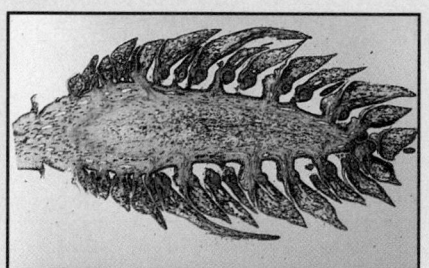

Immature female pine cone

Procedure

1. Examine prepared slides of male and female pine cones first with a hand lens and then under the low power of a microscope.

2. Make a sketch of each type of pine cone, and label the structures that you recognize.

3. Examine a prepared slide of a pine ovule under the low power of a compound microscope. Compare what you see with the photo above.

4. Draw a pine ovule, and label the following structures: scale, ovule, egg, pollen tube (if visible).

Analysis

1. **Compare** and **Contrast** the structure and contents of male and female pine cones.

2. **Critical Thinking Applying Information** It takes 15 months for a pine pollen tube to grow through the wall of a pine ovule. How would you describe the rate of pollen-tube growth in pines?

Life Cycle of a Conifer

Most gymnosperms are conifers, a group that includes pines. You can trace the stages in the life cycle of a pine in **Figure 8.** In pines, as in all plants, a diploid zygote results from sexual reproduction. The zygote develops into an embryo, which then becomes dormant (inactive). The embryo and the surrounding tissues form a seed. When their seeds are mature, seed cones open, and the seeds fall out. A seed of most pines has a wing that causes it to spin like the blade of a helicopter. Thus, pine seeds often travel some distance from their parent tree.

When conditions are favorable for growth, the embryos grow into new sporophytes. An adult pine tree produces both male and female cones. Spores form by meiosis, which occurs inside immature cones. The spores grow into gametophytes, which produce eggs and sperm by mitosis. After pollination, a pollen tube begins to grow from each pollen grain toward the eggs inside an ovule. Fertilization occurs as a sperm fuses with an egg, forming a zygote that will grow into a new sporophyte.

Figure 8   Conifer life cycle. In conifers, a very large sporophyte that produces cones alternates with tiny gametophytes that form on the scales of cones.

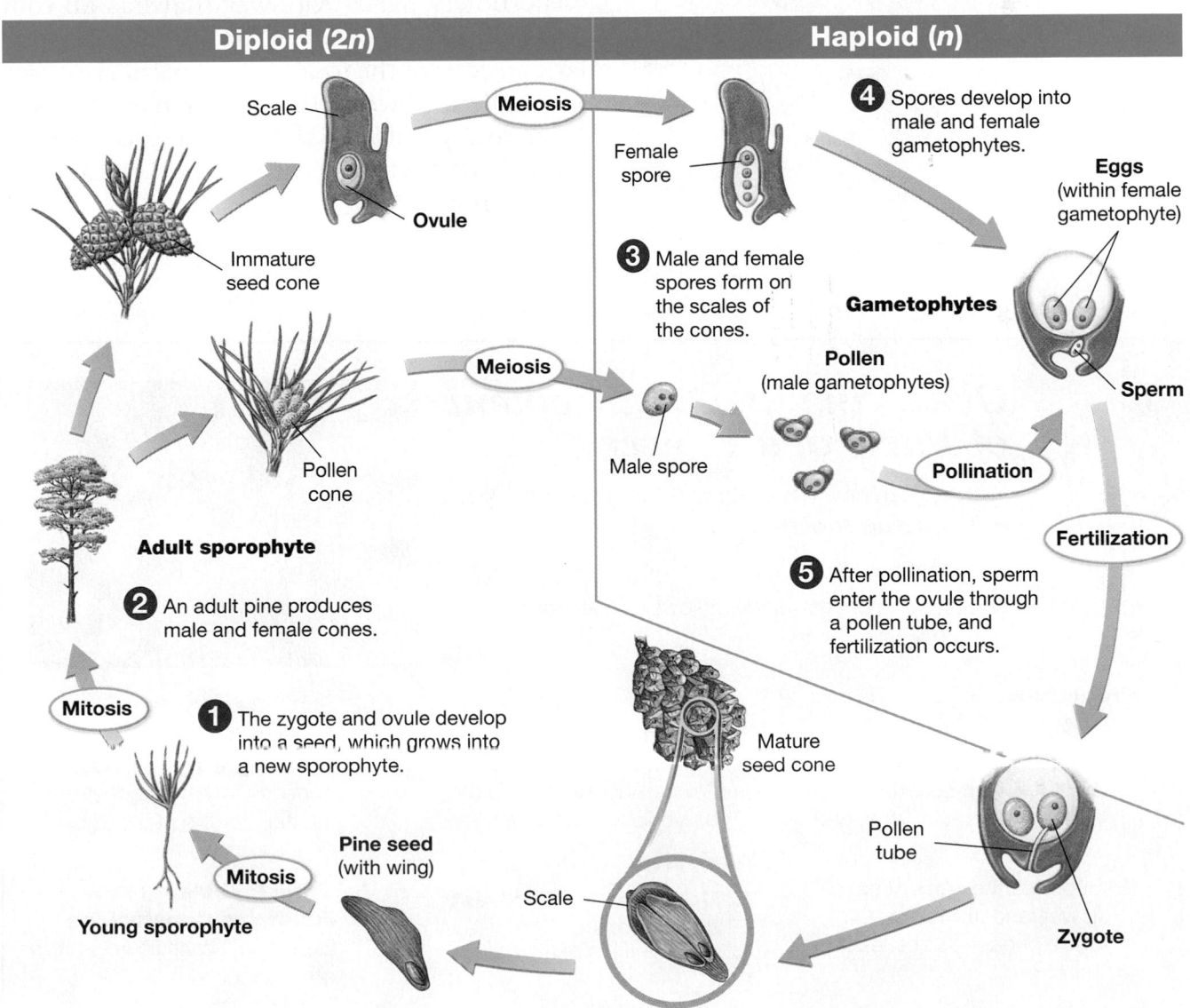

Flowers

In angiosperms, gametophytes develop within flowers. The basic structure of a flower is shown in **Figure 9.** Flower parts are arranged in four concentric whorls. The outermost whorl consists of one or more **sepals** *(SEE puhlz)*, which protect a flower from damage while it is a bud. The second whorl consists of one or more **petals,** which attract pollinators. The third whorl consists of one or more **stamens** *(STAY muhnz),* which produce pollen. Each stamen is made of a threadlike filament that is topped by a pollen-producing sac called an **anther.** The fourth and innermost whorl of a flower consists of one or more **pistils,** which produce ovules. Ovules develop in a pistil's swollen lower portion, which is called the ovary. Usually, a stalk, called the style, rises from the **ovary.** Pollen lands on and sticks to the stigma—the swollen, sticky tip of the style.

Flowers may or may not have all four of the basic flower parts. A flower that has all four parts is called a complete flower. Flowers that lack any one of the four types of parts are called incomplete flowers. If a flower has both stamens and pistils, it is called a perfect flower. Flowers that lack either stamens or pistils are called imperfect flowers.

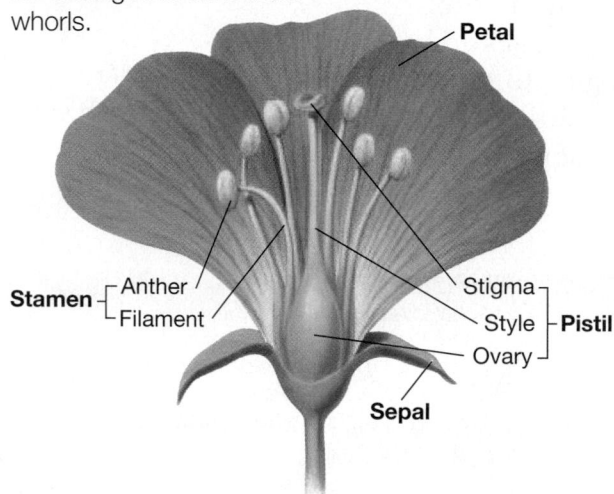

Figure 9   Basic flower structure. The four basic parts of a flower—sepals, petals, stamens, and pistils—are arranged in concentric whorls.

Observing the Arrangement of Parts of a Flower

You can see how the parts of flowers are arranged by dissecting flowers.

Materials

gloves, monocot flower, dicot flower, paper, tape

Procedure

1. Put on gloves. Examine a monocot flower and a dicot flower. Locate the sepals, petals, stamens, and pistil of each flower.

2. Separate the parts of each flower, and tape them to a piece of paper. Label each set of parts.

3. Count the number of petals, sepals, and stamens in each flower. Record this information below each flower.

Analysis

1. **Compare** and **Contrast** the appearance of the sepals and petals of each flower.

2. **Critical Thinking Forming a Hypothesis** For each flower, suggest a function for the petals based on their appearance.

3. **Critical Thinking Justifying Conclusions** Explain why each flower is from either a monocot or a dicot.

Flowers and Their Pollinators

Many flowers have brightly colored petals, sugary nectar, strong odors, and shapes that attract animal pollinators. Flowers are a source of food for pollinators such as insects, birds, and bats. For example, bees eat nectar and collect pollen, which is a rich source of protein they feed to their larvae. A bee gets coated with pollen as it visits a flower and then carries that pollen to other flowers.

Bees locate flowers by scent first and then by color and shape. Bee-pollinated flowers are usually blue or yellow and often have markings that show the location of nectar. Moths, which feed at night, tend to visit heavily scented white flowers, which are easy to find in dim light. Flies may pollinate flowers that smell like rotten meat.

Many flowers are not pollinated primarily by insects. Red flowers, for instance, may be pollinated by hummingbirds. Some large white flowers that open at night are pollinated by nighttime visitors—bats, as seen in **Figure 10.** Many flowers, such as those of grasses and oaks, are pollinated by wind. Wind-pollinated flowers are usually small and lack bright colors, strong odors, and nectar.

Figure 10 Bat pollination. This lesser long-nosed bat pollinates an organ pipe cactus as it feeds on the pollen of the plant's flowers.

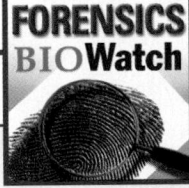

FORENSICS
BIOWatch

Telltale Pollen

Few living things are as short-lived as a flower. Delicate and lovely, its petals wither and die in just a few days. Yet flowers contain some of the toughest structures found in nature: pollen grains.

A thick, decay-resistant wall surrounds each tiny pollen grain. Remarkably, each flowering plant species has its own unique pollen. As shown in the scanning electron micrograph above right, under the microscope, pollen grains may have unusual shapes or may be covered with distinctive ridges, knobs, or spikes. Because no two plant species have identical pollen, investigators often can use pollen to link an object or a suspect to a scene of a crime.

Floral Footprints

Different areas have different kinds of pollen-producing plants. Dirt, mud, or dust collected from a person's clothing, hair, or shoes may contain pollen that can reveal a suspect's past whereabouts or link him or her to a specific location. Pollen recovered from the air filters of a vehicle can show where the vehicle might have traveled. In imported goods, the presence of certain pollens can verify the country of origin.

Drug Detection

Pollen from marijuana (*Cannibis* spp.) found trapped in dust filters, air ducts, or cracks in the floors can confirm that marijuana was grown or processed in a certain location.

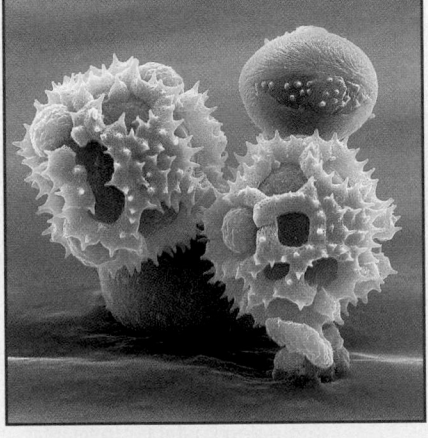

The mixture of different pollen types found in a sample can also be highly specific, much like a fingerprint. Law enforcement agencies can use this evidence to link plant-based drugs found on different people or at different places. Investigators can also use pollen evidence to pinpoint the region where a drug crop was grown.

Life Cycle of an Angiosperm

Figure 11 summarizes the life cycle of an angiosperm. Following fertilization, the zygote and the tissues of the ovule develop into a seed, which grows into a new sporophyte. The adult sporophytes of angiosperms produce spores by meiosis. These spores grow into gametophytes. The female gametophytes grow inside the ovules, which develop within the ovary of a pistil. The male gametophytes, or pollen grains, are produced in the anther of a stamen. A pollen grain contains two sperm cells. One sperm fuses with the egg, forming the zygote. The other sperm fuses with the haploid nuclei of two other cells produced by meiosis. The fusing of three haploid (*n*) cells forms a triploid (*3n*) cell that develops into endosperm. This is a process called **double fertilization.**

Figure 11   Angiosperm life cycle. In angiosperms, a large sporophyte alternates with a tiny gametophyte.

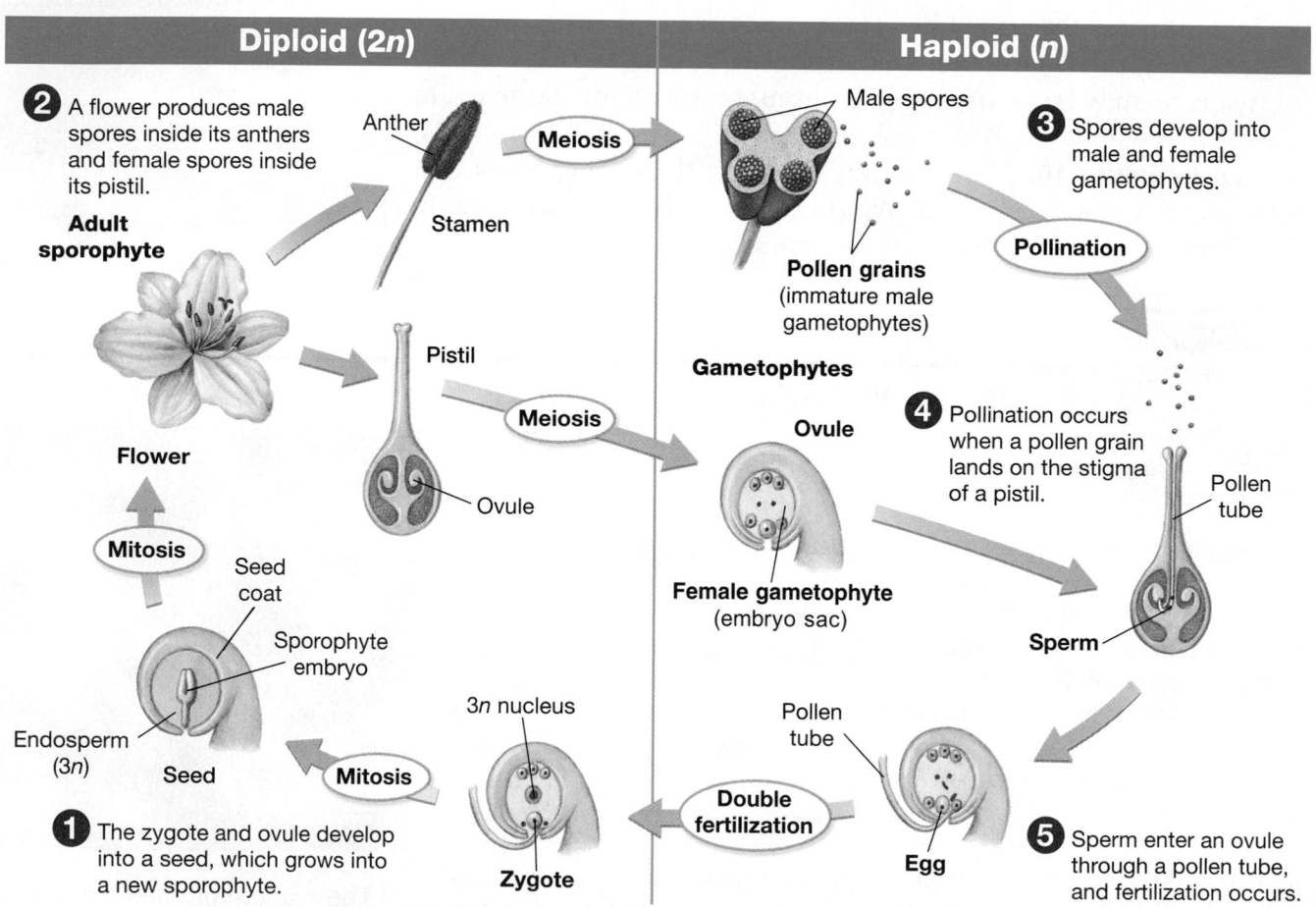

Diploid (2n)

2 A flower produces male spores inside its anthers and female spores inside its pistil.

Anther

Stamen

Meiosis

Adult sporophyte

Pistil

Meiosis

Ovule

Flower

Mitosis

Seed coat

Sporophyte embryo

Endosperm (3n)

Seed

Mitosis

3n nucleus

Zygote

1 The zygote and ovule develop into a seed, which grows into a new sporophyte.

Haploid (n)

Male spores

3 Spores develop into male and female gametophytes.

Pollen grains (immature male gametophytes)

Pollination

Gametophytes

Ovule

4 Pollination occurs when a pollen grain lands on the stigma of a pistil.

Pollen tube

Female gametophyte (embryo sac)

Sperm

Pollen tube

Double fertilization

Egg

5 Sperm enter an ovule through a pollen tube, and fertilization occurs.

Section 2 Review

1 **Distinguish** pollen grains from ovules.

2 **Describe** the function of each part of a seed.

3 **Summarize** the life cycle of a conifer.

4 **Describe** the main events in the life cycle of an angiosperm?

5 **Critical Thinking Relating Concepts** How is each part of a flower suited to its function?

6 **Standardized Test Prep** In angiosperms, pollen is produced in sacs called
A sepals.   C pistils.
B anthers.   D ovules.

Asexual Reproduction

Vegetative Reproduction

Most plants are able to reproduce asexually. The new individuals that result from asexual reproduction are genetically the same as the parent plant. Plants reproduce asexually in a variety of ways that involve nonreproductive parts, such as stems, roots, and leaves. The reproduction of plants from these parts is called **vegetative reproduction.** Many of the structures by which plants reproduce vegetatively are modified stems, such as runners, bulbs, corms, rhizomes, and tubers. **Table 1** describes these structures.

Vegetative reproduction is faster than sexual reproduction in most plants. A single plant can spread rapidly in a habitat that is ideal for its growth by reproducing vegetatively. Therefore, a mass of hundreds or even thousands of individuals, such as a stand of grasses or ferns, may have come from one individual. To learn about one unique method of vegetative reproduction in one plant, look at *Up Close: Kalanchoë,* on the next two pages.

Objectives

- **Describe** several types of vegetative reproduction in plants.
- **Distinguish** sexual reproduction in kalanchoës from asexual reproduction in kalanchoës.
- **Recommend** several ways to propagate plants.

Key Terms

vegetative reproduction
plant propagation
tissue culture

Table 1 Stems Modified for Vegetative Reproduction

Name	Description	Examples
Runner	Horizontal, above-ground stem	Airplane plant, Bermuda grass
Bulb	Very short, stem with thick, fleshy leaves; only in monocots	Onion, daffodil, tulip
Corm	Very short, thickened, underground stem with thin, scaly leaves	Gladiolus, crocus
Rhizome	Horizontal underground stem	Iris, fern, sugar cane
Tuber	Swollen, fleshy, underground stem	Potato, caladium

Up Close

Kalanchoë

- **Scientific name:** *Kalanchoë daigremontiana*
- **Size:** Grows from 30 cm (about 1 ft) to 1 m (3.3 ft) tall
- **Range:** Native to southwestern Madagascar; cultivated worldwide
- **Habitat:** Semiarid tropical grassland with moist summers and well-drained, fertile soil
- **Importance:** Kalanchoës (kal an KOH eez) are grown as indoor potted plants and as outdoor perennials in warm climates.

External Structures

Leaves The fleshy leaves are bluish green, with purple markings and saw-toothed margins. Leaf blades range from 12 to 25 cm (about 5 to 10 in.) long. Leaves are arranged in pairs that are opposite one another.

Flowers A cluster of flowers forms on a flowering stalk that grows from the end of a stem. The flowers are bell-shaped and about 2.5 cm (1 in.) long. Flower parts occur in fours. Each flower produces many tiny seeds.

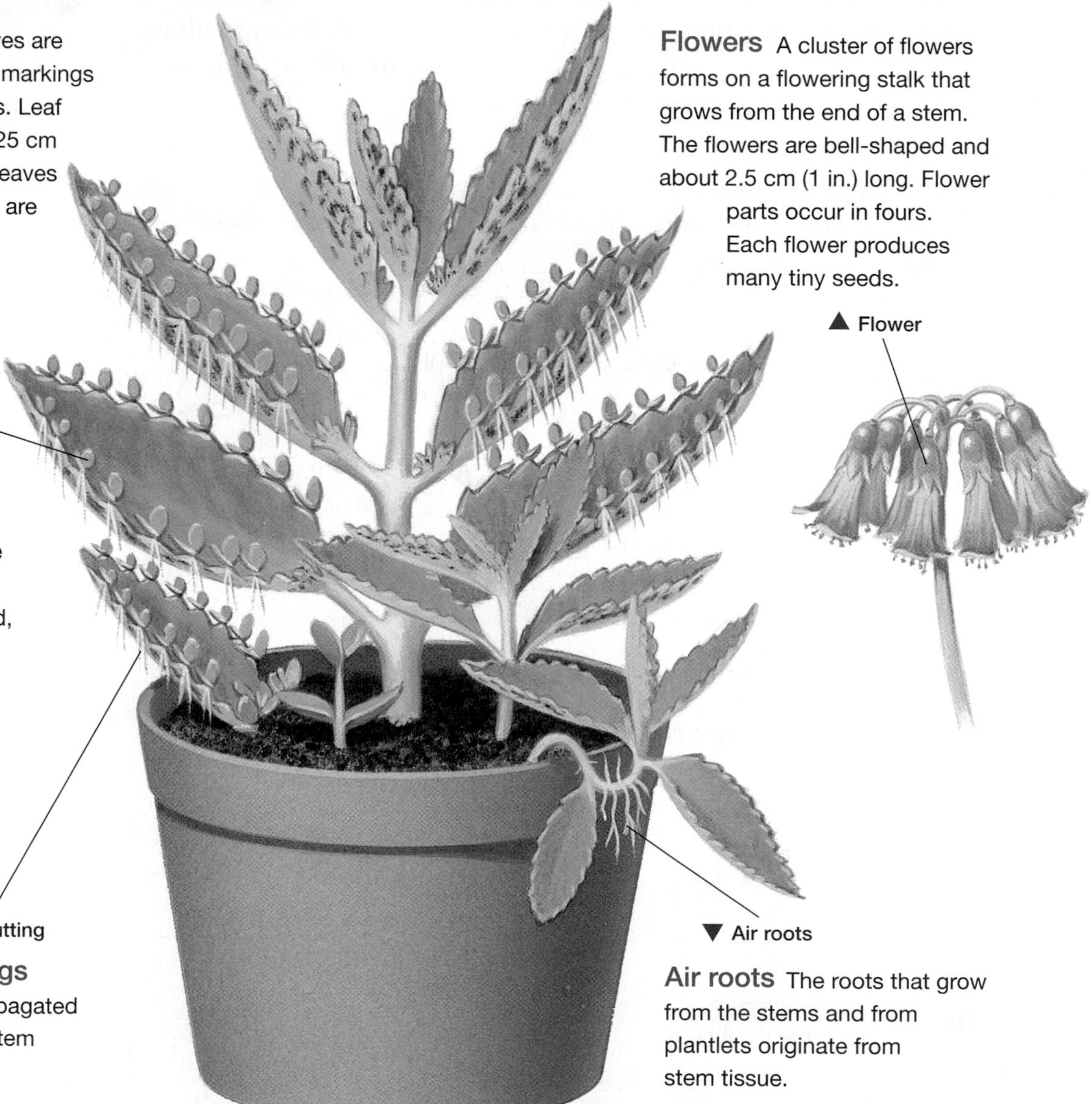

▲ Flower

▼ Plantlet

Plantlets Tiny new plants develop along leaf margins. These plantlets are a means of vegetative reproduction. When a plantlet falls to the ground, it grows into a new plant.

▼ Leaf cutting

Stem and leaf cuttings Kalanchoës are often propagated vegetatively by planting stem and leaf cuttings.

▼ Air roots

Air roots The roots that grow from the stems and from plantlets originate from stem tissue.

Leaf structure Kalanchoës are succulents, which means they have fleshy leaves and stems that store water. A kalanchoë leaf shows how some succulents are adapted for conserving water. A thick cuticle covers the leaf, and the epidermis (outer layer of cells) consists of several layers of cells. Relatively few, very small stomata dot the leaf surfaces.

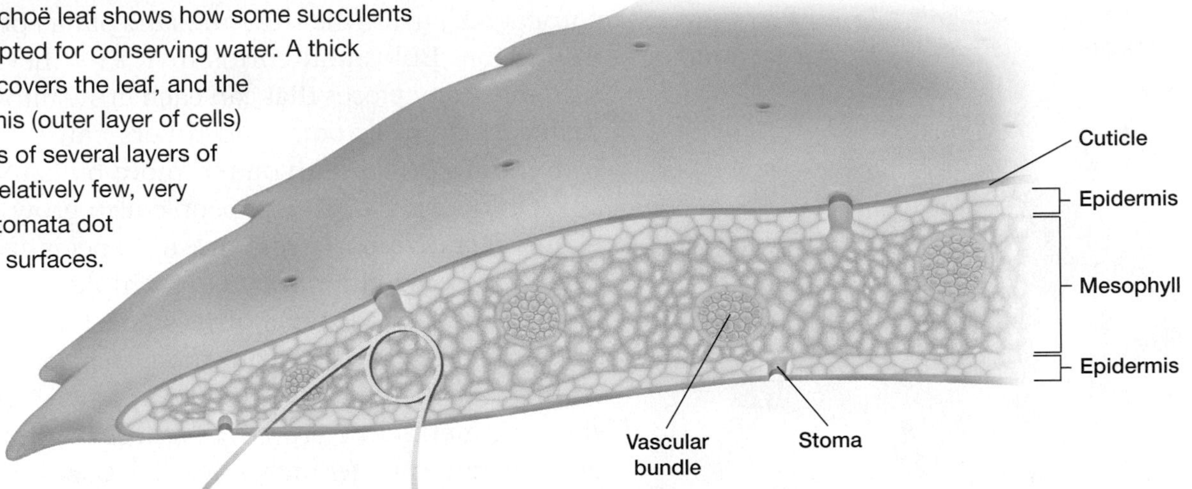

Cuticle

Epidermis

Mesophyll

Epidermis

Vascular bundle

Stoma

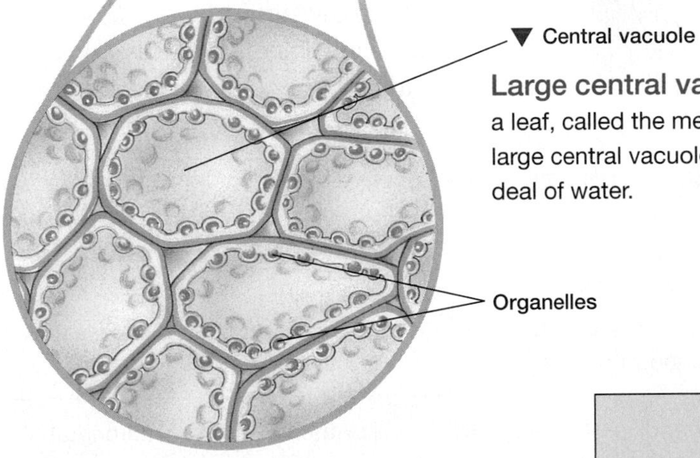

▼ Central vacuole

Large central vacuole The cells inside a leaf, called the mesophyll cells, have a large central vacuole that can hold a great deal of water.

Organelles

Mesophyll cells

Mesophyll cell

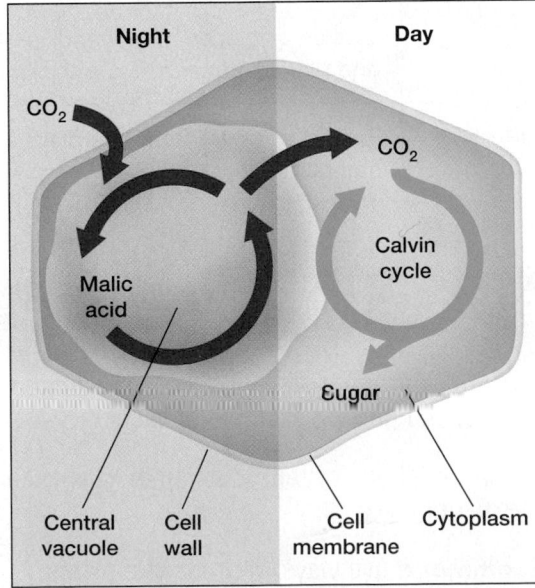

Night

Day

CO_2

CO_2

Calvin cycle

Malic acid

Sugar

Central vacuole

Cell wall

Cell membrane

Cytoplasm

CAM photosynthesis Kalanchoës belong to the Crassulaceae family, a group of succulent plants that are adapted to hot climates. Photosynthesis in kalanchoës involves a process called crassulacean acid metabolism (CAM).

The stomata of CAM plants open only at night, unlike those of other plants. At night, the plants fix carbon dioxide by using it to make malic acid. The malic acid is stored in the large central vacuoles of the mesophyll cells. In daytime, the stomata remain closed, which prevents water loss. Carbon dioxide is released from malic acid during the day and used by the Calvin cycle to make sugar.

Plant Propagation

Figure 12 Budding pears. A bud from a desirable variety of pears is attached to a stem of another pear species. The bud will grow into a branch that produces the desirable variety of pears.

People grow plants for many purposes, such as for food, to beautify homes, or to sell. Most field crops, such as cereal grains, vegetables, and cotton, are grown from seed. Many other plants are grown from vegetative parts. Growing new plants from seed or from vegetative parts is called **plant propagation.**

Plants are often propagated using the structures the plants produce for vegetative reproduction. Bulbs and corms divide as they grow, forming many pieces that can each grow into a new plant. Rhizomes, roots, and tubers can be cut or broken into pieces with one or more buds that can grow into new shoots. But people also grow plants from vegetative parts that are not specialized for vegetative reproduction. For example, pieces of plants, such as the stems of ivys and the leaves of African violets, are cut from the parent plant. The cuttings are then used to grow new plants. **Figure 12** shows a method of propagating trees called budding. In another technique called **tissue culture,** pieces of plant tissue are placed on a sterile medium and used to grow new plants. **Table 2** summarizes some of the methods of vegetative plant propagation that are widely used to grow plants.

Table 2 Methods of Vegetative Plant Propagation

Method	Description	Examples
Budding and grafting	Small stems from one plant are attached to larger stems or roots of another plant.	Grape vines, hybrid roses, fruit and nut trees
Taking cuttings	Leaves or pieces of stems or roots are cut from one plant and used to grow new individuals.	African violets, ornamental trees and shrubs, figs
Tissue culture	Pieces of tissue from one plant are placed on a sterile medium and used to grow new individuals.	Orchids, potatoes, many houseplants

Section 3 Review

1. **Describe** four types of vegetative reproduction in plants, and give an example of each.

2. **Classify** methods of reproduction in kalanchoës as sexual or asexual.

3. **Recommend** five ways to propagate plants.

4. **Critical Thinking Justifying Conclusions** Why would someone choose to propagate a particular plant for commercial purposes by using vegetative structures instead of seed?

5. **Standardized Test Prep** Bermuda grass reproduces asexually by means of horizontal, aboveground stems called

   A corms.   **C** tubers.

   B rhizomes.   **D** runners.

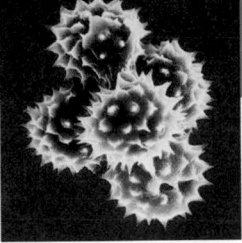

Key Concepts

1 Sexual Reproduction in Seedless Plants

- In mosses, the "leafy" green gametophytes are larger than the sporophytes, which consist of a bare stalk and a spore capsule.
- In the life cycle of a fern, the sporophytes are much larger than the gametophytes. The thin, green, heart-shaped gametophytes produce both sperm and eggs.
- Nonvascular plants and seedless vascular plants need water for fertilization because sperm must swim to eggs.

2 Sexual Reproduction in Seed Plants

- The tiny gametophytes of seed plants develop from spores that remain within sporophyte tissues. Male gametophytes are pollen grains. Female gametophytes develop inside ovules.
- A seed contains an embryo, which is a new sporophyte, and a supply of nutrients for the embryo. The cotyledons of an embryo help transfer nutrients to the embryo. A seed coat covers and protects a seed.
- In gymnosperms, male and female gametophytes develop in separate cones on the sporophytes. After fertilization, ovules develop into seeds, which grow into new sporophytes.
- Flowers have four types of parts—petals, sepals, stamens, and pistils. Petals attract pollinators. Sepals protect buds and may also attract pollinators. Pollen forms in the anthers of stamens. Seeds develop in the ovary of a pistil.
- In angiosperms, male and female gametophytes develop in the flowers of the sporophytes. After fertilization, ovules develop into seeds, which grow into new sporophytes.

3 Asexual Reproduction

- Vegetative reproduction is the growth of new plants from nonreproductive plant parts, such as stems, roots, and leaves.
- Kalanchoës are succulents that are often grown as potted plants and readily reproduce either vegetatively or by seeds.
- People often grow plants from their vegetative structures. This is called vegetative propagation.

Key Terms

Section 1

archegonium (530)
antheridium (530)
sorus (532)

Section 2

pollen grain (534)
ovule (534)
pollination (534)
pollen tube (534)
seed coat (535)
cotyledon (535)
sepal (538)
petal (538)
stamen (538)
anther (538)
pistil (538)
ovary (538)
double fertilization (540)

Section 3

vegetative reproduction (541)
plant propagation (544)
tissue culture (544)

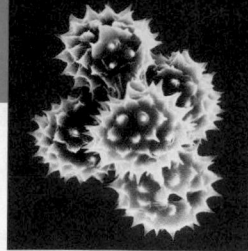

Understanding Key Ideas

1. Mosses and liverworts thrive in a moist environment because they need _____ for reproduction.
- **a.** bees
- **b.** birds
- **c.** water
- **d.** wind

2. The life cycle of a moss differs from the life cycle of a fern in that
- **a.** the gametophyte is absent in ferns.
- **b.** the sporophyte is absent in mosses.
- **c.** moss spores do not form on leaves.
- **d.** the gametophytes of mosses are green.

3. In angiosperms, the zygote and the first cell of the endosperm form by
- **a.** mitosis.
- **b.** meiosis.
- **c.** pollination.
- **d.** double fertilization.

4. Vegetative reproduction has *not* occurred when a new plant grows from a
- **a.** leaf.
- **b.** root.
- **c.** stem.
- **d.** seed.

5. Which of the following structures do kalanchoës produce for vegetative reproduction?
- **a.** seeds
- **b.** plantlets
- **c.** flowers
- **d.** bulbs

6. **BIOWatch** How can forensic botany help link a suspect to a location?

7. What is the function of the fruits in which seeds mature? (**Hint:** See Chapter 23, Section 1.)

8. Which of the following structures is *not* used to propagate dicots vegetatively?
- **a.** tubers
- **b.** rhizomes
- **c.** seeds
- **d.** stem cuttings

9. Look at the flower in the photograph below. It is the flower of the unicorn plant. How is this flower probably pollinated? Justify your answer.

10. **Concept Mapping** Make a concept map that explains how plants reproduce. Try to include the following terms in your map: *archegonium, antheridium, egg, sperm, ovule, zygote, stamen, anther, pistil, ovary, fertilization, spore,* and *vegetative reproduction.*

Critical Thinking

11. Evaluating Conclusions All nonvascular plants require a film of water for sperm to swim through and fertilize eggs. Therefore, many people conclude that nonvascular plants are not able to survive in very dry climates, such as deserts. Is this a valid conclusion? Justify your answer.

12. Justifying Conclusions A classmate has found a plant whose flowers lack petals and have many stamens. Your classmate tells you that the plant is wind-pollinated. Justify this conclusion.

13. Evaluating Methods You are asked to grow a large number of identical potted plants for a florist. The plants can be grown from either seeds or cuttings. Which method of plant propagation would you use? Justify your choice.

Alternative Assessment

14. Career Connection Plant Breeder Use the media center or Internet to find out about the field of plant breeding. Write a report on your findings.

Standardized Test Prep

Understanding Concepts

Directions (1–5): **For *each* question, write on a separate sheet of paper the letter of the correct answer.**

1 What is the purpose of the archegonium in mosses and ferns?
  A. produce eggs
  B. produce sperm
  C. develop into a cone
  D. develop into a seed

2 In seed plants, what structure transfers sperm from a pollen grain directly to an egg in an ovule?
  F. endosperm
  G. pollen tube
  H. pollinator
  I. seed coat

3 What type of plant produces a seed with a single cotyledon?
  A. dicot
  B. gymnosperm
  C. monocot
  D. nonvascular

4 Which part of a flower produces eggs?
  F. petal
  G. pistil
  H. sepal
  I. stamen

5 What conclusion can be drawn about a plant whose flowers are large and colorful?
  A. The plant must be self-pollinated.
  B. The plant must be wind-pollinated.
  C. The plant must be pollinated by insects.
  D. The plant must be a gymnosperm.

For questions involving life cycles, draw as much of the life cycle as you can remember. Looking at such a model may help you understand the question better and help you determine the correct answer.

Directions (6): **For the following question, write a short response.**

6 Plant breeders are scientists who work to improve plants that have commercial value. Estimate how plant breeders could breed plants with the most desirable flower color.

Reading Skills

Directions (7): **Read the passage below. Then answer the question.**

Use of pesticides could reduce the number of plants in a geographic area. Pesticides can kill insect pollinators such as bees and flies. Reducing the number of insect pollinators can result in a reduced number of flowering plants, which depend on insect pollination.

7 What other way could a flowering plant become pollinated if there are no insects?
  F. convert to asexual reproduction
  G. develop more stamens
  H. grow moving structures
  I. rely on birds, mammals, or wind

Interpreting Graphics

Directions (8): **Base your answer to question 8 on the diagram below.**

Plant Seed

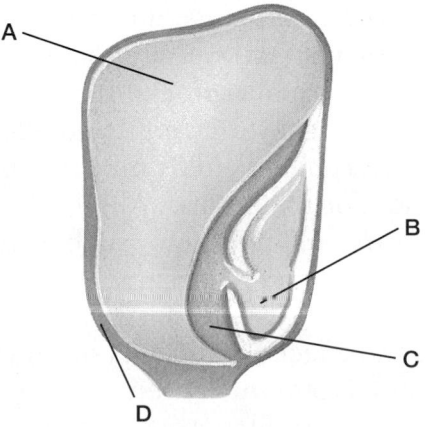

8 Which structure is the embryonic root?
  A. *A*   **C.** *C*
  B. *B*   **D.** *D*

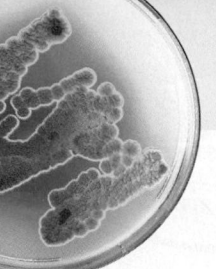

Exploration Lab

Observing the Effects of Nutrients on Vegetative Reproduction

SKILLS
- Using scientific processes
- Observing
- Graphing and analyzing data

OBJECTIVES
- **Identify** the structures of duckweed.
- **Compare** vegetative reproduction of duckweed in different nutrient solutions.

MATERIALS
- safety goggles
- lab apron
- duckweed culture
- 5 Petri dishes
- stereomicroscope or hand lens
- glass-marking pen
- beakers
- pond water
- Knop's solution
- 0.1% fertilizer solution
- distilled water

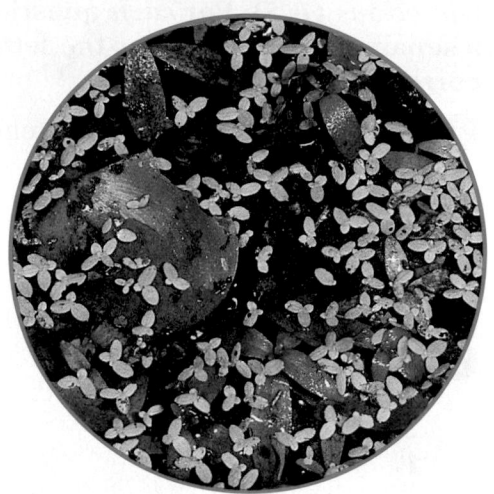

ChemSafety

CAUTION: Always wear safety goggles and a lab apron to protect your eyes and clothing.

CAUTION: Do not touch or taste any chemicals. Know the location of the emergency shower and eyewash station and how to use them. If you get a chemical on your skin or clothing, wash it off at the sink while calling to the teacher. Notify the teacher of a spill. Spills should be cleaned up promptly, according to your teacher's directions.

CAUTION: Glassware is fragile. Notify the teacher of broken glass or cuts. Do not clean up broken glass or spills with broken glass unless the teacher tells you to do so.

Before You Begin

Duckweed is a common aquatic plant. Like many flowering plants, duckweed reproduces readily by **vegetative reproduction,** which is a type of **asexual reproduction.** As individual plants grow, they divide into smaller individuals. Several individuals may remain joined together, forming a mat. All plants require certain **mineral nutrients,** such as nitrogen, phosphorus, and potassium, for the growth of vegetative parts. In this lab, you will investigate the effect of nutrients on the vegetative reproduction of duckweed.

1. Write a definition for each boldface term in the paragraph above.
2. Based on the objectives for this lab, write a question you would like to explore about vegetative reproduction in duckweed.

Procedure

PART A: Make Observations

1. Place a duckweed plant in a Petri dish. Then place a few drops of water on the plant.
2. Observe the duckweed plant with a stereomicroscope or a hand lens. Sketch what you see. Label the structures that you recognize.

PART B: Design an Experiment

3. Work with members of your lab group to explore one of the questions written for step 2 of **Before You Begin.** To explore the question, design an experiment that uses the materials listed for this lab.

You Choose

As you design your experiment, decide the following:

a. what question you will explore

b. what your hypothesis will be

c. what solutions to test

d. how much of each solution to use

e. how many individuals to use for each test

f. what your control will be

g. how you will judge which solution is the best

h. what data to record in your data table

4. Write a procedure for your experiment. Make a list of all the safety precautions you will take. Have your teacher approve your procedure and safety precautions before you begin the experiment.

PART C: Conduct Your Experiment

5. Put on safety goggles and a lab apron.

6. Set up your experiment. **CAUTION: Nutrient solutions are mild eye irritants. Avoid contact with your skin and eyes.** Complete step 8.

7. Conduct your experiment and collect data for two weeks.

PART D: Cleanup and Disposal

8. Dispose of solutions, broken glass, and duckweed in the designated waste containers. Do not pour chemicals down the drain or put lab materials in the trash unless your teacher tells you to do so.

9. Clean up your work area and all lab equipment. Return lab equipment to its proper place. Wash your hands thoroughly before you leave the lab and after you finish all work.

Analyze and Conclude

1. Summarizing Results Compare the appearance of plants growing in each nutrient solution with that of the plants in distilled water. Explain your observations.

2. Analyzing Data In which Petri dish did the greatest amount of growth (increase in numbers) take place?

3. Analyzing Results In which Petri dish did the least amount of growth take place?

4. Evaluating Hypotheses Did the results you observed agree with your hypothesis? If not, how are they different?

5. Recognizing Patterns As the number of new duckweed plants in a particular group increased, what happened to the group of plants?

6. Graphing Data Make a graph of your data. Label the *y*-axis "Number of plants," and the *x*-axis "Days." Use a different color to represent each solution you tested.

7. Drawing Conclusions What factors regulate the rate of vegetative reproduction in duckweed?

8. Evaluating Methods Why are the new duckweed plants produced by vegetative reproduction genetically the same as the parent plant?

9. Further Inquiry Write a new question about vegetative reproduction in duckweed that could be explored with another investigation.

? Do You Know?

Do research in the library or media center to answer these questions:

1. What do the flowers of duckweeds look like?

2. What is the smallest species of flowering plants, and how small are they?

Use the following Internet resources to explore your own questions about duckweed.

internet connect

www.scilinks.org
Topic: Duckweed
Keyword: HX4062

SCILINKS. Maintained by the National Science Teachers Association

A kokerboom, or quiver tree, *Aloe dichotoma*

25 Plant Structure and Function

Quick Review

Answer the following without referring to earlier sections of your book.

1. **Explain** what causes the adhesion and cohesion of water. *(Chapter 2, Section 2)*

2. **Describe** the organization of a plant cell. *(Chapter 3, Section 3)*

3. **Summarize** the steps in photosynthesis. *(Chapter 5, Section 2)*

4. **State** the relationship between stomata and guard cells. *(Chapter 23, Section 1)*

5. **Define** the terms *cuticle, stoma, xylem, phloem,* and *meristem. (Chapter 23, Section 1)*

Did you have difficulty? *For help, review the sections indicated.*

Reading Activity

Before you read this chapter, survey the sub-titles, headings, captions, and words in boldface type. Try to identify the purpose of this chapter. As you read this chapter, create a descriptive reading organizer.

internet connect

www.scilinks.org
National Science Teachers Association *sci*LINKS Internet resources are located throughout this chapter.

*sci*LINKS. Maintained by the National Science Teachers Association

• The structure of the quiver tree enables it to survive in the harsh southern African climate. For generations, the native San people have used the tree's bark and branches to make arrow quivers.

The Vascular Plant Body

Objectives

- **Identify** the three kinds of tissues in a vascular plant's body, and state the function of each.

- **Compare** the structures of different types of roots, stems, and leaves.

- **Relate** the structures of roots, stems, and leaves to their functions.

Key Terms

dermal tissue
ground tissue
epidermis
cork
vessel
sieve tube
cortex
root hair
root cap
herbaceous plant
vascular bundle
pith
heartwood
sapwood
petiole
mesophyll

Tissues

Like your body, a plant's body is made of tissues that form organs. In vascular plants, there are three types of tissue systems—the dermal tissue system, ground tissue system, and vascular tissue system. As you have read, vascular tissue forms strands that conduct water, minerals, and organic compounds throughout a vascular plant. **Dermal tissue** forms the protective outer layer of a plant. **Ground tissue** makes up much of the inside of the nonwoody parts of a plant, including roots, stems, and leaves. **Figure 1** shows how these three tissues are arranged in a nonwoody dicot. The tissues are organized a little differently in other types of vascular plants.

Each type of tissue contains one or more kinds of cells that are specialized to perform particular functions. As you read further in this chapter, you will learn that some specialized plant cells lack organelles found in other plant cells. In fact, some plant cells cannot perform their functions until they have lost most of their organelles.

Figure 1   Plant tissues

The leaves, stems, and roots of a vascular plant contain all three kinds of plant tissues.

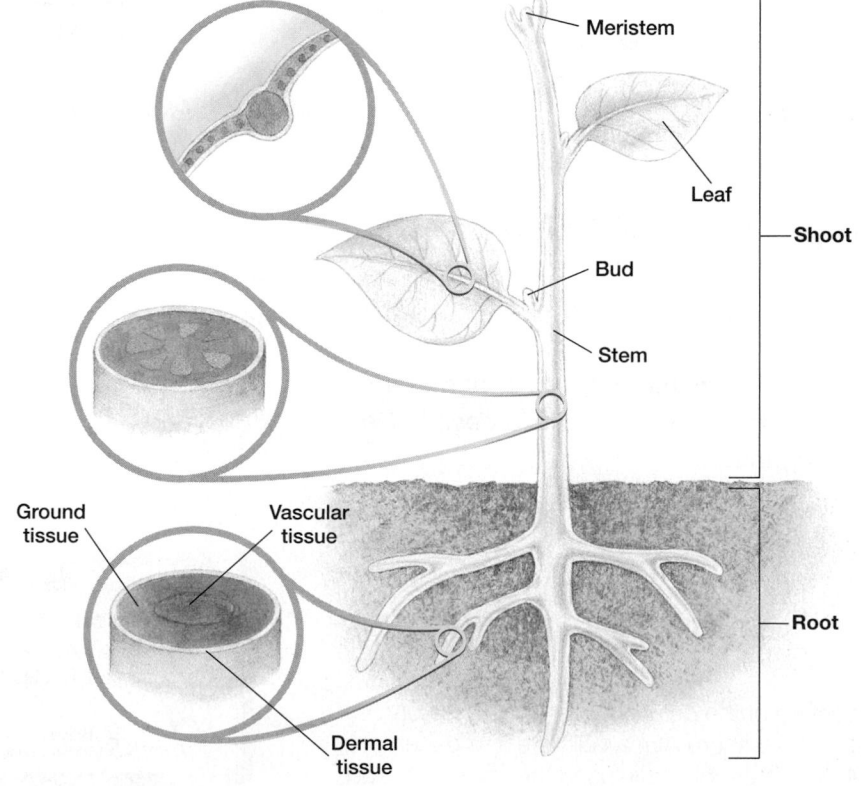

Figure 2 Dermal tissue.
Epidermis is one type of dermal tissue. Hairlike outgrowths of the epidermis, like the ones on this African violet leaf, help reduce water loss from the leaves of some plants.

Dermal Tissue System

Dermal tissue covers the outside of a plant's body. In the nonwoody parts of a plant, dermal tissue forms a "skin" called the **epidermis.** The word *epidermis* comes from the Greek words *epi,* meaning "upon," and *derma,* meaning "skin." The epidermis of most plants is made up of a single layer of flat cells. A waxy cuticle, which prevents water loss, coats the epidermis of the stems and leaves. Often, the cells of the epidermis have hairlike extensions or other structures, as **Figure 2** shows. Extensions of the epidermal cells on leaves and stems often help to slow water loss. Extensions of the epidermal cells on root tips help increase water absorption. The dermal tissue on woody stems and roots consists of several layers of dead cells that are referred to as **cork.** Cork cells contain a waterproof chemical and are not covered by a waxy cuticle. In addition to protection, dermal tissue also functions in gas exchange and in the absorption of mineral nutrients.

Ground Tissue System

Ground tissue makes up much of the inside of most plants. Most ground tissue consists of thin-walled cells that remain alive and keep their nucleus after they mature. In addition, ground tissue contains some thick-walled cells. Look for these different types of cells in the ground tissue shown in **Figure 3.** Ground tissue has different functions, depending on where it is located in a plant. The ground tissue in leaves, which is packed with chloroplasts, is specialized for photosynthesis. The ground tissue in stems and roots functions mainly in the storage of water, sugar, and starch. Throughout the body of a plant, ground tissue also surrounds and supports the third kind of plant tissue—vascular tissue.

Figure 3 Ground tissue.
A variety of cells make up the ground tissue visible in this cross section of a wheat stem. Thin-walled cells make up most of the ground tissue. Thick-walled cells strengthen the stem.

Magnification: 122×

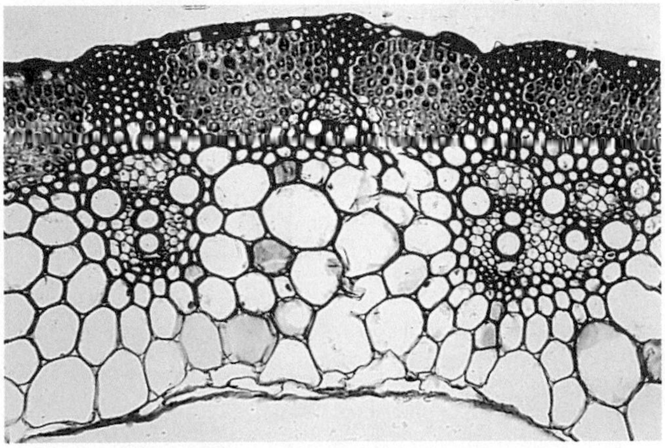

Vascular Tissue Systems

Plants have two kinds of vascular tissue—xylem and phloem. Both xylem and phloem contain strands of cells that are stacked end to end and act like tiny pipes, as you can see in **Figure 4.** These strands of cells act as a plumbing system, carrying fluids and dissolved substances throughout a plant's body.

Xylem

Xylem has thick-walled cells that conduct water and mineral nutrients from a plant's roots through its stems to its leaves. The conducting cells in xylem must lose their cell membrane, nucleus, and cytoplasm before they can conduct water. At maturity, all that is left of these cells is their cell walls. One type of xylem cell found in all vascular plants is called a tracheid *(TRAY kee ihd)*. Tracheids are narrow, elongated, and tapered at each end. Water flows from one tracheid to the next through pits, which are thin areas in the cell walls. Gnetophytes and flowering plants also have a second type of xylem cell, which makes up conducting strands called **vessels.** The vessel cells are wider than tracheids and have large perforations in their ends. The perforations allow water to flow more quickly between vessel cells.

Phloem

Phloem contains cells that conduct sugars and other nutrients throughout a plant's body. The conducting cells of phloem have a cell wall, a cell membrane, and cytoplasm. These cells either lack organelles or have modified organelles. The conducting strands in phloem are called **sieve tubes.** Pores in the walls between neighboring sieve-tube cells connect the cytoplasms and allow substances to pass freely from cell to cell. Beside the sieve tubes are rows of companion cells, which contain organelles. Companion cells carry out cellular respiration, protein synthesis, and other metabolic functions for the sieve-tube cells.

Figure 4 Xylem and phloem

Both xylem and phloem contain strands of tubular conducting cells that are stacked end to end like sections of pipe.

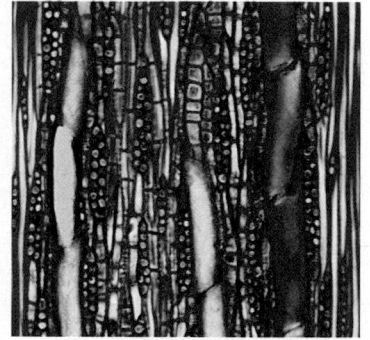

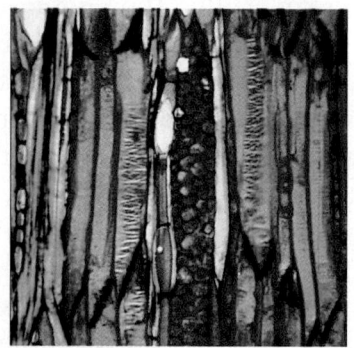

The conducting cells of xylem are tracheids and vessel cells. Tracheids are slender and have tapered ends. Vessel cells are larger and form vessels.

The conducting cells of phloem are sieve-tube cells, which form sieve tubes. Companion cells lie next to the sieve-tube cells.

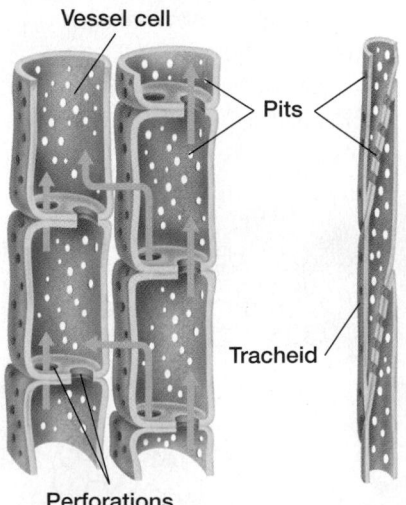

Vessel cell

Pits

Tracheid

Perforations

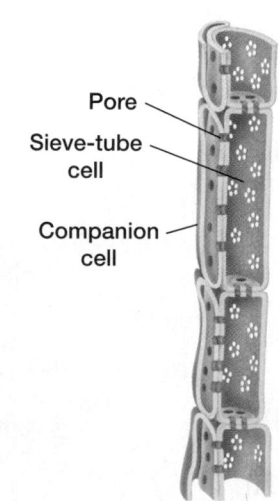

Pore

Sieve-tube cell

Companion cell

Roots

Most plants are anchored to the spot where they grow by roots, which also absorb water and mineral nutrients. In many plants, roots also function in the storage of organic nutrients, such as sugar and starch. Many dicots, such as carrots and radishes, have a large central root from which much smaller roots branch. This type of root system is called a *taproot* system. In contrast, most monocots, such as grasses, have a highly branched, fibrous root system, shown in **Figure 5.** Some plants have roots that grow from aboveground stems or leaves. Such roots are called adventitious *(ad ven TIH shuhs)* roots. The prop roots of corn and the aerial roots of orchids are examples of adventitious roots.

As **Figure 6** shows, a root has a central core of vascular tissue that is surrounded by ground tissue. The ground tissue surrounding the vascular tissue is called the **cortex.** Roots are covered by dermal tissue. An epidermis covers all of a root except for the root tip. The epidermal cells just behind a root tip often produce **root hairs,** which are slender projections of the cell membrane. Root hairs greatly increase the surface area of a root and its ability to absorb water and mineral nutrients. A mass of cells called the **root cap** covers and protects the actively growing root tip. A layer of cork replaces the epidermis in the older sections of a root. Many plants have roots that become woody as they get older. Layers of xylem replace the ground tissue in woody roots.

Figure 5 Fibrous root system. A fibrous roots system is made up of many roots that are about the same size.

Figure 6 Root structure. Roots have characteristic external and internal structures.

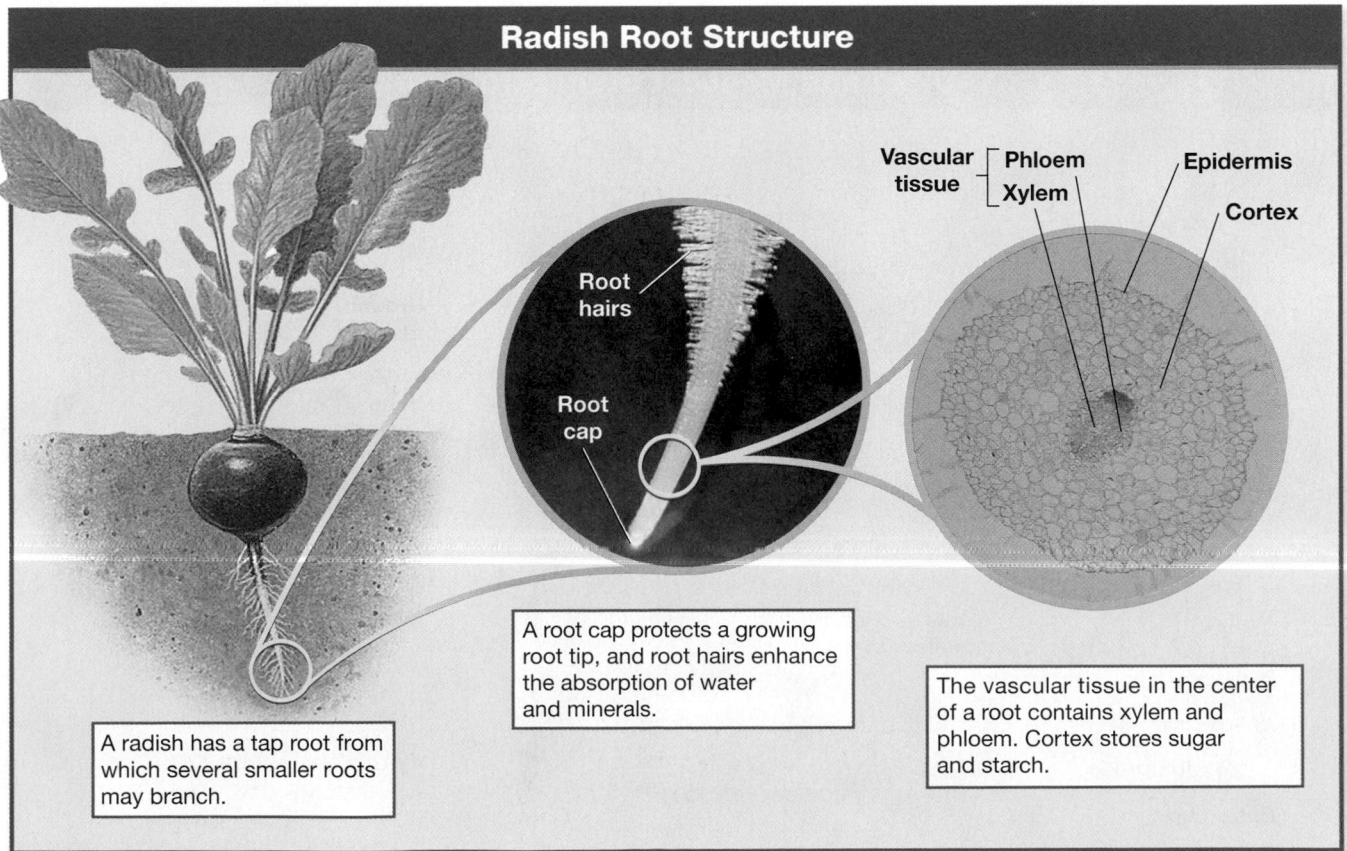

Radish Root Structure

Root hairs

Root cap

Vascular tissue
Phloem
Xylem
Epidermis
Cortex

A radish has a tap root from which several smaller roots may branch.

A root cap protects a growing root tip, and root hairs enhance the absorption of water and minerals.

The vascular tissue in the center of a root contains xylem and phloem. Cortex stores sugar and starch.

Stems

The shoots of most plants consist of stems and leaves. Stems support the leaves and house the vascular tissue, which transports substances between the roots and the leaves. Many plants have stems that are specialized for other functions. For example, the stems of cactuses store water. Potatoes are stems that are specialized for nutrient storage and for asexual reproduction.

Leaves are attached to a stem at points called nodes. The space between two nodes is called an internode. Buds that can grow into new branches are also located at the nodes on a stem. Look for these structures in **Figures 7** and **8.** Other features of a stem depend on whether the stem is woody or nonwoody.

Nonwoody Stems

A plant with stems that are flexible and usually green is called a **herbaceous** *(huhr BAY shuhs)* **plant.** Herbaceous plants include violets, clovers, and grasses. As Figure 7 shows, the stems of herbaceous plants contain bundles of xylem and phloem called **vascular bundles.** The vascular bundles are surrounded by ground tissue. In monocot stems, such as that of corn, illustrated in Figure 7, the vascular bundles are scattered in the ground tissue. In dicot stems, however, the vascular bundles are arranged in a ring. The ground tissue outside the ring of vascular bundles is called the cortex. The ground tissue inside the ring is called the **pith.** Herbaceous stems are covered by an epidermis. Stomata in the epidermis enable the stems to exchange gases with the outside air.

Real Life

Xylem forms wood's characteristic markings, or grain.

Wood is usually cut parallel to the axis of a tree trunk, producing straight-grained lumber. In some trees, however, xylem vessels form interesting patterns.

Recognizing Patterns
Examine several wooden objects. Decide whether they have straight grain or another type of grain.

Figure 7 Herbaceous stems. The vascular bundles are arranged differently in dicots and monocots.

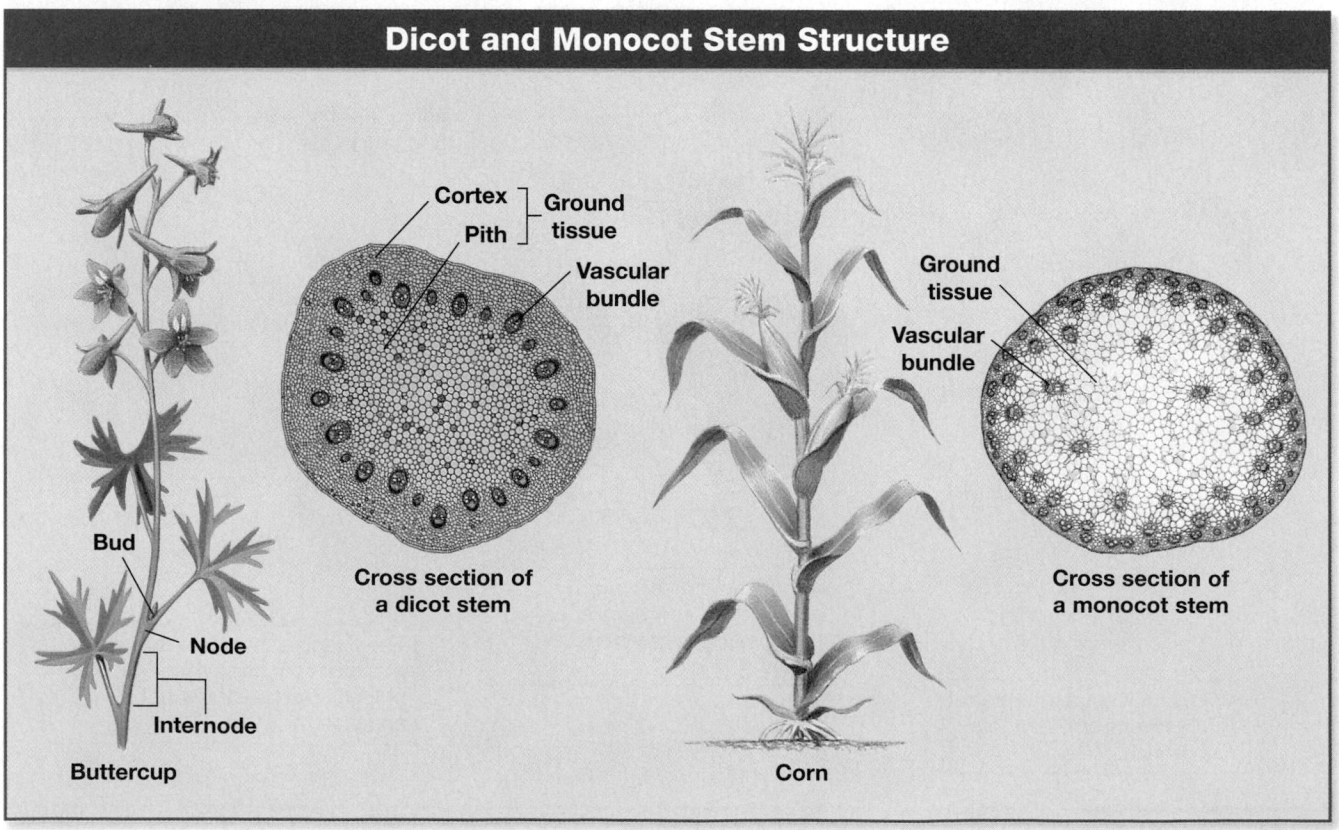

Dicot and Monocot Stem Structure

Cortex
Pith — Ground tissue

Vascular bundle

Cross section of a dicot stem

Bud

Node

Internode

Buttercup

Ground tissue

Vascular bundle

Cross section of a monocot stem

Corn

Woody Stems

Trees and shrubs, such as pines, oaks, roses, and hollies, have woody stems. As Figure 8 shows, woody stems are stiff and nongreen. Buds, which produce new growth, are found at the tips and at the nodes of woody stems. They exchange gases through pores in their bark. A young woody stem has a central core of pith and a ring of vascular bundles, which fuse into solid cylinders as the stem matures. Layers of xylem form the innermost cylinder and are the major component of wood. A cylinder of phloem lies outside the cylinder of xylem. Woody stems are covered by cork, which protects them from physical damage and helps prevent water loss. Together, the layers of cork and phloem make up the bark of a woody stem. A mature woody stem contains many layers of wood and is covered by a thick layer of bark. The wood in the center of a mature stem or tree trunk is called **heartwood.** The xylem in heartwood, which can no longer conduct water, provides support. **Sapwood,** which lies outside the heartwood, contains vessel cells that can conduct water.

Figure 8 Woody stems

Woody stems are typically stiff and nongreen.

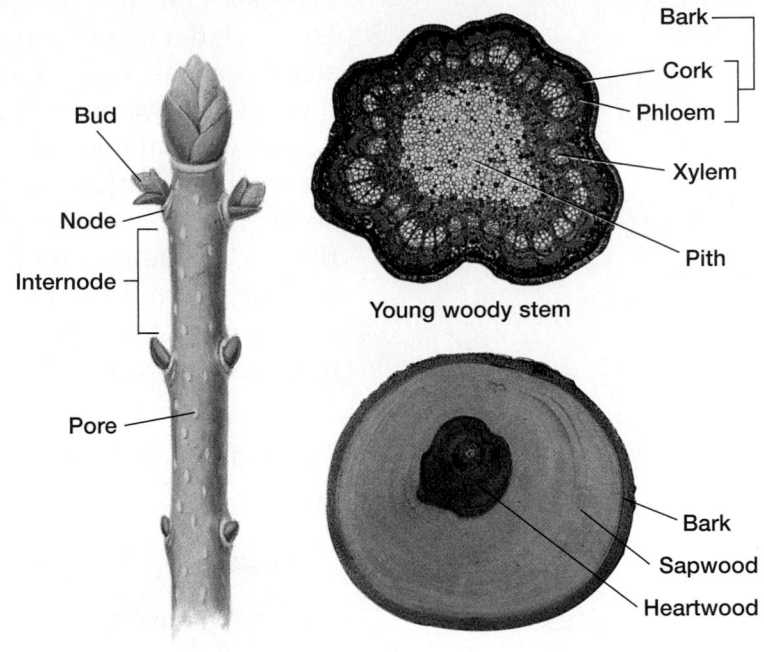

Young woody stem

Comparing the Structures of Roots and Stems

You can use a microscope to see differences in the internal structure of roots and stems.

Materials

compound microscope, prepared slide of a cross section of the following: dicot root, monocot root, dicot stem, monocot stem

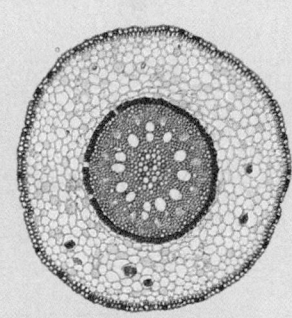

Cross section of a corn root

Procedure

1. View cross sections of dicot and monocot roots with a compound microscope. For each, draw and label what you see under low power. Then look at the vascular tissue in each root under high power. Draw what you see in each root, and label the xylem and phloem.

2. View cross sections of dicot and monocot stems with a compound microscope. For each, draw and label what you see under low power. Then look at a vascular bundle in each stem under high power. Draw each vascular bundle, and label the xylem and phloem.

Analysis

1. **Compare and contrast** the location of xylem and phloem in roots and stems.

2. **Compare and contrast** the arrangement and structure of the vascular bundles in monocot and dicot stems.

3. **Describe** the relationship between the structure and function of vascular tissue.

Leaves

Leaves are the primary photosynthetic organs of plants. Most leaves have a flattened portion, called the blade, that is often attached to a stem by a stalk called the **petiole** *(PEHT ee ohl)*. A leaf blade may be divided into two or more sections called leaflets, as shown in **Figure 9.** Leaves with an undivided blade are called simple leaves. Leaves with two or more leaflets are called compound leaves. Leaflets reduce the surface area of a leaf blade. Many plants have highly modified leaves that are specialized for particular purposes. For example, the spines of a cactus and the tendrils of a garden pea are modified leaves. Cactus spines are specialized for protection and water conservation, while garden-pea tendrils are specialized for climbing.

Figure 9 Simple and compound leaves

Most leaves consist of a flattened blade and a petiole that attaches to a stem.

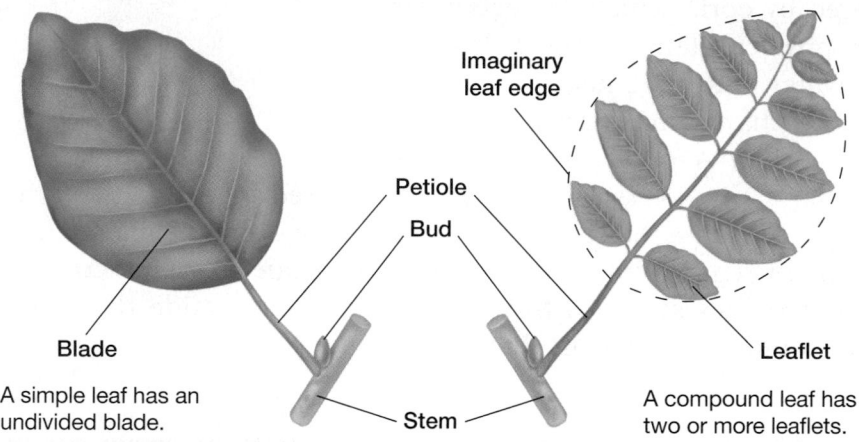

Imaginary leaf edge

Petiole

Bud

Blade

A simple leaf has an undivided blade.

Stem

Leaflet

A compound leaf has two or more leaflets.

QUICK LAB

Observing the Structures Inside a Leaf

With a microscope, you can see how a leaf is put together.

Materials

prepared slide of a leaf cross section, compound microscope

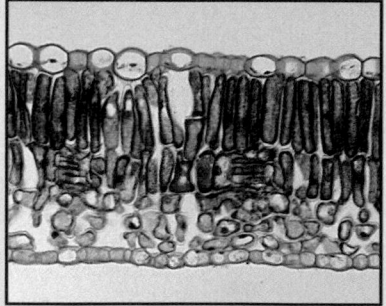

Cross section of a lilac leaf (530×)

Procedure

1. View a cross section of a leaf under low power of a compound microscope. Then switch to high power.

2. Use Figure 10 in this chapter to identify the following structures: stoma, guard cells, upper and lower epidermis, palisade layer, spongy layer, and vein.

Analysis

1. **Describe** a stoma, and relate the function of a stoma to your description.

2. **Describe** the location and contents of the veins.

3. **Critical Thinking Relating Concepts** How do the location and structure of the palisade and spongy layers help a leaf perform photosynthesis?

A leaf is a mass of ground tissue and vascular tissue covered by epidermis, as **Figure 10** shows. A cuticle coats the upper and lower epidermis. Both xylem and phloem are found in the veins of a leaf. Veins are extensions of vascular bundles that run from the tips of roots to the edges of leaves. In leaves, the ground tissue is called **mesophyll** (MEHS oh fihl). Mesophyll cells are packed with chloroplasts, where photosynthesis occurs. The chlorophyll in chloroplasts makes leaves look green.

Most plants have leaves with two layers of mesophyll. One or more rows of closely packed, columnar cells make up the palisade layer, which lies just beneath the upper epidermis. A layer of loosely packed, spherical cells, called the spongy layer, lies between the palisade layer and the lower epidermis. The spongy layer has many air spaces through which gases can travel. Stomata, the tiny holes in the epidermis, connect the air spaces to the outside air.

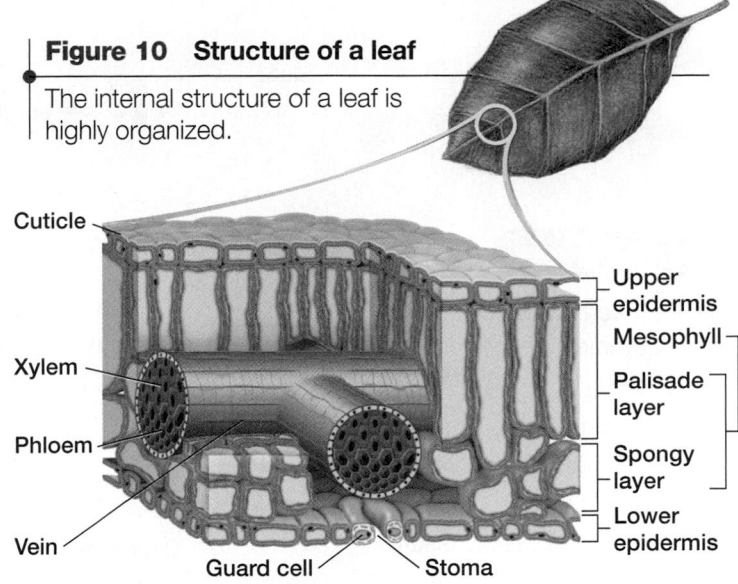

Figure 10   Structure of a leaf

The internal structure of a leaf is highly organized.

Cuticle
Xylem
Phloem
Vein
Guard cell
Stoma
Upper epidermis
Mesophyll
Palisade layer
Spongy layer
Lower epidermis

Exploring Further

C₃ Versus C₄ Leaves

Internal leaf structure varies among plants, depending on how they carry out photosynthesis. In photosynthesis, carbon dioxide from the air is added to organic molecules. This process, called carbon fixation, occurs during the Calvin cycle. Because the product of carbon fixation by the Calvin cycle is a three-carbon compound, plants that use *only* the Calvin cycle to fix carbon are called C_3 plants. The leaf structure shown in Figure 10 is typical of a C_3 plant. More than 90 percent of all plants are C_3 plants.

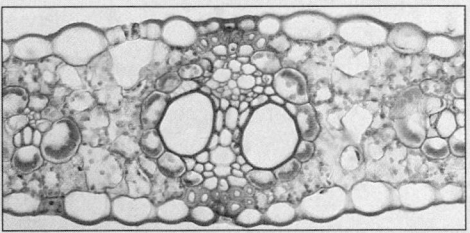

Cross section of a corn leaf (584×)

Plants such as corn and sugar cane use another chemical process to fix carbon before the Calvin cycle. The first products of this process are four-carbon compounds, so these plants are called C_4 plants. The leaves of C_4 plants have a different internal structure, as shown in the photo of a corn leaf above. C_4 plants are plentiful in the tropics because they conduct photosynthesis efficiently in high heat and intense light.

Section 1 Review

1 **Name** the three kinds of tissue that make up a plant, and list two functions for each.

2 **Summarize** the differences in the structure of the stems of monocots and dicots.

3 **Critical Thinking Forming Hypotheses** How do the structures of roots, stems, and leaves help them to carry out their functions?

4 **Standardized Test Prep** In which part of a plant would you find closely packed, columnar cells that contain many chloroplasts?

A woody stem

B nonwoody stem

C leaf

D root

- **Relate** transpiration to the movement of water up a plant.
- **Describe** how guard cells regulate the rate of transpiration.
- **Recognize** several distinguishing features of sugar maple trees.
- **Describe** the process of translocation in a plant.

Key Terms

transpiration
source
sink
translocation

Movement of Water

You know that water and mineral nutrients move up from a plant's roots to its leaves through xylem. However, some trees have leaves that are more than 100 m (328 ft) above the ground. How do plants manage to get water so high? Simply put, water is pulled up through a plant as it evaporates from the plant's leaves, as **Figure 11** shows.

Step ❶ Recall that the surfaces of leaves are covered with many tiny pores, the stomata. When the stomata are open, water vapor diffuses out of a leaf. This loss of water vapor from a plant is called **transpiration.** In most plants, more than 90 percent of the water taken in by the roots is ultimately lost through transpiration.

Step ❷ The xylem contains a column of water that extends from the leaves to the roots. The cohesion of water molecules causes water molecules that are being lost by a plant to pull on the water molecules still in the xylem. This pull extends through the water in the xylem. Water is drawn upward in the same way liquid is drawn through a siphon. As long as the column of water in the xylem does not break, water will keep moving upward as transpiration occurs.

Step ❸ Roots take in water from the soil by osmosis. This water enters the xylem and replaces the water lost through transpiration.

Figure 11

BIO graphic

Water Movement in Plants

Transpiration drives the movement of water through a plant.

❶ Water vapor exits the leaves through stomata by transpiration.

❷ The loss of water creates a pull that draws water up through the xylem.

❸ Water drawn into the roots from the soil by osmosis moves up the stem.

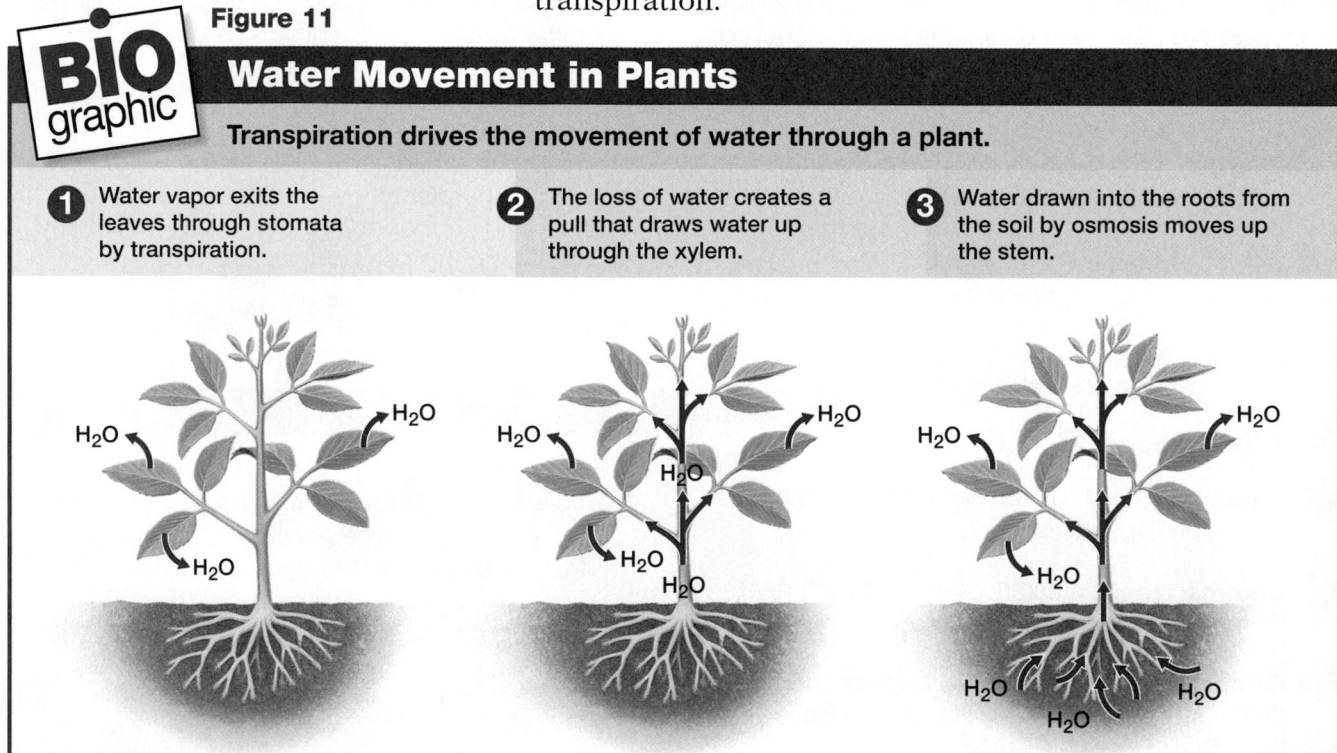

Figure 12 Control of stomatal opening

Changes in the shape of guard cells cause stomata to open or close.

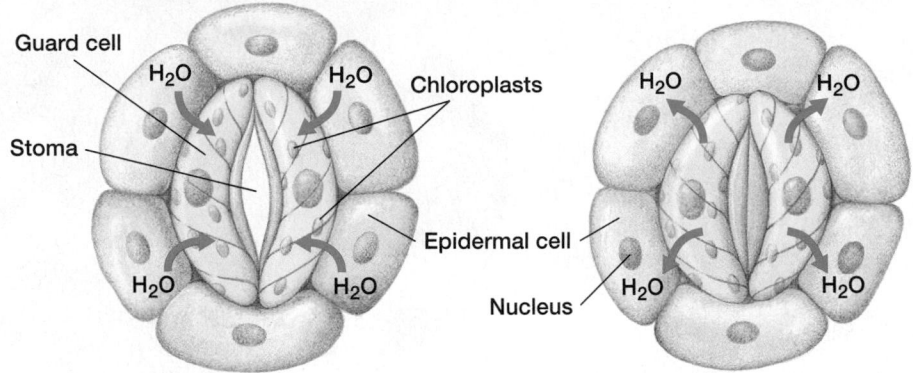

1. A stoma opens as the guard cells take in water, lengthen, and bow apart.

2. A stoma closes as the guard cells lose water, shorten, and come together.

Guard Cells and Transpiration

A stoma is surrounded by a pair of guard cells that are shaped like two cupped hands. Changes in water pressure within the guard cells cause the stoma to open or close, as shown in **Figure 12.** When the guard cells take in water, they swell. However, extra cellulose strands in their cell walls permit the cells to increase in length but not in diameter. As a result, guard cells that take in water bend away from each other, opening the stoma and allowing transpiration to proceed. When water leaves the guard cells, they shorten and move closer to each other, closing the stoma and stopping transpiration. Thus, the loss of water from guard cells for any reason causes stomata to close, stopping further water loss. This is an example of homeostasis in action.

internet connect

www.scilinks.org
Topic: Transpiration
Keyword: HX4178

*SC*LINKS. Maintained by the National Science Teachers Association

Inferring the Rate of Transpiration

Background

The graph shows the rate of water movement in a plant during high humidity and during low humidity. The rate of water movement indicates the rate of transpiration. Use the graph to answer the questions below.

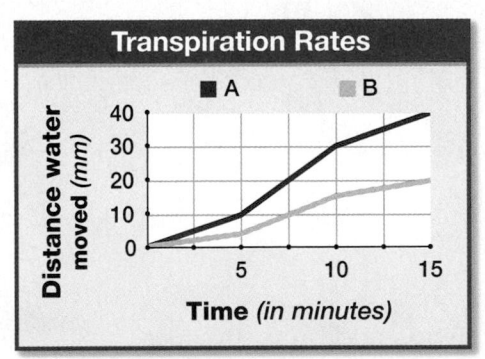

Transpiration Rates

■ A   ■ B

Distance water moved *(mm)*

Time *(in minutes)*

Analysis

1. **Determine** how far water had moved after 10 minutes under the condition represented by curve A.

2. **Critical Thinking Analyzing Results** After 15 minutes, how much farther had water moved under condition A than under condition B?

3. **Critical Thinking Recognizing Relationships** Which curve indicates a lower transpiration rate? Explain your reasoning.

4. **Critical Thinking Drawing Conclusions** Which curve shows the transpiration rate during low humidity? Justify your answer.

Up Close

Sugar Maple

- **Scientific name:** *Acer saccharum*
- **Size:** 12 to 37 m (40 to 121 ft) tall, canopy up to 14 m (46 ft) wide, trunk up to 1 m (3 ft) in diameter
- **Range:** Northeastern United States and adjacent regions of southeastern Canada
- **Habitat:** Northern temperate forests
- **Importance:** The wood of sugar maples is used to make furniture, musical instruments, and flooring. Their sap is made into maple syrup and maple sugar.

External Structures

Spring · Summer · Winter

Flowers Maple flowers appear before the leaves in early spring. The flowers lack petals and thus are incomplete. Most are imperfect (either male or female) as well.

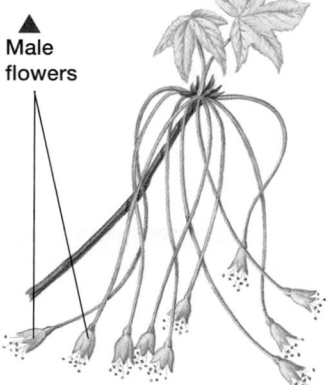

▲ Male flowers

Seeds and fruits Seeds mature inside winged fruits called samaras. The wings enable the dispersal of seeds by wind.

▲ Samara

Fall

Twigs and buds The twigs are green at first, and then they change to a glossy reddish brown. The reddish brown buds are conical, pointed, and about 4 mm (0.2 in.) long.

Bark Young sugar maples have smooth, light gray bark. The bark of older trees becomes dark gray or brown, rough, and deeply furrowed.

Leaves The leaves range from 7.5 to 15 cm (3 to 6 in.) across. Most leaves have five sharply toothed lobes. The main veins branch from a single point, like the fingers on a hand. In the fall, the leaves change to bright yellow, orange, or red, and fall from the tree.

Internal Structures

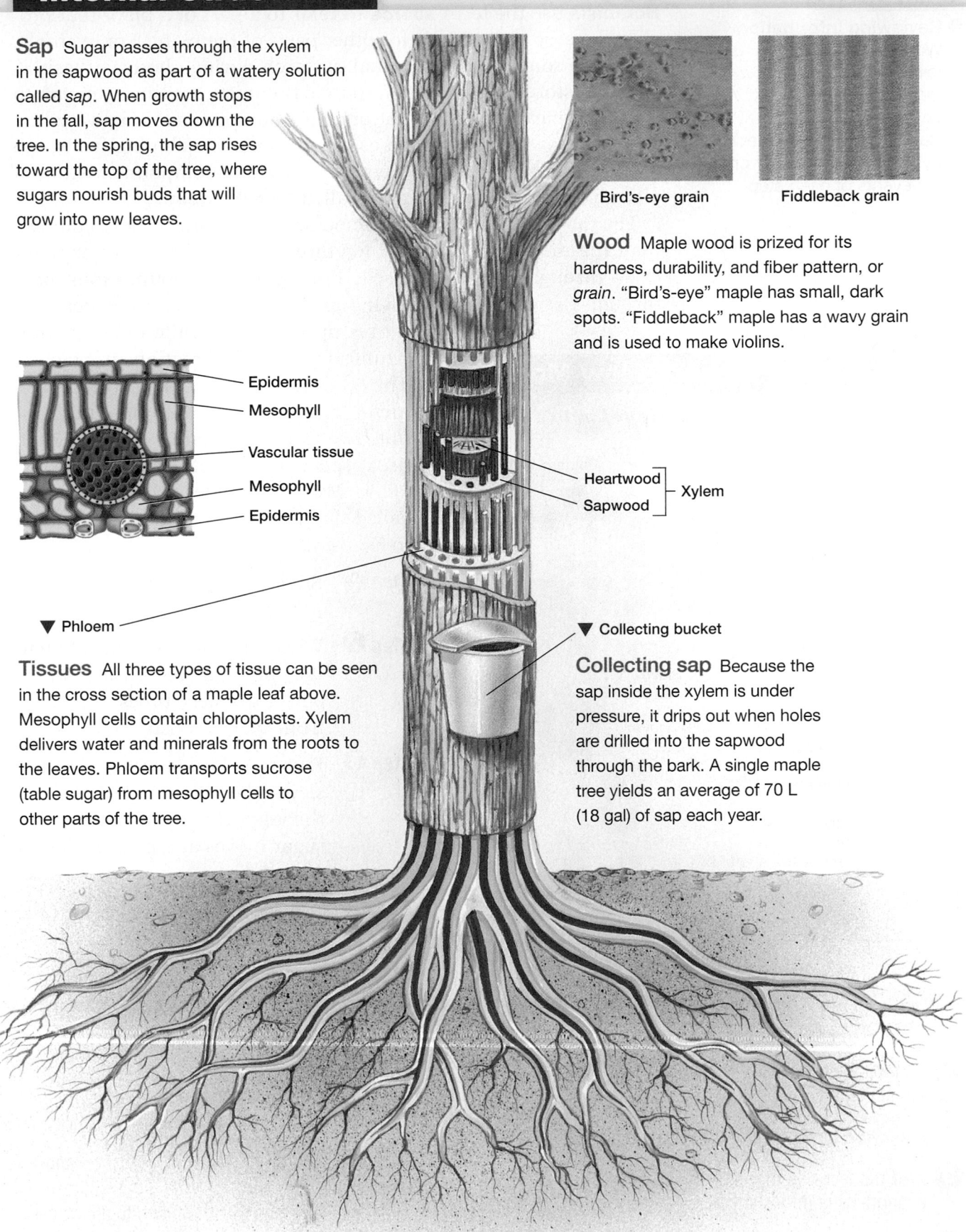

Sap Sugar passes through the xylem in the sapwood as part of a watery solution called *sap*. When growth stops in the fall, sap moves down the tree. In the spring, the sap rises toward the top of the tree, where sugars nourish buds that will grow into new leaves.

Bird's-eye grain

Fiddleback grain

Wood Maple wood is prized for its hardness, durability, and fiber pattern, or *grain*. "Bird's-eye" maple has small, dark spots. "Fiddleback" maple has a wavy grain and is used to make violins.

Epidermis
Mesophyll

Vascular tissue

Mesophyll

Epidermis

Heartwood
Sapwood
Xylem

▼ Phloem

▼ Collecting bucket

Tissues All three types of tissue can be seen in the cross section of a maple leaf above. Mesophyll cells contain chloroplasts. Xylem delivers water and minerals from the roots to the leaves. Phloem transports sucrose (table sugar) from mesophyll cells to other parts of the tree.

Collecting sap Because the sap inside the xylem is under pressure, it drips out when holes are drilled into the sapwood through the bark. A single maple tree yields an average of 70 L (18 gal) of sap each year.

Movement of Organic Compounds

Organic compounds move throughout a plant within the phloem. Botanists use the term **source** to refer to a part of a plant that provides organic compounds for other parts of the plant. For example, a leaf is a source because it makes starch during photosynthesis. A root that stores sugar is also a source. Botanists use the term **sink** to refer to a part of a plant that organic compounds are delivered to. Actively growing parts, such as root tips and developing fruits, are examples of sinks. The movement of organic compounds within a plant from a source to a sink is called **translocation.**

The movement of organic compounds in a plant is more complex than the movement of water for three reasons. First, water flows freely through empty xylem cells, but organic compounds must pass through the cytoplasm of living phloem cells. Second, water only moves up in xylem, while organic compounds move in all directions in phloem. Third, water can diffuse through cell membranes but organic compounds cannot. The German botanist Ernst Münch proposed a model of translocation in 1924. Münch's model, which is often called the pressure-flow model, is shown in **Figure 13.**

Step ❶ Sugar from a source enters phloem cells by active transport.

Step ❷ When the sugar concentration in the phloem increases, water enters the sieve tubes in phloem from xylem by osmosis.

Step ❸ Pressure builds up inside the sieve-tube cells and pushes sugar through the sieve tubes. The sugar moves at a rate as high as 100 cm/h.

Step ❹ Sugar moves from phloem cells into a sink by active transport.

Figure 13

BIOgraphic

The Pressure-Flow Model

Translocation is described by this model.

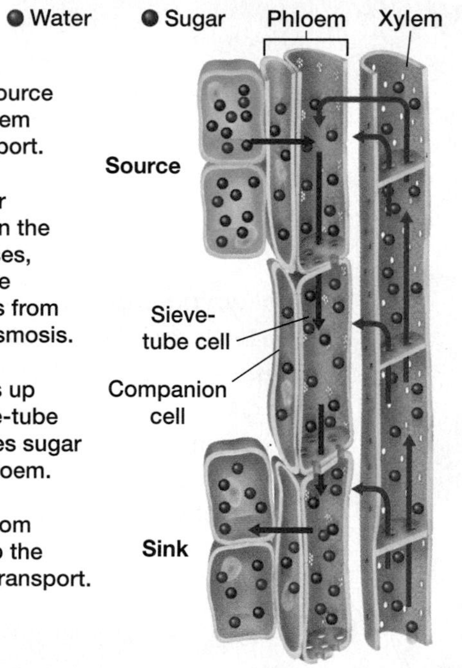

● Water   ● Sugar   Phloem   Xylem

❶ Sugar from a source enters the phloem by active transport.

Source

❷ When the sugar concentration in the phloem increases, water enters the sieve-tube cells from the xylem by osmosis.

Sieve-tube cell

❸ Pressure builds up inside the sieve-tube cells and pushes sugar through the phloem.

Companion cell

❹ Sugar moves from the phloem into the sink by active transport.

Sink

Section 2 Review

❶ Relate the process of transpiration to the movement of water through a plant.

❷ Describe how guard cells regulate transpiration.

❸ List the features that make sugar maples economically important.

❹ Describe how translocation occurs in plants.

❺ Critical Thinking Predicting Outcomes
When the soil is dry and it is very hot, how can a plant reduce its water loss? Explain your answer.

❻ **Standardized Test Prep** In plants, active transport is used to move

A sugar into the phloem.   C water into the phloem.

B sugar into the xylem.   D water into the xylem.

Key Concepts

1 The Vascular Plant Body

- A vascular plant's body contains three kinds of tissue systems—dermal tissue, ground tissue, and vascular tissue.

- Dermal tissue covers a plant. A thin layer of epidermis covers nonwoody parts. Several layers of cork cover woody parts.

- Ground tissue is specialized for photosynthesis in leaves and for storage and support in stems and roots.

- Vascular tissue conducts water, minerals, and organic compounds throughout the plant.

- Xylem contains vessels, which are made up of cells that conduct water only after they lose their cytoplasm. Water flows between cells through pits and perforations in their cell walls.

- Phloem contains sieve tubes, which are made up of cells that are still living. Substances pass between the cells through pores.

- Roots have a central core of vascular tissue that is surrounded by ground tissue and epidermal tissue. Root hairs on root tips increase the surface area available for absorption.

- Nonwoody stems contain bundles of xylem and phloem embedded in ground tissue. Woody stems have an inner core of xylem surrounded by a cylinder of phloem.

- Leaves are a mass of ground tissue and vascular tissue covered by epidermis. The ground tissue cells conduct photosynthesis. Gases are exchanged through the stomata in the epidermis.

2 Transport in Plants

- Transpiration, the loss of water from a plant's leaves, creates a pull that draws water up through xylem from roots to leaves.

- Guard cells control water loss by closing a plant's stomata when water is scarce. Thus, they also regulate the rate of transpiration.

- The sugar maple is a commercially valuable tree. Its hard wood is used to make furniture, musical instruments, and other products. Its sap is made into maple syrup and maple sugar.

- Organic compounds are pushed through the phloem from a source to a sink in a process called translocation.

Key Terms

Section 1

dermal tissue (552)
ground tissue (552)
epidermis (553)
cork (553)
vessel (554)
sieve tube (554)
cortex (555)
root hair (555)
root cap (555)
herbaceous plant (556)
vascular bundle (556)
pith (556)
heartwood (557)
sapwood (557)
petiole (558)
mesophyll (559)

Section 2

transpiration (560)
source (564)
sink (564)
translocation (564)

Understanding Key Ideas

1. Which of the following is *not* a function of ground tissue?
   a. storage
   b. support
   c. photosynthesis
   d. protection

2. Which of the following phrases describes the structure of a monocot stem?
   a. contains a ring of vascular bundles that surrounds a core of ground tissue
   b. contains a core of vascular tissue that is surrounded by a ring of ground tissue
   c. contains several layers of xylem that are surrounded by a ring of phloem
   d. contains vascular bundles that are scattered throughout the ground tissue

3. The stem in the photograph below is probably adapted for
   a. climbing.   c. nutrient storage.
   b. water storage.   d. conserving water.

4. The column of water in a plant's xylem can remain unbroken because of the
   a. cohesion of water molecules.
   b. repulsion between water molecules.
   c. strong walls in the xylem.
   d. stiff fibers in the bark.

5. Guard cells swell and become longer when
   a. carbon dioxide moves out of the cells.
   b. water moves out of the cells.
   c. water moves into the cells.
   d. oxygen moves into the cells.

6. **Exploring Further** How are C_3 leaves different from C_4 leaves?

7. **Up Close** What is the range and habitat of the sugar maple?

8. Organic compounds move through phloem
   a. by diffusion from a sink to a source.
   b. by diffusion from the leaves to the roots.
   c. by active transport within a sieve tube.
   d. because of the pressure created by the movement of water into the sieve tubes.

9. How is osmosis involved in translocation in a vascular plant? (**Hint:** See Chapter 4, Section 1.)

10. **Concept Mapping** Make a concept map that describes the organization of the vascular plant body. Try to include the following terms in your map: *cork, dermal tissue, epidermis, ground tissue, mesophyll, phloem, sieve tubes, tracheids, vascular tissue, vessels,* and *xylem.*

Critical Thinking

11. **Inferring Function** When a plant wilts, its stomata close. How does wilting help a plant maintain homeostasis?

12. **Evaluating Results** Some herbicides (weed killers) contain a chemical that breaks down waxy substances. Explain why such a chemical might be useful in a herbicide.

13. **Analyzing Information** Trace the path of a water molecule through a vascular plant, from the water molecule's entry into a root hair to its exit between two guard cells. Identify and describe the function of each structure the water molecule would encounter on its journey.

Alternative Assessment

14. **Organizing Information** Construct clay models of cross sections through a C_3 leaf and a C_4 leaf. Include the epidermis, mesophyll, veins, and guard cells in the model of each leaf. Use the models to explain to your class each leaf's structure and the function of its parts.

Standardized Test Prep

Understanding Concepts

Directions (1–4): **For *each* question, write on a separate sheet of paper the letter of the correct answer.**

1 Where would xylem cells that no longer conduct water be found in a plant?
  A. cortex   **C.** pith
  B. heartwood   **D.** sapwood

2 What is the function of guard cells?
  F. They open and close stomata, regulating the rate of transpiration.
  G. They cover the actively growing root tip and protect it from damage.
  H. They increase the surface area of a root and its ability to absorb water.
  I. They contain dozens of chloroplasts, the main site of photosynthesis in leaves.

3 How does the change from open stomata to closed stomata affect a plant's ability to carry out photosynthesis?
  A. It prevents the escape of glucose.
  B. It has no effect on photosynthesis.
  C. It blocks the entry of carbon dioxide.
  D. It exposes more chlorophyll to sunlight.

4 Water vapor lost through what structure drives the movement of water through a plant?
  F. buds
  G. leaves
  H. nodes
  I. root hairs

Directions (5): **For the following question, write a short response.**

5 Analyze why plants that have a fibrous root system rather than a taproot system would be more likely to prevent soil erosion on a steep hillside.

Test **TIP**

When faced with similar answers, define the answer choices and then use those definitions to narrow down the choices on a multiple-choice question.

Reading Skills

Directions (6): **Read the passage below. Then answer the question.**

Many plants that live in low-nitrogen habitats, such as acidic peat bogs, have a unique way of obtaining nitrogen. Instead of obtaining nitrogen compounds from soil through the roots, they trap and digest insects using specialized leaves. Examples of such plants are the Venus flytrap, the pitcher plant, and the sundew.

6 Why would having specialized leaves be more effective for these plants than having specialized roots?
  A. Their source of energy is located above ground.
  B. Their source of water is located above ground.
  C. Their source of nitrogen is located above ground.
  D. Their source of carbon dioxide is located above ground.

Interpreting Graphics

Directions (7): **Base your answer to question 7 on the diagram below.**

Leaf Cells Under Two Conditions

Condition 1   Condition 2
Guard cell
Stoma
Epidermal cell

7 What caused the change from condition 1 to condition 2?
  F. Sugar moved into the stoma.
  G. Sugar moved into the guard cells.
  H. Water moved out of the guard cells.
  I. Water moved out of the epidermal cells.

Skills Practice Lab

Separating Plant Pigments

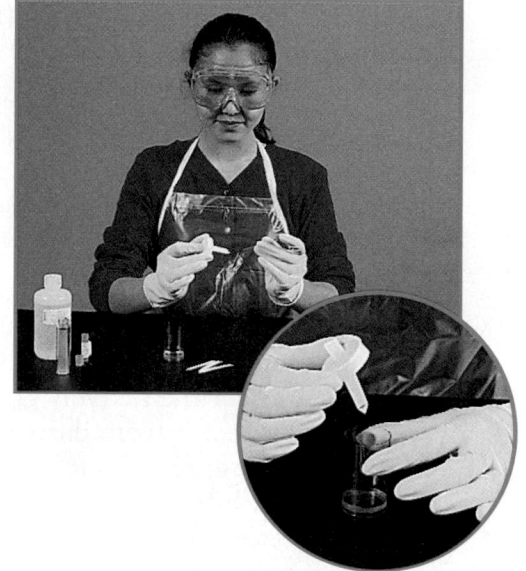

SKILLS
- Performing paper chromatography
- Calculating

OBJECTIVES
- **Separate** the pigments that give a leaf its color.
- **Calculate** the R_f value for each pigment.
- **Describe** how paper chromatography can be used to study plant pigments.

MATERIALS
- safety goggles
- lab apron
- strip of chromatography paper
- scissors
- metric ruler
- pencil
- capillary tube
- drop of simulated plant pigments extract
- 10 mL graduated cylinder
- 5 mL of chromatography solvent
- chromatography chamber

ChemSafety

CAUTION: Always wear safety goggles and a lab apron to protect your eyes and clothing.

CAUTION: Do not touch or taste any chemicals. Know the location of the emergency shower and eyewash station and how to use them. If you get a chemical on your skin or clothing, wash it off at the sink while calling to the teacher. Notify the teacher of a spill. Spills should be cleaned up promptly, according to your teacher's directions.

CAUTION: Glassware is fragile. Notify the teacher of broken glass or cuts. Do not clean up broken glass or spills with broken glass unless the teacher tells you to do so.

Before You Begin

Pigments produce colors by reflecting some colors of light and absorbing or transmitting others. Pigments can be removed from plant tissues using **solvents,** chemicals that dissolve other chemicals. The pigments can then be separated from the solvent and from each other by using **paper chromatography.** The word *chromatography* comes from the Greek words *chromat,* which means "color," and *graphon,* which means "to write." The **R_f** is the ratio of the distance that a pigment moves relative to the distance that a solvent moves. Since the R_f for a compound is constant, scientists can use it to identify compounds. In this lab, you will learn how to use paper chromatography to separate a mixture of pigments.

1. Write a definition for each boldface term in the previous paragraph and for each of the following terms: chlorophyll a, chlorophyll b, carotene, xanthophyll.

2. Make a data table similar to the one below.

DATA TABLE				
Band no.	Color	Pigment	Migration (in mm)	R_f value
1 (top)				
2				
3				
4				
Solvent				

3. Based on the objectives for this lab, write a question you would like to explore about plant pigments or paper chromatography.

Procedure

PART A: Making a Chromatogram

1. Put on safety goggles and a lab apron. Use scissors to cut the bottom end of a strip of chromatography paper to a tapered end. **CAUTION: Sharp or pointed objects may cause injury. Handle scissors carefully.**

2. Draw a faint pencil line 1 cm above the pointed end of the paper strip, as shown in the photo on the facing page. Use a capillary tube to apply a tiny drop of the simulated plant pigments extract on the center of the line.

3. Pour 5 mL of chromatography solvent into a chromatography chamber. Pull the chromatography paper through the opening of the cap, and adjust the length of the strip so that a small portion of the tip end is immersed in the solvent. DO NOT immerse the pigment in the solvent.

4. Place the cap over the chromatography chamber. Carefully bend the end of the strip of chromatography paper over the cap, as shown in the photograph on the facing page. Be sure that the strip does not touch the walls of the chamber.

5. Remove the strip from the chromatography chamber when the solvent nears the top of the chamber (within 5–7 minutes).

6. With a pencil, mark the position of the uppermost end of the solvent and the farthest distance each pigment moved. Measure the distance that the solvent and each pigment moved. Record your observations and measurements in your data table. Tape or glue your chromatogram to your lab report. Label the pigment colors.

7. Use the formula below to calculate and record the R_f for each pigment.

$$R_f = \frac{Distance\ substance\ (pigment)\ traveled}{distance\ solvent\ traveled}$$

PART B: Cleanup and Disposal

8. Dispose of chromatography paper, solutions, and broken glass in the designated waste containers. Do not pour chemicals down the drain or put lab materials in the trash unless your teacher tells you to do so.

9. Clean up your work area and all lab equipment. Return lab equipment to its proper place. Wash your hands thoroughly before you leave the lab and after you finish all work.

Analyze and Conclude

1. Summarizing Results Describe what happened to the simulated plant pigments during the lab.

2. Analyzing Data How do your R_f values compare with those of your classmates?

3. Inferring Conclusions What is a chromatogram?

4. Further Inquiry Write a new question about plant pigments that could be explored with another investigation.

On the Job

Paper chromatography is used to separate a variety of chemicals from living tissue extracts. Do research to discover how paper chromatography is used in research laboratories. For more about careers, visit **go.hrw.com** and type in the keyword **HX4 Careers.**

Bristlecone pine "flagged" by the wind

26 Plant Growth and Development

✔ Quick Review

Answer the following without referring to earlier sections of your book.

1. **Identify** the requirements for photosynthesis and cellular respiration. *(Chapter 5, Sections 2 and 3)*

2. **Define** *meristem, vascular tissue, xylem,* and *phloem*. *(Chapter 23, Section 1)*

3. **Identify** the parts of a seed. *(Chapter 23, Section 1)*

4. **Describe** the structures of stems and roots. *(Chapter 25, Section 1)*

Did you have difficulty? *For help, review the sections indicated.*

📖 Reading Activity

Before you read this chapter, write a short list of all the things you already know about plant growth and development. Then list things you want to know about plant growth and development.

internet connect.

www.scilinks.org
National Science Teachers Association *sci*LINKS Internet resources are located throughout this chapter.

sci**LINKS**. Maintained by the National Science Teachers Association

● Exposed to extreme cold and deprived of water, trees such as this one that grow near the tree line in the mountains are stunted. Like flags, they "stream" away from the prevailing wind.

How Plants Grow and Develop

- **Compare** seed germination in beans and corn.
- **Contrast** annuals, biennials, and perennials.
- **Explain** how primary and secondary growth are produced.
- **Describe** several traits of bread wheat.
- **Contrast** development in plants and animals.

Key Terms

germination
perennial
annual
biennial
primary growth
secondary growth
apical meristem
cork cambium
vascular cambium
annual ring

Seeds Sprout

A seed contains a plant embryo that is in a state of suspended animation. Some embryos can remain in suspended animation inside a seed for thousands of years. Seeds sprout with a burst of growth in response to certain changes in the environment. These changes, such as rising temperature and increasing soil moisture, usually signal the start of favorable growing conditions.

Many seeds must be exposed to cold or to light before they can sprout. The seed coats of other seeds must be damaged before they can sprout. Exposure to fire, passing through the digestive system of an animal, and falling on rocks are several natural ways that seed coats are damaged. A seed cannot sprout until water and oxygen penetrate the seed coat. When water enters a seed, the tissues in the seed swell, and the seed coat breaks. If enough water and oxygen are available after the seed coat breaks, the young plant, or seedling, begins to grow.

Germination

A plant embryo resumes its growth in a process called **germination.** The first sign of germination is the emergence of the embryo's root. What happens next varies somewhat from one type of plant to another, as you can see in **Figure 1.** The young shoots of some plants, such as beans, form a hook. The hook protects the tip of the shoot from injury as it grows through the soil.

Figure 1   Seed germination

Beans and corn show two characteristic patterns of seed germination.

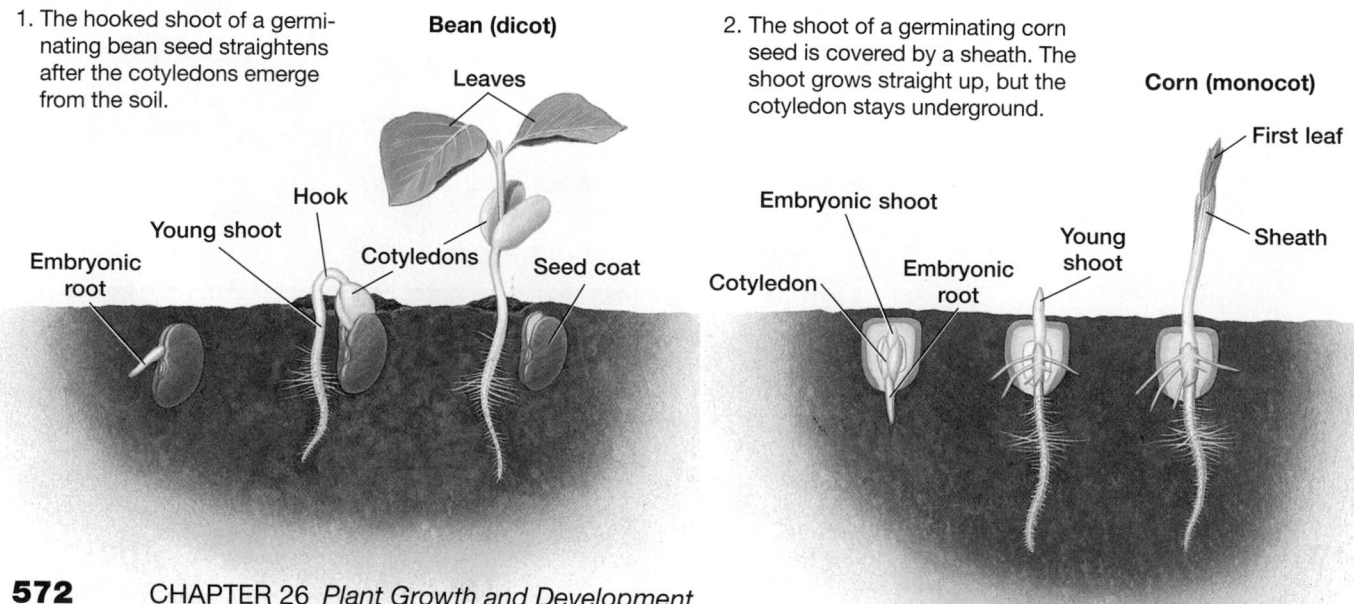

1. The hooked shoot of a germinating bean seed straightens after the cotyledons emerge from the soil.

Bean (dicot)

Leaves

Hook

Young shoot

Embryonic root

Cotyledons

Seed coat

2. The shoot of a germinating corn seed is covered by a sheath. The shoot grows straight up, but the cotyledon stays underground.

Corn (monocot)

First leaf

Embryonic shoot

Sheath

Cotyledon

Embryonic root

Young shoot

The young shoots of other plants, such as corn, have a protective sheath around their shoots. In some plants, such as beans and marigolds, the cotyledons emerge from the soil and unfold above-ground. In other plants, such as corn and peas, the cotyledons remain underground. After the shoot of a seedling emerges, its roots and shoots continue to grow throughout its life.

Plant Life Spans

As you have read in a previous chapter, bristlecone pines are the oldest known trees. They are estimated to be about 5,000 years old. In contrast, some plants live for only a few weeks. Depending on how long it lives, a plant can be classified as one of three basic types: perennial, annual, or biennial.

Perennials Many herbaceous plants and all woody plants are perennials. A **perennial** is a plant that lives for several years. Most perennials reproduce many times during their life span. Others, like the herbaceous perennial shown in **Figure 2,** reproduce only once before they die. Chrysanthemums, daffodils, and irises are familiar herbaceous perennials. These plants store nutrients for the next season's growth in fleshy roots or underground stems. The aboveground shoots of herbaceous perennials often die after each season of growth. Trees, shrubs, and many vines are woody perennials. Some woody perennials drop their leaves each year. Plants that drop all of their leaves each year, such as elms, maples, and grapevines, are known as deciduous (*dee SIHJ oo uhs*) plants. Those that drop a few leaves at a time throughout the year, such as firs, pines, and junipers, are called evergreens.

Annuals Sunflowers, beans, corn, and many weeds are annuals. An **annual** is a plant that completes its life cycle (grows, flowers, and produces fruits and seeds) and then dies within one growing season. Virtually all annuals are herbaceous plants. Most annuals grow rapidly when conditions are favorable. Individual plants can become quite large if they get enough water and nutrients.

Biennials Carrots, parsley, and onions are biennials. A **biennial** is a flowering plant that takes two growing seasons to complete its life cycle. During the first growing season, biennials produce roots and shoots. The shoots consist of a short stem and a rosette (circular cluster) of leaves. The roots store nutrients. In the second growing season, a biennial plant uses the stored nutrients to produce a flowering stalk. The plant dies after flowering and producing fruits and seeds.

Figure 2   A herbaceous perennial. Century plants live for many years but reproduce only once. Like the dried-up plant on the left, this flowering century plant will die when its seeds are mature.

Meristems

Plants grow by producing new cells in regions of active cell division called meristems. Almost all plants grow in length by adding new cells at the tips of their stems and roots. Growth that increases the length or height of a plant is called **primary growth.** Many plants also become wider as they grow taller. Growth that increases the width of stems and roots is called **secondary growth.** After new cells are formed by cell division, they grow and undergo differentiation. Recall from your reading that differentiation is the process by which cells become specialized in form and function.

Primary Growth

Apical *(AP ih kuhl)* **meristems,** which are located at the tips of stems and roots, produce primary growth through cell division. As shown in **Figure 3,** apical meristems are regions of small, undifferentiated cells. To better understand how primary growth occurs in most plants, imagine a stack of dishes. As you add more dishes to the top, the stack grows taller but not wider. Similarly, the cells in the apical meristems of most plants add more cells to the tips of a plant's body. New cells are added through cell division. The cells then lengthen. Thus, primary growth makes a plant's stems and roots get longer without becoming wider. To learn about primary growth in a monocot, look at *Up Close: Bread Wheat* later in this section.

The tissues that result from primary growth are called *primary tissues.* The new cells produced by apical meristems differentiate into the primary dermal, ground, and vascular tissues of roots, stems, and leaves. Some of the cells produced by the root apical meristem also become part of the root cap. These cells replace cells that are worn away as the root pushes through the soil.

🖳 internet connect ▤

www.scilinks.org
Topic: Primary Growth in Plants
Keyword: HX4148

SCiLINKS® Maintained by the National Science Teachers Association

Figure 3 Apical meristems

Both shoot tips and root tips contain apical meristems, where cell division occurs.

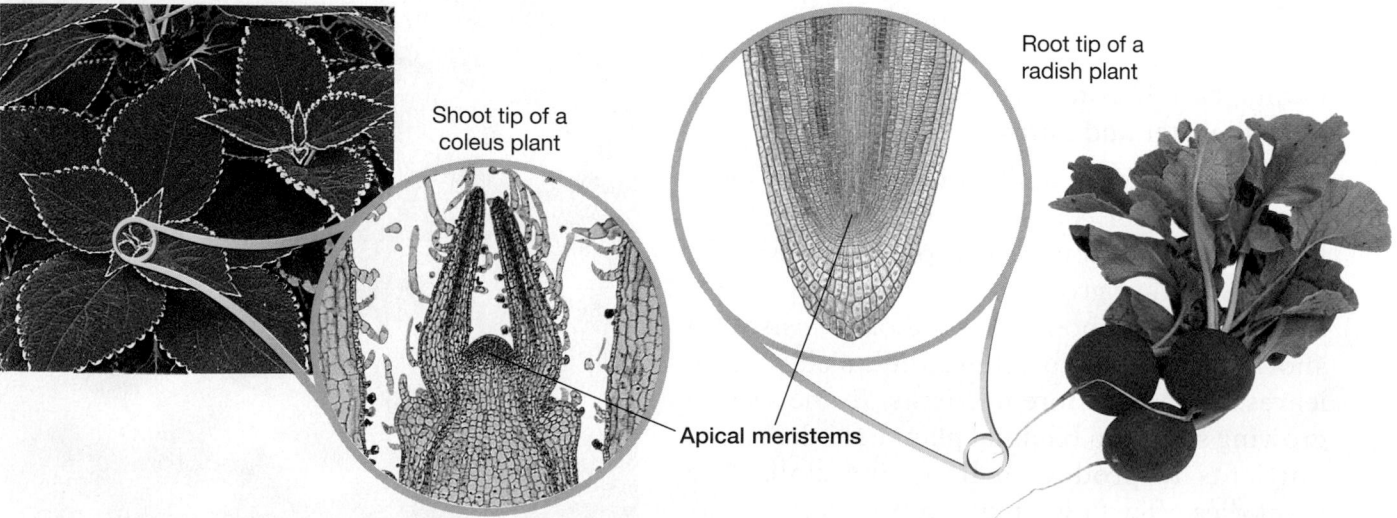

Shoot tip of a coleus plant

Root tip of a radish plant

Apical meristems

Secondary Growth

Secondary growth occurs in parts of many herbaceous plants, such as in carrot roots. However, it is most dramatic in woody plants. Secondary growth is produced by cell division in two meristems, which form thin cylinders near the outside of woody stems and roots. One meristem, called the **cork cambium** *(KAM bee uhm)*, lies within the bark and produces cork cells. The other meristem, called the **vascular cambium,** lies just under the bark and produces vascular tissues. The tissues that result from secondary growth are called secondary tissues. **Figure 4** shows how woody stems develop.

Step ❶ A young woody stem has a ring of vascular bundles between the cortex and the pith. Each vascular bundle contains primary xylem and primary phloem.

Step ❷ Vascular cambium develops between the primary xylem and the primary phloem in each vascular bundle. Secondary phloem is produced toward the outside of the stem. Secondary xylem is produced toward the inside of the stem. The cork cambium forms when the epidermis is stretched and broken as the stem grows in diameter.

Step ❸ Eventually, the vascular bundles merge into solid cylinders. No cortex or primary phloem remains. The cork, cork cambium, and secondary phloem make up the bark. The vascular cambium and secondary xylem lie inside the bark. Thick layers of secondary xylem, or wood, often form rings. Since one new ring is usually formed each year, the rings are called **annual rings.**

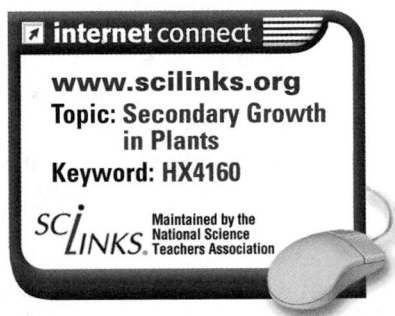

internet connect

www.scilinks.org
Topic: Secondary Growth in Plants
Keyword: HX4160

*SCI*LINKS. Maintained by the National Science Teachers Association

Figure 4

BIO graphic — Development of a Woody Stem

The wood in a woody stem results from secondary growth.

❶ Initially, the stem is covered by epidermis and contains cortex, pith, and a ring of vascular bundles with primary xylem and phloem.

❷ A vascular cambium forms between the xylem and phloem in each vascular bundle. Cork cambium forms under the epidermis.

❸ In a mature stem, the vascular cambium adds new layers of secondary xylem and phloem each year.

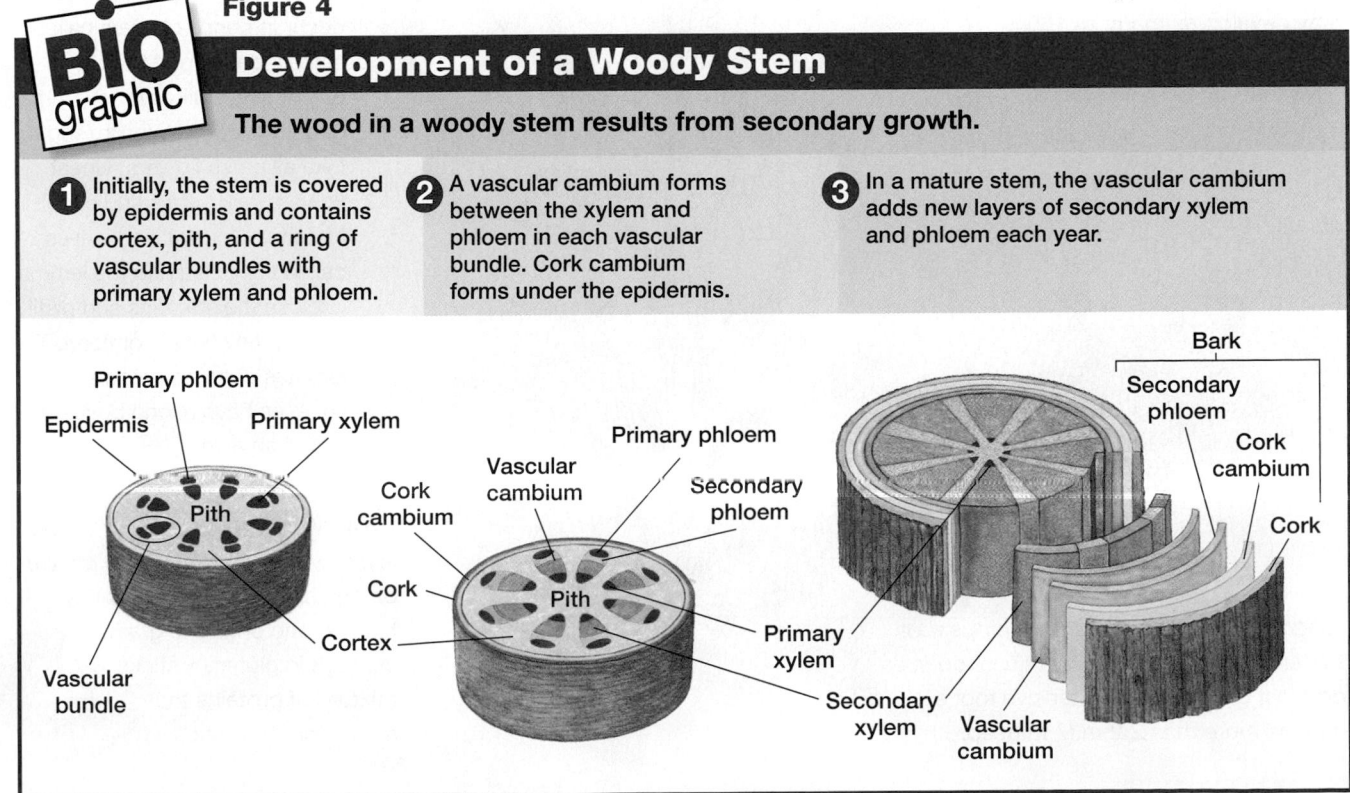

Up Close

Bread Wheat

- **Scientific name:** *Triticum aestivum*
- **Size:** 0.3 to 0.8 m (1 to 2.5 ft) tall
- **Range:** Agricultural regions worldwide
- **Habitat:** Cultivated fields in temperate and subtropical grasslands
- **Importance:** Wheat is the principal staple food in temperate regions of the world. The grains of *Triticum aestivum* are usually ground into flour that is used to make bread.

External Structures

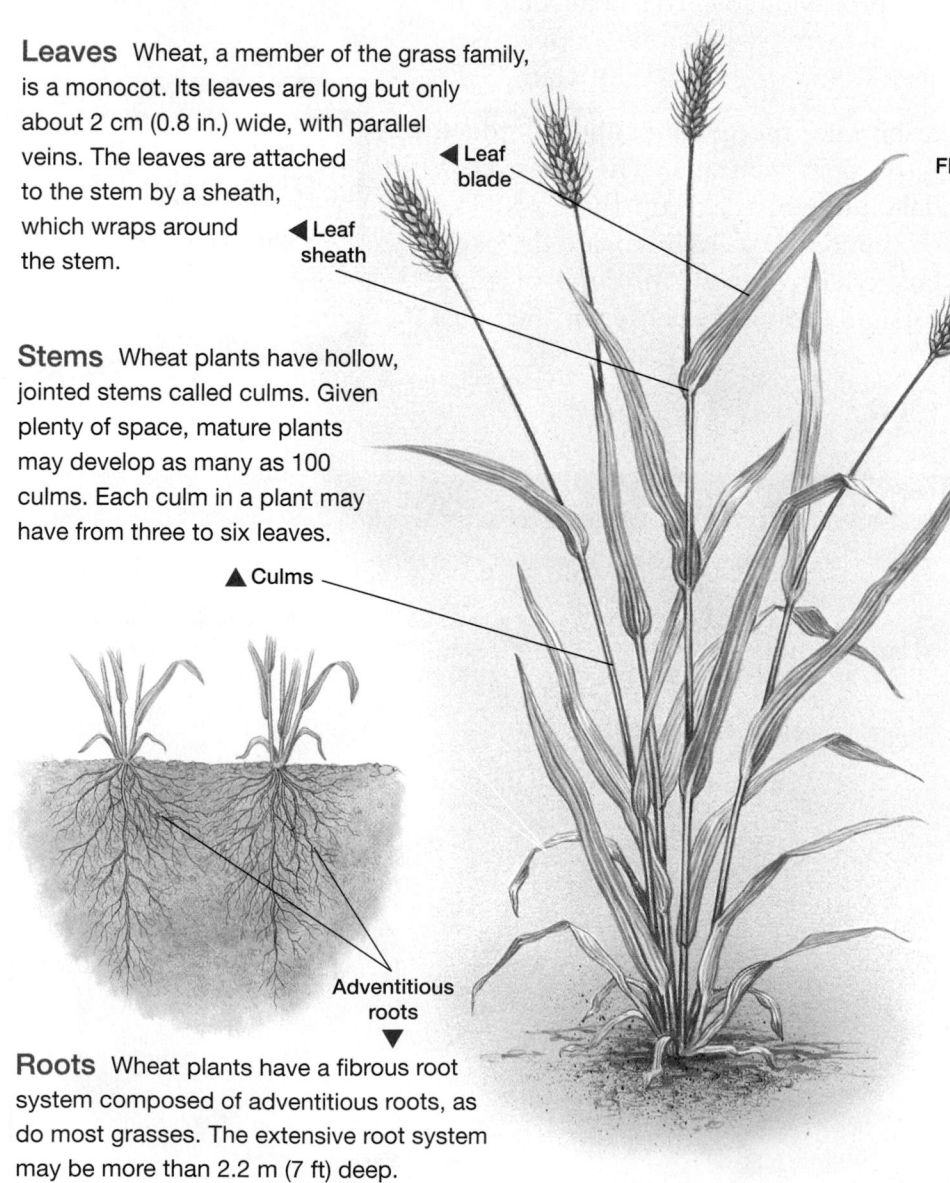

Leaves Wheat, a member of the grass family, is a monocot. Its leaves are long but only about 2 cm (0.8 in.) wide, with parallel veins. The leaves are attached to the stem by a sheath, which wraps around the stem.

◀ Leaf blade

◀ Leaf sheath

Stems Wheat plants have hollow, jointed stems called culms. Given plenty of space, mature plants may develop as many as 100 culms. Each culm in a plant may have from three to six leaves.

▲ Culms

Adventitious roots ▼

Roots Wheat plants have a fibrous root system composed of adventitious roots, as do most grasses. The extensive root system may be more than 2.2 m (7 ft) deep.

Awn

Palea
Anther
Stigma
Lemma

Floret

▼ Flower spike

Flowers The flowers, which occur in dense clusters called spikes, develop at the top of each culm. Spikes range from 5 to 13 cm (2 to 5 in.) in length. Like all grass flowers, wheat flowers lack petals and sepals. Instead, two modified leaves called the palea and the lemma enclose the stamens and pistil of each tiny flower, or floret. The lemmas of some bread-wheat varieties have a long bristle called an awn.

Fruit A kernel, or grain, of wheat is a one-seeded fruit with a crease on one side and a brush of tiny hairs at one end. The grains are high in gluten, a sticky mixture of proteins that make dough elastic.

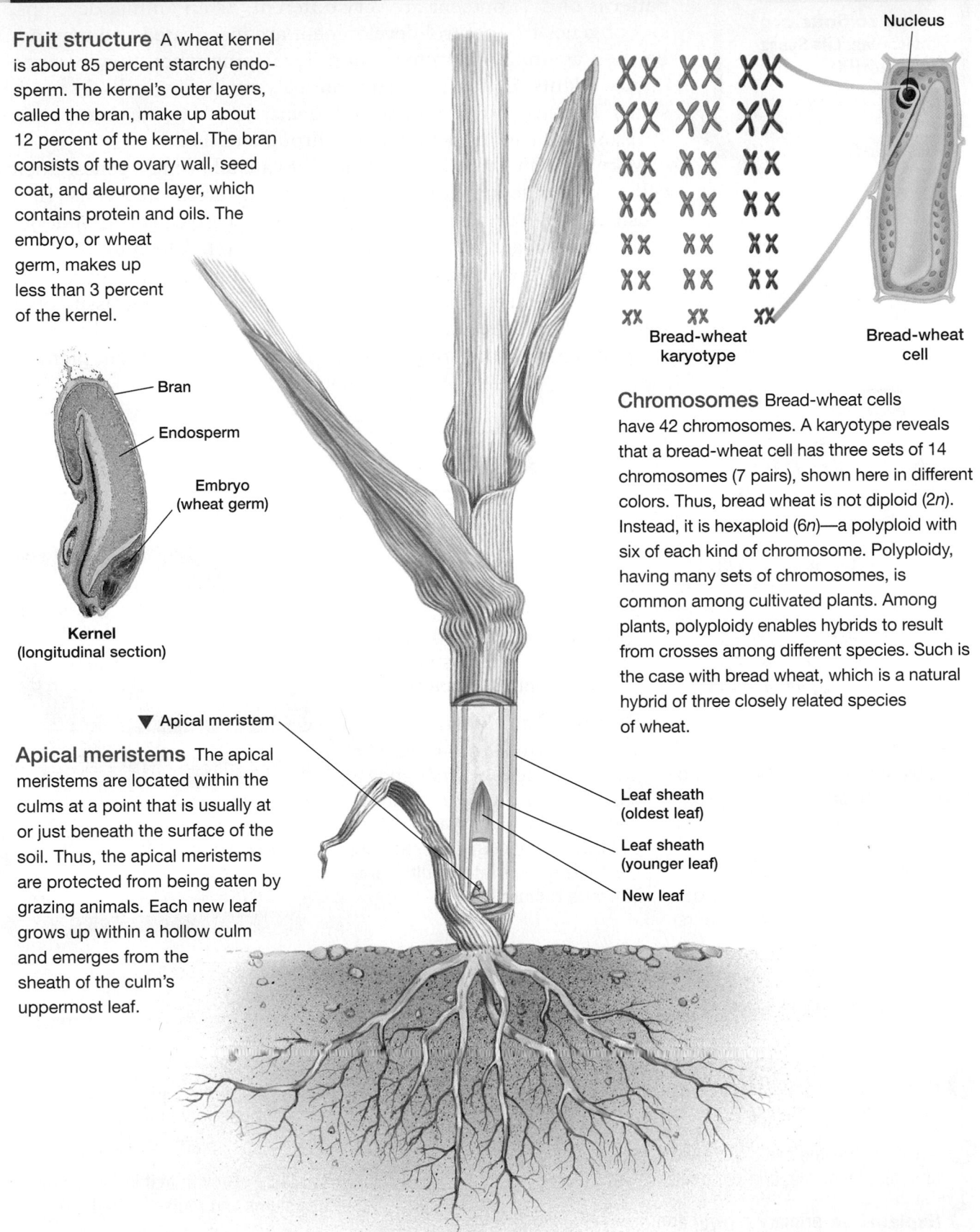

Internal Structures

Fruit structure A wheat kernel is about 85 percent starchy endosperm. The kernel's outer layers, called the bran, make up about 12 percent of the kernel. The bran consists of the ovary wall, seed coat, and aleurone layer, which contains protein and oils. The embryo, or wheat germ, makes up less than 3 percent of the kernel.

Bran

Endosperm

Embryo
(wheat germ)

Kernel
(longitudinal section)

▼ Apical meristem

Apical meristems The apical meristems are located within the culms at a point that is usually at or just beneath the surface of the soil. Thus, the apical meristems are protected from being eaten by grazing animals. Each new leaf grows up within a hollow culm and emerges from the sheath of the culm's uppermost leaf.

Nucleus

Bread-wheat karyotype

Bread-wheat cell

Chromosomes Bread-wheat cells have 42 chromosomes. A karyotype reveals that a bread-wheat cell has three sets of 14 chromosomes (7 pairs), shown here in different colors. Thus, bread wheat is not diploid ($2n$). Instead, it is hexaploid ($6n$)—a polyploid with six of each kind of chromosome. Polyploidy, having many sets of chromosomes, is common among cultivated plants. Among plants, polyploidy enables hybrids to result from crosses among different species. Such is the case with bread wheat, which is a natural hybrid of three closely related species of wheat.

Leaf sheath
(oldest leaf)

Leaf sheath
(younger leaf)

New leaf

Plant Development

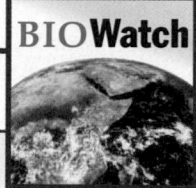
Genes guide the development of both plants and animals, but their patterns of development are very different. As an animal develops, sets of genes that control development are inactivated and may not be used again. Most animals stop developing when they become adults. Plants, in contrast, continuously make new cells in meristems. These cells differentiate and replace or add to existing tissues. Thus, a plant continues to develop throughout its life.

Many cells in a mature plant can activate all of their genes. Such cells can divide and form masses of undifferentiated cells. In a sense, they can reverse their development. These cells can undergo differentiation and develop into a mature plant. A technique called *tissue culture* is used to grow new plants from tissue that can reverse its development. A tissue culture is prepared by placing tissue on a sterile nutrient medium. Masses of undifferentiated cells that form grow into plants that are genetically identical to the parent plant.

BIOWatch

Plant Tissue Culture

Tissue culture is used to propagate orchids, houseplants, and fruit plants. Thousands of cultures can be made from a single plant. Tissue culture can also be used to produce plants with new characteristics.

Protoplast Fusion

Protoplast fusion has been used to produce hybrid petunias, potatoes, and carrots. A protoplast is a plant cell that has had its cell walls removed by enzymes. Certain chemicals or an electrical shock can cause two protoplasts to fuse, as the photo at right shows. If the protoplasts came from genetically different plants, a hybrid cell results. The hybrid is then placed in a tissue culture and grown into an adult plant.

Genetic Engineering

Tissue culture is also an essential part of producing genetically engineered plants. First, foreign genes are inserted into a plant's cells. The genetically altered cells are then grown into adult plants in tissue culture.

Magnification: 810×

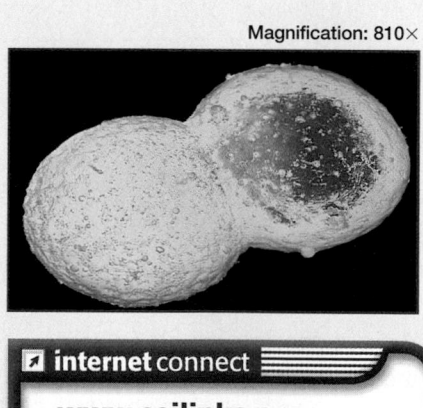

Section 1 Review

1 **Compare** and contrast the germination of bean seeds and corn seeds.

2 **Summarize** the basic differences between annuals, biennials, and perennials.

3 **Explain** how primary growth and then secondary growth produce a woody stem.

4 **Critical Thinking Analyzing Methods** In what ways does plant development differ from animal development?

5 **Standardized Test Prep** Growth that increases the width of a plant's stems and roots is called
  A germination.   **C** primary growth.
  B differentiation.   **D** secondary growth.

Regulating Growth and Development

Nutrients

Like all multicellular organisms, plants grow by adding new cells through cell division. Plants must have a steady supply of the raw materials they use to build new cells in order to grow. Plants need only two raw materials—carbon dioxide and water—to make all the carbohydrates in their tissues. As you learned earlier in this book, carbon dioxide and water are needed for photosynthesis.

Like animals, plants also need oxygen for cellular respiration. Although the green parts of a plant produce oxygen during photosynthesis, most of the oxygen used by leaves and stems comes from the air. Roots, which usually do not carry out photosynthesis, get oxygen from the air spaces between soil particles. If the soil around a plant's roots becomes compacted or saturated with water, it may not provide enough oxygen for the roots, and the plant could die.

However, carbon dioxide, water, and oxygen do not satisfy all of a plant's needs for raw materials. Plants also require small amounts of at least 14 **mineral nutrients,** which are elements absorbed mainly as inorganic ions. **Table 1** lists the six mineral nutrients needed in the greatest amounts for healthy plant growth and describes the importance of each nutrient. Commercial fertilizers may contain most of these mineral nutrients.

Objectives

- **Identify** the major nutrients plants need to grow.
- **Describe** how plant hormones control plant growth.
- **Relate** environmental factors to plant growth.

Key Terms

mineral nutrient
auxin
hormone
apical dominance
tropism
photoperiodism
dormancy

Table 1 Major Mineral Nutrients Required by Plants

Nutrient	Importance
Nitrogen	Part of proteins, nucleic acids, chlorophylls, ATP, and coenzymes; promotes growth of green parts
Phosphorus	Part of ATP, ADP, nucleic acids, phospholipids of cell membranes, and some coenzymes
Potassium	Needed for active transport, enzyme activation, osmotic balance, and stomatal opening
Calcium	Part of cell walls; needed for enzyme activity and membrane function
Magnesium	Part of chlorophyll; needed for photosynthesis and activation of enzymes
Sulfur	Part of some proteins and coenzyme A; needed for cellular respiration

Hormonal Control of Growth

For centuries, people have known that plants bend strongly toward a light source as their shoots elongate. In the 1920s, the Dutch biologist Frits Went hypothesized that a chemical produced in the shoot tip of a grass causes this bending response. Went named the growth-promoting chemical that causes stems to bend **auxin** *(AWK sin)*. The steps in Went's experiment are summarized in **Figure 5.**

Step ❶ Went removed the tip of an oat shoot and placed the tip on an agar block. Auxin diffused from the tip into the block.

Step ❷ Went then transferred the agar block to the cut end of a shoot, which was followed by growth of the shoot.

Step ❸ When Went placed an agar block with auxin on either side of cut shoots, the shoots grew in the opposite direction.

Step ❹ As a control, Went placed an agar block without auxin on the cut end of other shoots. These shoots did not grow.

Auxin

Auxin is one of many plant hormones. The word *hormone* comes from the Greek word *horman*, meaning "to set in motion." A **hormone** is a chemical that is produced in one part of an organism and transported to another part, where it causes a response. Auxin causes plant cell walls to become more acidic, which allows the cells

Figure 5

BIOgraphic

The Steps in Went's Experiment

Auxin apparently caused oat seedlings to elongate and bend toward light.

❶ Auxin diffused from the cut tip of an oat shoot into an agar block.

❷ Application of the agar block with the auxin to a second shoot was followed by growth.

❸ Cut shoots grew *away* from contact with agar with auxin.

❹ Application of agar without auxin was not followed by growth of the oat shoot.

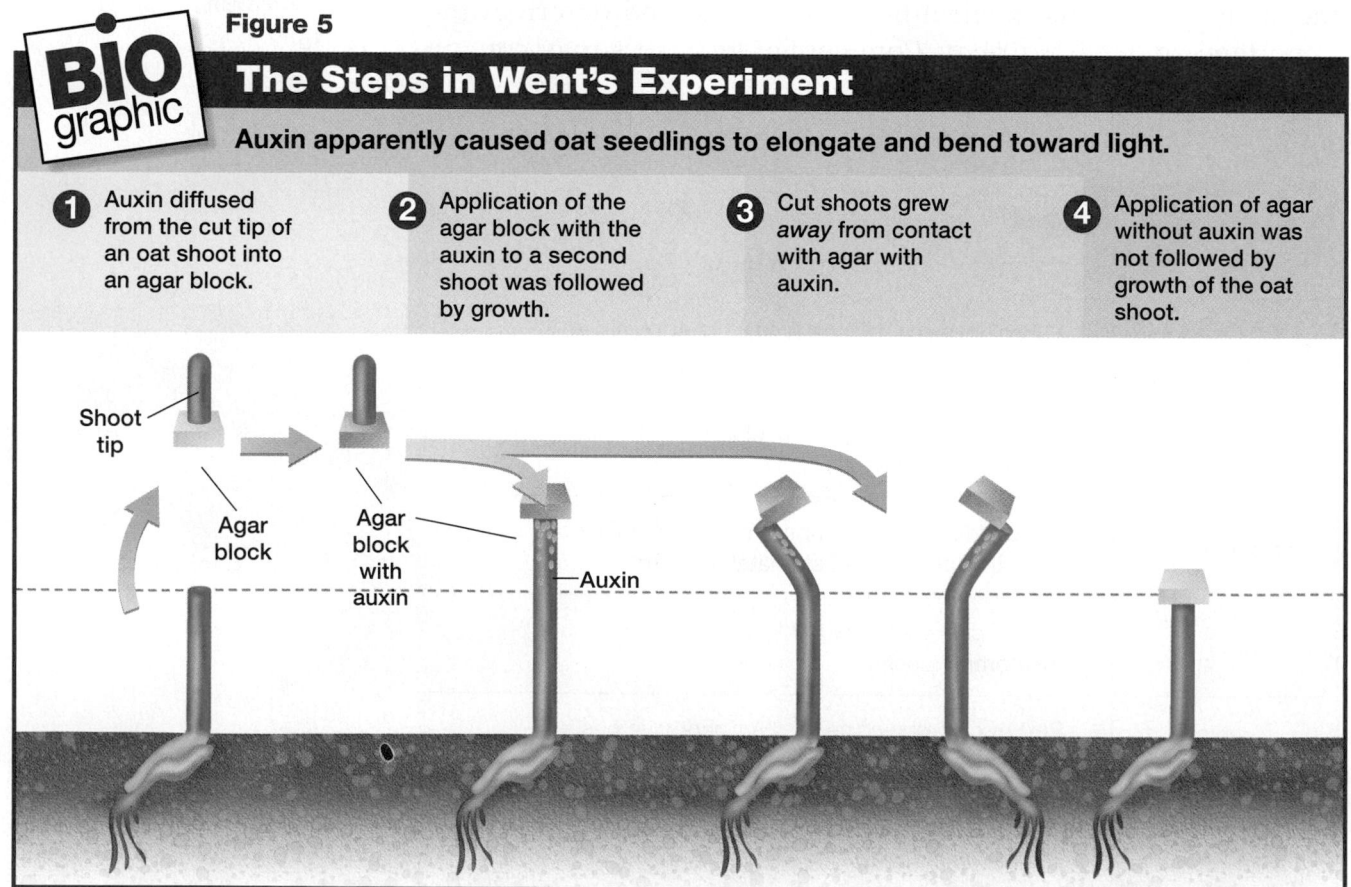

Shoot tip

Agar block

Agar block with auxin

Auxin

to elongate as they grow. Auxin accumulates on the dark side of a stem. As a result, the cells on the dark side of a stem elongate more than the cells on the light side. The difference in elongation causes the stem to grow toward the light. Auxin also inhibits the growth of the buds along a stem. This inhibition is called **apical dominance.** Cutting off the tip of a stem removes the source of auxin and enables the other buds to grow. That is why pruning the stems of a plant makes the plant become bushier.

Hormones in Agriculture

More than a century ago, citrus farmers discovered that they could cause citrus fruits to ripen by storing them in a room heated by a kerosene stove. The ripening was caused by ethylene, which is a gaseous organic compound produced when kerosene is incompletely burned. Most plant tissues produce ethylene. Today, ethylene is used to promote the ripening of tomatoes, bananas, and other fruits that are harvested before they ripen. Ethylene also loosens the fruit of cherries, blackberries, and blueberries, making it easier to harvest these crops mechanically.

Gibberellins *(jihb uhr EHL ihnz)* are produced in developing shoots and seeds. They stimulate stem elongation, fruit development, and seed germination. Gibberellins are used to enlarge Thompson seedless grapes. Other seedless fruits treated with gibberellins include apples, cucumbers, mandarin oranges, and peaches.

Cytokinins *(sie toh KIE nihnz)*, which are produced in root tips, stimulate cell division and slow the aging of some plant organs. Cytokinins are sprayed on cut flowers to keep them fresh and on fruits and vegetables to extend their shelf life. Cytokinins are added to tissue-culture media to cause undifferentiated cells to form shoots.

☐ internet connect

www.scilinks.org
Topic: Plant Hormones
Keyword: HX4140

*SCi*LINKS. Maintained by the National Science Teachers Association

Investigating the Effects of Ethylene on a Plant

QUICK LAB

You can use a ripe apple to see one of the effects of ethylene on plants.

Materials

4 L glass jars with lids (2), 2 plants in 5 cm pots, small ripe apple

Procedure

1. **Place** a plant inside one of the jars. Tightly secure the lid.

2. **Place** the other plant and the apple inside the other jar. Tightly secure the lid.

3. **Observe** both jars for several days. Record what you see.

Analysis

1. **Describe** any changes in the plant in each jar.

2. **Critical Thinking Drawing Conclusions** A ripe apple gives off ethylene gas. Based on your observations, how does ethylene affect a plant?

Environmental Influences on Growth

Because most plants are anchored in one spot, they cannot move from an unfavorable environment to a more favorable one as animals do. Instead, plants respond to their environment by adjusting the rate and pattern of their growth. For example, a plant that receives plenty of water and mineral nutrients may grow much faster and larger than it would if it received very little water and mineral nutrients. Also, a plant grown in full sun may grow much faster and larger than it would if it were grown in the shade or indoors. So the availability of light and nutrients affect the rate of plant growth. Many of a plant's responses to environmental stimuli, however, are triggered by the hormones that regulate plant growth.

Tropisms

A **tropism** *(TROH piz uhm)* is a response in which a plant grows either toward or away from a stimulus. Auxin is responsible for producing tropisms. **Figure 6** shows examples of three common types of tropisms. *Phototropisms* are responses to light. Responses to gravity are called *gravitropisms*. A *thigmotropism* is a response to touch. If a plant grows toward a stimulus, the response is called a positive tropism. If a plant grows away from the direction of the stimulus, the response is called a negative tropism. Thus, a shoot that grows up out of the ground shows both positive phototropism (growing toward the light) and negative gravitropism (growing away from the pull of gravity).

Figure 6 Tropisms

Tropisms are growth responses that occur either toward or away from a stimulus.

The bending of an amaryllis toward the light is a positive phototropism.

The upward growth of shoots is a negative gravitropism; the downward growth of roots is a positive gravitropism.

The coiling of grapevine tendrils around a wire is a thigmotropism.

Interpreting Annual Rings

DATA LAB

Background

The annual rings of a woody stem provide important clues to annual variations in the rainfall an area receives over time. Thick rings form in years with heavy rainfall. Relatively thin rings form in dry years. Use the photo at right to answer the following questions.

A

B

Analysis

1. **Critical Thinking**
   Interpreting Data What do the annual rings indicate about the climate where this plant grew?

2. **Critical Thinking**
   Drawing Conclusions Which ring, A or B, indicates a year when this plant received more rainfall?

3. **Critical Thinking**
   Making Predictions How will the annual rings of a nearby tree of the same age and species compare with those of this tree?

Photoperiodism

Certain plants bloom in the spring and others bloom in the summer or fall. Some plants bloom as soon as they reach a mature size. In many plants, seasonal patterns of flowering and other aspects of growth and development are caused by changes in the length of days and nights. The response of a plant to the length of days and nights is called **photoperiodism.**

Most plants can be categorized as one of three types in reference to photoperiodism. A plant that responds when days become shorter than a certain number of hours is said to be a short-day plant. A plant that responds when days become longer than a certain number of hours is called a long-day plant. Plants whose growth and development are not affected by day length are known as day-neutral plants. It is really the length of the nights, however, rather than the length of the days that controls photoperiodism, as **Figure 7** shows. Knowledge of photoperiodism is very important to the nursery and floral industries. The length of days and nights is controlled artificially in greenhouses where plants such as poinsettias and chrysanthemums are grown. Thus, commercial growers force the plants to produce flowers at times of the year when they ordinarily would not. This makes it possible for poinsettias to be available for Christmas and chrysanthemums to be available year round.

Figure 7 Flowering and photoperiodism. Long-day plants flower when nights are short. Short-day plants flower when nights are long. If a flash of light interrupts a long night, long-day plants flower and short-day plants do not.

Figure 8   Bud dormancy.
Thick scales cover the dormant buds on this twig from an apple tree.

Responses to Temperature

Temperature affects growth and development in many plants. For example, most tomato plants will not produce fruit if the nighttime temperatures are too high. Many plants that flower in early spring will not produce flowers until exposed to cold temperatures for a certain number of hours. Most deciduous woody plants drop their leaves in the fall in response to cooler temperatures and shorter periods of daylight. Thick, protective scales develop around their buds, as **Figure 8** shows. After a period of low temperatures, the buds begin growing into new leaves or sections of woody stem.

Dormancy is the condition in which a plant or a seed remains inactive, even when conditions are suitable for growth. Many plants and seeds remain dormant until they have been exposed to low temperatures for several weeks. Dormancy helps plants to survive by keeping buds from growing and seeds from germinating during warm spells before winter has ended.

Analyzing the Effect of Cold on Seed Germination

Background

In some plants, a period of low temperatures is needed to break seed dormancy. The graph at right shows how being stored at a low temperature (4°C) affected the ability of apple seeds to germinate. Use the graph to answer the following questions.

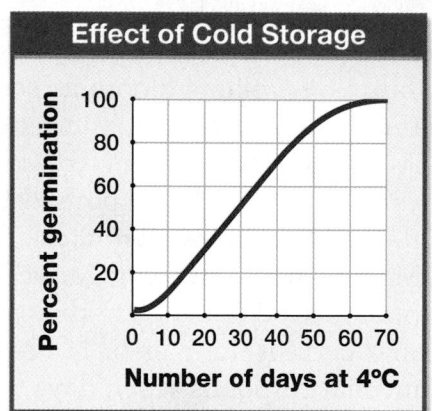

Analysis

1. **Summarize** the overall effect of cold temperatures on the germination of apple seeds.

2. **Calculate** the number of weeks that apple seeds must be stored at 4°C for at least 80 percent of the seeds to germinate.

3. **Critical Thinking Interpreting Graphs** What percentage of apple seeds germinate after storage at 4°C for 20 days?

4. **Critical Thinking Predicting Patterns** What percentage of apple seeds will germinate after being stored at 4°C for 80 days?

Section 2 Review

1 **List** the six mineral nutrients that plants require in the greatest amounts.

2 **Explain** how auxin causes a stem to grow toward a light source.

3 **Describe** an example of a negative gravitropism.

4 **Critical Thinking Predicting Outcomes** Why is it an advantage for plant growth and development to be regulated by environmental stimuli?

5 Standardized Test Prep When a vine responds to the touch of a fence wire by growing a tendril around the wire, the vine is exhibiting

A gravitropism.   **C** phototropism.

B thigmotropism.   **D** photoperiodism.

Key Concepts

1 How Plants Grow and Develop

- Germination is the resumption of growth by the embryo in a seed. Water and oxygen must penetrate the seed coat before germination can occur.

- Annuals complete their life cycle in one growing season. Biennials complete their life cycle in two growing seasons. Perennials live several years and may reproduce many times.

- Apical meristems located at the tips of shoots and roots produce primary growth. The tissues that result from primary growth are known as primary tissues.

- Secondary growth increases a plant's stem and root width. In woody stems, secondary growth is produced by the cork cambium and vascular cambium, two meristems near the outside of the stem.

- Bread wheat is a cereal grass with long leaves, hollow stems, and clusters of tiny flowers. The fruits, or grains, are usually ground into flour that is used to make bread.

- Plants develop throughout their lives. Plant development is reversible. Many mature plant cells can divide to form masses of undifferentiated cells, which can develop into a new plant.

2 Regulating Plant Growth and Development

- Plants need at least 14 mineral nutrients for growth. They also need carbon dioxide and water for photosynthesis and oxygen for cellular respiration.

- Hormones regulate plant growth and development. Auxin is a hormone that causes shoots to elongate and inhibits the growth of lateral buds.

- Plants modify their growth in response to the direction of light, gravity, and touch. Such growth responses are called tropisms.

- Seeds and many mature plants survive periods of unfavorable environmental conditions by becoming dormant.

- In many plants, seasonal patterns of flowering and other aspects of growth and development are caused by changes in the length of days and nights.

Key Terms

Section 1

germination (572)
perennial (573)
annual (573)
biennial (573)
primary growth (574)
secondary growth (574)
apical meristem (574)
cork cambium (575)
vascular cambium (575)
annual ring (575)

Section 2

mineral nutrient (579)
auxin (580)
hormone (580)
apical dominance (581)
tropism (582)
photoperiodism (583)
dormancy (584)

Understanding Key Ideas

1. Plant and animal development differ in that plant development
  a. is continuous and reversible.
  b. stops after a plant reaches maturity.
  c. is not affected by environment.
  d. is controlled by genes that cannot be reactivated.

2. Which of the following is *not* a raw material needed for plant growth?
  a. carbon dioxide
  b. water
  c. oxygen
  d. vitamins

3. Auxin causes cells to
  a. have less flexible cell walls.
  b. elongate more as they grow.
  c. bend toward light.
  d. develop lateral buds.

4. Which of the following caused the different growth patterns of the two pots of mums shown below?
  a. apical meristem
  b. auxin
  c. apical dominance
  d. all of the above

5. The response of a plant to the length of days and nights is
  a. gravitropism.
  b. photoperiodism.
  c. phototropism.
  d. thigmotropism.

6. Major mineral nutrients required by plants include all of the following except
  a. nitrogen. **c.** phosphorus.
  b. lead. **d.** potassium.

7. How do cytokinins and gibberellins affect plant growth, and how are these hormones used in agriculture?

8. Up Close How does the chromosome number of bread wheat differ from that of diploid organisms?

9. 🌐 **BIOWatch** How is tissue culture used to produce hybrid varieties of plants?

10. 🔲 **Concept Mapping** Construct a concept map that describes growth in vascular plants. Try to include the following terms in your map: *meristems, apical meristem, primary growth, primary tissues, cork cambium, secondary growth, cork, secondary phloem, secondary xylem,* and *vascular cambium.*

Critical Thinking

11. Applying Information Carrots are biennials. What role does the root of a carrot plant play in the plant's second year of growth?

12. Forming Reasoned Opinions Why is it possible for people to grow new plants from pieces of leaf, stem, or root tissue in which the cells have already undergone differentiation?

Alternative Assessment

13. Finding and Communicating Information Use the media center or Internet resources to learn how commercial growers produce plants such as poinsettias and chrysanthemums that flower at times when they would not flower in nature. Summarize your findings in a written report.

14. Career Connection Agronomist Agronomists study soil management and crop production. Write a report that includes a job description, training required, kinds of employers, growth prospects, and starting salary.

Standardized Test Prep

Understanding Concepts

Directions (1–4): For *each* question, write on a separate sheet of paper the letter of the correct answer.

1 The emergence of what part of the embryo is the first sign of germination in a bean?
  A. cotyledon   **C.** root
  B. hooked shoot   **D.** sheathed shoot

2 What term describes plants that live for several years or more?
  F. annuals   **H.** perennials
  G. biennials **I.** terminals

3 Which of the following is a tissue whose main function is to provide cells for growth at the tips of a plant?
  A. apical meristems
  B. cork cambium
  C. root cap
  D. vascular cambium

4 In which tissues does cell division that increases the diameter of a woody stem occur?
  F. apical meristems
  G. cork cambium
  H. primary tissues
  I. vascular cambium

Directions (5–6): For *each* question, write a short response.

5 Some seeds are soaked in acid before they are packaged and sold to farmers. What is the purpose of this treatment?

6 Dormancy is the condition in which seeds or buds do not grow even when environmental conditions are favorable for growth to occur. How does dormancy contribute positively to a plant's survival?

Test TIP

If you are not sure about the spelling of certain words when answering a short-response question, look at the question itself to see if the word appears in the question.

Reading Skills

Directions (7): Read the passage below. Then answer the question.

In an experiment, a student placed a green banana in each of 10 plastic bags. The student also placed a ripe pear in five of the bags and then sealed all of the bags. The bananas in the bags without pears took longer to ripen than the bananas in the bags with pears.

7 What could explain these experimental results?
  A. Pears cannot ripen without the presence of bananas.
  B. Bananas cannot ripen without the presence of pears.
  C. Bananas give off ethylene, which promotes ripening in the pears.
  D. Ripe pears give off ethylene, which promotes ripening in the bananas.

Interpreting Graphics

Directions (8): Base your answer to question 8 on the diagram below.

Hormonal Control of Growth in an Oat Shoot

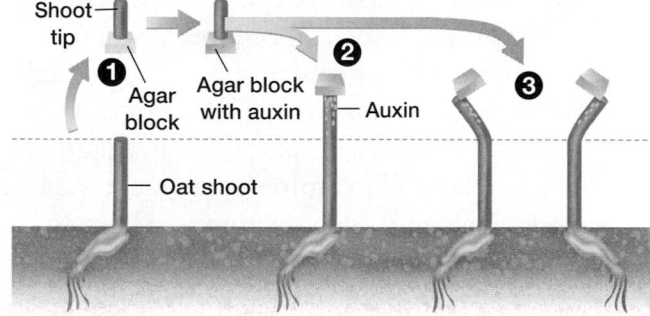

8 The diagram above summarizes an experimental investigation. What happened in step 2?
  F. The agar block stopped further growth of the shoot.
  G. Auxin from the agar block stimulated the shoot to grow.
  H. The shoot grew straight up in response to the agar block.
  I. The shoot produced auxin in response to the agar block.

Skills Practice Lab

Comparing Bean and Corn Seedlings

SKILLS
- Comparing
- Drawing
- Relating

OBJECTIVES
- **Observe** the structures of bean seeds and corn kernels.
- **Compare** and **Contrast** the development of bean embryos and corn embryos as they grow into seedlings.

MATERIALS
- 6 bean seeds soaked overnight
- stereomicroscope
- 6 corn kernels soaked overnight
- scalpel
- paper towels
- 2 rubber bands
- 150 mL beakers (2)
- glass-marking pen
- metric ruler

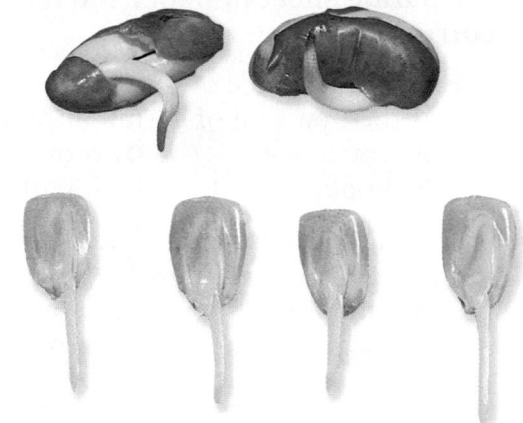

Before You Begin

A **seed** contains an inactive plant **embryo.** A plant embryo consists of one or more **cotyledons,** an embryonic shoot, and an embryonic **root.** Seeds also contain a supply of nutrients. In **monocots,** the nutrients are contained in the **endosperm.** In **dicots,** the nutrients are transferred to the cotyledons as seeds mature. A seed **germinates** when the embryo begins to grow and breaks through the protective **seed coat.** The embryo then develops into a young plant, or **seedling.** In this lab, you will examine bean seeds and corn kernels and then germinate them to observe the development of their seedlings.

1. Write a definition for each boldface term in the paragraph above.

2. Based on the objectives for this lab, write a question you would like to explore about seedling development.

Procedure

PART A: Observing Seed Structure

1. Remove the seed coat of a bean seed, and separate the two fleshy halves of the seed.

2. Locate the embryo on one of the halves of the seed. Examine the bean embryo with a stereomicroscope. Draw the embryo, and label the parts you can identify.

3. ◆ Examine a corn kernel, and locate a small light-colored oval area.
CAUTION: Sharp or pointed objects may cause injury. Handle scalpels carefully. Use a scalpel to cut the kernel in half along the length of this area.

4. Locate the corn embryo, and examine it with a stereomicroscope. Draw the embryo, and label the parts you can identify.

PART B: Observing Seedling Development

5. Fold a paper towel in half as shown in the photo at the top of the next page. Set five corn kernels on the paper towel. Roll up the paper towel, and put a rubber band around the roll. Stand the roll in a beaker with 1 cm of water in the bottom. Add water to the beaker as needed to keep the paper towels wet, but do not allow the corn kernels to be covered by water.

6. Repeat step 5 with five bean seeds.

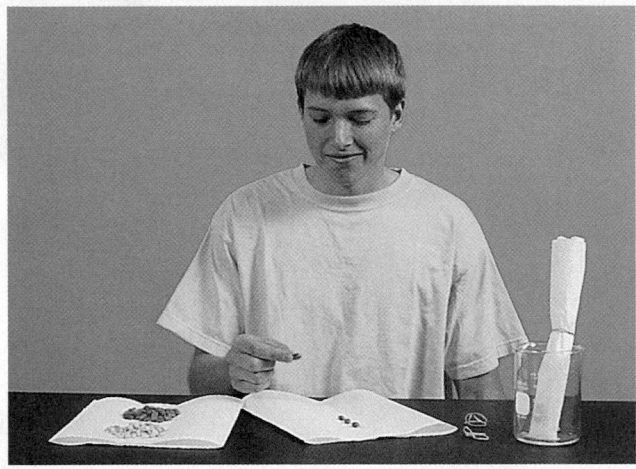

Analyze and Conclude

1. **Relating Concepts** Corn and beans are often cited as representative examples of monocots and dicots, respectively. Relate the seed structure of each to the terms *monocotyledon* and *dicotyledon*.

2. **Summarizing Results** What parts of a plant embryo were observed in all seedlings on the third day?

3. **Drawing Conclusions** In which part or parts of bean seedlings and corn seedlings do the seedlings grow in length? Explain.

4. **Forming Hypotheses** How are the tender young shoots of bean seedlings and corn seedlings protected as the seedlings grow through the soil?

5. **Evaluating Viewpoints** Defend the following statement: There are both similarities and differences in seed structure and seedling development in beans and corn.

6. **Further Inquiry** Write a new question about seedling development that could be explored with another investigation.

7. After three days, unroll the paper towels and examine the corn and bean seedlings. Use a glass-marking pen to mark the roots and shoots of the developing seedlings. Starting at the seed, make a mark every 0.5 cm along the root of each seedling. And again starting at the seed, make a mark every 0.5 cm along the stem of each seedling.

8. Draw a corn seedling and a bean seedling in your lab report. Label the parts of each seedling. Also show the marks you made on each seedling, and indicate the distance between the marks.

9. Using a fresh paper towel, roll up the seeds, place the rolls in the beakers, and add fresh water to the beakers.

10. After two more days reexamine the seedlings. Measure the distance between the marks. Repeat step 8.

PART C: Cleanup and Disposal

11. Dispose of seeds, broken glass, and paper towels in the designated waste containers. Do not put lab materials in the trash unless your teacher tells you to do so.

12. Clean up your work area and all lab equipment. Return lab equipment to its proper place. Wash your hands thoroughly before you leave the lab and after you finish all work.

On the Job

Plant physiology is the study of the processes that occur in plants. Do research to discover where plant physiologists work and what types of research are currently being conducted in the field of plant physiology. For more about careers, visit **go.hrw.com** and type in the keyword **HX4 Careers**.

UNIT 7

Exploring Invertebrates

Beekeeper examining a honeycomb frame full of honey

in perspective

Honeybees

Yesterday... Humans have raised bees for their honey and their beeswax for thousands of years. American colonists brought bees from England and early pioneers took them along on the journey west. **What are the characteristic features of insects?**

French beekeeper in the early 1700s

Today... Honeybees are valued today not only as a source of honey but also for their role as pollinators of crops such as almonds, cherries, apples, and pears. Migratory beekeepers carry bees from state to state, renting their bees to farmers in time to ensure a successful harvest. **Discover how a beehive is a highly organized social system.**

Honeybee

Tomorrow... In the United States, bee-dependent crops are worth $10 billion a year. Scientists continuc to study different varieties of bees in an effort to find those that are the most efficient pollinators.

Honeybee research

internet connect

www.scilinks.org
Topic: Honeybees
Keyword: HX4099

SCLINKS. Maintained by the National Science Teachers Association

Moon jellyfish

27 Introduction to Animals

✔ Quick Review

Answer the following without referring to earlier sections of your book.

1. **Define** the term *metabolism*. *(Chapter 1, Section 2)*

2. **Describe** the process of diffusion. *(Chapter 4, Section 1)*

3. **Summarize** how gametes are formed. *(Chapter 6, Section 1)*

4. **State** the relationship of a phylum to a kingdom. *(Chapter 14, Sections 1 and 2.)*

5. **Define** the terms *tissue* and *organ*. *(Chapter 19, Section 3)*

Did you have difficulty? *For help, review the sections indicated.*

Reading Activity

Before you begin to read, survey the chapter, noting the red headings at the tops of pages and the blue subheadings. Use these heads to make an outline of the chapter, and leave space after each heading. Fill in important facts as you read.

Looking Ahead

Section 1

Characteristics of Animals
General Features of Animals
Body Symmetry
Internal Body Cavity
Body Segmentation
Kinds of Animals

Section 2

Animal Body Systems
Tissues and Organs
Reproductive Strategies

🔲 **internet** connect

www.scilinks.org
National Science Teachers Association *sci*LINKS Internet resources are located throughout this chapter.

*SCi*LINKS. Maintained by the National Science Teachers Association

● The beautiful moon jellyfish, *Aurelia arita,* is found in the Atlantic and Pacific Oceans. People used to call jelly-fishes *sea lungs* because their pulsating movement seems similar to breathing.

Characteristics of Animals

Objectives

- **Identify** the features that animals have in common.

- **Distinguish** radial symmetry from bilateral symmetry.

- **Summarize** the importance of a body cavity.

- **Identify** how scientists determine evolutionary relationships among animals.

Key Terms

blastula
ectoderm
endoderm
mesoderm
body plan
asymmetrical
radial symmetry
bilateral symmetry
cephalization
coelom
acoelomate
pseudocoelomate
coelomate
phylogenetic tree

General Features of Animals

Humans have long marveled at and depended on animals. When Linnaeus first classified animals in the mid-1700s, he counted 4,236 kinds. Since then, many new animal species have been identified, and the count is now over a million. We share the planet with a fantastic array of animal forms, and our lives intersect in many ways. For example, if it were not for insects, such as the honeybee, many of our crops would not be pollinated. The prices of insect-pollinated fruits and vegetables would increase dramatically. And items such as apples and cucumbers might not be available at your local supermarket.

Think about a snail, a lizard, a hawk, an elephant—all are quite different. But the features these four animals share are just as important as their differences in size, shape, and behavior. Like all animals, they exhibit the following features—heterotrophy, mobility, multicellularity, diploidy, and sexual reproduction. They also exhibit blastula formation, cells organized into tissues, and the absence of a cell wall. These features are a legacy inherited from their common ancestor.

Heterotrophy

Animals are heterotrophs—that is, they cannot make their own food. Some animals that live in the ocean remain in one place and consume tiny particles of food that they filter from passing sea water. But most animals, such as those shown in **Figure 1,** move from place to place searching for food. Once food is located, it is eaten and then digested in a cavity inside the animal's body.

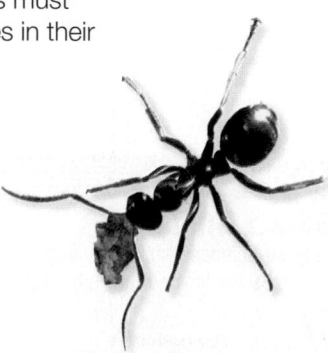

Figure 1 Heterotrophy. Unlike plants, animals must seek out food sources in their environment.

Figure 2   Mobility. A tiger finds food by roaming over its territory, which usually ranges from 10 to 30 mi^2.

Mobility

Animals are unique among living things in being able to perform rapid, complex movements. Animals move by means of muscle cells, specialized cells that are able to contract with considerable force. Animals can swim, crawl, walk, run, and even fly. In fact, flight has evolved four times among animals, in insects, pterosaurs (extinct reptiles from the time of the dinosaurs), birds, and bats. The tiger shown in **Figure 2** moves through its environment searching for food. The size of the tiger's home range depends on the availability of prey species.

Multicellularity

All animals are multicellular. Some are too small to be seen clearly with the naked eye, like *Daphnia*, shown in **Figure 3.** Others, such as some enormous whales, are larger than a city bus. In spite of their differences in body size, there is little difference in the size of most of the cells that make up these animals. The cells on the skin of your hand are roughly the same size as the cells in the heart of a whale or in the wing muscle of a hummingbird.

Figure 3   Life in a drop of water. *Daphnia*, commonly called water fleas, belong to a group of animals known as crustaceans.

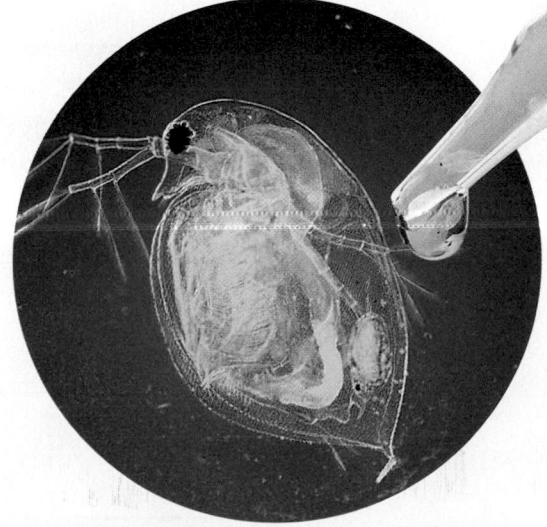

Diploidy

With few exceptions, animals are diploid *(DIP loyd)*, meaning adults have two copies of each chromosome, one inherited from their father and one from their mother. Only their gametes (egg and sperm) are haploid. In contrast, many plants have four or more copies of each chromosome, while fungi often have only one. A great advantage of diploidy is that it permits an animal to exchange genes between the two copies of a set of chromosomes, creating new combinations of genes.

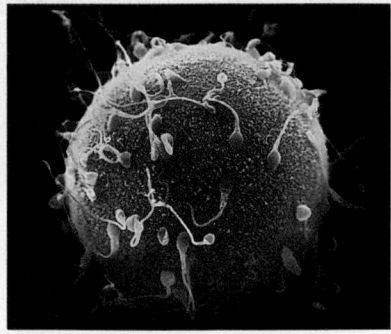

Magnification: 620×

Figure 4 Egg and sperm.
This human egg is surrounded
by sperm, only one of which
will fertilize the egg.

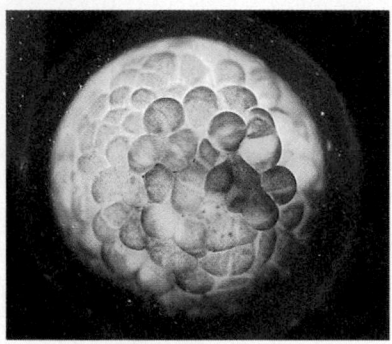

Figure 5 Blastula. The
individual cells of a blastula
form a sphere.

Sexual Reproduction

Almost all animals reproduce sexually by producing gametes, as do many plants, fungi, and protists. As shown in **Figure 4,** the females' egg cells are much larger than the males' sperm cells. Unlike the egg cells, the sperm cells of animals have a flagella and are highly mobile.

Absence of a Cell Wall

Among the cells of multicellular organisms, only animal cells lack rigid cell walls. The absence of a cell wall has allowed animals mobility that other multicellular organisms do not have. You may not realize this, but there are cells moving about in your body all the time. Cells called *macrophages*, for example, act as mobile garbage collectors, crawling over tissues and removing debris. A cell with a rigid cell wall could never perform such a task.

Blastula Formation

In all animals except sponges, the zygote (fertilized egg cell) undergoes cell divisions that form a hollow ball of cells called a **blastula** *(BLAS tyoo luh)*, shown in **Figure 5.** Cells within the blastula eventually develop into three distinct layers of cells—**ectoderm, endoderm,** and **mesoderm.** These layers are called the primary tissue layers because they give rise to all of the tissues and organs of the adult body. (A few simple invertebrates, such as *Hydra* and their kin, develop only two tissue layers, endoderm and ectoderm.) **Table 1** lists the three primary tissue layers and summarizes the body tissues and organs to which they give rise. Note that the table includes some organs, such as the urinary bladder, found only in vertebrates. The organs of vertebrates are complex structures containing cells that arise from more than one primary tissue layer. For example, the digestive system is formed primarily from endoderm and mesoderm.

Table 1 Origin of Animal Tissues and Organs	
Primary tissue layer	**Gives rise to**
Ectoderm	Outer layer of skin; nervous system; sense organs, such as eyes
Endoderm	Lining of digestive tract; respiratory system; urinary bladder; digestive organs; liver; many glands
Mesoderm	Most of the skeleton; muscles; circulatory system; reproductive organs; excretory organs

Tissues

The cells of all animals except sponges are organized into structural and functional units called tissues. Recall that tissues are groups of cells with a common structure that work together to perform a specific function. For example, the cells of adipose *(AD uh pohs)* tissue, shown in **Figure 6,** are specialized for storing fat. The cells of muscle tissue are specialized to contract, producing movement. The cells of nerve tissue are specialized to conduct signals.

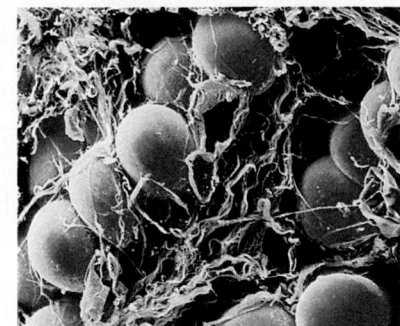

Figure 6 Adipose tissue. These adipose cells are specialized for storing fat, and each one contains a fat droplet. Together the cells make up adipose tissue.

Magnification: 230×

Exploring Further

From Zygote to Gastrula

Just as you did, this sea urchin began life as a zygote but quickly developed into an embryo made up of layers of tissue—endoderm, ectoderm, and mesoderm. A variation of this pattern of development is found in all animals except sponges and is evidence that animal life arose from a common ancestor. The illustrations at right show the sea urchin's pattern of development in its first hours of life.

Cell numbers increase
During a process called *cleavage,* the cells of the zygote divide, doubling the number of cells with each division. However, the mass of the developing embryo does not increase, and the cells formed are progressively smaller. After about 3 hours, the zygote has become a solid ball of cells. Cell division continues until a blastula is formed.

Cells change locations
During a second process, called *gastrulation,* the blastula begins to collapse inward. At the same time, its cells move about, changing their location within the blastula. The cells begin to vary in size and form three primary tissue (germ) layers. Gastrulation is complete, and the zygote is now an embryo.

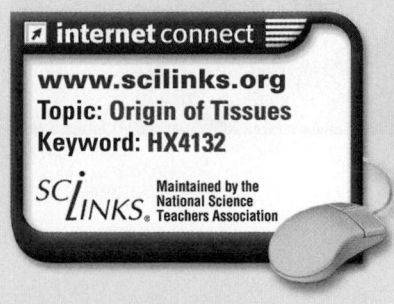

internet connect

www.scilinks.org
Topic: Origin of Tissues
Keyword: HX4132

SCLINKS. Maintained by the National Science Teachers Association

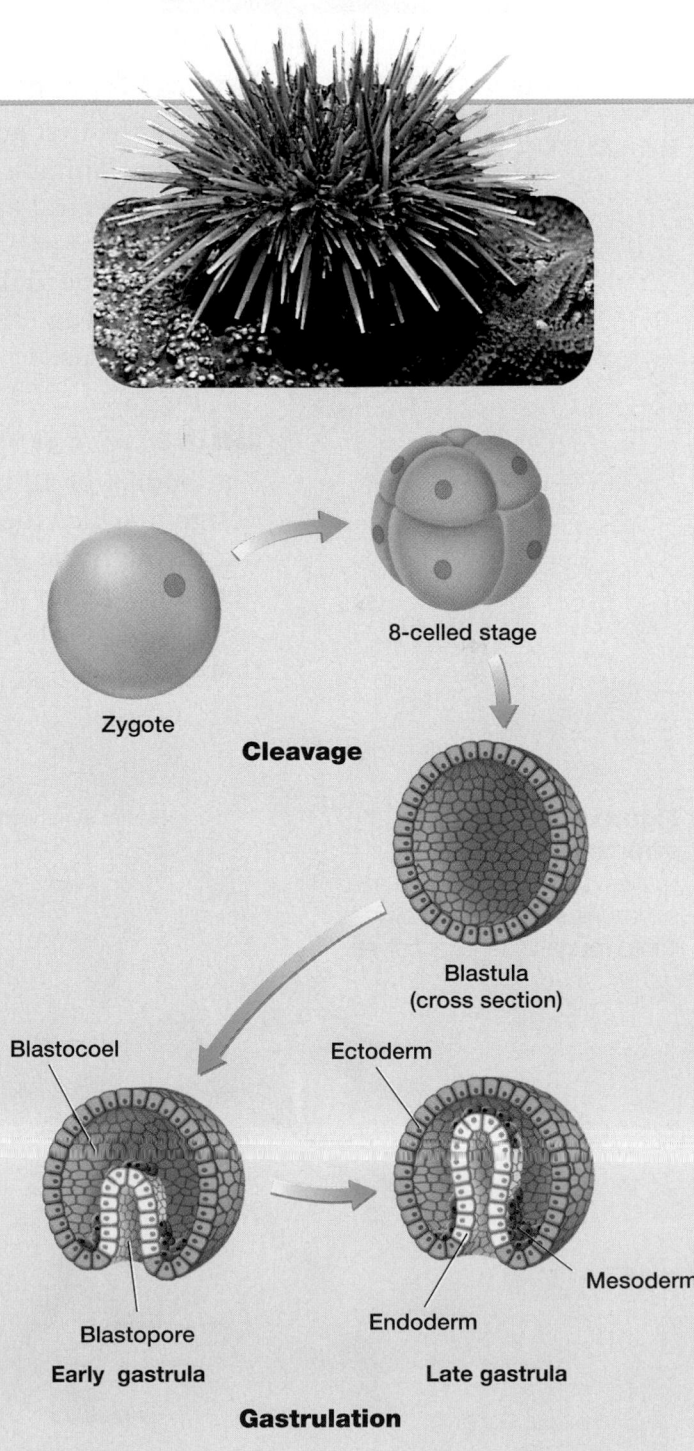

Zygote

8-celled stage

Cleavage

Blastula (cross section)

Blastocoel

Ectoderm

Blastopore

Endoderm

Mesoderm

Early gastrula

Late gastrula

Gastrulation

Body Symmetry

All animals have their own particular **body plan,** a term used to describe an animal's shape, symmetry, and internal organization. An animal's body plan results from a pattern of development programmed into the animal's genes by natural selection. Sponges, such as the one shown in **Figure 7,** have the simplest body plan of all animals. Sponges are **asymmetrical,** or irregular in shape, and sometimes their shape depends on where they are growing. The body plans of virtually all other animals show a definite body shape and symmetry.

Figure 7 Asymmetry. Animals that grow in an irregular pattern, such as this sponge, show asymmetry.

Radial Symmetry

Some of the first animals to evolve in the ancient oceans had radial symmetry. Animals with **radial symmetry** have body parts arranged around a central axis, somewhat like the spokes around a bicycle wheel. A plane passing through the central axis divides the organism into roughly equal halves, as shown in **Figure 8.** Today's radially symmetrical animals are aquatic. Most move slowly or drift in ocean currents.

Bilateral Symmetry

The bodies of all other animals show **bilateral symmetry,** a body design in which there are distinct right and left halves. A plane passing through the animal's midline divides the animal into mirror image halves, as shown in **Figure 9.** There is a dorsal (top) and a ventral (bottom) surface plus an anterior (front) end and a posterior (back) end.

Figure 8 Radial symmetry. Each of the planes passing through a sea anemone's central axis divides it into roughly identical halves.

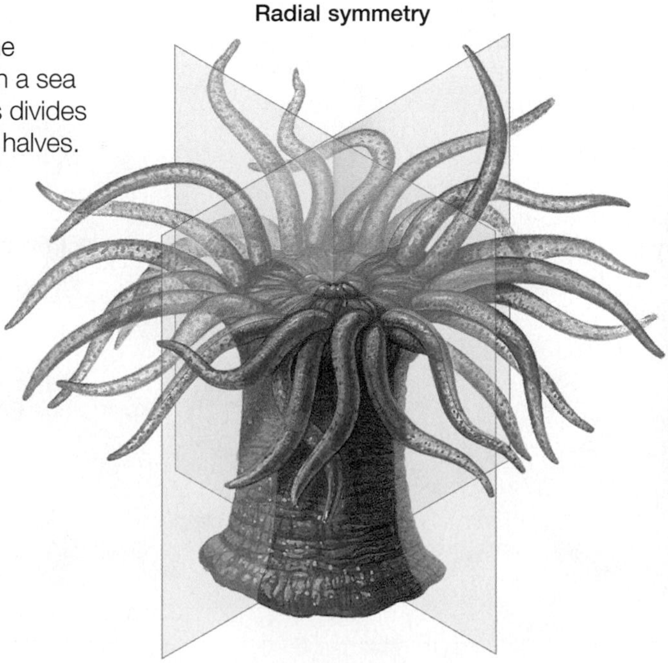

Radial symmetry

Bilateral symmetry was a major evolutionary change in animals because it enabled different parts of the body to become specialized in different ways. For example, most bilaterally symmetrical animals have evolved an anterior concentration of sensory structures and nerves, a process called **cephalization** *(SEF uhl lih ZAY shuhn)*. Animals with cephalic ends, or heads, are often active and mobile. With sensory organs concentrated in the front, such animals can more easily sense food and danger.

Dorsal

Anterior

Posterior

Ventral

Figure 9   Bilateral symmetry. The body parts of this gray squirrel are arranged so that it has a right and a left half that are mirror images of each other.

Recognizing Symmetry

QUICK LAB

You can use the letters of the alphabet to better understand the nature of symmetry.

Materials

envelope containing letters of the alphabet

Procedure

1. Spread the letters on the table in front of you so you can see all of them.

2. Sort the letters into groups based on their symmetry, using the terms *asymmetry*, *radial symmetry*, and *bilateral symmetry*. For example, the letters *A* and *T* show bilateral symmetry. The letter *J* is asymmetrical.

Analysis

1. **Propose** a definition for each kind of symmetry you found in the letters.

2. **List** any letters you found difficult to classify, explaining why it was difficult to classify these letters.

3. **Identify** the letters that show the same kind of symmetry as sponges.

4. **Identify** two or three animals that you might be familiar with that have the same kind of symmetry as the letter *M*.

5. **Critical Thinking Evaluating Methods** What are some strengths and weaknesses of using symmetry as a way of classifying or describing organisms?

Internal Body Cavity

Bilaterally symmetrical animals have one of three basic kinds of internal body plans, each illustrated in **Figure 10.** The body plan may include a body cavity, or **coelom** *(SEE luhm),* a fluid-filled space found between the body wall and the digestive tract (gut). This space is lined with cells that come from mesoderm. Animals with no body cavity are called **acoelomates** *(ay SEEL oh mayts).* The space between an acoelomate's body wall and gut is completely filled with tissues. Other animals, called **pseudocoelomates** *(SOO doh seel oh mayts),* have a body cavity located between the mesoderm and endoderm. Their body cavity is called a pseudocoelom (false coelom).

Coelomates have a true coelom, a body cavity located entirely within the mesoderm. Because the mesodermal layer lines the body wall and wraps around the gut, the gut and other internal organs of coelomates are suspended within the coelom.

A true coelom provides an internal space where mesoderm and endoderm can be in contact with each other during embryonic development. This aided the evolution of complex organs made of more than one type of tissue. Because these internal organs are suspended in a fluid-filled coelom, they are protected from the movement of surrounding muscles. Thus, an animal can move about without damaging the organs or interfering with their function.

Figure 10   Three body plans

Bilaterally symmetrical animals have one of three basic body plans.

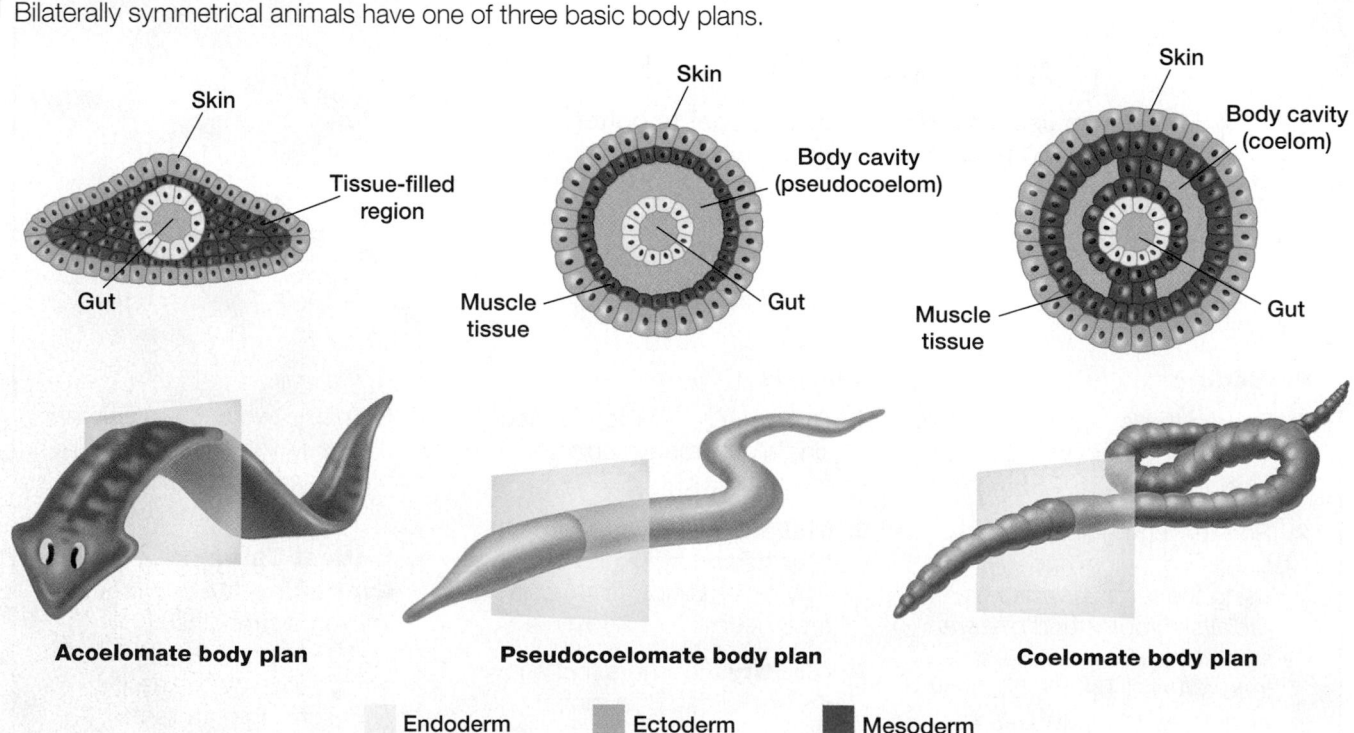

Acoelomate body plan | Pseudocoelomate body plan | Coelomate body plan

Endoderm   Ectoderm   Mesoderm

Body Segmentation

Segmented animals are composed of a series of repeating, similar units called *segments*. Segmentation underlies the organization of all "advanced" animals and is easy to observe in some animals, such as earthworms. Crustaceans (lobsters and their kin), spiders, and insects also show some degree of body segmentation, but it may be difficult to observe during their adult stage. For example, segments are not as apparent in the adult moth in **Figure 11,** but they are clearly visible in its immature caterpillar stage.

In vertebrates, segments are not visible externally, but there is evidence of segmentation in a vertebrate embryo. Vertebrate muscles develop from repeated blocks of tissue called *somites (SOH mietz),* and the vertebral column (backbone) consists of a stack of very similar vertebrae.

In segmented worms, each segment can move independently, permitting great flexibility and mobility. Therefore, a segmented worm's long body can move in ways that are often quite complex. For example, consider an earthworm as it crawls along a flat surface. It lengthens some parts of its body while shortening others. Many scientists think segmentation evolved as an adaptation that permitted more-efficient burrowing.

In highly segmented animals, such as earthworms, each segment repeats many of the organs in the adjacent segment. As a result, an injured animal can still perform vital life functions. Segments are not totally independent of each other, however. Materials pass from one segment to another through a circulatory system that connects them. Nerves also connect each segment to a brain that coordinates the body's movements. Segmentation also offers evolutionary flexibility. A small change in an existing segment can produce a new type of segment with a different function. For example, some segments are modified for feeding, while others are modified for reproduction.

Figure 11   Segmentation. This caterpillar shows clearly defined segments. In its adult form—a cecropia moth—the segments are more difficult to observe.

Kinds of Animals

Kingdom Animalia contains about 35 major divisions called *phyla* (singular, *phylum*), depending on how certain organisms are classified. The animals in the various phyla show an extraordinary range of body forms, internal body systems, and behaviors.

To visually represent the relationships among various groups of animals, scientists often use a type of branching diagram called a phylogenetic tree. A **phylogenetic tree,** such as the one in **Figure 12,** shows how animals are related through evolution. Clues to animal relationships can be found in the fossil record and by comparing the anatomy and physiology of living animals. Clues are also found by comparing patterns of development in animal embryos. The most direct evidence of evolutionary relationships, however, comes from comparing the DNA in the genes of various animal species. For example, scientists long debated whether giant pandas are related more closely to bears or to raccoons. Comparing their DNA showed that pandas are more closely related to bears than to raccoons.

Figure 12 Evolutionary milestones. This phylogenetic tree shows one hypothesis of the evolutionary relationships among nine of the major animal phyla. The circled numbers indicate important milestones in the evolution of the animal body, as listed in the table below.

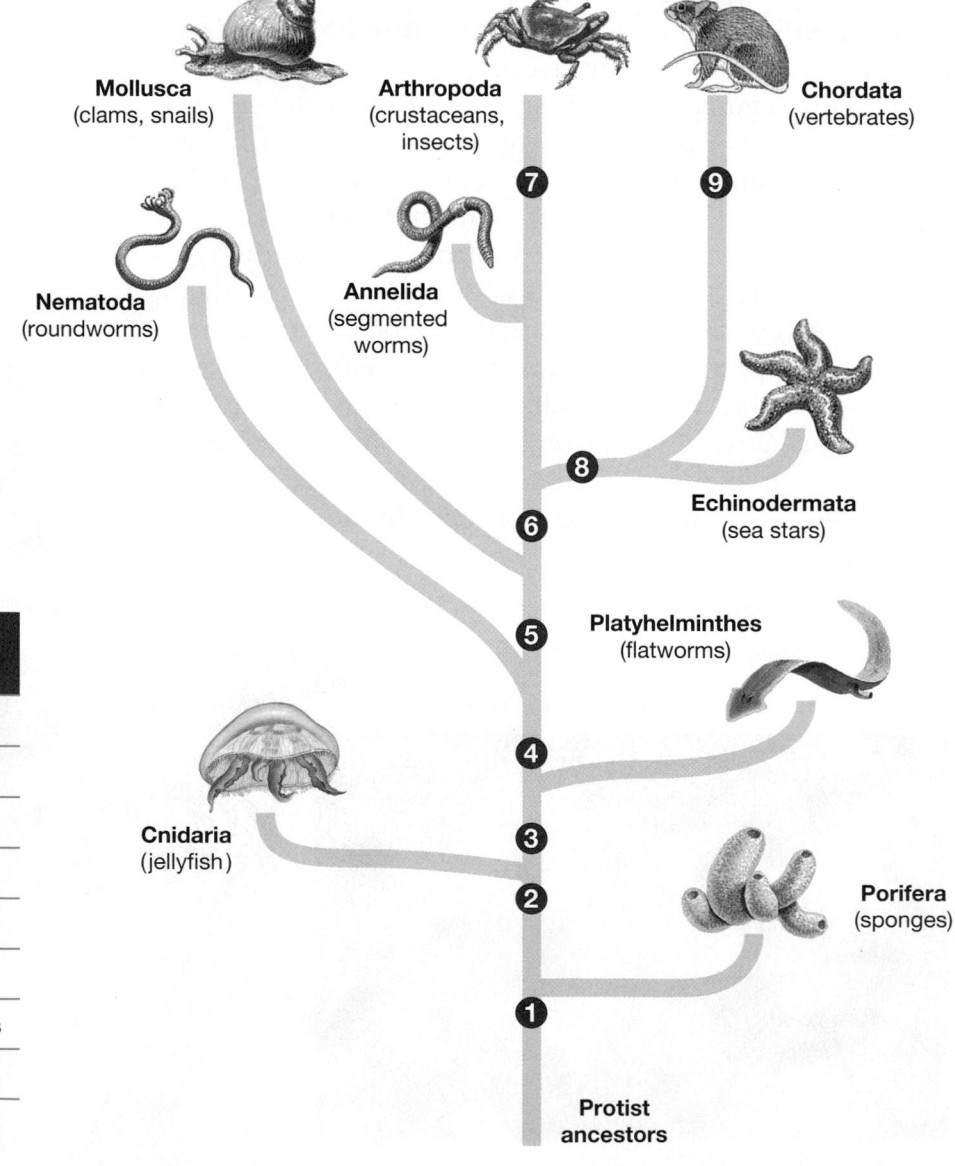

Evolutionary milestones	
1	Multicellularity
2	Tissues
3	Bilateral symmetry
4	Body cavity
5	Coelom
6	Segmentation
7	Jointed appendages
8	Deuterostomes
9	Notochord

The animal kingdom is often divided into two groups: invertebrates (animals without a backbone) and vertebrates (animals with a backbone). As you study these two groups, you will discover how a series of key evolutionary innovations in the body plan of animals has led to today's animals. The first animals to evolve did not have body cavities. After body cavities appeared, segmentation evolved. Later, jointed appendages (such as legs) evolved. Backbones evolved even later. Scientists consider these innovations (shown in Figure 12 as numbered milestones) when they attempt to determine evolutionary relationships among different groups of animals. Watch for these evolutionary stages and milestones as you continue studying invertebrates in later chapters in this book.

Exploring the Animal Kingdom

Background

You can find out more about the animal phyla by referring to the section "A Six-Kingdom System" in the Appendix. Turn to this section and locate the information for kingdom Animalia. Follow the procedure below to evaluate this information.

Procedure

1. Read the introductory paragraph for kingdom Animalia. Then quickly skim over the information presented. Do not read it word for word, but observe how the information is divided into sections.

2. Choose one phylum, and read all of the information about it.

3. As you read, notice how color is used and what types of information are given, for example, number of species found and habitat.

Analysis

1. List at least three types of information you found for the phylum you read about.

2. Analyze how color is used to distinguish between the different entries on a page.

3. Propose a way that you might use this information when studying about a particular animal phylum.

Section 1 Review

1 Describe each of the eight features animals have in common.

2 Summarize the difference between radial symmetry and bilateral symmetry.

3 Compare the body plans of acoelomates, pseudocoelomates, and coelomates.

4 Critical Thinking   Relating Concepts
In 1994, Western scientists first observed the Vietnamese saola, a hooted mammal. The saola was proved to be related to wild cattle and buffalo. How do you think scientists identified the saola's closest relatives?

5 Standardized Test Prep Which characteristic is found in all animals?

A multicellularity   **C** body cavity

B tissues   **D** body segmentation

Animal Body Systems

Tissues and Organs

As you go about your day, you make decisions that involve thinking—what to wear to school, how to solve a problem, where to sit during lunch. But a lot that happens during your day does not require your thought. You digest your food, you balance yourself as you walk, and your heart beats. These and thousands of other functions are carried out by your body. But how does your body, or the body of any animal, carry out these tasks?

Simple animals like sponges carry out the many tasks of living with little specialization in the cells of their body. More-complex animals have evolved tissues and organs that are specialized to perform specific functions. Six important functions of these tissues and organs are digestion, respiration, circulation, conduction of nerve impulses, support, and excretion.

Digestion

Single-celled organisms and sponges digest their food within their body cells. This means that their food source cannot be larger than their individual cells. All other animals digest their food extracellularly (outside of their body cells) within a digestive cavity, as shown in **Figure 13.** Digestive enzymes that are released into the cavity begin the breakdown of food, permitting the animal to prey on organisms larger than its body cells. Simple animals, such as the

Figure 13   Extracellular digestion

A hydra has a gastrovascular cavity, while a roundworm has a digestive tract in which food travels in one direction only.

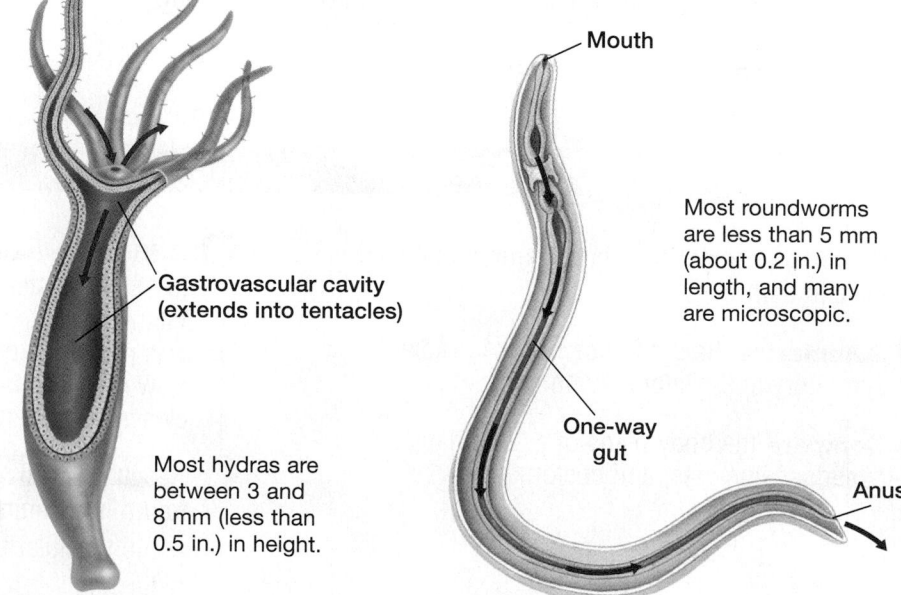

Gastrovascular cavity (extends into tentacles)

Most hydras are between 3 and 8 mm (less than 0.5 in.) in height.

Mouth

Most roundworms are less than 5 mm (about 0.2 in.) in length, and many are microscopic.

One-way gut

Anus

hydra and flatworms, have a **gastrovascular cavity,** a digestive cavity with only one opening. There can be no specialization within a gastrovascular cavity because every cell is exposed to all stages of food digestion.

Other animals have a digestive tract (gut) with two openings, a mouth and an anus. The anus is the opening through which undigested food leaves the body. In a digestive tract, food moves in one direction, from mouth to anus. Unlike a gastrovascular cavity, a one-way digestive system allows for specialization. For example, there may be a section of gut for food storage, a section for breaking down food into small pieces, and a section for the chemical digestion of food. Eventually the food is broken down into molecules small enough to pass through the lining of the gut and into the bloodstream.

Respiration

In simple animals, such as jellyfish, oxygen gas and carbon dioxide gas are exchanged directly with the environment by diffusion. The uptake of oxygen and the release of carbon dioxide, called **respiration,** can take place only across a moist surface, such as the damp skin of an earthworm. In larger, more complex animals, simple diffusion cannot provide for adequate gas exchange. Most large animals have specialized respiratory structures.

Some aquatic (and a few terrestrial) animals respire with **gills,** very thin projections of tissue that are rich in blood vessels. Gills provide a large surface for gas exchange. **Figure 14** shows the gills of a mud puppy. Gills are not suitable for most terrestrial animals because gills do not function unless they are kept moist. A variety of respiratory organs, such as lungs, have evolved in many terrestrial animals that allow them to respire on dry land.

Figure 14 Gills. The feathery gills of this mud puppy are supported by its watery environment.

Figure 15 Circulatory systems

In an open circulatory system, fluid leaves the circulatory vessels, but in a closed circulatory system, blood remains in the blood vessels.

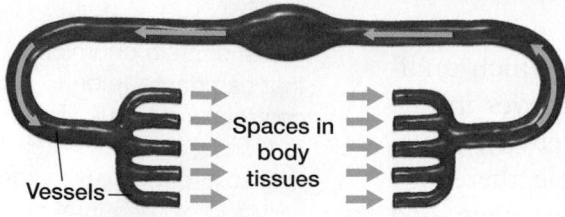

Open circulatory system

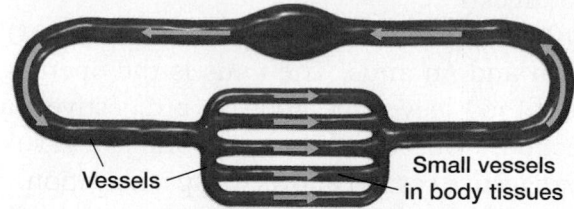

Closed circulatory system

Circulation

In simple animals, body cells are exposed to either the external environment or the gastrovascular cavity. None of the body cells are far away from sources of oxygen or nourishment. More-complex animals, however, have tissues that are several cell layers thick. Many of the cells in this tissue are not close enough to the surface of the cell layer to exchange materials directly with the environment. Oxygen and nutrients must be transported to these body cells by a circulatory system.

Two types of circulatory systems are shown in **Figure 15.** In an **open circulatory system,** a heart pumps fluid containing oxygen and nutrients through a series of vessels out into the body cavity. There the fluid washes across the body's tissues, supplying them with oxygen and nutrients. The fluid collects in open spaces in the animal's body and flows back to the heart. In a **closed circulatory system,** a heart pumps blood through a system of blood vessels. These blood vessels form a network that permits blood flow from the heart to all of the body's cells and back again. The blood remains in the vessels and does not come in direct contact with the body's tissues. Instead, materials pass into and out of the blood by diffusing through the walls of the blood vessels.

Conduction of Nerve Impulses

Nerve cells (neurons) are specialized for carrying messages in the form of electrical impulses (conduction). These cells coordinate the activities in an animal's body, enabling the animal to sense and respond to its environment. Members of all of the major animal phyla except sponges have nerve cells. **Figure 16** shows the arrangement of nerve cells in three animals—a hydra, a flatworm, and a grasshopper. The simplest arrangements of nerves are found in animals like hydras and jellyfishes. All of their nerve cells are similar and are linked to one another in a web called a nerve net. There is little coordination among the nerve cells in a nerve net.

Bilaterally symmetric animals have clusters of neurons called *ganglia*. The ganglia at the anterior end of the animal body became

Reviewing Information

Work with a partner to review the body functions in this section. Review each function by asking each other questions that include the boldface terms. For example, ask, "What is the difference between a gastrovascular cavity and a one-way gut?"

Figure 16 Nervous systems

The hydra has a simple nerve net, while the flatworm and the grasshopper have more-complex nervous systems.

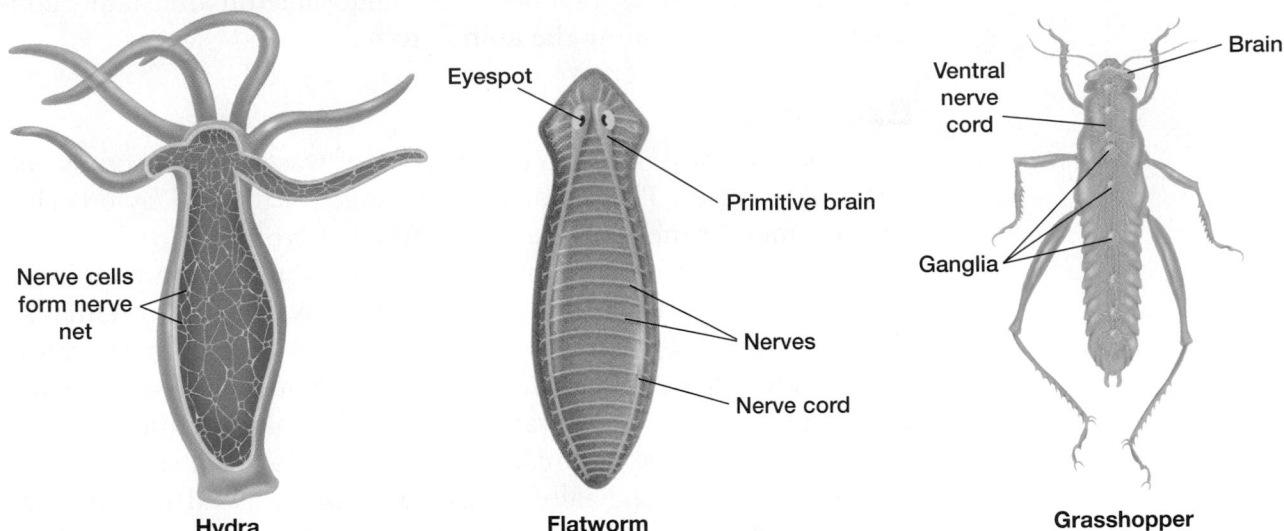

Hydra — Nerve cells form nerve net

Flatworm — Eyespot, Primitive brain, Nerves, Nerve cord

Grasshopper — Ventral nerve cord, Brain, Ganglia

larger and more complex, forming a brain-like structure, as seen in the flatworm. More-complex invertebrates, such as the grasshopper, have brains with sensory structures, such as eyes, associated with them. These cephalized animals—that is, animals with heads—could interact with their environments in more-complex ways than could other animals.

Support

An animal's skeleton provides a framework that supports its body. It is also vital to animal movement. All animals move using the same force: the contraction (shortening) of muscle tissue against a framework such as the skeleton provides.

Hydrostatic skeleton Many soft-bodied invertebrates have a hydrostatic skeleton. A **hydrostatic skeleton** consists of water that is contained under pressure in a closed cavity, such as a gastrovascular cavity or a coelom. Imagine a balloon filled with water. The water presses against the balloon, supporting it. If pressure is applied to the balloon in any place, the water must shift, altering the shape of the balloon. The hydrostatic skeleton of a hydra, shown in **Figure 17,** is formed by its gastrovascular cavity. Other soft-bodied invertebrates, such as earthworms, have a fluid-filled coelom that serves as a hydrostatic skeleton. In both cases, muscle forces exerted against the hydrostatic skeleton aid movement.

Exoskeleton Other invertebrates, such as insects, clams, and crabs, have a type of skeleton known as an exoskeleton. An **exoskeleton** is a rigid external skeleton that encases the body of an animal. An exoskeleton supports movement in a different manner than a hydrostatic skeleton does. The muscles of animals with exoskeletons are attached to the inside of the skeleton, which provides a surface for them to pull against. Exoskeletons also protect an organism's soft internal parts.

Magnification: 70×

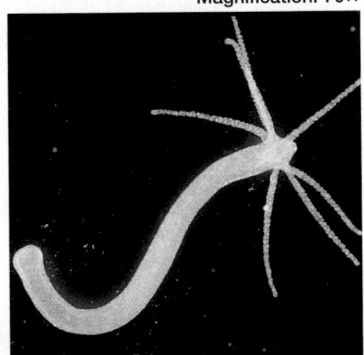

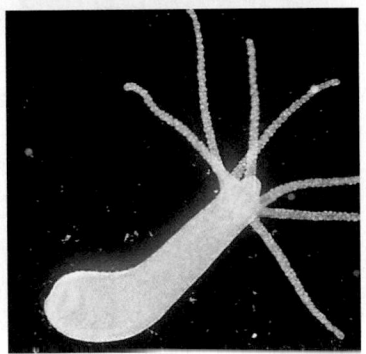

Figure 17 Hydrostatic skeleton. When the hydra closes the opening to its gastrovascular cavity and contracts muscles in its body wall, its body elongates. When water is released from the cavity and other muscles contract, the hydra's body shortens.

Endoskeleton An **endoskeleton** is composed of a hard material, such as bone, embedded within an animal. The most familiar type of endoskeleton is that of humans and other vertebrates. As with exoskeletons, muscles attached to the endoskeleton alternately contract and relax, enabling the animal to move.

Excretion

The term *excretion* refers to the removal of wastes produced by cellular metabolism. These waste products leave the cell by crossing the cell membrane and are then removed from the body. Some waste products are highly toxic and will poison an organism if not removed. The most important of these toxic wastes is ammonia. As ammonia forms, it dissolves in body fluids, becoming more dilute and thus less toxic. Simple aquatic invertebrates and some fishes excrete ammonia into the water through their skin or gills by diffusion. This is effective, but results in the loss of a lot of water.

Other animals, especially terrestrial animals, need to minimize water loss. A variety of ways have evolved in these animals by which wastes are removed from the body. One way is by converting ammonia to nontoxic chemicals, like urea. As the excretory system eliminates these wastes, water and other useful substances are returned to the body. Thus, eliminating metabolic wastes is linked to maintaining the body's water balance. For example, a mammal's kidneys filter fluid from the blood. The kidneys then concentrate the metabolic wastes filtered from the fluid and excrete them as concentrated urine. Simultaneously, the kidneys regulate the water content of the body by making the urine more or less dilute as necessary.

MATH LAB

Calculating Filtration Rate in the Human Kidney

Background

The human kidney filters fluid from the blood at the rate of approximately 125 mL per minute. However, only a small percentage of this fluid is excreted as urine—adult humans normally excrete between 1.5 and 2.3 L of urine a day.

Analysis

1. **Calculate** how many milliliters of fluid the human kidneys filter each hour.

2. **Calculate** how many milliliters of fluid the kidneys filter each day.

3. **Critical Thinking Analyzing Data** Convert your answer in item 2 from milliliters to liters. For help, see "Math and Problem-Solving Skills: SI Measurement" in the Appendix. To better visualize the quantity of fluid represented by your answer, think about the volume of fluid contained in a 1 L bottle of soda.

4. **Critical Thinking Predicting Outcomes** What would happen if the kidneys could not return water to the body?

Reproductive Strategies

While the reproductive system of an individual animal is not essential to its survival, reproduction is necessary if the species is to survive. There are two types of reproduction in animals, asexual and sexual.

Asexual Reproduction

Reproduction that does not involve the fusion of two gametes is called asexual reproduction. A sponge, for example, can reproduce by fragmenting its body. Each fragment grows into a new sponge. Some species of sea anemone reproduce by pulling themselves in half, forming two new adult anemones, as shown in **Figure 18.**

An unusual method of asexual reproduction is parthenogenesis *(pahr thuh noh JEN uh sis)*, in which a new individual develops from an unfertilized egg. Parthenogenesis is common among insects. In honeybees, for example, a queen bee mates only once and stores the sperm. Although she has mated, the queen bee has the ability to lay unfertilized eggs that develop by parthenogenesis into male bees, called drones. Female bees develop by sexual reproduction, when the queen releases stored sperm to fertilize her eggs. A few species of fishes, amphibians, and lizards reproduce by parthenogenesis. Animals that reproduce asexually are usually able to also reproduce sexually.

Sexual Reproduction

In sexual reproduction, a new individual is formed by the union of a male and a female gamete. Gametes are produced in the sex organs. The testes *(TEHS teez)* produce the male gametes (sperm), and the ovaries produce the female gametes (eggs). Some species of animals, called **hermaphrodites** *(huhr MAF roh dietz)*, have both testes and ovaries. Each individual functions as both a male and a female. But a hermaphrodite's sperm and eggs are usually produced at different times, so self-fertilization does not occur.

Figure 18   Asexual reproduction. This pink-tipped surf anemone is in the process of pulling into two halves, each of which will be an adult sea anemone.

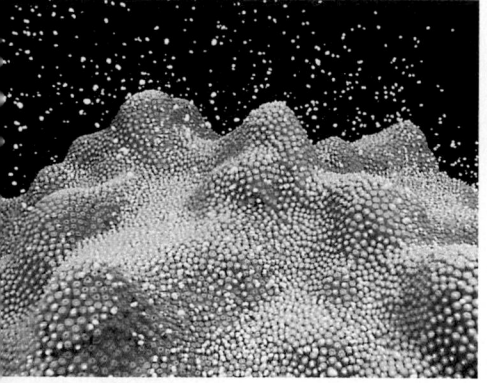

Figure 19 External fertilization. Coral on the Great Barrier Reef, off the coast of Australia, release billions of gametes in a mass spawning that occurs once a year.

Figure 20 Sea turtle. This green sea turtle digs a nest in the sand, where she deposits her eggs.

Among animals that rarely meet members of their own species, hermaphrodites can have a reproductive advantage. A hermaphrodite's chances of reproductive success are increased because it can either fertilize eggs or have its own eggs fertilized. Many simple invertebrates, including slugs and earthworms, and some fishes are hermaphrodites.

When sperm are released during sexual reproduction, their moving flagella propel them toward the egg. Most aquatic animals simply release the male and female gametes near one another in the water, where fertilization occurs. This method is called **external fertilization** because the egg is fertilized outside of the female's body. Often large numbers of gametes are released during external fertilization, but only a small percentage of the resulting fertilized eggs will survive to develop into adults. **Figure 19** shows the mass release of gametes by coral polyps on a coral reef.

External fertilization is not practical for animals that live on land because gametes dry out quickly when exposed to air. Most terrestrial animals reproduce sexually by means of internal fertilization. In **internal fertilization,** the union of the sperm and egg occurs within the female's body. The male places semen (*SEE muhn*), a fluid containing sperm and fluid secretions, directly into the female's body. In this way, fertilization takes place in a moist environment, and the gametes are protected from drying out.

Once fertilization occurs, developing eggs must be kept moist. This is not a problem for aquatic animals whose eggs are covered by a jellylike coat. The eggs of terrestrial animals need more protection. Their eggs have a shell that protects them from drying out and provides a degree of protection from physical damage. Some animals, such as the sea turtle shown in **Figure 20,** place their fertilized eggs in a safe place and leave them. Other animals remain with their eggs to protect them. In most mammals and a few other species, the eggs develop internally, and living young emerge from their mother's body.

Section 2 Review

1 **Summarize** the functions of the six body systems discussed.

2 **Describe** how a gastrovascular cavity differs from a one-way gut.

3 **Compare** open and closed circulatory systems.

4 **Describe** how asexual reproduction differs from sexual reproduction.

5 **Critical Thinking Justifying Conclusions** Which method of fertilization, external or internal, is more practical for most terrestrial animals? Justify your answer.

6 **(Standardized Test Prep)** Which two body systems in most animals are involved in taking up oxygen from the environment and transporting oxygen to body cells?

A digestive and respiratory

B respiratory and circulatory

C circulatory and nervous

D nervous and excretory

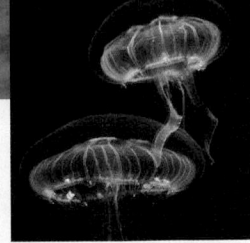

Study ZONE

Key Concepts

1 Characteristics of Animals

- All animals share these general features: heterotrophy, mobility, multicellularity, sexual reproduction, diploidy, the absence of a cell wall, cells organized as tissues, and blastula formation.

- Animals with radial symmetry have body parts arranged around a central axis. Animals with bilateral symmetry have a distinct right and left half, and most display cephalization.

- Animals have one of three basic body plans: acoelomate, pseudocoelomate, and coelomate.

- Segmentation in body structure underlies the organization of all advanced animals.

- There are about 35 animal phyla, which contain an extraordinary range of body forms and body systems.

- Scientists classify animals using several different types of data, which include comparing anatomy and physiology, patterns of development, and DNA.

- The animal kingdom is divided in two groups: vertebrates and invertebrates.

2 Animal Body Systems

- Body systems are specialized to carry out different tasks.

- Simple animals have a gastrovascular cavity with only one opening, while more-complex animals have a one-way gut.

- Simple animals exchange gases directly through their skin. More-complex aquatic animals use gills, while terrestrial animals use a variety of respiratory organs, such as lungs.

- In an open circulatory system, circulatory fluid leaves the vessels and enters the body cavity. In a closed circulatory system, blood remains in the vessels.

- While simple animals have little coordination among their nerve cells, complex animals have nerve cords and a brain with associated sensory structures.

- For most animals, eliminating wastes is linked to maintaining the correct water balance in their body.

- Asexual reproductive methods include fragmentation, splitting in two, and parthenogenesis. In sexual reproduction, male and female gametes combine to form a new individual.

Key Terms

Section 1

blastula (596)
ectoderm (596)
endoderm (596)
mesoderm (596)
body plan (598)
asymmetrical (598)
radial symmetry (598)
bilateral symmetry (598)
cephalization (599)
coelom (600)
acoelomate (600)
pseudocoelomate (600)
coelomate (600)
phylogenetic tree (602)

Section 2

gastrovascular cavity (605)
respiration (605)
gill (605)
open circulatory system (606)
closed circulatory system (606)
hydrostatic skeleton (607)
exoskeleton (607)
endoskeleton (608)
hermaphrodite (609)
external fertilization (610)
internal fertilization (610)

Understanding Key Ideas

1. Which of the following organisms do *not* have cells organized into tissues?
a. hydras
b. sea urchins
c. sponges
d. sea anemones

2. An animal's body plan includes all of the following except
a. internal organization.
b. shape.
c. size.
d. symmetry.

3. The presence of a true body cavity (coelom) allows
a. direct exchange of oxygen and carbon dioxide with the environment.
b. specialization of the gut.
c. cephalization.
d. bilateral symmetry.

4. Which of the following does *not* have nerve cells?
a. hydra
b. jellyfish
c. sponge
d. flatworm

5. Skeletal systems provide all of the following except
a. protection for an animal's soft parts.
b. absorption of nutrients from food.
c. a framework for supporting the body.
d. a framework for muscles to pull against.

6. Which of the following is *not* true of kidneys?
a. They remove oxygen from the body.
b. They remove wastes from the bloodstream.
c. They produce concentrated urine.
d. They help balance the body's water content.

7. How do kidneys function to reduce water loss in terrestrial animals?

8. **Exploring Further** Summarize the process of gastrulation, relating it to the formation of tissue layers and the development of tissues and organs.

9. Explain how the terms *diploid* and *haploid* apply to what you have learned in this chapter. (**Hint:** See Chapter 6, Section 1.)

10. **Concept Mapping** Construct a concept map that describes the characteristics of animals. Try to include the following terms in your concept map: *heterotrophic, multicellular, cell specialization, body plan, radial symmetry, bilateral symmetry, body system, open circulatory system, closed circulatory system,* and *sexual reproduction.*

Critical Thinking

11. **Forming Reasoned Opinions** Radially symmetric animals, such as hydras, are not found on land. However, bilaterally symmetrical animals live on land and in water. Propose a hypothesis to explain why radially symmetrical animals are best suited to aquatic life.

12. **Evaluating Hypotheses** Defend the position that cephalization gives terrestrial animals an advantage as they seek food.

13. **Justifying Conclusions** Support the argument that segmentation makes an animal more evolutionarily flexible.

Alternative Assessment

14. **Career Connection** Zookeeper Zookeepers are responsible for the day-to-day care of animals in their charge. Research a zookeeper's responsibilities, and write a report that includes a job description, training required, kinds of employers, growth prospects, and starting salary.

15. **Finding and Communicating Information** To learn what role public aquariums play in educating the public about aquatic life, develop a presentation designed to attract additional donations.

Understanding Concepts

Directions (1–4): **For *each* question, write on a separate sheet of paper the letter of the correct answer.**

1 Which of the following is a defining characteristic of heterotrophic organisms?
   A. They are capable of flight.
   B. They obtain food from their environment.
   C. They remain in one place throughout their lifetime.
   D. They make their own food from inorganic materials.

2 Which of the following cells in animals has two copies of each chromosome?
   F. a diploid cell
   G. a gamete cell
   H. a germ cell
   I. a haploid cell

3 What developmental process leads to the formation of tissue layers?
   A. asexual reproduction
   B. evolution
   C. fertilization
   D. gastrulation

4 Where does gas exchange take place in many aquatic phyla?
   F. gills
   G. gut
   H. kidneys
   I. lungs

Directions (5–6): **For *each* question, write a short response.**

5 Analyze how the excretory system of an animal helps maintain homeostasis.

6 Differentiate between the meaning of the terms *gastrovascular cavity* and *digestive tract (gut).*

Test TIP

When using a diagram to answer questions, carefully study each part of the figure as well as any lines or labels used to indicate parts of the figure.

Reading Skills

Directions (7): **Read the passage below. Then answer the question.**

A student used modeling clay to construct a cross-section model that showed the formation of primary tissue layers in one kind of animal. She used yellow clay to represent endoderm, blue clay to represent ectoderm, and red clay to represent mesoderm. The model consisted of a round outer ring of blue clay, a round middle ring of red clay, and an inner ring of yellow clay. She left a gap of space between the red and yellow rings and a round space inside the yellow ring.

7 What type of basic body plan did her model represent?
   A. acoelomate body plan
   B. asymmetrical body plan
   C. coelomate body plan
   D. pseudocoelomate body plan

Interpreting Graphics

Directions (8): **Base your answer to question 8 on the diagram below.**

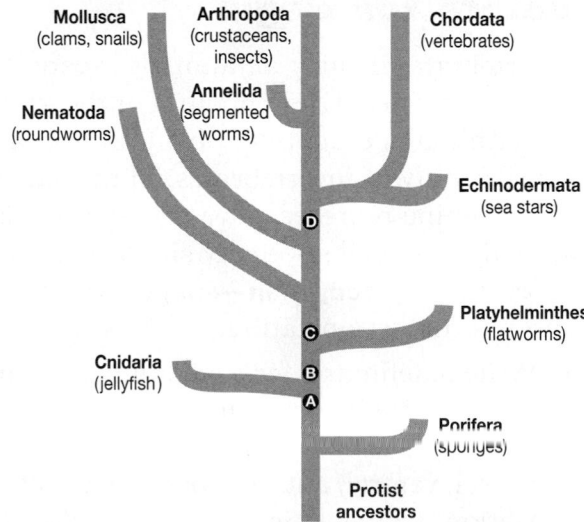

Phylogenetic Tree of Animal Phyla

8 Letters *A–D* represent evolutionary milestones. Which one is the presence of tissues?
   F. *A*   **H.** *C*
   G. *B*   **I.** *D*

Skills Practice Lab

Surveying Invertebrate Diversity

SKILLS
- Observing
- Comparing

OBJECTIVES
- **Observe** the similarities and differences among groups of invertebrates.
- **Relate** the structural adaptations of invertebrates to their evolution.

MATERIALS
- safety goggles
- lab apron
- preserved or living specimens of invertebrates
- prepared slides of sponges, hydras, planarians, and nematodes
- compound microscopes
- hand lenses or stereomicroscopes
- probes

ChemSafety

CAUTION: Always wear safety goggles and a lab apron to protect your eyes and clothing.

CAUTION: Do not touch or taste any chemicals. Know the location of the emergency shower and eyewash station and how to use them. If you get a chemical on your skin or clothing, wash it off at the sink while calling to the teacher. Notify the teacher of a spill. Spills should be cleaned up promptly, according to your teacher's directions.

CAUTION: Glassware is fragile. Notify the teacher of broken glass or cuts. Do not clean up broken glass or spills with broken glass unless the teacher tells you to do so.

Before You Begin

Invertebrates include all animals except those with backbones. Every phylum of the kingdom Animalia except the phylum Chordata consists only of invertebrates. In this lab, you will examine representatives of eight phyla of animals. You will see many similarities and differences in **body plan**—shape, symmetry, and internal organization.

1. Write a definition for each boldface term in the paragraph above and for the following terms: radial symmetry, bilateral symmetry, dorsal, ventral, anterior, posterior, cephalization, segmentation.

2. Describe the three basic body plans found in animals.

3. Make a data table similar to the one on the right to record observations about invertebrates.

4. Based on the objectives for this lab, write a question you would like to explore about the characteristics of invertebrates.

DATA TABLE				
Phylum	Symmetry	Body plan	Other	Examples

Procedure

PART A: Conducting a Survey

1.  Put on safety goggles and a lab apron.

2. Visit each invertebrate station, and examine the specimens there. Answer the questions, and record observations in your data table.

3. Sponges Examine each specimen.

   a. Describe the shape of a sponge.

   b. What do you think is the role of the many holes, or pores, in a sponge?

   c. Examine a prepared slide of a sponge with a compound microscope. What do you notice about the organization of the cells in sponges?

4. Cnidarians Examine each specimen.

   a. Divide the cnidarian specimens into two groups. What feature did you use to make your division?

   b. How many body openings does a cnidarian have?

   c. Examine a prepared slide of a hydra. What do you notice about the organization of the cells in cnidarians?

5. Flatworms and Roundworms Examine each specimen.

   a. How does a flatworm differ from a roundworm in external appearance?

   b. Do any of the worms appear to be segmented? Explain.

   c. Examine prepared slides of planarians and nematodes. How many body openings does each have?

6. Mollusks Examine each specimen.

   a. In what ways do the mollusks differ in external appearance?

   b. Which group of mollusks has the most noticeable "feet"?

7. Annelids Examine each specimen.

   a. How are an earthworm and a leech similar? How are they different?

   b. Describe any differences you see in the segments of the annelid worm.

8. Arthropods Examine each specimen.

   a. What characteristic do you observe in all arthropod appendages?

   b. How does the number of walking legs differ among these arthropods?

9. Echinoderms Examine each specimen.

   a. The word *echinoderm* means "spiny skin." Why is this name appropriate?

   b. What does an echinoderm's ventral surface look like?

PART B: Cleanup and Disposal

10. Dispose of broken glass in the designated waste containers. Do not put lab materials in the trash unless your teacher tells you to do so.

11. Wash your hands thoroughly before you leave the lab and after you finish all work.

Analyze and Conclude

1. Summarizing Data Which animal phyla show cephalization, and which do not?

2. Recognizing Patterns What type of symmetry is found with cephalization?

3. Recognizing Patterns What characteristics do annelids and arthropods share?

4. Analyzing Methods Were you able to identify the type of body plan found in all of the specimens? Explain.

5. Further Inquiry Write a new question about invertebrates that could be explored with another investigation.

On the Job

Parasitology is the study of parasites. Do research to discover how parasitologists help protect you and your pets from diseases caused by parasites. For more about careers, visit **go.hrw.com** and type in the keyword **HX4 Careers.**

Azure vase sponge

28 Simple Invertebrates

✓ Quick Review

Answer the following without referring to earlier sections of your book.

1. **Describe** the process of osmosis. *(Chapter 4, Section 1)*

2. **Identify** the organisms collectively called plankton. *(Chapter 17, Section 3 and Chapter 21, Section 2)*

3. **Define** the term *colonial organization*. *(Chapter 19, Section 1)*

4. **Distinguish** between ectoderm, mesoderm, and endoderm. *(Chapter 27, Section 1)*

5. **Distinguish** between acoelomates, pseudo-coelomates, and coelomates. *(Chapter 27, Section 1)*

Did you have difficulty? *For help, review the sections indicated.*

📖 Reading Activity

Before you read this chapter, write a short list of all the things you know about the members of each of the following phyla: sponges, cnidarians, flatworms, and roundworms. Then write a list of the things that you want to know about these simple invertebrates. Save your list, and to assess what you have learned, see how many questions you can answer after reading this chapter.

● Since ancient times, people have used sponges for a variety of functions. The ancient Greeks placed sponges in their helmets for padding, and the Romans used them for paintbrushes and mops. The beautiful azure vase sponge ranges in color from intense blue to pinkish-purple.

Looking Ahead

Section 1

Sponges
The Simplest Animals
Sponge Diversity
Reproduction

Section 2

Cnidarians
Two Body Forms
Hydrozoans
Scyphozoans
Anthozoans

Section 3

Flatworms and Roundworms
Flatworms
Roundworms

🔲 **internet** connect

www.scilinks.org
National Science Teachers Association *sci*LINKS Internet resources are located throughout this chapter.

*sci*LINKS® Maintained by the National Science Teachers Association

Figure 1 Sponge. The small openings in this sponge's body are ostia. The larger openings are oscula.

The Simplest Animals

Sponges are so unlike other animals that early naturalists classified them as plants. It wasn't until the late 1700s that scientists using improved microscope technology began studying sponges closely. Scientists then realized that sponges are animals. The bodies of most sponges completely lack symmetry and consist of little more than masses of specialized cells embedded in a gel-like substance called mesohyl *(MEHZ oh hil).* You could say that a sponge's body is somewhat like chopped fruit in gelatin. The chopped fruit represents the specialized cells, and the gelatin represents the mesohyl.

Sponge cells are not organized into tissues and organs. However, they do have a key property of all animal cells—cell recognition. A simple lab experiment can demonstrate that sponge cells can recognize other sponge cells. A living sponge can be passed through a fine silk mesh, causing the individual cells to separate. On the other side of the mesh, the individual sponge cells will recombine to form a new sponge.

Sponges have a body wall penetrated by tiny openings, or pores, called **ostia** *(AHS tee uh),* through which water enters. The name of the phylum, Porifera, refers to this system of pores. Sponges also have larger openings, or **oscula,** through which water exits. You can see the many oscula of the sponge shown in **Figure 1.** Sponges are also **sessile** *(SEHS uhl).* Early in their lives, sponges attach themselves firmly to the sea bottom or some other submerged surface, like a rock or coral reef. They remain there for life. Sponges can have a diameter as small as 1 cm (0.4 in.) or as large as 2 m (6.6 ft).

Most sponges are bag-shaped and have a large internal cavity. One or more oscula (singular, osculum) are located in the top of

Evolutionary Milestone

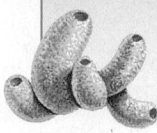

1 Multicellularity

The bodies of all animals, including sponges (phylum Porifera), are multicellular—made of many cells. Although the sponge is composed of several different cell types, these cells show only a small degree of coordination with each other.

Figure 2 Sponge interior

Water enters the sponge through many small pores (ostia) in its body wall and exits through the osculum, an opening at the top of the sponge.

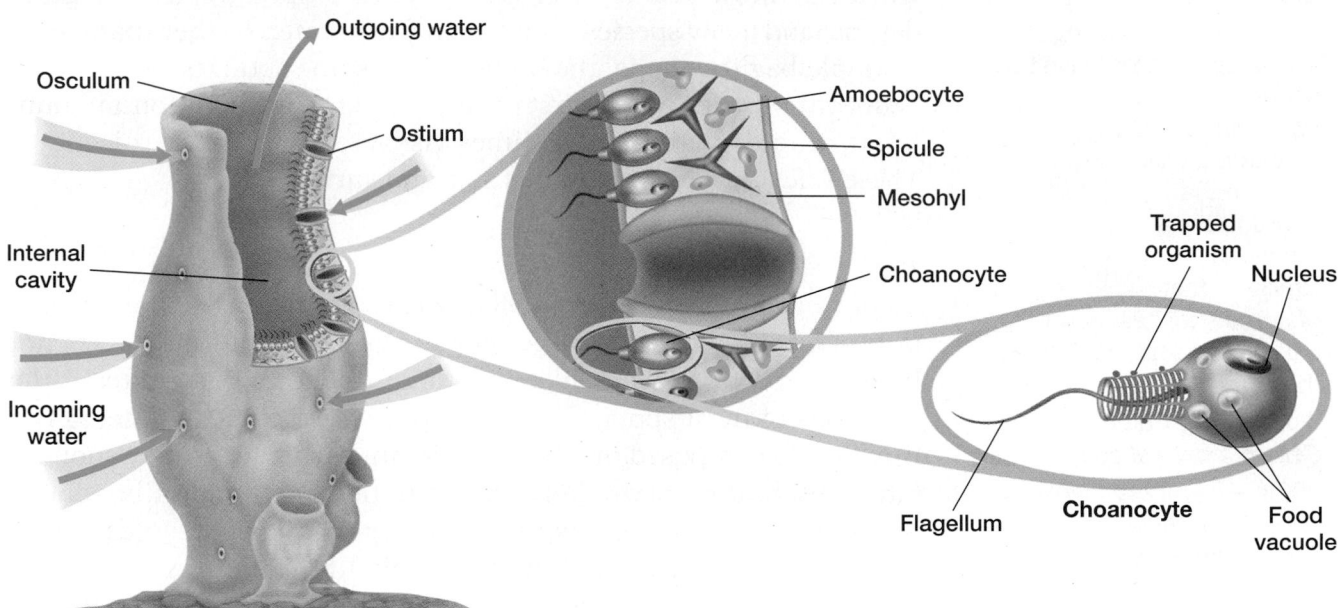

the body wall, as shown in **Figure 2.** Lining the internal cavity of a sponge is a layer of flagellated cells called **choanocytes** *(koh AN oh siets),* or collar cells. The flagella of these cells extend into the body cavity. As the flagella beat, water is drawn in through the pores in the body wall. The water is driven through the body cavity before it exits through the osculum.

As sea water passes through the sponge's body cavity, the collar cells function as sieves. These cells trap plankton and other tiny organisms in the small hairlike projections on the collar. The trapped organisms are then pulled into the interior of the collar cells, where they are digested intracellularly (within the cell). As sea water leaves the sponge, wastes are carried away in it.

How do the other sponge cells, such as those in the body wall, survive if the collar cells take in all of the food? The collar cells release nutrients into the mesohyl where other specialized cells, called *amoebocytes (uh MEE boh siets),* pick up the nutrients. **Amoebocytes** are sponge cells that have irregular amoeba-like shapes. They move about the mesohyl, supplying the rest of the sponge's cells with nutrients and carrying away their wastes.

Protistan Ancestors

The choanocytes of sponges very closely resemble a kind of protist called a choanoflagellate, shown in **Figure 3.** Ancient choanoflagellates are thought by many scientists to be the ancestors of sponges. Other free-swimming colonial flagellates closely resemble sponge larvae, however, and some scientists believe organisms similar to these other flagellates were the true ancestors of sponges.

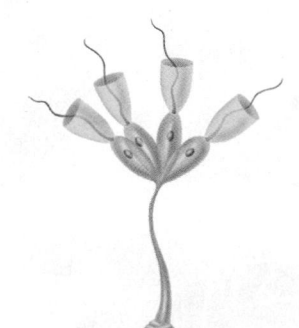

Figure 3 Choanoflagellate. Ancient choanoflagellates similar to the one shown above may be the ancestors of sponges.

Sponge Diversity

As any snorkeler can tell you, brilliantly colored sponges abound in warm, shallow sea waters. Other marine sponges live at great depths, and a few species even live in fresh water. Rather than being a simple baglike shape, the body wall of some sponges, such as the azure vase sponge on the first page of this chapter, may contain hundreds of folds that are sometimes visible as fingerlike projections. These folds increase a sponge's size and surface area.

Sponge Skeletons

To prevent the sponge from collapsing in on itself, the sponge body is supported by a skeleton. A sponge's skeleton, however, does not have a fixed framework like a human skeleton does. Instead, the skeletons of most sponges are composed of spicules. A **spicule** is a tiny needle composed of silica or calcium carbonate. A few sponges have skeletons composed of a resilient, flexible protein fiber called **spongin**. Some sponges contain both spongin and spicules. These supporting structures are found throughout the mesohyl.

Taxonomists group sponges into three types based on the composition of their skeletons. Calcareous sponges have spicules composed of calcium carbonate. Glass sponges and demosponges have spicules made of silica. Some demosponges also contain spongin. In some species the spongin is reinforced with spicules of silica. The three classes of sponges are shown in **Figure 4.**

Real Life

What is a luffa sponge?
A luffa sponge really isn't a sponge at all but a gourd. When dried, the fibrous material found in the gourd forms a "skeleton" similar to that of some sponges, and it can be used for many of the same purposes.

Comparing Structures
Obtain a natural sponge and a luffa sponge, and compare the nature of their "skeletons."

Figure 4   Three types of sponges

Sponges have skeletons made of spicules, spongin, or both.

Calcareous sponge

Magnification: 2403×

Glass sponge

Magnification: 203×

Demosponge

Magnification: 153×

Reproduction

Sponges can reproduce asexually. A remarkable property of sponges is that they regenerate when they are cut into pieces. Each bit of sponge, however small, will grow into a complete new sponge. As you might suspect, sponges frequently reproduce by shedding fragments, each of which develops into a new individual. Sponges also reproduce by budding. A third form of asexual reproduction occurs in some freshwater sponges. When living conditions become harsh (cold or very dry), some freshwater sponges form **gemmules** (*JEHM yools*), clusters of amoebocytes encased in protective coats. Sealed in with ample food, the cells survive even if the rest of the sponge dies. When conditions improve, the cells grow into a new sponge.

Sexual reproduction is also common among sponges. Most sponges are hermaphrodites, meaning they produce both eggs and sperm. Since eggs and sperm are produced at different times, self-fertilization is avoided. In most species of sponges, sperm cells from one sponge enter another sponge through its pores, as shown in **Figure 5.** Collar cells on the receiving sponge's interior pass the sperm into the mesohyl, where the egg cells reside, and fertilization occurs. The fertilized eggs develop into larvae and leave the sponge. After a brief free-swimming stage, the larvae attach themselves to an object and develop into new sponges.

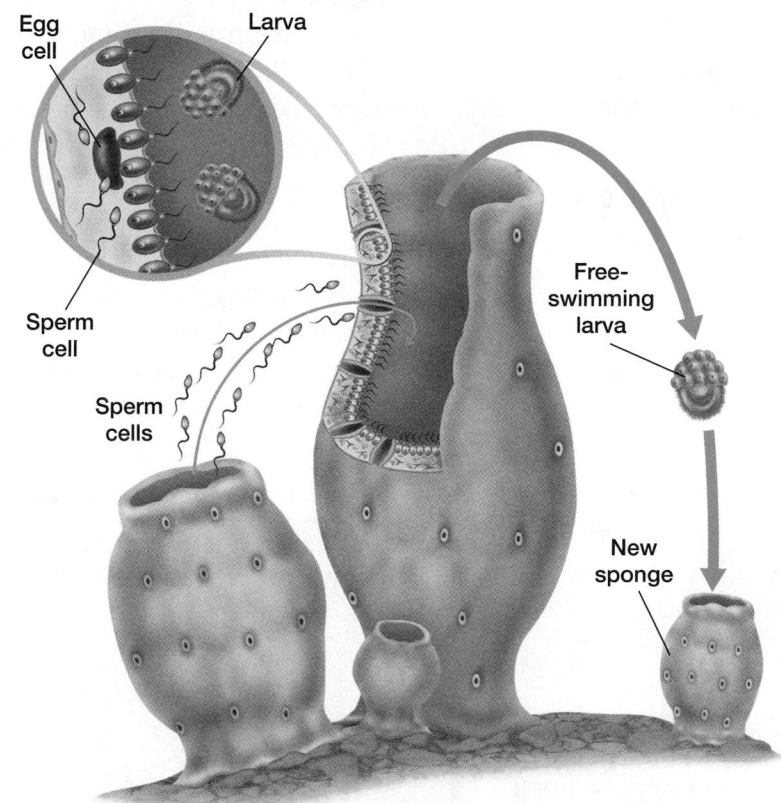

Figure 5 Sexual reproduction in sponges

In most species of sponges, sperm from one sponge fertilize eggs from another sponge.

Egg cell

Larva

Sperm cell

Sperm cells

Free-swimming larva

New sponge

Section 1 Review

1 **Draw** a simple sketch of a sponge body plan, and label all the parts you include.

2 **Summarize** how a sponge feeds and distributes nutrients.

3 **Describe** the three types of sponge skeletons.

4 **Compare** asexual and sexual reproduction in sponges.

5 **Critical Thinking Forming Hypotheses** What advantage might there be to a free-swimming larval stage in sponges?

6 **Critical Thinking Determining Factual Accuracy** Evaluate this statement: Sponges have two cell layers, mesohyl and collar cells.

7 **Standardized Test Prep** What is one function of choanocytes in a sponge?

A supporting the body **C** distributing nutrients

B fertilizing eggs **D** circulating water

Objectives

- **Describe** the two cnidarian body forms.
- **Summarize** how cnidocytes function.
- **Summarize** the life cycle of *Obelia*.
- **Compare** three classes of cnidarians.
- **Compare** asexual and sexual reproduction in cnidarians.

Key Terms

medusa
polyp
cnidocyte
nematocyst
basal disk
planula

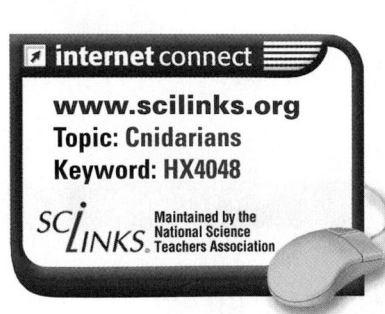

internet connect

www.scilinks.org
Topic: Cnidarians
Keyword: HX4048

SCI LINKS Maintained by the National Science Teachers Association

Two Body Forms

As the fragile bell of a jellyfish moves rhythmically through the water or the flowerlike sea anemone sways gently in the ocean currents, it's easy to be caught up in the mystery and beauty of these animals. But don't be deceived by their allure, for jellyfish and sea anemones are carnivores that can inflict a vicious sting. Along with hydras and corals, these animals belong to the phylum Cnidaria (*nih DAIR ee uh*). Cnidarians have two basic body forms, as shown in **Figure 6,** and both show radial symmetry. **Medusa** (*muh DOO suh*) forms are free-floating, jellylike, and often umbrella-shaped. **Polyp** (*PAHL ihp*) forms are tubelike and are usually attached to a rock or some other object. A fringe of tentacles surrounds the mouth, located at the free end of the body. Many cnidarians exist only as medusas, while others exist only as polyps. Still others alternate between these two phases during the course of their life cycle.

The cnidarian body has two layers of cells, as illustrated by the hydra in **Figure 7.** The outer layer derives from ectoderm, and the inner layer derives from endoderm. As in the sponge, there is a middle layer of mesoglea. But cnidarians differ from sponges in that cnidarians' cells are arranged into tissues.

Figure 6 Cnidarian body forms

The two body forms of cnidarians—medusa and polyp—consist of the same body parts arranged differently.

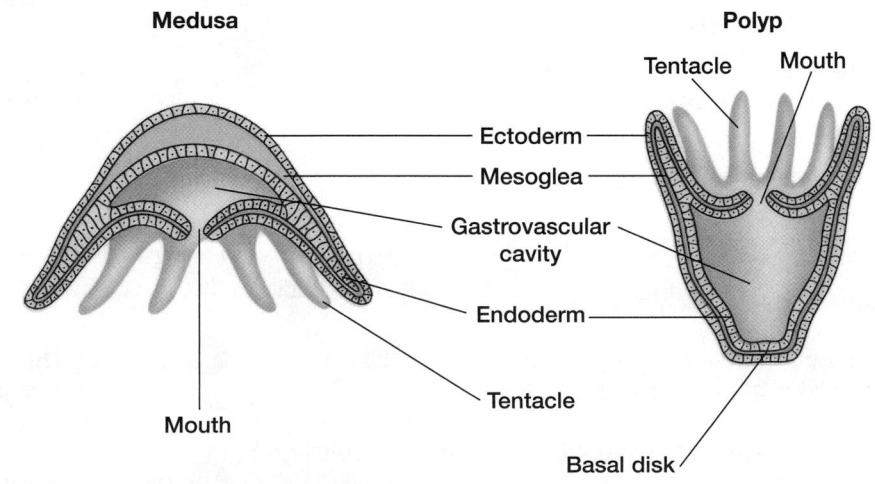

Medusa

Polyp

Tentacle Mouth

Ectoderm

Mesoglea

Gastrovascular cavity

Endoderm

Tentacle

Mouth

Basal disk

Evolutionary Milestone

2 Tissues

The cnidarian body plan is more complex than that of a sponge—it contains specialized tissues that carry out particular functions. The tissues, however, are not organized into organs.

Cnidocytes

Flexible fingerlike tentacles surround the opening to the gastrovascular cavity of cnidarians. Located on the tentacles are stinging cells called **cnidocytes** *(NIH doh siets)*, also shown in Figure 7. Cnidocytes are the distinguishing characteristic of the animals in the phylum Cnidaria. Within each cnidocyte is a small barbed harpoon called a **nematocyst** *(nehm AAH toh sihst)*. Nematocysts are used for defense and to spear prey. Some nematocysts contain deadly toxins, while others contain chemicals that stun but do not kill. When triggered, the nematocyst explodes forcefully and sinks into the cnidarian's prey. The captured prey is then pushed into the cnidarian's gastrovascular cavity by the tentacles.

Extracellular Digestion

In cnidarians, digestion begins extracellularly (outside the cell), in the gastrovascular cavity. Enzymes break food down into small fragments. Then cells lining the cavity engulf the fragments, and digestion is completed intracellularly. This allows cnidarians to feed on organisms larger than their own individual cells.

Figure 7   Cnidarian body plan

Like all cnidarians, this hydra is composed of tissues derived from endoderm and ectoderm.

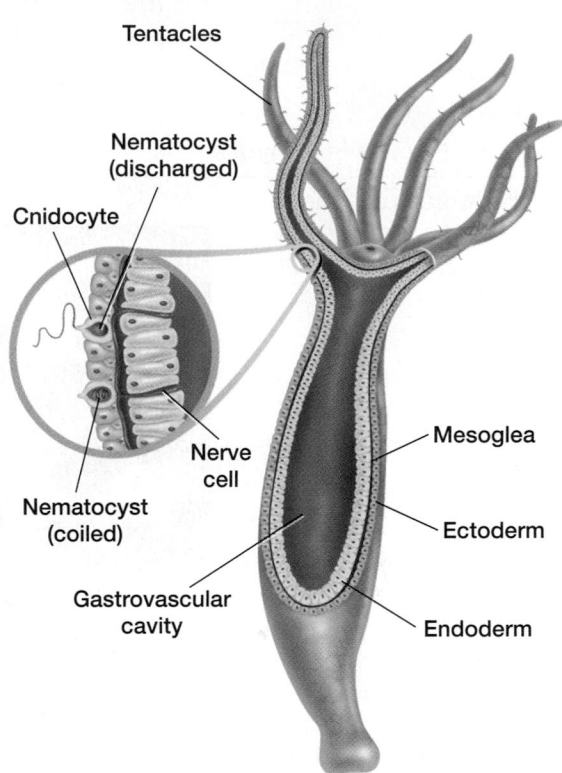

Tentacles
Nematocyst (discharged)
Cnidocyte
Nerve cell
Nematocyst (coiled)
Gastrovascular cavity
Mesoglea
Ectoderm
Endoderm

Estimating Size Using a Microscope

You can use the microscope to estimate the size of cnidarians that are too small to measure directly.

Materials

transparent millimeter ruler, compound microscope with low-power objective or a dissecting microscope, prepared slide of a medusa or polyp

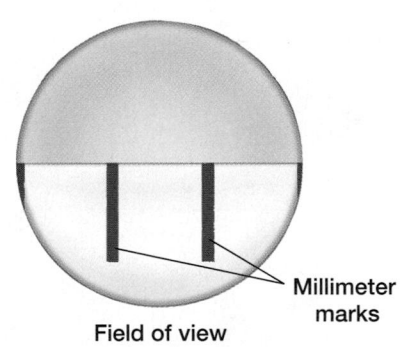

Millimeter marks

Field of view

Procedure

1. Identify the millimeter marks along the edge of the ruler.

2. With the microscope on low power (4× or lower), place the ruler on the stage and focus on the millimeter marks.

3. Adjust the ruler so that one edge lies across the diameter of the field, as shown above. Then measure the diameter of the field of view in millimeters.

4. Remove the ruler, and place the prepared slide on the stage. Identify the tentacles, gastrovascular cavity, and mouth.

5. Estimate the length and width of your organism as a ratio of the width of the field of view. For example, the length of your organism may appear to cover about two-thirds of the field of view.

Analysis

1. **Calculate** the size of your organism in millimeters by multiplying the ratio you found in step 5 by the width of the field of view you found in step 3.

2. **Describe** the body plan of the organism you viewed using terms from step 4.

Hydrozoans

The most primitive cnidarians are members of class Hydrozoa. Most species of hydrozoans are colonial marine organisms whose life cycle includes both polyp and medusa stages. Freshwater hydrozoans are less common, but are familiar to many people because they are often studied in school laboratories.

Figure 8   Freshwater hydra

This tiny hydra is attached to the leaf of a small aquatic plant. One way a hydra can move is by tumbling.

Magnification: 34×

Basal disk

Freshwater Hydrozoa

The abundant freshwater genus *Hydra* is unique among hydrozoans because it has no medusa stage and exists only as a solitary polyp. Hydras live in quiet ponds, lakes, and streams. They attach to rocks or water plants by means of a sticky secretion they produce in an area of their body called the **basal disk.** Hydras can glide around by decreasing the stickiness of the material secreted by their basal disk. Sometimes hydras move by tumbling, as shown in **Figure 8.** To tumble, the hydra bends its body over and touches the surface it is attached to with its tentacles. Then it pulls its basal disk free, flipping it over to the other side of its tentacles. The basal disk then reattaches, and the hydra returns to an upright position. Most hydras are brown or white, like the one in Figure 8. Others appear green because of the algae living within their cells.

Marine Hydrozoa

Marine hydrozoans are typically far more complex than freshwater hydrozoans. Often many individuals live together, forming colonies. The cells of the colony lack the interdependence that characterizes the cells of multicellular organisms. However, they often exhibit considerable specialization. For example, the colonial Portuguese man-of-war (genus *Physalia*) incorporates both medusas and polyps. A gas-filled float (probably a highly modified polyp) allows *Physalia* to float on the surface of the water. Dangling below the float are tentacles that can reach 60 m (about 197 ft). These tentacles are used to stun and entangle prey. Their nematocysts are tipped with powerful neurotoxins (nerve poisons) that are dangerous and may be fatal to humans. *Physalia,* shown in **Figure 9,** has other specialized polyps and medusas, each carrying out a different function, such as feeding or sexual reproduction.

Reproduction in Hydrozoans

Most hydrozoans are colonial organisms whose polyps reproduce asexually by forming small buds on the body wall. The buds develop into polyps that eventually separate from the colony and begin living independently. Many hydrozoans are also capable of sexual reproduction. Some species of *Hydra* are hermaphrodites, but in most species the sexes are separate.

The genus *Obelia* is typical of many marine colonial hydrozoans. *Obelia* lives in colonies that form when one polyp asexually produces buds that do not separate from it. Eventually, there are numerous polyps attached to one stem, forming the colony. The *Obelia* colony shown in **Figure 10** is branched like deer antlers, with various polyps attached to the branched stalks. The reproductive polyps give rise asexually to male and female medusas. These medusas leave the polyps and grow to maturity in the ocean waters.

During sexual reproduction, the medusas release sperm or eggs into the water. The gametes fuse and produce zygotes that develop into free-swimming, ciliated larvae called **planulae** *(PLAN yoo lee)*. The planulae eventually settle on the ocean bottom and develop into new polyps. Each polyp gives rise to a new colony by asexual budding, and the life cycle is repeated.

Figure 9 *Physalia.* A single Portuguese man-of-war colony can contain 1,000 individual medusas and polyps.

Figure 10 **Reproduction in *Obelia***

In *Obelia*'s life cycle, the medusa stage (sexual) and the polyp stage (asexual) alternate.

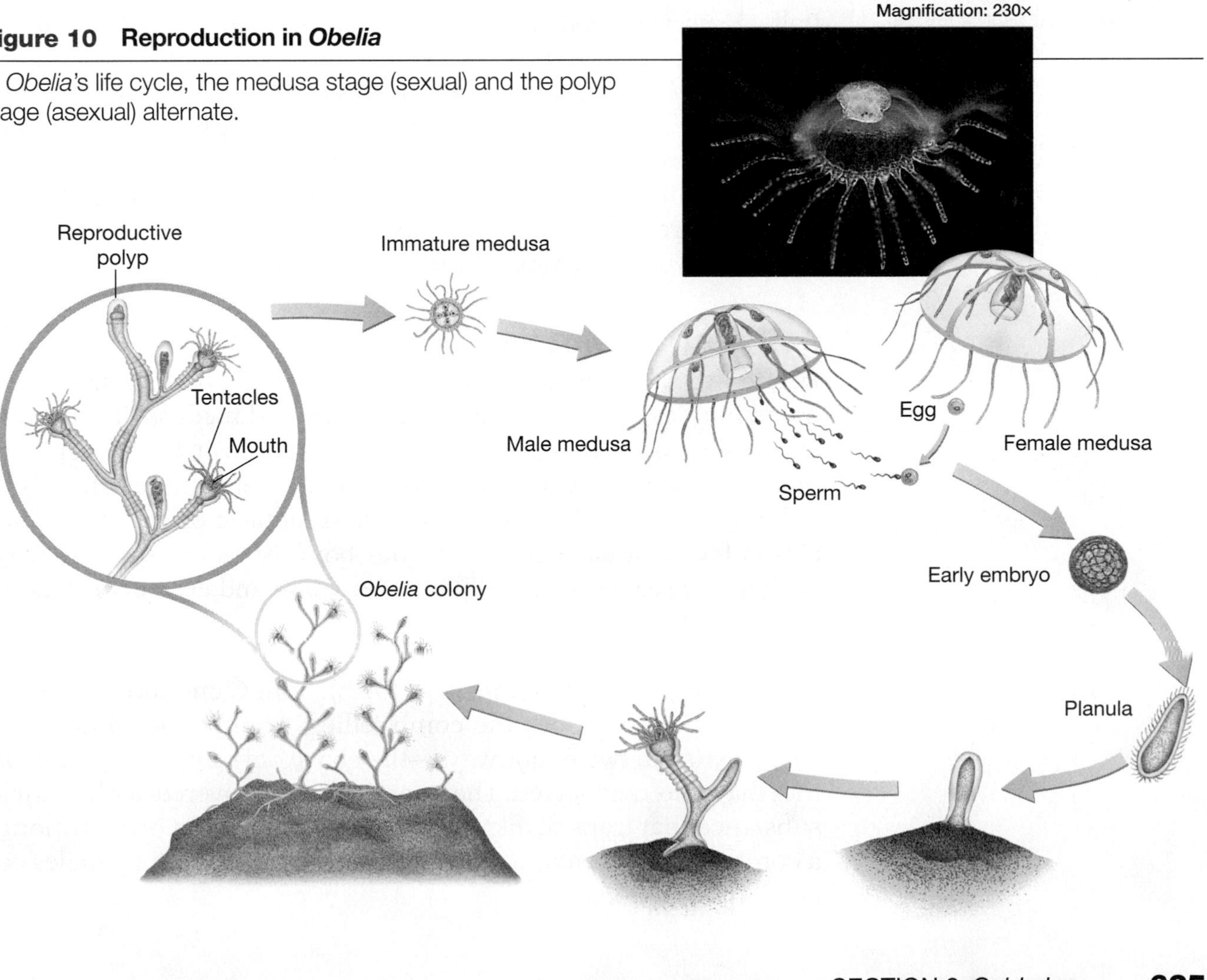

Magnification: 230×

Reproductive polyp

Immature medusa

Tentacles

Mouth

Male medusa

Egg

Female medusa

Sperm

Obelia colony

Early embryo

Planula

Figure 11 **Marine jellyfish**
Aurelia. *Aurelia* polyps are
about the size of hydras. The
free-swimming medusas
range from 10 cm (3.9 in.) to
25 cm (9.8 in.) in diameter.

Polyps

Medusa

Scyphozoans

Cnidarians belonging to the class Scyphozoa *(sie fuh ZOH uh)* are
the organisms usually referred to as true jellyfish. Scyphozoans are
active predators that ensnare and sting prey with their tentacles.
The toxins contained within the nematocysts of some species are
extremely potent. Scyphozoans range in size from as small as a
thimble to as large as a queen-size mattress.

The jellyfish seen in the ocean are medusas, which reproduce sexually. However, most species of jellyfish also go through an inconspicuous polyp stage at some point in their life cycle. The stinging
nettle, *Aurelia,* shown in **Figure 11,** is one of the most familiar jelly-
fishes. *Aurelia*'s tiny polyps hang from rocky surfaces. Periodically
the polyps release young medusas into the water. The *Aurelia* life
cycle is similar to that of *Obelia,* pictured on the previous page. The
major difference is that *Aurelia* spends most of its life as a medusa,
while *Obelia* spends most of its life as a polyp.

Jellyfish Relatives

Related to the jellyfish are the cubozoans, or box jellies. As their
name implies, cubozoans have a cube-shaped medusa. Their polyp
stage is inconspicuous, and in some species, it has never been
observed. Most box jellies are only a few centimeters in height,
although some are 25 cm (10 in.) tall. A tentacle or group of tentacles is found at each corner of the "box." Stings of some species,
such as the sea wasp, can inflict severe pain and even death among
humans. The sea wasp lives in the ocean along the tropical northern
coast of Australia.

Other relatives are members of the phylum Ctenophora *(tehn AW
for uh),* which includes the comb jellies. Comb jellies differ from
true jellyfish in two major ways—they have only a medusa stage and
they have no cnidocytes. Their tentacles are covered with a sticky
substance that traps plankton, the comb jelly's main prey. Although
a comb jelly is only about 2.5 cm (1 in.) in diameter, its tentacles can
be 10 times as long.

Anthozoans

The largest class of cnidarians is class Anthozoa. Anthozoans exist only as polyps. The most familiar anthozoans are the brightly colored sea anemones and corals. Other members of this class are known by such fanciful names as sea pansies, sea fans, and sea whips.

Anthozoans, such as the sea anemone shown in **Figure 12,** typically have a thick, stalklike body topped by a crown of tentacles that usually occur in groups of six. Nearly all of the shallow-water species contain symbiotic algae, such as dinoflagellates. The anthozoans provide a place for the dinoflagellates to live in exchange for some of the food that the dinoflagellates produce. The brilliant color of most anthozoans is actually that of dinoflagellates living within it. Some anthozoans reproduce asexually by forming buds, but they also reproduce sexually by releasing eggs and sperm into the ocean, where fertilization occurs. The fertilized eggs develop into planulae that settle and develop into polyps.

Figure 12　Sea anemone. When threatened, the sea anemone quickly retracts its tentacles and compresses its body.

Sea Anemones

Sea anemones are a large group of soft-bodied polyps found in coastal areas all over the world. Many species are quite colorful, and most do not grow very large, only from 5 mm (0.2 in.) to 100 mm (4.0 in.) in diameter. Sea anemones feed on fish and other marine life that happen to swim within reach of their tentacles.

Sea anemones are highly muscular and relatively complex animals. When touched, most sea anemones retract their tentacles into their body cavity and contract into a tight ball. Sea anemones often reproduce asexually by slowly pulling themselves into two halves. This method of reproduction often results in large populations of genetically identical sea anemones.

Corals

Most coral polyps live in colonies called reefs, such as the one shown in **Figure 13.** Each polyp secretes a tough, stonelike outer skeleton of calcium carbonate that is cemented to the skeletons of its neighbors. (Some corals called soft corals do not secrete hard exoskeletons.) Only the top layer of a coral reef contains living coral polyps. When coral polyps die, their skeletons remain and provide a

Figure 13　Coral. This coral reef is made up of hundreds of thousands of individual coral polyps. When the polyps feed (inset), they extend their tentacles from the protection of their stony skeleton.

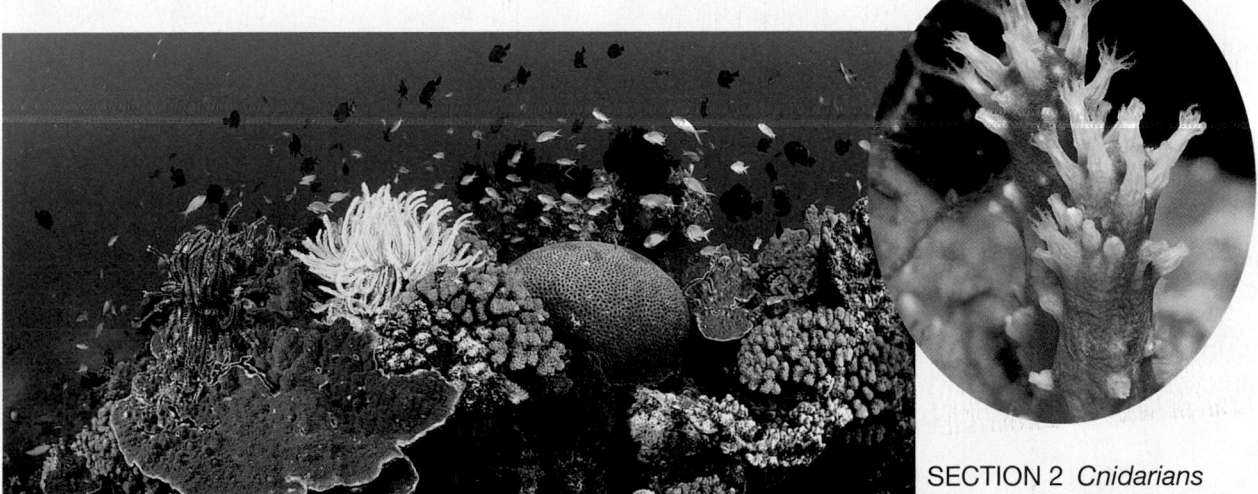

foundation for new coral polyps. Over thousands of years, these formations build up into coral reefs where hundreds of thousands of polyps live together on top of old skeletons. Coral reefs are found primarily in tropical regions of the world, where the ocean water is warm and clear, an environment that is ideal for the corals and the dinoflagellates that live inside them.

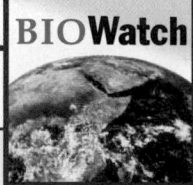

BIOWatch

Life on a Coral Reef

The diversity of life on a coral reef is rivaled only by that of a tropical rain forest. Algae growing on the corals and microscopic invertebrates help cement the corals together. Tube worms, mollusks, and other organisms donate their hard skeletons to the reef. A variety of small animals, including sponges, flatworms, shrimps, and crabs, find protection from predators in crevices in the reef. Sea anemones, hydroids, and feather duster worms anchor themselves to the reef and snare food from the surrounding water. Predators such as sea stars, octopuses, eels, and sharks patrol the reef in search of prey. As many as 3,000 species of animals may live on a single reef.

The Importance of Coral Reefs

Reef-building corals live symbiotically with photosynthetic algae. Because of their need for light, these corals only live in clear ocean waters less than 100 m deep.

Coral reefs directly benefit people by protecting coastlines from wave erosion and serving as resources for fisheries. Tourists who visit the reefs can be a significant source of income to nearby communities. Researchers are interested in reef-dwelling animals as sources of new medicines, including antibiotics and anticancer drugs. Today, coral reefs face a number of threats caused by human activity, including pollution of the waters surrounding

them. Some nations have established marine sanctuaries around their most vulnerable coral reefs to ensure the survival of these natural treasures.

internet connect

www.scilinks.org
Topic: Coral Reefs
Keyword: HX4051

SCLINKS. Maintained by the National Science Teachers Association

Section 2 Review

1 **Compare** the two body forms of cnidarians.

2 **Relate** cnidocytes and nematocysts to food gathering.

3 **Draw** and label the life cycle of *Obelia*.

4 **Summarize** the similarities and differences in the three classes of cnidarians described.

5 **Distinguish** between the two types of asexual reproduction found in cnidarians.

6 **Critical Thinking Forming Hypotheses** Some cnidarians are unique in exhibiting polyp and medusa forms. How might their two body forms give them an advantage over species that have only one body form?

7 **Standardized Test Prep** In cnidarians, digestion takes place

A only extracellularly.

B only intracellularly.

C in a gastrovascular cavity and intracellularly.

D in a digestive tract.

Flatworms and Roundworms

Flatworms

When you think of a worm, you probably visualize a creature with a long, tubular body, such as an earthworm. You might be less familiar with flatworms and roundworms. The flatworms are the largest group of acoelomate worms. Although the flatworm body plan is relatively simple, it is a great deal more complex than that of a sponge or cnidarian. Flatworms have a middle tissue layer, the mesoderm. And unlike sponges and cnidarians, the flatworm has tissues that are organized into organs.

The flatworm's body is bilaterally symmetrical and flat, like a piece of tape or ribbon. As a result, each cell in the animal's body lies very close to the exterior environment. This permits dissolved substances, such as oxygen and carbon dioxide, to pass efficiently through the flatworm's solid body by diffusion. In addition, portions of the flatworm's highly-branched gastrovascular cavity run close to practically all of its tissues. This gives each cell ready access to food molecules. Most flatworms have no respiratory or circulatory system.

Flatworms belong to phylum Platyhelminthes, which contains three major classes: Turbellaria, Cestoda, and Trematoda. They range in size from free-living forms less than 1 mm (0.04 in.) in length to parasitic intestinal tapeworms several meters long.

Objectives

- **Compare** the three classes of flatworms.
- **Summarize** the life cycle of a blood fluke.
- **Describe** the body plan of a roundworm.
- **Summarize** the life cycle of the roundworm *Ascaris*.

Key Terms

proglottid
fluke
tegument

Turbellaria

Almost all members of class Turbellaria are free-living marine flatworms, such as the one shown in **Figure 14.** However, marine flatworms are rarely studied by students because they are difficult to raise in captivity. Instead, students usually study a freshwater turbellarian such as *Dugesia*, one of a group of flatworms commonly called planarians. *Dugesia* is shown in *Up Close: Planarian*, on the following page.

Figure 14   Marine flatworm. Most free-living flatworms are marine species that swim with graceful wavelike movements.

Evolutionary Milestone

③ Bilateral Symmetry

Flatworms were likely the first bilaterally symmetrical animals, with left and right halves that mirror each other. Like all bilaterally symmetrical animals, flatworms have a distinct anterior (cephalic) end.

Up Close

Planarian

- **Scientific name:** *Dugesia* sp.
- **Size:** Average length of 3–15 mm (0.1–0.6 in.)
- **Range:** Worldwide
- **Habitat:** Cool, clear, permanent lakes and streams
- **Diet:** Protozoans and dead and dying animals

Dugesia feeding

Characteristics

Nervous System Sensory information gathered by the brain is sent to the muscles by two main nerve cords that are connected by cross branches. Light-sensitive structures called eyespots are connected to the brain. The eyespots are close to each other, giving *Dugesia* a cross-eyed appearance.

▲ Brain

Eyespot

Nerve cord

Pore

Tubule

▼ Excretory system

Flame cell

▲ Female reproductive system

▲ Male reproductive system

Pharynx

Mouth

Reproductive pore

▼ Intestine

Feeding *Dugesia,* a free-living flatworm, must extend its muscular pharynx out of its centrally located mouth in order to feed.

Reproduction *Dugesia* reproduces asexually in the summer by attaching its posterior end to a stationary object and stretching until it breaks in two, each of which will become a complete animal. Sexual reproduction also occurs. Individuals are hermaphrodites, and two individuals simultaneously transfer sperm to each other. Eggs of both individuals are fertilized and are released in clusters enclosed in a protective capsule. Several capsules are laid at a time, and the eggs inside hatch in 2 to 3 weeks.

Water Balance Because *Dugesia*'s body cells contain more solutes than fresh water does, water continually enters its body by osmosis. Excess water moves into a network of tiny tubules that run the length of *Dugesia*'s body. Side branches are lined with many flame cells, specialized cells with beating tufts of cilia that resemble a candle flame. The beating cilia draw water through pores to the outside of the worm's body.

Digestion The highly branched intestine enables nutrients to pass close to all of the flatworm's tissues. Nutrients are absorbed through the intestinal wall. Undigested food is expelled through the mouth.

Observing Planarian Behavior

Most bilaterally symmetrical organisms have sense organs concentrated in one end of the animal. You can observe how this arrangement affects the way they explore their environment.

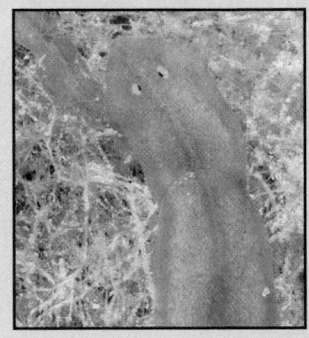

Materials

eyedropper, live culture of planaria, small culture dish with pond water, hand lens or dissecting microscope, forceps, and small piece of raw liver (3–7 cm)

Procedure

1. Using the tip of the eyedropper, place a planarian in the culture dish with pond water.

2. Using the hand lens or dissecting microscope, observe the planarian as it adjusts to its environment. Determine which end of the planarian contains sensory apparatus for exploring the environment.

3. Using forceps, place the liver in the pond water about 1 cm behind the planarian.

4. Observe the planarian's response. If the planarian approaches the liver, move the liver to a different position.

5. Continue observing the planarian for 5 minutes, moving the liver frequently.

Analysis

1. **Describe** the planarian's means of locomotion.

2. **Describe** how the planarian responded to the liver.

3. **Contrast** the feeding behavior of planarians with that of hydras, described earlier in this chapter.

4. **Critical Thinking Evaluating an Argument** Evaluate this statement: Bilateral symmetry gives planaria an advantage when feeding because sensory organs are concentrated in one end. Support your opinion with the observations you made on planaria.

Cestoda

Class Cestoda is made up of a group of parasitic flatworms commonly called tapeworms. Tapeworms use their suckers and a few hooklike structures, shown in **Figure 15,** to permanently attach themselves to the inner wall of their host's intestines. Food is then absorbed from the host's intestine directly through the tapeworm's skin. Tapeworms grow by producing a string of rectangular body sections called **proglottids** *(proh GLAHT ihds)* immediately behind their head. (Each proglottid is a complete reproductive unit, a fact that makes tapeworms easy to spread.) These sections are added continually during the life of the tapeworm. The long, ribbonlike body of a tapeworm may grow up to 12 m (40 ft) long.

Figure 15 Tapeworm

A tapeworm's body consists of a head and a series of proglottids.

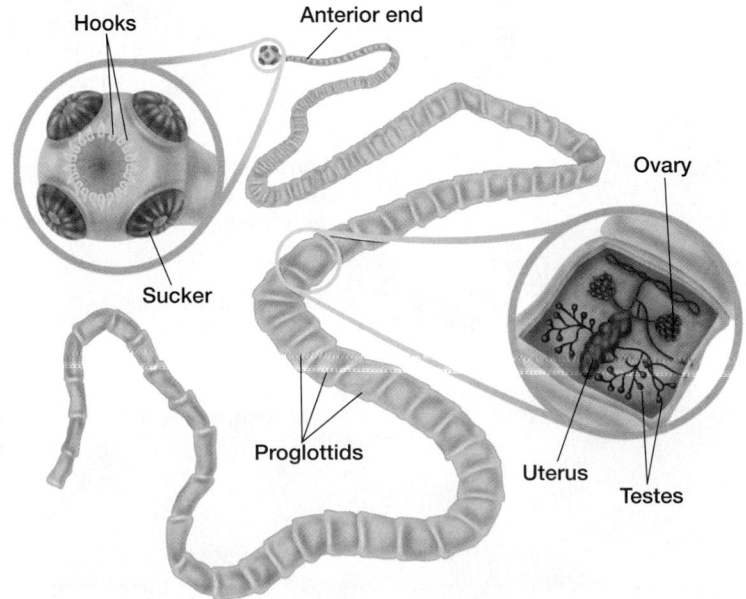

Most tapeworm infections occur in vertebrates, and about a dozen different kinds of tapeworms commonly infect humans. One of the tapeworms that infects humans is the beef tapeworm, *Taenia saginata*. Beef tapeworm larvae live in the muscle tissue of infected cattle, where they form enclosed fluid-filled sacs called cysts. Humans become infected when they eat infected beef that has not been cooked to a temperature high enough to kill the larvae.

Trematoda

The largest flatworm class, Trematoda, consists of parasitic worms called **flukes.** Some flukes are endoparasites, or parasites that live *inside* their hosts. Endoparasites have a thick protective covering of cells called a **tegument** that prevents them from being digested by their host. Other flukes are ectoparasites, or parasites that live on the *outside* of their hosts.

Flukes have very simple bodies with few organs. Flukes do not have well-developed digestive systems. Rather, they take their nourishment directly from their hosts. Flukes have one or more suckers that they use to attach themselves to their host. They use their muscular pharynx to suck in nourishment from the host's body fluids.

Most flukes have complex life cycles involving more than one host, one of which may be a human. Blood flukes of the genus *Schistosoma* are responsible for the disease schistosomiasis *(shihs tuh soh MIE uh sihs),* a major public health problem in the tropics. Infection occurs when people use or wade in water contaminated with *Schistosoma* larvae. The larval parasites bore through a person's skin and make their way to blood vessels in the intestinal wall. They block blood vessels, resulting in bleeding of the intestinal wall and damage to the liver. As shown in **Figure 16,** the life cycle of blood flukes includes a particular species of snail as an intermediate host.

Figure 16 *Schistosoma* **life cycle**

In the life cycle of blood flukes, snails are intermediate hosts and humans are final hosts.

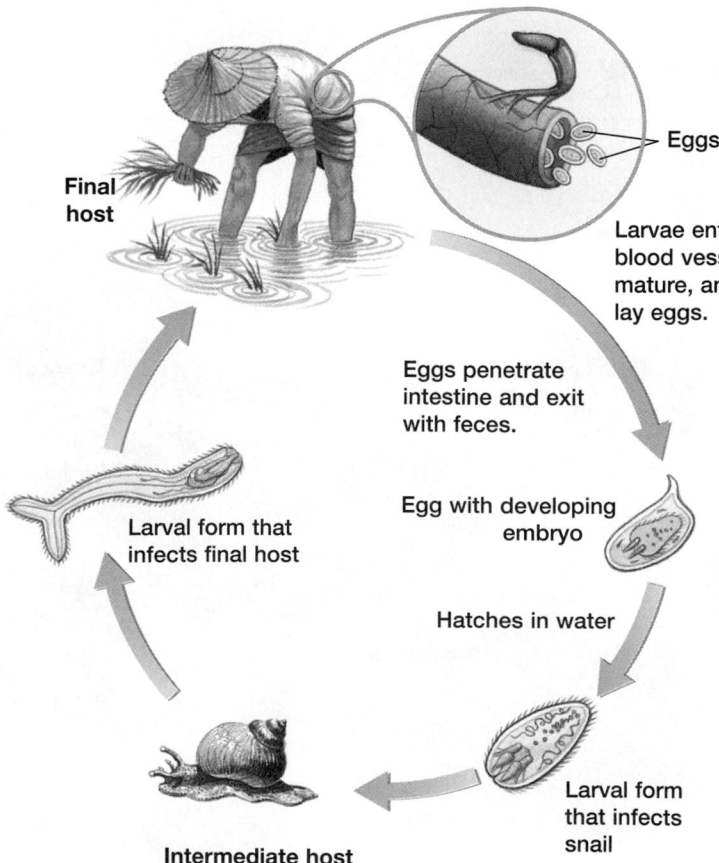

Final host

Eggs

Larvae enter blood vessels, mature, and lay eggs.

Eggs penetrate intestine and exit with feces.

Egg with developing embryo

Larval form that infects final host

Hatches in water

Larval form that infects snail

Intermediate host

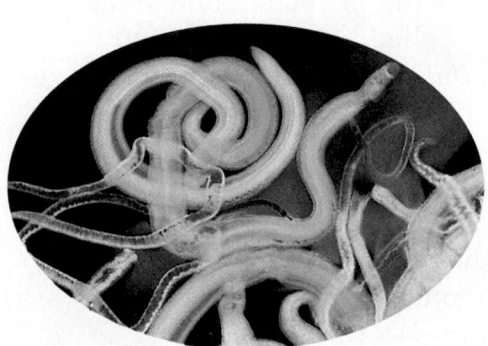

Adult male blood flukes are thick-bodied, while adult females are threadlike.

Roundworms

If you have a dog, you may be familiar with roundworms, some of which are canine parasites. Treatment for roundworms is a common reason for a trip to the vet, as shown in **Figure 17.** Roundworms (nematodes) are members of the phylum Nematoda and are characterized by the presence of a body cavity called a *pseudocoelom*. Movement of the fluid within the roundworm's pseudocoelom serves as a simple circulatory and gas exchange system. Oxygen and carbon dioxide move by diffusion into and out of the fluid. Nutrients from the digestive system also diffuse into the fluid and are distributed to the body cells.

Roundworms have long, cylindrical bodies and are the simplest animals to have a one-way digestive system. A flexible, thick layer of epidermis and cuticle form a protective cover and give the roundworm's body its shape. Beneath this cover, a layer of muscle extends along the length of the worm. These long muscles pull against the cuticle and the pseudocoelom (fluid-filled body cavity), whipping the worm's body from side to side. While some roundworms grow to be a foot or more in length, most are microscopic or only a few millimeters long. The vast majority of roundworms are free-living, active hunters.

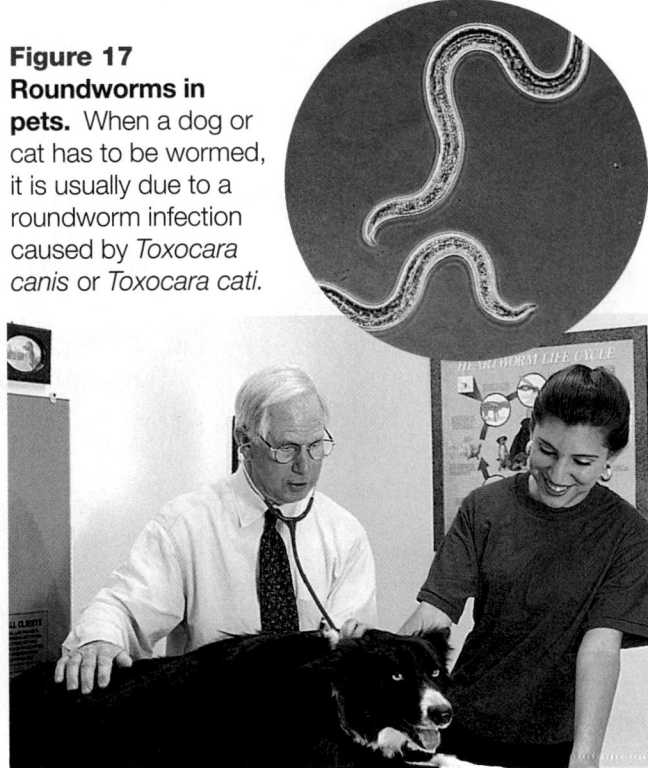

Magnification: 120×

**Figure 17
Roundworms in pets.** When a dog or cat has to be wormed, it is usually due to a roundworm infection caused by *Toxocara canis* or *Toxocara cati*.

Roundworm Infections

About 50 roundworm species are plant or animal parasites that cause considerable economic damage to crops and inflict terrible human suffering. Plant roundworms may attack any part of the plant—leaves, stem, roots—depending on the species. They feed on the living plant cells, causing wilting and withering of the plant. At least 14 species of roundworms infect humans. Three sources of human infection are *Ascaris lumbricoides*, *Trichinella spiralis*, and members of the genus *Necator*, commonly called hookworms.

The eggs of *Ascaris* are carried through human waste to the soil, where they can live for years. If ingested, the eggs enter the intestine, where they develop into larvae. The larvae bore through the blood vessels in the intestine and enter the bloodstream, which carries them to the lungs, causing respiratory distress. Some larvae may wander into the ducts of the pancreas or gallbladder, causing a blockage. Eventually, the larvae return

Evolutionary Milestone

4

Body Cavity

Roundworms have a pseudocoelom, a body cavity that forms between the gut and the body wall. All pseudocoelomates have a one-way gut in which food passes into the mouth and out of the anus.

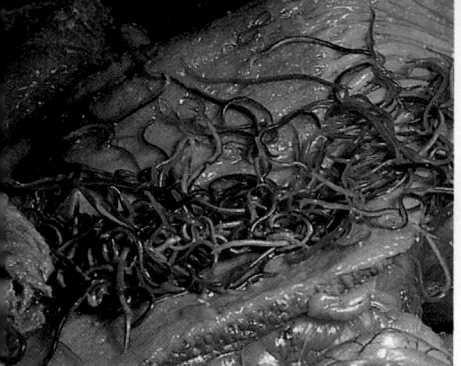

Figure 18 Roundworm
Ascaris. These adult *Ascaris* are in the stomach of a brown pelican.

to the intestine, where they mature and mate. Adult *Ascaris* may grow up to 0.3 m (1 ft) in length. **Figure 18** shows an *Ascaris* species whose final host is a bird.

Like *Ascaris, Trichinella* and *Necator* have complex life cycles that can involve a human host. *Trichinella* infects pigs and causes a serious disease called trichinosis *(trihk ih NOH sihs)* in humans. Infection with *Trichinella* can be avoided by not eating undercooked pork. Members of the genus *Necator* live mostly in the warm, moist soils of the tropics. Infection can occur when people step barefooted on soil containing hookworm larvae, which can enter the blood vessels if they penetrate the soles of the feet.

Identifying Parasites

Background

This graph shows how two drugs affect the release of eggs in a human infested with two parasites—*Schistosoma* and *Ascaris.* Drug 1 works by killing adult parasites in the intestines. Drug 2 works by killing adult parasites in the blood vessels. Use the graph and your knowledge of parasitic infections to answer the analysis questions.

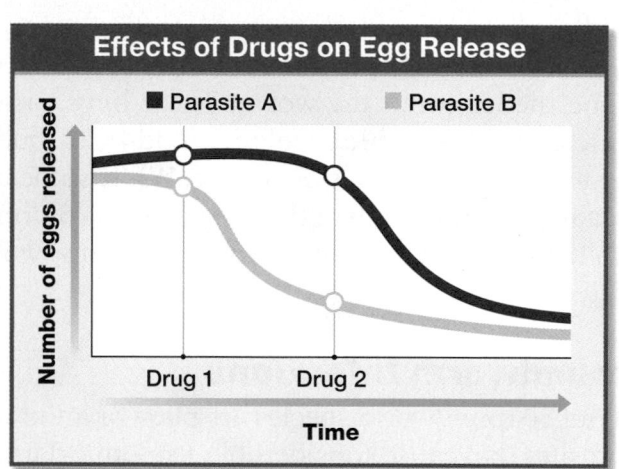

Analysis

1. **Describe** the response of the parasites to the two different drug treatments.

2. **Identify** the main human organs and tissues infected by the adult stages of *Schistosoma* and *Ascaris.* Use your textbook if necessary.

3. **Identify** which curve on the graph shows *Schistosoma* egg production and which shows *Ascaris* egg production.

4. **Critical Thinking Justifying Conclusions** Explain why you made the identifications you did in item 3.

5. **Critical Thinking Forming Hypotheses** *Schistosoma* spends part of its life cycle as a parasite of snails. Hypothesize a reason for an increase in the number of cases of schistosomiasis in villages near where hydroelectric dams have been built.

internet connect

www.scilinks.org
Topic: Roundworms
Keyword: HX4159

SC*LINKS* Maintained by the National Science Teachers Association

Section 3 Review

1. **Compare** the internal and external anatomy of a planarian with that of a parasitic flatworm.

2. **Summarize** in words or with a diagram the life cycle of a blood fluke.

3. **Summarize** the life cycle of *Ascaris.*

4. **Describe** a major innovation in body plan that first occurred in roundworms.

5. **Standardized Test Prep** Which organ system is missing in a planarian?
   A digestive   **C** nervous
   B respiratory   **D** reproductive

Key Concepts

1 Sponges

- Sponges lack symmetry and tissues.
- Sponges are sessile filter feeders that draw sea water through pores into an internal cavity, trapping tiny aquatic organisms.
- The sponge's supportive skeleton is composed of soft spongin fibers, hard spicules, or a combination of both.
- Sponges that reproduce sexually are usually hermaphrodites. Sponges also reproduce asexually.

2 Cnidarians

- Cnidarians are radially symmetrical, with bodies made up of tissue. Their body form may be a medusa or a polyp.
- Cnidocytes are stinging cells found in the tentacles of cnidarians. Harpoon-like nematocysts are located within the cnidocytes.
- Most hydrozoans are colonial organisms that reproduce asexually, though many forms can also reproduce sexually.
- Jellyfish are active predators, and some have extremely potent toxins within their nematocysts.
- Jellyfish spend most of their lives as medusas and usually reproduce sexually.
- Sea anemones and corals have thick, stalklike polyp bodies. Their life cycle includes no medusa form.

3 Flatworms and Roundworms

- Flatworms have flattened bodies that lack a body cavity. Most flatworms, such as planarians and marine flatworms, are free-living, but others, such as flukes and tapeworms, are parasites.
- Tapeworms are intestinal parasites that absorb food directly through their skin.
- Flukes are endoparasitic flatworms. They have a protective covering called a tegument that keeps them from being digested by their host.
- Roundworms have a pseudocoelom and a one-way gut. Most are free-living, but some are animal parasites.

Key Terms

Section 1

ostia (618)
oscula (618)
sessile (618)
choanocyte (619)
amoebocyte (619)
spongin (620)
spicule (620)
gemmule (621)

Section 2

medusa (622)
polyp (622)
cnidocyte (623)
nematocyst (623)
basal disk (624)
planula (625)

Section 3

proglottid (631)
fluke (632)
tegument (632)

Performance ZONE

Understanding Key Ideas

1. Which of the following is *not* a characteristic of sponges?
   a. body wall penetrated by many pores
   b. cells organized into tissues
   c. collar cells that trap food particles
   d. amoebocytes that transport food

2. What prevents self-fertilization among sponges?
   a. Gametes are released at different times.
   b. Few male sponges exist.
   c. Sponges are hermaphrodites.
   d. Encounters between members of the same species are rare.

3. A Portuguese man-of-war and a hydra are similar in that both
   a. are colonial.
   b. contain medusas and polyps.
   c. are hydrozoans.
   d. produce planulae.

4. Which sequence reflects the life cycle of *Obelia*?
   a. polyp → medusa → planula
   b. medusa → polyp → planula
   c. planula → medusa → polyp
   d. polyp → planula → medusa

5. Identify the function of the structure shown below.
   a. respiration
   b. water removal
   c. feeding
   d. digestion

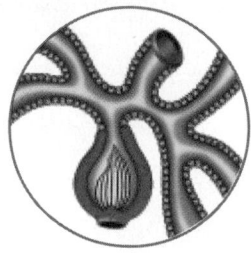

6. Sponges living today seem to have changed little compared to ancient sponges. How could evolutionary theory explain this?

7. How are parasitic flukes able to live when they no longer possess a well-developed digestive system?

8. Classify all of the organisms covered in this chapter as either acoelomate, pseudocoelomate, or coelomate. (**Hint:** See Chapter 27, Section 1.)

9. 🌐 **BIOWatch** What kinds of life-forms are supported by coral reefs, and why is it important for people to make sure that the reefs survive and develop?

10. 🗂 **Concept Mapping** Make a concept map that shows the major characteristics of sponges, cnidarians, flatworms, and roundworms. Include the following terms in your map: *sessile, choanocyte, spongin, medusa, polyp, cnidocyte, fluke, tegument,* and *proglottid.*

Critical Thinking

11. Applying Information Which animal would tend to have more water enter its body—a marine flatworm or a freshwater flatworm? Explain your answer.

12. Evaluating Conclusions A student concludes that infection with *Schistosoma* is more difficult to prevent than is infection with *Trichinella*. Evaluate this conclusion.

Alternative Assessment

13. Identifying Structures Make an anatomical drawing of the interior of a sponge, cnidarian, flatworm, or roundworm. Identify the species, and label at least 10 structures. Distribute copies of your drawing to your classmates.

14. Forming a Model In groups of three, research how one of the three different types of coral reefs—fringing, barrier, and atoll—is formed. Then build a model of your reef type or make a map showing where such reefs are located. Set up an exhibit of your work, and use a tape recording to create a "tour."

Standardized Test Prep

Understanding Concepts

Directions (1–4): **For *each* question, write on a separate sheet of paper the letter of the correct answer.**

1 What is a sponge's protein skeleton composed of?
   A. amoebocytes
   B. mesoglea
   C. spicules
   D. spongin

2 What are specialized stinging cells found in cnidarians called?
   F. choanocytes
   G. cnidocytes
   H. medusas
   I. polyps

3 Which of the following is an anthozoan?
   A. box jelly
   B. hydra
   C. jellyfish
   D. sea anemone

4 What term describes the covering that protects endoparasites from the actions of digestive enzymes?
   F. basal disk
   G. osculum
   H. proglittid
   I. tegument

Directions (5): **For the following question, write a short response.**

5 Individuals of a single species of sponge may vary in appearance depending on their environment. Factors that affect sponge shape include differences in the material on which they grow, availability of space, and the speed and temperature of water currents. How might these factors make the classification of sponges confusing?

Test TIP

If you find a particular question difficult, put a light pencil mark beside it and keep working. (Do not write in this book). As you answer other questions, you may find information that helps you answer the difficult question.

Reading Skills

Directions (6): **Read the passage below. Then answer the question.**

When locating sunken ships, some treasure seekers use dynamite to blast away portions of the ocean floor. Blasting disturbs the ocean floor, where many invertebrates live. It also can cause harm to invertebrates that are filter feeders.

6 How would blasting be harmful to filter feeders?
   A. The layers of ocean sediments are no longer perfectly horizontal.
   B. Large portions of the ocean floor may settle and destroy competitors.
   C. Particles stirred up by the blast fill the water, which interferes with filtering.
   D. Blasting disperses throughout the water the plankton that filter feeders eat.

Interpreting Graphics

Directions (7): **Base your answer to question 7 on the dichotomous key below.**

Key of Simple Animal Phyla

1. Cells are not organized into tissues	Phylum *W*
Cells are organized into tissues	Go to 2.
2. Tissues are not organized into organs	Phylum *X*
Tissues are organized into organs	Go to 3.
3. The body does not have a body cavity	Phylum *Y*
The body has a body cavity	Go to 4.
4. The body cavity is not a true coelom	Phylum *Z*
The body cavity is a true coelom	Go to a key of more complex animal phyla.

7 What is the name of Phylum *W*?
   F. Cnidaria
   G. Nematoda
   H. Platyhelminthes
   I. Porifera

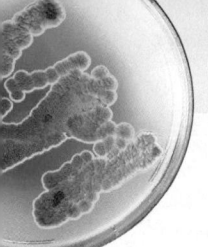

Exploration Lab

Observing Hydra Behavior

SKILLS
- Using scientific processes
- Observing

OBJECTIVES
- **Observe** a hydra finding and capturing prey.
- **Determine** how a hydra responds to stimuli.

MATERIALS
- silicone culture gum
- microscope slide
- 2 medicine droppers
- *Hydra* culture
- *Daphnia* culture
- concentrated beef broth
- filter paper cut into pennant shapes
- forceps
- stereomicroscope

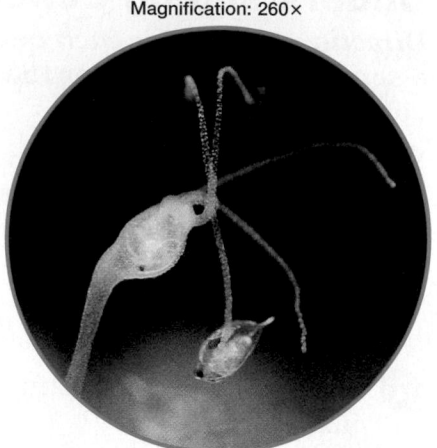

Magnification: 260×

Hydra feeding on *Daphnia*

Before You Begin

Cnidarians are carnivorous animals. A common cnidarian is **Hydra,** a freshwater organism that feeds on smaller freshwater animals, such as water fleas (*Daphnia*). Hydras find food by responding to stimuli, such as chemicals and touch. The way an animal responds to stimuli is called **behavior.** The tentacles of a cnidarian are armed with **nematocysts,** shown below, which are used in defense and in capturing prey. When a hydra receives stimuli from potential prey, its nematocysts spring out and harpoon or entangle the prey. In this lab, you will observe the feeding behavior of hydras to determine how they find and capture prey.

1. Write a definition for each boldface term in the paragraph above.

2. Based on the objectives for this lab, write a question you would like to explore about the feeding behavior of hydras.

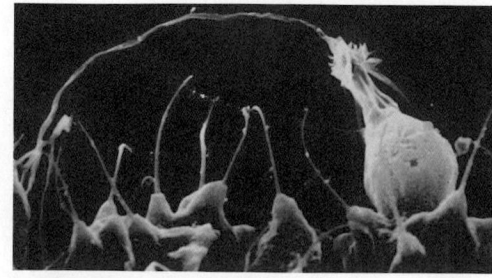

Nematocyst discharged

Procedure

PART A: Make Observations

1. To make an experimental pond for observing hydras, squeeze out a long piece of silicone culture gum. Arrange it to form a circular well on a microscope slide, as shown in the photograph below. **CAUTION: Glassware is fragile. Notify the teacher promptly of any broken glass or cuts.**

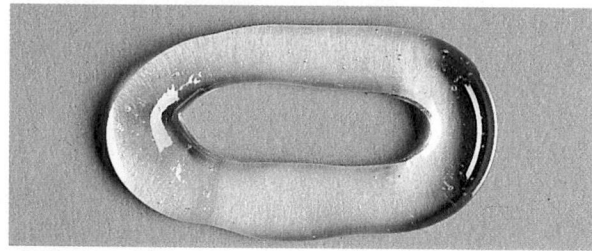

2. With a medicine dropper, gently transfer a hydra from its culture dish to the well on the slide, making sure the water covers the animal. **CAUTION: Handle hydras gently to avoid injuring them.** Allow the hydra to settle, then examine it under the high power of a stereomicroscope. Draw a hydra and label the body stalk, mouth, and tentacles.

PART B: Design an Experiment

3. Work with the members of your lab group to explore one of the questions written for step 2 of **Before You Begin.** To explore the question, design an experiment that uses the materials listed for this lab.

4. Write a procedure for your experiment. Make a list of all the safety precautions you will take. Have your teacher approve your procedure and safety precautions before you begin the experiment.

You Choose

As you design your experiment, decide the following:

a. what question you will explore

b. what hypothesis you will test

c. how to observe a hydra's feeding behavior

d. how to test a hydra's response to a stimulus, such as a chemical or a touch

e. what your test groups and controls will be

f. what to record in your data table

PART C: Conduct Your Experiment

5. Set up and carry out your experiment. **CAUTION: Handle hydras gently to avoid injuring them.**

6. Allow hydras to settle before exposing them to a test condition. If your hydra does not respond after a few minutes, obtain another hydra from the culture dish. Repeat your procedure.

PART D: Cleanup and Disposal

7. Dispose of lab materials and broken glass in the designated waste containers. Put hydras and daphnias in the designated containers. Do not put lab materials in the trash unless your teacher tells you to do so.

8. Clean up your work area and all lab equipment. Return lab equipment to its proper place. Wash your hands thoroughly before you leave the lab and after you finish all work.

Analyze and Conclude

1. Analyzing Results Describe a hydra's response to chemicals (beef broth).

2. Analyzing Results Describe a hydra's response to touch.

3. Drawing Conclusions How does a hydra detect its prey?

4. Justifying Conclusions Give evidence to support your conclusion about how hydras detect prey.

5. Inferring Conclusions Based on your observations, how do you think a hydra behaves when it detects a threat in its natural habitat?

6. Inferring Conclusions What happens to food that has not been digested by a hydra?

7. Inferring Conclusions How is a hydra adapted to a sedentary lifestyle?

8. Further Inquiry Write a new question about the behavior of hydras that could be explored with another investigation.

? Do You Know?

Do research in the library or media center to answer these questions:

1. What different kinds of food does a hydra eat?

2. How is the feeding method of a hydra different from that of a sponge?

Use the following Internet resources to explore your own questions about sponges and cnidarians.

internet connect

www.scilinks.org
Topic: Hydra
Keyword: HX4102

SCLINKS. Maintained by the National Science Teachers Association

Common octopus

CHAPTER
29 Mollusks and Annelids

✓ Quick Review

Answer the following without referring to earlier sections of your book.

1. **Describe** the process of diffusion. *(Chapter 4, Section 1)*
2. **Define** the term *plankton*. *(Chapter 17, Section 3)*
3. **Define** the terms *gill* and *hydrostatic skeleton*. *(Chapter 27, Section 2)*
4. **Distinguish** between an open and closed circulatory system. *(Chapter 27, Section 2)*

Did you have difficulty? *For help, review the sections indicated.*

📖 Reading Activity

As you read this chapter, draw and label a diagram of an idealized mollusk. Your mollusk should have all of the anatomical features described in the chapter (although not all mollusks possess all features).

Looking Ahead

Section 1

Mollusks
  A True Coelom
  Key Characteristics of Mollusks
  Body Plans of Mollusks

Section 2

Annelids
  The First Segmented Animals
  Annelid Groups

🔲 internet connect

www.scilinks.org
National Science Teachers Association *sci*LINKS Internet resources are located throughout this chapter.

sci**LINKS**. Maintained by the National Science Teachers Association

● The octopus is one of the most intelligent of the invertebrates. It can perform complex behaviors, including building its own house out of debris found on the ocean floor.

Objectives

- **Summarize** the evolutionary relationship between mollusks and annelids.
- **Describe** the key characteristics of mollusks.
- **Describe** excretion, circulation, respiration, and reproduction in mollusks.
- **Compare** the body plans and feeding adaptations of gastropods, bivalves, and cephalopods.

Key Terms

trochophore
visceral mass
mantle
foot
radula
nephridium
adductor muscle
siphon

A True Coelom

While most of the simple invertebrates you read about in the last chapter may be unfamiliar to you, chances are good that you have seen many mollusks and annelids. Snails, slugs, oysters, clams, scallops, octopuses, and squids are all mollusks. If you have seen an earthworm, then you know what an annelid is. While a snail may not seem to have much in common with an earthworm, these two very different-looking animals are related.

Mollusks and annelids were probably the first major groups of organisms to develop a true coelom. (Recall that in animals that have a true coelom, or body cavity, the gut and other internal organs are suspended from the body wall and cushioned by the fluid within the coelom.) Another feature shared by mollusks and annelids is a larval stage called a **trochophore** *(TRAHK oh fawr)*, which develops from the fertilized egg. In some species, the trochophore, shown in **Figure 1,** is free-swimming and propels itself through the water by movement of cilia on its surface. The presence of a trochophore larva in mollusks and annelids suggests that they share a common ancestor.

Members of the phylum Mollusca make up the second largest animal phyla, exceeded only by phylum Arthropoda. Mollusks are abundant in almost all marine, freshwater, and terrestrial habitats. There are more species of terrestrial mollusks than there are of terrestrial vertebrates. These mollusks often go unnoticed because people are not accustomed to looking for them. Seven classes of mollusks make up the phylum Mollusca. The three major classes are Gastropoda (snails and slugs), Bivalvia (clams, oysters, and scallops), and Cephalopoda (octopuses and squids).

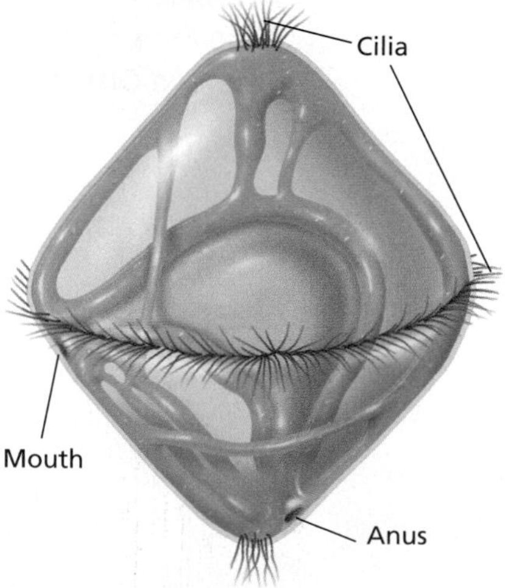

Cilia

Mouth

Anus

Figure 1 Trochophore larva

The microscopic trochophore larva has a belt of cilia that circles its body. The beating of the cilia propels the trochophore through the water.

Evolutionary Milestone

5 Coelom

A true coelom develops entirely within the mesoderm. Contact between the mesoderm and endoderm during the development of the embryo leads to the development of complex organs.

Key Characteristics of Mollusks

Despite their varied appearance, the members of the different groups of mollusks share a number of key characteristics.

1. **Body cavity.** The body cavity in mollusks is a true coelom, although in most species it is reduced to a small area immediately surrounding the heart.

2. **Symmetry.** Most mollusks exhibit bilateral symmetry.

3. **Three-part body plan.** The body of every mollusk has three distinct parts: the visceral mass, the mantle, and the muscular foot, as shown in **Figure 2.** The **visceral** *(VIS uhr uhl)* **mass** is a central section that contains the mollusk's organs. The mantle is wrapped around the visceral mass like a cape. The **mantle** is a heavy fold of tissue that forms the outer layer of the body. Finally, every mollusk has a muscular region called a **foot,** which is used primarily for locomotion.

4. **Organ systems.** Mollusks have organ systems for excretion, circulation, respiration, digestion, and reproduction.

5. **Shell.** Many mollusks have either one or two shells that serve as an exoskeleton, protecting their soft body. The shell is composed of protein that is strengthened by calcium carbonate, an extremely hard mineral.

6. **Radula.** All mollusks except bivalves have a **radula** *(RAJ uh luh),* a tongue-like organ located in their mouth. The radula, shown in Figure 2, has thousands of pointed, backward-curving teeth arranged in rows. When a mollusk feeds, it pushes its radula out of its mouth, and the teeth scrape fragments of food off rocks or plant matter. Mollusks that are predators use their radula for attacking their prey.

Figure 2 Three-part body plan

All mollusk bodies are composed of a visceral mass, a mantle, and a foot. Most mollusks also have a radula.

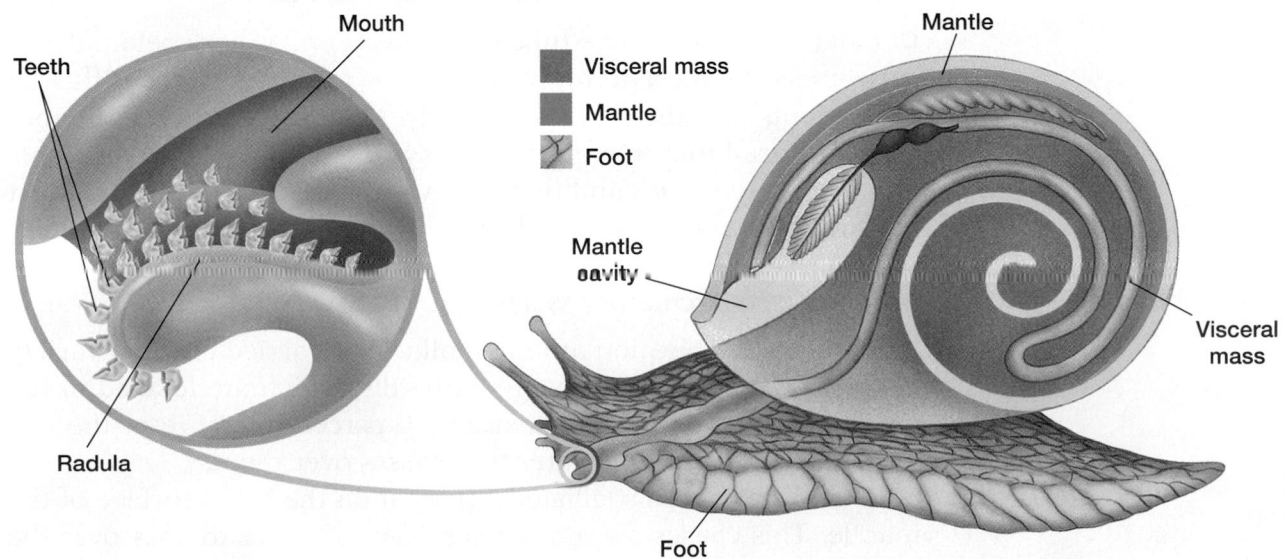

Mouth

Teeth

Visceral mass

Mantle

Foot

Mantle cavity

Mantle

Visceral mass

Radula

Foot

Figure 3 Mollusk body plan

Although mollusks vary greatly in body form, they all have complex organ systems.

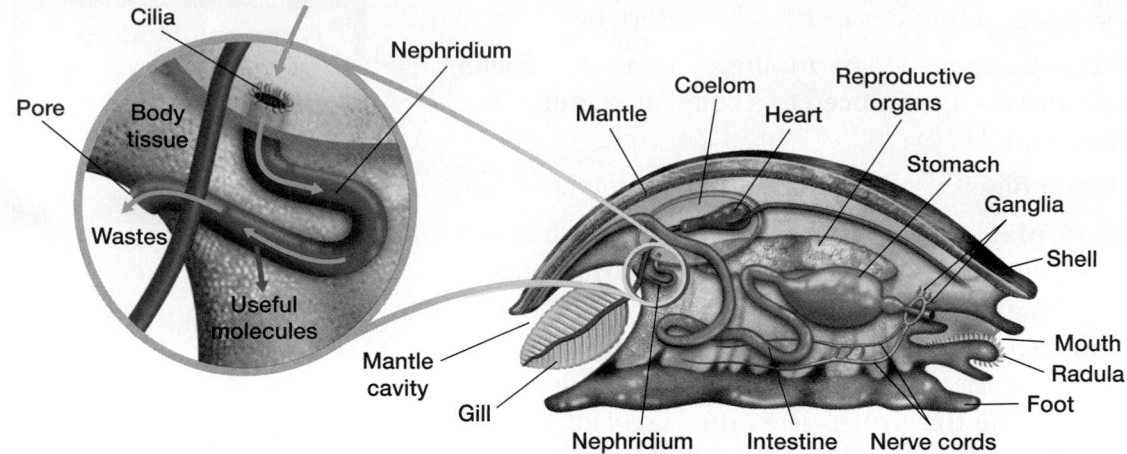

Organ Systems

Mollusks are the only coelomates without segmented bodies. Like the roundworms, mollusks have a one-way digestive system. Mollusks, however, are more complex than roundworms. As a group, mollusks are quite diverse, and no one mollusk can represent the phylum as a whole. **Figure 3** shows the basic mollusk body with organs that are characteristic of the phylum.

Excretion Mollusks are one of the earliest evolutionary lines to have developed an efficient excretory system. A mollusk's coelom is a collecting place for waste-laden body fluids. The beating of cilia pulls the fluid from the coelom into tiny tubular structures called **nephridia** *(nee FRIHD ee uh)*, also shown in Figure 3. The nephridia recover useful molecules (sugars, salts, and water) from the coelomic fluid. The recovered molecules are reabsorbed into the mollusk's body tissues. The remaining fluid waste leaves the mollusk's body through a pore that opens into the mantle cavity. Nephridia are found in all coelomate animals except arthropods and chordates.

Circulation The digestive tube of mollusks and other coelomates is surrounded by mesoderm, which acts as a barrier to the diffusion of nutrients into the cells of the body. Mollusks have a circulatory system. Recall that in a circulatory system, blood carries nutrients and oxygen to tissues and removes waste and carbon dioxide. Most mollusks have a three-chambered heart and an open circulatory system. Octopuses and squids are exceptions because they each have a closed circulatory system.

Respiration Respiration among mollusks is carried out in a variety of ways. Most mollusks respire with gills, which are located in the mantle cavity. Mollusk gills extract 50 percent or more of the dissolved oxygen from the water that passes over them.

In freshwater snails, ciliated gills beat on the inner surface of the mantle. This causes a continuous stream of water to pass over the gills. Most terrestrial snails have no gills. Instead, the thin membrane that lines their empty mantle cavity functions like a primitive lung. This membrane must be kept moist for oxygen to diffuse across it. Therefore, terrestrial snails, shown in **Figure 4,** are most active at night or after it rains when air has a high moisture content. During dry weather, a terrestrial snail pulls back into its shell and plugs the opening with a wad of mucus to keep water in. Sea snails also lack gills, and gas exchange takes place directly through their skin.

Reproduction Most species of mollusks have distinct male and female individuals, although some snails and slugs are hermaphrodites. Certain species of oysters and sea slugs are able to change from one sex to the other and back again. Many marine mollusks are moved from place to place as their trochophore larvae drift in the ocean currents. Octopuses, squids, freshwater snails, and some freshwater mussels have no free-swimming larvae. In these mollusks, the larval stage occurs within the egg, and a juvenile-stage mollusk hatches from the egg.

Figure 4 Terrestrial snails. Terrestrial snails are most active when the air is moist.

QUICK LAB

Modeling an Open Circulatory System

You can model an open circulatory system using simple items to represent the heart, blood vessels, blood, and body tissues of a living organism.

Materials

surgical tubing, 15 cm (about 6 in.) piece; clear plastic tubing, 15 cm (about 6 in.) and 7.5 cm (about 3 in.) pieces; shallow pan filled with water; eyedropper; food coloring

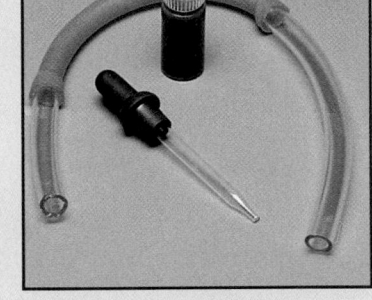

Procedure

1. Connect the surgical tubing to the two pieces of clear plastic tubing, as shown above.

2. Place the tubing into the tray filled with water. Allow the tubing to fill with water and rest on the bottom.

3. With the tubing still submerged, use an eyedropper to place two drops of food coloring into the short piece of clear plastic tubing.

4. With your thumb and index finger, squeeze along the piece of surgical tubing to pump the food coloring through the system.

5. As you continue to pump, observe the movement of food coloring.

Analysis

1. **Describe** what happened when you squeezed along the tubing.

2. **Identify** the structures represented by the pan of water, the surgical tubing, and the clear plastic tubing.

3. **Critical Thinking Evaluating Results** Evaluate your model's efficiency at pumping blood through the system.

4. **Critical Thinking Analyzing Methods** How does this model differ from a real circulatory system?

5. **Critical Thinking Analyzing Methods** How could you modify the model to make it more accurate?

Body Plans of Mollusks

The basic mollusk three-part body plan differs in each class of mollusks. As you read about the different classes of mollusks, you will see how the mollusk shell and foot are adapted for many different living conditions.

Gastropods

Gastropods—snails and slugs—are primarily a marine group that has successfully invaded freshwater and terrestrial habitats. They range in size from microscopic forms to the sea hare *Aplysia,* which reaches 1 m (almost 40 in.) in length. Most gastropods have a single shell. During the evolution of slugs and nudibranchs *(NOO dih branks),* the shell was lost completely. **Figure 5** shows three terrestrial tree snails and a nudibranch (sea slug). The foot of gastropods is adapted for locomotion. Terrestrial species secrete mucus from the base of their foot, forming a slimy path that they can glide along. Most gastropods have a pair of tentacles on their head with eyes often located at the tips.

Gastropods display varied feeding habits. Many are herbivores that scrape algae off rocks using their radula. Some terrestrial snails can be serious garden and agricultural pests, using their radula to saw off leaves. Sea slugs and many other gastropods are active predators. Whelks and oyster drills, for example, use their radula to bore holes in the shells of other mollusks. Then they suck out the soft tissues of their prey. In cone shells, such as the one shown in **Figure 6,** the radula is modified into a kind of poison-tipped harpoon that is shot into prey. The poison paralyzes the prey, which is then swallowed whole.

Throughout human history, snails have been a source of food for humans. Land snails belonging to the genus *Helix* are raised on snail farms and are consumed in great quantities. While freshwater snails are rarely eaten, a few marine species, such as conchs *(KAHNGKS),* are considered delicacies.

Tree snails

Sea slug

Figure 5　Snails and slugs. These Florida tree snails live where the air is moist enough to keep them from drying out. Sea slugs are often brilliantly colored, and most are under 15 cm (6 in.) in length.

Figure 6 **Cone shell.**
This cone shell searches the ocean bottom for prey. Once located, the prey is secured by the cone shell's radula and swallowed whole.

Bivalves

Most bivalves are marine, but some also live in fresh water. Many species of freshwater mussels are found throughout the rivers and lakes of North America and are important links in aquatic food chains. Oysters and mussels are important sources of food for humans. All bivalves have a two-part hinged shell. The valves, or shells, of a bivalve are secreted by the mantle. Two thick muscles, the **adductor muscles,** connect the valves. When these muscles are contracted, they cause the valves to close tightly. While most species of mollusks are sessile, some can move from place to place quite fast if necessary. For example, a swimming scallop opens and closes its valves rapidly. This pushes it along with the jets of water released when its valves snap shut.

Bivalves are unique among the mollusks because they do not have a distinct head region or a radula. A nerve ganglion above their foot serves as a simple brain. Most bivalves have some type of simple sense organs. For example, some bivalves have sensory cells located along the edge of their mantle that respond to light and touch.

Most bivalves are either male or female, but a few species are hermaphroditic. Bivalves reproduce sexually by releasing sperm and eggs into the water, where fertilization occurs. The fertilized eggs develop into free-swimming trochophore larvae. The larvae of a few freshwater mussels are brooded in a pouch within the mollusk's gills. The larvae are then released into the water, and they complete their larval stage as parasites on fish. This is a very unusual life cycle for a mollusk.

Most bivalves are filter feeders. Many, such as the clam illustrated in **Figure 7,** use their muscular foot to dig down into the sand. Once there, the cilia on their gills draw in sea water through hollow tubes called **siphons** (SIE fuhns). The water moves down one siphon tube, over the gills, and out the other siphon tube. The gills are used for feeding as well as respiration. A sticky mucus covers the gills, and as water moves over the gills, small marine organisms and organic material become trapped in the mucus. The cilia then direct the food-laden mucus to the bivalve's mouth.

WORD *Origins*

● The class name *Bivalvia* comes from the Latin *bi*, meaning "two," and *valva*, meaning "part of a door."

Figure 7 **Clam.** Many bivalves, like this clam, burrow into sand or mud and feed by drawing sea water in one siphon and expelling it out the other.

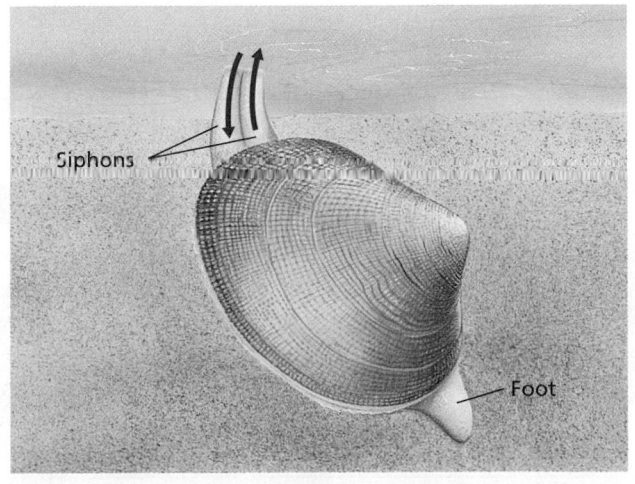

Siphons

Foot

Figure 8 Pearl oyster.
The best pearls come from oysters belonging to the genus *Pinctada.*

Like clams, oysters and scallops use their gills to filter food from the water. Oysters are permanently attached to rocks in the open water, where they feed. Scallops swim, and water passes over their gills as they move.

Many species of bivalves, such as the oyster in **Figure 8,** produce pearls. Pearls form when a tiny foreign object, such as a grain of sand, becomes lodged between the mollusk's mantle and shell. Bivalves respond to these irritants by coating them with thin sheets of nacre *(NAY kuhr),* also called mother-of-pearl. Nacre is the same hard, shiny substance that composes the inner shell surface. Successive layers of nacre are added until the foreign body is completely enclosed in the newly formed pearl.

While many bivalves form pearls, only a few species produce the beautifully colored nacre essential for gem-quality pearls. In fine pearls, the nacre contains tiny, overlapping mineral crystals. These crystals act like prisms, breaking up any light that falls on them into rainbows of color. This is what gives these pearls their iridescence. The mineral crystals found in the nacre of ordinary pearls are larger and do not reflect light as beautifully.

Eating Mollusks Safely

Since prehistoric times, mollusks have served as a human food resource. Today the annual worldwide harvest for bivalve mollusks alone amounts to an astounding 3 million metric tons (6,615,000,000 lb). But severe illness has been associated with eating mollusks. What causes these illnesses, and are mollusks really safe to eat?

Bivalves—oysters, mussels, clams, and scallops—are filter feeders. Any contaminant in a bivalve's environment circulates through and accumulates in its body. One source of contamination is water that is polluted with sewage, an ideal breeding ground for some bacteria and viruses. Because of this, harvesting is prohibited in certain areas. The primary danger is from eating infected bivalves raw. Contrary to popular opinion, hot sauce will not kill dangerous organisms that infect them. But it is now also evident that brief cooking, such as lightly steaming clams, does not destroy all pathogens either.

Follow the Guidelines

An estimated 20 million Americans eat oysters, and the yearly consumption of mollusks is increasing. To reduce the risk of illness from eating mollusks, the Food and Drug Administration (FDA) has developed guidelines for preparing them. If you observe these guidelines, the health risk from eating mollusks is very low. Avoid eating raw bivalves. Make your purchases from reputable sources to assure that the mollusks were not illegally harvested. Test for freshness by trying to move the shells sideways. If the

shells move, the mollusk is not fresh and should be discarded. Follow FDA recommendations for proper cooking times and temperatures. By following these guidelines, you can safely enjoy this delicious and healthful food choice.

Cephalopods

Squids, octopuses, cuttlefish, and nautiluses are all cephalopods. Most of their body is made up of a large head attached to tentacles (a foot divided into numerous parts), as shown in **Figure 9.** The tentacles are equipped with either suction cups or hooks for seizing prey. Squids have 10 tentacles, while octopuses have eight. The nautilus has 80–90 tentacles, although they are not nearly as long as those of the other cephalopods. Although cephalopods evolved from shelled ancestors, most modern cephalopods lack an external shell. The nautilus is the only living cephalopod species that still has an outer shell. Squids as well as the cuttlefish have a small internal shell. Cuttlefish "bones" are often attached to bird cages to provide calcium for canaries and other pet birds.

Cephalopods are the most intelligent of all invertebrates. They have a complex nervous system that includes a well-developed brain. Cephalopods are capable of exhibiting complex behaviors. Octopuses can easily be trained to distinguish between classes of objects, such as between a square and a cross, and they are the only invertebrates with this ability.

The structure of a cephalopod eye is similar in many ways to that of a vertebrate eye, and some species have color vision. The eyes of a squid can be very large. A giant squid that washed up on a beach in New Zealand in 1933 had eyes that were 40 cm (about 15.75 in.) across. At over 20 m (65 ft) in length, the giant squid is the largest of all invertebrates and has the largest eyes known in any animal.

Like most aquatic mollusks, cephalopods draw water into their mantle cavity and expel it through a siphon. In squids and octopuses, this system functions as a means of jet propulsion. When threatened, they quickly close their mantle cavity, causing water to shoot forcefully out of the siphon. Squids and octopuses can also

WORD Origins

● The name *cephalopod* comes from the Greek *kephalicos*, meaning "head," and *pous*, meaning "foot."

Figure 9   Cephalopods. Like all cephalopods, the squid is an active predator. Cuttlefish are agile swimmers that hunt at night, seeking small fishes and crustaceans. The nautilus swims with its coiled shell positioned over its head.

Nautilus

Squid

Cuttlefish

release a dark fluid that clouds the water and conceals the direction of their escape. The ink of the cuttlefish contains a reddish brown pigment called sepia. For centuries this ink was used by artists as a pigment and is found in many famous paintings.

All cephalopods are active marine predators. They feed on fish, mollusks, crustaceans, and worms. Once the prey has been snared by the tentacles, it is pulled to the mouth, where it is torn apart by strong, beaklike jaws. The cephalopod's radula then pulls the pieces into the mouth.

Analyzing the Molluscan Body Plan

Background

Mollusks share many common characteristics, yet there is great variety among the classes. The drawings on the right show how the shell (brown) and foot (green) vary in three classes of mollusks. Use the drawings to answer the analysis questions.

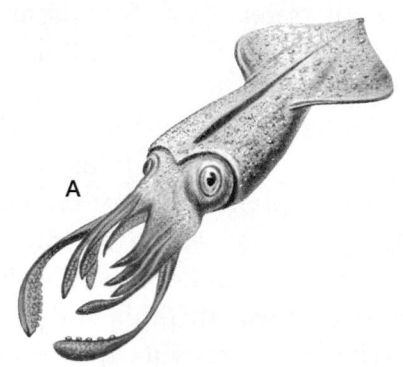

A

Analysis

1. **Determine** the class of mollusk *A*, mollusk *B*, and mollusk *C*.

2. **Compare** the shell modifications. Why might a shell suited to one mollusk be inappropriate for another?

3. **Critical Thinking Identifying Functions**
For each class shown, explain how the foot is useful for the animal's environment or kind of movement.

4. **Critical Thinking Predicting Outcomes**
Terrestrial snails and slugs are nearly identical except that slugs do not have a shell. Acidic forest soils are often poor in minerals, including calcium. Alkaline or neutral soils are rich in minerals. In which kind of soil would you be more likely to find a slug? Explain your answer.

B

C

Section 1 Review

1. **Identify** two characteristics that mollusks and annelids have in common.

2. **Summarize** six characteristics common to most groups of mollusks.

3. **Describe** how a nephridium functions in waste removal.

4. **Compare** the distinguishing features of each of the three major classes of mollusks.

5. **Critical Thinking Forming Hypotheses**
A chemical pollutant accidentally spills into a bay. One of the effects of this chemical is that it paralyzes cilia. The next day almost all of the oysters in the bay are dead. Develop a hypothesis that explains why the oysters died.

6. **Standardized Test Prep** A gastropod's radula is part of what organ system?
   A respiratory
   B circulatory
   C digestive
   D excretory

Annelids

The First Segmented Animals

You have probably heard the expression "a can of worms," which calls up an image of a lot of wiggly, wriggly creatures. An earthworm may come to mind, but there are many different species of segmented worms. Worms might not look like much, but this group of coelomates belongs to an ancient group, phylum Annelida. Annelid fossils can be found in rock that is 530 million years old. Scientists think that annelids evolved in the sea, where two-thirds of today's annelid species live. Most other annelid species are terrestrial earthworms. Annelids range in size from less than 1 mm (0.04 in.) long to more than 3 m (10 ft) long.

Annelids, such as the earthworm and fireworm shown in **Figure 10,** are easily recognized by their segments, which are visible as a series of ringlike structures along the length of their body. Each segment contains digestive, excretory, circulatory, and locomotor (movement) organs. Some of the segments are modified for specific functions, such as reproduction, feeding, or sensation. A well-developed **cerebral ganglion,** or primitive brain, is located in one anterior segment. The brain is connected to a nerve cord that runs along the underside of the worm's body.

Objectives

- **Identify** the major change in body plan that distinguishes annelids from mollusks.
- **Describe** the basic annelid body plan.
- **Describe** the annelid digestive system.
- **Compare** the three classes of annelids.

Key Terms

cerebral ganglion
septa
setae
parapodium

Figure 10 Annelids

The ringlike segments of this earthworm and marine fireworm identify them as annelids.

Earthworm

Fireworm

Evolutionary Milestone

6 Segmentation

Annelids were the first organisms to have a body plan based on repeated body segments. Segmentation underlies the body organization of all coelomate animals except mollusks.

Internal body walls, called **septa,** separate the segments of most annelids. Nutrients and other materials pass between the segments through the circulatory system. Sensory information is delivered by a nerve cord that connects nerve centers in the segments to the brain.

Characteristics of Annelids

In addition to segmentation, annelids share a number of other characteristics.

1. **Coelom.** The fluid-filled coelom is large and is located entirely within the mesoderm.

2. **Organ systems.** The organ systems of annelids show a high degree of specialization and include a closed circulatory system and excretory structures called nephridia. The gut has different regions that perform different functions in digestion.

3. **Bristles.** Most annelids have external bristles called **setae** *(SEET ee).* The paired setae located on each segment provide traction as the annelid crawls along. Some annelids, like the fireworm shown in Figure 10 on the previous page, also have fleshy appendages called **parapodia** *(par uh POH dee uh).*

QUICK LAB

Modeling a Closed Circulatory System

You can model a closed circulatory system using simple items to represent the heart, blood vessels, blood, and body tissues of a living organism.

Materials

clear plastic tubing, 30 cm (about 12 in.) piece; surgical tubing, 15 cm (about 6 in.) piece; shallow pan filled with water; eyedropper, food coloring

Procedure

1. Connect one end of the clear tubing to the surgical tubing. Submerge in the pan.

2. Use the eyedropper to insert several drops of food coloring into the surgical tubing. Close the tubing by attaching the two free ends together, as shown above.

3. With your thumb and index finger, squeeze along the piece of surgical tubing to pump the food coloring through the system.

4. Observe the food coloring as it moves through the tubing.

Analysis

1. **Describe** what happened when you pumped the food coloring through the system.

2. **Identify** what structures the pan of water, the surgical tubing, and the clear plastic tubing represent.

3. **Evaluate** your model's efficiency at pumping blood through the system.

4. **Critical Thinking Analyzing Methods** How could you modify the model to make it more accurate?

5. **Critical Thinking Inferring Relationships** If you did the Quick Lab in the previous section that models an open circulatory system, recall what happened in that system. Which model do you think exerted a greater pressure on the fluid in the tube?

Annelid Groups

Annelids differ in the number of setae (bristles) they have on each segment, and not all annelids have parapodia. These two external characteristics are used to classify annelids.

Marine Worms

Marine segmented worms are members of class Polychaeta *(PAHL ih keet uh),* the largest group of annelids. Polychaetes live in virtually all ocean habitats. They are often beautiful, showing unusual forms and iridescent colors. A distinctive characteristic of polychaetes is the pair of fleshy, paddle-like parapodia that occur on most of their segments. The parapodia, which usually have setae, are used to swim, burrow, or crawl. Parapodia also greatly increase the surface area of the polychaete's body, making gas exchange between the animal and the water more efficient.

Many polychaetes are burrowing species, but others live in protective tubes formed by the hardened secretions of glands located on their segments. Grains of sand or other foreign material may be cemented into the tube. Such tubeworms, like the feather duster shown in **Figure 11,** live with only their head stuck out of the tube. Featherlike head structures trap food particles from the water that passes over them. Other species of polychaetes feed by pumping water through their body. Free-swimming polychaetes, such as *Nereis* shown in **Figure 12,** are predators that use their strong jaws to feed on small animals.

Figure 11 Feather duster. Feather dusters filter-feed by trapping food particles in their featherlike head structures.

Figure 12 *Nereis*

Nereis, a polychaete worm, grasps its prey in its jaws, which open when it thrusts out its pharynx.

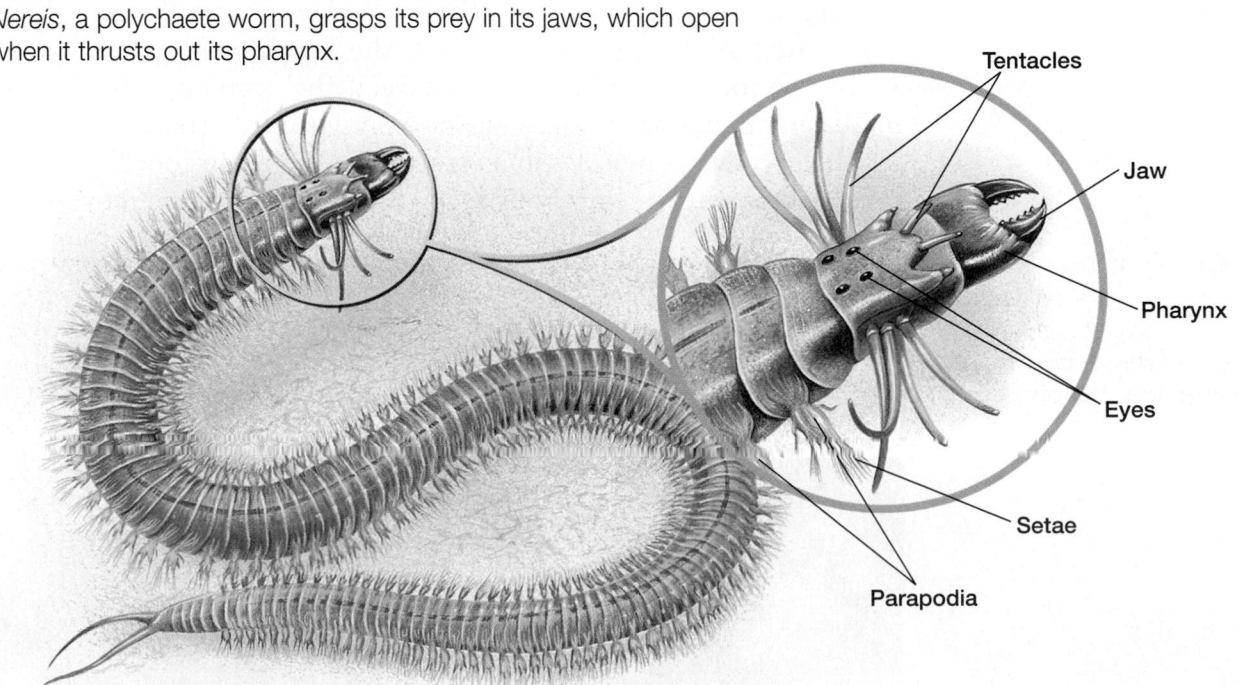

Tentacles

Jaw

Pharynx

Eyes

Setae

Parapodia

Earthworms

Earthworms and some related freshwater worms are members of the class Oligochaeta *(AHL ih goh KEET uh)*. Oligochaetes have no parapodia and only a few setae on each segment. Earthworms lack the distinctive head region of polychaetes and have no eyes. They do, however, have light-sensitive and touch-sensitive organs located at each end of their body. They have other sensory cells that detect moisture.

Earthworms, such as the ones in **Figure 13,** are highly specialized scavengers. They literally eat their way through the soil, consuming their own weight in soil every day. As they tunnel, earthworms take in organic matter and other materials using their muscular pharynx. The ingested soil moves through their one-way gut, down the esophagus and into a storage chamber called the *crop*. From here, the soil moves to an area called the *gizzard*. The grinding action of the gizzard crushes the soil particles together, breaking them down. The crushed material moves to the intestine, which extends to the posterior end of the earthworm's body. Digested food molecules are absorbed into the intestinal wall, and the remaining material passes out through the anus in a form called castings. The tunneling activity of earthworms allows air to penetrate the soil, and their castings fertilize it. Rich, organic soil may contain thousands of earthworms per acre.

Hydrostatic Skeleton

The fluid within the coelom of each body segment creates a hydrostatic skeleton that supports the segment. Each segment contains muscles that pull against this hydrostatic skeleton. Circular muscles wrap around the segment, while longitudinal muscles span its length. As shown in the *Up Close: Earthworm*, on the next page, when the circular muscles contract, the segment becomes longer. When the longitudinal muscles contract, the segment bunches up, increasing in diameter. An earthworm crawls by alternately contracting the two sets of muscles in its segments. The brain coordinates the muscular activity of each body segment, thus controlling movement.

Figure 13   Earthworms burrowing. Earthworms come to the surface only at night or during heavy rains. During dry or cold weather, they burrow deep into the soil and become inactive.

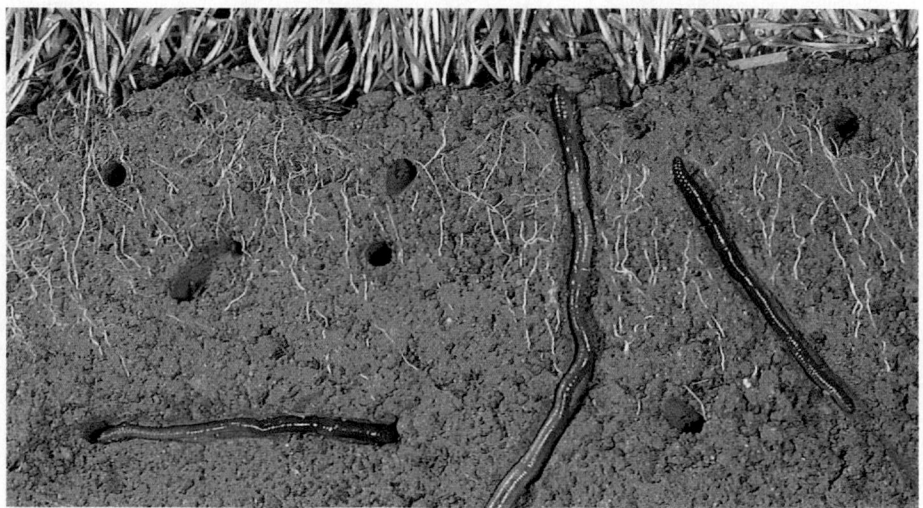

Up Close

Earthworm

- **Scientific name:** *Lumbricus terrestris*
- **Size:** Grows up to 30 cm (12 in.) long
- **Range:** Europe; eastern and northwestern North America
- **Habitat:** Damp soil
- **Diet:** Organic matter contained in soil

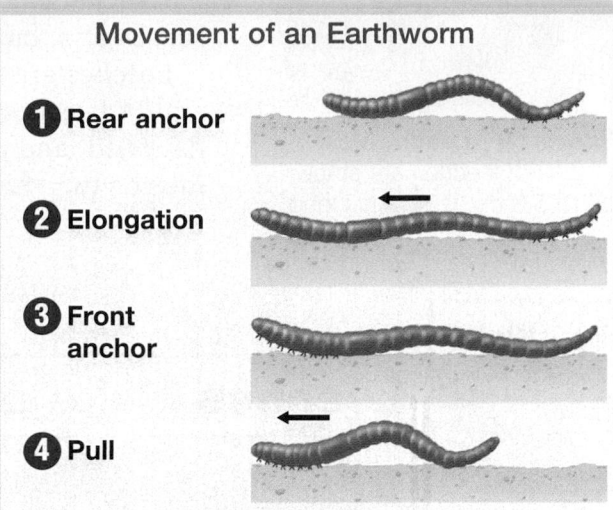

Movement of an Earthworm

1. **Rear anchor**
2. **Elongation**
3. **Front anchor**
4. **Pull**

Characteristics

Respiration Oxygen and carbon dioxide diffuse through the earthworm's skin. This exchange can take place only if the worm's skin is kept moist.

Digestion Earthworms "eat" soil, which is ground up in a thick, muscular gizzard. Food molecules pass across the walls of the intestine and are absorbed into the bloodstream.

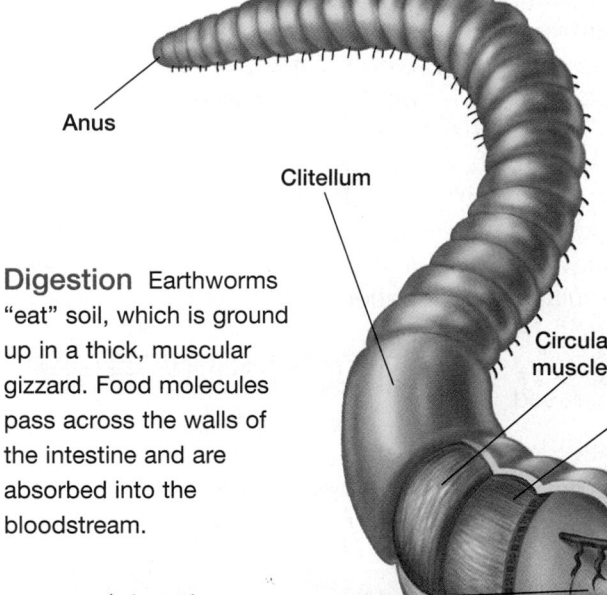

Anus

Clitellum

Circular muscles

▲ Longitudinal muscles

Dorsal blood vessel

Esophagus

Gizzard   Crop

Hearts

Pharynx

Mouth

▲ Intestine

Nephridium

Ventral blood vessel

Ventral nerve cord

Setae

◀ Reproductive organs

Segmental ganglion

▼ Cerebral ganglion (brain)

Movement As shown in the diagram, ❶ first the earthworm anchors several of its rear segments by sinking their setae into the ground. ❷ The worm then contracts the circular muscles in front of the anchored segments. This causes the anterior segments to elongate. ❸ Then the setae in front of the stretched region are anchored and the rear setae are released. ❹ The circular muscles relax and the longitudinal muscles contract, pulling the rear segments forward.

Reproduction Earthworms are hermaphrodites, each individual containing both sexes. Mating occurs when two earthworms join ventrally head to tail, exchanging sperm. During egg laying, the clitellum (a thickened, glandular ring of cells) of each worm secretes a mucous cocoon that encloses the fertilized eggs. Young worms emerge from the cocoon several weeks later.

Brain The brain coordinates the muscular activity of each body segment. It also processes sensory information from light-sensitive and touch-sensitive organs located at each end of the body.

Figure 14 Leech. Most species of leeches are small, 2.5 to 5.0 cm (1 to 2 in.) long.

Leeches

When people hear the word *leech,* they usually associate it with bloodsucking, and for a good reason. A leech, shown in **Figure 14,** has suckers at both ends of its body. Most species are predators or scavengers, but some are parasites of vertebrates and crustaceans.

Leeches are the only members of class Hirudinea *(hihr yoo DIHN ee uh).* Leeches lack both setae and parapodia. The body of a leech is flattened, and unlike other annelids, its segments are not separated internally.

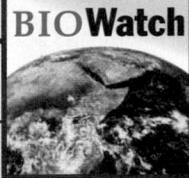 **BIO**Watch

Leeches Make a Comeback

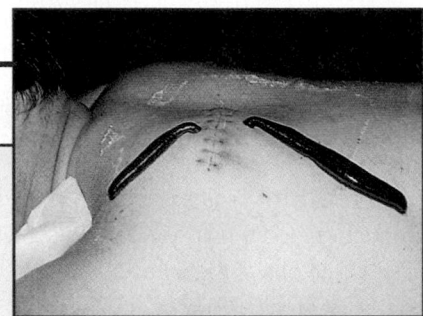

For many centuries, it was commonly believed that an excess amount of blood was the cause of a wide range of illnesses, from a fever or headache to severe heart disease. A standard treatment for these conditions was bloodletting using leeches. Physicians applied leeches to the patient's body, allowing the leeches to suck out the patient's "bad blood."

Use in Microsurgery

Although doctors no longer believe in "bad-blood," leeches are making a comeback in the field of health care. One use of leeches is during surgery to reattach severed limbs, fingers, or toes. In this type of operation, called microsurgery, the surgeon

uses tiny instruments and a microscope to reconnect tendons, blood vessels, and nerves. It is not possible to reconnect the smallest of the blood vessels, so circulation in the reattached part is usually poor. Often tissues in the region die, and the reattached part cannot heal. By applying leeches to suck out the accumulated blood, tissues remain healthy until new blood vessels grow and circulation is restored to normal. As a result, the success rate of surgery for reattachments has increased.

Other Applications

Leeches possess other useful qualities. Their saliva contains *anticoagulants*—substances that prevent blood from clotting—

and enzymes that can break up blood clots. It is not necessary to apply leeches to a patient to take advantage of these chemicals. Today these substances are produced through genetic engineering and have proven useful in the treatment of some heart patients.

internet connect

www.scilinks.org
Topic: Leeches
Keyword: HX4113

SCI*LINKS*® Maintained by the National Science Teachers Association

Section 2 Review

1. **Summarize** how you can tell if a wormlike organism is an annelid worm.

2. **Relate** an annelid's septa to its overall body plan.

3. **Describe** the major features of an earthworm's digestive system.

4. **Compare** the external appearance of marine annelids, earthworms, and leeches.

5. **Standardized Test Prep** What happens to a segment in an earthworm when the circular muscles in that segment contract?
   A It elongates.   C It increases in diameter.
   B It shortens.   D It bends to one side.

Key Concepts

1 Mollusks

- Many mollusks and annelids have a larval form called a trochophore.
- The mollusk body has three distinct parts: a visceral mass, a mantle, and foot.
- All mollusks except bivalves have a rasping tonguelike radula.
- Mollusks have a true coelom and well-developed organs.
- Most mollusks respire with gills but some respire with a primitive lung.
- Nephridia enable mollusks to recover the useful substances from their bodily wastes.
- Gastropods (snails and slugs) live in oceans, in fresh water, and on land.
- Bivalves (clams, oysters, and their kin) are aquatic and have hard shells called valves that protect their soft bodies.
- Gastropods and bivalves have an open circulatory system.
- Cephalopods (octopuses, squids, and their kin) have a well-developed head region, many tentacles, and a closed circulatory system. Most cephalopods have no external shell.

2 Annelids

- Annelids are coelomate worms that have segmented bodies and complex nervous systems.
- Annelids are classified according to the presence or absence of setae and parapodia.
- Annelids respire through their skin, and they have a closed circulatory system.
- Earthworms burrow through the soil, ingesting it as they crawl.
- Marine polychaetes have parapodia and setae. Some are active predators and others are filter feeders.
- Leeches lack parapodia and setae, and their segments are not separated internally. They may be aquatic or terrestrial, and some are parasites.

Key Terms

Section 1

trochophore (642)
visceral mass (643)
mantle (643)
foot (643)
radula (643)
nephridium (644)
adductor muscle (647)
siphon (647)

Section 2

cerebral ganglion (651)
septa (652)
seta (652)
parapodium (652)

Understanding Key Ideas

1. Both mollusks and annelids
   a. are coelomates.
   b. have at least a remnant of a shell.
   c. have no larval form.
   d. have a visceral mass.

2. Terrestrial snails respire
   a. with gills.
   b. with a primitive lung.
   c. through their skin.
   d. through their siphon.

3. Which of the following is *not* true of bivalves?
   a. They have a distinctive head region.
   b. They have sense organs.
   c. They have an open circulatory system.
   d. Most are filter feeders.

4. Cephalopods have all of the following characteristics *except*
   a. bilateral symmetry.
   b. a three-part body plan.
   c. an open circulatory system.
   d. a true coelom.

5. Annelids are divided into three classes. This classification is based on the number of setae and the presence or absence of
   a. segments.
   b. hearts.
   c. a gizzard.
   d. parapodia.

6. Blood in the circulatory system of an annelid
   a. flows into its body cavity.
   b. delivers carbon dioxide to its tissues.
   c. passes through gills.
   d. stays within its circulatory system.

7. Earthworm movement requires all of the following *except*
   a. circular muscles.
   b. secretion of mucus.
   c. muscle contractions.
   d. traction provided by setae.

8. ● **BIOWatch** What are some possible risks associated with cooking and eating bivalves purchased from an unknown vendor?

9. ● **BIOWatch** How can leeches be beneficial following microsurgery?

10. ⌗ **Concept Mapping** Make a concept map that shows the characteristics and diversity of the mollusks. Include the following words in your map: *foot, visceral mass, mantle, radula, siphons, gastropod, bivalve, cephalopod, open circulatory system,* and *closed circulatory system*.

Critical Thinking

11. **Recognizing Patterns** The gizzards of annelids that have returned to an aquatic environment over evolutionary time are smaller and less muscular than those of terrestrial annelids, such as earthworms. How do the differences between the two types of gizzards represent adaptations for aquatic and terrestrial feeding?

12. **Relating Concepts** Explain the significance of a coelom in the evolution of mollusks.

Alternative Assessment

13. **Finding and Communicating Information** Use the library or Internet resources to compile a list of mollusks used by humans as food sources. For each mollusk listed, tell where it is harvested and which part of the mollusk is eaten. Prepare a brochure to summarize your findings. Include a visual that informs the reader of the geographic location of the food source.

14. **Career Connection** **Worm Farmer** Research the field of growing segmented worms for use in research, as fishing bait, and for soil improvement. Your report should include a job description, training required, kinds of employers, growth prospects, and starting salary.

Standardized Test Prep

Understanding Concepts

Directions (1–4): **For** *each* **question, write on a separate sheet of paper the letter of the correct answer.**

1 What structure do mollusks use to extract useful molecules from coelomic fluid?
A. ganglia
B. nephridia
C. parapodia
D. radula

2 Which of the following is **not** a part of the mollusk's three-part body plan?
F. foot
G. mantle
H. radula
I. visceral mass

3 What structures connect the valves of bivalves?
A. adductor muscles
B. parapodia
C. siphons
D. visceral mass

4 What trait easily identifies an annelid?
F. body cavity
G. cephalization
H. nephridia
I. segmentation

Directions (5): **For the following question, write a short response.**

5 Classification of organisms has traditionally been based on physical similarities. However, gastropods, bivalves, and cephalopods are physically quite different. Given the physical differences of gastropods, bivalves, and cephalopods, how did they come to be grouped into the phylum Mollusca?

Test TIP

Whenever possible, highlight or underline numbers or words that are critical to the correct understanding of a question.

Reading Skills

Directions (6): **Read the passage below. Then answer the question.**

A particular bivalve mollusk called a shipworm can do extensive damage to ocean pier pilings (supports) by burrowing into them with its radula. As a science project, a student decides to determine if a new paint can reduce shipworm damage more than other paints do.

6 What variables must the student control in order for the experimental results to be considered valid?
A. All variables except for the type of paint must be identical.
B. All variables except for the thickness of paint must be identical.
C. All variables except for the type of pier pilings must be identical.
D. All variables except for the numbers of mollusks used must be identical.

Interpreting Graphics

Directions (7): **Base your answer to question 7 on the diagram below.**

Structure of a Mollusk

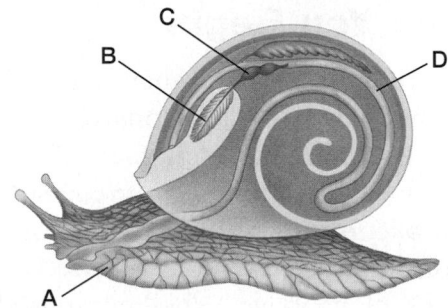

7 What structure is part of the respiratory system?
F. *A*
G. *B*
H. *C*
I. *D*

Skills Practice Lab

Observing Characteristics of Clams

SKILLS
- Observing
- Testing for the presence of a chemical

OBJECTIVES
- Observe the behavior of a live clam.
- Examine the structure and composition of a clam shell.

MATERIALS
- safety goggles
- lab apron
- live clam
- small beaker or dish
- eyedropper
- food coloring
- glass stirring rod
- clam shell
- Petri dish
- scalpel
- stereomicroscope
- 0.1 M HCl

ChemSafety

CAUTION: Always wear safety goggles and a lab apron to protect your eyes and clothing.

CAUTION: Do not touch or taste any chemicals. Know the location of the emergency shower and eyewash station and how to use them. If you get a chemical on your skin or clothing, wash it off at the sink while calling to the teacher. Notify the teacher of a spill. Spills should be cleaned up promptly, according to your teacher's directions.

CAUTION: Glassware is fragile. Notify the teacher of broken glass or cuts. Do not clean up broken glass or spills with broken glass unless the teacher tells you to do so.

Before You Begin

Clams are **mollusks,** and they have a two-part shell. The body of a clam consists of a visceral mass and a muscular **foot.** There is no definite head. Two tubes, an **incurrent siphon** and an **excurrent siphon,** extend from the body on the side opposite the foot. Like all mollusks, clams have a shell composed of **calcium carbonate.** A membrane called the **mantle** lines the shell and forms successive rings of shell as a clam grows. The **umbo** is the oldest part of a clam shell. In this lab, you will examine live clams and clam shells.

1. Write a definition for each boldface term in the paragraph.

2. Based on your objectives, write a question you would like to explore about clams.

Procedure

PART A: Observe a Live Clam

1. Put on safety goggles and a lab apron.

2. Place a live clam in a small beaker or shallow dish of water. Using an eyedropper, apply two drops of food coloring near the clam, as shown above.

3. Observe and record what happens to the food coloring.

4. Using a stirring rod, touch the clam's mantle. **CAUTION: Touch the clam gently to avoid injuring it.**

5. Observe and record the clam's response to touch.

PART B: Observe a Clam Shell

6. Examine the concentric growth rings on the shell. Locate the knob-shaped umbo on the shell. Count and record the number of growth rings on the clam shell.

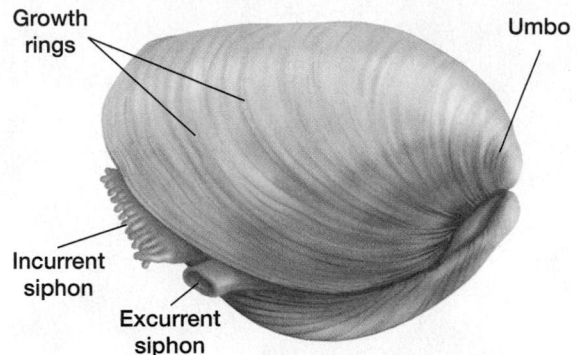

Growth rings

Umbo

Incurrent siphon

Excurrent siphon

7. Place the clam shell in a Petri dish. Use a scalpel to chip away part of the shell to expose its three layers. **CAUTION: Sharp or pointed objects may cause injury. Handle scalpels carefully.** View the shell's layers with a stereomicroscope. The outermost layer protects the clam from acids in the water. The innermost layer is mother-of-pearl, the material that forms pearls.

8. The middle layer of the shell contains crystals of calcium carbonate. To test for the presence of this compound, place one drop of 0.1 M HCl on the middle layer of the shell. **CAUTION: Hydrochloric acid is corrosive. Avoid contact with skin, eyes, and clothing. Avoid breathing vapors.** If calcium carbonate is present, bubbles of carbon dioxide will form in the drop. Record your observations.

PART C: Cleanup and Disposal

9. Dispose of solutions, broken glass, and pieces of clam shell in the waste containers designated by your teacher. Do not pour chemicals down the drain or put lab materials in the trash unless your teacher tells you to do so.

10. Clean up your work area and all lab equipment. Return live clams to the stock container. Return lab equipment to its proper place. Wash your hands thoroughly before you leave the lab and after you finish all work.

Analyze and Conclude

1. Analyzing Results Find the incurrent and excurrent siphons of the clam in the illustration on this page. Using this information, explain your observations in step 3.

2. Drawing Conclusions What is the purpose of a clam's shell?

3. Making Predictions Based on your observations, how do you think clams respond when they are touched or threatened in their natural habitat?

4. Forming a Hypothesis What does a clam take in from water that passes through its body?

5. Inferring Relationships Water that enters a clam's incurrent siphon passes over the clam's gills. How does this help the clam respire?

6. Further Inquiry Write a new question about clams that could be explored with another investigation.

On the Job

Geologists use hydrochloric acid to test rocks for the presence of calcium carbonate. Do research to discover the role that mollusks play in the formation of rocks. For more about careers, visit **go.hrw.com** and type in the keyword **HX4 Careers.**

European swallowtail

30 Arthropods

✔️ *Quick Review*

Answer the following without referring to earlier sections of your book.

1. **Summarize** the evolutionary movement of arthropods from the sea onto land. *(Chapter 12, Section 3)*

2. **Define** the term *chitin*. *(Chapter 22, Section 1)*

3. **Describe** the circulation of blood in an open circulatory system. *(Chapter 27, Section 2)*

4. **Define** the term *sessile*. *(Chapter 28, Section 1)*

Did you have difficulty? *For help, review the sections indicated.*

Reading Activity

As you read through this chapter, make an outline of the material presented. You may need to first review outlining methods, such as using Roman numerals, letters, and Arabic numerals. Review your outline to be sure that the main ideas are included.

📶 **internet** connect

www.scilinks.org
National Science Teachers Association *sci*LINKS Internet resources are located throughout this chapter.

*sci*LINKS® **Maintained by the National Science Teachers Association**

● The beautiful European swallowtail butterfly, *Papilio machaon,* produces one of two colors of pupa. Brown pupae are found on brown leaves or stems. Green pupae are found on green leaves or stems.

Features of Arthropods

Objectives

● **Summarize** the evolutionary relationship of arthropods and annelids.

● **Identify** the three subphyla of arthropods.

● **Describe** the characteristics of arthropods.

● **Describe** how growth occurs in arthropods.

Key Terms

appendage
thorax
cephalothorax
compound eye
molting
trachea
spiracle
Malpighian tubule

Jointed Appendages

Whether you are looking at a scorpion or a leaf-footed bug, as shown in **Figure 1,** when you see an arthropod, you will probably notice its appendages. An **appendage** is a structure that extends from the arthropod's body wall. Unlike the parapodia and setae of annelids, arthropod appendages have joints that bend. The phylum name, Arthropoda, literally means "joint foot." A variety of jointed appendages are found in arthropods, including legs for walking, antennae for sensing the environment, and mouthparts for sucking, ripping, and chewing food.

Arthropods almost certainly share a distant common ancestor with the annelid worms. Like annelids, arthropods have a coelom and a segmented body. Arthropod fossils, some as much as 600 million years old, are among the oldest, best-preserved fossils of multicellular animals. Among the most numerous of the early arthropods were the now-extinct trilobites, which lived in the sea. Like modern arthropods, trilobites had segmented bodies and jointed appendages, and they were the first animals to have eyes capable of forming images. Trilobites became extinct about 250 million years ago. The first terrestrial arthropods were probably scorpions similar to the modern scorpion shown in Figure 1.

Figure 1 Arthropods

This leaf-footed bug and hairy desert scorpion belong to phylum Arthropoda.

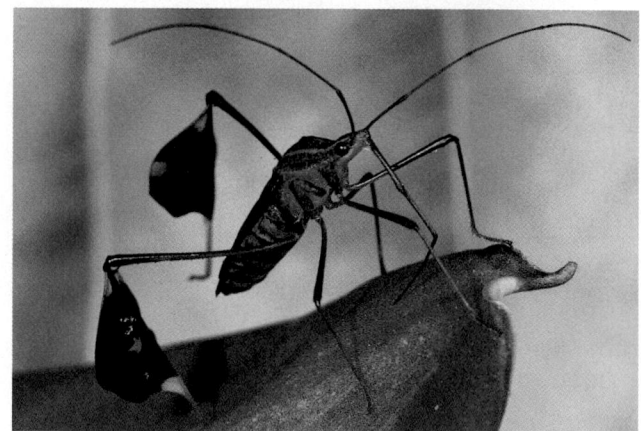

Leaf-footed bug

Scorpion

Evolutionary Milestone

7 Jointed appendages

Arthropods were the first animals to have jointed appendages. Joints permit powerful movement, aiding locomotion. Jointed appendages became specialized in many different ways, helping to create the vast diversity seen among the arthropods.

Arthropod Diversity

If a prize were given for sheer numbers, it would go to the arthropods. The total number of arthropod species exceeds that of all other kinds of animals combined. There may be 5,000,000 or more species of arthropods. There are more species of beetles alone than there are of vertebrates. Scientists estimate that 10^{18} arthropods are alive at any one moment! The great majority of arthropods are small, about 1 mm (0.04 in.) in length. Some parasitic mites are only 80 µm (0.003 in.) long. The largest arthropods are gigantic crabs 3.6 m (12 ft) across, found in the sea near Japan.

Living arthropods are traditionally divided into two groups, arthropods with jaws and arthropods with fangs or pincers. As shown in **Figure 2,** arthropods with jaws belong to either subphylum Uniramia *(yoo nuh RAY mee uh)* or to subphylum Crustacea *(kruhs TAY shuh)*. Arthropods with fangs or pincers belong to subphylum Chelicerata *(chuh LIS uh rahd uh)*. Each of these three subphyla represents a distinct evolutionary line.

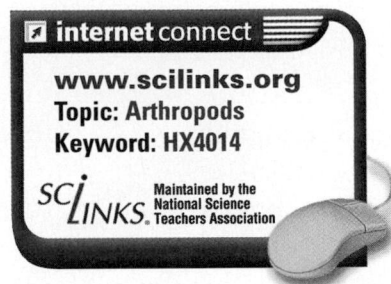

internet connect

www.scilinks.org
Topic: Arthropods
Keyword: HX4014

SCI*LINKS*. Maintained by the National Science Teachers Association

Figure 2 Phylogenetic tree

This phylogenetic tree shows the relationships among the three arthropod subphyla which form two groups: those with jaws and those with fangs or pincers.

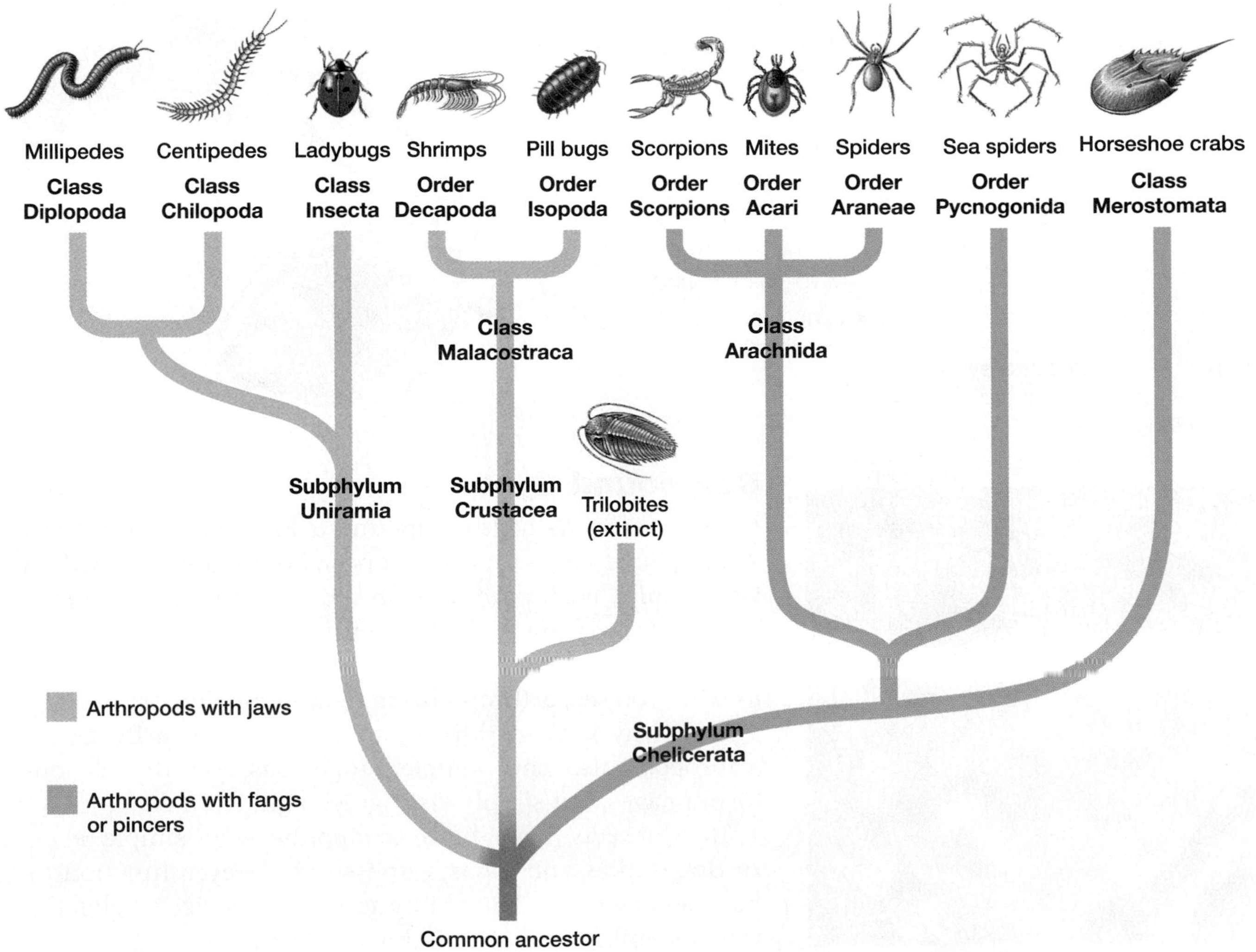

Millipedes — Class Diplopoda
Centipedes — Class Chilopoda
Ladybugs — Class Insecta
Shrimps — Order Decapoda
Pill bugs — Order Isopoda
Scorpions — Order Scorpions
Mites — Order Acari
Spiders — Order Araneae
Sea spiders — Order Pycnogonida
Horseshoe crabs — Class Merostomata

Class Malacostraca

Class Arachnida

Subphylum Uniramia

Subphylum Crustacea

Trilobites (extinct)

Arthropods with jaws

Arthropods with fangs or pincers

Subphylum Chelicerata

Common ancestor

Arthropod Body Plan

While arthropods may be quite different in appearance, they share a number of internal and external features, which are summarized in **Figure 3.** There is great variation in appearance among arthropod species, and not every species has every feature listed. However, these features are characteristic of the phylum as a whole.

Segmentation

In arthropods, individual body segments often exist only during the larval stage. For example, when you look at a butterfly larva (a caterpillar), you can easily see that it has many segments. However, if you look closely at an adult butterfly, you will see only three body regions. In most arthropods the many body segments fuse during development to form three distinct regions—the head, the **thorax** (midbody region), and the abdomen. In some arthropods, such as the crab shown at the top in Figure 3, the head is fused with the thorax to form a body region called the **cephalothorax.**

Figure 3 Arthropod characteristics. These eight characteristics are typical of arthropods, although not all arthropods show each characteristic.

Characteristics of Arthropods

- Jointed appendages
- Segmentation
- Distinct head, often with compound eyes
- Exoskeleton
- Respiration by gills, tracheae, or book lungs
- Open circulatory system
- Excretion through Malpighian tubules
- Wings on many arthropods

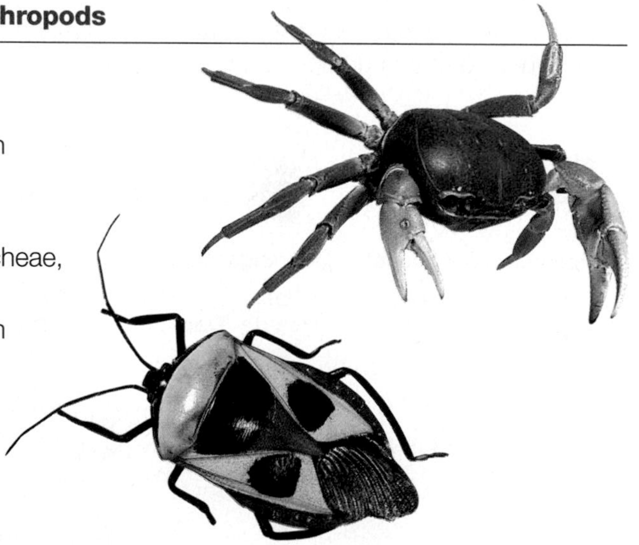

Figure 4 Compound eyes. The compound eye of this house fly is made of 800 or more individual units.

Magnification: 22×

Compound Eyes

Many arthropods have compound eyes, shown in **Figure 4.** A **compound eye** is an eye composed of multiple individual visual units, each with its own lens and retina. The brain receives input from each of the units, and then composes an image of an object. While the image formed is not as clear as what you see, arthropods see motion much more quickly. This is why it is so difficult to sneak up on a fly. Some arthropods also have simple, single-lens eyes that do not form images, but simply distinguish light from dark.

Most insects have both compound and simple eyes. In dragonflies and locusts, these simple eyes function as horizon detectors. The ability to see the horizon helps the insect stabilize its position during flight.

Exoskeleton

The outer layer of the arthropod body is a rigid exoskeleton (often called a shell) composed primarily of chitin. The exoskeleton is thin and flexible where the joints of the appendages are located. Muscles attached to the interior surfaces of the exoskeleton can pull against it, causing the animal's joints to bend. As shown in **Figure 5,** many arthropods can use their jointed appendages to perform complex movements. While chitin is tough, it is brittle and breaks easily. As arthropods increase in size, their exoskeletons must become thicker to withstand the pull of larger muscles without breaking. However, an increase in thickness of the exoskeleton adds weight, restricting the size arthropods can reach.

The exoskeleton of the different arthropod groups varies greatly in thickness. If you have ever attempted to swat a large insect, you know that its exoskeleton can be difficult to crush. Crustaceans, for example, have a thick, relatively inflexible exoskeleton. In comparison, the exoskeleton of other insects and some arachnids is fairly soft and flexible. Regardless of the nature of an arthropod's exoskeleton, it provides protection from injury and helps to prevent water loss.

Figure 5 Jointed appendages. The joints in the legs of this praying mantis permit it to perform many complex movements, such as manipulating prey.

QUICK LAB

Evaluating Jointed Appendages

To understand the importance of jointed appendages, test your range of movement without and with bending your joints.

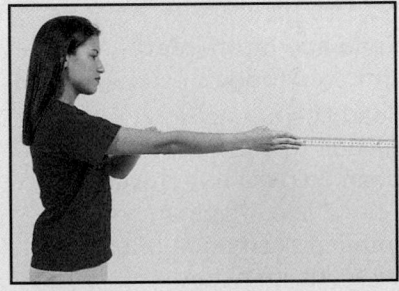

Materials

meterstick, paper, and pencil

Procedure

1. Work in pairs, and assign one person to be the test subject and one person to record the data.

2. The test subject extends one arm straight out in front of the body. The subject then places a meterstick along the inside of the arm, as shown in the illustration. The elbow should not be bent.

3. The recorder measures and records the distance along the meterstick that the test subject can reach with extended (not bent) fingers.

4. The test subject now tries to increase the range of movement by bending the fingers only. The recorder measures and records the closest and farthest distance along the meterstick that can be reached.

5. The test subject now tries to increase the range of motion by bending the elbow. The recorder measures and records the closest and farthest distance along the meterstick that can be reached.

Analysis

1. **Describe** how eating breakfast might be different if you did not have joints on your fingers and at your elbows.

2. **Predict** the advantages an animal with jointed appendages has over an animal without jointed appendages when capturing and consuming food.

3. **Predict** the advantages for an arthropod that has sense organs (eyes and odor detectors) on the ends of jointed appendages.

Figure 6 Molting. This green cicada is emerging from its old exoskeleton, which it leaves behind as a colorless ghost of itself.

Molting

A tough exoskeleton protects an arthropod from predators and helps prevent water loss. But an exoskeleton cannot grow larger, so an arthropod cannot simply grow bigger, as many other animals do. Imagine blowing up a balloon inside a soft drink can—after a certain point, the balloon cannot get any bigger. Arthropods have the same problem. In a process called **molting,** or ecdysis *(EHK duh sihs),* they shed and discard their exoskeletons periodically. Molting is triggered by the release of certain hormones. Just before molting, a new exoskeleton forms beneath the old one. When the new exoskeleton is fully formed, the old one breaks open. The arthropod emerges in its new, still-soft exoskeleton, as shown in **Figure 6.** The new exoskeleton hardens within a few hours or a few days, depending on the species.

Respiration

The majority of terrestrial arthropods respire through a network of fine tubes called **tracheae** *(TRAY kee ee),* as shown in **Figure 7.** Air enters the arthropod's body through structures called **spiracles** and passes into the tracheae, delivering oxygen throughout the body. Valves that control the flow of air through the spiracles and prevent water loss were a key adaptation for the first arthropods that invaded land more than 400 million years ago.

Figure 7 Tracheal system of a beetle

A complex series of hollow tubes called tracheae run through the bodies of most terrestrial arthropods.

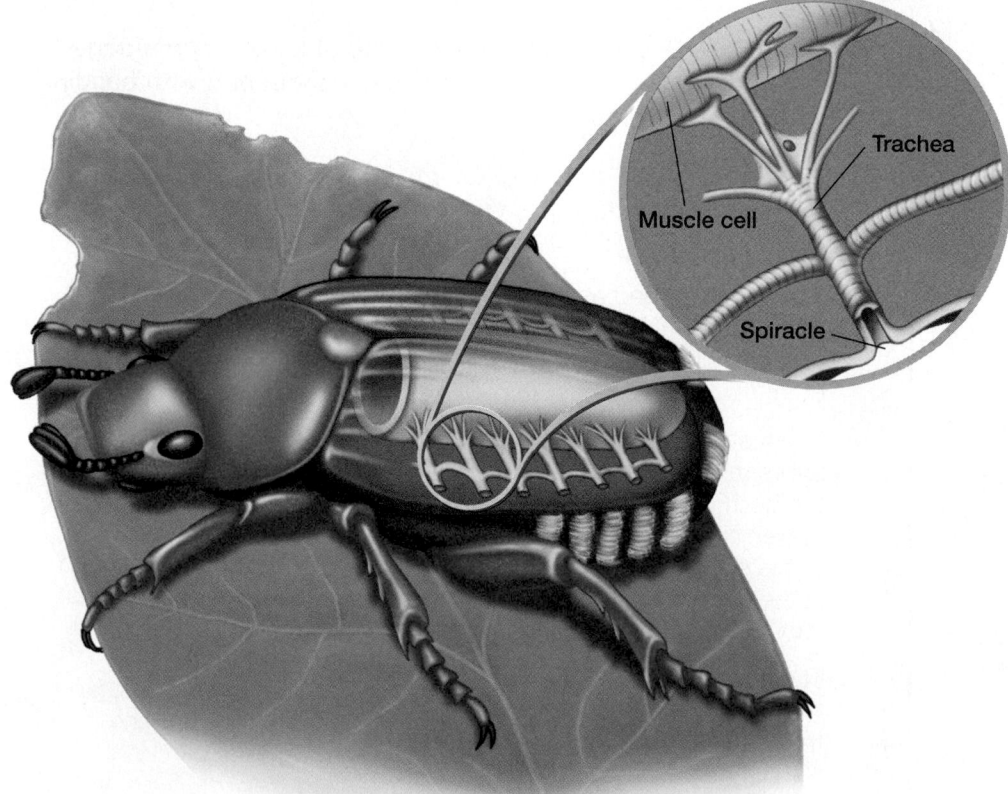

Trachea

Muscle cell

Spiracle

Excretion

Terrestrial arthropods have a unique excretory system that efficiently conserves water and eliminates metabolic wastes. This system is composed of excretory units called Malpighian *(mal PIHG ee uhn)* tubules. **Malpighian tubules** are slender, fingerlike extensions from the arthropod's gut that are bathed by blood. Water and small dissolved particles in the blood move through the tubules and into the arthropod's gut. As this fluid moves through the gut, most of the water, valuable ions, and metabolites from the fluid are reabsorbed into the arthropod's body tissues. Metabolic wastes remain in the gut and eventually leave the body through the anus. You can see the Malpighian tubules on the grasshopper in *Up Close: Grasshopper,* in Section 3 of this chapter.

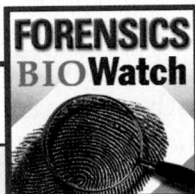

FORENSICS BIOWatch

Crawly Clues

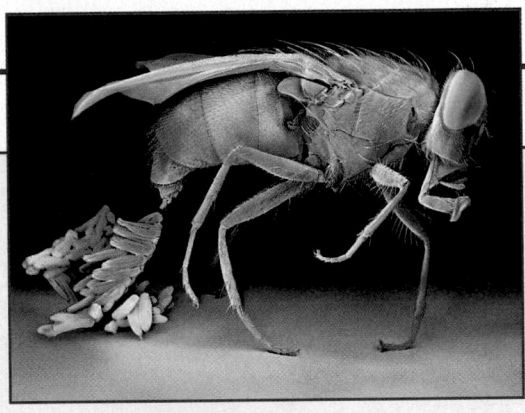

Entomologists are scientists who study the taxonomy, structure, and behavior of insects. Forensic entomologists apply their knowledge of insects to help solve crimes or resolve legal issues.

Postmortem Parade

Forensic medical entomologists are often asked to determine the time of a crime victim's death. To many insects, a victim's body is a banquet table. Blowflies (family Calliphoridae) and flesh flies (family Sarcophagidae) feed upon decaying human flesh. After the moment of death, certain species will appear at predictable intervals in a kind of grim postmortem parade. Blowflies, for example, typically infest a corpse beginning about two days after death. Other insect may come later.

Each species passes through each stage of metamorphosis at a precise rate. Insects develop most quickly in warmer weather. An experienced entomologist can sometimes identify the insect species based on the larval and pupal stages collected from the corpse. In some cases, the entomologist will raise the eggs, larvae, and pupae to adulthood to determine the species.

Location

Sometimes the kinds of insects found on a corpse can help determine the location of death. For example, blowflies are usually found in urban areas. A blowfly-infested corpse found along a rural road might have been killed in the city, and then moved to the countryside.

Section 1 Review

1 **State** the evolutionary relationship of the arthropods and the annelids.

2 **Describe** the three externally visible characteristics common to all arthropods.

3 **Summarize** how compound eyes function.

4 **Describe** how arthropods grow.

5 **Critical Thinking Relating Concepts** Draw a concept map of the three subphyla of arthropods, with two examples of each subphylum.

6 **Standardized Test Prep** Oxygen is delivered throughout an arthropod's body by the
A Malpighian tubules.   C exoskeleton.
B blood.   D tracheae.

Spiders and Other Arachnids

Objectives

- **Summarize** the characteristics of arachnids.

- **Identify** the internal and external characteristics of brown recluse spiders.

- **Compare** spiders, ticks, and mites.

- **Identify** the health threats posed by some arachnids.

Key Terms

chelicera
pedipalp
spinneret

Arachnid Modifications

Perhaps no other group of animals is more disliked and feared by humans than the arachnids—spiders, scorpions, ticks, mites, and daddy longlegs. While it is true that some spiders and scorpions are highly venomous, in general these creatures do more good than harm. For example, many spiders are major predators of insect pests, and gardeners usually welcome them. Arachnids *(uh RAK nihdz)* form the largest class in subphylum Chelicerata. Two minor classes, marine horseshoe crabs and sea spiders, also belong to this subphylum. The members of subphylum Chelicerata have mouthparts called **chelicerae** *(kuh LIS uh ree)* that are modified into pincers or fangs, as shown in **Figure 8.**

The arachnid body is made up of a cephalothorax and an abdomen. There are no antennae, and the first pair of appendages are chelicerae. The second pair of appendages are **pedipalps,** which are modified to catch and handle prey. (The pedipalps are sometimes specialized for sensory or even reproductive functions.) Following the pedipalps are four pairs of appendages called walking legs.

All arachnids except some mites are carnivores, and most are terrestrial. Since arachnids do not have jaws, they are able to consume only liquid food. To do so, the arachnid first injects its prey with powerful enzymes that cause the prey's tissues to liquefy. Then the arachnid sucks the liquid food into its stomach.

Figure 8 Chelicerae. The baboon spider's pointed black chelicerae (fangs) and its two pair of pedipalps are clearly seen in this close-up of its head region.

Spiders

The chelicerae of spiders are modified into fangs. Poison glands located in the spider's anterior end secrete a toxin through these fangs. The toxin kills or paralyzes the prey. The spider then injects enzymes into the prey that digest its tissues, and the spider sucks up the liquid food. Spiders are important predators of insects in almost every terrestrial ecosystem. Only two species of spiders living in the United States, the black widow and brown recluse, are dangerous to humans. Not all spiders build beautiful webs as the orb-builders do. Most spiders can secrete sticky strands of silk from appendages called **spinnerets** located at the end of the abdomen. Tubes located on some spinnerets do not produce silk. Instead, they excrete a sticky substance that the spider can use to make some silk strands adhesive.

Up Close

Brown Recluse Spider (Violin Spider)

- **Scientific name:** *Loxosceles reclusa*
- **Size:** Length of females, up to 10 mm (0.5 in.); males are smaller
- **Range:** South-central United States, from central Texas to Alabama, north to southern Ohio
- **Habitat:** Dark, dry sheltered sites outdoors or indoors
- **Diet:** Small insects

Characteristics

Cephalothorax Six eyes, in pairs, form a semicircle around the front of the cephalothorax. Two chelicerae and two pedipalps are located next to the mouth. Four pairs of walking legs attach to the cephalothorax, which is marked on top with a distinctive violin shape.

Abdomen The abdomen contains most of the spider's organs. Spinnerets located here are used to spin small, irregular webs.

Reproduction During mating, the male uses its pedipalps modified into sperm storage organs to insert sperm into the female's body. The female lays an average of 20–50 eggs inside a silk cocoon that she spins and hangs in her web.

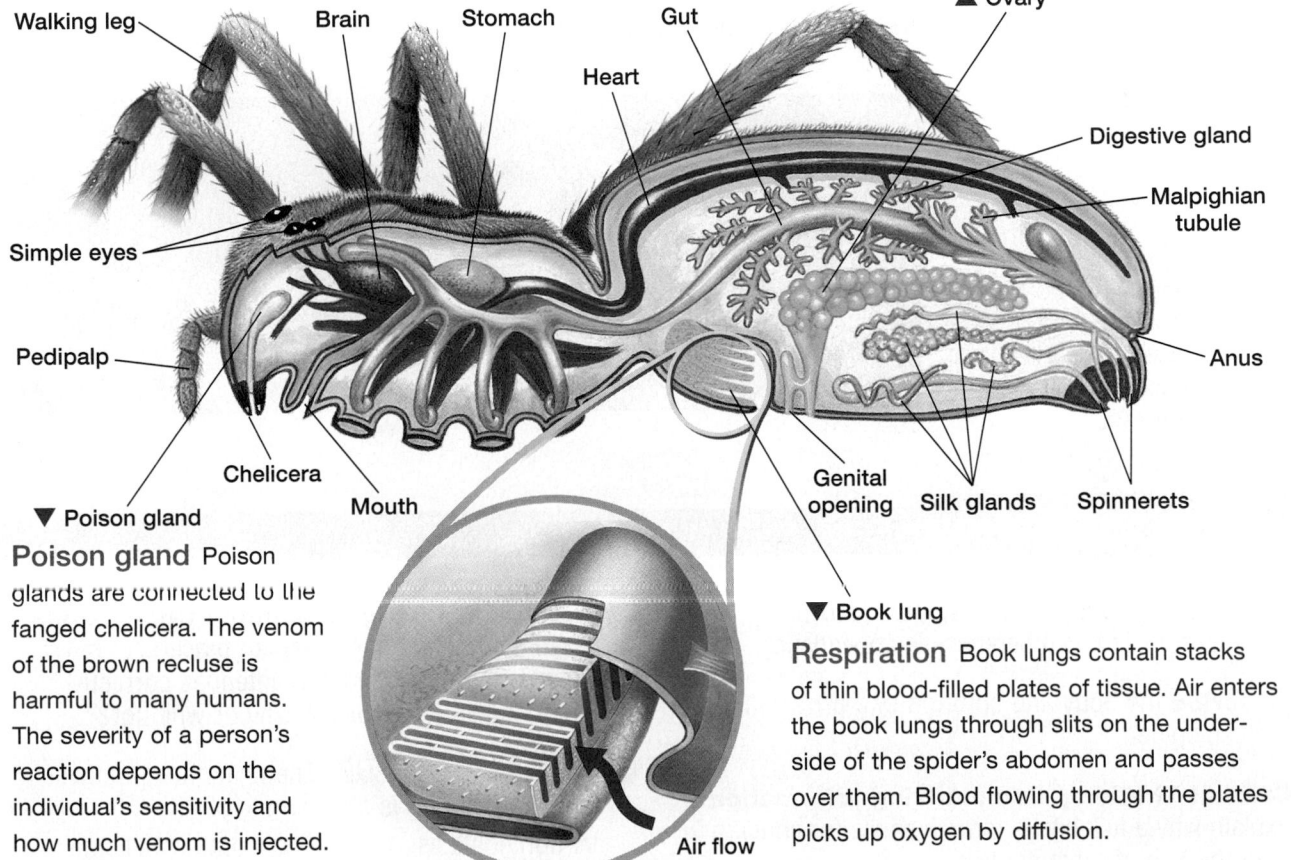

Walking leg · Brain · Stomach · Gut · Heart · ▲ Ovary · Simple eyes · Digestive gland · Malpighian tubule · Pedipalp · Anus · Chelicera · Mouth · Genital opening · Silk glands · Spinnerets · ▼ Poison gland · ▼ Book lung · Air flow

Poison gland Poison glands are connected to the fanged chelicera. The venom of the brown recluse is harmful to many humans. The severity of a person's reaction depends on the individual's sensitivity and how much venom is injected.

Respiration Book lungs contain stacks of thin blood-filled plates of tissue. Air enters the book lungs through slits on the underside of the spider's abdomen and passes over them. Blood flowing through the plates picks up oxygen by diffusion.

Scorpions and Mites

Two other familiar groups of arachnids are scorpions and mites. Like spiders, they have chelicerae and pedipalps, but these structures are modified differently.

Scorpions

Scorpions have long, slender, segmented abdomens that end in a venomous stinger used to stun their prey. The stinger-tipped abdomen is usually folded forward over the rest of the scorpion's body, a trait that makes scorpions instantly recognizable. The pedipalps of scorpions are large, grasping pincers, which are used not for defense but for seizing food and during sexual reproduction.

Mites

Mites are by far the largest group of arachnids. Some mites, including chiggers and ticks, are well known to humans because of their irritating bites. They are easily recognizable because their head, thorax, and abdomen are fused into a single, unsegmented body. Most adult mites, such as the one shown in **Figure 9,** are quite small, typically less than 1 mm (0.04 in.) long, but ticks grow larger. Many aquatic mites are herbivores, while terrestrial mites are usually predators.

Most mites are not harmful, but some are plant and animal pests. While feeding, plant mites may pass viral and fungal infections to the plant. Blood-sucking ticks attach themselves to a host, often a human. Lyme disease is spread by bites from infected deer ticks, like the one shown in Figure 9.

Figure 9 Mites. The house dust mite is a major cause of allergies in humans. The bite of an infected deer tick can cause Lyme disease.

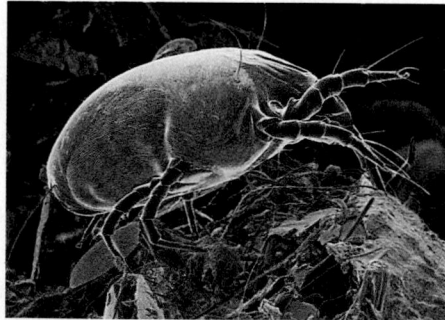

Dust mite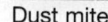

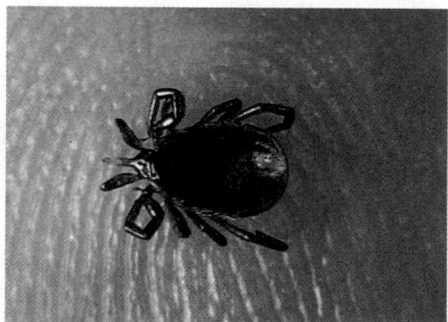

Deer tick

Section 2 Review

1. **Compare** the body plan of spiders, scorpions, and mites, including differences in appendages.

2. **Describe** the body and coloring of a brown recluse spider.

3. **Critical Thinking Summarizing Information** Explain why a tick bite is more a cause for concern than the bite of most spiders.

4. **Critical Thinking Evaluating Conclusions** Is an animal that has grasping pincers, a segmented body, and two antennae correctly identified as a scorpion? Why or why not?

5. **Standardized Test Prep** The relationship between a tick and a dog is an example of
   A homeostasis.   C commensalism.
   B parasitism.   D mutualism.

Insects and Their Relatives

Insect Diversity

Anyone who has ever been on a picnic in a wooded area does not have to be told that insects are numerous. Ants, mosquitoes, gnats, flies, bees, crickets—they all want to join in while the cicadas sing in the background. These animals all belong to the arthropod subphylum Uniramia, an enormous group of mostly terrestrial arthropods that have chewing mouthparts called **mandibles** (jaws). Uniramians consist of three classes: Insecta (insects), Diplopoda (millipedes), and Chilopoda (centipedes).

The insects are by far the largest group of organisms on Earth, with more than 700,000 named species. Most scientists agree that there may be several million insect species in existence, with most of the undiscovered species living in the tropics. As shown in **Figure 10,** more than 50 percent of all named animal species are insects. More than 90 percent of these species belong to one of the four orders shown in **Table 1.** To read about other orders of insects, see "A Six-Kingdom System of Classification" in the Appendix of this book.

Objectives

- **Describe** the characteristics of insects.
- **Compare** complete and incomplete metamorphosis.
- **Identify** the external and internal structures of the Eastern Lubber grasshopper.
- **Compare** millipedes and centipedes with insects.

Key Terms

mandible
metamorphosis
chrysalis
pupa
nymph
caste

Figure 10   Species of insects

The predominance of insects, especially beetles (Coleoptera), in the living world is illustrated by the blue section of this pie chart.

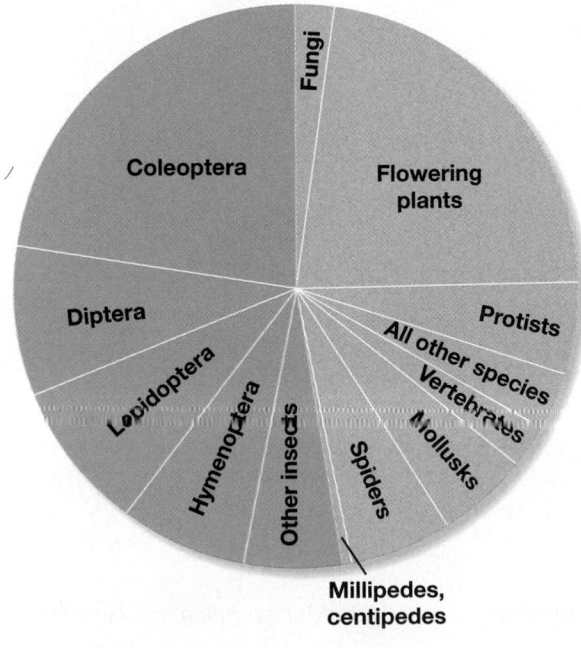

Millipedes, centipedes

Table 1 Four Orders of Insects

Order	Examples		Number of species
Coleoptera "shield winged"	Beetles, weevils		350,000
Diptera "two winged"	Flies, mosquitoes		120,000
Lepidoptera "scale winged"	Butterflies, moths		120,000
Hymenoptera "membrane winged"	Ants, wasps, bees		100,000

Insect Body Plan

Insects are primarily a terrestrial group, and aquatic insects probably had terrestrial ancestors. Although the great majority of insects are small (some are only a few centimeters in length), others are much larger. The African Goliath beetle, for example, exceeds 10 cm (4 in.) in length. Generally, the larger insects live in tropical areas. Despite great variation in their size, all insects share the same general body plan, made up of three body sections.

1. **Head.** Located on an insect's head are mandibles, specialized mouthparts, and one pair of antennae. The mandibles and mouthparts of different insect species are adapted for eating different foods, as shown in **Figure 11.** In addition, an insect's head usually has a relatively large pair of compound eyes and a pair of antennae. Like the mouthparts, antennae vary greatly in size and shape.

2. **Thorax.** The thorax is composed of three fused segments. Attached to the thorax are three pairs of jointed walking legs. Some insects, such as fleas, lice, and silverfish, lack wings, but other adult insects have one or two pairs attached to the thorax.

3. **Abdomen.** The abdomen is composed of 9 to 11 segments. In adult insects, there are no wings or legs attached to the abdomen.

Turn the facing page to learn more about one particular insect, the grasshopper, in *Up Close: Eastern Lubber Grasshopper.*

Study TIP

● **Organizing Information**
The numbered list on this page lists three important characteristics of insects. Use this information to draw a concept map that summarizes insect characteristics.

Figure 11   Insect mouthparts

The mouthparts of the different insect species are adapted for different functions.

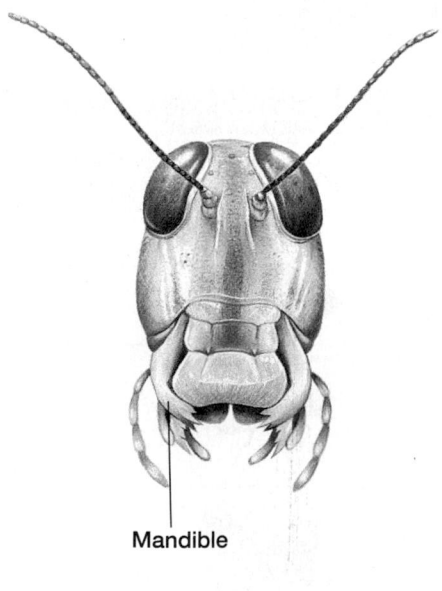

Grasshopper
(adapted for biting and chewing)

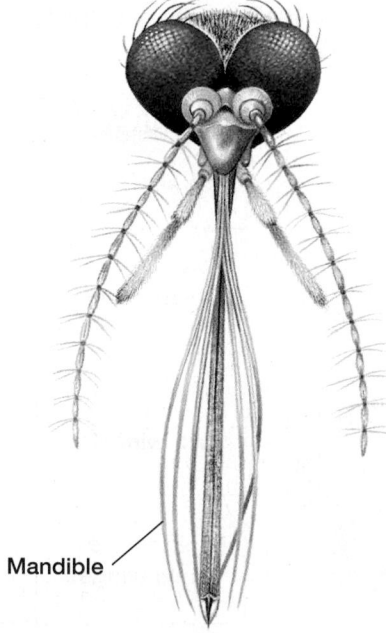

Mosquito
(adapted for piercing and sucking)

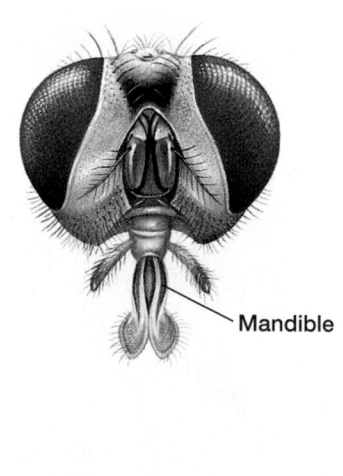

Fly
(adapted for sponging and lapping)

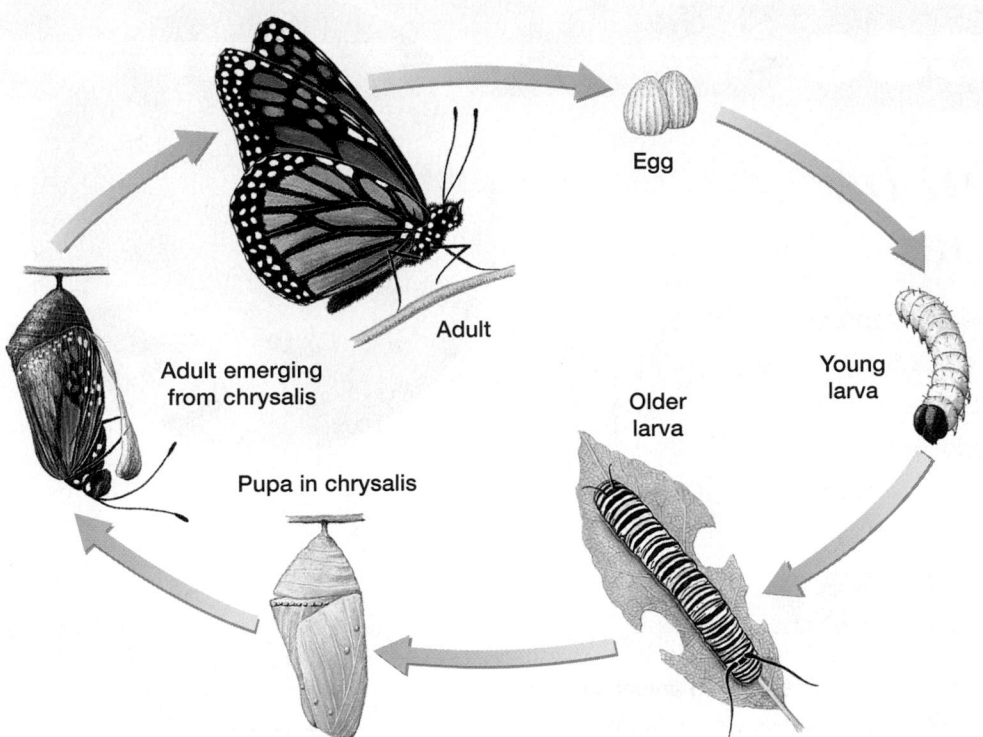

Egg

Adult

Adult emerging from chrysalis

Young larva

Older larva

Pupa in chrysalis

Insect Life Cycle

The life cycles of most insects are complex, and often several molts are required before the adult stage is reached. During the last molt, the young insect undergoes a dramatic physical change called **metamorphosis.**

Complete Metamorphosis Almost all insect species undergo "complete" metamorphosis, as shown in **Figure 12.** In complete metamorphosis, the wingless, wormlike larva encloses itself within a protective capsule called a **chrysalis** *(KRIHS uh lihs).* Here, it passes through a **pupa** stage, in which it changes into an adult.

A complete metamorphosis is a complex life cycle. The larvae can, however, exploit different habitats and food sources than adults. For example, the larvae of nectar-drinking butterflies are caterpillars that eat leaves! This ecological separation of young from adults eliminates competition. This increases the chance of survival for each phase of the life cycle.

Incomplete Metamorphosis A smaller number of species develop into adults in a much less dramatic incomplete metamorphosis, as shown in **Figure 13.** In these species, the egg hatches into a juvenile, or **nymph** *(NIHMF),* that looks like a small, wingless adult. After several molts, the nymph develops into an adult.

Figure 13 Incomplete metamorphosis. The nymph passes through several molts before it becomes an adult.

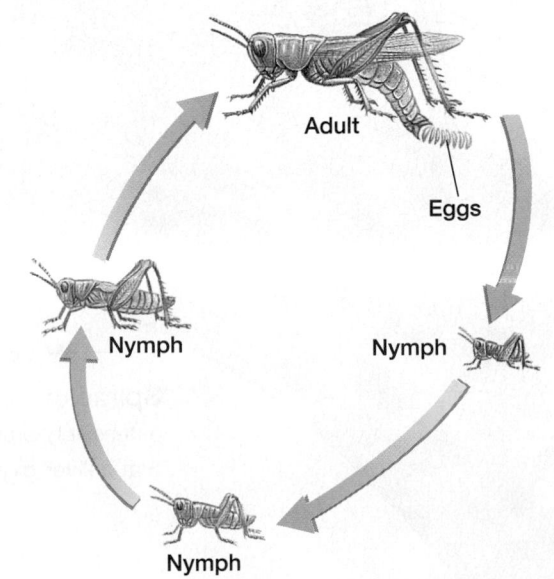

Adult

Eggs

Nymph

Nymph

Nymph

Up Close

Eastern Lubber Grasshopper

- **Scientific name:** *Romalea microptera*
- **Size:** 5 cm (2 in.) to 6.5 cm (2.6 in.) in length
- **Range:** Eastern United States
- **Habitat:** Fields and meadows
- **Diet:** Grasses and other leafy vegetation

External Structures

Thorax The thorax is composed of three fused segments, each with a pair of legs. The front two pairs are walking legs. The rear pair is larger jumping legs.

During mating season, males "sing" to potential mates by rubbing a row of pegs on a jumping leg against ridges on a forewing.

Wings Grasshoppers have a pair of leathery forewings that protect the more delicate flying wings (This grasshopper is flightless).

▲ Forewing

▲ Rudimentary flying wing

Jumping leg

Ovipositor

Head Two antennae contain sense organs for touch and smell. On each side of the head is a very large compound eye. Located high on the forehead are three light-detecting ocelli.

◄ Head

▲ Thorax

Antenna

Compound eye

Ocellus

Labrum
Mandible
Maxilla

Labium   ▼ Mouthparts

Walking leg

Abdomen

▼ Spiracles

Mouthparts The stiff upper labrum and lower labium (lips) hold a leaf or blade of grass in place while the mandibles (jaws), assisted by maxillas (graspers), tear off pieces of the plant.

Spiracles Spiracles admit air to the extensively branching system of tracheae that deliver oxygen throughout the body.

Internal Structures

Reproductive system The female collects the male's sperm in a storage pouch called a seminal receptacle. Later, the female digs a hole using two pairs of pointed ovipositors. As she releases the eggs into the hole, they are fertilized by the stored sperm.

Circulatory system A long blood vessel with a series of muscular "hearts" runs along the grasshopper's back. Blood is pumped out of the open system and bathes the body tissues directly before returning to the heart.

▲ Dorsal blood vessel

Hearts

Gastric ceca

Flying wing

Brain

▲ Seminal receptacle

Mouth

Anus

Salivary gland

Midgut

Gizzard (within gut)

▼ Crop

Ganglia

Malpighian tubules

▼ Nerve cord

Nervous system The nervous system is composed of a major ventral nerve cord with ganglia located in each body segment. Three fused ganglia in the head serve as the brain.

Digestive system Chewed food enters a storage pouch called a crop and passes to the gizzard, where it is shredded and crushed. Food is digested in the midgut, and food molecules pass through the midgut wall into the fluid of the coelom. This fluid eventually enters the circulatory system.

Figure 14   Insect flight.
This stop-action series shows how this insect's wings move during flight.

Flight

Insects were the first animals to have wings. For more than 100 million years, until flying reptiles appeared, insects were the only flying organisms. Flight, illustrated in **Figure 14,** was a great evolutionary innovation. Flying insects were able to reach previously inaccessible food sources and to escape quickly from danger.

An insect's wings develop from saclike outgrowths of the body wall of the thorax. The wings of adult insects are composed entirely of chitin, strengthened by a network of tubes called veins (which carry air and a bloodlike substance). In most insects, the power stroke of the wing during flight is downward, and it is produced by strong flight muscles. When at rest, most insects fold their wings over their abdomen, but a few insects are unable to do this. Dragonflies, for example, keep their wings outstretched when they rest beside a pond. Most insects have two pairs of wings. A few groups of insects, such as fleas and lice, are wingless.

In most insects only one pair of wings is functional for flight. In some species, the second pair of wings serves another purpose. For example, in grasshoppers and beetles, the forewings act as protective wing covers. In flies, the hindwings are modified into knoblike structures that help control stability during flight.

Analyzing the Effects of Pesticide Use

Background

In nature, insect pests are usually kept in balance by the presence of predators, including other insects. The use of some pesticides can upset this balance, as shown in the graph below. Examine the graph, and answer the analysis questions.

Analysis

1. **Identify** the years during which the two insect populations appear to maintain stability in relation to each other. Justify your answer with data from the graph.

2. **Describe** the relationship between the two insect species before year 4.

3. **Describe** the changes in the two populations after the use of a pesticide.

4. **Compare** the annual changes in population size of the pest species before and after the use of a pesticide.

5. **Critical Thinking Developing Hypotheses** Propose a hypothesis that might explain the dramatic changes that occur in the insect populations after the use of pesticides.

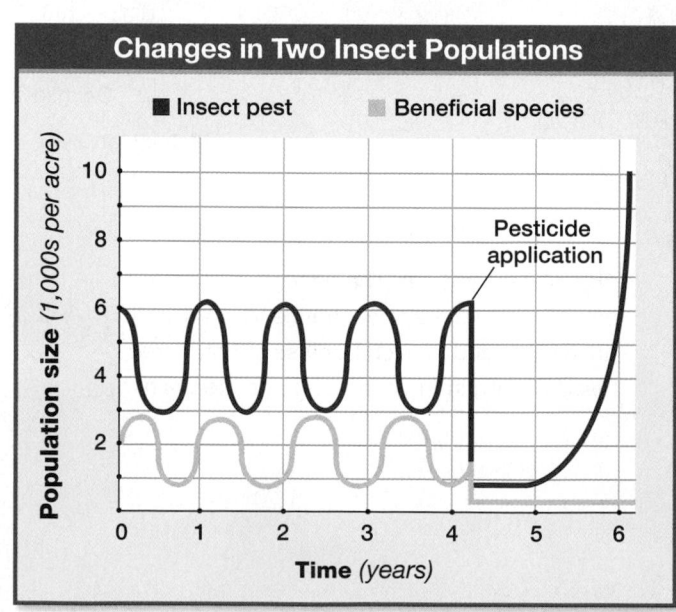

Social Insects

Two orders of insects, Hymenoptera (ants, bees, and wasps) and Isoptera (termites), have elaborate social systems. These insects often live in highly organized societies of genetically related individuals. Within these insect societies, there is a marked division of labor, with different kinds of individuals performing specific functions. The role played by an individual in a colony is called its **caste.** Caste is determined by a combination of heredity; diet, especially as a larva; hormones; and *pheromones,* chemical substances used for communication. In the termite colony shown in **Figure 15,** for example, small, active members called workers gather the food, raise the young, and excavate tunnels. Other, larger termites, called soldiers, defend the colony with their immense jaws. Both workers and soldiers are sterile. Reproduction is a function of only the queen and king.

Figure 15 Termite colony. Most of the members of this termite colony are unable to reproduce. The queen, with her enormous abdomen, is the egg-laying machine of the colony.

Insect Relatives

Centipedes and millipedes, shown in **Figure 16,** have similar bodies. Each has a head region followed by numerous similar segments. Each segment bears one or two pairs of legs. Centipedes have one pair of legs per segment and can have up to 173 segments. Modern millipedes have from 11 to 100 or more body segments, and most millipede segments have two pairs of legs. While centipedes are carnivores, most millipedes are herbivores.

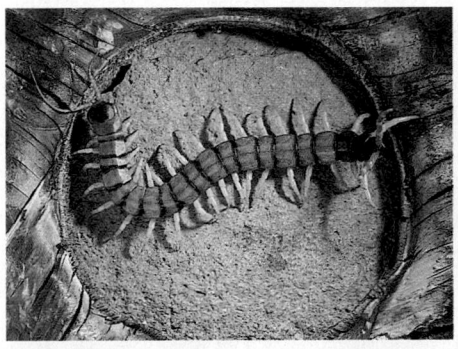

Centipede

Millipede

Figure 16 Centipedes and millipedes. Centipedes are carnivorous predators, while millipedes are herbivores that feed on decayed vegetation.

Section 3 Review

1 **Relate** the Eastern Lubber grasshopper's body plan to that of a typical insect.

2 **Compare** the life cycles of grasshoppers and butterflies.

3 **Identify** the distinguishing characteristics of insects, millipedes, and centipedes.

4 **Critical Thinking Forming Hypotheses** Based on the information given in Table 1, what characteristic is key to determining an insect's classification? Support your answer.

5 **Standardized Test Prep** A grasshopper's antennae contain sense organs for

A touch and smell.   **C** hearing and vision.

B smell and hearing.   **D** vision and touch.

Objectives

- **Summarize** how crustaceans and insects are similar and dissimilar.
- **Describe** the body plan of decapods.

Key Terms

nauplius
krill

Crustacean Habitats

Just as insect species dominate on land, crustaceans abound in the world's oceans. Their great numbers have earned them the nickname "the insects of the sea." Many are microscopic creatures that drift as plankton in the ocean currents. While primarily marine, members of subphylum Crustacea are also found in fresh water and in a few terrestrial habitats. Crustaceans include crabs, lobsters, crayfish, shrimps, barnacles, water fleas *(Daphnia)*, and pill bugs.

Many crustaceans have a distinctive larval form called a **nauplius** *(NAW plee uhs)*. The nauplius, shown in **Figure 17,** has three pairs of branched appendages. Like insects, the nauplius undergoes a series of molts before it takes on its adult form.

Adult crustaceans also have mandibles, as insects do. But crustaceans differ from insects in a number of important respects, as summarized in **Table 2.**

Terrestrial Crustaceans

Only a few crustacean groups have successfully invaded terrestrial habitats. The most widespread group of terrestrial crustaceans is composed of the pill bugs and sow bugs. They live among leafy ground litter found in gardens and woods. Pill bugs and sow bugs belong to a group called isopods and are the only crustaceans that are truly terrestrial. Another group, the sand fleas, includes several thousand species typically found along beaches. In addition, a few species of land crabs live in damp areas. Land crabs are only partly adapted to terrestrial living. They are active primarily at night, when the air is more moist. Their life cycle is tied to the ocean, where the larvae live until maturity.

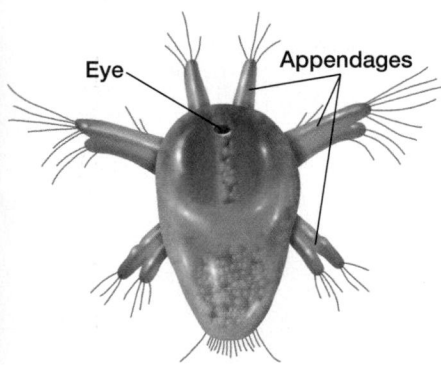

Figure 17   Nauplius.
A microscopic, free-swimming nauplius larva is a developmental stage of almost all crustaceans.

Eye   Appendages

Table 2 Comparison of Crustaceans and Insects		
Characteristic	**Crustaceans**	**Insects**
Nature of appendages	Most are branched at the end	Unbranched at the end
Antennae	Two pairs	One pair
Chewing appendages	Usually three pairs	One pair
Location of appendages	Cephalothorax and abdomen	Head and thorax
Respiration	Gills	Tracheal system

Aquatic Crustaceans

Crustaceans are a major food source for humans and some animals. The members of some orders of crustaceans are quite small. Common are fairy shrimps, water fleas, ostracods, and tiny copepods *(KOH puh pahds)*. Copepods are among the most abundant multicellular organisms on Earth and are a key food source in the marine food chain. Another small marine crustacean, *Euphausia superba,* swarms in huge groups and is known by its common name, **krill.** Krill, shown in **Figure 18,** are the chief food for many marine species.

Figure 18   Krill. Found in icy Antarctic waters, krill are the favorite food of many marine animals.

Decapods

Large marine crustaceans such as shrimps, lobsters, and crabs, along with the freshwater crayfish shown in **Figure 19,** have five pairs of legs and are often referred to as decapods. Almost one quarter of all crustaceans are decapods. The head and thorax of decapods are fused into a single cephalothorax, which is covered on top by a protective shield called a carapace.

In crayfish and lobsters, the anterior pair of legs are modified into large pincers called chelipeds *(KEE luh pehdz)*. Appendages called swimmerets are attached to the underside of the abdomen and are used in swimming and in reproduction. Flattened, paddle-like appendages called uropods are at the end of the abdomen. Many decapods have a telson, or tail spine. Decapods can propel themselves through the water by forcefully flexing their abdomen.

Figure 19   Crayfish

Like all decapods, the crayfish has five pairs of legs.

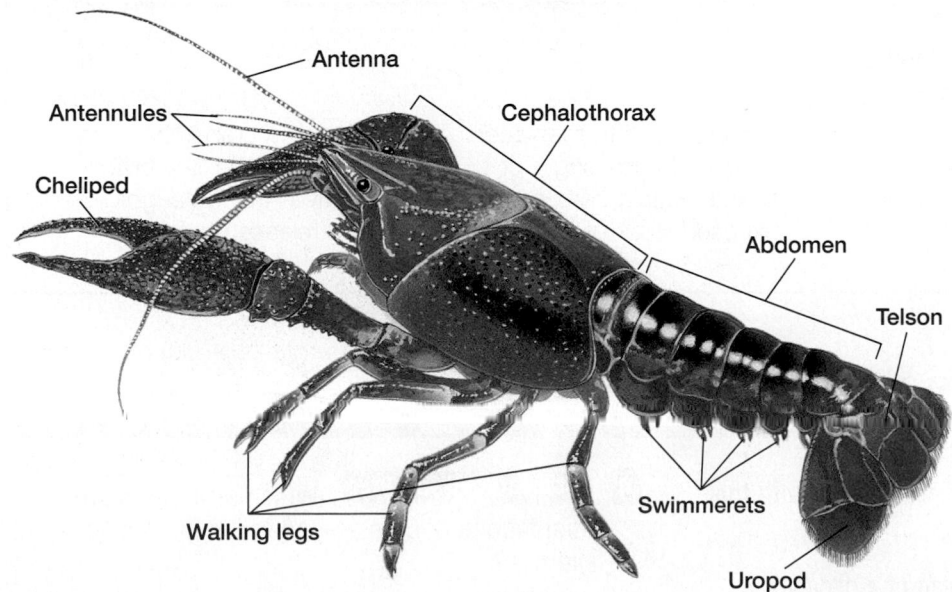

Sessile Crustaceans

Barnacles are a group of crustaceans that are sessile as adults. Free-swimming larvae attach themselves to a rock, post, or some other submerged object, where they remain. Hard plates that can open and close protect the barnacle's body. When feeding, barnacles extend their jointed feeding appendages (legs) through the open plates. Their feathery legs stir food from the water into the barnacles' mouth. Unlike most crustaceans, barnacles are hermaphrodites. However, they do not usually fertilize their own eggs.

DATA LAB

Relating Molting to Mortality Rates

Background

During the soft-shell stage that follows molting, many crustaceans die of disease or are eaten by predators. The bar graph below shows the percent mortality for crabs over a 9-month period. Study the data, and answer the analysis questions.

Analysis

1. **Summarize** what the data in the graph tell you about crab mortality.

2. **Summarize** what the graph shows about molting in crabs.

3. **Describe** the relationship between the mortality rates and molting periods of crabs.

4. **Critical Thinking Developing Hypotheses** Propose a hypothesis that explains the relationship between the percent of crabs molting and mortality rates.

5. **Critical Thinking Making Predictions** Most states have laws that require crab fishers to return molting crabs to the water. How might the length of time a molting crab is exposed to air or how roughly a crab is handled affect whether the crab survives being caught and released?

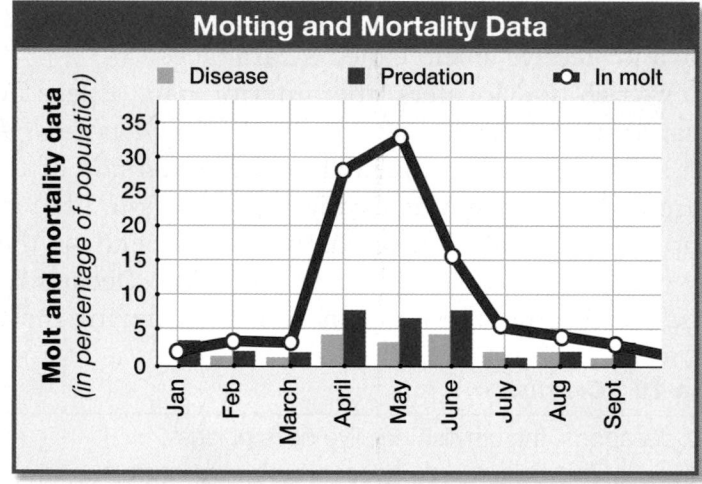

Section 4 Review

1 **Compare** the body plan of a crustacean with that of an insect.

2 **Describe** the external body plan of a decapod.

3 **Critical Thinking Making Predictions** Pill bugs respire with gills. How might this affect the distribution of pill bugs in an ecosystem?

4 **Standardized Test Prep** Which structures are adaptations of crustaceans to life in aquatic biomes?
   A antennae.
   B gills.
   C walking legs.
   D mandibles.

Key Concepts

1 Features of Arthropods

- All arthropods have a coelom, a segmented body, and jointed appendages that are modified to perform different functions.
- Arthropods have an exoskeleton made of chitin, which they discard periodically in a process called molting.
- Arthropods are grouped into three subphyla: Chelicerata, Uniramia, and Crustacea.

2 Spiders and Other Arachnids

- Members of subphylum Chelicerata have mouthparts, called chelicerae, that are modified into fangs or pincers.
- Spiders have a head and a cephalothorax, no antennae, six or eight pairs of simple eyes, a pair of fangs and pedipalps, and four pairs of walking legs.
- Scorpions have pedipalps modified into large, grasping pincers. A stinger is located at the end of their abdomen.
- Mites have body parts that are fused to form an unsegmented body. Many mites transmit diseases.

3 Insects and Their Relatives

- Insects make up more than half of all named animal species.
- All insects have a body plan with three body sections (head, thorax, and abdomen), three pairs of legs (all attached to the thorax), and one pair of antennae.
- The life cycles of insects are complex and involve a process of change called metamorphosis, during which larvae change into the adult insects.

4 Crustaceans

- Crustaceans have a distinctive larval form called a nauplius.
- Copepods and krill, which are tiny marine crustaceans, are the chief food of many marine species.
- Most crustaceans have branched appendages, two pairs of antennae, three chewing appendages, walking legs attached to the thorax, and gills. Like insects, crustaceans have jaws called mandibles.

Key Terms

Section 1

appendage (664)
thorax (666)
cephalothorax (666)
compound eye (666)
molting (668)
trachea (668)
spiracle (668)
Malpighian tubule (669)

Section 2

chelicera (670)
pedipalp (670)
spinneret (670)

Section 3

mandible (673)
metamorphosis (675)
chrysalis (675)
pupa (675)
nymph (675)
caste (679)

Section 4

nauplius (680)
krill (681)

Performance ZONE

CHAPTER REVIEW

Understanding Key Ideas

1. What evidence suggests that arthropods are closely related to annelids?
   a. Arthropods and annelids have gills.
   b. Both groups have marine species.
   c. Segmentation is present in both groups.
   d. Arthropods have vestigial parapodia.

2. Which is a feature of the arthropod body plan?
   a. a hydrostatic support system
   b. pharyngeal slits
   c. an exoskeleton
   d. a nonsegmented body

3. Arthropods molt because
   a. their body grows faster than their shell.
   b. of damage to their exoskeleton.
   c. their exoskeleton cracks and lets in water.
   d. their hard exoskeleton cannot grow larger.

4. In adult insects
   a. the abdomen has wings.
   b. there are two pairs of antennae.
   c. the legs are attached to the thorax.
   d. the first appendages are chelicerae.

5. Which of the following sequences shows a complete metamorphosis?
   a. egg → larva → pupa → adult
   b. egg → larva → adult
   c. egg → young juvenile → older juvenile
   d. egg → pupa → winged juvenile → adult

6. Millipedes and centipedes differ in that millipedes
   a. are terrestrial and segmented.
   b. have one pair of legs on each segment.
   c. have poisonous fangs.
   d. are herbivores.

7. Which of the following is *not* a crustacean?
   a. lobster   c. copepod
   b. scorpion   d. water flea

8. Copepods are said to be the most important animals on Earth because they are
   a. a critical link in the marine food chain.
   b. found in both the ocean and fresh water.
   c. accomplished predators.
   d. easier to collect and study than other arthropods.

9. **BIOWatch** Why might a forensic entomologist need to raise adult insects from eggs or immature forms?

10. **Concept Mapping** Construct a concept map that outlines the three major groups of arthropods and that gives the characteristics for each group. Try to include the following terms in your concept map: *appendages, cephalothorax, tracheae, spiracles, chelicerae, pedipalps, complete metamorphosis, chrysalis, pupa,* and *nauplius.*

Critical Thinking

11. **Recognizing Logical Inconsistencies** A neighbor commented that there was an increased number of insects around her house and that she was killing every spider she saw. How might these actions affect the number of insects around a house?

Alternative Assessment

12. **Forming a Model** Using papier mâché or some other material, make a model of a grasshopper or a spider. Decide in advance how many features of insects you must show to make an adequate model. Present your model to the class, and describe the structures that you modeled.

13. **Communicating Information** Use the media center or Internet sources to learn more about diseases transmitted by arthropods. Develop a brochure that presents your findings. The brochure should discuss the species that transmit the disease, as well as the symptoms and treatment of the disease.

Standardized Test Prep

Understanding Concepts

Directions (1–4): For *each* question, write on a separate sheet of paper the letter of the correct answer.

1 Which of the following structures are used by arachnids to catch and handle prey?
A. chelicerae   **C.** pedipalps
B. nephridia   **D.** spiracles

2 Through what structures do insects respire?
F. lungs   **H.** gills
G. ganglia   **I.** spiracles

3 The wings of the atlas moth, *Attacus atlas,* look much like the head of a snake. Why might it benefit a flying insect to look like a snake?
A. Its prey might think the insect is a snake and run away from it.
B. Snakes might come near the insect and protect it from predators.
C. Predators might think the insect is a snake and therefore not attack it.
D. The markings act like camouflage and allow it to hide on the head of a snake.

4 On what body part are a spider's spinnerets located?
F. abdomen
G. cephalothorax
H. pedipalps
I. thorax

Directions (5): For the following question, write a short response.

5 An unknown arthropod has three body segments, one pair of antennae, and three pairs of jointed legs. What kind of arthropod is it? Explain how you arrived at this conclusion.

Test TIP

Choose your answer to a question based both on what you already know and any information presented in the question.

Reading Skills

Directions (6): Read the passage below. Then answer the question.

If levels of vegetation were reduced in an environment, a species of arthropod that undergoes complete metamorphosis would likely have a greater advantage over one that undergoes incomplete metamorphosis. This is because, in complete metamorphosis, the larvae have different food sources from the adults. Thus, grasshoppers would likely not do as well as butterflies during a drought.

6 Why might this difference help more butterflies survive than grasshoppers?
A. Incomplete metamorphosis is slowed while complete metamorphosis speeds up during a drought.
B. One stage of the butterfly's lifecycle would not be affected, while both stages of the grasshopper's would suffer.
C. Grasshoppers would begin to undergo complete metamorphosis and would not be able to compete with butterflies.
D. The larvae of butterflies and the nymph of grasshoppers would perish, but both adult arthropods would not be affected.

Interpreting Graphics

Directions (7): Base your answer to question 7 on the diagram below.

Grasshopper External Structure

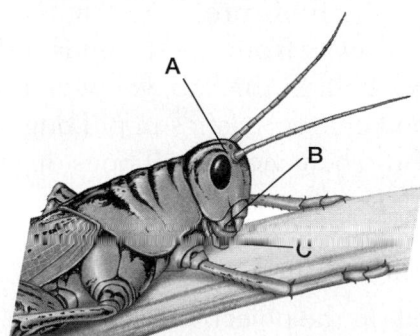

7 What are structures *B* and *C* used for?
F. inhaling and exhaling air
G. biting and chewing leaves
H. catching and handling prey
I. sponging and lapping liquids

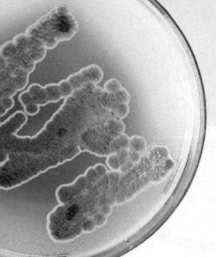

Exploration Lab

Observing Pill Bug Behavior

SKILLS
- Using scientific methods
- Observing

OBJECTIVES
- **Identify** arthropod characteristics in a pill bug.
- **Observe** the behavior of pill bugs on surfaces with different textures.
- **Infer** the adaptive advantages of pill bug behaviors.

MATERIALS
- 4 adult pill bugs
- 2 Petri dishes
- stereomicroscope or hand lens
- blunt probe
- fabrics with different textures
- scissors
- transparent tape
- clock or watch with second hand

Before You Begin

Pill bugs live in moist terrestrial environments, such as under rocks and logs. Like other **crustaceans,** pill bugs respire with gills and have hard outer shells and jointed appendages. They respond to a **stimulus,** such as light, moisture, or touch, by moving toward or away from the stimulus or by curling into a ball. In this lab, you will look for arthropod characteristics in pill bugs and observe the behavior of pill bugs on surfaces with different textures.

1. Write a definition for each boldface term in the paragraph above.

2. Based on the objectives for this lab, write a question you would like to explore about pill bug characteristics and behavior.

Procedure

PART A: Make Observations

1. Place a pill bug in a Petri dish, and observe it with a stereomicroscope or hand lens. Observe it from a dorsal viewpoint as well as from the side. List the characteristics that tell you the pill bug is an arthropod.

2. Touch the pill bug with a blunt probe. **CAUTION: Touch pill bugs gently to avoid injuring them.** Record your observations.

PART B: Design an Experiment

3. Work with the members of your lab group to explore one of the questions written for step 2 of **Before You Begin.** To explore the question, design an experiment that uses the materials listed for this lab.

4. Write a procedure for your experiment. Make a list of all the safety precautions you will take. Have your teacher approve your procedure and safety precautions before you begin the experiment.

5. ◆ Set up and carry out your experiment. **CAUTION: Sharp or pointed objects can cause injury. Handle scissors carefully.**

PART C: Cleanup and Disposal

6. ◆ Dispose of fabric scraps and broken glass in the designated waste containers. Put pill bugs in the designated container. Do not put lab materials in the trash unless your teacher tells you to do so.

7. ◆ Clean up your work area and all lab equipment. Return lab equipment to its proper place. Wash your hands thoroughly before you leave the lab and after you finish all work.

Analyze and Conclude

1. Analyzing Methods Why did you test several pill bugs in this investigation instead of just one pill bug?

2. Analyzing Results Did all of your pill bugs show a similar pattern of movement? Explain.

3. Graphing Results Make a graph of your data. Plot the average time spent on the material on the y-axis and the type of material on the x-axis.

4. Analyzing Results Rank the fabrics according to the total amount of time spent on them by the pill bugs.

5. Drawing Conclusions Which fabric texture do pill bugs seem to prefer?

6. Inferring Conclusions How is a pill bug's response to disturbances an advantage?

7. Inferring Conclusions How is being able to detect surface texture helpful to pill bugs in their natural habitat?

8. Further Inquiry Write a new question about pill bugs that could be explored with another investigation.

 Do You Know?

Do research in the library or media center to answer these questions:

1. What are the advantages and disadvantages of biological pest control when compared with the use of chemical pesticides?

2. What role do crustaceans play in the lives of humans?

Use the following Internet resources to explore your own questions about arthropods.

🖙 internet connect

www.scilinks.org
Topic: Biological Pest Control
Keyword: HX4021

SCLINKS. Maintained by the National Science Teachers Association

Should the genes for insect resistance be transferred into crops?

Insect-Resistant Crops

Each year, Americans use hundreds of millions of tons of pesticides to protect their homes, crops, forests, livestock, and pets from insects and other pests. A drawback of pesticides is that they remain effective only for a short time because pests acquire resistance to the effects of pesticides. Pest control has become a race between the scientists who are trying to invent new pesticides and the ever-evolving pests. And because of pests' adaptability—some insects have evolved resistance to a new insecticide within a single growing season—pests seem to have the edge.

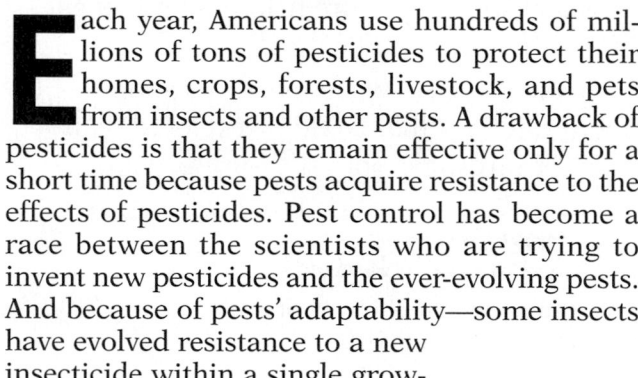

Pesticides from Bacteria

One pesticide with many advantages is the *Bt* toxin, which is derived from the common soil bacterium *Bacillus thuringiensis*. This bacterium produces a toxin that binds to a receptor protein in the digestive system of some insects, killing the insect. But only a few groups of insects are susceptible, including beetles, butterflies, flies, and ants. *Bt* toxin does not harm most beneficial insects, spiders and other arthropod predators of insects, or vertebrates, including humans.

Bt toxin has two other things going for it. First, different strains of the bacterium make different toxins that are effective against different pests. This gives farmers a choice of weapons to use, depending on which insects are attacking their crops. It also reduces the likelihood that resistance will evolve, because the odds that an insect will be resistant to more than one toxin are small. Second, *Bt* toxin breaks down into harmless byproducts within 2 or 3 days of being sprayed on a field. As a result, pests are exposed

Boll weevil
Boll weevils cause extensive damage to United States cotton crops each year.

Harvesting cotton
Seed pods (called bolls) contain fibers that are used to make fabric.

to less of the toxin, and the probability of their evolving resistance is reduced.

Genetic Engineering of Insect Resistance

Through genetic engineering, the genes for *Bt* toxins have been inserted into several crop plants, including corn, cotton, and potatoes. These plants now produce their own insecticides. During the 1990s, the Environmental Protection Agency (EPA), the federal agency that regulates pesticides, approved the sale of crops genetically engineered to produce *Bt* toxin. Since then, the area of farmland planted with the crops has grown rapidly.

Not everyone agrees with the EPA's decision that the crops are safe. Critics also note that the question is not whether resistance will become widespread but how fast that will happen. They point out that the genetically engineered plants produce the toxin continuously, which increases evolutionary pressure on the pests to adapt.

Supporters of the EPA's decision note that the farmers who buy the seeds are required to plant part of their land with nonengineered crops. This land will serve as a "refuge" on which susceptible

insects can live. Supporters reason that if any *Bt*-resistant insects appear, their resistance traits will be swamped when they interbreed with the nonresistant individuals from the refuge. Finally, because some pests attack both corn and cotton, the EPA prohibited the planting of genetically engineered corn and cotton in the same state. This should reduce the pests' exposure to *Bt* toxin, supporters say. ■

Analyzing STS Issues

Science and Society

1 What are alternatives to synthetic chemical pesticides? Use library resources or the Internet to research integrated pest management. How does it differ from traditional methods of controlling pests? What are some of its shortcomings?

2 Should genetically engineered crops be labeled? Current regulations do not require labeling of genetically engineered crops that are sold in stores. Many consumer advocates contend that all such crops should be labeled so that consumers know what they are eating. Research both sides of this issue, and then write a logical, persuasive essay arguing either for or against labeling.

3 What is biological control? Biological control is an alternative to chemical pesticides. Using library resources or the Internet, research the topic of biological control. What is biological control? Describe one example of biological control.

Technology: Genetic Engineering

4 How do scientists genetically engineer plants? Use library resources or the Internet to find the two main methods for inserting genes into plant cells. Why do scientists want to add genes for herbicide resistance to plants? What are the benefits to plants such as cotton?

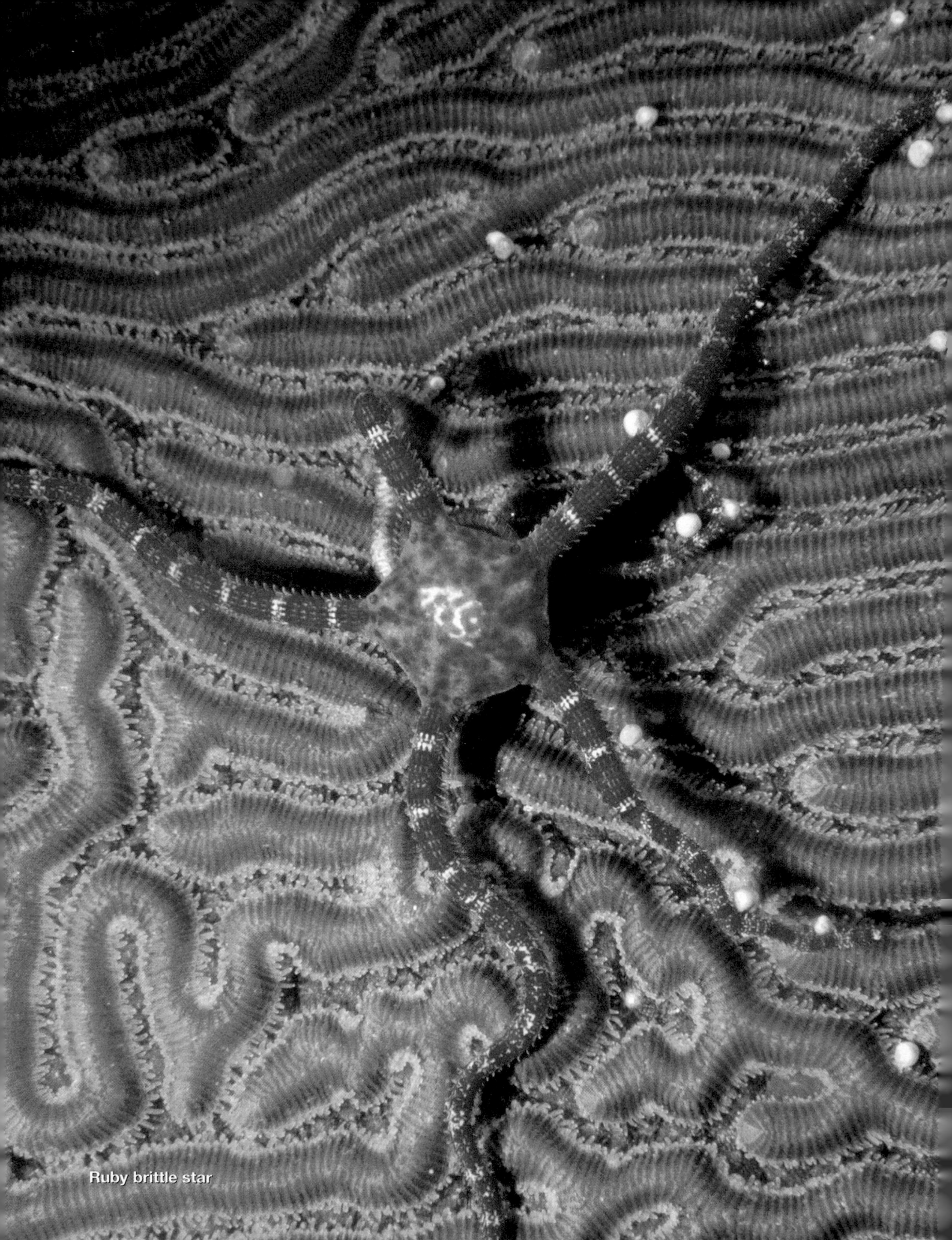

Ruby brittle star

31 Echinoderms and Invertebrate Chordates

✔️ *Quick Review*

Answer the following without referring to earlier sections of your book.

1. **Distinguish** invertebrates from vertebrates. *(Chapter 19, Section 3)*

2. **Define** the terms *gastrula* and *blastula*. *(Chapter 27, Section 1)*

3. **Distinguish** between radial symmetry and bilateral symmetry. *(Chapter 27, Section 1)*

Did you have difficulty? *For help, review the sections indicated.*

Reading Activity

Before you begin to read this chapter, write down all of the key words for both sections of the chapter. Then, write a definition next to each word that you have heard of. As you read the chapter, write definitions next to the words that you did not previously know, and modify as needed your original definitions of words familiar to you.

Looking Ahead

Section 1

Echinoderms
Animal Development
Modern Echinoderms
Echinoderm Diversity

Section 2

Invertebrate Chordates
The Chordate Skeleton
Invertebrate Chordates

🔲 **internet** connect

www.scilinks.org
National Science Teachers Association *sci*LINKS Internet resources are located throughout this chapter.

SCI**LINKS**® Maintained by the National Science Teachers Association

---

● Ruby brittle stars are echinoderms that are commonly found among corals and sponges in deeper parts of coral reefs. Brittle stars have radial symmetry, five flexible arms and a hard internal skeleton. They are the fastest moving echinoderms.

Echinoderms

- **Compare** the developmental pattern found in protostomes with that found in deuterostomes.

- **Describe** the major characteristics of echinoderms.

- **Summarize** how the sea star's water vascular system functions.

blastopore
protostome
deuterostome
ossicle
water-vascular system
skin gill

Animal Development

If you have been to a saltwater aquarium, you're sure to have seen echinoderms, which are spiny invertebrates that live on the ocean bottom. How could echinoderms like the brittle star shown on the first page of this chapter be related to animals such as chordates, which are primarily vertebrates? The answer lies in their early development. As an embryo develops, it goes through a gastrula stage. As shown in **Figure 1,** a gastrula has an opening to the outside called the **blastopore.** In acoelomate animals, the mouth develops from or near the blastopore. This pattern of development also occurs in some coelomate animals, such as annelids, mollusks, and arthropods. Animals with mouths that develop from or near the blastopore are called **protostomes.**

Some animals follow a different pattern of development. In phylums Echinodermata and Chordata, the anus—not the mouth—develops from or near the blastopore. (The mouth forms later, on another part of the embryo.) Animals with this pattern of development are called **deuterostomes,** also shown in Figure 1. If you know the origin of these two terms, it's easy to remember the differences between the two developmental patterns. The term *protostome* is from the Greek *protos,* meaning "first," and *stoma,* meaning "mouth." The prefix *deutero-* is from the Greek *deuteros,* meaning "second." In deuterostomes, the anus develops first and the mouth develops second.

Figure 1   Embryonic development

The development of an animal embryo follows one of two patterns.

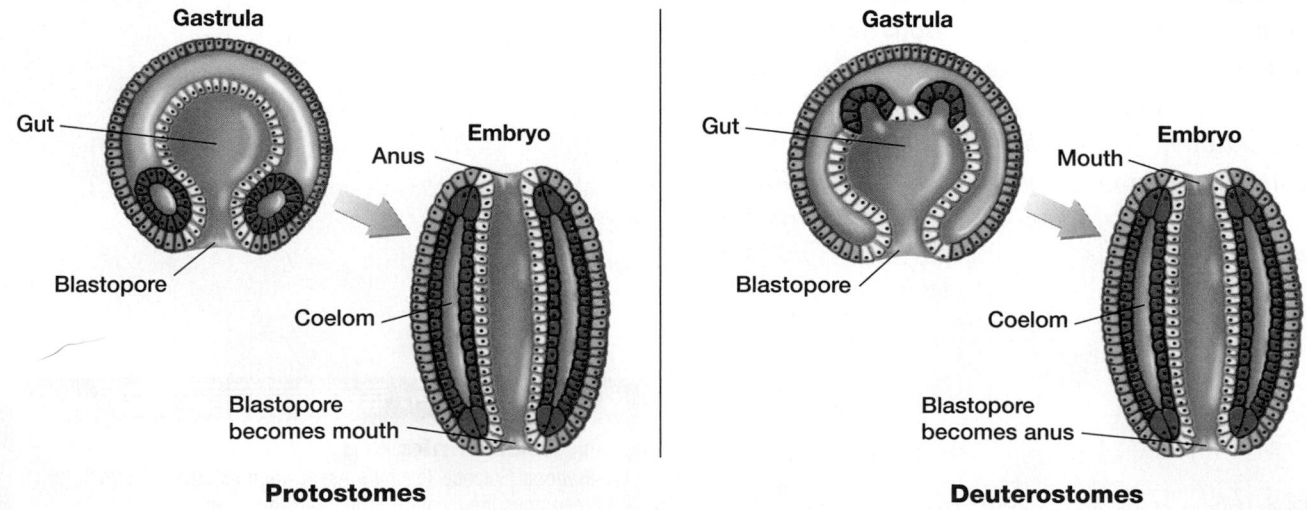

Protostomes

Deuterostomes

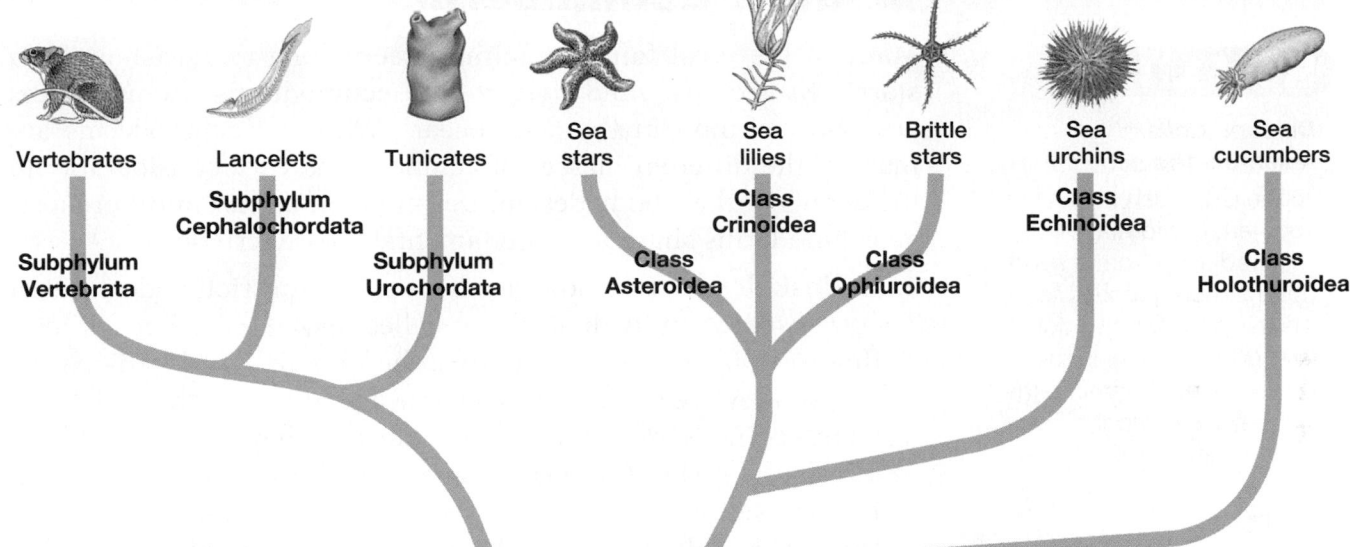

Figure 2 Evolution of chordates and echinoderms. This phylogenetic tree shows the relationship of the major chordate and echinoderm groups.

The first deuterostomes were marine echinoderms that evolved more than 650 million years ago. They were also the first animals to develop an endoskeleton. Today, most people are familiar with echinoderms known as "starfish," which are not really fish and are more properly called sea stars. In addition to sea stars, many other animals commonly seen along the sea shore—sea urchins, sand dollars, and sea cucumbers—are echinoderms. All are marine, and all are radially symmetrical as adults.

Chordates, as well as a few other small phyla, are also deuterostomes. (Humans and all other vertebrates are chordates.) Like the echinoderms, chordates have an internal skeleton. This developmental similarity unites these seemingly dissimilar animal phyla. It also leads scientists to believe that chordates and echinoderms derived from a common ancestor, as shown in the phylogenetic tree in **Figure 2.** The identity of the ancestral deuterostome is not known. The fossil record indicates that echinoderms, such as the sea lily in **Figure 3,** were abundant in the ancient seas.

Evolutionary Milestone

8 Deuterostomes

Echinoderms are coelomates that have a deuterostome pattern of embryo development. The same pattern of development occurs in the chordates.

Figure 3 Fossil sea lily. Sea lilies such as the one preserved by this fossil were plentiful in the ancient oceans.

Modern Echinoderms

Many of the most familiar animals seen along the seashore—sea stars, sea urchins, sand dollars—are echinoderms. Echinoderms are also common in the deep ocean. While all echinoderms are marine, the different classes of echinoderms vary considerably in the details of their body design. Despite their apparent differences, all echinoderms share four fundamental characteristics.

1. **Endoskeleton.** Echinoderms have a calcium-rich endoskeleton composed of individual plates called **ossicles.** When ossicles first form in young echinoderms, they are enclosed in living tissue, so they are a true endoskeleton. Even though the ossicles of adult echinoderms appear to be external, they are covered by a thin layer of skin (although sometimes the skin is worn away). In adult sea stars and in many other echinoderms, a large number of these plates are fused together. The fused plates function much like an arthropod exoskeleton. They provide sites for muscle attachment and shell-like protection. In most echinoderms, the plates of the endoskeleton bear spines that project upward through their skin.

2. **Five-part radial symmetry.** All echinoderms are bilaterally symmetrical as larvae. During their development into adults, the larvae's body plan becomes radially symmetrical. Most adult echinoderms, such as the one shown on the left in **Figure 4,** have a five-part body plan with arms that radiate from a central point. However, the number of arms can vary. Echinoderms have no head or brain. Instead, the nervous system consists of a central ring of nerves with branches extending into each of the arms. Although echinoderms are capable of complex response patterns, each arm acts more or less independently. Many species, including sea stars, can regenerate a new arm if a portion of an arm is lost. In some species of sea stars, a complete animal can regenerate from an arm connected to a portion of the central disk. However, a complete sea star cannot regenerate from an arm alone.

Figure 4 Five-part body plan. The echinoderm five-part body plan is easily seen in this colorful African species. Other sea stars, such as the sunstar, have more than five arms.

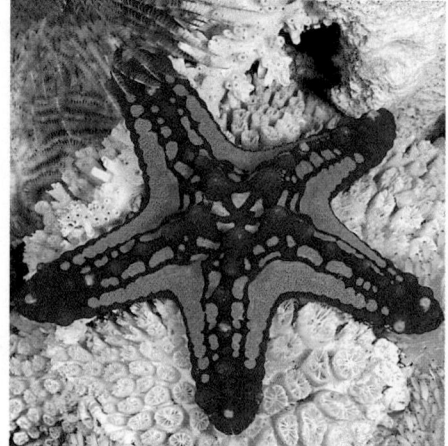

3. **Water-vascular system.** Echinoderms have a water-filled system of interconnected canals and thousands of tiny hollow tube feet called a **water-vascular system.** In some echinoderms, such as the sea star, the tube feet extend outward through openings in the ossicles. In some species, each tube foot has a sucker at its tip. Many echinoderms use their tube feet to crawl across the sea floor. The water-vascular system also functions in feeding and gas exchange. A sea star can use the hundreds of tube feet on its arms to pull the valves of a bivalve open. Some gas exchange and waste excretion takes place through the thin walls of the tube feet.

4. **Coelomic circulation and respiration.** The echinoderm body cavity functions as a simple circulatory and respiratory system. Particles, including respiratory gases, move freely throughout the large, fluid-filled coelom. Many echinoderms have skin gills that aid respiration and waste removal. **Skin gills,** shown in **Figure 5,** are small, fingerlike projections that grow among the echinoderm's spines. These projections create an increased surface area through which respiratory gases can be exchanged. Skin gills also function as excretory structures, and wastes that accumulate in them are released into the surrounding water.

You can learn more about the structure of one particular echinoderm, the sea star, in *Up Close: Sea Star,* on the following page.

Figure 5 Skin gills. An echinoderm's skin gills function as both respiratory and excretory organs.

🖅 internet connect ▤

www.scilinks.org
Topic: Echinoderms
Keyword: HX4065

SC*i*LINKS. Maintained by the National Science Teachers Association

---

Determining How Predators Affect Prey

Background

Sea stars can be very effective predators, and they frequently eat mollusks. The chart at right shows the relative number of two species of mollusks before and after the introduction of a predatory sea star. Study the chart, and answer the Analysis questions.

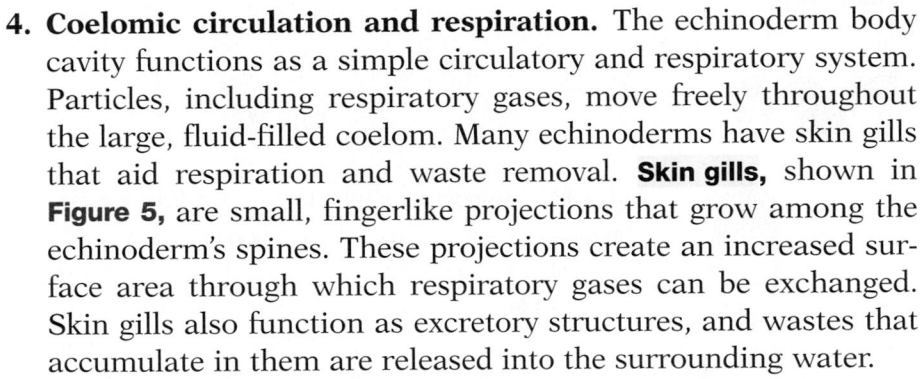

Sea Star Predation of Mollusks

Species A ■ Species B

Percentage of original population

Sea stars introduced

Time

Analysis

1. **Compare** the relative sizes of the two mollusk populations before the introduction of the sea star.

2. **Identify** the preferred prey of the sea star, and use the data presented in the graph to support your answer.

3. **Critical Thinking Analyzing Data** When the sea star began preying on the nonpreferred species, the preferred species had dropped to what percent of its original population?

4. **Critical Thinking Inferring Relationships** What factors might cause

the sea star to begin consuming a nonpreferred species, even when its preferred prey is still present?

5. **Critical Thinking Predicting Outcomes** Predict the relative abundance of the two species of mollusks if the sea star remains in the area indefinitely.

Up Close

Sea Star

- **Scientific name:** *Asterias vulgaris*
- **Size:** Typically from 15 to 30 cm (6 to 12 in.) in diameter
- **Range:** East coast of North America
- **Habitat:** Intertidal; often on hard, rocky surfaces
- **Diet:** Slow-moving or sessile species, including mollusks, crustaceans, polychaetes, and corals

Characteristics

Water-vascular system
Water enters and leaves the system through pores in the madreporite *(ma druh PAWR it),* a sievelike structure that filters out large particles. Water then moves into the ring canal and passes into the radial canals.

Tube feet

▲ Radial canal

Ray (arm)

Ring canal

Tube feet Hundreds of tube feet extend from the bottom of the radial canals, and each foot is connected to a waterfilled sac called an ampulla *(am PUHL uh).* When water is pumped from the sacs into the tube feet, they expand outward. Suckers on the ends of the tube feet attach firmly to solid surfaces. When muscles force water back into the ampulla, the tube feet shorten, pulling *Asterias* forward.

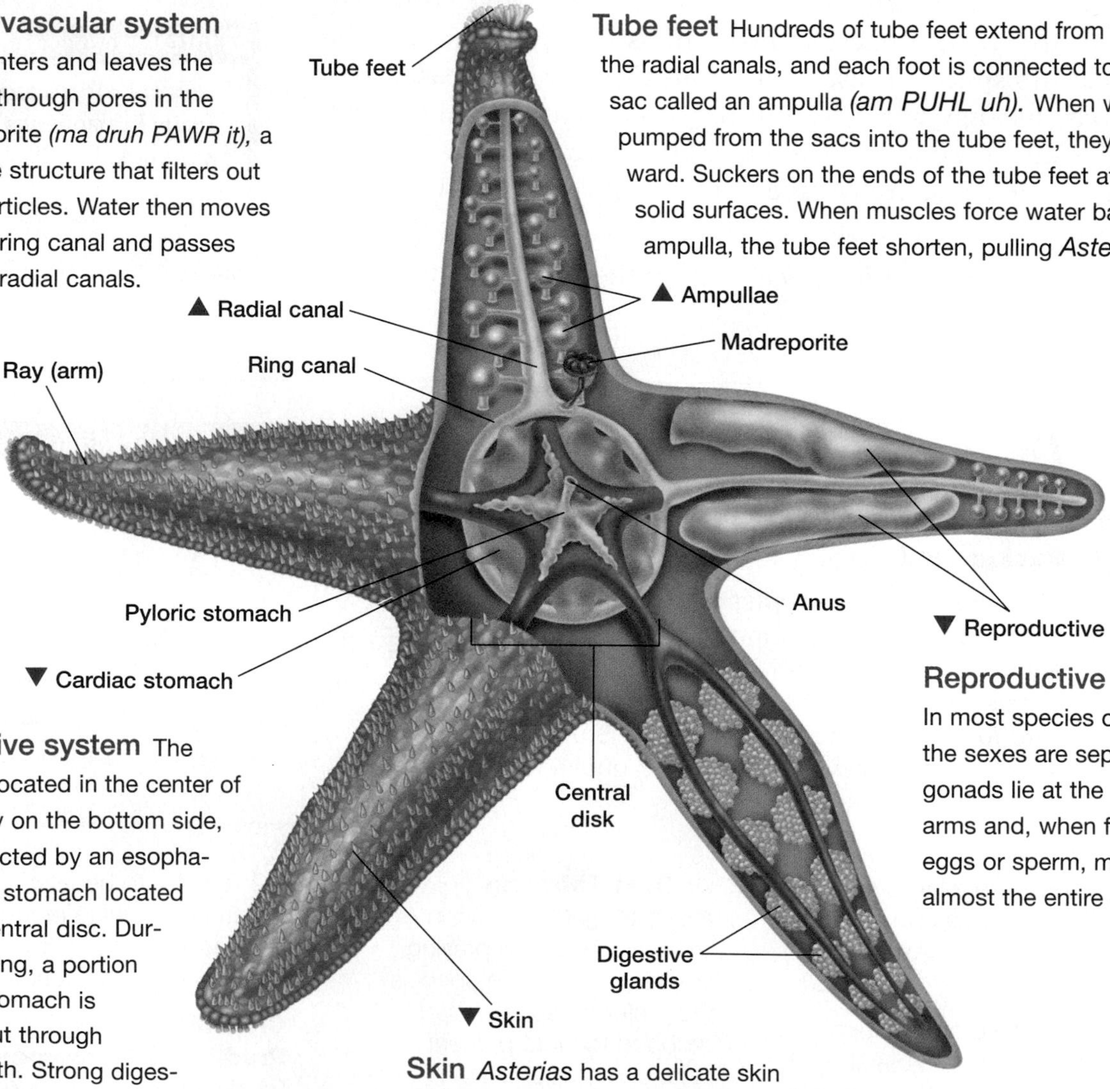

▲ Ampullae

Madreporite

Anus

▼ Reproductive organs

Reproductive system
In most species of *Asterias* the sexes are separate. The gonads lie at the base of the arms and, when filled with eggs or sperm, may occupy almost the entire arm.

Pyloric stomach

▼ Cardiac stomach

Central disk

Digestive system The mouth, located in the center of the body on the bottom side, is connected by an esophagus to a stomach located in the central disc. During feeding, a portion of the stomach is thrust out through the mouth. Strong digestive juices liquefy the prey, which is then ingested.

Digestive glands

▼ Skin

Skin *Asterias* has a delicate skin stretched over an endoskeleton of spiny plates.

Echinoderm Diversity

Echinoderms are one of the most numerous of all marine phyla. In the past, they were even more plentiful than they are now. There are more than 20 extinct classes of echinoderms and an additional six classes of living members. As you saw on the phylogenetic tree that appeared earlier in this section, the living classes include sea stars, sea lilies, brittle stars, sea urchins, and sea cucumbers. The recently discovered sea daisy does not appear on the phylogenetic tree because its relationship to the other echinoderms is not fully understood.

Figure 6 Sea star. This sea star is using its tube feet to pry open the shell of a clam. Then it will feed on the clam's soft tissues.

Sea Stars

Sea stars are the echinoderms most familiar to people. Almost all species of sea stars are carnivores, and they are among the most important predators in many marine ecosystems. For example, the crown-of-thorns sea star eats coral polyps. In 1 year, a single crown-of-thorns can consume up to 6 m^2 of a reef. Over time, this sea star can destroy an entire coral reef ecosystem. Other sea stars prey on bivalve mollusks, whose shells they pull open with their powerful tube feet, as shown in **Figure 6.**

The ossicles of many species of sea stars produce pincerlike structures called pedicellaria *(ped uh suh LAH ree uh)*. Pedicellaria contain their own muscles and nerves, and they snap at anything that touches them. This action prevents small organisms from attaching themselves to the surface of the sea star.

Brittle Stars

The sea star's relatives, the brittle stars and sea baskets, make up the largest class of echinoderms. Brittle stars have slender branched arms that they move in pairs to row along the ocean floor. Their arms break off easily, a fact that gives brittle stars their name. Brittle stars and sea baskets live primarily on the ocean bottom, and they usually hide under rocks or within crevices in coral reefs. Although a few species are predators, most brittle stars are filter feeders or feed on food in the ocean sediment.

Figure 7 Feather star. The feathery arms of these feather stars are adapted for filter feeding.

Sea Lilies and Feather Stars

The sea lilies and feather stars are the most ancient and primitive living echinoderms. They differ from all other living echinoderms because their mouth is located on their upper, rather than lower, surface. Sea lilies are sessile and are attached to the ocean floor by a stalk that is about 60 cm (23 in.) long. Feather stars, shown in **Figure 7,** use hooklike projections to attach themselves directly to the ocean bottom or a coral reef. They sometimes crawl or swim for short distances.

Sand dollar

Figure 8 Sea urchin and sand dollar. Sea urchins usually live on rocky ocean bottoms, while sand dollars live on sandy ocean bottoms.

Sea urchin

Sea Urchins and Sand Dollars

The sea urchins and sand dollars, shown in **Figure 8,** lack distinct arms but have the basic five-part body plan seen in other echinoderms. Both sea urchins and sand dollars have a hard, somewhat flattened endoskeleton of fused plates covered with spines protruding from it. The spines provide protection and, in some species of sea urchins, contain a venom that causes a severe burning sensation. In some other species of sea urchin, a specialized type of pedicellarium contains a toxin used to paralyze prey. Sea urchins are found on the ocean bottoms while sand dollars live in sandy areas along the sea coast.

Sea Cucumbers

Sea cucumbers are soft-bodied, sluglike animals without arms. They differ from other echinoderms in that their ossicles are small and are not fused together. Because of this, the sea cucumber's long, cylindrical body is soft. Often the body has a tough, leathery exterior. The sexes of most sea cucumbers are separate, but some species are hermaphrodites.

Sea cucumbers feed by trapping tiny organisms present in the sea water. Their mouth, located at one end of the body, is surrounded by several dozen tube feet modified into tentacles. The tentacles are covered with a sticky mucus that entraps plankton. Periodically, the sea cucumber draws its tentacles into its mouth and cleans off the plankton and mucus. The tentacles are then coated with a fresh supply of mucus. When threatened, a sea cumber has an unusual means of defending itself. As shown in **Figure 9,** the sea cucumber can release a number of sticky threads from its anus to entrap its attacker.

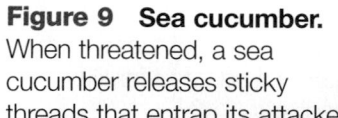

Figure 9   Sea cucumber. When threatened, a sea cucumber releases sticky threads that entrap its attacker.

Sea Daisies

In 1986, a new class of echinoderm was discovered: strange disk-shaped little animals called sea daisies. Less than 1 cm (0.39 in.) in diameter, these creatures were first found in deep waters off the coast of New Zealand. Only a few species are known. Sea daisies have five-part radial symmetry but no arms. Their tube feet are located around the edges of the disk rather than along the radial lines, like they are in other echinoderms.

BIOWatch
Monitoring Water Quality

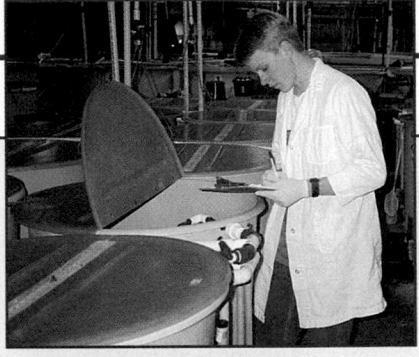

Recording water temperature in a bioassay tank

If you were swimming or fishing in coastal waters, you likely would not be able to detect the presence of toxic chemicals in the water, the sediments, or the sea life. To help protect humans and marine organisms, scientists have developed several tests to monitor marine environments for potential health hazards. Since sea urchin sperm and eggs are very sensitive to many pollutants, they are used in one of these tests, known as the sea urchin fertilization bioassay. (A bioassay is the use of a living organism or cell culture to test for the presence of a substance.)

Using Sea Urchins

Samples of ocean water, sediment, and industrial wastes that are discharged into the ocean are collected regularly from different sites. Then they are tested under controlled conditions in a lab. In this bioassay, sea urchin sperm and eggs are mixed together with the collected samples. After a short waiting period, scientists compare the fertilization success rate in the collected water samples with the fertilization success rate found in control water samples. If the test samples show a lower fertilization rate, scientists conclude that toxic contaminants are present.

Taking Action

What happens when the test indicates the presence of contaminants? More specific tests may be run to determine exactly what contaminants are present. If the toxicity can be traced to runoff from a factory or sewage treatment plant, the plant may be forced to clean its waste before discharging it. Sediments may have to be removed or decontaminated. In the future, it may be possible to clean up some pollutants by using plants that have the ability to remove toxic chemicals from the water they are growing in. The use of this process, known as phytoremediation, in marine environments is an exciting new area of research.

internet connect

www.scilinks.org
Topic: Bioassay
Keyword: HX4019

SC**LINKS**® Maintained by the National Science Teachers Association

Section 1 Review

1 **Summarize** why echinoderms are considered to be more closely related to tunicates, lancelets, and vertebrates than to other animals.

2 **Summarize** the four major echinoderm characteristics.

3 **Describe** how the sea stars use their water-vascular system to move along the sea floor.

4 **Standardized Test Prep** In an echinoderm, the functions of a circulatory system are carried out by the

A ossicles.

B pedicellariae.

C coelom.

D madreporite.

Invertebrate Chordates

The Chordate Skeleton

The second major group of deuterostomes are the **chordates.** Chordates have a very different kind of endoskeleton from that of echinoderms. The chordate endoskeleton is completely internal. During the development of the chordate embryo, a stiff rod called the **notochord** develops along the back of the embryo. Using muscles attached to this rod, early chordates could swing their backs from side to side, enabling them to swim through the water. The development of an internal skeleton was an important step that led to the evolution of vertebrates. The endoskeleton, which muscles attach to, made it possible for animals to grow large and to move quickly.

Other Chordate Characteristics

Chordates also share three other characteristics. They have a single, hollow, dorsal nerve cord with nerves attached to it that travel to different parts of the body. Chordates also have a series of **pharyngeal pouches.** Pharyngeal pouches of aquatic chordates develop in the wall of the pharynx and develop into the gill structure later in the animals' development. In terrestrial chordates, the pharyngeal pouches develop into different structures, such as the parathyroid gland and the inner ear. Another chordate characteristic is a postanal tail, which is a tail that extends beyond the anus. All chordates have all four of these characteristics at some time in their life, even if it is only briefly as embryos. **Figure 10** shows these chordate characteristics as seen in the body of an adult lancelet.

Figure 10 Lancelet interior. Adult lancelets possess all of the characteristics of chordates.

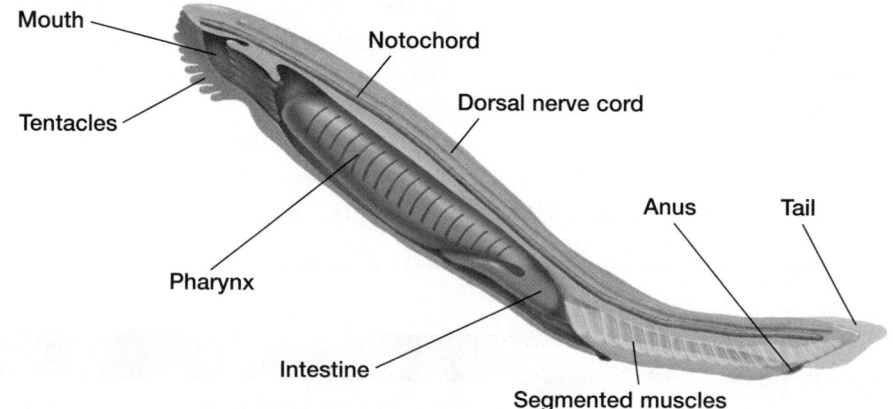

Mouth
Notochord
Tentacles
Dorsal nerve cord
Anus Tail
Pharynx
Intestine
Segmented muscles

Evolutionary Milestone

9 Notochord

Tunicates, lancelets, and all the vertebrates belong to phylum Chordata (chordates). Chordates are coelomate animals that have a flexible, dorsal rod called a **notochord.** In vertebrate chordates, the notochord is replaced during embryonic development by a vertebral column (backbone).

Invertebrate Chordates

Phylum Chordata is divided into three subphyla. The vast majority of chordate species belong to subphylum Vertebrata, which you will study in the next unit. Two other subphyla, Urochordata (the tunicates) and Cephalochordata (the lancelets), contain a small number of species. Because members of these two subphyla are chordates that do not have backbones, they are called **invertebrate chordates.**

Tunicates

Only the free-swimming tunicate larvae have a nerve cord, notochord, and postanal tail. These features are lost during the larvae's transformation into adulthood. However, adult tunicates, shown in **Figure 11,** retain their pharyngeal slits. Most adult tunicates are sessile, filter-feeding marine animals. A tough sac, called a tunic, develops around the adult's body and gives tunicates their name. Cilia beating within the tunicate cause water to enter the incurrent siphon. The water circulates through the tunicate's body, passes though its pharyngeal slits, and leaves the body through the excurrent siphon. As water passes through the slits in the pharynx, food is filtered from it and passed into the stomach. Undigested food passes to the anus, which empties into the excurrent siphon.

All tunicates are hermaphrodites, and some are also able to reproduce asexually by budding. While some tunicates are solitary, budding can result in colonies of identical tunicates.

Study TIP

● **Organizing Information**
Make a table to organize information about invertebrate chordates. Across the top, write the headings *Adult characteristics* and *Larval characteristics.* Along the sides, write *Lancelet* and *Tunicate.* Add information to the table as you read Section 2.

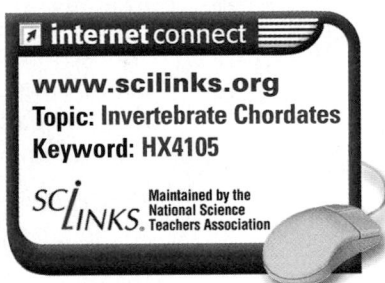

☑ internet connect

www.scilinks.org
Topic: Invertebrate Chordates
Keyword: HX4105

SC*LINKS.* Maintained by the
National Science
Teachers Association

Figure 11   Adult tunicate

Pharyngeal slits are the only chordate characteristic retained by adult tunicates.

Adult tunicate, dorsal view

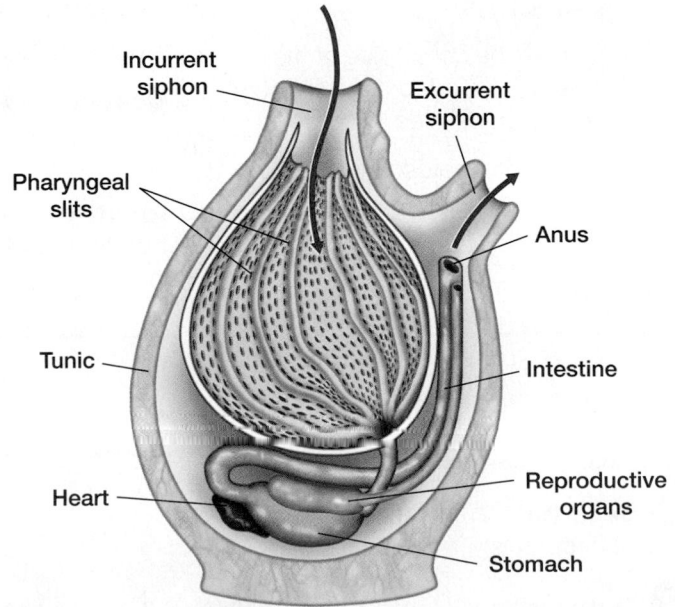

Tunicate interior

Figure 12 Lancelet.
Lancelets bury their tails in sediment.

Lancelets

Lancelets, shown in **Figure 12,** receive their name from their bladelike shape. Although lancelets may resemble fish, they are not fish. Lancelet fossils have been found in rocks over 550 million years— lancelets are much older than any fish species is. Lancelets are scaleless chordates only a few centimeters long. The lancelets' V-shaped bundles of muscles are arranged in a series of repeating segments. Lancelets are found worldwide in shallow ocean water. They spend most of their time with their mouths protruding from mud or sand. The beating of cilia that line the front end of their digestive tract draws water through the mouth and pharynx and out the pharyngeal slits. Lancelets feed on microscopic protists that they filter out of the water. Unlike tunicates, the sexes are separate in lancelets.

QUICK LAB

Comparing the Structures of the Notochord and Nerve Cord

The notochord and hollow nerve cord are two important characteristics of all chordates. While both are located on an animal's dorsal side, they differ in size, structure, and location. You can compare the two when viewing a cross section of an adult lancelet.

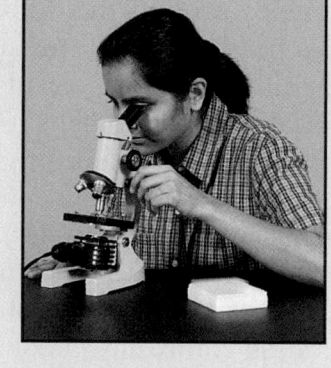

Materials

compound microscope, prepared slide of the cross section of an adult lancelet

Procedure

1. Place a prepared slide of a cross section of an adult lancelet under the microscope.

2. Locate the dorsal side of the specimen, and turn the slide so the dorsal side is on top.

3. Locate the notochord and hollow nerve cord. If visible, locate the intestine.

4. Sketch the specimen, and label the dorsal and ventral sides, the notochord, the nerve cord, and the intestines.

Analysis

1. **Describe** the structure, location, and size of the nerve cord and the notochord.

2. **Identify** the kind of symmetry observed in the adult lancelet.

3. **Compare** the lancelet's symmetry with the symmetry of adult echinoderms.

4. **Critical Thinking Forming Hypotheses** In vertebrate chordates, the notochord becomes a backbone that encases the nerve cord. Why might this arrangement be an advantage to an animal?

Section 2 Review

1 Describe the characteristics common to all chordates.

2 Summarize why tunicates and lancelets are classified as invertebrate chordates.

3 Compare the chordate characteristics found in adult tunicates with those found in adult lancelets.

4 Standardized Test Prep The pharyngeal slits of a tunicate play a role in what system?

 A skeletal   C nervous
 B reproductive   D digestive

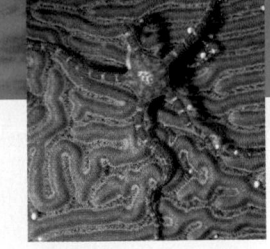

Key Concepts

1 Echinoderms

- During embryonic development, a protostome's mouth develops from the blastopore. In a deuterostome, the anus forms from the blastopore and the mouth forms later from a different opening.

- Because echinoderms and chordates are both deuterostomes, scientists believe that both groups were derived from a common ancestor.

- Echinoderms lack a head or brain. Their nervous system consists of a central ring of nerves with branches extending into each of the five parts of its body plan.

- Echinoderms share four characteristics: an endoskeleton composed of ossicles; five-part radial symmetry; a water-vascular system; and coelomic circulation and respiration.

- In many echinoderm species, respiration and waste removal are performed by skin gills.

- Echinoderms are a diverse group consisting of seven classes: sea stars, brittle stars, sea lilies, feather stars, sea urchins and sand dollars, sea cucumbers, and sea daisies.

2 Invertebrate Chordates

- At some point in their lives, all chordates have a notochord, a dorsal nerve chord, pharyngeal pouches, and a postanal tail.

- Phylum Chordata includes invertebrate and vertebrate chordates.

- Invertebrate chordates do not have a backbone (vertebral column). Two invertebrate subphyla are Urochordata (tunicates) and Cephalochordata (lancelets).

- Tunicate larvae have a nerve cord, notochord, pharyngeal pouches, and a postanal tail. As adults, they lose most of these characteristics.

- Lancelets retain their notochord, dorsal nerve cord, pharyngeal pouches, and postanal tail into adulthood.

Key Terms

Section 1

blastopore (692)
protostome (692)
deuterostome (692)
ossicle (694)
water-vascular system (695)
skin gill (695)

Section 2

chordate (700)
notochord (700)
pharyngeal pouch (700)
invertebrate chordate (701)

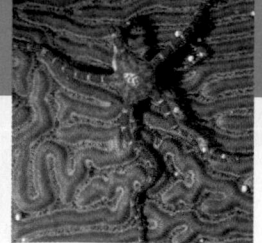

Performance ZONE CHAPTER REVIEW

Understanding Key Ideas

1. The phylum characterized by radial symmetry and a water-vascular system is
   a. Chordata.
   b. Arthropoda.
   c. Echinodermata.
   d. Cnidaria.

2. The presence of which of the following characteristics is typical among echinoderms?
   a. nauplius larvae
   b. a notochord
   c. an exoskeleton
   d. a water-vascular system

3. Which of the following pairs is a correct match between an adult echinoderm and one of its characteristics?
   a. sea cucumber: leathery epidermis
   b. sand dollar: five arms
   c. sea lily: free-swimming
   d. sea star: sessile

4. Which echinoderm group includes species that specialize in hunting bivalves?
   a. sea cucumbers
   b. sand dollars
   c. sea stars
   d. sea lilies

5. Animals with a notochord, a dorsal nerve cord, and pharyngeal slits are members of the phylum
   a. Echinodermata.
   b. Chordata.
   c. Annelida.
   d. Cnidaria.

6. Which of the following do adult chordates and adult echinoderms have in common?
   a. a nonsegmented body
   b. an internal skeleton
   c. a water-vascular system
   d. bilateral symmetry

7. In what way was the notochord an important development in the evolution of vertebrates?

8. **BIOWatch** Once preliminary tests determine that coastal waters are polluted, what kinds of actions might be taken to reduce the pollution?

9. Gas exchange is essential to life. Compare how gas exchange occurs in most crustaceans, insects, and echinoderms. (Hint: See Chapter 30, Sections 1 and 4.)

10. **Concept Mapping** Construct a concept map that outlines the development and body plan of echinoderms and invertebrate chordates. Try to include the following terms in your map: *blastopore, protostome, deuterostome, water-vascular system, skin gills, notochord, nerve cord,* and *pharyngeal slits.*

Critical Thinking

11. **Evaluating Conclusions** Sea cucumbers and sea lilies are relatively sessile. Their larvae, however, are capable of swimming. Explain how swimming larvae provide an advantage for these echinoderms.

12. **Forming Reasoned Opinions** A scientist collects several specimens of an unidentified animal. After conducting an in-depth study of the mysterious species, the scientist observes that they have tube feet, an endoskeleton, and a protostome pattern of embryonic development. Why is the classification of these organisms difficult?

13. **Inferring Relationships** Explain why the larval form of an organism can be valuable in determining relationships among species.

Alternative Assessment

14. **Finding and Communicating Information** Use the media center or Internet resources to learn more about the crown-of-thorns sea star, or *Acanthaster planci,* which is abundant on the Great Barrier Reef. Find out why this echinoderm poses such a threat to its environment, and prepare a report on your findings. Include steps being taken by the Australian government.

Standardized Test Prep

Understanding Concepts

Directions (1–4): For *each* question, write on a separate sheet of paper the letter of the correct answer.

1 Which of the following patterns of development is characteristic of echinoderms and lancelets?
A. acoelomate
B. deuterostome
C. protostome
D. pseudocoelomate

2 A scientist comes across a colony of tunicates in which all of the individuals are identical to one another. What conclusion might the scientist draw about the colony?
F. These tunicates are mobile and feed by trapping their prey as a group.
G. The colony came about when two different species of tunicates mated.
H. The tunicates in the colony arose from asexual reproduction by budding.
I. These tunicates are hermaphrodites and reproduced sexually to form the colony.

3 What pincerlike structures do the ossicles of *Asterias* produce?
A. ampullae   C. pedicellaria
B. madreporites   D. tube feet

4 What structure coordinates the movements of a sea star?
F. mantle   H. radula
G. nerve ring   I. tube foot

Directions (5): For the following question, write a short response.

5 Differentiate between a protostome and a deuterostome.

Test TIP

Take time to read each question completely on a standardized test, including all of the answer choices. Consider each answer choice before determining which one is correct.

Reading Skills

Directions (6): **Read the passage below. Then answer the question.**

Most animal phyla that exist today probably originated during a relatively short time during the late Precambrian and early Cambrian periods. Scientists have found many echinoderm fossils from the Cambrian period, but they have found few fossils of other species that must have lived then.

6 What might explain the very large number of fossilized echinoderms and the relatively small number of other animal fossils from the Cambrian period?
A. Endoskeletons of echinoderms are rigid and are easily fossilized.
B. Other animals live on the sea floor where sedimentary rock forms.
C. Scientists have incorrectly identified most of the fossils from that period.
D. Fossils of other animals from that period would be found in deeper layers of rock.

Interpreting Graphics

Directions (7): **Base your answer to question 7 on the diagram below.**

Adult Tunicate

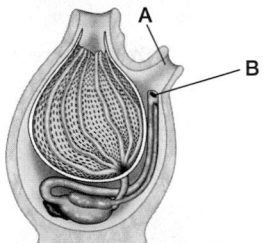

7 What is the significance of the relationship between structures *A* and *B*?
F. *B* collects food from the water that enters the body through *A*.
G. *B* extracts oxygen from the water that enters the body through *A*.
H. *B* releases undigested food into the water that leaves the body through *A*.
I. *B* releases gametes into the water that leaves the body through *A*.

Skills Practice Lab

Analyzing Sea Star Anatomy

SKILLS
- Observing
- Collecting data
- Inferring

OBJECTIVES
- **Observe** anatomical structures of an echinoderm.
- **Infer** function of body parts from structure.

MATERIALS
- disposable gloves
- preserved sea star
- dissection tray
- dissection scissors
- hand lens
- dissecting microscope
- forceps
- blunt probe
- sharp probe
- dissection pins

ChemSafety

CAUTION: Always wear safety goggles and a lab apron to protect your eyes and clothing.

CAUTION: Do not touch or taste any chemicals. Know the location of the emergency shower and eyewash station and how to use them. If you get a chemical on your skin or clothing, wash it off at the sink while calling to the teacher. Notify the teacher of a spill. Spills should be cleaned up promptly, according to your teacher's directions.

CAUTION: Glassware is fragile. Notify the teacher of broken glass or cuts. Do not clean up broken glass or spills with broken glass unless the teacher tells you to do so.

Before You Begin

Sea stars are members of the phylum Echinodermata, a group of invertebrates that also includes sand dollars, sea urchins, and sea cucumbers. **Echinoderms** share four main characteristics: an **endoskeleton, five-part radial symmetry,** a **water-vascular system,** and circulation and respiration through their **coelom.**

1. Write a definition for each boldface term above.
2. Create a data table like the one shown to the right.
3. Based on the objectives for this lab, write a question you would like to explore about sea star anatomy.

DATA TABLE		
Function of Sea Star Structures		
Structure	**Observations**	**Inferred function**
Madreporite		
Spine		
Skin gill		

Procedure

PART A: External Anatomy

1. **CAUTION: Put on safety goggles, a lab apron, and protective gloves.** As you observe the sea star body structures, record your observations and your inference of each structure's function in the table. On a separate sheet of paper or in your lab notebook, draw and label the sea star and the structures that you observe.

2. Using forceps, hold a preserved sea star under running water to gently but thoroughly remove excess preservative. Then place the sea star in a dissecting tray.

3. Refer to the diagram of a sea star in **Up Close** in this chapter to locate the madreporite on the upper surface of the sea star.

4. Use a hand lens to observe the sea star's spines. Are they distributed in any recognizable pattern? Are they exposed or covered by tissue? Are they movable or fixed?

5. Use the dissecting microscope to look for small skin gills, If any are present, describe their location and structure.

6. Examine the sea star's lower surface. Find the mouth, and use forceps or a probe to gently move aside any soft tissues. What structures are found around the mouth?

7. Locate the tube feet. Describe their distribution. Using a dissecting microscope, observe and then draw a single tube foot.

PART B: Internal Anatomy

8. ◆ **CAUTION: Scissors, probes, and pins are sharp. Use care not to puncture your gloves or injure yourself or others.** Using scissors and forceps, carefully the cut the body wall away from the upper surface of one of the sea star's arms. Start near the end of the arm and work toward the center.

9. Find the digestive glands in the arm you have opened. Then, locate the short branched tube that connects the digestive glands to the pyloric stomach,

10. Cut the tube that connects the digestive glands to the stomach, and move the digestive glands out of the arm. Look for the reproductive organs.

11. Locate the two rows of ampullae that run the length of the arm.

12. Carefully remove the body wall from the upper surface of the central region of the sea star. Locate the pyloric stomach and the cardiac stomach.

13. Remove the stomachs and find the ring canal and the radial canals. In which direction does water move through these canals?

14. ◆ Dispose of sea stars and sea star body parts in the waste container designated by your teacher. Do not put lab materials in the trash unless your teacher tells you to do so.

15. ◆ Clean up your work area and all lab equipment. Return lab equipment to its proper place. Wash your hands thoroughly before you leave the lab and after you finish all work.

Analyze and Conclude

1. **Analyzing Results** What type of symmetry is found in the sea star?

2. **Inferring Relationships** What is the relationship between the ampullae and the tube feet?

3. **Making Predictions** How does a sea star use its stomach during feeding?

4. **Making Predictions** If the ring canals and radial canals did not function properly, how would this affect the sea star's ability to move and feed?

5. **Further Inquiry** Write a new question about echinoderms that could be explored with another investigation.

Use the following Internet resources to explore your own questions about echinoderms.

◢ **internet** connect

www.scilinks.org
Topic: Echinoderms
Keyword: HX4065

*SCi*LINKS. Maintained by the
National Science Teachers Association

UNIT 8 Exploring Vertebrates

Only a few wild tigers remain in the world.

in perspective

The Tiger

Yesterday... Native to Asia, tigers once lived throughout much of the southern half of the continent. Tigers were killed in large numbers as hunters killed them for their skins and humans cleared the forests in which they lived. **What characteristics of tigers are representative of other mammals?**

Prince of Wales, Edward VII, on a hunting expedition in India

Today... Tigers everywhere are at critically low numbers. Poaching, the illegal killing of tigers for their bones, skin, and other body parts, continues in many countries. In India, Nepal, and Russia, protected areas have been set aside to help preserve tiger habitat. **Discover how a mammal's teeth provide clues about its diet.**

Tiger skin confiscated by the police in Delhi, India

Tomorrow... The long-term survival of tigers may depend on breeding programs in zoos and nature reserves. In some areas, more rigorous enforcement of anti-poaching laws is helping to protect wild populations. **Find out to which order of mammals tigers belong.**

Bengal tiger cubs

internet connect

www.scilinks.org
Topic: Tigers
Keyword: HX4176

SCLINKS Maintained by the National Science Teachers Association

Excavated mammoth skeleton

32 Introduction to Vertebrates

✔ Quick Review

Answer the following without referring to earlier sections of your book.

1. **Summarize** how mass extinctions have affected life on Earth. *(Chapter 12, Section 2)*

2. **Define** the terms *vertebrate* and *continental drift*. *(Chapter 12, Section 3)*

3. **Distinguish** between a carnivore and a herbivore. *(Chapter 16, Section 2)*

4. **Define** the terms *bilateral symmetry, cephalization*, and *coelom*. *(Chapter 27, Section 1)*

5. **Summarize** the function of genes. *(Chapter 10, Section 1)*

6. **Explain** the importance of phylogenetic relationships. *(Chapter 13, Section 2)*

Did you have difficulty? *For help, review the sections indicated.*

📖 Reading Activity

As you read the first section of this chapter, complete a reading organizer to describe the series of adaptations in body structure for agnathans, ostracoderms, acanthodians, and placoderms.

● Mammoths were large mammals with tusks and a trunk similar to those of modern elephants. Now extinct, mammoths lived throughout North America starting about 1.7 million years ago.

Vertebrates in the Sea and on Land

Objectives

- **Identify** the key characteristics of vertebrates.

- **Describe** two adaptations found in early fishes.

- **Identify** the relationship of fishes to amphibians.

- **Summarize** the key adaptations of amphibians for life on land.

Key Terms

vertebra
agnathan
acanthodian
cartilage

Adaptations of Vertebrates

If you go to a zoo, many of the animals you see—the lions, elephants, snakes, turtles, and birds—are vertebrates, just as people are. Vertebrates are chordates with a backbone. Even though these vertebrates appear different, they are very much alike internally, indicating that in the distant past they had a common ancestor. Present-day differences in vertebrates reflect their different evolutionary paths.

What are the internal similarities shared by vertebrates? First, vertebrates are chordates with a backbone. They take their name from the individual segments, called **vertebrae** *(VUR tuh bree)* (singular, *vertebra*), that make up the backbone. **Figure 1** shows a typical vertebrate and its endoskeleton. In most vertebrates, the backbone completely replaces the notochord found in invertebrate chordates.

The backbone provides support for and protects a dorsal nerve cord. It also provides a site for muscle attachment. These functions paved the way for the development of an internal skeleton that allowed vertebrates to grow larger than their invertebrate ancestors. In addition to a backbone, vertebrates have a bony skull that encases and protects their brain.

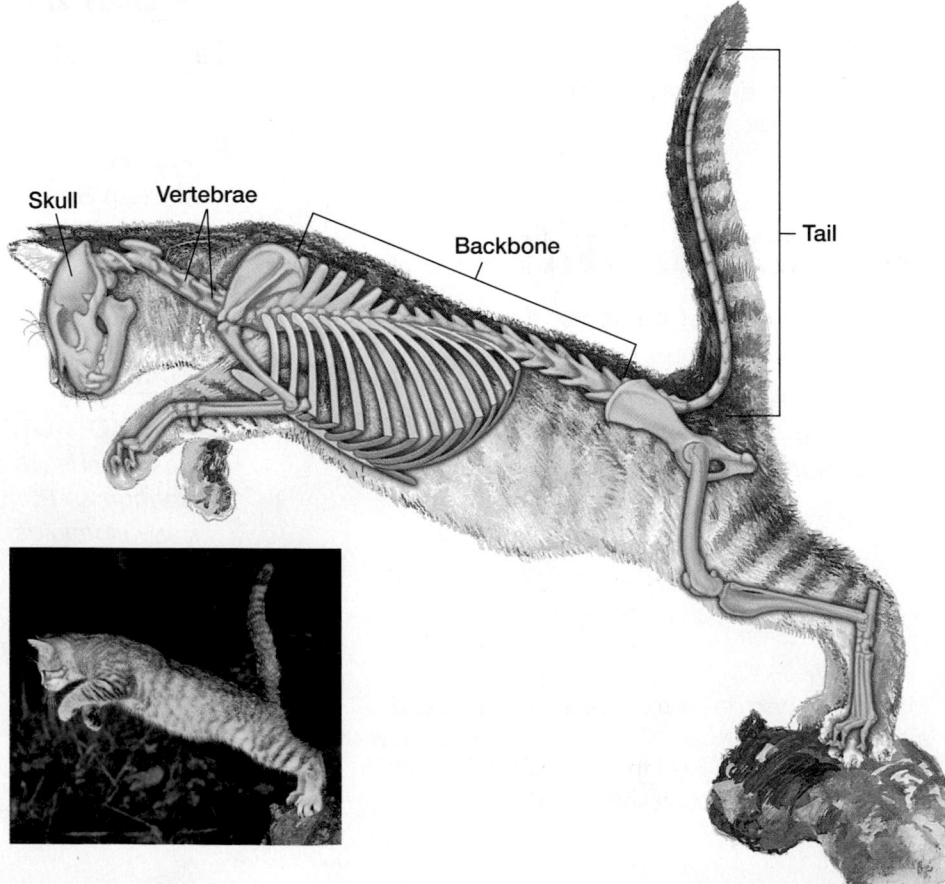

Skull   Vertebrae   Backbone   Tail

Figure 1   Vertebrate.
This cat's skeleton includes a skull and a backbone composed of vertebrae.

Vertebrates share a number of other characteristics, including segmentation, bilateral symmetry, and two pairs of jointed appendages, such as limbs or fins. They exhibit cephalization and have complex brains and sense organs. All vertebrates have a true coelom and a closed circulatory system with a chambered heart.

The tissues of vertebrates compose organs. In turn, the organs compose organ systems. Vertebrate organ systems tend to be more complex than the organ systems found in invertebrates. For example, recall the grasshopper's digestive system, seen in *Up Close* in an earlier chapter. Its digestive tract is divided into several specialized regions. However, the only associated structures are salivary glands. In contrast, the cat's digestive system, shown in **Figure 2,** consists of a digestive tract, together with two accessory organs, the liver and the pancreas. These complex organs secrete enzymes that aid digestion. Figure 2 shows the major organ systems of a typical vertebrate.

Real Life

How many vertebrae do giraffes have in their neck?

Nearly all vertebrates, including giraffes, have seven neck vertebrae. At almost 6 m (20 ft) tall, the giraffe is the tallest of all vertebrates, and its seven cervical (neck) vertebrae are greatly elongated.

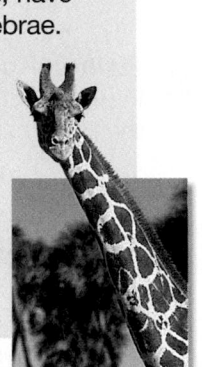

Figure 2 Major organ systems

A vertebrate's organ systems perform a variety of functions.

Vertebrate Organ Systems

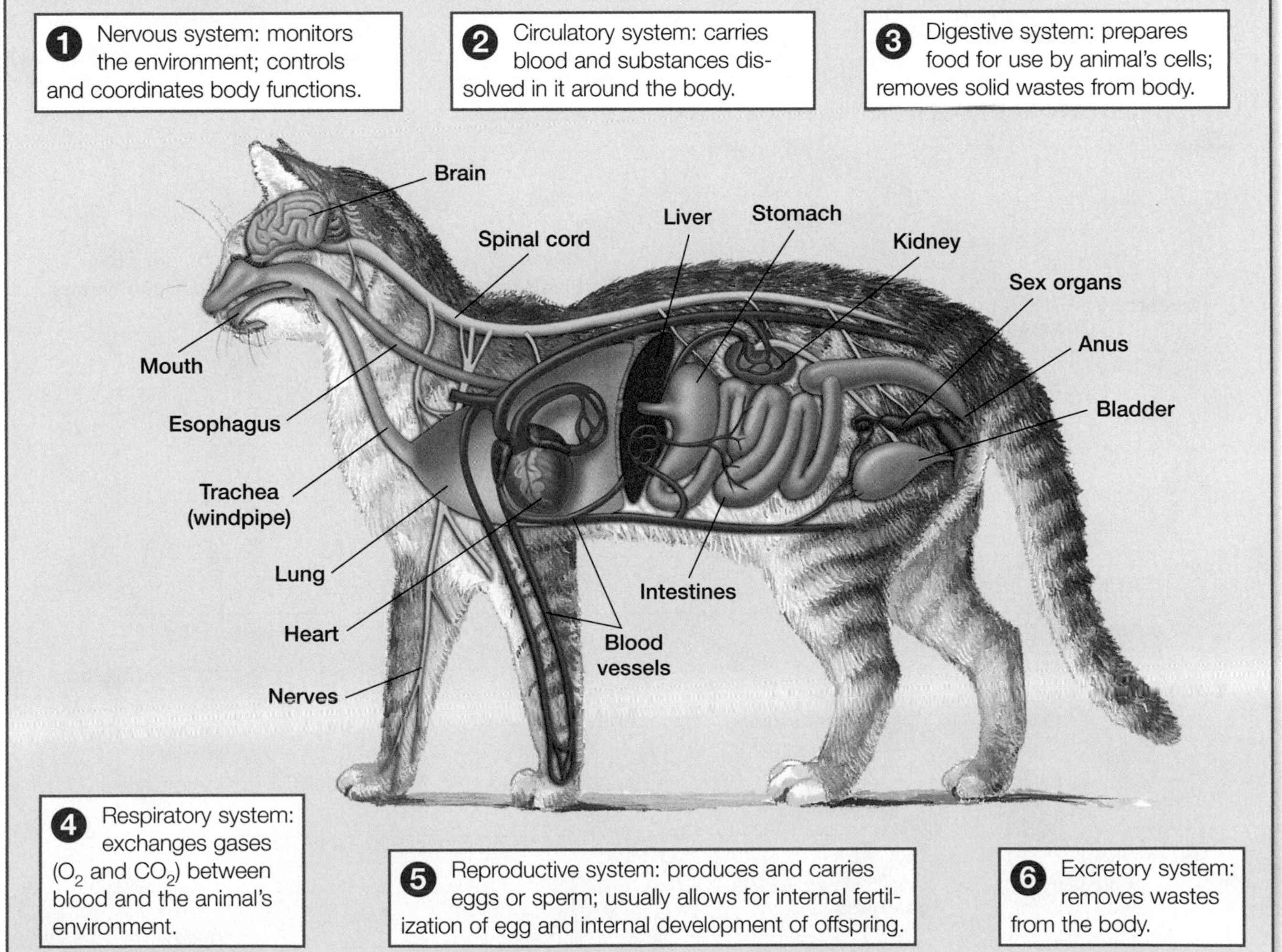
1 Nervous system: monitors the environment; controls and coordinates body functions.
2 Circulatory system: carries blood and substances dissolved in it around the body.
3 Digestive system: prepares food for use by animal's cells; removes solid wastes from body.
Brain
Spinal cord
Liver
Stomach
Kidney
Sex organs
Anus
Bladder
Mouth
Esophagus
Trachea (windpipe)
Lung
Heart
Nerves
Intestines
Blood vessels
4 Respiratory system: exchanges gases (O_2 and CO_2) between blood and the animal's environment.
5 Reproductive system: produces and carries eggs or sperm; usually allows for internal fertilization of egg and internal development of offspring.
6 Excretory system: removes wastes from the body.

The First Vertebrates

The first chordates evolved about 550 million years ago. At that time, many different groups of organisms appeared in the shallow seas that covered a large portion of Earth's continents. According to the fossil record, the first vertebrates appeared about 50 million years later. The first vertebrates were fishes similar to the one shown in **Figure 3.** Unlike most of the fishes you are familiar with, the earliest fishes, called **agnathans,** had neither jaws nor paired fins. But agnathans did have a backbone, which provided a central axis for muscle attachment. As their muscles pulled against the backbone, the agnathans propelled themselves along the ocean bottom.

Within another 50 million years, jawless fishes had diversified into a great variety of species. The major group was the ostracoderms *(ahs TRAK uh durms)*, which had primitive fins and massive plates of bony tissue on their body. Jawless fishes dominated the oceans for about 100 million years, until they were replaced by new kinds of fishes that were hunters. **Figure 4** shows one hypothesis about evolutionary relationships among fishes.

Figure 3   Ancient fish. This fossil fish is typical of the early jawless fishes.

Figure 4   Evolution of fishes

This simplified phylogenetic tree shows one hypothesis about the relationship between the early fishes and the fishes and amphibians that evolved later.

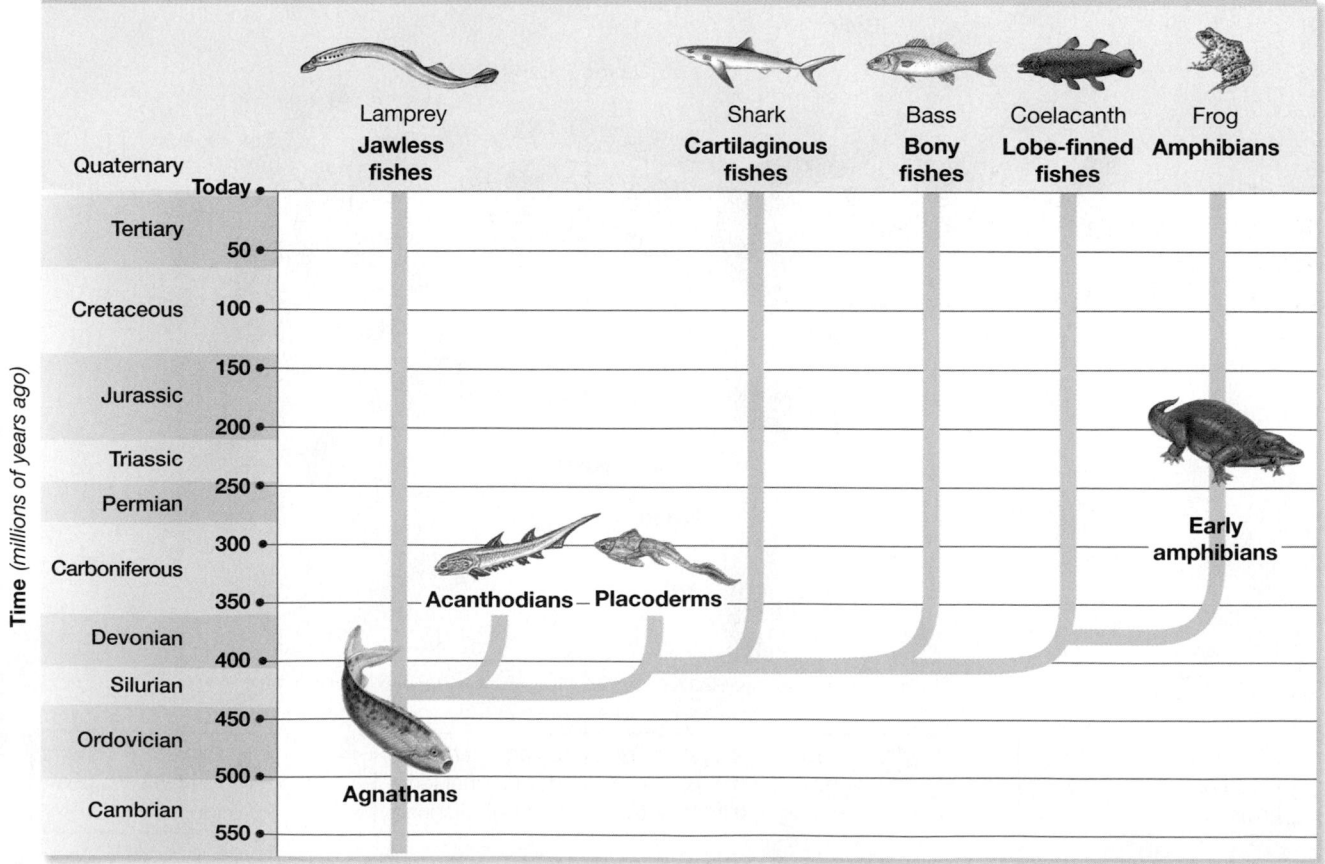

Evolution of Fishes

To survive as predators in the water, fishes must meet two important challenges. The first challenge is that of pursuing prey through the water. Fishes are fast and agile swimmers. The bodies of fishes are streamlined and many are flattened sideways, which makes it easy for them to move through the water. Fishes also have paired fins supported by spines, which provide fine control of movement.

The second challenge is that of grasping prey once it is within reach. About 430 million years ago, the **acanthodians** *(uh kan THOH dee uhns)*, or spiny fishes, appeared. Acanthodians had strong jaws with jagged, bony edges that served as teeth, enabling them to hold onto prey. The development of jaws in fishes was a key evolutionary innovation. As shown in **Figure 5,** jaws are thought to have evolved from gill arch supports made of **cartilage** *(KAHRT lihj)*, a lightweight, strong, flexible tissue.

The spiny fishes had internal skeletons of cartilage, although some fossils indicate that their skeletons also contained some bone. Their scales also contained small plates of bone. The presence of bone in the spiny fishes foreshadowed the much larger role that bone would play in their descendants.

About 20 million years after the acanthodians appeared, the placoderms evolved. Placoderms *(PLAK uh durms)* were jawed fishes with massive heads armored with bony plates. By the end of the Devonian period, almost all of the early fishes, including the placoderms, had disappeared. After dominating the seas for almost 50 million years, they were replaced by swifter swimmers—the sharks and bony fishes.

Figure 5   Evolution of jaws

The development of jaws was a key evolutionary innovation.

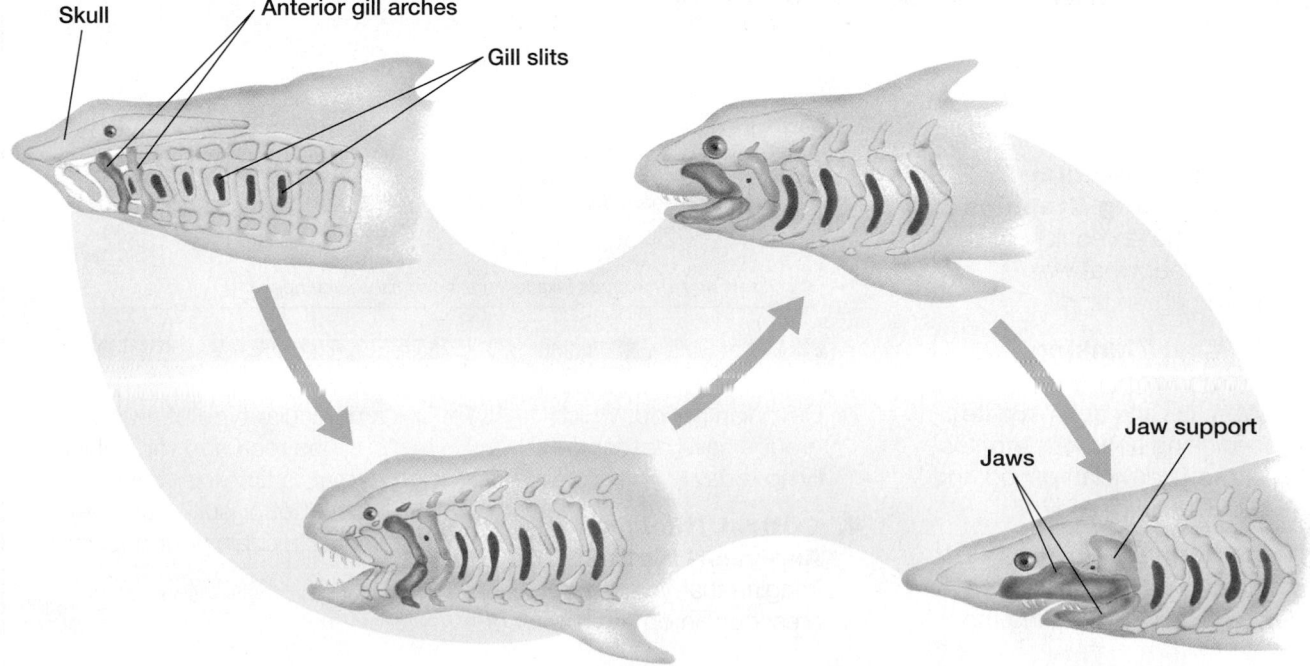

Skull

Anterior gill arches

Gill slits

Jaws

Jaw support

Sharks and Bony Fishes

By the end of the Devonian period, almost all of the early fishes had disappeared. At about the same time (400 million years ago), sharks and bony fishes appeared. Sharks and bony fishes, which are superior swimmers, thrived in the Devonian seas. Most have streamlined bodies that are well adapted for rapid movement through the water. Both sharks and bony fishes also have an assortment of movable fins that greatly aid their swimming. However, their evolutionary pathways differ. Sharks, such as the extinct *Hybodus*, shown in **Figure 6,** have a skeleton composed of cartilage. Calcium carbonate (the mineral oyster shells are made of) is deposited in the outer layers of the cartilage. A thin layer of bone covers this reinforced cartilage. The result is a very light but strong skeleton. Bony fishes, on the other hand, have a skeleton made of bone, which is heavier and less flexible than cartilage. Bony fishes have a swim bladder, which compensates for this increased weight. This gas-filled sac buoys them in the water, just as an air-filled balloon buoys a swimmer.

Figure 6 *Hybodus.* Early sharks, such as *Hybodus*, were among the first vertebrates to have jaws.

Using Timelines and Phylogenetic Trees

Background

Use the timeline in Section 2 of the chapter titled "History of Life on Earth" and the phylogenetic tree in Figure 4 in this chapter to answer the analysis questions. Tell which graphic you used to answer each question.

Analysis

1. **Critical Thinking Interpreting Graphics** When the jawed fishes appeared, what was occurring on land?

2. **Critical Thinking Interpreting Graphics** What are the approximate beginning and ending dates of the Ordovician period and the Silurian period?

3. **Critical Thinking Interpreting Graphics** Of the fishes living during the Devonian period, which groups have descendants living today?

4. **Critical Thinking Analyzing Methods** Imagine that you are giving a presentation on the history of a particular type of music, such as rock and roll. Which format, a timeline or a tree, would better suit your presentation? Explain your answer.

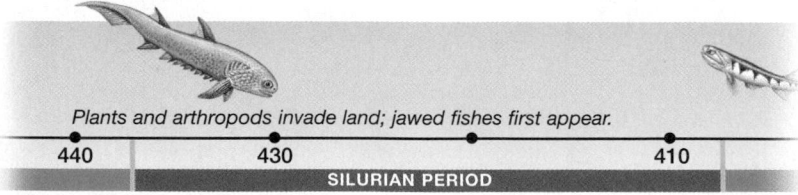

Age *(in millions of years ago)*

Plants and arthropods invade land; jawed fishes first appear.

| 440 | 430 | | 410 |

SILURIAN PERIOD

Modern Fishes

Today there are more fishes than any other group of vertebrates, both in terms of numbers of individuals and numbers of species. If you consider that water covers three-fourths of Earth's surface and that fishes live in virtually every aquatic habitat, this fact is not surprising. Today's fishes belong to one of three major groups: the agnathans, the cartilaginous *(kart'l AJ uh nuhs)* fishes, and the bony fishes. The agnathans (hagfishes and the lampreys) resemble the early jawless fishes.

The first cartilaginous fishes (sharks) and the bony fishes evolved at about the same time, 400 million years ago. These two groups of fishes likely evolved from the same early, jawless fishes that gave rise to the acanthodians and the placoderms. The shark's relatives, the skates and rays, evolved later.

Bony fishes are the fishes most familiar to us. Bony fishes make up about 95 percent of modern fish species. Because bony fishes are adapted to many different environments, they vary greatly in size, color, and shape. **Figure 7** shows three examples of bony fishes.

Figure 7 Bony fishes

Bony fishes live in a wide variety of habitats, and there is great variation in their appearance.

Russian sturgeon

Blue-faced angelfish

Eastern mud minnow

Evolution of Amphibians

The first group of vertebrates to live on land were the amphibians, which appeared about 370 million years ago. Amphibians probably share a common ancestor with modern lungfishes and other lobe-finned fishes. As shown in **Figure 8,** the pattern of bones in an amphibian's limbs bears a strong resemblance to that of a lobe-finned fish. While several species of lobe-finned fishes exist today, those species thought to be ancestral to amphibians are extinct.

The "Age of Amphibians"

Although amphibians first appeared in the Devonian period, they increased greatly in numbers during the Carboniferous period. During this time, which began what biologists call the "age of amphibians," the number of amphibian families increased from 14 to about 34. (Recall that a family is a taxonomic category composed of one or more genera.) By the late Carboniferous period, much of what was to become North America was covered by low-lying tropical swamplands. Amphibians thrived in this moist environment, sharing it with early reptiles.

In the Permian period that followed, amphibians reached their greatest diversity, increasing to 40 families. In the early Permian period, a remarkable change occurred among amphibians—many of them began to leave the marshes for dry uplands. By the middle Permian, 60 percent of all amphibian species were living in dry environments. *Eryops*, illustrated in **Figure 9,** was typical of amphibians of this period and was well adapted for life on land.

Figure 8   From fin to limb

As shown by the colors below, the pattern of limb bones in an early amphibian bears a distinct resemblance to the pattern of fin bones in a lobe-finned fish.

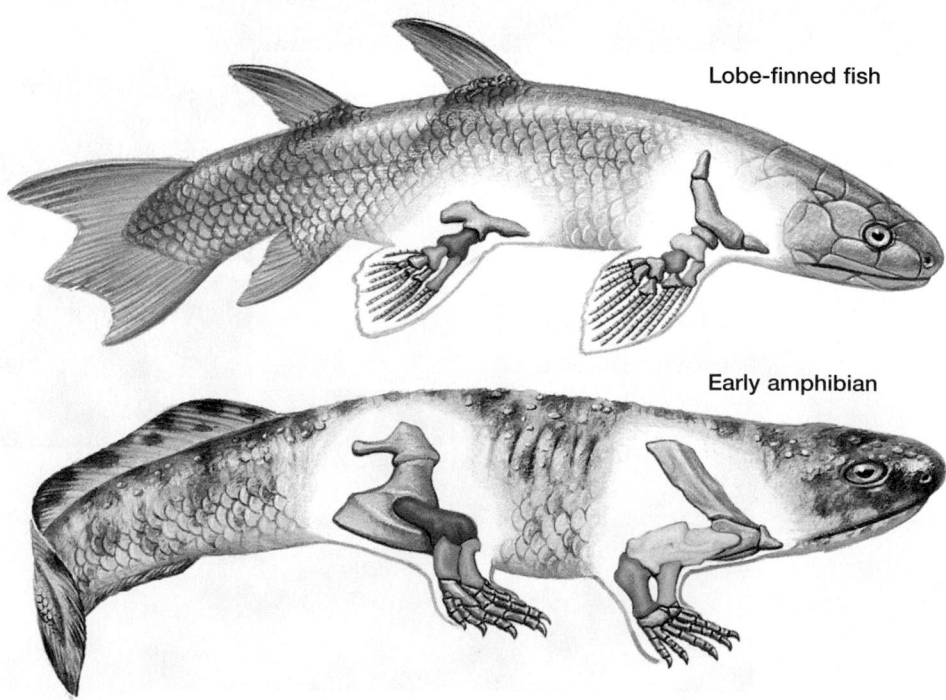

Lobe-finned fish

Early amphibian

Adaptations for Life on Land

Life on land is quite different from life in the water. Thus, a number of major adaptations allowed some species to successfully invade land.

1. **Legs.** Legs support the body's weight as well as allow movement from place to place.

2. **Lungs.** The delicate structure of a fish's gills depends on water for support. On land, lungs carry out gas exchange.

3. **Heart.** Walking on land requires a greater expenditure of energy than swimming. Land animals tend to have higher metabolic rates than aquatic animals. This, in turn, requires greater amounts of oxygen. The structure of the vertebrate heart allows oxygen to be delivered to the body efficiently.

Modern amphibians have such adaptations, but many groups are not well-adapted to dry land. For example, amphibian eggs are not watertight, and modern amphibians must seek out water or damp areas in which to reproduce.

Figure 9 *Eryops. Eryops* grew to 1.8 m (6 ft) long and probably lived much like a modern alligator, moving in and out of freshwater habitats.

Comparing the Surface Areas of Gills

6 cm

7 vertical cuts

Background

Air contains more oxygen than water, so why do fish die when removed from water? To understand what happens to gills when they are removed from water, follow the procedure below.

Materials

cellophane wrap, scissors, toothpick, ruler, container of water

Procedure

1. Cut an 8 × 6 cm piece of cellophane wrap. Use the piece of wrap and a toothpick to make a model of fish gills, as shown in the drawing.

2. Measure the approximate length and width of the model. Calculate the area of the model by using the formula $a = l \times w$.

3. Submerge the model in water and allow it to float at the top. Notice any change in shape of the model.

4. Grasp the model by the toothpick and pull it gently through the water. Then remove the model and place it on the table.

5. Without rearranging the model in any way, measure its approximate length and width. Calculate the area of the model again.

Analysis

1. **Summarize** any difference you observed between the gill model in the water and the wet model out of the water.

2. **Compare** the area obtained in step 2 with that obtained in step 5. If the areas are different, identify which was larger.

3. **Critical Thinking Analyzing Data** Consider what you know about the requirements for gas exchange across a gill's surface. Do the data you obtained suggest a reason why fishes cannot live out of water? Explain your answer.

4. **Critical Thinking Comparing Structures** Your model is two-dimensional. To calculate the surface area of an actual gill, you would need to know another measurement. What is that measurement, and why is it important?

Modern Amphibians

The middle Permian period marked the peak of amphibian success. By the end of this period a new kind of vertebrate, a reptile called a *therapsid*, had become common and began to replace the amphibians. By the end of the Triassic period that followed, there were only 15 families of amphibians, including the first species of frogs. And by the Jurassic period, only three groups of amphibians remained—frogs, salamanders, and caecilians. The age of amphibians was over.

All of today's amphibians are descendants of the amphibians that survived into the Jurassic period. They are found in aquatic and moist habitats throughout the temperate and tropical regions of the world. Frogs and toads make up the largest, and probably the most familiar, group of modern amphibians. Salamanders and newts are far less numerous. Caecilians account for less than 1 percent of today's amphibian species. **Figure 10** shows a representative sample of today's amphibians.

Figure 10 Amphibians

Today's amphibians are descendants of the three amphibian groups that survived into the Jurassic period.

Bullfrog

Spotted salamander

Woodhouse's toad

Yellow-striped caecilian

Section 1 Review

1 **Describe** two characteristics that distinguish vertebrates from invertebrates.

2 **Discuss** two adaptations that enabled early fishes to dominate the oceans.

3 **Relate** the structure of the limbs of a lobe-finned fish to the evolution of amphibians.

4 **Critical Thinking Evaluating a Hypothesis** Evaluate this statement: Amphibians are not fully adapted for life on land.

5 **Standardized Test Prep** Fossil evidence indicates that the first fishes lacked

A a backbone.   C fins.

B jaws.   D muscles.

Terrestrial Vertebrates

Early Reptiles

To understand the problem that water loss can be, all you need to do is stay outside on a hot day. It won't be long before you are sweating and thinking about a cool drink. If you don't increase your fluid intake and cool off, you may start to feel dizzy and nauseated. Fluid loss is a problem for all **terrestrial** animals, that is, animals that live on land. The adaptations that permitted amphibians to live on land further developed in reptiles. Two very important adaptations for terrestrial life evolved in reptiles. Reptiles were the first animals to have skin and eggs that are both almost watertight, and they differ from amphibians in this respect.

When reptiles first evolved, about 320 million years ago, Earth was entering a long, dry period. Early reptiles were better suited to these conditions than amphibians were, and the reptiles quickly diversified. Within 50 million years, reptiles had replaced amphibians as the dominant terrestrial vertebrates. **Figure 11** shows one hypothesized evolutionary relationship of terrestrial vertebrates.

Objectives

- **Summarize** why dinosaurs became the dominant land vertebrates.
- **Contrast** ectotherms with endotherms.
- **Identify** the dinosaurlike and the birdlike features of *Archaeopteryx*.
- **Summarize** why mammals replaced dinosaurs.

Key Terms

terrestrial
thecodont
Pangaea
ectothermic
endothermic
therapsid

Figure 11   Evolution of terrestrial vertebrates

This simplified phylogenetic tree shows one hypothesis of the relationship between early reptiles and the vertebrates that evolved from them.

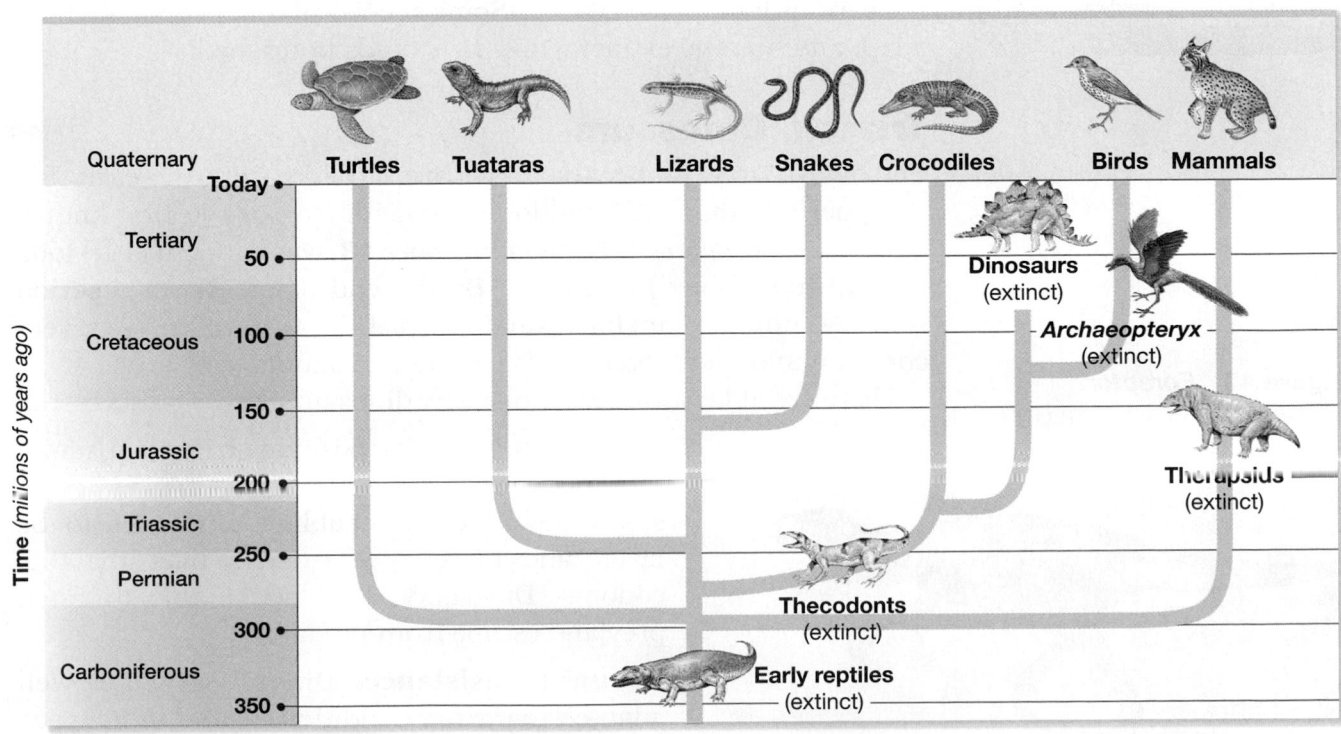

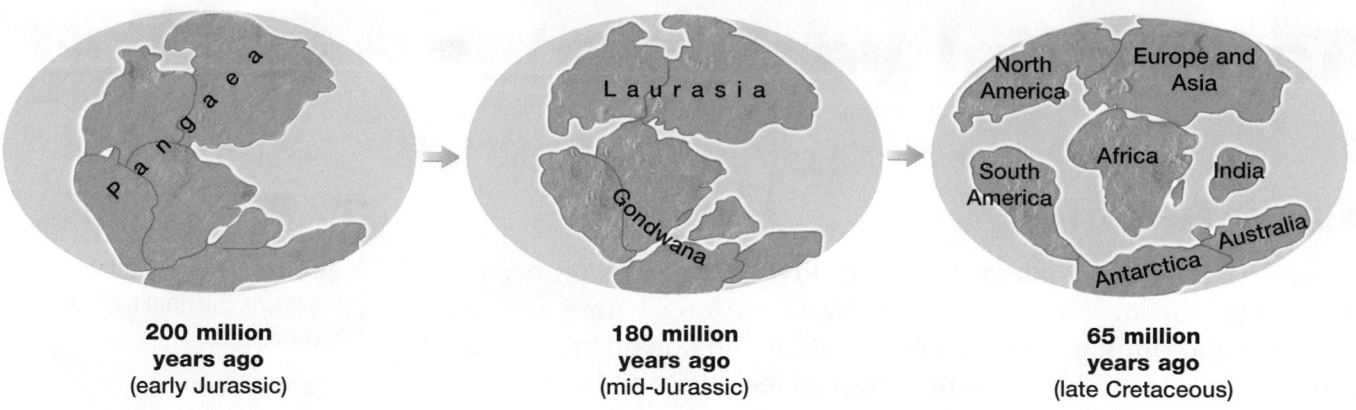

| 200 million years ago (early Jurassic) | 180 million years ago (mid-Jurassic) | 65 million years ago (late Cretaceous) |

Figure 12 Breakup of Pangaea. The supercontinent Pangaea began to break up during the Jurassic period. By the late Cretaceous period today's continents began to be recognizable.

internet connect

www.scilinks.org
Topic: Pangaea
Keyword: HX4135

SC*i*LINKS Maintained by the National Science Teachers Association

Evolution of Dinosaurs

Beginning about 235 million years ago, dinosaurs dominated life on land for roughly 150 million years. They evolved from the **thecodonts,** an extinct group of crocodile-like reptiles. During their long history, dinosaurs changed a great deal, in part because the world they inhabited changed. Thus, dinosaurs represent a long parade of change and adaptation—animals that lived at different times and were adapted to very different environments.

One factor that affected dinosaur evolution was the movement of the continents, which radically altered Earth's climate. When the dinosaurs first appeared, all of Earth's landmasses were joined in a single supercontinent called **Pangaea** *(pan GEE uh),* shown in **Figure 12.** There were few mountain ranges over this enormous stretch of land, and the interior was dry. Coastal climates were much the same all over the world—quite warm, with a dry season followed by a very wet rainy season. As Pangaea broke apart, the climates of the various landmasses varied. Some species of dinosaurs could not adapt and became extinct, while new kinds flourished.

Triassic Dinosaurs

The oldest known dinosaur fossils are in rocks from the early Triassic period, about 235 million years ago. One of the first known dinosaurs, *Eoraptor*, illustrated in **Figure 13,** was a 30 cm (1 ft) long bipedal (two-footed) carnivore. By the end of the Triassic period some 22 million years later, small, carnivorous dinosaurs were very common and had largely replaced the thecodonts.

There are at least three reasons why dinosaurs were so successful.

1. **Leg structure.** Legs positioned directly under the body provided good support for the dinosaur's body weight, enabling dinosaurs to be faster and more agile runners than the thecodonts. Dinosaurs were better able to catch prey and escape from predators.

2. **Drought resistance.** Dinosaurs were well adapted to the dry conditions found in Pangaea during the late Triassic period.

Figure 13 *Eoraptor.* *Eoraptor* was only 30 cm (1 ft) long—about the size of a chicken.

Figure 14 Sauropod

The sauropod *Brachiosaurus* stood 12.5 m (41 ft) tall and reached up to 23 m (75 ft) in length. It weighed 81,000 kg (89 tons), more than 14 African elephants or 1,500 students.

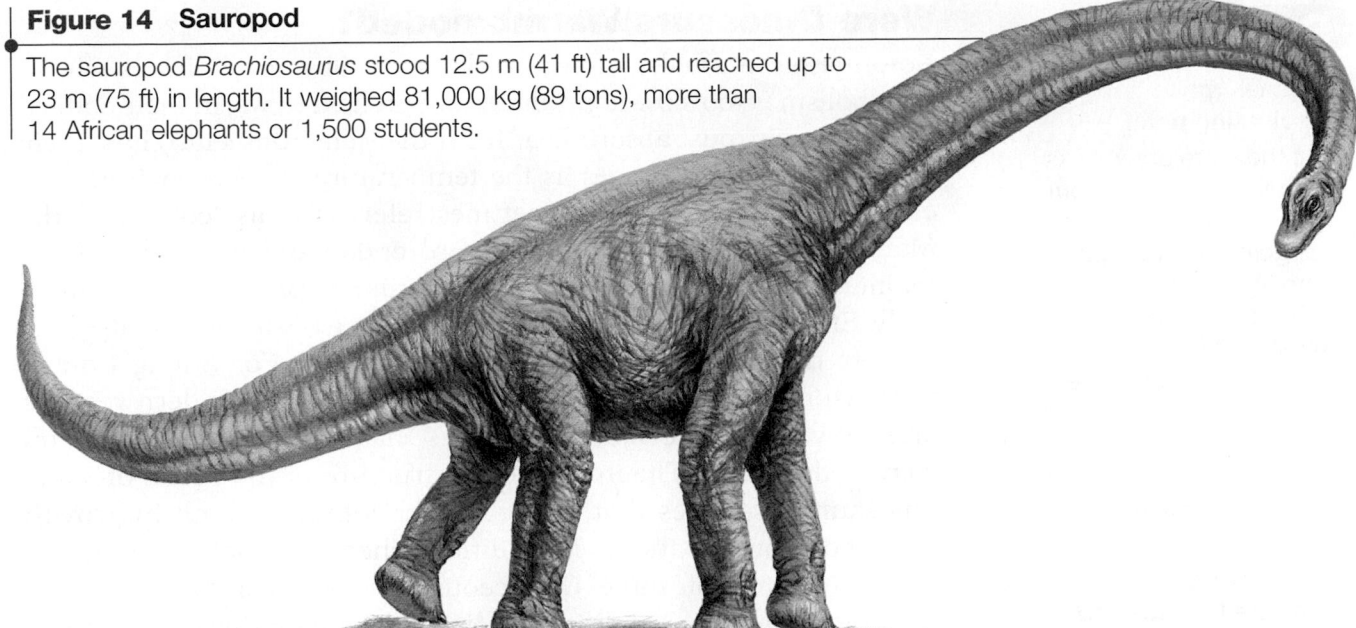

3. Extinction of other animal species. At the end of the Triassic period a large meteorite struck northeastern Canada. (Its impact site, the Manicouagan Crater, is still visible today.) This event might have been responsible for the great loss of animal diversity that occurred at the end of the Triassic period. Thecodonts and many other species became extinct, but the dinosaurs survived.

☑ **internet** connect ▤▤▤

www.scilinks.org
Topic: Comparing Dinosaurs
Keyword: HX4049

SC*LINKS* Maintained by the National Science Teachers Association

The Jurassic Period— The Golden Age of Dinosaurs

The Jurassic period is considered the golden age of dinosaurs because of the variety and abundance of dinosaurs that lived during this time. They included the largest land animals of all time, the sauropods (*SAWR oh pawdz*). As **Figure 14** shows, sauropods had enormous, barrel-shaped bodies, heavy columnlike legs, and very long necks and tails. Sauropods were the dominant herbivores of the Jurassic period.

By the late Jurassic period, the carnivorous theropods (*THEHR uh pawdz*) were common. Theropods stood on two powerful legs and had short arms. Their large heads were equipped with sharp teeth, and each foot had sickle-shaped claws used for ripping open prey. This anatomy was well suited for rapid running and quick, slashing attacks. Theropods preyed on the large herbivorous dinosaurs and were the dominant terrestrial predators until the end of the Cretaceous period. **Figure 15** shows a representative theropod.

Figure 15 Theropod. The theropod *Velociraptor* was about 1.8 m (6 ft) long, about the height of a large human.

Were Dinosaurs Warmblooded?

Ectothermic animals, such as today's reptiles, are animals whose metabolism is too slow to produce enough heat to warm their bodies. Such animals must absorb heat from their environment. Thus, their body temperature changes as the temperature of their environment changes. Ectotherms are sometimes referred to as "coldblooded." Mammals and birds, in contrast, are **endothermic** animals. They maintain a high, constant body temperature by producing heat internally. Endotherms are sometimes referred to as "warmblooded."

Were dinosaurs ectotherms or endotherms? For a long time it was assumed that dinosaurs were ectotherms, as modern reptiles are. However, new evidence indicates that at least some dinosaurs were endotherms. The microscopic structure of the bones of some dinosaurs resembles that of modern endotherms, both in growth rate and growth pattern. In addition, chemical analysis indicates that the bones of some Cretaceous dinosaurs probably formed under endothermic conditions. A lively debate on this question will probably continue for some time.

The Cretaceous Extinction

Toward the end of the Cretaceous period, sea levels began to fall and the climate began to cool. Many kinds of dinosaurs began to decrease in number. Then, 65 million years ago, all dinosaurs abruptly disappeared from the fossil record. No one knows for certain why this happened. Most scientists now agree that the major contributing cause was the impact on Earth of one or more meteorites or asteroids, the largest of which may have been 8 to 16 km (5 to 10 mi) in diameter. A crater 320 km (200 mi) wide off the coast of Mexico's Yucatán peninsula is very likely the site where this meteorite or asteroid collided with Earth approximately 65 million years ago, at the end of the Cretaceous period. A thin line of sediment marks the end of the Cretaceous period in rocks throughout the world. This sediment is rich in iridium, a mineral rare in Earth's crust but common in meteorites.

Scientists think that such an impact would have thrown large amounts of material from Earth's surface into the atmosphere. This material would have blocked out sunlight for a considerable period of time and created a prolonged, worldwide cold period. The endothermic birds and mammals, which were relatively small and insulated with feathers and fur, survived. The smaller ectothermic reptiles and amphibians also survived. But the dinosaurs did not survive.

Scientists are not sure why the dinosaurs did not survive this period of intense cold. Disease or competition from mammals might have led to their extinction. However, if most Cretaceous dinosaurs were endotherms, as some scientists now think they were, the dinosaurs would not have been able to survive the cold. They lacked the insulation of birds and mammals and could not lower their activity level as smaller reptiles and amphibians could.

Evolution of Birds

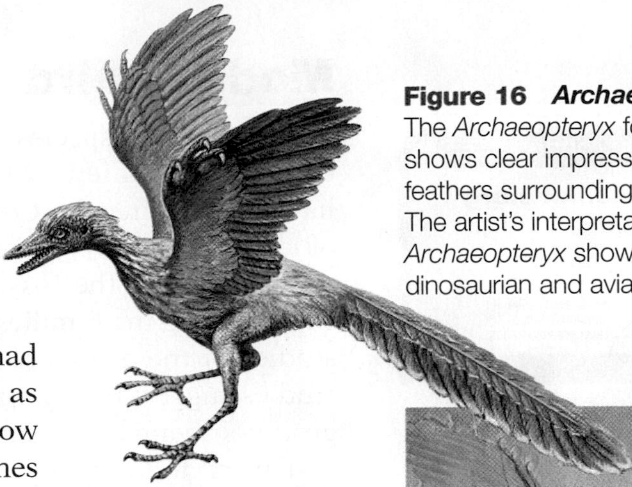

The earliest known bird is *Archaeopteryx* (meaning "ancient wing"), illustrated in **Figure 16.** *Archaeopteryx* was about the size of a crow and shared many features with small theropods. For example, it had a long reptilian tail. This is a feature of a dinosaur, not of a bird. *Archaeopteryx* had no breastbone to anchor its flight muscles, as do modern birds. And unlike the hollow bones of modern birds, *Archaeopteryx*'s bones were solid. It also had forelimbs similar to those of a dinosaur. Because of these dinosaurlike features, several *Archaeopteryx* fossils were originally classified as dinosaurs.

The impressions of feathers on some *Archaeopteryx* fossils raised questions about their classification as dinosaurs. *Archaeopteryx* fossils appear avian (birdlike) due to the feathers on their wings and tails. *Archaeopteryx* also had other avian features, notably a fused collar bone—the wishbone—which dinosaurs did not have. Today, most biologists agree that *Archaeopteryx* is very closely related to the small dinosaur *Compsognathus*. Some biologists go so far as to classify *Archaeopteryx* as a "feathered dinosaur." Recent fossil discoveries in China, such as the fossil shown in **Figure 17,** support this opinion. However, many biologists continue to classify birds in a separate class, Aves, because of their distinct features.

The fossil record now reveals a diverse collection of toothed birds with hollow bones and breastbones adapted for flight. By the early Cretaceous period, only 15 million years after *Archaeopteryx* lived, a variety of birds with many of the features of modern birds had evolved.

Figure 16 *Archaeopteryx.* The *Archaeopteryx* fossil below shows clear impressions of feathers surrounding the wings. The artist's interpretation of *Archaeopteryx* shows both its dinosaurian and avian features.

Figure 17 A feathered dinosaur? This dinosaur fossil, *Protoarchaeopteryx,* found in China in 1996, shows impressions of feathers along its back.

Modern Birds

internet connect

www.scilinks.org
Topic: Birds
Keyword: HX4024

SC**LINKS** Maintained by the
National Science
Teachers Association

There are more species of birds than of any other terrestrial vertebrate. All but a few of the modern orders of birds are thought to have arisen after the Cretaceous extinction. Since the impressions of feathers are rarely fossilized and since modern birds have hollow delicate bones, the fossil record of birds is incomplete. Relationships among the families of modern birds are mostly inferred from studies of the degree of DNA similarity among living birds. These studies suggest that the ostrich and its relatives belong to the oldest group of living birds. Ducks, geese, and waterfowl likely arose next and were followed by a diverse group of woodpeckers, parrots, swifts, and owls. Next came the songbirds, which include 60 percent of today's bird species. The most recent birds to appear were the more specialized orders, including many birds of prey, flamingos, and penguins. **Figure 18** shows representative modern birds.

Birds live in a wide variety of environments and differ greatly in appearance. While most birds have wings and can fly, the ostrich has wings and does not fly. The penguin's wings have been modified to be used as flippers that propel it swiftly through ocean waters. Striking colors often found in male birds help them attract a mate. All birds eat frequently because they have a high metabolic rate and typically store little body fat. The beaks of birds are adapted to the type of food the bird eats. Most birds consume small, energy-rich meals of fruits, seeds, worms, or insects. Hawks and owls are predators; they eat rabbits, rodents, and other small mammals, and sometimes snakes and lizards. Gulls and pelicans eat fish. Vultures feed on dead animals.

Figure 18   Modern birds

Although birds vary greatly in appearance, they all have feathers, a key defining characteristic.

Ostrich

Yellow warbler

American kestrel

Snow goose

Modern Reptiles

Of the 16 orders of reptiles known to have existed, only four remain today. Representatives of these orders are shown in **Figure 19.** The turtles have changed very little in structure since the time of the dinosaurs. The vast majority of living reptiles belong to the second group to evolve—snakes and lizards. Tuataras belong to the third group of surviving reptiles to evolve.

The fourth line of living reptiles, the crocodilians—crocodiles and their relatives, including the familiar alligators—appeared on Earth much later than the first three groups. Crocodilians have changed very little in more than 200 million years. Like dinosaurs, crocodilians are descendants of the thecodonts.

In some ways, such as the structure of their heart, crocodilians resemble birds far more than they resemble other living reptiles. And crocodilians are the only living reptiles that care for their young. What does this mean in terms of their relationships to other vertebrate species? Today, many biologists think that birds are direct descendants of the dinosaurs. If this is true, then crocodilians and birds are more closely related to dinosaurs, and to each other, than they are to other living reptiles. This close evolutionary relationship may account for some of their similarities.

Study *TIP*

● **Reading Effectively**

Locate the thecodonts in Figure 11. Notice how the lines leading to the crocodilians and birds stem from the thecodonts. This suggests that crocodiles and birds come from a different lineage than the other reptile groups.

Figure 19   Reptiles

Of the four living orders of reptiles, the order that includes crocodiles and alligators is the most recent. Crocodiles and their close relatives differ in several ways from other reptiles.

American alligator

Snapping turtle

Boa constrictor

Tuatara

Figure 20 *Eozostrodon.* Only 12 cm (5 in.) in total length, *Eozostrodon* is typical of the small, early nocturnal mammals.

Evolution of Mammals

The first mammals appeared about 220 million years ago, just as the dinosaurs were evolving from thecodonts. It is most likely that mammals were descendants of the **therapsids,** an extinct order of animals that were probably endotherms. Mammals are the only vertebrates that have fur and mammary glands. Early mammals, such as *Eozostrodon*, illustrated in **Figure 20,** were small—about the size of mice. They were insect-eating tree dwellers that were active at night.

For 155 million years, while the dinosaurs flourished, mammals were a minor group that changed little. Only five orders of mammals arose in that time, and their fossils are scarce. In the Cretaceous extinction, most animal species larger than a small dog disappeared. The smaller reptiles, including lizards, turtles, crocodilians, and snakes, survived. Mammals and birds also survived.

Diversification of Mammals

Just as after earlier mass extinctions, the stage was again set for a great evolutionary play. But the world's climate was no longer dry, as it was after the Triassic extinction. The adaptations that served reptiles well in dry climates were no longer so important. Mammals and birds became the dominant vertebrates on land. In the Tertiary period, mammals rapidly diversified, taking over many of the ecological roles once dominated by dinosaurs. Mammals reached their greatest diversity in the late Tertiary period, about 15 million years ago. At that time, tropical conditions existed over much of Earth. During the last 15 million years, world climates have changed, and the area covered by tropical habitats has decreased. As a result, the number of mammalian species has declined.

Ice Age Mammals Today, all very large land animals are mammals. However, many large land mammals existed during the last ice age (about 2 million to 10,000 years ago). At that time, many species of enormous mammals such as the Irish elk, illustrated in **Figure 21,** roamed Earth.

Figure 21 Irish elk. The Irish elk was actually a species of large deer that lived throughout Europe, northern Africa, and northern Asia.

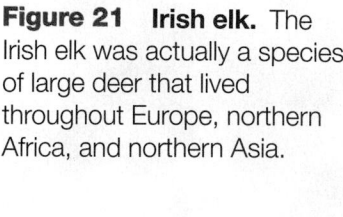

Giant ground sloths lived then and weighed three tons, as large as a modern elephant. Enormous vegetarian cave bears populated Europe during the ice ages. Large lionlike saber-toothed cats hunted with jaws that opened an incredible 120 degrees to allow the cat to drive its huge upper pair of saber teeth into prey. The shoulders of Irish elk, shown in Figure 21, were over 2.1 m (7 ft) high, and the span of the antlers could be up to 3.65 m (12 ft). All of these giant mammals are now extinct. Today only elephants approach the size of the enormous mammoths and giant camels that roamed Earth then.

Later Mammals Today, there are more than 4,500 species of mammals, and they inhabit virtually every habitat on Earth. They are found in jungles, deserts, and on polar ice. They live in fresh-water streams, and deep in the ocean. They even populate the air—bats are among the most successful of all mammal groups. Mammals range in size from the blue whale, which may weigh as much as 136,000 kg (150 tons), to the tiny Kitti's hognose bat, weighing about 1.5 grams (less than 0.1 oz). Most mammals are covered with a dense coat of fur, but some have little fur. Whales have only a few hairs.

internet connect

www.scilinks.org
Topic: Early Mammals
Keyword: HX4063

SCLINKS. Maintained by the National Science Teachers Association

FORENSICS BIOWatch

If Bones Could Talk

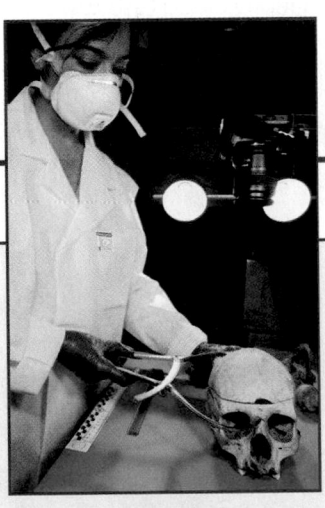

When a murder victim has been dead for weeks, months, or years, clues can be few and far between. In these cases, homicide investigators often seek the aid of forensic physical anthropologists. Physical anthropology is the scientific study of human variation, classification, and evolution. Forensic anthropologists use their knowledge of skeletal anatomy and biology to help solve crimes.

Silent Witness

Long after soft body tissue has decayed, bones remain, bearing silent witness to their owner's fate. In the hands of an experienced forensic anthropologist, telltale evidence will often emerge. If a victim suffered a violent death (for example, a hammer blow to the head), an anthropologist can analyze signs of physical injury. Using a microscope, the anthropologist can see precise details of tool marks, which may suggest a likely murder weapon. Such details can also help investigators distinguish saw marks, for example, from postmortem damage caused by a gnawing rodent. Based on the condition of the body, forensic anthropologists can also estimate the time that has passed since death.

Who Is It?

Many times a criminal investigation will seek to determine the victim's identity. Skeletal remains can reveal the victim's age, sex, and height. Based on patterns of muscle attachment, a forensic anthropologist can even determine whether the victim was right- or left-handed. Computer-assisted modeling techniques let anthropologists create a possible "face" that fits the underlying bone structure of the skull. A law enforcement agency or a member of the public may recognize this image. The anthropologist can also adjust the image to "age" the subject, a technique often used when investigating child kidnappings.

Modern Mammals

The first group of mammals to evolve laid eggs, as did their early ancestors. The only surviving direct descendants of these early mammals are two species of echidna *(ee KIHD nuh)* and the duck-bill platypus, shown in **Figure 22.** These egg-laying mammals are in a group of their own called the monotremes.

Mammals other than monotremes show one of two patterns of development. Both of these groups give birth to live young, but their embryos develop in different ways.

1. **Marsupials.** Marsupial *(mahr SOO pee uhl)* mammals include kangaroos, also shown in Figure 22, opossums, koalas, and wombats. Their young are born at a very immature stage and complete their development in their mother's pouch (called a marsupium).

2. **Placentals.** Placental mammals include dogs, cats, horses, sheep, gorillas, humans, and most of the other mammals you are familiar with. Placental mammals, such as the goat seen in Figure 22, develop within their mother's body and are nourished by an organ called the placenta. In a later chapter, you will learn more about the mammals.

Figure 22 Mammals.
There are three types of mammals—placentals, marsupials, and monotremes—each represented below.

Domestic goat

Red kangaroo

Duck-bill platypus

Section 2 Review

1. **Describe** two key adaptations that allowed reptiles to live and reproduce on land.

2. **Summarize** the factors that contributed to the dinosaurs' success on land.

3. **Compare** ectothermic and endothermic animals.

4. **Relate** changes in the world's climate to the increase in numbers of mammals.

5. **Critical Thinking Forming Reasoned Opinions** Evaluate the change in classification of *Archaeopteryx* from dinosaur to bird.

6. **Standardized Test Prep** At the end of the Cretaceous period 65 million years ago, natural selection resulted in the diversification of

   A mammals.   **C** theropods.

   B sauropods.   **D** thecodonts.

Evolution of Primates

Characteristics of Primates

Fossil evidence indicates that small, insect-eating mammals with large eyes and small, sharp teeth lived about 80 million years ago, during the age of dinosaurs. These ancient mammals were the ancestors of the first primates. A **primate** is a member of the mammalian order Primates, which includes prosimians, monkeys, apes, and humans. The first primates evolved about 50 million years ago. These animals had two features that enabled them to stalk and capture prey in the branches of trees.

1. **Grasping hands and feet** The grasping hands and feet of primates enable them to cling to their mothers when they are young, grip limbs, hang from branches, and seize food.

2. **Forward orientation of the eyes** Unlike the eyes of their ancestors, which were located on the sides of the head, the eyes of primates are positioned at the front of the face. This forward placement of the eyes produces overlapping binocular vision that enables the primate brain to judge distance more precisely (depth perception). The ability to judge depth is very important for an animal that leaps from branch to branch high above the ground. Some other mammals have binocular vision, but only primates have both binocular vision and grasping hands.

Prosimians

According to the fossil record, the modern primates that most closely resemble early primates are the prosimians. A **prosimian** is a member of a group of mostly night-active primates that live in trees. Modern prosimians include lorises, lemurs, and tarsiers, such as the one shown in **Figure 23.** Fossil evidence indicates that prosimians were common about 38 million years ago in North America, Europe, Asia, and Africa.

Objectives

- **Name** two unique features of primates.
- **Contrast** prosimians with monkeys.
- **Distinguish** monkeys from apes.
- **Describe** the evolutionary relationship between humans and apes.
- **Identify** the evidence that indicates human ancestors walked upright before their brains enlarged.

Key Terms

primate
prosimian
diurnal
hominid

Figure 23 Prosimian.
The tarsier, *left*, has bendable, clawed fingers and toes and forward-facing eyes, which are key adaptations for life in the trees.

Nonhuman Primates

About 36 million years ago, a revolution occurred in how primates live. Many primate species became diurnal. **Diurnal** *(die UR nuhl)* animals are active during the day, and they sleep at night. The evolution of a diurnal pattern gave primates more opportunities to feed and enabled them to better detect predators. Modern day-active primates include monkeys and apes.

Monkeys

Feeding mainly on fruits and leaves, monkeys were among the first primates to have opposable thumbs. An opposable thumb—such as your own—stands out at an angle from the other fingers and can be bent inward toward them to hold an object. This gives the hand a greatly increased level of ability to manipulate objects.

Apes

Apes, which share a common ancestor with monkeys, first appeared about 30 million years ago. Modern apes include gibbons, orangutans, gorillas, and chimpanzees. Apes have larger brains with respect to their body size than monkeys, and none of the apes have tails.

The phylogenetic tree shown in **Figure 24** represents the theoretical evolutionary relationships between modern apes and humans. DNA analysis has shown that the genes of humans and chimpanzees are remarkably similar. Indeed, human and chimpanzee DNA nucleotide sequences may differ by as little as 5 percent. For example, the 287 amino acids that make up two kinds of chains of protein in human hemoglobin are identical to the amino acids in chimpanzee hemoglobin.

Figure 24 Phylogenetic tree of apes and humans. The evolutionary group that led to modern gibbons diverged earlier than other primate groups.

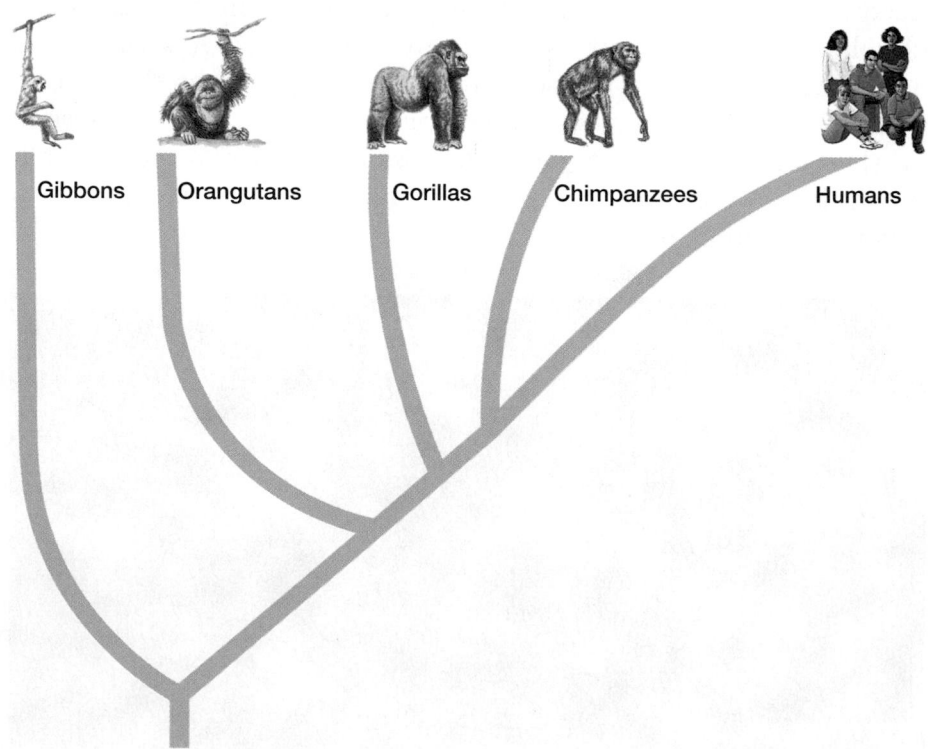

Gibbons   Orangutans   Gorillas   Chimpanzees   Humans

Early Hominids

Hominids *(HAHM ih nihds)* are primates that walk upright on two legs. Hominids are members of the group that led to the evolution of humans. According to the fossil record, hominids first appeared on Earth about 5 million to 7 million years ago. All early primates walked on all four limbs. What, then, caused hominids to stand up and walk on two legs? Hypotheses about this aspect of human evolution differ. Fifteen million years ago, the world's climate began to cool, and the great forests of Africa were largely replaced by savannas (treeless plains). Natural selection may have favored primate forms adapted to living on the ground.

Australopithecines

The early hominids best represented by fossil finds belong to the group known as australopithecines *(aw stray loh PIHTH uh seenz)*. Australopithecines belong to the genus *Australopithecus*. **Figure 25** shows Lucy, the fossil remains of an australopithecine that lived more than 3 million years ago. Their fossil remains indicate that australopithecines walked upright on two legs; thus they are classified as hominids. As **Table 1** shows, the skeletons of australopithecines differ from those of modern apes in several ways.

The brains of australopithecines had a slightly greater volume, relative to body weight, than the brain of an ape. Some australopithecine species weighed about 18 kg (40 lb) and were approximately 1.1 m (3.5 ft) tall, about the size of a small chimpanzee. Other australopithecine species were larger, weighing more than 45 kg (100 lb) and standing more than 1.5 m (5 ft) tall. Their brains, 400 cm³ in volume, were generally as large as those of modern chimpanzees. Australopithecine brains were much smaller, however, than the brains of modern humans, which are about 1,350 cm³ (83 in.³).

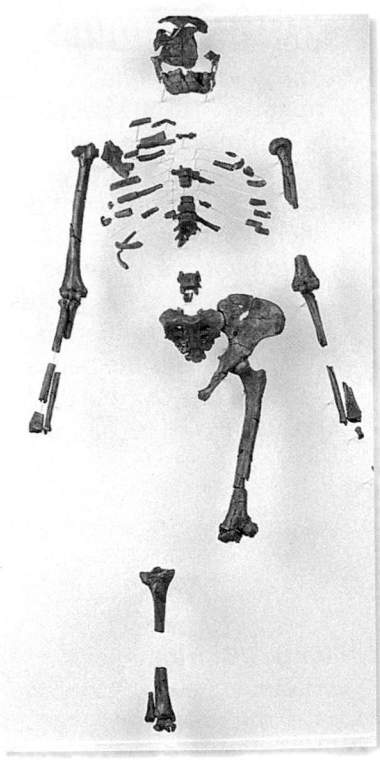

Figure 25   Lucy. Lucy's leg bones indicate that she walked upright. She stood about 1 m (3 ft) tall.

Table 1 Comparison of Gorilla and Australopithecine Skeletons

	Gorilla	Australopithecine
	Skull atop C-shaped spine; spinal cord exits near rear of skull	Skull atop S-shaped spine; spinal cord exits at bottom of skull
	Arms longer than legs; arms and legs used for walking	Arms shorter than legs; only legs used for walking
	Tall and narrow pelvis, allowing the body weight to shift forward	Bowl-shaped pelvis, centering the body weight over the legs
	Femurs (thighbones) angled away from pelvis when walking upright	Femurs angled inward, directly below body to carry its weight

Gorilla   ■ **Skull**   ■ **Spine**   ■ **Arms**   ■ **Pelvis**   ■ **Femurs**   **Australopithecine**

Other Early Hominids

In the past few years, scientists have discovered fossils of hominids that lived at the same time or earlier than *Australopithecus*. The relationship between early hominid species, shown in **Figure 26,** is uncertain. These early fossils indicate that the hominid line is an old one, extending back past 6 million years. The discovery in 2002 of a 6 million to 7 million year old fossil in central Africa, is creating lively debate among scientists about the origins of the hominids. This new species, *Sahelanthropus tchadensis*, appears at the same time more primitive and more modern than the australopithecines. Some scientists think that *Sahelanthropus* represents a species at the root of the family tree of hominids. The modern features of this very old fossil also suggest to some scientists that australopithecines may not have been ancestral humans, but rather a group of species that left no descendants.

Until more early hominid fossils are discovered, the historical picture remains too unclear to draw any firm conclusion about true lines of descent. What is clear is that over the last 6 million years, a variety of different hominids have existed, with more than one species living at one time. As new fossils are found, our picture of the root of the human phylogenetic tree may come into sharper focus.

Figure 26 Hominid evolution. Several species of hominids arose and then died out.

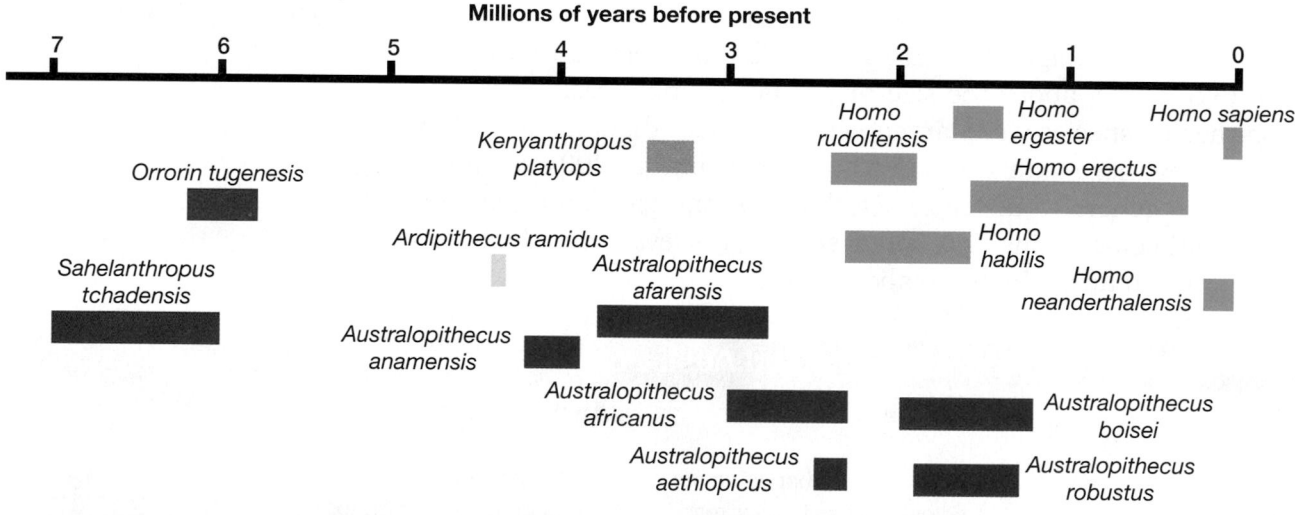

Millions of years before present

Section 3 Review

1. **Describe** two unique features of primates.

2. **Describe** one characteristic that distinguishes prosimians from monkeys and one characteristic that distinguishes monkeys from apes.

3. **Identify** and explain the evidence that closely links humans to chimpanzees.

4. **Critical Thinking Comparing Structures** Summarize the differences between the skeleton of an ape and the skeleton of an australopithecine.

5. **Standardized Test Prep** What evidence would help scientists establish whether or not a primate was bipedal?

   A skull size **C** femur weight

   B pelvis shape **D** number of ribs

The Genus *Homo*

Homo habilis

Our genus, *Homo*, is composed of at least three species. Of these three, modern humans, *Homo sapiens sapiens*, is the only surviving species. The first members of the genus *Homo* appeared on Earth more than 2 million years ago. In the early 1960s, hominid bones were discovered near the site where *Australopithecus* fossils had been found. Stone tools were scattered among the bones. Scientists disagreed about whether this fossil was that of an early human or an australopithecine. Reconstruction of the bones of the skull indicated that the volume of the brain was about 640 cm³ (39 in.³). This is larger than the australopithecine brain volume of 400 to 550 cm³ (24 to 34 in.³).

Because of its association with tools, this hominid was named *Homo habilis*. The Latin word *homo* means "man," and the Latin word *habilis* means "handy." Bones discovered in 1987 indicate that *Homo habilis* stood about 1.2 m (4 ft) tall. Fossils indicated that *Homo habilis* lived in Africa for about 500,000 years and then became extinct.

Objectives

- **Compare** *H. habilis* with australopithecines.
- **Describe** the characteristics of *Homo erectus*.
- **Describe** the evidence that suggests that *H. sapiens* evolved in Africa.
- **Compare** Neanderthals with modern humans.

Key Terms

None

Figure 27 **Richard Leakey and *Homo habilis* skull**

Homo habilis lived in eastern Africa.

Homo habilis had a brain volume of about 700 cm³ on the average and many of the characteristics of modern human skulls.

Richard Leakey (seated, far right) and his co-workers found one of the most complete *Homo habilis* skulls to date.

735

Homo erectus

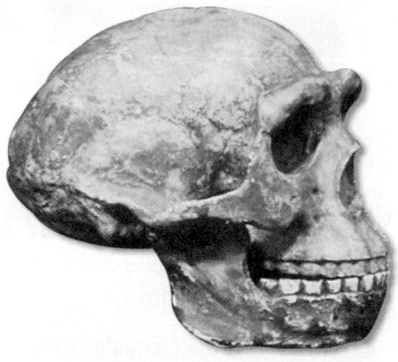

Figure 28 Homo erectus.
Homo erectus had large brow ridges and smaller teeth than *Homo habilis*.

Our understanding of *Homo habilis* is limited because it is based on only a few fossil specimens. In contrast, scientists have found many fossils of the species that replaced *Homo habilis*—the species *Homo erectus*.

The first *Homo erectus* specimen found, called Java Man, was discovered in 1891 on the island of Java, in southeast Asia. *Homo erectus* was larger than *Homo habilis*—about 1.5 m (5 ft) tall. *Homo erectus* also had a large brain of about 1,000 cm³ (60 in.³). Java Man may be more than about 700,000 years old. Similar fossils found in China are thought to be as old as 500,000 years. In 1976, a *Homo erectus* skull that may be 1.5 million years old was discovered in eastern Africa. **Figure 28** shows a fossil *Homo erectus* skull.

Homo erectus evolved in Africa and migrated into Asia and Europe, as shown in **Figure 29.** *Homo erectus* probably lived in groups of 20 to 50 individuals. They hunted large animals, used fire, and made both stone and bone tools. *Homo erectus* survived for more than 1 million years. The species disappeared about 200,000 years ago, as early modern humans emerged. Most scientists think that *Homo erectus* was the direct ancestor of our species, *Homo sapiens*.

Figure 29   Hominid migration

Homo erectus evolved in Africa and migrated to Asia and Europe. Later, *Homo sapiens* appeared in Africa and migrated to Europe and Asia.

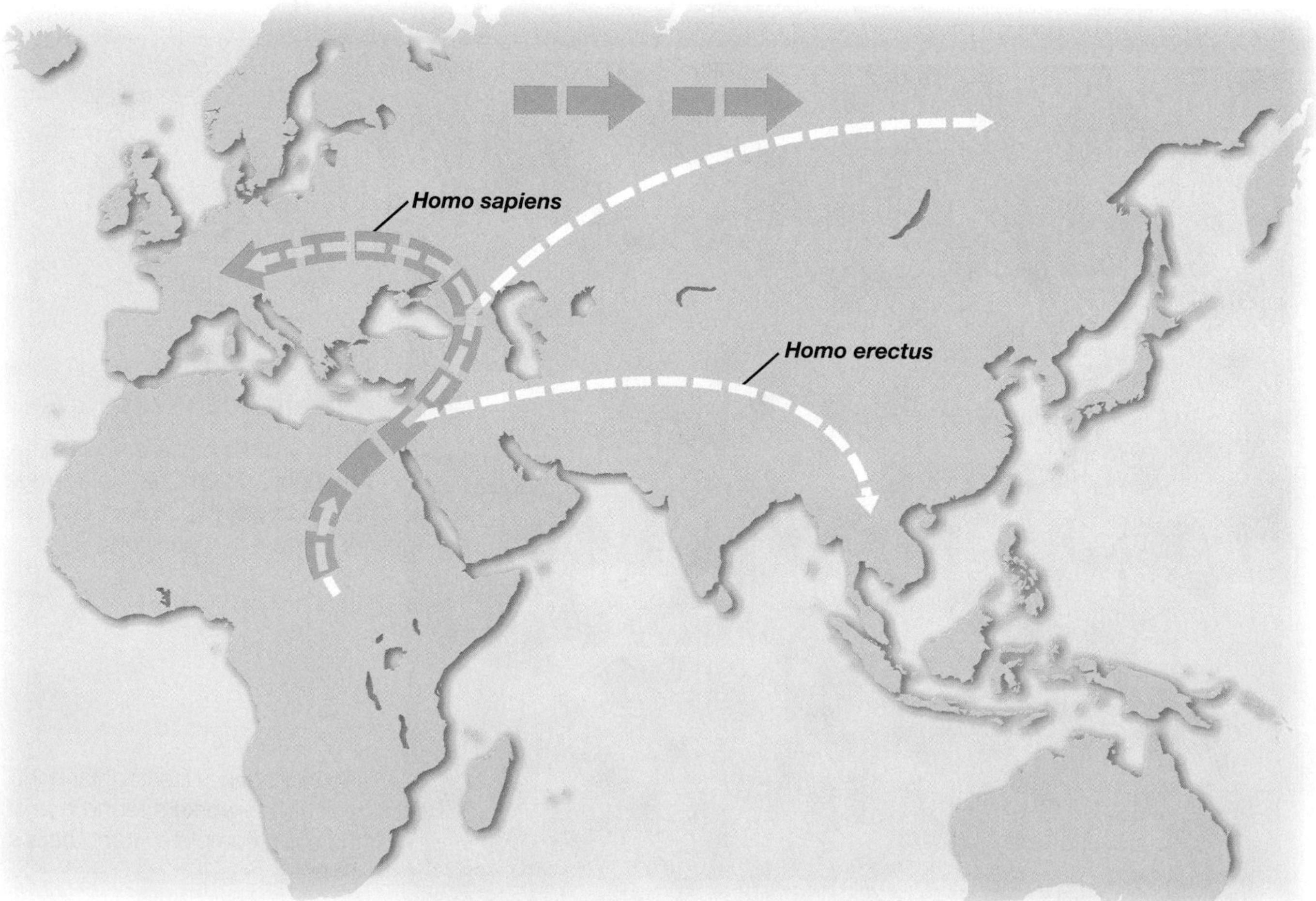

Homo sapiens

Homo erectus

Homo sapiens

Homo sapiens is the only surviving species of the genus *Homo*. The name *Homo sapiens* is from the Latin *homo*, meaning "man," and *sapiens*, meaning "wise." *Homo sapiens* is a newcomer to the hominid family. *Homo sapiens* has not existed as long as *Homo erectus* did. Early *Homo sapiens* left behind many fossils and artifacts, including the first known paintings, such as the one shown in **Figure 30.**

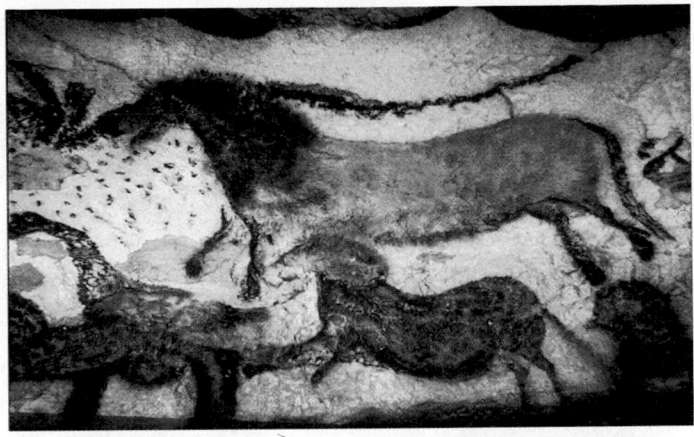

Figure 30 Cave painting. This thundering herd of horses was painted on the wall of a cave in Lascaux, France. The cave paintings may be about 20,000 years old.

African Origins?

In the past, some scientists proposed that independent *Homo erectus* groups living in Africa, Europe, and Asia interbred and that *Homo sapiens* thus arose as a new species at the same time all over Earth. Most scientists, however, argue that *Homo sapiens* appeared in one place (Africa) and then migrated to Europe and Asia, where the species gradually replaced *Homo erectus*.

Exploring Further

DNA Dating

To answer the question of when and where *Homo sapiens* evolved, researchers are studying the DNA of modern humans. Recall that DNA tends to accumulate more and more mutations as time passes. Therefore, individuals that share a recent common ancestor have relatively similar DNA. In contrast to this, individuals who share only a *distant* common ancestor have less similar DNA. Thus individuals in older human populations should be more genetically different from each other than individuals in newer populations are.

Mitochondrial DNA

Human DNA has been looked at in two ways that shed light on the issue. The first approach focuses on mitochondrial DNA (mDNA). Humans inherit mDNA from only their mother—it is found in the egg cell. mDNA is thus passed down unchanged from generation to generation. Researchers began by comparing the mDNA of individuals from different ethnic backgrounds. Based on their data, the researchers hypothesize that all modern humans share a common ancestor that lived about 170,000 years ago. According to fossil evidence, this was about the time *Homo sapiens* appeared in Africa.

The researchers used the mDNA data to generate a human "family tree." When analyzed in this way, the data show a distinct branch that arose 52,000 years ago between Africans and non-Africans. This supports the hypothesis that some modern *Homo sapiens* left Africa and spread to Europe and Asia. They thus retraced the path taken by *Homo erectus* many years before.

Y Chromosomal DNA

Researchers have also analyzed DNA on the Y chromosome. Like mDNA, DNA on the Y chromosome does not undergo recombination during meiosis. Thus DNA on the Y chromosomes is passed down unchanged in males from one generation to the next. Researchers analyzed DNA variations in more than 1,000 European males. They found that 80 percent of the males studied share a single pattern of DNA. This suggests that modern Europeans share a relatively recent common ancestor. The data indicate the shared pattern arose about 40,000 to 50,000 years ago. This piece of evidence suggests that modern humans came to Europe at that time.

Taken together, the analysis of both mDNA and Y chromosome DNA indicate that modern humans arose only once, in Africa.

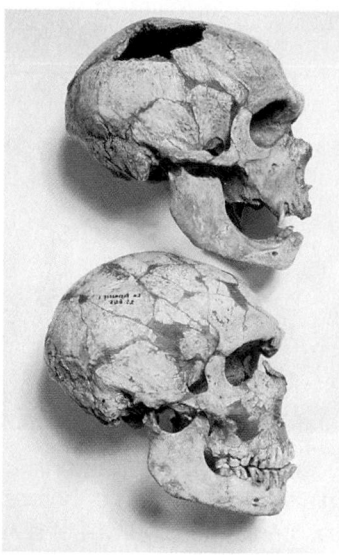

Figure 31 Neanderthals.
Neanderthals were distinguished by massive ridges over the eyes.

Homo sapiens in Europe

Members of the species *Homo sapiens* first appeared in Europe about 130,000 years ago. The fossils of a group of *Homo sapiens* called Neanderthals (*nee AN dur TALZ*—the *h* is silent) were found in 1856 in the Neander Valley of Germany. The Neanderthals were short and powerfully built. Their skulls were massive, as shown in **Figure 31,** with protruding faces and heavy, bony brow ridges. The average Neanderthal brain was slightly larger than that of modern humans, although there is much overlap in size. Many scientists that study human evolution now classify Neanderthals as a separate human species, *Homo neanderthalensis*.

Neanderthals became more and more abundant in Europe and Asia, and by 70,000 years ago, they had become fairly common. Neanderthals cared for their injured and sick. They also commonly buried their dead and often placed food, weapons, and even flowers with the bodies. Such attention to the dead suggests that they may have believed in a life after death. Neanderthals were the first hominids to show evidence of abstract thinking.

Modern *Homo sapiens*

About 34,000 years ago, the European Neanderthals were abruptly replaced by *Homo sapiens* of modern appearance. These early humans are thought to have evolved first in Africa and then migrated to Europe and Asia. Early modern humans lived by hunting. They had complex patterns of social organization, and they probably used language. They spread across Siberia and reached North America at least 12,000 years ago, when a land bridge connected Siberia and Alaska.

The ability of humans to make and use tools effectively has been important to the success of humans. Humans use symbolic language and can shape concepts out of experience. Written language has enabled humans to transmit concepts from one generation to another. Humans can also reason abstractly, applying existing knowledge to new situations. Humans also have cultural evolution. Through culture, we have found ways to change our environment to suit our needs, rather than changing ourselves in response to the environment. This is both an exciting potential and an enormous responsibility.

Section 4 Review

1 **Describe** how *Homo habilis* differs from australopithecines.

2 **Describe** the evidence that identifies *Homo erectus* as the first human species to have left Africa.

3 **Analyze** the two hypotheses of the origin of *Homo sapiens*.

4 **Contrast** Neanderthals with modern humans.

5 **(Standardized Test Prep)** What evidence have scientists used to estimate when and where modern humans evolved?
  A mDNA
  B cave paintings
  C analysis of language
  D articles buried with the dead

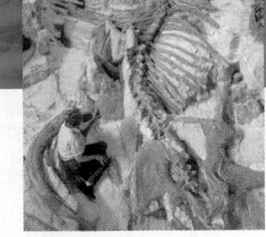

Key Concepts

1 Vertebrates in the Sea and on Land

- Vertebrates have an endoskeleton that includes a backbone, bilateral symmetry, cephalization, a coelom, a closed circulatory system, and a chambered heart.
- The first vertebrates, jawless fishes, evolved about 500 million years ago.
- Spiny fishes with jaws appeared 430 million years ago.
- Cartilaginous sharks and bony fishes evolved about 400 million years ago. Bony fishes account for more than 90 percent of today's fishes.
- Amphibians, which appeared about 370 million years ago, were the first group of vertebrates to live on land.

2 Terrestrial Vertebrates

- Reptiles evolved about 320 million years ago and have watertight skin and a watertight egg.
- Dinosaurs dominated the land for 150 million years.
- Modern birds may be the descendants of a group of small theropods, a type of dinosaur.
- The first mammals appeared 220 million years ago and took over many ecological roles once dominated by dinosaurs.

3 Evolution of Primates

- The first primates evolved 50 million to 60 million years ago.
- Primates have grasping hands and binocular vision.
- DNA evidence indicates that humans are more closely related to chimpanzees than to any other primate species.
- Early hominids walked upright before their brains became significantly larger.

4 The Genus *Homo*

- *Homo habilis* had a brain much larger than that of australopithecines and used tools.
- *Homo erectus* evolved in Africa about 1.5 million years ago and may have been the direct ancestor of *Homo sapiens*.
- *Homo sapiens* probably evolved in Africa and migrated to the rest of the world.

Key Terms

Section 1
vertebra (712)
agnathan (714)
acanthodian (715)
cartilage (715)

Section 2
terrestrial (721)
thecodont (722)
Pangaea (722)
ectothermic (724)
endothermic (724)
therapsid (728)

Section 3
primate (731)
prosimian (731)
diurnal (732)
hominid (733)

Section 4
None

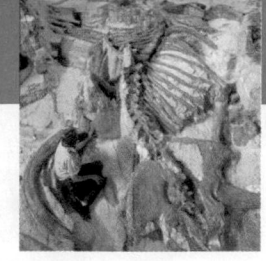

Understanding Key Ideas

1. To survive as predators, fish had to adapt to which of the following two challenges?
 a. low oxygen conditions, shallow seas
 b. avoiding predators, grasping prey
 c. pursuing prey in water, grasping prey
 d. filtering invertebrates from the water, avoiding predators

2. Which of the following is *not* an amphibian adaptation for life on land?
 a. legs
 b. lungs
 c. watertight skin
 d. more efficient heart

3. The first fully terrestrial vertebrates were
 a. frogs.
 b. reptiles.
 c. dinosaurs.
 d. mammals.

4. As Pangaea broke apart,
 a. dinosaurs became more alike.
 b. mammals became dominant.
 c. dinosaur species flourished.
 d. world climates changed.

5. Mammals are thought to have survived the climatic changes that caused the extinction of dinosaurs because the bodies of mammals
 a. have little insulation.
 b. are ectothermic.
 c. were relatively small.
 d. lacked insulation.

6. Which sequence reflects the order in which the major groups of vertebrates are thought to have evolved?
 a. bony fishes → reptiles → amphibians
 b. bony fishes → amphibians → reptiles
 c. amphibians → mammals → reptiles
 d. birds → mammals → reptiles

7. What evidence convinced scientists that *Archaeopteryx* should be classified as a bird rather than a small dinosaur?
 a. teeth
 b. hollow bones
 c. hard-shelled eggs
 d. fused collarbone

8. What evidence is there that crocodiles and birds are likely to be more closely related to each other than to other living reptiles?

9. **BIOWatch** List three causes of death that could be discovered through the work of forensic anthropologist.

10. **Concept Mapping** Construct a concept map that shows the relationships among the various living and nonliving vertebrates. Include the following terms in your concept map: *jawless fishes, spiny fishes, armored fishes, cartilaginous fishes, bony fishes, amphibians, thecodonts, dinosaurs, reptiles, therapsids, mammals, Archaeopteryx,* and *birds.*

Critical Thinking

11. **Recognizing Relationships** Describe the positioning of the eyes in the first primates. How did this positioning of the eyes improve the primates' ability to navigate their environment?

12. **Evaluating Hypotheses** Some stone tools are more than 2 million years old. Evaluate the hypothesis that they must have been made by a species from the genus *Homo.*

Alternative Assessment

13. **Recognizing Relationships** Vertebrates share several characteristics. Develop an illustrated guide that informs readers about these characteristics.

14. **Forming a Model** Make a series of models or create a mural that shows either the evolution of jaws or the evolution of limbs. Create a written guide or an audiotape that guides the viewer through your work.

Standardized Test Prep

Understanding Concepts

Directions (1–5): **For** *each* **question, write on a separate sheet of paper the letter of the correct answer.**

1 Which of the following is **not** considered a key characteristic of vertebrates?
A. backbone
C. skull
B. jaws
D. vertebrae

2 What were the earliest fishes called?
F. acanthodians
G. agnathans
H. thecodonts
I. therapsids

3 Which of the following are night-active primates?
A. ectoderms
B. endoderms
C. prosimians
D. thecodonts

4 Apes have larger brains than monkeys. What do apes lack that monkeys have?
F. bipedalism
G. depth perception
H. opposable thumbs
I. tail

5 Which of the following describes reptiles but **not** birds?
A. They are ectotherms.
B. They are endotherms.
C. They have feathers.
D. They have watertight eggs.

Directions (6): **For the following question, write a short response.**

6 Analyze the ways in which the adaptations of reptiles to land are similar to the adaptations of plants to land.

Test *TIP*

When a question uses a word you don't know, try to determine the meaning of the word by breaking it down into smaller parts and inferring the meaning of each part.

Reading Skills

Directions (7): **Read the passage below. Then answer the question.**

Fossil evidence collected in the 1980s in Arctic Alaska suggests that some dinosaurs lived year-round in areas of freezing temperatures and total darkness during the winter months. This evidence of "Arctic dinosaurs" contradicts the hypothesis that the extinction of dinosaurs was due to a climate change produced by a meteor collision sending debris into the Earth's atmosphere.

7 Why would this evidence contradict the hypothesis?
F. All dinosaurs had inadequate insulation from cold and could not slow down their metabolism to survive the cold.
G. The meteor that hit the Earth would have made the Arctic regions even colder, killing off dinosaurs there.
H. Dinosaurs that could survive the Arctic climate would not go extinct when debris blocked the sun and cooled the Earth.
I. A period of intense cold that killed other dinosaurs would cause Arctic dinosaurs to evolve and become warm blooded.

Interpreting Graphics

Directions (8): **Base your answer to question 8 on the drawings below.**

Two Extinct Animals

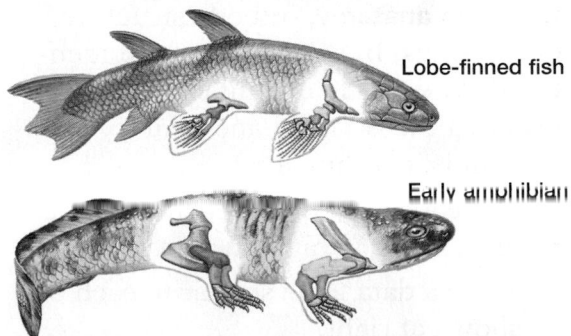

Lobe-finned fish

Early amphibian

8 What are the bones in the drawings an example of?
A. vertebrae
C. vestigial structures
B. gill arches
D. homologous structure

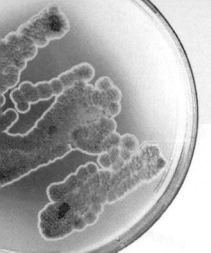

Exploration Lab

Comparing Hominid Skulls

SKILLS
- Measuring
- Comparing anatomical features

OBJECTIVES
- **Identify** differences and similarities between the skulls of apes and the skulls of humans.
- **Identify** differences and similarities between the fossilized skulls of hominids.
- **Classify** the features of hominid skulls as apelike, humanlike, or intermediate.

MATERIALS
- metric ruler
- protractor

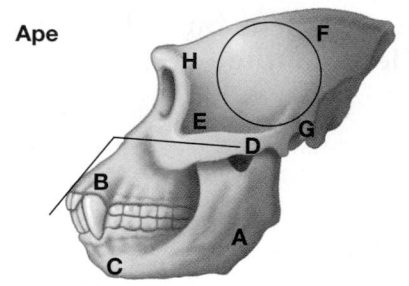

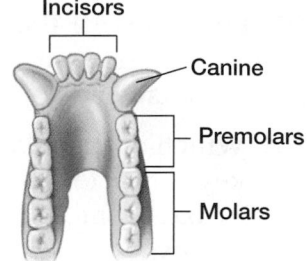

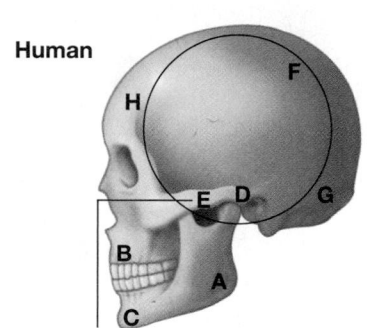

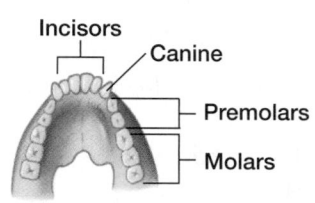

Before You Begin

Modern **apes** and humans share a **common ancestor.** Much of our understanding of human evolution is based on the study of the fossilized remains of **hominids.** By studying fossilized bones and identifying similar and dissimilar structures, scientists can infer the **anatomy,** or body structure, of a species. In this lab, you will identify differences and similarities between the skulls of apes, early hominids, and humans.

1. Write a definition for each boldface term in the paragraph above.

2. Make a data table similar to each one shown at right.

3. Based on the objectives for this lab, write a question you would like to explore about human evolution.

DATA TABLE 1						
Name	Cranial capacity (cm³)	Lower face area (cm²)	Brain area (cm²)	Jaw angle (degrees)	Brow ridge	Teeth
Ape						
Human						

DATA TABLE 2				
Name				
Australopithecus robustus				
Australopithecus africanus				
Homo erectus				
Neanderthal				

How to Interpret the Features of a Skull

Cranial capacity: Use the circles drawn on the skulls to estimate brain volume, or cranial capacity. Measure the radius of each circle in centimeters. Then cube this number, and multiply the result by 1,000 to calculate the approximate life-size cranial capacity in cubic centimeters.

Lower face area: Measure *A* to *B* and *C* to *D* in centimeters for each skull. Multiply these two numbers together, and multiply the product by 40 to approximate the life-size lower face area in square centimeters.

Brain area: Measure *E* to *F* and *G* to *H* in centimeters for each skull. Multiply these two numbers and multiply the product by 40 to approximate the life-size brain area in square centimeters.

Jaw angle: Note the two lines that come together near the nose of each skull. Use a protractor to measure the inside angle made by the lines and to determine how far outward the jaw projects.

Brow ridge: Note the presence or absence of a bony ridge above the eye sockets.

Teeth: Count the number of each kind of teeth in the lower jaw.

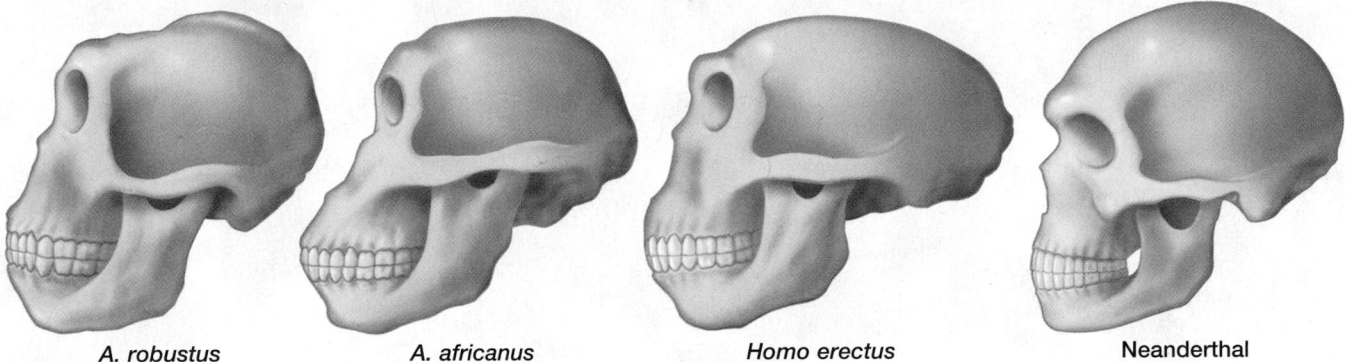

A. robustus   A. africanus   Homo erectus   Neanderthal

Procedure

PART A: Ape Skulls and Human Skulls

1. Examine the diagrams of the skull and jaw of an ape and a human. Look for similarities and differences between the features listed in the chart "How to Interpret the Features of a Skull." Record your observations and measurements for each feature listed in Data Table 1.

PART B: Fossil Hominids

2. Examine the four fossil hominid skulls. On the hominid skulls, observe and measure four features that are listed in the chart "How to Interpret the Features of a Skull." Use the human skull as a model for taking measurements. Record your observations and measurements in Data Table 2.

3. Compare your data for the hominids with your data for the modern ape and human. Classify each feature of the hominid skulls as being apelike, humanlike, or intermediate by writing an *A*, *H*, or *I* next to your observation or measurement for that feature.

4. Using your data, predict the order in which the hominids shown here evolved.

Analyze and Conclude

1. **Summarizing Results** How did skull structure change as hominids evolved?

2. **Drawing Conclusions** Which fossil skull is most apelike? most humanlike?

3. **Further Inquiry** Write a new question about human evolution that could be explored with another investigation.

On the Job

Anthropology is the scientific study of humans. Do research to learn about a famous anthropologist, such as Louis or Mary Leakey. For more about careers, visit **go.hrw.com** and type in the keyword **HX4 Careers.**

Leopard frog

33 Fishes and Amphibians

✓ Quick Review

Answer the following without referring to earlier sections of your book.

1. **Describe** the process of osmosis. *(Chapter 4, Section 1)*
2. **Define** the term *gills*. *(Chapter 27, Section 2)*
3. **State** the function of the excretory systems. *(Chapter 27, Section 2)*
4. **Distinguish** between external and internal fertilization. *(Chapter 27, Section 2)*
5. **Define** the term *metamorphosis*. *(Chapter 30, Section 3)*

Did you have difficulty? *For help, review the sections indicated.*

📖 Reading Activity

Take a break after reading each section of this chapter, and closely study the figures in the section. Reread the figure captions, and for each one write out a question that can be answered by referring to the figure and its caption. Refer to your list of figures and questions as you review the concepts addressed in the chapter before you complete the Performance Zone chapter review.

● The Northern Leopard frog, *Rana pipiens,* has become scarce in areas where the populations had once been dense. This amphibian is now being bred in protected habitats in those areas to increase its numbers.

Looking Ahead

🔲 **internet** connect

www.scilinks.org
National Science Teachers Association *sci*LINKS Internet resources are located throughout this chapter.

*sci*LINKS® Maintained by the National Science Teachers Association

Objectives

- **Describe** the characteristics of modern fishes.
- **Summarize** how fish obtain oxygen.
- **Summarize** how blood circulates through a fish.
- **Contrast** how marine and freshwater fishes balance their salt and water content.
- **Describe** two methods of reproduction in fishes.

Key Terms

gill filament
gill slit
countercurrent flow
nephron

Key Characteristics of Modern Fishes

What makes a goldfish instantly recognizable as a fish? You might name characteristics such as its fins, gills, scales, and typical fish shape as traits that contribute to the goldfish's "fishiness." But some fishes don't look quite so fishy. This is because the term *fish* refers to any member of one of three general categories of vertebrates: Agnatha (jawless fishes), Chondrichthyes (cartilaginous fishes), and Osteichthyes (bony fishes). The great diversity of fishes found today reflects various adaptations that enable fishes to live in the oceans and fresh waters around the world. Fishes vary in size from whale sharks longer than a moving van to gobies no larger than your fingernail. Despite the variation seen among fishes, shown in **Figure 1,** all share certain key characteristics.

1. **Gills.** Fishes normally obtain oxygen from the oxygen gas dissolved in the water around them. They do this by pumping a great deal of water through their mouths and over their gills.

2. **Single-loop blood circulation.** Blood is pumped from the heart to the capillaries in the gills. From the gills, blood passes to the rest of the body and then returns to the heart. (Lungfishes, which have a double-loop circulation, are an exception.)

3. **Vertebral column (backbone).** All fishes have an internal skeleton made of either cartilage or bone, with a vertebral column surrounding the spinal cord. The brain is fully encased within a protective covering called the *skull* or *cranium*.

To learn about one common fish, read *Up Close: Yellow Perch* later in this chapter.

Figure 1   Fish diversity. While these three fish appear quite different externally, they share a number of characteristics.

Sea horse

Stingray

Trout

Gills

One challenge faced by all animals is the need to get enough oxygen for cellular respiration. Sponges, cnidarians, many flatworms and roundworms, and some annelids obtain oxygen by diffusion through the body surface. Other marine invertebrates, such as mollusks, arthropods, and echinoderms have gills, which are specialized respiratory organs. Fishes also respire with gills.

If you look closely at the face of a swimming fish, you will notice that as it swims, the fish continuously opens and closes its mouth, as if it were trying to eat the water. What looks like eating is actually breathing. The major respiratory organ of a fish is the gill, shown in **Figure 2.** Gills are made up of rows of **gill filaments**—fingerlike projections through which gases enter and leave the blood. The gill filaments hang like curtains between a fish's mouth and cheeks. At the rear of the cheek cavity is an opening called a **gill slit.** When a fish "swallows," water is forced over the gills and out through the gill slits.

This swallowing procedure is the core of a great change in gill design shown by fishes—countercurrent flow, also shown in Figure 2. In **countercurrent flow,** water passes over the gills in one direction as blood flows in the opposite direction through capillaries in the gills. Countercurrent flow ensures that oxygen diffuses into the blood over the entire length of the capillaries in the gills. Due to this arrangement, the gills of bony fishes are extremely efficient respiratory organs. Fish gills can extract up to 85 percent of the oxygen in the water passing over them.

Figure 2   Fish gill structure. Countercurrent flow increases the gill's efficiency.

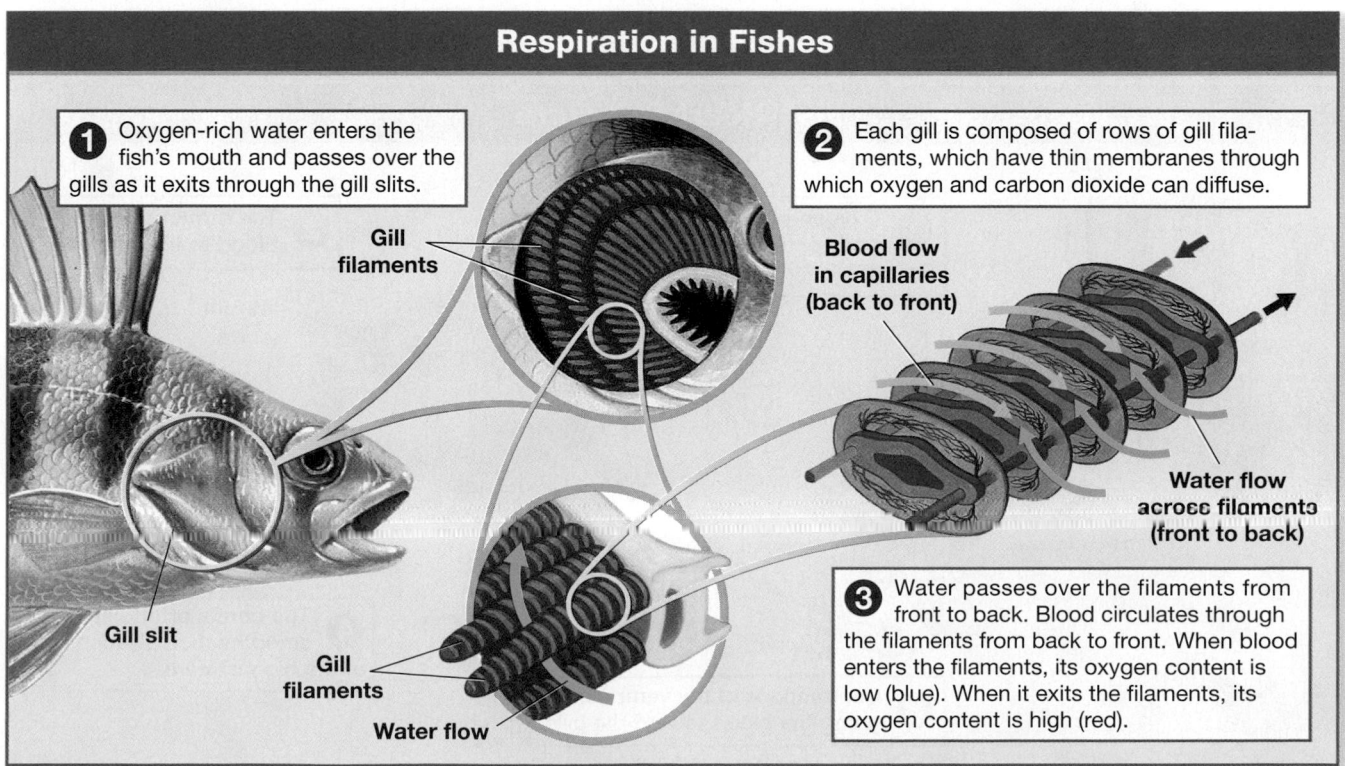

Respiration in Fishes

❶ Oxygen-rich water enters the fish's mouth and passes over the gills as it exits through the gill slits.

❷ Each gill is composed of rows of gill filaments, which have thin membranes through which oxygen and carbon dioxide can diffuse.

Gill filaments

Blood flow in capillaries (back to front)

Water flow across filaments (front to back)

❸ Water passes over the filaments from front to back. Blood circulates through the filaments from back to front. When blood enters the filaments, its oxygen content is low (blue). When it exits the filaments, its oxygen content is high (red).

Gill slit

Gill filaments

Water flow

Circulation of Blood

WORD Origins

The term *sinus venosus* is Latin, as are many anatomical terms. The Latin word *sinus* means "bend," and the Latin word *venosus* refers to veins. The sinus venosus is a bent collecting chamber that conducts blood into the heart.

Chordates that were ancestral to the vertebrates had a simple tubular "heart." This structure was little more than a specialized zone of one artery that had more muscle tissue than the other arteries had. When a tubular heart contracts, blood is pushed in both directions, and circulation is not very efficient.

For an organism with gills, such as a fish, a simple tubular heart is not an adequate pump. The tiny capillaries in a fish's gills create resistance to the flow of blood, so a stronger pump is needed to overcome this resistance. In fishes, the tubular heart of early chordates has been replaced with a simple chamber-pump heart, shown in **Figure 3.** The chamber-pump heart can be thought of as a tube with four chambers in a row.

1. **Sinus venosus** *(SIE nuhs vuh NOH suhs).* This collection chamber acts to reduce the resistance of blood flow into the heart.

2. **Atrium.** Blood from the sinus venosus fills this large chamber, which has thin, muscular walls.

3. **Ventricle.** The third chamber is a thick-walled pump with enough muscle tissue to contract strongly, forcing blood to flow through the gills and eventually to the rest of the body.

4. **Conus arteriosus** *(KOH nuhs ahr TIHR ee oh suhs).* This chamber is a second pump that smoothes the pulsations and adds still more force.

The fish heart represents one of the great evolutionary changes found in vertebrates—a heart that pumps fully oxygenated blood through a single circulatory loop to the body's tissues.

Figure 3   Fish chamber-pump heart. These four steps show how blood flows through the heart of a fish.

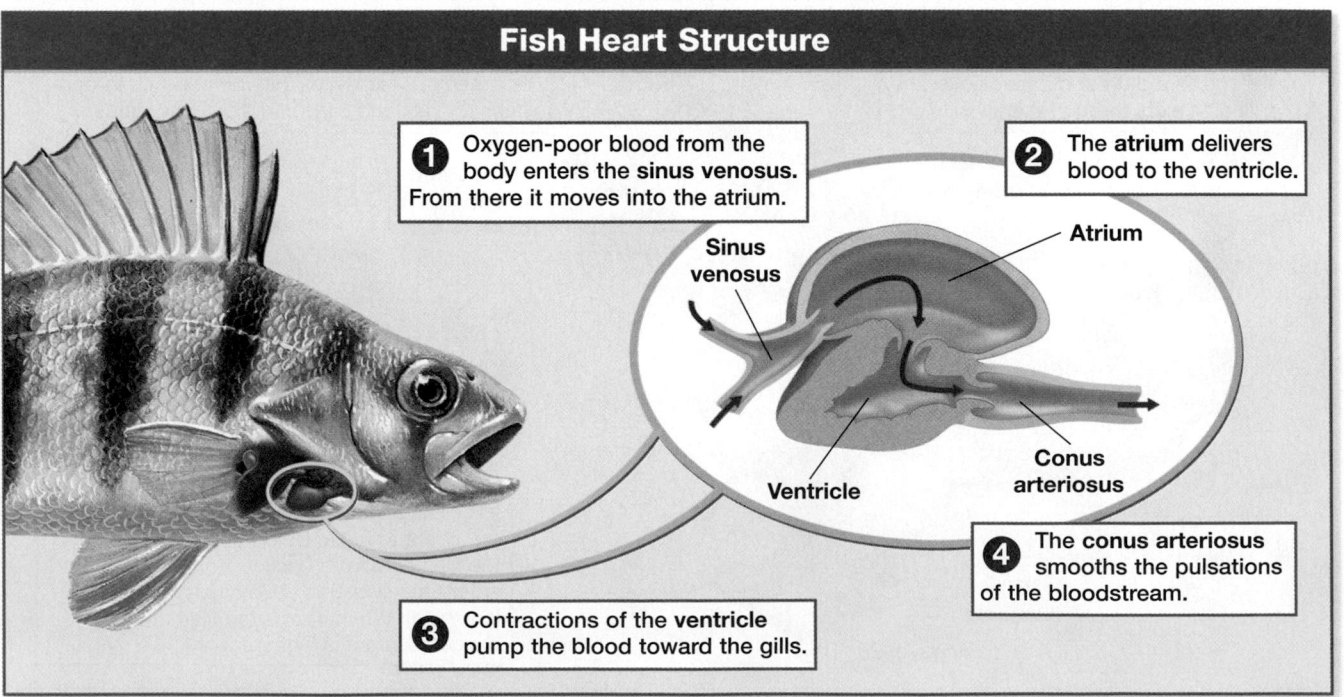

Fish Heart Structure

1 Oxygen-poor blood from the body enters the **sinus venosus.** From there it moves into the atrium.

2 The **atrium** delivers blood to the ventricle.

Sinus venosus

Atrium

Ventricle

Conus arteriosus

4 The **conus arteriosus** smooths the pulsations of the bloodstream.

3 Contractions of the **ventricle** pump the blood toward the gills.

Kidneys

Vertebrates have a body that is about two-thirds water, and most will die if the amount of water in their body falls much lower than this. Therefore, minimizing dehydration (water loss) has been a key evolutionary challenge facing all vertebrates. Even some fishes must cope with the problem of water loss. If this seems strange to you, remember that osmosis causes a net movement of water through membranes toward regions of higher ion concentration.

The ion (salt) concentration of sea water is three times that of the tissues of a marine bony fish. As a result, these fishes lose water to the environment through osmosis. To make up for the water they lose, marine bony fishes drink sea water and actively pump excess ions out of their body. Freshwater fishes have the opposite problem. Because their bodies contain more ions than the surrounding water, they tend to take in water through osmosis. The additional water dilutes their body salts, so freshwater fishes regain salts by actively taking them in from their environment.

Although the gills play a major role in maintaining a fish's salt and water balance, another key element is a pair of kidneys. Kidneys are organs made up of thousands of nephrons. **Nephrons** are tubelike units that regulate the body's salt and water balance and remove metabolic wastes from the blood. Excess water and bodily wastes leave the kidneys in the form of a fluid called *urine*. Marine fishes excrete small amounts of urine and rid their bodies of ammonia largely through their gills. Freshwater fishes excrete large amounts of dilute urine.

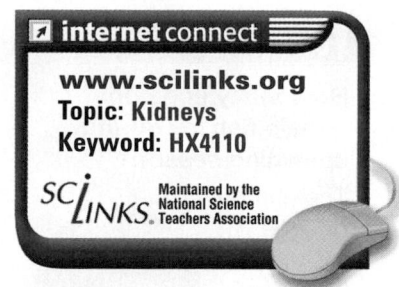

internet connect

www.scilinks.org
Topic: Kidneys
Keyword: HX4110

SC*LINKS* Maintained by the National Science Teachers Association

Analyzing Ion Excretion in Fish

Background

A few species of fish, such as adult salmon, are able to move between salt-water and freshwater environments. The graph at right shows the excreted ion concentration of a fish as it travels from one body of water to another. Examine the graph, and answer the analysis questions.

Analysis

1. **Determine** if the fish is losing or gaining ions by excretion as it travels.

2. **Critical Thinking Inferring Conclusions** Is the fish traveling from fresh to salt water or from salt to fresh water?

3. **Summarize** the reasoning you used to answer item 2.

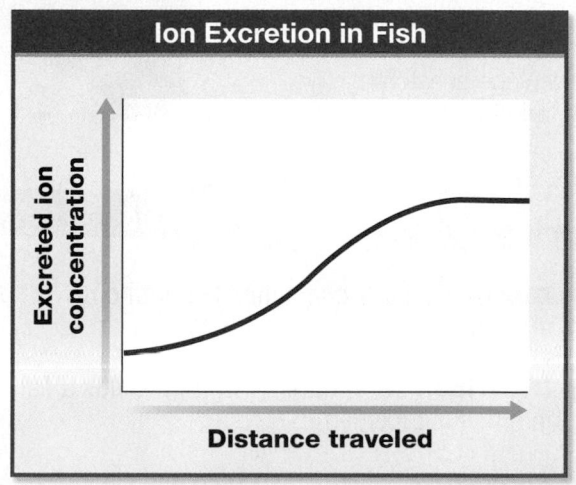

Ion Excretion in Fish

Excreted ion concentration

Distance traveled

Reproduction

The sexes are separate in most fishes, and generally fertilization takes place externally. In a process called *spawning*, male and female gametes are released near one another in the water, as shown in **Figure 4.** A yolk sac within each egg contains nutrients the developing embryo will need for growth. The yolk sac remains attached to the hatchling fish but is quickly used up. Then, the growing fish must seek its own food. More likely than not, it will become food for some larger animal. Many species of fishes release large numbers of eggs that are fertilized in a single spawning season. This practice helps ensure that some individuals will survive to maturity.

The eggs of sharks, skates, and rays are fertilized inside the female's body. During mating, the male uses two organs called *claspers* to insert sperm into the female. In most species, the eggs develop inside the female and the young are born live. A few species of sharks lay eggs.

Figure 4   Fish spawning.
These salmon spawn in shallow river waters. Thousands of eggs are released in a single mating, but only a small percentage of hatchlings live until adulthood.

Section 1 Review

① **Discuss** the key characteristics found in all fishes.

② **Describe** how countercurrent flow aids a fish in obtaining oxygen.

③ **Summarize** why the fish heart and circulatory system are considered important evolutionary changes.

④ **Contrast** reproduction in sharks with that of most other fishes.

⑤ **Critical Thinking Forming a Hypothesis**
A student removes Fish A from a saltwater aquarium and Fish B from a freshwater aquarium. By mistake, the student returns each fish to the wrong aquarium. The next day, both fish are dead. Form a hypothesis that explains why.

⑥ **Standardized Test Prep** Which chamber of a fish's heart generates most of the force that pumps blood through the body?

A atrium

B ventricle

C sinus venosus

D conus arteriosus

Today's Fishes

Jawless Fishes

Perhaps the most unusual fishes found today are the surviving jawless fishes, the lampreys and hagfishes. These primitive creatures have changed little over the past 330 million years. Little is known about hagfishes, which is not surprising when you consider where they live—on the ocean floor at depths as great as 1,700 m (about 1 mi). Lampreys are better understood and are found in both salt and fresh water. Interestingly, all species of lampreys must return to fresh water to reproduce, suggesting that their ancestors lived in fresh water.

Lampreys and hagfishes have scaleless, eel-like bodies with multiple gill slits and unpaired fins. Their skeletons are made of cartilage, a strong fibrous connective tissue, and both kinds of fishes retain their notochord into adulthood. Hagfishes, such as the one shown in **Figure 5,** are scavengers of dead and dying animals on the ocean bottom. Because of this behavior, they are sometimes called the "vultures of the sea." When threatened, a hagfish can produce huge quantities of slime from its roughly 200 slime glands. Recently, biologists have discovered that hagfishes are far more numerous than once thought and play a vital role in the ecology of the oceans.

Most lampreys, such as the one shown in Figure 5, are parasitic on other living fishes. A lamprey has a suction-cup-like structure around its mouth that it uses to attach itself to its host. After attachment, the lamprey gouges out a wound with its rough tongue, feeding on blood and bits of flesh from the wound.

Objectives

- **Distinguish** between the three general categories of modern fishes.
- **Describe** the major external and internal characteristics of the yellow perch.
- **Summarize** features of bony fishes.

Key Terms

lateral line
operculum
swim bladder
teleost

Figure 5 Hagfish and lamprey

These two modern jawless fishes have changed little over the past 330 million years.

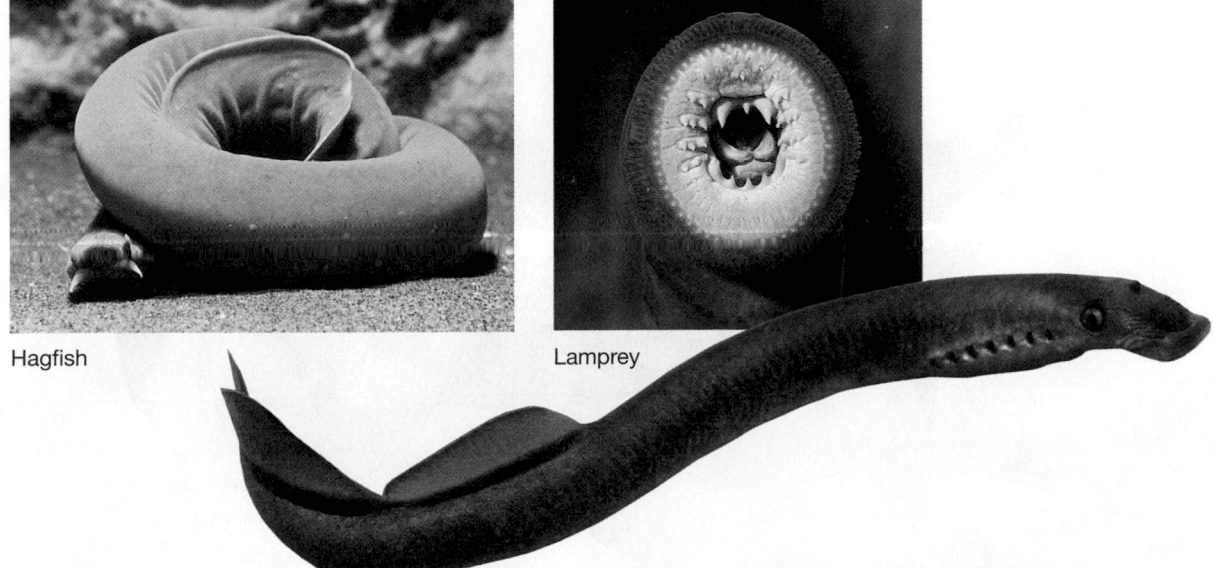

Hagfish

Lamprey

Cartilaginous Fishes

Sharks, skates, and rays are cartilaginous *(KAHRT'l AJ uh nuhs)* fishes. Their skeletons are made of cartilage strengthened by the mineral calcium carbonate (the material oyster shells are made of). Calcium carbonate is deposited in the outer layers of cartilage and forms a thin layer that reinforces the cartilage. The result is a very light but strong skeleton.

The shark's light, streamlined body allows it to move quickly through the water in search of prey. Its skin contains cone-shaped placoid scales, which give the skin a rough texture. As you can see in **Figure 6,** the shark's scales and teeth are quite similar in structure. This is because the teeth are actually modified scales. The teeth are arranged in 6 to 10 rows along the shark's jaw. The teeth in front are pointed and sharp and are used for biting and cutting. Behind the front teeth, rows of immature teeth are growing. When a front tooth breaks or is worn down, a replacement tooth moves forward. One shark may use more than 20,000 teeth during its lifetime. This system of tooth replacement guarantees that the teeth being used are always new and sharp.

Two smaller groups of cartilaginous fishes, the skates and rays, have flattened bodies that are well adapted to life on the sea floor. Rays are usually less than 1 m (3.3 ft) long, while skates are typically smaller. However, the giant manta ray may be up to 7 m (23 ft) wide. Most species of skates and rays have flattened teeth that are used to crush their prey, mainly small fishes and invertebrates.

Figure 6   Shark scales and teeth. The shark's skin feels like sandpaper because it is covered with toothlike scales. The teeth, which are modified scales, are similar in structure but are much larger.

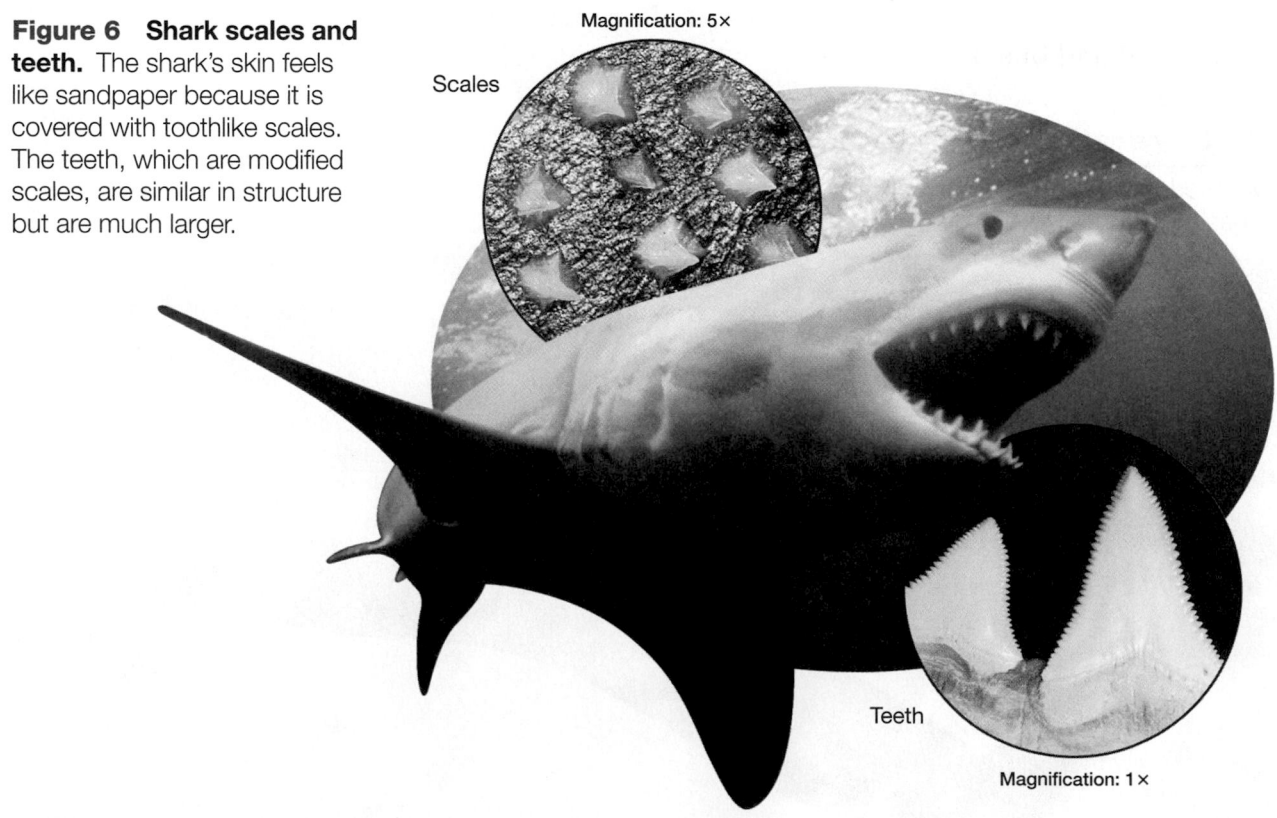

Magnification: 5×

Scales

Teeth

Magnification: 1×

Bony Fishes

Jawless and cartilaginous fishes are not as diverse as bony fishes, which are the most numerous of all the fishes. In addition to a strong, internal skeleton made completely of bone, bony fishes have a series of unique structural adaptations that contribute greatly to their success.

1. **Lateral line system.** Bony fishes have a fully developed lateral line system. The **lateral line,** shown in **Figure 7,** is a sensory system that extends along each side of a bony fish's body. As moving water presses against the fish's sides, nerve impulses from ciliated sensory cells within the lateral line permit the fish to perceive its position and rate of movement. For example, a trout moving upstream to spawn uses its lateral line system to obtain the sensory information it needs to orient its head upstream.

The lateral line system also enables a fish to detect a motionless object by the movement of water deflected by that object. The way that a fish detects an object with its lateral line and the way that you hear music with your inner ear are quite similar. Both processes share the same basic mechanism—sensory cells with cilia detect vibrations and send this information to the brain.

Study TIP

● **Organizing Information**
Use the information on this page and the next four pages to draw a concept map that summarizes the characteristics of bony fishes. On your concept map, include information from *Up Close: Yellow Perch*.

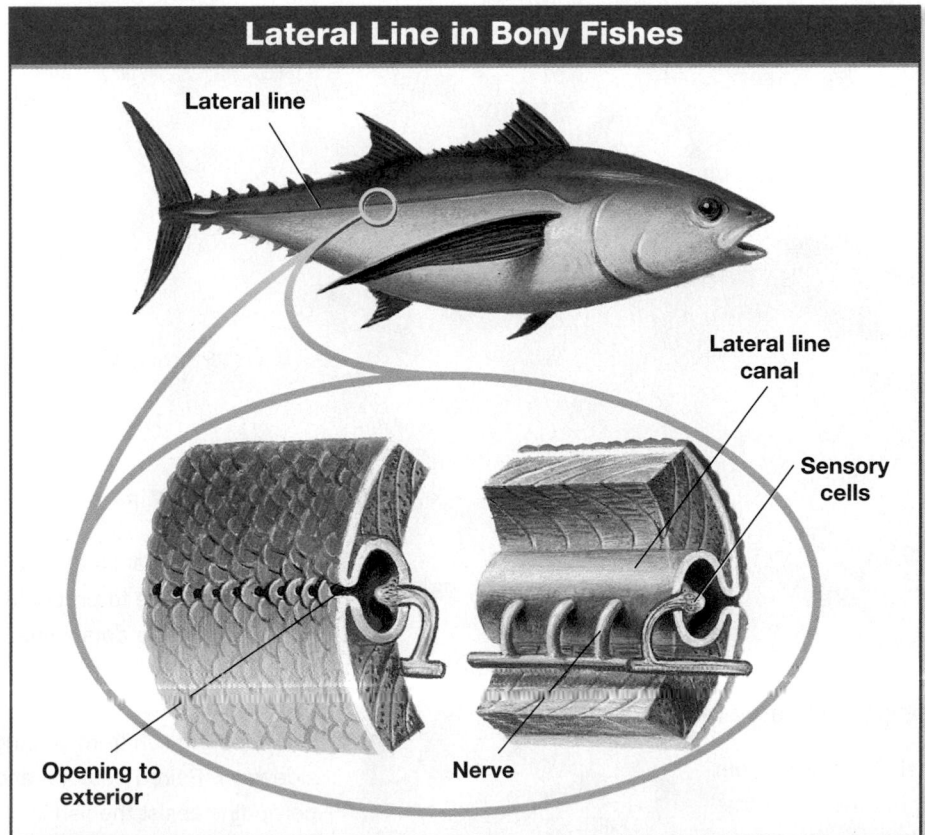

Lateral Line in Bony Fishes

Lateral line

Lateral line canal

Sensory cells

Opening to exterior

Nerve

Figure 7   Lateral line.
The lateral line contains sensory cells that help a fish perceive its position in the water.

Up Close

Yellow Perch

- **Scientific name:** *Perca flavescens*
- **Size:** About 0.3 m (1 ft) long and up to 2.3 kg (5 lb)
- **Range:** Found in lakes and rivers from the Great Lakes to the Atlantic coast and as far south as South Carolina
- **Habitat:** Lives concealed among vegetation or submerged tree roots
- **Diet:** Feeds on insect larvae, crustaceans, and other fishes

External Structures

Lateral line The lateral line is a sense organ that detects vibrations in water that are caused by currents or pressure waves. The perch uses this sensory information to direct its movement as it swims and to detect objects in its environment, including predators and prey.

▲ Lateral line

Posterior dorsal fin

Caudal fin

Anterior dorsal fin

Opercula Movements of the opercula draw water into the perch's mouth. The water then moves over the gills, where oxygen and carbon dioxide are exchanged. Then the water is forced out through the gill slits.

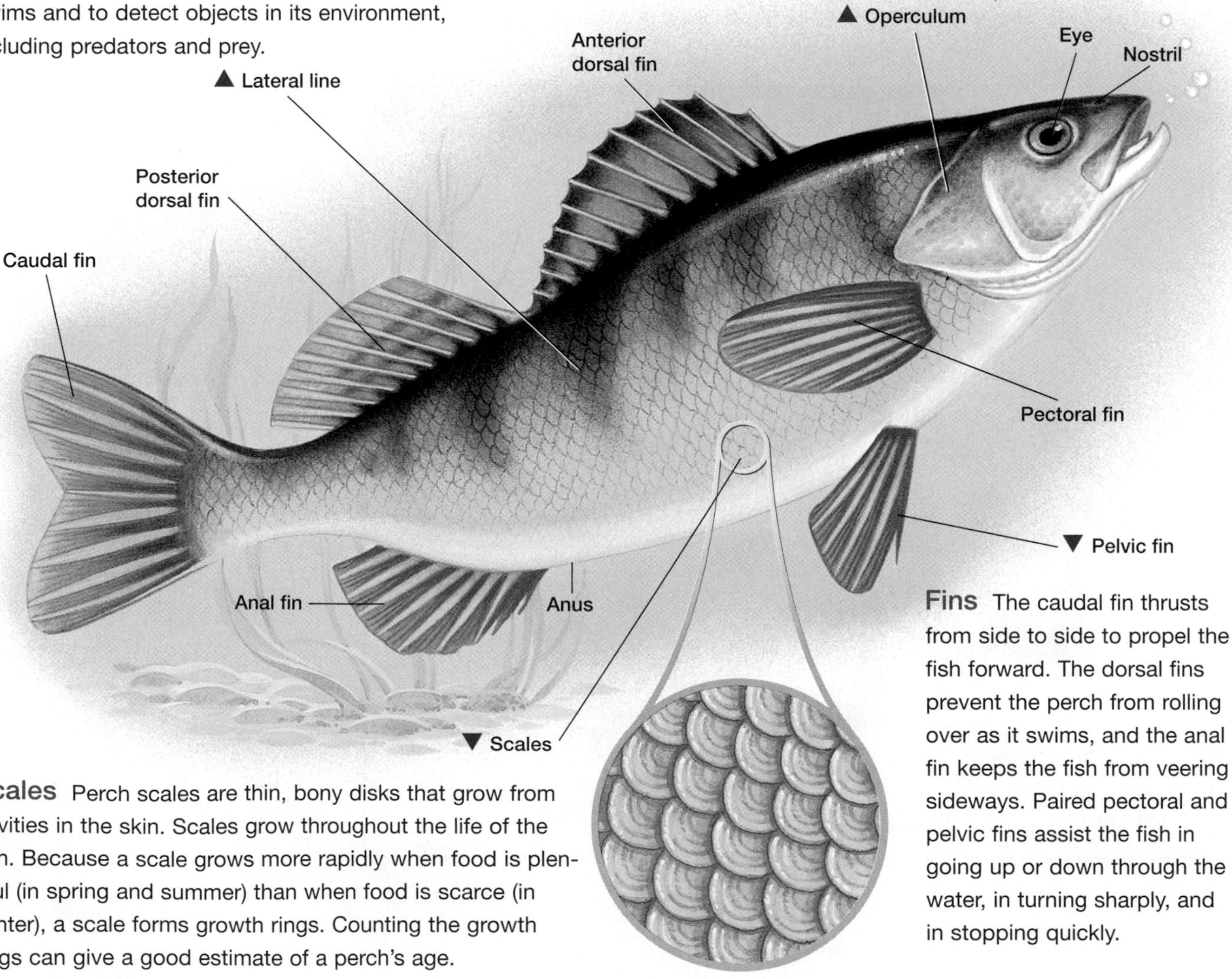

▲ Operculum

Eye

Nostril

Pectoral fin

Anal fin

Anus

▼ Pelvic fin

▼ Scales

Fins The caudal fin thrusts from side to side to propel the fish forward. The dorsal fins prevent the perch from rolling over as it swims, and the anal fin keeps the fish from veering sideways. Paired pectoral and pelvic fins assist the fish in going up or down through the water, in turning sharply, and in stopping quickly.

Scales Perch scales are thin, bony disks that grow from cavities in the skin. Scales grow throughout the life of the fish. Because a scale grows more rapidly when food is plentiful (in spring and summer) than when food is scarce (in winter), a scale forms growth rings. Counting the growth rings can give a good estimate of a perch's age.

Reproductive organs Yellow perch produce gametes during their breeding season in the spring. The female lays strings of eggs that are fertilized externally. In warm water, the young hatch within days and grow quickly. In cold water, the development of the eggs may take much longer.

Brain The optic lobe receives information from the eyes, and the olfactory bulbs receive information from chemical-sensing cells. The cerebrum processes this and other information. The cerebellum coordinates muscle activity, and the medulla oblongata controls the function of many internal organs.

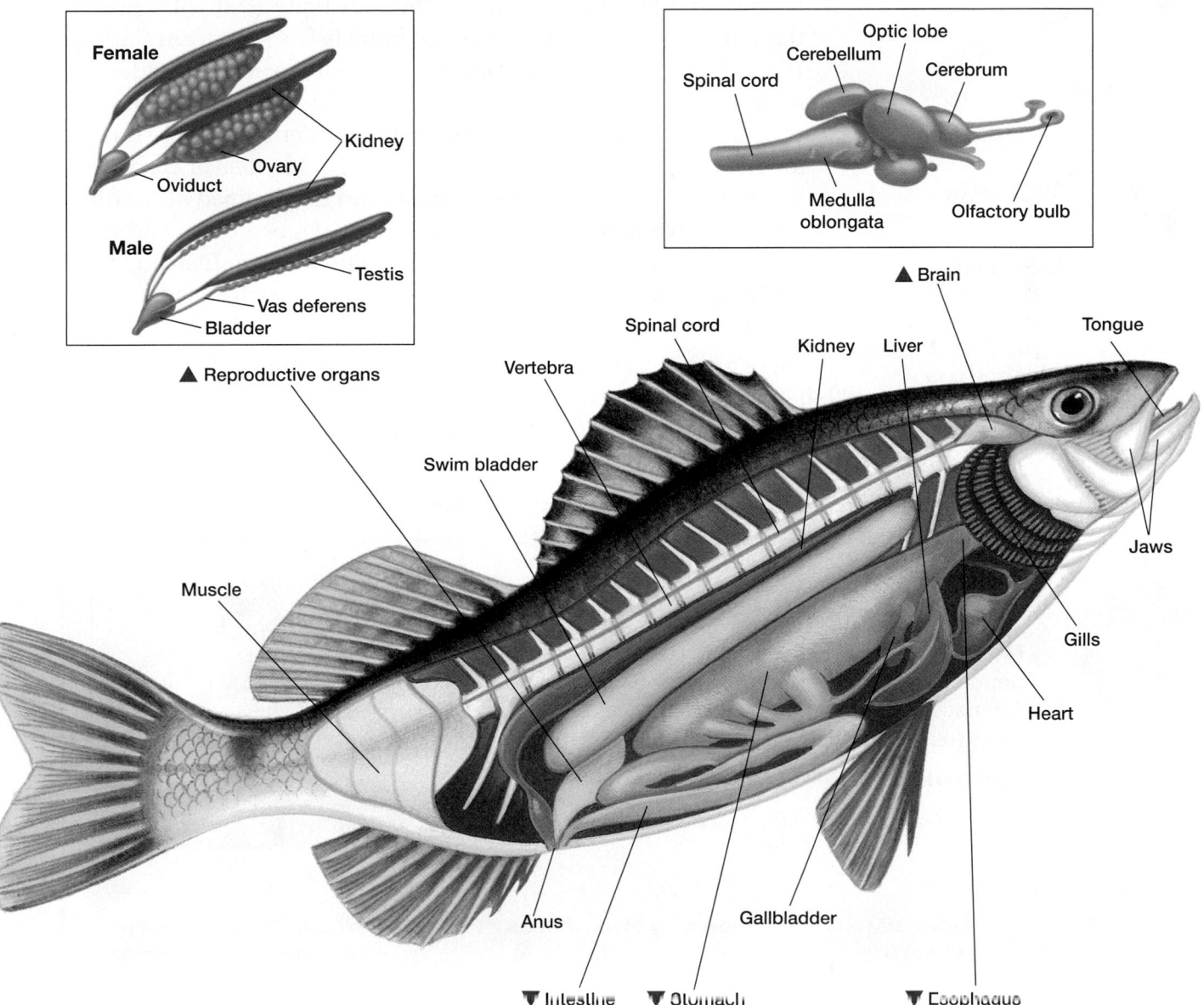

Female
Kidney
Ovary
Oviduct
Male
Testis
Vas deferens
Bladder

▲ Reproductive organs

Optic lobe
Cerebellum
Cerebrum
Spinal cord
Medulla oblongata
Olfactory bulb

▲ Brain

Spinal cord
Vertebra
Kidney
Liver
Tongue

Swim bladder
Jaws

Muscle
Gills

Heart

Anus
Gallbladder

▼ Intestine   ▼ Stomach   ▼ Esophagus

Digestive system The digestive system reflects a basic arrangement of structures found in all vertebrates. Food enters the mouth and passes from the esophagus to the stomach. The liver secretes bile, and the pancreas secretes enzymes into a short intestine. The bile and enzymes help break down the food. Absorption of digested food occurs through the inner lining of the intestine. Undigested material exits through the anus.

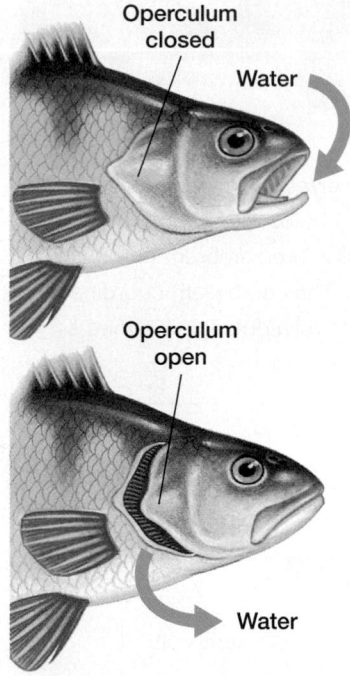

Figure 8 Operculum.
When a fish's mouth opens, water enters and the opercula close over the gills. When the mouth closes, the opercula open and water moves across the gills and out of the fish.

2. Gill cover. Most bony fishes have a hard plate, an **operculum,** that covers the gills on each side of the head. Movements of certain muscles and of the opercula, shown in **Figure 8,** permit a bony fish to draw water over the gills, which enables the fish to take in oxygen. By using this mechanism, most bony fishes can move water over their gills while remaining stationary in the water. A bony fish doesn't have to swim forward with its mouth open to move water over its gills. This ability to respire without swimming enables a bony fish to conserve energy that can be spent chasing after prey and escaping from predators.

3. Swim bladder. The density of the fish body is slightly greater than that of sea water. How then do bony fishes keep from sinking? Bony fishes contain a special gas sac called a **swim bladder.** By adjusting the gas content of the swim bladder, bony fishes can regulate their buoyancy. As the swim bladder fills, the fish rises, and as it empties, the fish sinks. The swim bladder of early bony fishes was connected to their throat, and they gulped air to fill it. The swim bladder of modern bony fishes, shown on the previous page, does not have a direct passage to the mouth. Instead, gas is exchanged between the bloodstream and the swim bladder. This permits the fish to maintain or change its depth in the water.

There are two groups of bony fishes, the ray-finned fishes and the lobe-finned fishes. The yellow perch described in *Up Close: Yellow Perch* on the previous two pages is a common type of ray-finned fish.

Modeling the Action of a Swim Bladder

Most fish use a swim bladder to regulate their depth in water. As gas enters the swim bladder, the fish rises in the water. As gas is expelled, the fish sinks to a lower depth.

Materials

100 mL beaker or small glass; cold, clear, carbonated soft drink; 2 very dry raisins

Procedure

1. Fill a 100 mL beaker with a cold, carbonated soft drink.

2. Drop two raisins into the beaker, and observe what happens over the next 5 minutes.

Analysis

1. Describe what happened after you dropped the raisins into the soft drink.

2. Forming Hypotheses Develop a hypothesis to explain your observations.

3. Critical Thinking Analyzing Results How does the lifting of the raisins differ from the use of a swim bladder to control buoyancy?

4. Critical Thinking Forming Reasoned Opinions Think about the energy you would have to expend to keep yourself in one position under water. What advantage might a swim bladder provide to a fish?

Ray-Finned Bony Fishes

Ray-finned bony fishes, such as the ones shown in **Figure 9,** comprise the vast majority of living fishes. Their fins are supported by bony structures called *rays*. **Teleosts** *(TEL ee ahsts),* such as the yellow perch you saw in *Up Close,* are the most advanced of the ray-finned bony fishes. Teleosts have highly mobile fins, very thin scales, and completely symmetrical tails. About 95 percent of all living fish species are teleosts.

Lobe-Finned Bony Fishes

Only seven species of lobe-finned fishes survive today. One species is the coelacanth *(SEE luh kanth),* shown in **Figure 10,** and the other six species are all lungfishes. The lobe-finned fishes have paired fins that are structurally very different from the fins of ray-finned fishes. In many lobe-finned fishes, each fin consists of a long, fleshy, muscular lobe that is supported by a central core of bones. These bones are connected by joints, like the joints between the bones in your hand. Bony rays are found only at the tips of each lobed fin. Muscles within each lobe can move the bony rays independently of each other.

Scientists have debated whether the direct ancestor of amphibians was a coelacanth or a lungfish. However, recent evidence has led biologists to believe that it was neither. The ancestor of the amphibians most likely was a third type of lobe-finned fish that is now extinct.

Figure 9 Pacific bluefin tuna

At sexual maturity, Pacific bluefin tuna can weigh about 136 kg (300 lb), although some grow larger.

Figure 10 Coelacanth.
Coelacanths were thought to have been extinct for millions of years, until one was caught off the coast of Africa in 1938. Coelacanths can reach up to nearly 3 m (9.8 ft) in length.

Section 2 Review

1. **Compare** the three categories of modern fishes.

2. **Summarize** the role of the operculum in fish respiration.

3. **Summarize** how the swim bladder can be viewed as an energy-saving mechanism.

4. **Describe** the digestive process in a yellow perch.

5. **Relate** a yellow perch's lateral line system to the human ear.

6. **Critical Thinking Evaluating Conclusions** An unidentified species of fish has rough skin, several rows of teeth, and no opercula. Based on these characteristics, a student infers that the fish has a swim bladder. Explain why you agree or disagree with this conclusion.

7. **Standardized Test Prep** The mouth of a lamprey is specialized for
   A straining plankton.
   B chewing seaweed.
   C scavenging dead animals.
   D parasitizing other fish.

Amphibians

Objectives

- **Summarize** the characteristics of modern amphibians.
- **Compare** the three orders of living amphibians.
- **Describe** the major external and internal characteristics of the leopard frog.

Key Terms

lung
pulmonary vein
septum

Key Characteristics of Modern Amphibians

The next time you see a frog, consider that it is a surviving member of an ancient amphibian group, the first vertebrates to walk on land. The croaking and peeping of frogs, such as the one shown in **Figure 11,** make it difficult not to notice them, but their smaller, quieter relatives live nearby, hidden in damp habitats. Class Amphibia contains three orders of living amphibians: order Anura (frogs and toads), order Urodela (salamanders and newts), and order Apoda (caecilians). Most amphibians share five key characteristics.

1. **Legs.** The evolution of legs was an important adaptation for living on land. Frogs, toads, salamanders, and newts have four legs. Caecilians lost their legs during the evolutionary course of adapting to a burrowing existence.

2. **Lungs.** Although larval amphibians have gills, most adult amphibians breathe with a pair of lungs. Lungless salamanders are an exception.

3. **Double-loop circulation.** Two large veins called *pulmonary veins* return oxygen-rich blood from the lungs to the heart. The blood is then pumped to the tissues at a much higher pressure than in the fish heart.

4. **Partially divided heart.** The atrium of the amphibian heart is divided into left and right sides, but the ventricle is not. A mixture of oxygen-rich and oxygen-poor blood is delivered to the amphibian's body tissues.

5. **Cutaneous respiration.** Most amphibians supplement their oxygen intake by respiring directly through their moist skin. Cutaneous respiration ("skin breathing") limits the maximum body size of amphibians because it is efficient only when there is a high ratio of skin surface area to body volume.

Lungs

Although air contains about 20 times as much oxygen as sea water does, gills cannot function as respiratory organs when out of water. Thus, one of the major challenges that faced the first land vertebrates was that of obtaining oxygen from air. The evolutionary solution to this challenge was the lung.

A **lung** is an internal, baglike respiratory organ that allows oxygen and carbon dioxide to be exchanged between the air and the bloodstream. The amount of oxygen a lung can absorb depends on its internal surface area. The greater the surface area, the greater the amount of oxygen that can be absorbed. In amphibians, the

Figure 11   Spring peeper.
In some areas, the call of the spring peeper is one of the first signs of spring.

lungs are hardly more than sacs with folds on their inner membrane that increase their surface area, as shown in **Figure 12.** With each breath, fresh oxygen-rich air is drawn into the lungs. There it mixes with a small volume of air that has already given up most of its oxygen. Because of this mixing, the respiratory efficiency of lungs is much less than that of gills. Because there is much more oxygen in air than there is in water, however, lungs do not have to be as efficient as gills. Many amphibians also obtain oxygen through their thin, moist skin.

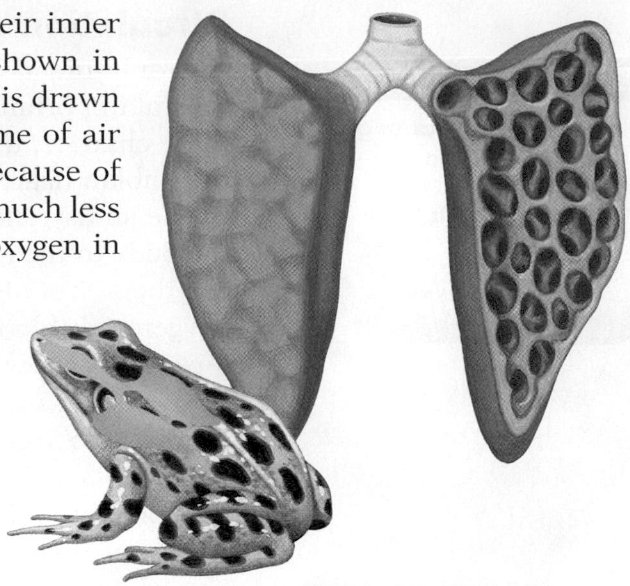

Double-Loop Circulation

As amphibians evolved and became active on land, their circulatory system changed, resulting in a second circulatory loop. This change allowed more oxygen to be delivered to their muscles. **Figure 13** compares the single-loop circulation of most fishes with the double-loop circulation of amphibians (also found in lungfishes). Notice that amphibians have a pair of blood vessels not found in fishes, the pulmonary veins. The **pulmonary veins** carry oxygen-rich blood from the amphibian's lungs to its heart. The heart pumps the oxygen-rich blood to the rest of the body. The advantage of this arrangement is that oxygen-rich blood can be pumped to the amphibian's tissues at a much higher pressure than it can in fishes. (Recall that in fish, blood is pumped through the gills before reaching the body organs. As a result, much of the force of the heartbeat is lost.)

Figure 12 Amphibian lungs. The lungs of an amphibian are sacs with a folded internal membrane that provides a large surface for gas exchange.

Figure 13 Circulatory loops

Circulation in fishes involves a single loop. Amphibians have a second loop that goes from the heart to the lungs and back to the heart.

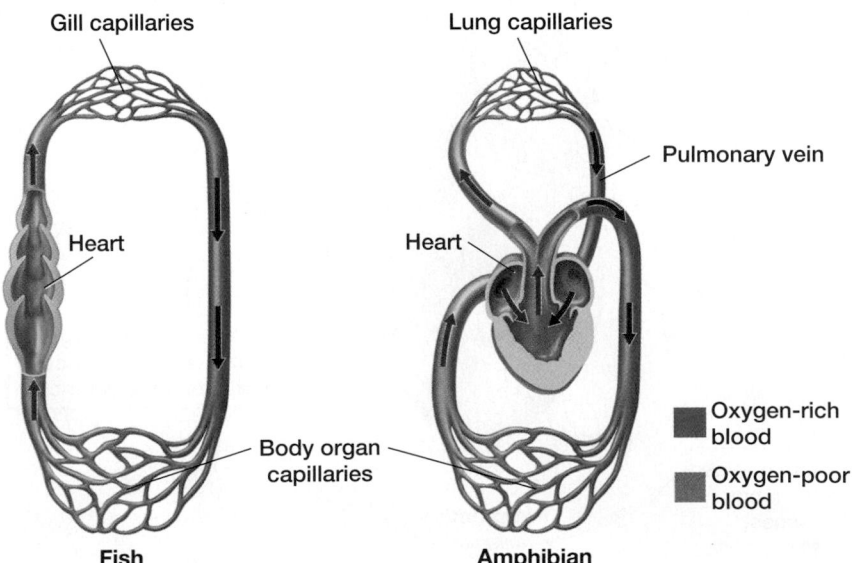

Gill capillaries

Lung capillaries

Pulmonary vein

Heart

Heart

Body organ capillaries

■ Oxygen-rich blood

■ Oxygen-poor blood

Fish

Amphibian

Circulation of Blood

Not only did the path of circulation in amphibians change, but several important changes occurred in the heart. As you read about these changes, use **Figure 14** to trace the flow of blood through the amphibian heart.

The sinus venosus continues to deliver oxygen-poor blood from the body to the right side of the heart, as shown in step 1. (You can see the sinus venosus on the left side of Figure 14.) In addition, oxygen-rich blood from the lungs enters the left side of the heart directly, as shown in step 2.

A dividing wall known as the **septum** separates the amphibian atrium into right and left halves. You cannot see the septum in Figure 14 as it is beneath the conus arteriosus. The septum prevents the complete mixing of oxygen-rich and oxygen-poor blood as each enters the heart. As shown in step 3, both types of blood empty into a single ventricle, where some mixing of oxygen-rich and oxygen-poor blood occurs. However, due to the anatomy of the ventricle, the two streams of blood remain somewhat separate, as shown in step 4. Oxygen-rich blood tends to stay on the side that directs blood toward the body. Oxygen-poor blood tends to stay on the side that directs blood toward the lungs.

A number of amphibians have a spiral valve that divides the conus arteriosus. The spiral valve also helps to keep the two streams of blood separate as they leave the heart. Even so, some oxygen-poor blood is delivered to the body's tissues. Recall, however, that amphibians also obtain oxygen through their skin. This additional oxygen partly offsets the limitations of their circulatory system.

Figure 14   Amphibian heart. These four steps show how blood flows through the heart of an amphibian.

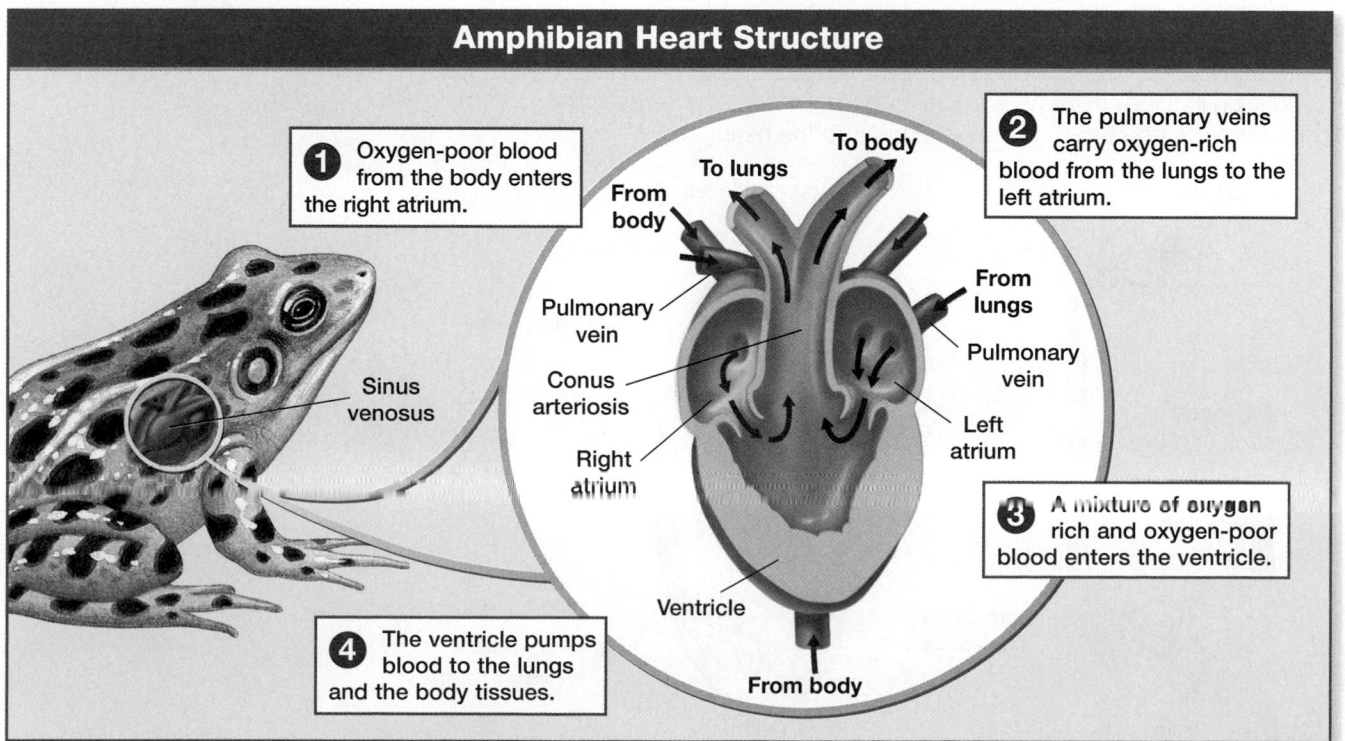

Amphibian Heart Structure

1 Oxygen-poor blood from the body enters the right atrium.

2 The pulmonary veins carry oxygen-rich blood from the lungs to the left atrium.

3 A mixture of oxygen rich and oxygen-poor blood enters the ventricle.

4 The ventricle pumps blood to the lungs and the body tissues.

Sinus venosus

Pulmonary vein

Conus arteriosis

Right atrium

From body

To lungs

To body

From lungs

Pulmonary vein

Left atrium

Ventricle

From body

Frogs and Toads

The order Anura is made up of frogs and toads that live in environments ranging from deserts to rain forests, valleys to mountains, and ponds to puddles. Adult anurans are carnivorous, eating a wide variety of small prey. Some species have a sticky tongue that they extend rapidly to catch their prey. The frog body, particularly its skeleton, is adapted for jumping, and its long muscular legs provide the power. To learn about the leopard frog, see *Up Close: Leopard Frog* on the next two pages. Toads, such as the one shown in **Figure 15,** are very similar to frogs but have squat bodies and shorter legs. Their skin is not smooth like that of a frog but is covered with bumps.

Reproduction in Frogs

Like most living amphibians, frogs depend on the presence of water to complete their life cycle. The female releases her eggs into the water and a male's sperm fertilize them externally. After a few days, the fertilized eggs hatch into swimming, fishlike larval forms called *tadpoles*. Tadpoles breathe with gills and feed mostly on algae. After a period of growth, the body of the tadpole changes into that of an adult frog. The rate at which tadpoles develop depends on the species and the availability of food. This process of dramatic physical change, called *metamorphosis*, is shown in **Figure 16.**

Figure 15   Toad. Toads like this common Asian toad have dry, bumpy skin and relatively short legs.

Figure 16   Frog life cycle

The transition of a larval frog (tadpole) to an adult involves a complex series of external and internal body changes.

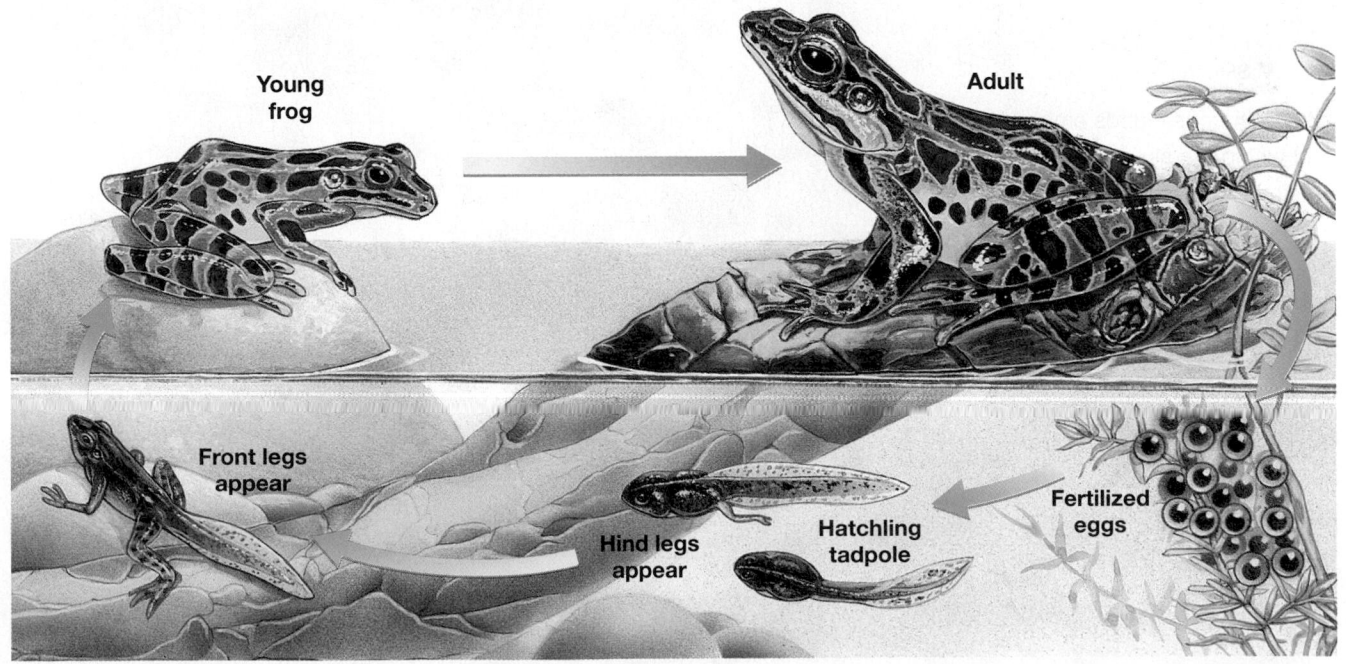

Young frog

Adult

Front legs appear

Hind legs appear

Hatchling tadpole

Fertilized eggs

Up Close

Leopard Frog

- **Scientific name:** *Rana pipiens*
- **Size:** Body length (legs excluded) of 5–9 cm (2–3.5 in.)
- **Range:** From northern Canada to southern New Mexico and from eastern California to the Atlantic coast
- **Habitat:** Lives in the short grass of meadows and around ponds
- **Diet:** Feeds on crickets, mosquitoes, and other small animals

External Structures

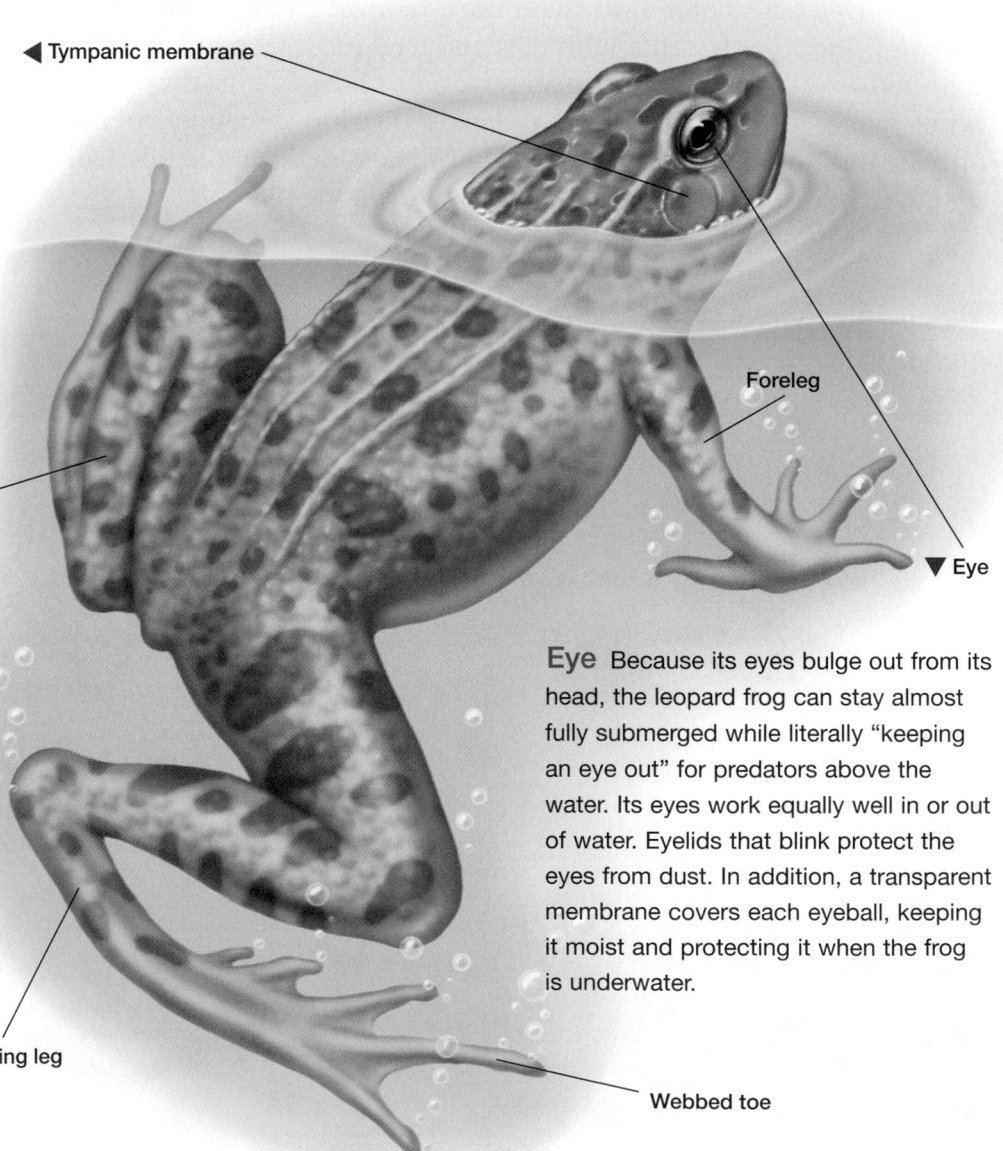

Tympanic membrane

When sound causes the tympanic membrane (eardrum) to vibrate, a tiny bone transmits the vibrations to the inner ear. There, ciliated sensory cells (similar to those found in the lateral line of a fish) detect the vibrations and help the frog maintain balance. Leopard frogs hear well in both water and air.

◀ Tympanic membrane

Foreleg

▼ Skin

▼ Eye

Skin Mucous glands embedded within the skin supply a lubricant that keeps the skin moist, a necessity for respiration. Unlike those of many frogs and toads, the leopard frog's skin glands do not secrete poisonous or foul-tasting substances. Instead, the leopard frog must rely on its protective coloration and speed to evade predators.

Eye Because its eyes bulge out from its head, the leopard frog can stay almost fully submerged while literally "keeping an eye out" for predators above the water. Its eyes work equally well in or out of water. Eyelids that blink protect the eyes from dust. In addition, a transparent membrane covers each eyeball, keeping it moist and protecting it when the frog is underwater.

Jumping leg

Webbed toe

Internal Structures

Brain The frog's brain differs from the fish's brain in that its components are more complex. For example, the larger, more complex cerebrum of a frog is able to process a wider assortment of sensory information than the cerebrum of a fish can.

Cerebrum Optic lobe Cerebellum

Olfactory lobe Medulla oblongata

Skeleton The skeletal system of the leopard frog (and all other modern frogs) has only nine vertebrae and no ribs. The rest of the vertebrae are fused into a single bone (urostyle). When a frog is sitting, the urostyle points upward, which gives the frog its characteristic humped back. When a frog jumps the long hind legs extend, which produces a powerful thrust forward.

Tongue and jaw The tongue flicks out with great speed, curls around the prey, and returns to the mouth. Two large teeth that project from the roof of the mouth impale struggling prey. In addition the upper jaw is lined with small, sharp teeth that prevent the prey from escaping. Food is not chewed but swallowed whole.

◄ Brain

▲ Tongue

Teeth

Esophagus

Sacral vertabra Kidney

Urostyle

◄ Pelvic girdle

Lung

Heart

Intestine Stomach Liver

Urinary bladder

▼ Reproductive organs

▼ Cloaca

Cloaca Undigested material passes into the cloaca, a chamber that opens to the outside of the body. Urine from the kidneys travels to the bladder and then passes into the cloaca, as do gametes from the reproductive organs. All of these materials exit the body through the cloacal opening.

Male **Female**

Mature ovary

Oviduct

Testis Kidney

Ureter

Cloaca

Reproductive organs Prior to breeding, the reproductive organs of male and female leopard frogs produce enormous numbers of gametes. The female releases a large cluster of eggs into the water. The male then discharges his sperm over them, fertilizing them externally.

Salamanders and Caecilians

Salamanders have elongated bodies, long tails, and smooth, moist skin. There are about 369 species of salamanders, all belonging to the order Urodela. They typically range from 10 cm to 0.3 m (4 in. to 1 ft) in length. However, giant Asiatic salamanders of the genus *Andrias* grow as long as 1.5 m (5 ft) and weigh up to 41 kg (90 lb). Because salamanders need to keep their skin moist, most are unable to remain away from water for long periods, although some salamander species manage to live in dry areas by remaining inactive during the day.

Figure 17 Salamander. This four-toed salamander has deposited its eggs in a damp, mossy environment.

Salamanders lay their eggs in water or in moist places, as shown in **Figure 17.** Fertilization is usually external. A few species of salamanders practice a type of internal fertilization in which the female picks up a sperm packet that has been deposited by the male and places it in her cloaca. Unlike frog and toad larvae, salamander larvae do not undergo a dramatic metamorphosis. The young that hatch from salamander eggs are carnivorous and resemble small versions of the adults, except that the young usually have gills. A few species of salamanders, such as the North American mudpuppy and the Texas spring salamander, never lose their larval characteristics. They retain their external gills as adults.

Caecilians

Caecilians (order Apoda) are a highly specialized group of tropical, burrowing amphibians with small, bony scales embedded in their skin. They feed on small invertebrates found in soil. These legless, wormlike animals, shown in **Figure 18,** grow to about 0.3 m (1 ft) long, although some species can be up to 1.2 m (4 ft) long. During breeding, the male deposits sperm directly into the female. Depending on the species, the female may bear live young or lay eggs that develop externally. Caecilians are rarely seen, and scientists do not know a lot about their behavior.

Figure 18 Caecilian. Like most caecilians, this one from Colombia, South America, burrows beneath the soil and is rarely seen.

Section 3 Review

1. **Summarize** how amphibians take in oxygen.

2. **Contrast** the single-loop circulation of fish with the double-loop circulation of amphibians.

3. **Compare** the external characteristics of each order of amphibians.

4. **Compare** reproduction and development in frogs and salamanders.

5. **Relate** the tongue of the leopard frog to its feeding habits.

6. **Explain** why it is difficult to "sneak up" on a frog.

7. **Standardized Test Prep** In a frog's heart, the blood that enters the left atrium
   A comes from the lungs.   C then goes to the lungs.
   B comes from the body.   D then goes to the body.

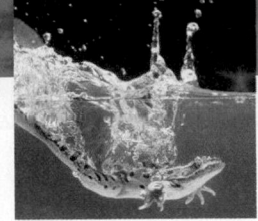

Key Concepts

1 The Fish Body

- All fishes have gills and a backbone, and they circulate oxygen-rich blood from their gills directly to body tissues.

- Countercurrent flow maximizes the amount of oxygen that can be extracted from water.

- The four-chambered fish heart collects oxygen-poor blood from the body and pumps it through the gills where it receives oxygen. Oxygen-rich blood then circulates to the rest of the body.

- A fish relies on its gills and a pair of kidneys to regulate its salt and water balance.

- Most fishes fertilize their eggs externally as males and females release their gametes near one another in the water.

2 Today's Fishes

- Hagfishes and lampreys are the only surviving jawless fishes.

- Sharks have light, highly streamlined bodies well suited for rapid swimming, which makes them swift and efficient predators.

- Bony fishes are the most diverse and abundant group of fishes.

- Bony fishes have an internal skeleton made completely of bone, a swim bladder, a lateral line sensory system, and a set of gill covers called *opercula*.

3 Amphibians

- Most amphibians have legs, breathe with lungs and through their skin, and have two circulatory loops.

- Most amphibians supplement their oxygen intake through cutaneous respiration—respiration through their moist skin.

- An amphibian lung is basically an air sac with a large surface area for gas exchange.

- The amphibian heart pumps oxygen-poor blood to the lungs and receives oxygen-rich blood from the lungs. The oxygen-rich blood is then pumped to the body.

- Salamanders are semiaquatic predators with tails, and caecilians are legless amphibians specialized for burrowing.

Key Terms

Section 1

gill filament (747)
gill slit (747)
countercurrent flow (747)
nephron (749)

Section 2

lateral line (753)
operculum (756)
swim bladder (756)
teleost (757)

Section 3

lung (758)
pulmonary vein (759)
septum (760)

Understanding Key Ideas

1. Which of the following is *not* a key characteristic of fishes?
  a. vertebral column
  b. gills
  c. single-loop circulation
  d. tympanic membrane

2. Which of the following shark characteristics is *not* an adaptation for predation?
  a. streamlined body
  b. internal fertilization of eggs
  c. sharp, replaceable teeth
  d. lightweight skeleton

3. Yellow perch and sharks share all of the following characteristics *except*
  a. gills.
  b. an internal skeleton.
  c. a single-loop circulatory system.
  d. a swim bladder.

4. A shark's skeleton is
  a. composed of cartilage.
  b. composed of bone.
  c. very dense.
  d. quite rigid.

5. Most adult amphibians respire
  a. through their skin.
  b. through their skin and gills.
  c. through their lungs.
  d. through their skin and lungs.

6. Which of the following is *not* a characteristic of amphibians?
  a. lungs
  b. heart with two ventricles
  c. cutaneous respiration
  d. double-loop circulation

7. Which of the following characteristics of leopard frogs is *not* an adaptation for avoiding predators?
  a. fast, flicking tongue
  b. skeleton adapted for jumping
  c. spotted, greenish-brown skin
  d. position of the eyes

8. How do tadpoles differ from frogs?
  a. Tadpoles have gills; frogs do not.
  b. Tadpoles are carnivorous; frogs are herbivorous.
  c. Frogs show body symmetry; tadpoles do not.
  d. Frogs live in water; tadpoles live in damp vegetation.

9. Compare amphibian metamorphosis with insect metamorphosis. (**Hint:** See Chapter 30, Section 3.)

10. **Concept Mapping** Construct a concept map describing the characteristics of jawless, cartilaginous, and bony fishes. Try to include the following terms in your concept map: *gills, countercurrent flow, cartilage, operculum,* and *teleosts*.

Critical Thinking

11. **Inferring Relationships** Explain how marine and freshwater fishes differ in the way they maintain their salt and water balance.

12. **Distinguishing Relevant Information** A student is writing a paper on the evolution of the heart. Which of the following terms do not pertain to her topic? Explain. *Sinus venosus, pulmonary veins, septum, osmotic balance, atrium, operculum, conus arteriosus,* and *cutaneous respiration*.

Alternative Assessment

13. **Forming a Model** Construct a model that shows how water passes over the gills of a bony fish. Then explain in writing why countercurrent flow increases respiratory efficiency.

14. **Finding and Communicating Information** Use the media center or Internet resources to learn more about amphibians that live in your area. Create an illustrated reference table that includes their scientific and common names and information about size, habitat, diet, and population size.

Standardized Test Prep

Understanding Concepts

Directions (1–4): For *each* question, write on a separate sheet of paper the letter of the correct answer.

1 What organ in bony fishes senses pressure changes in water?
A. gill
B. lateral line
C. operculum
D. septum

2 What organ in bony fishes regulates buoyancy?
F. atrium
G. conus arteriosus
H. lateral line
I. swim bladder

3 A newspaper article reports that some carp in a local pond are approximately 50 years old. How can this claim be verified?
A. by counting the number of growth rings on a scale from one of the fish
B. by comparing the size of the fish with younger fish from the same pond
C. by comparing the color of the fish with younger fish from the same pond
D. by counting the layers of dark and light layers in a cross section of one fish

4 What group of amphibians is legless?
F. caecelians
G. lampreys
H. skates
I. toads

Directions (5): For the following question, write a short response.

5 Assess why you would **not** expect the digestive system of a tadpole to function like that of an adult frog.

Test *TIP*

When using a diagram to answer questions, carefully study each part of the figure as well as any lines or labels used to indicate parts of the figure.

Reading Skills

Directions (6): Read the passage below. Then answer the question.

A scientist writes the following description of a vertebrate circulatory system: Oxygen-rich blood is pumped from the capillaries to a small chamber in the heart. From the small chamber, the blood is pumped to a larger chamber in the heart where it mixes with some oxygen-poor blood. The blood is then pumped to the rest of the body. When it returns from the body, the oxygen-poor blood enters another small chamber in the heart. From that chamber the blood is pumped to the larger chamber and then on to the capillaries.

6 In which type of environment would you be most likely find an animal that has this kind of circulatory system?
A. coral reef
B. deep ocean
C. moist habitat on land
D. icy habitat on a glacier

Interpreting Graphics

Directions (7): Base your answer to question 7 on the diagram below.

Vertebrate Circulatory Systems

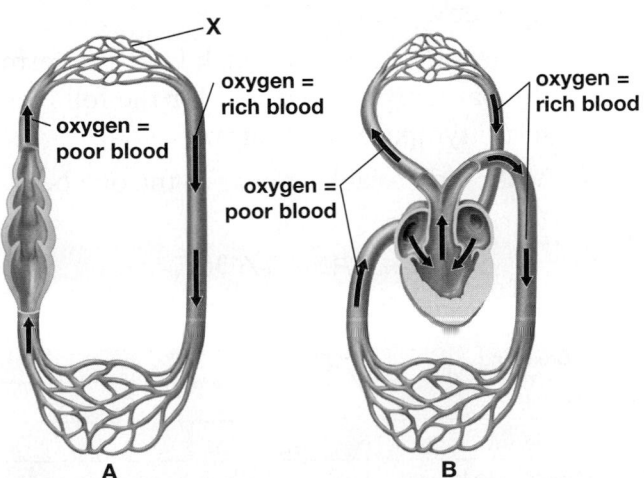

7 Where are the capillaries labeled *X* located?
F. in the brain
G. in the gills
H. in muscles
I. in the lungs

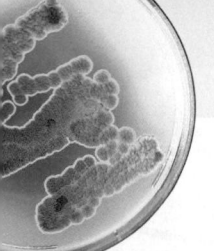

Exploration Lab

Observing a Live Frog

SKILLS
- Observing
- Relating

OBJECTIVES
- **Examine** the external features of a frog.
- **Observe** the behavior of a frog.
- **Explain** how a frog is adapted to life on land and in water.

MATERIALS
- live frog in a terrarium
- live insects (crickets or mealworms)
- 600 mL beaker
- aquarium half-filled with dechlorinated water

Before You Begin

Frogs, which are **amphibians**, are adapted for living on land and in water. For example, a frog's eyes have an extra eyelid called the **nictitating membrane**. This eyelid protects the eye when the frog is underwater and keeps the eye moist when the frog is on land. The smooth skin of a frog acts as a respiratory organ by exchanging oxygen and carbon dioxide with the air or water. The limbs of a frog enable it to move both on land and in water. In this lab, you will examine a live frog in both a terrestrial environment and an aquatic environment.

1. Write a definition for each boldface term in the paragraph above and for the following term: tympanic membrane.

2. Make a data table similar to the one below.

DATA TABLE

Behavior/structure	Observations
Breathing	
Eyes	
Legs	
Response to food	
Response to noise	
Skin	
Swimming behavior	

3. Based on the objectives for this lab, write a question you would like to explore about frogs.

Procedure

PART A: Observing a Frog

1. Observe a live frog in a terrarium. Closely examine the external features of the frog. Make a drawing of the frog. Label the eyes, nostrils, tympanic membranes, front legs, and hind legs.

2. Watch the frog's movements as it breathes air into and out of its lungs. Record your observations.

3. Look closely at the frog's eyes, and note their location. Examine the upper and lower eyelids as well as a third transparent eyelid called a *nictitating membrane*. Describe how the eyelids move.

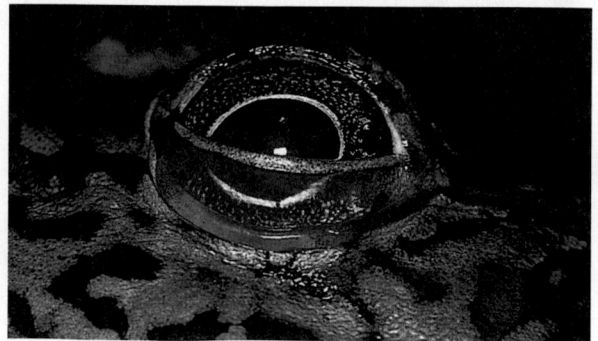

4. Study the frog's legs, and note the difference between the front and hind legs.

5. Place a live insect, such as a cricket or a mealworm, into the terrarium. Observe how the frog reacts.

6. Tap the side of the terrarium farthest from the frog, and observe the frog's response.

7. Place a 600 mL beaker in the terrarium. **CAUTION: Handle live frogs gently. Frogs are slippery! Do not allow a frog to injure itself by jumping from a lab table to the floor.** Carefully pick up the frog, and examine its skin. How does it feel? Now place the frog in the beaker. Cover the beaker with your hand, and carry it to a freshwater aquarium. Tilt the beaker, and gently lower it into the water until the frog swims out.

8. Watch the frog float and swim. Notice how the frog uses its legs to swim. Also notice the position of the frog's head. As the frog swims, bend down to view the underside of the frog. Then look down on the frog from above. Compare the color on the dorsal and ventral sides of the frog.

PART B: Cleanup and Disposal

9. Dispose of broken glass in the designated waste containers. Put live animals in the designated containers. Do not pour chemicals down the drain or put lab materials in the trash unless your teacher tells you to do so.

10. Clean up your work area and all lab equipment. Return lab equipment to its proper place. Wash your hands thoroughly before you leave the lab and after you finish all work.

Analyze and Conclude

1. **Summarizing Information** How does a frog use its hind legs for moving on land and in water?

2. **Recognizing Relationships** How does the position of a frog's eyes benefit the frog while it is swimming?

3. **Analyzing Data** What features of an adult frog provide evidence that it has an aquatic life and a terrestrial life?

4. **Analyzing Methods** Were you able to determine in this lab how a frog hears? Explain.

5. **Inferring Conclusions** What can you infer about a frog's field of vision from the position of its eyes?

6. **Forming Hypotheses** How is the coloration on the dorsal and ventral sides of a frog an adaptive advantage?

7. **Further Inquiry** Write a new question about frogs that could be explored with another investigation.

On the Job

Herpetology is the study of reptiles and amphibians. Do research to discover how herpetologists are working with the Declining Amphibian Task Force (FROGLOG) to solve the mystery of the worldwide decline in amphibian populations. For more about careers, visit **go.hrw.com** and type in the keyword **HX4 Careers**.

A reptile emerging from an amniotic egg

CHAPTER

34 Reptiles and Birds

 Quick Review

Answer the following without referring to earlier sections of your book.

1. **Describe** the process of molting in arthropods. *(Chapter 30, Section 1)*

2. **Distinguish** between ectotherms and endotherms. *(Chapter 32, Section 2)*

3. **Summarize** the evolutionary relationships between reptiles and birds. *(Chapter 32, Section 2)*

4. **Relate** countercurrent flow to the efficiency of the fish gill. *(Chapter 33, Section 1)*

Did you have difficulty? *For help, review the sections indicated.*

 Reading Activity

Write down the title of this chapter and the titles of its three sections on a piece of paper or in your notebook. Leave a few blank lines after each section title. Then, write down what you think you will learn in each section. Save your list, and after you finish reading this chapter, check off everything that you learned that was on your list.

Looking Ahead

Section 1

The Reptilian Body
Key Characteristics of Reptiles
Water Retention
Respiration
Reproduction

Section 2

Today's Reptiles
Lizards and Snakes
Other Orders of Reptiles

Section 3

Characteristics and Diversity of Birds
Key Characteristics of Birds
Adaptations of Birds

---

internet connect

www.scilinks.org
National Science Teachers Association *sci*LINKS Internet resources are located throughout this chapter.

sciLINKS. Maintained by the National Science Teachers Association

● Ancient amphibians gave rise to a new group of animals that were able to lay eggs on dry land—the reptiles. The amniotic egg is a significant evolutionary milestone for land-dwelling animals, such as reptiles and birds.

CHAPTER 34 *Reptiles and Birds* **771**

The Reptilian Body

Key Characteristics of Reptiles

Many people react with fear or repulsion when they see a snake slither across a yard or field. But snakes and their reptile relatives are important members of most ecosystems, and they kill large numbers of insect pests and small rodents. It's true that some reptiles—venomous snakes and crocodilians (crocodiles and alligators)—are dangerous. Most reptiles, however, live quietly and go about their business, preferring to avoid humans.

Members of class Reptilia live throughout the world in a wide variety of habitats, except in the coldest regions, where it is impossible for ectotherms to survive. Reptiles share certain fundamental characteristics, features they retain from the time when reptiles replaced amphibians as the dominant terrestrial vertebrates. **Figure 1** summarizes these key features.

Reptiles have a strong, bony skeleton, and most have two pairs of limbs, although snakes and some lizards are legless. The legs of reptiles are positioned more directly under their body than are the limbs of amphibians. Thus, reptiles can move more easily on land than amphibians can. Unlike amphibians, reptiles have toes with claws, which are used for climbing and digging. Claws also enable reptiles to get a good grip on the ground, allowing many reptiles to run quickly for short distances.

The nervous system of a reptile is very similar to that of an amphibian. Like their dinosaur ancestors, modern reptiles have a brain that is small in relation to their body. For example, an alligator about 2.5 m (8 ft) long has a brain that is about the size of a walnut. Despite this small brain size, reptiles are capable of complex behaviors, including elaborate courtship.

Figure 1 Characteristics of living reptiles. This male anole is extending his dewlap, a display used during courtship and when defending territory.

Key Features of Reptiles

- Strong, bony skeleton and toes with claws
- Ectothermic metabolism
- Dry, scaly skin, almost watertight
- Amniotic eggs, almost watertight
- Respiration through well-developed lungs
- Ventricle of heart partly divided by a septum
- Internal fertilization

Ectothermic Metabolism

Reptiles' ectothermic metabolism is too slow to generate enough heat to warm their bodies, so they must absorb heat from their surroundings. As a result, a reptile's body temperature is largely determined by the temperature of its environment. Many reptiles regulate their temperature behaviorally, by basking in the sun to warm up or seeking shade to cool down. **Figure 2** shows that a lizard can maintain a relatively constant body temperature throughout the day by moving between sunlight and shade. At very low temperatures, most reptiles become sluggish and unable to function. Intolerance of cold generally limits their geographical range and, in temperate climates, forces them to remain inactive through the winter.

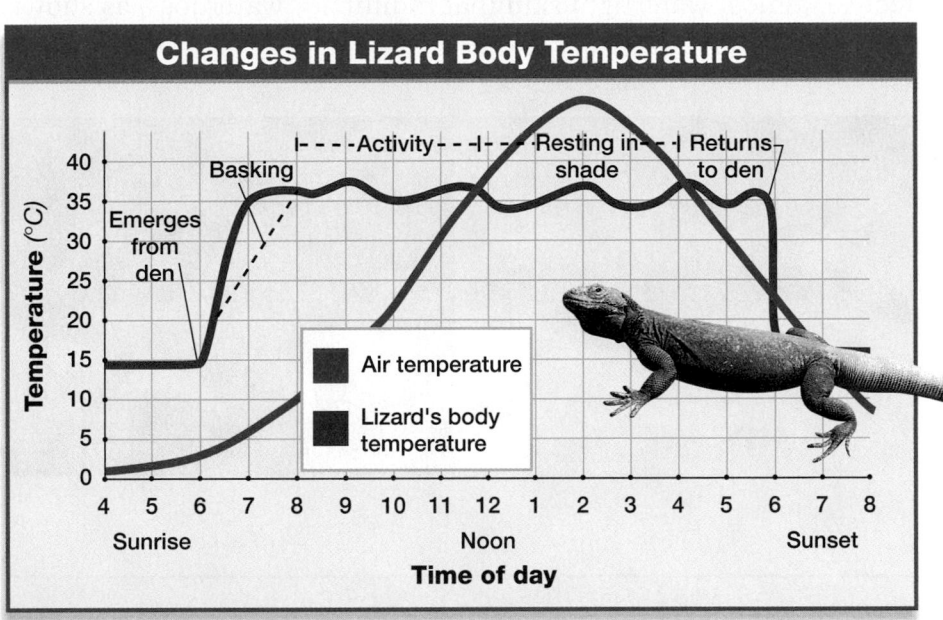

Figure 2 Body temperature in a lizard. A lizard may regulate its body temperature by moving repeatedly between sun and shade.

Identifying Ectotherms

Background

The body temperature of all animals changes during the course of a day. How it changes can help you identify an animal as an ectotherm or an endotherm.

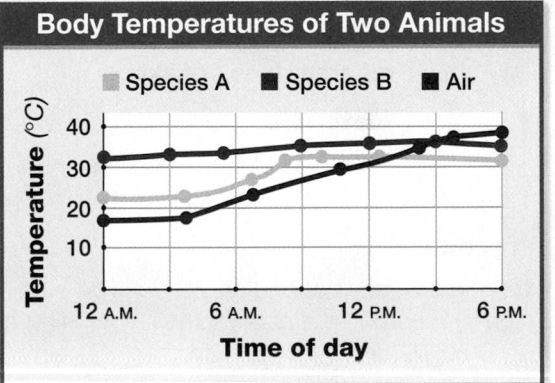

Analysis

1. **Analyze** the data and determine which animal species, A or B, is most likely an ectotherm. Explain your reasoning.

2. **Identify** the time of day the animal you identified as an ectotherm reaches its lowest body temperature.

3. **Identify** the time of day the animal you identified as an ectotherm reaches its highest body temperature.

4. **Propose** a reason why the ectotherm's body temperature is highest at this time.

5. **Predict** what the endotherm's graph line would look like if it were extended to show body temperature between 6 P.M. and midnight.

Water Retention

Study TIP

● **Reviewing Information**
Reread the bulleted list of key features of reptiles in Figure 1. Then write them down on a separate piece of paper, leaving room to write notes about each characteristic. As you read, summarize how the information relates to a particular characteristic.

Amphibians such as frogs cannot be considered fully terrestrial because they lose too much water through their skin. Amphibians must stay moist to avoid dehydration, and their method of reproduction requires a moist environment. Reptiles have evolutionary adaptations that free them from the water requirements of amphibians.

Watertight Skin

Terrestrial animals face a serious problem of water loss as water evaporates through their skin. Modern reptiles have evolved a skin made of light, flexible scales. These scales overlap and form a protective, almost watertight skin that minimizes water loss, as shown in **Figure 3.**

Figure 3 Reptilian scales. The scales of a reptile's skin form a tight seal that retains moisture within the reptile's body.

Modeling Watertight Skin

QUICK LAB

Scales make a reptile's skin almost watertight. This is one of reptiles' adaptations to terrestrial life. You can use grapes to model and compare water loss in different types of skin.

Materials

forceps, 2 grapes, balance, Petri dish

Procedure

1. Find the mass of one grape, and record it in a data table. Then place the grape in an open Petri dish.

2. Using forceps, peel the skin from the second grape. Find and record the mass of the peeled grape. Then place it in the same Petri dish, but do not let the two grapes touch.

3. Wait 15 minutes, and then find and record the mass of each grape again.

Analysis

1. **Calculate** the difference between the original and final masses of each grape.

2. **Propose** an explanation for any changes in mass you observed.

3. **Determine** which grape represents an amphibian's skin and which represents a reptile's skin.

4. **Describe** how a watertight skin is an adaptation to terrestrial life. Include information you have learned in this lab in your explanation.

Watertight Eggs

For a reptile living on dry land, reproduction presents another serious water-loss problem. Without a watery environment, both sperm and eggs will dry out. A reptile's fertilized eggs need a moist environment in which to develop. As you will read later in this chapter, the first problem is overcome by internal fertilization.

The nature of a reptile's amniotic *(am nee AHT ic)* egg solves the second problem. An **amniotic egg** contains both a water supply and a food supply and is key to a reptile's success as a terrestrial animal. Because the egg's tough shell makes it essentially watertight, it does not dry out, even in very dry habitats. Most reptiles, all birds, and three species of mammals reproduce by means of amniotic eggs with shells. (Other mammals produce amniotic eggs, but the embryo develops within the female's uterus rather than within a shell. You will learn about the development of these eggs in a later chapter.) The formation of amniotic eggs with shells suggests that these three groups of animals evolved from a common ancestor.

BIOWatch

The Amniotic Egg

Both reptiles and birds have amniotic eggs, which are very much alike internally. Although a reptile's eggshell is leathery and a bird's is hard, both are almost watertight. However, the shells are porous enough to allow oxygen to enter the egg and carbon dioxide to leave. The shell and the albumen (egg white) lying beneath it protect and cushion the developing embryo. The albumen is also a source of protein and water for the embryo.

Within the egg, four specialized membranes—the amnion, the yolk sac, the allantois, and the chorion—play important roles in maintaining a stable environment in which the embryo can develop.

The amnion *(AM nee awn)* encloses the embryo within a watery environment. In a sense, this membrane creates a little pond that substitutes for the water in which amphibians lay their eggs. This watery enclosure also protects the embryo by cushioning it.

The yolk sac contains the yolk, the developing embryo's food supply. The embryo absorbs nourishment from the yolk through blood vessels connecting its gut and the yolk sac.

The allantois *(uh LAHN toh is)* is a sac that stores waste products from the embryo. It also serves as the embryo's organ for gas exchange. Blood vessels in its walls carry oxygen to and carbon dioxide from the embryo.

Surrounding the amnion, yolk sac, and allantois is a membrane called the chorion *(KAWR ee*

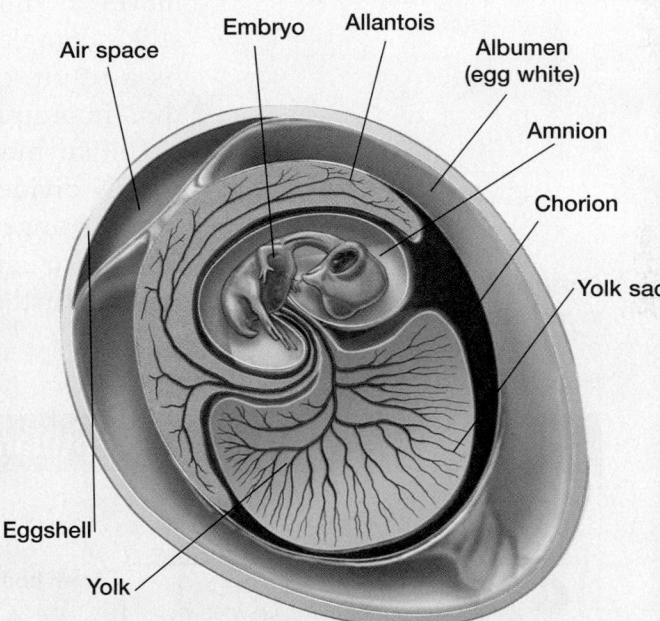

Air space · Embryo · Allantois · Albumen (egg white) · Amnion · Chorion · Yolk sac · Eggshell · Yolk

awn). The chorion allows oxygen to enter the egg and carbon dioxide to leave.

internet connect

www.scilinks.org
Topic: Amniotic Egg
Keyword: HX4005

SCiLINKS. Maintained by the National Science Teachers Association

---

Respiration

Because most reptiles are far more active than amphibians, they have greater metabolic requirements for oxygen. Their bodies meet this demand in several ways.

Lungs

A reptile's scaly skin does not permit gas exchange, so reptiles cannot use their skin as an additional respiratory surface, as many amphibians can. However, the lungs of most reptiles have many internal folds, as shown in **Figure 4.** These folds greatly increase the respiratory surface area of a reptile's lungs. In addition, reptiles have strong muscles attached to their rib cage. The action of these muscles helps to move air into and out of the lungs, increasing the lungs' efficiency.

Figure 4   Reptilian lungs. The lungs of reptiles contain numerous internal folds.

Heart

Recall that the ventricle of the amphibian heart is not divided by a septum. Oxygen-poor blood and oxygen-rich blood mix somewhat in the amphibian's ventricle. In most reptiles, however, the septum extends into the ventricle, partly dividing it into right and left halves, as shown in **Figure 5.** The septum enables a much better, but still incomplete, separation of oxygen-rich and oxygen-poor blood. As a result, oxygen is delivered to the body cells more efficiently than in amphibians.

Unlike most reptiles, crocodilians have a heart with a completely divided ventricle that consists of two pumping chambers. This arrangement fully separates the lung circulation from the body circulation. Thus, the delivery of oxygen throughout the body is further improved in these animals.

Figure 5   Reptilian heart. In most reptiles, the ventricle of the heart is partly divided by a septum.

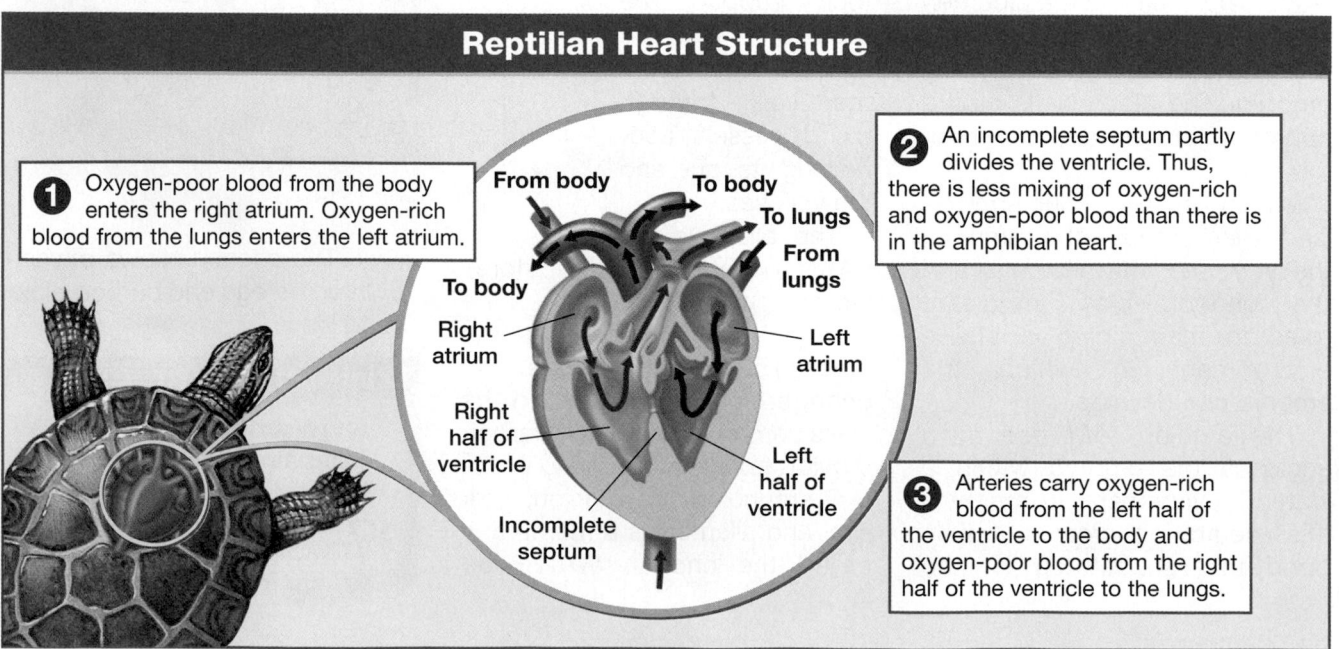

Reptilian Heart Structure

1 Oxygen-poor blood from the body enters the right atrium. Oxygen-rich blood from the lungs enters the left atrium.

2 An incomplete septum partly divides the ventricle. Thus, there is less mixing of oxygen-rich and oxygen-poor blood than there is in the amphibian heart.

3 Arteries carry oxygen-rich blood from the left half of the ventricle to the body and oxygen-poor blood from the right half of the ventricle to the lungs.

From body

To body

To lungs

From lungs

To body

Right atrium

Left atrium

Right half of ventricle

Left half of ventricle

Incomplete septum

Reproduction

Unlike the eggs of most amphibians, reptilian eggs are fertilized within the female, a process called internal fertilization. The male reptile introduces his semen directly into the female's body. The semen contains sperm and fluid secretions. Internal fertilization protects the gametes from drying out, even though the adult animals are fully terrestrial.

Many reptiles are **oviparous** *(oh VIHP urh uhs)*, meaning the young hatch from eggs, as shown in **Figure 6.** In most cases, the eggs are not protected by the parents. Most snakes and lizards, all turtles and tortoises, and all crocodilians are oviparous. All birds and three species of mammals are also oviparous.

Some species of snakes and lizards are **ovoviviparous,** which means the female retains the eggs within her body until shortly before hatching, or the eggs may hatch within the female's body. Although the embryos receive water and oxygen from the female, their nourishment comes from the yolk sac. The offspring of a snake, shown in Figure 6, are born able to fend for themselves. In ovoviviparous reptiles, the eggs are less vulnerable to predators.

WORD *Origins*

● The term *ovoviviparous* comes from three different Latin words: *ovum*, meaning "egg," *vivus*, meaning "alive," and *parere*, meaning "to bring forth or bear."

Figure 6 Reproduction. Sea turtles hatch from eggs buried on sandy beaches. The eggs of some species of snakes are incubated within the female's body, and the young are born live.

Sea turtles

Hatchling snakes

Section 1 Review

1. **Identify** seven characteristics of reptiles.

2. **Describe** how the ectothermic nature of reptiles influences their physical activity and feeding habits.

3. **Summarize** the skin and egg adaptations that allow reptiles to live on land.

4. **Critical Thinking Forming Reasoned Opinions** Data show that an animal's temperature changes over the course of a day. A student asserts that this proves the animal is an ectotherm. What must the student consider before making such a claim?

5. **Describe** how reptiles meet their need for more oxygen than amphibians require.

6. **Standardized Test Prep** If a lizard's internal temperature sensors detect a decrease in body temperature, the lizard can maintain homeostasis by

   A speeding up its metabolism.

   B slowing its metabolism.

   C basking in the sunshine.

   D resting in the shade.

Today's Reptiles

Lizards and Snakes

You've probably walked by a snake or lizard without even knowing it was there. Most are quiet, and their coloration often conceals them from view. Even if you visited the jungles of South America, you might not notice an anaconda unless it moved. What's an anaconda? It's the world's largest snake, frequently reaching 5 m (about 16 ft) in length. The largest anaconda ever found was twice that long. Very large anacondas have been known to prey on jaguars. After such a meal, the anaconda may not eat again for up to a year.

Snakes and lizards belong to order Squamata. A distinguishing characteristic of this order is a lower jaw that is only loosely connected to the skull. This allows the mouth to open wide enough to accommodate large prey and explains how an anaconda can swallow a jaguar. This ability is a contributing factor to the success of snakes and most lizards as predators.

Lizards

Common lizards include iguanas, chameleons, geckos, anoles, and horned lizards (often mistakenly called "horny toads"). A few species of lizards are herbivores, but most are carnivores. Most lizards are small, measuring less than 30 cm (1 ft) in length, but lizards that belong to the monitor family can be quite large. The Komodo dragon of Indonesia, shown in **Figure 7,** is the largest monitor lizard. It can be up to 3 m (10 ft) in length and weigh up to 125 kg (275 lb). The tail of some species of lizards, such as the gecko shown in Figure 7, breaks off easily when seized by a predator, allowing the lizard to escape. Lizards can regenerate a new tail, but it does not have any vertebrae in it.

Gecko

Figure 7 Lizards. Geckos are small reptiles, rarely exceeding 24 cm (10 in.) in length. The Komodo dragon is the world's largest lizard.

Komodo dragon

Snakes

Snakes probably evolved from lizards during the Cretaceous period. The close relationship between lizards and snakes is reflected in their many similarities. In fact, it is often difficult to distinguish the legless species of lizards from snakes. Snakes lack movable eyelids and external ears, as do several species of lizards. Also, both snakes and lizards molt periodically, shedding their outer layers of skin.

Body Structure The skeleton of snakes is unique. Most snakes have no trace of a pectoral girdle (the supporting bones for the bones of the forelimbs), which is found even in legless lizards. The snake's jaw is very flexible because it has five points of movement. (Your jaw, in contrast, has only one movement point.) One of these points is the chin, where the halves of the lower jaw are connected by an elastic ligament. This ligament permits the lower jaw to spread apart when a large meal is being swallowed. The African egg-eating snake, shown in **Figure 8,** can swallow eggs that are much larger than its head in a process that can take an hour or more.

Feeding While many snakes simply seize their prey and swallow it whole, some snakes use other methods to subdue their prey. Many very large snakes, such as anacondas, boas, and pythons, are constrictors, as are some smaller species, such as king snakes. Constrictors wrap their body around their prey, gradually squeezing tighter and tighter until the prey suffocates. The snakes then swallow their prey whole, even if the prey is very large. Like all snakes, constrictors have no teeth that are suited for cutting and chewing.

Some snakes kill their prey with venom (poison). Of the 13 or more commonly recognized families of snakes, only four are venomous: (1) cobras, kraits, and coral snakes; (2) sea snakes; (3) adders and vipers; and (4) rattlesnakes, water moccasins, and copperheads. In most venomous snakes, modified salivary glands produce a venom that is injected into the victim through grooved or hollow teeth. The African boomslang and twig snakes produce venom but do not inject it. Instead, they bite their prey with fangs located at the back of their mouth. Grooved teeth direct the venom into their victim's wound. You can read more about the biology of snakes in *Up Close: Timber Rattlesnake,* on the following pages.

Figure 8 Snake feeding. Snakes have flexible jaws that allow them to swallow prey much larger than their head.

Up Close

Timber Rattlesnake

- **Scientific name:** *Crotalus horridus*
- **Size:** Typically 90–150 cm (36–60 in.) long; maximum 189 cm (74 in.)
- **Range:** Eastern and central United States, from northern New York to northern Florida and west, to central Texas
- **Habitat:** Prefers thick brush, dense woodland, or swamp
- **Diet:** Primarily small mammals

External Structures

Rattle The rattle typically consists of 5 to 7 interlocking rings made of keratin, a protein. When shaken, it produces a rattling sound that serves as a warning. Contrary to popular belief, the snake does not add a rattle each year. Instead, each time the snake sheds its skin during molting, a new ring is added to the base of the rattle. The more rapidly the snake grows, the more rattles it accumulates during a given time. This is why the number of rattles a snake has increases with the size of the snake.

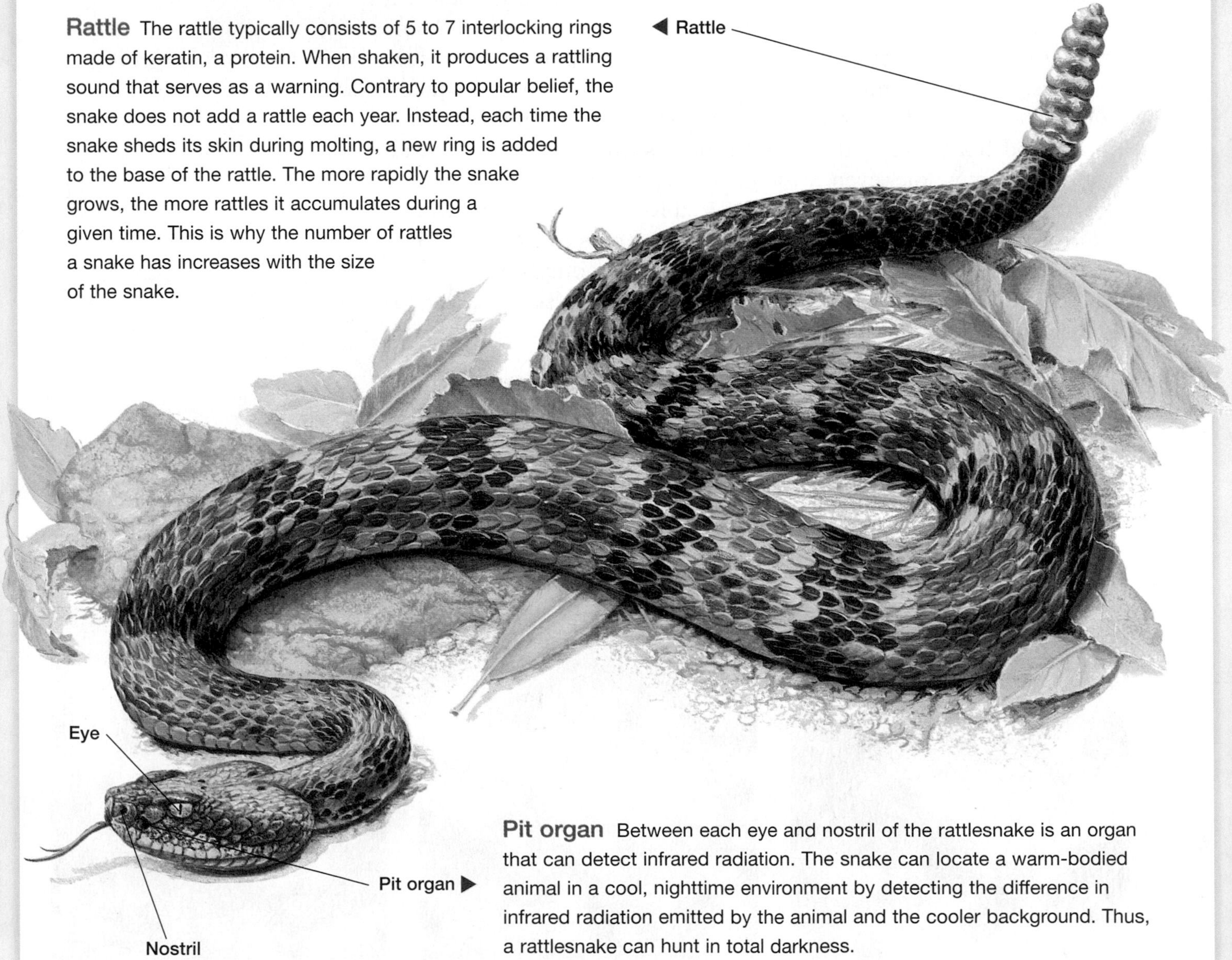

◀ Rattle

Eye

Pit organ ▶

Nostril

Pit organ Between each eye and nostril of the rattlesnake is an organ that can detect infrared radiation. The snake can locate a warm-bodied animal in a cool, nighttime environment by detecting the difference in infrared radiation emitted by the animal and the cooler background. Thus, a rattlesnake can hunt in total darkness.

Internal Structures

Venom glands The timber rattlesnake has hollow upper front teeth, or fangs. When the rattlesnake strikes, these hinged fangs swing forward from the roof of the mouth and inject venom deep into the prey. The venom contains hemotoxins, proteins that attack the circulatory system, destroying red blood cells and causing internal hemorrhaging. Modified salivary glands in the upper jaw produce the venom.

Jacobson's organs Flicking its forked tongue into the air, the rattlesnake takes in chemical samples from the environment. These chemicals are transferred to two depressions in the roof of the mouth called Jacobson's organs, which detect the odor of the chemicals. The snake uses these organs to follow the scent trail of prey.

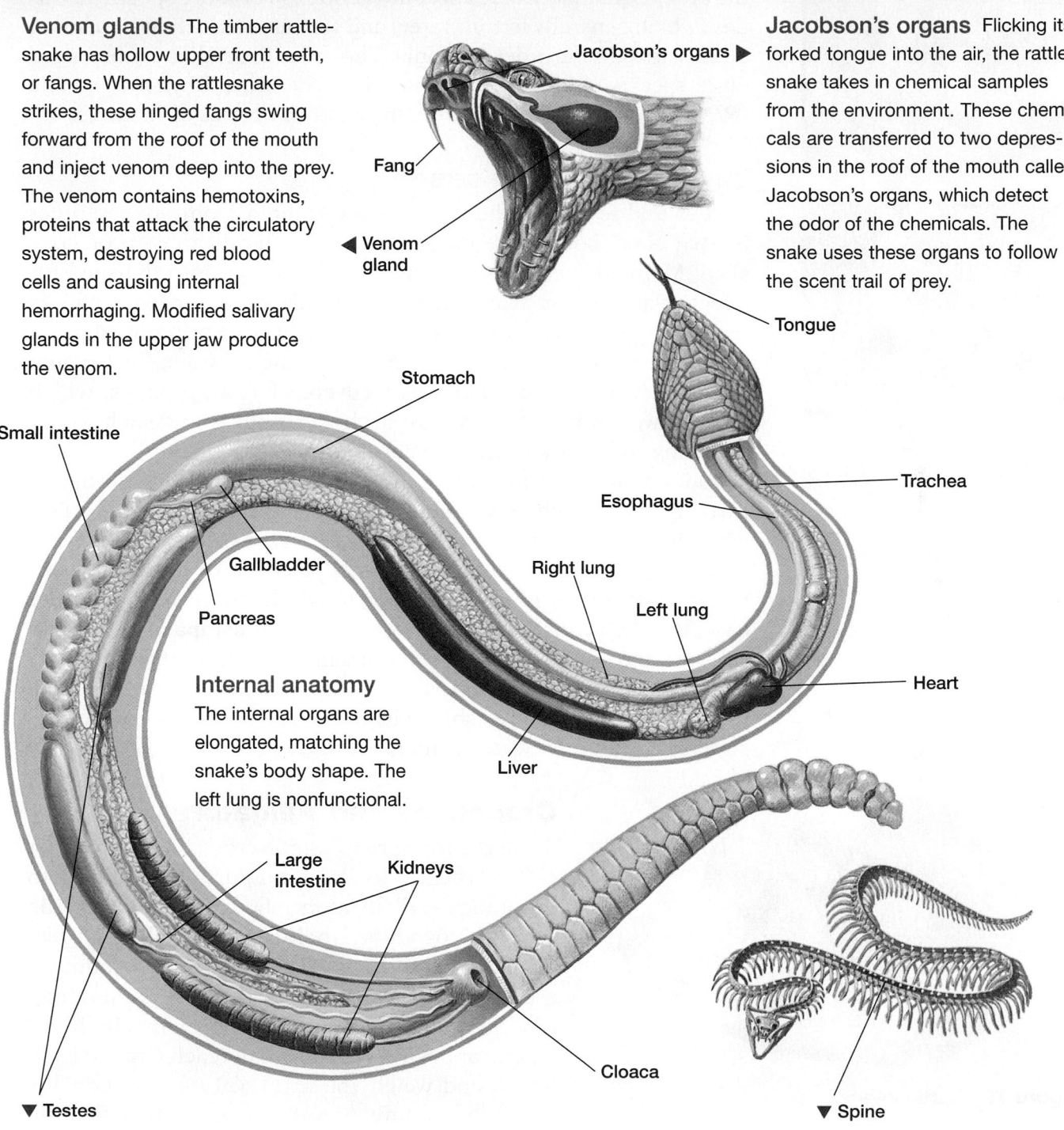

Jacobson's organs

Fang

Venom gland

Tongue

Stomach

Small intestine

Trachea

Esophagus

Gallbladder

Right lung

Pancreas

Left lung

Internal anatomy
The internal organs are elongated, matching the snake's body shape. The left lung is nonfunctional.

Heart

Liver

Large intestine

Kidneys

▼ Testes

Cloaca

▼ Spine

Reproductive structures This male rattlesnake produces sperm in his testes. Female timber rattlesnakes are ovoviviparous. A female carries her fertilized eggs in her body while they develop. Each egg has a thin membrane through which water and oxygen pass from the mother to the embryo. All nourishment is provided by the egg's yolk. After the eggs hatch in the mother's body, the live young are ejected and must fend for themselves.

Spine The rattlesnake's spine is made up of several hundred vertebrae, each with its own pair of attached ribs. It provides the framework for thousands of muscles that manipulate not only the skeleton but also the snake's skin, causing the overlapping scales to extend or lie flat.

Other Orders of Reptiles

The remaining orders of living reptiles contain far fewer species than the order Squamata does. There are about 250 or more species of turtles (which generally live in water) and tortoises (which live on land), all classified in the order Chelonia. The order Crocodilia is composed of 25 species of large, aquatic reptiles. The order Rhynchocephalia (*RING koh seh FAY lee uh*) contains only two species of tuataras.

Turtles and Tortoises

Turtles and tortoises, shown in **Figure 9,** differ from other reptiles in that their bodies are encased within a hard, bony, protective shell. Many of them can pull their head and legs into the shell for effective protection from predators. While most tortoises have a dome-shaped shell, water-dwelling turtles have a streamlined, disk-shaped shell that permits rapid maneuvering in water. Turtles and tortoises lack teeth but have jaws covered by sharp plates, which form powerful beaks. Many are herbivores but some, such as the snapping turtle, are aggressive carnivores.

Today's turtles and tortoises differ little from the earliest known turtle fossils, which date to more than 200 million years ago. This evolutionary stability may reflect the adaptive aspects of their basic shell-covered body structure. The shell is made of fused plates of bone covered with horny shields or tough, leathery skin. In either case, the shell consists of two basic parts. The **carapace** is the dorsal (top) part of the shell, and the **plastron** is the ventral (bottom) portion. The vertebrae and ribs of most species are fused to the inside of the carapace, as shown in **Figure 10.** The shell provides the support for all muscle attachments in the torso.

Figure 9 Turtle and tortoise. Like other sea turtles, this green sea turtle (top) spends virtually its entire life in the sea. The Galápagos tortoise (bottom) spends its life on land.

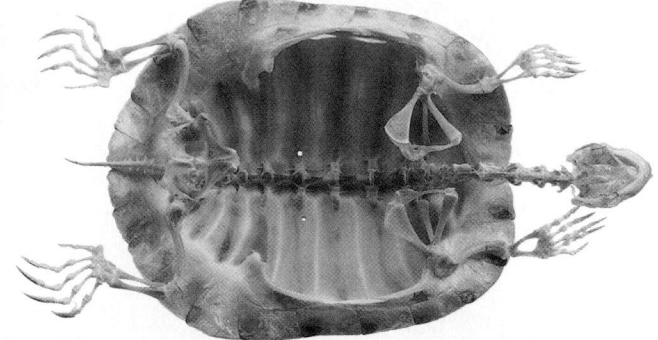

Figure 10 Turtle interior. In this ventral view, a turtle's plastron has been removed to show the relationship of the vertebral column, ribs, pelvis, and pectoral girdle to the carapace.

Crocodiles and Alligators

Of all the living reptiles, the crocodilians are most closely related to the dinosaurs. In addition to crocodiles and alligators, shown in **Figure 11,** the order Crocodilia includes the alligator-like caimans and the long-snouted gavial. Crocodilians are aggressive carnivores. Some are quite large. American alligators can reach 5.5 m (18 ft) in length, and Nile crocodiles can reach 6 m (20 ft) in length and weigh 750 kg (1,650 lb). Crocodilians generally capture prey by stealth, often floating just beneath the water's surface near the shore. When an animal comes to the water to drink, the crocodilian explodes out of the water and seizes its prey. The crocodilian then hauls the prey back into the water to be drowned and eaten. The bodies of crocodilians are well adapted for this form of hunting. Their eyes are high on the sides of the head, and their nostrils are on top of the snout. As a result, they can see and breathe while lying nearly submerged in the water. Crocodilians have a very strong neck and an enormous mouth studded

American alligator

Australian crocodiles

Figure 11   Crocodilians. In general, the snouts of alligators are shorter and broader than those of crocodiles.

with sharp teeth. A valve in the back of the mouth prevents water from entering the lungs when crocodilians feed underwater.

Unlike other living reptiles, crocodilians care for their young after hatching. For instance, a female American alligator builds a nest of rotting vegetation for her eggs. After the eggs hatch, the mother may tear open the nest to free the hatchlings. The young alligators remain under her protection for up to a year.

Tuataras

The two living species of tuataras are members of the genus *Sphenodon* and are native to New Zealand. *Sphenodon punctatus*, the more common species, is shown in **Figure 12.** Tuataras are lizardlike reptiles up to 70 cm (2 ft) long. Unlike most reptiles, tuataras are most active at low temperatures. They burrow or bask in the sun during the day and feed on insects, worms, and other small animals at night. Tuataras are sometimes called living fossils because they have survived almost unchanged for 150 million years. Since the arrival of humans in New Zealand about 1,000 years ago, the tuatara's range has diminished, and their numbers are declining.

Figure 12   Tuatara. Tuataras live on only a few small islands in New Zealand.

Section 2 Review

1 Describe the characteristics shared by lizards and snakes.

2 Describe the function of two different organs that help snakes locate their prey.

3 Summarize the ways turtles and tortoises differ from other reptiles.

4 Compare the parental care shown by alligators with that shown by most other reptiles.

5 Critical Thinking Recognizing Relationships How does the position of a crocodile's nostrils and eyes relate to its method of hunting?

6 Standardized Test Prep The pit organ of a rattlesnake is sensitive to

A airborne chemicals.

B faint sounds.

C ground vibrations.

D infrared radiation.

Characteristics and Diversity of Birds

Objectives

- **Summarize** the key characteristics of birds.

- **Describe** how a bird's feathers and bone structure aid flight.

- **Summarize** how a bird's lungs and heart are adapted for high efficiency.

- **Relate** the structure of a bird's feet and beak to its habits and diet.

Key Terms

contour feather
preen gland
down feather

internet connect

www.scilinks.org
Topic: Characteristics of Birds
Keyword: HX4036

SCiLINKS. Maintained by the National Science Teachers Association

Key Characteristics of Birds

Why do people use the expression "free as a bird"? Most likely it comes from a bird's ability to fly seemingly wherever it wishes. Through human history, the gift of flight has been celebrated in stories, poetry, and songs. But there is more to birds than flight; in fact, some species of birds can't fly.

The birds you see today are the modern members of class Aves. Unlike their reptilian relatives, birds usually lack teeth and have a tail that is greatly reduced in length. But they do retain some reptilian characteristics. For instance, birds lay amniotic eggs that are very similar to those of reptiles, and the feet and legs of birds are covered with scales. Other characteristics unique to birds distinguish them from all other animals. The most obvious is the presence of feathers and the modification of the forelimbs into wings. **Figure 13** lists some distinguishing features of birds. To learn more about the anatomy and habits of one bird, see *Up Close: Bald Eagle* later in this section.

Feathers

Feathers are modified reptilian scales that develop from tiny pits, called follicles, in the skin. Just as snakes and lizards replace their skin by molting, birds molt and replace their feathers. However, few birds shed all of their feathers at one time.

Birds have two main types of feathers: contour feathers and down feathers. **Contour feathers** cover the bird's body and give adult birds their shape. Specialized contour feathers, called *flight feathers*, are found on a bird's wings and tail. These feathers help provide lift for flight. As shown in **Figure 14,** a contour feather has many branches called barbs. Each barb has many projections, called

Figure 13 Characteristics of birds. Like most birds, this tern is well adapted to flight.

Characteristics of Birds
- Forelimbs modified into wings
- Body covered with feathers
- Lightweight bones
- Endothermic metabolism
- Super-efficient respiratory system
- Heart with completely divided ventricle

Figure 14   Contour feather

The structure of a contour feather helps create a smooth, aerodynamic surface, aiding flight.

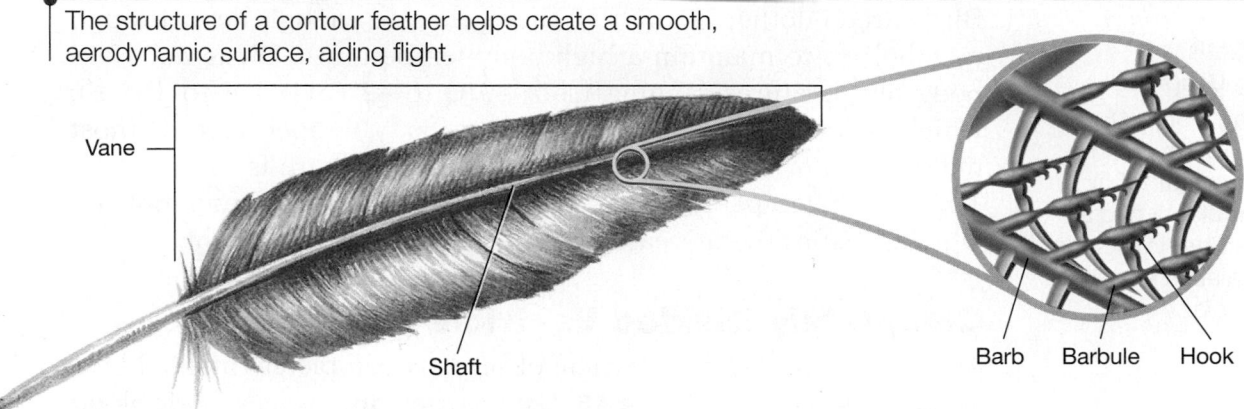

Vane

Shaft

Barb   Barbule   Hook

barbules, that are equipped with microscopic hooks. These hooks link the barbs to one another, giving the feather a continuous surface and a sturdy but flexible shape. With use, the connections become undone. When you see a bird pulling its feathers through its beak, it is relinking these connections. This process is called *preening*. Preening also serves another function. Most birds have a gland called a **preen gland** which secretes oil. When a bird preens, it spreads the oil over its feathers, cleaning and waterproofing them.

Down feathers cover the body of young birds and are found beneath the contour feathers of adults. Their soft, fluffy structure provides good insulation for the bird, helping the bird conserve body heat.

Feathers are important for other reasons too. Their coloration may be protective (as camouflage) or may be important in the selection of a mate. For example, the feathers of some birds allow them to blend in with their surroundings. In other species, the males develop special plumage during the breeding season.

Strong, Lightweight Skeleton

If you have ever picked up a bird, such as a parakeet, you may have been surprised at how light it was compared to a mammal of a similar size. This is because the bones of birds are thin and hollow. Many of the bones are fused, making a bird's skeleton more rigid than a reptile's. The fused sections form a sturdy frame that anchors muscles during flight. The power for flight (or for swimming underwater in the case of some birds, like penguins) comes from large breast muscles that can make up 30 percent of a bird's body weight. These muscles stretch from the wing to the breastbone. The breastbone is greatly enlarged and bears a prominent keel for muscle attachment, as illustrated in **Figure 15.** Muscles also attach to the fused collarbones (wishbone). No other living vertebrates have a keeled breastbone or fused collarbones.

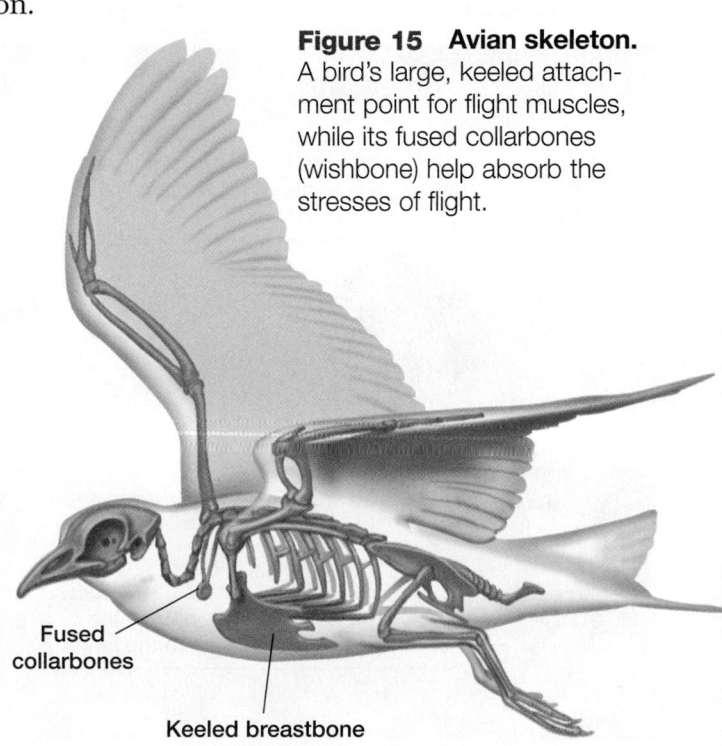

Figure 15   Avian skeleton. A bird's large, keeled attachment point for flight muscles, while its fused collarbones (wishbone) help absorb the stresses of flight.

Fused collarbones

Keeled breastbone

Endothermic Metabolism

Birds are endotherms; that is, they generate enough heat through metabolism to maintain a high body temperature. Birds maintain body temperatures ranging from 40°C to 42°C (104°F to 108°F), which is significantly higher than the body temperature of most mammals. For comparison, your body temperature is 37°C (98°F). These high temperatures are due to a high rate of metabolism, which satisfies the increased energy requirements of flight.

Completely Divided Ventricle

As in crocodilians, the ventricle of birds is completely divided by a septum, as shown in **Figure 16.** Oxygen-rich and oxygen-poor blood are kept separate, meaning that oxygen is delivered to the body cells more efficiently. The sinus venosus, which is a prominent part of the fish heart, is not a separate chamber of the heart in birds (or mammals). However, a small amount of tissue from it remains in the wall of the right atrium. This tissue is the point of origin of the heartbeat and is known as the heart's pacemaker.

Highly Efficient Lungs

Birds such as the geese shown in **Figure 17** use a considerable amount of energy when they fly. Since birds often fly for long periods of time, their cellular demand for energy exceeds that of even the most active mammals. How do birds get the energy they need?

Recall that reptiles meet their increased need for oxygen with lungs that have a larger surface area than the lungs of amphibians. But there is a limit to how much the efficiency of a lung can be improved just by increasing its surface area. Another way to

Figure 16   Avian heart.
A bird's heart has a complete septum.

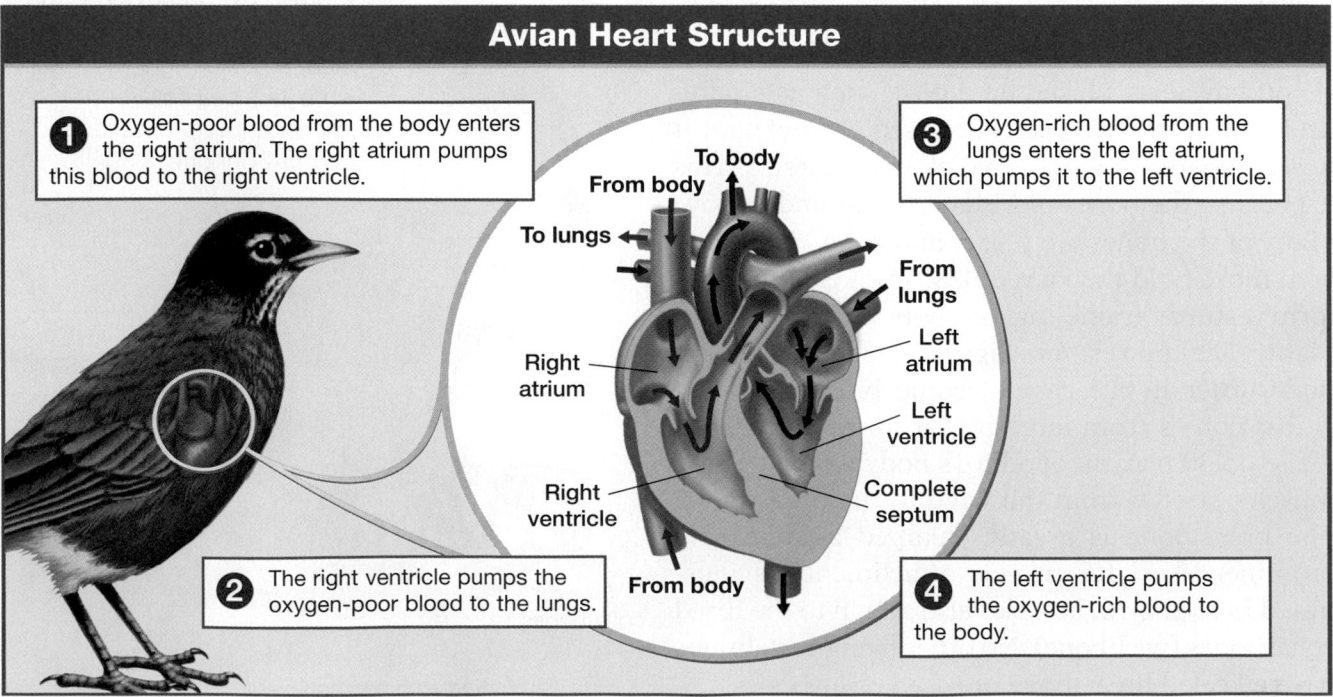

Avian Heart Structure

1 Oxygen-poor blood from the body enters the right atrium. The right atrium pumps this blood to the right ventricle.

3 Oxygen-rich blood from the lungs enters the left atrium, which pumps it to the left ventricle.

To body
From body
To lungs
From lungs
Right atrium
Left atrium
Left ventricle
Right ventricle
Complete septum
From body

2 The right ventricle pumps the oxygen-poor blood to the lungs.

4 The left ventricle pumps the oxygen-rich blood to the body.

increase the efficiency of a lung is to have air pass over its respiratory surface in one direction only, just as water flows over a fish's gills in one direction. This is what happens in birds. One-way air flow is possible in birds because they have air sacs connected to their lungs, as shown in **Figure 18.** There is no gas exchange in the air sacs. They simply act as holding tanks.

There are two important advantages to one-way air flow. First, the lungs are exposed only to air that is almost fully oxygenated, increasing the amount of oxygen transported to the body cells. Second, the flow of blood in the lungs runs in a different direction than the flow of air does. Unlike the flow of water and blood in fish gills, the flow of air and blood in bird lungs are not completely opposite (countercurrent). Nevertheless, the difference in direction does increase oxygen absorption.

These three characteristics—endothermic metabolism, a completely divided ventricle, and highly efficient lungs—provide the energy a bird needs for takeoff and sustained flight. They enable a hummingbird to flap its wings rapidly (20–80 beats per second) as it hovers by a flower. They also permit migrating birds to fly thousands of kilometers without stopping. One species of shorebirds called the lesser yellowlegs flies across the open ocean from Massachusetts to Martinique in the West Indies. Incredibly, some of these birds cover this distance of 3,220 km (about 2,000 mi) in less than 6 days. Note, however, that many birds, such as gulls and vultures, remain aloft for long periods of time using little energy. These birds take advantage of upward air movements that lift them.

Figure 17 Flight. These barnacle geese expend an enormous amount of energy during take off and flight.

Figure 18 Avian respiration. A single breath of air stays in a bird's respiratory system for two cycles of inhalation and exhalation.

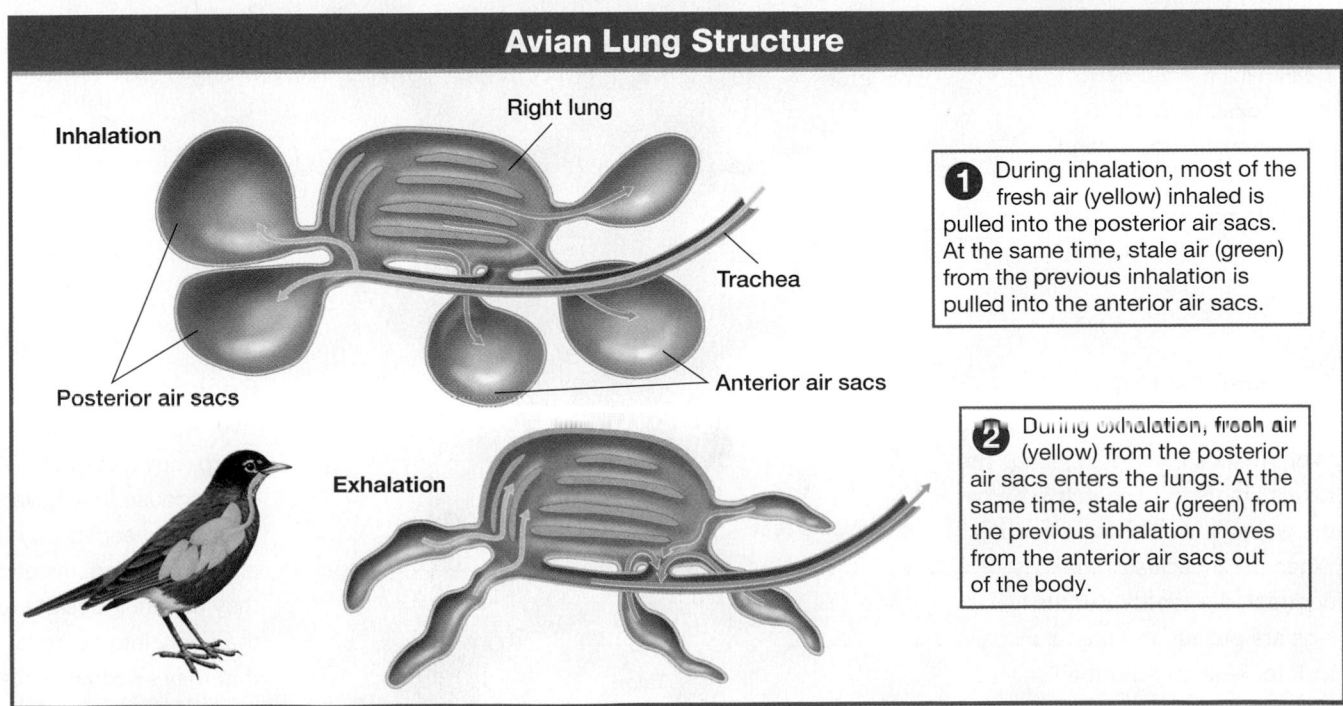

Avian Lung Structure

Inhalation

Right lung

Trachea

Posterior air sacs

Anterior air sacs

Exhalation

❶ During inhalation, most of the fresh air (yellow) inhaled is pulled into the posterior air sacs. At the same time, stale air (green) from the previous inhalation is pulled into the anterior air sacs.

❷ During exhalation, fresh air (yellow) from the posterior air sacs enters the lungs. At the same time, stale air (green) from the previous inhalation moves from the anterior air sacs out of the body.

Up Close

Bald Eagle

- **Scientific name:** *Haliaeetus leucocephalus*
- **Size:** Wingspan is typically over 2 m (6.5 ft), and body weight often exceeds 7 kg (15 lb)
- **Range:** Nearly all of North America, from Florida to northern Alaska
- **Habitat:** Forested areas near water that have tall trees for perching and nesting
- **Diet:** Fish, small mammals, birds, carrion

External Structures

Eyes Vision is a bald eagle's most important sense. The bird's keen eyesight allows it to see prey at great distances. Its visual acuity is 3–4 times higher than ours.

▲ Eye

Nostril

▼ Beak

▼ Feathers

Feathers The body of the bald eagle is covered with feathers everywhere except the feet and the beak, which are bare. Both sexes develop the characteristic white head and neck at maturity.

Grasping feet The bald eagle has large feet and talons—the hind talon may be 5 cm (2 in.) long. The talons are used to snatch fish from the water while the eagle is flying. When the muscles of the legs contract, the tendons in the lower legs are pulled, and the talons lock together around the fish.

◄ Grasping feet

Talon

Beak The beak is massive, with an elongated, sharp, downward-curving tip. Because they have no teeth, bald eagles do not chew their food. Instead, they use their beak to tear their prey into portions that they swallow whole.

Internal Structures

Brain In the ratio of brain size to body size, birds rank second among vertebrates, behind only mammals. The large cerebellum receives and integrates information from the muscles, eyes, and inner ears. This makes possible the precise control of movement and balance necessary for flight. The optic lobe is large because it processes input from the eagle's most important sense organs—the eyes. The cerebrum performs many functions, including evaluation of sensory information, control of behavior, and learning.

Cerebellum   Cerebrum

Brainstem   Optic lobe

▲ Brain

Excretory system The excretory system is efficient and lightweight. It does not store waste liquids in a bladder. Instead, the bald eagle (and other birds) converts its nitrogenous wastes to uric acid, which is concentrated into a harmless white paste. The uric acid travels to the cloaca and is eliminated.

Left lung

Left ovary

◀ Kidney

Small intestine

Large intestine

▼ Cloaca

Esophagus

Trachea

Air sac

Crop

Heart

Liver

Pancreas   ▼ Gizzard

Cloaca The cloaca is a common collecting chamber for the excretory, digestive, and reproductive systems. As fertilized eggs travel down the female eagle's oviduct, egg white, membranes, and the shell are added. The completed egg then passes into the cloaca and out of the female's body.

Digestive system Large meals are temporarily stored in the crop, the expandable lower portion of the esophagus. The food then passes into a two-chamber stomach. In the first chamber, stomach acids begin breaking down the food. The partially digested food is then passed to the second chamber, the gizzard, where it is ground and crushed. Undigested material is eliminated through the cloaca.

Adaptations of Birds

While there is great diversity among the 28 orders of birds, 60 percent of all bird species belong to order Passeriformes. These birds, also known as the songbirds, number approximately 5,300 species and are by far the largest group of terrestrial vertebrates. Birds are adapted for different ways of life, and you can tell a great deal about the habits and diet of a bird by examining its beak (bill), legs, and feet. Carnivorous birds such as hawks have curved talons for seizing prey and a sharp beak for tearing apart their meal. The beaks of ducks are flat for shoveling through water or mud, and their webbed feet enable them to swim. Finches are seed eaters, and their short, thick beak is adapted for crushing seeds while their curved toes enable them to cling to branches. Other birds, such as the penguins shown in **Figure 19,** are flightless, and their wings and feet are modified for swimming.

During the evolutionary history of birds, their beaks, legs, and feet have been adapted to the particular environment the birds live in, as shown in **Table 1.** Some birds are more highly specialized than others, and many birds are highly flexible in their eating habits. The song sparrow, for example, has a strong bill that it uses in winter to crack hard seeds. In summer, the sparrow uses its bill to catch soft-bodied insects.

Figure 19 Penguin. The penguin's wings are adapted for swimming rather than flying.

Table 1 Avian Adaptations		
Type of bird	**Beak adaptations**	**Foot adaptations**
Songbirds (e.g., cardinal, robin)	Seed-cracking: Short, thick, strong beak Insect-catching: Long, slender beak for probing	Perching: Toes can cling to branches; one toe points backward
Hummingbirds	Probing: Thin, slightly curved beak for inserting into flowers to sip nectar	Hovering: Legs so small the bird cannot walk on the ground; tiny feet

Type of bird	Beak adaptations	Foot adaptations
Woodpeckers	**Drilling:** Strong, chisel-like beak	**Grasping:** Feet with two toes pointing forward and two pointing backward
Parrots	**Cracking, tearing:** Short, stout, hooked beak used to crack seeds and nuts and to tear vegetation	**Climbing/grasping:** Strong toes, two pointing forward, two pointing backward; adapted for perching, climbing, and holding food
Birds of prey	**Tearing:** Curved, pointed beak for pulling apart prey	**Grasping:** Powerful, curved talons for seizing and gripping prey
Ducks	**Sieving:** Long, flattened, rounded bill	**Swimming:** Three toes linked by webs for improved swimming
Long-legged waders	**Fishing:** Long, slender, spear-shaped beak for fishing	**Wading:** Long legs; toes spread out over a large area to support bird on soft surfaces

Other Adaptations

There are many groups of birds, each of which is adapted to its particular living conditions. For example, gulls and terns have streamlined bodies that are adapted for flying over the water in search of fish. Owls' excellent low-light vision enables them to survive as nocturnal hunters. For a list of the orders of birds, see "Classification in Kingdoms and Domains" in the Appendix.

Calculating Average Bone Density

Background

Density is the ratio of the mass of an object to its volume. Several teams of students determined the density of bones from two different animals. You can use their data to practice calculating average bone density.

DATA TABLE

Bone type	Team 1	Team 2	Team 3	Team 4
Animal 1	1.6 g/cm³	1.0 g/cm³	1.2 g/cm³	1.4 g/cm³
Animal 2	2.3 g/cm³	1.8 g/cm³	1.8 g/cm³	2.1 g/cm³

1. **Add the densities of one bone type.** For example, if three bone samples have densities of 3.0, 3.1, and 2.9 g/cm³, their sum would be 9.0 g/cm³.

2. **Divide the sum of the densities by the number of samples.**

$$\text{Average density} = \frac{\text{sum of the densities}}{\text{number of samples}} = \frac{9.0 \text{ g/cm}^3}{3} = 3.0 \text{ g/cm}^3$$

Analysis

1. **Calculate** the average bone density for each of the two animals in the data table. Express your answer in grams per cubic centimeter.

2. **Critical Thinking Evaluating Methods** Why is it important to analyze several samples and obtain the average of your data?

3. **Critical Thinking Drawing Conclusions** Based on your answer to item 1, which of the two animals is more likely to be a bird?

Section 3 Review

1 Identify the adaptations of birds for flight.

2 Summarize how birds obtain the energy necessary for flight.

3 Relate the bald eagle's methods of hunting and feeding to its external body features.

4 Critical Thinking Evaluating Hypotheses A student examines a bird that has delicate, perching feet with long, slender toes. Its beak is small but slightly long and pointed. The student concludes that the bird is a seed-eating songbird. Do you agree? Explain your reasoning.

5 Standardized Test Prep Which structure is part of the excretory, digestive, and reproductive systems of a bird?

A kidney   C gizzard

B cloaca   D ovary

Key Concepts

1 The Reptilian Body

- Reptiles have a strong, bony skeleton.
- Reptiles are ectothermic.
- Reptiles have nearly watertight skin and eggs, both of which enable them to be terrestrial animals.
- Reptiles have paired lungs that have a greater surface area for gas exchange than the lungs of amphibians.
- Reptiles have a double-loop circulatory system. Most have a ventricle that is partly divided into right and left halves, resulting in incomplete separation of oxygen-rich and oxygen-poor blood.
- Reptilian fertilization is internal.

2 Today's Reptiles

- Snakes and lizards (order Squamata) share many characteristics, such as periodic molting, but snakes have no legs.
- The shells of turtles and tortoises (order Chelonia) are made of fused plates of bone covered with horny shields or leathery skin.
- Unlike other reptiles, crocodilians (order Crocodilia), care for their young after hatching. They also have a completely divided ventricle.
- There are only two species of tuataras (order Rhynchocephalia).

3 Characteristics and Diversity of Birds

- Birds are endotherms. Their high rate of metabolism helps them meet the large energy requirements for flight.
- A bird's contour feathers give the bird its shape and aid flight. Its down feathers provide insulation.
- The bones of birds are thin and hollow, and many of them are fused; all are adaptations for flight.
- One-way airflow through the lungs provides the large amounts of oxygen birds need for flight.
- The ventricle of the bird heart is completely divided by a septum.

Key Terms

Section 1

amniotic egg (775)
oviparous (777)
ovoviviparous (777)

Section 2

carapace (782)
plastron (782)

Section 3

contour feather (784)
preen gland (785)
down feather (785)

Performance ZONE

Understanding Key Ideas

1. Which is *not* an adaptation of reptiles for life on land?
   - **a.** watertight skin
   - **b.** external fertilization
   - **c.** amniotic egg
   - **d.** kidneys

2. The heart of most reptiles has
   - **a.** no septum.
   - **b.** a partly divided ventricle.
   - **c.** a fully divided ventricle.
   - **d.** two pumping chambers.

3. In reptiles, fertilization
   - **a.** is internal.
   - **b.** is external.
   - **c.** always occurs in water.
   - **d.** does not occur.

4. Snakes differ from lizards in that snakes do *not* have
   - **a.** lungs.
   - **b.** kidneys.
   - **c.** a pectoral girdle.
   - **d.** a flexible jaw.

5. Jacobson's organs are involved in the sense of
   - **a.** smell.
   - **b.** hearing.
   - **c.** sight.
   - **d.** touch.

6. The feathers of most birds are well adapted for
   - **a.** swimming and repelling water.
   - **b.** expelling heat and feeding.
   - **c.** flying and conducting heat.
   - **d.** flying and insulating.

7. The foot illustrated below is most likely of a bird adapted for
   - **a.** wading.
   - **b.** grasping.
   - **c.** perching.
   - **d.** swimming.

8. Which group of living reptiles is most closely related to birds?
   - **a.** snakes
   - **b.** turtles
   - **c.** rhynchocephalians
   - **d.** crocodilians

9. 🌐 **BIOWatch** Name the four membranes contained in an amniotic egg, and describe how they make the egg an independent life-support system.

10. 🔗 **Concept Mapping** Construct a concept map that describes the characteristics of both reptiles and birds. Include the following terms in your map: *ectotherm, endotherm, oviparous, ovoviviparous, scales, feathers, reptiles, snakes, lizards, tuataras, turtles, crocodilians, three-chambered heart,* and *four-chambered heart.*

Critical Thinking

11. **Predicting Outcomes** How might having a three-chambered heart, like that of most reptiles, affect a hummingbird in flight?

12. **Recognizing Logical Connections** How might a long period of parental care be related to the number of offspring an animal produces?

Alternative Assessment

13. **Being a Team Member and Communicating** Work with two or three of your classmates to find out what kinds of birds are common in your area. Select at least six birds to explore in depth and research the following: its habitat, its food, its beak and foot adaptations, and where it winters. Present the information you gathered in an illustrated guide. Make copies of your guide available for interested students.

14. **Organizing Information** Create a habitat in a terrarium for a small lizard, and observe the lizard's behavior. Make a labeled drawing of the environment you create, and keep a journal of your observations.

Standardized Test Prep

Understanding Concepts

Directions (1–4): **For *each* question, write on a separate sheet of paper the letter of the correct answer.**

1 What term describes the eggs of reptiles and birds?
A. amniotic
B. oviparous
C. ovoviviparous
D. externally fertilized

2 What is the function of the preen gland?
F. secreting oil
G. sensing chemicals
H. stimulating egg production
I. changing body temperature

3 When a piece of a lizard's tail breaks off, the separated portion may wiggle about forcefully. How might this adaptation be an advantage for the lizard?
A. The movement of the tail frightens the predator away.
B. Eating the tail provides the lizard with the energy it needs to escape.
C. Predators are distracted by the wiggling tail, giving the lizard time to escape.
D. The wiggling tail can injure a predator, preventing it from pursuing the lizard.

4 What are the two basic parts of a turtle's shell called?
F. keratin and cloaca
G. septum and amnion
H. chorion and allantois
I. carapace and plastron

Directions (5): **For the following question, write a short response.**

5 Many viviparous snakes and lizards live in cold climates. Evaluate why viviparity might be advantageous in such environments.

Test *TIP*

Scan the answer choices for words such as *never* and *always*. Such words often are used in incorrect statements because they are too broad.

Reading Skills

Directions (6): **Read the passage below. Then answer the question.**

Reptile and bird eggs contain the water, nutrients, and protection a delicate embryo needs during development. Unlike amphibian eggs, reptile and bird eggs are watertight. Because they do not dry out, the eggs can remain on dry land yet still provide the embryo with all of the water it needs. Though water cannot pass through the shell, oxygen and carbon dioxide can pass through.

6 Why is it important that egg shells be permeable to some gases?
A. Embryos use water vapor to maintain homeostasis.
B. Embryos need to take up oxygen for cellular respiration.
C. Embryos need to release carbon dioxide for photosynthesis.
D. Embryos use nitrogen gas to build proteins and nucleic acids.

Interpreting Graphics

Directions (7): **Base your answer to question 7 on the chart below.**

Normal Ranges of Body Temperature

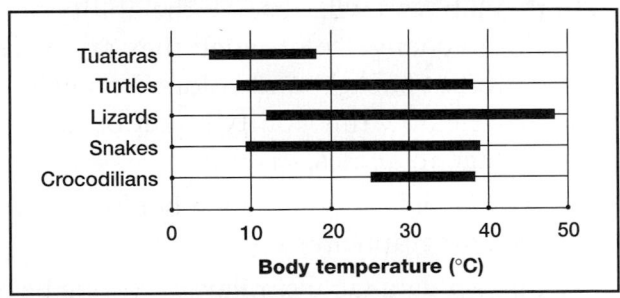

7 Which of these statements is supported by the data in the chart?
F. Turtles and snakes have similar body temperature ranges.
G. Lizards always have a higher body temperature than tuataras.
H. Crocodilians have a greater body temperature range than tuataras.
I. Some tuataras can have a higher body temperature than some crocodilians

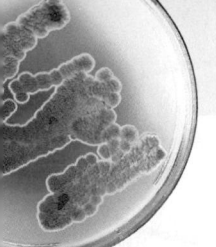

Exploration Lab

Observing Color Change in Anoles

SKILLS
- Using scientific methods
- Observing

OBJECTIVES
- **Observe** live anoles.
- **Relate** the color of an anole to the color of its surroundings.

MATERIALS
- glass-marking pencil
- 2 large, clear jars with wide mouths and lids with air holes
- 2 live anoles
- 6 shades each of brown and green construction paper, ranging from light to dark (2 swatches of each shade)

Before You Begin

Lizards are a group of **reptiles.** There are 250–300 species of anoles, lizards in the genus *Anolis*. Like chameleons, anoles can change color, ranging from brown to green. Anoles live in shrubs, grasses, and trees. Light level, temperature, and other factors, such as whether the animal is frightened or has eaten recently, can all affect the color of an anole. When anoles are frightened, they usually turn dark gray or brown and are unlikely to respond to other **stimuli.** Anoles generally change color within a few minutes. In this lab, you will observe the ability of anoles to change color when they are placed on different background colors. You will also determine how this ability might be an advantage to anoles.

1. Write a definition for each boldface term in the paragraph above.

2. Make a data table similar to the one below.

3. Based on the objectives for this lab, write a question you would like to explore about the color-changing behavior of anoles.

Procedure

PART A: Make Observations

1. Observe live anoles in a terrarium. Make a list of characteristics that indicate that anoles are reptiles.

2. Work with a partner to place anoles to be studied in separate glass jars.
CAUTION: Handle anoles gently, and follow instructions carefully. Anoles run fast and are easily frightened. Plan your actions before you start. By working efficiently, you can keep your anole from becoming overly frightened. Carefully pick up one anole by grasping it firmly but gently around the shoulders. Do not pick up anoles by their tail. Place the anole in a glass jar. Quickly and carefully place a lid with air holes on the jar.

3. When anoles become overly frightened, they remain dark. While you are designing your experiment, do not disturb your anoles, and let them recover from your handling.

PART B: Design an Experiment

4. Work with members of your lab group to explore one of the questions written for step 3 of **Before You Begin.** To explore the question, design an experiment that uses the materials listed for this lab.

DATA TABLE				
	Color 1		Color 2	
Anole	Change	Time	Change	Time
1				
2				

5. Write a procedure for your experiment. Make a list of all the safety precautions you will take. Have your teacher approve your procedure and safety precautions before you begin the experiment.

6. Set up and carry out your experiment.

PART C: Cleanup and Disposal

7. Dispose of construction paper and broken glass in the designated waste containers. Put anoles in the designated container. Do not put lab materials in the trash unless your teacher tells you to do so.

8. Clean up your work area and all lab equipment. Return lab equipment to its proper place. Wash your hands thoroughly before you leave the lab.

Analyze and Conclude

1. Summarizing Results Briefly state how the variable you tested influenced the color-changing behavior of anoles.

2. Evaluating Results Did any unplanned variables influence your data? (For example, was there a loud noise, or was a jar suddenly moved?)

3. Analyzing Methods How could your experiment be modified to improve the certainty of your results?

4. Analyzing Data Were there any inconsistencies in your data? (For example, two anoles reacted in different ways.) If so, offer an explanation for them.

5. Drawing Conclusions After considering your data, make a statement about color-changing behavior in anoles.

6. Further Inquiry Write a new question about anoles that could be explored with another investigation.

? Do You Know?

Do research in the library or media center to answer these questions:

1. What other behaviors are characteristic of anoles?

2. How is a chameleon different from an anole?

Use the following Internet resources to explore your own questions about lizards that change color.

internet connect

www.scilinks.org
Topic: Adaptations of Reptiles
Keyword: HX4003

SCiLINKS Maintained by the
National Science Teachers Association

Mountain lion

CHAPTER
35 Mammals

✓ Quick Review

Answer the following without referring to earlier sections of your book.

1. **Define** the term *metabolism*. (Chapter 1, Section 2)

2. **Compare** the diets of carnivores and herbivores. (Chapter 16, Section 2)

3. **Summarize** the difference between endothermic and ectothermic animals. (Chapter 32, Section 2)

4. **Describe** the breakup of Pangaea. (Chapter 32, Section 2)

Did you have difficulty? *For help, review the sections indicated.*

Reading Activity

After reading each section of this chapter, closely study the figures in the section. Reread the figure captions, and for each one, write a question that can be answered by referring to the figure and its caption. As you review the chapter, but before you complete the Performance Zone chapter review, refer to your list of figures and questions.

● The mountain lion, *Puma concolor*, has the widest distribution of any species of native mammals in the Western Hemisphere—from Canada throughout South America. It is now threatened or endangered in much of its range.

internet connect

www.scilinks.org
National Science Teachers Association *sci*LINKS Internet resources are located throughout this chapter.

SCI**LINKS**® Maintained by the National Science Teachers Association

The Mammalian Body

- **Describe** three functions of hair.
- **Relate** a mammal's teeth to its diet.
- **Summarize** how mammals maintain a high body temperature.
- **Relate** the characteristics of mammals to one mammal, the grizzly bear.
- **Describe** parental care in mammals.

Key Terms

hair
alveolus
mammary gland
weaning

Key Characteristics of Mammals

If you were to look out over an African landscape, you would notice the big mammals—lions, elephants, antelopes, and zebras. Your eye would not as readily pick out the many birds, snakes, lizards, and frogs that also live there. The fact that almost all of today's large, land-dwelling vertebrates are mammals makes them much more easily noticed. While most mammals are terrestrial, some, like whales, swim in the sea, and others—the bats—fly through the air. In spite of their differences, mammals share key characteristics that are summarized in **Figure 1.**

Mammals are well adapted for terrestrial living and are able to retain water more efficiently than reptiles. That's because the mammalian kidney has an exceptional ability to concentrate waste products in a small volume of urine. Later in this book, you will find out how the mammalian kidney functions.

Hair

Of all animal species, only mammals have hair. Even whales and dolphins, which appear to be hairless, have a few sensitive bristles on their snout. A **hair** is a filament composed mainly of dead cells filled with the protein keratin. The evolutionary origin of hair is unknown, but hair probably did not derive from reptilian scales.

The primary function of hair is insulation. Mammals, such as the polar bear shown in **Figure 1,** typically maintain body temperatures higher than the temperature of their surroundings. As a result, they tend to lose body heat. However, the dense coat of hair that covers most mammals holds heat in. Like other mammals that live in cold environments, the polar bear has a layer of fat under its skin that provides additional insulation from the cold. Humans have a sparse

Figure 1 Characteristics of mammals. Despite their vast external differences, mammals share a number of distinctive features.

Characteristics of Mammals

- Hair
- Diverse and specialized teeth
- Endothermic metabolism
- Mammary glands that produce milk

covering of hair and a limited amount of body fat; in most climates we need clothes to provide adequate insulation.

Hair has functions other than insulation. The coloration and pattern of a mammal's coat often enable the animal to blend in with its surroundings. A small brown mouse is almost invisible against the dark forest floor. The orange and black stripes of a Bengal tiger, shown in **Figure 2,** conceal it in the tall, orange-brown grass in which it hunts. Some animals, such as the arctic fox, show a seasonal change in the color of their coat from white in winter to brown in summer that provides protective coloration year round. The color of a mammal's coat may also be a clear signal; the black and white fur of a skunk, for instance, warns would-be predators to stay away.

In some animals, specialized hairs serve a sensory function. The whiskers of cats and dogs are stiff hairs that are very sensitive to touch. Mammals that are active at night or that live underground often rely on their whiskers for information about the environment. Other specialized hairs can be used as a defensive weapon. For example, when threatened, porcupines defend themselves by raising their sharp, barbed quills, also shown in Figure 2.

Figure 2 Functions of hair. This tiger's stripes and reddish fur help the tiger blend in with the surrounding grasses. Porcupines often use their quills for defense.

Evaluating the Insulation Value of Hair

QUICK LAB

When you are getting dressed on a cold day, why are you are more likely to choose a wool sweater than a cotton one? In this lab you will compare the insulating abilities of the animal fiber wool and the vegetable fiber cotton.

Materials

MBL or CBL system with appropriate software, temperature probe, 1 wool sock, beaker of ice, graph paper (optional), 1 cotton sock

Procedure

1. Set up an MBL/CBL system to collect data from a temperature probe at 6-second intervals for 100 data points.

2. Find and record the room temperature.

3. Insert the end of the probe into one thickness of wool sock. Then place the sock-covered probe into a beaker of ice. Collect temperature data for 10 minutes.

4. View the graph of your data. If possible, print out the graph. Otherwise, plot the graph on graph paper.

5. Repeat steps 1–4 using a cotton sock and fresh ice.

Analysis

1. **Analyze** your data and determine which sock was the better insulator.

2. **Summarize** why these results are of importance to mammals.

Mammalian Teeth

Unlike some other vertebrates whose teeth are constantly being lost and replaced, mammals usually have only two sets of teeth. The first set, commonly called baby teeth or milk teeth, is replaced by permanent teeth, which are not replaced if lost or damaged. Animals use their teeth in a variety of ways—to secure and chew food, for protection or as a threat signal, and to perform tasks, as when a beaver cuts down trees to make a dam.

In most mammals, four types of teeth can be recognized: incisors, canines, premolars, and molars. Each type of tooth performs a different function in eating. Incisors, the front teeth, are for biting and cutting. Behind them are canines used for stabbing and holding. Lining the jaw are the premolars and molars. As a mammal chews, its upper and lower molars fit together, crushing and grinding the food more thoroughly than a reptile's teeth can. The resulting smaller bits of food can be quickly digested, which permits a mammal to eat enough food to fuel its endothermic metabolism.

A mammal's teeth are specialized for the food it eats, and it is usually possible to determine a mammal's diet by examining its teeth. **Figure 3** shows the differences between the teeth of a coyote (a carnivore) and those of a deer (a herbivore). The coyote has long canine teeth that are suited for grasping prey, and its sharp molars can cut off pieces of flesh. In contrast, the deer's canines are small, and it uses its incisors to nip off selected pieces of plant material. The deer's premolars and molars are flat and covered with ridges that form a surface on which plant material can be ground.

Figure 3 Specialized teeth

A mammal's teeth provide clues about its diet.

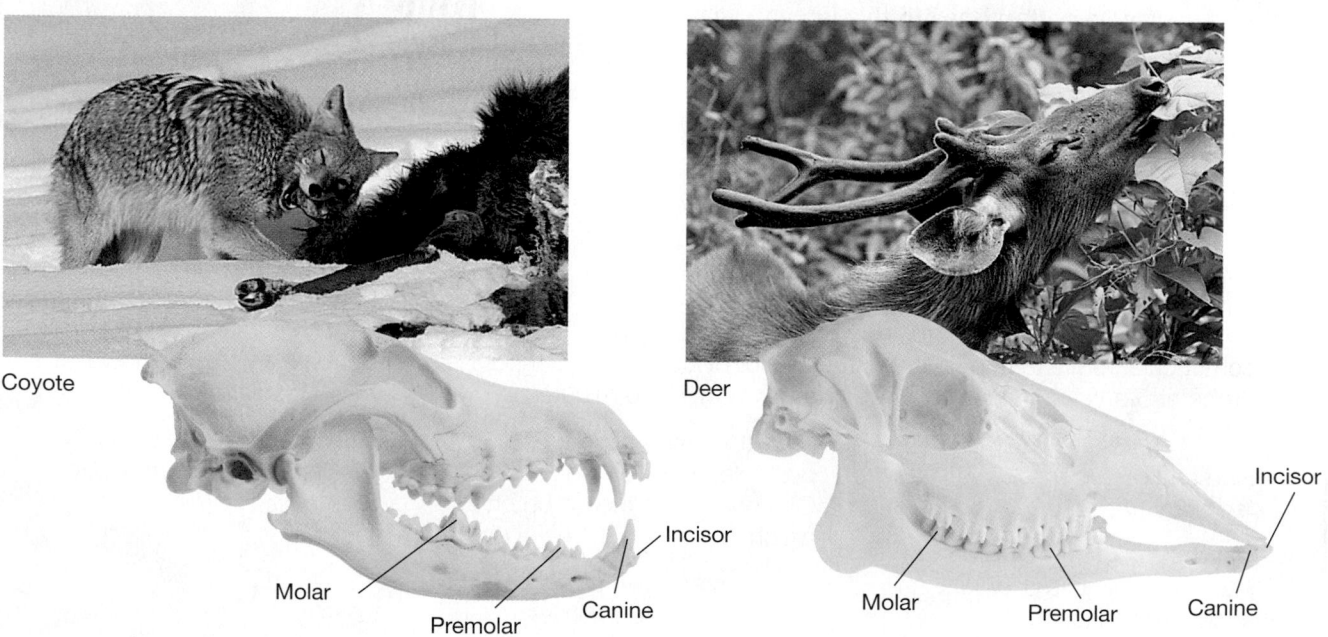

Body Temperature

Like birds, mammals are endotherms, generating heat internally through the rapid metabolism of food. Because a mammal's body temperature remains relatively constant regardless of the temperature of its surroundings, mammals can be active at any time of day or night. They also can live in very cold climates, where most ectothermic amphibians and reptiles cannot. In addition, endothermic metabolism permits mammals to sustain activities that require high levels of energy, such as running or flying long distances.

To maintain the high metabolic rate required by an endotherm, a mammal must eat about 10 times as much food as an ectotherm of similar size. Metabolizing this food requires a considerable amount of oxygen. Mammals, like birds, have respiratory and circulatory systems that are very efficient at acquiring and distributing oxygen.

Respiratory System

Mammalian lungs, shown in **Figure 4,** have a large internal surface area that aids the exchange of oxygen and carbon dioxide. Thus, mammalian lungs are much more efficient at obtaining oxygen from the air than are reptilian and amphibian lungs. Respiration in mammals is aided by the diaphragm, a sheet of muscle that separates the chest cavity from the abdominal cavity. When the diaphragm contracts, the chest cavity enlarges, drawing air into the lungs.

The lungs of mammals contain small, grape-shaped chambers called **alveoli** *(al VEE uh lie),* (singular, *alveolus),* also shown in Figure 4. Alveoli provide a very large respiratory surface area. In more active mammals, the alveoli are smaller and more numerous, further increasing the surface area for diffusion.

Heart and Circulatory System

Like crocodiles and birds, mammals have a four-chambered heart with a septum that completely divides the ventricle. The division of the ventricle creates two pumping chambers, one for each loop of the mammal's double-loop circulatory system. One chamber pumps oxygen-rich blood to the body, while the other pumps oxygen-poor blood to the lungs. Because the two do not mix, only oxygen-rich blood is delivered to the tissues, a condition vital for meeting the oxygen needs of endotherms.

You will learn more about the mammalian respiratory and circulatory systems later in this book. You can read more about the anatomy of one mammal in *Up Close: Grizzly Bear* on the following two pages.

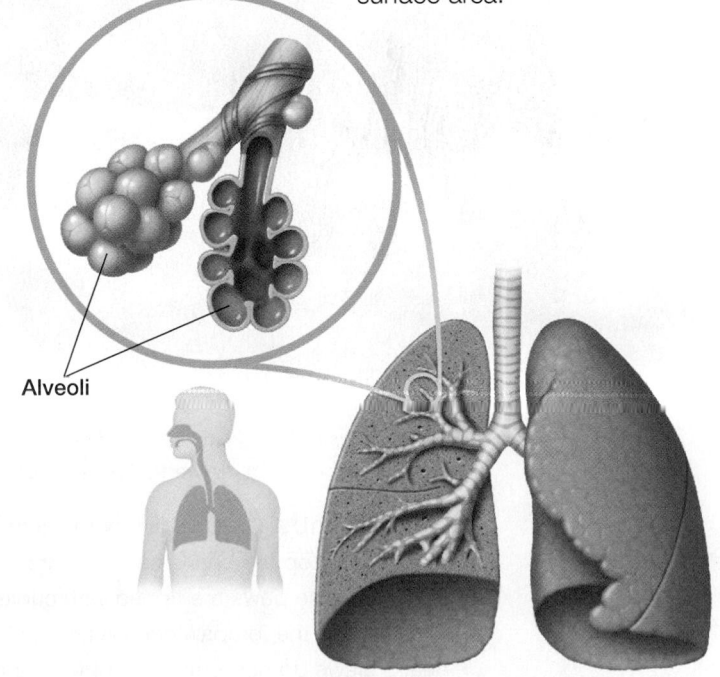

Figure 4 Mammalian lungs. The lungs of mammals contain clusters of alveoli that provide a very large internal surface area.

Alveoli

Grizzly Bear

Female with cub

- **Scientific name:** *Ursus arctos horribilis*
- **Size:** Males average 160 kg (350 lb) and can reach 1.2 m (4 ft) at the shoulder; females are smaller
- **Range:** Alaska and western Canada, with small populations in Washington, Idaho, Montana, and Wyoming
- **Habitat:** Tundra and mountainous forests and meadows
- **Diet:** Omnivorous; primary diet is vegetation; hunts insects, small mammals, and fish; eats carrion

External Structures

Fur Thick fur—ranging from yellowish-brown to black—covers the body. The name "grizzly" comes from silver-tipped hairs that are often sprinkled over the bear's head and back.

Senses Grizzlies have good hearing but relatively poor eyesight. They rely primarily on their excellent sense of smell to follow an odor trail or catch the scent of distant food.

▲ Fur

▲ Ear ▲ Eye

▼ Powerful limbs

Forepaw

Hind paw

Strong limbs Grizzlies are extremely strong and have great endurance. The hump atop the bear's back is a knot of strong muscles that power the forelimbs. The paws are tipped with curved claws up to 10 cm (4 in.) long. One swat of the forepaw can kill an adult moose or elk. Unlike a cat's, the bear's claws do not retract, and they are not adapted for climbing.

Internal Structures

Brain Like most mammals, the bear has a well-developed cerebrum, the portion of the brain where higher mental functioning occurs. The cerebellum, a center for motor coordination, is also large and connects directly to the portions of the cerebrum that govern motor activity.

Skull The long skull protects the bear's brain and serves as an anchor for the strong jaw muscles. Molars at the back of the jaw are rounded and have a wrinkled surface that is used for grinding up tough grasses and leaves.

Fat layer Grizzlies snooze away the winter in underground dens. During this time, the bear's metabolism slows, and its heart rate and breathing rate decrease. It does not eat or drink, obtaining all of its energy from a thick layer of stored fat.

Reproductive system Like all placental mammals, grizzlies nourish their embryos through a placenta. Mating occurs from May to June, but the fertilized eggs are not implanted in the uterus until late fall. Females reproduce every 2–4 years. One to four cubs about the size of a rat are born in late winter. The cubs suckle their mother's rich milk, and by the time they emerge from the den in spring, their weight may have increased 20-fold. Cubs usually remain with their mother at least 2 years.

Digestive system Although they eat large amounts of plant material, bears have no specialized structures, such as a multi-chambered stomach, for digesting cellulose. However, a bear's small and large intestines are relatively long, which helps break down hard-to-digest plant material. Bacteria in the large intestine also contribute to the digestion of plants.

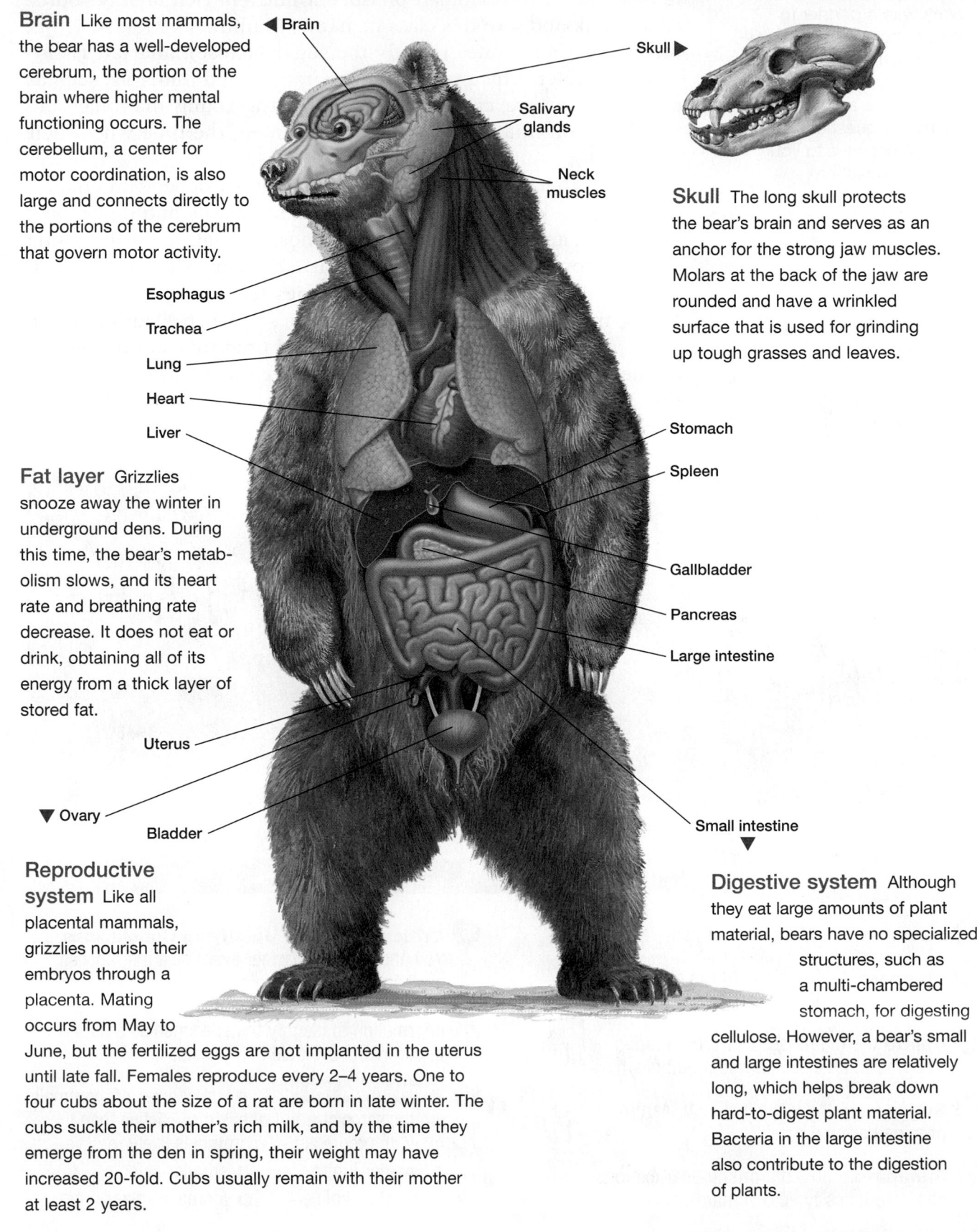

◄ Brain
Skull ►
Salivary glands
Neck muscles
Esophagus
Trachea
Lung
Heart
Liver
Stomach
Spleen
Gallbladder
Pancreas
Large intestine
Uterus
▼ Ovary
Bladder
Small intestine ▼

Parental Care

Study TIP

● **Reviewing Information**

Work with a partner to review the four characteristics of mammals. For each characteristic, both partners should write a question. Exchange questions, and find the answers to your partner's questions.

Mammals are unique among the vertebrates in the way that they nourish their young after birth. **Mammary glands** located on the female's chest or abdomen produce a nutrient-rich energy source called milk and give this class its name, Mammalia. Milk is rich in protein, carbohydrates (chiefly the sugar lactose), and fat. It also contains water, which prevents dehydration, and minerals, such as calcium, that are critical to early growth. Young mammals are nourished on milk from birth until **weaning,** the time when their mother stops nursing them.

Unlike other vertebrates, young mammals are dependent on their mother for a relatively long period, receiving milk and other food, protection, and shelter from her. For most animals, once the young can fend for themselves, the mother leaves them. The participation of the father in raising the young varies from species to species. Young mammals, such as the sea otter shown in **Figure 5,** often learn necessary survival skills during the time they are dependent on their mother. For primates, such as the chimpanzees in Figure 5, this early learning is especially important, and primates have the longest period of parental dependency of all mammals.

Figure 5 Parental care.
This young otter will spend 6 to 8 months with its mother. In contrast, this baby chimp will remain close to its mother for several years.

Section 1 Review

1. **Describe** three functions of hair.

2. **Compare** the functions of the different types of mammalian teeth.

3. **Relate** the mammal's heart and respiratory systems to its endothermic metabolism.

4. **Summarize** the ways in which mammals provide parental care.

5. **Summarize** how the grizzly bear exhibits characteristics typical of mammals.

6. **Critical Thinking Justifying Conclusions**
You and your lab partner examine a mammalian skull and jaw that contains only incisor teeth. Your partner concludes that you do not have enough information to identify the specimen as a herbivore or carnivore. Evaluate this conclusion.

7. **Standardized Test Prep** If two species of mammals are the same size but one is more active than the other, the more active mammal is likely to have
   A smaller lungs.   **C** smaller alveoli.
   B fewer alveoli.   **D** a smaller diaphragm.

Today's Mammals

Mammalian Diversity

Cave paintings, some perhaps 40,000 years old, show that human life has long been intertwined with the lives of animals. The vast majority of the animals shown in ancient cave paintings are mammals. While some of the mammals represented are extinct, the descendents of many (horses, hyenas, rhinoceroses, panthers, bison, and lions) still live. They are a part of the astounding array of mammals that share our planet with us. In their diversity of size, anatomy, and habitat, mammals surpass all other vertebrate groups. Mammals range in size from tiny shrews that weigh about 1.5 g (less than 0.1 oz) to gigantic blue whales that can weigh up to 136,000 kg (150 tons).

Consider also the differences between a bat and a whale, compared in **Figure 6.** These two mammals are adapted to live in very different environments. Bats fly and are active primarily at night, while whales are permanently aquatic. However, both groups face a similar challenge—how to navigate in an environment where visibility is often limited. Bats and some whales have a similar solution to this problem: they use echolocation, which works something like the sonar of a ship. In echolocation, animals emit high-frequency sound waves. As the waves travel, they strike objects in the environment, and a portion of each wave is reflected back to the animal. The brain interprets the reflected wave, or echo, revealing the object's size and location.

Objectives

- **Recognize** how mammals are adapted to different environments.
- **Compare** reproductive patterns in monotremes, marsupials, and placental mammals.
- **Relate** the distribution of monotremes and marsupials to the breakup of Pangaea.

Key Terms

placenta
gestation period
ungulate
cud

Comparison of Bats and Whales

Bats
- Forelimbs modified into wings and covered with leathery skin
- Body covered with hair
- Active at night; use echolocation to navigate

Whales
- Forelimbs flattened and paddle-shaped; no hind limbs
- Nearly hairless, streamlined body
- Communicate with sound; some use echolocation

Figure 6 Comparison of bats and whales. Although both are mammals, bats and whales are adapted to live in different environments.

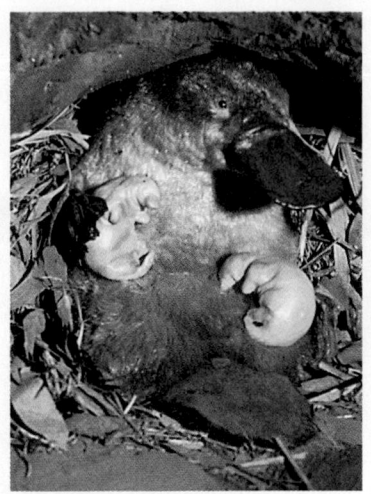

Figure 7   Platypus.
Although these platypus young hatched from eggs, they drink milk produced by their mother's mammary glands.

Reproduction

All mammals reproduce by internal fertilization: The male releases sperm into the female's reproductive tract where one or more eggs are fertilized. But there is some variation among mammals in how and where the fertilized egg develops. Present-day mammals are divided into three groups—monotremes, marsupials, and placental mammals—based on their pattern of development.

Monotremes

The monotremes, the most primitive of the mammals, are represented by only one order, Monotremata. All modern monotremes live in Australia, New Guinea, and South America. This distribution may be the result of the movement of these landmasses by continental drift.

The three living monotreme species—the duckbill platypus and two species of echidnas *(ee KIHD nuhz)*, or spiny anteaters—show a curious mix of characteristics. They have shoulders and forelimbs that are quite reptilian in appearance. Among living mammals, only monotremes reproduce by laying eggs. Their eggs, like reptile eggs, have leathery shells. Like birds, the female monotreme incubates her eggs with her body heat, and at hatching, the newborns are only partially developed. Also like birds and reptiles, monotremes have a cloaca, a common passageway for the digestive, reproductive, and urinary systems. No other mammals have a cloaca. Unlike other mammals, adult monotremes do not have true teeth. Because of these characteristics, monotremes may more closely resemble early mammals than do many other living mammal species.

Two features monotremes share with other mammals are hair and milk production. Unlike other mammals, however, a female monotreme does not have nipples and young monotremes do not nurse. Instead, the young lap up milk that oozes from glands located on their mother's belly.

The platypus, shown in **Figure 7,** inhabits lakes and streams in eastern and southern Australia. Its broad, flat tail and webbed front feet make it an excellent swimmer. The platypus uses its flat bill to probe for crustaceans, worms, and snails. Echidnas, shown in **Figure 8,** are found in parts of Australia and in New Guinea. They are terrestrial and have very strong, sharp claws and an elongated, beaklike snout which they use for burrowing and for digging out insects and other invertebrates.

Figure 8   Echidna. Echidnas are covered with sharp, barbless quills. They use their powerful claws and long, beaklike snout to dig for their main food, termites and ants.

Marsupials

Order Marsupialia includes not just the well-known kangaroos, but wombats, wallaroos, koalas, and the opossums shown in **Figure 9**. In marsupial mammals the young are born only days or weeks after fertilization—tiny and incompletely developed except for their front limbs. Without any parental help, the newborns crawl to their mother's nipples, usually located in a pouch (the marsupium) on her abdomen. There each newborn attaches itself to a nipple and continues its growth and development for several months. When the young marsupials are able to function on their own, they emerge from the pouch, although they continue to return there to nurse.

Today, many marsupial species are found in the Australian region—Australia, New Guinea, and a few nearby islands—where about half of mammalian species are marsupials. This limited distribution is the result of the breakup of Pangaea. About 70 million years ago, the Australian region separated from the continents of Antarctica and South America. As placental mammals had not yet reached the Australian region, the marsupials there developed in isolation.

Figure 9   Virginia opossum. At birth, these four young opossums looked like the pink newborn opossum, which is smaller than your thumb.

BIOWatch
Selecting Dairy Products

Unlike all other mammals, humans continue to drink milk (and eat products made from it) far into adulthood. Today, most of our dairy products are made from cow's milk, although goat milk products are becoming increasingly popular. In other parts of the world, milk from other mammals, such as water buffalo, sheep, horses, and even reindeer, is used as a food source.

Next time you are in the dairy section of a supermarket, take a look at what is available. Just deciding which milk to purchase—whole, skimmed, 2 percent, lactose reduced, flavored—may take some time. And then there is the array of yogurts, kefirs, cottage cheeses, hard cheeses, and string cheeses.

Read the Label

Although dairy products are nutritious and an excellent source of calcium, some are high in fat. This does not mean that you should never eat them, but you can make informed decisions about how much to eat. Begin by locating the information on the label that tells you the product's per-serving calorie and fat content.

Once you've made a purchasing decision, be certain to check the expiration date on the packaging. Consume perishable dairy products while they are still fresh. Dairy products require refrigeration, so do your other shopping first, and pick up perishable items last.

internet connect

www.scilinks.org
Topic: Dairy Products
Keyword: HX4056

SCILINKS. Maintained by the National Science Teachers Association

Placental Mammals

The young of placental mammals develop within the female's uterus, where they are nourished by nutrients from her blood. An organ called the **placenta** allows the diffusion of nutrients and oxygen from the mother's blood, across placental membranes, and into the blood of the developing fetus. Waste materials from the fetus diffuse in the opposite direction and are eliminated by the mother's excretory system.

The period of time between fertilization and birth is called the **gestation** *(jeh STAY shuhn)* **period.** Most placental mammals have a longer gestation period than marsupial mammals do, and their young are more completely developed at birth. Some placental mammals, like the foal in **Figure 10,** can stand and walk within a few hours of birth. Others, like the rabbits in Figure 10, are born blind, deaf, and helpless.

Placental mammals vary greatly in size, shape, diet, and habits. They live in a variety of habitats, from hot, moist rainforests to the frigid tundra. Although mammals share many similarities, species have different characteristics. For example, some placental mammals are adapted for running, leaping, swimming, or flying.

Alone among vertebrates, some placental mammals have hooves, horns, or antlers. Hooves, such as the zebra hoof shown in **Figure 11,** are specialized pads that cover the toes of many running mammals. Hooves are made of keratin, a versatile protein that is a component of many mammalian structures. The horns of sheep, cattle, and antelopes are composed of a core of bone surrounded by a sheath of keratin. This bony core is firmly attached to the skull, and the horn is never shed. (The horn of a rhinoceros is composed not of bone but of hairlike fibers of keratin that form a hard structure.)

Other placental mammals, such as deer and elk, grow and shed a set of antlers each year. Antlers, grown only by the male, are composed of bone but are not covered by a keratin sheath. While they

Figure 10   Young mammals

Newborn foals are on their feet nursing within a couple of hours, while young rabbits do not even open their eyes until they are about 10 days old.

Mare and foal

Baby rabbits

are growing, the antlers are covered by a thin layer of soft skin called velvet. When the antlers are fully grown, the velvet dries up and comes off. The male uses his antlers during breeding season to attract females and to combat other males. After the breeding season is over, the antlers are shed, and a new pair grows the next year. Because the male grows a larger pair of antlers each year, antler size gives an indication of a male deer's age.

Domestic Animals Domestic animals are animals that have been kept and bred by people for special purposes. Most domestic animals are placental mammals whose association with humans dates back at least 2,000 years. These animals include dogs, cats, cattle, horses, donkeys, mules, rabbits, sheep, goats, pigs, camels, llamas, and alpacas.

Different breeds of domestic mammals have been developed through selective breeding. For example, some breeds of goats produce more milk than others. Their milk is used to produce a variety of dairy products. Other goats, such as angora goats, are bred for their fine hair, which is spun into yarn. Some domestic mammals are hybrids of two different species. Mules, for example, are the offspring of a female horse (mare) and a male donkey. Like most hybrids, mules are sterile. The mule's strength, endurance, and surefootedness make it valuable as a pack animal and for chores such as plowing.

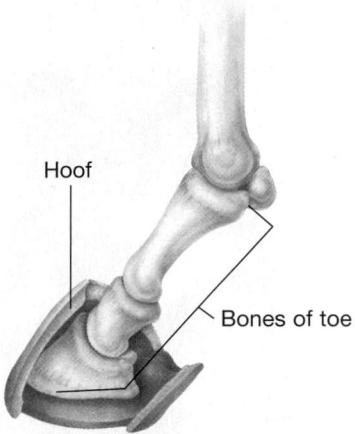

Figure 11 Hoof. A zebra's foot is modified so that the zebra walks on a single toe that is covered with a hoof made of keratin.

DATA LAB

Comparing Gestation Periods

Background

If you have ever raised gerbils or hamsters, you know that they can produce several litters of young each year. That's because they have a very short gestation period compared to other mammals. Use the table below to find out more about gestation periods.

Analysis

1. **Critical Thinking Recognizing Patterns** State a generalization about the relationship between the length of the gestation period and the number of offspring per pregnancy.

2. **Critical Thinking Forming Hypotheses** Propose a hypothesis to explain this relationship.

3. **Recommend** a way that the information in the table could be rearranged to show this relationship more clearly.

Gestation Periods in Mammals		
Mammal	**Gestation period**	**Offspring per pregnancy**
Vampire bat	210 days	1
Gerbil *	19–21 days	4–7
Human	about 265 days	1
Quarter horse	332–342 days	1
Black spider monkey	226–232 days	1
Grey squirrel	44 days usually	3
Rabbit *	about 31 days	3–6
Sperm whale	420–430 days	1
Arctic wolf	63 days	4–5

*More than two litters per year

Figure 12 Mouse

Figure 13 Bat

Figure 14 Shrew

Modern Placental Mammals

There are 19 orders of placental mammals, which include more than 90 percent of all mammal species. Terrestrial placental mammals inhabit all continents except Antarctica, and aquatic placental mammals inhabit all oceans. The following 12 orders contain the vast majority of mammal species.

Order ❶ Order Rodentia

Over 40 percent of all placental mammals are gnawing mammals called *rodents*. Rodents are distinguished from most other mammals by their teeth, which are specialized for gnawing. All rodents have two pairs of large, curving incisor teeth that grow continuously. As the rodent gnaws, the back side of the incisors wears away faster than the front, creating a sharp chisel edge on the teeth. The success of this group of herbivores can be attributed to their intelligence, small size, and rapid rate of reproduction. Most rodents, such as the mouse shown in **Figure 12,** are small, but a few are larger. For example, beavers can weigh between 18 and 43 kg (between 40 and 95 lb).

Order ❷ Order Chiroptera

This order is composed of bats, the only mammals capable of true flight. A bat's front limbs are modified into wings. The thumb, which is not attached to the wing, has a curved claw that can be used for clinging or grasping. Bats generally live in groups and are active only at night. During the day, bats hang upside down in caves or some other protected place, as shown in **Figure 13.** Most bats are carnivorous, using echolocation to find insects, which they catch while in flight. Other bats eat fruit or nectar from night-blooming flowers.

Order ❸ Order Insectivora

Insectivores are the mammals most similar to the ancestors of the placental mammals. As their name implies, these small mammals eat mainly insects, but their diet may also include fruit, small birds, and snakes, as well as other insectivores. Insectivores have an enormous appetite, and some, such as the shrew shown in **Figure 14,** must eat more than two times their body weight daily to fuel their extremely high metabolic rate.

Order ❹ Order Carnivora

Some of the best-known animals are the flesh-eating hunters called *carnivores*. Scientists generally divide this order into two subgroups: the cat family and the dog family, represented by the fox shown in **Figure 15.** Carnivores are strong and extremely intelligent, and they have keen senses of smell, vision, and hearing—characteristics that have enabled them to become successful hunters. Their long canine teeth are specialized for capturing prey and tearing flesh. Some members of this order are no longer carnivorous: raccoons and bears are omnivores, while pandas are herbivores.

Figure 15   Fox

Order ❺ Order Pinnipedia

This order of marine carnivores includes seals and sea lions that feed at sea but return to land to mate, rear their young, and rest. All four limbs are modified as flippers for swimming, and their bodies are streamlined for rapid movement through the water. An insulating layer of blubber protects them from the cold ocean waters. Most species live in large colonies called rookeries which are headed by a large male, such as the walrus shown in **Figure 16.**

Figure 16   Walrus

Order ❻ Order Primates

Humans belong to the order Primates. Other members of this order include the prosimians, which are active at night, and monkeys and apes, which are active during the day. Most nonhuman primates, like the monkey in **Figure 17,** are tree dwellers, and many of their characteristic features are adaptations for living in trees. Flexible, grasping hands and feet aid tree-dwelling primates in climbing. Monkeys and prosimians have a tail that is used for balance. The excellent depth perception of the primates is critical for those that live in the trees. Primates are extremely curious and their ability to learn is exceptional.

Figure 17   Monkey

Figure 18 Sheep

Figure 19 Rhinoceroses

Figure 20 Orca

Order ❼ Order Artiodactyla

Mammals belonging to this order and the following order, Perissodactyla, are classified as **ungulates** (*UNG gyoo lits*), mammals with hoofs. Ungulates walk, not on their entire foot as most vertebrates do, but with their weight supported by their hoof-covered toes. Most ungulates are herbivores that live together in herds. The young are well developed at birth and can move along with the herd within a day or two. An ungulate's main means of defense are the security of the herd and the ability to run very fast when danger approaches.

Artiodactyls have an *even* number of toes within their hooves. Many of these mammals have a stomach with a storage chamber called a *rumen*. Microbes in the rumen break down the cellulose in the plant material artiodactyls eat. Mammals with a rumen regurgitate partly digested food, called **cud,** rechew it, and swallow it again for further digestion. This order includes pigs, hippopotamuses, camels, deer, antelope, cattle, goats, and giraffes as well as sheep, such as the one shown in **Figure 18.**

Order ❽ Order Perissodactyla

Ungulates with an *odd* number of toes within their hooves are classified as perissodactyls. This order includes horses, zebras, tapirs, and rhinoceroses, such as the two shown in **Figure 19.**

Perissodactyls do not chew their cud. Instead of a rumen, they have a cecum, a pouch branching from their large intestine. The cecum contains microbes that digest the cellulose in their diet. Perissodactyls are far less numerous than artiodactyls.

Order ❾ Order Cetacea

Cetaceans are divided into two groups: the predatory toothed whales, dolphins, and porpoises, and the filter-feeding baleen whales. Whales, such as the orca shown in **Figure 20,** are probably descendants of land mammals that returned to the sea about 50 million years ago and have adapted to a fully aquatic life. Their streamlined bodies have front limbs modified into flippers, no hind limbs, and a broad, flat tail for swimming. A nostril called a blowhole is located on top of their head. Cetaceans are very intelligent animals that communicate with each other by making sounds that we hear as a series of clicks.

Order ⑩ Order Lagomorpha

This order is composed of rabbits and hares. Like rodents, they have one pair of long, continually growing incisors, but they also have an additional pair of peg-like incisors that grow just behind the front pair. Rabbits and hares have long hind legs and are specialized for hopping, as shown in **Figure 21.** Rabbits build nests that the female lines with fur. The young are born furless and their eyes are closed. Hares do not construct nests. Their young are born with fur and their eyes are open.

Order ⑪ Order Sirenia

These somewhat barrel-shaped marine animals include the dugongs and manatees, shown in **Figure 22.** Like whales, they have front limbs modified as flippers and no hind limbs. A flattened tail is used for propulsion through the tropical oceans, estuaries, and rivers where sirenians live, grazing on aquatic plants. Despite their appearance and habitat, sirenians are closely related to elephants and are often called *sea cows*.

Order ⑫ Order Proboscidea

The two living species of this order, the African elephant and the Indian elephant shown in **Figure 23,** are the largest land animals alive today. Their long, boneless trunk is really an elongated nose and upper lip and is used for a variety of tasks. An elephant's upper incisor teeth are modified into long ivory tusks. Elephants live in herds made up of a dominant male, a number of females, and young of varying ages.

Figure 21   Rabbit

Figure 22   Manatees

Figure 23   Indian elephant

The seven remaining orders of placental mammals are summarized in **Table 1.** These orders contain few species, but some are quite interesting. The hyrax, for example, looks as if it could be kin to rabbits or rodents. But traits such as its hooflike nails lead some biologists to think the hyrax is most closely related to elephants or perissodactyls.

Table 1 Orders of Placental Mammals

Order	Description
Edentata anteaters, armadillos	Toothless or with poorly developed teeth that lack enamel; found only in the Western Hemisphere
Macroscelidea elephant shrews	Ground-dwelling insect eaters; long, flexible snout; hop about somewhat like small kangaroos
Scandentia tree shrews	Omnivorous; small, squirrel-like mammals; long snout, sharp teeth; live mainly on the ground, despite their name
Pholidota pangolins (scaly anteaters)	Body covered with overlapping scales; no teeth; very long tongue for capturing ants
Hyracoidea hyraxes	Rabbitlike body; short ears; four hoofed toes on front feet; three hoofed toes on back feet
Dermoptera flying lemurs	Squirrel-like; glide on a sheet of skin stretching between their forelegs and hind legs
Tubulidentata aardvarks	Nocturnal; piglike body; big ears; long snout used to feed on ants and termites

Section 2 Review

1. **Compare** the reproductive patterns of monotremes, marsupials, and placental mammals.

2. **Relate** the location of modern-day marsupials to the breakup of Pangaea.

3. **Describe** how artiodactyls and perissodactyls are adapted for digesting plant material.

4. **Summarize** the ways in which aquatic mammals are adapted to life in the water.

5. **Standardized Test Prep** In which biome would you most likely find a mammal that uses baleen for filter feeding?

   A tundra   **C** deep ocean
   B savanna   **D** temperate grassland

CHAPTER HIGHLIGHTS

Key Concepts

1 The Mammalian Body

- Mammals are the only animals with hair. The primary function of hair is to insulate a mammal's body, though it can also provide camouflage or a clear signal through coloration, serve a sensory function, or be a defensive weapon.

- Mammals usually have two sets of teeth in their lifetime. Teeth in the second set are not replaced, even if lost or damaged.

- The four types of mammalian teeth are highly specialized: incisors are for biting and cutting; canines are for stabbing and holding; and premolars and molars crush and grind the food.

- Mammals are endotherms, generating heat internally through the rapid metabolism of food. Endothermy is made possible by highly efficient respiratory and circulatory systems.

- Mammals nurse their young with milk from the mammary glands of the female.

2 Today's Mammals

- In terms of anatomy and habitat, mammals are the most diverse of all vertebrate groups.

- All mammals reproduce by internal fertilization.

- Monotremes consist of three species that have a cloaca and lay eggs.

- Marsupials give birth to incompletely developed young that complete their development in the mother's pouch.

- Placental mammals nourish their unborn young in the uterus through the placenta.

- The different mammalian species have evolved a variety of adaptations that permit them to live in a wide range of habitats.

- Many aquatic mammals have a layer of blubber that insulates them from the cold. Their forelimbs are modified into flippers, and they have no hind limbs. A flattened tail aids in swimming.

- Ungulates have digestive systems modified for digesting cellulose. Even-toed ungulates chew their cud.

Key Terms

Section 1
hair (800)
alveolus (803)
mammary gland (806)
weaning (806)

Section 2
placenta (810)
gestation period (810)
ungulate (814)
cud (814)

Understanding Key Ideas

1. The primary function served by mammalian hair is
a. camouflage.
b. insulation.
c. defense.
d. sensory.

2. Which of the following is *not* a type of mammal tooth?
a. canine
b. baleen
c. molar
d. incisor

3. Endothermic metabolism permits mammals to do all of the following *except*
a. live in very cold climates.
b. generate heat internally.
c. run or fly for long distances.
d. eat less than ectotherms.

4. Care given by mammals to their young
a. begins after the young are weaned.
b. is similar to that given by reptiles.
c. involves nursing and teaching survival skills.
d. ends soon after the young are born.

5. Monotremes differ from marsupials in that monotremes
a. lay eggs.
b. are viviparous.
c. nourish unborn via the placenta.
d. do not have mammary glands.

6. Artiodactyls differ from perissodactyls in that artiodactyls
a. have a cecum.
b. do not have a rumen.
c. seldom live in herds.
d. have an even number of toes.

7. Which of the following is *not* a mammalian adaptation for aquatic living?
a. limbs modified as flippers
b. layer of blubber
c. streamlined body
d. keen sense of smell

8. Which of the following grizzly bear features is *not* typical of most mammals?
a. thick hair
b. specialized teeth
c. powerful claws
d. placental reproduction

9. **BIOWatch** What advice would you give to someone who wanted to include more dairy products in their diet?

10. **Concept Mapping** Make a concept map that describes the different methods of mammalian reproduction. Include the following terms in your map: *monotreme, marsupial, mammal, mammary gland, egg,* and *placenta.*

Critical Thinking

11. Drawing Conclusions How do hair and a high rate of metabolism help a mammal maintain homeostasis?

12. Applying Information A mammal must eat about 10 times as much food as an ectotherm of similar size. What role might the respiratory system play in this need?

13. Comparing Structures In what way are a mammal's hair and a bird's down feathers alike? How are they different?

14. Evaluating Conclusions Some mammal species must care for their young for many years before they reach maturity and can survive on their own. Can you conclude that all vertebrate young would benefit from this type of parental care? Explain why or why not.

Alternative Assessment

15. Recognizing Patterns Find out more about how plate tectonics can explain the pattern of distribution of mammals on Earth. Analyze and critique the theory, and write a report on your findings. Include a description of how the breakup of Pangaea led to the predominance of placental mammals on all the continents except Australia.

Standardized Test Prep

Understanding Concepts

Directions (1–4): **For *each* question, write on a separate sheet of paper the letter of the correct answer.**

1 Clusters of what structures greatly increase the respiratory surface of the lungs?
   A. alveoli
   B. mammary glands
   C. placentas
   D. ungulates

2 After what stage does a young animal no longer nurse?
   F. gestation
   G. hibernation
   H. insulating
   I. weaning

3 What term describes an animal with hooves?
   A. cetacean
   B. marsupial
   C. monotreme
   D. ungulate

4 Which statement describes the function of the mammalian heart?
   F. Two pumping chambers supply oxygen-rich blood to the lungs.
   G. Four pumping chambers supply oxygen-poor blood to the lungs.
   H. One pumping chamber supplies blood to the lungs and body organs.
   I. One pumping chamber supplies oxygen-poor blood to the lungs, and one pumping chamber supplies oxygen-rich blood to the body organs.

Directions (5): **For the following question, write a short response.**

5 What function does the placenta serve in placental mammals?

Test TIP

Before looking at the answer choices for a question, try to answer the question yourself.

Reading Skills

Directions (6): **Read the passage below. Then answer the question.**

   Mammals are the only animals with hair, but not all mammalian hair serves the same function. Hair serves some unusual functions in some mammals. For example, some mammals have specialized hairs that act as defensive weapons. Porcupine quills are stiff, sharp hair structures that can injure an attacker. Rhinoceros horns are formed from hair. Though biologists are not certain of the horn's function, mother rhinoceroses have been known to use their horns to protect their young from predators.

6 What is the primary function of hair?
   A. It attracts a mate.
   B. It insulates the body.
   C. It has a sensory function.
   D. It acts as a defensive weapon.

Interpreting Graphics

Directions (7): **Base your answer to question 7 on the diagram below.**

Anatomy of a Bear

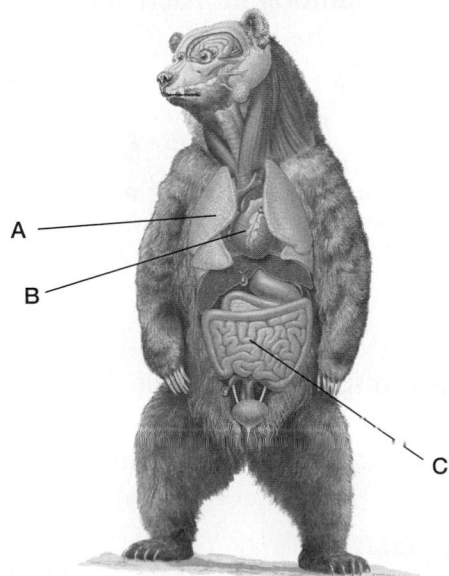

7 Which organisms in structure *C* contribute to the digestion of plant material?
   F. algae   **H.** tunicates
   G. bacteria   **I.** yeasts

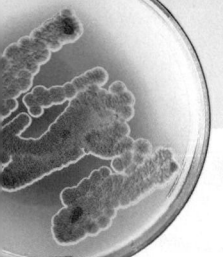

Exploration Lab

Exploring Mammalian Characteristics

SKILLS
- Observing
- Drawing
- Inferring

OBJECTIVES
- **Examine** distinguishing characteristics of mammals.
- **Infer** the functions of mammalian structures.

MATERIALS
- hand lens or stereomicroscope
- prepared slide of mammalian skin
- compound microscope
- mirror
- specimens or pictures of vertebrate skulls (some mammalian, some non-mammalian)

Magnification: 28×

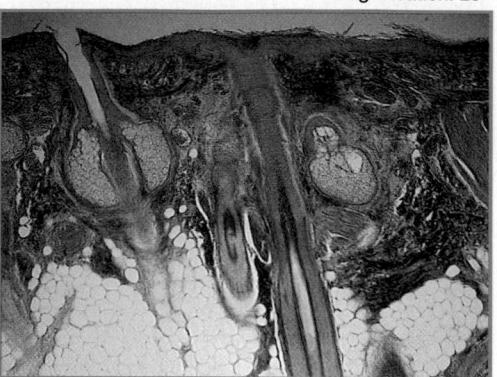

Mammalian skin (human scalp)

Before You Begin

Mammals are vertebrates with **hair, mammary glands,** a single lower jawbone, and specialized teeth. Other characteristics of mammals include **endothermy** and a four-chambered heart. Mammals also have **oil (sebaceous) glands** in their skin, and most have **sweat glands.** In this lab, you will examine some of the characteristics of mammals that distinguish them from other vertebrates.

1. Write a definition for each boldface term in the paragraph above.

2. Make a data table similar to the one below.

DATA TABLE				
Mammal	Incisors	Canines	Premolars	Molars
Human				

3. Based on the objectives for this lab, write a question you would like to explore about the characteristics of mammals.

Procedure

PART A: Examining Mammalian Skin

1. Use a hand lens to look at several areas of your skin, including areas that appear to be hairless. Record your observations.

2. Look at a prepared slide of mammalian skin under low power of a compound microscope. Notice the glands in the skin. Look for the oil (sebaceous) glands and the sweat glands. Draw and label an example of each type of gland.

PART B: Examining Mammalian Teeth and Skulls

3. Wash your hands thoroughly with soap and water. Use a mirror to look in your mouth. Identify the four kinds of mammalian teeth you see.

4. Count each kind of tooth on one side of your lower jaw. Multiply the number of each kind of tooth by four, and record these numbers in the appropriate columns of your data table. Wash your hands again before continuing.

5. Look at the skulls of several mammals. Identify the kinds of teeth in each skull. For each skull, find the number of each kind of tooth as you did in step 4.

6. Look at the skulls of several nonmammalian vertebrates, and compare nonmammalian teeth to mammalian teeth.

7. Compare the jaws of mammalian skulls to those of nonmammalian vertebrates. As you look at each skull, notice the structure

of the lower jawbone and how the upper jaw-bone and the lower jawbone connect.

PART C: Cleanup and Disposal

8. Dispose of broken glass in the waste container designated by your teacher.

9. Clean up your work area and all lab equipment. Return lab equipment to its proper place. Wash your hands thoroughly before you leave the lab and after you finish all work.

Analyze and Conclude

1. **Summarizing Information** List the characteristics that distinguish mammals from other vertebrates.

2. **Interpreting Graphics** Compare the amount of hair on humans to that on the mammals shown in the photographs below.

3. **Inferring Relationships** What role, if any, might hair or fur play in enabling mammals to be endotherms?

4. **Forming Hypotheses** Besides the role of hair you identified in item 3 above, what other roles do you think hair might play in mammals?

5. **Recognizing Patterns** Where are the oil (sebaceous) glands located in the skin of mammals?

6. **Forming Hypotheses** Do you think the mammals in the photos on the left below have more sweat glands or fewer sweat glands than humans? Explain.

7. **Comparing Structures** How is the mammalian jaw different from nonmammalian jaws?

8. **Inferring Conclusions** Based on the shape of your teeth, would you classify humans as carnivores (meat eaters), herbivores (plant eaters), or omnivores (meat and plant eaters)? Explain.

9. **Evaluating Conclusions** Justify the following conclusion: The kinds and shapes of a mammal's teeth can be used to determine its diet.

10. **Further Inquiry** Write a new question about the characteristics of mammals that could be explored with a new investigation.

On the Job

Comparative anatomy is the study of the anatomical similarities and differences between organisms. Do research to discover how comparative anatomy is used to hypothesize the relationships among the animal phyla. For more about careers, visit **go.hrw.com** and type in the keyword **HX4 Careers.**

Great egret and chicks

36 Animal Behavior

 ## Quick Review

Answer the following without referring to earlier sections of your book.

1. **Describe** the relationship of genes to inherited traits. *(Chapter 30, Section 1)*
2. **Relate** natural selection to adaptation. *(Chapter 32, Section 2)*

Did you have difficulty? *For help, review the sections indicated.*

 ## Reading Activity

Before you read this chapter, write a short list of all of the things you know about animal behavior. Then write a list of the things that you want to know about animal behavior. Save your list, and to assess what you have learned, see how many of your own questions you can answer after reading this chapter.

internet connect

www.scilinks.org
National Science Teachers Association *sci*LINKS Internet resources are located throughout this chapter.

SC*I*LINKS® **Maintained by the National Science Teachers Association**

● The great egret, *Casmerodius albus,* lays from three to five eggs. Young birds may stay with their parents for more than six weeks.

Evolution of Behavior

Objectives

- **Distinguish** between "how" and "why" questions about behavior.
- **Describe** how natural selection shapes behavior.
- **Compare** innate and learned behaviors.
- **Summarize** how behavior is influenced by both heredity and learning.

Key Terms

behavior
innate behavior
fixed action pattern
  behavior
learning
conditioning
reasoning
imprinting

What Is Behavior?

A squirrel buries a nut; a hungry baby cries. A frog jumps into a pond to avoid a predator, and a driver applies the brakes when approaching a red traffic light. These are all examples of behavior. A **behavior** is an action or series of actions performed by an animal in response to a stimulus. The stimulus might be something in the environment, such as a sound, a smell, a color, or another individual. The stimulus can also be related to the internal state of the animal, such as being hungry or cold. For example, when under threat, the lizard shown in **Figure 1** flares out the folds of skin around its head. This gives it a threatening appearance, which tells potential enemies to stay away.

Scientists studying behavior investigate two kinds of questions—"how" questions and "why" questions. "How" questions are about how a behavior is triggered, controlled, and performed. For instance, consider the squirrel, also shown in Figure 1. "How" questions about squirrel behavior might include "How does a squirrel select which nuts to bury?" "How does it choose where to bury the nut?" and "How does it remember where the nut is?"

However, answering "how" questions provides only a partial understanding of a behavior. Scientists also try to answer "why" questions, such as "Why do squirrels bury nuts?" "Why" questions concern the reasons a behavior exists and are really questions about the evolution of behavior. A study of animal behavior may therefore seek to identify the benefits of a particular behavior.

Figure 1 Animal behavior. This Australian frilled lizard and European red squirrel are engaging in behaviors typical of their species.

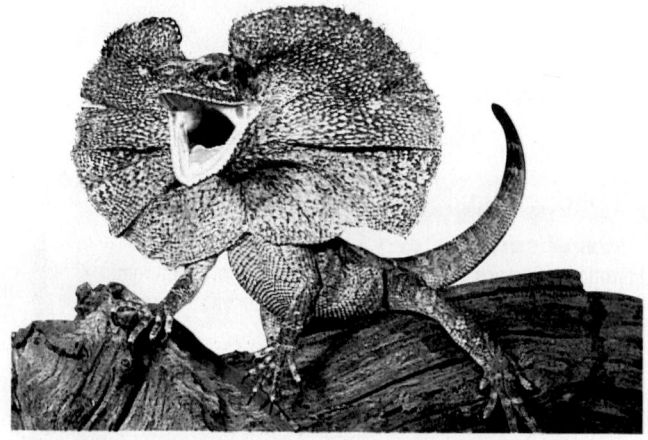

Natural Selection and Behavior

Recall that evolution by natural selection is a process by which populations change in response to their environment. Natural selection favors traits that improve the likelihood that an individual will survive to reproduce. Over time, traits that provide a reproductive advantage become more common. Traits that do not provide an advantage become less common and may disappear.

An understanding of natural selection can help answer a "why" question. A good example of this is seen in East African lions, which live in small groups called *prides*. Each pride contains several adult females, several young lions (called cubs), and one or more adult males. The adult males father all the cubs and defend the pride against other males. But a male or group of males usually can control a pride for only a couple of years. Then they are forced out by younger males who take over the pride. When this happens, the new males often kill all the young cubs in the pride, as shown in **Figure 2.** In contrast, male lions are usually quite tolerant of their own offspring, also shown in Figure 2.

To understand this behavior, we need to understand how the new males would benefit from it. The new males will control the pride for only a few years, so they have a very short time in which to reproduce. But female lions with cubs will not breed until their cubs are grown, which may take more than two years. If a female's cubs die, however, she will mate again almost immediately.

Why do the new males kill the cubs? One hypothesis suggests that the new males will father more cubs as a result of this behavior. Note, however, that this does not suggest that male lions are aware that they are killing the offspring of other males, or that they understand how they will benefit from this behavior.

Figure 2 Behavior in male lions. After taking over a pride, a male lion often kills the young cubs in the pride. However, the males are usually tolerant of their own cubs.

Individual Selection

You may have heard it said that a trait or behavior ensures the survival of the species. This once popular belief is now considered false. Most scientists now agree that natural selection favors traits that contribute to the survival and reproduction of *individuals*, not species. The actions of male lions support this idea. Cub-killing increases the already high death rate among cubs and actually reduces the likelihood that the species will survive. Because natural selection favors traits that benefit individuals, the male lions usually will behave in ways that are favorable for them, not for the pride as a whole.

internet connect

www.scilinks.org
Topic: Animal Behavior
Keyword: HX4008

SC*LINKS* Maintained by the National Science Teachers Association

Genetically Influenced Behavior

From years of observation and experimentation, biologists have learned that many kinds of animal behaviors are influenced by genes. Genetically programmed behavior is often called **innate behavior,** or more commonly, instinct. The orb spider, shown in **Figure 3,** builds her web the same way every time. There is little or no variation in what she does, and her female offspring will build their webs in the same manner without being taught. This type of innate behavior is called **fixed action pattern behavior** because the action always occurs the same way.

Demonstrating the Genetic Basis of Behavior

Nest building is an innate behavior exhibited by most birds, including African lovebirds, shown in **Figure 4.** These small parrots construct their nests from materials that they collect and carry back to the nest site. One species of lovebird, Fischer's lovebird, carries a single long strip of nesting material in its beak. A second, closely related species, the peach-faced lovebird, carries several short strips of nesting material tucked into the feathers near its tail.

Evidence that these behaviors have a genetic basis comes from studies in which the two types of lovebirds were interbred. As shown in Figure 4, the resulting hybrid birds showed nesting behaviors that resembled those of both parents. They chose medium-length strips of nesting material and tried to place the strips in the feathers near their tail. But the hybrid birds were rarely successful because they did not let go of the strips after placing them in their feathers. Eventually, some of the hybrid birds learned to carry the nesting material in their beak.

Figure 3 Fixed action pattern behavior. Like all web-building spiders, the orb spider is genetically predisposed to build her web the same way each time.

Figure 4 Nest building behavior

The hybrid offspring of two lovebird species show nest-building behaviors similar to those of both parents.

Fischer's lovebirds carry nesting material in their beak.

Peach-faced lovebirds carry nesting material tucked into the feathers near their tail.

Their hybrid offspring tuck nesting material into their feathers but never let go of it. As a result, the nesting material does not stay in place.

Learning and Behavior

Behaviors are influenced by genes, but to what degree can behaviors be modified by experience? The development of behaviors through experience is called **learning.** Learning can influence the expression of innate behavior and the expression of behaviors that are not innate. One simple kind of learning is habituation. In habituation, an animal learns to ignore a frequent, harmless stimulus. For example, birds may at first stay away from a garden that has a new scarecrow. But if the position of the scarecrow is not changed on a regular basis, the birds learn to ignore it and go into the garden unafraid.

Classical Conditioning

A more complex type of learning is **conditioning,** or learning by association. One of the most famous studies of conditioning was Russian psychologist Ivan Pavlov's work with dogs, carried out in the late 1890s and early 1900s. When Pavlov presented meat powder (a stimulus) to a hungry dog, the dog salivated—an innate response to food. At the same time the dog received the meat powder, Pavlov also presented the dog with a second, unrelated stimulus—a ringing bell. After repeated trials, the dog learned to associate the ringing bell with the meat powder and would salivate in response to the bell alone. The dog became conditioned to associate the ringing of the bell with a reward (meat powder). This type of conditioning, in which an animal comes to associate an unrelated response with a stimulus, is called *classical conditioning*.

Trial-and-Error Learning

Animals also learn by trial-and-error that performing a certain action will result in a reward or a punishment. For example, a dog may learn to avoid a particular cat after being scratched on the nose once or twice. When trial-and-error learning occurs under highly controlled conditions, it is called *operant conditioning*.

Operant conditioning was demonstrated in another famous set of experiments conducted by the American psychologist B. F. Skinner. Skinner studied learning in rats by placing them in a "Skinner box," illustrated in **Figure 5.** Once inside, the rat would explore the box. Occasionally, it would accidentally press a lever, and a pellet of food would appear. At first, the rat would ignore the lever and continue to move about, but it soon learned to press the lever to obtain food. This sort of trial-and-error learning is of major importance to most vertebrates, and it influences many behaviors essential to survival, such as searching for food.

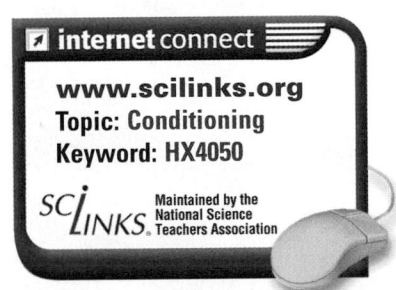

Study TIP

● **Recognizing Differences**
To understand the difference between classical and operant conditioning, think about how the behavior is learned. In classical conditioning, the stimulus has no relationship to an activity. In operant conditioning, there is a direct relationship.

🖉 **internet** connect

www.scilinks.org
Topic: Conditioning
Keyword: HX4050

SCI LINKS. Maintained by the National Science Teachers Association

Figure 5 Skinner box.
When placed in a Skinner box, this rat learned by trial-and-error to push a lever to receive a reward of food.

Reasoning

The ability to analyze a problem and think of a possible solution is called **reasoning.** Reasoning involves using experience to develop an insight into how to solve a new problem. The dog shown in **Figure 6** cannot think of a solution to the situation it is in. Humans and some other primates show the ability to reason. For example, in one experiment a chimpanzee was placed in a room with some boxes and a banana hung high overhead. Although it had not been in a similar situation, the chimpanzee stacked up the boxes to reach the banana, a behavior that required reasoning.

Figure 6 An inability to reason. Although the solution seems obvious to humans, this dog is unable to figure out how to reach the food.

There has been much research and debate over the ability of other animals, such as dolphins, some parrots, and even octopuses, to reason. So far, there is no clear evidence that other animals can reason. Many animals can, however, learn complex tasks. Some birds living in cities have learned to remove the foil covering from bottles of non-homogenized milk to reach the cream at the top. Japanese snow monkeys have learned to float grain on water to separate the grain from sand.

QUICK LAB

Recognizing Learned Behavior

Sow bugs must keep moist to survive. Follow the procedure below to see if sow bugs can learn to find moisture.

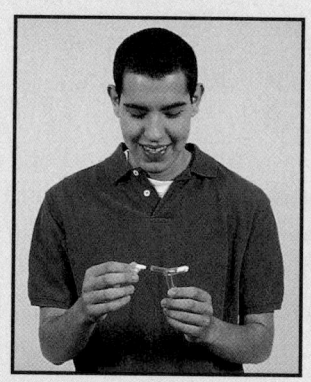

Materials

small wads of paper towel (one moist and one dry), T-maze made of two 6 cm (about 2.25 in.) pieces of 1.25 cm (0.5 in.) clear vinyl tubing, sow bug, blunt probe

Procedure

1. Place the moist paper wad in the open end of the left side of the T, and place the dry paper wad on the right side.

2. Place the sow bug at the bottom of the T. If it does not start to crawl, gently prod it with a blunt probe.

3. Observe what the sow bug does when it reaches the T section. Retrieve the sow bug and perform as many trials as time allows.

4. Keep a record of the results of each trial.

5. Using the same sow bug, repeat this procedure for three days.

Analysis

1. **Summarize** your sow bug's behavior, in writing or on a graph.

2. **Describe** any trend in behavior that you observed.

3. **Determine** if your sow bug modified its behavior through learning, using evidence to support your answer.

4. **Evaluate** the value of performing a final trial in which the T-maze contains two dry paper wads.

Genetic and Learned Aspects of Behavior

Do genes determine most behaviors, or do animals usually learn how to behave from experience? Over the last century, this topic has been debated. Some scientists argued that most behaviors are genetically programmed because different individuals in the same species act in the same ways. Other scientists claimed that behaviors are shaped by an animal's experiences. Most biologists who study animal behavior have come to think that animal behavior, particularly the complex behavior of vertebrates, has both genetic and learned components.

Imprinting

Learning that can occur only during a specific period early in the life of an animal and cannot be changed once it occurs is called **imprinting.** Imprinting is easily observed in young geese and ducks, which have no innate recognition of their mother. Instead, these birds are genetically programmed to follow the first moving object they see during a short period immediately after they hatch.

There is great survival value to this behavior, as the young must follow their mother as she leads them to water, helps them find food, and keeps them out of danger. However, the young will follow any object they see during this period just as they would their mother—including toy wagons, boxes, and balloons. Once the young birds imprint on an object, they prefer to follow it, even when given the opportunity to follow a member of their own species.

Konrad Lorenz, a Nobel Prize-winning pioneer in the study of animal behavior, observed imprinting when he raised a group of newly hatched goslings (young geese) by hand and found that they imprinted on him. **Figure 7** shows Konrad Lorenz leading his "family" of goslings. The goslings ability to imprint on an object during a sensitive period is not a learned behavior; it is programmed into their genes. However, the process of imprinting is a form of learned behavior. Thus learning determines the final shape of this genetically based behavior.

Figure 7 Imprinting. These goslings imprinted on Konrad Lorenz and followed him around just as if he were their mother.

Section 1 Review

1 Describe the difference between "how" and "why" questions in regard to animal behavior.

2 Summarize how cub killing by male lions supports the hypothesis that natural selection shapes behavior.

3 Distinguish between and give an example of innate and learned behavior.

4 Analyze the behaviors involved in imprinting.

5 Critical Thinking Forming Reasoned Opinions A friend is teaching his dog a new trick in which it is rewarded each time the trick is performed correctly. The friend says his method is called *classical conditioning*. Evaluate your friend's use of this term.

6 Standardized Test Prep When an Australian frilled lizard flares the folds of skin around its head in response to a threat, the lizard exhibits

A imprinting.   **C** classical conditioning.

B trial-and-error learning.   **D** an innate behavior.

Types of Behavior

Objectives

- **Discuss** six types of animal behavior.

- **Discuss** how animals use signals.

- **Summarize** how sexual selection can influence evolution.

Key Terms

sexual selection

Categories of Animal Behavior

As you sit on a park bench, a pigeon approaches, and you soon realize that it expects food. You toss out a bit of your sandwich and the pigeon eats it, immediately looking up for more. This urban pigeon has been conditioned to seek food from people. Has its human supplier been conditioned too?

Behavior is an animal's most immediate way of dealing with its environment. Because the environment is complex and can change rapidly, most animals have many different kinds of behavior, each suited to a particular situation. For instance, a squirrel may perform one kind of behavior when it finds a nut on the ground—it digs a hole. It performs a completely different behavior when a snake approaches—it runs for shelter—because digging a hole would not help it escape from the snake.

Like the squirrel, the musk oxen of the Arctic display many different types of behavior, some of them cooperative. When predators, usually wolves, appear, the adult musk oxen form a defensive circle around their young, as shown in **Figure 8.** The tight circle and the danger of injury from the adult's horns and hooves usually prevent a successful attack. While musk oxen can run, running would not protect the herd in the way that their group defense does.

Biologists have classified the behaviors animals perform into several broad categories. **Figure 9** shows some of these categories and gives an example of each.

Figure 8   Musk oxen.
When threatened, adult musk oxen form a defensive circle around their young.

Influences on Behavior

While many of the behaviors illustrated in Figure 9 may seem different from one another, they all tend to favor survival and reproductive success. To gain a better idea of the nature of animal behavior, let's examine one of these behaviors—foraging behavior—in detail. *Foraging* is finding and getting food.

Animals can be divided into two broad groups based on the range of food items each group consumes. Specialists feed primarily or exclusively on one kind of food. Some species of ants, for example, eat only spider eggs. Generalists, in contrast, consume many different kinds of food. For example, some insects eat the leaves of a wide variety of plants. Generalists are typically less efficient than specialists at

Figure 9 Animal behavior.
Although their methods differ, all animals engage in at least some of the behaviors shown below.

Foraging behavior	Migratory behavior	Defensive behavior
Locate, obtain, and consume food	**Move to a more suitable environment as seasons change**	**Protect from predators**

A raccoon searches along streams and ponds for fish, frogs, crayfish, and small rodents. It also hunts for insects and fruit in woodlands.

Monarch butterflies migrate thousands of kilometers, from the United States to central Mexico.

When threatened, a hognose snake turns onto its back and plays dead.

Territorial behavior	Courtship behavior	Parental care
Protect a resource for exclusive use	**Attract a mate**	**Ensure survival of young**

Like many wild cats, this young cheetah claws on trees, leaving a scent that marks its territory.

During its breeding season, the male stickleback fish develops a bright coloring and builds an elaborate nest to attract a female.

This robin is feeding an insect to its offspring.

feeding on any one type of food. However, generalists have the advantage of being able to collect more than one kind of food. Which approach to foraging is better? When one kind of food source is plentiful, specialists forage more successfully. But when food sources are diverse and no particular one of them is more common, generalists find more to eat.

For predators, food typically comes in a variety of sizes. Larger food items contain more energy. But larger items are harder to capture, and they are usually less abundant.

Foraging thus involves a trade-off between a food's energy content and its availability. Animals tend to feed on prey that maximize their energy intake per unit of foraging time. This approach is called optimal foraging. Natural selection has favored the evolution of foraging behaviors of this sort. Sometimes, however, animals will consume foods that are low sources of energy. Often this is because those foods supply an important nutrient. The location of the food source may also allow the consumer to avoid being captured by some other predator.

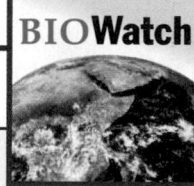

BIOWatch

Tracking Animal Movements by Satellite

The migrations of whales, birds, butterflies, bats, and other animals are among the most fascinating examples of animal behavior. Some animals migrate thousands of kilometers between the same two places every year. To track these movements, biologists have turned to satellites for help.

How Satellite Tracking Works

As shown on the right, a small transmitter containing electronic circuitry, a battery, and an antenna is attached to an animal. For birds and other small animals, these parts must be kept as light as possible. Satellites orbiting about 850 km (530 mi) above Earth pick up the radio signals produced by the transmitter and relay the signals to a central computer on the ground. The computer analyzes the information and calculates the animal's location. By connecting to

this computer on the Internet, biologists all over the world can get information on the movements of the animals they are interested in tracking.

Why Use Satellites?

Satellite tracking allows biologists to track animals that would be impossible to follow on foot or in a vehicle. For example, biologists used satellite tracking to trace the winter migration of Swainson's hawks. Over the past few decades, the number of these hawks has been declining in some areas of North America. Satellite tracking showed that the hawks spend the winter in a specific region in central Argentina. By traveling to that region, researchers discovered that thousands of the hawks were being killed accidentally as a result of the use of pesticides. The Argentinian government and the pesticide manufacturers are now working together to protect the hawks.

Swainson's hawk

Horned lizard

Communication

You approach an unfamiliar dog and it begins to bark. You know that if you go closer, the dog might bite you, so you stop and talk to it. The dog continues to bark but not so aggressively, and it begins to wag its tail. You and the dog have each responded to a signal given by the other.

A signal can be a sound, posture, movement, color, scent, or facial expression. These signals are sent and received through all of the senses familiar to us—sight, hearing, smell, touch, and taste. Animals use signals to influence the behavior of other animals. Because they face a variety of social situations in which communication is needed, animals usually have several different signals, each suited to a different situation, as shown in **Figure 10.**

Natural selection has shaped animal signals so that they reach the intended receiver efficiently and stimulate a response. To be transmitted efficiently, a signal must be able to travel through the environment from sender to receiver. A signal must also be recognizable to the receiver, or it won't have any effect on behavior. Consider the loud mating call emitted by the male túngara frog. The call carries a long distance, reaching even far-off females. At night, when túngara frogs are active, a loud call is the best way to communicate. Visual signals, such as colors and movements, would be noticeable from only a short distance away and would not be nearly as effective at attracting a mate.

Figure 11 Primate communication. Vervet monkeys have distinct calls that identify different kinds of predators.

Primate Communication

Among animals, vocal communication may be most developed in the primates. Some primates, such as those shown in **Figure 11,** have a "vocabulary" of calls that allows individuals to communicate the identity of specific predators, such as eagles, leopards, and snakes. Chimpanzees and gorillas can be taught by humans to recognize and use a large number of symbols to communicate abstract concepts. Chimpanzees and gorillas cannot talk, however, because they are physically unable to produce the sounds of speech. Some researchers believe that chimpanzees can combine symbols they have learned in meaningful ways. But chimpanzees cannot rearrange symbols to form a new sentence with a different meaning. That is a very complex task, which only humans can perform.

In humans, language develops at a very early age. Infants begin to learn language by trial-and-error during the "babbling baby" phase of childhood, at about six months of age. At first, infants all over the world babble the same consonant sounds, including sounds they have never heard. Soon, however, the infants pick out the sounds used by the people around them and repeat only those sounds. The other sounds begin to drop away and are forgotten. Children quickly and effortlessly learn a vocabulary of thousands of words, a feat no chimpanzee can do. This ability to learn language rapidly seems to be genetically influenced. But language is not the only form of human communication. Evidence suggests that odors and nonverbal signals (body language) may also be important.

QUICK LAB

Observing Territorial Behavior in Crickets

The chirp of a male cricket attracts females and warns other males to stay away from his territory. You can study chirping behavior by observing crickets in an aquarium.

Materials

5 male crickets, each marked with a different color; 5 unmarked female crickets; covered aquarium; slice of apple and of potato; small plastic jar; 5 cm (2 in.) square of cardboard

DATA TABLE

Cricket Behavior

Cricket	Apple	Potato	Jar	Tent	Female
Blue					
Yellow					
Red					
Green					
White					

Procedure

1. Place the crickets and food in an aquarium. Make two shelters by turning the plastic jar on its side and by folding the cardboard in half to form a tent-like structure.

2. Make a chart like the one above to record the behavior of the male crickets.

3. Observe the crickets for 10 minutes. Among the males, look for territorial (aggressive) behaviors—chirping, stroking others with antennae, pushing others away, etc.

4. For each observation of aggressive behavior, record the color of the aggressive male and where the behavior occurred—for example, next to the jar or the tent.

5. For each cricket, tally the number of times aggressive behavior was observed. Make a list that ranks each cricket, placing the cricket with the highest tally on top.

6. Then tally the numbers for where the behaviors occurred. Rank the locations.

Analysis

1. **Critical Thinking Analyzing Data** Were any crickets more aggressive than the others? Give evidence to support your answer.

2. **Describe** the circumstances in which most aggressive behavior occurred.

3. **Propose** a reason to explain your answer to item 2.

4. **Critical Thinking Forming Hypotheses** For each aggressive behavior you observed, form a hypothesis that explains its function.

Reproductive Behavior

When ready to mate, animals produce signals to communicate with potential partners. Each species usually produces a unique courtship signal that ensures that individuals do not mate with individuals of another species. For example, the flash patterns of fireflies differ between species. A female firefly recognizes a male of her own species by the pattern of his flashes, and she will ignore any male that has a different flash pattern. The chemical produced by a female silk moth attracts only males of her own species. Many species of insects, amphibians, and birds produce unique sounds or songs to attract mates. A white-crowned sparrow will respond to the song of another white-crowned sparrow, but it will totally ignore the song of an English sparrow.

During the breeding season, animals make several important "decisions" concerning mating and parenting. These include how many mates to have and how much time and energy to devote to rearing offspring. These decisions are all aspects of an animal's reproductive strategy, a set of behaviors that have evolved and that maximize reproductive success.

Mate Choice

Males and females usually differ in their reproductive strategies. In many animals, females do not mate with the first male they encounter. Instead, the female seems to evaluate the male before she decides whether to mate. This behavior, called *mate choice*, has been observed in many invertebrate and vertebrate species. Female túngara frogs, for instance, have been observed "shopping around" among calling males. A female will sit near a male and listen to his call for several minutes, then move on to another male and listen to his call. She may evaluate several males before choosing one with which to mate.

Figure 12 Widowbirds. During breeding season, the tail of the male widowbird, shown above, grows to more than three times the length of his body. At other times of the year, it is similar in length to that of the female, which is the bird on the left in the inset photograph.

Sexual Selection What characteristics do animals use in choosing a mate? When Charles Darwin considered this question more than a century ago, he made an important discovery about evolution. Darwin noticed that males often have extreme characteristics that they use in their courtship displays. Take, for example, widowbirds, shown in **Figure 12.** During the breeding season, the male widowbird grows an extremely long tail, up to five times longer than the female's. How did such differences between the sexes evolve? The long tail of the male widowbird cannot be essential for survival, since the female bird survives quite well with a much shorter tail.

Figure 13   Gorillas. On average, male gorillas are 50 percent larger than females. Full grown males can weigh as much as 200 kg (440 lb).

Darwin recognized that extreme traits, such as the male widowbird's tail, could have evolved if they helped males attract or acquire mates. He proposed the mechanism of sexual selection to account for such traits. **Sexual selection** is selection in which a mate is chosen based on a certain trait or set of traits. Thus, traits that increase the ability of individuals to attract or acquire mates appear with increased frequency. Even traits that have a negative effect on survival can evolve in this way, provided that their benefits to reproduction are high enough.

It is usually females that select mates based on their physical traits. One explanation for this tendency is that reproduction has a greater metabolic cost and requires a greater investment from females. For example, eggs are less numerous and much larger than sperm. (Humans' eggs are about 195,000 times larger!) In mammals, females are also responsible for gestation and milk production. Males may show mate choice as well if their parental involvement is high. This happens in crickets, where the sperm packet a male deposits makes up 30 percent of his body weight, contributes nutrition for the female, and helps her develop eggs.

Competition by Males What kinds of traits help males acquire mates? Because male animals usually compete among themselves for the chance to mate with a female, selection has often favored traits that make males more intimidating or better at combat. For example, in many animal species, such as the gorillas shown in **Figure 13,** the male is much larger than the female, and the largest male will have the most opportunities to mate. Other examples of traits related to male competition are antlers in deer and moose, horns in bighorn sheep, and manes in lions. The extreme traits of some male animals, such as large size and tusks in walrus, often permit males to assess each other. Males that are not physically well matched rarely engage in serious fights. Thus, some extreme traits may reduce conflict among males rather than increase it.

Competition among males can also take subtle forms. In some species, a male can gain a reproductive advantage over other males by interfering with their reproduction. For instance, in some species of worms, butterflies, and snakes, the male seals the female's reproductive tract after mating, thus preventing other males from mating with her.

Section 2 Review

1 Describe the function of six different animal behaviors.

2 Summarize in words and with examples the ways in which animals use signals.

3 Discuss how selection can account for the extreme traits found in the males of some species.

4 Standardized Test Prep A female firefly recognizes males of her own species because the males produce a specific

A chemical.

B flash pattern.

C sound.

D type of nest.

Study ZONE

CHAPTER HIGHLIGHTS

Key Concepts

1 Evolution of Behavior

- A behavior is an action or a series of actions performed in response to a stimulus.
- Natural selection favors behavioral traits that increase the likelihood of an individual's survival and reproduction.
- Genetically programmed behaviors are called innate behaviors, instincts, or fixed action pattern behaviors, and there is little or no variation in how they are performed.
- Learning is the modification of behavior by experience. Learning may occur by association with an unrelated stimulus (classical conditioning) or by trial-and-error (one type of which is operant conditioning).
- Reasoning is the ability to think of a possible solution to a problem.
- Many behaviors, especially complex behaviors, have both genetic and learned aspects.
- Learning determines the final shape of many genetically based behaviors, such as imprinting.

2 Types of Behavior

- Animal behaviors fall into several broad categories, which include parental care, courtship behavior, defensive behavior, foraging behavior, migratory behavior, and territorial behavior.
- Most animals use signals, often vocal or visual, to communicate with one another.
- Primates are unique among animals in using symbols to communicate.
- The human ability to learn language rapidly during childhood seems to be genetically influenced.
- By the mechanism of sexual selection, traits that increase the ability of an individual to attract a mate appear with increased frequency.

Key Terms

Section 1

behavior (824)
innate behavior (826)
fixed action pattern behavior (826)
learning (827)
conditioning (827)
reasoning (828)
imprinting (829)

Section 2

sexual selection (836)

Performance ZONE

CHAPTER REVIEW

Understanding Key Ideas

1. A male lion that takes over a pride may kill all the young cubs in order to ensure the survival and reproduction of
a. the females.
b. the pride.
c. himself.
d. his siblings.

2. The nest-building behavior of Fischer's lovebirds is
a. learned.
b. a result of operant conditioning.
c. innate.
d. gradually learned.

3. Which of the following best represents classical conditioning?
a. the feeding behavior shown by rats in Skinner boxes
b. a male túngara frog calling to a female
c. a dog salivating at the sound of a bell
d. a primate giving a warning signal to members of its troop

4. Parental care is performed in order to
a. ensure survival of young.
b. protect parents from predators.
c. protect a resource for exclusive use.
d. locate, obtain, and consume food.

5. The purpose of foraging behavior is to
a. ensure survival of young.
b. protect individuals from predators.
c. protect a resource for exclusive use.
d. locate, obtain, and consume food.

6. Which of the following behaviors is least likely to be associated with food resources?
a. territorial   **c.** foraging
b. migratory   **d.** defensive

7. Which of the following best demonstrates the mechanism of sexual selection?
a. the bright colors of a poisonous frog
b. the ability of a lion to run fast
c. the long tail of a male widowbird
d. the web-spinning behavior of a spider

8. **🌐 BIOWatch** What advantages does satellite tracking of animals provide over other methods, such as radio collars?

9. Summarize the mechanism of natural selection, and explain why behaviors are just as important to survival and reproduction as physical features are. (**Hint:** See Chapter 13, Section 1.)

10. **⌁ Concept Mapping** Construct a concept map describing animal behavior. Use the following terms in your map: *behavior, stimulus, innate behavior, fixed action pattern behavior, learned behavior, conditioning, reasoning, imprinting,* and *sexual selection.*

Critical Thinking Skills

11. **Evaluating an Argument** "A child's behavior closely resembles the behavior of its parents. Therefore, most human behaviors are genetically controlled." Explain the logical flaw in this argument.

12. **Recognizing Relationships** In many bird species, the male is more brightly colored and heavily plumed than the female. Explain this in terms of sexual selection and evolution.

Alternative Assessment

13. **Finding and Communicating Information** Use the media center or Internet resources to learn more about imprinting. Work with a small group to find examples of imprinting, and develop a class presentation on the subject.

14. **Summarizing Information** At times, some species of whales seem to deliberately swim into waters that are shallow. Often they end up stranded on the beach, where they cannot survive. Using the Internet or other resources, search for information you can use to answer the "how" and "why" questions about this phenomenon. Present your findings in a report to your class.

Understanding Concepts

Directions (1–4): **For *each* question, write on a separate sheet of paper the letter of the correct answer.**

1 What is learning by association?
 A. assuming   **C.** imprinting
 B. conditioning   **D.** reasoning

2 The orb spider builds her web in exactly the same way every time. What is this an example of?
 F. learned behavior
 G. random behavior
 H. inherited behavior
 I. fixed action pattern behavior

3 What is the ability to analyze a problem and think of a possible solution?
 A. assuming
 B. conditioning
 C. imprinting
 D. reasoning

4 How can a behavior that results in the death of one offspring evolve by natural selection?
 F. The behavior has no adaptive value.
 G. The behavior protects all the young in a group.
 H. The behavior leads to the death of all the young in a group.
 I. The behavior helps more of one individual's offspring to survive.

Directions (5): **For the following question, write a short response.**

5 Captured male widowbirds whose tails have been cut mate with half as many females as wild males with longer tails. How can a scientist design an experiment to determine whether female widowbirds prefer males with longer tails?

Test TIP

When using experimental data to answer a question, determine the constants, variables, and control before answering the question.

Reading Skills

Directions (6): **Read the passage below. Then answer the question.**

Imprinting is learning that can occur only during a specific period early in the life of an animal and cannot be changed once it occurs. Many birds, such as ducks and geese, imprint on their mother shortly after they hatch. Goslings can even imprint on humans. When they do, they will follow the human around in single file just as if he or she were their mother.

6 Why might natural selection have favored the evolution of imprinting behavior?
 A. Goslings that can recognize a human are more likely to find food.
 B. Adult geese that associate with humans are more likely to reproduce.
 C. Adult geese that travel in single file as a group are more likely to avoid predators.
 D. Goslings that follow their mother are more likely to find food and avoid danger.

Interpreting Graphics

Directions (7): **Base your answer to question 7 on the diagram below.**

Fisher's Lovebird and Peach-faced Lovebird

Fischer's lovebirds carry nesting material in their beak.

Peach-faced lovebirds carry nesting material tucked into the feathers near their tail.

7 If the nest-building behaviors of these two kinds of lovebirds are innate, how might the hybrid offspring raised in captivity try to carry nesting materials?
 F. balanced on their backs
 G. carried in their claws
 H. kicked along the ground
 I. tucked near their tail feathers

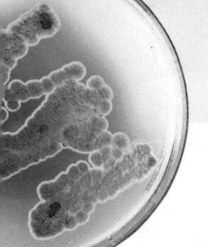

Exploration Lab

Studying Nonverbal Communication

SKILLS
- Observing
- Analyzing
- Graphing

OBJECTIVES
- **Recognize** that posture is a type of nonverbal communication.
- **Observe** how human posture changes during a conversation.
- **Determine** the relationship of gender to the postural changes that occur during a conversation.

MATERIALS
- stopwatch or clock with a second hand
- paper
- pencil

Before You Begin

People communicate nonverbally with their **posture,** or body position. The position of the body while standing is called the **stance.** In an **equal stance,** the body weight is supported equally by both legs. In an **unequal stance,** more weight is supported by one leg than by the other. In this lab, you will observe and analyze how stance changes during conversations between pairs of people who are standing.

1. Write a definition for each boldface term in the paragraph above.

2. Make a data table similar to the one below. The sample data entered in row 1 show how to enter data. Do not copy these data.

DATA TABLE

	Gender		15-s intervals		
Pairs	Involved	Observed	15 s	30 s	45 s
1	F, M	M	U, W	E	E
2					
3					

3. Based on the objectives for this lab, write a question you would like to explore about nonverbal communication.

Procedure

PART A: Observing Behavior

1. Work in a group of two or three to observe conversations between pairs of people. Each conversation must last between 45 seconds and 5 minutes. One person in your group should be the timekeeper and the other group members should record data. **Note:** Be sure that your subjects are unaware they are being observed.

2. Observe at least three conversations. Record the genders of the two participants in each conversation and the gender of the one person whose posture you observe. **Note:** Be sure that the timekeeper accurately clocks the passage of each 15-second interval.

3. For each 15-second interval, record all of the changes in stance by the person you are observing. For example, note every time your subject shifts from an equal stance to an unequal stance, or vice versa. To record the stance simply, you may write *E* to identify an equal stance and *U* to identify an unequal stance.

4. If the subject assumes an unequal stance, also record the number of weight shifts from one foot to the other. Indicate a weight shift simply by writing *W*.

5. When a conversation ends, write down whether the pair departed together or separately. To record this, write *T* to indicate departing together or *S* to indicate departing separately.

6. After you have completed each observation, tally the total number of weight shifts within each 15-second block. **IMPORTANT!** Retain data only for conversations that last at least 45 seconds. If a conversation ends before you have collected data for 45 seconds, observe another conversation.

PART B: Analyzing Data

7. After all observations have been completed, combine the data from all of the groups in your class. Analyze the data, without regard to gender.

   a. Determine the most common stance during the first 15 seconds of a conversation, the middle 15 seconds, and the last 15 seconds. Make a bar graph to summarize the class data.

   b. Find the average number of weight shifts in the beginning, middle, and end intervals. Make a bar graph to summarize the class data.

8. Repeat step 7, but analyze the data according to gender this time.

9. Compile the data and make bar graphs for each of the following: males talking with a male, males talking with a female, females talking with a male, and females talking with a female. Compare these graphs with the ones you made in Step 7.

Analyze and Conclude

1. Analyzing Results Which stance was used most often during a conversation?

2. Recognizing Relationships Which behavior most often signals that a conversation is about to end: stance change or weight shift?

3. Drawing Conclusions Do males and females differ in their departure signals? Justify your conclusion.

4. Forming Hypotheses What do you think might be an adaptive significance of a departure signal?

5. Forming Reasoned Opinions What other behaviors you observed were forms of nonverbal communication? Justify your answer.

6. Further Inquiry Write a new question about animal behavior that could be explored with a new investigation.

On the Job

Psychology is the study of human and animal behavior. Do research to discover how psychology is used to treat behavioral disorders in humans or in pets. For more about careers, visit **go.hrw.com** and type in the keyword **HX4 Careers.**

UNIT 9 Exploring Human Biology

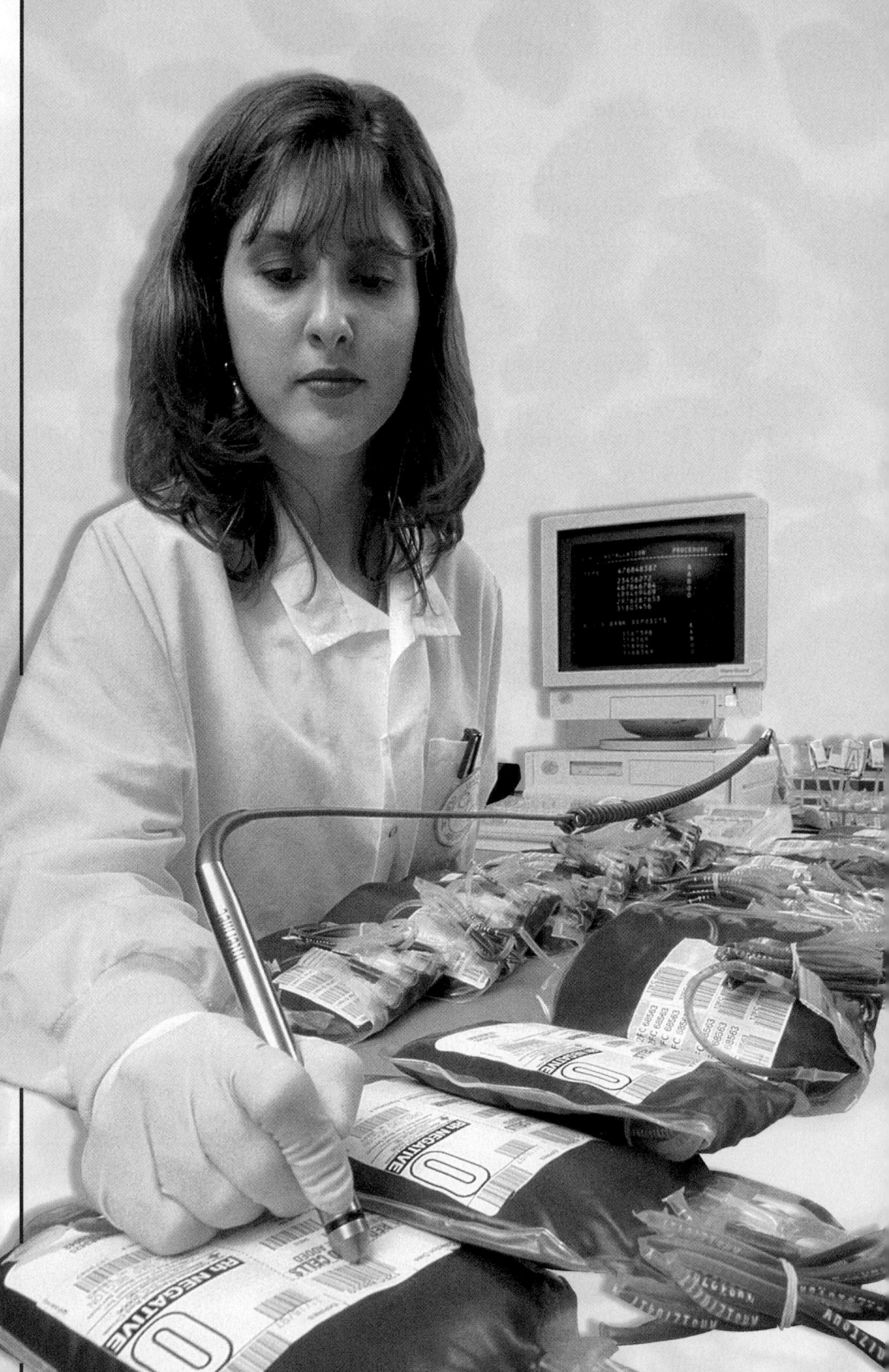

Donated blood saves thousands of lives every year.

in perspective

Circulation

Colleagues of William Harvey, a 17th century English anatomist, considered his conclusion that blood flows one way in a continuous system of vessels to be a strange idea. **Read to learn how valves prevent blood from flowing backwards.** In the 1940s, Dr. Charles Drew, the first African-American to earn an M.D. at Columbia University, developed commercial procedures for the safe transfusion of plasma, the liquid component of blood.

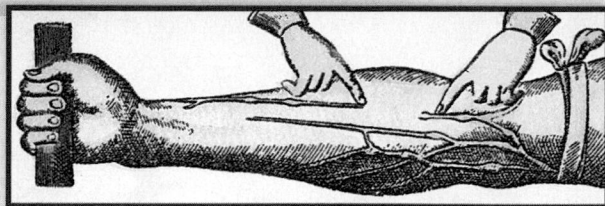

William Harvey's theory of blood circulation

Dr. Charles Drew

Today...
Scientists know that lifestyle choices—for example, what we eat and how much we exercise—affect the health of our circulatory system. Eating a healthful diet and exercising regularly will reduce the likelihood of hypertension (commonly called high blood pressure) and lower the risk of heart attack. **What are the dangers of hypertension?**

Moderate exercise

Tomorrow...
Blood is the river of life. Researchers are currently exploring artificial substitutes for red blood cells. **What are the components of blood?**

Artificial blood research

🖉 internet connect

www.scilinks.org
Topic: **Blood**
Keyword: **HX4025**

SC**i**LINKS. Maintained by the
National Science
Teachers Association

Hurdler

CHAPTER
37 Introduction to Body Structure

✓ Quick Review

Answer the following without referring to earlier sections of your book.

1. **Describe** the role of ATP in metabolism. *(Chapter 5, Section 1)*

2. **Summarize** the two stages of cellular respiration. *(Chapter 5, Section 1)*

3. **Distinguish** between tissues, organs, and organ systems. *(Chapter 20, Section 3)*

Did you have difficulty? *For help, review the sections indicated.*

📖 Reading Activity

Before you read this chapter, survey the subtitles, headings, captions, and words in boldface type. Try to identify the purpose of this chapter. As you read each section, create a descriptive reading organizer.

☑ internet connect

www.scilinks.org
National Science Teachers Association *sci*LINKS Internet resources are located throughout this chapter.

SCI*LINKS* Maintained by the
National Science Teachers Association

● An activity as complicated as jumping hurdles while running full speed requires a tremendous amount of energy and the coordination of many body systems.

Body Organization

- **Identify** four levels of structural organization within the human body.
- **Analyze** the four kinds of body tissues.
- **List** the body's major organ systems.
- **Evaluate** the importance of endothermy in maintaining homeostasis.

Key Terms

epithelial tissue
nervous tissue
connective tissue
muscle tissue
body cavity

Levels of Structural Organization

The human body contains more than 100 trillion cells and more than 100 kinds of cells. How do these cells work together? The body is structurally organized into four levels: cells, tissues, organs, and organ systems. Recall that a tissue is a group of similar cells that work together to perform a common function. The cell types of the body are grouped by function into four basic kinds of tissues: epithelial, nervous, connective, and muscle tissues. These tissues, shown in **Figure 1,** are the building blocks of the human body.

Four Kinds of Tissues

Epithelial tissue There are many different kinds of epithelial *(ehp ih THEE lee uhl)* tissue. **Epithelial tissue** lines most body surfaces, and it protects other tissues from dehydration and physical damage. An epithelial layer is usually no more than a few cells thick. These cells are typically flat and thin, and they contain only a small amount of cytoplasm. Epithelial tissue is constantly being replaced as cells die.

Figure 1   Body tissues

Cells of the body are grouped into different kinds of tissues.

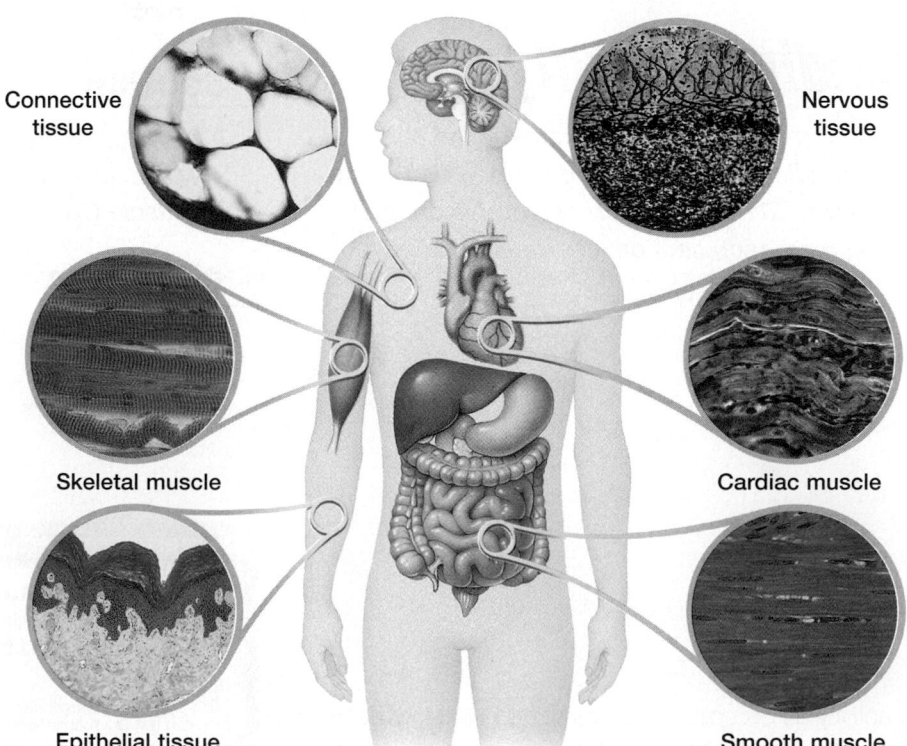

Connective tissue

Nervous tissue

Skeletal muscle

Cardiac muscle

Epithelial tissue

Smooth muscle

Nervous Tissue The nervous system is made of **nervous tissue.** Nervous tissue consists of nerve cells and their supporting cells. Nerve cells carry information throughout the body. You will learn more about the nervous system in Chapter 41.

Connective Tissue Various types of **connective tissue** support, protect, and insulate the body. Connective tissue includes fat, cartilage, bone, tendons, and blood. Some connective tissue cells, such as those in bone, are densely packed. Others, such as those found in blood, are farther apart from each other.

Muscle Tissue Three kinds of **muscle tissue** enable the movement of body structures by muscle contraction. The three kinds of muscle tissues are skeletal muscle, smooth muscle, and cardiac muscle.

1. **Skeletal muscle.** Skeletal muscle is called voluntary muscle because you can consciously control its contractions. Skeletal muscles move bones in the trunk and limbs.

2. **Smooth muscle.** Smooth muscle is called involuntary muscle because you cannot consciously control its slow, long-lasting contractions. Smooth muscles line the walls of blood vessels and hollow organs. Some contract only when stimulated by signal molecules; others contract spontaneously.

3. **Cardiac muscle.** Cardiac muscle is involuntary and is found in the heart. The powerful, rhythmic contractions of cardiac muscle pump blood to all body tissues. Groups of cardiac cells contract all at once, stimulating adjacent groups of cells to contract.

Stem Cells

Every human starts life as a single fertilized egg, which rapidly divides into a small cluster of cells. After about 5 days, a small ball of a few hundred cells is formed, which encloses a mass of embryonic stem cells. These early, undifferentiated cells will give rise to all of the types of cells of the developing body. Embryonic stem cells are immortal—that is, they divide indefinitely. And embryonic stem cells are not yet specialized. Indeed any embryonic stem cell is capable of becoming any type of tissue found in the adult body.

Because they can develop into any tissue, embryonic stem cells offer the possibility of repairing damaged tissues. Stem cell therapy in mice has been shown to repair heart muscle and to produce functional nerve cells in the brain. The use of human embryonic stem cells is very controversial. Because obtaining embryonic stem cells destroys an early embryo, therapeutic use of embryonic stem cells raises serious ethical issues.

Adults also have stem cells. Stem cells in bone marrow produce different types of blood cells. The adult brain contains stem cells that develop into new nerve cells. Adult stem cells are not as versatile as embryonic stem cells, and they are not immortal. Most stop reproducing after fewer than 100 cell divisions. Scientists are now at work on several therapeutic applications of adult stem cells.

Organg Systems

Body organs are made of combinations of two or more types of tissues working together to perform a specific function. The heart, for example, contains cardiac muscle tissue and connective tissue, and the heart is stimulated by nervous tissue. Each organ belongs to at least one organ system, which is a group of organs that work together to carry out major activities or processes. The different organs in an organ system interact to perform a certain function, such as digestion. The digestive system is composed of the mouth, throat, esophagus, stomach, intestines, liver, gallbladder, and pancreas. Some organs function in more than one organ system. The pancreas, for example, functions in both the digestive system and the endocrine system. **Table 1** lists the body's major organ systems.

Table 1 Major Organ Systems of the Body

System	Major structures	Functions
Circulatory	Heart, blood vessels, blood (cardiovascular) lymph nodes and vessels, lymph (lymphatic)	Transports nutrients, wastes, hormones, and gases
Digestive	Mouth, throat, esophagus, stomach, liver, pancreas, small and large intestines	Extracts and absorbs nutrients from food; removes wastes; maintains water and chemical balances
Endocrine	Hypothalamus, pituitary, pancreas and many other endocrine glands	Regulates body temperature, metabolism, development, and reproduction; maintains homeostasis; regulates other organ systems
Excretory	Kidneys, urinary bladder, ureters, urethra, skin, lungs	Removes wastes from blood; regulates concentration of body fluids
Immune	White blood cells, lymph nodes and vessels, skin	Defends against pathogens and disease
Integumentary	Skin, nails, hair	Protects against injury, infection, and fluid loss; helps regulate body temperature
Muscular	Skeletal, smooth, and cardiac muscle tissues	Moves limbs and trunk; moves substances through body; provides structure and support
Nervous	Brain, spinal cord, nerves, sense organs	Regulates behavior; maintains homeostasis; regulates other organ systems; controls sensory and motor functions
Reproductive	Testes, penis (in males); ovaries, uterus, breasts (in females)	Produces gametes and offspring
Respiratory	Lungs, nose, mouth, trachea	Moves air into and out of lungs; controls gas exchange between blood and lungs
Skeletal	Bones and joints	Protects and supports the body and organs; interacts with skeletal muscles, produces red blood cells, white blood cells, and platelets

Body Cavities

The body contains four large fluid-filled spaces, or **body cavities,** that house and protect the major internal organs. Within the body cavities, shown in **Figure 2,** organs are suspended in fluid that supports their weight and prevents them from being deformed by body movements. These organs are also protected by bones and muscles. For example, your heart and lungs are protected by the rib cage and the sternum inside the thoracic *(thoh RAS ik)* cavity. Your brain, encased within the cranial *(KRAY nee uhl)* cavity, is protected by the skull. Your digestive organs, located in the abdominal cavity, are protected by the pelvis and abdominal muscles. Your spinal cord is protected by the vertebrae that surround the spinal cavity.

Endothermy

Like all mammals, humans are endotherms. Humans maintain a fairly constant internal temperature of about 37°C (98.6°F). Your body uses a great deal of energy to maintain a stable internal condition. For example, a large percentage of the energy you consume in food is devoted to maintaining your body temperature. You would not survive very long if your temperature fell much below the normal range. Very high temperatures, such as occur with fever, are also dangerous because they can inactivate critical enzymes. Your body maintains a constant temperature due to the flow of blood through blood vessels just under the skin. To release heat to the air, blood flow is increased to these vessels. To retain heat, blood is shunted away from the skin. As an endotherm, you can remain active at external temperatures that would slow the activity of ectotherms. Endothermy enables you to sustain strenuous activity, such as exercise, for a long time.

To maintain homeostasis, the body's organ systems must function smoothly together. The nervous system and the endocrine system operate on negative feedback with other organ systems. This promotes stability throughout the body. In addition to temperature regulation, homeostasis involves adjusting metabolism, detecting and responding to environmental stimuli, and maintaining water and mineral balances.

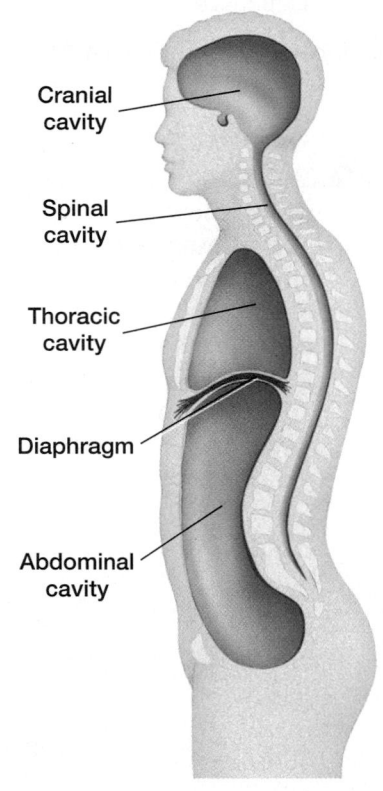

Figure 2   Body cavities.
Many organs and organ systems are encased in protective body cavities.

Cranial cavity

Spinal cavity

Thoracic cavity

Diaphragm

Abdominal cavity

Section 1 Review

1 **Summarize** the four levels of structural organization in the body.

2 **List** the four different kinds of body tissues, and give an example of each kind.

3 **Describe** the relationship between organs and organ systems.

4 **Relating Concepts** How is endothermy advantageous to humans?

5 **Critical Thinking Inferring** Why should fever be controlled during an illness?

6 (Standardized Test Prep) In which part of the body would you most likely find flat, thin cells that contain only a small amount of cytoplasm?
 A bone
 B cardiac muscle
 C digestive tract lining
 D skeletal muscle

Skeletal System

Objectives

- **Distinguish** between the axial skeleton and the appendicular skeleton.
- **Analyze** the structure of bone.
- **Summarize** the process of bone development.
- **List** two ways to prevent osteoporosis.
- **Identify** the three main classes of joints.

Key Terms

axial skeleton
appendicular skeleton
bone marrow
periosteum
Haversian canal
osteocyte
osteoporosis
joint
ligament

The Skeleton

What keeps your body from collapsing like a limp noodle? An internal skeleton of bones shapes and supports your body. Your skeleton provides protection for internal organs and, along with muscles, enables movement with a versatile system of levers and joints. Muscles pull against bones at joints, moving the limbs and the trunk. Your skeleton is made mostly of bone, a type of hard connective tissue that is constantly being formed and replaced. The human skeleton, shown in **Figure 3,** contains 206 individual bones. Of these, 80 bones form the **axial skeleton,** which includes bones of the skull, spine, ribs, and sternum. The other 126 bones, including those of the arms, legs, pelvis, and shoulder, form the **appendicular** *(ap uhn DIHK yoo luhr)* **skeleton.**

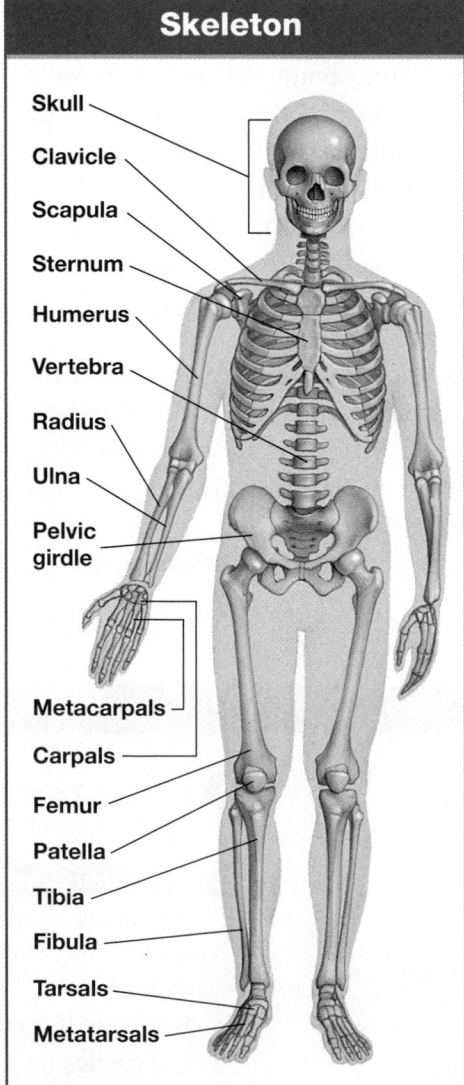

Skeleton

Skull
Clavicle
Scapula
Sternum
Humerus
Vertebra
Radius
Ulna
Pelvic girdle
Metacarpals
Carpals
Femur
Patella
Tibia
Fibula
Tarsals
Metatarsals

Figure 3   Skeleton. Bones of the appendicular skeleton "hang" from bones of the axial skeleton (purple). Immature humans have 33 vertebrae. During development, vertebrae of the sacrum and coccyx fuse so that there are 26 separate bones.

Axial Skeleton

The most complex part of the axial skeleton is the skull. Of the 29 bones in the skull, 8 bones form the cranium, which encases the brain. The skull also contains 14 facial bones, 6 middle-ear bones, and a single bone that supports the base of the tongue. The skull is attached to the top of the spine, or backbone, which is a flexible, curving column of 26 vertebrae that supports the center of the body. Curving forward from the middle vertebrae are 12 pairs of ribs, which form a protective rib cage around the heart and lungs.

Appendicular Skeleton

The appendicular skeleton forms the appendages or limbs—the shoulders, arms, hips, and legs. The arms and legs are attached to the axial skeleton at the shoulders and hips, respectively. The shoulder attachment, called the pectoral girdle, contains two large, flat shoulder blades, or scapulas, and two slender, curved collarbones, or clavicles. The clavicles connect the scapulas to the upper region of the sternum and hold the shoulders apart. This arrangement enables full rotation of the arms about the shoulder. The hip attachment, called the pelvic girdle, contains two large pelvic bones. The pelvic bones distribute the weight of the upper body evenly down the legs.

WORD Origins

● The word *periosteum* is from the Greek *peri*, meaning "around," and *osteon*, meaning "bone."

Structure of Bone

As shown in **Figure 4,** bones are made of a hard outer covering of compact bone surrounding a porous inner core of spongy bone. Compact bone is a dense connective tissue that provides a great deal of support. Spongy bone is a loosely structured network of separated connective tissue. Some cavities in spongy bone are filled with a soft tissue called **bone marrow.** Red bone marrow begins the production of all blood cells and platelets. The hollow interior of long bones is filled with yellow bone marrow. Yellow bone marrow consists mostly of fat, which stores energy. Bones are surrounded and protected by a tough exterior membrane called the **periosteum** *(pair ee AHS tee uhm).* The periosteum contains many blood vessels that supply nutrients to bones.

Figure 4   Structure of bone

Many bones contain bone marrow, blood vessels, and both compact and spongy bone tissue.

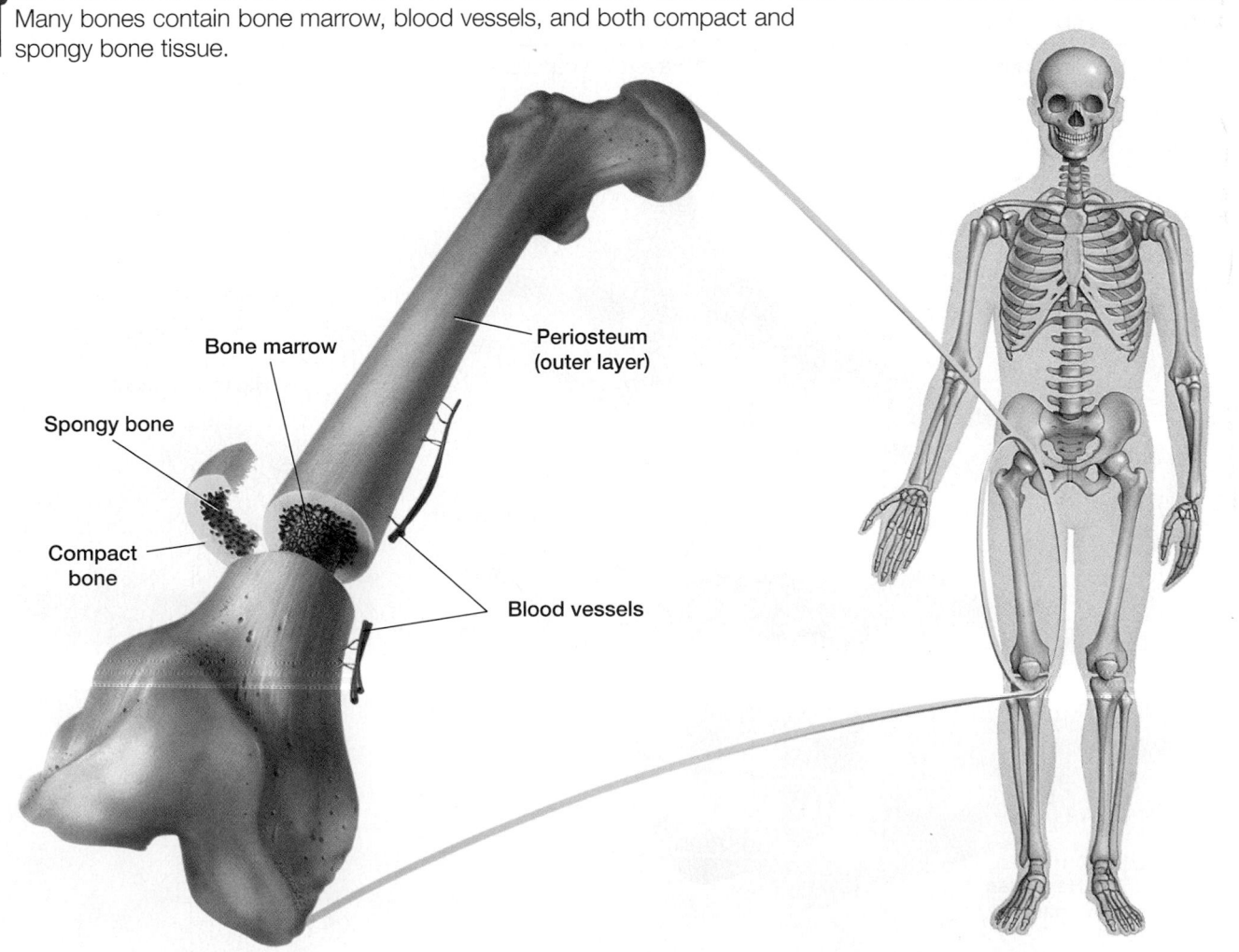

Spongy bone

Bone marrow

Periosteum (outer layer)

Compact bone

Blood vessels

Growth of Bones

Real Life

Years after an unsolved murder is committed, the victim's bones may hold clues to the crime.

Forensic anthropologists have solved many cases by analyzing bones and other human remains found at crime scenes.

Finding Information
Read some accounts of crimes solved by forensic anthropologists to learn how these scientists analyze bones.

In early development, the skeleton is made mostly of cartilage, a type of connective tissue that serves as a template for bone formation. During development, most cartilage is gradually replaced by bone as minerals are deposited. Deposits of calcium and other minerals harden bones, enabling them to withstand stress and provide support. In compact bone, new bone cells are added in layers around narrow, hollow channels called **Haversian canals.** Haversian canals extend down the length of a bone, and they contain blood vessels that enter the bone through the periosteum.

As shown in **Figure 5,** layers of new bone cells form several concentric rings around Haversian canals. These rings form columns that enable the bone to withstand tremendous amounts of stress. Eventually, bone cells called **osteocytes** *(AHS tee oh siets)* become embedded within the bone tissue. Osteocytes maintain the mineral content of bone. The blood vessels that run through each Haversian canal supply the osteocytes with nutrients needed to maintain bone cells. Bones continue to thicken and elongate through adolescence as bone cells replace cartilage. Bone elongation occurs at the ends of long bones. Cartilage degenerates as new bone cells are added, causing bones to lengthen.

Figure 5   Compact bone

In compact bone, concentric rings of bone surround Haversian canals.

Bone marrow

Haversian canal

Osteocytes

Periosteum

Vein

Artery

Haversian canals

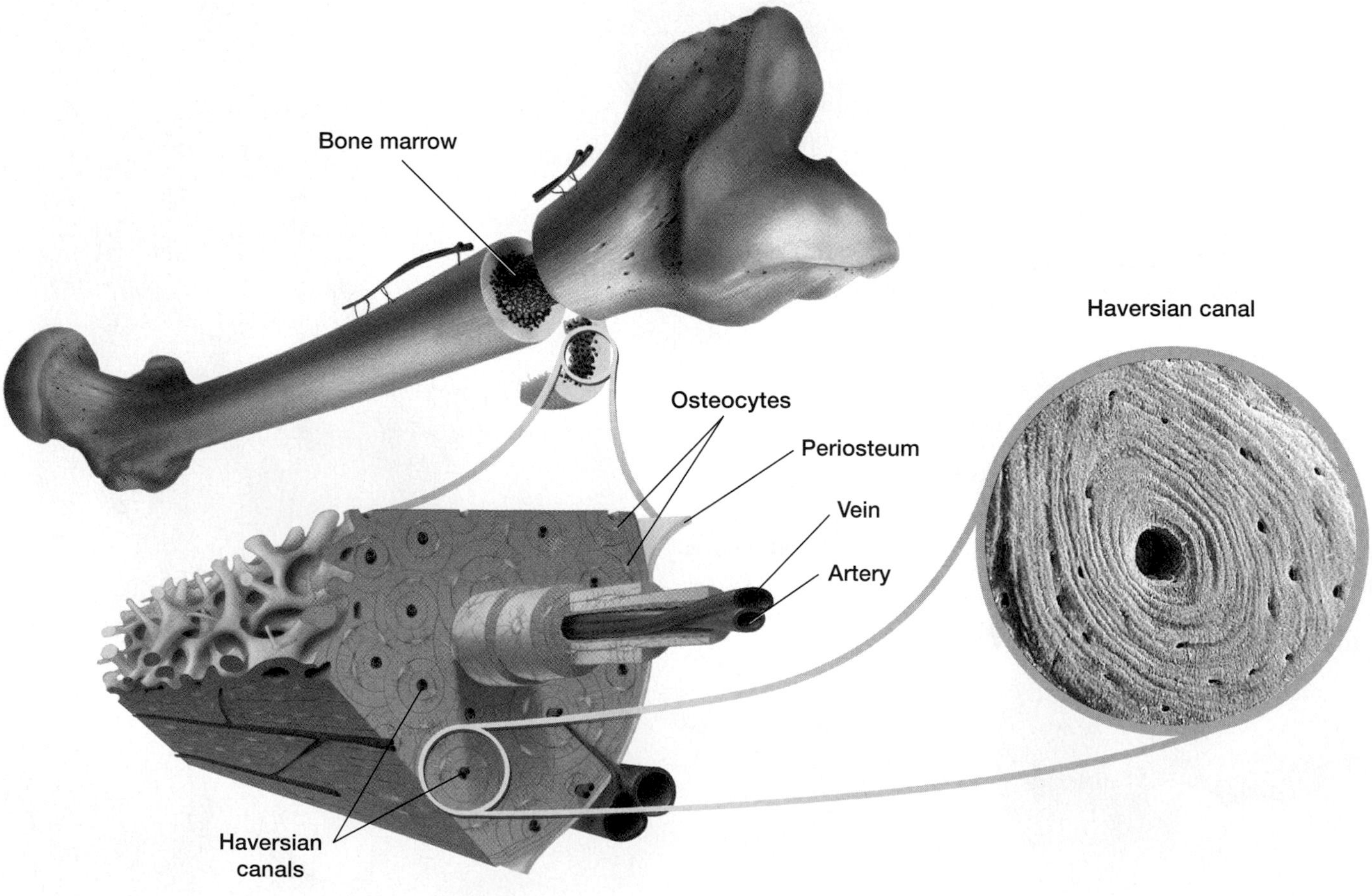

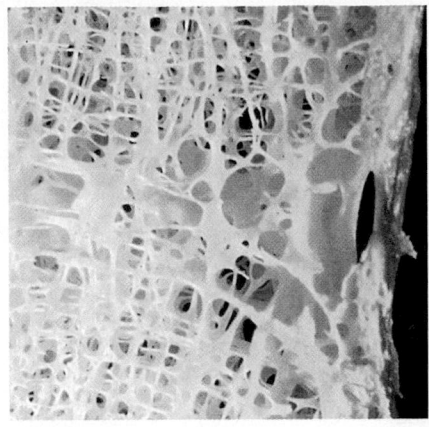

Normal bone

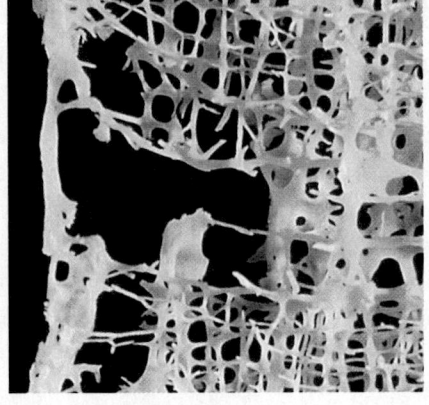

Bone in osteoporosis

Dowager's hump

Osteoporosis

In young adults, the density of bone usually remains constant. However, around the age of 35, bone replacement gradually becomes less efficient and some bone tissue is lost. Severe bone loss, as shown in **Figure 6,** can lead to a condition called **osteoporosis** *(ahst ee oh puh ROH sihs),* which means "porous bone." Bones affected by osteoporosis become brittle and are easily fractured.

Although both women and men lose bone tissue as they age, more women than men are affected by osteoporosis. Because women's bones are usually smaller, women cannot afford to lose as much bone tissue as men. In addition, the production of sex hormones decreases after menopause. This decrease in hormone production has been linked to an increased rate of bone loss in women following menopause.

You can take action now to prevent future osteoporosis. Building strong bones now will make you less likely to be affected by osteoporosis later in life. A mineral-rich diet that includes dairy products, green leafy vegetables, whole grains, and legumes such as those shown in **Figure 7,** together with regular exercise throughout your life will help maintain bone density.

Figure 6 Effects of osteoporosis. Compare the density of a normal bone with that of a bone weakened by osteoporosis. One sign of osteoporosis is the familiar "dowager's hump" of the back, caused by curvature of the spine.

Figure 7 Preventing osteoporosis. A proper diet and regular exercise help reduce the risk of osteoporosis.

Joints

A **joint** is where two bones meet. Pads of cartilage cushion the ends of the bones of a joint, enabling the joint to withstand great pressure and stress. The bones of a joint are held together by strong bands of connective tissue called **ligaments.** Ligaments not only help stabilize joints but also prevent joints from moving too far in any one direction. Many sports-related injuries to ligaments are caused by an impact that forces a joint to overextend. Injury occurs because the impact exceeds the tension that the ligaments can withstand.

Three Main Types of Joints

The skeletal system contains three main types of joints that enable varying degrees of movement: immovable joints, slightly movable joints, and freely movable joints. Examples of the three types of joints are shown in **Figure 8.**

Immovable Joints Tight joints that permit little or no movement of the bones they join are called immovable joints. The cranial bones of the skull are joined by sutures, a type of immovable joint in which the bones are separated by only a thin layer of connective tissue.

Slightly Movable Joints Joints that permit limited movement of the bones they join are called slightly movable joints. For example, the vertebrae of the spine are joined by cartilaginous joints, which are a kind of slightly movable joint in which a bridge of cartilage connects the bones. Slightly movable joints are also located between bones of the rib cage.

Freely Movable Joints In joints that permit movement, the direction of bone movement is determined by the structure of the joint. Joints that permit the most movement are called freely movable joints. Some kinds

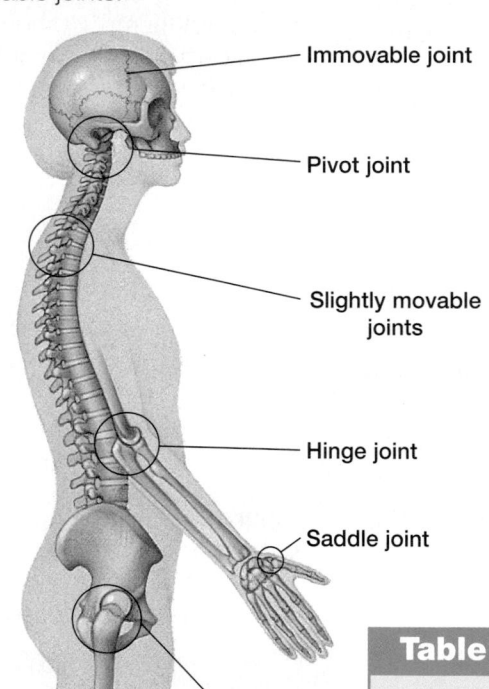

Figure 8 Types of joints.
The body contains immovable, slightly movable, and freely movable joints.

- Immovable joint
- Pivot joint
- Slightly movable joints
- Hinge joint
- Saddle joint
- Ball-and-socket joint
- Gliding joint

Table 2 Movable Joints		
Joint	**Type of movement**	**Examples**
Ball-and-socket joint	All types	Shoulders and hips
Pivot joint	Rotation	Top of spine (turning of head)
Hinge joint	Bending and straightening	Elbows, knuckles of fingers and toes
Gliding joint	Sliding motion	Wrists and ankles
Saddle joint	Rotation, bending, and straightening	Base of thumbs

Figure 9 Knee. The knee is an example of a freely movable joint.

Muscle

Cartilage

Ligaments

Fibula
(bone)

Tendon

Patella
(kneecap)

Tibia
(bone)

of freely movable joints are listed in **Table 2.** The structure of one freely movable joint, the knee, is shown in **Figure 9.**

Disorders of Joints

Recall that ligaments hold the bones of a joint together. A lining of tissue that surrounds the joint secretes a lubricating fluid that reduces friction at the ends of the bones. When a disease afflicts the bones, connective tissue, or lubricating tissues in a freely movable joint, the joint's ability to move may be greatly impaired.

Rheumatoid arthritis is a painful inflammation of freely movable joints. This condition occurs when cells of the immune system attack the tissues around joints, severely damaging the joints. Symptoms of rheumatoid arthritis include stiffening and swelling of the joints. Osteoarthritis is a similar disorder that causes the degeneration of cartilage that covers the surfaces of bones. As the cartilage wears away, the bones rub together, causing pain.

Section 2 Review

1 **Distinguish** between the axial skeleton and the appendicular skeleton.

2 **Differentiate** between compact bone and spongy bone.

3 **Describe** how bones elongate in development.

4 **List** the three main types of joints, and give an example of each type.

5 **Analyzing Information** Why are women more likely than men to develop osteoporosis?

6 **Critical Thinking Relating Concepts**
The bones of a newborn baby are made mostly of cartilage. Why is that an advantage during childbirth?

7 **Standardized Test Prep** What effect does regular exercise have on the skeletal system?

A reduces bone mass

B leads to osteoporosis

C maintains bone density

D makes bones more porous

Muscular System

- **Describe** the action of muscle pairs in moving the body.

- **Relate** the structure of a skeletal muscle to the muscle's ability to contract.

- **Describe** how energy is supplied to muscles for contraction.

Key Terms

tendon
flexor
extensor
actin
myosin
myofibril
sarcomere

Muscles and Movement

Every time you move, you use your muscles. Walking and running, for example, require precisely timed and controlled contractions of many skeletal muscles. When you lift a heavy object, the total force produced by muscle contractions in your arm must overcome the weight of the object. Muscles in your jaw contract and enable you to chew food with your teeth. Even when you are idle, many skeletal muscles, including those in your back and neck, remain partially contracted to maintain balance and posture.

Movement of the Skeleton

Muscles can move body parts because muscles are attached to bones of the skeleton. As shown in **Figure 10,** most skeletal muscles are attached to bones by strips of dense connective tissue called **tendons.** One attachment of the muscle, the origin, is a bone that remains stationary during a muscle contraction. The muscle pulls against the origin. The other attachment, the insertion, is the bone that moves when the muscle contracts. Movement occurs when a muscle contraction pulls the muscle's insertion toward its origin.

Skeletal muscles are generally attached to the skeleton in opposing pairs. One muscle in a pair pulls a bone in one direction, and the other muscle pulls the bone in the opposite direction. In the limbs, each opposing pair of muscles includes a flexor muscle and an extensor muscle, as shown in Figure 10. A **flexor** muscle causes a joint to bend. An **extensor** muscle causes a joint to straighten.

Figure 10   Muscle pair

Pairs of opposing muscles work together to move bones at joints.

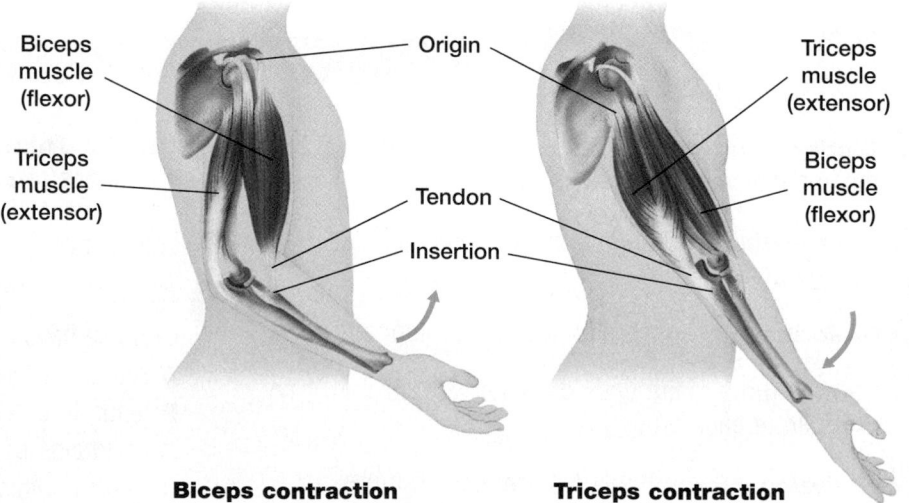

Biceps muscle (flexor)

Triceps muscle (extensor)

Origin

Tendon

Insertion

Triceps muscle (extensor)

Biceps muscle (flexor)

Biceps contraction   **Triceps contraction**

Figure 11   Skeletal muscle

In skeletal muscle, contraction occurs within the sarcomeres of muscle fibers.

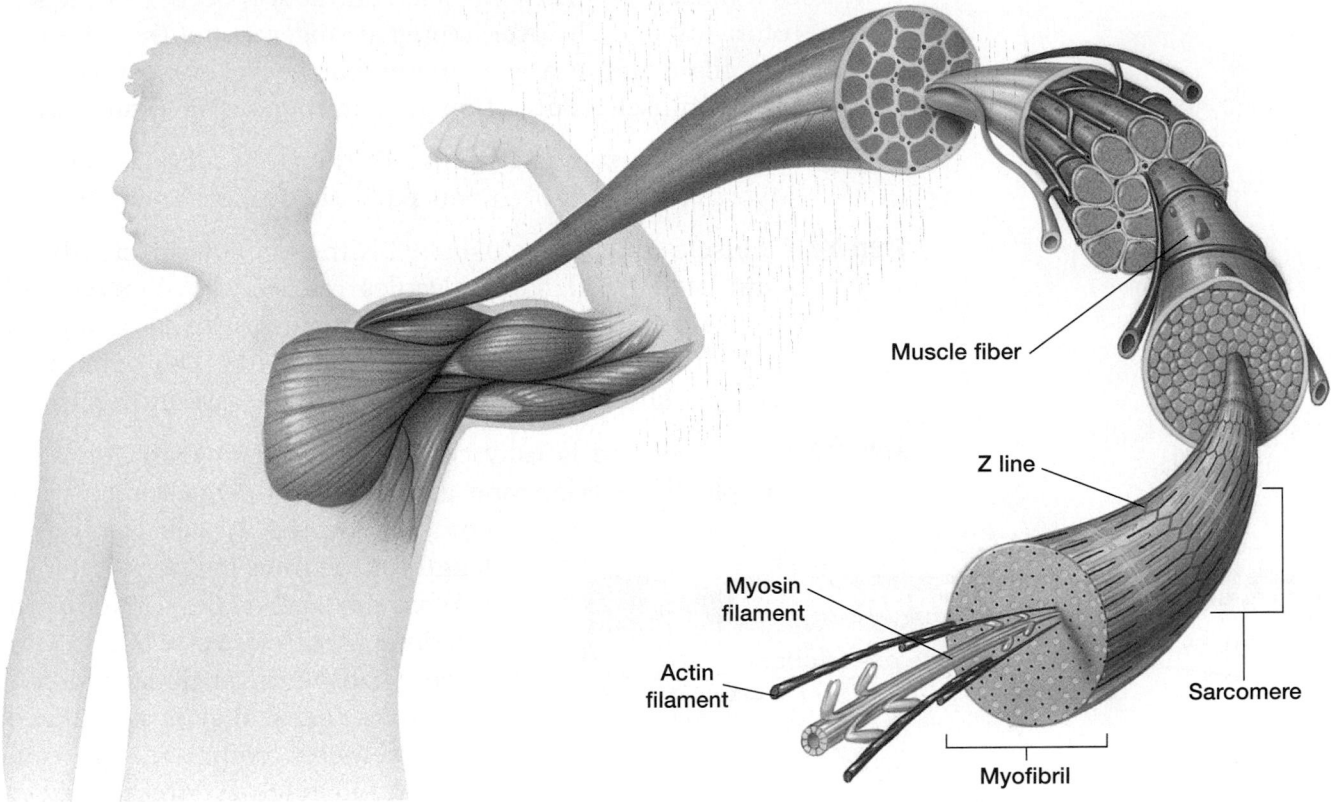

Muscle fiber

Z line

Myosin filament

Actin filament

Sarcomere

Myofibril

Muscle Structure

Muscles contain some connective tissue, which holds muscle cells together and provides elasticity. Muscle tissue also contains large amounts of contractile protein filaments. These protein filaments, called **actin** and **myosin** *(MIE oh sihn),* enable muscles to contract. Actin and myosin are usually found in the cytoskeleton of eukaryotic cells, but they are far more abundant in muscle cells. Other characteristics of muscle tissue include the ability to stretch or expand and the ability to respond to stimuli, such as signal molecules released by nerve cells.

Skeletal muscle tissue consists of many parallel elongated cells called muscle fibers. As shown in **Figure 11,** each muscle fiber contains small cylindrical structures called **myofibrils** *(mie oh FIE bruhlz).* Myofibrils have alternating light and dark bands that produce a characteristic striated, or striped, appearance when viewed under a microscope. In the center of each light band is a structure called a Z line, which anchors actin filaments. The area between two Z lines is called a **sarcomere** *(SAHR koh mihr).* Thus, a myofibril is a grouping of sarcomeres linked end to end. Each sarcomere contains overlapping thin and thick protein filaments that move and interact with each other. The thin filaments are actin, and the thick filaments are myosin. The filaments run parallel to one another along the length of the sarcomere. The dark bands that occur in the middle of the sarcomere are regions where the thick filaments and the thin filaments overlap.

internet connect

www.scilinks.org
Topic: Muscle Structure
Keyword: HX4126

SCiLINKS Maintained by the
National Science
Teachers Association

Muscle Contraction

How does a muscle contract? Muscle contraction occurs in the sarcomeres of myofibrils. The overlapping arrangement of the thick and thin protein filaments in a sarcomere enables muscle contraction. The events of a muscle contraction are summarized in **Figure 12.**

Step ❶ Before a muscle is stimulated, the sarcomere is relaxed. Myosin and actin filaments partially overlap one another.

Step ❷ A muscle contraction usually begins when a muscle fiber is stimulated by signal molecules released by a nerve cell. This causes myosin and actin filaments to "slide" along one another so that they overlap even more. The sarcomere becomes shorter as the Z lines are pulled closer together.

Step ❸ The sarcomere is fully contracted, and myosin and actin completely overlap one another. This shortening of sarcomeres occurs down the entire length of the muscle fiber.

What determines the force of contraction? A muscle exerts the greatest force when all of its fibers are contracted. When a fiber is stimulated, its sarcomeres contract. The total amount of force a muscle exerts depends on how often muscle fibers are stimulated and how many muscle fibers contract.

How is the force of muscular contraction controlled? As different numbers of fibers in a muscle contract at one time, the total force generated by contraction varies. For example, the total amount of force needed to lift a pencil is much less than the force needed to lift a brick. Thus, fewer muscle fibers in your arm contract when you lift a pencil than when you lift a brick.

The set of muscle fibers activated by a nerve cell is called a motor unit. Every time a nerve cell activates its motor unit, all the fibers in that unit contract. Muscles that require a finer degree of control, such as muscles that move the fingers, have only a few muscle fibers in each motor unit. Large muscles, such as muscles in the leg, have several hundred muscle fibers in each motor unit.

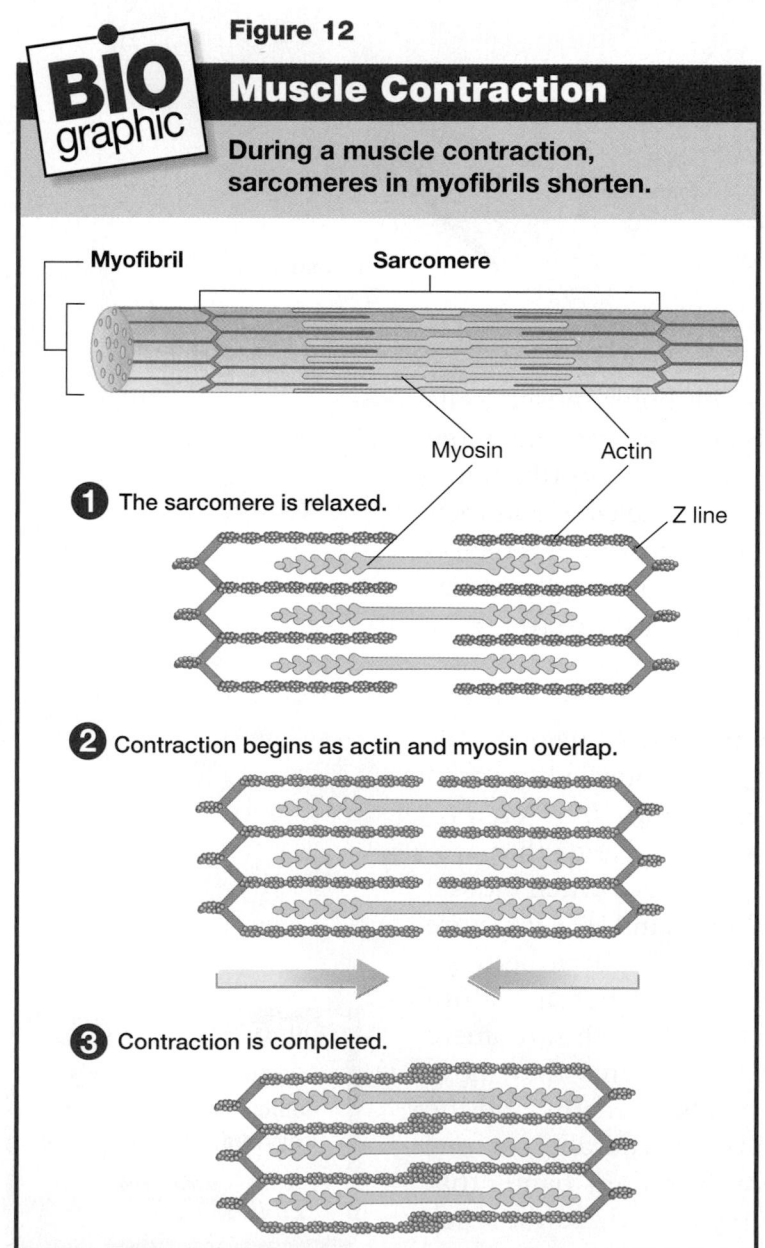

Figure 12

BIOgraphic

Muscle Contraction

During a muscle contraction, sarcomeres in myofibrils shorten.

Myofibril — Sarcomere

Myosin — Actin

Z line

❶ The sarcomere is relaxed.

❷ Contraction begins as actin and myosin overlap.

❸ Contraction is completed.

Interaction of Myosin and Actin

How do actin and myosin cause sarcomeres to shorten during a muscle contraction? Myosin filaments have long, finger-like projections with an enlarged "head" at one end. Actin filaments contain many sites to which myosin can bind during a muscle contraction. Stimulation of a muscle fiber leads to the exposure of these binding sites on actin filaments. As shown in **Figure 13,** the myosin heads attach to the binding sites on actin filaments and then rotate, causing myosin to move relative to actin.

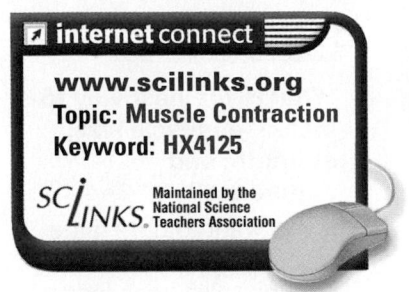

internet connect

www.scilinks.org
Topic: Muscle Contraction
Keyword: HX4125

SCI LINKS Maintained by the National Science Teachers Association

Step ❶ Muscle contraction begins as a myosin head attaches to an exposed binding site on an actin filament.

Step ❷ The myosin head rotates, causing the actin filament to "slide" against the myosin filament. This sliding causes the filaments to overlap one another.

Step ❸ ATP is used as the myosin head detaches and snaps back into its original position. The myosin head reattaches to actin at a binding site farther along the actin filament. When the myosin heads cannot move farther, they release momentarily and reposition themselves to grab the actin and pull again. Thus, the myosin heads "walk" along actin filaments, essentially "stepping" at each available binding site. This grabbing and pulling action is repeated, causing the sarcomere to shorten as the Z lines are pulled closer together.

A lot of energy is needed to power a muscle contraction. ATP is used each time a myosin head moves from one binding site on an actin filament to another. Without ATP, myosin heads would remain attached to actin filaments, keeping the muscle contracted. ATP is also used to regulate calcium ions in muscle cells. Calcium ions are needed for binding sites to be exposed on actin filaments. Without calcium ions and ATP, a muscle could not contract.

Figure 13

BIO graphic

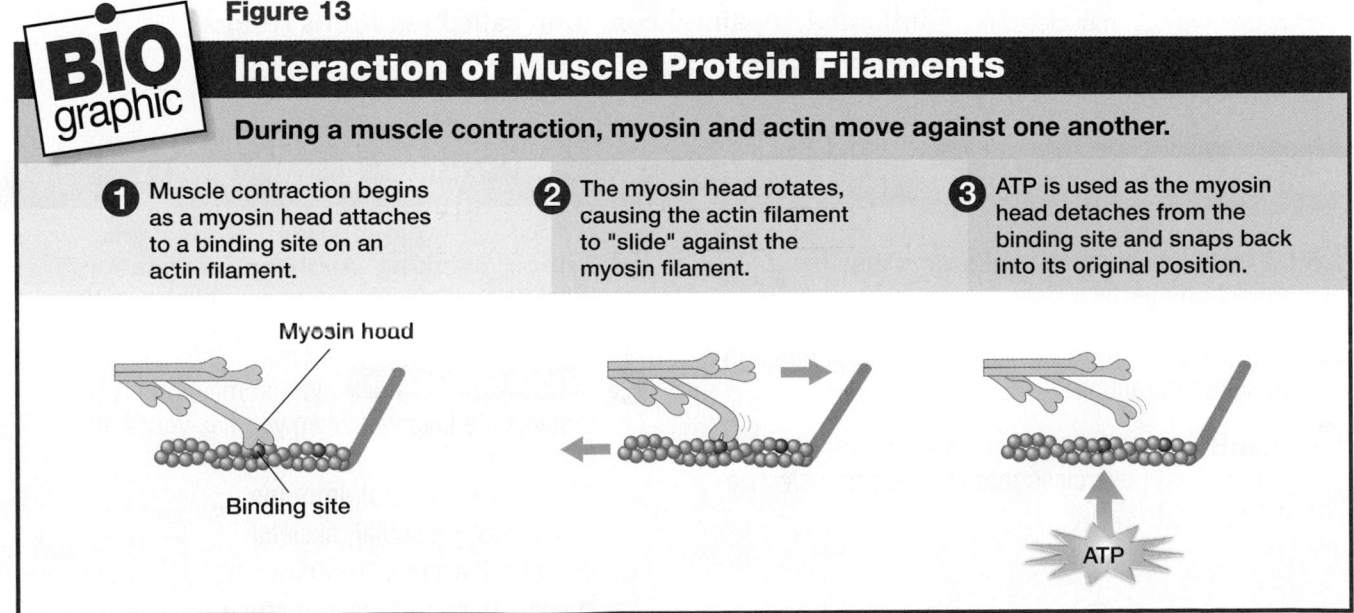

Interaction of Muscle Protein Filaments

During a muscle contraction, myosin and actin move against one another.

❶ Muscle contraction begins as a myosin head attaches to a binding site on an actin filament.

❷ The myosin head rotates, causing the actin filament to "slide" against the myosin filament.

❸ ATP is used as the myosin head detaches from the binding site and snaps back into its original position.

Myosin head

Binding site

ATP

Aerobic and Anaerobic Energy Pathways

The ATP used to power contractions is usually supplied by aerobic respiration. During prolonged exercises, such as a long-distance walk, oxygen is consumed at a sustainable, steady rate, and aerobic respiration yields most of the ATP. However, during brief, intense activities, such as sprinting and weight lifting, anaerobic processes take over. Most of the ATP used in such activities comes from glycolysis, as the oxygen available to muscle cells rapidly decreases.

When both anaerobic and aerobic energy pathways become insufficient for muscle contraction, muscles can use only glycogen as an energy source. As glycogen is used up, the body begins to use fat as an energy source. When ATP consumption exceeds ATP production, muscle fatigue and soreness may result, leaving muscle fibers unable to recover from contraction.

Exercise and Fitness

Consistent aerobic exercise makes the heart pump more efficiently and thus increases the energy available to muscles as a result of improved blood circulation. More oxygen is extracted by the body with each breath, increasing the oxygen supply to muscles. More ATP is available for muscle contractions, thereby reducing muscle fatigue. The increase in muscle efficiency results in greater endurance, or the ability to continue exercising.

Resistance exercises, such as weight lifting, shown in **Figure 14,** can increase muscle size and strength. Resistance exercises are mostly anaerobic, so they do not usually improve the uptake of oxygen to muscles. Muscle mass is increased by resistance training. The amount of tension and the rate of exercise are both important factors. However, the short-term demands of such strength training do not cause the circulatory changes that increase endurance.

Excessive exercise or failure to warm up properly can lead to muscle injury. Muscles can tear if they are stretched too far during strenuous exercise. If excessive stress causes tendons to become inflamed, a painful condition called tendinitis results.

Figure 14   Resistance exercise. Weight lifting and other resistance exercises primarily involve anaerobic energy pathways in muscles.

Section 3 Review

1 **Describe** how muscle pairs work together to move body parts.

2 **Compare** the roles of thick and thin filaments in muscle contraction.

3 **Identify** the energy pathway that is primarily involved with exercises that increase muscle size and strength.

4 **Critical Thinking Applying Information** What causes muscle cramping after rigorous exercise or a repeated movement?

5 **Standardized Test Prep** Which main organ systems are involved when you flex your arm at the elbow?

A muscular, skeletal, immune

B nervous, muscular, skeletal

C skeletal, excretory, nervous

D endocrine, muscular, immune

Skin, Hair, and Nails

Skin

The skin, which makes up about 15 percent of your total body weight, is the largest organ of the body. Many specialized structures are found in the skin, which along with the hair and nails, forms the integumentary system. The skin protects the body from injury, provides the first line of defense against disease, helps regulate body temperature, and prevents the body from drying out through evaporation. As shown in **Figure 15,** the skin is made mostly of connective tissue and layers of epithelial tissue. The two primary layers of skin are the epidermis and the dermis.

Epidermis

The **epidermis** is the outermost layer of the skin. About as thick as this page, the epidermis is made of several layers of epithelial cells. The part of the epidermis you see when you look in a mirror is a thin layer of flattened, dead cells that contain keratin. **Keratin** is a protein that makes skin tough and waterproof. The cells of the epidermis are continuously damaged by the environment. They are scraped, ripped, worn away by friction, and dried out because of moisture loss. Your body deals with this damage not by repairing the cells, but by replacing them.

Objectives

- **Analyze** the structure and function of the epidermis.
- **Describe** how the dermis helps the body maintain homeostasis.
- **Summarize** how hair and nails are formed.
- **Identify** various types of skin disorders.

Key Terms

epidermis
keratin
melanin
dermis
hair follicle
subcutaneous tissue
sebum

Figure 15   Structure of skin

Skin has two distinct layers that contain many blood vessels, nerve cells, muscles, hairs, and glands.

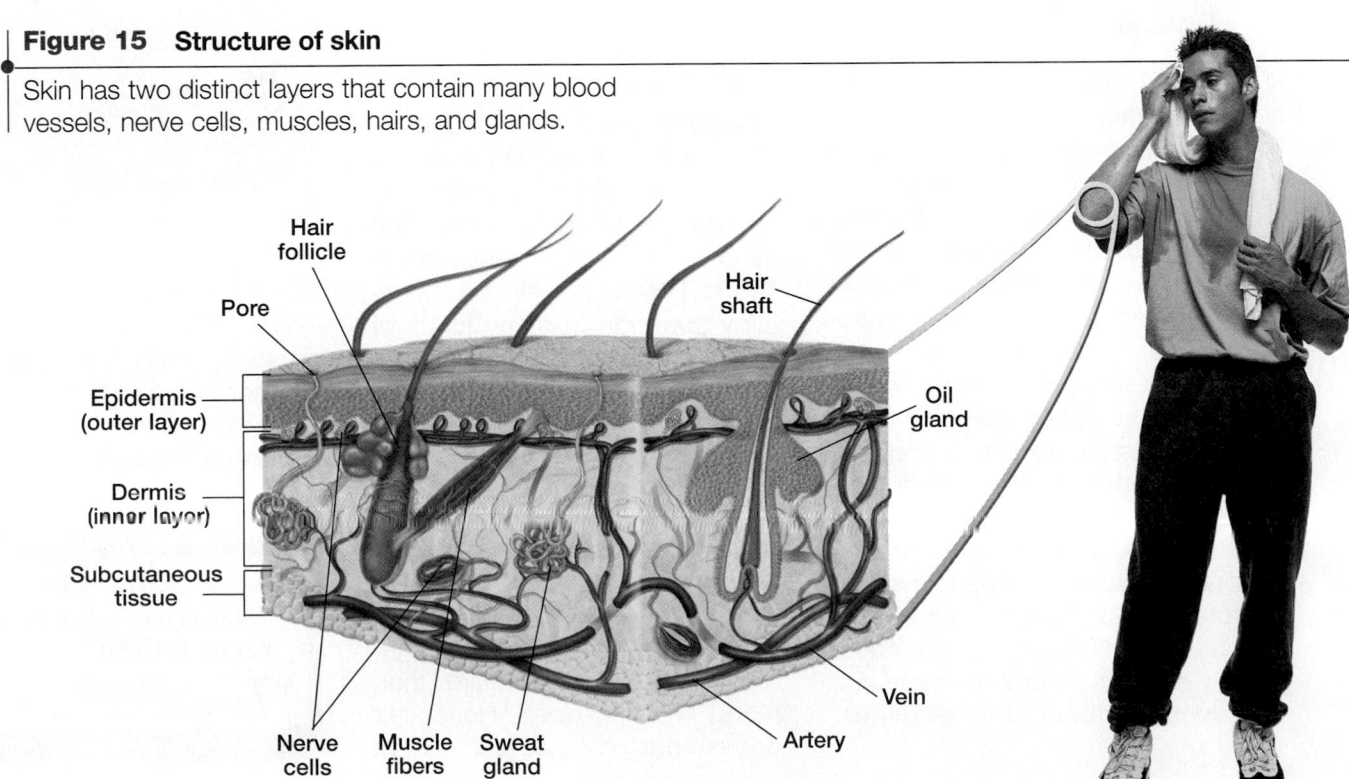

Hair follicle

Pore

Hair shaft

Epidermis (outer layer)

Oil gland

Dermis (inner layer)

Subcutaneous tissue

Nerve cells

Muscle fibers

Sweat gland

Artery

Vein

The outermost cells of the skin are continually shed and replaced by a layer of actively dividing cells at the base of the epidermis. As new skin cells form, they migrate upward and produce large amounts of keratin. These cells are shed about a month after they reach the surface. The inner layer of the epidermis also contains cells that produce the pigment melanin. **Melanin** *(MEHL uh nihn)* ranges in color from yellow to reddish brown to black, and it helps determine skin color. People with more melanin tend to have darker skin, and people with less melanin usually have lighter skin. Melanin also absorbs ultraviolet (UV) radiation, protecting the skin from exposure to sunlight. Exposure to UV radiation increases the production of melanin. This is why some people become "tan" after exposure to sunlight. However, UV radiation has been shown to cause skin cancer, especially in people with light skin. Thus, you should avoid excessive exposure to sunlight and wear sunscreen when outdoors.

Dermis

The **dermis** is the functional layer of skin that lies just beneath the epidermis. Connective tissue in the dermis makes the skin tough and elastic. The dermis contains many nerve cells, blood vessels, **hair follicles,** and specialized skin cells. Sensations of touch,

FORENSICS BIOWatch

Courtroom Science

In the United States, medical evidence is often used in criminal trials. Hairs left at a crime scene can provide enough information to make or break a case. To the unaided eye, the differences between hairs amount to no more than color, coarseness, and whether the hairs are curly or straight. Using a microscope, however, scientists can distinguish more than two dozen characteristics in a single hair. Experts compare hairs found at a crime scene with hairs of suspects. These comparisons can rule out a suspect when the hairs are different, but they usually are not enough to convict a suspect without other evidence.

DNA Fingerprinting

Hair consists of cells that contain DNA. Thus, scientists can get additional evidence from hair through DNA fingerprinting. First used as forensic evidence in the 1980s, DNA fingerprinting is now fairly common in many kinds of criminal trials. By 1997, DNA fingerprinting had been used in more than 50,000 cases in the United States.

Drug Testing

Hair can also provide evidence of drug use. Hair collects drugs that are delivered to hair follicles by blood. Hair provides a longer record of drug exposure than blood or urine does. Head hair grows an average of about 1 cm

Forensic scientist

(0.4 in.) per month; a 10 cm (4 in.) length of hair gives evidence of a person's drug use during the last 10 months. Such evidence has been introduced in cases involving illegal drug use.

internet connect

www.scilinks.org
Topic: Forensic Analysis
Keyword: HX4087

SCLINKS Maintained by the National Science Teachers Association

temperature, and pain originate in nerve cells within the dermis. The dermis also contains tiny muscles that are attached to hair follicles in your skin. When you get cold, these muscles contract and pull the hair shafts upright, helping to insulate the body. These muscles also cause goose bumps on the skin's surface.

Temperature Regulation A network of blood vessels in the dermis provides nourishment to the living cells of the skin. These blood vessels also help regulate body temperature by either radiating heat into the air or conserving heat. If your body gets too hot, blood vessels just under the skin dilate so that blood flows near the skin's surface, releasing heat from the body. This is why people with light complexions turn slightly red during strenuous exercise, as shown in **Figure 16.** If your body gets too cold, the blood vessels constrict, reducing heat loss.

Sweat glands in the dermis also help remove excess body heat. The evaporation of sweat from the skin's surface removes heat more efficiently than the dilation of blood vessels. Most sweat is about 99 percent water and 1 percent dissolved salts and acids. Certain sweat glands located in body areas with dense hair, such as the armpits, also secrete proteins and fatty acids. Because these substances provide a rich food source for bacteria, stale sweat often releases the offensive odor of bacterial waste products.

Figure 16 Cooling mechanism. Flushed (reddened) skin is a sign that the body is overheated and is releasing heat from the skin.

Subcutaneous Tissue

Subcutaneous tissue, located beneath the skin just under the dermis, is a layer of connective tissue made mostly of fat. Subcutaneous tissue acts as a shock absorber, provides additional insulation to help conserve body heat, and stores energy. Subcutaneous tissue also anchors the skin to underlying organs. The thickness of subcutaneous tissue varies in different parts of the body. For example, the eyelids have very little, while the buttocks and thighs may have a lot. The pads of subcutaneous tissue in the soles of your feet may be more than 6 mm (0.25 in.) thick.

Hair and Nails

Hair and nails are derived from the epidermis. Hair follicles produce individual hairs, which help protect and insulate the body. Shown in **Figure 17,** hair is made mostly of dead, keratin-filled cells. A shaft of hair grows up from the hair follicle and through the skin's surface. Each hair on your head grows for several years. Then the follicle enters a resting phase for several months, and the hair is eventually shed. Hair color is primarily determined by the presence of the pigment melanin. Blonde hair and red hair typically contain less melanin than brown hair and black hair.

Nails are produced by specialized epidermal cells located in the light, semicircular area at the base of each nail. These cells become filled with keratin as they are pushed outward by new cells. Nails protect the tips of the fingers and toes and continue to grow throughout life.

Figure 17 Hair. An average human head contains about 100,000 hairs.

Skin Disorders

The skin is continuously exposed to damaging factors such as insect bites, microorganisms, and ultraviolet radiation. Injuries such as scrapes and blisters are often minor and usually heal rapidly without permanent scarring. Burns, however, can be very serious and can result in permanent scarring or even death. Some skin disorders are the result of changes that occur within the body over time.

Figure 18 Skin cancer. In its early stages, a carcinoma may look like a wart. A malignant melanoma often looks like a mole that changes in size, shape, or color.

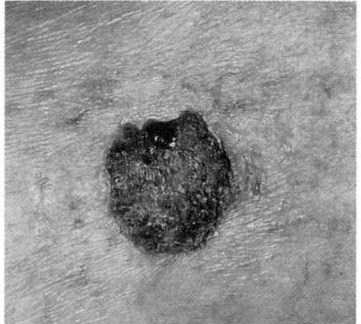

Carcinoma

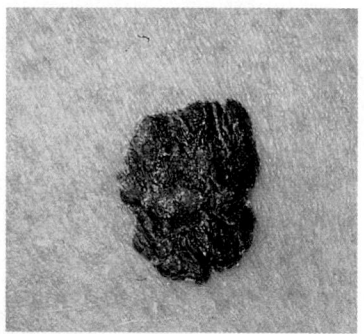

Malignant melanoma

Acne

The most common skin problem for teenagers is acne *(AK nee)*, a chronic inflammatory condition that involves the skin's oil-producing glands. Oil glands in the dermis release **sebum** *(SEE buhm)*, an oily secretion that lubricates the skin. Sebum is released through ducts, or pores, into nearby hair follicles. These oil glands are especially active during adolescence. Acne is caused by excessive secretion of sebum, which blocks pores with oil, dirt, and bacteria. Makeup and other cosmetic products can contribute to clogging. As a result, the surrounding tissue becomes infected and inflamed, and the pores accumulate pus, producing pimples. Serious acne may need to be treated using antibiotics. Although acne cannot be prevented, it can usually be managed with proper skin care.

Skin Cancer

Skin cancer can result from genetic mutations caused by overexposure to UV radiation. The most common types of skin cancer are carcinomas *(kahr sih NOH mahz)*, which originate in skin cells that do not produce pigments. If they are detected early, carcinomas can be treated. A small percentage of skin cancers are caused by mutations that occur in pigment-producing skin cells. These cancers, called malignant melanomas *(mehl uh NOH mahz)*, grow very quickly and spread easily to other parts of the body. About 8 out of 10 skin cancer deaths are from malignant melanomas. A carcinoma and a malignant melanoma are shown in **Figure 18.** You can reduce the risk of skin cancer by avoiding overexposure to either natural or artificial UV radiation and by using protective sunscreens.

Section 4 Review

1. **Describe** the structure of the epidermis.

2. **List** two ways that the dermis helps regulate body temperature.

3. **Summarize** how nails are formed.

4. **Identify** the most common cause of skin cancer and how it can be avoided.

5. **Critical Thinking Recognizing Relationships** Why is a third-degree burn, which destroys the epidermis and dermis of the skin, such a serious injury?

6. **Standardized Test Prep** When your body temperature becomes too low, blood vessels just under your skin
   A dilate.
   B constrict.
   C release sweat.
   D raise hair shafts.

Study ZONE CHAPTER HIGHLIGHTS

Key Concepts

1 Body Organization

- Cells are grouped into four types of body tissues: epithelial tissue, nervous tissue, connective tissue, and muscle tissue.
- Body organs contain several types of body tissues.
- Organs are grouped into organ systems in which organs interact to perform a certain function, such as digestion.
- Endothermy enables the body to maintain homeostasis at all times, regardless of the temperature outside the body.

2 Skeletal System

- The skeleton supports the body, provides protection for internal organs, and enables movement.
- The 206 bones of the skeleton are divided into the axial skeleton and the appendicular skeleton.
- Bones are made of hard compact bone surrounding porous spongy bone.
- Early in development, the skeleton is mostly cartilage. Bones harden as calcium and other mineral deposits build up.
- Bones thicken and elongate as development continues.
- Three kinds of joints fasten bones together: immovable joints, slightly movable joints, and freely movable joints.

3 Muscular System

- Muscles are attached to bones by tendons.
- Muscle pairs move parts of the body by pulling on bones.
- Sarcomeres shorten during muscle contraction.
- During muscle contraction, actin and myosin interact.
- Energy is required for muscles to contract.

4 Skin, Hair, and Nails

- The skin consists of two layers: the epidermis and the dermis.
- Subcutaneous tissue anchors skin to underlying organs.
- Hair and nails are derived from the epidermis.
- Most skin disorders are caused by damage to the epidermis.

Key Terms

Section 1

epithelial tissue (846)
nervous tissue (847)
connective tissue (847)
muscle tissue (847)
body cavity (849)

Section 2

axial skeleton (850)
appendicular skeleton (850)
bone marrow (851)
periosteum (851)
Haversian canal (852)
osteocyte (852)
osteoporosis (853)
joint (854)
ligament (854)

Section 3

tendon (856)
flexor (856)
extensor (856)
actin (857)
myosin (857)
myofibril (857)
sarcomere (857)

Section 4

epidermis (861)
keratin (861)
melanin (862)
dermis (862)
hair follicle (862)
subcutaneous tissue (863)
sebum (864)

Performance ZONE

Understanding Key Ideas

1. The thoracic cavity contains
 a. the spinal cord.
 b. the brain.
 c. organs of the respiratory system.
 d. organs of the reproductive system.

2. Which of the following is a function of the skeletal system?
 a. support
 b. protection
 c. movement
 d. All of the above

3. Resistance exercises
 a. decrease muscle endurance.
 b. decrease muscle strength.
 c. increase muscle size.
 d. increase the number of muscle cells.

4. Shortening of sarcomeres causes
 a. muscles to contract.
 b. Z lines to move apart.
 c. muscles to relax.
 d. None of the above

5. The dermis helps regulate body temperature by producing
 a. sweat.
 b. acne.
 c. oil.
 d. sebum.

6. The risk of developing skin cancer is increased by
 a. eating oily foods.
 b. using sunscreen.
 c. exercising.
 d. sunbathing.

7. What factors contribute to acne, shown in the photograph below?

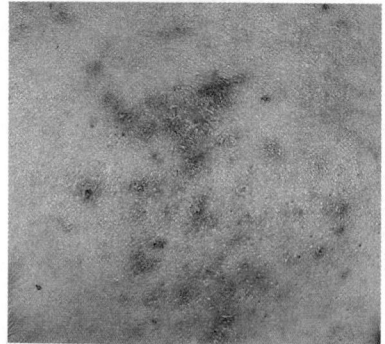

8. Give an example of a ball-and-socket joint and a hinge joint. What type of movement is permitted by each of these joints?

9. **BIOWatch** What are some ways in which forensic science can help in identifying a criminal? Which method produces the most compelling evidence?

10. **Concept Mapping** Make a concept map that illustrates the body's four levels of structural organization. Try to include the following terms in your map: *muscle tissue, connective tissue, epithelial tissue, nervous tissue, organ,* and *organ system.*

Critical Thinking

11. **Inferring Relationships** Relate the events of muscle contraction to cellular respiration.

12. **Inferring Relationships** Young thoroughbred horses that are raced too early in life have an increased risk of breaking the bones in their legs. Using this information, what can you infer about bone development in horses?

13. **Predicting Outcomes** Oil glands in the skin secrete a substance that helps kill bacteria. What might happen if you washed your skin too often?

Alternative Assessment

14. **Relating Structure to Function** Use a compound light microscope to compare prepared slides of bone, muscle, and epithelial tissue. Write a short report that describes the structural differences between cells. Relate the structure of cells to their functions.

15. **Communicating** Find out about the causes, symptoms, and treatment of muscular disorders such as muscular dystrophy and Lou Gehrig's disease. Summarize your findings in an oral report.

Standardized Test Prep

Understanding Concepts

Directions (1–4): For *each* question, write on a separate sheet of paper the letter of the correct answer.

1 What do the bones of the arms and legs form?
  A. the appendicular skeleton
  B. the axial skeleton
  C. the pectoral girdle
  D. the vertebrae

2 Which sequence identifies the levels of organization found in the body?
  F. muscle cell → muscular system → muscle → muscle tissue
  G. muscle cell → muscle tissue → muscle→ muscular system
  H. muscular system → muscle tissue → muscle cell → muscle
  I. muscle → muscle cell → muscle tissue → muscular system

3 What is the bone at the end of a muscle attachment that does **not** move during muscle contraction called?
  A. flexor
  B. insertion
  C. origin
  D. tendon

4 Which of the following is made mostly of connective tissue?
  F. dermis
  G. epidermis
  H. hair
  I. keratin

Directions (5): For the following question, write a short response.

5 Red blood cells are produced in red bone marrow. How are red blood cells transported to the rest of the body?

Reading Skills

Directions (6): **Read the passage below. Then answer the question.**

The incidence of skin cancer is different in different parts of the world. Skin cancer is more common in fair-skinned people and in places with intense sunshine, such as Arizona and Hawaii. It is very common in Australia because many residents are fair-skinned people of European descent who have little natural melanin to protect them from sun.

6 What types of causes result in skin cancer?
  A. It is a result of only genetic causes.
  B. It is a result of only environmental causes.
  C. It is a result of both genetic and environmental causes.
  D. It is a result of neither genetic nor environmental causes.

Interpreting Graphics

Directions (7): **Base your answer to question 7 on the diagram below.**

Sarcomere Under Two Conditions

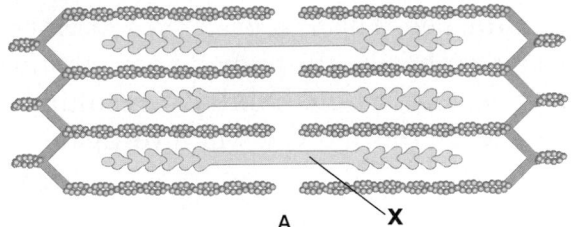

A   X

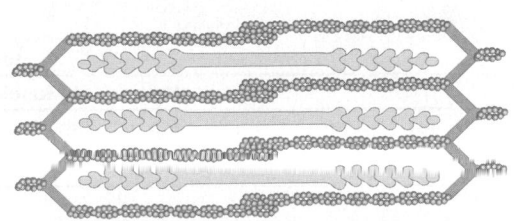

B

7 When a sarcomere changes from condition *A* to condition *B*, what happens to the muscle cell that contains it?
  F. It contracts.   **H.** It rotates.
  G. It relaxes.   **I.** It slides.

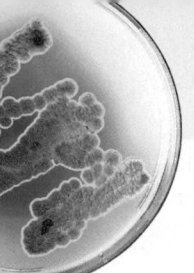

Exploration Lab

Analyzing the Work of Muscles

SKILLS
- Using scientific methods
- Data collection
- Data interpretation

OBJECTIVE
- **Relate** muscles to the work they do.
- **Observe** the effects of fatigue

MATERIALS
- watch with second hand
- graph paper
- spring hand grips

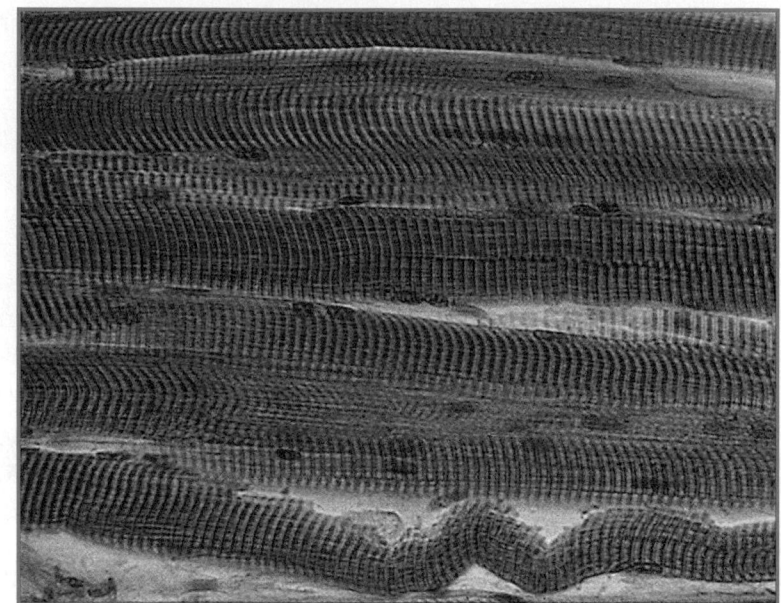

Striated muscle

Before You Begin

Muscles are attached to bones. As muscles contract, they move the bones to which they are attached. This is a basic type of work accomplished by the human body. As muscles are used, lactic acid builds up, resulting in **fatigue.** In this lab you will investigate how fatigue affects the amount of work that muscles can do.

1. Write a definition for the bold face term in the preceding paragraph.

2. Create a data table like the one below.

DATA TABLE													
for Muscle Contractions													
Number of Muscle Contractions in 10-Second Intervals													
Time	1st 10 sec.	2nd 10 sec.	3rd 10 sec.	4th 10 sec.	5th 10 sec.	6th 10 sec.	7th 10 sec.	8th 10 sec.	9th 10 sec.	10th 10 sec.	11th 10 sec.	12th 10 sec.	13th 10 sec.	14th 10 sec.
Trial 1													
Trial 2													
Trial 3													
Trial 4													

Procedure

1. Perform this investigation with a partner. Designate one laboratory partner to observe and record while the other performs the experiment.

2. Hold the spring hand grips in your left hand if you are righthanded, or in your right hand if you are lefthanded. Squeeze the grips rapidly and as hard as possible at a steady pace, until complete fatigue is experienced in the muscles of your hand and forearm.

3. The recorder should count and record the number of squeezes for every 10 seconds.

4. Allow the experimenter to rest for one minute and repeat the procedure for two more trials.

5. Record the data in your table. Some spaces may be left blank.

6. Switch roles with your partner and repeat the procedure.

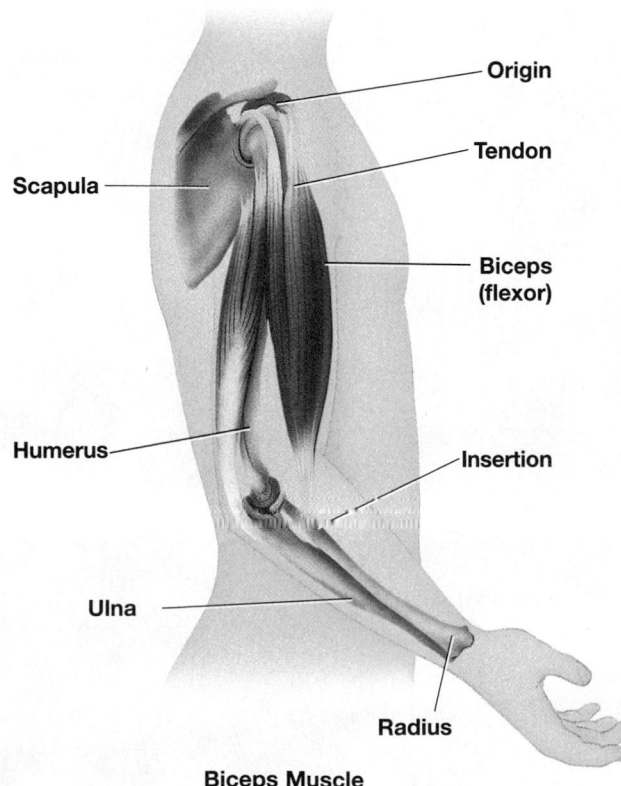

Biceps Muscle

Labels: Scapula, Humerus, Ulna, Radius, Origin, Tendon, Biceps (flexor), Insertion

Analyze and Conclude

1. **Summarizing Results** Plot the results of the three trials on a graph. The X axis should be used for time in seconds, and the Y axis for the number of muscle contractions.

2. **Analyzing Data** Account for the differences in the amount of work done by the muscles during the three trials.

3. **Drawing Conclusions** What is the relationship between the work muscles can do and fatigue?

4. **Predicting Patterns** How does the work done in the muscles of your hands and arms relate to the work done by the muscle of your heart?

5. **Further Inquiry** Compare the charts and the graphs of the athletes and nonathletes.

? Do You Know?

Do research in the library or media center to answer these questions:

1. Some bacteria produce lactic acid. Explain how lactic acid from bacteria can be used as a food preservative.

2. Some sports drink labels claim to include ingredients that neutralize lactic acid. Why would they make this claim? Can a sports drink eliminate lactic acid from muscles?

Use the following Internet resources to explore your own questions about muscle contraction.

internet connect

www.scilinks.org
Topic: Muscle Contraction
Keyword: HX4125

SCiLINKS. Maintained by the National Science Teachers Association

Teen playing soccer

38 Circulatory and Respiratory Systems

✔ *Quick Review*

Answer the following without referring to earlier sections of your book.

1. **Define** *homeostasis.* *(Chapter 1, Section 1)*
2. **Define** the terms *diffusion* and *osmosis.* *(Chapter 4, Section 1)*
3. **Summarize** the role of oxygen in aerobic respiration. *(Chapter 5, Section 3)*
4. **Describe** cardiac muscle, smooth muscle, epithelial tissue, and connective tissue. *(Chapter 37, Section 1)*

Did you have difficulty? *For help, review the sections indicated.*

Reading Activity

Before you begin to read this chapter, write down all of the key words for each of the three sections in the chapter. Then, write a definition next to each word that you have heard of. As you read the chapter, write definitions next to the words that you did not previously know, and modify as needed any definitions of words you know.

internet connect

www.scilinks.org
National Science Teachers Association *sci*LINKS Internet resources are located throughout this chapter.

Maintained by the
National Science Teachers Association

● Sports that require running, such as soccer, provide aerobic exercise. Aerobic exercise increases the body's use of oxygen and causes breathing rate and heart rate to increase.

The Circulatory System

- **List** five types of molecules that are transported by the cardiovascular system.

- **Differentiate** between arteries, capillaries, and veins.

- **Relate** the function of the lymphatic system to the functions of the cardiovascular and immune systems.

- **Relate** each component of blood to its function.

- **Summarize** how a person's blood type is determined.

Key Terms

cardiovascular system
artery
capillary
vein
valve
lymphatic system
plasma
red blood cell
anemia
white blood cell
platelet
ABO blood group system
Rh factor

Transport and Distribution

Regardless of your activities—whether you are roller-blading, swimming, singing, reading, or just sleeping—your body transports nutrients, hormones, and gases, and it gets rid of wastes. Two body systems play major roles in these functions. The circulatory system, which includes the cardiovascular and lymphatic systems, transports these materials to different parts of the body. The respiratory system exchanges gases with the environment—it takes in oxygen, O_2, and releases carbon dioxide, CO_2.

The human **cardiovascular system,** shown in **Figure 1,** functions like a network of highways. The cardiovascular system connects the muscles and organs of the body through an extensive system of vessels that transport blood, a mixture of specialized cells and fluid. The heart, a muscular pump, propels blood through the blood vessels.

Different kinds of molecules move through the cardiovascular system:

1. Nutrients from digested food are transported to all cells in the body through the blood vessels of the cardiovascular system.

2. Oxygen from the lungs, where the oxygen is taken in, is transported to all cells through blood vessels.

3. Metabolic wastes, such as carbon dioxide, are transported through blood vessels to the organs and tissues that excrete them.

4. Hormones, substances which help coordinate many activities of the body, are transported through blood vessels.

5. The cardiovascular system also distributes heat more or less evenly in order to maintain a constant body temperature. For example, in a warm environment, blood vessels in the skin relax to allow more heat to leave the body. In a cold environment, blood vessels constrict, conserving heat by diverting blood to deeper tissues. This diversion of blood prevents heat from escaping the body.

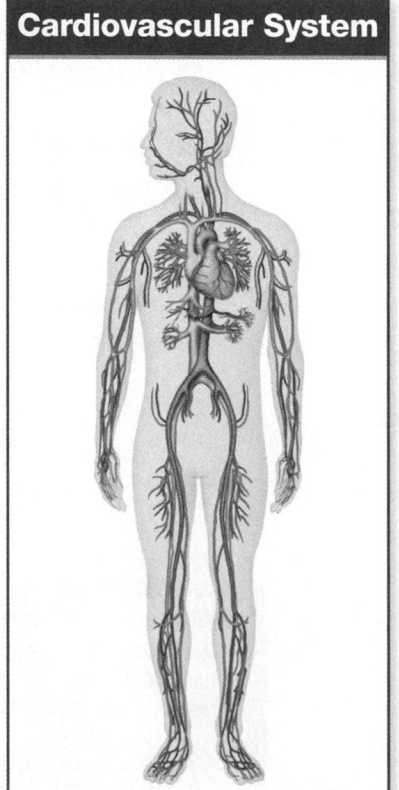

Cardiovascular System

Figure 1 Blood vessels, blood, and a heart.
The cardiovascular system transports materials throughout the body and distributes heat.

Blood Vessels

Blood circulates through the body through a network of vessels. **Arteries** *(AHRT uh reez)*, shown in **Figure 2,** are blood vessels that carry blood away from the heart. Blood passes from the arteries into a network of smaller arteries called arterioles *(ahr TIHR ee ohls)*. Eventually, blood is pushed through to the capillaries.

Capillaries are tiny blood vessels that allow the exchange of gases, nutrients, hormones, and other molecules in the blood. The molecules are exchanged with the cells of the body. From the capillaries, the blood flows into small vessels called venules *(VEHN yools)*. From the venules, blood empties into larger vessels called veins *(vaynz)*. **Veins** are blood vessels that carry the blood back to the heart.

Arteries

With each contraction, the heart forcefully ejects blood into arteries. To accommodate each forceful pulse of blood, an artery's wall expands and then returns to its original size. Elastic fibers in the walls of arteries allow arteries to expand.

The wall of an artery is made up of three layers of tissue, as shown in Figure 2. The innermost layer is a thin layer of epithelial tissue called the endothelium. The endothelium is made up of a single layer of cells. Surrounding the endothelium is a layer of smooth muscle tissue with elastic fibers. Finally, a protective layer of connective tissue with elastic fibers wraps around the smooth muscle tissue. Just as a balloon expands when you blow more air into it, the elastic artery expands when blood is pumped into it.

Magnification: 1,150×

Figure 2   Blood vessels

Blood vessels transport blood and allow for the exchange of substances.

Endothelium
Smooth muscle
Connective tissue

Arteriole
(connects arteries
to capillaries)

Capillaries
(exchange gases, nutrients,
wastes, and hormones)

Venule
(connects veins
to capillaries)

Artery
(carries blood away from the heart)

Vein
(returns blood to the heart)

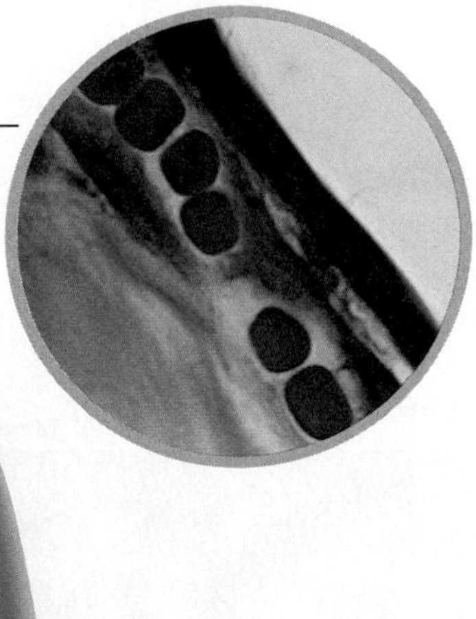

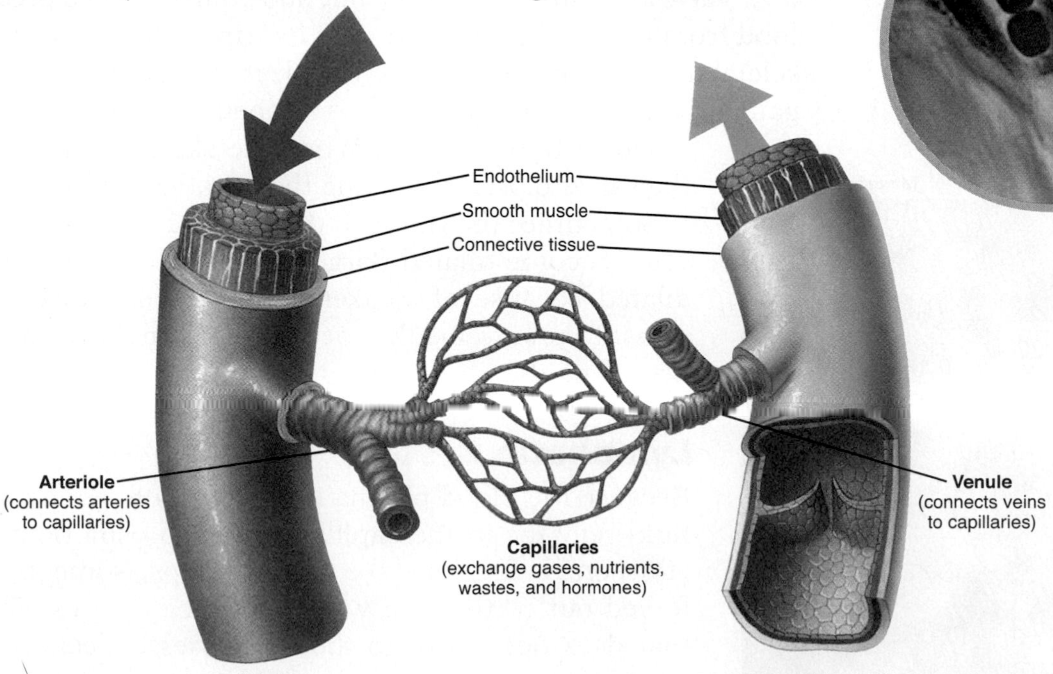

Capillaries

No cell in your body is more than a few cell diameters away from a capillary. At any moment, about 5 percent of your blood is in capillaries. In capillaries, gases, nutrients, hormones and other molecules are transferred from the blood to the body's cells. Carbon dioxide and other wastes are transferred from the body's cells to the capillaries.

The extensive back-and-forth traffic in the capillaries is possible because of two key properties. Capillary walls are only one cell thick, so gas and nutrient molecules easily pass through their thin walls. Capillaries are also very narrow, with an internal diameter of about 8 μm (0.0003 in.)—a diameter only slightly larger than the diameter of a red blood cell. Thus, blood cells passing through a capillary slide along the capillary's inner wall, as shown in the photo in Figure 2. This tight fit makes it easy for oxygen and carbon dioxide to diffuse to and from red blood cells through the capillaries.

Veins

The walls of veins consist of a much thinner layer of smooth muscle, than the walls of arteries. They are farther from the heart pump and exposed to lower pressures. Veins do not receive the pulsing pressure that arteries do.

As shown in Figure 2, veins also differ from arteries in that they are larger in diameter. A large blood vessel offers less resistance to blood flow than a narrower one, so the blood can move more quickly through large veins. The largest veins in the human body are about 3 cm in diameter—about the same diameter as your thumb.

Most veins have one-way valves. A **valve** is a flap of tissue that ensures that the blood or fluid that passes through does not flow back. Valves in veins, such as the one shown in **Figure 3,** prevent the blood from flowing backward during its trip to the heart. When the skeletal muscles in your arms and legs contract, they squeeze against the veins, causing the valves to open and thus, allowing the blood to flow through. When the skeletal muscles relax, the valves close, preventing the backflow of blood.

Sometimes the valves in the veins become weak and the veins become dilated (larger in diameter). Veins that are dilated because of weakened valves are called varicose veins. Dilated veins that occur in the anal area are called hemorrhoids.

Lymphatic System

Because the blood plasma is rich in proteins, most of the fluid remains in the capillaries due to osmotic pressure. However, every time the heart pumps, some fluids are forced out of the thin walls of the capillaries. The fluid that does not return to the capillaries collects in spaces around the body's cells. The fluid that collects around the

Figure 3   Valves in veins.
Valves are most abundant in the veins of the arms and legs, where the upward flow of blood is opposed by gravity.

Magnification: 122×

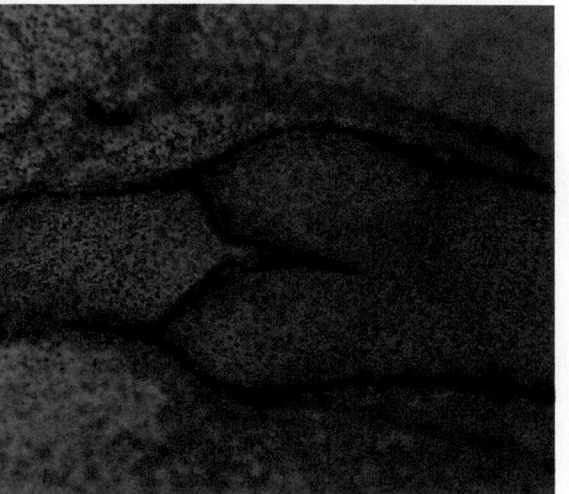

Mapping the Valves in Veins

By applying pressure to your arm, you can locate the valves in the veins of your arm.

Materials

nontoxic felt-tip pen

Procedure

1. Have a classmate make a fist and extend his or her arm, with the hand palm up and slightly below elbow level. Locate a prominent vein on the inside of the forearm. Using one finger, press down on the vein at a point near the wrist to block the blood flow.

2. Gently place a second finger along the vein about 5 cm from the first finger (toward the elbow). Release the second finger, but not the first. The vein should refill partway. Mark this point, which indicates the location of a valve, with a pen. You may have to try more than one vein to locate a valve.

Analysis

1. **Identify** the direction blood flows in the vein you chose.

2. **Propose** why the subject must make a fist and hold his or her arm slightly down.

3. **Infer** what effect standing in one place for long periods of time might have on the veins in the legs.

---

cells is picked up by the lymphatic system and returned to the blood supply.

The **lymphatic system** collects and recycles fluids leaked from the cardiovascular system and is involved in fighting infections. As shown in **Figure 4,** the lymphatic system is made up of a network of vessels called lymphatic vessels and tiny bean-shaped structures called lymph nodes. Lymph tissue is also located in various places throughout the body, including the thymus, tonsils, spleen, and bone marrow.

Lymphatic vessels carry the leaked fluid, called lymph, back to two major veins in the neck. Similar to veins, lymphatic vessels contain valves that prevent the backflow of the fluid. The fluid is pushed through the lymphatic vessels when the skeletal muscles in the arms and legs contract.

The lymphatic system also acts as a key element in the immune system. Immune cells in the lymph nodes and lymphatic organs help defend the body against bacteria, viruses, other infecting microbes, and cancerous cells. Lymph nodes, which are concentrated in the armpits, neck, and groin, sometimes get tender and swell when they are actively fighting infection and filled with white blood cells. Health-care professionals are trained to detect certain types of infections by feeling for the lymph node swellings on the body.

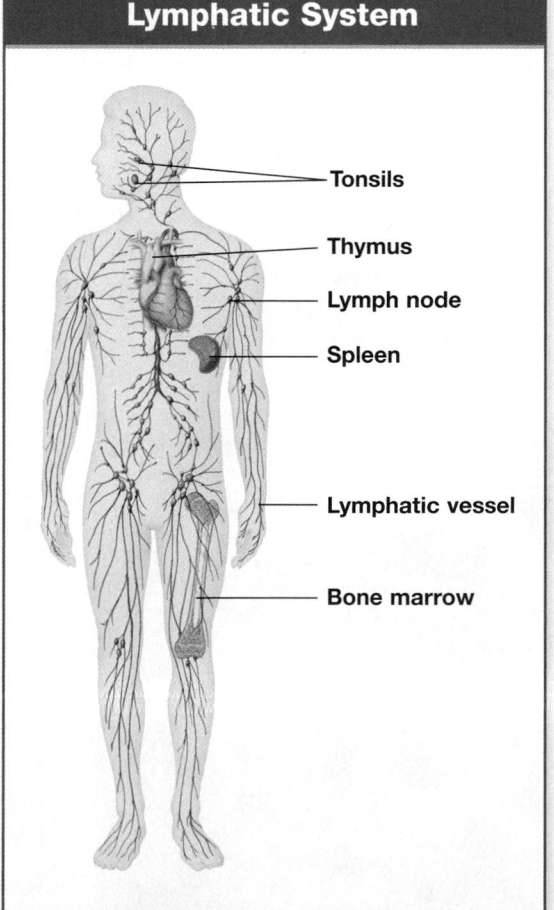

Lymphatic System

- Tonsils
- Thymus
- Lymph node
- Spleen
- Lymphatic vessel
- Bone marrow

Figure 4 Lymphatic tissues. Lymphatic tissues are located throughout the body.

Components of Blood

Blood has been called the river of life because it is responsible for transporting so many substances throughout the body. In life-threatening situations, a person's blood volume is carefully monitored, as shown in **Figure 5.** Typically, blood appears to us as a red, watery fluid. Blood is composed of water, but it also contains a variety of molecules dissolved or suspended in the water, as well as three kinds of cells.

Plasma

About 60 percent of the total volume of blood is **plasma,** the liquid portion of blood. Plasma is made of 90 percent water and 10 percent solutes. The solutes include metabolites, wastes, salts, and proteins.

Water Water in the plasma acts as a solvent. It carries other substances.

Metabolites and Wastes Dissolved within the plasma are glucose and other nutrient molecules. Vitamins, hormones, gases, and nitrogen-containing wastes are also found in plasma.

Salts (Ions) Salts are dissolved in the plasma as ions. The chief plasma ions are sodium, chloride, and bicarbonate. The ions have many functions, including maintaining osmotic balance and regulating the pH of the blood and the permeability of cell membranes.

Proteins Plasma proteins, the most abundant solutes in plasma, play a role in maintaining the osmotic balance between the cytoplasm of cells and that of plasma. Water does not move by osmosis from the plasma to cells because the plasma is rich in dissolved proteins. Some of the plasma proteins are essential for the formation of blood clots. Other proteins called antibodies help the body fight disease.

Some plasma proteins help thicken the blood. The thickness of blood determines how easily it flows through blood vessels. Other plasma proteins serve as antibodies, defending the body from disease. Still other plasma proteins, called clotting proteins or blood-clotting factors, play a major role in blood clotting. When blood is collected for clinical purposes, the blood-clotting factors are removed from the blood and stored for later use.

Blood Cells and Cell Fragments

About 40 percent of the total volume of blood is cells and cell fragments that are suspended in the plasma. There are three principal types of cells in human blood: red blood cells, white blood cells, and platelets.

Red blood cells Most of the cells that make up blood are **red blood cells**—cells that carry oxygen. Each milliliter of human blood contains about 5 million red blood cells. Red blood cells are also called erythrocytes (eh RIHTH roh seyets).

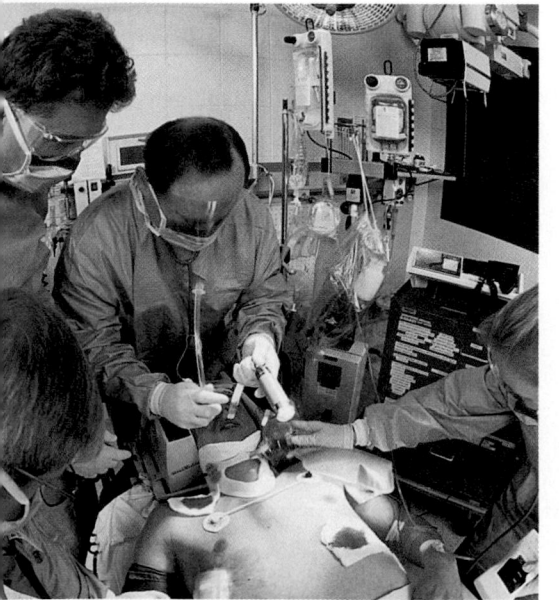

Figure 5 **The river of life.** The loss of too much blood can create a life-threatening situation.

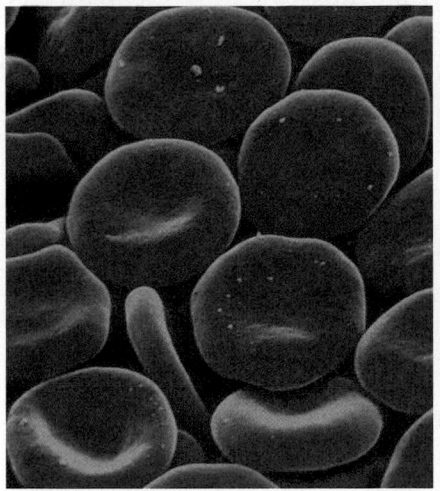

Red blood cells

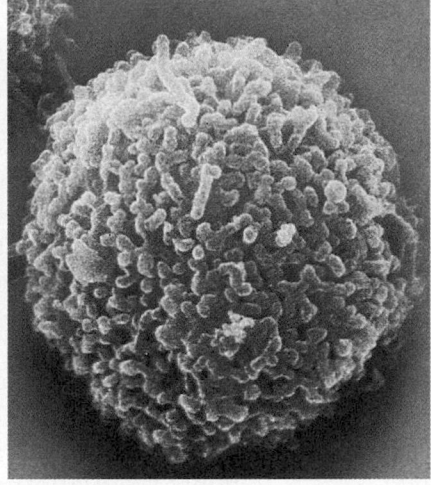

White blood cell

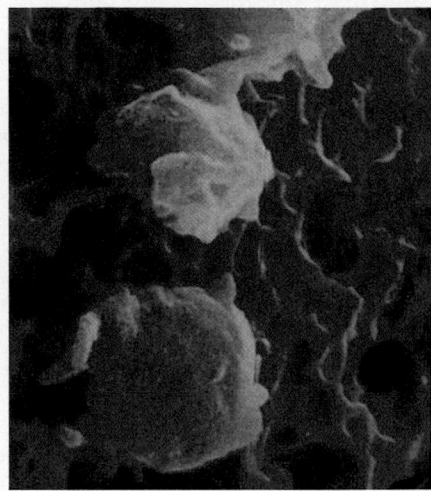

Platelets

Figure 6   Three kinds of blood cells. Red blood cells transport oxygen and some carbon dioxide. White blood cells help defend the body against disease. Platelets are involved in blood clotting.

Most of the interior of a red blood cell is packed with hemoglobin. Hemoglobin is an iron-containing protein that binds oxygen in the lungs and transports it to the tissues of the body. Mature red blood cells do not have nuclei and therefore cannot make proteins or repair themselves. Red blood cells have a biconcave shape, as shown in **Figure 6,** and a short life span (about 4 months). New red blood cells are produced constantly by stem cells, specialized cells in bone marrow.

An abnormality in the number or function of red blood cells can result in anemia. **Anemia** *(uh NEE mee uh)* is a condition in which the oxygen-carrying ability of the blood is reduced. Anemia may result from blood loss or nutritional deficiencies.

White Blood Cells There are only 1 or 2 white blood cells, or leukocytes *(LOO koh sites),* for every 1,000 red blood cells. **White blood cells** are cells whose primary job is to defend the body against disease. White blood cells, shown in Figure 6, are larger than red blood cells and contain nuclei.

There are many different kinds of white blood cells, each with a different immune function. For example, some white blood cells take in and then destroy bacteria and viruses. Other white blood cells produce antibodies, proteins that mark foreign substances for destruction by other cells of the immune system.

Platelets In certain large cells in bone marrow, bits of cytoplasm are regularly pinched off. These cell fragments, called **platelets** *(PLAYT lihts),* are shown in Figure 6. Platelets play an important role in the clotting of blood. If a hole develops in a blood vessel wall, rapid action must be taken by the body, or blood will leak out of the system and death could occur.

When circulating platelets arrive at the site of a broken vessel, they assume an irregular shape, get larger, and release a substance that makes them very sticky. The platelets then attach to the protein fibers on the wall of the broken blood vessel and eventually form a sticky clump that plugs the hole.

internet connect

www.scilinks.org
Topic: White Blood Cells
Keyword: HX4190

SCi
*L*INKS. Maintained by the National Science Teachers Association

Figure 7 Blood-clotting cascade

The release of enzymes from platelets at the site of a damaged blood vessel initiates a "clotting cascade."

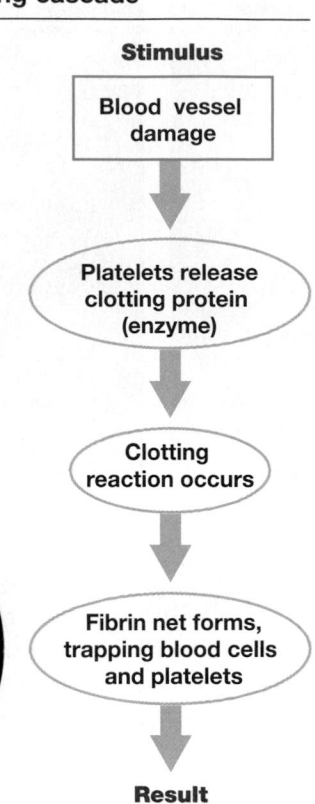

Stimulus

Blood vessel damage

Platelets release clotting protein (enzyme)

Clotting reaction occurs

Fibrin net forms, trapping blood cells and platelets

Result

Blood clot

Fibrin net   Blood cells

For wounds such as an open cut, the platelets release a clotting enzyme that activates a series of chemical reactions. Eventually, a protein called fibrin is formed. The fibrin threads form a net, trapping blood cells and platelets, as shown in **Figure 7.** The net of fibrin and platelets develops into a mass, or clot, that plugs the blood vessel hole. A mutation in a gene for one of the blood-clotting proteins causes hemophilia, a blood clotting disorder.

Blood Type

Occasionally, an injury or disorder is so serious that a person must receive blood or blood components from another person. The blood types of the recipient, the person receiving the blood, and that of the donor, the person giving the blood, must match. Blood type is genetically determined by the presence or absence of a specific complex carbohydrate found on the surface of red blood cells.

One system used to type blood is the **ABO blood group system.** Under this system, the primary blood types are A, B, AB, and O. The letters *A* and *B* refer to complex carbohydrates on the surface of red blood cells that act as antigens, substances that can provoke an immune response.

Blood Witness

Blood is the most common evidence in the world of criminal justice. Serology is the scientific study of blood and its components. A forensic serologist analyzes blood, semen, saliva, and other body fluids to help solve crimes.

A Link Between Suspect and Victim

Many criminals attempt to clean up a violent crime scene, but blood often remains behind. Investigators may find blood under a victim's fingernails, on automobile upholstery, on carpet or clothing, or sometimes even in a household drain. A forensic serologist first determines if the substance found is blood, then verifies that it is human blood.

In a violent crime, blood can reveal the identities of both the victim and the criminal. Red blood cells contain telltale antigens attached to their surfaces. Blood plasma contains proteins that serve as antibodies. Using antibodies, serologists can determine the blood type of a sample.

Positive Identification

For decades, ABO blood typing was the primary tool of forensic serologists. One problem, however, is that blood evidence from two or more individuals can combine. For example, a type

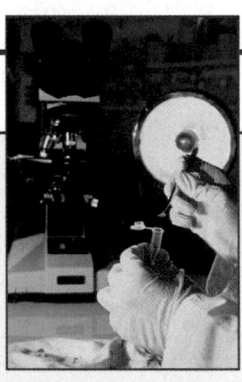

AB sample might actually consist of blood from a type A victim and a type B assailant. ABO blood typing evidence could *exclude* a suspect whose blood type did not match the evidence. But it could not positively *identify* a suspect.

Today blood is commonly used in DNA testing, which results in highly accurate identification. DNA evidence can also indicate familial relationships, which can help investigators discover that the assailant is related to the victim.

Table 1 Blood Types

Blood type	Antigen on the red blood cell	Antibodies in plasma	Can receive blood from	Can donate blood to
A	A	B	O, A	A, AB
B	B	A	O, B	B, AB
AB	A, B	Neither A nor B	O, A, B, AB	AB
O	Neither A nor B	A, B	O	O, A, B, AB

As summarized in **Table 1,** people with type A blood have the A antigen on their red blood cells. People with type B blood have the B antigen. People with type AB blood have both the A and the B antigen, while those with type O blood have neither antigen.

Antibodies are defensive proteins made by the immune system. People with type A blood produce antibodies against the B antigen, even if they have never been exposed to it. In these people, type B red blood cells clump and can block blood flow. For this reason, blood transfusion recipients must receive blood that is compatible with their own.

People with type AB blood are universal recipients (they can receive A, B, AB, or O blood) because they do not have anti-A or anti-B antibodies. Type O individuals are universal donors (they can donate blood to those with A, B, AB or O blood) because their blood cells do not carry A or B antigens and therefore do not react with either anti-A or anti-B antibodies.

Rh Factor

Another important antigen on the surface of red blood cells is called **Rh factor,** which was originally identified in rhesus monkeys. People who have this protein are said to be Rh+, and those who lack it are Rh−. When an Rh− mother gives birth to an Rh+ infant, the Rh− mother begins to make anti-Rh antibodies. The mother's antibodies may be passed to an Rh+ fetus in a future pregnancy, which can lead to fetal death.

Real Life

RhoGAM is a blood product that can suppress the ability to respond to Rh+ red blood cells.

It is given to an Rh− woman who is pregnant with an Rh+ fetus to prevent her from developing antibodies that would harm her baby.

Section 1 Review

1. **Name** the system that transports nutrients, oxygen, wastes, hormones, and heat.

2. **Compare** the structures and functions of arteries, capillaries, and veins.

3. **Describe** the role of the lymphatic system.

4. **Summarize** the functions of water, red blood cells, white blood cells, and platelets.

5. **Predict** the blood types that would be safe for a person with type A blood to receive during a transfusion.

6. **Standardized Test Prep** Which antigens are on the red blood cells of a person with type O blood?
   A 0 antigens
   B Both A and B antigens
   C Either A or B antigens
   D Neither A nor B antigens

Objectives

- **Differentiate** the pulmonary circulation loop from the systemic circulation loop.
- **Summarize** the path that blood follows through the heart.
- **Name** the cluster of heart cells that initiates contraction of the heart.
- **Describe** three ways to monitor the health of the circulatory system.
- **Name** two vascular diseases, and identify factors that contribute to their development.

Key Terms

atrium
ventricle
vena cava
aorta
coronary artery
sinoatrial node
blood pressure
pulse
heart attack
stroke

A Muscular Pump

Blood vessels allow for the movement of blood to all cells in the body. The pumping action of the heart, however, is needed to provide enough pressure to move blood throughout the body. The heart is made up mostly of cardiac muscle tissue, which contracts to pump blood.

Two Separate Circulatory Loops

As shown in **Figure 8,** the human heart has two separate circulatory loops. The right side of the heart is responsible for driving the pulmonary *(PUHL muh nehr ee)* circulation loop, which pumps oxygen-poor blood through the pulmonary arteries to the lungs. Gas exchange—the release of carbon dioxide and pick up of oxygen—occurs in the lungs. The oxygenated blood is then returned to the left side of the heart through pulmonary veins.

The left side of the heart is responsible for driving the systemic circulation loop, which pumps oxygen-rich blood through a network of arteries to the tissues of the body. Oxygen-poor blood is then returned to the right side of the heart through the veins.

Figure 8   Simplified diagram mapping

The pulmonary circuit transports blood between the heart and lungs; the systemic circuit transports blood between the heart and the rest of the body.

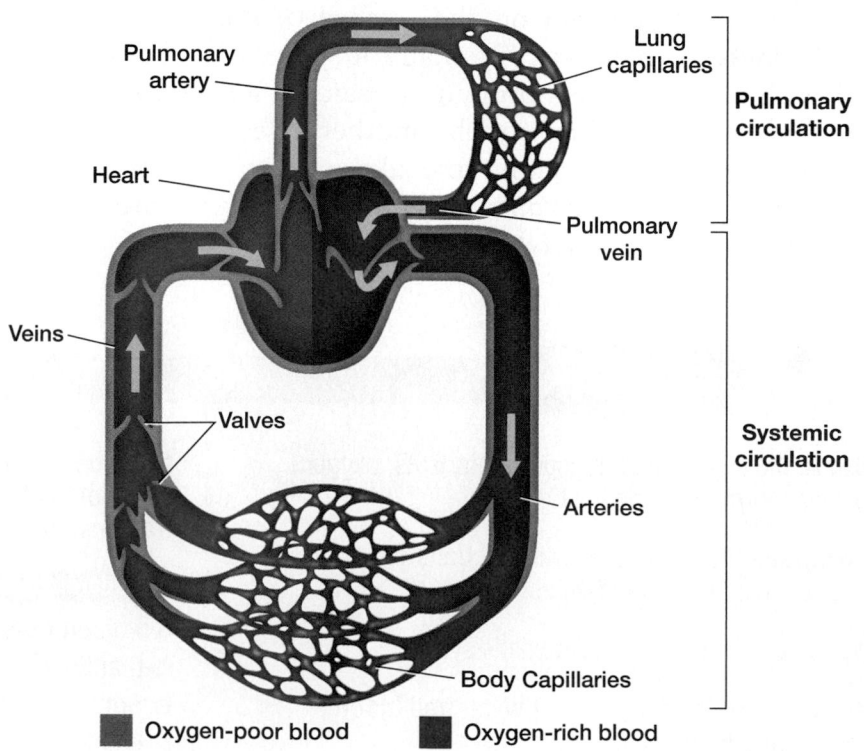

Pulmonary artery

Lung capillaries

Pulmonary circulation

Heart

Pulmonary vein

Veins

Valves

Systemic circulation

Arteries

Body Capillaries

■ Oxygen-poor blood   ■ Oxygen-rich blood

Circulation of Blood

As shown in **Figure 9,** the heart has a wall that divides the right and left sides of the heart. At the top of the heart are the left and right atria *(AY tree uh)*. The **atria** (singular, *atrium*), are chambers that receive blood returning to the heart. Below the atria are the left and right **ventricles,** thick-walled chambers that pump blood away from the heart. A series of one-way valves in the heart prevent blood from moving backward. Figure 9 summarizes the path blood follows through the heart:

❶ Two large veins called the inferior **vena cava** and superior vena cava collect all of the oxygen-poor blood from the body. The venae cavae empty blood directly into the right atrium of the heart.

❷ The blood from the right atrium moves into the right ventricle.

❸ As the right ventricle contracts, it sends the blood into the pulmonary arteries.

❹ The pulmonary arteries carry the blood to the right and left lungs. At the capillaries of the lungs, oxygen is picked up and carbon dioxide is unloaded.

❺ The freshly oxygenated blood returns from the lungs to the left side of the heart through the pulmonary veins, which empty the blood directly into the left atrium.

❻ From the left atrium, the blood is pumped into the left ventricle.

Study *TIP*

● **Interpreting Graphics**
In human anatomy, the terms *left* and *right* always refer to the left and right from the perspective of the subject. This will help you understand why the terms *left* and *right* appear reversed in anatomical drawings, such as that of the heart in Figures 8 and 9.

Figure 9   Blood flow through the heart

The arrows trace the path of blood as it travels through the heart.

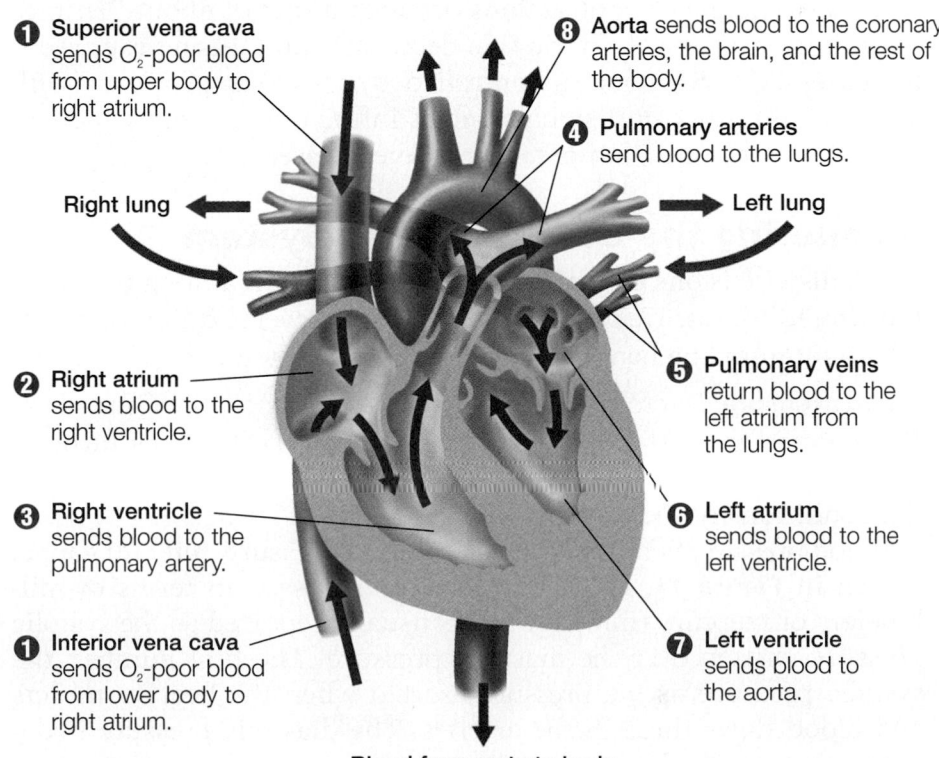

❶ Superior vena cava sends O$_2$-poor blood from upper body to right atrium.

❽ Aorta sends blood to the coronary arteries, the brain, and the rest of the body.

❹ Pulmonary arteries send blood to the lungs.

Right lung ← → Left lung

❷ Right atrium sends blood to the right ventricle.

❺ Pulmonary veins return blood to the left atrium from the lungs.

❸ Right ventricle sends blood to the pulmonary artery.

❻ Left atrium sends blood to the left ventricle.

❶ Inferior vena cava sends O$_2$-poor blood from lower body to right atrium.

❼ Left ventricle sends blood to the aorta.

Blood from aorta to body

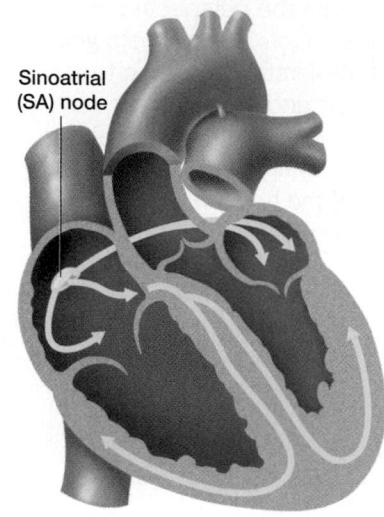

Figure 10 Electrical regulation of the heart. The SA node, or pacemaker, fires ahead of each heart contraction. The wave of contraction spreads across both atria and delays for an instant before it travels to the ventricles.

Sinoatrial (SA) node

❼ After a slight delay that permits the left atrium to empty completely, the left ventricle contracts. The walls of the left ventricle are muscular, so the left ventricle's contraction is forceful.

❽ The blood then enters one of the largest arteries of the body, the **aorta** *(ay OHR tuh).*

The first arteries to branch from the aorta are the **coronary** *(KOHR uh neh ree)* **arteries,** which carry freshly oxygenated blood to the heart muscle. Other arteries also branch from the aorta and carry oxygen-rich blood to all parts of the body.

After delivering oxygen to the cells of the body and picking up carbon dioxide, the cycle continues when blood returns to the heart through the inferior or superior venae cavae.

Initiating Contraction

Contraction of the heart is initiated by a small cluster of cardiac muscle cells called the **sinoatrial** *(SIE noh ay tree uhl)* **node,** which is embedded in the upper wall of the right atrium. The cells that make up the sinoatrial node (SA node, for short) act as the pacemaker of the heart. These cells "fire" an electrical stimulus in a regular rhythm. Each stimulus is followed immediately by a contraction that travels quickly in a wave and causes both atria to contract almost simultaneously, as shown in **Figure 10.**

The wave of contraction spreads from the atria to the ventricles, but almost one-tenth of a second passes before the ventricles start to contract. The delay permits the atria to finish emptying blood into the ventricles before the ventricles contract simultaneously. The wave of contraction is conducted rapidly over both ventricles by a network of fibers in the heart.

On average, heart contractions occur at a rate of about 72 times per minute. During sleep the rate decreases, and during exercise it increases. The SA node is controlled by two sets of nerves with antagonistic (opposite) signals and is influenced by many factors, including hormones, temperature, and exercise.

Monitoring the Cardiovascular System

Heart disease is one of the leading causes of death among people in the United States. Health professionals use several different methods to monitor the health of the circulatory system.

Blood pressure Doctors routinely measure patients' blood pressure. **Blood pressure** is the force exerted by blood as it moves through blood vessels. Blood pressure readings provide information about the conditions of the arteries.

Blood pressure is measured with a blood pressure cuff and gauge, shown in **Figure 11.** Blood pressure is expressed in terms of millimeters of mercury (mm Hg) and is usually reported as the systolic pressure written over the diastolic pressure. The first number, the systolic pressure, is the pressure exerted when the heart contracts and blood flows through the arteries. The diastolic pressure is the pressure exerted when the heart relaxes.

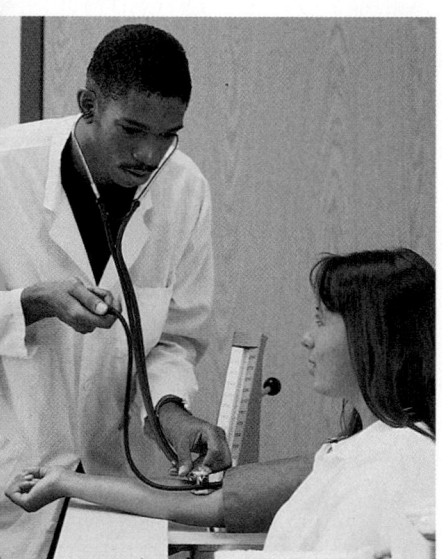

Figure 11 Monitoring blood pressure. Blood pressure is measured with a blood pressure cuff, a stethoscope, and a mercury (Hg) column gauge.

Normal blood pressure is less than 120 for systolic pressure and less than 80 for diastolic pressure. An example of a normal reading would be written 115/75 mm Hg. These figures indicate the blood is pushing against the artery walls with a pressure of 115 mm Hg as the heart contracts and 75 mm Hg as the heart rests.

Many Americans suffer from a condition called high blood pressure, or hypertension. High blood pressure places a strain on the walls of the arteries and increases the chance that a vessel will burst. Left untreated, hypertension can lead to heart damage, brain damage, or kidney failure. Regular aerobic activity can help people to maintain a healthy blood pressure. Hypertension can be easily diagnosed and usually can be controlled by medicine, diet, and exercise.

Electrocardiograms (ECGs or EKGs) A common way to monitor the heart's function is to measure the tiny electrical impulses produced by the heart muscle when it contracts. Because the human body is composed mostly of water with dissolved ions, it conducts electrical currents. A small portion of the heart's electrical activity reaches the body surface. As shown in **Figure 12,** an instrument called an electrocardiograph uses special sensors to detect the electrical activity. A recording of these electrical impulses is called an electrocardiogram, abbreviated as ECG or EKG. In one normal heartbeat, three successive electrical-impulse waves are recorded, as shown in Figure 12.

Heart Rate It takes only a watch with a second hand to measure your pulse. Your **pulse** is a series of pressure waves within an artery caused by the contractions of the left ventricle. A person's pulse is an indicator of his or her heart rate—how fast or slow the heart is beating. Each time the blood surges from the aorta, the elastic walls in the blood vessels expand and stretch. This rhythmical expansion can be felt as a pulse in areas where the vessels near the surface of the skin. The number of pulses counted per minute represents the number of heartbeats per minute. The most common site for taking a pulse is at a radial artery, on the thumb side of each wrist. The average pulse rate ranges from 70 to 90 beats per minute for adults.

Figure 12 Monitoring heart contractions. The electrical changes with each heart contraction can be detected with an electrocardiograph. The characteristic up-and-down waves are analyzed to assess the health of the heart.

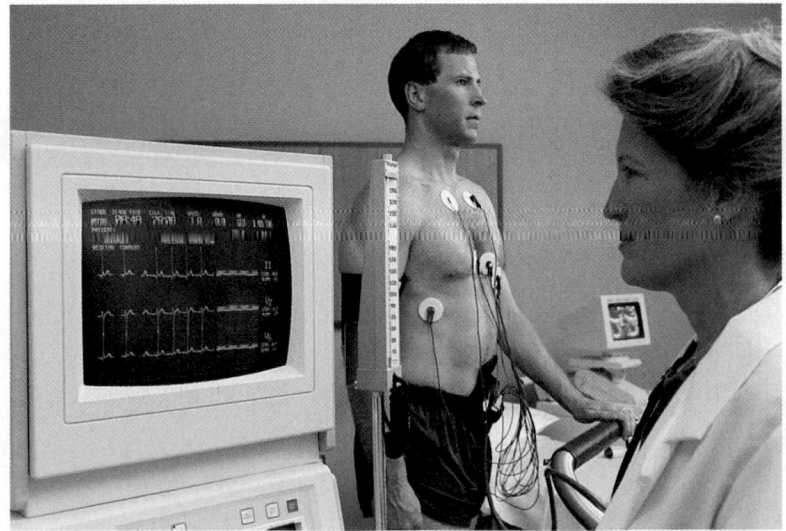

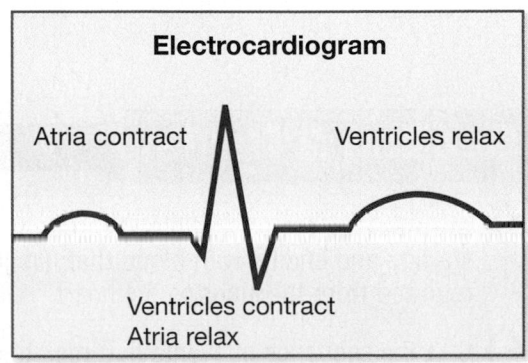

Electrocardiogram

Atria contract   Ventricles relax

Ventricles contract
Atria relax

Diseases of the blood vessels serving the heart and brain are leading causes of premature death and disability in the United States. When either the heart or the brain does not get enough blood, parts of the organ die. A **heart attack** occurs when an area of the heart muscle stops working and dies. When an area of the brain dies the result is a **stroke.** Death or varying degrees of disability may result. Factors that contribute to heart attacks and strokes are cigarette smoking, lack of physical activity, diets high in saturated fats, and unmanaged stress.

Exploring Further

What Is a Heart Attack?

You may feel a sharp, crushing, squeezing pain in your chest. You may have mild pain in your jaws or down an arm. You may break into a cold sweat and feel nauseated. Some of these symptoms occur in almost 2 million Americans each year when they experience a heart attack. Some people experience almost no symptoms.

Why Do Heart Attacks Occur?
Heart attacks usually happen when the arteries that deliver oxygen to the heart, the coronary arteries, become blocked. Heart cells begin to die very quickly without blood. If a large part of the heart is affected, the victim can die immediately or within a few days or weeks.

Blockage of Arteries
A blood clot formed somewhere else in the body can break loose and flow to the heart or to the brain, where it blocks the flow of blood. Blood flow is also blocked by the buildup of fatty deposits, including cholesterol, a condition called atherosclerosis *(ath uhr oh skluh ROH sihs).*

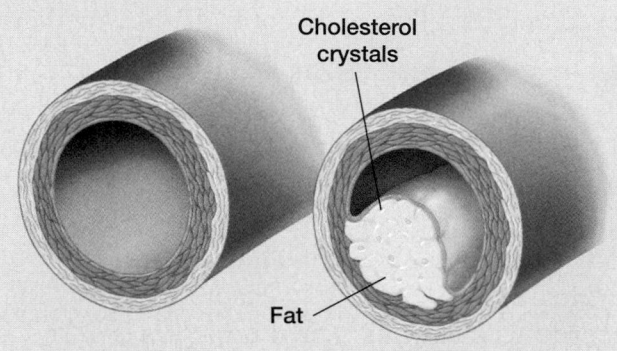

Cholesterol crystals

Fat

Normal artery   **Artery with fatty deposits**

When calcium is deposited in the fatty buildup, the condition is called arteriosclerosis *(ahr tihr ee oh skluh ROH sihs),* or hardening of the arteries. Hardened arteries cannot expand to handle the volume of blood that enters every time the heart contracts. Pressure builds up in the artery and causes the heart to work harder.

Prevention
High blood pressure, high cholesterol levels, and cigarette smoking are all controllable risk factors in heart disease. Not smoking at all, early diagnosis and treatment of high blood pressure, regular medical checkups, a healthy diet, and regular exercise can all help prevent a heart attack or decrease the severity of one.

Section 2 Review

1 **Summarize** the path of blood through the body starting and ending with blood that has just returned from the lungs to the heart.

2 **List** the sequence of events that results in atrial and ventricular contraction.

3 **Describe** the function of the SA node.

4 **Identify** three ways that the condition of the cardiovascular system can be monitored.

5 **Differentiate** between a heart attack and a stroke.

6 **Standardized Test Prep** When the right ventricle contracts, it pumps blood to the
A lungs.   **C** right atrium.
B aorta.   **D** rest of the body.

The Respiratory System

Gas Exchange

A person can go without water for a few days and without food for more than a week. But if a person stops breathing for more than a few minutes, he or she will die. Breathing is the means by which your body obtains and releases gases. Oxygen is used by your cells to completely oxidize glucose and then make ATP, the main energy currency in your cells. Without oxygen, your body cannot obtain enough energy from food to survive. Excess carbon dioxide produced as a waste product of aerobic respiration is toxic to cells and must be removed.

The Path of Air

Breathing is only one part of gas exchange. The gases must be transported by the cardiovascular system and then exchanged at the cells. All of the organs and tissues that function in this exchange of gases make up the respiratory system, as shown in **Figure 13.**

A breath of air enters the respiratory system through the nose or mouth. Air is made up of many gases. About 21 percent of air is oxygen gas. Hairs in your nose filter dust and particles out of the air. Tissues that line the nasal cavity moisten and warm the air.

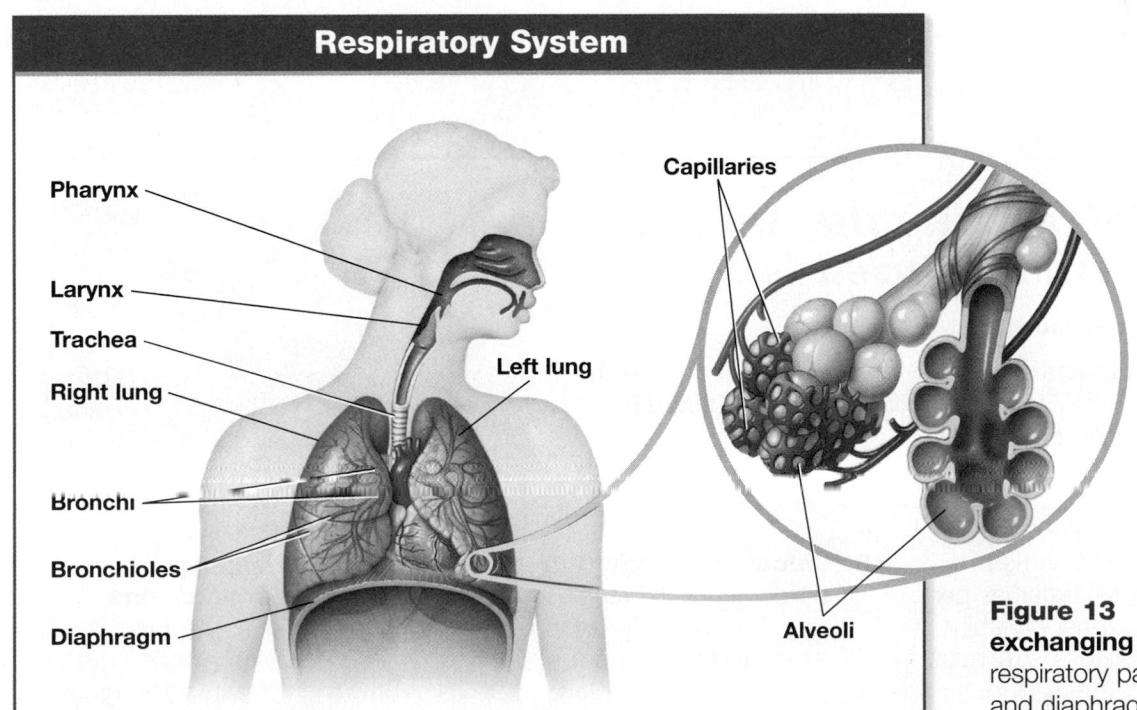

Respiratory System

Pharynx

Larynx

Trachea

Right lung

Bronchi

Bronchioles

Diaphragm

Left lung

Capillaries

Alveoli

Figure 13 Taking in and exchanging gases. The respiratory passages, lungs, and diaphragm make up the respiratory system.

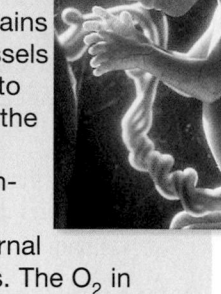
From the nose, air passes through a muscular tube in the upper throat called the **pharynx** *(FAIR ingks)*, which serves as a passageway for air and food. The air continues on through a passageway for air, called the **larynx** *(LAIR ingks)*, or voice box located in the neck. A flap of tissue, the epiglottis *(ehp uh GLAHT ihs)*, covers the opening to the larynx when you swallow food and liquids. This prevents food and liquids from passing into your lungs.

From the larynx the air passes into the **trachea** *(TRAY kee uh)*, a long, straight tube in the chest cavity. The trachea, or windpipe, divides into two smaller tubes, the **bronchi** *(BRAHNG kie)*, which lead to the lungs. Within the lung, the bronchi (singular, *bronchus*) divide into smaller and smaller tubes called bronchioles *(BRAHNG kee ohls)*. The smallest bronchioles end in clusters of air sacs called **alveoli** *(al VEE uh lie)*, where gases are actually exchanged. As shown in Figure 13, each of the 300 million small alveoli is surrounded by a jacket of capillaries. Alveoli increase the surface area of your lungs to as much as 42 times the surface area of your body.

The cells that line the bronchi and trachea secrete mucus that traps foreign particles in the air. The mucus is directed upward by cilia to the epiglottis, where the mucus is swallowed and digested. Microbes in the mucus are destroyed by acids and enzymes in the stomach.

Lungs

The lungs, which are among the largest organs in the body, are suspended in the chest cavity, bounded on the sides by the ribs and on the bottom by the diaphragm *(DIE uh fram)*. The **diaphragm** is a powerful muscle spanning the rib cage under the lungs, and it aids in respiration. A double membrane surrounds both lungs. The outermost membrane is attached to the wall of the thoracic cavity, and the inner membrane is attached to the surface of the lungs. Between both membranes is a small space filled with fluid.

Calculating the Amount of Air Respired

Background

Most adults take in about 0.5 L of air with each breath. The normal breathing rate is about 8 to 15 breaths per minute.

Analysis

1. **Calculate** the volume of air in liters an adult breathes per minute if his or her breathing rate is 15 breaths per minute.

2. **Calculate** the volume of air in liters an adult breathes per hour if his or her breathing rate is 15 breaths per minute.

3. **Critical Thinking Inferring Conclusions** The breathing rate of an infant is about 40 breaths per minute. Why might infants have higher respiratory rates than adults?

Breathing

Air is drawn into and pushed out of the lungs by the mechanical process known as breathing. Breathing occurs because of air pressure differences between the lungs and the atmosphere, as shown in **Figure 14.** To draw air into the lungs, a process called inhalation, the rib muscles contract. This draws the rib cage up and out, and the diaphragm contracts, moving downward. The volume of the chest cavity increases, which reduces the air pressure within the cavity below the atmospheric pressure. Because air flows from a high pressure area to a low pressure area, air is drawn into the lungs.

Normal exhalation (breathing out) is a passive process. The rib cage and diaphragm muscles relax, which returns the rib cage and diaphragm to their resting position. The relaxation of these muscles decreases the volume in the chest cavity and increases the air pressure in the lungs. Because the air pressure is now higher in the lungs than in the atmosphere, air is forced out—from a high pressure area to a low pressure area.

Breathing Rate

You took your first breath within moments of being born. Since then, you have repeated the process more than 200 million times. What controls how fast or slow you breathe? Receptors in the brain and cardiovascular system continually monitor the levels of oxygen and carbon dioxide in the blood. The receptors enable the body to automatically regulate oxygen and carbon dioxide concentrations by sending nerve signals to the brain. The brain responds by sending nerve signals to the diaphragm and rib muscles in order to speed or slow the rate of breathing.

It may surprise you to know that carbon dioxide levels have a greater effect on breathing than do oxygen levels. For example, if the concentration of carbon dioxide in your blood increases, such as during exercise, you respond by breathing more deeply, ridding your body of excess carbon dioxide. When the carbon dioxide level drops, your breathing slows. Factors such as stress, pain, and fear also influence breathing rate.

The signals that travel from the breathing center of the brain are not subject to voluntary control. You cannot simply decide to stop breathing indefinitely. You can hold your breath for a while, but even if you lose consciousness your respiratory control center will take over and force your body to breathe.

Figure 14   Inhalation and exhalation

The diaphragm and the muscles between the ribs are involved in the movement of the chest cavity during breathing.

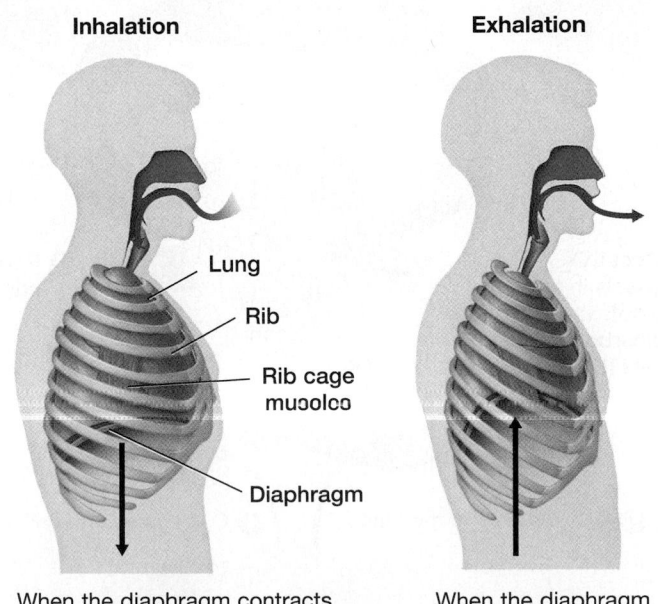

Inhalation

Exhalation

Lung

Rib

Rib cage muscles

Diaphragm

When the diaphragm contracts, it moves down and air rushes in.

When the diaphragm relaxes, it moves up and air is forced out.

Gas Transport

Breathing is the first step to getting oxygen to the trillions of cells in your body. When oxygen molecules move from the air into your alveoli, their journey has just begun, as shown in **Figure 15.** As oxygen passes into the plasma of the bloodstream, it is picked up by red blood cells that carry an iron-containing protein called hemoglobin.

Oxygen Transport

Each hemoglobin molecule contains four atoms of iron. The iron atoms in the hemoglobin give red blood cells their red color. The iron atoms bind reversibly with oxygen. Reversible binding means that at the appropriate time, the oxygen can be released elsewhere in the body and be taken up by the cells that need it.

Figure 15 summarizes the path of oxygen and carbon dioxide through the body:

❶ Oxygen from the outside air reaches the lungs.

❷ The oxygen diffuses from the alveoli to the pulmonary capillaries. At the high oxygen levels that occur in the blood within the lungs, most hemoglobin molecules carry a full load of oxygen.

❸ The oxygen-rich blood then travels to the heart. The heart pumps the blood to the tissues of the body.

❹ Oxygen diffuses into the cells for use during aerobic respiration. In the tissues, oxygen levels are lower. This causes the hemoglobin to release its oxygen.

❺ In tissues, the presence of carbon dioxide produced by cellular respiration makes the blood more acidic and causes the hemoglobin molecules to assume a different shape, one that gives up oxygen more easily. The carbon dioxide diffuses from the cells to the blood.

❻ Most of the carbon dioxide travels to the heart as bicarbonate (HCO_3^-) ions.

❼ The heart pumps the blood to the lungs. In the lungs, carbon dioxide is released in its gaseous form to the alveoli.

❽ The carbon dioxide is expelled.

Figure 15   O$_2$ and CO$_2$ transport in the blood.
Hemoglobin molecules inside red blood cells transport oxygen, while most carbon dioxide is transported as bicarbonate ions in the plasma.

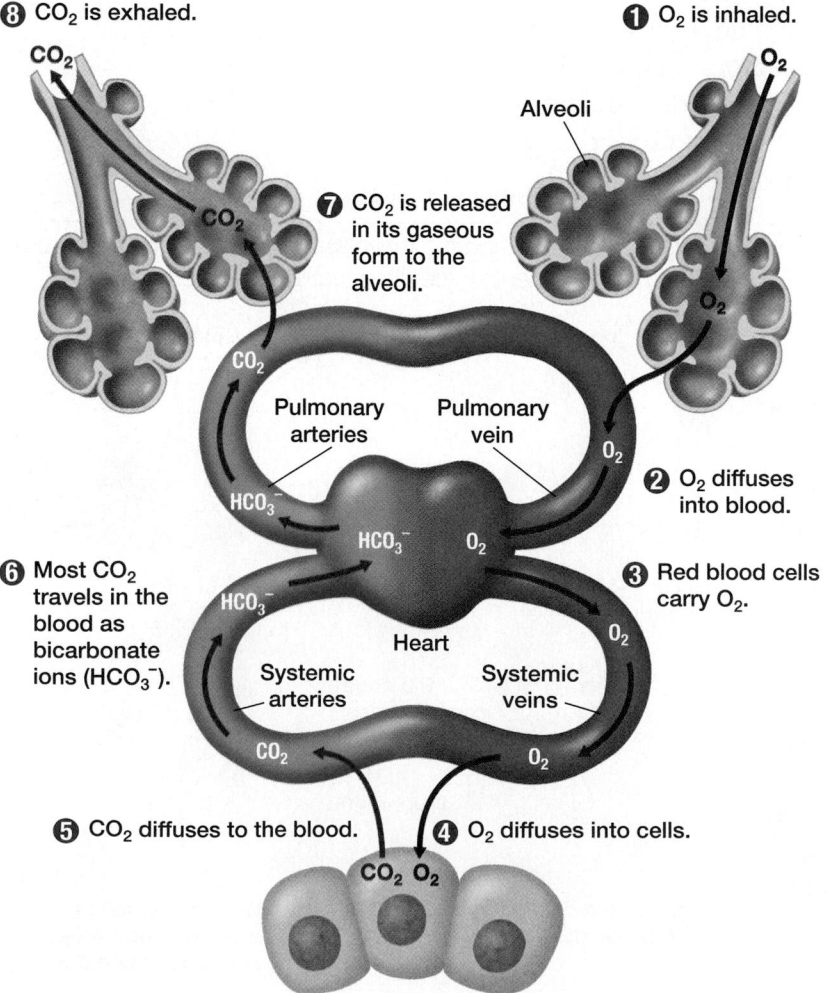

❽ CO_2 is exhaled.

❶ O_2 is inhaled.

Alveoli

❼ CO_2 is released in its gaseous form to the alveoli.

Pulmonary arteries

Pulmonary vein

❷ O_2 diffuses into blood.

❻ Most CO_2 travels in the blood as bicarbonate ions (HCO_3^-).

Heart

❸ Red blood cells carry O_2.

Systemic arteries

Systemic veins

❺ CO_2 diffuses to the blood.

❹ O_2 diffuses into cells.

Cells

Carbon Dioxide Transport

At the same time that the red blood cells are unloading oxygen to tissues, they are also taking in carbon dioxide from the tissues. Carbon dioxide is carried by the blood in three forms:

1. About 7 percent of CO_2 is dissolved in the blood plasma.

2. About 23 percent of CO_2 is attached to hemoglobin molecules inside red blood cells.

3. The majority of CO_2, 70 percent, is carried in the blood as bicarbonate ions.

How is CO_2 carried as bicarbonate ions? In the presence of a certain enzyme, carbon dioxide combines with water to form carbonic acid, H_2CO_3. This is shown in the equation below. The carbonic acid then breaks up to form a bicarbonate ion, HCO_3^-, and a hydrogen ion, H^+:

$$H_2O + CO_2 \rightleftharpoons H_2CO_3 \rightleftharpoons HCO_3^- + H^+$$

Thus, most of the CO_2 travels in the blood as bicarbonate ions. The hydrogen ions make the blood more acidic. When the blood reaches the lungs, the series of reactions is reversed:

$$HCO_3^- + H^+ \rightleftharpoons H_2CO_3 \rightleftharpoons H_2O + CO_2$$

A bicarbonate ion combines with a hydrogen ion to form carbonic acid, which in turn forms CO_2 and water. The CO_2 diffuses out of the capillaries into the alveoli and is exhaled into the atmosphere.

QUICK LAB

Modeling the Role of Bicarbonate in Homeostasis

You can use pH indicator paper, water, and baking soda to model the role of bicarbonate ions in maintaining blood pH levels in the presence of carbon dioxide.

Materials

two 250 mL beakers, 250 mL distilled water, 2.8 g baking soda, glass stirring rod, 4 strips of wide-range pH paper, 2 drinking straws

Procedure

1. Label one beaker A and another B. Fill each beaker halfway with distilled water.

2. Add 1.4 g of baking soda to beaker B, and stir well.

3. Test and record the pH of the contents of each beaker.

4. *Gently* blow through a straw, into the water in beaker A.

Test and record the pH of the resulting solution.

5. Repeat step 4 for beaker B.

Analysis

1. **Describe** what happened to the pH in the two beakers during the experiment.

2. **State** the chemical name for baking soda.

3. **Propose** the chemical reaction that might have caused a change in pH in beaker A.

4. **Summarize** the effect the baking soda had on the pH of the solution in beaker B after blowing.

5. **Critical Thinking Applying Information** Relate what happened in beaker B to what occurs in the bloodstream.

Respiratory Diseases

Respiratory diseases affect millions of Americans. A chronic pulmonary—or lung—disease is one for which there is no cure.

Asthma

Asthma *(AZ muh)* is a chronic condition in which the bronchioles of the lungs become inflamed, because of their sensitivity to certain stimuli in the air. The bronchial walls tighten and extra mucus is produced, causing the airways to narrow. In severe asthma attacks, the alveoli may swell enough to rupture. Stressful situations and strenuous exercise may trigger an asthma attack. Left untreated, asthma can be deadly. Fortunately, prescribed inhalant medicines may help to stop an asthma attack by expanding the bronchioles. People of all ages can have asthma.

Emphysema

Emphysema *(ehm fuh SEE muh)* is a chronic pulmonary disease resulting from a chemical imbalance that destroys elastic fibers in the lungs. Normally, these elastic fibers allow the lungs to expand and contract. Emphysema begins with the destruction of alveoli. Damage to the alveoli is irreversible and results in constant fatigue and breathlessness. Severely affected individuals must breathe from tanks of oxygen in order to live. Smoking is the cause of up to 90 percent of emphysema cases. Emphysema affects millions of lives annually.

Lung Cancer

Lung cancer is one of the leading causes of death in the world today. As shown in **Figure 16,** cancer is a disease characterized by abnormal cell growth. In the United States alone, about 28 percent—155,000—of all cancer deaths each year are attributed to lung cancer. Smoking is the major cause of lung cancer. Once cancer is detected, the affected lung is sometimes removed surgically. Even with such drastic measures, lung cancer usually is not curable. About 15 percent of lung cancer victims live more than 5 years after diagnosis.

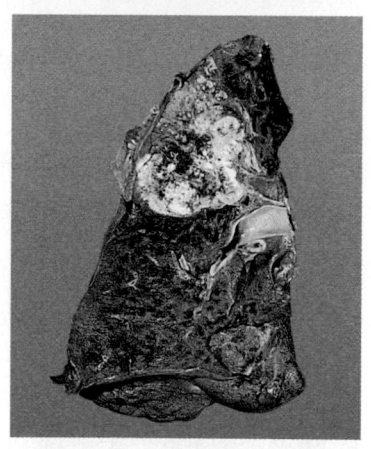

Figure 16 Healthy lungs and cancerous lungs. The top photo shows a healthy lung. The bottom photo shows a cancerous lung which contains cells that divide uncontrollably.

Section 3 Review

1. **Sequence** the path a breath of air follows through the respiratory system. (Begin with air that enters through the nose or mouth.)

2. **State** the direction that the diaphragm and rib cage move to cause inhalation.

3. **Name** the main factor that regulates the rate of breathing.

4. **Evaluate** the role that bicarbonate plays in transporting carbon dioxide in the blood.

5. **Critical Thinking Justifying Conclusions** Would a person with emphysema have trouble climbing stairs? Explain.

6. **Standardized Test Prep** Gases are exchanged between the blood and inhaled air in the
   A larynx.   C trachea.
   B alveoli.   D bronchi.

Key Concepts

1 The Circulatory System

- The human cardiovascular system is made up of blood vessels, blood, and the heart, which together function to transport materials, remove wastes, and distribute heat.

- Arteries carry blood away from the heart. Materials are exchanged at the capillaries. Veins contain valves and carry blood back to the heart. Fluids not returned to the capillaries are picked up by lymphatic vessels.

- Blood consists of plasma (water, metabolites, wastes, salts, and proteins), red blood cells, white blood cells, and platelets.

- Blood types are defined by the presence or absence of complex carbohydrates on the surface of red blood cells.

2 The Heart

- The right side of the heart receives oxygen-poor blood from the body and circulates it to the lungs. In the lungs, gases are exchanged. The left side of the heart receives oxygenated blood from the lungs and circulates it to the rest of the body.

- Atria receive blood entering the heart. Ventricles pump blood away from the heart.

- Contraction of the heart is initiated by the sinoatrial node. The health of the cardiovascular system can be monitored by measuring blood pressure, electrical impulses, and pulse rate.

- Blockages in blood vessels can lead to heart attacks or strokes.

3 The Respiratory System

- A series of tubes and bunched air sacs (alveoli) take in oxygen and remove carbon dioxide.

- Breathing is caused by pressure changes within the chest cavity.

- The concentration of carbon dioxide in the blood is the most critical factor affecting a person's breathing rate and depth.

- Oxygen is transported to tissues by combining with hemoglobin molecules inside red blood cells. Most carbon dioxide is transported to the lungs as bicarbonate ions.

- Asthma, emphysema, and lung cancer limit lung function.

Key Terms

Section 1
cardiovascular system (872)
artery (873)
capillary (873)
vein (873)
valve (874)
lymphatic system (875)
plasma (876)
red blood cell (876)
anemia (877)
white blood cell (877)
platelet (877)
ABO blood group system (878)
Rh factor (879)

Section 2
atrium (881)
ventricle (881)
vena cava (881)
aorta (882)
coronary artery (882)
sinoatrial node (882)
blood pressure (882)
pulse (883)
heart attack (884)
stroke (884)

Section 3
pharynx (886)
larynx (886)
trachea (886)
bronchus (886)
alveolus (886)
diaphragm (886)

Understanding Key Ideas

1. Lymphatic vessels
- **a.** transport blood.
- **b.** return fluid to the blood.
- **c.** produce antibodies.
- **d.** control blood clotting.

2. What type of blood contains A antibodies (but *not* B antibodies) in the plasma and lacks Rh antigens?
- **a.** AB negative
- **b.** A positive
- **c.** B negative
- **d.** O positive

3. Blood in the pulmonary veins is
- **a.** oxygen-rich.
- **b.** iron-poor.
- **c.** oxygen-poor.
- **d.** calcium-rich.

4. The diaphragm contracts and the pressure in the chest cavity decreases during
- **a.** bronchitis.
- **b.** exhalation.
- **c.** inhalation.
- **d.** asthma attacks.

5. Breathing rate will automatically increase when
- **a.** blood pH is high.
- **b.** the amount of carbon dioxide in the blood increases.
- **c.** blood acidity decreases.
- **d.** hemoglobin is unloaded.

6. Which organ receives the richest oxygen supply from blood returning from the lungs?
- **a.** stomach
- **b.** brain
- **c.** heart
- **d.** kidney

7. Which is *not* a factor that contributes to chronic coronary disease?
- **a.** a diet rich in fat
- **b.** unmanaged stress
- **c.** cigarette smoking
- **d.** vigorous exercise

8. **BIOWatch** Why does DNA testing of blood lead to a more reliable match of blood to suspect than does blood typing?

9. **Exploring Further** One of the effects of aspirin is that it thins the blood. Why is aspirin sometimes prescribed for people at risk for heart attack?

10. **Concept Mapping** Make a concept map that outlines the path of blood through the body. Try to include the following terms in your map: *artery, capillary, vein, lymphatic system, pulmonary circulation, systemic circulation, atrium, ventricle, aorta,* and *vena cava.*

Critical Thinking

11. Forming Reasoned Opinions The frequency of blood clots and heart attacks is much lower among the Inuit, the nomadic hunters of the North American Arctic, than it is among other North Americans. This difference is credited to fish oils in the Inuit diet that cause blood platelets to be more slippery. How might slippery platelets affect the clotting ability of the Inuit's blood?

12. Evaluating Results As altitude increases, the atmosphere becomes thinner. When a runner who trained at sea level competes at a location 500 m above sea level, how will his or her performance compare with his or her training performance?

13. Inferring Function How is body temperature regulated by blood vessel diameter?

14. Finding Information Use the media center or Internet resources to find out which foods are recommended as foods that prevent heart disease.

Alternative Assessment

15. Career Connection Respiratory Therapist Research the field of respiratory therapy, and write a report on your findings. Your report should include a job description, training required, kinds of employers, growth prospects, and starting salary.

Standardized Test Prep

Understanding Concepts

Directions (1–5): **For *each* question, write on a separate sheet of paper the letter of the correct answer.**

1 Which part of the heart is the pacemaker?
A. coronary sinus
B. inferior vena cava
C. left ventricle
D. sinoatrial node

2 In what organ is carbon dioxide released in its gaseous form?
F. diaphragm   H. lung
G. heart   I. muscle

3 Which of the following is a disease in which the elastic fibers in the alveoli are destroyed?
A. arteriosclerosis
B. asthma
C. atherosclerosis
D. emphysema

4 Which chamber of the heart sends blood to the lungs?
F. left atrium   H. left ventricle
G. right atrium   I. right ventricle

5 Which of the following is a factor that contributes to heart attacks and strokes?
A. unmanaged stress
B. regular physical activity
C. avoiding cigarette smoke
D. diet high in unsaturated fats

Directions (6): **For the following question, write a short response.**

6 What role do surface markers play in blood typing?

Test TIP

When using a graph to answer a question, read the graph's title and the labels on the graph's axes. For graphs that show a change in some variable over time, keep in mind that the steepness and direction of a curve indicate the relative rate of change at a given point in time.

Reading Skills

Directions (7): **Read the passage below. Then answer the question.**

Dr. Charles Drew (1904–1950) was an African-American physician who worked at Columbia University in New York. Among his many accomplishments was the development of a way to keep blood plasma fresh. He separated the plasma from the red blood cells and froze these parts of blood separately. Dr. Drew also helped save many lives by starting the first blood bank.

7 Which component of blood allows it to carry oxygen to tissues in the body?
F. plasma
G. platelets
H. red blood cells
I. white blood cells

Interpreting Graphics

Directions (8): **Base your answer to question 8 on the graph below.**

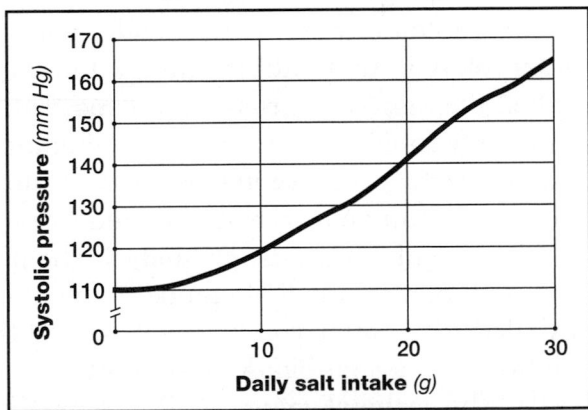

Daily Salt Intake and Blood Pressure in Humans

8 What conclusion can be drawn from the chart?
A. A person can reduce hypertension by consuming more salt.
B. A daily intake of 10 g or less is associated with a risk to health.
C. Raising one's systolic pressure leads to a greater appetite for salt.
D. Increasing one's salt intake leads to an increased systolic pressure.

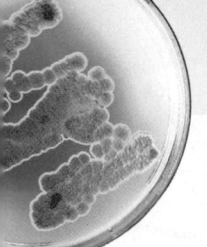

Exploration Lab

Determining Lung Capacity

SKILLS
- Measuring
- Organizing data
- Comparing

OBJECTIVES
- **Measure** your tidal volume, vital capacity, and expiratory reserve volume.
- **Determine** your inspiratory reserve capacity and lung capacity.

- **Predict** how exercise will affect tidal volume, vital capacity, and lung capacity.

MATERIALS
- spirometer
- spirometer mouthpiece

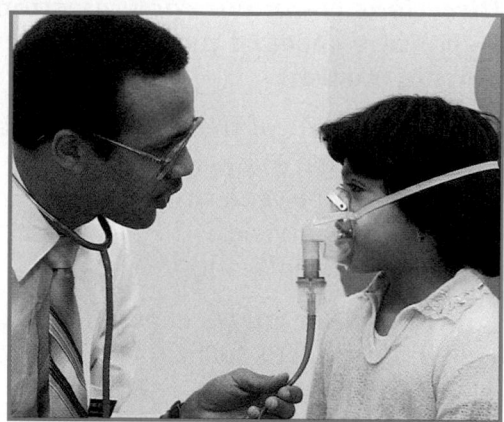

Before You Begin

Lung capacity is the total volume of air that the lungs can hold. The lung capacity of an individual is influenced by many factors, such as gender, age, strength of diaphragm and chest muscles, and disease.

During normal breathing, only a small percentage of your lung capacity is inhaled and exhaled. The amount of air inhaled or exhaled in a normal breath is called the **tidal volume.** An additional amount of air, called the **inspiratory reserve volume,** can be forcefully inhaled after a normal inhalation. The **expiratory reserve volume** is the amount of air that can be forcefully exhaled after a normal exhalation. **Vital capacity** is the maximum amount of air that can be inhaled or exhaled. Even after you have exhaled all the air you can, a significant amount of air called the **residual volume** still remains in your lungs.

In this lab, you will determine your lung capacity by using a **spirometer,** which is an instrument used to measure the volume of air exhaled from the lungs.

1. Write a definition for each boldface term in the paragraph above.

2. Make a data table similar to the one shown.

DATA TABLE	
Tidal volume	
Expiratory reserve volume	
Inspiratory reserve volume	
Vital capacity	
Estimated residual volume	
Estimated lung capacity	

3. Based on the objectives for this lab, write a question about breathing that you would like to explore.

Procedure

PART A: Measuring Volume

1. Place a clean mouthpiece in the end of a spirometer. **CAUTION: Many diseases are spread by body fluids, such as saliva. Do NOT share a spirometer mouthpiece with anyone.**

2. To measure your tidal volume, first inhale a normal breath. Then exhale a normal breath into the spirometer through the mouthpiece. Record the volume of air exhaled in your data table.

3. To measure your expiratory reserve volume, first inhale a normal breath and then exhale normally. Then forcefully exhale as much air as possible into the spirometer. Record this volume.

4. To measure your vital capacity, first inhale as much air as you can, and then forcefully exhale as much air as you can into the spirometer. Record this volume.

PART B: Calculating Lung Capacity

The table below contains average values for residual volumes and lung capacities for young adults.

Residual Volumes and Lung Capacities		
	Males	**Females**
Residual volume*	1,200 mL	900 mL
Lung capacity*	6,000 mL	4,500 mL

*Athletes can have volumes 30–40% greater than the average for their gender.

5. Inspiratory reserve volume (IRV) can be calculated by subtracting tidal volume (TV) and expiratory reserve volume (ERV) from vital capacity (VC). The formula for this calculation is as follows:

$$IRV = VC - TV - ERV$$

Use the data in your data table and the equation above to calculate your estimated inspiratory reserve volume.

6. Lung capacity (LC) can be calculated by adding residual volume (RV) to vital capacity (VC). The formula for this calculation is as follows:

$$LC = VC + RV$$

Use the data in your data table and the table above to calculate your estimated lung capacity.

PART C: Cleanup and Disposal

7. Dispose of your mouthpiece in the designated waste container.

8. Clean up your work area and all lab equipment. Return lab equipment to its proper place. Wash your hands thoroughly before you leave the lab and after you finish all work.

Analyze and Conclude

1. Interpreting Data How does your expiratory reserve volume compare with your inspiratory reserve volume?

2. Interpreting Tables How does the residual volume and lung capacity of an average young adult female compare with those of an average young adult male?

3. Analyzing Data How did your tidal volume compare with that of others?

4. Recognizing Relationships Why was the value you found for your lung capacity an estimated value?

5. Analyzing Methods Why didn't you measure inspiratory reserve volume directly?

6. Inferring Conclusions Why would males and athletes have greater vital capacities than females?

7. Justifying Conclusions Use data from your class to justify the conclusion that exercise increases lung capacity.

8. Further Inquiry Write a new question that could be explored with another investigation.

On the Job

Spirometry is the use of a spirometer to study respiratory function. Nurses and respiratory therapists use spirometers to evaluate patients with respiratory diseases. Do research to discover how spirometry is used to distinguish different respiratory diseases. For more about careers, visit **go.hrw.com** and type in the keyword **HX4 Careers**.

Healthier lifestyles can lower the risk of diabetes, while new treatments help reduce the damage caused by the disease.

Controlling Diabetes

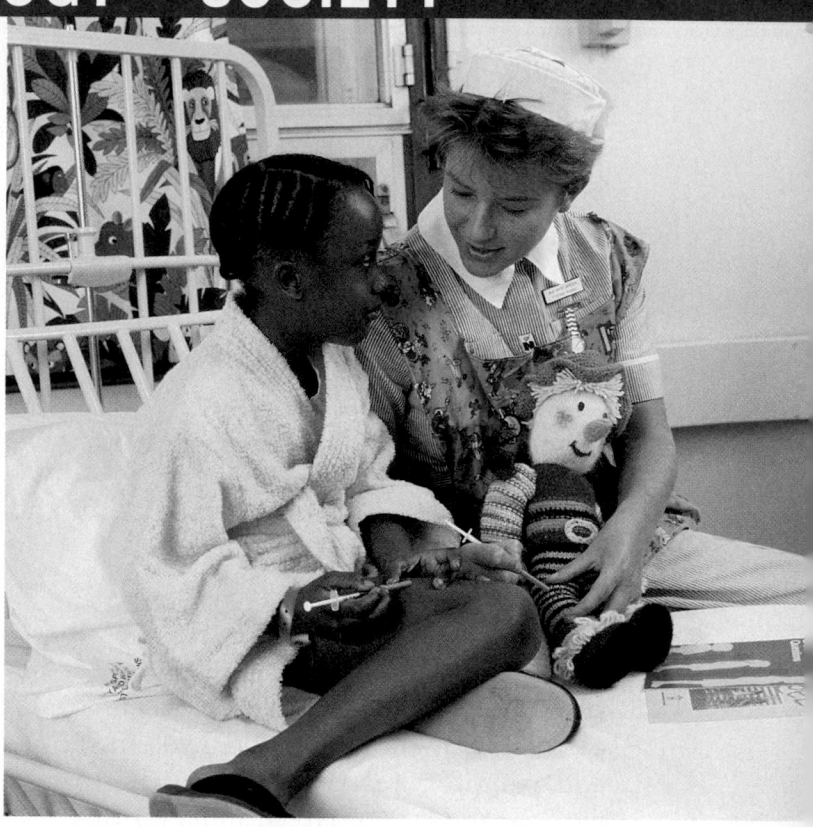

It is the fifth leading cause of death in the United States, killing more than 200,000 people each year. But almost one-third of the people who have this disease—more than 5 million individuals—are unaware they are sick. During the 7 to 10 years that typically elapse between the onset of the illness and diagnosis, the heart, eyes, kidneys, and nerves in the hands and feet can suffer irreversible damage. In later stages, the disease can cause blindness, kidney failure, heart disease, stroke, and infections so severe that often the affected limb must be amputated.

The disease is diabetes mellitus. It affects nearly 17 million Americans and is responsible for more than $100 billion a year in medical expenses, disability payments, and other costs. Although all racial and ethnic groups suffer from diabetes, it is much more common among African Americans, Native Americans, and Hispanics.

With a combination of drugs, diet, and exercise, many diabetics can prevent the development of complications and delay the progression of the disease—although no cure yet exists. Even better news is that people at risk for diabetes can take some simple steps to dramatically lower their likelihood of becoming sick. The bad news is that, despite these advances in treatment and prevention, the number of diabetics is rising fast in the United States and throughout the world.

What Is Diabetes?

Diabetes is a disease of the endocrine system in which the body loses the ability to regulate the amount of glucose in the blood. It stems from a defect in the body's production, use, or transport of insulin, a hormone secreted by the pancreas

Smart choices
A healthful diet helps reduce the likelihood of diabetes.

Young diabetic
Diabetes affects both young and old.

that stimulates cells to take in glucose. Blood glucose levels, normally tightly controlled, rise abnormally high, and the excess glucose is excreted in the urine.

There are two forms of diabetes. Type I diabetes usually begins suddenly when the person is a child or young adult. The immune system attacks the insulin-producing cells of the pancreas. What stimulates this attack remains a mystery, although some recent research suggests a virus may be responsible. People with Type I diabetes cannot make insulin and must take daily injections of insulin to survive.

By contrast, in Type II diabetes, the pancreas either stops producing enough insulin or the body's cells become insensitive to insulin's effects, taking in less glucose. In both cases, chronically high blood glucose levels result. Between 90 and 95 percent of diabetics have this form of the disease, which can often be controlled by diet and exercise instead of insulin injections.

Type II diabetes results from a combination of genetic and environmental causes. Susceptibility to the disease runs in families—the odds of becoming sick are higher if a parent or sibling suffers from the disease. Your lifestyle is just as important, since the disease is usually brought on by an environmental risk factor. The environmental risk factors are a diet high in fat and sugar but low in fiber, lack of exercise, high blood pressure, and obesity (weighing more than 20 percent over your ideal body weight).

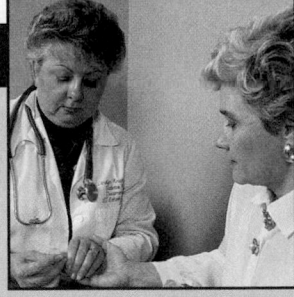

Career

Community Health Educator

Diabetes educator

Profile

Community health educators counsel individuals and groups on health practices designed to prevent disease and promote good health.

Job Description

Persons who work in community health education may specialize in a particular disease (such as diabetes) and its effects and treatments. Many community health educators are employed by state and local governments, public health clinics, social service agencies, or resident care facilities. Employment is expected to grow rapidly as the population ages and public interest in health education continues to increase.

Science/Math Career Preparation

Biology	Biochemistry
Microbiology	Psychology
Chemistry	Sociology

Although it's more often diagnosed in people over 40, Type II diabetes results from damage that accumulates over years or decades. However, Type II diabetes is becoming more common in younger adults and children. Eating a healthful diet, getting plenty of exercise, and avoiding obesity will reduce your likelihood of diabetes and can pay off in other ways, such as reducing your risk of heart disease, stroke, and some kinds of cancers. ∎

Analyzing STS Issues

Science and Society

1 **What are some other lifestyle diseases?** Diabetes is often called a lifestyle disease because it frequently results from how one chooses to live—diet, activity level, and so forth. Using library resources or the Internet, research some other lifestyle diseases, such as certain kinds of cancer and heart disease. What are the risk factors for these diseases? How have the frequencies of these diseases changed over the last century?

2 **What are healthy lifestyle choices to prevent diabetes?** When it comes to health, little

steps make a big difference. What type of diet can help prevent diabetes? What other measures can be taken to maintain a healthy lifestyle?

Technology: Biomedical Devices

3 **How might new technology further improve diabetes treatment?** Over the last few decades, new methods for treating diabetes have come on the market. These include new ways of delivering insulin and better methods for monitoring blood glucose levels. Research a new technology for treating diabetes that is being developed or tested. Then write a short report describing what you learned.

Villi in human intestine (1,280×)

39 Digestive and Excretory Systems

✔ Quick Review

Answer the following without referring to earlier sections of your book.

1. **Differentiate** between carbohydrates, fats, and proteins. *(Chapter 2, Section 3)*

2. **Compare** saturated and unsaturated fatty acids. *(Chapter 2, Section 3)*

3. **Describe** the role of enzymes in chemical reactions. *(Chapter 2, Section 4)*

4. **Summarize** the function of cellular respiration. *(Chapter 5, Section 3)*

5. **Discuss** the balance of water and salt in vertebrates. *(Chapter 33, Section 1)*

Did you have difficulty? *For help, review the sections indicated.*

📖 Reading Activity

Before you begin to read, survey the chapter, noting the red headings at the tops of pages and the blue subheadings. Use these heads to make an outline of the chapter, leaving space after each heading. Fill important facts in as you read.

internet connect

www.scilinks.org
National Science Teachers Association *sci*LINKS Internet resources are located throughout this chapter.

*sci*LINKS. Maintained by the National Science Teachers Association

● The multiple, protruding villi of the small intenstine greatly increase the surface area through which nutrients are absorbed.

Your Body's Need for Food

Objectives

- **Identify** five nutrients found in foods.

- **Relate** the role of carbohydrates, proteins, lipids, vitamins, minerals, and water in maintaining a healthy body.

- **Describe** each of the parts of the USDA food guide pyramid.

- **Name** one health disorder associated with high levels of saturated fats in the diet.

Key Terms

nutrient
digestion
calorie
vitamin
mineral

Food

Your body uses energy to move, to grow, and even to lie still and sleep. The amount of energy you need depends on many factors, including your age, your sex, your rate of growth, and your level of physical activity. Different activities use different amounts of energy, as shown in **Figure 1.**

You obtain energy from the nutrients in the foods and beverages you consume. A **nutrient** is a substance required by the body for energy, growth, repair, and maintenance. Nutrients in food and beverages include carbohydrates, lipids, proteins, vitamins, and minerals. Each nutrient plays a different role in keeping your body healthy. Water is essential for life and for maintaining health.

The large molecules in food must be broken down in order to be absorbed into the blood and carried to cells throughout the body. The process of breaking down food into molecules the body can use is called **digestion.** Your cells then break the chemical bonds of the digested food molecules and use the energy that is released to make ATP during the process of cellular respiration.

The energy available in food is measured by using a unit called a calorie. A **calorie** is the amount of heat energy required to raise the temperature of 1 g of water 1°C (1.8°F). The greater the number of calories in a quantity of food, the more energy the food contains. Because a calorie represents a very small amount of energy, nutritionists use a unit called the Calorie (with a capital C), which is equal to 1,000 calories. On food labels and throughout this book, the word *calories* represents Calories (1,000 calories).

Figure 1   Energy required for common activities. Quiet activities require just a little more energy than what it takes to keep you alive. Strenuous activities require more energy.

Energy and Building Materials

Each nutrient plays a different role in maintaining a healthy body. Carbohydrates, proteins, and lipids are involved in providing both energy and building materials to the cells.

Carbohydrates

Carbohydrates that exist as single sugar molecules are called monosaccharides or simple carbohydrates. Carbohydrates made of two or many sugar molecules linked together by chemical bonds are called complex carbohydrates. Complex carbohydrates must be digested (broken down) into simple sugars before cells can use their energy.

Many foods contain carbohydrates, as shown in **Figure 2.** Glucose, fructose, and other simple sugars are found in fruits, honey, and onions. Glucose, a simple sugar, is used by cells for energy, and it can be directly absorbed into your bloodstream. Table sugar contains sucrose, two simple sugars linked together. Starches are long chains of sugars found in cereal grains and in vegetables such as potatoes, beans, and corn. Cellulose is a major component of plant cell walls and is found in all foods that come from plants. Cellulose, which is a major part of fiber, does not provide energy because we do not have the enzymes to digest it. However, cellulose aids in human digestion by stimulating the walls of the digestive tract to secrete mucus, which helps pass food through the digestive tract.

If excess carbohydrates are consumed, they are stored as the carbohydrate glycogen in the liver and in some muscle tissue. Glycogen can later be broken back down into glucose when the body needs energy. The remainder of the excess glucose is converted to fat and stored in fatty tissue.

Study TIP

● **Organizing Information**
Make a table to organize information about food nutrients and water. Across the top, write the headings *Carbohydrates, Proteins, Lipids, Vitamins, Minerals,* and *Water.* Along the left side, write *Functions, Food sources,* and *Additional comments.* Add information to the table as you read Section 1.

Figure 2 Nutrients in food

Although most foods contain a mix of nutrients, some foods are richer than others in a specific nutrient.

Carbohydrate-rich foods

(Carbohydrates contain 4 calories per gram.) Breads, pasta, grains, cereals, potatoes, fruits

Protein-rich foods

(Proteins contain 4 calories per gram.) Fish, eggs, poultry, beef pork, nuts, legumes, milk, cheese, tofu

Fat-rich foods

(Fats contain 9 calories per gram.) Milk, cheese, meats, butter, olives, avocados, fried foods, oils, chips

Proteins

The digestive products of proteins—amino acids—are normally used by the body for making other protein molecules, such as enzymes and antibodies. When more protein is eaten than is needed by the cells, the amino acids are used for energy or converted to fat. The body requires 20 different amino acids to function. A child's or teen's body can make 10 of the amino acids from other amino acids. The other 10, called essential amino acids, must be obtained directly from food. Most animal products, such as eggs, milk, fish, poultry, and beef, contain all of the essential amino acids. No single plant food contains all of the essential amino acids. But eating certain combinations of two or more plant products can supply all the essential amino acids. Adults must get eight essential amino acids from food.

The guidelines for healthy eating are summarized in the USDA (U.S. Department of Agriculture) food guide pyramid, as shown in **Figure 3.** The pyramid lists the daily number of servings needed from each food group to obtain a variety of nutrients in your diet.

Lipids

Lipids, organic compounds that are insoluble in water, are used to make steroid hormones and cell membranes and to store energy. Fats are lipids that store energy in plants and animals. Fats are

Figure 3   The USDA food guide pyramid

Food groups at the bottom of the pyramid should be eaten in greater amounts than those at the top.

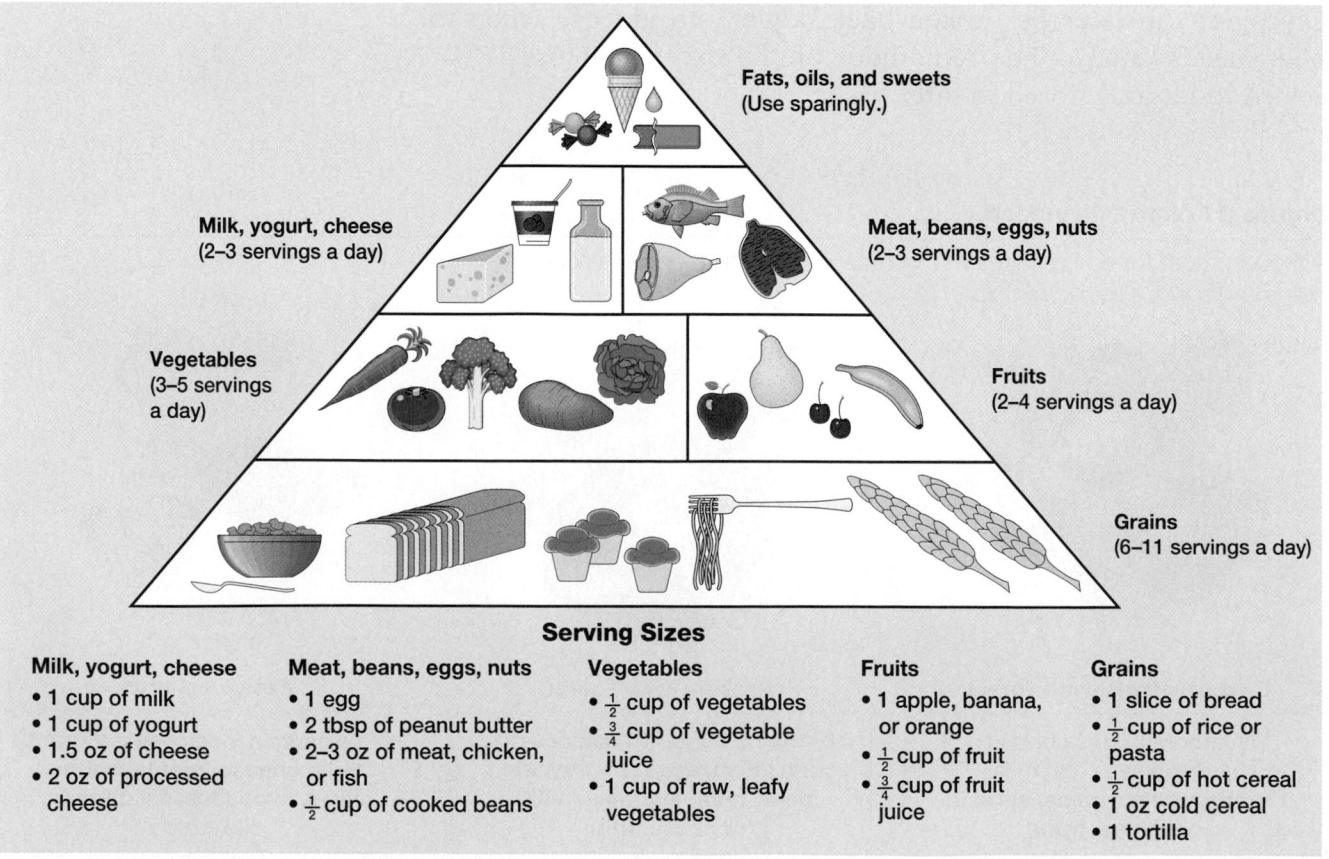

Serving Sizes

Milk, yogurt, cheese	Meat, beans, eggs, nuts	Vegetables	Fruits	Grains
• 1 cup of milk • 1 cup of yogurt • 1.5 oz of cheese • 2 oz of processed cheese	• 1 egg • 2 tbsp of peanut butter • 2–3 oz of meat, chicken, or fish • $\frac{1}{2}$ cup of cooked beans	• $\frac{1}{2}$ cup of vegetables • $\frac{3}{4}$ cup of vegetable juice • 1 cup of raw, leafy vegetables	• 1 apple, banana, or orange • $\frac{1}{2}$ cup of fruit • $\frac{3}{4}$ cup of fruit juice	• 1 slice of bread • $\frac{1}{2}$ cup of rice or pasta • $\frac{1}{2}$ cup of hot cereal • 1 oz cold cereal • 1 tortilla

stored around organs and act as padding and insulation. Fats also act as solvents for fat-soluble vitamins.

Although lipids are essential nutrients, too much fat in the diet is known to harm several body systems. For example, a diet high in saturated fats is linked to high blood-cholesterol levels, which in turn may be connected to cardiovascular diseases. It is recommended that a person limits his or her consumption of saturated fats and that most of the fats in the diet be unsaturated.

Energy-Giving Nutrient	Percentage of Daily Calories
Carbohydrates	45 to 65
Protein	10 to 15
Fats	25 to 35

Balancing Nutrients and Energy

Regardless of their source, the excess calories you eat will be stored as either glycogen or body fat, and you will gain weight. If you use more calories than you take in, additional energy will be obtained from your body's energy stores, and you will lose weight. Your diet and overall activity level determine in part whether you store excess calories as glycogen or as fat. **Figure 4** summarizes what percentage of the day's total calories should come from each nutrient.

Obesity is described as being more than 20 percent heavier than your ideal body weight. Obesity significantly increases an individual's risk of diabetes, heart disease, osteoarthritis, and many other disorders. Regular physical activity is important in maintaining energy balance.

Figure 4 A balanced meal. The percentages of the day's total calories that should come from each nutrient are shown. About half of your day's calories should come from foods high in complex carbohydrates.

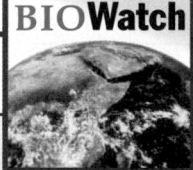

BIOWatch

Obesity and Health

The National Center for Health Statistics reported that by the year 2000, 31 percent of adults in the United States were obese. Medically, obesity is defined as having a body mass index (BMI) of 30 or higher. To calculate your BMI, first multiply your weight in pounds by 703, and then divide by your height in inches squared.

$$BMI = \frac{weight\ in\ pounds \times 703}{(height\ in\ inches)^2}$$

For example, someone who is 6'0" (72 inches) tall and who weighs 189 pounds has a body mass index of $(189 \times 703) \div (72 \times 72) = 25.6$.

Unfortunately, being obese often has a serious consequence— diabetes. Late-onset diabetes affected 14 million Americans in 2000, and 80 percent of those so affected were obese. Like the rate of obesity, the rate of late-onset diabetes is soaring, up 40 percent in the last 10 years.

Late-onset diabetes is a disorder in which the hormone insulin is unable to induce the body's cells to take up glucose from the blood. Glucose-starved tissues consume their own protein, and waste away. Diabetes is the leading cause of kidney failure, blindness, and amputation in adults.

What is the link between obesity and late-onset diabetes? Researchers have recently found that fat cells produce a hormone called *resistin* that prompts cells to resist insulin. Mice given resistin by researchers lose much of the ability to respond to insulin and fail to take up blood sugar. Drugs that inhibit resistin restore insulin's effectiveness. Researchers hypothesize that resistin blocks the same glucose-transporting molecules in the plasma membrane that insulin activates.

Vitamins, Minerals, and Water

Vitamins, minerals, and water are required in our diets. They do not provide energy, but they contribute to many different functions, including regulating the reactions that release energy.

Vitamins Many different **vitamins,** organic substances that occur in many foods in small amounts, are necessary in trace amounts for the normal metabolic functioning of the body.

Vitamins dissolve in either water or fat, as summarized in **Table 1.** Fat-soluble vitamins—vitamins A, D, E, and K—can be stored in body fat. Excessive amounts of vitamins A and D can be toxic. Excess water-soluble vitamins—vitamin C and the B vitamins—are excreted in urine and must be replenished by the diet.

Table 1 Vitamins

Vitamin	Food sources	Role	Effects of deficiency
Water-soluble			
Vitamin B$_1$ (thiamin)	Most vegetables, nuts, whole grains, organ meats	Assists in carbohydrate metabolism, helps nerves and heart to function properly	Digestive disturbances, impaired senses
Vitamin B$_2$ (riboflavin)	Fish, poultry, cheese, yeast, green vegetables	Needed for healthy skin and tissue repair, assists in carbohydrate metabolism	Blurred vision, cataracts, cracking of skin, lesions of intestinal lining
Vitamin B$_3$ (niacin)	Whole grains, fish, poultry, tomatoes, legumes, potatoes	Keeps skin healthy, assists in carbohydrate metabolism	Mental disorders, diarrhea, inflamed skin
Vitamin B$_{12}$ (cobalamin)	Meat, poultry, milk, dairy products	Needed for formation of red blood cells	Reduced number of red blood cells
Vitamin C (ascorbic acid)	Citrus fruits, strawberries, potatoes	Needed for wound healing, healthy gums and teeth	Swollen and bleeding gums, loose teeth, slow-healing wounds
Fat-soluble			
Vitamin A (retinol)	Butter, eggs, liver, carrots, green leafy vegetables, sweet potatoes	Keeps eyes and skin healthy, needed for strong bones and teeth	Infections of urinary and digestive systems, night blindness
Vitamin D (cholecalciferol)	Salmon, tuna, fish liver oils, fortified milk, cheese	Assists in calcium uptake by the gut, needed for strong bones and teeth	Bone deformities in children, loss of muscle tone
Vitamin E (tocopherol)	Many foods, especially wheat and other vegetable oils, olives, whole grains	Protects cell membranes from damage by reactive oxygen compounds (free radicals)	Reduced number of red blood cells, nerve tissue damage in infants
Vitamin K (menadione)	Leafy green vegetables, liver, cauliflower	Needed for normal blood clotting	Bleeding caused by prolonged clotting time

Minerals Different minerals are required to maintain a healthy body. **Minerals** are naturally occurring inorganic substances that are used to make certain body structures and substances, for normal nerve and muscle function, and to maintain osmotic balance. Some minerals are essential for enzyme function. Minerals are not produced by living organisms. Minerals must be replaced on a daily basis because they are soluble in water. Teeth and bones require the minerals calcium and phosphorus. Iron is required for transporting oxygen. Magnesium, calcium, sodium, potassium, and zinc help regulate function of the nerves and muscles.

Trace elements, such as those listed in **Table 2,** are minerals present in the body in small amounts. Humans usually obtain adequate amounts of the required trace elements directly from the plants they eat or indirectly from animals that have eaten plants.

Water You can survive only a few days without water, though you can live several weeks without food. Water is used by the body as a medium to transport gases, nutrients, and waste products. Water also plays a role in regulating body temperature. Two-thirds of the body's weight is water.

Table 2 Trace Elements

Trace element	Food sources	Role
Iodine	Seafood, plants grown in iodine-rich soil, iodized table salt	Synthesis of thyroid hormones
Cobalt	Leafy vegetables, liver, kidney	Synthesis of vitamin B_{12}
Zinc	Meat, shellfish, dairy products	Synthesis of digestive enzymes, proper immune function
Molybdenum	Legumes, cereals, milk	Protein synthesis
Manganese	Whole grains, nuts, legumes	Hemoglobin synthesis, urea formation
Selenium	Meat, seafood, cereal grains	Preventing chromosome breakage

Section 1 Review

1 **Predict** four nutrients that would be found in a serving of green beans.

2 **Compare** the functions of carbohydrates and proteins in maintaining a healthy body.

3 **Describe** the type of information the USDA food guide pyramid provides.

4 **Evaluate** the roles vitamins, minerals, and water play in maintaining a healthy body.

5 **Critical Thinking Applying Information** Your friend wants to feed her elderly grandmother more food in order to keep her healthy. Is this a good idea? Explain.

6 **Standardized Test Prep** One of the functions of lipids in the body is to
 A enhance enzyme activity.
 B make glycogen.
 C make steroid hormones.
 D make proteins.

Objectives

- **Relate** the four major functions of the digestive system to the processing of food.

- **Summarize** the path of food through the digestive system and the major digestive processes that occur in the mouth, stomach, small intestine, and large intestine.

- **Describe** how nutrients are absorbed from the digestive system into the bloodstream or lymphatic system.

- **Identify** the role of the pancreas and liver in digestion.

Key Terms

amylase
esophagus
pepsin
lipase
villus
colon

Breaking Down Food

Imagine you just ate your favorite meal. What happens to that food? Before your body can use the nutrients in the food you eat, the large food molecules must be broken down physically and chemically. The process of breaking down food into molecules the body can use is called digestion. The digestive system is the body system that is involved in the taking in and processing of food for use by your body cells. The digestive system takes in food, breaks it down into molecules small enough for the body to absorb, and gets rid of undigested molecules and waste.

As shown in **Figure 5,** the digestive system is made up of a long, winding tube, the digestive tract, that begins at the mouth and winds through the body to the anus. Food travels more than 8 m (26 ft) through your digestive tract. The digestive tract includes the mouth, pharynx, esophagus, stomach, small intestine, large intestine, and rectum. Although the liver and pancreas *(PAN kree uhs)* are not part of the digestive tract, they deliver secretions into the digestive tract through ducts (tubes).

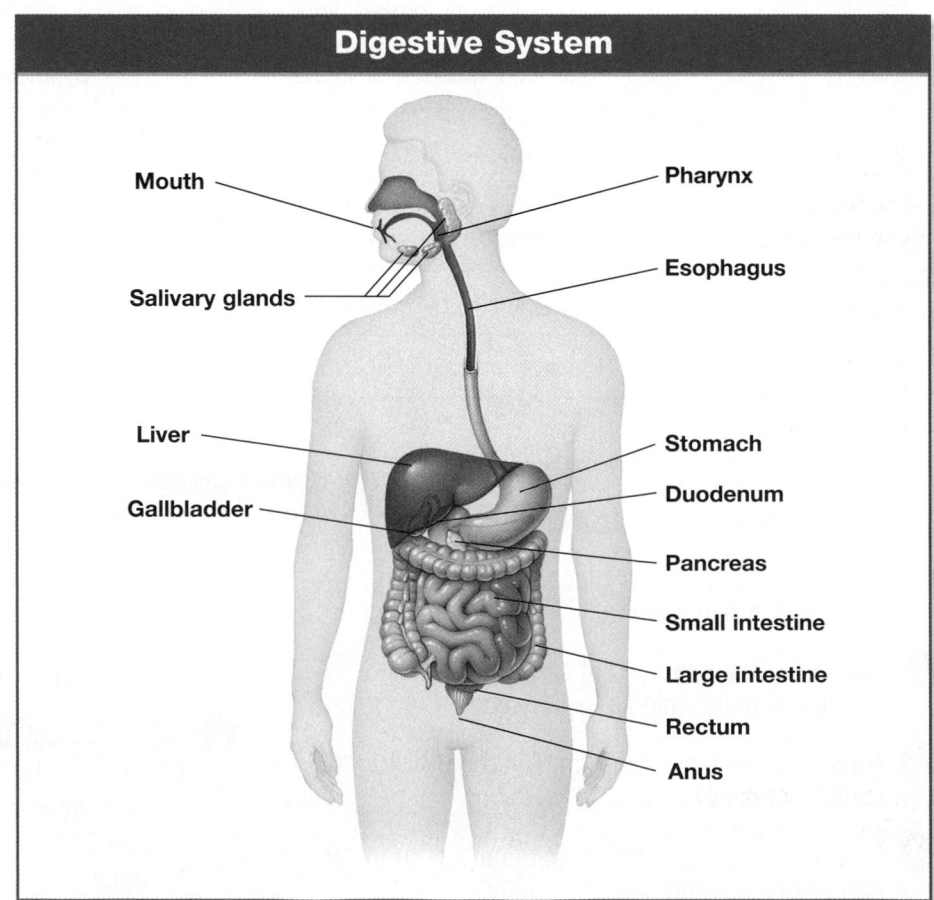

Digestive System

Mouth — Pharynx
Salivary glands — Esophagus
Liver — Stomach
Gallbladder — Duodenum
Pancreas
Small intestine
Large intestine
Rectum
Anus

Figure 5 Processing food.
The digestive system breaks down food into individual nutrient molecules that can be absorbed into the bloodstream.

Starting Digestion

The digestion of food begins as soon as the food is ingested. The teeth rip and chew food into shreds, and the tongue mixes the pieces with a watery solution called saliva. Taste buds on the tongue are sensitive to certain chemicals in the food. Saliva is secreted into the mouth by three pairs of salivary glands, shown in Figure 5. Saliva moistens and lubricates the food so that it can be swallowed more easily.

Saliva also contains **amylases** *(AM uh lay sehs)*, enzymes that begin the breakdown of carbohydrates such as starch, into monosaccharides (single sugars). The mechanical action of chewing and the chemical action of amylase are both part of the digestion of carbohydrates in the mouth.

Notice in **Figure 6** that the structure of our teeth helps in the breakdown of food. The two front teeth, the incisors, cut food. The cuspids, or canines, shred food. The back teeth, the molars, crush and grind food.

After passing through the region in the back of the throat called the pharynx *(FAIR ihnks)*, the food triggers a swallowing response. The action of swallowing moves the epiglottis (a flap of cartilage) over the opening of the trachea—the tube that leads to the lungs. This action prevents food from entering the trachea and eventually the lungs. Instead, food enters the esophagus *(ih SAHF uh guhs)*.

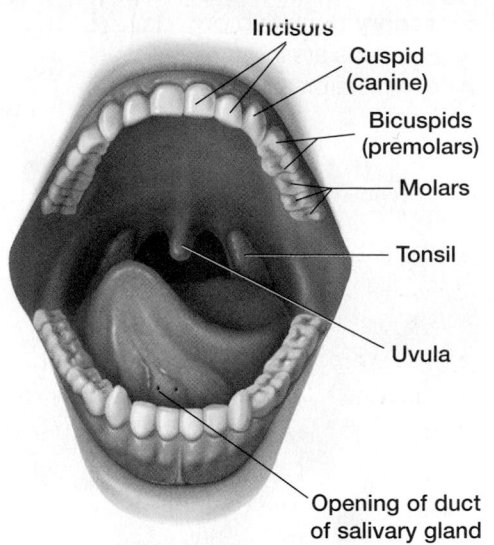

Figure 6 Teeth and digestion. Canine and incisor teeth are used for cutting and tearing food. The molars are used to grind food.

The Esophagus

The **esophagus** is a long tube that connects the mouth to the stomach. No digestion takes place in the esophagus. Its role is to act as a kind of descending elevator, moving food down to the stomach.

The esophagus is about 25 cm (10 in.) long. The lower two-thirds of the esophagus is wrapped in sheets of smooth muscle. Food does not simply fall into the stomach; it is pushed down, as shown in **Figure 7.** Successive rhythmic waves of smooth muscle contraction in the esophagus, called peristaltic *(pehr uh STAHL tihk)* contractions, or peristalsis, move the food toward the stomach. Peristalsis can be thought of as waves moving through the muscle with the area where the wave is passing causing the muscle to narrow. It takes about 5 to 10 seconds for food to pass down the esophagus and into the stomach.

The Stomach

Food exits the esophagus and enters the stomach through a muscular valve called a sphincter *(SFIHNGK tuhr)*. The sphincter prevents acid-soaked food in the stomach from making its way back into the esophagus. The stomach is a saclike organ located just beneath the diaphragm. Besides temporarily storing food, the stomach, shown in **Figure 8** on the next page, also mechanically breaks down food and chemically unravels and breaks down proteins.

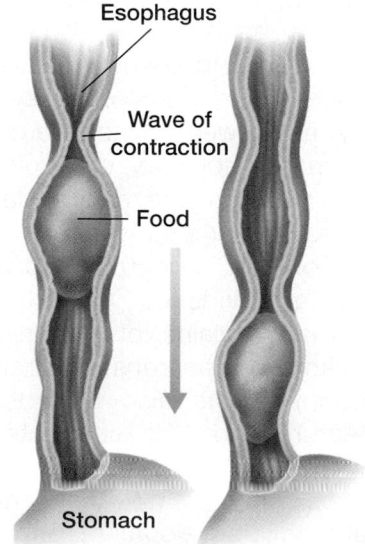

Figure 7 Peristalsis moves food. Food is pushed down the esophagus and toward the stomach by waves of smooth muscle contractions in the wall of the esophagus.

Figure 8 The stomach and accessory digestive organs. Many organs are involved in the complete breakdown of nutrients.

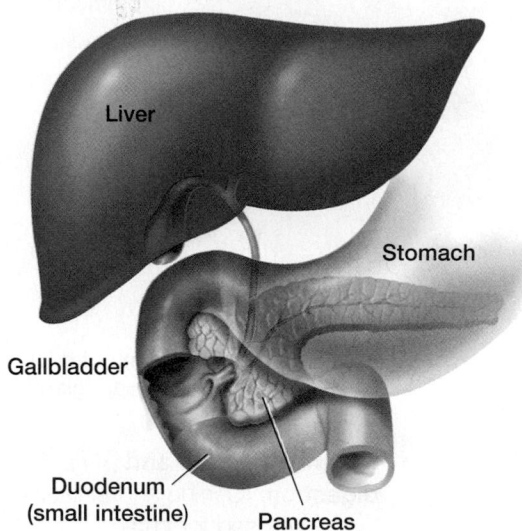

Liver

Stomach

Gallbladder

Duodenum (small intestine)

Pancreas

When food enters the stomach, gastric juice is secreted by the cells that line the inside of the stomach. Gastric juice is a combination of hydrochloric *(HIE droh klawr ihk)* acid (HCl) and pepsin. The acid breaks the bonds in proteins and unfolds large protein chains into single protein strands. **Pepsin,** a digestive enzyme secreted by the stomach, cuts the single protein strands into smaller chains of amino acids. Pepsin is effective only in an acidic environment.

The stomach mixes its contents by using peristaltic waves. Swallowed food can spend from 2 to 6 hours in the stomach. Your stomach secretes about 2 L (2.11 qt) of HCl every day, which creates a solution about 3 million times more acidic than your bloodstream. The hormone gastrin regulates the synthesis of HCl. Thus, HCl is made only when the pH in the stomach is higher than about 1.5.

A coating of mucus protects the lining of the stomach from gastric acid. Bicarbonate in the stomach helps neutralize digestive fluids. Blood circulation in the stomach lining also helps protect stomach tissues.

FORENSICS BIOWatch

Dental Records

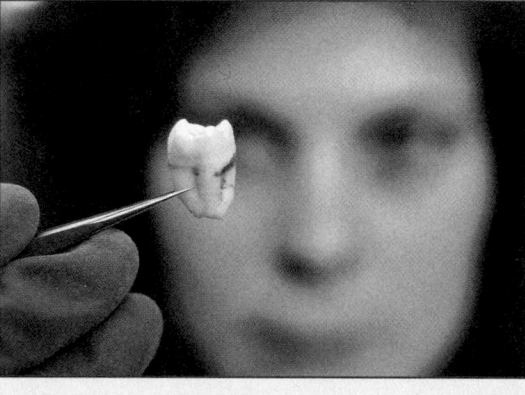

The next time you eat an apple, pause between bites to look at the apple. You've left bite marks, patterns that match your teeth—and no one else's. If the apple were found at a crime scene, a forensic odontologist could use the bite mark evidence to search for a suspect.

Forensic odontologists use dental evidence to identify human remains of missing or unknown persons. Forensic odontologists also use dental features to link suspects to crimes. Odontology is the scientific study of the teeth, their anatomy, development, and diseases. Forensic odontologists compare crime scene evidence with the dental records of a possible suspect or victim.

Identifying the Dead
When an unidentified body is found, forensic odontologists examine the teeth and the jaw. Tooth enamel is the hardest biological substance known. Teeth are a reliable source of evidence, lasting long after most of the body has decayed. The size of the teeth and the amount of wear help odontologists estimate a victim's age.

Forensic odontologists play an important role in the identification of victims of catastrophic events, such as airplane crashes, fires, or terrorist attacks. Disasters such as these result in many casualties and the bodies of victims are often badly damaged. After odontologists have completed their work, law enforce-

ment officials can match forensic data with known dental records, X rays, or photographs.

Linking a Suspect to a Crime
Sometimes human bite marks are found on the body of the victim of a crime. Forensic odontologists analyze bite marks in cases of assault or homicide. Computer modeling can enable odontologists to match a bite mark to a dental mold of a suspect's mouth. The odontologist may also collect saliva from the bite wound for later DNA testing.

The Small Intestine

Food passes from the stomach into the small intestine when a sphincter between the two organs opens. The small intestine is a coiled tubular organ about 6 m (19.8 ft) long that is continuous with the stomach and that functions mainly in the digestion and absorption of nutrients. The word *small* refers to the small diameter of the small intestine as compared with the diameter of the large intestine—not to its length. Peristalsis mixes the food, which remains in the small intestine for about 3 to 6 hours.

The first part of the small intestine, the duodenum *(doo oh DEE nuhm)*, receives secretions from the pancreas, liver, and gallbladder, as shown in Figure 8. Cells that line the small intestine and the pancreas secrete digestive enzymes involved in completing the digestion of carbohydrates into monosaccharides, proteins into amino acids, and lipids into fatty acids and glycerol.

Before fats can be digested by pancreatic enzymes called **lipases** *(LIE pays uhs)*, the fats must first be treated with bile, a greenish fluid produced by the liver. Bile breaks up fat globules into tiny fat droplets, a process called emulsification *(ee MUHL suh fih kay shuhn)*. The gallbladder, a green muscular sac attached to the liver, concentrates and stores bile until it is needed in the small intestine.

Most absorption (passage of nutrients to the blood or lymph) occurs in the small intestine. The lining of the small intestine is covered with fine fingerlike projections called **villi** (singular form, villus). Villi, shown in **Figure 9,** are too small to see with the naked eye. In turn, the cells covering each villus have projections on their outer surface called microvilli. The villi and microvilli greatly increase the area available for absorption of nutrients. Sugars and amino acids enter capillaries in the villi and are carried in the blood to the liver for further metabolism. Fatty acids and glycerol enter lymphatic vessels in the villi and eventually enter the bloodstream.

Figure 9   Villi in the small intestine

Inside each villus are capillaries and lymphatic vessels where nutrients enter the bloodstream.

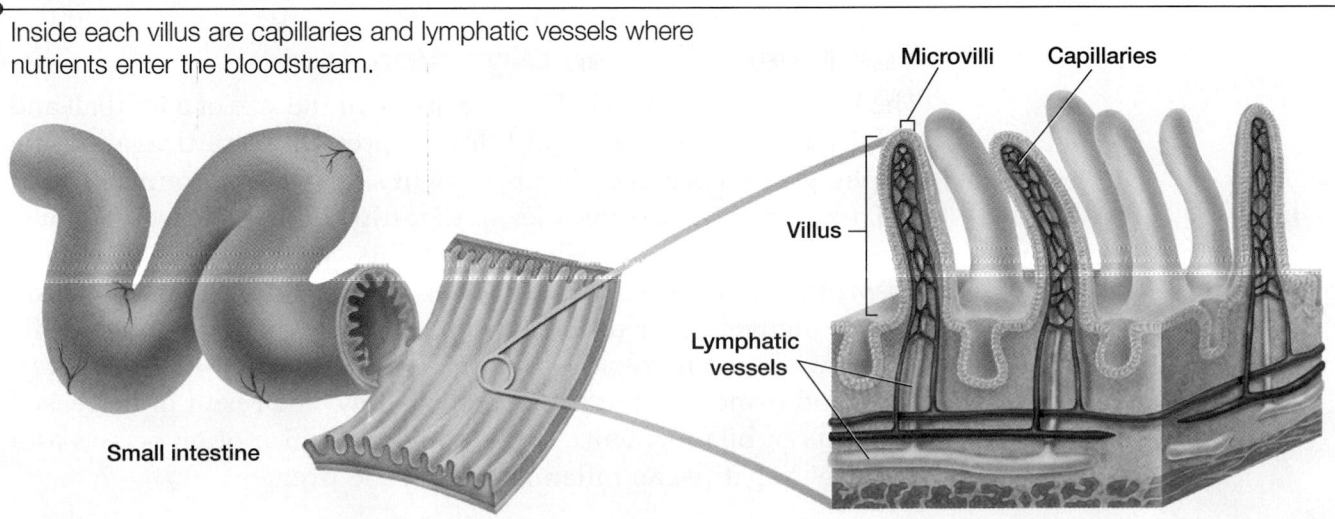

Small intestine

Microvilli

Capillaries

Villus

Lymphatic vessels

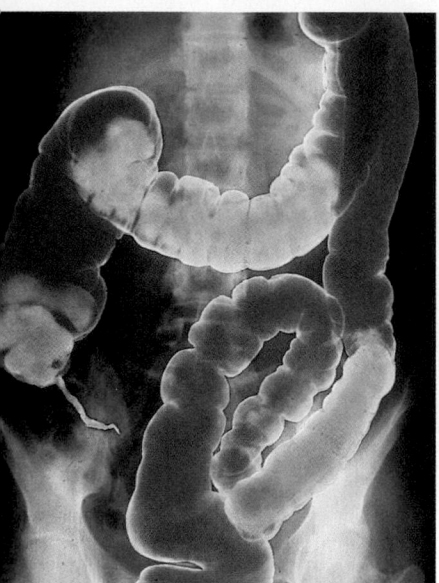

Figure 10 X ray of the large intestine (colon). Stretched out, the large intestine, which appears orange, is about 1 m (3.3 ft) long.

The Large Intestine

All components of food that are not for energy production (for example, cellulose) are considered wastes. The wastes move into the large intestine. The large intestine, or **colon** *(KOH luhn)*, shown in **Figure 10,** is much shorter than the small intestine. However, the diameter of the large intestine is about three times larger than the diameter of the small intestine. The colon is not coiled like the small intestine. Instead, it is composed of three relatively straight segments. No digestion takes place in the colon.

The volume of material that flows through the digestive system each day is large—about 10 L of food, drink, and secretions enter the small intestine. But the amount of material that leaves the body as waste is small. This is because almost all of the fluids and solids (about 90–95 percent) are absorbed during their passage through the small intestine. Mostly mineral ions and water are absorbed through the wall of the large intestine.

Most of the colon's contents are dead cells, mucus, digestive secretions, bacteria, and yeast. A thriving colony of bacteria live in the human colon. These microbes synthesize many compounds that your body needs and cannot get easily from the food you eat, including vitamin K and several B vitamins. In addition, bacteria aid in transforming and compacting the undigested materials into the final waste product, feces.

The final segment of the large intestine is the rectum. Solids in the colon pass into the rectum as a result of peristalsis in the large intestine. From the rectum, the solid feces are eliminated from the body through the anus. Undigested material passes through the large intestine and is expelled through the anus in 12 to 24 hours.

Balancing water absorption in the intestine is important. Wastes rushed through the large intestine before the remaining water is absorbed result in diarrhea (watery feces). When food remains in the colon for long periods of time, which causes much water to be absorbed, constipation (hard feces) results. Hard feces are difficult to pass.

The Liver's Role in Digestion

The human liver, shown in Figure 8, is about the size of a football and weighs more than 1.4 kg (3.1 lb). It presses upward against the diaphragm and occupies the upper right side of the abdominal cavity.

The liver plays several roles in digestion. The liver secretes bile, which aids in the emulsification of fats. Bile also promotes the absorption of fatty acids and the fat-soluble vitamins A, D, E, and K. Bile pigments (the products of hemoglobin breakdown) give bile a yellowish green color. Jaundice, a condition in which the eyes, skin, and urine become abnormally yellow, is a result of increased amounts of bile pigments in the blood. Jaundice often occurs as a result of hepatitis, an inflammation of the liver.

The Liver's Role in Metabolism

Digested food molecules in the bloodstream are transported to the liver. The liver stabilizes blood sugar by converting extra sugar to glycogen for storage. The liver then breaks down the glycogen when it is needed for energy. The liver also modifies amino acids. Fat-soluble vitamins and iron are stored in the liver. The liver monitors the production of cholesterol and detoxifies poisons. If the liver is unable to change a substance's harmful form, it stores it. In this way, toxins, including heavy metals and pesticides, accumulate in the liver.

The liver can also be damaged by viral infections, chronic drug or alcohol use, and traumatic injury. As a result of any of these, healthy liver cells are destroyed and replaced by scar tissue. The scarring of the liver is called cirrhosis *(suh ROH sihs)*.

🔲 internet connect ▤

www.scilinks.org
Topic: Poisons
Keyword: HX4142

SCLINKS® Maintained by the National Science Teachers Association

---

QUICK LAB

Modeling the Function of Bile

You can use a detergent and cooking oil to simulate the effect bile has on breaking up (emulsifying) fats as part of digestion.

Materials

two 250 mL beakers, water, cooking oil, dish detergent, stirring rod, graduated cylinder

Procedure

1. Label one beaker *A* and one beaker *B*. Fill each beaker halfway with water.

2. Add 10 mL of the cooking oil to each beaker.

3. While stirring, slowly add 10 drops of dish detergent to beaker *B* only.

Analysis

1. **Describe** how oil reacts with the water.

2. **Describe** what happened to the oil when the dish detergent was added.

3. **Compare** the effect of dish detergent on oil with the effect of bile on fats.

4. **Critical Thinking Inferring Conclusions** Do the detergents and bile increase or decrease the surface area of oil? In the case of bile, how does this help the digestive process?

Section 2 Review

1 Summarize the path a piece of cheese pizza would follow through the digestive system.

2 Relate the role of the mouth, stomach, small intestine, and large intestine in the digestion of a piece of cheese pizza.

3 Locate the area of the digestive system where nutrients are absorbed into the bloodstream.

4 State how the liver and pancreas are involved in digestion.

5 Critical Thinking Applying Information A person has a small intestine that has villi but a reduced number of microvilli. Would you expect this person to be underweight or overweight? Explain.

6 Standardized Test Prep The enzyme pepsin is involved in the digestion of

A starches.   C monosaccharides.

B fats.   D proteins.

Objectives

● **Identify** major wastes produced by humans and the organ or tissue where they are eliminated from the body.

● **Relate** the role of nephrons to the filtering of blood in the kidneys.

● **Summarize** how nephrons form urine.

● **Describe** the path of urine through the human urinary system.

● **Predict** how kidney damage might affect homeostasis and threaten life.

Key Terms

excretion
urea
nephron
urine
ureter
urinary bladder
urethra

Water and Metabolic Wastes

Cleaning up, though not always a pleasant chore, must be done to maintain a healthy living environment. In the same way, our bodies must get rid of wastes to maintain health. Food residues are eliminated from the body in the form of feces. Other wastes produced as a result of metabolic reactions that occur in the body must also be eliminated. For example, water and carbon dioxide are produced during cellular respiration. During the metabolism of proteins and nucleic acids, a toxic nitrogen-containing waste, ammonia, is formed.

The body must remove wastes. It must also maintain osmotic balance and stable pH by either excreting or conserving salts and water. **Excretion** is the process that rids the body of toxic chemicals, excess water, salts, and carbon dioxide while maintaining osmotic and pH balance.

The organs involved in excretion are shown in **Figure 11.** Carbon dioxide (and some water vapor) is transported to your lungs by the circulatory system and excreted every time you exhale. Excess water is excreted through the skin in sweat and through the kidneys in urine. In the liver, ammonia is converted to a much less toxic nitrogen waste called **urea** *(yoo REE uh),* which is then carried by the bloodstream to the kidneys, where it is removed from the blood.

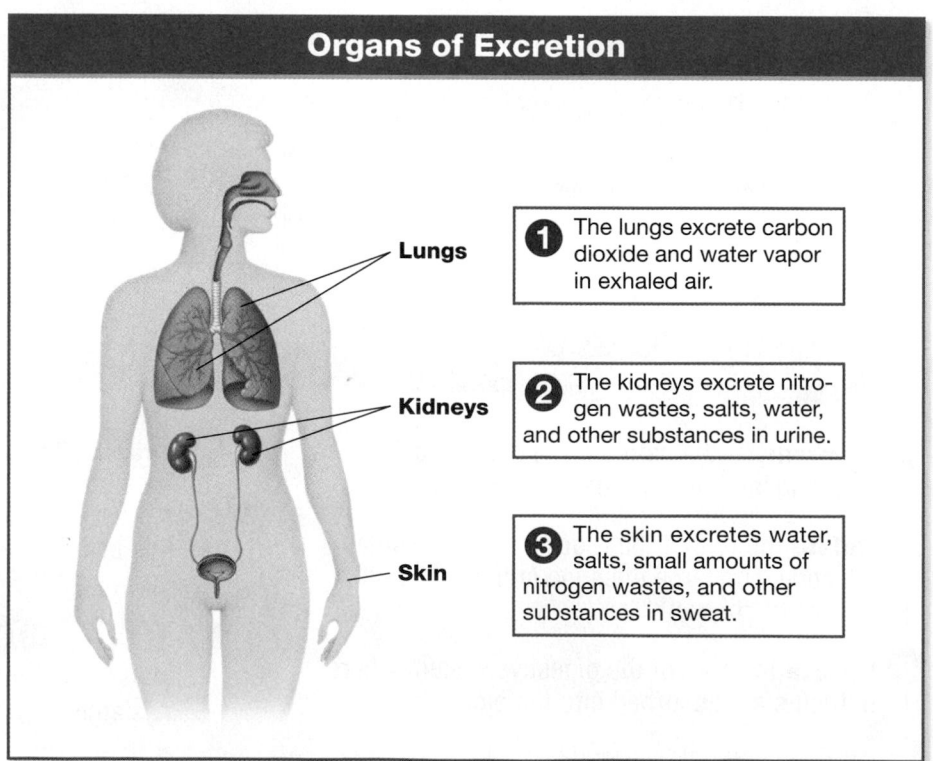

Organs of Excretion

Lungs

❶ The lungs excrete carbon dioxide and water vapor in exhaled air.

Kidneys

❷ The kidneys excrete nitrogen wastes, salts, water, and other substances in urine.

Skin

❸ The skin excretes water, salts, small amounts of nitrogen wastes, and other substances in sweat.

Figure 11 Organs of excretion. The lungs, the kidneys, and the skin all function as excretory organs. The main excretory products are carbon dioxide, water, and nitrogen wastes (urea).

The Kidneys

The kidneys are extremely important organs because of their role in regulating the amount of water and salts contained in blood plasma. The kidneys are a pair of bean-shaped, reddish brown organs located in the lower back. Each kidney is the size of a small fist. The body has to maintain a certain level of salts in the blood plasma and in the fluid surrounding cells, or serious harm to the body's cells and organ systems can result.

Blood Filters

Each kidney is a complex organ composed of roughly 1 million microscopic blood-filtering units called nephrons *(NEHF rahns)*, as shown in **Figure 12. Nephrons** are tiny tubes in the kidneys. One end of a nephron is a cup-shaped capsule surrounded by a tight ball of capillaries that filters wastes from the blood, retains useful molecules, and produces urine. Three different phases occur as the blood flows through a nephron: filtration, reabsorption and secretion.

Filtration Filtration begins at the cup-shaped capsule called Bowman's capsule. Within each Bowman's capsule an arteriole enters and splits into a fine network of capillaries called a glomerulus *(gloh MEHR yoo luhs)*. The glomerulus acts as a filtration device. The blood pressure inside the capillaries forces a fluid composed of water, salt, glucose, amino acids, and urea into the hollow interior of the Bowman's capsule. This fluid is called filtrate. Blood cells, proteins, and other molecules too large to cross the membrane remain in the blood.

Figure 12 Kidneys and nephrons

The kidneys filter out toxins, urea, water, and mineral salts from the blood as fluid passes through the microscopic filtering units called nephrons.

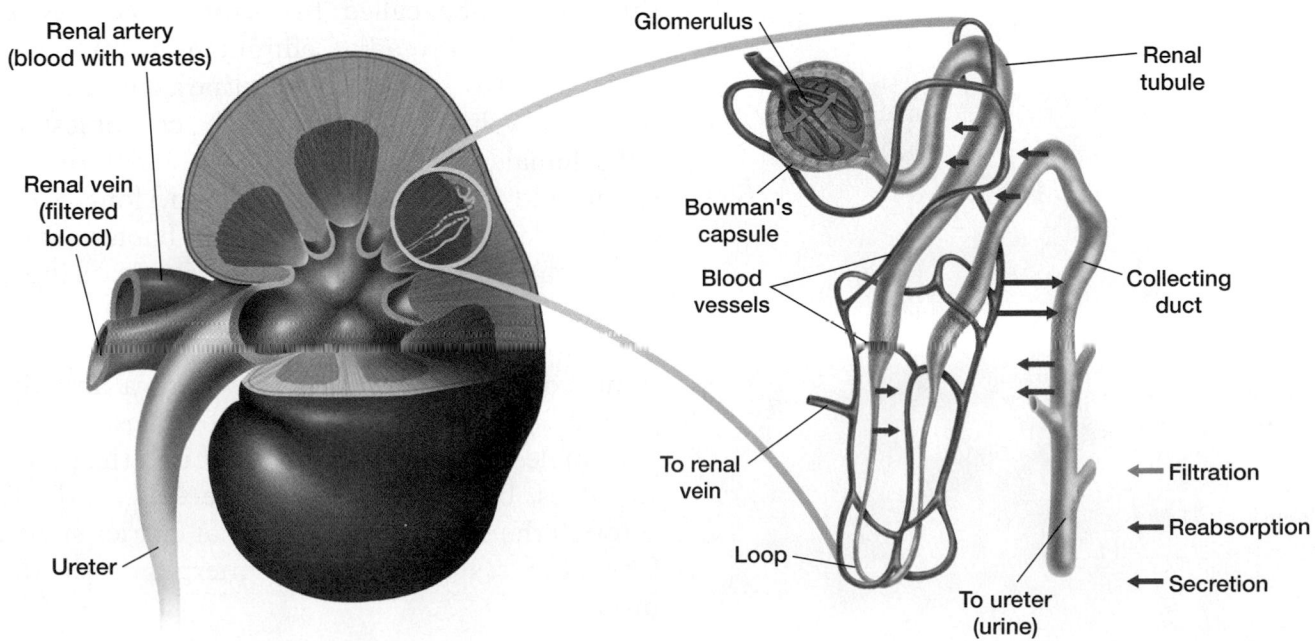

Reabsorption and Secretion Reabsorption begins when the filtrate passes from the Bowman's capsules into the renal tubules—long, narrow tubes connected to Bowman's capsules. Renal tubules bend at their center, which forms a loop. As the filtrate passes through the renal tubules, the tubules extract from the filtrate a variety of useful molecules, including glucose, ions, and some water. These substances reenter the bloodstream through capillaries that wrap around the tubule. This arrangement prevents these molecules from being eliminated from the body in the urine. Some substances can pass from the blood into the filtrate in a process called secretion.

Urine Formation The **urine** that is excreted from the body is formed from the water, urea, and various salts that are left after the absorption and secretion processes. Collecting ducts receive fluid from several nephron segments and empty the urine into areas of the kidneys that lead to the ureters. **Ureters** *(yoo REET uhrs)* are tubes that carry the urine from the kidney to the urinary bladder. The collecting duct removes much of the water from the filtrate that passes through it. As a result, human urine can be very concentrated. In fact, it can be as much as four times more concentrated than blood plasma is.

Elimination of Urine

The ureters, shown in **Figure 13,** have smooth muscle in their walls. The slow, rhythmic contractions of this muscle move the urine through the ureters. The ureters direct the urine into the **urinary bladder,** a hollow, muscular sac that stores urine. The urinary bladder gradually expands as it fills. The average urinary bladder can hold up to about 0.6 L (0.63 qt) of urine. The urinary bladders of males tend to be larger than those of females.

Muscular contractions of the bladder force urine out of the body. Urine leaves the bladder and exits the body through a tube called the **urethra** *(yoo REE thruh).* A healthy adult eliminates from about 1.5 L (1.6 qt) to 2.3 L (2.4 qt) of urine a day, depending on the volume of fluid he or she consumes.

In females the urethra lies in front of the vagina and is only about 2.5 cm (1 in.) long. Such a short length makes it easy for bacteria and other pathogens to invade the female urinary system, which explains why females are more prone to urinary infections than males are. There is no connection between the urethra and the genital (reproductive) system in females.

In males the urethra passes through the penis. In males, both sperm and urine exit the body through the urethra. The tube that carries sperm from the testes eventually merges with the urethra.

Figure 13 The organs of urinary excretion. Urine exits the kidneys by way of two ureters that empty into a storage organ called the urinary bladder. Urine exits the body through the urethra.

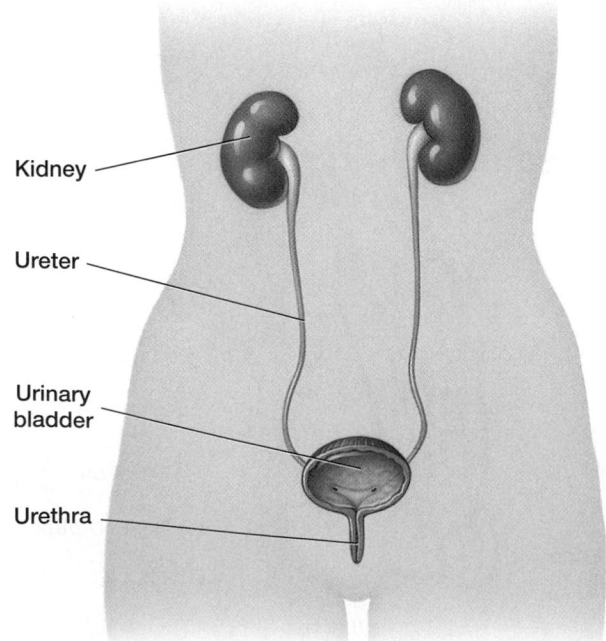

Kidney

Ureter

Urinary bladder

Urethra

The elimination of urine from the body through the urethra is called urination. When the bladder fills with urine, stretch receptors in the bladder's wall send nerve impulses to the spinal cord. In response, the spinal cord returns impulses to the bladder and urethra simultaneously. These impulses cause contraction of the bladder's muscular walls and relaxation of the rings of muscle closing off the urethra. The bladder then empties its contents through the urethra. In older children and adults, the brain overrides this urination reflex, which delays the release of urine until a convenient time.

Damage to the Kidneys

Because of the vital role played by the kidneys in maintaining homeostasis, diseases affecting these organs may eventually threaten life. If one kidney is lost in an accident or by disease, the other may enlarge and do the work of both. Nephrotic syndrome refers to a number of signs and symptoms that result from damage to the glomeruli, which leads over time to kidney failure. The most common causes of kidney failure are infection, diabetes, high blood pressure, and damage to the kidneys by the body's own immune system.

Because of their function in excretion, kidneys often are exposed to hazardous chemicals that have entered the body through the lungs, skin, or gastrointestinal tract. Household substances that, in concentration, can damage kidneys include paint, varnishes, furniture oils, glues, aerosol sprays, air fresheners, and lead. When kidneys fail, toxic wastes, such as urea, accumulate in the plasma, and blood-plasma ion levels increase to dangerous levels. If both kidneys fail, there are only two treatment options.

Kidney Dialysis Kidney dialysis, also called hemodialysis *(HEE moh die AL uh sihs),* is a procedure for filtering the blood by using a dialysis machine, as shown in **Figure 14.** A dialysis machine, just as the nephrons in the kidney, sorts small molecules in the blood, keeping some and discarding others. Dialysis machines are sometimes used for kidneys that are damaged but either will eventually heal or be replaced by a kidney transplant.

Kidney Transplants A more permanent solution to kidney failure is transplantation of a kidney from a healthy donor. A major problem with kidney transplants is common to all organ transplants— rejection of the transplanted organ by the recipient's immune system. Recall that the cells of your body have "self-markers," or antigens, on their surfaces that identify the cells to your immune system so it will not attack them. The combination of these antigens displayed on your body's cells is as unique as your fingerprints.

Figure 14 Hemodialysis.
Hemodialysis has prolonged the lives of many people with damaged or diseased kidneys. The dialysis machine functions like a kidney in that it filters urea and excess ions from the blood.

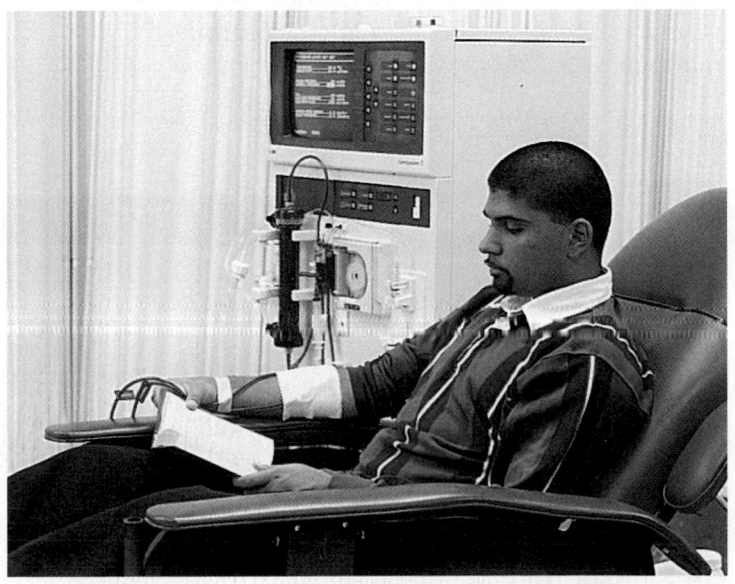

Only identical twins have the same set of antigens. The more closely related two individuals are, the more likely they are to have common antigens. This is why tissue transplants are more likely to succeed if the donor and recipient are closely related. But even in close matches, there is some chance of transplant rejection. To reduce chances of rejection, the recipient is treated with drugs that suppress the activity of the immune system.

Exploring Further

Kidney Dialysis

People whose kidneys are damaged cannot filter their blood. Kidney dialysis is one option for artificially filtering the blood. In kidney dialysis, tubes called catheters are surgically inserted into an artery and a vein, usually on a forearm. The catheters are equipped with valves. Every few days the catheters are connected to a dialysis machine, as shown to the right.

Blood Is Filtered

Blood passes from the patient's artery into the dialysis machine. Inside the machine, the blood travels through many hollow tubes, each of which is surrounded by a thin, permeable membrane. Waste materials and ions that have accumulated in the person's blood diffuse through the membrane into a fluid that has the same makeup as normal blood plasma and is free of wastes. The filtered blood is then returned to the person's vein.

Dialysis is not a permanent solution to kidney failure. A single healthy kidney can meet all of the homeostatic needs of the body, but no dialysis machine can. Dialysis patients must carefully manage their salt, protein, and water intake

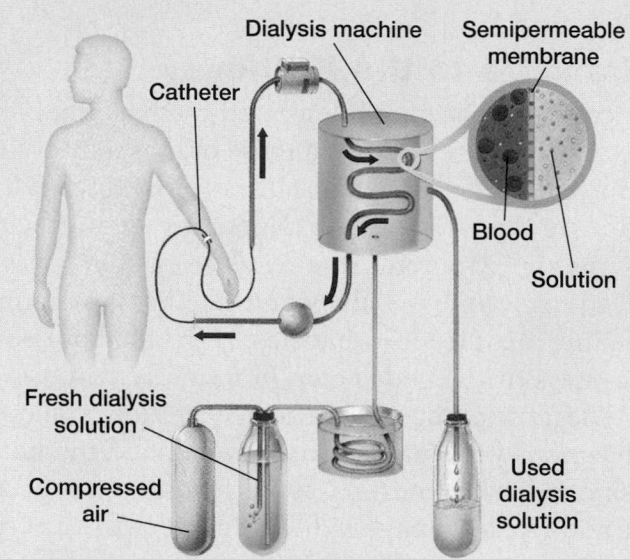

because the dialysis machine cannot regulate these blood components as well as the kidney can.

internet connect

www.scilinks.org
Topic: Kidney Dialysis
Keyword: HX4108

*SCI*LINKS® Maintained by the National Science Teachers Association

Section 3 Review

1 Identify how the carbon dioxide in your body is produced and excreted.

2 Relate the following terms to the formation of urine: filtration, reabsorption, and secretion.

3 Name the liquid stored inside the collecting duct of a nephron.

4 Summarize how urine is stored and eliminated from the body.

5 Critical Thinking Applying Information
A doctor has just informed a patient that his urine contains a high sugar concentration. Explain why this may indicate damaged kidneys.

6 Standardized Test Prep The substances that are removed from the body by the excretory system are carried to the kidneys by the
A nervous system.   **C** circulatory system.
B respiratory system.   **D** digestive system.

Key Concepts

1 Your Body's Need for Food

- Food and beverages provide the nutrients and water required by the body for growth, energy, repair, and maintenance.

- Carbohydrates and lipids provide most of the body's energy. Proteins are normally used for making other proteins.

- The USDA food guide pyramid graphically summarizes the daily recommended servings from each food group.

- Vitamins enhance the activity of enzymes and regulate the release of energy. Minerals are used to make certain body structures and substances, for normal nerve and muscle function, to maintain osmotic balance, and for enzyme function.

- Water acts as a lubricant, solvent, and coolant, and as a support medium for cells and tissues.

2 Digestion

- Teeth break down food into smaller pieces. Amylase begins the breakdown of starch to sugars. The stomach stores and mechanically breaks down food. Stomach acid and pepsin chemically break down proteins.

- Most chemical digestion occurs in the small intestine with the help of secretions from the pancreas, liver, and gallbladder.

- Usable compounds are absorbed into capillaries or lymphatic vessels in villi. Compounds not absorbed are eventually excreted as feces.

- The liver releases bile, helps to maintain blood sugar levels, and detoxifies poisons.

3 Excretion

- The skin, lungs, and kidneys are specialized to excrete wastes.

- Nephrons in the kidneys filter wastes from the blood. Most of the water, some of the salts, and all of the sugar and amino acids in the filtrate are reabsorbed into the blood-stream. The water, urea, and salts that remain in the nephron are eliminated as urine.

- Kidney dialysis and organ transplants are treatment options when both kidneys fail.

Key Terms

Section 1

nutrient (900)
digestion (900)
calorie (900)
vitamin (904)
mineral (905)

Section 2

amylase (907)
esophagus (907)
pepsin (908)
lipase (909)
villus (909)
colon (910)

Section 3

excretion (912)
urea (912)
nephron (913)
urine (914)
ureter (914)
urinary bladder (914)
urethra (914)

Performance ZONE — CHAPTER REVIEW

Understanding Key Ideas

1. The primary function of carbohydrates is to
 a. break down molecules.
 b. aid in digestion.
 c. supply the body with energy.
 d. regulate the flow of acid.

2. Food from the _____ food group should be eaten in the greatest abundance.
 a. grains
 b. fats, oils, and sweets
 c. vegetables
 d. milk, yogurt, and cheese

3. The body needs vitamins because they
 a. supply energy.
 b. activate enzymes.
 c. function as enzymes.
 d. act as hormones.

4. The _____ are involved in excretion.
 a. kidneys and stomach
 b. liver and pancreas
 c. pancreas and kidneys
 d. kidneys and lungs

5. During secretion in the kidney, substances move from
 a. the filtrate to the blood.
 b. the urethra to the bladder.
 c. the blood to the filtrate.
 d. the bladder to the urethra.

6. Identifying Information Identify the food groups represented by *A* and *B*. Indicate the number of servings that should be eaten daily for *A* and *B*.

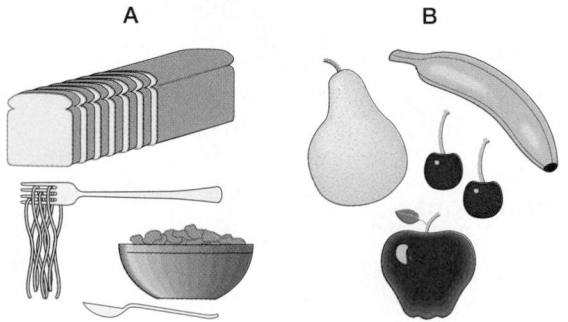

A   B

7. Which substance is *not* a waste eliminated from the body through the kidneys?
 a. urea   **c.** salts
 b. water   **d.** oxygen

8. BIOWatch A suspect leaves a half-eaten sandwich at a crime scene. How would the sandwich be useful to a forensic odontologist?

9. Exploring Further Summarize how a kidney dialysis machine filters blood.

10. Concept Mapping Make a concept map that shows how nutrients are digested. Try to include the following words in your map: *carbohydrates, proteins, lipids, enzymes, saliva, pancreas, stomach, bile, liver, small intestine,* and *large intestine*.

Critical Thinking

11. Evaluating Conclusions The length of the looped tubule in a nephron varies among mammal species. A friend believes the looped tubules of mammals that live in the water would be shorter than those found in humans. Do you agree or disagree? Explain.

12. Recommending Information Calcium needs an acidic environment for optimal absorption. What kinds of foods would you recommend be combined with calcium-rich foods in order to maximize absorption?

13. Relating Concepts Describe the symbiotic role of bacteria in the human intestine.

Alternative Assessment

14. Communicating Write an article that discusses diuretics (substances that increase urine excretion) for your school or local newspaper. Emphasize diuretics that most people have heard of, such as the caffeine in coffee and soft drinks.

Standardized Test Prep

Understanding Concepts

Directions (1–4): For *each* question, write on a separate sheet of paper the letter of the correct answer.

1 What biological molecules make up proteins?
  A. amino acids
  B. fatty acids
  C. glycerol
  D. monosaccharides

2 Which of the following correctly pairs the enzyme with the food molecule it digests?
  F. amylase, fat
  G. lipase, fat
  H. lipase, starch
  I. pepsin, starch

3 Through what structures in the small intestine must nutrients pass in order to enter the blood stream?
  A. gastric pits
  B. glomeruli
  C. nephrons
  D. villi

4 What is the main function of dietary fiber?
  F. to provide energy
  G. to maintain osmotic balance
  H. to provide materials for making enzymes
  I. to help food pass through the digestive tract

Directions (5–6): For *each* question, write a short response.

5 A friend believes that a vegetarian diet would decrease his intake of saturated fat and cholesterol. Do you agree with his idea? Why or why not?

6 For the following set of terms, choose the term that does not belong and explain why it does not belong: nephron, villi, glomerulus, renal tubule.

Test TIP

Sometimes only a portion of a graph or table is needed to answer a question. Focus only on the necessary information to avoid confusion.

Reading Skills

Directions (7): Read the passage below. Then answer the question.

A food is considered to be a good source of a vitamin or mineral if it provides at least 10 percent of the daily value of that vitamin or mineral. A nutrition facts label on a box of cereal states that one-half cup of the cereal contains 15% of the daily value of vitamin A, 0% of the daily value of vitamin C, 10% of the daily value of vitamin D, 50% of the daily value of iron, and 8% of the daily value of zinc. Adding half a cup of milk increases vitamin A by 5% and zinc by 2%.

7 What can you conclude about a breakfast that includes half a cup of cereal and half a cup of milk?
  A. It is a good source of vitamin A, vitamin C, vitamin D, iron, and zinc.
  B. It is a good source of vitamin A, vitamin D, iron, and zinc but not of vitamin C.
  C. It is a good source of vitamin A, vitamin D, and iron but not of vitamin C and zinc.
  D. It is a good source of vitamin A and vitamin D but not of vitamin C, iron, and zinc.

Interpreting Graphics

Directions (8): Base your answer to question 8 on the table below.

Food Label

Nutrition Facts	Amount/serving	% DV*	Amount/serving	% DV*
Serv. size 2 oz (56 g/$\frac{1}{8}$ box) Servings per container 8	Total fat 1 g	1%	Total carb. 43 g	14%
	Sat. fat 0 g	0%	Dietary fiber 2 g	8%
	Cholesterol 0 mg	0%	Sugars 3 g	
Calories 210 Fat Cal. 10	Sodium 0 mg	0%	Protein 6 g	

*Percent Daily Values (DV) are based on a 2,000 Calorie diet. Vitamin A 0% • Vitamin C 0% • Calcium 2% • Iron 10% Thiamin 30% • Riboflavin 10% • Niacin 15%

8 Approximately what percentage of the calories in this food come from fats?
  F. 1 percent
  G. 5 percent
  H. 10 percent
  I. 14 percent

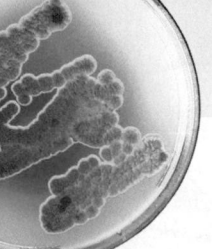

Exploration Lab

Demonstrating Lactose Digestion

SKILLS
- Using scientific methods
- Observing
- Comparing

OBJECTIVES
- **Describe** the relationship between enzymes and the digestion of food molecules.
- **Evaluate** the ability of a milk-treatment product to promote lactose digestion.
- **Infer** the presence of lactose in milk and foods that contain milk.

MATERIALS
- milk-treatment product (liquid)
- toothpicks
- depression slides
- droppers
- whole milk
- glucose solution
- glucose test strips

ChemSafety

CAUTION: Always wear safety goggles and a lab apron to protect your eyes and clothing.

CAUTION: Do not touch or taste any chemicals. Know the location of the emergency shower and eyewash station and how to use them. If you get a chemical on your skin or clothing, wash it off at the sink while calling to the teacher. Notify the teacher of a spill. Spills should be cleaned up promptly, according to your teacher's directions.

CAUTION: Glassware is fragile. Notify the teacher of broken glass or cuts. Do not clean up broken glass or spills with broken glass unless the teacher tells you to do so.

Before You Begin

People with a condition known as **lactose intolerance** often experience stomach and intestinal pain, bloating, and diarrhea when they eat foods that contain milk. These symptoms result from an inability to digest lactose, a sugar found in milk. **Lactose** is a disaccharide made of one glucose unit and one galactose unit. Lactose molecules are broken down into glucose and galactose molecules during **digestion.** People who cannot digest lactose do not produce **lactase,** the digestive enzyme that aids the breakdown of lactose. In this lab, you will investigate a milk-treatment product that is designed to aid lactose digestion.

1. Write a definition for each boldface term in the paragraph above.

2. List at least 10 foods that contain milk.

3. Make a data table similar to the one below.

4. Based on the objectives for this lab, write a question you would like to explore about enzymes and digestion.

DATA TABLE		
Solution	**Result (+ or −)**	**Interpretation**

Procedure

PART A: Design an Experiment

1. Read the information sheet that comes with the milk-treatment product. Discuss with your lab group what the product is and what it does. Write a summary of your discussion for your lab report.

2. Work with the members of your lab group to explore one of the questions written for step 4 of **Before You Begin.** To explore the question, design an experiment that uses the materials listed for this lab.

You Choose

As you design your experiment, decide the following:

a. what question you will explore

b. what hypothesis you will test

c. what your controls will be

d. how much milk and milk-treatment product to use for each test

e. how to determine whether lactose was broken down

f. what data to record in your data table

3. Write the procedure for your group's experiment. Make a list of all the safety precautions you will take. Have your teacher approve your procedure and safety precautions before you begin the experiment.

4. Set up your group's experiment, and collect data.

PART B: Cleanup and Disposal

5. Dispose of solutions, broken glass, and glucose test strips in the designated waste containers. Do not pour chemicals down the drain or put lab materials in the trash unless your teacher tells you to do so.

6. Clean up your work area and all lab equipment. Return lab equipment to its proper place. Wash your hands throughly before you leave the lab and after you finish all work.

Analyze and Conclude

1. **Summarizing Information** What are the milk-treatment product's ingredients?

2. **Recognizing Relationships** What is the relationship between lactose and lactase?

3. **Analyzing Methods** What role did the glucose solution play in your experiment?

4. **Drawing Conclusions** What does the milk-treatment product do to milk?

5. **Analyzing Conclusions** How do your results justify your conclusion?

6. **Evaluating Methods** Why should you test the milk-treatment product with glucose test strips?

7. **Analyzing Results** What do you infer from the results of this lab about treatments for other medical problems resulting from enzyme deficiencies?

8. **Forming Reasoned Opinions** As a person grows older, will he or she be more likely or less likely to develop lactose intolerance? Explain your answer.

9. **Predicting Patterns** Do you think lactose intolerance might be inherited? Explain your answer.

10. **Further Inquiry** Write a new question about enzymes and digestion that could be explored with another investigation.

? Do You Know?

Do research in the library or media center to answer these questions:

1. What are some other food-treatment products that contain digestive enzymes?

2. Why does the improper breakdown of certain food molecules cause symptoms such as stomach pain, gas, and diarrhea?

Use the following Internet resources to explore your own questions about lactose intolerance.

internet connect

www.scilinks.org
Topic: Lactose Intolerance
Keyword: HX4111

SCiLINKS. Maintained by the National Science Teachers Association

White blood cell attacking bacteria (6,480×)

 Quick Review

Answer the following without referring to earlier sections of your book.

1. **Identify** the role of receptor proteins in cellular communication. *(Chapter 4, Section 2)*
2. **Explain** the relationship between HIV and AIDS. *(Chapter 21, Section 1)*
3. **Differentiate** between antibodies and antigens. *(Chapter 38, Section 1)*

Reading Activity

Write down the title of this chapter and the titles of its four sections on a piece of paper or in your notebook. Leave a few blank lines after each section title. Then write down what you think you will learn in each section. Save your list, and after you finish reading this chapter, check off everything that you learned that was on your list.

▪ **internet** connect

www.scilinks.org
National Science Teachers Association *sci*LINKS Internet resources are located throughout this chapter.

*sci*LINKS. Maintained by the National Science Teachers Association

● A white blood cell (macrophage) ingests bacteria as part of the immune system's response to infection. The macrophage degrades bacterial proteins into peptides, which then form antigens on the cell's surface.

Nonspecific Defenses

Objectives

● **Describe** how skin and mucous membranes defend the body.

● **Compare** the inflammatory response with the temperature response.

● **Identify** proteins that kill or inhibit pathogens.

● **Analyze** the roles of white blood cells in combating pathogens.

Key Terms

pathogen
mucous membrane
inflammatory response
histamine
complement system
interferon
neutrophil
macrophage
natural killer cell

Two Lines of Nonspecific Defenses

Some animals, including turtles, clams, and armadillos, defend themselves with their hard armor shells. However, even armor will not protect against the most dangerous enemies that they or the human body faces—harmful bacteria, viruses, fungi, and protists. You, as well as most animals, survive because your body's immune system defends against these pathogens. A **pathogen** is a disease-causing agent. The immune system consists of cells and tissues found throughout the body. The body uses both nonspecific and specific defense mechanisms to detect and destroy pathogens, thereby preventing or reducing the severity of infection.

First Line of Nonspecific Defenses

The body's surface defenses are nonspecific, meaning they do not target specific pathogens. Your skin is the first of your immune system's nonspecific defenses against pathogens. Skin acts as a nearly impenetrable barrier to invading pathogens, keeping them outside the body. This barrier is reinforced with chemical weapons. Oil and sweat make the skin's surface acidic, inhibiting the growth of many pathogens. Sweat also contains the enzyme lysozyme, which digests bacterial cell walls.

Mucous membranes cover some body surfaces that come into contact with pathogens. **Mucous** *(MYOO kuhs)* **membranes** are layers of epithelial tissue that produce a sticky, viscous fluid called mucus. Mucous membranes line the digestive system, nasal passages, lungs, respiratory passages, and the reproductive tract. Like the skin, mucous membranes serve as a barrier to pathogens and produce chemical defenses. Cells lining the bronchi and bronchioles in the respiratory tract secrete a layer of mucus that traps pathogens before they can reach the warm, moist lungs, which are an ideal breeding ground for microorganisms. Cilia on cells of the respiratory tract continually sweep mucus toward the opening of the esophagus. Mucus then can be swallowed, sending pathogens to the stomach, where they are digested by acids and enzymes.

Skin and mucous membranes work to prevent any pathogens from entering the body. Occasionally these defenses are penetrated. You take pathogens into your body when you breathe, because many microbes and microbial spores are suspended in the air. Other pathogens may be present in the food you eat. Pathogens can also enter through wounds or open sores. When invaders reach deeper tissue, a second line of nonspecific defenses takes over.

Second Line of Nonspecific Defenses

What happens when pathogens break through your body's first line of defense? When the body is invaded, four important nonspecific defenses take action: the inflammatory response; the temperature response; special proteins that kill or inhibit pathogens; and white blood cells, which attack and kill pathogens.

Inflammatory response Injury or local infection, such as a cut or a scrape, causes an inflammatory response. An **inflammatory response** is a series of events that suppress infection and speed recovery. Imagine that a splinter has punctured your finger, creating an entrance for pathogens, as shown in **Figure 1.** Infected or injured cells in your finger release chemicals, including histamine. **Histamine** (*HIHST uh meen*) causes local blood vessels to dilate, increasing blood flow to the area. Increased blood flow brings white blood cells to the infection site, where they can attack pathogens. This also causes swelling and redness in the infected area. The whitish liquid, or pus, associated with some infections contains white blood cells, dead cells, and dead pathogens.

Temperature response When the body begins its fight against pathogens, body temperature increases several degrees above the normal value of about 37°C (98.6°F). This higher temperature is called a fever, and it is a common symptom of illness that shows the body is responding to an infection. Fever is helpful because many disease-causing bacteria do not grow well at high temperatures. Although fever may slow the growth of bacteria, very high fever is dangerous because extreme heat can destroy important cellular proteins. Temperatures greater than 39°C (103°F) are considered dangerous, and those greater than 41°C (105°F) can be fatal.

Figure 1   Inflammatory response

When pathogens penetrate your body, an inflammatory response is triggered.

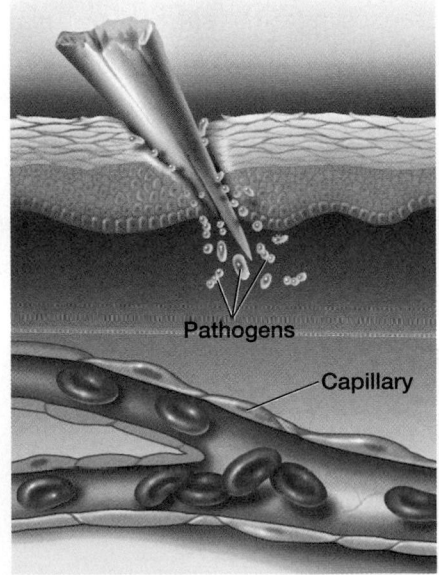

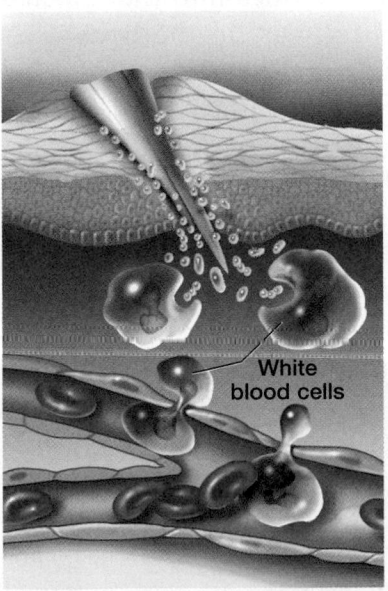

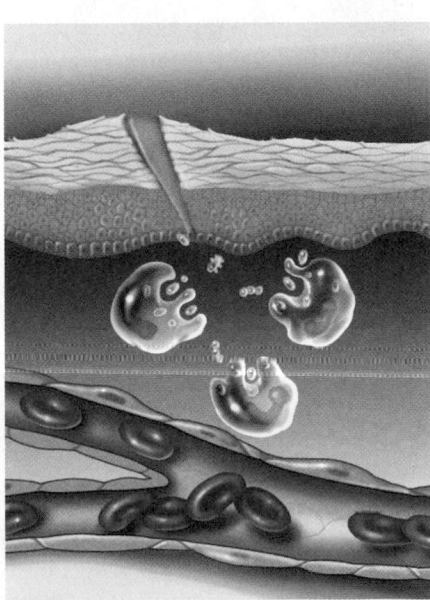

1. When the skin is punctured, pathogens enter the body.

2. Blood flow to the area increases, causing swelling and redness.

3. White blood cells attack and destroy the pathogens.

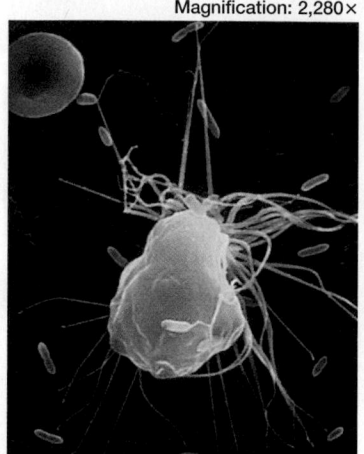

Magnification: 2,280×

Figure 2   Macrophage.
Cytoplasmic extensions of this macrophage (yellow) are capturing bacteria (blue).

Proteins Various proteins also provide nonspecific defenses. One defense mechanism, called the **complement system,** consists of about 20 different proteins. Complement proteins circulate in the blood and become active when they encounter certain pathogens. Then some of these proteins form a membrane attack complex (MAC), a ring-shaped structure. The MAC punches a hole in the cell membrane, causing the cell to leak and die. Another nonspecific defense is **interferon** *(in tuhr FEER ahn),* a protein released by cells infected with viruses. Interferon causes nearby cells to produce an enzyme that prevents viruses from making proteins and RNA.

White blood cells The most important counterattacks in the second line of nonspecific defenses are carried out by three kinds of white blood cells: neutrophils, macrophages, and natural killer cells. These cells patrol the bloodstream, wait within the tissues for pathogens, and then attack the pathogens. Each kind of cell uses a different mechanism to kill pathogens.

1. **Neutrophils.** A **neutrophil** *(NOO truh fihl)* is a white blood cell that engulfs and destroys pathogens. The most abundant type of white blood cell, neutrophils engulf bacteria and then release chemicals that kill the bacteria—and themselves. Neutrophils can also squeeze between cells in the walls of capillaries to attack pathogens at the site of an infection.

Magnification: 14,250×

2. **Macrophages.** White blood cells called **macrophages** *(MA kroh fay jez),* shown in **Figure 2,** ingest and kill pathogens they encounter. They also clear dead cells and other debris from the body. Most macrophages travel through the body in blood, lymph, and fluid between cells. Macrophages are concentrated in particular organs, especially the spleen and lungs.

3. **Natural killer cells.** A **natural killer cell** is a large white blood cell that attacks cells infected with pathogens. Natural killer cells destroy an infected cell by puncturing its cell membrane. Water then rushes into the infected cell, causing the cell to swell and burst. One of the body's best defenses against cancer, natural killer cells can detect and kill cancer cells, as shown in **Figure 3,** before a tumor can develop.

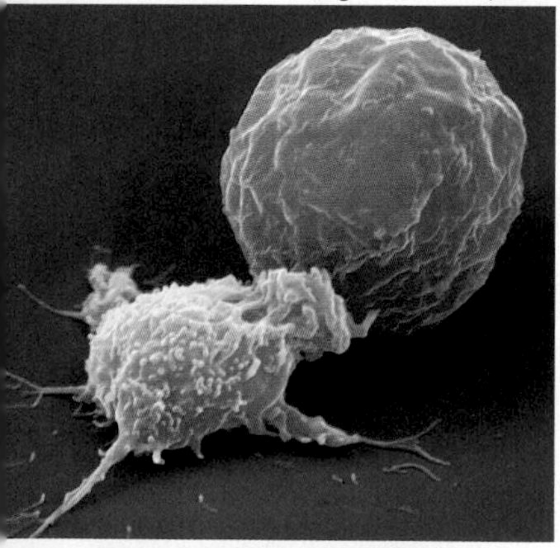

Figure 3   Natural killer cell.
This natural killer cell (yellow) is attacking a cancer cell (pink).

Section 1 Review

1. **Describe** how the inflammatory and temperature responses help defend against infection.

2. **Identify** the role of white blood cells in the second line of nonspecific defenses.

3. **Critical Thinking Relating Concepts**
Explain why taking a drug that reduces fever might delay rather than speed up your recovery from an infection.

4. **Standardized Test Prep** In the inflammatory response, local blood vessels dilate when infected or injured cells release
   A interferon.
   B histamine.
   C mucus.
   D complement proteins.

Immune Response

Specific Defenses

What happens when pathogens occasionally overwhelm your body's nonspecific defenses? Pathogens that have survived the first and second lines of nonspecific defenses still face a third line of specific defenses—the immune response. The immune response consists of an army of individual cells that rush throughout the body to combat specific invading pathogens. The immune response is not localized in the body, nor is it controlled by a single organ. It is more difficult to evade than the nonspecific defenses.

Cells Involved in the Immune Response

White blood cells are produced in bone marrow and circulate in blood and lymph. Of the 100 trillion or so cells in your body, about 2 trillion are white blood cells. Four main kinds of white blood cells participate in the immune response: macrophages, cytotoxic T cells, B cells, and helper T cells. Each kind of cell has a different function. Macrophages consume pathogens and infected cells. **Cytotoxic** (*sie toh TAHKS ihk*) **T cells** attack and kill infected cells. **B cells** label invaders for later destruction by macrophages. **Helper T cells** activate both cytotoxic T cells and B cells. Macrophages can attack any pathogen. B cells and T cells, however, respond only to pathogens for which they have a genetically programmed match. These four kinds of white blood cells interact to remove pathogens from the body.

Recognizing Invaders

To understand how the third line of defenses works, imagine that you have just come down with influenza—the flu. You have inhaled influenza virus particles, but they were not all trapped by mucus in the respiratory tract. The virus has begun to infect and kill your cells. At this point, macrophages begin to engulf and destroy the virus.

An infected body cell will display antigens of an invader on its surface. An **antigen** (*AN tih jihn*) is a substance that triggers an immune response. Antigens typically include proteins and other parts of viruses or pathogen cells. Antigens are present on the surface of the infected body cell. White blood cells of the immune system are covered with receptor proteins that respond to infection by binding to specific antigens on the surfaces of the infecting microbes. These receptors recognize and bind to antigens that match their particular shape, as shown in **Figure 4.**

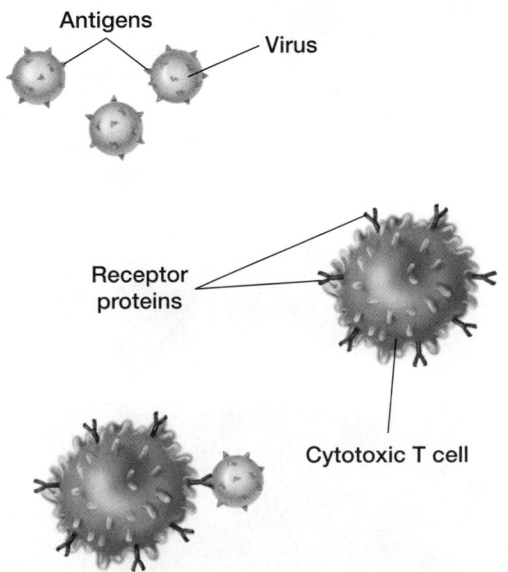

Figure 4 Antigens. Some cells of the immune system have receptor proteins that bind to specific antigens.

The Immune Response Has Two Main Parts

Two distinct processes work together in an immune response. One is the B cell response, a defense that aids the removal of extracellular pathogens from the body. The other is the T cell response, a defense that involves the destruction of intracellular pathogens by cytotoxic T cells. Both the T cell response and the B cell response are regulated by helper T cells. Both responses, which happen simultaneously, are summarized in **Figure 5**.

Figure 5

BIOgraphic

Immune Response

The immune response involves several kinds of white blood cells.

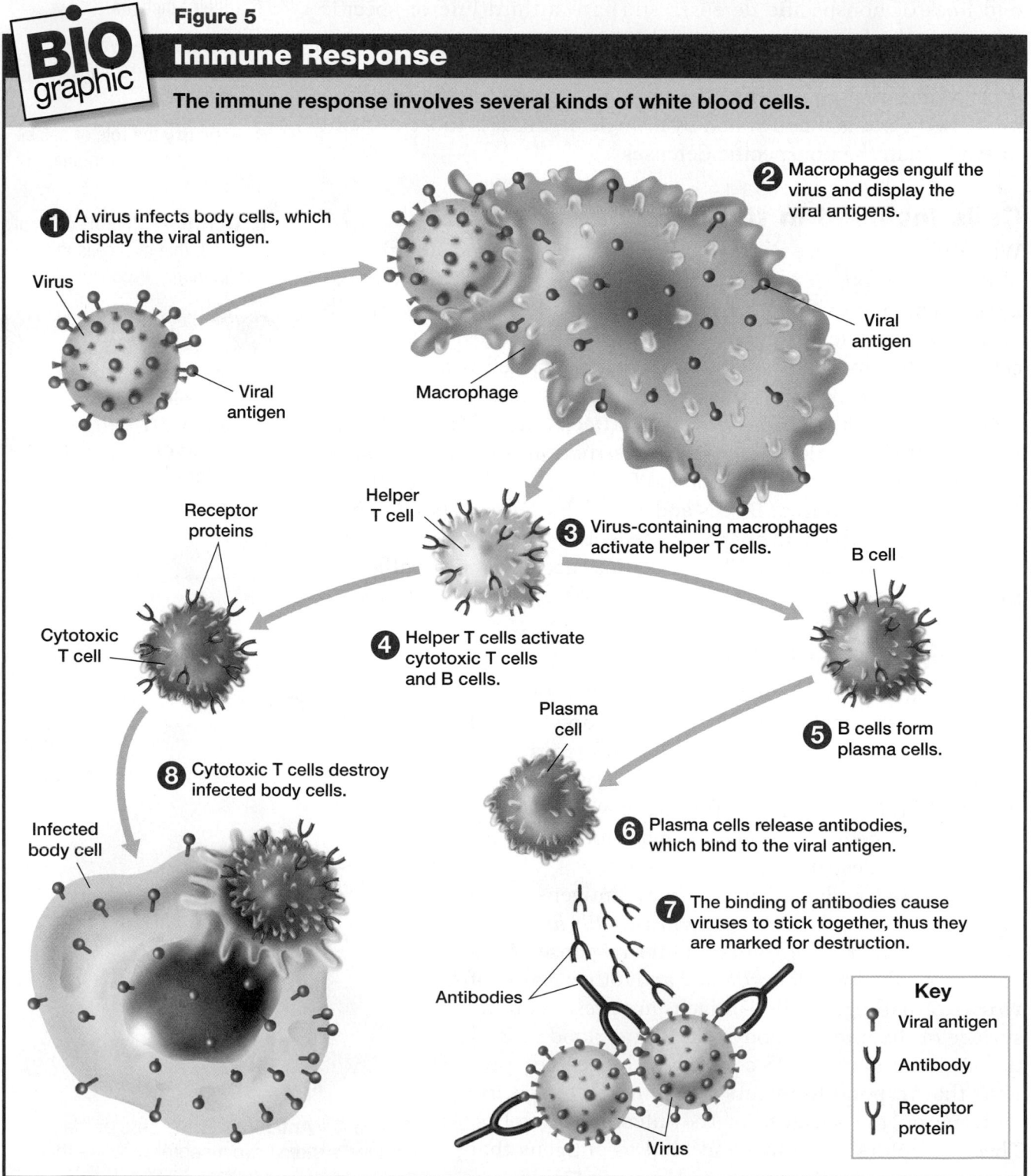

1 A virus infects body cells, which display the viral antigen.

Virus

Viral antigen

2 Macrophages engulf the virus and display the viral antigens.

Viral antigen

Macrophage

Helper T cell

3 Virus-containing macrophages activate helper T cells.

B cell

Receptor proteins

Cytotoxic T cell

4 Helper T cells activate cytotoxic T cells and B cells.

Plasma cell

5 B cells form plasma cells.

8 Cytotoxic T cells destroy infected body cells.

Infected body cell

6 Plasma cells release antibodies, which bind to the viral antigen.

7 The binding of antibodies cause viruses to stick together, thus they are marked for destruction.

Antibodies

Virus

Key

🌑 Viral antigen

Y Antibody

Y Receptor protein

Step ❶ When a virus infects body cells, the infected cells display the viral antigen on their surfaces.

Step ❷ Macrophages engulf the virus and display the viral antigens on their surfaces.

Step ❸ Receptor proteins on helper T cells bind to the viral antigen displayed by the macrophages. The macrophages release a protein called interleukin-1 *(ihn tuhr LOO kihn).*

Step ❹ Interleukin-1 activates helper T cells, but helper T cells do not attack pathogens directly. Instead, helper T cells activate cytotoxic T cells and B cells. Stimulation by interleukin-1 causes helper T cells to release interleukin-2. Interleukin-2 stimulates further division of helper T cells and cytotoxic T cells, amplifying the body's response to the infection.

Step ❺ Interleukin-2 released by helper T cells also activates B cells. Activated B cells divide and develop into plasma cells. **Plasma cells** are cells that release Y-shaped antibodies into the blood. An **antibody** is a defensive protein produced upon exposure to a specific antigen, which can bind to that antigen.

Step ❻ Plasma cells divide repeatedly and make large numbers of antibodies. Plasma cells release antibodies into the bloodstream where they attach to the viruses. Antibodies bind to the viral antigen and mark the virus for destruction.

Step ❼ The binding of antibodies cause viruses and antigens to stick together, forming clumps that can be easily identified and destroyed by macrophages.

Step ❽ Activated cytotoxic T cells destroy infected cells by puncturing their cell membranes. Your body makes millions of different T cells, each with receptor proteins that bind to a specific antigen. Receptor proteins on cytotoxic T cells bind to the viral antigen displayed by infected cells. For example, any of your body's cells that bear traces of an influenza virus will be destroyed by cytotoxic T cells with receptor proteins that bind to the antigen of that virus.

Section 2 Review

❶ **List** the different kinds of white blood cells involved in the immune response.

❷ **Describe** how white blood cells recognize and bind to pathogens.

❸ **Compare** the roles of B cells and T cells in the immune response.

❹ **Recognizing Relationships** Explain the role of helper T cells in the immune response.

❺ **Critical Thinking Predicting Outcomes** How would an enzyme that destroys interleukins affect the immune response?

❻ **Standardized Test Prep** Which cells produce antibodies and release them into the blood?
  A cytotoxic T cells   **C** plasma cells
  B helper T cells   **D** macrophages

Disease Transmission and Prevention

Objectives

- **List** five ways diseases can be transmitted to humans.
- **Summarize** Koch's postulates for identifying pathogens.
- **Analyze** how the body produces immunity to pathogens.
- **Describe** how vaccines produce immunity to pathogens.

Key Terms

Koch's postulates
immunity
vaccination
vaccine
antigen shifting

Disease Transmission

In general, you can get infectious diseases in any of five different ways: through person-to-person contact, air, food, water, and animal bites. Diseases transferred from person to person are considered contagious, or communicable. For example, when a person sneezes, droplets of saliva and mucus carrying pathogens are expelled from the mouth and nose, as shown in **Figure 6.** If another person breathes these droplets, the pathogens can infect that person. People directly transmit some diseases by kissing, shaking hands, touching sores, or having sexual contact. People can also transmit diseases indirectly through objects contaminated with pathogens, such as drinking glasses, toys, plumbing, and needles used to inject drugs or in tatooing.

By minimizing exposure to pathogens, you can decrease your chances of becoming ill. For example, to prevent illnesses caused by bacteria found in foods that contain animal products, these foods should always be cooked thoroughly. Utensils and other surfaces that foods touch should be sanitized.

Detecting Disease

The German physician Robert Koch (1843–1910) established a procedure for diagnosing causes of infection. Koch determined that bacteria cause anthrax, a disease that afflicts cattle, sheep, goats, and humans. Anthrax is a serious disease although it is not passed from person to person. In an experiment, Koch isolated bacteria from a cow with anthrax and then infected a healthy cow with the bacteria. The healthy cow developed anthrax and had the same bacteria that the first cow had. In his research, Koch developed the following four-step procedure, known as **Koch's postulates,** as a guide for identifying specific pathogens.

1. The pathogen must be found in an animal with the disease and not in a healthy animal.
2. The pathogen must be isolated from the sick animal and grown in a laboratory culture.
3. When the isolated pathogen is injected into a healthy animal, the animal must develop the disease.
4. The pathogen should be taken from the second animal and grown in a laboratory culture. The cultured pathogen should be the same as the original pathogen.

Figure 6   Disease transmission. When a person sneezes, pathogens are expelled from the mouth and nose.

Long-Term Protection

The specific immune response is very powerful, and it can be a long-lasting defense. After an immune response, some B cells and T cells become memory cells that continue to patrol your body's tissues. Some memory cells provide lifelong protection against previously encountered pathogens. If a pathogen ever appears again, memory cells activate antibody production against that pathogen. As shown in **Figure 7,** a second exposure to the same pathogen causes a sharp increase in antibody concentration. This enables macrophages to destroy the pathogen before you become ill. You are said to be "immune," or resistant, to the disease caused by that pathogen.

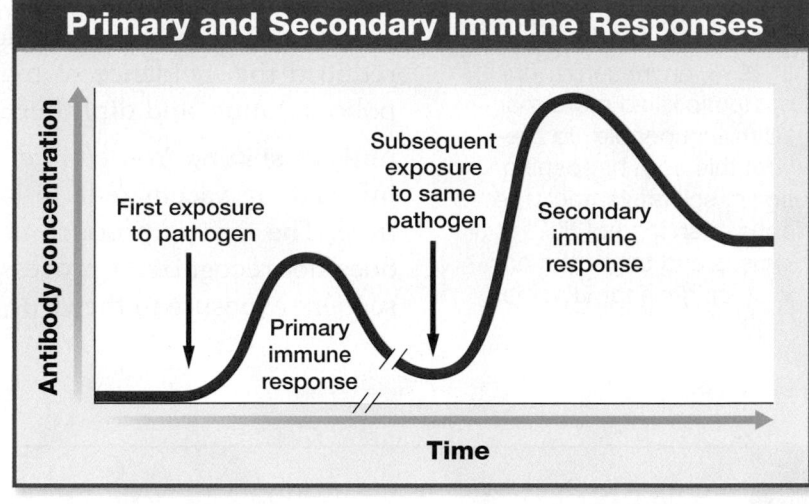

Primary and Secondary Immune Responses

First exposure to pathogen

Primary immune response

Subsequent exposure to same pathogen

Secondary immune response

Antibody concentration

Time

Figure 7 Immune responses. The first time you are exposed to a pathogen, your immune system responds normally. If you become exposed to the same pathogen again, antibody production increases quickly.

Resistance to Disease

Resistance to a particular disease is called **immunity.** It has long been observed that individuals who recover from an infectious disease develop an immunity to that disease. This knowledge preceded the development of immunology, a branch of science that deals with antigens, antibodies, and immunity. Immunologists study the body's defenses and ways to help protect against disease.

In 1796, an English doctor named Edward Jenner performed an experiment that marks the beginning of immunology. Smallpox, which is caused by a virus, was a common and deadly disease then. Jenner observed that milkmaids who had contracted cowpox, a mild form of smallpox, rarely became infected with smallpox. Jenner hypothesized that cowpox produced protection against smallpox. To test his hypothesis, Jenner infected healthy people with cowpox. As Jenner had predicted, many of the people he infected never developed smallpox, even though they had been exposed to the virus. We now know that smallpox and cowpox are caused by two similar viruses. The cowpox infection caused an immune response that later prevented smallpox infection in Jenner's patients.

Vaccination Jenner's procedure of injecting the cowpox virus to produce resistance to smallpox is called vaccination. **Vaccination** *(vak sih NAY shuhn)* is a medical procedure used to produce immunity. You have probably been to the doctor for vaccination to guard against various diseases. Modern vaccination usually involves an injection, or "shot," of a vaccine under the skin. A **vaccine** *(vak SEEN)* is a solution that contains a dead or weakened pathogen or genetic material from a pathogen.

A vaccine triggers an immune response against the pathogen without symptoms of infection. For several days after you are vaccinated, your immune system develops antibodies and memory cells against the pathogen. You develop a long-lasting immunity to the disease. In 1977, smallpox became the first infectious disease to be eradicated from the public by vaccination. Vaccination has also reduced the incidence of many other diseases, including measles, polio, tetanus, and diphtheria.

Antigen shifting You can get the flu even if you have already been infected or vaccinated. Influenza viruses constantly mutate over time. The viruses produce new antigens that your immune system does not recognize, a process known as **antigen shifting.** With subsequent exposure to the virus, your body must make new antibodies.

Simulating Antigen Activity

Using simulated blood, you can see what happens when antigens encounter specific antibodies.

Materials

safety goggles, disposable gloves, lab apron, 2 blood-typing trays, simulated blood (types AB and O), simulated anti-A and anti-B blood-typing serums, 4 toothpicks

Procedure

1. Put on safety goggles, disposable gloves, and a lab apron.

2. Place 3–4 drops of type AB simulated blood into each well in a clean blood-typing tray. **CAUTION: Use only simulated blood provided by your teacher.**

3. Add 3–4 drops of anti-A blood-typing serum to one well. Stir the mixture for 30 seconds using a toothpick. Add 3–4 drops of anti-B blood-typing serum to the other well. Use a new toothpick to stir the mixture. Look for clumps separating from the mixtures.

4. Repeat steps 2 and 3 using simulated type O blood.

5. Dispose of your materials according to your teacher's directions. Clean up your work area and wash your hands.

Analysis

1. **Determine** which blood type has antigens that are recognized by the blood-typing sera.

2. **Evaluating Results** What does clumping of the blood mixtures indicate?

3. **Predicting Outcomes** What would happen if you did the same experiment using type A blood and type B blood?

Section 3 Review

1. **List** two ways that diseases can be transmitted between people.

2. **Summarize** Koch's postulates for identifying specific pathogens.

3. **Describe** how vaccination produces immunity.

4. **Critical Thinking Relating Concepts** Explain why you cannot get many diseases more than once.

5. **Standardized Test Prep** Smallpox is caused by a
   A virus.   **C** fungus.
   B bacterium.   **D** protist.

Disorders of the Immune System

Autoimmune Diseases

The ability of your immune system to distinguish cells and antigens of your body from foreign cells and antigens is crucial to the fight against pathogens. In some people, the immune system cannot distinguish between the body's antigens and foreign antigens, causing an autoimmune disease. In an **autoimmune disease,** the body launches an immune response against its own cells, attacking body cells as if they were pathogens. The immune system cannot distinguish between antigens of "self" and "nonself." This effect may be caused by the inappropriate production of antibodies specific to the antigens of body cells.

Autoimmune diseases affect organs and tissues in various areas of the body. For example, multiple sclerosis *(skleh ROH sihs)* usually strikes people between the ages of 20 and 40. Multiple sclerosis (MS) is generally thought to be an auto-immune disease. In people with multiple sclerosis, the immune system attacks and gradually destroys insulating material surrounding nerve cells in the brain, in the spinal cord, and in the nerves leading from the eyes to the brain. This impairs and may eventually stop the functioning of these nerve cells. Multiple sclerosis causes problems with vision, speech, and coordination. **Table 1** lists and describes several autoimmune diseases.

Objectives

- **Describe** several auto-immune diseases.
- **Summarize** how HIV disables the immune system.
- **List** five ways HIV is transmitted.
- **Identify** causes of an allergic reaction.

Key Terms

autoimmune disease
AIDS
HIV
CD4
allergy

Table 1 Autoimmune Diseases		
Disease	**Areas affected**	**Symptoms**
Graves' disease	Thyroid gland	Weakness, irritability, heat intolerance, increased sweating, weight loss, insomnia
Multiple sclerosis (MS)	Nervous system	Weakness, loss of coordination, problems with vision and speech
Rheumatoid arthritis	Joints	Severe pain, fatigue, disabling inflammation of joints
Systemic lupus erythematosus (SLE)	Connective tissue, joints, kidneys	Facial skin rash, painful joints, fever, fatigue, kidney problems, weight loss
Type I diabetes	Insulin-producing cells in pancreas	Increased blood glucose level, excessive urine production, problems with vision, weight loss, fatigue, irritability

HIV Infection

Before 1981, **AIDS,** or acquired immunodeficiency syndrome, was unknown. Between 1981 and 2000, more than 448,000 Americans died of AIDS. Since then, the total number of people living with HIV in the United States has increased to more than 850,000. AIDS is a disease caused by **HIV,** or the human immunodeficiency virus.

Many scientists think HIV evolved from a virus similar to one that infects nonhuman primates in Africa. A mutation enables HIV to recognize a receptor protein called **CD4** on some human cells. HIV, shown in **Figure 8,** enters white blood cells by binding to CD4. HIV usually invades helper T cells, which begin to produce HIV soon after infection. As helper T cells die, the immune system gradually weakens and becomes overwhelmed by pathogens that it would normally detect and destroy. The body becomes susceptible to other diseases, called opportunistic infections, that generally cause illness only in people with weakened immune systems.

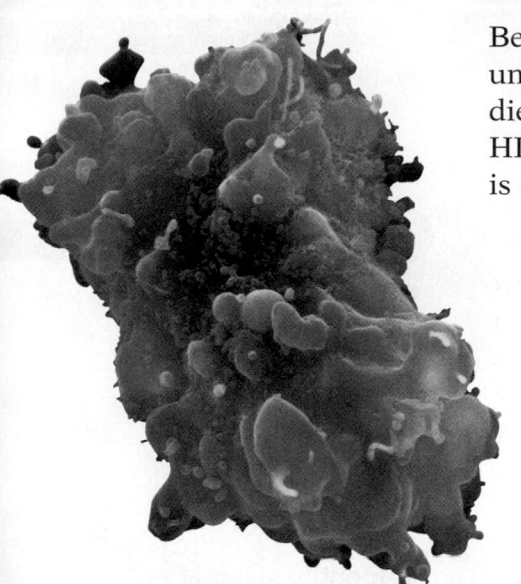

Figure 8   HIV. Small HIV particles (purple) surround a helper T cell (orange).

Testing for HIV

Antibodies to HIV can be detected in blood. Someone whose blood contains antibodies to HIV is said to be HIV positive. A diagnosis of AIDS may be made based on several criteria, including a helper T cell count less than 200 cells/mL of blood. **Figure 9** shows how the number of helper T cells may decline over time in an HIV-positive person.

The time between HIV infection and the onset of AIDS can exceed 10 years, and this time period is increasing as new treatments for HIV infection are developed. A person with HIV may feel and appear healthy but can infect other people. In the United States, the number of deaths caused by AIDS has dropped from more than 38,000 in 1996 to about 22,000 in 1997, and to about 15,000 in 2000. This decrease does not reflect a decline in HIV infection, but rather more effective drug therapies, which postpone onset of the disease.

Figure 9   Onset of AIDS. The graph at right shows the decline over time in the number of helper T cells in a person infected with HIV.

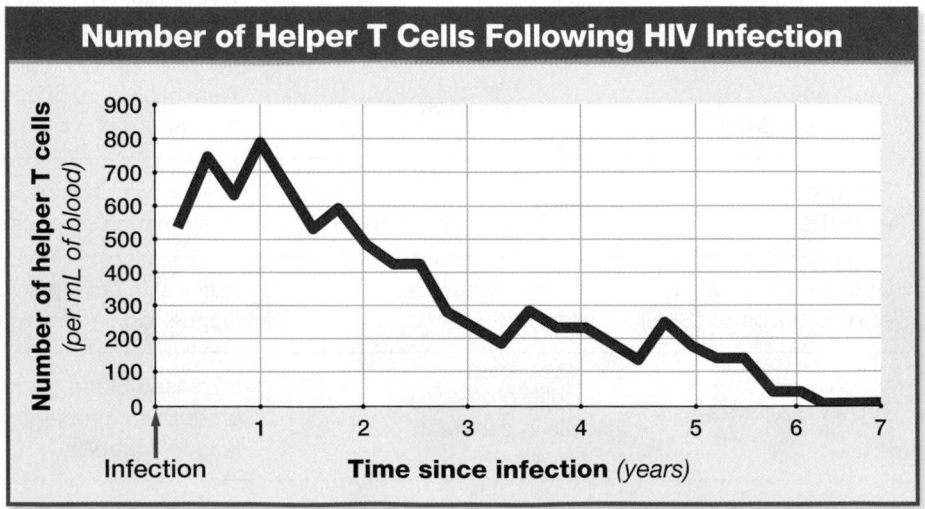

Number of Helper T Cells Following HIV Infection

Number of helper T cells *(per mL of blood)*

Time since infection *(years)*

Infection

Transmission of HIV

You can become infected with HIV if you come in contact with body fluids—including the blood—of an infected person. The most common method of HIV transmission is through sexual contact. Use of a latex condom during sexual contact reduces but does not eliminate the risk of getting or spreading HIV. Many people infected with HIV do not know they are infected. The only sure way to prevent HIV infection by sexual contact is through abstinence (the conscious decision to refrain from sexual activity).

HIV can be passed between drug users who share a hypodermic needle because HIV-infected blood often remains in the needle or syringe. Several years ago, many people became infected with HIV after receiving transfusions of HIV-contaminated blood. This is very unlikely now because blood made available for transfusion is tested for HIV. In addition, pregnant or nursing mothers can pass HIV to their infants through blood and breast milk.

HIV is not transmitted through the air, on toilet seats, by kissing or handshaking, or by any other medium where HIV-infected white blood cells could not survive. Although HIV has been found in saliva, tears, and urine, these body fluids usually contain too few HIV particles to cause an infection. Mosquitoes and ticks do not transmit HIV because they do not carry HIV-infected white blood cells.

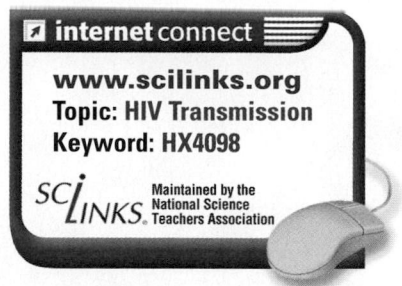

internet connect

www.scilinks.org
Topic: HIV Transmission
Keyword: HX4098

SCI LINKS Maintained by the National Science Teachers Association

DATA LAB

Tracking the Spread of AIDS

Background

The graph below shows the total AIDS cases reported in the United States between 1996 and 2001. Use the graph to answer the following questions:

Analysis

1. **Describe** how the number of people with AIDS has changed since 1996.

2. **Inferring Relationships** Is the number of Americans infected with HIV most likely greater than or less than the number of people with AIDS? Explain why.

3. **Evaluating Data** The graph indicates that the number of new AIDS cases reported each year has decreased since 1996. Suggest a possible reason for this decline.

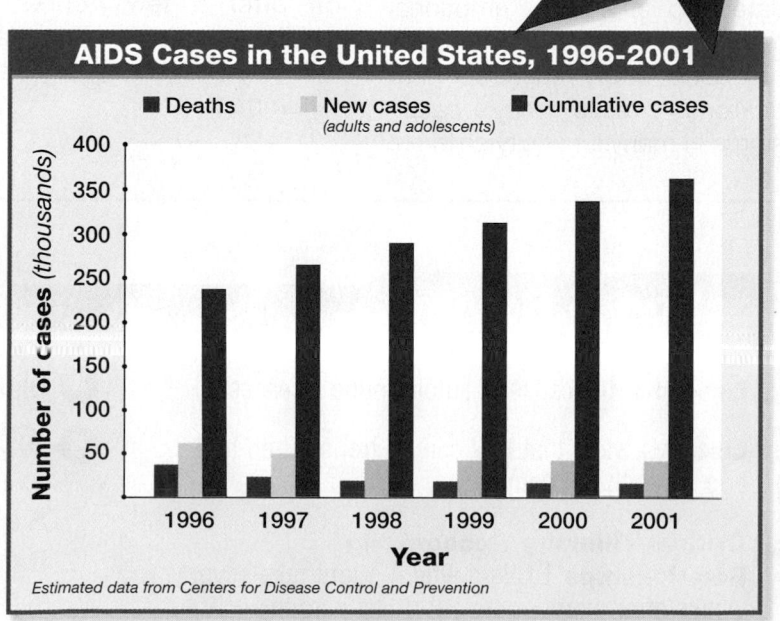

AIDS Cases in the United States, 1996-2001

■ Deaths   ■ New cases (adults and adolescents)   ■ Cumulative cases

Estimated data from Centers for Disease Control and Prevention

Allergic Reactions

Many health problems are caused by inappropriate responses of the immune system. One example is an allergic reaction. An **allergy** is the body's inappropriate response to a normally harmless antigen. Allergy-causing antigens include pollen, the feces of dust mites, fungal spores, and substances found in some foods and drugs. Most allergic reactions are merely uncomfortable. Cells exposed to allergy-causing antigens release histamine. Histamine causes swelling, redness, increased mucus production, runny nose, itchy eyes, and nasal congestion. Most allergy medicines contain antihistamines, which are drugs that prevent the action of histamine. Severe allergic reactions, such as asthma, can be life threatening if they are not treated immediately.

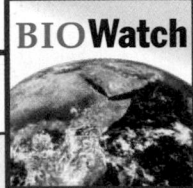
BIO Watch

Asthma

Asthma is an inflammation of the respiratory tract often caused by an allergic reaction to substances in the air. Asthma affects about 15 million Americans and causes more than 5,000 deaths each year. Inner-city residents get asthma three times as often as people who live outside cities. In some cities, the death rate from asthma is eight times the national average. Some scientists think increased asthma rates in inner-city residents is related to pollution, emotional stress, and limited access to health care. One study suggests that cockroach feces may cause asthma in many inner-city children.

Asthma Attack
During an asthma attack, the respiratory passages become inflamed and swollen. Then mucus collects in the lungs, restricting airflow. Finally, muscles that surround the bronchial tubes tighten, causing shortness of breath.

Treating Asthma
Asthma sufferers can take medicines that increase airflow by relaxing bronchial-tube muscles, but their effects wear off after a few hours. Other medicines provide long-lasting relief by preventing or reducing inflammation.

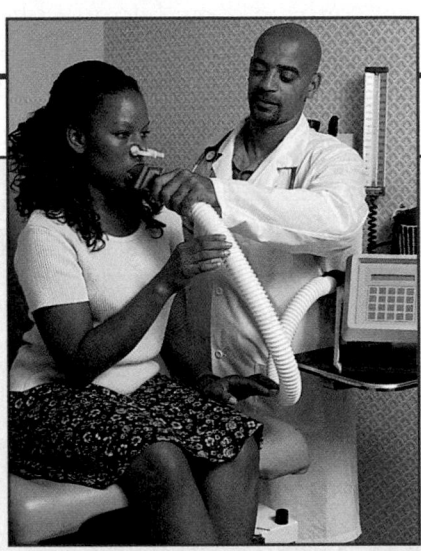
Measuring lung capacity

internet connect

www.scilinks.org
Topic: Asthma
Keyword: HX4015

SCI LINKS. Maintained by the National Science Teachers Association

Section 4 Review

1 **Describe** the cause of autoimmune diseases.

2 **List** two ways that HIV can be transmitted and two ways that it cannot.

3 **Critical Thinking Recognizing Relationships** Explain why it might take several weeks after exposure to HIV for a person's HIV antibody test to be positive.

4 **Distinguish** between HIV infection and AIDS.

5 **Standardized Test Prep** One common symptom of an allergic reaction to airborne antigens is
A a weakened immune response.
B opening nasal passages.
C reduced mucus production.
D itchy eyes.

Key Concepts

1 Nonspecific Defenses

- Skin and mucous membranes act as barriers to pathogens.
- The inflammatory response increases blood flow to an infected area, while the temperature response inhibits bacterial growth.
- Complement proteins form a membrane attack complex (MAC). Interferon stimulates cells and inhibits viruses.
- Neutrophils, macrophages, and natural killer cells use different methods to attack and destroy invading pathogens.

2 Immune Response

- Receptors on white blood cells bind to specific antigens.
- The T cell response is a defense in which cytotoxic T cells destroy pathogens.
- The B cell response is a defense in which antibodies mark pathogens for destruction by white blood cells.

3 Disease Transmission and Prevention

- Diseases are transmitted to humans through person-to-person contact, air, food, water, and animal bites.
- Biologists use Koch's postulates to identify pathogens.
- Memory cells can produce long-term immunity to pathogens.
- Vaccination produces long-term immunity to pathogens.
- Antigen shifting makes the immune response of memory cells ineffective.

4 Disorders of the Immune System

- In an autoimmune disease, the immune system attacks body cells as if they were pathogens.
- HIV, the virus that causes AIDS, invades helper T cells, causing them to produce more HIV particles and eventually die.
- HIV is transmitted by HIV-infected white blood cells in body fluids, through sexual contact or by the sharing of a hypodermic needle with an infected person.
- An allergic reaction is an inappropriate response to normally harmless antigens.

Key Terms

Section 1

pathogen (924)
mucous membrane (924)
inflammatory response (925)
histamine (925)
complement system (926)
interferon (926)
neutrophil (926)
macrophage (926)
natural killer cell (926)

Section 2

cytotoxic T cell (927)
B cell (927)
helper T cell (927)
antigen (927)
plasma cell (929)
antibody (929)

Section 3

Koch's postulates (930)
immunity (931)
vaccination (931)
vaccine (931)
antigen shifting (932)

Section 4

autoimmune disease (933)
AIDS (934)
HIV (934)
CD4 (934)
allergy (936)

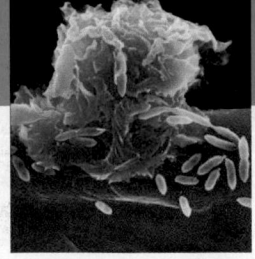

Understanding Key Ideas

1. Robert Koch
  a. treated smallpox patients.
  b. established a four-step procedure for identifying pathogens.
  c. perfected vaccination.
  d. identified complement proteins.

2. Flu vaccinations are given each year because
  a. influenza viruses mutate often.
  b. influenza is caused by bacteria.
  c. very few memory cells are produced.
  d. macrophages cannot engulf flu viruses.

3. HIV can be transmitted by
  a. sexual contact.   **c.** shaking hands.
  b. mosquito bites.   **d.** vaccination only.

4. Rheumatoid arthritis is an example of
  a. an allergic reaction.
  b. an autoimmune disease.
  c. an AIDS-related infection.
  d. a bacterial infection.

5. HIV disables the immune system by
  a. blocking the action of macrophages.
  b. destroying helper T cells.
  c. activating production of B cells.
  d. All of the above

6. For each pair of terms, explain the differences in their meanings.
  a. macrophage, neutrophil
  b. immunity, vaccine
  c. allergy, histamine

7. Name three types of white blood cells, and explain their roles in the immune system.

8. How do cytotoxic T cells recognize antigens?

9. 🌐 **BIOWatch** What symptoms are usually associated with an asthma attack?

10. ⌗ **Concept Mapping** Make a concept map that describes the immune response. Include the following terms in your map: *pathogen, macrophage, helper T cell, cytotoxic T cell, B cell, plasma cell,* and *antibody.*

11. The graph below shows the decrease in the number of helper T cells in a person with AIDS. How many months after infection did the onset of AIDS occur?

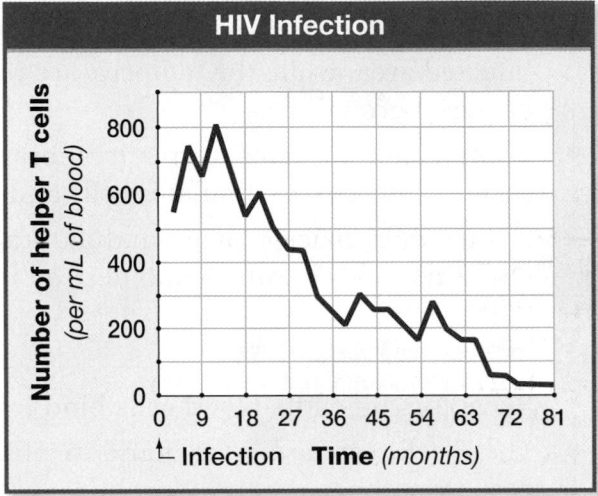

Critical Thinking

12. **Analyzing Information** Plasma cells contain a large Golgi apparatus and large amounts of rough endoplasmic reticulum. How is the presence of these organelles related to the function of plasma cells?

13. **Inferring Relationships** People who are severely burned often die from infection. Given what you know about disease transmission, explain why this is common.

Alternative Assessment

14. **Summarizing Information** Use the media center or the Internet to research three different vaccines. Make a large chart or table on poster board listing the pathogens they protect against, their effectiveness, side effects, and boosters required, if any. Present your chart to your class.

15. **Career Focus** **Immunologist** Research the field of immunology, and write a report on your findings. Your report should include a job description, education and training required, kinds of employers, growth prospects, and starting salary.

Standardized Test Prep

Understanding Concepts

Directions (1–4): **For** *each* **question, write on a separate sheet of paper the letter of the correct answer.**

1 Which of the following is a nonspecific defense?
   A. antibody response
   B. the B cell response
   C. the inflammatory response
   D. the T cell response

2 Which of the following functions is a role that mucous membranes play in the immune system?
   F. causing blood clots
   G. producing antibodies
   H. activating helper T cells
   I. secreting mucous, which traps pathogens

3 B cells and cytotoxic T cells are stimulated by interleukin-2. What kind of cell releases interleukin-2?
   A. helper T cells
   B. macrophages
   C. natural killer cells
   D. neutrophils

4 Which of the following statements is true of plasma cells?
   F. They engulf pathogens.
   G. They produce antibodies.
   H. They result from cytotoxic T cells.
   I. They are directly stimulated by interleukin-1.

Directions (5–6): **For** *each* **question, write a short response.**

5 Under what circumstances can a child be born with HIV?

6 Differentiate between a helper T cell and a cytotoxic T cell.

Test *TIP*

To help you learn the stages of the immune response, make a note card describing each stage, mix the cards up, and practice reordering the stages.

Reading Skills

Directions (7): **Read the passage below. Then answer the question.**

Scientific research into treatments and a cure for AIDS is an ongoing process. One important development is a clinical trial of a potential HIV vaccine conducted in 15 U. S. medical centers. Also, a new medication, which combines two proven medications (3Tc and AZT) into one, may help people with HIV live longer, healthier lives. Another medication, alitretinoin, treats lesions of Kaposi's sarcoma.

7 How could an HIV vaccine prevent people from developing AIDS?
   A. The vaccine would treat lesions of Kaposi's sarcoma and other symptoms.
   B. The vaccine would cure an infected person by destroying all of his HIV viruses.
   C. The vaccine would cause the body to produce HIV-fighting cells before a person is infected with the virus.
   D. The vaccine would prevent a person with HIV from developing AIDS, by preventing the virus from mutating.

Interpreting Graphics

Directions (8): **Base your answer to question 8 on the diagram below.**

Immune Response to a Virus

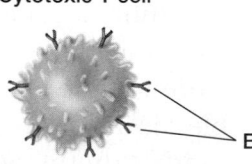

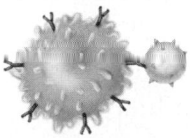

8 Why can structures *A* and *B* interact with each other?
   F. Both of them are "nonself."
   G. They have matching shapes.
   H. Both of them are viral proteins.
   I. They are produced by the same cells.

Skills Practice Lab

Simulating Disease Transmission

SKILLS
- Modeling
- Organizing and analyzing data

OBJECTIVES
- **Simulate** the transmission of a disease.
- **Determine** the original carrier of the disease.

MATERIALS
- safety goggles
- lab apron
- disposable gloves
- dropper bottle of unknown solution
- large test tube
- indophenol indicator

ChemSafety

CAUTION: Always wear safety goggles and a lab apron to protect your eyes and clothing.

CAUTION: Do not touch or taste any chemicals. Know the location of the emergency shower and eyewash station and how to use them. If you get a chemical on your skin or clothing, wash it off at the sink while calling to the teacher. Notify the teacher of a spill. Spills should be cleaned up promptly, according to your teacher's directions.

CAUTION: Glassware is fragile. Notify the teacher of broken glass or cuts. Do not clean up broken glass or spills with broken glass unless the teacher tells you to do so.

Before You Begin

Communicable diseases are caused by **pathogens** and can be transmitted from one person to another. You can become infected by a pathogen in several ways, including by drinking contaminated water, eating contaminated foods, receiving contaminated blood, and inhaling infectious **aerosols** (droplets from coughs or sneezes). In this lab, you will simulate the transmission of a communicable disease. After the simulation, you will try to identify the original infected person in the closed class population.

1. Write a definition for each boldface term in the paragraph above.

2. Make data tables similar to the ones shown at right.

3. Based on the objectives for this lab, write a question you would like to explore about disease transmission.

DATA TABLE 1	
Round number	Partner's name

DATA TABLE 2			
Name of infected person	Names of infected person's partners		
	Round 1	Round 2	Round 3

Procedure

PART A: Simulate Disease Transmission

1. Put on safety goggles, a lab apron, and gloves.

2. You will be given a dropper bottle of an unknown solution. When your teacher says to begin, transfer 3 dropperfuls of your solution to a clean test tube.

3. Select a partner for Round 1. Record the name of this partner in Data Table 1.

4. Pour the contents of one of your test tubes into the other test tube. Then pour half the solution back into the first test tube. You and your partner now share any pathogens either of you might have.

5. On your teacher's signal, select a new partner for Round 2. Record this partner's name in Data Table 1. Repeat step 4.

6. On your teacher's signal, select another new partner for Round 3. Record this partner's name. Repeat step 4.

7. Add one dropperful of indophenol indicator to your test tube. "Infected" solutions will stay colorless or turn light pink. "Uninfected" solutions will turn blue. Record the results of your test.

PART B: Trace the Disease Source

8. If you are infected, write your name and the name of your partner in each round on the board or on an overhead projector. Mark your infected partners. Record all the data for your class in Data Table 2.

9. To trace the source of the infection, cross out the names of the uninfected partners in Round 1. There should be only two names left. One is the name of the original disease carrier. To find the original disease carrier, place a sample from his or her dropper bottle in a clean test tube, and test it with indophenol indicator.

10. To show the disease transmission route, make a diagram similar to the one below. Show the original disease carrier and the people each disease carrier infected.

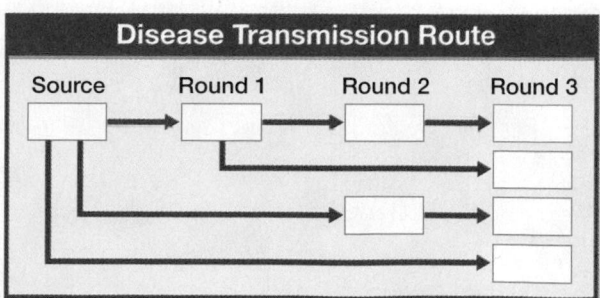

Disease Transmission Route

Source	Round 1	Round 2	Round 3

PART C: Cleanup and Disposal

11. Dispose of solutions and broken glass in the designated waste containers. Do not pour chemicals down the drain unless your teacher tells you to do so.

12. Clean up your work area and all lab equipment. Return lab equipment to its proper place. Wash your hands thoroughly before you leave the lab and after you finish all work.

Analyze and Conclude

1. **Interpreting Data** After Round 3, how many people were "infected"? Express this number as a percentage of your class.

2. **Relating Concepts** What do you think the clear fluids each student started with represent? Explain why.

3. **Drawing Conclusions** Can someone who does not show any symptoms of a disease transmit that disease? Explain.

4. **Further Inquiry** Write a new question about disease transmission that could be explored with another investigation.

On the Job

Public health officials, such as food inspectors, study and work to prevent the spread of diseases in human populations. Do research to find out how public health officials trace the origin of communicable diseases. For more about careers, visit **go.hrw.com** and type in the keyword **HX4 Careers.**

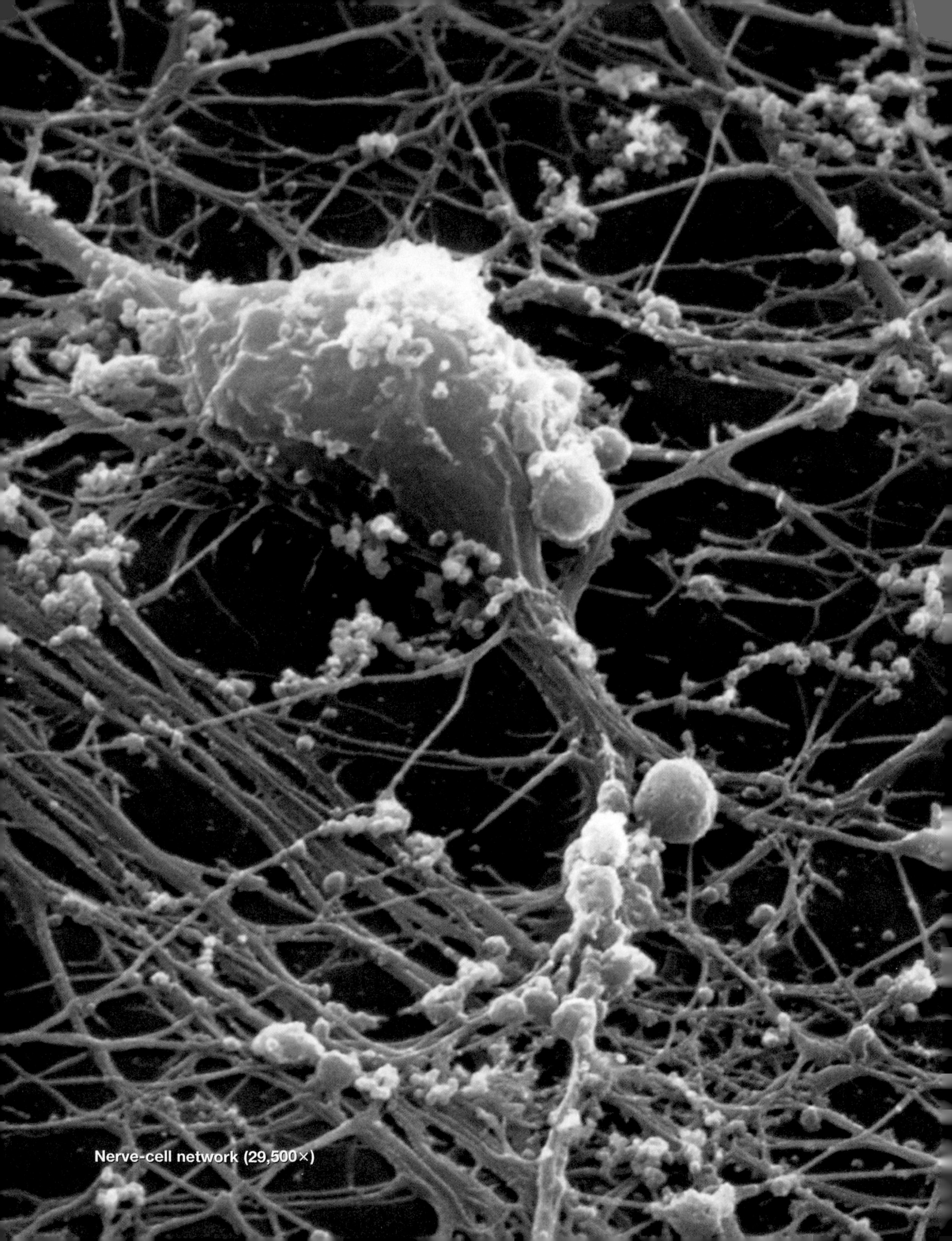

Nerve-cell network (29,500×)

41 Nervous System

✓ Quick Review

Answer the following without referring to earlier sections of your book.

1. **Describe** the importance of ion channels in cell transport. *(Chapter 4, Section 1)*

2. **Identify** the role of sodium-potassium pumps in cells. *(Chapter 4, Section 2)*

3. **Distinguish** between endocytosis and exocytosis. *(Chapter 4, Section 2)*

4. **List** three functions of receptor proteins. *(Chapter 4, Section 2)*

Did you have difficulty? *For help, review the sections indicated.*

📖 Reading Activity

Copy the following statements in your notebook:

1. *Addiction is a purely psychological response to drug use.*

2. *Reflexes occur before the brain is aware of danger.*

3. *Prolonged exposure to loud noise can cause permanent hearing loss.*

Before you read this chapter, write down whether you agree or disagree with each statement. After you have finished reading the chapter, decide whether or not you still agree with your first response.

● Nerve-cell networks like this one transmit thought, emotions, and sensations by conducting electro-chemical signals from cell to cell.

Looking Ahead

▣ internet connect

www.scilinks.org
National Science Teachers Association *sci*LINKS Internet resources are located throughout this chapter.

*sci*LINKS. Maintained by the National Science Teachers Association

Neurons and Nerve Impulses

Objectives

● **Analyze** the structure and function of neurons.

● **Describe** how the resting potential is established in a neuron.

● **Sequence** the steps of a nerve impulse.

● **List** the events that occur in synaptic transmission of a nerve impulse.

Key Terms

neuron
dendrite
axon
nerve
membrane potential
resting potential
action potential
synapse
neurotransmitter

Neurons

If your body used only chemical signals to send messages, your interaction with the environment would be slow. A quicker means of communication is needed, especially if your brain has an urgent message for the muscles in your legs, such as "Contract quickly! A speeding car is headed this way!" In addition to chemical signals, your nervous system uses electrical signals to send messages rapidly throughout your body.

The nervous system contains a complex network of nerve cells, or **neurons** *(NOO rahns)*. Neurons, such as those shown in **Figure 1,** are specialized cells that transmit information throughout the body. Neurons enable many important functions, such as movement, perception, thought, emotion, and learning.

Structure of Neurons

A neuron's unique structure enables it to conduct electrical signals called nerve impulses. Neurons communicate by transmitting nerve impulses to body tissues and organs, including muscles, glands, and other neurons. Neurons vary greatly in form, but a typical neuron is similar to the one shown in **Figure 2. Dendrites** *(DEHN driets),* which extend from the cell body of the neuron, are the "antennae" of the neuron. Dendrites receive information from other cells. The neuron's cell body collects information from dendrites, relays this information to other parts of the neuron, and maintains the general functioning of the neuron. An **axon** is a long membrane-covered extension of the cytoplasm that conducts nerve impulses. The ends of an axon are called axon terminals. When a neuron communicates with other cells, it does so at its axon terminals.

Nervous tissue consists mostly of neurons and their supporting cells. Bundles of axons are called **nerves.** The arrangement of axons in a nerve is similar to a telephone cable with many different communication channels, each carried by a separate wire. Nerves appear as fine, white threads when viewed with the unaided eye.

Insulated Neurons

Many neurons have a layer of insulation on their axon called a myelin *(MIE uh lihn)* sheath, as shown in Figure 2. Myelin is produced by supporting cells that surround the axon. The presence of myelin causes nerve impulses to move faster down the axon. The myelin sheath is interrupted at intervals, leaving gaps called nodes of Ranvier

Figure 1 Two neurons. An average adult human brain contains about 100 billion neurons.

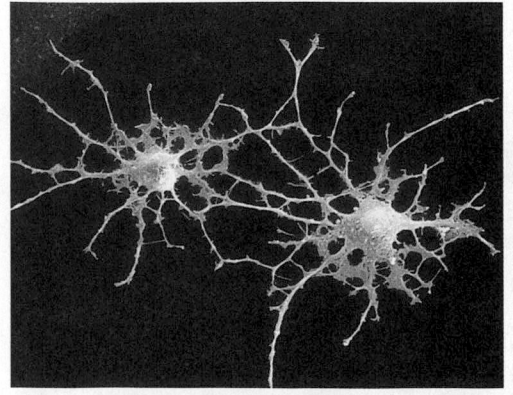

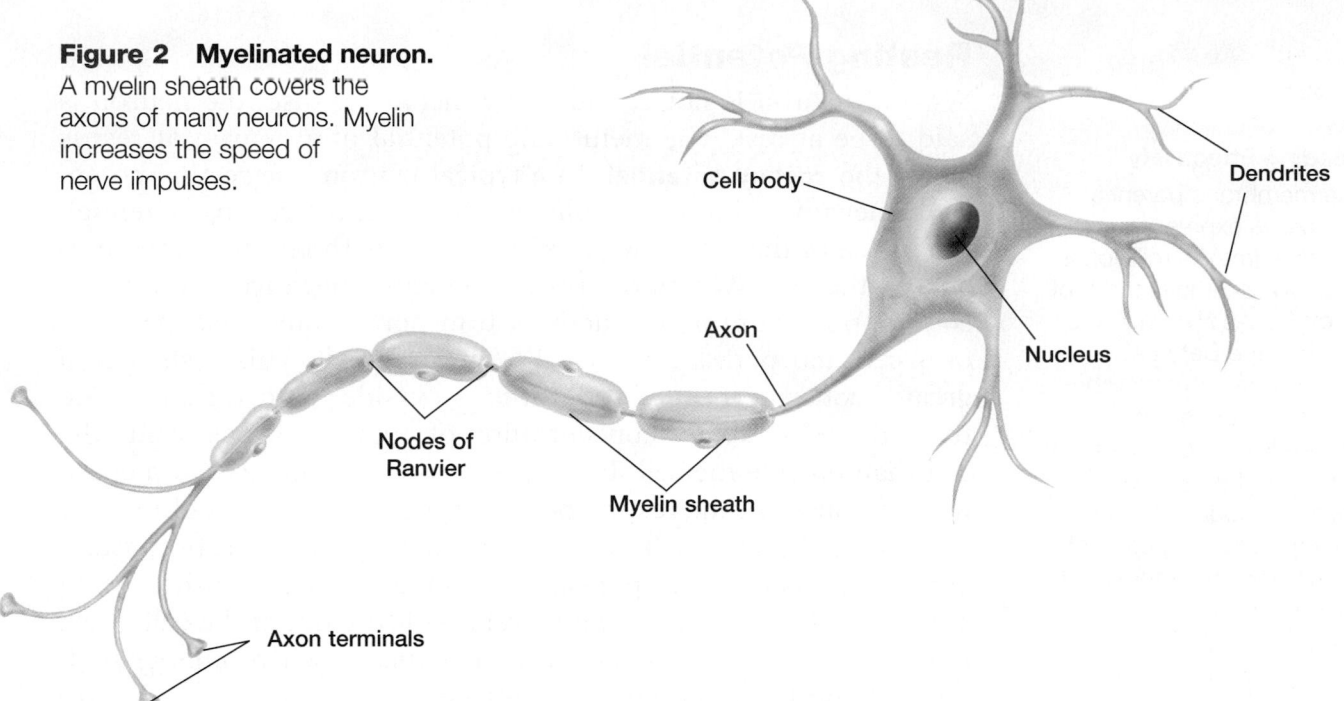

Figure 2   Myelinated neuron.
A myelin sheath covers the
axons of many neurons. Myelin
increases the speed of
nerve impulses.

Cell body

Dendrites

Axon

Nucleus

Nodes of
Ranvier

Myelin sheath

Axon terminals

(RAHN vee ay), where the axon membrane is exposed to the surrounding fluid. Conduction of nerve impulses is faster in myelinated axons because nerve impulses "jump" from node to node as
they move down the axon. Myelin is especially beneficial in neurons
that must transmit information very rapidly, such as those involved
with quick movement.

The speed of impulse conduction is also related to axon diameter.
A large-diameter axon conducts impulses faster than a small-
diameter axon, assuming both axons are either myelinated or
unmyelinated.

internet connect

www.scilinks.org
Topic: Neurons
Keyword: HX4129

SCI*LINKS* Maintained by the
National Science
Teachers Association

Neuron Function

All cells have an electrical charge on the inner surface of the cell
membrane that is different from the electrical charge of the fluid
outside the cell. The difference in electrical charge across the cell
membrane, called the **membrane potential,** results from the movement of ions into and out of the cell. This movement depends on
the relative concentration of ions inside and outside the cell, the
ability of the ions to diffuse across the cell membrane, and the electrical charge of the ions. The membrane potential is expressed as
voltage, like that of a battery.

Ions diffuse across a neuron's cell membrane by passing through
proteins that act as ion channels. Each type of channel allows
only specific ions to pass. Certain channels are voltage-gated—that
is, whether they are open or closed depends on the membrane
potential. Even a small change in the membrane potential can
affect the permeability of the cell membrane to certain ions. As
these ions diffuse into or out of the neuron, they in turn affect the
membrane potential.

Study TIP

● **Reading Effectively**

The membrane potential of a cell is expressed in millivolts (mV). A millivolt is equal to one-thousandth of a volt (V). As shown in the graph in the Data Lab below, a neuron's membrane potential can be positive or negative. The resting potential of an average neuron is about –70 mV. During an action potential, the membrane potential of the neuron reaches about +40 mV.

Resting Potential

When a neuron is not conducting a nerve impulse, the neuron is said to be at rest. The membrane potential of a neuron at rest is called the **resting potential.** In a typical neuron, the resting potential is negative, about –70 millivolts (mV). At the resting potential, the inside of the cell is negatively charged with respect to the outside of the cell. Why is the resting potential negative? Recall that sodium-potassium pumps actively transport sodium ions, Na^+, out of a cell and potassium ions, K^+, into the cell. This results in a greater concentration of sodium ions outside the cell than inside the cell, and a greater concentration of potassium ions inside the cell than outside the cell. In a neuron, voltage-gated sodium channels are closed at the resting potential. Thus, very few sodium ions can diffuse into the cell, despite their strong concentration gradient. Some voltage-gated potassium channels are open at the resting potential. Potassium ions can therefore diffuse out of the cell down their concentration gradient, carrying their positive charge with them. Neurons also contain negatively charged proteins that are too large to exit the cell.

Action Potential

When a neuron is conducting a nerve impulse, changes occur in the cell membrane of the neuron. A nerve impulse is also called an action potential. An **action potential** is a local reversal of polarity—from a negative charge to a positive charge—inside the neuron. An action potential moves down an axon like a flame burning down a fuse. The events of an action potential are summarized in **Figure 3.**

Analyzing Changes During a Nerve Impulse

DATA LAB

Background

The graph below illustrates changes that occur in the membrane potential of a neuron during an action potential. Use the graph to answer the following questions. Refer to Figure 3 as needed.

Analysis

1. **Determine** about how long an action potential lasts.

2. **State** whether voltage-gated sodium channels are open or closed at point *A*.

3. **State** whether voltage-gated potassium channels are open or closed at point *B*.

4. **Critical Thinking Recognizing Relationships** What causes the membrane potential to become less negative at point *A*?

5. **Critical Thinking Recognizing Relationships** What causes the membrane potential to become more negative at point *B*?

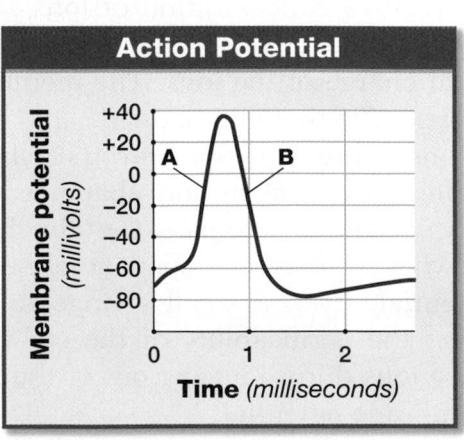

Action Potential

Figure 3

Conduction of a Nerve Impulse

An action potential moves rapidly down an axon.

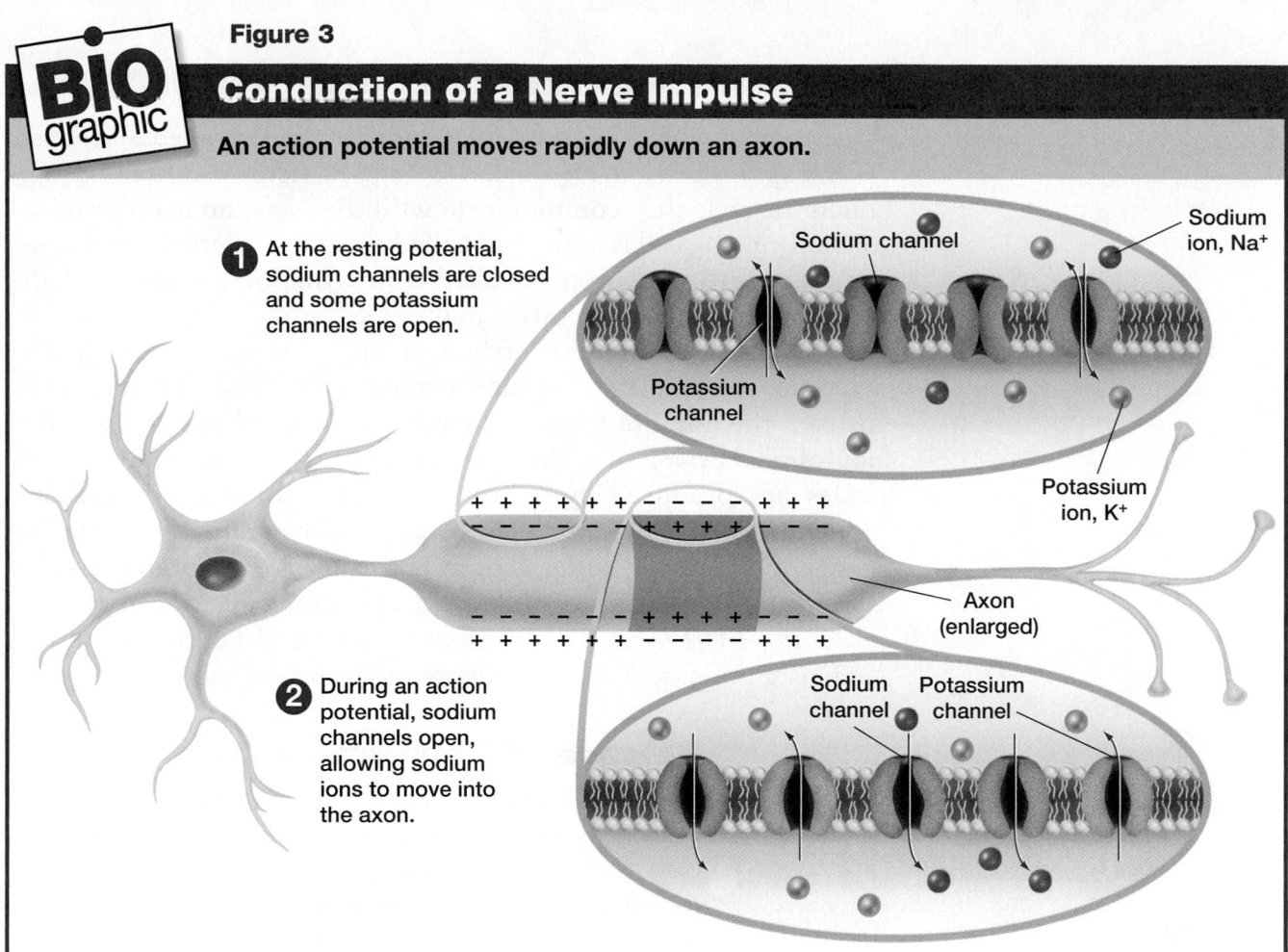

1 At the resting potential, sodium channels are closed and some potassium channels are open.

Sodium channel

Sodium ion, Na$^+$

Potassium channel

Potassium ion, K$^+$

Axon (enlarged)

2 During an action potential, sodium channels open, allowing sodium ions to move into the axon.

Sodium channel   Potassium channel

Step ❶ At the resting potential, the inside of the neuron is negatively charged with respect to the outside of the neuron. The neuron is ready to conduct an action potential.

Step ❷ An action potential begins when a stimulus, such as a signal molecule, causes a local change in the membrane potential to a more positive value. This change causes voltage-gated sodium channels to open, and sodium ions rapidly flow into the axon. For a brief moment, the membrane potential approaches about +40 mV as the inside of the axon becomes positively charged. This sudden local reversal of polarity begins a chain reaction that causes voltage-gated sodium channels to open down the entire length of the axon. As each sodium channel opens, sodium ions flow into the axon. The action potential conducts rapidly down the axon toward the axon terminals.

Voltage-gated sodium channels close immediately after the action potential has passed. Then additional voltage-gated potassium channels open, allowing potassium ions to flow out of the axon. As a result, the membrane potential becomes negative again immediately after the action potential. The resting potential is fully restored as sodium-potassium pumps reestablish the original concentrations of sodium ions and potassium ions inside and outside the axon. The neuron cannot conduct another action potential until that time.

Communication Between Neurons

A junction at which a neuron meets another cell is called a **synapse** *(SIHN aps)*, shown in **Figure 4.** At synapses, neurons usually do not touch the cells they communicate with. Between an axon terminal and a receiving cell is a tiny gap called a synaptic cleft. At a synapse, the transmitting neuron is called a presynaptic neuron, and the receiving cell is called a postsynaptic cell.

When a nerve impulse arrives at an axon terminal of a presynaptic neuron, the impulse cannot cross the synaptic cleft. Instead, the impulse triggers the release of signal molecules called **neurotransmitters** into the synaptic cleft. Neurotransmitter molecules are produced by neurons and are stored inside vesicles. There are many different neurotransmitters and several mechanisms of neurotransmitter action. For example, in human muscles the principal neurotransmitter is a chemical called acetylcholine *(as ee tihl KOH leen)*. The brain utilizes several neurotransmitters such as glutamate *(GLOO tuh mayt)* and dopamine.

Release of Neurotransmitter

A nerve impulse causes a presynaptic neuron to release neurotransmitter molecules into the synaptic cleft. When an action potential reaches an axon terminal of the presynaptic neuron, vesicles that contain neurotransmitter molecules fuse with the cell membrane. This releases neurotransmitter molecules into the synaptic cleft by exocytosis. Neurotransmitter molecules diffuse across the synaptic cleft and interact with the postsynaptic cell. As shown in **Figure 5,** neurotransmitter molecules bind to receptor proteins on the postsynaptic cell. In some cells, ion channels open when a neurotransmitter binds to these receptor proteins. Such channels are called chemical-gated ion channels; whether these channels are open or closed depends on the binding of a chemical—in this case a neurotransmitter molecule.

A neurotransmitter may either excite or inhibit the activity of the postsynaptic cell it binds to. For example, when the neurotransmitter opens chemical-gated ion channels, ions move across the cell membrane of the postsynaptic cell. This causes the membrane potential of the postsynaptic cell to change depending on the charge of the ions that move into or out of the cell. If positively charged ions enter a postsynaptic neuron, an action potential may be produced (excitation). On the other hand, if positively charged ions flow out of the postsynaptic neuron, or if negatively charged ions enter the neuron, an action potential may be suppressed (inhibition).

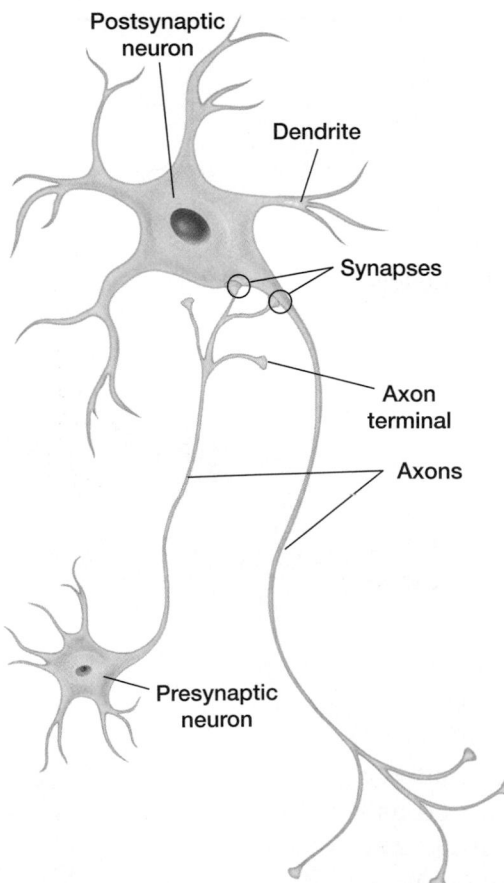

Figure 4 Synapse.
A synapse is a junction at which signals are transmitted between a neuron and another cell.

Postsynaptic neuron

Dendrite

Synapses

Axon terminal

Axons

Presynaptic neuron

Neurotransmitter molecules do not remain in the synaptic cleft indefinitely. Instead, most neurotransmitter molecules are cleared from the synaptic cleft very shortly after they are released. Many presynaptic neurons reabsorb neurotransmitter molecules and use them again. At other synapses, neurotransmitter molecules are broken down by enzymes or other chemicals. This happens, for example, at the synapses between neurons and skeletal muscle cells. The reuptake or breakdown of the neurotransmitter molecules ensures that their effect on postsynaptic cells is not prolonged.

Figure 5 Synaptic transmission

Neurotransmitter molecules are released from a presynaptic neuron, diffuse across the synaptic cleft, and interact with a postsynaptic cell.

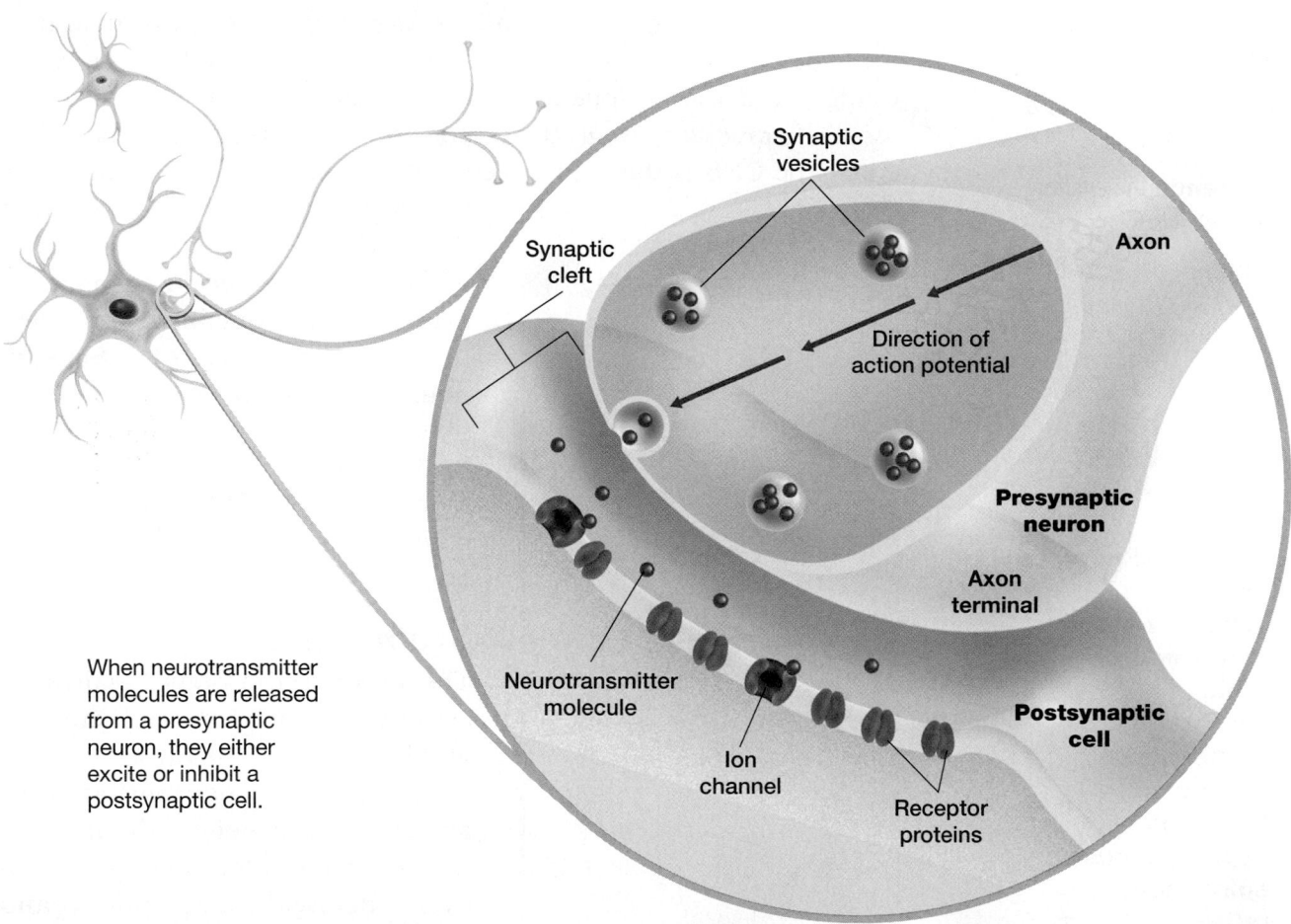

When neurotransmitter molecules are released from a presynaptic neuron, they either excite or inhibit a postsynaptic cell.

Section 1 Review

1 Describe the structure of a typical neuron.

2 Describe how the movement of ions across the cell membrane determines the membrane potential.

3 Summarize the events involved in the synaptic transmission of a nerve impulse.

4 Critical Thinking Inferring Relationships How does the membrane potential affect the permeability of a neuron's cell membrane?

5 Standardized Test Prep The junction at which a neuron communicates with another neuron or a muscle cell is called a

A myelin sheath.   **C** nerve.

B synapse.   **D** neurotransmitter.

Structures of the Nervous System

Central Nervous System

Neurons are the most important cells of the nervous system. The functions of the nervous system depend on the complex interaction between billions of neurons. Networks of neurons constantly gather, integrate, interpret, and respond to information about the body's internal state and environmental conditions. How are neurons organized in the nervous system? As shown in **Figure 6,** there are two main divisions of the nervous system—the central nervous system, shown in orange, and the peripheral nervous system, shown in purple. The **central nervous system** (CNS) consists of the brain and the spinal cord. The CNS is the control center of the body. The CNS interprets and responds to information from the environment and from within the body. The **peripheral nervous system** (PNS) contains sensory neurons and motor neurons. **Sensory neurons** send information from sense organs, such as the skin, to the CNS. **Motor neurons** send commands from the CNS to muscles and other organs.

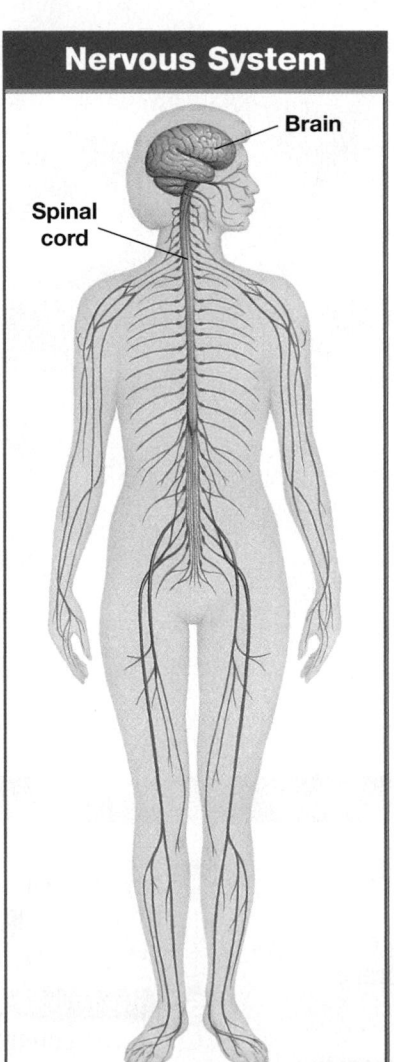

Nervous System

Brain

Spinal
cord

Figure 6 Nervous system.
The central nervous system (orange) consists of the brain and the spinal cord. The peripheral nervous system (purple) branches throughout the body.

Brain

The **brain** is the body's main processing center. Encased entirely within the skull, the brain contains about 100 billion neurons. An average adult brain weighs about 1.5 kg (3 lb). Thoughts, feelings, emotions, behavior, perception, and memories are controlled by your brain. Your brain also enables you to learn and process information, such as the text in this book. Scientists have determined the location of various functions in the brain. The brain consists of three major parts, shown in **Figure 7**—the cerebrum, the cerebellum, and the brain stem.

Cerebrum The **cerebrum** *(seh REE bruhm)* is the largest part of the brain. The capacity for learning, memory, perception, and intellectual function resides in the cerebrum. The cerebrum has a folded outer layer with many bumps and grooves. A long, deep groove down the center divides the cerebrum into right and left halves, or hemispheres. The cerebral hemispheres communicate through a connecting band of axons called the corpus callosum *(KOR puhs kuh LOH suhm)*. In general, the left cerebral hemisphere receives sensations from and controls movements of the right side of the body. The right cerebral hemisphere receives sensations from and controls movements of the left side of the body.

Most sensory and motor processing occurs in the cerebral cortex *(KOHR teks)*, the folded, thin (2–4 mm) outer layer of the cerebrum. The cerebral cortex contains about 10 percent of the brain's neurons. The folded outer surface of the cerebrum is the cerebral cortex, which has a large surface area. The cerebral cortex is primarily involved with the functioning of sensory systems.

Cerebellum The **cerebellum** *(ser uh BEL uhm)*, which is located at the posterior base of the brain, regulates balance, posture, and movement. The cerebellum smooths and coordinates ongoing movements, such as walking, by timing the contraction of skeletal muscles. The cerebellum integrates and responds to information about body position from the cerebrum and the spinal cord to control balance and posture.

Brain Stem At the base of the brain is the stalklike **brain stem**. The brain stem is a collection of structures leading down to the spinal cord and connecting the cerebral hemispheres with the cerebellum. The lower brain stem consists of the midbrain, the pons, and the

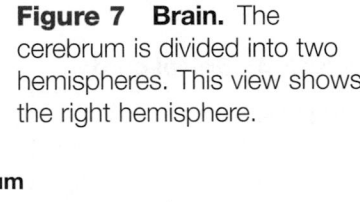

Real Life

Each year, nearly 250 bicyclists die because of brain injuries.
Wearing a bicycle helmet reduces the risk of head trauma by more than 70 percent.
Evaluating Viewpoints Should there be helmet laws for bicyclists just as there are seatbelt laws for automobile drivers?

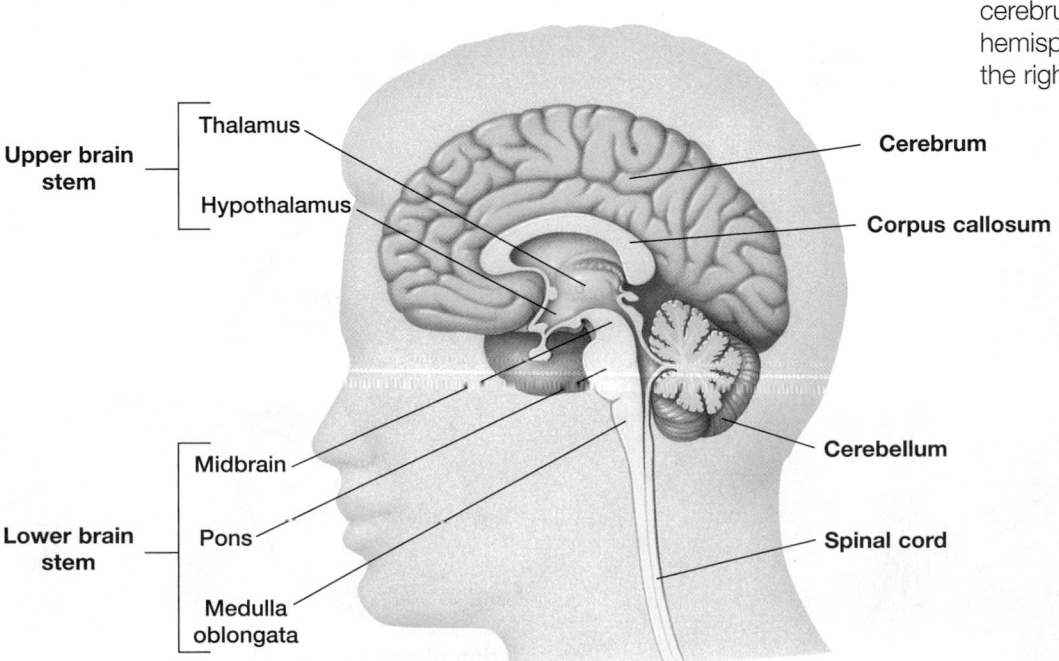

Figure 7 Brain. The cerebrum is divided into two hemispheres. This view shows the right hemisphere.

Upper brain stem
- Thalamus
- Hypothalamus

Lower brain stem
- Midbrain
- Pons
- Medulla oblongata

Cerebrum
Corpus callosum
Cerebellum
Spinal cord

● **Interpreting Graphics**

As you look at Figure 8, think of the spinal cord as a busy two-way highway with sensory traffic going north and motor traffic going south.

medulla oblongata *(mi DUHL uh ahb lahn GAHT uh)*. These structures relay information throughout the CNS and play an important role in homeostasis by regulating vital functions, such as heart rate, breathing rate, body temperature, and sleep.

The upper brain stem contains important relay centers that direct information to and from different parts of the brain. The **thalamus** *(THAL uh muhs)* is a critical site for sensory processing. Sensory information from all parts of the body converges on the thalamus, which relays the information to appropriate areas of the cerebral cortex. Below the thalamus, at the base of the brain, is the hypothalamus. The **hypothalamus,** along with the medulla oblongata, helps regulate many vital homeostatic functions, such as breathing and heart rate. The hypothalamus is responsible for feelings of hunger and thirst. It also regulates many functions of the endocrine system by controlling the secretion of many hormones.

The thalamus and hypothalamus are linked to some areas of the cerebral cortex by an extensive network of neurons called the limbic system. The limbic system includes structures of both the brain stem and the cerebrum. The limbic system has an important role in memory, learning, and various emotions, such as pleasure and anger.

Spinal Cord

The **spinal cord,** shown in **Figure 8,** is a dense cable of nervous tissue that runs through the vertebral column. The spinal cord extends from the medulla oblongata through the vertebrae to a level just below the ribs. The spinal cord links the brain to the PNS. The brain receives information that travels upward through the spinal cord. Through the spinal cord, the brain also sends commands that control the rest of the body. In addition to relaying messages, the spinal cord functions in reflexes. A **reflex** is a sudden, involuntary contraction of muscles in response to a stimulus.

Figure 8   Spinal cord

Spinal nerves have a dorsal root and a ventral root that diverge as they enter the spinal cord.

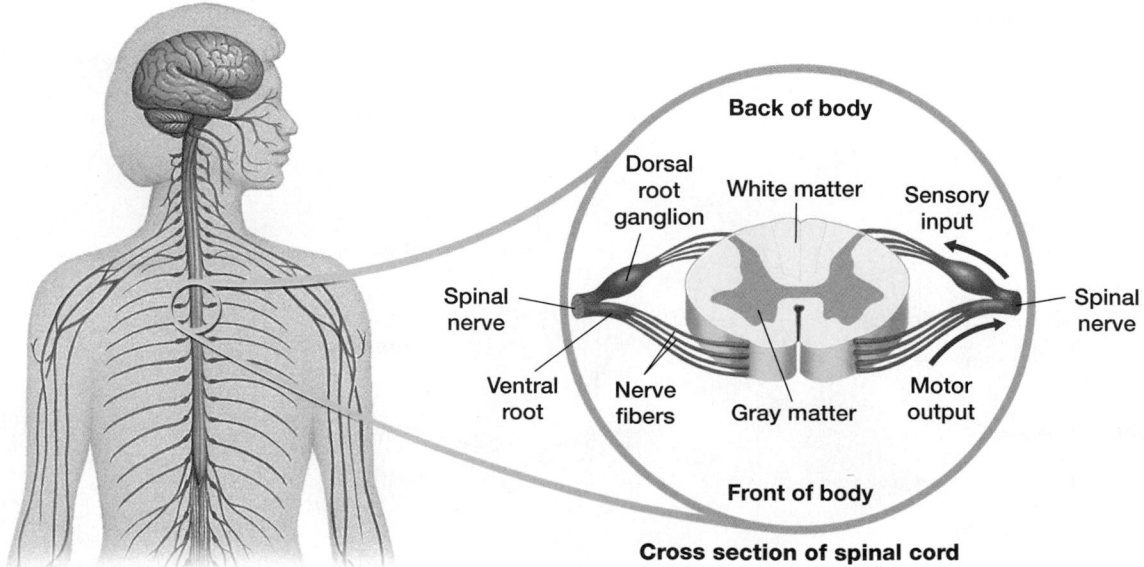

Back of body

Dorsal root ganglion   White matter   Sensory input

Spinal nerve

Spinal nerve

Ventral root   Nerve fibers   Gray matter   Motor output

Front of body

Cross section of spinal cord

The spinal cord is linked to the PNS through 31 pairs of spinal nerves. The spinal nerves, which branch from the spinal cord, carry information to and from the CNS. Spinal nerves in the upper part of the spinal cord branch into the arms and upper body, and spinal nerves in the lower part of the spinal cord branch into the legs and lower body. Each spinal nerve has a dorsal root and a ventral root. Dorsal roots contain sensory neurons, which carry information from areas of sensory input to the CNS. Ventral roots contain motor neurons, which carry motor responses from the CNS to muscles, glands, and other organs. As shown in Figure 8, dorsal and ventral roots come together to form the spinal nerves near the spinal cord.

The spinal cord contains a core of gray matter covered by a sheath of white matter, as shown in Figure 8. Gray matter contains the cell bodies of neurons, whereas white matter contains the axons of neurons. Included in the gray matter are **interneurons,** which link neurons to each other.

BIOWatch

Spinal Cord Injury

Unlike most other parts of the body, the spinal cord does not heal after an injury. Damaged neurons stop conducting nerve impulses at the site of injury, permanently paralyzing the legs or, in some cases, all four limbs.

Every year, spinal cord injuries—whether incurred in athletics or automobile accidents—leave nearly 15,000 Americans partially or totally paralyzed. In 1995, actor Christopher Reeve, shown in the photo at right, injured his spinal cord after falling headfirst from a horse. The fall broke vertebrae in Reeve's neck, paralyzing him from the neck down.

A treatment currently available for people with spinal cord injuries is an anti-inflammatory drug called methylprednisolone. If given within 8 hours after the spinal cord is injured, the drug can improve chances of recovery. Even with this drug, however, recovery is usually far from complete.

Stopping Cell Death

Cells continue to die near the site of a spinal cord injury for several weeks after injury occurs. Myelin-producing cells die, leaving neurons in the spinal cord unable to function. Some scientists think that stopping the death of these cells could help avoid paralysis. In experiments on rats, researchers have found that a cell-death inhibitor improves the rats' ability to use their hind legs after a spinal cord injury. Researchers are investigating other cell-death inhibitors that could be used on humans.

Bridging the Gap

After the spinal cord is injured, damaged axons begin to regrow. However, their growth is inhibited by substances in the spinal cord. Peripheral nerves lack these substances, so the axons in these nerves can regrow quite well. To stimulate the growth of axons in the injured spinal cord, researchers have grafted pieces of peripheral nerves into the spinal cord. The nerve grafts provide tunnels for regrowing axons. Rats with such nerve grafts begin to show signs of recovery within 3 weeks. Within a year, they can support their own weight. Similar grafts have not yet been tried on humans.

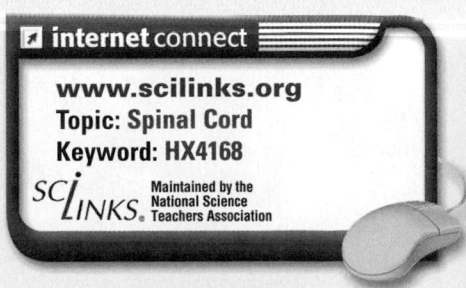

internet connect

www.scilinks.org
Topic: **Spinal Cord**
Keyword: **HX4168**

SCi LINKS. Maintained by the National Science Teachers Association

Peripheral Nervous System

Study TIP

● **Interpreting Graphics**

As you look at Figure 9, notice that in a spinal reflex, motor neurons stimulate muscles in the same region in which the stimulus that caused the reflex originated.

The peripheral nervous system connects the brain and the spinal cord to the rest of the body. In addition to the 31 pairs of spinal nerves, 12 pairs of cranial nerves connect the brain with areas in the head and neck. The PNS contains two principal divisions—the sensory division and the motor division. The sensory division directs sensory information to the central nervous system. The motor division carries out responses to sensory information. The motor division of the PNS consists of two independent systems—the somatic nervous system and the autonomic nervous system.

Somatic Nervous System

Most motor neurons that stimulate skeletal muscles are under our conscious control. These neurons are part of the somatic nervous system. Some activity in the somatic nervous system, such as spinal reflexes, is involuntary. A spinal reflex is a self-protective motor response. Spinal reflexes are extremely rapid because they usually involve the spinal cord but do not involve the brain.

The knee-jerk reflex, shown in **Figure 9,** is an example of a spinal reflex. When the ligament below your kneecap is tapped, your lower leg suddenly kicks forward. Tapping the ligament stimulates a sensory neuron, shown in red. The sensory neuron sends a nerve impulse to the spinal cord and excites a motor neuron, shown in green, which causes the quadriceps to contract. This causes the leg to extend rapidly. The sensory neuron also stimulates an interneuron, shown in blue. The interneuron inhibits a motor neuron that would normally cause the hamstrings to contract, allowing the hamstrings to relax.

Figure 9   Knee-jerk reflex

When the ligament below the patella is tapped, the quadriceps contracts, the hamstrings relax, and the leg rapidly extends.

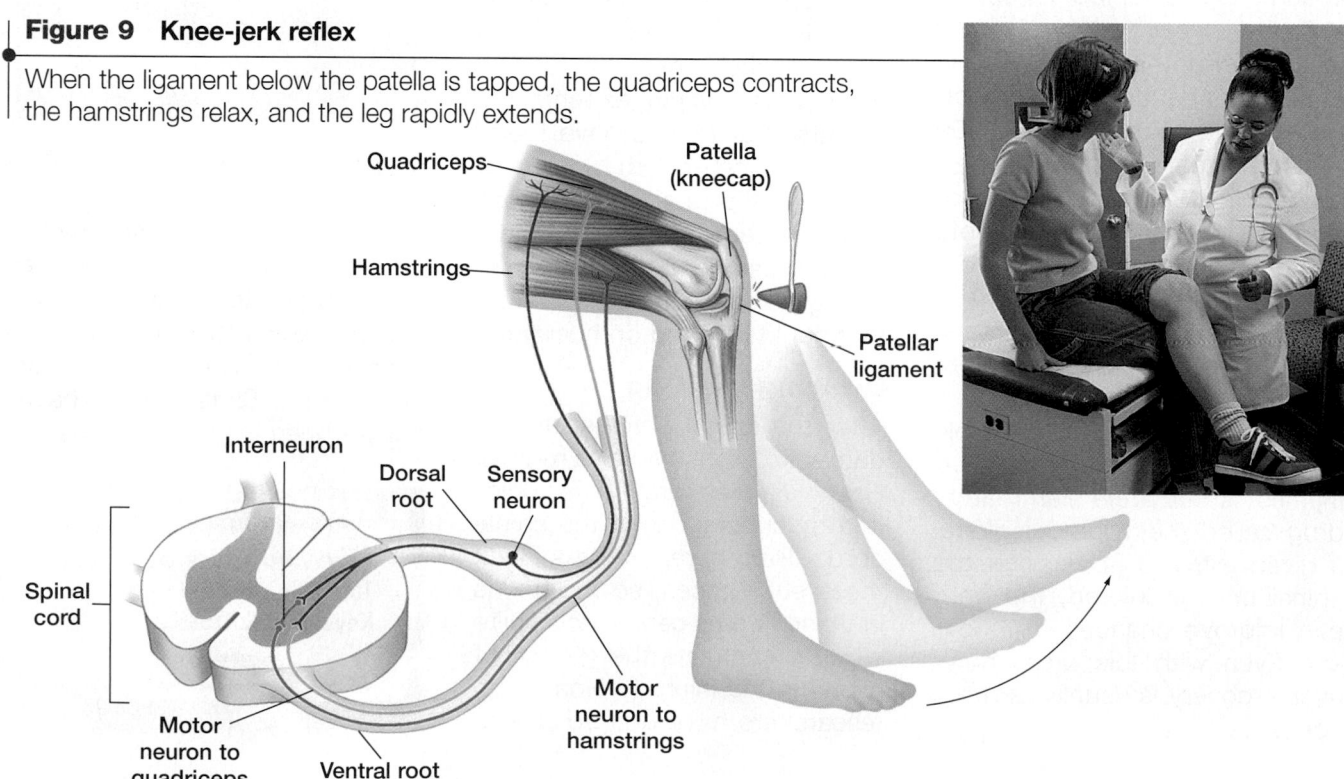

Autonomic Nervous System

Peripheral motor neurons that regulate smooth muscles do not require our conscious control. These neurons are part of the autonomic nervous system, which regulates heart rate and blood flow by controlling contractions of cardiac muscle in the heart and smooth muscle lining the walls of blood vessels. It also controls muscles in the digestive, urinary, respiratory, and reproductive systems, as well as the secretions of many glands.

Two divisions of the autonomic nervous system—the parasympathetic division and the sympathetic division—maintain stability in the body by counterbalancing each other's effects. The parasympathetic division is most active under normal conditions. It keeps your body functioning even when you are not active. For example, you continue to breathe when you fall asleep.

The sympathetic division dominates in times of physical or emotional stress. It controls the "fight-or-flight" response that you experience during a stressful situation, such as "nervousness" when taking a pop quiz. The sympathetic division increases blood pressure, heart rate, and breathing rate. It also directs blood flow toward your heart and skeletal muscles. Effects of the autonomic nervous system are summarized in **Table 1.**

internet connect

www.scilinks.org
Topic: Autonomic
Nervous System
Keyword: HX4017

SCILINKS. Maintained by the National Science Teachers Association

Table 1 Physiological Effects of the Autonomic Nervous System		
Organ	**Effect of sympathetic division**	**Effect of parasympathetic division**
Eyes	Pupils dilate	Pupils constrict
Heart	Heart rate increases	Heart rate decreases
Lungs	Bronchioles dilate	Bronchioles constrict
Intestines	Gastric secretions decrease	Gastric secretions increase
Blood vessels	Blood vessels dilate	Little or none

Section 2 Review

1. **Name** the two main divisions of the nervous system, and state their general functions.

2. **Compare** the functions of the cerebellum and the brain stem.

3. **Distinguish** between dorsal roots and ventral roots of the spinal cord.

4. **Name** the division of the autonomic nervous system that is more active under normal conditions.

5. **Critical Thinking Comparing Functions** Why is a spinal reflex more rapid than a voluntary movement?

6. **Standardized Test Prep** A sudden stretch of the quadriceps muscle triggers the knee-jerk reflex, which maintains homeostasis by causing the

   A patella to elongate.

   B hamstrings to contract.

   C quadriceps to elongate.

   D quadriceps to contract.

Sensory Systems

Objectives

- **List** five types of sensory receptors and the stimuli to which they respond.

- **Identify** sites of sensory processing in the brain.

- **Analyze** the structure of the eye and its role in the visual system.

- **Describe** how the ear detects sound and helps maintain balance.

- **Compare** the senses of taste and smell.

Key Terms

sensory receptor
retina
rod
cone
optic nerve
cochlea
semicircular canal

Perception of Stimuli

The perception of everything you respond to in the environment, such as the horn of a passing car or cold rain on your face, is made possible by sensory systems. Sensory systems are essential to survival, and they enable us to experience both pleasurable and painful stimuli. Sensory systems help maintain homeostasis by constantly adjusting body conditions to respond to changes in the environment. This requires the integration of the peripheral nervous system and the central nervous system. The sensory division of the PNS collects information about sensory stimuli in and around the body. The sensory information is sent to the brain, which processes the information and, if necessary, generates a motor response to the stimuli.

How does the nervous system detect sensory stimuli? Specialized neurons called **sensory receptors** detect sensory stimuli and then convert the stimuli to electrical signals, in the form of nerve impulses, that can be interpreted by the brain. Although sensory receptors are located throughout the body, they are most concentrated in the sense organs—the eyes, ears, nose, mouth, and skin. **Table 2** lists several types of sensory receptors and some of their locations.

Sensory Receptors

Mechanoreceptors throughout the body respond to physical stimuli— such as pressure and tension—that cause distortion or bending of tissue. These stimuli alter the electrical activity of mechanoreceptors. Many mechanoreceptors are found in the skin, and they are concentrated in very sensitive areas, including the face, hands, fingertips, and neck. Pain receptors, which respond to potentially harmful stimuli—such as intense heat or cold and tissue damage—are responsible for painful sensations. Pain is a very important sensation

Table 2 Types of Sensory Receptors		
Receptor type	**Stimuli**	**Locations**
Thermoreceptors	Temperature change	Skin, hypothalamus
Pain receptors	Tissue damage	All tissues and organs except the brain
Mechanoreceptors	Movement, pressure, tension	Skin, ears, muscles
Photoreceptors	Light	Eyes
Chemoreceptors	Chemical	Tongue, nose

Figure 10 **Processing sites and lobes of the cerebral cortex**

Specific areas of the cerebral cortex control different functions of the body.

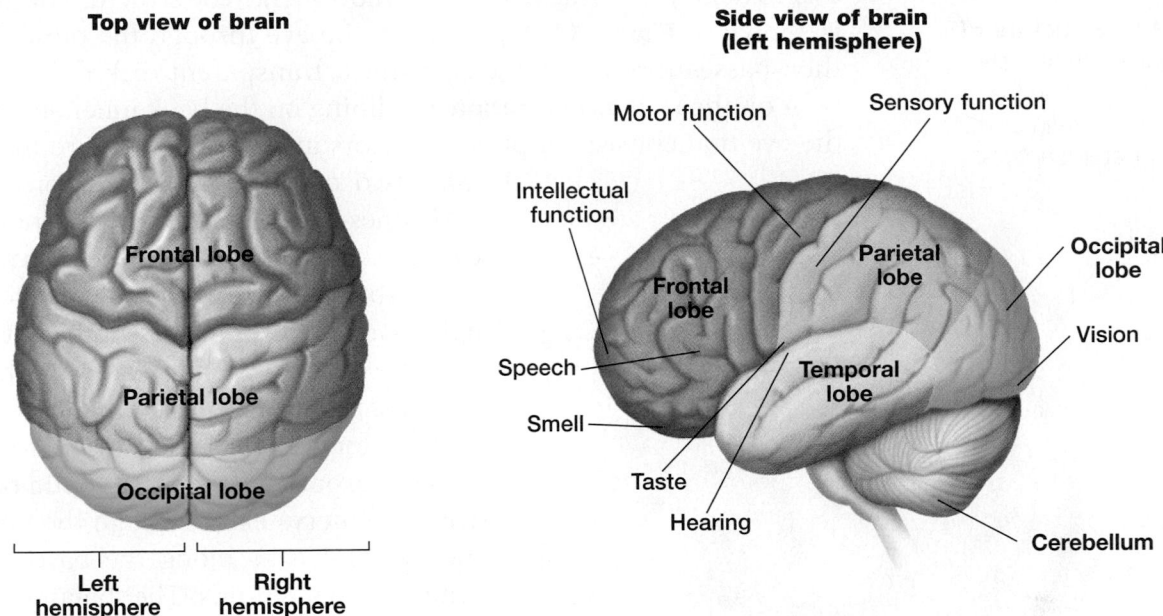

Top view of brain

Frontal lobe

Parietal lobe

Occipital lobe

Left hemisphere — Right hemisphere

Side view of brain (left hemisphere)

Motor function
Sensory function
Intellectual function
Parietal lobe
Occipital lobe
Frontal lobe
Vision
Speech
Temporal lobe
Smell
Taste
Hearing
Cerebellum

because it informs you that something is wrong in your body. Many self-protective responses, such as reflexes, are initiated by pain receptors. *Thermoreceptors*, located in the skin and hypothalamus, detect changes in temperature. Thermoreceptors play an important role in homeostasis, helping to keep the body temperature within its normal range.

Sensory receptors are located throughout the body, and sensory input from these receptors enters the central nervous system in an organized fashion. Sensory stimuli that originate in the lower body enter the lower part of the spinal cord. Sensory stimuli that originate in the upper body enter the upper part of the spinal cord and the brain.

Processing of Sensory Information

Recall that the cerebral cortex contains a large percentage of the brain's neurons. Many of the neurons in the cerebral cortex are responsible for processing incoming sensory information from the sense organs. The thalamus relays information from the sense organs to certain regions of the cerebral cortex. As shown in **Figure 10,** deep grooves divide the cerebral hemispheres into four general areas, or lobes: the occipital *(ahk SIP ih tuhl)* lobe, the parietal *(puh RIE uh tuhl)* lobe, the temporal lobe, and the frontal lobe. Sensory neurons from the different sense organs come together at certain regions in the cerebral cortex. For example, most visual processing takes place in the occipital lobe, located at the back of the head. Similarly, processing of sound is carried out within the temporal lobe.

Eyes

Humans have very good eyesight. Our eyes enable us to see in color and to distinguish fine details and movement. The structure of the eye is shown in **Figure 11.** Light enters the eye through the pupil. Light then passes through the lens, a thick, transparent disk that focuses light on the retina. The **retina** is a lining on the back inner surface of the eye that consists of photoreceptors and neurons. The retina contains two types of photoreceptors—rods and cones—which convert light energy to electrical signals that can be interpreted by the brain. **Rods** respond best to dim light. **Cones** respond best to bright light and enable color vision. The retina also contains many other neurons that process visual information. The axons of some of these neurons make up the **optic nerve.** The optic nerve exits through the back of the eye and runs along the base of the brain to the thalamus. The thalamus then relays visual information to the occipital lobe of the cerebral cortex, where the information is processed.

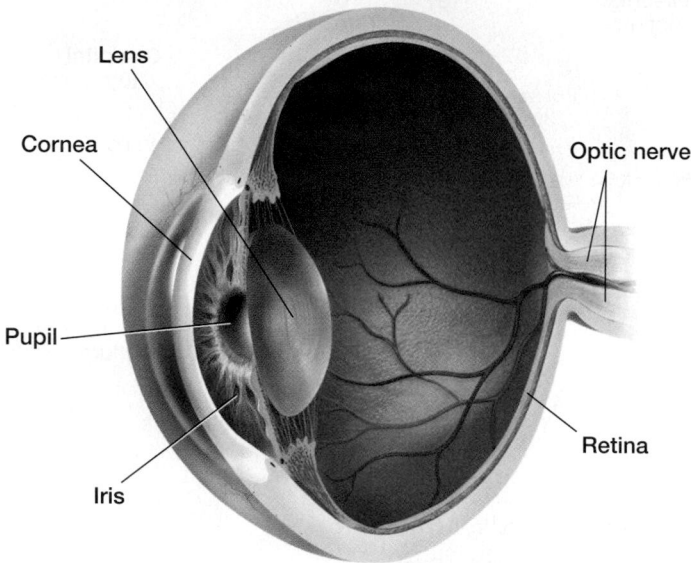

Figure 11 Structure of eye. Light enters the eye through the pupil and is focused on the retina, which contains photoreceptors.

Lens
Cornea
Optic nerve
Pupil
Retina
Iris

QUICK LAB

Demonstrating Your Blind Spot

The blind spot in your visual field corresponds to the site where the optic nerve exits the back of the eye. There are no photoreceptors at the site where the optic nerve exits. Use the procedure below to demonstrate your blind spot.

Materials

unlined 3 × 5 index card, pencil

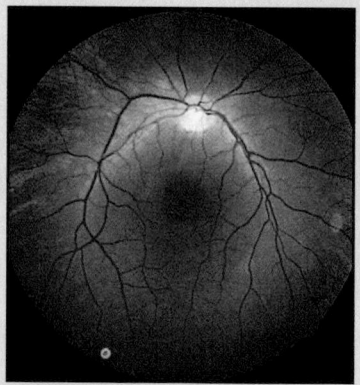

Retina

Procedure

1. On the index card, draw an *X* about 1 in. from the left side of the card. Draw an *O* about the same size 3 in. to the right of the *X*.

2. Hold your index card in front of you at arm's length. Close your right eye and stare at the *O* with your left eye. Slowly move the card toward you while continuing to stare at the *O* until the *X* disappears from view.

Analysis

1. Name the two kinds of photoreceptors found in the retina.

2. Propose why you cannot see images that fall on the site where the optic nerve exits the eye.

3. Critical Thinking Relating Concepts What is the relationship between the structure of the retina and the disappearance of the *X* on the index card?

Ears

How do your ears enable you to hear? Your ears convert the energy in sound waves to electrical signals that can be interpreted by your brain. **Figure 12** shows the structure of the ear. Sound waves enter the ear through the ear canal and strike the tympanic *(tim PAN ik)* membrane, or eardrum, causing it to vibrate. Behind the eardrum, three small bones of the middle ear—the hammer, anvil, and stirrup—transfer the vibrations to a fluid-filled chamber within the inner ear. This chamber, called the **cochlea** *(KAHK lee uh)*, is coiled like a snail's shell, and it contains mechanoreceptors called hair cells. Hair cells rest on a membrane that vibrates when waves enter the cochlea. Waves of different frequencies cause different parts of the membrane to vibrate and thus stimulate different hair cells. When hair cells are stimulated, they generate nerve impulses in the auditory nerve. The impulses travel to the brain stem through the auditory nerve. The thalamus then relays the information to the temporal lobe of the cerebral cortex, where the auditory information is processed.

Keeping Your Balance

The ears not only enable you to hear but also help you maintain equilibrium. The **semicircular canals** are fluid-filled chambers in the inner ear that contain hair cells. Clusters of these hair cells respond to changes in head position with respect to gravity. When your head moves, the hair cells are stimulated according to the magnitude and direction of the fluid's movement, and they send electrical signals to the brain. Signals generated by the hair cells enable the brain to determine the orientation and position of the head.

WORD *Origins*

● The word *cochlea* is from the Greek *kochlias*, meaning "snail shell."

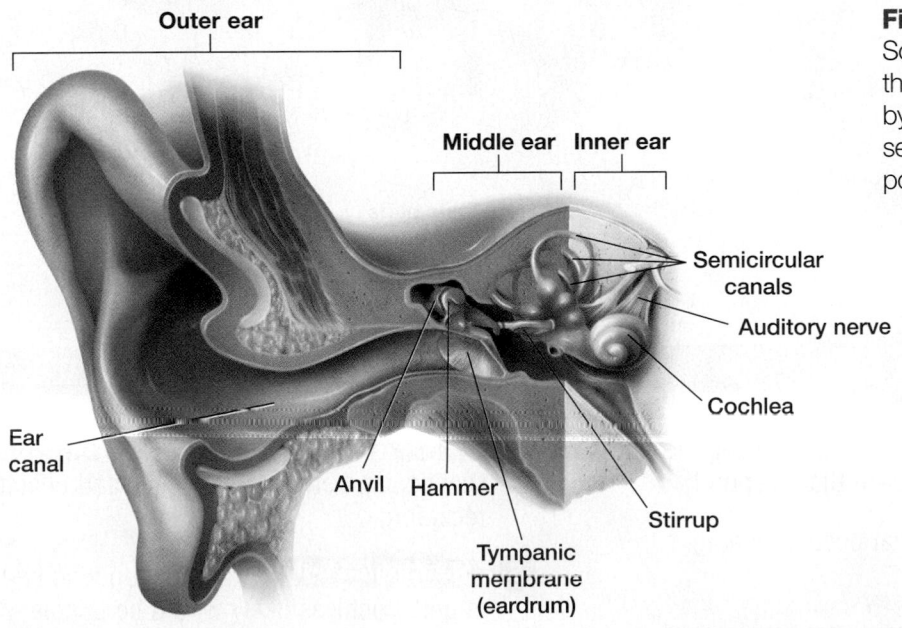

Figure 12 Structure of ear. Sound waves are transmitted to the inner ear and are detected by mechanoreceptors. The semicircular canals detect the position of the head.

Chemical Senses

Embedded within the surface of the tongue are 2,000–5,000 taste buds. Most taste buds are located within small projections on the surface of the tongue. A taste bud, shown in **Figure 13,** is a cluster of 50–100 taste cells. Taste cells are chemoreceptors that detect at least four basic chemical substances: sugars (sweet), acids (sour), alkaloids (bitter), and salts (salty). Each taste cell is generally sensitive to all tastes but is most sensitive to only one of them. A taste bud is stimulated when food molecules dissolved in saliva bind to taste cells. Taste cells generate electrical signals that can be interpreted by the brain.

Chemoreceptors that detect odors, called olfactory *(ahl FAK tuh ree)* receptors, are located in the roof of the nasal passage. Chemicals in the air stimulate olfactory receptors, which generate electrical signals that can be interpreted by the brain. Your sense of smell affects your enjoyment of food. When you have a bad cold and your nose is stuffed up, your food may seem to have little taste.

Figure 13   Location and structure of taste buds

A taste bud is a cluster of taste cells surrounding a taste pore.

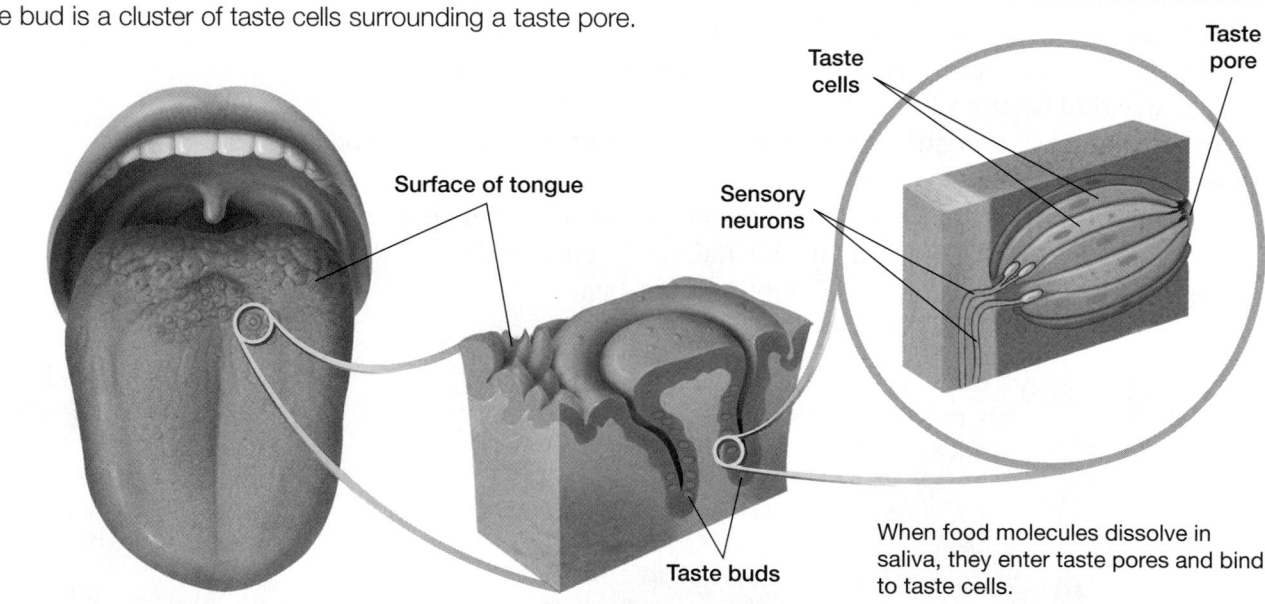

Surface of tongue

Taste cells

Sensory neurons

Taste pore

Taste buds

When food molecules dissolve in saliva, they enter taste pores and bind to taste cells.

Section 3 Review

1 **List** two different types of sensory receptors and the kinds of stimuli to which they respond.

2 **Sequence** the events that occur when light enters the eye.

3 **Describe** how sound waves are transmitted through the ear.

4 **Critical Thinking Comparing Structures** Distinguish between taste cells and olfactory receptors.

5 **Standardized Test Prep** A person who had defects in both cochleas likely would be unable to detect

A odors.

B colors.

C sounds.

D tastes.

Drugs and the Nervous System

Psychoactive Drugs

Many different kinds of drugs are available to the public. Advertisements tell you about pain relievers, antacids, cough syrups, and other medications that can help you feel better. Drugs can prevent, treat, or cure many different illnesses. However, drugs, whether legal or illegal, can also be misused or abused.

In the broadest sense, a drug is a chemical that alters body structures or biological functions. Drugs that alter the functioning of the central nervous system are known as **psychoactive drugs.** Many medications, such as those prescribed by doctors to treat mental disorders, contain psychoactive drugs. Caffeine, found in coffee and soft drinks, is also a psychoactive drug. Alcohol, marijuana, and cocaine are examples of commonly abused psychoactive drugs. Psychoactive drugs also include many other substances, such as inhalants. Many psychoactive drugs produce physiological dependence and addiction. Abuse of psychoactive drugs can damage the body, and in some cases, can result in death. **Table 3** lists several classes of commonly abused psychoactive drugs.

Objectives

- **Identify** types of psychoactive drugs, and describe their effects.
- **Describe** how drug addiction develops.
- **Describe** effects of commonly abused drugs on the nervous system.

Key Terms

psychoactive drug
addiction
tolerance
withdrawal
stimulant
depressant

Table 3 Psychoactive Drugs of Abuse

Drug	Examples	Psychoactive effects	Risks associated with use
Depressants	Barbiturates (sedatives), tranquilizers, alcohol	Decreased activity of the central nervous system	Drowsiness, depression, brain or nerve damage, coma, respiratory failure
Stimulants	Cocaine, crack, nicotine, amphetamines	Increased activity of the central nervous system	Aggressive behavior, paranoia, cardiac arrest, high blood pressure, brain damage
Inhalants	Nitrous oxide, ether, paint thinner, glue, cleaning fluid, aerosols	Disorientation, confusion, memory loss	Brain damage, kidney and liver damage, respiratory failure
Hallucinogens	LSD, PCP, MDMA (ecstasy), peyote (mescaline), psilocybe mushroom	Sensory distortion, anxiety, hallucinations, numbness	Depression, paranoia, aggressive behavior
THC	Marijuana, hashish	Short-term memory loss, impaired judgment	Lung damage, loss of motivation
Narcotics	Heroin, morphine, codeine, opium	Feeling of well-being, sedation, impaired sensory perception, impaired reflexes	Coma, respiratory failure

Drug Addiction and Neuron Function

Figure 14 Coca plant. Cocaine is derived from the coca plant, *Erythroxylon coca.*

Addiction is a physiological response caused by use of a drug that alters the normal functioning of neurons and synapses. Once a neuron or synapse has been altered by a drug, it cannot function normally unless the drug is present. With repeated exposure to a drug, a person addicted to the drug develops tolerance to the drug. **Tolerance** is a characteristic of drug addiction in which increasing amounts of the drug are needed to achieve the desired sensation. **Withdrawal** is a set of emotional and physical symptoms caused by removal of the drug from the body. The severity of drug addiction is evident in recovering addicts who experience withdrawal when they stop taking an addictive drug. Symptoms of withdrawal may include vomiting, headache, depression, and seizures. Withdrawal from barbituates, and withdrawal in cases of severe alcohol addiction, can cause death and should be supervised by a doctor.

A Model of Drug Addiction

Cocaine is a highly addictive stimulant found in the leaves of the coca plant, *Erythroxylon coca,* shown in **Figure 14.** A **stimulant** is a drug that generally increases the activity of the central nervous system. Despite being illegal, cocaine is still used by many people.

Recall that in synaptic transmission, neurotransmitter molecules are released from a presynaptic neuron and bind to receptor proteins on a postsynaptic cell. Some neurotransmitter molecules are

Figure 15

Action of Cocaine

Cocaine alters the function of dopamine-producing neurons in the limbic system.

1 **Normal synapse**
Dopamine is reabsorbed by the presynaptic neuron.

2 **Synapse with cocaine**
Cocaine blocks the reabsorption of dopamine.

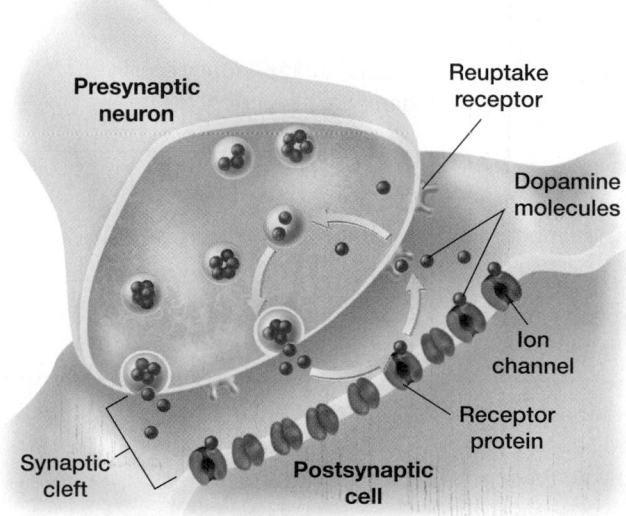

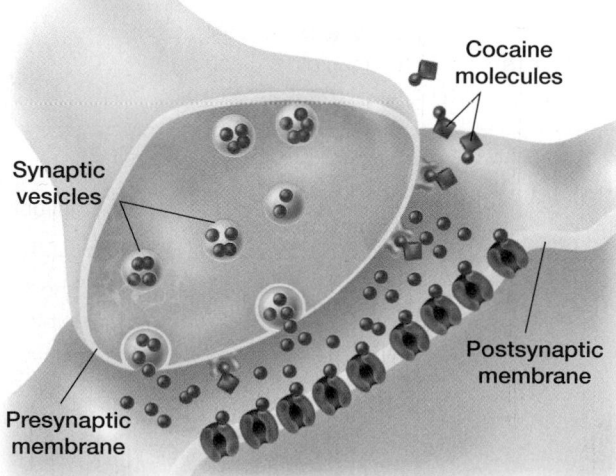

reabsorbed by presynaptic neurons after they have been released into the synaptic cleft. Cocaine is an example of a drug that interferes with a presynaptic neuron's ability to reabsorb, or reuptake, neurotransmitter molecules. Cocaine affects dopamine *(DOH pah meen)* neurons in the limbic system, which plays an important role in the sensation of pleasure. The mechanism of cocaine action is summarized in **Figure 15.**

Step ❶ At a normal synapse, reuptake receptors move molecules of dopamine in the synaptic cleft back into the presynaptic neuron.

Step ❷ Cocaine blocks the reuptake of dopamine molecules by interfering with these reuptake receptors.

Step ❸ As a result, excess dopamine remains in the synaptic cleft, overstimulating the postsynaptic cell. Overstimulation produces an intense feeling of exhilaration and well-being. Because the post synaptic cell has been overstimulated, the number of dopamine receptors will decrease over time.

Step ❹ If cocaine is removed from the synaptic cleft, the number of dopamine molecules returns to normal. This level is now too low to adequately stimulate the postsynaptic cell because it has fewer receptor proteins. Addiction occurs because more cocaine must be taken to maintain adequate stimulation of the postsynaptic cell.

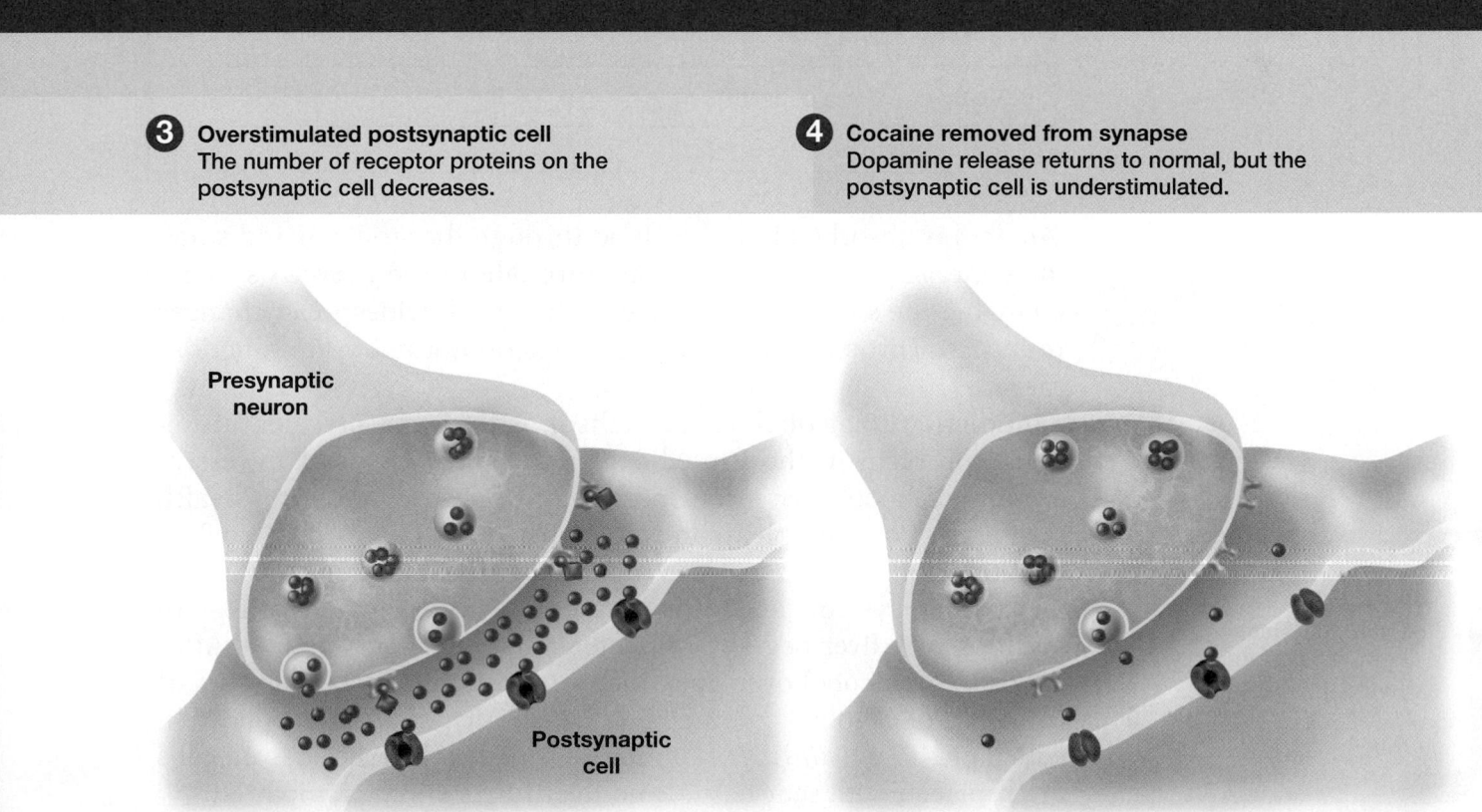

❸ **Overstimulated postsynaptic cell**
The number of receptor proteins on the postsynaptic cell decreases.

❹ **Cocaine removed from synapse**
Dopamine release returns to normal, but the postsynaptic cell is understimulated.

Presynaptic neuron

Postsynaptic cell

Figure 16 Breath test.
Law enforcement officials use
a device that detects the level
of alcohol vapors in the breath
to estimate the BAC of drunk-
driving suspects.

Alcohol

Of all the psychoactive drugs, alcohol (ethanol) is one of the most widely used and abused. Alcohol, found in wine, beer, and liquor, is a depressant that produces a sense of well-being when taken in small amounts. A **depressant** is a drug that generally decreases the activity of the central nervous system. As more alcohol is consumed, reaction time increases, and coordination, judgment, and speech become impaired. This produces a state of intoxication known as being "drunk." Drunkenness results as the blood-alcohol concentration (BAC) increases. BAC can be measured by a breath test, illustrated in **Figure 16,** that detects the level of alcohol vapors in the breath. **Table 4** shows the effects of alcohol at various concentrations in the blood.

Table 4 Effects of Blood Alcohol Concentration	
BAC*	**Condition**
0.02–0.04	Slight impairment and sedation
0.05–0.06	Slight impairment of coordination; increased reaction time
0.07–0.09	Slurred speech; blurred vision; intoxication
0.10–0.15	Severe intoxication; impaired coordination, vision, and balance
0.15–0.30	Dizziness; confusion; inability to walk; extremely severe intoxication
0.30–0.50	Unconsciousness
0.50–0.60	Coma or death

*in mg of alcohol per mL of blood

Alcohol is absorbed into the blood through the stomach and small intestine. Alcohol affects neurons throughout the nervous system, changing the shape of receptor proteins. Such widespread changes in receptor proteins have various effects on normal brain functioning.

Addiction to alcohol, or alcoholism, is the most prevalent drug-abuse problem in the United States. People who drink excessive amounts of alcohol over long periods of time develop serious health problems. For example, many alcoholics do not eat properly when drinking heavily. This can lead to malnutrition, abnormalities in the circulatory system, and inflammation of the stomach lining. In addition, the liver begins to use alcohol as an energy source. After exposure to alcohol over time, the liver accumulates fat deposits. If drinking of alcohol continues, a potentially fatal liver condition called cirrhosis *(sih ROH sis)* may develop. In a cirrhotic liver, cells are replaced with scar tissue, and liver functioning is impaired.

internet connect
www.scilinks.org
Topic: Blood Alcohol Concentration
Keyword: HX4026
SCiLINKS. Maintained by the National Science Teachers Association

Nicotine

About 50 million Americans smoke cigarettes despite convincing evidence that smoking causes mouth cancer, heart disease, lung cancer, and emphysema. So why do people continue to smoke? Many smokers say they would like to stop smoking but find the habit too difficult to overcome. They are addicted to nicotine, a drug in cigarette smoke.

Figure 17 Tobacco plant.
Tobacco leaves are dried and crushed and are then smoked in cigarettes, cigars, and pipes. Tobacco is also chewed and snuffed.

Effects of Nicotine

Nicotine is the highly addictive stimulant found in the leaves of the tobacco plant, *Nicotiana tabacum*, shown in **Figure 17.** Nicotine is extremely toxic; a dose of only 60 mg is lethal in humans. Tobacco leaves are dried or crushed and are then smoked in cigarettes, cigars, and pipes. Tobacco is also chewed and snuffed.

Nicotine quickly enters the bloodstream and circulates through the body. In the brain, nicotine mimics the action of the neurotransmitter acetylcholine. Scientists have extensively studied the behavior of the brain when exposed to nicotine. Nicotine binds to brain cells at specific sites usually reserved for acetylcholine. These sites are the central controls of the brain—mechanisms the brain uses to adjust levels of many of its activities. Like twisting the dial on a central control, the binding of nicotine to these sites produces many changes. After a while, the smoker's body makes adjustments, and systems almost return to normal—as long as the smoker keeps smoking. Take away the nicotine, however, and all those adjustments throw everything out of balance all at once. The only way to keep things "normal" is to keep smoking. The smoker is addicted.

Effects of Tobacco

Smokers get more than nicotine from cigarette smoke. Inhaled smoke contains hundreds of toxic and mutagenic chemicals that pass through the mouth, air passages, and lungs. These chemicals, also called tars, are produced by burning tobacco. Because tars and other chemicals in tobacco smoke are powerful mutagens, smoking causes lung cancer. Almost all cases of lung cancer, a major cause of death in the United States, are attributed to smoking.

In the United States, smoking-related illnesses cause more than 400,000 deaths each year. Smoking is associated with cancer of the mouth and larynx, and smoking may increase the risk of cancer of the pancreas and bladder. Smoking is also a major contributor to often-fatal respiratory disorders, such as emphysema. The tars in smoke irritate mucous membranes in the mouth, nose, and throat. They accumulate in the lungs and paralyze cilia that move debris from the lungs. Tars also blacken lung tissue and decrease breathing capacity. People who are exposed to secondhand smoke are at risk for the same diseases as people who smoke. Women who smoke during pregnancy are more likely to have miscarriages or to give birth to stillborn babies.

Real Life

Is smokeless tobacco harmful?
The use of smokeless tobacco, such as chewing tobacco, causes cancers of the lips, mouth, and gums. When chewing tobacco is placed between the cheek and gum, nicotine and other chemicals are absorbed into the bloodstream.
Finding Information
Find out about mouth cancers caused by tobacco.

Drugs of Abuse

Narcotics are extremely addictive psychoactive drugs that relieve pain and induce sleep. Some of the most potent narcotics are derived from the poppy plant, *Papaver somniferum*, shown in **Figure 18.** The sap that oozes from the cut seed pod forms a thick, gummy substance called opium. Drugs derived from opium, called opiates or narcotics, include codeine *(KOH deen)*, morphine, and heroin, a more potent form of morphine. Codeine is widely prescribed by doctors for pain relief. Morphine is one of the most effective pain-relieving drugs used today. Heroin addiction and abuse are among the most serious illegal-drug problems in society.

Recall that pain receptors throughout the body detect painful stimuli. As uncomfortable as it may feel, pain plays a very important role in the body. Pain notifies you that body tissues have been injured or damaged. Imagine how your body would look and function today if you did not have the ability to sense pain. Pain begins as a signal at damaged nerve endings. Nerve impulses generated by pain receptors travel to the spinal cord toward the brain. After reaching the spinal cord, a pain signal is suppressed by a class of neurotransmitters called enkephalins *(ihn KEHF uh lihnz)*. When enkephalins bind to neurons in the spinal cord, they prevent pain signals from reaching the brain.

Narcotics mimic the action of enkephalins by binding to the same receptor proteins in the spinal cord. These receptor proteins are called opiate receptors because scientists observed opiates binding to them before enkephalins were ever discovered. Narcotics also affect the limbic system, producing a feeling of well-being.

Marijuana

In addition to alcohol and tobacco, marijuana, though illegal, is a widely consumed drug. Marijuana comes from various species of the hemp plant, *Cannabis,* shown in **Figure 19.** Hashish also comes from the hemp plant. The active ingredient in marijuana and hashish is commonly known as THC. When marijuana is smoked, it may cause disorientation, impaired judgment, short-term memory loss, and general loss of motivation. Scientists continue to research the effects of THC on the nervous system.

Figure 18 Opium poppy.
Opium is a narcotic derived from the poppy plant, *Papaver somniferum*.

Figure 19 Hemp. Marijuana is produced from the hemp plant, *Cannabis.*

Section 4 Review

1 **Describe** how tolerance to a drug develops.

2 **Summarize** how cocaine produces addiction.

3 **Distinguish** between stimulants and depressants. Give an example of each.

4 **Critical Thinking Recognizing Relationships** What do all psychoactive drugs have in common?

5 **Critical Thinking Applying Information** Why is drug addiction considered a physiological condition?

6 **Standardized Test Prep** Cocaine interferes with the normal functions of the limbic system by blocking
  A reuptake of dopamine.
  B release of enkephalins.
  C sensory perception.
  D synaptic transmission.

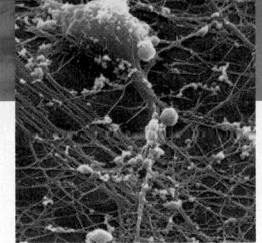

Key Concepts

1 Neurons and Nerve Impulses

- Neurons are specialized cells that rapidly transmit information as electrical signals throughout the body.
- At the resting potential, the inside of a neuron is negatively charged with respect to the outside of the neuron.
- An action potential moves rapidly down an axon.
- Synaptic transmission involves the release of neurotransmitters at synapses.

2 Structures of the Nervous System

- The central nervous system consists of the brain and spinal cord.
- The brain contains three major parts: the cerebrum, the cerebellum, and the brain stem.
- The spinal cord links the brain to the peripheral nervous system, which branches throughout the body.

3 Sensory Systems

- Sensory receptors detect various sensory stimuli.
- Photoreceptors in the eyes convert light into electrical signals that are interpreted by the brain.
- The ear converts sound into electrical signals that are interpreted by the brain.
- The semicircular canals monitor the position of the head.
- Taste and smell are related chemical senses.

4 Drugs and the Nervous System

- Psychoactive drugs affect the central nervous system.
- Drug addiction involves physiological changes in neurons.
- Alcohol is an addictive depressant that widely affects the central nervous system.
- Nicotine is an addictive stimulant found in tobacco products.

Key Terms

Section 1

neuron (944)
dendrite (944)
axon (944)
nerve (944)
membrane potential (945)
resting potential (946)
action potential (946)
synapse (948)
neurotransmitter (948)

Section 2

central nervous system (950)
peripheral nervous system (950)
sensory neuron (950)
motor neuron (950)
brain (950)
cerebrum (951)
cerebellum (951)
brain stem (951)
thalamus (952)
hypothalamus (952)
spinal cord (952)
reflex (952)
interneuron (953)

Section 3

sensory receptor (956)
retina (958)
rod (958)
cone (958)
optic nerve (958)
cochlea (959)
semicircular canal (959)

Section 4

psychoactive drug (961)
addiction (962)
tolerance (962)
withdrawal (962)
stimulant (962)
depressant (964)

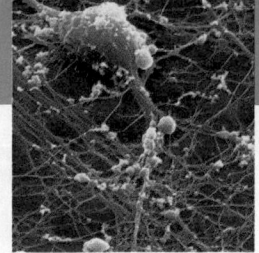

Understanding Key Ideas

1. A myelin sheath on the axon of a neuron
   a. covers the axon completely.
   b. decreases the rate of impulse conduction.
   c. increases the rate of impulse conduction.
   d. has no effect on impulse conduction.

2. When a neuron is at the resting potential,
   a. the inside is positively charged.
   b. the outside is negatively charged.
   c. the inside is negatively charged.
   d. None of the above

3. During an action potential,
   a. sodium ions flow into a neuron.
   b. sodium ions flow out of a neuron.
   c. potassium ions flow into a neuron.
   d. there is no movement of ions.

4. In a spinal reflex, the signal travels
   a. immediately to the brain.
   b. to the spinal cord and out to a muscle.
   c. only through sensory neurons.
   d. only through motor neurons.

5. Drug addiction is considered a physiological condition because addictive drugs
   a. can be purchased illegally.
   b. must be injected.
   c. alter the functioning of neurons.
   d. are used in social settings.

6. The diagram below shows the brain of a fish. How is the cerebrum of the fish brain different from that of a human brain? What do the large olfactory bulbs of the fish brain indicate about the relative importance of the sense of smell to the fish?

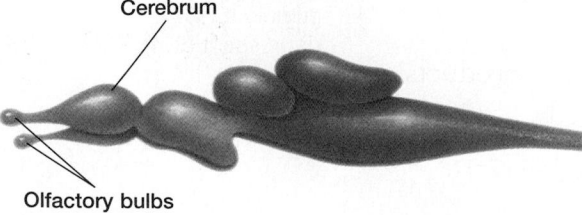

Cerebrum

Olfactory bulbs

7. Action potentials travel in only one direction along a neuron—toward the axon terminals and away from the cell body. What structures of the neuron ensure that this pattern is always followed?

8. List three ways that the binding of a neurotransmitter to a receptor protein on a postsynaptic cell could cause changes in the cell. (Hint: See Chapter 4, Section 2.)

9. **BIOWatch** Although neurons in the spinal cord do not grow and regenerate when they have been injured, scientists have developed ways to help prevent paralysis. Describe two of these ways.

10. **Concept Mapping** Make a concept map that describes the structures and functions of the nervous system. Try to include the following terms: *spinal cord, brain, neuron, nerve, synapse,* and *neurotransmitter*.

Critical Thinking

11. **Inferring Relationships** People who suffer from vertigo feel dizzy and disoriented in certain situations. What is the relationship between vertigo and the semicircular canals?

12. **Recognizing Relationships** Suggest a possible defect of the retina or of retinal cells that would cause colorblindness, a condition in which a person cannot distinguish between certain colors.

Alternative Assessment

13. **Summarizing Information** Research the causes and symptoms of various disorders of the nervous system caused by the degeneration of neurons. Some disorders include Alzheimer's disease, Parkinson's disease, and multiple sclerosis. Find out how these disorders affect the nervous system. List the types of drugs or other methods used to treat these disorders.

Standardized Test Prep

Understanding Concepts

Directions (1–3): For *each* question, write on a separate sheet of paper the letter of the correct answer.

1 Which of the following is a sensory receptor that is stimulated by light?
  A. cochlea
  B. cone
  C. interneuron
  D. optic nerve

2 If the sodium–potassium pump of a neuron failed, what effect would this likely have on the neuron's function?
  F. The concentrations of positive and negative ions would cause the neuron to be negative inside.
  G. Voltage-gated potassium channels and voltage-gated sodium channels would no longer function.
  H. The neuron could not conduct another action potential until the resting potential was fully restored.
  I. The concentration of sodium ions would be higher outside the cell and that of potassium ions would be higher inside.

3 Which part of the ear contains mechanoreceptors?
  A. cochlea
  B. eardrum
  C. hammer
  D. stirrup

Directions (4): For the following question, write a short response.

4 During an epileptic seizure, many neurons in the brain produce large bursts of action potentials, causing the body to become rigid and to jerk or convulse. From what you know about the brain's control of muscles and posture, how might you explain these symptoms?

Test TIP

When using a diagram to answer a question, look in the image for evidence that supports your potential answer.

Reading Skills

Directions (5): **Read the passage below. Then answer the question.**

Depression affects several million Americans. Symptoms of depression include withdrawal, anger, poor communication, sadness, and indifference to surroundings. Depression may be triggered by the loss of a friend or relative, a major disappointment at work, prescription drugs, prolonged illness, alcohol or drug withdrawal, or hormones. Treatments include counseling, several types of drugs, and exercise.

5 Which structure is likely a target for drugs that treat depression?
  F. cerebellum
  G. motor neuron
  H. neurotransmitter
  I. spinal cord

Interpreting Graphics

Directions (6): **Base your answer to question 6 on the diagram below.**

Neuron

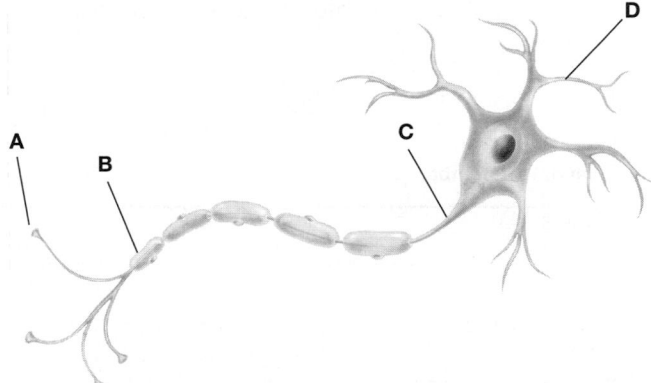

6 Which structure increases the speed at which the axon conducts action potentials?
  A. *A*
  B. *B*
  C. *C*
  D. *D*

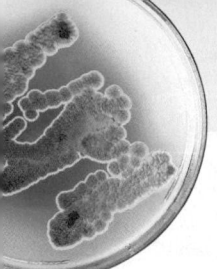

Exploration Lab

Calculating Reaction Times

SKILLS
- Measuring
- Calculating

OBJECTIVES
- **Determine** human reaction times.
- **Design** an experiment that measures changes in reaction times.

MATERIALS
- meterstick

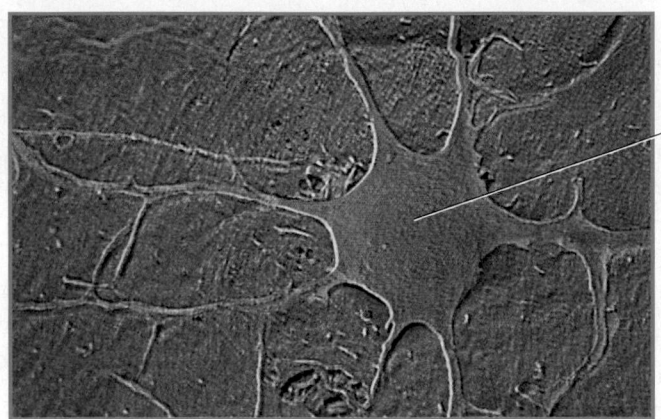

Cell body

Neuron

Before You Begin

When you want to move your hand, your brain must send a message all the way to the muscles in your arms. How long does that take? In this exercise, you will work with a partner to see how quickly you can react. In this lab, you will investigate reaction times and design an experiment to investigate influences on reaction times.

1. Create a Data Table to record reaction times.

2. Write a hypothesis about an influence on reaction times. (For example: "People who have eaten breakfast have faster reaction times than people who have not eaten breakfast.")

Procedure

PART A: Calculating Reaction Times

1. Sit in a chair and have a partner stand facing you while holding a meterstick in a vertical position.

DATA TABLE		
Hand: trial number	Subject 1 reaction time (s)	Subject 2 reaction time (s)
Left: 1		
Left: 2		
Left: 3		
Left: average		
Right: 1		
Right: 2		
Right: 3		
Right: average		

2. Hold your thumb about 3 cm from your fingers near the bottom end of the stick. The meterstick should be positioned to fall between your thumb and fingers.

3. Tell your partner to let go of the meterstick without warning.

4. When your partner releases the meter stick, catch the stick by pressing your thumb and fingers together. Your partner should be ready to catch the top of the meterstick if it begins to tip over.

5. Record the number of centimeters the stick dropped before you caught it. The distance that the meterstick falls before you catch it can be used to evaluate your reaction time.

6. Repeat the procedure several times, and calculate the average number of centimeters.

7. Try this procedure with your other hand.

8. Close your eyes and have your partner say "now," when the stick is released.

9. Exchange places with your partner, and repeat the procedure.

PART B: Designing Your Own Experiment

10. Work with the members of your lab group to explore one of the hypotheses written in the **Before You Begin** section of this lab.

You Choose

As you design your experiment, decide the following:

a. what hypothesis you will explore

c. how you will test the hypothesis

d. what the controls will be

e. how many trials to perform

f. what data to record in your data table.

---

11. Write a procedure for your experiment. Make a list of all the safety precautions you will take. Have your teacher approve your procedure and safety precautions before you begin the experiment.

12. Set up your group's experiment and collect data.

Analyze and Conclude

1. **Summarizing Results** What was your fastest reaction time?

2. **Analyzing Data** How does your reaction time when using your dominant hand compare with your reaction time when using your other hand?

3. **Drawing Conclusions** Why may each hand have a different reaction time? Why may each person have a different reaction time? Compared to earlier trials, was the reaction time in step 8 faster, slower or the same? If the time was faster or slower hypothesize a reason for the difference.

4. **Predicting Patterns** Compile the data gathered by each pair in your class. Can you identify any trends in the data? (For example, do males and females have the same average reaction times?)

5. **Further Inquiry** Write a new question about reaction times that could be explored in another investigation.

? Do You Know?

Do research in the library or media center to answer these questions:

1. How can athletes improve their reaction times?

2. What factors can influence how fast a person reacts to a stimulus?

Use the following Internet resources to explore your own questions about neurons.

internet connect

www.scilinks.org
Topic: Neurons
Keyword: HX4129

SCiLINKS. Maintained by the National Science Teachers Association

Gail Devers

✓ Quick Review

Answer the following without referring to earlier sections of your book.

1. **Describe** the function and structure of glycogen. *(Chapter 2, Section 3)*

2. **Describe** the action of enzymes. *(Chapter 2, Section 4)*

3. **Summarize** the location and function of receptors. *(Chapter 3, Section 2)*

4. **Summarize** the role of DNA and mRNA in protein synthesis. *(Chapter 10, Section 1)*

5. **Describe** the role of the sympathetic nervous system. *(Chapter 41, Section 2)*

Did you have difficulty? *For help, review the sections indicated.*

Reading Activity

Before you read this chapter, write a short list of all of the things you know about glands and hormones. Then write a list of the things that you want to know about glands and hormones. Save your lists, and to assess what you have learned, see how many of your own questions you can answer after reading this chapter.

• Severe weight loss, shaking fits, and loss of vision in one eye were symptoms experienced in 1988 by American sprinter Gail Devers. She was later diagnosed with Grave's disease, a condition in which excessive amounts of hormones are made by the thyroid gland. Devers went on to win the gold medal in the 100 m dash in two Summer Olympics.

internet connect

www.scilinks.org
National Science Teachers Association *sci*LINKS Internet resources are located throughout this chapter.

SCiLINKS. Maintained by the National Science Teachers Association

Hormones

Coordination of Activities

Reacting to fear, growing taller, and developing male or female characteristics are all activities in the body that are partially regulated by hormones *(HAWR mohnz)*. **Hormones** are substances secreted (released) by cells that act to regulate the activity of other cells in the body.

You—and the cyclist shown in **Figure 1**—make hormones that keep your body functioning properly. The functions of hormones include the following:

1. Regulating growth, development, behavior, and reproduction

2. Coordinating the production, use, and storage of energy

3. Maintaining homeostasis (temperature regulation, metabolism, excretion, and water and salt balance)

4. Responding to stimuli from outside the body

Hormones act as chemical messengers, carrying instructions that cause cells to change their activities. For example, some hormones cause the cells of the heart to increase the rate at which the heart is beating. In the past, it was thought that hormones, once secreted from a cell, had to be transported through the bloodstream to reach the cells they were to act on. Today, we know that some hormones act directly on adjacent cells without traveling through the blood.

The instructions a hormone carries are determined by both the hormone itself and the cell it affects. For example, a hormone may instruct a cell to make a specific protein or to activate a specific enzyme. The same hormone can also instruct a different cell to alter the permeability of its cell membrane or even to release another hormone. Hormones can instruct muscle cells to relax and nerve cells to fire action potentials.

Each hormone is very specific about which types of cells can receive its instructions. Each hormone acts like a key that opens a lock on or inside the cell. A hormone will act only on cells with the right lock. The locks, as discussed later in this chapter, are receptors on or inside the cell.

Figure 1 Hormones and homeostasis. Combining activities, such as water balance and temperature regulation, requires coordination. Such coordination is maintained by hormones.

Endocrine Glands and Tissues

A gland is an organ whose primary function is to secrete materials into other regions of the body. **Endocrine** *(EN doh krihn)* **glands** are ductless organs that secrete hormones directly into either the bloodstream or the fluid around cells (extracellular fluid). In addition to endocrine glands, several other organs contain cells that secrete hormones. These organs include the brain, stomach, small intestine, kidney, liver, and heart.

All of the endocrine glands and hormone-secreting tissues collectively make up the endocrine system, shown in **Figure 2.** The endocrine system coordinates all of the body's sources of hormones.

Some organs, such as the pancreas, are both endocrine and exocrine *(EHKS oh krihn)* glands. Exocrine glands deliver substances through ducts (tubelike structures). The ducts transport the substances to specific locations inside and outside the body. Sweat glands, mucous glands, salivary glands, and other digestive glands are examples of exocrine glands. The exocrine part of the pancreas produces digestive enzymes and delivers them to the small intestine through ducts. The endocrine part of the pancreas secretes two hormones into the bloodstream that regulate blood glucose levels.

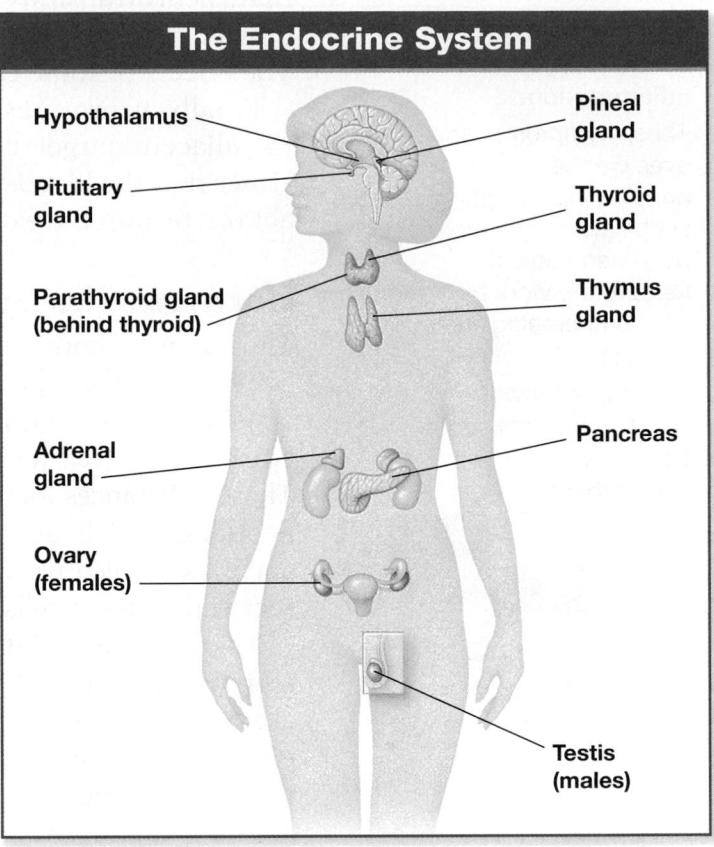

The Endocrine System

Hypothalamus

Pituitary gland

Parathyroid gland (behind thyroid)

Adrenal gland

Ovary (females)

Pineal gland

Thyroid gland

Thymus gland

Pancreas

Testis (males)

Figure 2 Coordinating the body's activities.
Endocrine glands are located throughout the human body. In addition to the organs shown above, many other organs secrete hormones.

Hormones and Neurotransmitters Are Chemical Messengers

As you learned in the previous chapter, the nervous system is also involved in coordinating the body's activities. The endocrine system and nervous system interact in this shared role. However, with some exceptions, each system acts through different chemical messengers and in different ways.

The chemical messengers of the nervous system are known as neurotransmitters, while the chemical messengers of the endocrine system are called hormones. Some nerve cells, however, are capable of secreting hormones, and several chemicals serve as both hormones in the endocrine system and neurotransmitters in the nervous system. For example, epinephrine is both a neurotransmitter and a hormone. When secreted from a nerve cell, epinephrine conveys messages to other neurons. When secreted from an endocrine cell in the adrenal gland, epinephrine acts as a "fight-or-flight" hormone.

Real Life

Prostaglandins can cause pain, fever, and inflammation.

These symptoms, however, can be treated. Many nonprescription drugs, such as aspirin, ibuprofen, naproxen sodium, and ketoprofen, work by inhibiting prostaglandin production.

Finding Information
What symptoms can occur with long-term use of these drugs?

Another difference between the endocrine and nervous systems is that neurotransmitters are fast-acting and usually short-lived messengers. Hormones are usually slower-acting and longer-lived. The effects of some can last for days, weeks, or even years.

Finally, neurotransmitters are released from nerve cells directly to adjacent target cells. Endocrine cells can release hormones either into the bloodstream, where they travel to the cell they are to act on, or into the extracellular fluid to act on nearby cells.

Hormonelike Substances

The human body has many substances that regulate cellular activities much as hormones do. These substances were not initially considered hormones because they are not secreted into the bloodstream. However, most scientists today classify them as hormones. These substances include a large number of chemicals called neuropeptides, which are secreted by the nervous system, as well as chemicals called prostaglandins *(prahs tuh GLAN dihnz)*, which are secreted by most cells.

There are several different groups of neuropeptides. Enkephalins *(ehn KEHF uh lihnz)*, which were discussed in the previous chapter, are a group of neuropeptides that inhibit pain messages traveling toward the brain. Endorphins *(ehn DOHR fihnz)*, which are thought to regulate emotions, influence pain, and affect reproduction, are another important group of neuropeptides. Unlike neurotransmitters, enkephalins and endorphins tend to affect many cells near the nerve cells that release them.

Prostaglandins are modified fatty acids that have a variety of functions. They tend to accumulate in areas where tissues are disturbed or injured. There are dozens of different prostaglandins, and they produce a variety of effects. For example, some prostaglandins cause the constriction of blood vessels. The constricted blood vessels in turn affect blood pressure and body temperature. Other prostaglandins cause blood vessels to dilate, producing inflammation. A headache may result when blood vessels swell and their walls press against nerves in the brain.

Section 1 Review

1 **Describe** four ways in which hormones coordinate the activities of the body.

2 **Name** the type of gland that secretes substances into the bloodstream or extracellular fluid.

3 **Differentiate** the actions and chemical messengers in the endocrine system from those in the nervous system.

4 **Recognizing Relationships** Compare endorphins to neurotransmitters.

5 **Standardized Test Prep** Antidiuretic hormone is secreted when the body becomes dehydrated. How might this hormone help maintain water balance?

A by stimulating an appetite for salt

B by causing profuse sweating

C by promoting water reuptake from urine

D by causing the release of dilute urine

How Hormones Work

Target Cells

After hormones are released from the cell in which they are made, they bind to and act only on target cells. A **target cell** is a specific cell that a hormone binds to and acts on (carries the message to). Imagine what would happen if hormones were not specific. All the cells in the body would respond to the hormone, resulting in uncoordinated activities, such as activation of many enzymes.

A hormone recognizes a target cell because the target cell has specific receptors. The receptors are located either on the surface of the target cell (on the cell membrane) or inside the cell (in the cytoplasm or nucleus). Recall that a receptor is a protein to which a molecule binds. A hormone's shape matches that of a particular receptor protein much like a key fits into a lock, as shown in **Figure 3.** Thus, a hormone binds only to cells that have a particular receptor protein, ignoring all other cells.

Types of Hormones

Most hormones are classified as either amino-acid-based or steroid hormones. **Amino-acid-based hormones** are hormones made of amino acids (either a single modified amino acid or a protein composed of 3 to 200 amino acids). Most amino-acid-based hormones are water soluble. **Steroid hormones** are lipid hormones that the body makes from cholesterol. Steroid hormones are fat soluble.

Objectives

- **Relate** how hormones act only on specific cells.
- **Summarize** how amino-acid-based hormones function.
- **Summarize** how steroid and thyroid hormones function.
- **Relate** how negative feedback is used to regulate hormone levels.

Key Terms

target cell
amino-acid-based
  hormone
steroid hormone
second messenger
negative feedback

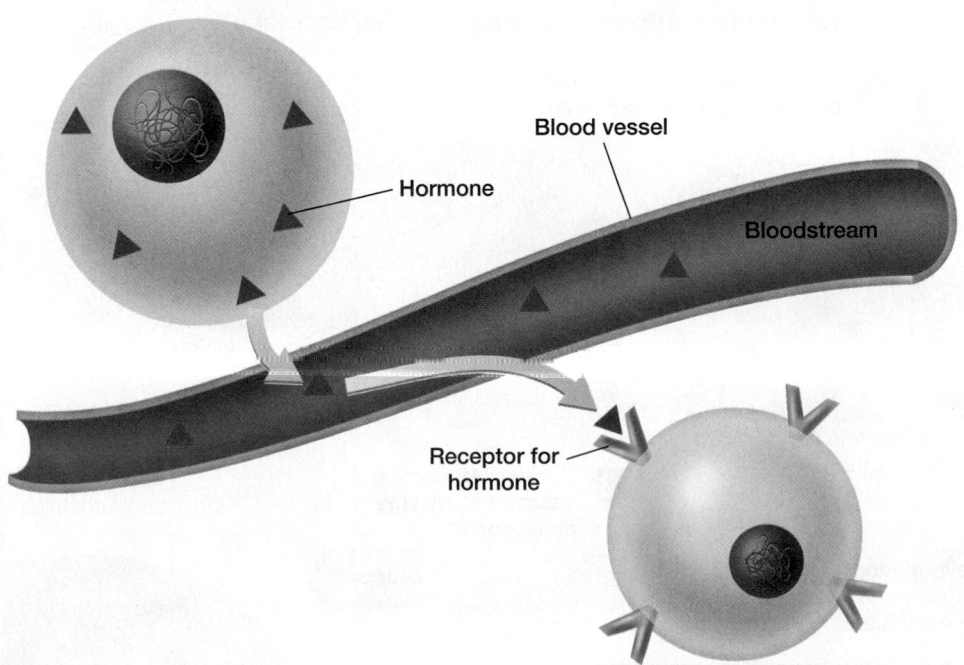

Figure 3   Hormones act on target cells. Hormones travel in the blood or in the fluid around cells to reach their target cells. The binding of the hormone with its receptor signals the target cell to change its activity.

Receptors

When a hormone binds to a specific receptor on a target cell, the hormone brings the target cell a message. What happens after the hormone binds, however, depends on the type of hormone.

Amino-Acid-Based Hormones

Amino-acid-based hormones are not fat soluble, and most bind to cell membrane receptors, as shown in **Figure 4.**

Step ❶ When an amino-acid-based hormone binds to a receptor, the shape of the receptor changes.

Step ❷ This change in shape eventually activates a **second messenger,** a molecule that passes the message from the first messenger (the hormone) to the cell. For example, when glucagon, a hormone secreted by the pancreas, binds to a receptor, an enzyme is activated that converts ATP to a second messenger called cyclic AMP (cAMP).

Step ❸ The second messenger then activates or deactivates certain enzymes in a cascade fashion. That is, one enzyme activates another enzyme, which activates yet another, and so on. In the case of glucagon, the second messenger cAMP activates a series of enzymes that breaks down glycogen into glucose.

Step ❹ Eventually the activity of the target cell is changed by the final enzyme in the cascade—even though the hormone never enters the cell!

Figure 4

BIO graphic How Amino-Acid-Based Hormones Work

Most amino-acid-based hormones, such as glucagon, bind to cell-membrane receptors, which activate second messengers that relay the hormone's message.

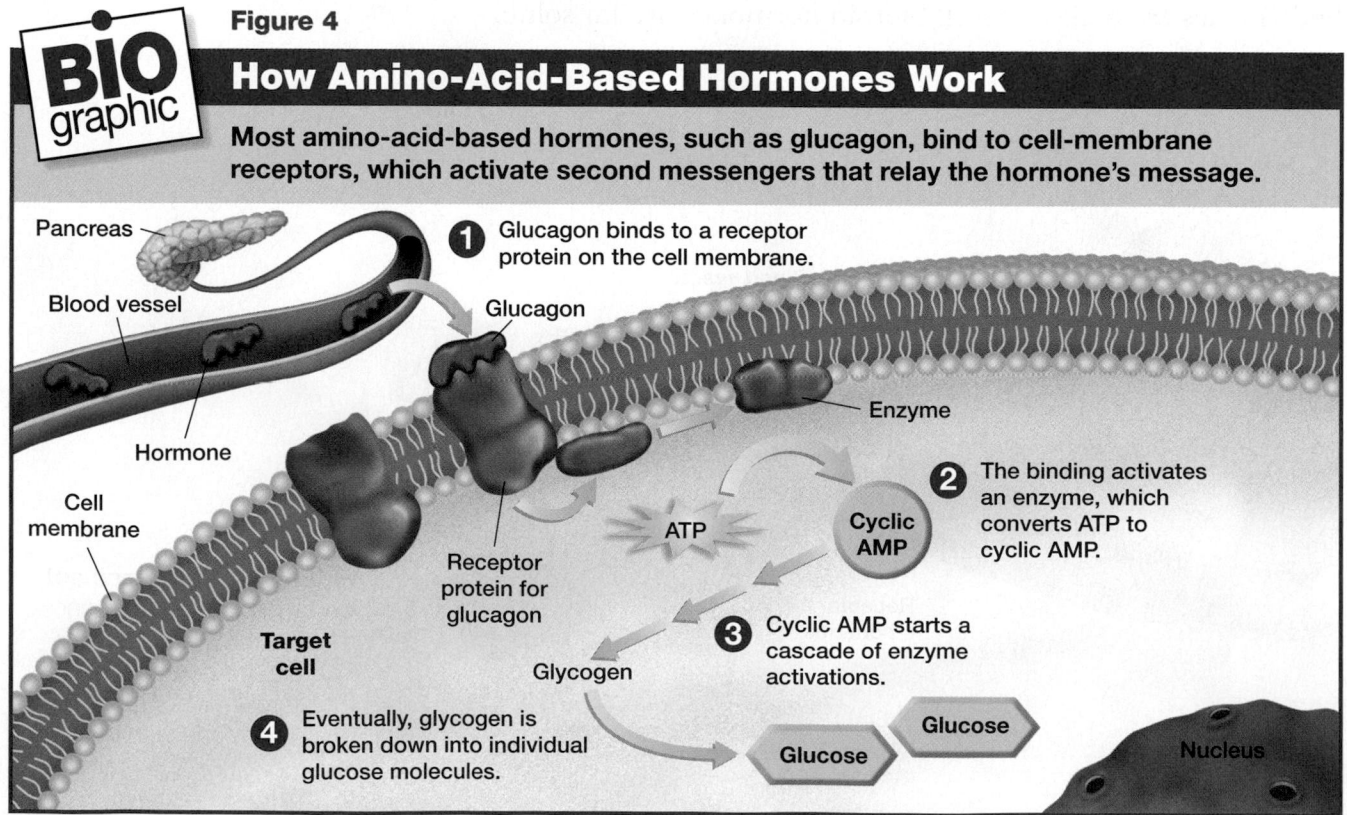

Pancreas

Blood vessel

Hormone

Cell membrane

Target cell

Receptor protein for glucagon

Glucagon

❶ Glucagon binds to a receptor protein on the cell membrane.

Enzyme

❷ The binding activates an enzyme, which converts ATP to cyclic AMP.

ATP

Cyclic AMP

❸ Cyclic AMP starts a cascade of enzyme activations.

Glycogen

❹ Eventually, glycogen is broken down into individual glucose molecules.

Glucose

Glucose

Nucleus

Steroid and Thyroid Hormones

Because steroid and thyroid hormones are fat soluble, they readily pass through the cell membranes of their target cells. Steroid hormones bind to receptors located in a target cell's cytoplasm or its nucleus; thyroid hormones bind to receptors in a target cell's nucleus.

Cortisol is a steroid hormone made in the adrenal glands and released in response to stressful situations, such as the one shown in **Figure 5.** How steroid hormones such as cortisol work is summarized in **Figure 6.**

Step ❶ The hormone diffuses through the cell membrane and binds to its receptor. The hormone and receptor form a hormone-receptor complex in the cytoplasm.

Step ❷ The hormone-receptor complex enters the nucleus of the cell and binds to DNA.

Step ❸ Depending on the hormone and the target cell, the binding either activates or inactivates a gene. That is, either the gene is transcribed and the resulting mRNA is translated into protein, or transcription and translation are inhibited.

Step ❹ The target cell's activities are altered. For example, cortisol stimulates the making of enzymes that break down proteins and fats into glucose.

If the receptor for a steroid or thyroid hormone is located in the nucleus, the hormone enters the nucleus and binds to the receptor there. The hormone-receptor complex then binds to and affects the DNA in the same manner as it does with receptors in the cytoplasm.

Figure 5 Steroid hormones and stress. Stressful situations, such as taking an exam, result in the release of the steroid hormone cortisol.

Figure 6

BIOgraphic

How Steroid Hormones Work

The steroid hormone–receptor complex binds to DNA in the nucleus and activates or inactivates transcription and translation of a gene.

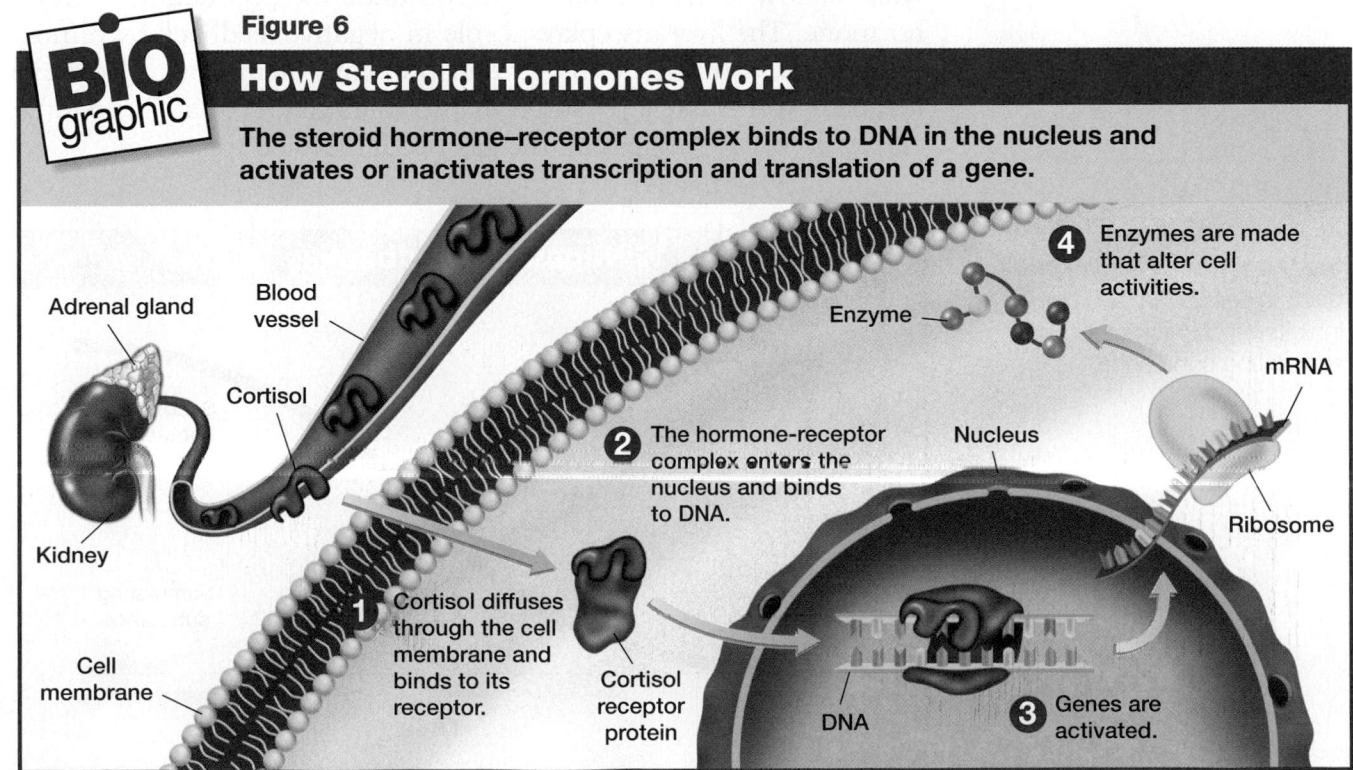

Adrenal gland

Blood vessel

Cortisol

Kidney

Cell membrane

❶ Cortisol diffuses through the cell membrane and binds to its receptor.

Cortisol receptor protein

❷ The hormone-receptor complex enters the nucleus and binds to DNA.

Nucleus

DNA

❸ Genes are activated.

❹ Enzymes are made that alter cell activities.

Enzyme

mRNA

Ribosome

Feedback Mechanisms

The human body makes more than 40 hormones, and the body must regulate the release of the hormones. Nerve impulses alone can increase or decrease secretion of some hormones. For example, a baby nursing on a mother's breast stimulates the release of the hormone oxytocin, which in turn stimulates the release of milk from the mother's mammary glands.

Recall that homeostasis is essential in all living things. The endocrine system plays an important role in homeostasis. Different hormones moving through the bloodstream affect specific target tissues, and the amounts of various hormones must be maintained within a very narrow range.

In many cases the level of a hormone in the blood turns production of the hormone off and on through feedback mechanisms. Feedback mechanisms detect the amount of hormones in circulation or the amount of other chemicals produced because of hormone action. The endocrine system then adjusts the amount of hormones being made or released.

If high levels of a hormone stimulate the output of even *more* hormone, the regulation is called positive feedback. For example, the hormone that stimulates egg release also regulates the female hormone estrogen. A rise in estrogen levels, however, will stimulate the release of *more* of the regulatory hormone.

In humans, the release of most hormones is regulated through negative feedback, as shown in **Figure 7.** In **negative feedback,** a change in one direction stimulates the control mechanism to counteract further change in the same direction. For example, high levels of a hormone inhibit the production of more hormone, whereas low levels of a hormone stimulate the production of more hormone. The liver also plays a role in negative feedback by removing hormones from the blood and breaking them down. Negative feedback works like a person trying to maintain a certain speed in a car by pressing or releasing the gas pedal (accelerator).

Figure 7 Negative feedback. In negative feedback, a secondary substance inhibits production of its initial stimulating substance.

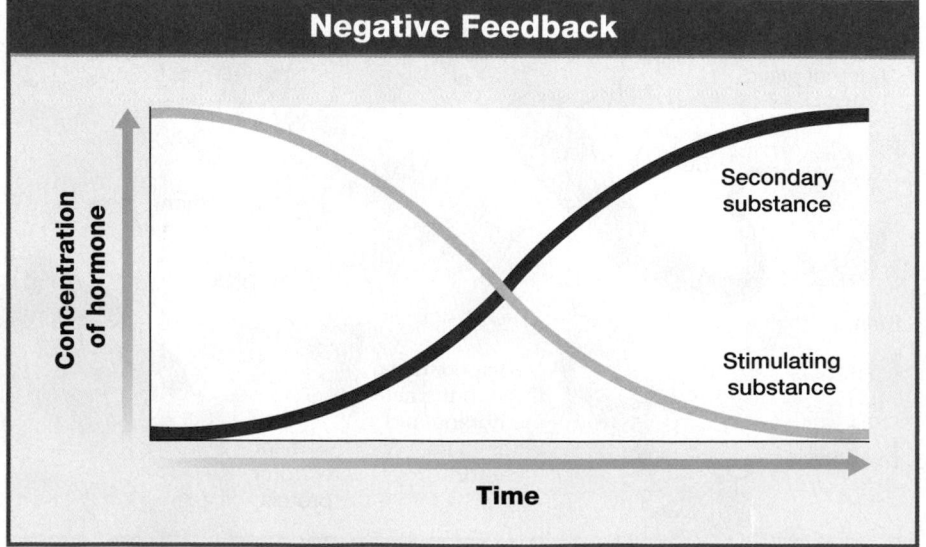

Negative Feedback

Concentration of hormone

Secondary substance

Stimulating substance

Time

Anabolic Steroids Are Dangerous

Many athletes use anabolic (protein-building) steroids and other hormone therapies to increase the size of their muscles and improve their performance. The unnatural use of steroids disrupts the feedback mechanisms that regulate hormone concentrations in the body.

Which Steroid Hormones Are Used?

The steroids used by athletes include synthetic hormones that mimic the male sex hormone testosterone. Many precursors to testosterone (such as androstenedione) are also used. Testosterone is secreted during puberty, when it stimulates many of the characteristics associated with being a man. For example, hair grows on the face, the underarms, and the pubic area; the voice deepens; and bigger muscles develop in the arms, legs, shoulders, and elsewhere.

Do Steroids Really Improve an Athlete's Performance?

When athletes inject steroids, they are trying to stimulate the production of proteins in the muscle cells as a way of increasing muscle mass and strength. In large doses, steroids can promote increases in mass, strength, and endurance.

Many Side Effects Are Associated with the Use of Steroids.

There are many side effects that accompany steroid use. When steroids are used before the skeleton matures completely, they stop the bones from growing. The body never reaches adult height and may look distorted. Liver cancer and other liver disorders may also result from steroid use. Some males who use anabolic steroids develop enlarged breasts and shriveled testes. Females who use these chemi-

cals may develop facial hair, deepening of the voice, and male-pattern baldness. Finally, the virus that causes AIDS can be transmitted if shared needles are used to inject the steroids. *The long-term risks to health are often greater than any benefits from the use of steroids.*

internet connect

www.scilinks.org
Topic: Anabolic Steroids
Keyword: HX4006

SCILINKS Maintained by the National Science Teachers Association

Section 2 Review

1 Name the structures found on or inside cells that allow hormones to recognize their target cells.

2 Relate how an amino-acid-based hormone changes a cell's activity.

3 Relate how a steroid or thyroid hormone changes a cell's activity.

4 Analyzing Graphics Use **Figure 7** to describe how hormone levels are regulated by negative feedback.

5 Standardized Test Prep *X* and *Y* are hormones. *X* stimulates the secretion of *Y*, which exerts negative feedback on the cells that secrete *X*. What happens when the blood level of *Y* decreases?

A Less *X* is secreted.

B More *X* is secreted.

C Secretion of *X* stops.

D Less *Y* is secreted.

The Major Endocrine Glands

Endocrine Control

Feedback mechanisms fine-tune the levels of hormones in circulation, but two endocrine glands control the initial release of many hormones. The hypothalamus *(HIE poh THAL uh muhs)* and the pituitary *(pi TOO uh tehr ee)* gland, shown in **Figure 8,** together serve as a major control center for the rest of the endocrine system.

The Hypothalamus

The **hypothalamus** is the area of the brain that coordinates the activities of the nervous and endocrine systems. It controls many body functions, including body temperature, blood pressure, and emotions. The hypothalamus receives information about external and internal conditions from other brain regions. The hypothalamus responds to these signals from the nervous system as well as to blood concentrations of circulating hormones. The hypothalamus responds by issuing instructions—in the form of hormones—to the pituitary gland.

Figure 8   The hypothalamus and pituitary gland

Many hormones are released in a cascade starting with the release of hormones from the hypothalamus.

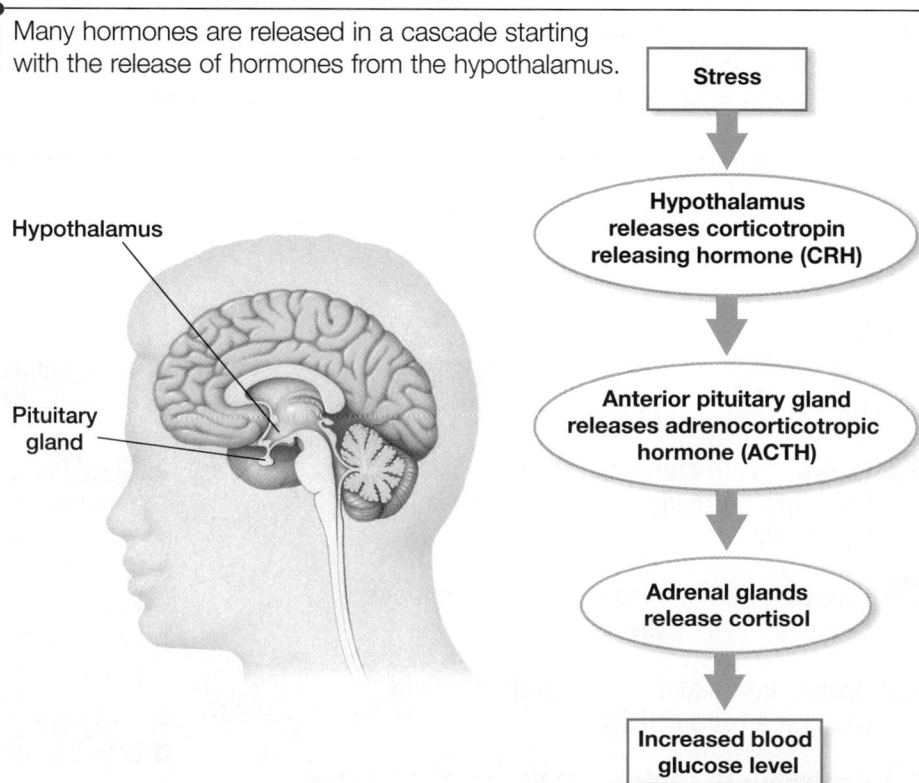

The Pituitary Gland

As shown in Figure 8, the **pituitary gland** is an endocrine gland suspended from the hypothalamus by a short stalk. The pituitary gland secretes many hormones, including some that control endocrine glands elsewhere in the body.

The nerve cells in the hypothalamus make at least six hormones that are released into a special network of blood vessels between the hypothalamus and the pituitary gland. Some of these hormones are "releasing" hormones, which cause the front part of the pituitary gland, the anterior pituitary, to make and then release a corresponding pituitary hormone. "Inhibiting" hormones signal the anterior pituitary to stop secretion of one of its hormones.

Certain pituitary hormones travel to a distant endocrine gland and cause the gland to begin producing its particular hormone. One example of this cascade of events is shown in Figure 8. Other pituitary hormones act directly on organs and tissues that are not endocrine glands, as summarized in **Table 1.**

The nerve cells of the hypothalamus also have axons that extend to the back part of the pituitary gland, the posterior pituitary. The nerve cells in the hypothalamus make two hormones that are stored in the posterior pituitary and released when needed: oxytocin *(ahks ih TOH sihn)* and antidiuretic hormone (ADH or vasopressin). Oxytocin triggers milk ejection during nursing and uterine contractions during childbirth. ADH causes the kidneys to form more-concentrated urine, thereby conserving water in the body.

Table 1 Hormones Secreted by the Pituitary Gland

Hormone	Target tissue	Effects
Adrenocorticotropic hormone (ACTH)	Adrenal glands	Stimulates the release of cortisol and other steroid hormones from the adrenal cortex
Follicle-stimulating hormone (FSH)	Ovaries and testes	Regulates the development of male and female gametes
Luteinizing hormone (LH)	Ovaries and testes	Stimulates the release of an egg (ovulation) from an ovary; stimulates secretion of sex hormones from ovaries and testes
Prolactin	Mammary glands	Stimulates milk production in breasts
Growth hormone (GH)	Many tissues	Stimulates protein synthesis and bone and muscle growth
Thyroid-stimulating hormone (TSH)	Thyroid gland	Stimulates synthesis and release of the thyroid hormones by the thyroid gland
Antidiuretic hormone (ADH)	Kidneys, blood vessels	Stimulates reabsorption of water from the kidney; constricts blood vessels
Oxytocin	Mammary glands, uterus	Stimulates uterine contractions and milk secretion

The Thyroid and Parathyroid Glands

As shown in **Figure 9,** the thyroid gland is an endocrine gland shaped like a shield of armor. It is located just below the Adam's apple in the front of the neck. The name *thyroid* comes from the Greek word *thyros*, which means "shield."

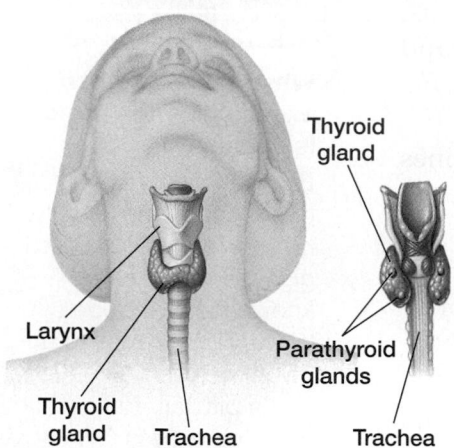

Figure 9 **The thyroid and parathyroid glands.** The thyroid gland, located in the neck, is wrapped around the trachea. The parathyroid glands are located on the back of the thyroid gland.

Regulating Metabolism and Development

The thyroid gland makes and releases thyroid hormones. Thyroid hormones regulate the body's metabolic rate and promote normal growth of the brain, bones, and muscles during childhood. Thyroid hormones also affect reproductive functions and maintain mental alertness in adults.

Thyroid hormones are modified amino acids produced by the addition of iodide to the amino acid tyrosine. If iodide salts are lacking in the diet, the thyroid gland becomes greatly enlarged. An enlarged thyroid gland, like the one shown in **Figure 10,** is called a goiter *(GOY tuhr)*. Goiters resulting from iodide deficiency are now rare in the United States because iodide is added to commercially available table salt.

The underproduction of thyroid hormones is known as hypothyroidism. In childhood hypothyroidism, an underproduction of thyroid hormones can cause permanently stunted growth, mental retardation, or both. In adults, hypothyroidism can cause a lack of energy, dry skin, and weight gain. Overproduction of thyroid hormones, or hyperthyroidism, can cause nervousness, sleep disorders, an irregular heart rate, and weight loss.

Regulating Calcium Levels

A high level of calcium in the blood stimulates the thyroid gland to produce a hormone called calcitonin. Calcitonin causes calcium to be deposited in bone tissue rapidly, lowering the blood-calcium level. Calcium is used for different purposes. For example, calcium ions are required for muscle contraction and for the release of certain substances from cells.

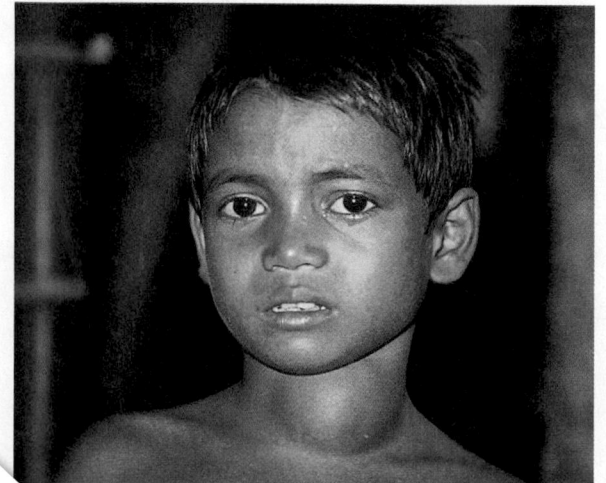

Figure 10 **Goiter.** Goiters result from a lack of iodide in the diet or improper functioning of the thyroid gland.

Parathyroid hormone (PTH) is a hormone that is produced by four parathyroid glands attached to the back part of the thyroid gland, as shown in Figure 9. PTH is made and released in response to a falling level of calcium in the blood. PTH acts in three ways to raise calcium levels. First, it stimulates bone cells to break down bone tissue and release calcium into the blood. Second, it causes the kidneys to reabsorb calcium ions from urine. Third, PTH leads to activation of vitamin D, which is necessary for calcium absorption by the intestine.

The Adrenal Glands

The body has two **adrenal glands,** which are endocrine organs located above each kidney. Each almond-size adrenal gland is actually two glands in one, as seen in **Figure 11:** an inner core, called the adrenal medulla, and an outer shell, called the adrenal cortex.

Immediate Response to Stress

The adrenal medulla acts as a warning system in times of stress by releasing the "fight-or-flight" hormones **epinephrine** *(ehp uh NEHF rihn)* and **norepinephrine** (formerly called adrenaline and noradrenaline, respectively). The effects of these hormones, which prepare the body for action in emergencies, are identical to the effects of the sympathetic nervous system in response to a stressful situation, but longer lasting. In stressful situations the fight-or-flight hormones increase heart rate, blood pressure, blood glucose level, and blood flow to the heart and lungs.

Longer-Term Response to Stress

The adrenal cortex makes several hormones, including cortisol and aldosterone. The adrenal cortex hormones provide a slower, more long-term response to stress than epinephrine and norepinephrine. Cortisol makes more energy available to the body. For example, cortisol causes the body to increase the level of blood glucose and to break down proteins for energy. A high level of cortisol, such as occurs when the body is under stress for a long period of time, suppresses the immune system. Artificial derivatives of this hormone, such as prednisone *(PREHD nih sohn)*, are widely used as anti-inflammatory drugs.

Aldosterone *(al DAHS tuh rohn)* helps reabsorb sodium ions from the fluids removed by the kidneys so that these ions are not lost in the urine. In contrast, aldosterone stimulates the kidneys to secrete potassium ions into the urine. When the aldosterone level is too low, potassium ions in the blood may accumulate to a dangerous level. The overall effect of aldosterone to prolonged stress is that the volume of blood is increased, which raises blood pressure.

Figure 11   The adrenal gland

Each adrenal gland has two parts—the adrenal medulla and the adrenal cortex—which secrete different hormones.

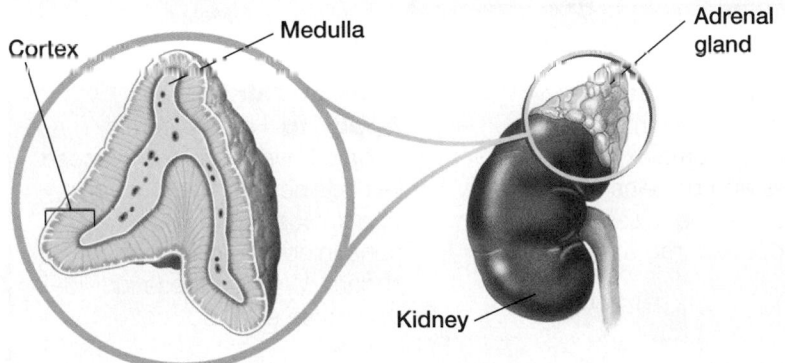

The Pancreas and Other Organs

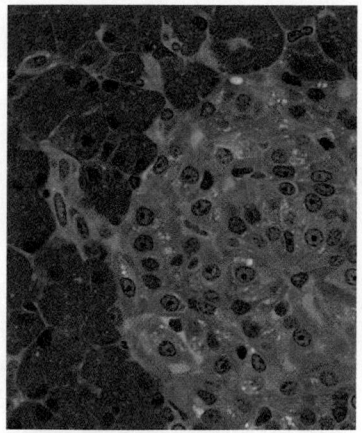

Figure 12 Islets of Langerhans. Islets of Langerhans are clusters of cells in the pancreas. The lighter-stained cells produce glucagon. The darker-stained cells produce insulin.

In addition to those mentioned so far, several other organs and glands produce hormones. For example, the stomach, small intestine, thymus, kidney, liver, and heart all contain endocrine cells. Recall that the stomach and small intestine secrete hormones, such as gastrin, that regulate the release of acids and digestive enzymes.

Regulating Blood Glucose Levels

The pancreas contains clusters of specialized cells, called the islets *(IE litz)* of Langerhans *(LAHNG uhr hahns)*, shown in **Figure 12.** Two hormones made by the islets interact to control the level of glucose in the blood. **Insulin** is a hormone that lowers blood glucose levels by promoting the accumulation of glycogen in the liver. Insulin also stimulates muscle cells to take up glucose and convert it into glycogen. **Glucagon** has the opposite effect of insulin—it raises blood glucose levels. Glucagon causes liver cells to release glucose that was stored as glycogen.

Analyzing Blood Glucose Regulation

Background

Eating simple sugars causes glucose to enter the bloodstream faster than eating complex carbohydrates or proteins. The rise in sugar levels triggers the secretion of insulin, which decreases blood glucose levels.

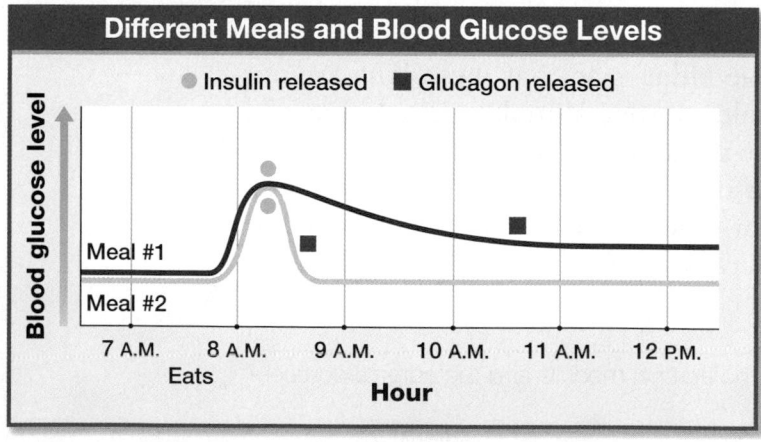

Different Meals and Blood Glucose Levels

● Insulin released   ■ Glucagon released

Meal #1

Meal #2

Analysis

1. **Identify** which meal causes a faster rise in blood glucose.

2. **Critical Thinking**
   Inferring Determine which meal has complex carbohydrates and proteins that allow glucose to be released into the bloodstream more slowly.

3. **Critical Thinking**
   Applying Hypoglycemic people have low blood glucose levels. They are often advised to eat six small meals a day containing little or no simple sugars. Why are these individuals given such advice?

Diabetes *(die uh BEET eez)* **mellitus** *(MEH liet uhs)* is a serious disorder in which cells are unable to obtain glucose from the blood, resulting in high blood glucose levels. The kidneys excrete the excess glucose, and water follows, resulting in excessive volumes of urine and persistent thirst. Because cells cannot take up glucose, they use the body's supply of fats and proteins for energy. The fat breakdown results in acidic products that accumulate in the blood, leading to low blood pH, coma, and, in extreme cases, death.

There are two kinds of diabetes mellitus. About 10 percent of affected individuals suffer from Type I diabetes, and 90 percent suffer from Type II diabetes. Type I diabetes is a hereditary autoimmune disease. The immune system attacks the islets of Langerhans, causing low insulin levels. Type I diabetes usually is treated with daily injections of insulin. It usually develops before age 20.

People with Type II diabetes may have normal levels of insulin in their blood. Their fat cells may produce a hormone that blocks insulin activation of glucose transport. This makes insulin ineffective. Type II diabetes often develops in people over age 40 due to obesity and an inactive lifestyle. Type II diabetes is usually treated with diet and exercise and, sometimes, medication other than insulin.

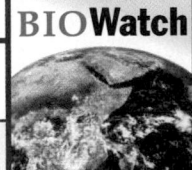

BIOWatch

Hormones and Body Fat

People with very little body fat, including many long-distance runners and gymnasts, often have disrupted reproductive systems. Very thin women may stop having menstrual periods, and very thin men may have lower testosterone levels.

During puberty, girls accumulate body fat before their first menstrual period. If they are very thin, their first period may be delayed by a year or more. Scientists are searching for hormones that tie a person's reproductive state to his or her body-fat content.

Hormone Made by Fat Cells

In 1994, researchers discovered that fat cells secrete a hormone called leptin that helps control metabolism.

When injected into young female mice, leptin causes the mice to reach sexual maturity sooner. Mutant female mice that cannot make leptin do not produce eggs and thus are infertile. If the mutant mice are injected with leptin, they begin to produce eggs and can become pregnant.

The more body fat he or she has, the more leptin in his or her blood. Leptin is involved in regulating body weight. Scientists are unsure how leptin controls human reproduction. Cells in the ovaries and hypothalamus have leptin receptors.

Female Hormones

Some women suffer severe bone loss (osteoporosis). Women with more body fat tend to have stronger bones and are at lower risk for osteoporosis after menopause. Estrogens, female sex hormones, help maintain strong bones in women. Secretion of estrogens decreases during and after menopause. The bones eventually become less dense and break more easily. After menopause, a woman's ovaries and adrenal cortex secrete small amounts of the "male" hormone testosterone, which is converted to estrogen by enzymes.

🖅 **internet** connect

www.scilinks.org
Topic: Hormones and Body Fat
Keyword: HX4101

SC*i*INKS. Maintained by the National Science Teachers Association

Regulating Reproduction

The ovaries and the testes, which also produce gametes, secrete hormones that regulate reproduction. The ovaries secrete estrogens and progesterone, and the testes produce testosterone. These hormones affect the formation of gametes and control sexual behavior and cycles. They also stimulate the development of secondary sex characteristics, such as breast size, hair growth, and muscle development.

Regulating Daily Rhythms

The pineal *(PIHN ee uhl)* gland is a pea-sized gland located in the brain. The pineal gland secretes the hormone melatonin, which is a modified form of the amino acid tryptophan.

Melatonin seems to be released by the human pineal gland as a response to darkness. Therefore, the pineal gland is thought to be involved in establishing daily biorhythms, such as the one shown in **Figure 13.** The pineal gland may also play a role in mood disorders such as seasonal affective disorder (SAD) syndrome and in a variety of aspects of sexual development.

Figure 13   A daily biorhythm. The daily variation in body temperature is an example of a biorhythm thought to be influenced by melatonin.

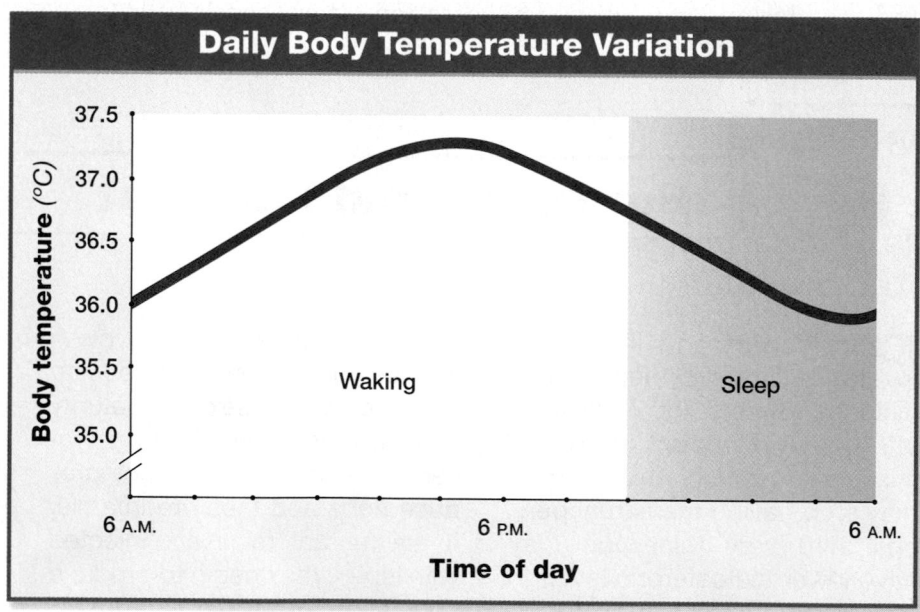

Section 3 Review

1 **Explain** why the hypothalamus and pituitary gland are considered the major control center of the endocrine system.

2 **Evaluate** the consequences of an underproduction of thyroid hormones during childhood.

3 **Compare** the effects of glucagon and insulin on blood glucose levels.

Identify the functions of reproductive ~rmones.

5 **Critical Thinking Applying** A classmate states that hormones from the adrenal medulla, but not from the adrenal cortex, are secreted in response to stress. Do you agree? Explain.

6 **Standardized Test Prep** Cortisol exerts negative feedback on the hypothalamic cells that release CRH. Which of the following results from a rise in the blood level of cortisol?
A Blood glucose levels fall. C Stress levels rise.
B Less ACTH is released.   D More CRH is released.

Key Concepts

1 Hormones

- Hormones are chemical messengers secreted by cells that act to regulate the activity of other cells. Ductless glands called endocrine glands make most of the body's hormones.
- Hormones are usually slower-acting but longer-lasting than neurotransmitters.
- Similar to hormones, endorphins, enkephalins, and prostaglandins act on nearby cells to regulate cellular activities.

2 How Hormones Work

- Amino-acid-based hormones bind to cell-membrane receptors, activating a second messenger. The second messenger then activates or deactivates enzymes in a cascade fashion.
- Steroid and thyroid hormones bind to receptors inside the cell. The hormone-receptor complex binds to DNA in the nucleus and turns genes either on or off.
- Most hormones are regulated by negative feedback.

3 The Major Endocrine Glands

- The hypothalamus and pituitary gland serve as the major control center for the release of many hormones.
- The thyroid hormones regulate metabolism and development. Calcitonin and parathyroid hormone regulate blood calcium levels.
- The inner medulla of the adrenal glands produces the fight-or-flight hormones. The outer cortex of the adrenal glands produces cortisol, aldosterone, and other steroid hormones.
- The pancreas secretes insulin and glucagon, which are involved in regulating blood glucose levels.
- In diabetes mellitus, an individual's cells are unable to take up glucose from the blood. The cause is either an abnormally low level of insulin or insulin resistance.
- Hormones secreted by the ovaries and testes regulate reproductive functions.
- Melatonin is thought to regulate daily body rhythms.

Key Terms

Section 1

hormone (974)
endocrine gland (975)

Section 2

target cell (977)
amino-acid-based hormone (977)
steroid hormone (977)
second messenger (978)
negative feedback (980)

Section 3

hypothalamus (982)
pituitary gland (983)
adrenal gland (985)
epinephrine (985)
norepinephrine (985)
insulin (986)
glucagon (986)
diabetes mellitus (987)

Performance ZONE

Understanding Key Ideas

1. Steroid hormones
 a. bind to cell-membrane receptors.
 b. bind to mRNA.
 c. eventually form hormone-receptor complexes that bind to DNA.
 d. never enter cells.

2. The _____ interact to control the secretion of other hormones.
 a. pancreas and thyroid gland
 b. hypothalamus and pineal gland
 c. adrenal gland and pancreas
 d. hypothalamus and pituitary gland

3. Thyroid hormones
 a. slow growth.
 b. inhibit insulin production.
 c. promote sperm production.
 d. control metabolic activities.

4. What adrenal cortex hormone acts, at high levels, to reduce inflammation?
 a. calcitonin   **c.** prostaglandin
 b. aldosterone   **d.** cortisol

5. Insulin leads to
 a. higher blood glucose levels.
 b. lower blood glucose levels.
 c. release of additional insulin.
 d. glycogen breakdown.

6. Which of the following endocrine glands secretes melatonin and is believed to be involved in establishing biorhythms?
 a. pituitary gland   **c.** pineal gland
 b. thyroid gland   **d.** adrenal gland

7. Summarizing Information Identify the endocrine gland labeled A, and name two hormones it makes.

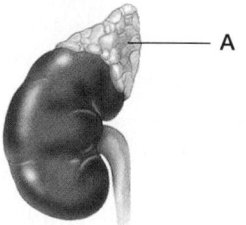

—— A

8. 🌐**BIOWatch** Summarize the major side effects experienced by athletes who use anabolic steroids.

9. 🌐**BIOWatch** If having more body fat lowers a woman's risk of developing osteoporosis, why don't doctors advise their female patients to gain weight?

10. 🔲 **Concept Mapping** Make a concept map that describes the endocrine system. Try to include the following terms in your map: *hypothalamus, pituitary gland, thyroid gland, hormones, adrenal glands,* and *target cell.*

Critical Thinking

11. Inferring Relationships Describe the importance of "fit" between a receptor protein and a hormone.

12. Applying Information Before iodide was added to table salt, goiters were common among people living in inland regions but rare among people living in coastal areas. Why do you think this was so?

13. Distinguishing Relevant Information During a medical examination, a person is found to be unable to move glucose, stored as glycogen, from the liver into the blood. Further tests show that glucagon levels are normal, as is the structure of the hormone. Why do you think glucagon is unable to carry out its function in this case?

Alternative Assessment

14. Career Focus Endocrinologist Research endocrinology, and write a report on your findings. Your report should include a job description, training required, kinds of employers, growth prospects, and starting salary.

15. Evaluating Information Interview several coaches at your school to determine their attitudes toward steroid testing. Write an article that discusses your findings and explains how steroids affect the body.

Standardized Test Prep

Understanding Concepts

Directions (1–4): For *each* question, write on a separate sheet of paper the letter of the correct answer.

1 What are the chemical messengers of the endocrine system?
- **A.** blood cells
- **B.** carbohydrates
- **C.** hormones
- **D.** neurons

2 Which of the following is true of exocrine glands?
- **F.** function only after puberty
- **G.** include the brain and liver
- **H.** release products through ducts
- **I.** release products into the bloodstream

3 How may amino-acid-based hormones use cyclic AMP?
- **A.** as a coenzyme
- **B.** as a receptor
- **C.** as a second messenger
- **D.** as a target cell

4 What leads to an increase in hormone levels when hormone levels rise above normal?
- **F.** feedback inhibition
- **G.** negative feedback
- **H.** neutral feedback
- **I.** positive feedback

Directions (5–6): For *each* question, write a short response.

5 Analyze the relationship between transcription factors and steroid and thyroid hormones.

6 How are hormones and neurotransmitters alike and different?

Test TIP

When using a graph to answer a question, be sure to study the graph carefully before choosing a final answer. Some of the answer choices may be based on common misinterpretations of graphs.

Reading Skills

Directions (7): **Read the passage below. Then answer the question.**

The pineal gland secretes the hormone melatonin, which is a modified form of the amino acid tryptophan. The daily variation in body temperature is an example of a biorhythm thought to be influenced by melatonin. Melatonin seems to be released as a response to darkness. Soon after the melatonin is released, the person goes to sleep.

7 What type of disorder might be treated by using melatonin supplements?
- **A.** attention deficit
- **B.** bulimia
- **C.** depression
- **D.** insomnia

Interpreting Graphics

Directions (8): **Base your answer to question 8 on the chart below.**

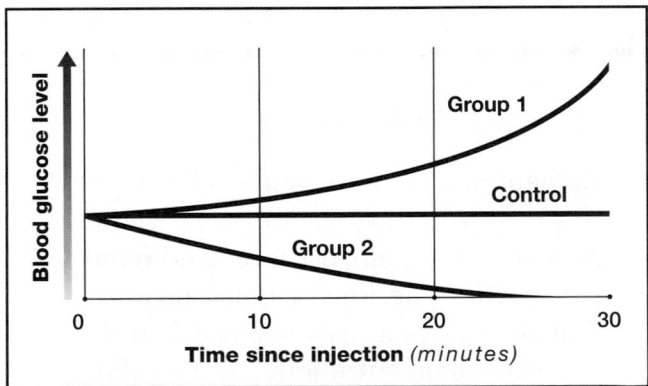

Glucose Levels in Rats Injected with Hormones

8 The chart shows the effects of hormone injections on blood glucose levels in rats. Rats in groups 1 and 2 were injected with saline containing a hormone. Rats in the control group were injected with only saline. Which hormone was likely contained in the injection given to rats in group 1?
- **F.** calcitonin
- **G.** glucagon
- **H.** insulin
- **I.** oxytocin

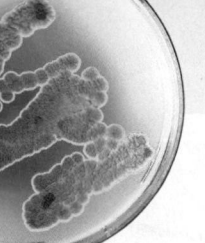

Exploration Lab

The Effect of Epinephrine on Heart Rate

SKILLS
- Using scientific methods
- Graphing
- Calculating

OBJECTIVES
- **Determine** the heart rate of *Daphnia*.
- **Observe** the effect of the hormone epinephrine on heart rate in *Daphnia*.
- **Determine** the threshold concentration for the action of epinephrine on *Daphnia*.

MATERIALS
- medicine droppers
- *Daphnia*
- *Daphnia* culture water
- depression slides
- petroleum jelly
- coverslips
- compound microscope
- watch or clock with second hand
- paper towels
- 100 mL beaker
- 10 mL graduated cylinders
- epinephrine solutions (0.001%, 0.0001%, 0.00001%, and 0.000001%)

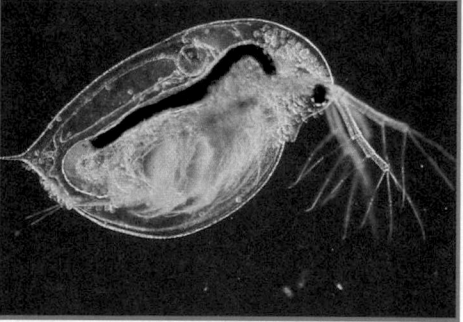

Daphnia

ChemSafety

CAUTION: Always wear safety goggles and a lab apron to protect your eyes and clothing.

CAUTION: Do not touch or taste any chemicals. Know the location of the emergency shower and eyewash station and how to use them. If you get a chemical on your skin or clothing, wash it off at the sink while calling to the teacher. Notify the teacher of a spill. Spills should be cleaned up promptly, according to your teacher's directions.

CAUTION: Glassware is fragile. Notify the teacher of broken glass or cuts. Do not clean up broken glass or spills with broken glass unless the teacher tells you to do so.

Before You Begin

Epinephrine is a hormone released in response to stress. It increases blood pressure, blood glucose level, and **heart rate** (HR). The lowest concentration that stimulates a response is called the **threshold concentration.** In this lab, you will observe the effect of epinephrine on HR using the crustacean *Daphnia*. Epinephrine affects the HR of *Daphnia* and humans in similar ways.

1. Write a definition for each boldface term in the paragraph above.

2. Make a data table similar to the one above.

3. Based on the objectives for this lab, write a question you would like to explore about the action of hormones.

	DATA TABLE				
Solution	HR (beats/s) Trial 1 (A)	HR (beats/s) Trial 2 (B)	HR (beats/s) Trial 3 (C)	Average HR (beats/s) [(A+B+C)/3]	Average HR (beats/min)

Procedure

PART A: Observing Heart Rate in *Daphnia*

1. **Caution: Do not touch your face while handling microorganisms.**
   Use a clean medicine dropper to transfer one *Daphnia* to the well of a clean depression slide. Place a dab of petroleum jelly in the well. Add a coverslip. Observe with a compound microscope under low power.

2. Count the *Daphnia's* heartbeats for 10 seconds. Divide this number by 10 to find the HR in beats/s. Record this number under Trial 1 in your data table. Turn off the microscope light, and wait 20 seconds. Repeat the count for Trials 2 and 3.

3. After calculating the average HR in beats/s, calculate the HR in beats/min by using the following formula: HR (in beats/min) = Average HR (in beats/s) × 60 s/min.

PART B: Design an Experiment

4. Work with the members of your lab group to explore one of the questions written for step 3 of **Before You Begin.** To explore the question, design an experiment that uses the materials listed for this lab.

You Choose

As you design your experiment, decide the following:

a. what question you will explore

b. what hypothesis you will test

c. how many *Daphnia* to use

d. what your controls will be

e. what concentrations of epinephrine to test

f. how many trials to perform

g. what data to record in your data table

---

5. Write a procedure for your experiment. Make a list of all the safety precautions you will take. Have your teacher approve your procedure and safety precautions before you begin the experiment.

PART C: Conduct Your Experiment

6.  Put on safety goggles, gloves, and a lab apron.

7. To add a solution to a prepared slide, first place a drop of the solution at the edge of the coverslip. Then place a piece of paper towel along the opposite edge to draw the solution under the coverslip. Wait 1 minute for the solution to take effect.

8. Set up your group's experiment, and collect data. **Caution: Epinephrine is toxic and is absorbed through the skin.**

PART D: Cleanup and Disposal

9. Dispose of solutions and broken glass in the designated waste containers. Place treated *Daphnia* in a "recovery container." Do not pour chemicals down the drain or put lab materials in the trash unless your teacher tells you to do so.

10. Clean up your work area and all lab equipment. Return lab equipment to its proper place. Wash your hands thoroughly before you leave the lab and after you finish all work.

Analyze and Conclude

1. Summarizing Results Make a graph of your group's data. Plot "Epinephrine concentration (%)" on the *x*-axis. Plot "Average heart rate (beats/min)" on the *y*-axis.

2. Analyzing Data Which solutions affected the heart rate of *Daphnia*?

3. Drawing Conclusions What was the threshold concentration of epinephrine?

4. Predicting Patterns Based on the information you have and on your data, predict how epinephrine concentration would affect human heart rates.

5. Further Inquiry Write a new question about hormones that could be explored with another investigation.

? Do You Know?

Do research in the library or media center to answer these questions:

1. What is anaphylactic shock?

2. Why is epinephrine used to treat anaphylactic shock?

Use the following Internet resources to explore your own questions about hormones.

internet connect

www.scilinks.org
Topic: Hormones
Keyword: HX4100

SCILINKS. Maintained by the National Science Teachers Association

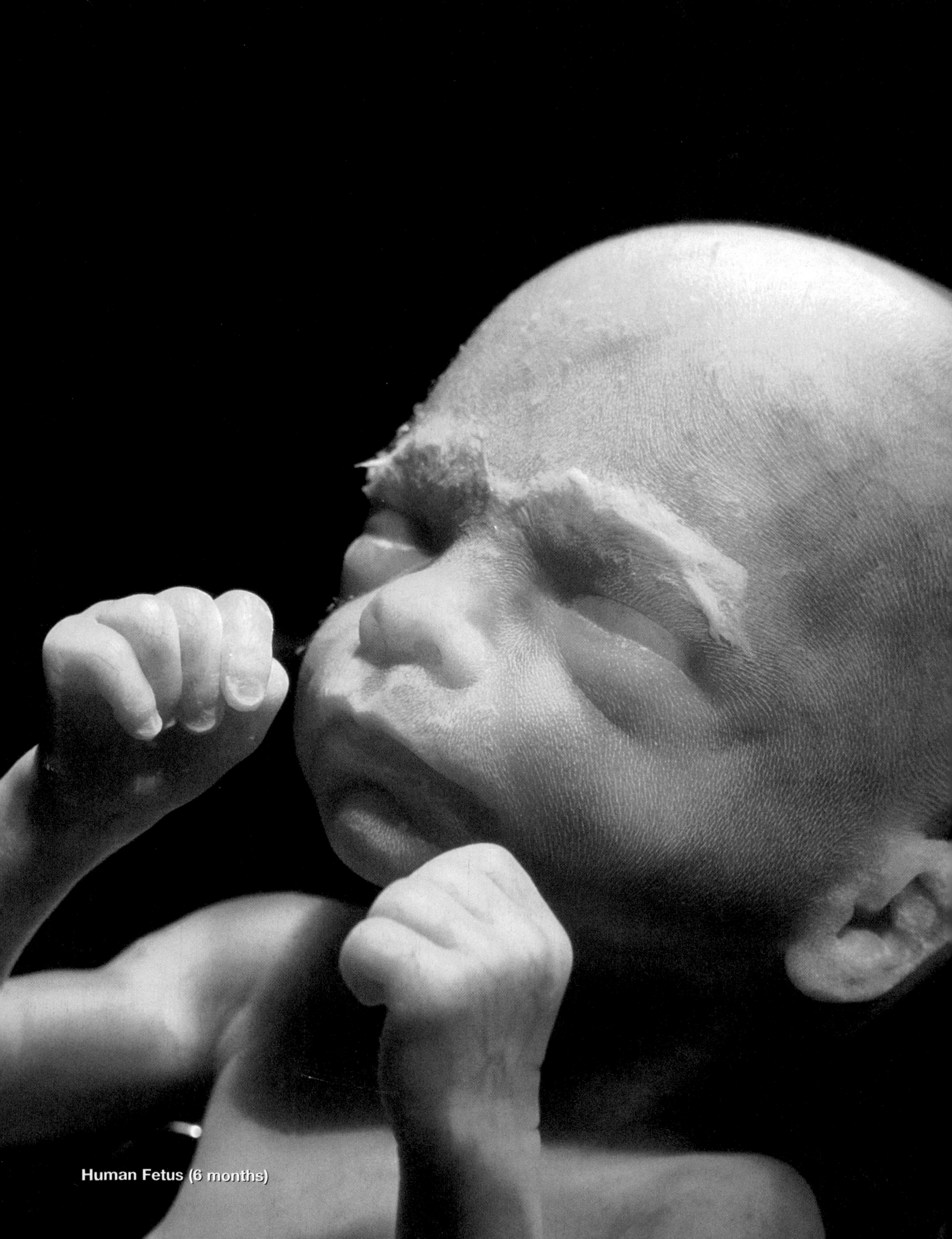

Human Fetus (6 months)

43 Reproduction and Development

✓ Quick Review

Answer the following without referring to earlier sections of your book.

1. **Sequence** the phases of meiosis. *(Chapter 7, Section 1)*

2. **List** the stages of animal development in the zygote. *(Chapter 27, Section 1)*

3. **Describe** the functions of sex hormones. *(Chapter 42, Section 3)*

4. **Summarize** the role of the hypothalamus in the endocrine system. *(Chapter 42, Section 3)*

Did you have difficulty? *For help, review the sections indicated.*

Reading Activity

Take a break after reading each section of this chapter, and closely study the figures in the section. Reread the figure captions, and for each one, write out a question that can be answered by referring to the figure and its caption. Refer to your list of figures and questions as you review the concepts addressed in the chapter before you complete the Performance Zone chapter review.

☑ internet connect

www.scilinks.org
National Science Teachers Association *sci*LINKS Internet resources are located throughout this chapter.

*sci*LINKS. Maintained by the National Science Teachers Association

● At six months, the human fetus is about 35.6 cm (14 in.) long and has a mass of about 908 g (weighs about 2 lb). The fetus has some brain wave activity, and its eyes will open soon.

Male Reproductive System

Objectives

- **Describe** how sperm are produced.
- **Identify** the major structures of the male reproductive system.
- **Relate** the structure of a sperm cell to its functions.
- **Sequence** the path taken by sperm as they leave the body.

Key Terms

testes
seminiferous tubules
epididymis
vas deferens
seminal vesicles
prostate gland
bulbourethral glands
semen
penis

The Testes

What are the roles of a human male in sexual reproduction? Recall that sexual reproduction involves the formation of a diploid zygote from two haploid sex cells, or gametes, through fertilization. The roles of a male in sexual reproduction are to produce sperm cells—the male gametes—and to deliver the sperm cells to the female reproductive system to fertilize an egg cell—the female gamete.

Where are sperm produced? Two egg-shaped **testes** *(TEHS teez)*, or testicles, are the gamete-producing organs of the male reproductive system. The testes are located in the scrotum *(SKROHT uhm)*, an external skin sac. The testes first form inside the abdominal cavity then move down into the scrotum either before or shortly after birth. The normal body temperature of 37°C (98°F) is too high for sperm to complete development. In the scrotum the temperature is about 3°C lower than it is in the rest of the body, making it an ideal location for sperm production.

Production of Sperm

The testes begin to produce sperm during the adolescent stage of development known as puberty *(PYOO buhr tee)*. As shown in **Figure 1,** each testis contains hundreds of compartments packed with many tightly coiled tubules, called **seminiferous** *(sehm uh NIHF uhr uhs)* **tubules.** Sperm cells are produced through meiosis in the lining of the seminiferous tubules. Thus, human sperm cells contain only 23 chromosomes (the haploid number) instead of the usual 46 chromosomes (the diploid number) found in other body cells. Two hormones released by the anterior pituitary regulate the functioning of the testes. Luteinizing hormone (LH) stimulates secretion of the sex hormone testosterone. Follicle-stimulating hormone (FSH), along with testosterone, stimulates sperm production in the seminiferous tubules. Cells located between the seminiferous tubules secrete testosterone.

Figure 1 Testes

The testes produce sperm cells.

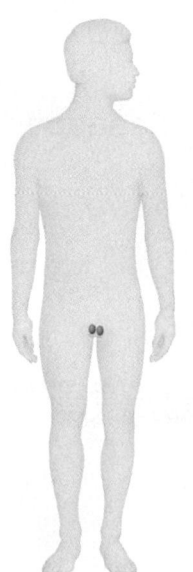

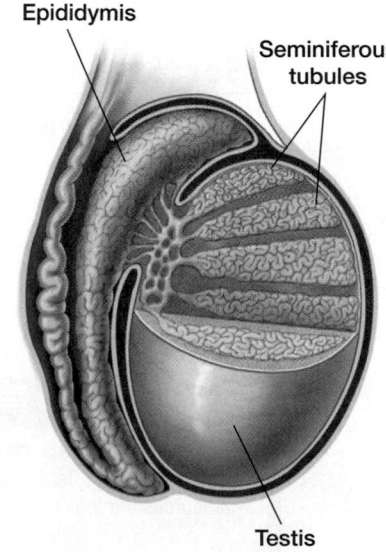

Epididymis

Seminiferous tubules

Testis

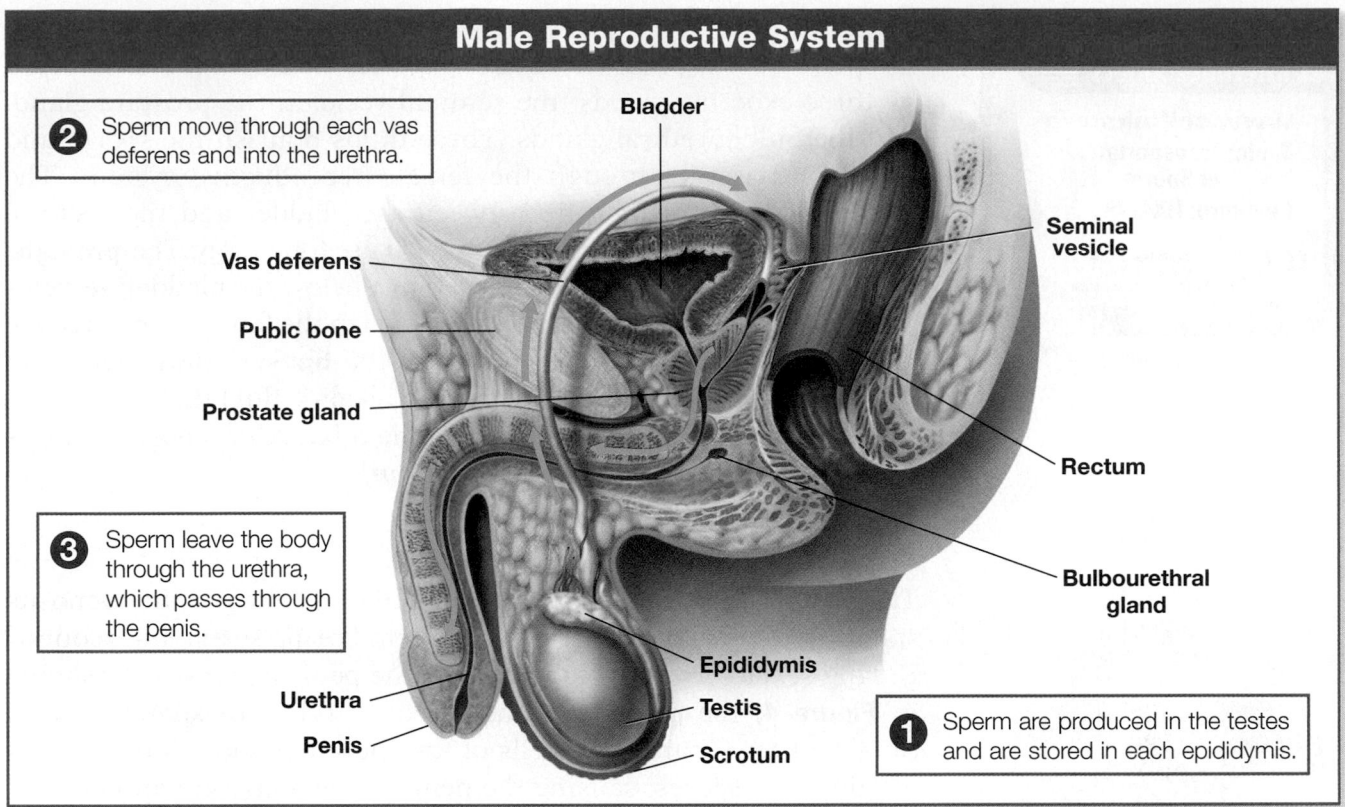

Male Reproductive System

2 Sperm move through each vas deferens and into the urethra.

Bladder

Seminal vesicle

Vas deferens

Pubic bone

Prostate gland

Rectum

3 Sperm leave the body through the urethra, which passes through the penis.

Bulbourethral gland

Epididymis

Urethra

Testis

Penis

Scrotum

1 Sperm are produced in the testes and are stored in each epididymis.

Maturation and Storage of Sperm

A typical adult male produces several hundred million sperm cells each day. After being produced in the seminiferous tubules, the sperm travel through a series of long tubes though they are not yet capable of swimming. Sperm then enter a long coiled tube called the **epididymis** *(ehp uh DIHD ih mihs)*, shown in Figure 1. Within each epididymis, the sperm mature and become capable of moving.

The epididymis is also the site where most of the sperm are stored. From the epididymis, some sperm move to another long tube, the **vas deferens** *(vas DEHF uh rehnz)*. Sperm move through the vas deferens and into the urethra, as shown by the arrows in **Figure 2.** Sperm leave the body by passing through the urethra, the same duct through which urine exits the body.

Structure of Mature Sperm

As shown in **Figure 3,** a mature sperm cell consists of a head with very little cytoplasm, a midpiece, and a long tail. Enzymes at the tip of the head help the sperm cell penetrate an egg cell during fertilization. The midpiece contains many mitochondria that supply sperm with the energy needed to propel themselves through the female reproductive system. The tail of a sperm cell is a powerful flagellum that whips back and forth, enabling the sperm cell to move. ATP produced in the mitochondria power the whiplike movements of the tail. During fertilization, only the head of a sperm enters an egg, so a father's mitochondria are not passed to offspring.

Figure 2 Male reproductive system. The arrows indicate the path taken by sperm cells from the testes as they exit the body.

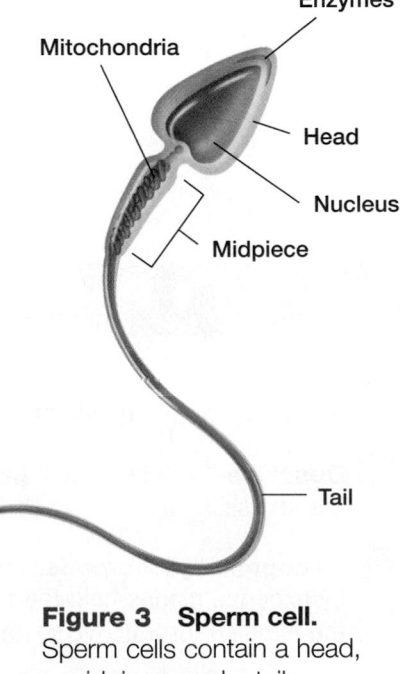

Enzymes

Mitochondria

Head

Nucleus

Midpiece

Tail

Figure 3 Sperm cell. Sperm cells contain a head, a midpiece, and a tail.

Semen

As sperm cells move into the urethra, they mix with fluids secreted by three exocrine glands: the seminal vesicles, the prostate gland, and the bulbourethral glands. These fluids nourish the sperm and aid their passage through the female reproductive system. The **seminal vesicles,** which lie between the bladder and the rectum, produce a fluid rich in sugars that sperm use for energy. The **prostate** *(PRAHS tayt)* **gland,** which is located just below the bladder, secretes an alkaline fluid that neutralizes the acids in the female reproductive system. Before semen leaves the body, the **bulbourethral** *(buhl boh yoo REE thruhl)* **glands** also secrete an alkaline fluid that neutralizes traces of acidic urine in the urethra. The mixture of these secretions with sperm is called **semen** *(SEE muhn).*

Delivery of Sperm

The urethra passes through the **penis,** the male organ that deposits sperm in the female reproductive system during sexual intercourse. During sexual arousal, blood flow to the penis increases. As shown in **Figure 4,** the penis contains three cylinders of spongy tissue. Small spaces separate the cells of the spongy tissue. Blood collects within these spaces, causing the penis to become rigid and erect.

Sperm exit the penis through ejaculation *(ee jak yoo LAY shun),* the forceful expulsion of semen. During ejaculation, muscles around each vas deferens contract, moving sperm into the urethra. Muscles at the base of the penis force semen out of the urethra.

After the semen is deposited in the female reproductive system, sperm swim until they encounter an egg cell or until they die. If sperm are unable to reach an egg, fertilization does not occur. Covering the penis with a thin rubber sheath called a condom helps prevent fertilization during sexual intercourse. Abstaining from sexual intercourse is the surest way to prevent fertilization.

About 3.5 mL of semen, containing 300–400 million sperm, is expelled during ejaculation. Because most sperm die in the female reproductive system, fertilization usually requires a high sperm count. Males with fewer than 20 million sperm per mL of semen are generally considered sterile.

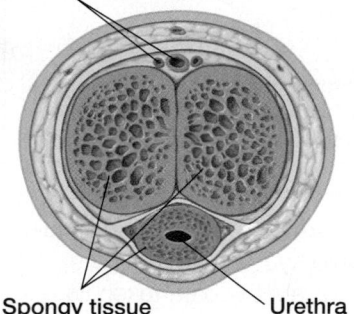

Blood vessels

Spongy tissue   Urethra

Figure 4   Cross section of penis. The penis contains three cylinders of spongy tissue. When the spaces in these cylinders fill with blood, the penis becomes erect.

Section 1 Review

1 **Sequence** the path that mature sperm take from the testes to the outside of the body.

2 **Describe** the role of each part of a mature sperm cell.

3 **Recognizing Relationships** How do secretions by exocrine glands help the delivery of sperm to the female reproductive system?

4 **Describe** the functions of the testes.

5 **Inferring Relationships** If a male's left vas deferens is blocked, how is his sperm count affected? Explain your answer.

6 **Standardized Test Prep** A fluid containing sugars that sperm use for energy is secreted by the
  A seminal vesicles.   **C** vas deferens.
  B prostate gland.   **D** epididymis.

Female Reproductive System

The Ovaries

Each month, the female reproductive system prepares for a possible pregnancy by producing a mature egg cell—the female gamete. After sperm have been deposited and fertilization has occurred, the role of the male in reproduction is complete. If pregnancy occurs, the female reproductive system will nourish and protect the fertilized egg through nine months of development.

Production of Eggs

Two egg-shaped ovaries, shown in **Figure 5,** are located within the abdominal cavity. The **ovaries** *(OH vuh reez)* are the gamete-producing organs of the female reproductive system. Females are born with all of the egg cells they will ever produce. At birth, the ovaries contain about 2 million immature egg cells that already have begun the first division of meiosis. Like sperm cells, egg cells contain 23 chromosomes (the haploid number) because eggs also are formed through meiosis.

After meiosis begins, egg cells become stalled in prophase of the first meiotic division. When a female reaches puberty, the increased production of sex hormones enables meiosis to resume. However,

Objectives

- **Describe** how eggs are produced.
- **Identify** the major structures of the female reproductive system.
- **Analyze** the events of the ovarian and menstrual cycles.

Key Terms

ovary
ovum
fallopian tube
uterus
vagina
ovarian cycle
ovulation
follicle
corpus luteum
menstrual cycle
menstruation

Figure 5   Female reproductive organs

The ovaries produce egg cells. The uterus nurtures the fetus during pregnancy.

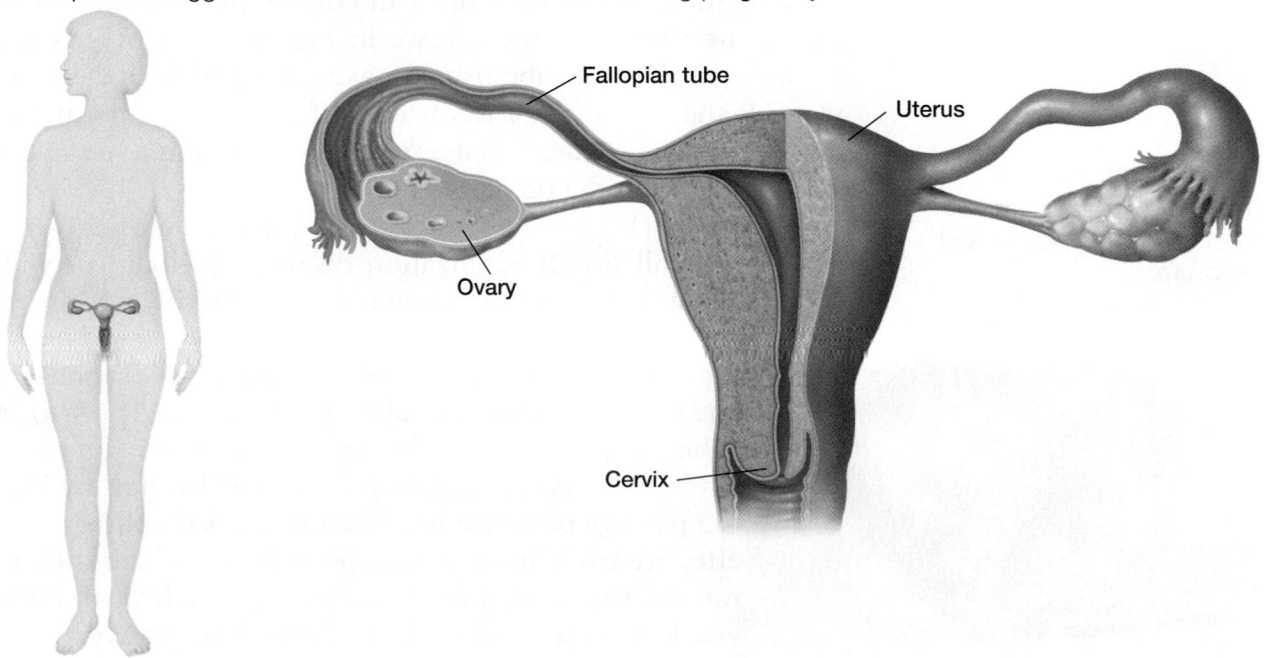

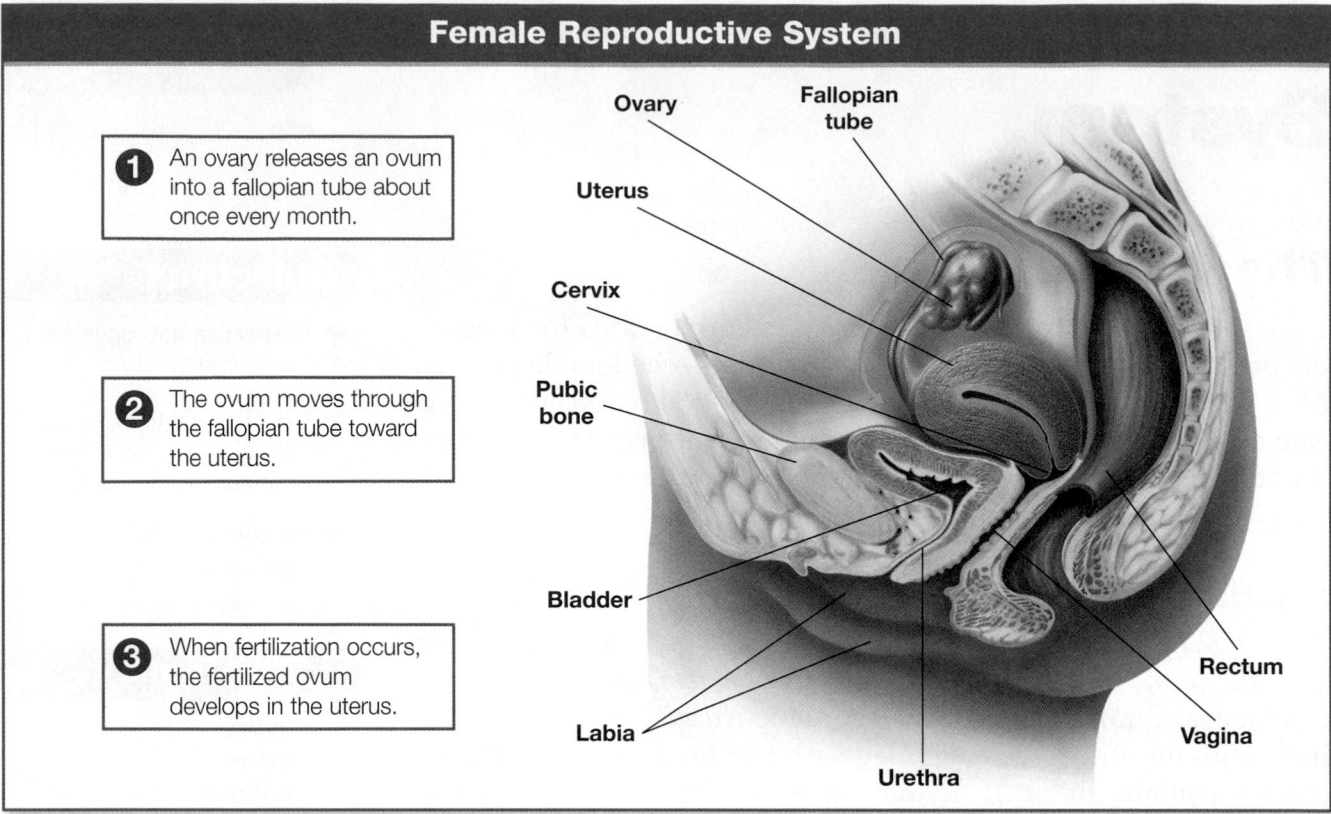

Female Reproductive System

1 An ovary releases an ovum into a fallopian tube about once every month.

2 The ovum moves through the fallopian tube toward the uterus.

3 When fertilization occurs, the fertilized ovum develops in the uterus.

Ovary
Fallopian tube
Uterus
Cervix
Pubic bone
Bladder
Labia
Urethra
Rectum
Vagina

Figure 6   Female reproductive system. The arrows indicate the path taken by an ovum from an ovary to the uterus.

normally only one immature egg cell matures each month. In the lifetime of a female, only 300–400 egg cells will mature. When an egg cell matures, it is called an **ovum** *(OH vuhm)*.

Structures of the Female Reproductive System

An ovum is released from an ovary about every 28 days. Cilia sweep the ovum into a fallopian tube. Each **fallopian** *(fuh LOH pee uhn)* **tube** is a passageway through which an ovum moves from an ovary toward the uterus. Smooth muscles lining the fallopian tubes contract rhythmically, moving the ovum down the tube and toward the uterus, as shown by the arrows in **Figure 6.** An ovum's journey through a fallopian tube usually takes three to four days to complete. If the ovum is not fertilized within 24–48 hours, it dies. An ovum, shown in **Figure 7,** is many times larger than a sperm cell and can be seen with the unaided eye.

The **uterus** *(YOO tuh ruhs)* is a hollow, muscular organ about the size of a small fist. If fertilization occurs, development will take place in the uterus. During sexual intercourse, sperm are deposited inside the **vagina** *(vuh JIE nuh)*, a muscular tube that leads from the outside of the female's body to the entrance to the uterus, called the *cervix (SUR vihks)*. A soft rubber cap called a *diaphragm (DIE uh fram)* can be used to cover the cervix and help prevent fertilization by blocking the passage of sperm into the uterus. A diaphragm is more effective when used with a sperm-killing chemical, or spermicide. During childbirth, a baby passes through the cervix and leaves the mother's body through the vagina.

Figure 7   Ovum. Notice the great difference in size between the sperm and the ovum.

The Ovum

The ovaries prepare and release an ovum in a series of events collectively called the **ovarian cycle**. The release of an ovum from an ovary is called **ovulation** *(ahv yoo LAY shuhn)*. The ovum is then swept into the fallopian tube and begins to move toward the uterus, awaiting fertilization. Although the duration of the ovarian cycle varies from female to female, the cycle generally spans about 28 days.

Phases of the Ovarian Cycle

Follicular phase The ovarian cycle has two distinct phases: the follicular phase *(fuh LIK yoo luhr)* and the luteal phase. These phases are regulated by hormones released by the hypothalamus and the anterior pituitary. The events of the ovarian cycle are summarized in **Figure 8.** In an ovary, egg cells mature within follicles. A **follicle** *(FAHL i kuhl)* is a cluster of cells that surrounds an immature egg cell and provides the egg with nutrients. During the follicular phase of the ovarian cycle, hormones regulate the completion of an egg cell's maturation. The follicular phase, which marks the beginning of the ovarian cycle, begins when the anterior pituitary releases follicle-stimulating hormone (FSH) and luteinizing hormone (LH) into the bloodstream. FSH causes the follicle to develop. The follicle produces estrogen, a sex hormone that aids in the growth of the follicle.

Figure 8   Ovarian cycle

During the ovarian cycle, ovulation occurs about every 28 days.

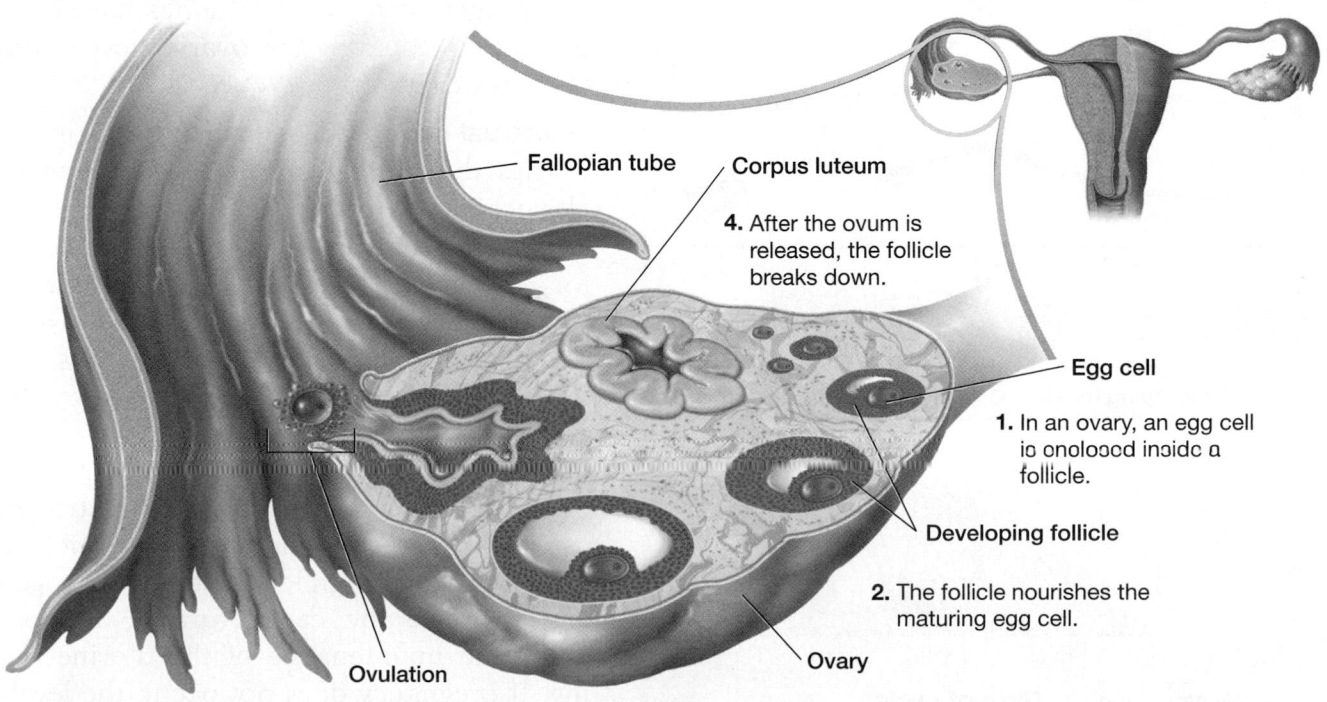

Fallopian tube

Corpus luteum

4. After the ovum is released, the follicle breaks down.

Egg cell

1. In an ovary, an egg cell is enclosed inside a follicle.

Developing follicle

2. The follicle nourishes the maturing egg cell.

Ovary

Ovulation

3. At ovulation, the ovum (mature egg) bursts from the follicle into the fallopian tube.

Ovulation At first the small increase in the level of estrogen prevents further release of FSH and LH from the anterior pituitary. This is caused by a negative feedback mechanism. But as the follicle approaches maturity, it begins to secrete large amounts of estrogen. The anterior pituitary responds to this high level of estrogen by greatly increasing secretion of LH. This increase in LH secretion is caused by a positive feedback mechanism. This surge of LH causes the egg cell to complete the first meiotic division, and it causes the follicle and the ovary to rupture. When the follicle bursts, ovulation occurs, as shown in Figure 8.

Luteal phase The luteal *(LOOT ee uhl)* phase of the ovarian cycle follows the follicular phase, as shown in **Figure 9.** After ovulation occurs, LH causes the cells of the ruptured follicle to grow, forming a corpus luteum. A **corpus luteum** *(KOHR puhs LOOT ee uhm)* is a yellowish mass of follicular cells that functions like an endocrine gland. LH causes the corpus luteum to secrete estrogen and progesterone, another sex hormone. Estrogen and progesterone inhibit the release of FSH and LH. This prevents the development of new follicles during the luteal phase.

Figure 9 **Ovarian and menstrual cycles.** The ovarian cycle is regulated by hormones produced by the hypothalamus and the pituitary gland. The menstrual cycle is regulated by hormones produced by the follicle and the corpus luteum.

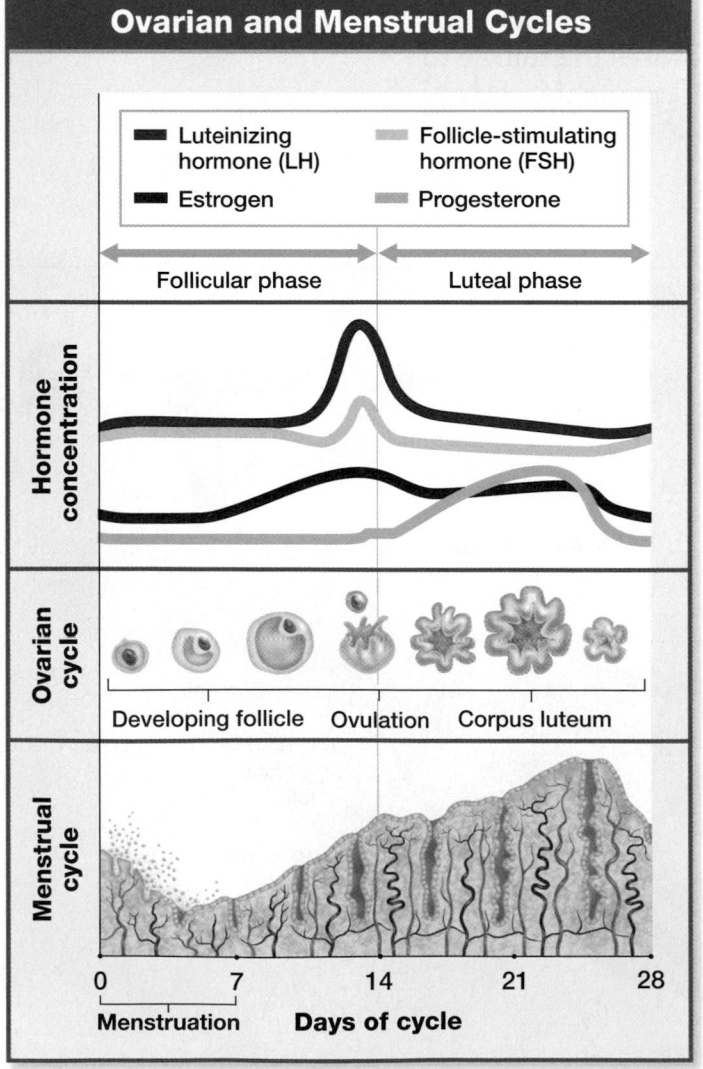

Preparation for Pregnancy

Progesterone signals the body to prepare for fertilization. If fertilization occurs, the corpus luteum continues to produce progesterone for several weeks. If fertilization does not occur, production of progesterone slows and eventually stops, marking the end of the ovarian cycle. Prescription drugs containing relatively large doses of synthetic estrogen and progesterone-like hormones have been designed to disrupt the ovarian cycle and prevent ovulation.

Menstrual cycle While changes occur in the ovaries during the ovarian cycle, changes also occur in the uterus, as shown in Figure 9. The series of changes that prepare the uterus for a possible pregnancy each month is called the **menstrual** *(MEN struhl)* **cycle.** The menstrual cycle lasts about 28 days.

The menstrual cycle is influenced by the changing levels of estrogen and progesterone during the ovarian cycle. Prior to ovulation, increasing levels of estrogen cause the lining of the uterus to thicken. After ovulation, high levels of both estrogen and progesterone cause further development and maintenance of the uterine lining. If pregnancy does not occur, the levels of estrogen and progesterone decrease. This decrease causes the lining of the uterus to shed, marking the end of the menstrual cycle. The end of the menstrual cycle coincides with the end of the luteal phase of the ovarian cycle.

Menstruation

When the lining of the uterus is shed, blood vessels break and bleeding results. A mixture of blood and discarded tissue then leaves the body through the vagina. This process, called **menstruation** *(men STRAY shuhn)*, usually occurs about 14 days after ovulation. At the end of the ovarian and menstrual cycles, estrogen and progesterone levels are low. Negative feedback of estrogen and progesterone thus causes the pituitary to again begin to produce FSH and LH, starting the cycles again.

Women eventually stop menstruation, usually between the ages of 45 and 55. After this event, called *menopause*, a woman no longer ovulates and thus moves out of the childbearing phase of her life. During menopause, many women experience symptoms, such as hot flashes, caused by a decrease in estrogen production. Estrogen, which can be taken to relieve symptoms of menopause, is a widely used prescription drug in the United States.

Analyzing Hormone Secretions

Background

The ovarian and menstrual cycles are regulated by hormones secreted by the hypothalamus, the pituitary gland, and the ovaries. Feedback mechanisms play a major role in these cycles. Use Figure 9 and the explanation in the text to answer the following questions.

Ovulation

Analysis

1. **Identify** the hormones that are secreted in large amounts prior to ovulation.

2. **Describe** the effect of estrogen production on the secretion of LH.

3. **Critical Thinking Analyzing Concepts** What type of feedback mechanism causes a decrease in the secretion of LH and FSH during the luteal phase?

4. **Critical Thinking Analyzing Concepts** What type of feedback mechanism causes the surge of LH secretion during the follicular phase?

Section 2 Review

1 **Describe** the functions of ovarian follicles.

2 **Compare** the regulatory roles of LH and FSH.

3 **Recognizing Relationships** What causes the lining of the uterus to thicken and then to be shed during the menstrual cycle?

4 **Relating Concepts** How could the maturation of an egg cell be halted in the ovary?

5 **Standardized Test Prep** When do the egg cells of a human female begin meiosis?
A before she is born   **C** during ovulation
B at the start of puberty   **D** during menstruation

Objectives

- **Sequence** the events of fertilization, cleavage, and implantation.

- **Summarize** the three trimesters of pregnancy.

- **Describe** the effects of drug use on development.

Key Terms

cleavage
blastocyst
implantation
gestation
pregnancy
embryo
placenta
fetus

Fertilization

If sperm are present in the female reproductive system within a few days after ovulation, fertilization may occur. To fertilize an ovum, a sperm cell must swim to a fallopian tube, where fertilization usually occurs. During fertilization, a sperm cell penetrates an ovum by releasing the enzymes at the tip of its head. These enzymes break down the jellylike outer layers of the ovum. The head of the sperm enters the ovum, and the nuclei of the ovum and sperm fuse together. This produces a diploid cell called a *zygote*.

Cleavage and Implantation

In the first week after fertilization, the zygote undergoes a series of internal divisions known as **cleavage**, as shown in **Figure 10.** Cleavage produces many smaller cells—first two cells, then four, then eight, and so on—within the zygote. Cleavage continues as the zygote moves through the fallopian tube toward the uterus. By the time it reaches the uterus, the zygote is a hollow ball of cells called a **blastocyst** (*BLAS toh sist*). About six days after fertilization, the blastocyst burrows into the lining of the uterus in an event called **implantation.** There it will undergo development, eventually forming a living human.

Figure 10 Early zygote development

Fertilization, cleavage, and implantation occur after ovulation.

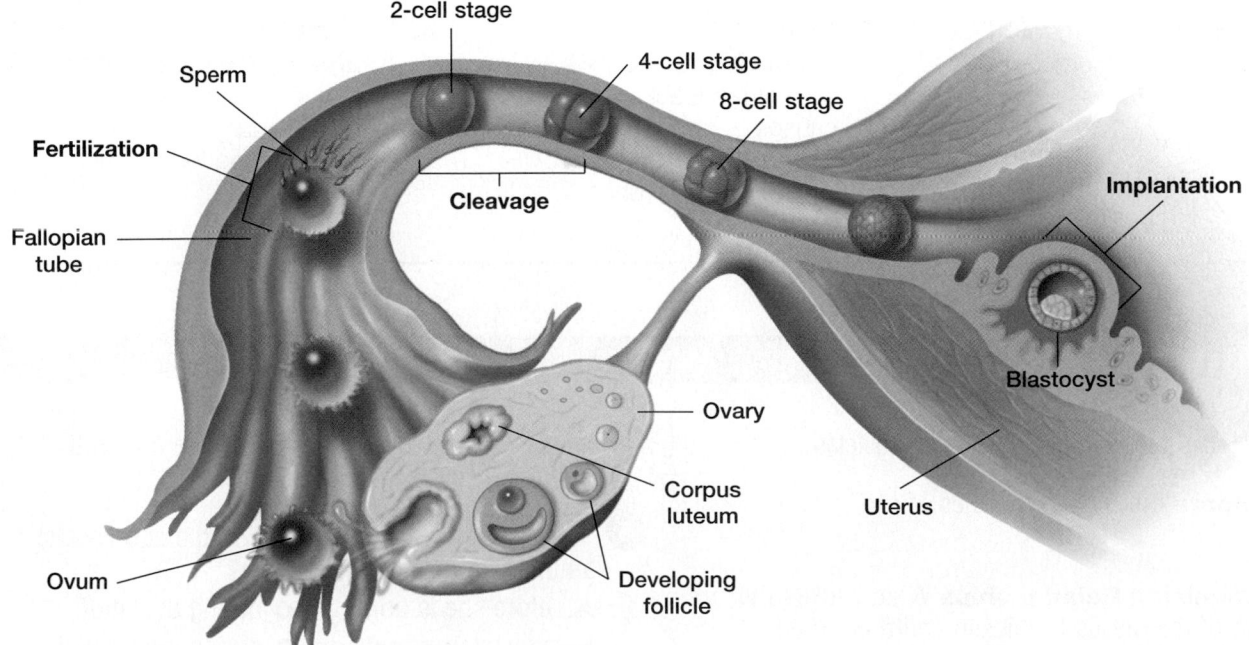

Pregnancy

Development begins with a single diploid cell from which billions of other cells arise. The uterus provides protection and nourishment during development. Human development takes about 9 months— a period known as **gestation** *(jes TAY shuhn)*, or pregnancy. The 9 months of **pregnancy** are often divided into three trimesters, or 3-month periods. For the first 8 weeks of pregnancy, the developing human is called an **embryo** *(EHM bree oh)*.

First Trimester

Supportive membranes The most crucial events of development occur very early in the first trimester. In the second week after fertilization—shortly after implantation—the embryo grows rapidly. Membranes that will protect and nourish it also develop. One of these membranes, the amnion *(AM nee ahn)*, encloses and protects the embryo. Another membrane, the chorion *(KOHR ee ahn)*, interacts with the uterus to form the placenta.

The **placenta** *(plah SEHN tah)* is the structure through which the mother nourishes the embryo. As shown in **Figure 11,** the mother's blood normally never mixes with the blood of the embryo. Instead, nutrients in the mother's blood diffuse through the placenta and are carried to the embryo through blood vessels in the umbilical *(uhm BIL i kuhl)* cord.

The waste products of the embryo also pass through the placenta into the mother's blood. Most other substances, including drugs and pathogens, can also diffuse through the placenta. Thus, if the mother ingests any harmful substances, the embryo is also affected. For example, alcohol use by pregnant women, especially during early pregnancy, is a leading cause of birth defects. Fetal alcohol

Figure 11   Placenta

The developing human is nourished through the placenta.

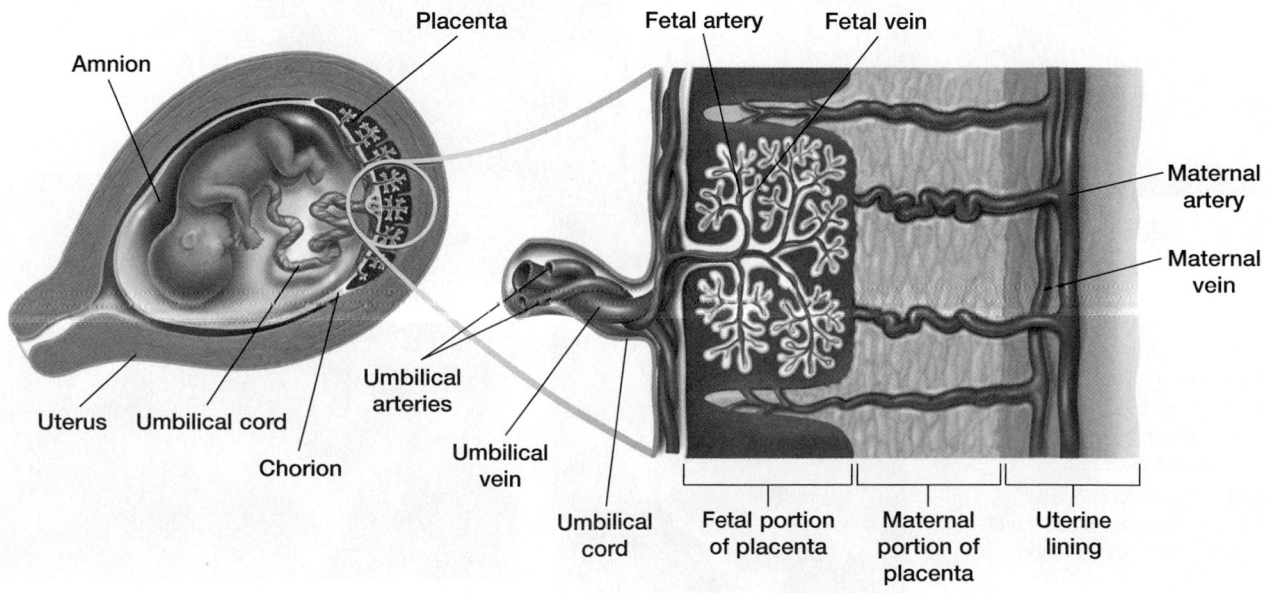

syndrome is a collection of birth defects in which a baby may have a deformed face and often severe mental, behavioral, and physical retardation. Women should abstain from alcohol and avoid all unnecessary drugs throughout pregnancy.

Development of embryo As the placenta forms, the inner cells of the blastocyst form the three primary tissue layers—endoderm, mesoderm, and ectoderm. By the end of the third week, blood vessels and the gut begin to develop, and the embryo is about 2 mm (0.08 in.) long. In the fourth week, the arms and legs also begin to form, and the embryo more than doubles in length to about 5 mm (0.2 in.). By the end of the fourth week, all of the major organs begin to form, and the heart begins to beat.

During the second month, the final stage in embryonic development takes place. The arms and legs take shape. Within the body cavity, the major internal organs, including the liver and pancreas, are evident. By the end of the second month, the embryo is about 22 mm (0.9 in.) long and weighs about 1 g (0.036 oz).

Development of fetus From the eighth week of pregnancy until childbirth, the developing human is called a **fetus** *(FEET uhs)*. By the end of the first trimester, the sex of the fetus can be distinguished. A fetus has recognizable body features, and its organ systems have begun to form, as shown in **Figure 12.**

Figure 12   Development of fetus

A fetus has recognizable body characteristics.

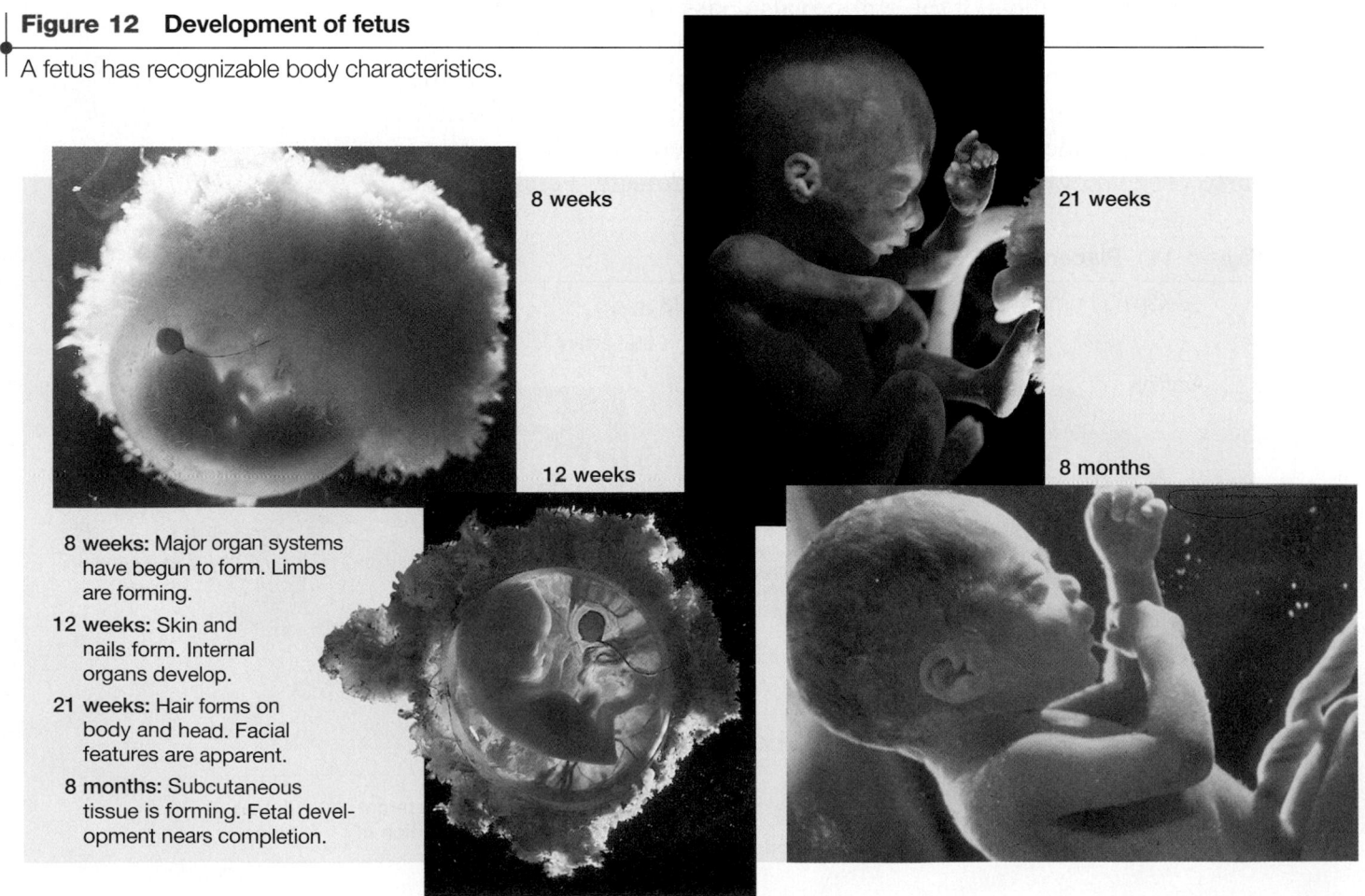

8 weeks

21 weeks

12 weeks

8 months

8 weeks: Major organ systems have begun to form. Limbs are forming.

12 weeks: Skin and nails form. Internal organs develop.

21 weeks: Hair forms on body and head. Facial features are apparent.

8 months: Subcutaneous tissue is forming. Fetal development nears completion.

Second and Third Trimesters

During the second and third trimesters, the fetus grows rapidly as its organs become functional. By the end of the third trimester, the fetus is able to exist outside the mother's body. After about 9 months of development, the fetus leaves the mother's body in a process called *labor*, which usually lasts several hours. During labor, the walls of the uterus contract, expelling the fetus from the uterus, as shown in **Figure 13.** The fetus leaves the mother's body through the vagina. The placenta and the umbilical cord are expelled after the baby is born. After birth, development is far from complete. Physical growth and neurological development continue for years after birth.

Figure 13   Childbirth

During childbirth, the fetus exits the mother's body through the vagina.

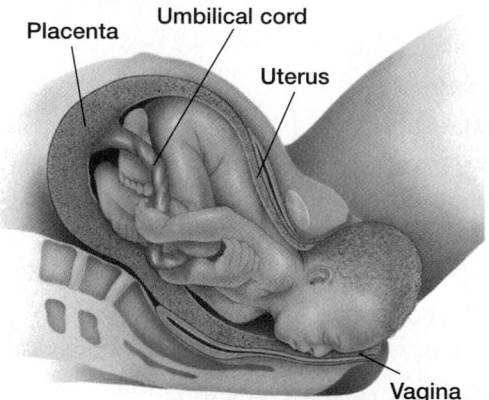

Placenta
Umbilical cord
Uterus
Vagina

BIOWatch

Ultrasound Imaging

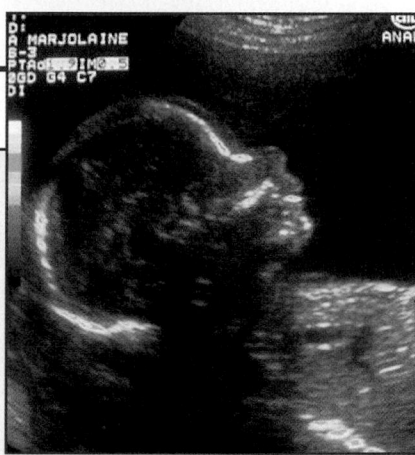

Ultrasound image of fetus

Since the 1960s, several methods have been developed for looking inside the body without surgery. One such method is ultrasound imaging. To make an ultrasound image, a physician holds a small probe against a patient's skin. The probe emits high-frequency sound waves, which produce echoes when they reflect off structures in the patient's body. The probe detects the echoes, which are converted to an image on a video screen.

Medical Uses of Ultrasound

Ultrasound measurements of the size of the embryo or fetus can indicate its age, while other signs indicate its health. For example, heart motions can usually be seen by 7 weeks. In addition, many fetal abnormalities can be diagnosed using ultrasound.

Advantages of Ultrasound

The biggest advantage of ultrasound imaging is its safety. Unlike X rays, ultrasound does not involve ionizing radiation, which can cause mutations. Ultrasound imaging has no known harmful effects.

internet connect

www.scilinks.org
Topic: Ultrasound
Keyword: HX4180

SCiLINKS. Maintained by the National Science Teachers Association

Section 3 Review

1 **Summarize** the events in development that occur in the embryo's first month.

2 **Describe** the function of the placenta.

3 **Describe** fetal alcohol syndrome.

4 **Relating Concepts** Why are some drugs harmful when they are taken during pregnancy?

5 **Predicting Outcomes** What might happen if more than one egg were released from the ovaries prior to fertilization?

6 **Standardized Test Prep** Following fertilization, cleavage begins while the zygote is in

A an ovary.   C the placenta.

B the uterus.   D a fallopian tube.

Sexually Transmitted Diseases

Objectives

- **Identify** the causes and symptoms of several bacterial STDs.

- **Identify** the causes and symptoms of some viral STDs.

- **Compare** the treatment and cure rates of viral STDs with those of bacterial STDs.

Key Terms

gonorrhea
syphilis
chlamydia
pelvic inflammatory
 disease
genital herpes

STDs

Recall that disease-causing pathogens are transmitted in many ways. Pathogens present in body fluids, such as semen, can be passed from one person to another through sexual contact. Diseases spread by sexual contact are called *sexually transmitted diseases,* or *STDs.* Both viruses and bacteria can cause STDs. **Table 1** lists several types of STDs. Abstinence is the only sure way to protect yourself from contracting an STD.

Bacterial STDs

Most STDs that are caused by bacteria can be successfully treated and cured with antibiotics. Unfortunately, the early symptoms of most bacterial STDs are very mild and often are not detected. Early detection and treatment are necessary to prevent serious consequences that can result from infection. For example, untreated bacterial STDs can cause sterility in both men and women. Three major bacterial STDs are gonorrhea, syphilis, and chlamydia.

Gonorrhea *(gahn uh REE uh)* is a bacterial STD that causes painful urination and a discharge of pus from the penis in males. In females, gonorrhea sometimes causes a vaginal discharge but more

Table 1 Sexually Transmitted Diseases		
Disease	**Symptoms**	**Pathogen**
AIDS	Immune-system failure and susceptibility to opportunistic infections	Human immunodeficiency virus (HIV)
Chlamydia	Painful urination and penile discharge in males; vaginal discharge and abdominal pain in females	*Chlamydia trachomatis* (bacterium)
Genital herpes	Painful blisters on genital region, thighs, or buttocks and flulike symptoms	Herpes simplex virus (HSV)
Genital warts	Warts on genital or anal region	Human papilloma virus (HPV)
Gonorrhea	Painful urination and penile discharge in males; vaginal discharge and abdominal pain in females	*Neisseria gonorrhoeae* (bacterium)
Hepatitis B	Flulike symptoms and yellowing of skin	Hepatitis B virus
Syphilis	Chancre on penis in males; chancre in vagina or on cervix in females; fever and rash	*Treponema pallidum* (bacterium)

often has no symptoms. In males, untreated gonorrhea can spread to the vas deferens, epididymis, or testes. In females, it can spread to the fallopian tubes and cause pain and scarring that may lead to infertility. Some strains of gonorrhea are resistant to commonly used antibiotics, such as penicillin.

Syphilis *(SIHF uh lihs)* is a serious bacterial STD that usually begins with the appearance of a small, painless ulcer called a *chancre (SHAHN kuhr)* 2–3 weeks after infection. In males, the chancre usually appears on the penis. In females, the chancre may form inside the vagina or on the cervix. If syphilis is not treated, it may cause fever, swollen lymph glands, or a rash like the one shown in **Figure 14** a few weeks after infection. These symptoms disappear without treatment. Years later, however, syphilis may cause destructive lesions on the nervous system, blood vessels, bones, and skin. A pregnant woman infected with syphilis can also transmit the disease to the fetus. As a result, the fetus may be stillborn or suffer serious damage to organ systems.

Chlamydia *(kluh MIHD ee ah)* is the most common bacterial STD in the United States. The symptoms of chlamydia are similar to those of a mild case of gonorrhea: painful urination in males and vaginal discharge in females. Like gonorrhea, chlamydia often is not detected. Chlamydia, even more than gonorrhea, is likely to cause scar tissue in infected fallopian tubes, leading to infertility, or the inability to become pregnant.

Pelvic inflammatory disease One of the most common causes of infertility in women is **pelvic inflammatory disease,** or **PID.** PID is a severe inflammation of the uterus, ovaries, fallopian tubes, or abdominal cavity that results from a bacterial STD that has gone untreated. **Figure 15** shows the damage that PID can cause in the fallopian tubes. Most cases of PID are the result of gonorrhea or chlamydia infections.

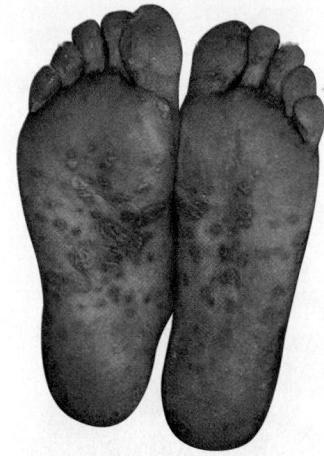

Figure 14 Syphilis.
A rash such as this one is a symptom of the second stage of syphilis. Even at this stage, syphilis can be cured by treatment with antibiotics.

Figure 15 Pelvic inflammatory disease

Most cases of PID result from gonorrhea or chlamydia infections.

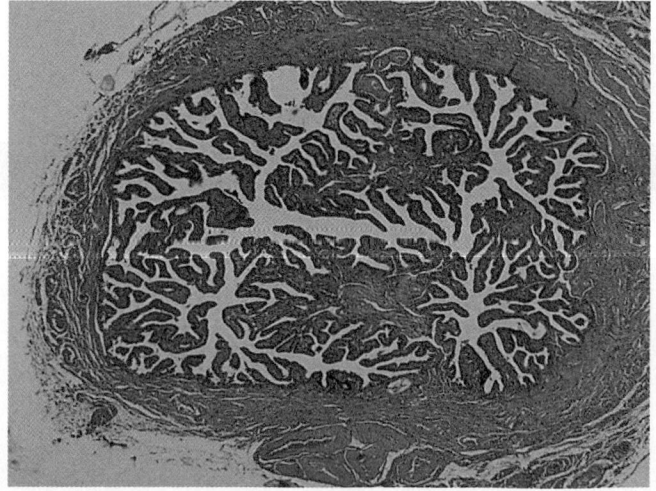

A normal fallopian tube has a highly folded lining and many spaces through which gametes can pass.

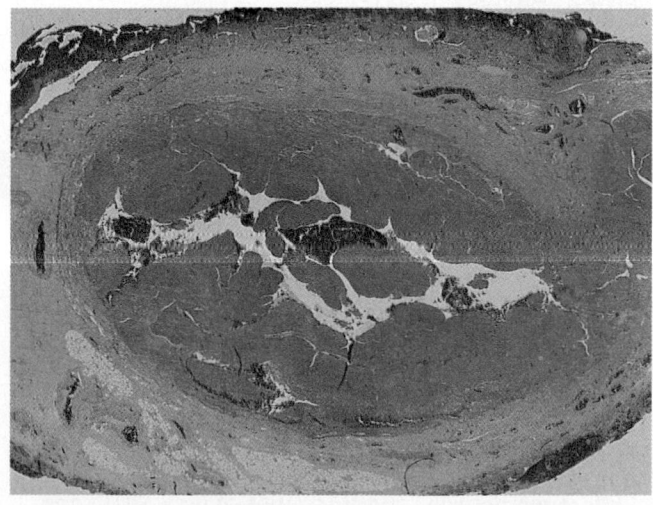

In a fallopian tube scarred by PID, many of these spaces have become blocked with tissue.

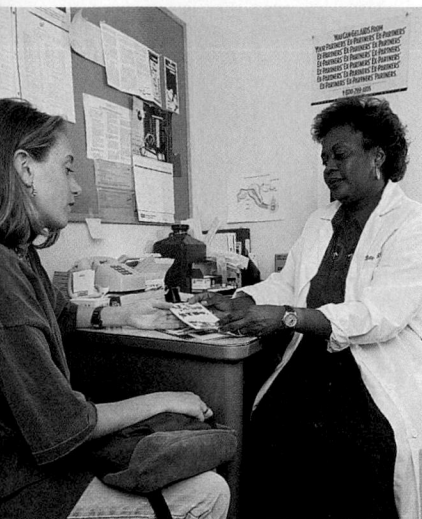

Figure 16 Counseling.
Counseling is available to
people with HIV.

Viral STDs

Because viruses are not affected by antibiotics, STDs that are caused by viruses cannot be treated and cured with antibiotics. AIDS and genital herpes are two common viral diseases that are transmitted through sexual contact. Other common viral STDs include genital warts and hepatitis B.

AIDS is a fatal disease caused by the human immunodeficiency virus (HIV). Transmission through sexual contact is the most common way that people become exposed to HIV. HIV destroys the immune system of infected individuals by attacking white blood cells. People with AIDS generally die from opportunistic infections that persist only in people with weakened immune systems.

The number of HIV infections among teenagers and young adults has increased dramatically over the last decade. AIDS is now the leading killer of African-American men between the ages of 25 and 44. More than 460,000 people in the United States have already died from AIDS. The number of young adults in the United States with AIDS has increased drastically over the last 15 years. While the number of new AIDS cases reported has decreased each year since 1993 due to improved drug treatments, new HIV infections have not decreased, and are most frequent among young adults—like the young woman shown in **Figure 16.** Researchers are trying to develop new treatments for AIDS.

Genital herpes is caused by herpes simplex virus (HSV). About 70 percent of genital herpes infections are caused by HSV-2. The rest are caused by HSV-1, which more commonly causes cold sores, or fever blisters, around and inside the mouth.

Symptoms of genital herpes include periodic outbreaks of painful blisters in the genital region, as shown in **Figure 17,** and flu-like aches and fever. Antiviral drugs can temporarily eliminate the blisters caused by genital herpes, but they cannot eliminate HSV from the body. Although genital herpes is not life threatening, it can have serious consequences. Women with genital herpes have a greater risk of developing cervical cancer. Like HIV, herpes simplex virus can be passed from mother to fetus during pregnancy or birth. Infants infected with HSV may suffer severe damage to their nervous system or even die as a result of the infection.

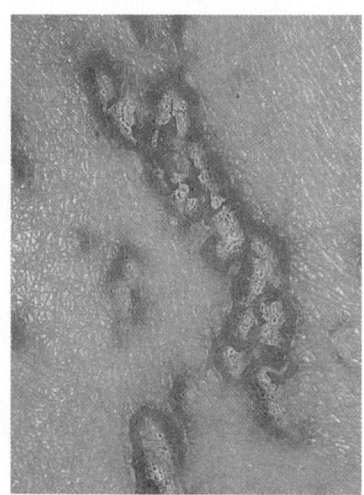

**Figure 17 Genital
herpes.** Blisters caused by
genital herpes may appear on
or near the genitalia.

Section 4 Review

1 **Name** three common STDs caused by bacteria. Why is early detection of these diseases important?

2 **Describe** how HIV weakens the immune system of an infected individual.

3 **Recognizing Differences** What is the main difference between the treatment of viral STDs and the treatment of bacterial STDs?

4 **List** three symptoms of genital herpes.

5 **Applying Information** How can you best protect yourself from contracting a sexually transmitted disease?

6 **Standardized Test Prep** A sexually transmitted disease that can be treated with antibiotics is

A genital herpes.   **C** hepatitis B.

B syphilis.   **D** AIDS.

CHAPTER HIGHLIGHTS

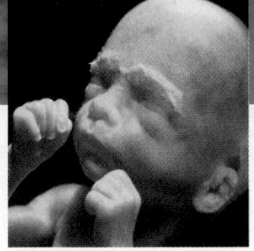

Key Concepts

1 Male Reproductive System

- Sperm cells are produced by meiosis in the testes.
- Sperm mature and are stored in each epididymis.
- A mature sperm cell consists of a head, a midpiece, and a long, powerful tail.
- Sperm move through the epididymis and the vas deferens and exit the body through the urethra.

2 Female Reproductive System

- Egg cells are produced by meiosis in the ovaries.
- An egg cell matures in a 28-day ovarian cycle.
- The menstrual cycle prepares the uterus for pregnancy.
- The menstrual and ovarian cycles are regulated by hormones.

3 Development

- After fertilization, cleavage and implantation occur.
- The human gestation period is about 9 months.
- The most crucial events of development occur during the first trimester of pregnancy.
- The mother nourishes the fetus through the placenta.
- Primary tissue layers develop into organs and tissues.
- The fetus leaves the mother's body during labor.

4 Sexually Transmitted Diseases

- Many STDs are caused by bacteria and viruses.
- Syphilis is a severe bacterial STD that can have destructive effects on the nervous system, bones, and skin if untreated.
- Gonorrhea and chlamydia are common bacterial STDs that can scar the fallopian tubes and lead to infertility.
- AIDS is a viral STD in which HIV destroys immune system cells, leaving the body vulnerable to opportunistic infections.
- Genital herpes is a viral STD that causes blistering.

Key Terms

Section 1

testes (996)
seminiferous tubules (996)
epididymis (997)
vas deferens (997)
seminal vesicles (998)
prostate gland (998)
bulbourethral glands (998)
semen (998)
penis (998)

Section 2

ovary (999)
ovum (1000)
fallopian tube (1000)
uterus (1000)
vagina (1000)
ovarian cycle (1001)
ovulation (1001)
follicle (1001)
corpus luteum (1002)
menstrual cycle (1002)
menstruation (1003)

Section 3

cleavage (1004)
blastocyst (1004)
implantation (1004)
gestation (1005)
pregnancy (1005)
embryo (1005)
placenta (1005)
fetus (1006)

Section 4

gonorrhea (1008)
syphilis (1009)
chlamydia (1009)
pelvic inflammatory disease (1009)
genital herpes (1010)

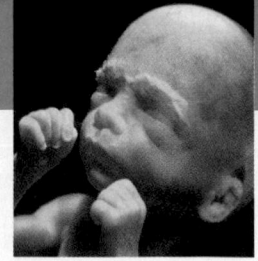

Performance ZONE

CHAPTER REVIEW

Understanding Key Ideas

1. The correct pathway of sperm is from the
   a. testes to vas deferens to epididymis.
   b. epididymis to urethra to vas deferens.
   c. testes to epididymis to vas deferens.
   d. urethra to vas deferens to testes.

2. Which of the following is *not* a function of the female reproductive system?
   a. production of gametes
   b. nourishment of the fetus
   c. maturation of eggs
   d. secretion of FSH

3. Which of the following sexually transmitted diseases *cannot* be treated with antibiotics?
   a. genital herpes   c. gonorrhea
   b. syphilis   d. chlamydia

4. Which of the following is *not* true for human development?
   a. Alcohol and drugs taken during pregnancy may harm the embryo or fetus.
   b. Crucial events occur during the first trimester of pregnancy.
   c. Drugs and alcohol taken during pregnancy cannot cause birth defects.
   d. Normal development may be affected by viral diseases.

5. A symptom associated with the earliest stage of syphilis is
   a. painful urination.
   b. blisters in the genital area.
   c. fever blisters and cold sores.
   d. a painless chancre.

6. The diagram below shows a mature human sperm cell. Explain the roles of the structures labeled A, B, and C in the sperm's ability to fertilize an egg.

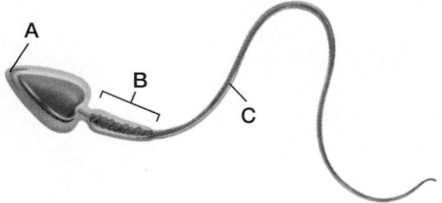

7. **Recognizing Patterns** Describe the role of feedback mechanisms in the maturation of an egg cell during the ovarian cycle of an adult female.

8. How do birth control pills, which contain synthetic estrogen and progesterone-like hormones, prevent pregnancy?

9. **BIOWatch** Describe two different medical uses for ultrasound imaging. What can be learned about a fetus by using this technology? Find out other ways ultrasound imaging is used in medicine.

10. **Concept Mapping** Make a concept map that describes the ovarian and menstrual cycles. Include the following terms in your map: *ovary, fallopian tube, uterus, ovarian cycle, follicle, ovulation, corpus luteum,* and *menstrual cycle.* Include additional terms in your map as needed.

Critical Thinking

11. **Making Inferences** A man interested in fathering children wants to know his sperm count. He finds out that he has a sperm count of fewer than 60 million sperm in a 3.5 mL sample of semen. If you were his physician, what would you tell him about the results of the test?

12. **Predicting Outcomes** What do you think would happen if more than one sperm were able to fertilize an egg?

13. **Inferring Relationships** Why should a pregnant woman eat many healthy foods?

Alternative Assessment

14. **Interpreting Information** Research some causes of and treatments for infertility for both men and women. What are the benefits and shortcomings of each type of treatment? Summarize your findings in a written report.

Standardized Test Prep

Understanding Concepts

Directions (1–5): **For *each* question, write on a separate sheet of paper the letter of the correct answer.**

1 What structure is a tube in which sperm mature?
  A. epididymis
  B. seminiferous tubules
  C. urethra
  D. vas deferens

2 Which of the following is true of the follicular phase of the ovarian cycle?
  F. stops estrogen production
  G. ends when ovulation occurs
  H. starts when fertilization occurs
  I. occurs when LH levels drop to zero

3 How are semen and sperm related?
  A. Both are stored in the bladder.
  B. Sperm and semen are both gametes.
  C. Semen is made up of fluids and sperm.
  D. Both are produced in the prostate gland.

4 When does an embryo develop endoderm, mesoderm, and ectoderm?
  F. during cleavage
  G. during fertilization
  H. during gastrulation
  I. during implantation

5 Where does fertilization usually take place?
  A. cervix   **C.** fallopian tubes
  B. epididymis   **D.** uterus

Directions (6): **For the following question, write a short response.**

6 In the 1960s, many women who took a tranquilizer called *thalidomide* early in pregnancy gave birth to babies with serious limb defects. Other women who took the drug later in pregnancy gave birth to normal babies. What does this tell you about the pattern of fetal development?

Test TIP

When analyzing a graph, pay attention to its title. The title should tell you what is plotted on the graph and provide some context for the data.

Reading Skills

Directions (7): **Read the passage below. Then answer the question.**

The first *in vitro* fertilization baby, Louise Brown, was born in England in July 1978. During an *in vitro* fertilization procedure, an egg is taken from the female, sperm is taken from the male, and fertilization occurs externally. The zygote is released into the woman's uterus two to six days later so that implantation can proceed.

7 Where does the fertilized egg finally implant once it is released into the woman's uterus?
  F. in the ovaries
  G. in the corpus luteum
  H. in the fallopian tubes
  I. in the lining of the uterus

Interpreting Graphics

Directions (8): **Base your answer to question 8 on the graph below.**

Hormone Concentrations During the Ovarian Cycle

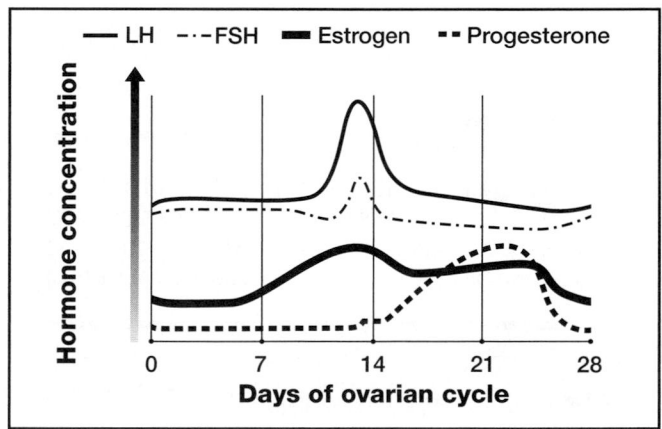

8 Approximately when during the ovarian cycle does ovulation occur?
  A. day 7
  B. day 14
  C. day 21
  D. day 28

Skills Practice Lab

Observing Embryonic Development

SKILLS

- Observing
- Comparing and contrasting
- Making drawings
- Drawing conclusions

OBJECTIVES

- **Identify** the stages of early animal development.
- **Describe** the changes that occur during early development.
- **Compare** the stages of human embryonic development with those of echinoderm embryonic development.

MATERIALS

- prepared slides of sea star development, including
- unfertilized egg
- zygote
- 2-cell stage
- 4-cell stage
- 8-cell stage
- 16-cell stage
- 32-cell stage
- 64-cell stage
- blastula
- early gastrula
- middle gastrula
- late gastrula
- compound light microscope
- paper and pencil

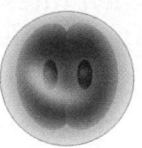

2-cell stage

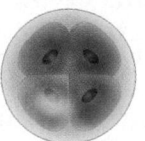
4-cell stage

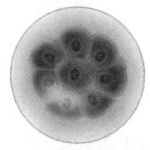
8-cell stage

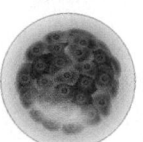
64-cell stage

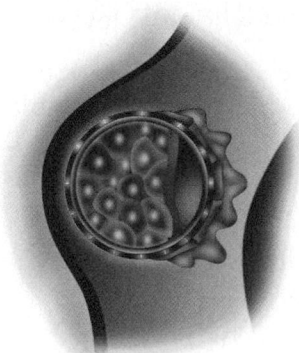
Blastocyst

Before You Begin

Most members of the animal kingdom begin life as a single cell—the fertilized egg, or **zygote.** The early stages of development are quite similar in different species. Cleavage follows fertilization. During cleavage, the zygote divides many times without growing. The new cells migrate and form a hollow ball of cells called a **blastula.** The cells then begin to organize into the three primary germ layers: endoderm, mesoderm, and ectoderm. During this process, the developing organism is called a **gastrula.**

1. Write a definition for each boldface term in the preceding paragraph.

2. Based on the objectives for this lab, write a question you would like to explore about embryonic development.

Procedure

1. Obtain a set of prepared slides that show star eggs at different stages of development. Choose slides labeled unfertilized egg, zygote, 2-cell stage, 4-cell stage, 8-cell stage, 16-cell stage, 32-cell stage, 64-cell stage, blastula, early gastrula, middle gastrula, late gastrula, and young sea star larva. (Note: *Blastula* is the general term for the embryonic stage that results from cleavage. In mammals, a blastocyst is a modified form of the blastula.)

2. Examine each slide using a compound light microscope. Using the microscope's low-power objective first, focus on one good example of the developmental stage listed on the slide's label. Then switch to the high-power objective, and focus on the image with the fine adjustment.

3. In your lab report, draw a diagram of each developmental stage that you examine (in chronological order). Label each diagram with the name of the stage it represents and the magnification used. Record your observations as soon as they are made. Do not redraw your diagrams. Draw only what you see; lab drawings do not need to be artistic or elaborate. They should be well organized and include specific details.

4. Compare your diagrams with the diagrams of human embryonic stages shown at left.

5. Clean up your materials and wash your hands before leaving the lab.

Analyze and Conclude

1. **Summarizing Results** Compare the size of the sea star zygote with that of the blastula. At what stage does the embryo become larger than the zygote?

2. **Analyzing Data** What is the earliest stage in which all of the cells in the embryo no longer look exactly alike? How do cell shape and size change during successive stages of development?

3. **Drawing Conclusions** From your observations of changes in cellular organization, why do you think the blastocoel (the space in the center of the hollow sphere of cells of a blastula) is important during embryonic development?

4. **Predicting Patterns** How are the symmetries of a sea star embryo and a sea star larva different from the symmetry of an adult sea star? Would you expect to see a similar change in human development? What must happen to the sea star gastrula before it becomes a mature sea star?

5. **Further Inquiry** How do your drawings of sea star embryonic development compare with those of human embryonic development? Based on your observations, in what ways do you think sea star embryos could be used to study early human development?

? Do You Know?

Do research in the library or media center to answer these questions:

1. How do identical twins develop?
2. How do fraternal twins develop?
3. What is *in vitro* fertilization?

Use the following Internet resources to explore your own questions about embryonic development and cloning.

internet connect
www.scilinks.org
Topic: Cloning
Keyword: HX4047

SC*LINKS* Maintained by the National Science Teachers Association

What Is an Active Reader?

Active reading means thinking and interacting with the text before, during, and after reading it. You might be wondering, "Isn't everyone *thinking* while they are reading?" Not necessarily. Many students read without consciously directing their thinking. Active readers are different—they have developed habits of thinking to use *before, during,* and *after* they read. They use these habits to successfully reach their goal: to understand what they're reading about and remember it well enough to be able to pass the test.

Before You Read

Active readers begin by getting familiar with the reading assignment. This is called previewing. The word *preview* means "to look before." Looking over the chapter before you read helps you predict what you will be learning. It can also help you recall what you already know about a subject—this can help you learn more material and remember it longer.

Scan the Horizon

An active reader will turn to the first page of a chapter, ask questions, and use the information on the page to find answers. Here are some questions to ask.

1. **What do I know?** Questions at the beginning of each chapter in the **Quick Review** prompt you to remember vital information from previous chapters.

2. The **Reading Activity** will help you more fully engage the text. Many Reading Activities encourage you to survey your own knowledge of the chapter's topic. All Reading Activities will help you become more aware of the information that will be presented in the chapter.

3. **What's here?** If you're in an unfamiliar city, you might use a road map to find your way around. In this textbook, the section titled **Looking Ahead** is your map, preparing you for what's coming up. Take time to really look at the organization of the chapter. This organization tells you the focus and the purpose of the chapter.

Get to Know the Neighborhood

Now you're ready to begin the first section. Start by recognizing the elements that are there to help you:

- **Objectives** tell you what you should be able to do after reading the section. Read the objectives before you read the section.
- **Key Terms** are the lesson's key vocabulary words listed in order of appearance. They appear in **bold** black print, highlighted in yellow, within the lesson.

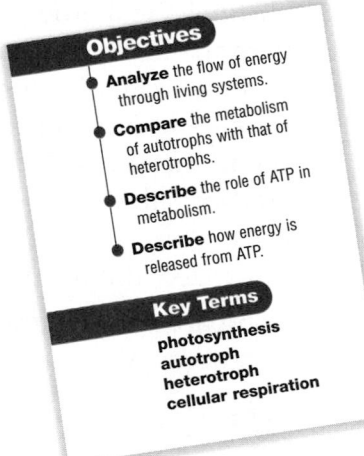

Find Your Way

As you look over the chapter, you will notice text headings in **bold** red print and text subheadings in **bold** blue print. Each red heading is a main idea for the lesson. Red headings are always located at the top of a page, making it easy for you to scan the chapter and find the main ideas. Smaller, blue headings are words or phrases. These subheadings identify important topics that are covered under each red heading. As you read, notice that bold blue print is used for references to nearby photos, illustrations, and tables (ex. **Figure 1, Table 2**).

As You Read

As they read, active readers remain self-aware. They ask themselves questions about how well they understand what they are reading. As you read the lesson, try asking yourself questions.

Monitor your understanding.
- What did I just learn from that paragraph?
- Do I need to read this paragraph again?
- What can I say about that illustration?
- Can I respond to the Objectives now?

Express what you know.
- What can I say about what I just learned?

Ask more questions.
- What is it that I want to know now?
- How does this affect my life?

Study Tips along the way will give you tips for improvement. Look for the following types of Study Tips as you read.

Reading Effectively strategies for finding meaning, understanding content, and identifying main ideas

Organizing Information tips on how to organize what you're learning

Comparing and Contrasting recognizing similarities and differences

Reviewing Information how to review—and remember—what you're reading

Interpreting Graphics suggestions for understanding graphs and illustrations

After You Read

When finished reading, active readers take a few moments to assess their overall level of understanding. Ask yourself:

- What can I say about each illustration or figure?
- Did I learn what I expected to learn?
- What do I know about each lesson title, Objective, heading, subheading, figure, or Key Term in the text?
- What can I tell others about what I know?

Check Your Comprehension

At the end of each section, **Section Review** questions will let you see how much you have learned. Each Section Review question measures how well you mastered an Objective at the beginning of the section. If you cannot answer a Section Review question, go back to the section of text that covers the Objective and re-read the information on that page.

Are You Ready for the Test?

When you have finished reading the chapter, turn to the **Study Zone** page at the end of the chapter to review the **Key Concepts** (main ideas) and the **Key Terms.** If you understand the Key Concepts and Key Terms, then you may answer the questions in the **Performance Zone** to see how well you've learned the material.

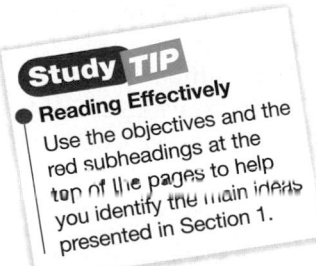

Study TIP

Reading Effectively
Use the objectives and the red subheadings at the top of the pages to help you identify the main ideas presented in Section 1.

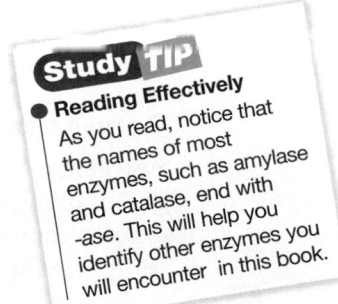

Study TIP

Reading Effectively
As you read, notice that the names of most enzymes, such as amylase and catalase, end with -ase. This will help you identify other enzymes you will encounter in this book.

WORD Origins
The word *filial* is from the Latin *filialis*, meaning "of a son or daughter." Knowing this makes it easier to remember that the F generations refer to any generation following the parental (P) generation.

What's the Best Way to Study?

Although there are many ways to study, *concept mapping* may help you understand ideas by showing you their connections to other ideas. A concept map not only identifies the major concepts from a chapter or your class notes but also shows the relationships between the concepts, much as a road map illustrates how highways and other roads are linked to cities.

Identifying Concepts

Suppose you have just finished reading a section of a chapter and you would like to make a concept map as a study aid. How do you begin? First you need to identify the concepts in that section. For instance, examine the following words:

- pool
- grass
- tree
- water
- sky
- playing
- biking
- raining
- thinking

All of these words are concepts, which usually form a picture in your mind. Now read the following series of words:

- the
- to
- has
- when
- was
- be
- with
- can

Are they concepts? No. They do not form a picture in your mind. They are linking words. Linking words play an important role in concept mapping; you use them to connect concepts in your map. See if you can identify the concepts and linking words in the concept map shown in **Figure A** on the facing page. As you can see, the concepts are listed in boxes, and the linking words and lines are used to connect

the concepts. Notice that one linking word can be used to link several concepts. You can also use arrows to help link concepts more clearly.

Organizing Concepts

Some concepts are more general and include other concepts. These general concepts will be the main ideas in your map. Determine the main idea for the following list of concepts:

- Tokyo
- Mexico City
- Seoul
- New York
- Bombay

They are cities, of course. Now determine the main idea for the following list:

- car
- bus
- train
- bicycle
- truck

Each of these concepts is an example of a vehicle.

In a concept map, the main concept should go at the top. The more specific concepts and examples should go below. Capitalize the first letter of the main concept. Write the other concepts using all lowercase letters. For example, in the concept map shown in Figure A, the main concept is "Biology." The more specific concepts listed below it in the map, such as "taxonomy," describe biology. Study this concept map before you try one yourself. Then read the following paragraph and make a list of any important words you think should be defined.

What is life? What is the difference between living and nonliving things? If you were in a wilderness area, it would be easy for you to pick out the living and nonliving things. The animals and plants are the living organisms. Organisms are made of substances organized into living systems. The rocks, air, water, and soil you see are nonliving. They contribute substances to the living organisms.

Connecting Ideas

Two general ideas prevail in the preceding paragraph: living and nonliving. You could make two separate concept maps, the main concept of one being "Living" and the other being "Nonliving,"

> ### Study TIP
> ● **Organizing Information**
> You can use concept maps as a study guide to help you organize and review information.

Figure A

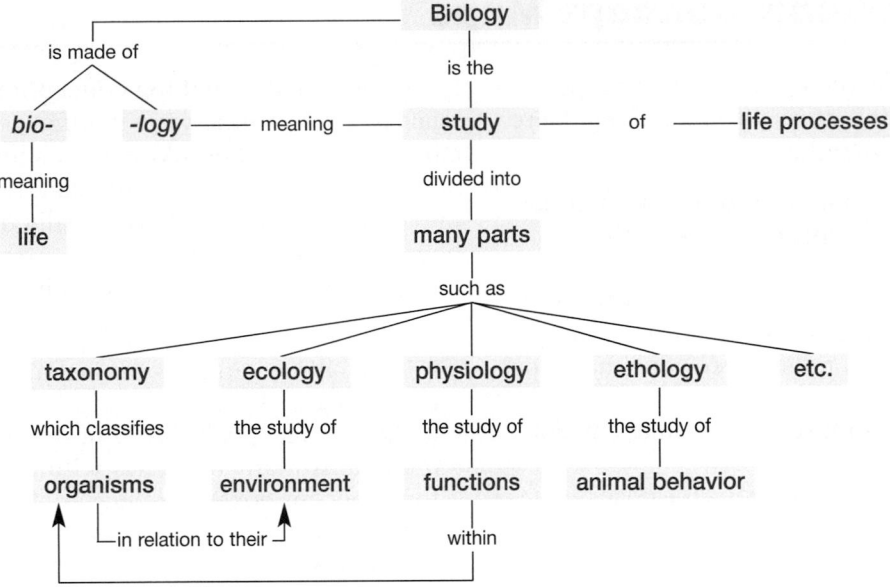

or you could make one map and use "Natural things" as the main concept. Natural things can be living or nonliving. Therefore, the concepts "living" and "nonliving" are parts of the main concept "Natural things" and should be placed below it in the map.

You have learned that living things are organisms and that they are made of substances that are organized into living systems. Living things are different from but related to nonliving things. So you should make the connection between living and nonliving things near the top of the concept map. The rest of the map should relate living and nonliving things. Nonliving things contribute substances that are used by living systems.

Listing Concepts

The next step is to put the concepts in order from the most general to the most specific. You can do this by writing the concepts on pieces of paper. Some concepts will share the same rank and be equally specific. Remember that the examples are the most specific and will be at the bottom of the map.

Now begin to rearrange the concepts you have written on the pieces of paper. Start with the most general; get the main idea. Then, if there are two or more concepts that are equally specific, place them on the same level. For example, "Natural

things" is the main idea. Then "living" and "nonliving" are placed under "Natural things" on the same row or level.

Now continue to lay out all the other concepts under the subconcepts in the first row or level until you have used them all. You can rearrange your pieces of paper any time, so keep pushing them around as if they were pieces of a jigsaw puzzle until you have arranged them the way you think they belong.

Linking Concepts

Now make the connections between the concepts. Use lines to connect the concepts, and write linking words on the lines to show or tell why they are connected. Use linking words for all the lines connecting all the concepts. Glue or tape down your concept papers if you want to make the map permanent, or use a separate piece of paper to draw a sketch showing the way you have arranged the concepts.

Now you have the completed concept map shown in **Figure B** on the following page. If you had a choice between reading the paragraph or looking at this map, you would probably agree that the map shows the concepts more clearly. This map gives you the main idea more quickly, and it is easier to understand all the ideas because their relationships to other ideas are shown.

Features of Good Concept Maps

Remember, practice is the key to good concept mapping. You will get better as you go along. Here are some things to remember:

- A concept map does not have to be symmetrical. It can have more concepts on one side than on the other.
- There are no "perfect" concept maps, only maps that come closer to the meanings of the concepts. As the mapmaker, you must make your map work for you.
- Do not put more than three words in a concept box.
- Do not have more than four concept boxes in a row without branching out.
- Connect every pair of concepts with linking words. Use as few linking words as possible.

If the relationships you have made between any two concepts are wrong, your teacher will help you sort out your misconception. Even if your relationships are absolutely correct, maps made by your classmates may be different. These maps could be equally correct, even though they may look nothing like yours. Everyone thinks a little bit differently, and as a result, other people may see different relationships between certain concepts.

As you practice making concept maps, your teacher will examine your linking words more closely. Because the linking words and the lines between concept boxes relate concepts, your linking words will tell you if you really understand how concepts are connected.

A concept map should always have the following characteristics:

- It is two dimensional—not just a list of concepts connected by lines.
- It shows concepts in order of importance.
- It contains many branches with no more than four concept boxes in a row and no more than three words in each concept box.
- It contains only concepts in the boxes and only linking words on the lines.

Evaluating Your Skills

For the first map you make on your own, think about something you know very well. Do you play a team sport or an individual sport? Do you have a hobby? Do you enjoy a particular kind of music? Whatever topic you choose, use it as the main concept for your concept map. This will be more fun and easier because you know this topic so well.

Figure B

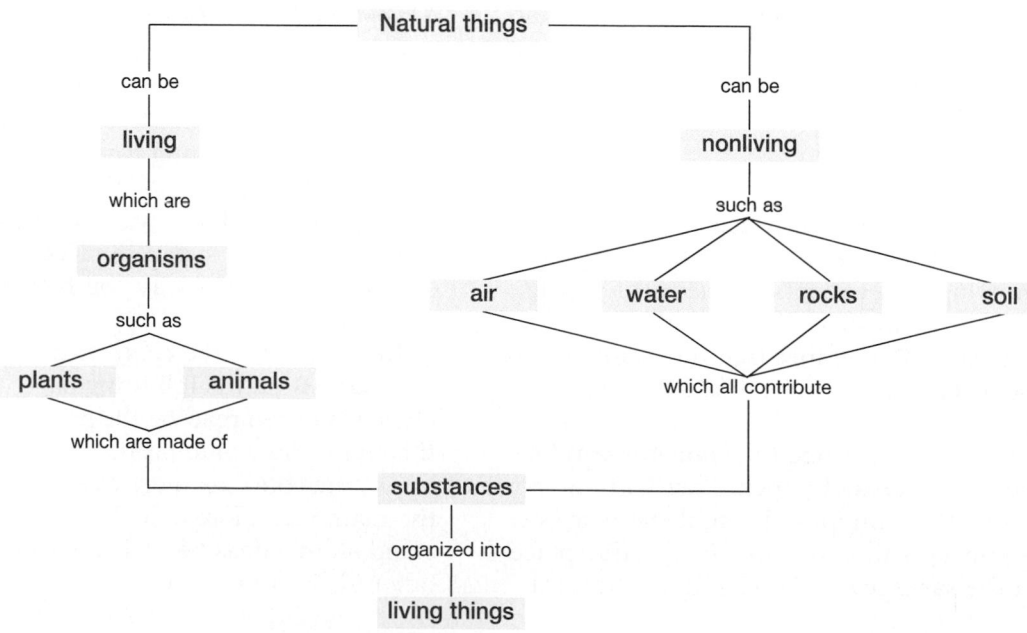

Determining the Meanings of Words

The challenge of understanding a new word can often be simplified by carefully examining the word. Many words can be divided into three parts: a prefix, root, and suffix. The **prefix** consists of one or more syllables placed in front of a **root.** The root is the main part of the word. The **suffix** consists of one or more syllables at the end of a root. Prefixes and suffixes modify or add to the meaning of the root. A knowledge of common prefixes, roots, and suffixes can give you clues to the meaning of unfamiliar words and can help make learning new words easier. For example, each of the word parts in **Table 1** can be combined with the root *derm* to form a word.

Table 2 lists word prefixes and suffixes commonly used in biology. Each word part is followed by its usual meaning, an example of a word in which it is used, and a definition of that word. Examine the definition and the example. Decide whether each word part in the first column is a prefix or suffix, depending on how the word part is used in the example.

Table 1 Word Parts

Prefix	Root	Suffix
hypo-	derm	*-ic*
pachy-	derm	
	derm	*-atology*
	derm	*-atologist*
	derm	*-atitis*

Use Table 1 to form five words using the root *derm.* Then use the list of word parts and their definitions to write what you think is each word's meaning. An example is shown below.

Example:	**Dermatologist**
derm (root):	skin
-logy (suffix):	the study of
-ist (suffix):	someone who practices or deals with something
dermatologist:	someone who studies or deals with skin

Table 2 Word Prefixes and Suffixes

Prefix or suffix	Definition	Example
a-	not, without	asymmetrical: not symmetrical
ab-	away, apart	abduct: move away from the middle
-able	able	viable: able to live
ad-	to, toward	adduct: move toward the middle
amphi-	both	amphibian: type of vertebrate that lives both on land and in water
ante-	before	anterior: front of an organism
anti-	against	antibiotic: substance, such as penicillin, capable of killing bacteria
arche-	ancient	*Archaeopteryx: a* fossilized bird
arthro-	joint	arthropod: jointed-limbed organism belonging to the phylum Arthropoda
auto-	self, same	autotrophic: able to make its own food
bi-	two	bivalve: mollusk with two shells
bio-	life	biology: the study of life
blast-	embryo	blastula: hollow ball stage in the development of an embryo
carcin-	cancer	carcinogenic: cancer-causing
cereb-	brain	cerebrum: part of the vertebrate brain
chloro-	green	chlorophyll: green pigment in plants needed for photosynthesis

Table 2 Word Prefixes and Suffixes

Prefix or suffix	Definition	Example
chromo-	color	chromosome: structure found in eukaryotic cells that contains DNA
chondro-	cartilage	Chondrichthyes: cartilaginous fish
circ-	around	circulatory: system for moving fluids through the body
-cide	kill	insecticide: a substance that kills insects
co-, con-	with, together	conjoined twins: identical twins physically joined by a shared portion of anatomy at birth
-cycle	circle	pericycle: layer of plant cells
cyt-	cell	cytology: the study of cells
de-	remove	dehydration: removal of water
derm-	skin	dermatology: study of the skin
di-	two	diploid: full set of chromosomes
dia-	through	dialysis: separating molecules by passing them through a membrane
ecol-	dwelling, house	ecology: the study of living things and their environments
ecto-	outer, outside	ectoderm: outer germ layer of developing embryo
-ectomy	removal	appendectomy: removal of the appendix
endo-	inner, inside	endoplasm: cytoplasm within the cell membrane
epi-	upon, over	epiphyte: plant growing upon another plant
ex-, exo-	outside of	exobiology: the search for life elsewhere in the universe
gastro-	stomach	gastropod: type of mollusk
-gen	type	genotype: genes in an organism
-gram	write or record	climatogram: depicting the annual precipitation and temperature for an area
hemi-	half	hemisphere: half of a sphere
hetero-	different	heterozygous: different alleles inherited from parents
hist-	tissue	histology: the study of tissues
homeo-	the same	homeostasis: maintain a constant condition
hydro-	water	hydroponics: growing plants in water instead of soil
hyper-	above, over	hypertension: blood pressure higher than normal
hypo-	below, under	hypothalamus: part of the brain located below the thalamus
-ic	of or pertaining to	hypodermic: pertaining to under the skin
inter-	between, among	interbreed: breed within a family or strain
intra-	within	intracellular: inside a cell
iso-	equal	isogenic: having an identical set of genes
-ist	someone who practices or deals with something	biologist: someone who studies life
-logy	study of	biology: the study of life
macro-	large	macromolecule: large molecule, such as DNA or proteins

Table 2 Word Prefixes and Suffixes

Prefix or suffix	Definition	Example
mal-	bad	malnourishment: poor nutrition
mega-	large	megaspore: larger of two types of spores produced by some ferns and flowering plants
meso-	in the middle	mesoglea: jellylike material found between outer and inner layers of coelenterates
meta-	change	metamorphosis: change in form
micro-	small	microscopic: too small to be seen with unaided eye
mono-	one, single	monoploid: one set of alleles
morph-	form	morphology: study of the form of organisms
neo-	new	neonatal: newborn
nephr-	kidney	nephron: functional unit of the kidneys
neur-	neuron	neurotransmitter: chemical released by a neuron
oo-	egg	oogenesis: gamete formation in female diploid organisms
org-	living	organism: living thing
-oma	swelling	carcinoma: cancerous tumor
orth-	straight	orthodontics: the practice of straightening teeth
pachy-	thick	pachyderm: thick-skinned animal, such as an elephant
para-	near, on	parasite: organism that lives on and gets nutrients from another organism
path-	disease	pathogen: disease-causing agent
peri-	around	pericardium: membrane around the heart
photo-	light	phototropism: bending of plants toward light
phyto-	plants	phytoplankton: plankton that consists of plants
poly-	many	polypeptide: sequence of many amino acids joined together to form a protein
-pod	foot	pseudopod: false foot that projects from the main part of an amoeboid cell
pre-	before	prediction: a forecast of events before they take place
-scope	instrument used to see something	microscope: instrument used to see very small objects
semi-	partially	semipermeable: allowing some particles to move through
-some	body	chromosome: structure found in eukaryotic cells that contains DNA
sub-	under	substrate: molecule on which an enzyme acts
super-, supra-	above	superficial: on or near the surface of a tissue or organ
syn-	with	synapse: junction of a neuron with another cell
-tomy	to cut	appendectomy: operation in which the appendix is removed
trans-	across	transformation: the transfer of genetic material from one organism to another
ur-	referring to urine	urology: study of the urinary tract
visc-	organ	viscera: internal organs of the body

It is your responsibility to protect yourself and other students by conducting yourself in a safe manner while in the laboratory. You can avoid accidents in the laboratory by following directions, handling materials carefully, and taking your work seriously. Read the following general safety guidelines below before attempting to do work in the laboratory. Make sure you understand all safety guidelines before entering the laboratory. If necessary, ask your teacher for clarification of laboratory rules and procedures.

General Guidelines for Laboratory Safety

- **Only perform experiments specifically assigned by your teacher.** Do not attempt any laboratory procedure without your teacher's direction, and do not work alone in the laboratory.

- **Familiarize yourself with the investigation and all safety precautions before entering the lab.** Be aware of the potential hazards of the required materials and procedures. Before you begin, ask your teacher to explain any parts of an investigation that you do not understand.

- **Before beginning work, tie back long hair, roll up loose sleeves, and put on any required personal protective equipment as directed by your teacher.** Avoid or confine loose clothing that could knock things over, catch on fire, or absorb chemical solutions. Nylon and polyester fabrics burn and melt more readily than does cotton. Do not wear open-toed shoes, sandals, or canvas shoes in the laboratory.

- **Always wear a lab apron and safety goggles.** Wear this equipment even if you are not working on an experiment. Laboratories contain chemicals that can damage your clothing, skin, and eyes. If your safety goggles cloud up or are uncomfortable, ask your teacher for help. Lengthening the strap slightly, washing the goggles with soap and warm water, or using an anti-fog spray may help the problem.

- **No contact lenses are allowed in the lab.** Even if you are wearing safety goggles, chemicals could get between contact lenses and your eyes and cause irreparable eye damage. If your doctor requires that you wear contact lenses instead of glasses, then you should wear eye-cup safety goggles—similar to goggles worn for underwater swimming—in the lab. Ask your doctor or your teacher how to use eye-cup safety goggles to protect your eyes.

- **Know the location of all safety and emergency equipment used in the laboratory.** Ask your teacher where the nearest eyewash stations, safety blankets, safety shower, fire extinguisher, first-aid kit, and chemical spill kit are located.

- **Immediately report any accident, incident, or hazard—no matter how trivial—to your teacher.** Any incident involving bleeding, burns, fainting, chemical exposure, or ingestion should also be reported immediately to the school nurse or to a physician.

- **In case of fire, alert your teacher and leave the lab.** Standard fire-safety procedures should be followed.

- **Do not have or consume food or drink in the lab.** Do not store or eat food in the laboratory.

- **Do not fool around in the lab.** Take your lab work seriously and behave appropriately in the laboratory. Be aware of your classmates' safety as well as your own at all times.

- **Do not apply cosmetics in the lab.** Some hair-care products and nail polish are highly flammable.

- **Keep your work area neat and uncluttered.** Have only books and other materials that are needed to conduct the experiment in the laboratory.

- **Clean your work area at the conclusion of each lab period as directed by your teacher.** Broken glass, chemicals, and other laboratory waste products should be disposed of in separate special containers. Dispose of waste materials as directed by your teacher.

- **Wash your hands with soap and hot water after each lab period.** Wash your hands at the conclusion of each lab period and before leaving the laboratory to avoid contamination.

Key to Safety Symbols and Their Precautions

Before you begin working in the laboratory, familiarize yourself with the following safety symbols, which are used throughout this textbook, and guidelines that you should follow when you see these symbols.

 Eye Safety

- **Wear approved safety goggles as directed.** Safety goggles should always be worn in the laboratory, especially when you are working with a chemical or solution, a heat source, or a mechanical device.

- **In case of eye contact, do the following:** Go to an eyewash station immediately and flush your eyes (including under the eyelids) with running water for at least 15 minutes. Hold your eyelids open with your thumb and fingers, and roll your eyeball around. While doing this, have another student notify your teacher.

- **Do not wear contact lenses in the lab.** Chemicals can be drawn up under a contact lens and into the eye. If you must wear contacts prescribed by a physician, tell your teacher. You must also wear approved eye-cup safety goggles to help protect your eyes.

- **Do not look directly at the sun through any optical device or lens system, and do not reflect direct sunlight to illuminate a microscope.** Such actions concentrate light rays to an intensity that can severely burn your retinas, possibly causing blindness.

 Hand Safety

- **Do not cut objects while holding them in your hand.** Dissect specimens in a dissecting tray.

- **Wear protective gloves when working with an open flame, chemicals, solutions, or wild or unknown plants.**

 Safety with Gases

- **Do not inhale any gas or vapor unless directed to do so by your teacher.** Never inhale pure gases.

- **Handle materials that emit vapors or gases in a well-ventilated area.** This work should be done in an approved chemical fume hood.

 Sharp-Object Safety

- **Use extreme care when handling all sharp and pointed instruments, such as scalpels, sharp probes, and knives.**

- **Do not use double-edged razor blades in the laboratory.**

- **Do not cut objects while holding them in your hand.** Cut objects on a suitable work surface. Always cut in a direction away from your body.

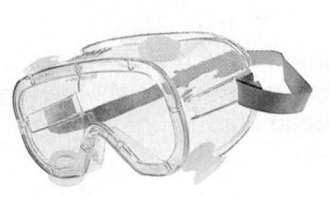

 Animal Care and Safety

- **Do not approach or touch any wild animals.** When working outdoors, be aware of poisonous or dangerous animals in the area.

- **Always get your teacher's permission before bringing any animal (including pets) into the school building.**

- **Handle animals only as directed by your teacher.** Mishandling or abusing any animal will not be tolerated.

 Heating Safety

- **Be aware of any source of flames, sparks, or heat (open flames, electric heating coils, hot plates, etc.) before working with flammable liquids or gases.**

- **When heating chemicals or solutions in a test tube, do not point the test tube toward anyone.**

- **Avoid using open flames.** If possible, work only with hot plates that have an "On-Off" switch and an indicator light. Do not leave hot plates unattended. Do not use alcohol lamps. Turn off hot plates and open flames when they are not being used.

- **Know the location of laboratory fire extinguishers and fire-safety blankets.**

- **Use tongs or appropriate insulated holders when handling heated objects.** Heated objects often do not appear to be hot. Do not pick up an object with your hand if it could be warm.

- **Keep flammable substances away from heat, flames, and other ignition sources.**

 Hygienic Care

- **Keep your hands away from your face and mouth while working in the lab.**

- **Wash your hands thoroughly before leaving the laboratory.**

- **Remove contaminated clothing immediately.** If you spill caustic substances on your skin or clothing, use the safety shower or a faucet to rinse. Remove affected clothing while under the shower, and call to your teacher. (It may be temporarily embarrassing to remove clothing in front of your classmates, but failure to rinse a chemical off your skin could result in permanent damage.)

- **Launder contaminated clothing separately.**

- **Use the proper technique demonstrated by your teacher when handling bacteria or other microorganisms.** Treat all microorganisms as if they are pathogens. Do not open Petri dishes to observe or count bacterial colonies.

- **Return all stock and experimental cultures to your teacher for proper disposal.**

 Glassware Safety

- **Inspect glassware before use; do not use chipped or cracked glassware.** Use heat-resistant glassware for heating materials or storing hot liquids.

- **Do not attempt to insert glass tubing into a rubber stopper without specific instruction from your teacher.**

- **Immediately notify your teacher if a piece of glassware breaks. Do not attempt to clean up broken glass.**

 Proper Waste Disposal

- **Clean and sanitize all work surfaces and personal protective equipment after each lab period as directed by your teacher.**

- **Dispose of all sharp objects (such as broken glass) and other contaminated materials (biological or chemical) in special containers as directed by your teacher.**

Electrical Safety

- **Do not use equipment with frayed electrical cords or loose plugs.**

- **Fasten electrical cords to work surfaces using tape.** This will prevent tripping and will ensure that equipment cannot fall off the table.

- **Do not use electrical equipment near water or with wet hands or clothing.**

- **Hold the rubber cord when you plug in or unplug equipment.** Do not touch the metal prongs of the plug, and do not unplug equipment by pulling on the cord.

 ## Clothing Protection

- **Wear an apron or laboratory coat at all times in the laboratory to prevent chemicals or chemical solutions from contacting skin or clothes.**

 ## Plant Safety

- **Do not ingest any plant part used in the laboratory (especially commercially sold seeds).** Do not touch any sap or plant juice directly. Always wear gloves.

- **Wear disposable polyethylene gloves when handling any wild plant.**

- **Wash hands thoroughly after handling any plant or plant part (particularly seeds). Avoid touching your face and eyes.**

- **Do not inhale or expose yourself to the smoke of any burning plant.** Smoke contains irritants that can cause inflammation in the throat and lungs.

- **Do not pick wildflowers or other plants unless directed by your teacher.**

 ## Chemical Safety

- **Always wear safety goggles, gloves, and a lab apron or coat when working with any chemical or chemical solution to protect your eyes and skin.**

- **Do not taste, touch, or smell any chemicals or bring them close to your eyes, unless specifically instructed by your teacher.** If you are directed by your teacher to note the odor of a substance, do so by waving the fumes toward you with your hand. Do not pipette any chemicals by mouth; use a suction bulb as directed by your teacher.

- **Know the location of the emergency lab shower and eyewash and how to use them.** If you get a chemical on your skin or clothing, wash it off at the sink while calling to your teacher.

- **Always handle chemicals or chemical solutions with care.** Check the labels on bottles, and observe safety procedures. Do not return unused chemicals or solutions to their original containers. Return unused reagent bottles or containers to your teacher.

- **Do not mix any chemicals unless specifically instructed by your teacher.** Otherwise harmless chemicals can be poisonous or explosive if combined.

- **Do not pour water into a strong acid or base.** The mixture can produce heat and splatter.

- **Report any spill immediately to your teacher.** Spills should be cleaned up promptly as directed by your teacher.

 **ChemSafety**

CAUTION: Always wear safety goggles and a lab apron to protect your eyes and clothing.

CAUTION: Do not touch or taste any chemicals. Know the location of the emergency shower and eyewash station and how to use them. If you get a chemical on your skin or clothing, wash it off at the sink while calling to the teacher. Notify the teacher of a spill. Spills should be cleaned up promptly, according to your teacher's directions.

CAUTION: Glassware is fragile. Notify the teacher of broken glass or cuts. Do not clean up broken glass or spills with broken glass unless the teacher tells you to do so.

Laboratory Skills: Using a Compound Light Microscope

Parts of the Compound Light Microscope

- The **eyepiece** magnifies the image, usually 10×.

- The **low-power objective** further magnifies the image, up to 4×.

- The **high-power objectives** further magnify the image, from 10× to 43×.

- The **nosepiece** holds the objectives and can be turned to change from one objective to another.

- The **body tube** maintains the correct distance between the eyepiece and the objectives. This is usually about 25 cm (10 in.), the normal distance for reading and viewing objects with the unaided eye.

- The **coarse adjustment** moves the stage up and down in large increments to allow gross positioning and focusing of the objective lens.

- The **fine adjustment** moves the stage slightly to bring the image into sharp focus.

- The **stage** supports a slide that contains the viewed specimen.

- The **stage clips** secure the slide in position for viewing.

- The **diaphragm** (not labeled), located under the stage, controls the amount of light allowed to pass through the object being viewed.

- The **light source** provides light for viewing the image. It can be either a light reflected with a mirror or an incandescent light from a small lamp. Never use reflected direct sunlight as a light source.

- The **arm** supports the body tube.

- The **base** supports the microscope.

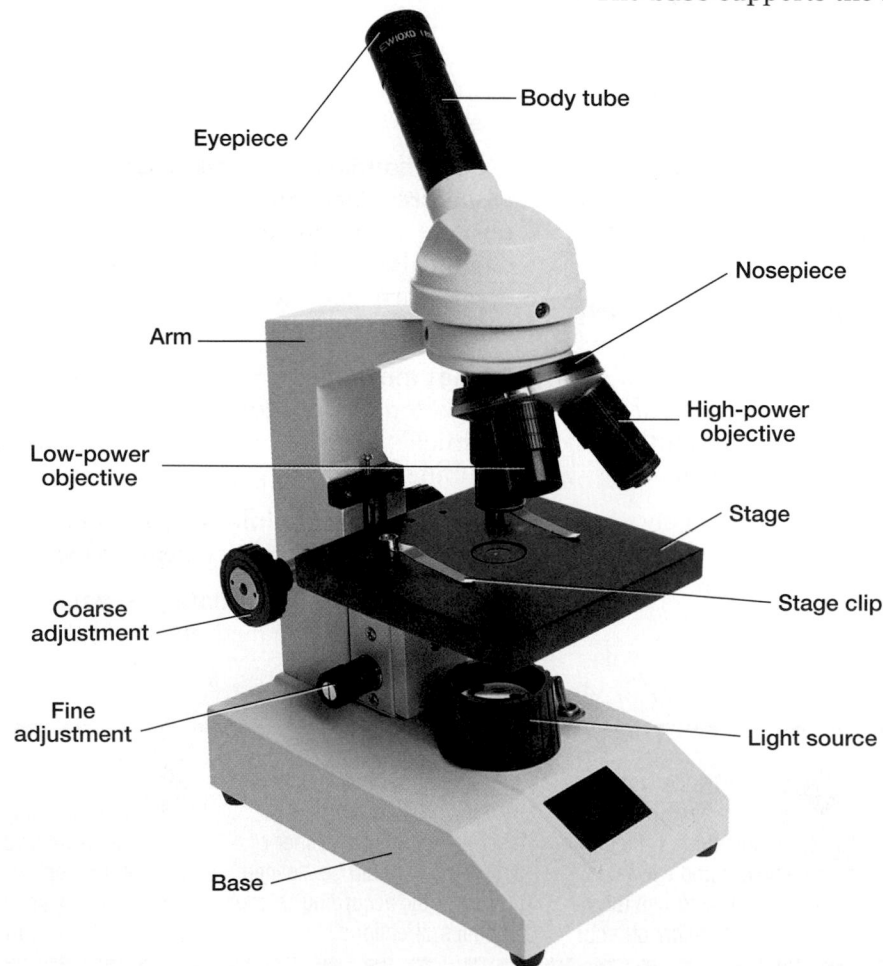

Body tube

Eyepiece

Nosepiece

Arm

High-power objective

Low-power objective

Stage

Coarse adjustment

Stage clip

Fine adjustment

Light source

Base

Proper Handling and Use of the Compound Light Microscope

1. Carry the microscope to your lab table using both hands, one supporting the base and the other holding the arm of the microscope. Hold the microscope close to your body.

2. Place the microscope on the lab table at least 5 cm (2 in.) from the edge of the table.

3. Check to see what type of light source the microscope has. If the microscope has a lamp, plug it in, making sure that the cord is out of the way. If the microscope has a mirror, adjust it to reflect light through the hole in the stage.

   CAUTION: If your microscope has a mirror, do not use direct sunlight as a light source. Using direct sunlight can damage your eyes.

4. Adjust the revolving nosepiece so that the low-power objective is aligned with the body tube.

5. Place a prepared slide over the hole in the stage, and secure the slide with the stage clips.

6. Look through the eyepiece, and move the diaphragm to adjust the amount of light that passes through the specimen.

7. Now look at the stage at eye level. Slowly turn the coarse adjustment to raise the stage until the objective almost touches the slide. Do not allow the objective to touch the slide.

8. While looking through the eyepiece, turn the coarse adjustment to lower the stage until the image is in focus. Never focus objectives downward. Use the fine adjustment to achieve a sharply focused image. Keep both eyes open while viewing a slide.

9. Make sure that the image is exactly in the center of your field of vision. Then switch to the high-power objective. Focus the image with the fine adjustment. Never use the coarse adjustment at high power.

10. When you are finished using the microscope, remove the slide. Clean the eyepiece and objectives with lens paper, and return the microscope to its storage area.

Procedure for Making a Wet Mount

1. Use lens paper to clean a glass slide and coverslip.

   CAUTION: Glass slides and coverslips break easily. Handle them carefully. Notify your teacher if you break a slide or coverslip.

2. Place the specimen that you wish to observe in the center of the slide.

3. Using a medicine dropper, place one drop of water on the specimen.

4. Position the coverslip so that it is at the edge of the drop of water and at a 45° angle to the slide. Make sure that the water runs along the edge of the coverslip.

5. Lower the coverslip slowly to avoid trapping air bubbles.

6. If a stain or solution will be added to a wet mount, place a drop of the staining solution on the microscope slide along one side of the coverslip. Place a small piece of paper towel on the opposite side of the coverslip.

7. As the water evaporates from the slide, add another drop of water by placing the tip of the medicine dropper next to the edge of the coverslip, just as you would if adding stains or solutions to a wet mount. If you have added too much water, remove the excess by using the corner of a paper towel as a blotter. Do not lift the coverslip to add or remove water.

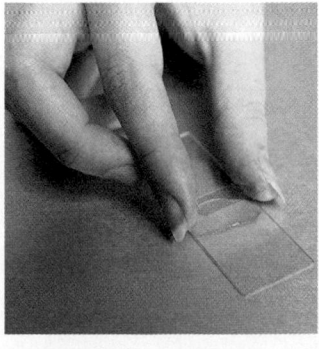
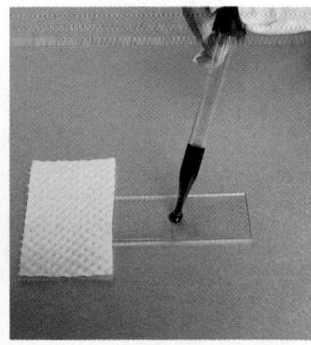

Laboratory Skills
Determining Mass and Temperature

Reading a Balance

A single-pan balance, such as the one shown at right, has one pan and three or four beams. The scale of measure for each beam depends on the model of the balance. When an object is placed on the pan, the riders are moved along the beams until the mass on the beams equals the mass of the object in the pan.

Measuring Mass

When determining the mass of a chemical or powder, use weighing or filter paper. Determine the mass of the paper, and subtract that mass from the total mass. Use the following procedure for determining the mass of objects:

1. Make sure the balance is on a level surface and the pan is allowed to move freely. Position all the riders at zero. If the pointer does not come to rest in the middle of the scale, calibrate the balance using the adjustment knob (usually located under and to the left of the pan).

   CAUTION: Never place a hot object or chemical directly on a balance pan.

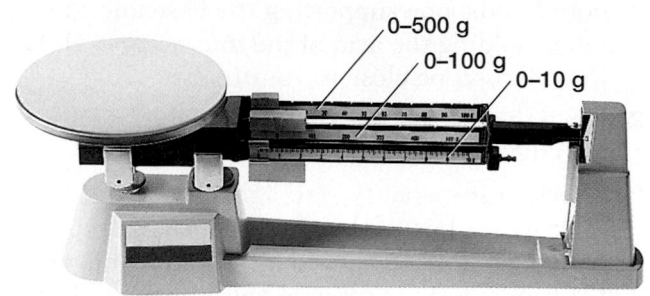

0–500 g
0–100 g
0–10 g

2. Place the object on the pan.
3. Move the largest rider along the beam to the right until it is at the last notch that does not move the pointer below the zero point in the middle of the scale.
4. Follow the same procedure with the next rider.
5. Move the smallest rider until the pointer rests at zero in the middle of the scale.
6. Add up the readings on all the beams to determine the mass of the object.

Practice Exercises

1. **Determine** the mass of each of the following items using a single-pan balance:
   a. an empty 250 mL beaker
   b. 250 mL beaker filled with 100 mL of water
   c. 250 mL beaker filled with 100 mL of vegetable oil
   d. a house key
   e. a small book
   f. a paper clip or small safety pin

2. **Determine** the mass of each object represented by the balance readings shown.

a.

b.

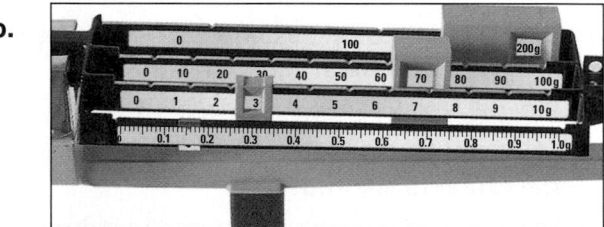

c.

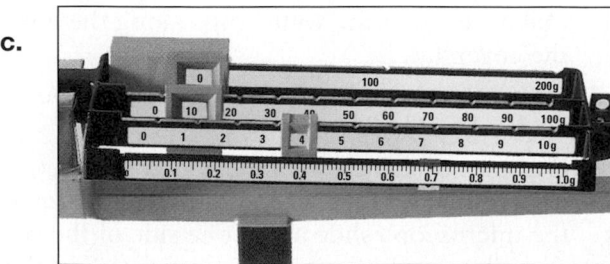

Reading a Thermometer

Many laboratory thermometers are the bulb-type shown below. The sensing bulb of the thermometer is filled with a colored liquid (alcohol) that expands when heated. When the liquid expands, it moves up the stem of the thermometer through the capillary tube. Thermometers usually measure temperature in degrees Celsius (°C).

Measuring Temperature

Use the following procedure when measuring the temperature of a substance.

1. Carefully lower the bulb of the thermometer into the substance. The stem of the thermometer may rest against the side of the container, but the bulb should never rest on the bottom where heat is being applied. If the thermometer has an adjustable clip for the side of the container, the thermometer can be suspended in the liquid.

> **CAUTION: Do not hold a thermometer in your hand while measuring the temperature of a heated substance.**

2. Gently rotate the thermometer in the clip. Watch the rising colored liquid in the capillary tube. When the liquid in the capillary tube stops rising, note the whole-degree increment nearest the top of the liquid column. If your thermometer is marked in tenths of a degree, count the increments up or down from the nearest whole degree to obtain your reading. For example, if the top of the colored liquid column is closest to the 51°C mark but somewhat above it, as shown below, what is the accurate temperature reading? Because the top of the column about one-half of a degree above the 51°C mark, the temperature is 51.5°C. Add one-half of a degree to 51°C to obtain your reading.

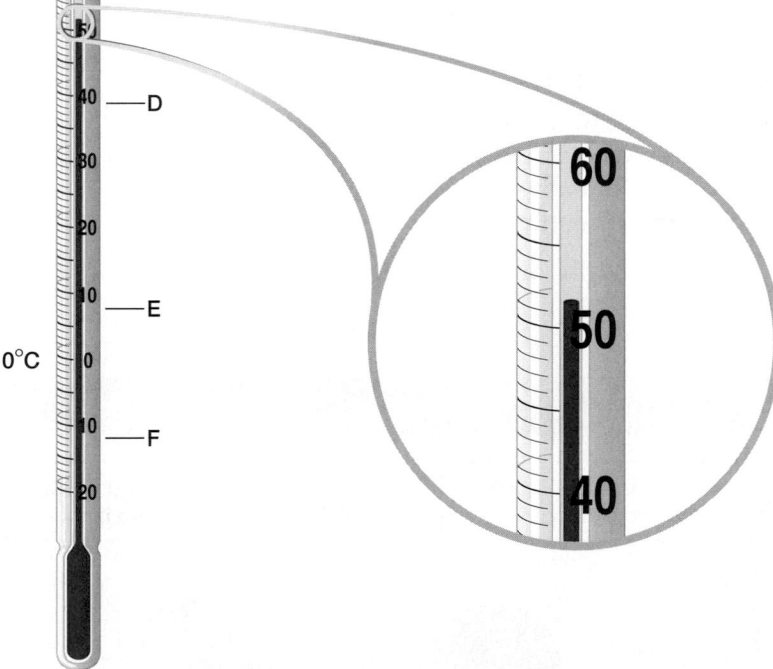

Practice Exercises

Use the thermometer shown above to answer the following questions:

1. Identify the scale used for this thermometer.

2. Determine whether this thermometer is marked only in whole degrees or in tenths of degrees.

3. Estimate the temperature reading on this thermometer.

4. SKILL Interpreting Variables What would be the temperature reading if the top of the column were resting at each of the following points?

a. *A*   **d.** *D*

b. *B*   **e.** *E*

c. *C*   **f.** *F*

SI Units

Scientists throughout the world use the metric system. The official name of the metric system is the Système International d'Unités, or the International System of Measurements. It is usually referred to simply as SI. Most measurements in this book are expressed in metric units. You will always use metric units when you take measurements in the lab.

SI Prefixes

SI is a decimal system; that is, all relationships between SI units are based on powers of 10. Most units have a prefix that indicates the relationship of that unit to a base unit. For example, the SI base unit for length is the meter. A meter equals 100 centimeters (cm), or 1,000 millimeters (mm). A meter also equals 0.001 kilometer (km). **Table 1** summarizes the prefixes used in SI units.

Table 1 SI Prefixes

Prefix	Symbol	Factor of base unit
giga-	G	1,000,000,000
mega-	M	1,000,000
kilo-	k	1,000
hecto-	h	100
deka-	da	10
deci-	d	0.1
centi-	c	0.01
milli-	m	0.001
micro-	μ	0.000001
nano-	n	0.000000001
pico-	p	0.000000000001

Conversion Factors

Conversion between SI units requires a conversion factor. For example, to convert from meters to centimeters, you need to know the relationship between meters and centimeters.

$$1 \text{ cm} = 0.01 \text{ m} \quad \text{or} \quad 1 \text{ m} = 100 \text{ cm}$$

If you need to convert 15.5 centimeters to meters, you could do either of the following:

$$15.5 \text{ cm} \times \frac{1 \text{ m}}{100 \text{ cm}} = 0.155 \text{ m}$$

or

$$15.5 \text{ cm} \times \frac{0.01 \text{ m}}{1 \text{ cm}} = 0.155 \text{ m}$$

Sizes of Objects

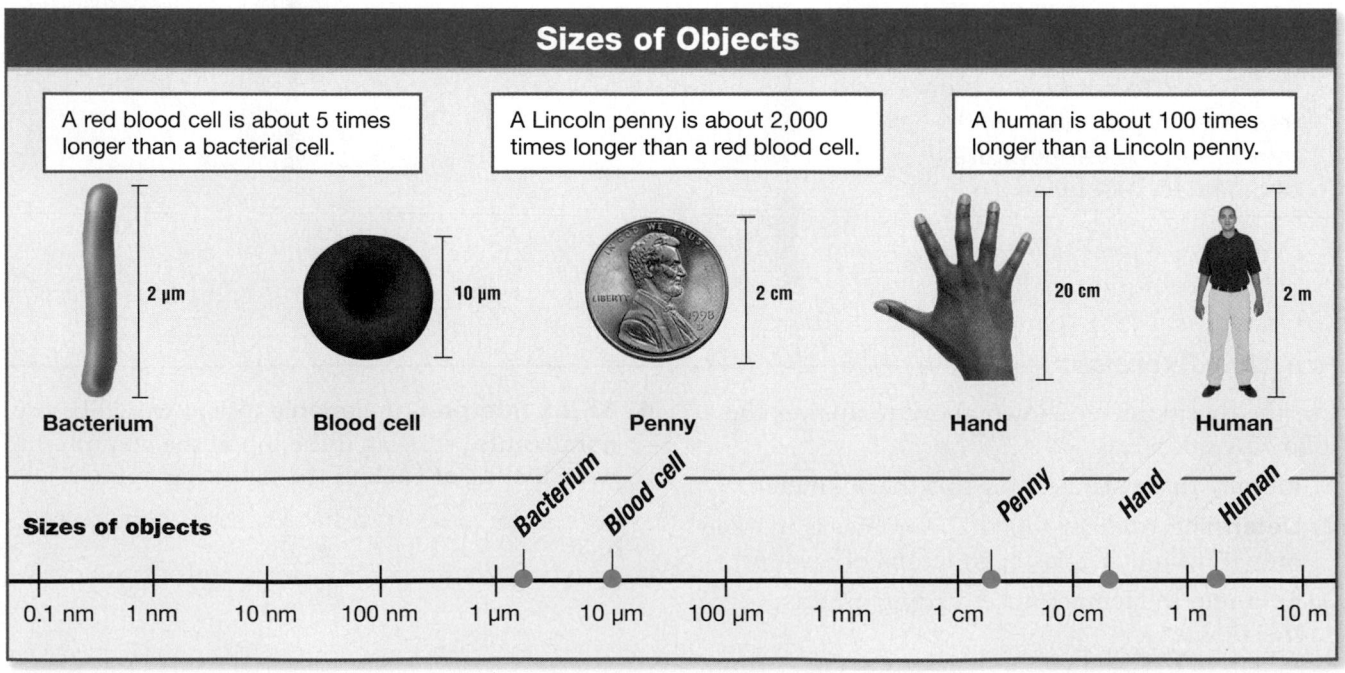

A red blood cell is about 5 times longer than a bacterial cell.

A Lincoln penny is about 2,000 times longer than a red blood cell.

A human is about 100 times longer than a Lincoln penny.

Bacterium — 2 μm
Blood cell — 10 μm
Penny — 2 cm
Hand — 20 cm
Human — 2 m

Sizes of objects

0.1 nm | 1 nm | 10 nm | 100 nm | 1 μm | 10 μm | 100 μm | 1 mm | 1 cm | 10 cm | 1 m | 10 m

Base Units

In this book, you will see three fundamental quantities represented by base units in SI: mass, length, and time. The base units of these quantities are the kilogram (kg), the meter (m), and the second (s). These quantities, their abbreviations, and their equivalent measurements are listed in **Table 2.**

Derived Units

Other important quantities, such as area (m²) and liquid volume (m³), are expressed in derived units. A derived unit is a combination of one or more base units. Like base units, derived units can be expressed using SI prefixes. These quantities are listed in **Table 3.**

Table 2 Conversions for SI Base Units
Mass: unit = kilogram (kg)
1 kilogram (kg) = 1,000 g
1 gram (g) = 0.001 kg
1 milligram (mg) = 0.001 g
1 microgram (μg) = 0.000001 g
Length: unit = meter (m)
1 kilometer (km) = 1,000 m
1 meter (m) = 100 cm
1 centimeter (cm) = 0.01 m
1 millimeter (mm) = 0.001 m
1 micrometer (μm) = 0.000001 m
Time: unit = second (s)
1 minute (min) = 60 s
1 hour (h) = 3,600 s = 60 min
1 day (d) = 24 h

Table 3 Conversions for SI Derived Units
Area: unit = square meter (m²)
1 square kilometer (km²) = 100 ha
1 hectare (ha) = 10,000 m²
1 square meter (m²) = 10,000 cm²
1 square centimeter (cm²) = 100 mm²
Liquid volume: unit = cubic meter (m³)
1 cubic meter (m³) = 1 kL
1 kiloliter (kL) = 1,000 L
1 liter (L) = 1,000 mL
1 milliliter (mL) = 0.001 L
1 cubic centimeter (cm³) = 1 mL
Mass density: unit = kilograms per cubic meter (kg/m³)
Temperature: unit = degrees Celsius (°C)
Velocity: unit = meters per second (m/s)

Temperature

In SI, the Celsius scale is used to express temperature. In the Celsius scale, 0°C (32°F) is the freezing point of water, and 100°C (212°F) is the boiling point of water. You can use the temperature scale shown below to convert between the Celsius scale and the Fahrenheit scale, which is commonly used in the United States. You can also use the following equation to convert between degrees Celsius (T_C)

and degrees Fahrenheit (T_F):

$$T_F = \frac{9}{5} T_C + 32$$

For example, to convert 0°C to degrees Fahrenheit, do the following:

$$T_F = \frac{9}{5}(0°C) + 32°F = 0 + 32°F = 32°F$$

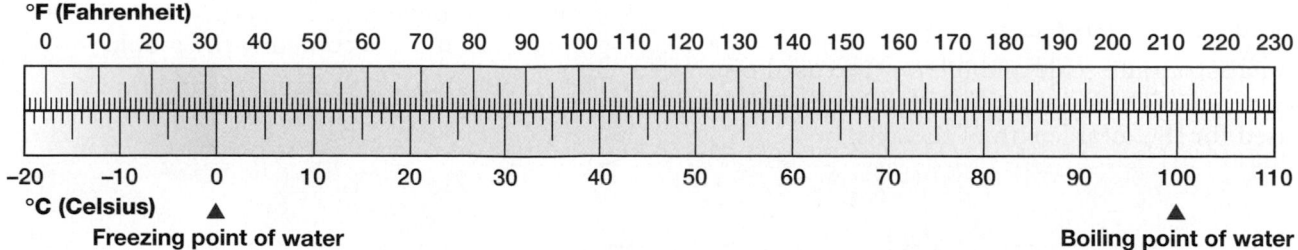

°F (Fahrenheit)

°C (Celsius)

Freezing point of water

Boiling point of water

Math and Problem-Solving Skills

Graphing

Line Graphs

Line graphs, such as the one shown at right, are most often used to compare or relate one or more sets of data that show continuous change. In the graph shown at right, both "Daily salt intake" and the "Systolic pressure" are the **variables,** or sets of data that are being compared.

In graphs, values are assigned to the independent variable. In this case, "Daily salt intake" is the **independent variable.** "Systolic pressure" is called the **dependent variable** because blood pressure is affected by salt intake, according to the graph. Each set of data—the independent and dependent variables—is called a data pair.

Another way to think about independent and dependent variables is to think about the amount of sleep you get. You know that how alert or tired you feel often depends on the number of hours of sleep that you had the night before. The amount of sleep is the independent variable; your alertness is the dependent variable. Studying biology, you will see many examples of dependent and independent variables represented in graphs.

When you are making a line graph using data pairs, first organize data pairs into a table. Then

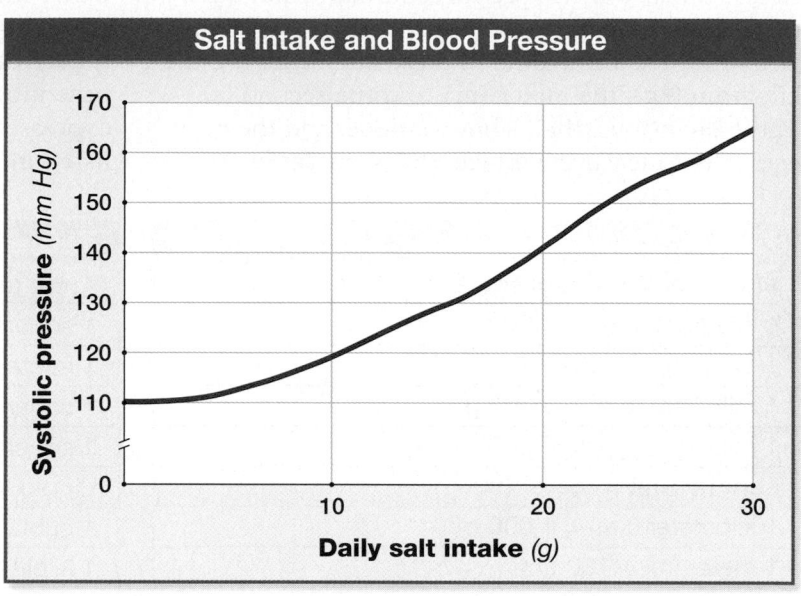

Salt Intake and Blood Pressure

draw the horizontal and vertical axes of your graph. Be sure to label each axis, including units where appropriate. Refer to your data table to determine the scale and interval of each axis. Make sure that the scale and interval of each axis are consistent. Plot each data pair on the graph. Then connect the plotted data points to make a line, or curve. Finally, give the graph a title that clearly indicates the relationship between the data shown by the graph.

Characteristics of Line Graphs

Important characteristics of line graphs include the following:

- The independent variable is graphed on the horizontal *(x)* axis.
- The dependent variable is graphed on the vertical *(y)* axis.
- Both axes are labeled.
- An appropriate scale and interval are used on each axis. The same scale and interval must be used for the total length of the axis.

- Reasonable starting points are used for each axis.
- The data pairs are plotted as accurately as possible.
- The title of the graph accurately reflects the data presented.
- If more than one set of data is presented on a graph, a key must accompany the graph.
- The graph is easy to understand and interpret.

Bar Graphs

Sometimes it is not appropriate to use a line graph to represent data. A bar graph, such as the one shown at right, is appropriate for data that are not continuous. A bar graph is a good indicator of trends if the data are taken over a sufficiently long period of time. For example, studying color variations in moths requires that data be collected over a long period of time. Even after years of study, predictions can still be difficult to make with certainty. Notice that a bar graph is also useful in comparing multiple sets of data, such as those for the light and dark moths found in the woods near Birmingham and Dorset.

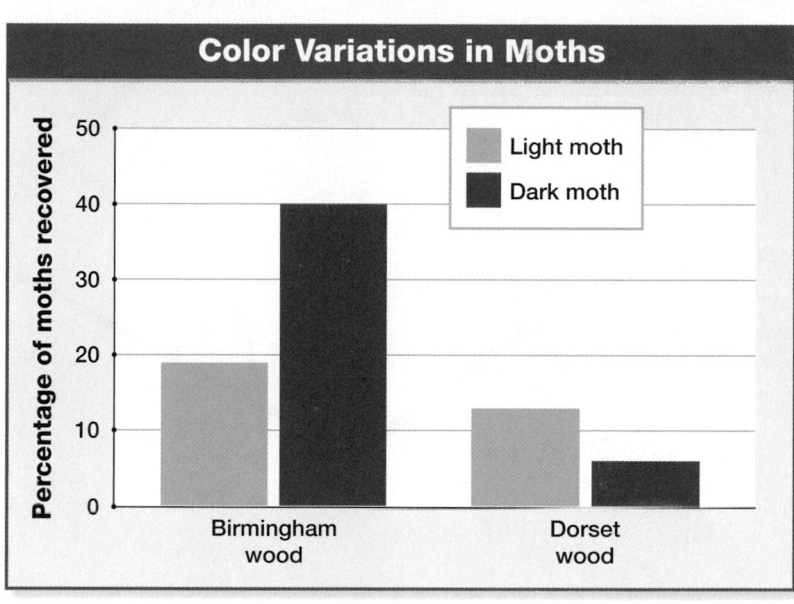

Characteristics of Bar Graphs

Important characteristics of bar graphs include the following:

- An appropriate scale is used on each axis.
- Reasonable starting points are used for each axis.
- A key accompanies the graph.

- The axes are labeled.
- Data are accurately plotted.
- The title of the graph accurately reflects the data presented.
- The graph is easy to understand and interpret.

Using Graphs to Make Predictions

Graphs show trends in data that may not be obvious from a data chart or table. Examine the graph of salt intake and blood pressure on the facing page. Do you notice any trends? Do you think you can conclude that salt intake affects blood pressure? The process of going beyond the data points in a graph to determine a relationship between the data is called **extrapolation.** The further we extrapolate, the less certain we can be of our predictions.

We can also use graphs to link information quickly between two sets of data pairs. For example, we can read the graph to determine that when the subject takes in 10 g of salt daily, the subject's systolic blood pressure is about 120 mm Hg.

Linking information between two sets of data pairs in this way is called **interpolation.** Interpolation will help you identify relationships between sets of data in data pairs.

● **Interpreting Graphics**
When reading graphs, identify the dependent and independent variables. Then see if you can determine a relationship between the two variables by interpolation.

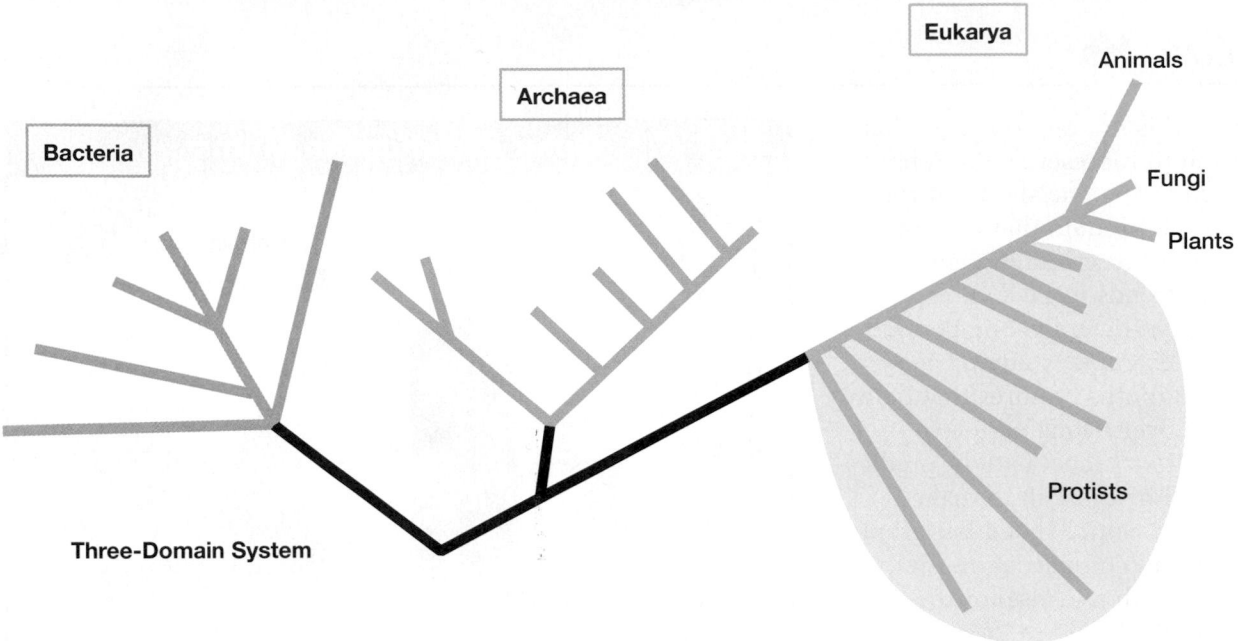

Bacteria

Archaea

Eukarya

Animals

Fungi

Plants

Protists

Three-Domain System

The classification system used in this book is based on both the commonly recognized six-kingdom system and the newer, three-domain system. A **kingdom** is a group of related phyla, and a **domain** is a group of related kingdoms. In the three-domain system illustrated above, all living things are grouped into three distinct domains based on similarities in their nucleic acid sequences. Two of the three domains consist of prokaryotes and one consists of eukaryotes.

The domain **Bacteria** is thought to be the oldest of the three domains. The domain Bacteria consists of the single kingdom Eubacteria—prokaryotic microbes commonly called *bacteria*. A second prokaryotic domain is **Archaea.** Archaea is composed of the single kingdom Archaebacteria—prokaryotic microbes called *archaebacteria*. Although both are prokaryotic microbes, archaebacteria and bacteria differ greatly.

The third domain, **Eukarya,** is composed of all of the eukaryotic organisms. In the four kingdoms within the domain Eukarya are found the animals, plants, fungi, and protists (the kingdoms Animalia, Plantae, Fungi, and

Protista). The phylogenetic tree below shows the division of organisms into six kingdoms.

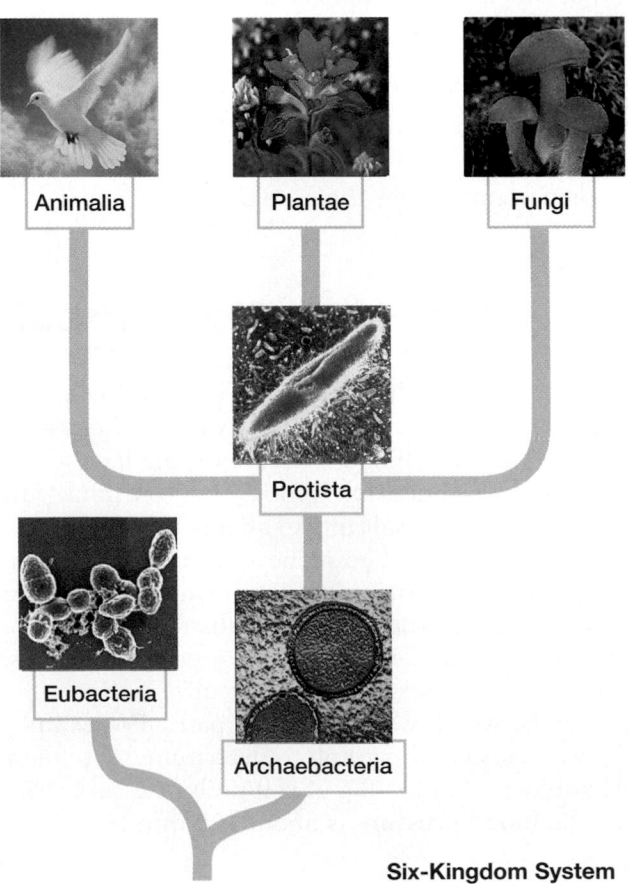

Animalia

Plantae

Fungi

Protista

Eubacteria

Archaebacteria

Six-Kingdom System

The information on the following pages is organized according to kingdoms. Not every phylum or group in each kingdom is discussed, however, and the classification of some groups is controversial. For example, biologists do not agree how the kingdoms Eubacteria, Archaebacteria, and Protista should be divided into phyla.

In this book, these kingdoms are divided into convenient and commonly recognized groups. Biologists also disagree about the number of species in various groups. Unless stated otherwise, the numbers given represent approximate numbers of living, named species.

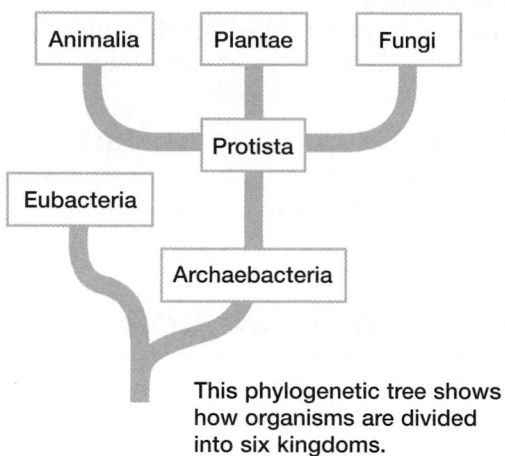

This phylogenetic tree shows how organisms are divided into six kingdoms.

Kingdom Eubacteria

More than 4,000 species
Typically unicellular; prokaryotic; without membrane-bound organelles; nutrition mainly heterotrophic (by absorption), but some are photosynthetic or chemosynthetic; reproduction usually by fission or budding.

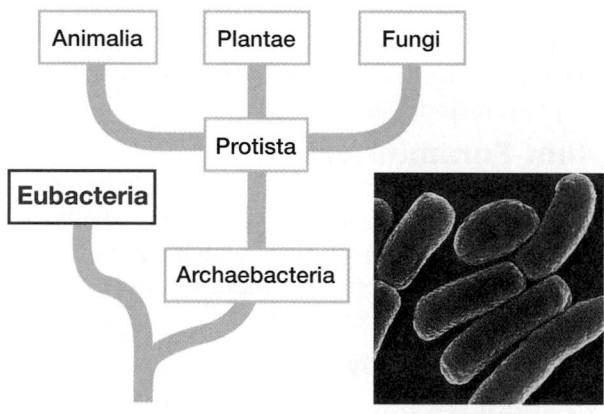

Escherichia coli

Cyanobacteria
Photosynthetic; surrounded by a pigmented covering; common on land and in the ocean; probably ancestors of chloroplasts in some protists.
Examples: *Anabaena, Oscillatoria, Spirulina*

Chemoautotrophs
Ancient bacteria that can grow without sunlight or other organisms; derive energy from reduced gases—ammonia (NH_3), methane (CH_4), hydrogen sulfide (H_2S); play critical roles in Earth's nitrogen cycles; includes nitrobacteria and sulfur bacteria.
Examples: *Nitrosomonas, Nitrobacter*

Enterobacteria
Typically rigid, rod-shaped, heterotrophic bacteria; can be aerobic or anaerobic; have flagella; responsible for many serious diseases of plants and humans.
Examples: *Escherichia coli, Salmonella typhimurium*

Pseudomonads
Straight or curved rods with flagella at one end; strict aerobes; common in soil; many are plant pathogens.
Example: *Pseudomonas aeruginosa*

Spirochaetes
Long, spiral cells; flagella originating at each end; responsible for several serious diseases.
Examples: *Treponema pallidum, Borrelia burgdorferi*

Actinomycetes
Filamentous bacteria that are often mistaken for fungi; spore-producing; sources of antibiotics including streptomycin, tetracycline, and chloramphenicol; cause diseases including dental plaque, leprosy, and tuberculosis.
Example: *Mycobacterium tuberculosis*

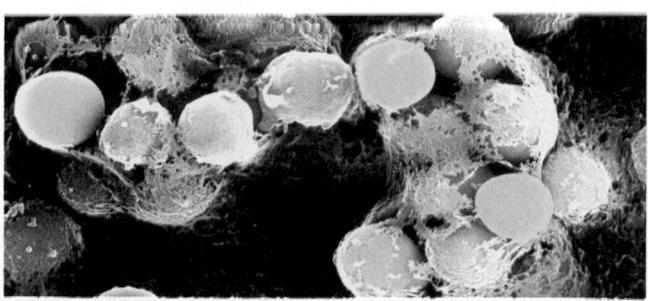

Streptococcus

Rickettsias

Parasitic bacteria found within the cells of vertebrates and arthropods; cause serious diseases.
Example: *Rickettsia rickettsii*

Gliding and budding bacteria

Rod-shaped cells; secrete slimy polysaccharides; often aggregate into gliding masses; live mainly in soil.
Example: *Myxobacteria*

Kingdom Archaebacteria

Fewer than 100 described species
Includes anaerobic and aerobic bacteria adapted to extreme environments; prokaryotic; differ from eubacteria in structure of cell wall and cell membrane; similarities to eukaryotes suggest that archaebacteria are more closely related to eukaryotes than to eubacteria; asexual reproduction only. There is evidence they also exist freely in oceans and soil but have not yet been cultured in labs.

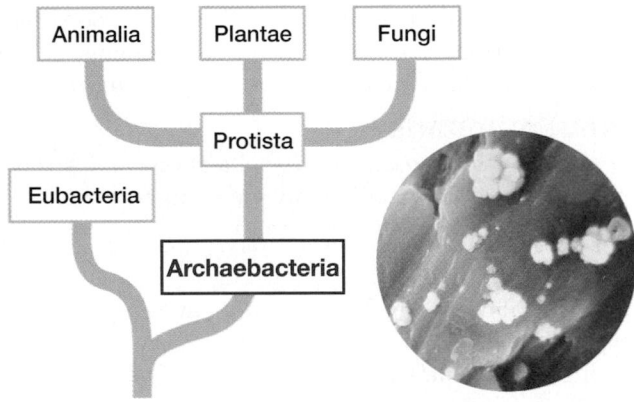

Acidianus brierleyi

Methanogens

Anaerobic methane producers; most species use carbon dioxide as a carbon source; inhabit soil, swamps, and the digestive tracts of animals, particularly grazing mammals such as cattle; produce nearly 2 trillion kilograms (2 billion tons) of methane gas annually.
Example: *Methanobrevibacter ruminatium*

Thermoacidophiles

Inhabit hot, acidic environments; can tolerate high temperatures; require sulfur; mostly anaerobic.
Example: *Sulfolobus solfataricus*

Extreme halophiles

Inhabit environments with very high salt content (salinity 15 to 20 percent), including the Dead Sea and the Great Salt Lake; many aerobic; gram-negative.
Example: *Halobacteroides holobius*

Kingdom Protista

About 43,000 species
Includes eukaryotes that are not plants, fungi, or animals; the most structurally diverse kingdom; both unicellular and multicellular; membrane-bound nucleus; nearly all have chromosomes, mitochondria, and internal compartments; many have chloroplasts; most have cell walls; reproduce sexually and asexually; aquatic or parasitic; many live in soil.

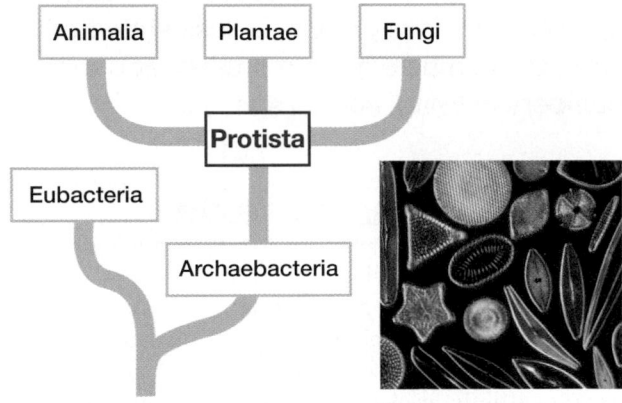

Diatoms

Phylum Rhizopoda

About 300 species
Unicellular and heterotrophic; amorphously shaped cells that move using cytoplasmic extensions called pseudopods; includes amoebas.

Phylum Foraminifera

About 300 species
Unicellular and heterotrophic; marine; have shells of organic material with pores through which many cytoplasmic threads project; includes forams.
Example: *Heterostegina depressa*

Phylum Chlorophyta

About 7,000 species
Unicellular, colonial, and multicellular; photosynthetic; contain chlorophylls *a* and *b;* contain chloroplasts similar to those of plants; scientists think plants descended from this group; includes green algae.
Examples: *Chlamydomonas, Chorella, Oedogonium, Spirogyra, Ulva, Volvox*

Phylum Rhodophyta

About 4,000 species
Almost all are multicellular; photosynthetic; most are marine; contain chlorophyll *a* and phycobilins; chloroplasts probably evolved from symbiotic cyanobacteria; includes red algae.
Example: *Porphyra*

Phylum Phaeophyta

About 1,500 species

Multicellular and photosynthetic; nearly all are marine; contain chlorophylls *a* and *c* and fucoxanthin, which is the source of their brownish color; includes brown algae.

Examples: *Fucus, Laminaria, Postelsia, Sargassum*

Phylum Bacillariophyta

More than 11,500 species

Unicellular and photosynthetic; secrete a unique shell made of opaline silica that resembles a box with a lid; chloroplasts resemble those of brown algae; contain chlorophylls *a* and *c* and fucoxanthin; includes diatoms.

Phylum Dinoflagellata

More than 2,100 species

Unicellular; heterotrophic and autotrophic species; mostly marine; body enclosed within two cellulose plates; contain chlorophylls *a* and *c* and carotenoids; includes dinoflagellates.

Examples: *Gonyaulax, Noctiluca*

Phylum Euglenophyta

About 1,000 species

Unicellular; both photosynthetic and heterotrophic species; asexual; most live in fresh water; chloroplasts are similar to those of green algae and are thought to have evolved from the same symbiotic bacteria; includes euglenoids.

Example: *Euglena*

Phylum Kinetoplastida

About 3,000 species

Mostly unicellular; heterotrophic; all have at least one flagellum; includes zoomastigotes.

Examples: *Giardia, Leishmania, Trypanosoma*

Phylum Ciliophora

About 8,000 species

Very complex single cells; heterotrophic; have rows of cilia and two types of cell nuclei; includes ciliates.

Examples: *Didinium, Paramecium, Stentor, Vorticella*

Phylum Acrasiomycota

About 70 species

Heterotrophic; amoeba-shaped cells that aggregate into a moving mass called a slug when they are deprived of food; cells within the slug retain their membranes and do not fuse; a slug produces spores that form new amoebas elsewhere; includes cellular slime molds.

Example: *Dictyostelium*

Phylum Myxomycota

About 500 species

Heterotrophic; individuals stream along as a multi-nucleate mass of cytoplasm; can give rise to spores that start a new individual in a more favorable environment; includes plasmodial slime molds.

Example: *Physarum*

Phylum Oomycota

About 580 species

Heterotrophic; unicellular parasites or decomposers; cell walls composed of cellulose, not chitin as in fungi; includes water molds, white rusts, and downy mildews.

Example: *Phytophthora*

Phylum Apicomplexa

About 3,900 species

Unicellular; heterotrophic; nonmotile; spore-forming parasites of animals; have complex life cycles; asexual and sexual reproduction; includes sporozoans.

Examples: *Plasmodium, Toxoplasma*

Kingdom Fungi

About 77,000 species

Eukaryotic, terrestrial heterotrophs with nutrition by absorption; all but yeasts are multicellular; body is typically composed of filaments (called hyphae) and is multinucleate, with incomplete divisions (called septae) between cells; cell walls made of chitin; about 17,000 species (known as deuteromycetes) without a sexual stage.

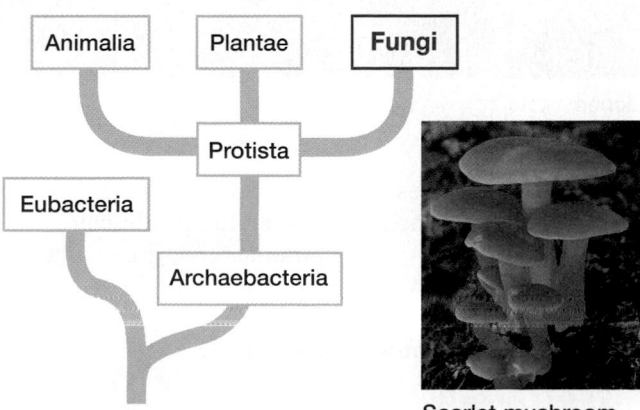

Scarlet mushroom

Phylum Zygomycota

About 665 species

Usually lack septae; fusion of hyphae leads to formation of zygote, which divides by meiosis when it germinates; terrestrial or parasitic; includes bread molds.

Examples: *Pilobolus, Rhizopus*

Phylum Ascomycota

About 30,000 species

Hyphae usually have perforated septae; fusion of hyphae leads to formation of densely interwoven mass that contains characteristic microscopic reproductive structures called asci (singular, ascus); many fungi formally grouped as Fungi Imperfecti are now grouped here; terrestrial, marine, and freshwater species; includes brewer's and baker's yeasts, molds, morels, and truffles.

Examples: *Neurospora, Saccharomyces*

Phylum Basidiomycota

About 16,000 species

Hyphae usually have incomplete septae; reproduction is typically sexual; fusion of hyphae leads to the formation of densely interwoven reproductive structure (mushroom) with characteristic microscopic structures called basidia (singular, basidium); includes mushrooms, toadstools, shelf fungi, rusts, and smuts.

Fungal Associations

About 20,000 species

Fungi form symbiotic associations with plants, green algae, and cyanobacteria.

Lichen

Lichens

About 15,000 species

Mutualistic relationships between fungi (almost always ascomycetes) and cyanobacteria, green algae, or both; the photosynthetic partners actually live among the hyphae of the fungus; the fungus derives energy from its photosynthetic partners.

Mycorrhizae

About 5,000 species

Mutualistic relationships between fungi and the roots of plants; 80 percent of all plants have mycorrhizae associated with their roots; the plant provides sugars to the fungi; in return, the fungi serve as accessory roots, greatly increasing the surface area available for the absorption of nutrients.

Kingdom Plantae

About 280,000 species

Multicellular; eukaryotic; mostly autotrophic; mostly terrestrial organisms containing tissues and organs; cell walls with cellulose; contain chlorophylls *a* and *b* in plastids; life cycle is alternation of generations.

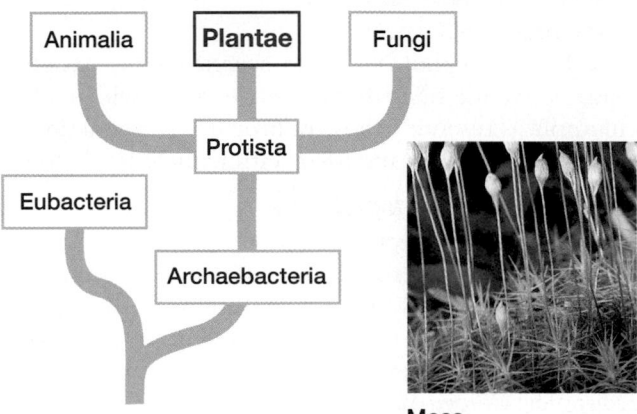

Moss

Phylum Bryophyta

About 10,000 species

Nonvascular plants; gametophytes are larger than sporophytes; sporophytes grow on gametophytes; have simple conducting tissue; lack roots, stems, and leaves; includes mosses.

Example: *Sphagnum*

Phylum Hepatophyta

About 6,000 species

Nonvascular plants; gametophytes are larger than sporophytes; sporophytes grow on gametophytes; lack stomata, roots, stems, and leaves; includes liverworts.

Example: *Marchantia*

Phylum Anthocerophyta

About 100 species

Nonvascular plants; gametophytes are larger than sporophytes; sporophytes grow on gametophytes; sporophytes have stomata; lack roots, stems, and leaves; includes hornworts.

Example: *Anthoceros*

Phylum Pterophyta

About 11,000 species

Seedless vascular plants; sporophytes are larger than gametophytes; sporophytes have roots, stems, and leaves that produce spores on their lower surfaces; gametophytes are small, flat, and independent; includes ferns.

Example: *Salvinia*

Phylum Lycophyta

About 1,000 species
Seedless vascular plants; sporophytes are larger than gametophytes; sporophytes produce spores in cones, resemble moss gametophytes, and have roots, stems, and leaves; gametophytes are small, flat, and independent; includes club mosses.
Examples: *Lycopodium, Selaginella*

Phylum Sphenophyta

15 species
Seedless vascular plants; sporophytes are larger than gametophytes; sporophytes produce spores in cones and have roots, leaves, and jointed stems; gametophytes are small, flat, and independent; includes horsetails.
Example: *Equisetum*

Phylum Psilotophyta

Several species
Seedless vascular plants; sporophytes are larger than gametophytes; sporophytes produce spores in sporangia at tips of stems and have roots and stems but no leaves; gametophytes are small, flat, and independent; includes whisk ferns.
Example: *Psilotum*

Phylum Coniferophyta

About 550 species
Gymnosperms, seed plants that produce naked seeds; sporophytes are mostly evergreen trees or shrubs with needlelike or scalelike leaves; male and female gametophytes are microscopic and develop from spores produced within cones on sporophytes; includes pines, spruces, firs, larches, and yews.
Examples: *Pinus, Taxus*

Phylum Cycadophyta

About 100 species
Gymnosperms, seed plants that produce naked seeds; sporophytes are evergreen trees and shrubs with palmlike leaves; male and female gametophytes are microscopic and develop from spores produced within cones on separate sporophytes; includes cycads.
Example: *Cycas*

Phylum Ginkgophyta

1 species
Gymnosperm, seed plant that produces naked seeds; sporophyte is a deciduous tree with fan-shaped leaves and fleshy seeds; male and female gametophytes are microscopic and develop from spores produced by separate sporophytes; includes *Ginkgo biloba*.

Phylum Gnetophyta

About 70 species
Gymnosperms, seed plants that produce naked seeds; sporophytes are shrubs or vines with some angiosperm characteristics; male and female gametophytes are microscopic and develop from spores produced within cones on sporophytes; includes gnetophytes.
Examples: *Ephedra, Welwitschia*

Phylum Anthophyta

About 250,000 species
Angiosperms, seed plants that produce seeds within a fruit; sporophytes are trees, shrubs, herbs, or vines that produce flowers; male and female gametophytes are microscopic and develop from spores produced within the reproductive structures of a flower; includes flowering plants.
Examples: *Aster, Prunus, Quercus, Zea*

Class Monocotyledones

About 70,000 species
Embryos have one cotyledon; flower parts in multiples of three; leaf veins parallel; vascular bundles scattered through stem tissue; includes grasses, sedges, lilies, irises, palms, and orchids.

Class Dicotyledones

About 180,000 species
Embryos have two cotyledons; flower parts in multiples of two, four, or five; leaves with netlike veins; vascular bundles in stems are arranged in rings; includes daisies, roses, maples, and elms.

Rose

Kingdom Animalia

More than 1 million species
Multicellular, eukaryotic, heterotrophic organisms; nutrition mainly by ingestion; most have specialized tissues, and many have complex organs and organ systems; no cell walls or chloroplasts; sexual reproduction predominates; both aquatic and terrestrial forms.

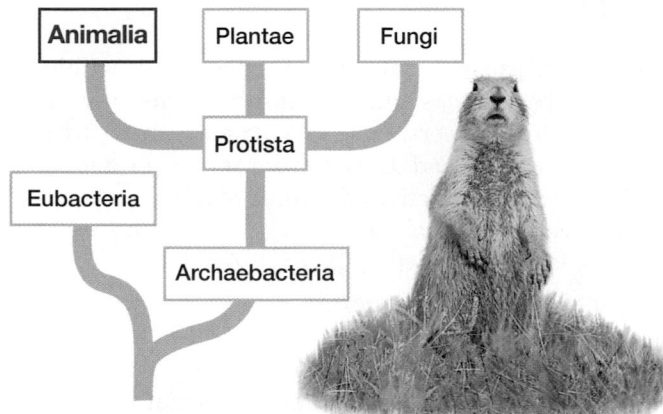

Prairie dog

Phylum Porifera

About 9,000 species
Asymmetrical; lack tissues and organs; body wall consists of two cell layers, penetrated by numerous pores; internal cavity is lined with unique food-filtering cells called choanocytes; sexual and asexual reproduction; mostly marine; includes sponges.

Phylum Cnidaria

About 10,000 species
Radially symmetrical; most have distinct tissues; baglike body of two cell layers; gelatinous; marine and freshwater species.

Class Hydrozoa
About 2,700 species
Most have both polyp and medusa stages in life cycle; includes hydras.
Examples: *Hydra, Obelia, Physalia*

Class Scyphozoa
About 200 species
Exclusively marine; medusa stage dominant; includes jellyfish.
Example: *Aurelia*

Class Anthozoa
About 6,200 species
Marine; solitary or colonial; medusa stage absent; includes sea anemones, corals, and sea fans.

Phylum Ctenophora

About 100 species
Radially symmetrical; transparent, gelatinous bodies resembling jellyfish; marine; includes comb jellies.

Phylum Platyhelminthes

About 20,000 species
Bilaterally symmetrical acoelomates; body flat and ribbonlike, without true segments; organs present; three germ layers; includes flatworms.

Class Turbellaria
More than 3,000 species
Mostly free-living aquatic or terrestrial forms; includes planarians.
Example: *Dugesia*

Class Cestoda
About 1,500 species
Specialized internal parasites; no digestive system; body sections called proglottids; hooked scolex for attaching to host; includes tapeworms.
Example: *Taenia saginata*

Class Trematoda
About 6,000 species
Internal parasites, with mouth at anterior end; often have complex life cycle with alternation of hosts; cause disease in humans and other animals; includes flukes.
Examples: *Schistosoma, Chlonorchis sinensis*

Phylum Nematoda

More than 12,000 species
Tiny, parasitic, unsegmented worms; long, slender body; pseudocoelomates; includes roundworms.
Examples: *Ascaris, Trichinella spiralis, Necator, Toxocara canis, Toxocara cati*

Phylum Mollusca

About 110,000 species
Soft-bodied animals with a true coelom; three-part body consisting of foot, visceral mass, and mantle; protostomes; most have a unique rasping tongue (radula); terrestrial, freshwater, and marine.

Cuttlefish

Class Polyplacophora
About 600 species
Elongated body and reduced head; similar to ancestral mollusk form; includes chitons.

Class Gastropoda
About 80,000 species
Visceral mass twisted during development; head, distinct eyes, and tentacles usually present; includes gastropods, such as snails, slugs, and whelks.

Class Bivalvia
10,000 species
Two shells connected by a hinge; no radula; large, wedge-shaped foot; includes bivalves, such as oysters, clams, and scallops.

Class Cephalopoda
More than 600 species
Foot modified into tentacles; includes cephlapods, such as squids, octopuses, nautilus, and cuttlefish.

Phylum Annelida
About 12,000 species
Serially segmented worms; bilaterally symmetrical; protostomes.

Class Polychaeta
About 8,000 species
Fleshy outgrowths called parapodia extend from segments; many bristles (setae); marine; includes feather dusters.
Example: *Nereis*

Class Oligochaeta
About 3,100 species
Head not well developed; no parapodia; few setae; terrestrial and freshwater forms; includes earthworms.

Class Hirudinea
About 600 species
Body flattened; no parapodia; usually suckers at both ends; many are external parasites; includes leeches.

Phylum Arthropoda
About 1 million species
Segmented bodies with paired, jointed appendages; bilaterally symmetrical; chitinous exoskeleton; protostomes; aerial, terrestrial, and aquatic forms.

Subphylum Chelicerata
Distinguished by absence of antennae and presence of chelicerae; all appendages unbranched; four pairs of walking legs; two body regions (cephalothorax and abdomen); predominantly terrestrial.

Class Arachnida
About 57,000 species
Terrestrial; use book lungs and tracheae for respiration; four pairs of legs; includes spiders, scorpions, ticks, and mites.

Class Merostomata
5 species
Cephalothorax covered by protective "shell"; sharp spike on tail; includes horseshoe crabs.

Class Pycnogonida
About 1,000 species
Small marine predators or parasites; usually four pairs of legs; includes sea spiders.

Subphylum Crustacea
About 35,000 species
Two pairs of antennae, mandibles, and appendages with two branches; predominantly aquatic.

Crab

Class Malacostraca
About 20,000 species
Typically five pairs of legs; two pairs of antennae; most are aquatic; includes crayfish, lobsters, crabs, shrimp, sow bugs, and krill.

Subphylum Uniramia
Antennae, mandibles, and unbranched appendages.

Class Insecta
About 750,000 species
Head, thorax, and abdomen; three pairs of legs, all attached to thorax; usually two pairs of wings.

Order Coleoptera: includes beetles, ladybugs, and weevils.
Order Diptera: includes flies, mosquitoes, gnats, and midges.
Order Lepidoptera: includes butterflies and moths.
Order Hymenoptera: includes bee ants, wasps, hornets, and ichneumon fly. Beetle
Order Hemiptera: includes water striders, water boatmen, back swimmers, bedbugs, squash bugs, stink bugs, and assassin bugs.
Order Homoptera: includes cicadas, aphids, leaf hoppers, and scale insects.
Order Orthoptera: includes grasshoppers, cockroaches, walking sticks, praying mantises, and crickets.
Order Odonata: includes dragonflies and damsel flies.
Order Neuroptera: includes ant lions and lacewings.

Order Thysanura: includes silverfish, bristletails, and firebrats.
Order Anoplura: includes sucking lice.
Order Isoptera: includes termites.
Order Ephemeroptera: includes mayflies.
Order Siphonaptera: includes fleas.
Order Dermaptera: includes earwigs.

Class Chilopoda

About 2,500 species
Body flattened and consisting of 15–170 or more segments; one pair of legs attached to each segment; includes centipedes.

Class Diplopoda

About 10,000 species
Elongated body of 15–200 segments; two pairs of legs per segment; herbivorous; includes millipedes.

Phylum Echinodermata

About 6,000 species
Deuterostomes; adults radially symmetrical with five-part body plan; most forms have water vascular system with tube feet for locomotion; marine.

Class Asteroidea

About 1,500 species
Body usually with five arms and double rows of tube feet on each arm; mouth directed downward; includes sea stars.

Sea star

Class Ophiuroidea

About 2,000 species
Usually with five slender, delicate arms or rays; includes brittle stars and basket star.

Class Crinoidea

About 600 species
Mouth faces upward and is surrounded by many arms; includes sea lilies and feather stars.

Class Echinoidea

About 900 species
Body spherical, oval, or disk-shaped; arms lacking but five-part body plan still apparent; includes sea urchins and sand dollars.

Class Holothuroidea

About 1,500 species
Elongated, thickened body with tentacles around the mouth; includes sea cucumbers.

Phylum Chordata

About 42,500 species
Bilaterally symmetrical; deuterostomes; coelom present; have notochord, dorsal nerve cord, pharyngeal slits, and tail; aquatic and terrestrial.

Subphylum Urochordata

About 1,250 species
Saclike covering, or tunic, in adults; larvae are free-swimming and have nerve cord and notochord; marine; includes tunicates.

Subphylum Cephalochordata

23 species
Small and fishlike with a permanent notochord; filter feeders; includes lancelets.

Subphylum Vertebrata

About 40,000 species
Most of the notochord is replaced by a spinal column composed of vertebrae that protect the dorsal nerve cord; recognizable head containing a brain.

Jawless fishes

63 species
Freshwater or marine eel-like fishes without true jaws, scales, or paired fins; cartilaginous skeleton; includes lampreys and hagfish.

Cartilaginous fishes

About 850 species
Fishes with jaws and paired fins; gills present; no swim bladder; cartilaginous skeleton; includes sharks, rays, and skates.

Bony fishes

About 18,000 species
Freshwater and marine fishes with gills attached to gill arch; jaws and paired fins; bony skeleton; most have swim bladder; includes rayfins, such as tuna, sea horse, trout, perch, sturgeon, and angelfish; and lobefins, such as coelacanth and lungfish.

Class Amphibia

About 4,200 species
Freshwater or terrestrial; gills present at some stage; skin often slimy and lacking scales; eggs typically laid in water and fertilized externally.

Order Anura: includes frogs and toads.
Order Urodela: includes salamanders and newts.
Order Apoda: includes caecilians.

Class Reptilia

About 7,000 species
Terrestrial or semiaquatic vertebrates; breathe using lungs at all stages; body covered by scales; most species lay amniotic eggs covered with a protective shell; fertilization internal.

Order Squamata: includes lizards and snakes.
Order Chelonia: includes turtles and tortoises.

Order Crocodilia: includes alligators, crocodiles, gavials, and caimans.

Order Rhynchocephalia: includes tuataras.

Class Aves

About 9,000 species

Body covered with feathers; forelimbs modified into wings; four-chambered heart; endothermic; lay shelled, amniotic eggs.

Order Passeriformes: includes robins, bluebirds, sparrows, warblers, and thrushes.

Order Apodiformes: includes swifts and hummingbirds.

Order Piciformes: includes woodpeckers, sapsuckers, flickers, and toucans.

Order Psittaciformes: includes parrots, parakeets, macaws, and cockatoos.

Order Charadriiformes: includes snipes, sandpipers, plovers, gulls, terns, auks, puffins, and ibises.

Tern

Order Columbiformes: includes pigeons and doves.

Order Falconiformes: includes hawks, falcons, eagles, kites, and vultures.

Order Gaviiformes: includes loons.

Order Gruiformes: includes cranes, coots, gallinules, and rails.

Order Anseriformes: includes ducks, geese, and swans.

Order Strigiformes: includes owls.

Order Ciconiiformes: includes herons, bitterns, egrets, storks, spoonbills, and ibises.

Order Pelecaniformes: includes pelicans, cormorants, and gannets.

Order Galliformes: includes pheasants, turkeys, quails, partridges, and grouse.

Order Procellariiformes: includes albatrosses and petrels.

Order Cuculiformes: includes cuckoos and roadrunners.

Order Caprimulgiformes: includes goatsuckers, whippoorwills, and nighthawks.

Order Coraciiformes: includes kingfishers.

Order Sphenisciformes: includes penguins.

Order Struthioniformes: includes ostriches.

Order Apterygiformes: includes kiwis.

Class Mammalia

About 4,400 species

Hair on at least part of body; young nourished with milk secreted by mammary glands; endothermic; breathe with lungs.

Order Rodentia: includes squirrels, woodchucks, mice, rats, muskrats, and beavers.

Order Chiroptera: includes bats.

Bat

Order Insectivora: includes moles and shrews.

Order Marsupialia: includes opossums, kangaroos, koalas, and wallabies.

Order Carnivora: includes bears, weasels, mink, otters, skunks, lions, tigers, and wolves.

Order Primates: includes monkeys, lemurs, gibbons, orangutans, gorillas, chimpanzees, and humans.

Order Artiodactyla: includes hippopotamuses, camels, llamas, deer, giraffes, cattle, sheep, and goats.

Order Cetacea: includes whales, porpoises, and dolphins.

Order Lagomorpha: includes rabbits, hares, and pikas.

Order Pinnipedia: includes seals, sea lions, and walruses.

Order Edentata: includes armadillos, sloths, and anteaters.

Order Macroscelidea: includes elephant shrews.

Order Perissodactyla: includes tapirs, rhinoceroses, horses, and zebras.

Order Scandentia: includes tree shrews.

Order Hyracoidea: includes hyraxes.

Order Pholidota: includes pangolins.

Order Sirenia: includes sea cows, dugongs, and manatees.

Order Monotremata: includes duckbill platypus and spiny anteaters.

Order Dermoptera: includes flying lemurs.

Order Proboscidea: includes elephants.

Order Tubulidentata: includes aardvark.

Polar bear

Periodic Table

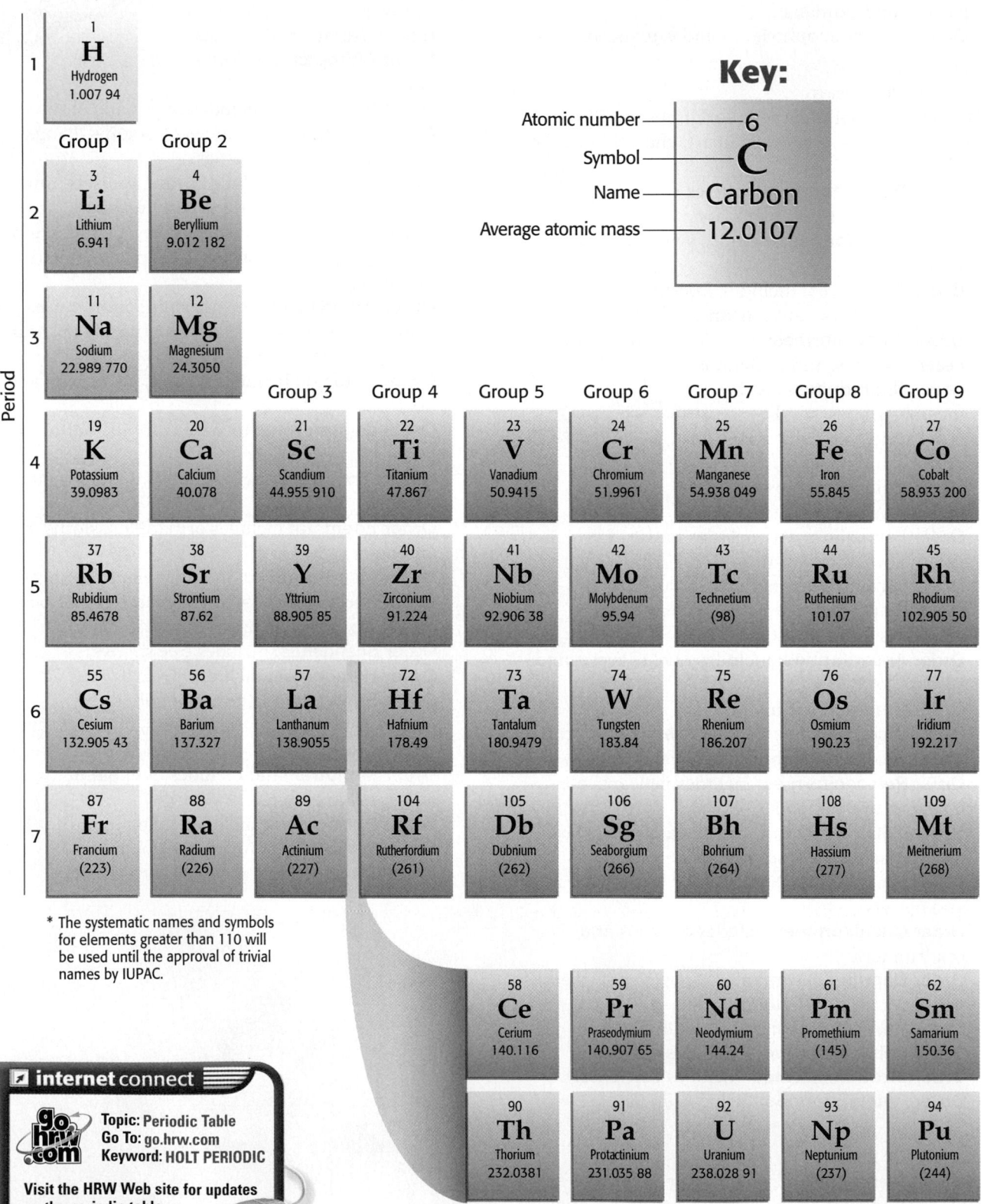

Key:

Atomic number — 6
Symbol — **C**
Name — Carbon
Average atomic mass — 12.0107

Period	Group 1	Group 2		Group 3	Group 4	Group 5	Group 6	Group 7	Group 8	Group 9
1	1 **H** Hydrogen 1.007 94									
2	3 **Li** Lithium 6.941	4 **Be** Beryllium 9.012 182								
3	11 **Na** Sodium 22.989 770	12 **Mg** Magnesium 24.3050								
4	19 **K** Potassium 39.0983	20 **Ca** Calcium 40.078		21 **Sc** Scandium 44.955 910	22 **Ti** Titanium 47.867	23 **V** Vanadium 50.9415	24 **Cr** Chromium 51.9961	25 **Mn** Manganese 54.938 049	26 **Fe** Iron 55.845	27 **Co** Cobalt 58.933 200
5	37 **Rb** Rubidium 85.4678	38 **Sr** Strontium 87.62		39 **Y** Yttrium 88.905 85	40 **Zr** Zirconium 91.224	41 **Nb** Niobium 92.906 38	42 **Mo** Molybdenum 95.94	43 **Tc** Technetium (98)	44 **Ru** Ruthenium 101.07	45 **Rh** Rhodium 102.905 50
6	55 **Cs** Cesium 132.905 43	56 **Ba** Barium 137.327		57 **La** Lanthanum 138.9055	72 **Hf** Hafnium 178.49	73 **Ta** Tantalum 180.9479	74 **W** Tungsten 183.84	75 **Re** Rhenium 186.207	76 **Os** Osmium 190.23	77 **Ir** Iridium 192.217
7	87 **Fr** Francium (223)	88 **Ra** Radium (226)		89 **Ac** Actinium (227)	104 **Rf** Rutherfordium (261)	105 **Db** Dubnium (262)	106 **Sg** Seaborgium (266)	107 **Bh** Bohrium (264)	108 **Hs** Hassium (277)	109 **Mt** Meitnerium (268)

* The systematic names and symbols for elements greater than 110 will be used until the approval of trivial names by IUPAC.

58 **Ce** Cerium 140.116	59 **Pr** Praseodymium 140.907 65	60 **Nd** Neodymium 144.24	61 **Pm** Promethium (145)	62 **Sm** Samarium 150.36
90 **Th** Thorium 232.0381	91 **Pa** Protactinium 231.035 88	92 **U** Uranium 238.028 91	93 **Np** Neptunium (237)	94 **Pu** Plutonium (244)

internet connect

go.hrw.com

Topic: Periodic Table
Go To: go.hrw.com
Keyword: HOLT PERIODIC

Visit the HRW Web site for updates on the periodic table.

Legend

- Hydrogen
- Semiconductors
  (also known as *metalloids*)

Metals
- Alkali metals
- Alkaline-earth metals
- Transition metals
- Other metals

Nonmetals
- Halogens
- Noble gases
- Other nonmetals

Group 13	Group 14	Group 15	Group 16	Group 17	Group 18
					2 **He** Helium 4.002 602
5 **B** Boron 10.811	6 **C** Carbon 12.0107	7 **N** Nitrogen 14.0067	8 **O** Oxygen 15.9994	9 **F** Fluorine 18.998 4032	10 **Ne** Neon 20.1797
13 **Al** Aluminum 26.981 538	14 **Si** Silicon 28.0855	15 **P** Phosphorus 30.973 761	16 **S** Sulfur 32.065	17 **Cl** Chlorine 35.453	18 **Ar** Argon 39.948

Group 10	Group 11	Group 12						
28 **Ni** Nickel 58.6934	29 **Cu** Copper 63.546	30 **Zn** Zinc 65.409	31 **Ga** Gallium 69.723	32 **Ge** Germanium 72.64	33 **As** Arsenic 74.921 60	34 **Se** Selenium 78.96	35 **Br** Bromine 79.904	36 **Kr** Krypton 83.798
46 **Pd** Palladium 106.42	47 **Ag** Silver 107.8682	48 **Cd** Cadmium 112.411	49 **In** Indium 114.818	50 **Sn** Tin 118.710	51 **Sb** Antimony 121.760	52 **Te** Tellurium 127.60	53 **I** Iodine 126.904 47	54 **Xe** Xenon 131.293
78 **Pt** Platinum 195.078	79 **Au** Gold 196.966 55	80 **Hg** Mercury 200.59	81 **Tl** Thallium 204.3833	82 **Pb** Lead 207.2	83 **Bi** Bismuth 208.980 38	84 **Po** Polonium (209)	85 **At** Astatine (210)	86 **Rn** Radon (222)
110 **Ds** Darmstadtium (281)	111 **Uuu**[*] Unununium (272)	112 **Uub**[*] Ununbium (285)	113 **Uut**[*] Ununtrium (284)	114 **Uuq**[*] Ununquadium (289)	115 **Uup**[*] Ununpentium (288)			

A team at Lawrence Berkeley National Laboratories reported the discovery of elements 116 and 118 in June 1999. The same team retracted the discovery in July 2001. The discovery of elements 113, 114, and 115 has been reported but not confirmed.

63 **Eu** Europium 151.964	64 **Gd** Gadolinium 157.25	65 **Tb** Terbium 158.925 34	66 **Dy** Dysprosium 162.500	67 **Ho** Holmium 164.930 32	68 **Er** Erbium 167.259	69 **Tm** Thulium 168.934 21	70 **Yb** Ytterbium 173.04	71 **Lu** Lutetium 174.967
95 **Am** Americium (243)	96 **Cm** Curium (247)	97 **Bk** Berkelium (247)	98 **Cf** Californium (251)	99 **Es** Einsteinium (252)	100 **Fm** Fermium (257)	101 **Md** Mendelevium (258)	102 **No** Nobelium (259)	103 **Lr** Lawrencium (262)

The atomic masses listed in this table reflect the precision of current measurements. (Values listed in parentheses are the mass numbers of those radioactive elements' most stable or most common isotopes.)

Forensics Lesson

An Introduction to Forensic Science

A police takes a plaster cast of a tire track from the mud of a crime scene. A botanist uses a microscope to identify the source of pollen from the clothing of a crime suspect. A rescue worker sifts through debris after a fire, looking for the bodies of victims. In each of these cases, a small amount of evidence will be used to piece together a story or an identity.

Background

When most people think about forensic science, they recall crime dramas from television, in which a crime scene investigator seizes the smallest bits of evidence and ties them together to reconstruct the crime, enabling the identification of even the most cunning criminal.

Many of the scientists who work in forensics are involved in law enforcement. But there's much more to forensics than that. Basically, forensics is the making of knowledge and information available in a public forum, such as a court of law. A forensic scientist is a person who applies scientific knowledge and techniques to the investigation of evidence for the purpose of identification of a person or to establish a sequence of events that took place in the past.

For example, the evidence associated with a crime may be samples of hair, fibers, paint, glass, soil, blood, or plant material. Forensics scientists from different areas of specialization analyze evidence using a variety of different techniques, depending on the type of evidence.

You are familiar with some forensic techniques. For example, fingerprint analysis and blood typing have been in use for many years. The use of DNA to identify a person is now commonplace in the courtroom. Other forensic areas of specialization may be less familiar.

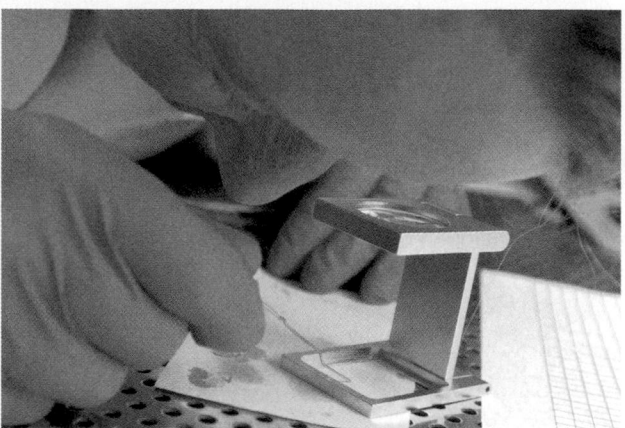

Figure 1 Document examination

Areas of Specialization in Forensics

Document Examination

Document examiners attempt to determine the authenticity of documents. Such determinations, such as the one shown in **Figure 1,** focus on handwriting or the machine (such as a typewriter, copier, or fax) used to produce the document, the inks used, and the material on which it is written.

Researchers, are now looking at ways to analyze the language patterns in a document, including particular words, sentence construction, and verb tenses, in order to help identify the author of the document.

Latent Fingerprints Research

Fingerprints at a crime scene, such as the one shown in **Figure 2,** are vital to many forensic investigations. The old method of dusting for fingerprints is very time consuming, however, and investigators may miss prints.

New methods are being developed that allow an entire room, or even an entire house, to be scanned for fingerprints in a matter of hours. One method uses fumes, produced when dried cyanoacrylate glue is

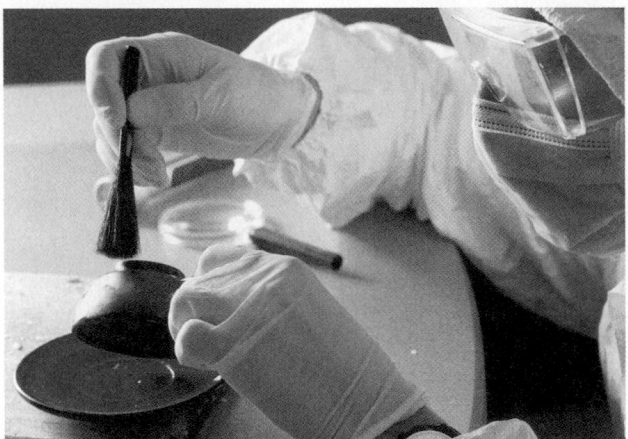
Figure 2 Fingerprint analysis

burned, to make latent, or hidden, fingerprints visible.

Trace Evidence Examination

Glass fragments, paint chips, and gunpowder residue are examples of trace evidence—physical evidence recovered from a crime scene or from the body of a victim or a suspect. Analyzing very small samples, such as the one shown in **Figure 3,** usually involves both physical and chemical tests, and may destroy the sample. New research is underway to find ways to analyze very small samples. Some methods being investigated are sophisticated types of chromatography, laser scanning, fluorescent imaging, and use of plasma devices. A trace evidence researcher, needs to have a broad background in chemistry, physics, and materials science.

Forensic Dentistry

If a dead body has been badly burned or is decomposed, identification of the remains may be very difficult. In such cases, the only part of the body left intact may be the jaws and teeth. A forensic dentist can identify human remains based on the dental records, if they exist, of possible victims. The forensic dentist may perform a dental exam during an autopsy, which may include X rays and charts of the teeth and the skull.

In other cases, the forensic dentist may examine human bite marks on a victim for comparison with the tooth patterns of a suspect. The examination of dental injuries is another task that may be required of the forensic dentist.

Firearms Analysis

A gun leaves a set of scratches as unique as a fingerprint on each bullet fired from the gun. Because these scratches are unique to each gun, it is possible to match a particular gun to a fired bullet or to tell if two bullets were fired from the same gun. A firearms analyst, such as the one shown in **Figure 4,** compares scratch patterns in an attempt to match the bullet to a particular gun. Automated analysis using a computer database is the goal of research in this area. The scratch pattern from a bullet would be scanned into a computer for comparison with the database of known weapons.

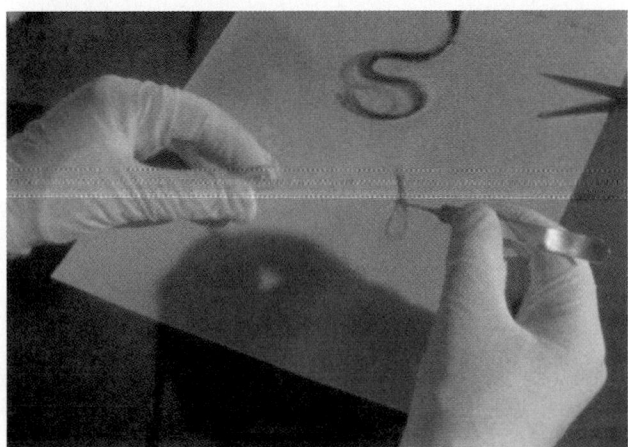
Figure 3 Trace evidence analysis

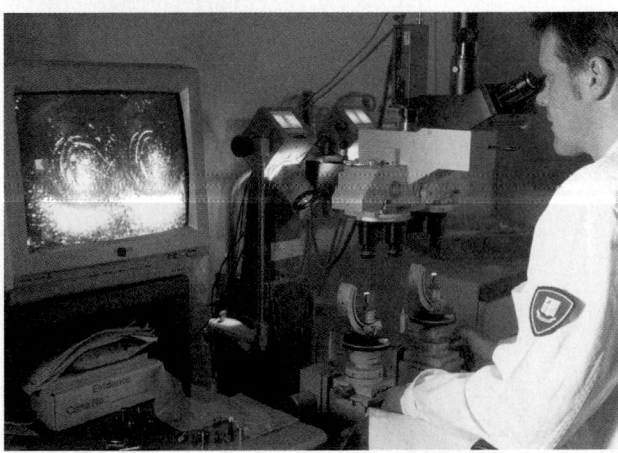
Figure 4 Firearms analysis

Hot Topic: Forensic mtDNA Analysis

One of the newest areas of forensic science involves the analysis of mitochondrial DNA (mtDNA), a type of DNA that is not found in the nuclei of cells. Rather, mtDNA occurs in mitochondria, small structures about the size of a bacterium that are found in the cell material surrounding the nucleus. Mitochondria, such as the ones shown in **Figure 5,** exist in large numbers in every cell, numbering from several to a thousand or more. Mitochondria have their own DNA, and each mitochondrion contains several identical copies of it. A person's mitochondria and thus their mitochondrial DNA are always inherited from the mother, whereas nuclear DNA is inherited from both parents.

One advantage of using mtDNA for forensic analysis comes from the number of mitochondria in each cell. This abundance allows a very small sample to provide a large mtDNA sample. Another advantage is that mtDNA exists in all types of cells and therefore any type of body tissue can be used, including bone.

One disadvantage is that, although the total size of mtDNA is small compared to nuclear DNA, an entire mitochondrion must be analyzed. This must be done because the number of variations in mtDNA is not large and the variations are very subtle.

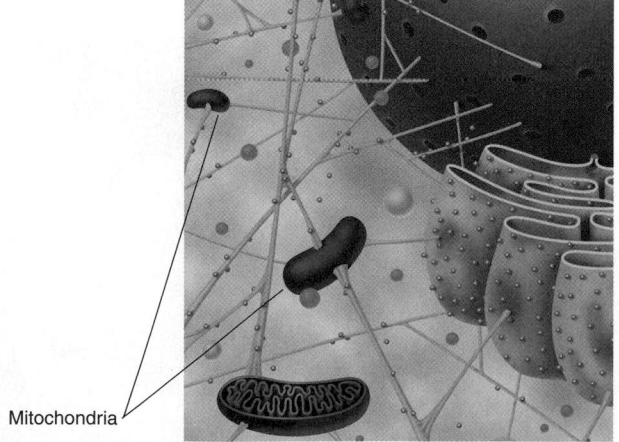

Mitochondria

Figure 5 Mitochondria

Figure 6 The Russian royal family

Forensic Genealogy

Forensic investigators can use mtDNA analysis to aid in the identification of human remains. A person may be identified if their mtDNA can be compared to that from another person descended along maternal lines. One such investigation used the mtDNA obtained from the female human remains in a grave in Russia thought to be that of the last Russian czar, his wife, the czarina, and one of their daughters, shown in **Figure 6.** The mtDNA from Prince Phillip of the British royal family was used for the comparison because Prince Phillip is a direct descendant of the czarina's mother.

The mtDNA samples matched and thus the members of the Russian royal family were identified more than seventy years after they were killed during the Russian revolution. A related analysis proved that a woman claiming to be the missing princess Anastasia, another daughter of the czar, was not related.

Topic Questions

1. Describe in one sentence what a forensic scientist does.
2. What new method is being developed for determining the author of a document?
3. How can a firearms analyst tell what gun a certain bullet was fired from?
4. Where is mtDNA found, and how does it differ from the DNA in the nucleus?

Forensics Lesson

Identification

Biological evidence is highly prized because it is not subject to the same sources of error as other types of evidence about identity. Eyewitnesses often get confused or make mistakes. Other types of physical evidence may show how the crime was committed, but may not offer much information about who committed it.

Analysis of blood or DNA evidence, on the other hand, can give investigators very specific information that can positively identify, or eliminate, a suspect as having been at the scene of the crime. Biological evidence such as this has its limitations, too. Contamination of samples can easily invalidate results, and laboratory specialists must be very careful in their analysis, as the samples they study are often very small. But since blood and DNA analysis methods were first developed, they have been important tools of identification for forensic scientists.

Blood Typing

Human blood can be one of four types—A, B, AB, or O—depending on which kinds of molecules, called antigens, are attached to the outer cell membrane of the red blood cells. These blood antigen molecules come in two forms: A antigens and B antigens.

A person's blood is type A if only A antigens are present on the red blood cells. The blood is type B if it has only B antigens. If both kinds of antigens are present, the blood type is AB. If neither kind of antigen is present, the blood is type O.

Determining a person's blood type is important not only for identification, but also to ensure that the person receives the correct blood type if a blood transfusion is needed. Giving a person the wrong blood type in transfusion can be fatal.

How Identification by Blood Type Works

Identification of blood type is very simple because red cells clump together when incompatible blood types are mixed. This clumping, called *agglutination*, is shown in **Figure 1.** *Serum* is the watery component of blood. The serum of type-A blood contains antibodies against the antigens of the cells in type-B blood. These antibodies, called *agglutinins*, attack the type-B red cells and cause them to clump together. Conversely, the serum from type-B blood contains antibodies that cause type-A red cells to clump. So, if A-serum is added to a blood sample and clumping occurs, the sample contains B antigens. However, because the sample could be type-B or type-AB, a test with B-serum is needed. If a second sample of the same blood forms clumps when combined with B-serum, A antigens are also present, and the blood is type-AB. The red blood cells in type-O blood (which carry neither A nor B antigens) will not form clumps when serum from type-A blood, -B, or -AB blood is added. Type O blood or serum (which contains antibodies to both A and B antigens) will, however, cause clumping of red blood cells in type-A, -B, or -AB whole blood.

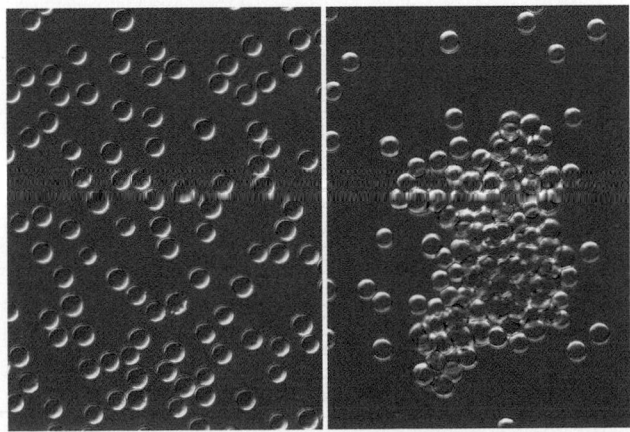

Figure 1 Normal blood cells (left); agglutinated blood cells (right)

DNA Fingerprinting

With the exception of identical twins, no two people have the same DNA sequence. Scientists estimate that the genes of a human consist of over three billion nucleotide pairs (also called "base pairs"). Each nucleotide is made up of a phosphate group bonded to a deoxyribose sugar molecule, which is bonded to one of only four nitrogenous bases—adenine, guanine, cytosine, and thymine. A person's entire genetic code consists of a very long sequence of these base pairs in a certain arrangement, which varies from person to person.

Given the extreme length of a DNA molecule, you may think that comparing DNA from one person to DNA of another person is very difficult. In fact, the process is fairly simple. Scientists have discovered certain enzymes in bacteria, called restriction enzymes, that digest—that is, break—DNA chains in specific places. A specific restriction enzyme digests a DNA strand only where certain nucleotide sequences occur. These nucleotide sequences are found in all human DNA, but their placement along the strand varies from person to person.

How Identification by DNA Works

When a restriction enzyme is added to a DNA sample, the products will be DNA fragments of various sizes, depending on where the target sequence of nucleotides occurred. These fragments—called *restriction fragments*—can be separated according to size by a process called *electrophoresis*, shown in **Figure 2.** During electrophoresis, DNA fragments migrate through a gel toward an electrode at various speeds according to their electrical charges (which are, in turn, determined by the size of the fragments). The result is a pattern, called a *DNA fingerprint*, which is unique to one person's DNA. You can see in **Figure 2** that the patterns for samples II and III match, while the pattern for sample I is different.

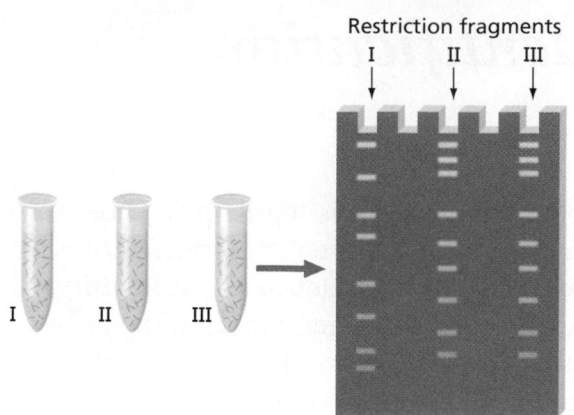

Figure 2 DNA-restriction enzyme mixtures (left); restriction fragments in gel (right)

DNA from a different person will produce a different DNA fingerprint. This is because some of the restriction fragments of the second person's DNA differ in size from those of the first person's DNA. An unknown DNA sample can be matched with a known sample of a person's DNA to an accuracy of 1 in 10 billion people. Due to its accuracy, DNA fingerprinting is often used to determine identity in criminal cases and to establish hereditary relationships, such as paternity.

Topic Questions

1. Explain why the blood cells of type B blood clump together when it is mixed with type A blood.

2. A student carrying out blood typing finds that the blood sample forms clumps when type A serum is added. The sample forms no clumps when type B serum is added. What is the type of the blood sample?

3. How does DNA determine most of the unique characteristics of a single person? Can a person have characteristics that are not determined by DNA? Explain and give two examples.

4. What characteristics of DNA fragments determine their separation during electrophoresis?

Forensics Skills Lab

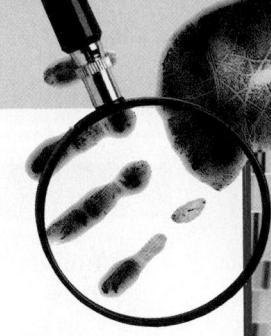

Blood Typing

SKILLS

- Experimenting
- Measuring
- Predicting
- Organizing and Analyzing Data

OBJECTIVES

- **Determine** the ABO and Rh blood types of unknown simulated blood samples.

MATERIALS

- blood samples, of unknown type from four subjects, simulated
- serum, anti-A, simulated (blood-typing)
- serum, anti-B, simulated (blood-typing)
- serum, anti-Rh, simulated (blood-typing)
- blood typing trays (4)
- toothpicks (12)
- wax pencil

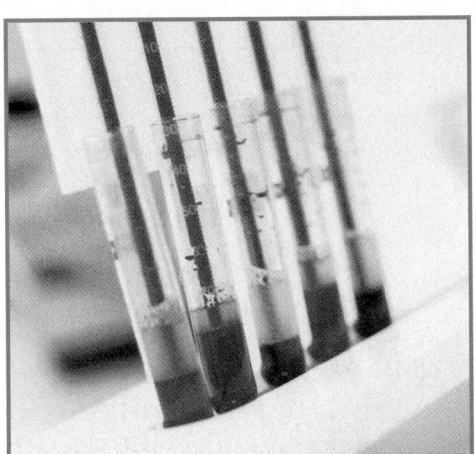

ChemSafety

⚠ CAUTION: Always wear safety goggles and a lab apron to protect your eyes and clothing.

⚠ CAUTION: Do not touch or taste any chemicals. Know the location of the emergency shower and eyewash station and how to use them. If you get a chemical on your skin or clothing, wash it off at the sink while calling to the teacher. Notify the teacher of a spill. Spills should be cleaned up promptly, according to your teacher's directions.

⚠ CAUTION: Glassware is fragile. Notify the teacher of broken glass or cuts. Do not clean up broken glass or spills with broken glass unless the teacher tells you to do so.

Background

Blood is one of the most common kinds of physical evidence at a crime scene where someone was injured. One of the first things a lab technician in a forensics lab must learn is how to determine ABO blood type from a blood sample. In this lab, you will determine the blood type of samples of simulated human blood from four different people.

Blood typing is performed using *antiserum*, blood serum that contains specific antibodies. For ABO blood typing, antibodies against the A and B antigens are used. These antibodies are called *anti-A* and *anti-B agglutinins*. If clumping—called *agglutination*—occurs in the test blood only when it is exposed to anti-A serum, the blood contains the A antigen and

is said to be type A. If clumping occurs in the test blood only when it is exposed to anti-B serum, the blood contains the B antigen and is said to be type B. If agglutination occurs with anti-A and separately, anti-B sera, the blood is type AB, which has both A and B antigens. If no agglutination occurs with either serum type, the blood type is O. This information is summarized in **Table 1.**

Another type of marker protein on the surface of red blood cells is the Rh factor, so named because it was originally identified in rhesus monkeys. People whose blood contains the Rh factor are said to be Rh positive (Rh+). People whose blood does not contain the Rh factor are Rh negative (Rh–).

Blood Type	Serum antibodies	Agglutination reaction of red blood cells below with blood serum in "Blood Type" column			
		A	B	AB	O
A	Anti-B	–	+	+	–
B	Anti-A	+	–	+	–
AB	None	–	–	–	–
O	Anti-A Anti-B	+	+	+	–

A person with Rh– blood has no antibodies to Rh+ blood unless the person was exposed to Rh+ blood at an earlier age. No agglutination occurs the first time an Rh– person receives a blood transfusion from an Rh+ person. Agglutination can occur, however, the second time the Rh– person receives Rh+ blood. In addition to testing for ABO blood type, it is also important to test blood for transfusion for its Rh factor.

PART A: Typing Blood

Procedure

1. Put on safety goggles and a lab apron. With a wax pencil, label each of four blood-typing trays as follows:

   Tray 1—Mr. Thomas, Tray 2—Ms. Chen, Tray 3—Mr. Juarez, Tray 4—Ms. Brown.

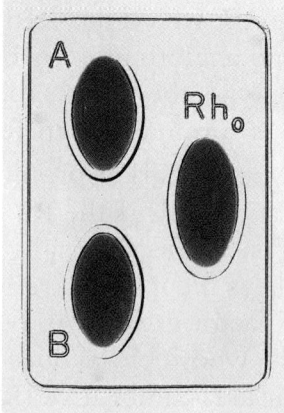

Figure 1 Blood-typing tray

2. **CAUTION: Use ONLY the simulated blood provided by your teacher. Do not use real blood.** Place 3 to 4 drops of Mr. Thomas's simulated blood in each of the A, B, and Rh wells of Tray 1 as shown in **Figure 1.**

3. Place 3 to 4 drops of Ms. Chen's simulated blood in each of the A, B, and Rh wells of Tray 2.

4. Place 3 to 4 drops of Mr. Juarez's simulated blood in each of the A, B, and Rh wells of Tray 3.

5. Place 3 to 4 drops of Ms. Brown's simulated blood in each of the A, B, and Rh wells of Tray 4.

6. Add 3 to 4 drops of the simulated anti-A serum to each A well on the four trays.

7. Add 3 to 4 drops of the simulated anti-B serum to each B well on the four trays.

8. Add 3 to 4 drops of the simulated anti-Rh serum to each Rh well on the four trays.

9. Use separate toothpicks to stir each sample of serum and blood. Record your observations in your lab report in a table like **Data Table A.** Indicate an agglutination reaction with a + and no reaction with a –. Also record your observations of each test. *Note: A positive test is indicated by obvious clumping of the red blood cells. Rh+ blood samples will undergo an agglutination reaction when exposed to anti-Rh serum.*

DATA TABLE A

	Anti-A serum	Anti-B serum	Anti-Rh Serum	Blood type	Observations
Agglutination reaction of blood samples with blood serum					
Tray 1: Mr. Thomas					
Tray 2: Ms. Chen					
Tray 3: Mr. Juarez					
Tray 4: Ms. Brown					

PART B: Cleanup and Disposal

10. Dispose of paper and broken glass in the designated waste containers. Do not put lab materials in the trash unless your teacher tells you to do so.

11. Clean up your work area and all lab equipment. Return lab equipment to its proper place. Wash your hands thoroughly before you leave the lab.

Analyze and Conclude

1. **Applying Concepts** What factors determine the ABO blood types?

2. **Applying Concepts** What is the difference between an antigen and agglutinin?

3. **Applying Concepts** If Ms. Brown were serving as a blood donor, what ABO blood type(s) could receive her blood safely?

4. **Applying Concepts** Which person among the four represented by the simulated blood samples can receive donated blood from Ms. Chen? Explain your answer.

5. **Inferring Conclusions** People with type O blood who are also Rh– are commonly called universal donors. Explain why.

6. **Inferring Conclusions** A person with what blood type would be considered a universal recipient? Explain your answer.

Extension

1. **Further Inquiry**
The first baby with Rh+ blood born to a woman with Rh– blood usually has no health problems. The second Rh+ child, however, can be seriously threatened before birth if the mother produces antibodies against the Rh antigens of her baby. Find out why this happens and what treatment is given to babies in this situation to save their lives.

2. **Further Inquiry**
Find out what an emergency medical technician gives to a patient when the technician administers an emergency transfusion in the field. Why do technicians use this substance instead of blood for transfusing their patients?

Forensics Lab

DNA Fingerprinting

SKILLS
- Collecting data
- Identifying patterns
- Inferring
- Interpreting
- Analyzing data
- Predicting

OBJECTIVES
- **Perform** a restriction digestion of DNA samples.
- **Evaluate** the results of simulated DNA fingerprints.
- **Identify** a hypothetical burglar by analyzing simulated DNA fingerprints.

MATERIALS
- safety goggles
- lab apron
- gloves
- microcentrifuge tubes (6)
- microcentrifuge tube rack/float
- micropipet, 10 or 20 μL
- micropipet tips (30)
- DNA samples (5)
- *Pvu II* restriction enzyme
- *Pvu II* reaction buffer, 10×
- water bath, 37° C
- loading dye
- water bath, 65° C
- agarose gel on gel tray (0.8 percent)
- electrophoresis system, battery powered
- beaker, 500 mL
- graduated cylinder, 250 mL

- TBE running buffer, 1×, 350 mL
- gel staining tray
- DNA stain
- Lambda DNA/*Hind III* marker
- resealable plastic bag

ChemSafety

CAUTION: Always wear safety goggles and a lab apron to protect your eyes and clothing.

CAUTION: Do not touch or taste any chemicals. Know the location of the emergency shower and eyewash station and how to use them. If you get a chemical on your skin or clothing, wash it off at the sink while calling to the teacher. Notify the teacher of a spill. Spills should be cleaned up promptly, according to your teacher's directions.

CAUTION: Glassware is fragile. Notify the teacher of broken glass or cuts. Do not clean up broken glass or spills with broken glass unless the teacher tells you to do so.

Background

The police are investigating the burglary of a residence. As the police arrived, the masked burglar smashed a glass door with a chair and quickly escaped. Police found small bits of bloodstained fabric on some of the pieces of the glass door. Four suspects have been brought in, and DNA samples have been collected from each. The police lab thus has the DNA from the blood samples found at the crime scene and DNA samples taken from each of the four suspects.

In this lab, you will perform some of the experimental procedures involved in DNA fingerprinting and use your results to identify the burglar.

Procedure

PART A: Restriction Enzyme Digestion of DNA

In the first step of DNA fingerprinting, samples of known and unknown origin are obtained and then *digested,* or cut into small fragments, by a restriction enzyme. In this case, the samples would include DNA from the blood sample found at the scene (the burgler's DNA) and DNA from each of the suspects.

1. Put on safety goggles and a lab apron.

2. Obtain five microcentrifuge tubes.

3. Label tubes as follows: Crime Scene, Suspect #1, Suspect #2, Suspect #3, Suspect #4.

4. Set the micropipet to 2 µL and place a clean tip on the pipet.

5. Add 2 µL of *Pvu II* 10× reaction buffer to each tube. *Note: It may be helpful to place the pipet tip against the side of the micro-centrifuge tube when dispensing small volumes.*

6. Set the micropipet to 10 µL.

7. Using a fresh pipet tip for each sample, add 10 µL of each DNA sample to the appropriate tube.

8. Set the micropipet to 2 µL.

9. Add 2 µL of *Pvu II* restriction enzyme to each tube. Again, be sure to use a fresh micropipet tip for each sample.

10. Close each tube and snap the tubes in a downward motion with your wrist to force all of the reagents to the bottom of the tubes.

11. Incubate all five tubes at 37° C for 45–60 minutes.

12. Add 2 µL of loading dye to each tube.

13. Incubate all five tubes for 5 minutes at 65° C to stop the restriction enzyme activity.

14. Place the samples in the freezer if you will not be using them immediately. Make sure all samples for your group are stored together and labeled with your group name.

PART B: Gel Electrophoresis of DNA Samples

The next step of DNA fingerprinting is to separate the restriction fragments by gel electrophoresis. In gel electrophoresis, the digested DNA samples are loaded into wells on a jellylike slab called a *gel*. The gel is then exposed to an electrical current. DNA has a negative electrical charge, so the restriction fragments are attracted to the positive pole when an electric current is applied. Smaller fragments travel farther and faster through the gel than do longer ones. Recall that restriction fragments harvested from one person's DNA differ somewhat from the restriction fragments of another person. So the pattern made by restriction fragments migrating through the gel also differs from person to person. A special dye, called a *loading dye*, is added to the DNA samples for two reasons. The loading dye is heavy, which helps the samples to stay in the wells. The loading dye also runs slightly faster than the DNA samples and indicates when the gel has finished running.

15. Place the tray with the gel on the lab bench.

16. Set the pipet to 10 µL and place a new tip on the end of your micropipet.

17. Open the microtube containing the DNA marker and use the pipet to load 10 µL of the marker DNA into the well in Lane 1 of an agarose gel. To do this, place both elbows on the lab table, lean over the gel, and slowly lower the micropipet tip into the opening of the well before depressing the plunger. *Note: Do not jab the micropipet tip through the bottom of the well.*

18. Set the pipet to 15 µL and place a new tip on the end of your micropipet.

19. Open the microtube containing the crime scene sample and use the pipet to load 15 µL of the sample into the well in Lane 2 of an agarose gel.

20. Using a new micropipet tip for each tube, repeat step 19 for each of the remaining samples. Place each sample in a well as follows:

Lane 1: DNA marker

Lane 2: Crime scene DNA

Lane 3: Suspect #1 DNA

Lane 4: Suspect #2 DNA

Lane 5: Suspect #3 DNA

Lane 6: Suspect #4 DNA

21. Follow your teacher's instructions to carefully place the agarose gel (still in a gel-casting tray) in the chamber of an electrophoresis apparatus, such as the one shown in **Figure 1,** so that the wells are closest to the negative electrode.

22. Pour approximately 350 mL of 1× TBE running buffer into a beaker.

23. Gently and slowly pour the running buffer from the beaker into one side of the electrophoresis chamber until the gel is completely covered (approximately 1 to 2 mm above the top surface of the gel). *Note: Pouring too fast will rinse your DNA sample out of the wells. Be careful not to overfill the chamber with buffer.*

24. Place the cover on the electrophoresis chamber. Wipe off any spills around the electrophoretic apparatus before doing the next step.

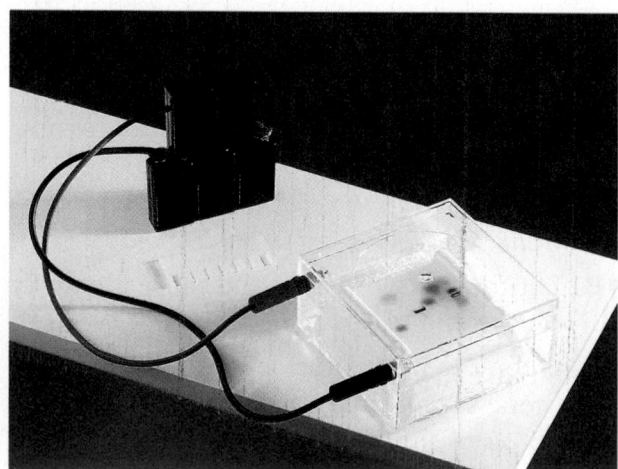

Figure 1 Electrophoresis set-up

25. ⚠ Connect five 9 V alkaline batteries as instructed by your teacher. **CAUTION: Do not touch both ends of the patch cords or both terminals on the battery pack at the same time.**

26. Connect the red (positive) patch cord to the red terminal on the chamber and the red terminal on the battery pack. Follow the same procedure with the black (negative) patch cord and the black terminals.

27. Observe the migration of the loading dye along the gel toward the red (positive) electrode.

28. Disconnect the battery pack when the loading dye band has run halfway off the end of the gel. *Note: If gels have been run overnight, this step will take place on the following day.*

29. Remove the cover from the electrophoresis chamber.

30. Carefully lift the gel tray (containing the gel) from the chamber onto a piece of paper towel. Notch one side of the gel so that you can identify the lanes.
Note: Agarose gels are very fragile, so use extreme care when handling the gel.

31. Use a metric ruler to measure the distance of the dye bands in Lane 6 (in mm) from each of the six sample wells. *Note: Be sure to measure from the center of the well to the center of the band.*

32. Place the gel in a resealable bag and add 1–2 mL of 1× TBE buffer and refrigerate the gel until the next lab period. If there is enough time (at least 45 minutes) left in the class period, proceed to step 33.

PART C: Staining the Gel

After the DNA fragments have been separated, the gel can be stained to visualize the banding pattern of each DNA fingerprint.

33. Gently place the gel in the staining tray.

34. Put on safety goggles and a lab apron. Pour approximately 100 mL of warm dilute stain into the staining tray so that it covers the gel.

35. Cover the tray and let the gel stain for approximately 30 minutes.

36. Carefully decant the used stain. *Note: Make sure the gel remains flat and does not move up against the corner. Decant the stain directly to a sink drain and flush with water.*

37. Add distilled water or tap water to the staining tray. Do not pour water directly onto the gel.

38. Allow the gel to de-stain either overnight or for 20 minutes. To de-stain in 20 minutes, gently rock the tray, and change the water several times. Overnight de-staining does not require a change of water.

39. View the gel against a white sheet of paper. Sketch the bands you see on the blank gel, which will look similar to the image in **Figure 2**.

40. Store the gel in a resealable plastic bag with 1–2 mL of TBE buffer.

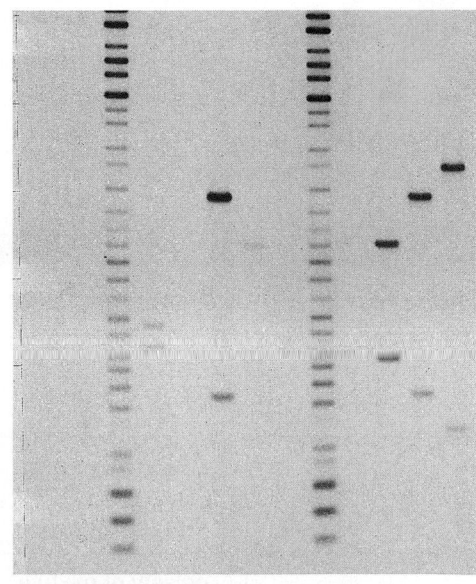

Figure 2 DNA fingerprint

PART D: Cleanup and Disposal

41. Dispose of paper and broken glass in the designated waste containers. Do not put lab materials in the trash unless your teacher tells you to do so.

42. Clean up your work area and all lab equipment. Return lab equipment to its proper place. Wash your hands thoroughly before you leave the lab.

Analyze and Conclude

1. Explaining Events Where are the smallest DNA fragments located on the gel? the largest? Explain why.

2. Explaining Events Why was loading dye added to each sample prior to loading the samples on the gel.

3. Examining Data Are any of the DNA fingerprints identical? If so, which ones?

4. Drawing Conclusions Based on your results, which suspect appears to be the burglar? Explain your answer.

5. Interpreting Information Explain why each suspect has a different banding pattern (or DNA fingerprint).

6. Making Predictions What do you think would happen if you placed your gel in the electrophoresis chamber with the wells containing DNA next to the red electrode instead of the black electrode?

7. Making Inferences From what types of crime scene evidence (other than blood) might police obtain DNA?

Extension

1. Research and Communications Research another way that DNA fingerprinting is used in society and present your results to your classmates in a short oral report.

2. Research and Communications Research the procedures involved in collecting evidence at a crime scene.

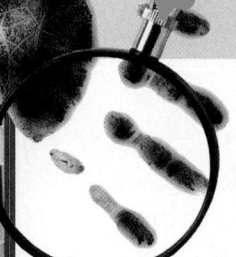

Forensics Lesson

Chromatography

Green paint has been thrown all over several walls of a local business. The police have several suspects and have found some green paint at each of the suspect's homes. How can they determine if any of the suspect samples match the paint thrown on the walls?

Background

Chromatography is a technique used to separate a mixture of different substances. It was originally developed in 1903 by a Russian botanist, Mikhail Tswett, who used this approach to separate colored plant pigments. *Chromatography* means "to write with colors" and comes from the Greek words *chroma*, color, and *graphein*, to write.

There are several types of chromatography, each one depending on the nature of the substance in a mixture. In most types of chromatography, polarity is the basic principle at work. A polar molecule has an uneven distribution of electric charge. Water is the most familiar example of a polar substance. Its uneven distribution of charge means that one side of a water molecule strongly attracts the opposite side of the water molecule next to it, as shown in **Figure 1.** This attraction between opposite charges means that polar

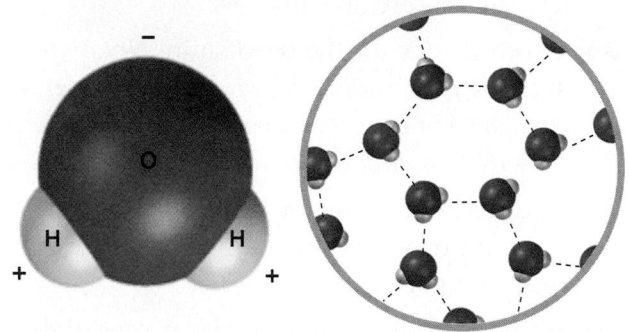

Figure 1 Water molecule (left); attraction between polar molecules (right)

substances—even different polar substances—have strong attraction to each other. On the other hand, nonpolar substances, such as oils, have a much stronger attraction for each other than for polar substances.

How Chromatography Works

All kinds of chromatography involve two systems, or *phases*, in contact with each other. The mobile phase is a solvent that dissolves some or all of the substances to be separated in the mixture. The stationary phase on which the mixture is placed is often some type of solid material such as chromatography paper or a thin layer of a gel coated onto a glass or plastic plate. **Table 1,** shows some examples of different kinds of chromatography.

TABLE 1: SOME TYPES OF CHROMATOGRAPHY			
Category	**Mobile phase**	**Stationary phase**	**Separating principle**
Paper chromatography	some solvent; can be water, methanol, etc.	paper	polarity of mixture components relative to mobile phase
Thin-layer chromatography	some solvent; can be water, methanol, etc.	silica gel plate	polarity of mixture components relative to mobile phase
Column chromatography	some solvent; can water, methanol, etc.	powdered adsorbent packed in a glass column	polarity of mixture components relative to mobile phase

As the mobile phase moves through or across the stationary phase, the stationary phase separates the components of the mixture. The result of the process is called a *chromatogram,* shown in **Figure 2,** so named for the colored bands produced when the separation involves a mixture of colored substances.

A paper chromatogram is produced by dabbing a sample of the mixture to be separated near one end of a piece of chromatography paper (the stationary phase). Another part of the chromatography paper is then wetted with the solvent. While the paper is in contact with the solvent, the solvent is wicked through the paper by capillary action.

The substances that make up the sample mixture are dissolved by the solvent and move along the paper with the solvent. As the solvent moves through the paper, the molecules of the various dissolved substances are attracted to both the solvent and the paper. The strengths of these attractions are determined by how polar the substance is. Because these attractions are different for each substance in a mixture, each substance moves through the paper at a different rate. The substances with a greater attraction to the solvent move faster and farther along the paper. Those with a greater attraction to the paper move more slowly, and consequently not as far as other substances.

Polar solvents, such as water, methanol, and acetic acid, tend to work best with substances that are also polar, such as water-based inks and dyes. A less polar solvent such as petroleum ether, cyclohexane, or methylene chloride will tend to work best with nonpolar or weakly polar substances, such as oil-based inks and dyes.

In addition, large molecules tend to move more slowly than small molecules because the large molecules cannot easily pass through the stationary phase material. The result of the differences in attraction and size cause the various substances to separate into distinct areas on the paper. The paper is then removed from the solvent and allowed to dry, producing a permanent record that can be analyzed or processed further.

How to Use Chromatography

Chromatography is often used as a tool for comparing the makeup of two or more mixtures. If the chromatogram is the same for each mixture, then the mixtures are probably the same. If the chromatograms are different, then the mixtures are probably different. Often, different samples to be tested are placed on the same piece of chromatography paper. This way, the conditions that create the chromatogram are the same for all the mixtures.

Topic Questions

1. In chromatography, what are the functions of the mobile and stationary phases?

2. Red dye can be made from a mixture of yellow dye and magenta (a purplish blue) dye. How could you determine if a particular red dye is made from a single dye or from a mixture of yellow and magenta dyes?

3. Two mixtures are compared using chromatography. Why is it desirable to place samples of both mixtures on the same paper when making the chromatogram?

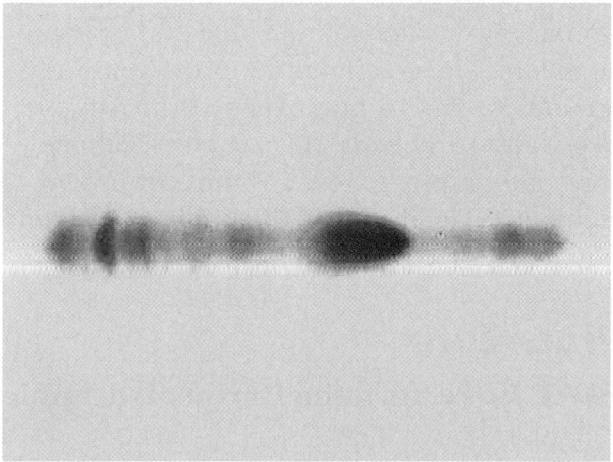

Figure 2 Chromatogram of pigments in a species of algae

Forensics Lab

The Counterfeit Drugs

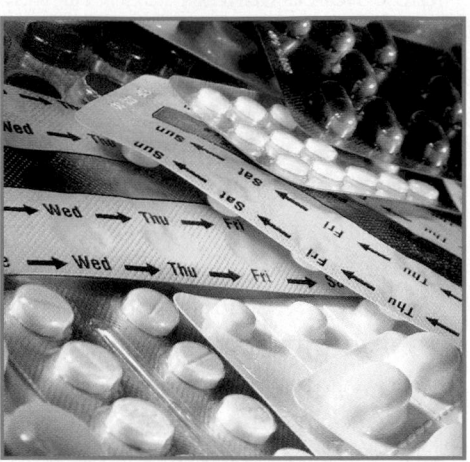

SKILLS
- Experimenting
- Inferring
- Interpreting
- Communicating

OBJECTIVE
- **Determine** by strip paper chromatography if a suspect dye contains a particular pigment.

MATERIALS
- distilled water
- pigment solutions to be tested
- strips of chromatography paper (2)
- tape
- beakers, small (2)
- eyedropper
- paper clips, small
- pencil
- scissors

CAUTION: Always wear safety goggles and a lab apron to protect your eyes and clothing.

CAUTION: Do not touch or taste any chemicals. Know the location of the emergency shower and eyewash station and how to use them. If you get a chemical on your skin or clothing, wash it off at the sink while calling to the teacher. Notify the teacher of a spill. Spills should be cleaned up promptly, according to your teacher's directions.

CAUTION: Glassware is fragile. Notify the teacher of broken glass or cuts. Do not clean up broken glass or spills with broken glass unless the teacher tells you to do so.

Background

A small pharmaceutical company has been accused of illegally manufacturing and exporting a counterfeit version of a popular blood pressure medication. The Food and Drug Administration (FDA) suspects that the company manufactures a generic version of the patented drug and labels it identically to the real drug.

The FDA, along with the Drug Enforcement Administration (DEA), has confiscated a sample of the drug as it was being loaded onto a ship. Officials need to determine whether or not their sample is the brand-name drug or the identical-looking, counterfeit version of the medication.

While the active chemical compound of the brand name drug and the counterfeit drug is the same, officials know that there is

a difference between the genuine and the counterfeit pills. While the pills appear to be identical, the pigments used in the brown color coating of the pills differ. The brown coating used in the counterfeit pills contains a mixture of water-soluble blue and orange pigments. The coating used in the genuine version contains a homogeneous brown dye, also water-soluble. In this lab, you will use chromatography to test a sample of brown pigment to determine if the pill it came from is genuine or counterfeit.

Procedure

PART A: Performing Chromatographic Analysis

1. 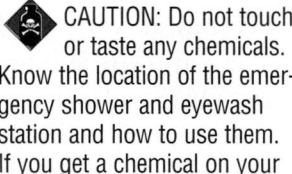 Put on safety goggles and a lab apron. Cut a length of chromatography paper equal to the depth

of the beaker you are using. You may check to make sure the length is correct by taping the top of the paper to a pencil and lowering the paper into the beaker, allowing the pencil to rest across the top of the beaker.

2. Once the paper is cut to the correct length, remove it from the beaker. Clip a paper clip to the bottom of the paper to keep the paper hanging straight down while it is in the beaker. (Keep the pencil attached to the top of the paper.) Using an eyedropper, dab some of the suspect dye onto the chromatography paper about 2 cm above where the water level will be in the bottom of the beaker.

3. Allow the dye to dry for a few minutes, and then repeat this spotting process several times (at the same place on the paper each time) to build up a concentration of pigment that will yield good results.

4. Now pour distilled water into the beaker to a depth of about 2 cm. It is important to have the test spot be 2 cm above the water level, and not immersed. Make sure that the dye to be tested does not come into contact with the water. Refer to **Figure 1** which shows the correct relationship between the dye spot and the water level.

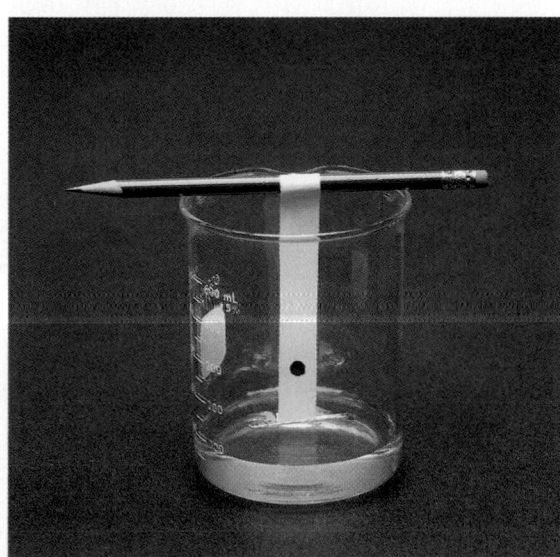

Figure 1 Paper chromatography setup

5. To conduct a chromatographic run, set up the beaker, paper, pencil, and water as shown in **Figure 1.** Allow the water to move up the paper by capillary action and separate the pigments in the dye. This process may take several minutes, so plan accordingly.

6. When the water migrates to within a few millimeters from the top of the paper, stop the run by removing the pencil and attached paper from the beaker. Empty the water from the beaker, place the pencil back across the top of the beaker, and allow the chromatogram to dry overnight.

PART B: Cleanup and Disposal

7. Dispose of paper and broken glass in the designated waste containers. Do not put lab materials in the trash unless your teacher tells you to do so.

8. Clean up your work area and all lab equipment. Return lab equipment to its proper place. Wash your hands thoroughly before you leave the lab.

Analyze and Conclude

1. **Explaining Events** Why do the pigments of a dye separate out on the chromatography paper?

2. **Analyzing Methods** Why is it important that tests in cases like this one be carried out such that the analysts do not know what products they are testing?

3. **Drawing Conclusions** Is the sample you tested from the genuine drug or the counterfeit drug? How can you tell?

Extension

1. **Research** Research and report on science-related careers in the FDA or DEA.

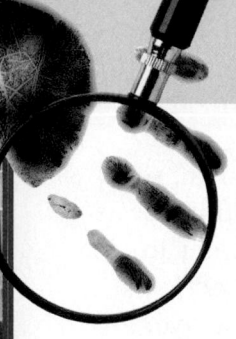

Forensics Lab

The Questionable Autograph

SKILLS
- Experimenting
- Inferring
- Interpreting
- Communicating

OBJECTIVES
- **Determine** by examination of pigments through thin layer chromatography if an autograph was written with a particular type of pen.

MATERIALS
- thin layer chromatography plates (2)
- teasing needle or pushpin
- single-hole puncher
- suspect autographs, and ink samples from an authentic pen
- spot plates (4)
- dropper bottles of distilled water and methanol
- open-ended capillary tubes (2)
- ruler
- graduated cylinder, 10 mL
- 2-oz. glass bottles with caps (2)

ChemSafety

CAUTION: Always wear safety goggles and a lab apron to protect your eyes and clothing.

CAUTION: Methanol is poisonous. Wear protective gloves. Avoid prolonged exposure to vapors and use in the hood or a well-ventilated area, as directed by the teacher. Keep methanol away from heat and flames, as it is flammable.

CAUTION: Do not touch or taste any chemicals. Know the location of the emergency shower and eyewash station and how to use them. If you get a chemical on your skin or clothing, wash it off at the sink while calling to the teacher. Notify the teacher of a spill. Spills should be cleaned up promptly, according to your teacher's directions.

CAUTION: Glassware is fragile. Notify the teacher of broken glass or cuts. Do not clean up broken glass or spills with broken glass unless the teacher tells you to do so.

Background

The United States Postal Service has been called on to investigate a report of mail fraud. Through an online auction, a sports fan has purchased a ticket stub autographed by his favorite ballplayer. Among the ballplayer's many commercial endorsements is one for a certain brand of pen—he agreed to use only that brand of pen to sign autographs. The suspicious sports fan thinks the signature on the ticket stub looks like it was made using an ordinary ballpoint pen.

The online auction company is about to ban the seller because of other complaints, and an investigation of mail fraud is under-

way. Postal inspectors need to determine whether or not the autograph was signed with the right kind of pen.

In this lab, you will use thin layer chromatography (TLC) to compare the ink from the signed ticket stub with ink from the type of pen the athlete would have used.

Because some inks are water soluble while others dissolve in methanol, you will need to conduct four thin layer chromatography (TLC) runs: two (one for the suspect ink and one for the authentic ink) using water as the solvent and another two of the same samples using methanol as the solvent.

Procedure

PART A: Performing Thin Layer Chromatography

1. Put on safety goggles, protective gloves, and a lab apron. Score two TLC plates, such as the one shown in **Figure 1,** in half lengthwise by scratching a line in the TLC gel coating with the teasing needle or push-pin, to allow two runs per plate.

2. To extract ink from an autograph, use a hole punch to punch about 25 holes in areas containing ink. Place half of these small disks into a spot plate well and then add a few drops of water to dissolve and remove the ink from the paper.

3. Place the rest of the paper disks containing autograph ink into another spot plate well, and add a few drops of methanol to dissolve and remove the ink from the paper. **CAUTION: Methanol is poisonous and flammable.**

4. Repeat steps 2 and 3 with ink samples from the authentic pen.

5. You will use one of the TLC plates to run the suspect and authentic inks side by side with water as the solvent, and the other TLC plate to run the suspect and authentic inks side by side with methanol as the solvent. Use a capillary tube to spot some of the extracted ink onto the TLC plates about 2 cm from one end. Allow the initial spot to dry, then repeat the process over the same location a few times to build up a concentration of pigment that will yield good results. Use a different capillary tube for each of the two ink samples.

6. Pour water to a depth of 1 cm in one of the beakers. Pour methanol to a depth of 1 cm in the other beaker. It is important that the test spots are 1 cm above the solvent level and not immersed or even touching the solvent initially.

7. Place the TLC plates, as upright as possible, into the beakers. The solvents will move up the plates and separate the pigments in the ink. Observe the process carefully, as it may take just a few minutes, or up to 30 minutes, for the solvent to move to the top of the plate. When the solvent level has risen to within 1 cm of the top of the plate, remove the plate from the bottle. Allow the TLC plates to dry overnight before examining them.

PART B: Cleanup and Disposal

8. Dispose of paper and broken glass in the designated waste containers. Do not put lab materials in the trash unless your teacher tells you to do so.

9. Clean up your work area and all lab equipment. Return lab equipment to its proper place. Wash your hands thoroughly before you leave the lab.

Analyze and Conclude

1. Are the pigments contained in the suspect and authentic inks similar?

2. If the two inks tested are identical, does this prove with certainty that the autograph is authentic?

3. What factors could affect the accuracy of your results? How could your procedure have been improved in order to control for these factors?

4. What additional tests could be conducted in order to make the results of the case more certain?

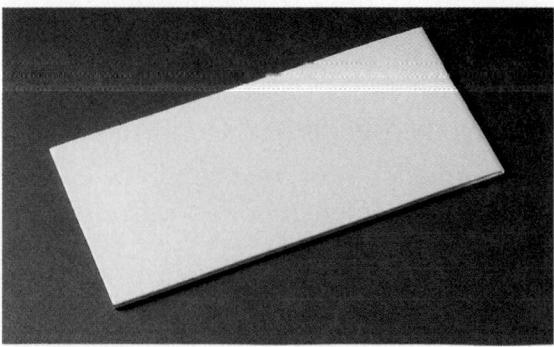
Figure 1 Thin-layer chromatography plate

Forensics Lesson

Determination of Density

You find a shard of glass and wonder if it is from a window or eyeglasses. A large, old necklace may be made of gold or of a metal far less valuable. One of the easiest to analyze properties of an object is its density, which may be a clue to its identity or source.

Background

Density is the ratio of the mass of an object to its volume. This ratio can be expressed as an equation, as shown below.

$$\text{density} = \text{mass/volume}$$

also written $\qquad d = m/V$

An object has a density characteristic of the material of which it is made. At standard temperature and pressure, for example, the density of water is 1.0. When a solid object is placed in fluid, the difference between its density and the fluid's density determines whether the object will sink or float.

- If the density of the object is *greater* than the density of the fluid, the object will sink. The volume of fluid displaced will be the same as the volume of the object, as shown in **Figure 1.** (The fluid displaced by the brick in **Figure 1** is shown as overflow in the smaller beaker. This overflow fluid volume is equal to the volume of the brick.)

- If the density of the object is *the same* as the density of the fluid, the object will neither sink nor float: it will remain suspended in the fluid. The volume of fluid displaced will be equal to the volume of the object.

- If the density of the object is *less* than the density of the fluid, the object will float. The volume of fluid displaced will have a weight equal to the weight of the object.

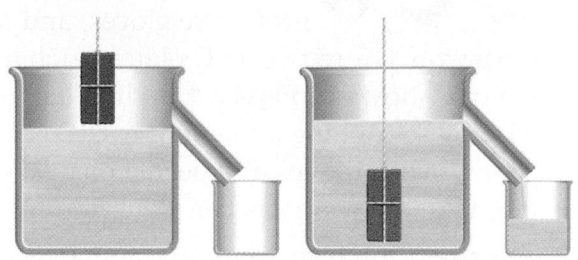

Figure 1 Object displacing its own volume of a fluid

Archimedes' Principle

Archimedes, a Greek who lived in the third century BCE, discovered an important fact about buoyant force, the upward force exerted on an object by the fluid in which it is immersed. Archimedes' principle states that for an object in a fluid, a buoyant force is exerted by the fluid on the object that is equal to the weight of the fluid displaced by the object. Archimedes' principle can be stated in a very simple equation, shown below.

$$\text{buoyant force} = \text{weight of displaced fluid}$$

Because of buoyant force, a submerged object will have an apparent weight that is less than its weight in air. (You experience buoyant force in a swimming pool and even in a bathtub.)

A floating object, on the other hand, appears to be weightless. This is because a floating object is less dense than the fluid, so only part of its volume displaces the fluid, and the weight of the displaced fluid is the same as the weight of the entire object.

Archimedes' principle can be used to determine the density of an object submerged in a fluid.

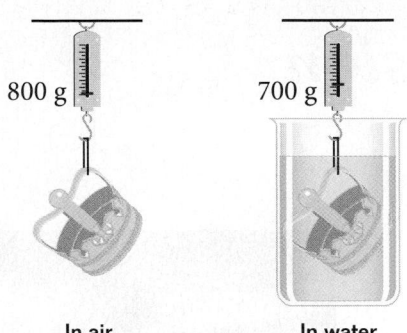

Figure 2 Weight of crown in air (left) and in water (right)

In air   In water

Density Determination of an Irregularly-Shaped Object

The density of a relatively large but irregularly shaped object, such as the crown shown in **Figure 2,** is most easily determined by using Archimedes' principle. In this method, the weight of an object is compared to its apparent weight when it is submerged in a fluid.

Density Determination of a Very Small Object

To determine the density of very small objects such as a tiny piece of glass or plastic, it is often more convenient and accurate to place the object in a fluid of known density to determine whether the object sinks or floats. The density of the fluid can then be adjusted by adding a second fluid that mixes with the first fluid but has a different density. The second fluid is added, dropwise, until the object being tested is suspended—neither floating nor sinking. At this point, the density of the fluid and the density of the object are the same.

Accurate results require careful recording of the number of drops of fluid added. Knowing the exact number of drops of each fluid added, and the density of each fluid, you can calculate the final volume of the mixtures using weighted averages. For this, a "weight" is given to the density of each fluid based on how much of each was present in the mixture at the point where it achieved the same density as the glass sample.

First, multiply the density of each fluid by the number of drops that were present. To compute the final density of the mixture, add the two weighted densities together and divide that by the total number of drops of fluid in the mixture.

For example, say you have glass sample that is suspended in a mixture. To make the mixture, you had added 30 drops of a fluid with a density of 2.50 g/mL, plus 5 drops of water, which has a density of about 0.998 g/mL. First, you weight the density of each fluid by multiplying it by the number of drops in your solution, and add these values together:

$$(30 \text{ drops} \times 2.50 \text{ g/mL})$$
$$+(5 \text{ drops} \times 0.998 \text{ g/mL})$$
$$= 80.0 \text{ drops} \cdot \text{g/mL}$$

To compute the density of the fluid mixture, divide this weighted total by the total number of drops:

$$\text{density} = \frac{80.0 \text{ drops} \cdot \text{g/mL}}{35 \text{ drops}} = 2.29 \text{ g/mL}$$

Because 2.29 g/mL is the final density of the mixture when the glass sample was suspended, we know 2.29 g/mL is also the density of the glass sample.

Topic Questions

1. What is the mathematical definition of density?

2. State Archimedes' principle.

3. An object placed in a fluid sinks and displaces fluid. What is the relationship between the volume of fluid displaced and the volume of the object?

4. If an object is submerged in a fluid, how is the weight of the object affected? Why?

5. Suppose that to determine the density of an object, you use the weighted-difference technique. What information is necessary for the determination to be successful? Explain.

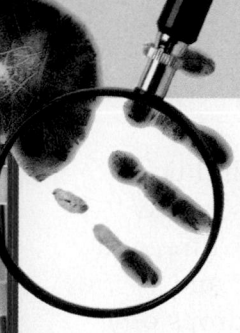

Forensics Lab

The Parking Lot Collision

SKILLS
- Designing experiments
- Experimenting
- Collecting data
- Inferring
- Interpreting
- Measuring

OBJECTIVES
- **Determine** the density of a small glass chip by the method of suspension.

MATERIALS
- safety goggles
- lab apron
- gloves
- dropper bottles of distilled water and saturated ZnI_2 solution
- glass sample
- test tube, 2.5 mL with cap
- amber bottle for liquid disposal and recovery

ChemSafety

 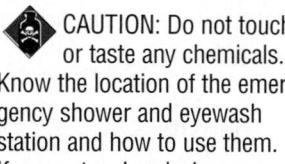 **CAUTION: Always wear safety goggles and a lab apron to protect your eyes and clothing.**

CAUTION: Do not touch or taste any chemicals. Know the location of the emergency shower and eyewash station and how to use them. If you get a chemical on your skin or clothing, wash it off at the sink while calling to the teacher. Notify the teacher of a spill. Spills should be cleaned up promptly, according to your teacher's directions.

CAUTION: Glassware is fragile. Notify the teacher of broken glass or cuts. Do not clean up broken glass or spills with broken glass unless the teacher tells you to do so.

Background

The high school campus police are following up on the report of a fender bender. A student said she was exiting the school when she saw another student back into the front of her car with a large truck, and then leave the scene. The police soon found the alleged perpetrator, who denied any involvement in the collision that left the front of the small car crushed, grill and headlights broken.

While talking to the driver of the truck, the officer noticed a shard of glass stuck in the soft rubber covering of the truck's bumper. When questioned, the driver of the truck claimed not to know the source of the glass, but announced confidently that she had not run into anyone's car. In this lab, you will use the suspension method to determine the density of the piece of glass retrieved from the bumper of the suspect's vehicle.

Procedure

PART A: Design an Experiment

1. Work with members of your lab group to design a procedure to test the density of the glass fragment in order to solve the case. If the glass is the same density as headlight glass, the incident and the suspect will require further investigation. You will use the method of density determination by suspension and with the lab materials provided. See "Forensics Lesson: Density" that precedes this lab, for hints.

2. Write a procedure for your experiment. Create a data table that clearly displays your calculations and results. Make a list of all the safety precautions you will take. Have your teacher approve your procedure and safety precautions before you begin the experiment.

PART B: Conduct Your Experiment

3. Put on safety goggles, protective gloves, and a lab apron.

4. **CAUTION; ZnI$_2$ is corrosive and is a poison. Do not taste or touch.** You will use two clear liquids of different density, saturated zinc iodide solution and distilled water. Glass densities are closer to the density of the saturated zinc iodide solution, so you will want to start by counting 25 drops of the zinc iodide solution into the test tube with the glass sample in it. Then add water dropwise, keeping track of the number of drops until suspension is achieved. Refer to **Figure 1** as a guide to achieving suspension of the glass fragment in the mixture.

5. **CAUTION: Glass samples have sharp edges. Handle them with care.** When the glass fragment is suspended, you will be able to determine the density of the mixture by calculating a weighted average from the densities of the two liquids (based on the number of drops added). If your technique and calculations are accurate, this calculated density of the mixture when the glass piece is suspended will be the same as the density of the glass piece.

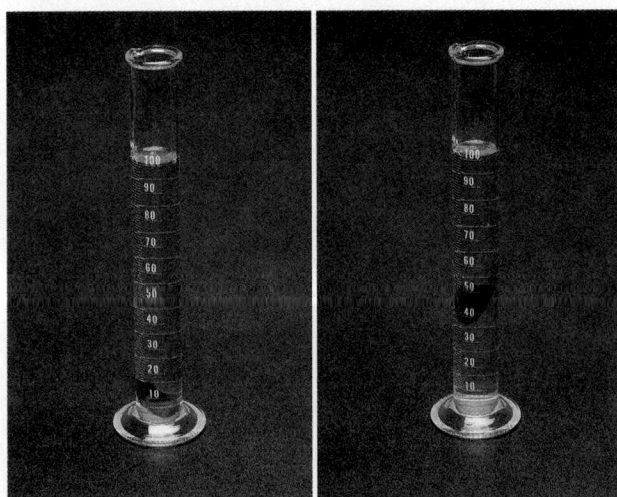

Figure 1 Glass denser than solution (left); glass same density as solution (right)

Substance	Characteristic density range
Water at 20°C	0.998 g/mL
Window-pane glass	2.47–2.56 g/mL
Headlight glass	2.47–2.63 g/mL
Ophthalmic glass	2.65–2.81 g/mL
Saturated ZnI$_2$ solution	2.73 g/mL

Density values you will need are listed in **Table 1.** If the suspect glass fragment in this case is determined to be headlight glass, investigation into the incident will continue.

PART C: Cleanup and Disposal

6. Dispose of paper and broken glass in the designated waste containers. Do not put lab materials in the trash unless your teacher tells you to do so.

7. Clean up your work area and all lab equipment. Return lab equipment to its proper place. Wash your hands thoroughly before you leave the lab.

Analyze and Conclude

1. What is the density of the glass sample? Show how you calculated your result.

2. What statement can you make based on the density of the glass sample?

3. What factors could affect the accuracy of your results? How could your procedure have been improved in order to control for these factors?

4. How would you interpret a value of 2.50 g/mL for the suspect glass chip?

Forensics Lab

The Sports Shop Theft

SKILLS
- Designing experiments
- Experimenting
- Collecting data
- Inferring
- Interpreting
- Measuring
- Organizing and analyzing data
- Communicating

OBJECTIVES
- **Determine** the comparative densities of two small glass chips by the method of suspension in a gradient column.

MATERIALS
- safety goggles
- lab apron
- gloves
- dropper bottles of distilled water (d = 0.998 g/mL) and saturated ZnI2 solution (d = 2.73 g/ml)
- size 0 cork stoppers
- markers
- crime scene and suspect glass samples
- glass tubing, about 25 cm in length, 1/4 in. inside diameter
- ring stand
- clamp
- graduated cylinder, 10 mL
- pipette
- small bottle to support the glass tubing
- amber bottle, for liquid disposal and recovery

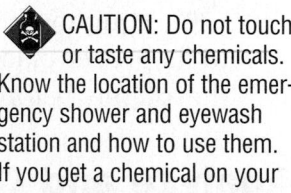

ChemSafety

CAUTION: Always wear safety goggles and a lab apron to protect your eyes and clothing.

CAUTION: Do not touch or taste any chemicals. Know the location of the emergency shower and eyewash station and how to use them. If you get a chemical on your skin or clothing, wash it off at the sink while calling to the teacher. Notify the teacher of a spill. Spills should be cleaned up promptly, according to your teacher's directions.

CAUTION: Glassware is fragile. Notify the teacher of broken glass or cuts. Do not clean up broken glass or spills with broken glass unless the teacher tells you to do so.

Background

The police are investigating the burglary of a sporting goods store. The thief broke a pane of glass in the door to gain access and headed straight for the store's valuable collection of jerseys and collectors-edition posters autographed by popular ball players. The investigating officer suspected that the job was planned—that the thief knew the layout of the store and where to find the most valuable items. The owner told the police that he had had to fire a young man the previous week. When the police paid a visit to the former employee, they asked to inspect the shoes in his closet. The employee denied that he had stolen any shoes, but that was not what interested the investigating officer. She found several shards of glass in one pair of sports shoes, and she asked where it came from. The former employee—now a suspect—said that he had accidentally stepped on his own eyeglasses, breaking the lenses.

In this lab, you will use the density column gradient method to compare the density of the glass from the shoes with the density of both ophthalmic glass (used for lenses) and of windowpane glass.

Procedure

PART A: Design an Experiment

1. Work with members of your lab group to design a procedure to test the comparative densities of the glass fragments in order to solve the case, using the method of density determination using a gradient column and with the lab materials provided. See "Forensics Lesson: Density" earlier in this lab appendix, for hints.

2. Write a procedure for your experiment. Create a data table that clearly displays your calculations and results. Make a list of all the safety precautions you will take. Have your teacher approve your procedure and safety precautions before you begin the experiment.

PART B: Conduct Your Experiment

3. Put on safety goggles, protective gloves, and a lab apron.

4. **CAUTION: Glass samples have sharp edges.** Handle them with care. You will use the density gradient column method for this analysis. This method uses glass tubing as a column, which will contain several layers of liquid, each of a different density. The densest layer will be found at the bottom and the least dense layer at the column's top. When an object is placed into a liquid and the substance neither sinks nor floats, but instead remains suspended, the density of the object and the liquid are the same. When dropped into the column, the suspect and crime scene glass samples will be suspended in the layers of the column that match their densities. If the two samples stop in the same density layer, that would indicate that they have the same density.

5. Your teacher will discuss with you the nature of the liquids to be used and how many layers your column should contain. See **Figure 1** for a diagram of the suspension column setup.

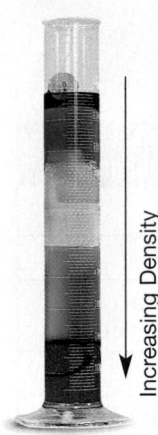

Increasing Density

Figure 1 Suspension column

6. **CAUTION; ZnI_2 is corrosive and is a poison. Do not taste or touch.** Different concentrations of ZnI_2 are commonly used in the column. After you prepare each layer in a graduated cylinder, pour the layer carefully down the side of the column so that the different layers are not mixed. Mark the top of each layer with a marker. Density values you will need are listed in **Table 1** in the lab "The Parking Lot Collision" that precedes this one.

PART C: Cleanup and Disposal

7. Dispose of paper and broken glass in the designated waste containers. Do not put lab materials in the trash unless your teacher tells you to do so.

8. Clean up your work area and all lab equipment. Return lab equipment to its proper place. Wash your hands thoroughly before you leave the lab.

Analyze and Conclude

1. What is the density of the glass sample? the crime scene sample?

2. What can you conclude about the case based on the density of the suspect glass sample as compared with the density of the crime scene sample?

3. What factors could affect the accuracy of your results? How could you have controlled for these factors?

Glossary

Pronunciation Key

Sound	As In	Phonetic Respelling
ahy	bat	(BAT)
ay	face	(FAYS)
ah	lock argue	(LAHK) (AHR gyoo)
ow	out	(OWT)
ch	chapel	(CHAP uhl)
eh	test	(TEHST)
ai	rare	(RAIR)
ee	eat feet ski	(EET) (FEET) (SKEE)
ih	bit	(BIHT)
ie	idea	(ie DEE uh)
y	ripe	(RYP)

Sound	As In	Phonetic Respelling
ihng	going	(GOH ihng)
k	card kite	(KAHRD) (KEYET)
ng	anger	(ANG guhr)
oh	over	(OH vuhr)
aw	dog horn	(DAWG) (HAWRN)
oy	foil	(FOYL)
u	pull	(PUL)
oo	pool	(POOL)
s	cell sit	(SEHL) (SIHT)
sh	sheep	(SHEEP)

Sound	As In	Phonetic Respelling
th	that thin	(THAT) (THIHN)
uh	cut	(CUHT)
ur	fern	(FURN)
y	yes	(YEHS)
yoo	globule	(GLAHB yool)
yu	cure	(KYUR)
z	bags	(BAGZ)
zh	treasure	(TREHZH uhr)
uh	medal pencil onion	(MEHD uhl) (PEHN suhl) (UHN yuhn)
uhr	paper	(PAY puhr)

A

abiotic factor an environmental factor that is not associated with the activities of living organisms (340)

ABO blood group system a system used to classify human blood by antigens found on the surface of red blood cells (878)

acanthodian an early fish; the earliest known vertebrate to have jaws (715)

acid any compound that increases the number of hydronium ions when dissolved in water; acids turn blue litmus paper red and react with bases and some metals to form salts (33)

acid rain precipitation that has a pH below normal and has an unusually high concentration of sulfuric or nitric acids, often as a result of chemical pollution of the air from sources such as automobile exhausts and the burning of fossil fuels (386)

acoelomate an animal that lacks a coelom, or body cavity (600)

actin a protein responsible for the contraction and relaxation of muscle (857)

action potential a sudden change in the polarity of the membrane of a neuron, gland cell, or muscle fiber that facilitates the transmission of electrical impulses (946)

activation energy the minimum amount of energy required to start a chemical reaction (39)

active site the site on an enzyme that attaches to a substrate (41)

active transport the movement of chemical substances, usually across the cell membrane, against a concentration gradient; requires cells to use energy (81)

adaptation the process of becoming adapted to an environment; an anatomical, physiological, or behavioral change that improves a population's ability to survive (279)

addiction a physiological or psychological dependence on a substance, such as alcohol or drugs (962)

adductor muscle the thick muscle that joins the two valves in mollusks and that causes the shell to open (647)

adhesion the attractive force between two bodies of different substances that are in contact with each other (31)

adrenal gland one of the two endocrine glands located above each kidney (985)

aerobic describes a process that requires oxygen (104)

aggregation a grouping of cells or other organisms (418)

agnathan a member of a class of primitive, jawless fishes (714)

AIDS acquired immune deficiency syndrome, a disease caused by HIV, an infection that results in an ineffective immune system (934)

allele one of the alternative forms of a gene that governs a characteristic, such as hair color (167)

allergy a physical response to an antigen, which can be a common substance that produces little or no response in the general population (936)

alternation of generations the alternation of sexual reproduction and asexual reproduction in certain plants and animals (463)

alveolus any of the tiny air cells of the lungs where oxygen and carbon dioxide are exchanged (886)

amino acid any one of 20 different organic molecules that contain a carboxyl and an amino group and that combine to form proteins (36)

amino acid–based hormone a hormone that is made up of simple amino acids, peptides, or proteins (977)

amniotic egg a type of egg that is produced by reptiles, birds, and egg-laying mammals and that contains a large amount of yolk; usually surrounded by a leathery or hard shell within which the embryo and its embryonic membranes develop (775)

amylase an enzyme that breaks down starches into sugars (907)

anaerobic describes a process that does not require oxygen (104)

anemia a condition in which the oxygen-carrying ability of red blood cells is reduced and the production of red blood cells decreases (877)

angiosperm a flowering plant that produces seeds within a fruit (514)

annual a plant that completes its life cycle, reproduces, and dies within one growing season (573)

annual ring in secondary xylem (wood), the growth ring formed in one season (575)

anther in flowering plants, the tip of a stamen, which contains the pollen sacs where grains form (538)

antheridium a reproductive structure that produces male sex cells in flowerless and seedless plants (530)

antibiotic a substance that can inhibit the growth of or kill some microorganisms (443)

antibody a protein that reacts to a specific antigen or that inactivates or destroys toxins (929)

anticodon a region of tRNA that consists of three bases complementary to the codon of mRNA (212)

antigen a substance that stimulates an immune response (927)

antigen shifting the production of new antigens by a virus as it mutates over time (932)

aorta the main artery in the body; it carries blood from the left ventricle to systemic circulation (882)

apical dominance the inhibition of lateral bud growth on the stem of a plant by auxin produced in the terminal bud (581)

apical meristem the growing region at the tips of stems and roots in plants (574)

appendage a structure that extends from the main body, such as a limb, tentacle, fin, or wing (664)

appendicular skeleton the bones of the arms and legs (850)

aquifer a porous rock that stores and allows the flow of groundwater (393)

Archaebacteria a classification kingdom made up of bacteria that live in extreme environments; differentiated from other prokaryotes by various important chemical differences (258)

archegonium a female reproductive structure of small, nonvascular plants that produces a single egg and in which fertilization and development take place (530)

artery a blood vessel that carries blood away from the heart to the body's organs (873)

arthropod a member of the phylum Arthropoda, which includes invertebrate animals such as insects, crustaceans, and arachnids; characterized by having segmented bodies and paired appendages (266)

ascus the spore sac where ascomycetes produce ascospores (487)

asexual reproduction reproduction that does not involve the union of gametes and in which a single parent produces offspring that are genetically identical to the parent (150)

asymmetrical irregular in shape; without symmetry (598)

atom the smallest unit of an element that maintains the properties of that element (28)

ATP adenosine triphosphate, an organic molecule that acts as the main energy source for cell processes; composed of a nitrogenous base, a sugar, and three phosphate groups (37)

atrium a chamber that receives blood that is returning to the heart (881)

autoimmune disease a disease in which the immune system attacks the organism's own cells (933)

autosome any chromosome that is not a sex chromosome (122)

autotroph an organism that produces its own nutrients from inorganic substances or from the environment instead of consuming other organisms (94)

auxin a plant hormone that regulates cell elongation (580)

axial skeleton the bones of the skull and vertebral column (850)

axon an elongated extension of a neuron that carries impulses away from the cell body (944)

B

B cell a white blood cell that matures in bones and makes antibodies (927)

bacillus a rod-shaped bacterium (443)

bacteriophage a virus that infects bacteria (192)

basal disk an area on cnidarians, such as hydras, jellyfish, and corals, that enables them to adhere to surfaces (624)

base any compound that increases the number of hydroxide ions when dissolved in water; bases turn red litmus paper blue and react with acids to form salts (33)

base pairing rules the rules stating that cytosine pairs with guanine and adenine pairs with thymine in DNA, and that adenine pairs with uracil in RNA (197)

basidium a structure that produces asexual spores in basidiomycetes (488)

behavior an action that an individual carries out in response to a stimulus or to the environment (824)

biennial a plant that has a two-year life cycle (573)

bilateral symmetry a condition in which two equal halves of a body mirror each other (598)

binary fission a form of asexual reproduction in single-celled organisms by which one cell divides into two cells of the same size (119)

binomial nomenclature a system for giving each organism a two-word scientific name that consists of the genus name followed by the species name (300)

biodiversity the number and variety of organisms in a given area during a specific period of time (341)

biogeochemical cycle the circulation of substances through living organisms from or to the environment (350)

biological magnification the accumulation of increasingly large amounts of toxic substances within each successive link of the food chain (391)

biological species a group of organisms than can reproduce only among themselves and that are usually contained in a geographic region (305)

biology the scientific study of living organisms and their interactions with the environment (6)

biomass organic matter that can be a source of energy; the total mass of the organisms in a given area (349)

biome a large region characterized by a specific type of climate and certain types of plant and animal communities (372)

biotic factor an environmental factor that is associated with or results from the activities of living organisms (340)

blastocyst the modified blastula stage of mammalian embryos (1004)

blastopore an opening that develops in the blastula (692)

blastula the stage of an embryo before gastrulation (596)

blood pressure the force that blood exerts on the walls of the arteries (882)

body cavity any cavity that houses organs, such as the thoracic, abdominal, or pelvic cavity (849)

body plan an animal's shape, symmetry, and internal organization (598)

bone marrow soft tissue inside bones where red and white blood cells are produced (851)

brain the mass of nerve tissue that is the main control center of the nervous system (950)

brain stem the stemlike portion of the brain that connects the cerebral hemispheres with the spinal cord and that maintains the necessary functions of the body, such as breathing and circulation (951)

bronchus one of the two tubes that connect the lungs with the trachea (886)

budding asexual reproduction in which a part of the parent organism pinches off and forms a new organism (487)

bulbourethral gland one of the two glands in the male reproductive system that add fluid to the semen during ejaculation (998)

C

calorie the amount of energy needed to raise the temperature of 1 g of water 1°C; the Calorie used to indicate the energy content of food is a kilocalorie (900)

Calvin cycle a biochemical pathway of photosynthesis in which carbon dioxide is converted into glucose using ATP (102)

cancer a tumor in which the cells begin dividing at an uncontrolled rate and become invasive (12)

capillary a tiny blood vessel that allows an exchange between blood and cells in tissue (873)

capsid a protein sheath that surrounds the nucleic acid core in a virus (435)

capsule in mosses, the part that contains spores; in bacteria, a protective layer of polysaccharides around the cell wall (443)

carapace a shieldlike plate that covers the cephalothorax of some crustaceans (782)

carbohydrate any organic compound that is made of carbon, hydrogen, and oxygen and that provides nutrients to the cells of living things (34)

carbon fixation the synthesis of organic compounds from carbon dioxide, such as in photosynthesis (102)

carnivore an animal that eats other animals (346)

carotenoid a class of pigments that are present mostly in plants and that aid in photosynthesis (98)

carrier protein a protein that transports substances across a cell membrane (80)

carrying capacity the largest population that an environment can support at any given time (322)

cartilage a flexible and strong connective tissue (715)

caste a group of insects in a colony that have a specific function (679)

cell in biology, the smallest unit that can perform all life processes; cells are covered by a membrane and have a nucleus and cytoplasm (7)

cell cycle the life cycle of a cell; in eukaryotes, it consists of a cell-growth period in which DNA is synthesized and a cell-division period in which mitosis takes place (125)

cell membrane a phospholipid layer that covers a cell's surface and acts as a barrier between the inside of a cell and the cell's environment (56)

cell theory the theory that states that all living things are made up of cells, that cells are the basic units of organisms, that each cell in a multicellular organism has a specific job, and that cells come only from existing cells (55)

cell wall a rigid structure that surrounds the cell membrane and provides support to the cell (57)

cellular respiration the process by which cells produce energy from carbohydrates; atmospheric oxygen combines with glucose to form water and carbon dioxide (95)

central nervous system the brain and the spinal cord; its main function is to control the flow of information in the body (950)

central vacuole a large cavity or sac that is found in plant cells or protozoans and that contains air or partially digested food (66)

centromere the region of the chromosome that holds the two sister chromatids together during mitosis (119)

cephalization the concentration of nerve tissue and sensory organs at the anterior end of an organism (599)

cephalothorax in arachnids and some crustaceans, the body part made up of the head and the thorax (666)

cereal any grass that produces grains that can be used for food, such as rice, wheat, corn, oats, or barley (518)

cerebellum a posterior portion of the brain that coordinates muscle movement and controls subconscious activities and some balance functions (951)

cerebral ganglion one of a pair of nerve-cell clusters that serve as a primitive brain at the anterior end of some invertebrates, such as annelids (651)

cerebrum the upper part of the brain that receives sensation and controls movement (951)

chelicera in arachnids, either of a pair of appendages used to attack prey (670)

chitin a carbohydrate that forms part of the exoskeleton of arthropods and other organisms, such as insects, crustaceans, fungi, and some algae (482)

chlamydia a bacterial sexually transmitted disease marked by painful urination and vaginal discharge (1009)

chlorofluorocarbons hydrocarbons in which some or all of the hydrogen atoms are replaced by chlorine and fluorine; used in coolants for refrigerators and air conditioners and in cleaning solvents; their use is restricted because they destroy the ozone (abbreviation, CFC) (387)

chlorophyll a green pigment that is present in most plant cells, that gives plants their characteristic green color, and that reacts with sunlight, carbon dioxide, and water to form carbohydrates (98)

chloroplast an organelle found in plant and algae cells where photosynthesis occurs (66)

choanocyte any of the flagellate cells that line the cavities of a sponge (619)

chordate an animal that at some stage in its life cycle has a dorsal nerve, a notochord, and pharyngeal pouches; examples include mammals, birds, reptiles, amphibians, fish, and some marine lower forms (700)

chromatid one of the two strands of a chromosome that become visible during meiosis or mitosis (119)

chromosome in a eukaryotic cell, one of the structures in the nucleus that are made up of DNA and protein; in a prokaryotic cell, the main ring of DNA (119)

chrysalis the hard-shelled pupa of certain insects, such as butterflies (675)

cilium a hairlike structure arranged in tightly packed rows that projects from the surface of some cells (50)

cladistics a phylogenetic classification system that uses shared derived characters and ancestry as the sole criterion for grouping taxa (327)

cladogram a diagram that is based on patterns of shared, derived traits and that shows the evolutionary relationships between groups of organisms (309)

class a taxonomic category containing orders with common characteristics (302)

cleavage in biological development, a series of cell divisions that occur immediately after an egg is fertilized (1004)

climate the average weather conditions in an area over a long period of time (371)

clone an organism that is produced by asexual reproduction and that is genetically identical to its parent; to make a genetic duplicate (150)

closed circulatory system a circulatory system in which the heart circulates blood through a network of vessels that form a closed loop; the blood does not leave the blood vessels, and materials diffuse across the walls of the vessels (606)

cnidocyte a stinging cell of a cnidarian (623)

coccus a sphere-shaped bacterium (443)

cochlea a coiled tube that is found in the inner ear and that is essential to hearing (959)

codominance a condition in which both alleles for a gene are fully expressed (178)

codon in DNA, a three-nucleotide sequence that encodes an amino acid or signifies a start signal or a stop signal (211)

coelom a body cavity that contains the internal organs (600)

coelomate an animal that has a body cavity in which the internal organs are located (600)

coevolution the process in which long-term, interdependent changes take place in two species as a result of their interactions (362)

cohesion the force that holds molecules of a single material together (31)

colon a section of the large intestine (910)

colonial organism a collection of genetically identical cells that are permanently associated but in which little or no integration of cell activities occurs (418)

commensalism a relationship between two organisms in which one organism benefits and the other is unaffected (364)

community a group of species that live in the same habitat and interact with each other (340)

competition the relationship between species that attempt to use the same limited resource (365)

competitive exclusion the exclusion of one species by another due to competition (369)

complement system a system of proteins that circulate in the bloodstream and that combine with antibodies to protect against antigens (926)

complementary base pairing a characteristic of nucleic acids in which the sequence of bases on one strand is paired to the sequence of bases on the other (197)

compound a substance made up of atoms of two or more different elements joined by chemical bonds (29)

compound eye an eye composed of many light detectors separated by pigment cells (666)

concentration gradient a difference in the concentration of a substance across a distance (74)

conditioning the process of learning by association (827)

cone in animals, a photoreceptor within the retina that can distinguish colors and is very sensitive to bright light (958)

cone in plants, a seed-bearing structure (511)

connective tissue a tissue that has a lot of intracellular substance and that connects and supports other tissues (847)

consumer an organism that eats other organisms or organic matter instead of producing its own nutrients or obtaining nutrients from inorganic sources (345)

continental drift the hypothesis that states that the continents once formed a single landmass, broke up, and drifted to their present locations (268)

contour feather one of the most external feathers that cover a bird and that help determine its shape (784)

control group in an experiment, a group that serves as a standard of comparison with another group to which the control group is identical except for one factor (17)

convergent evolution the process by which unrelated species become more similar as they adapt to the same kind of environment (307)

cork the outer layer of bark of any woody plant (553)

cork cambium a layer of tissue under the cork layer where cork cells are produced (575)

coronary artery one of the two arteries that supply blood directly to the heart (882)

corpus luteum the structure that forms from the ruptured follicle in the ovary after ovulation; it releases hormones (1002)

cortex in plants, the primary tissue located in the epidermis; in animals, the outermost portion of an organ (555)

cotyledon the embryonic leaf of a seed (535)

countercurrent flow in fish gills, an arrangement whereby water flows away from the head and blood flows toward the head (747)

crossing-over the exchange of genetic material between homologous chromosomes during meiosis; can result in genetic recombination (144)

cud partly digested food that is regurgitated, rechewed, and reswallowed for further digestion by mammals that have a rumen (814)

cuticle a waxy or fatty and watertight layer on the external wall of epidermal cells (502)

cyanobacteria a bacterium that can carry out photosynthesis, such as a blue-green alga (258)

cystic fibrosis a fatal genetic disorder in which excessive amounts of mucus are secreted, blocking intestinal and bronchial ducts and causing difficulty in breathing (12)

cytokinesis the division of the cytoplasma of a cell; cytokinesis follows the division of the cell's nucleus by mitosis or meiosis (125)

cytoplasm the region of the cell within the membrane that includes the fluid, the cytoskeleton, and all of the organelles except the nucleus (56)

cytoskeleton the cytoplasmic network of protein filaments that plays an essential role in cell movement, shape, and division (56)

cytotoxic T cell a type of T cell that recognizes and destroys cells infected by virus (927)

D

decomposer an organism that feeds by breaking down organic matter from dead organisms; examples include bacteria and fungi (347)

dendrite a cytoplasmic extension of a neuron that receives stimuli (944)

density-dependent factor a variable affected by the number of organisms present in a given area (322)

density-independent factor a variable that affects a population regardless of the population density, such as climate (324)

deoxyribose a five-carbon sugar that is a component of DNA nucleotides (194)

dependent variable in an experiment, the variable that is changed or determined by manipulation of one or more factors (the independent variables) (17)

depressant a drug that reduces functional activity and produces muscular relaxation (964)

derived character a unique characteristic of a particular group of organisms (307)

dermal tissue the outer covering of a plant (552)

dermis the layer of skin below the epidermis (862)

detritivore a consumer that feeds on dead plants and animals (346)

deuterostome an animal whose mouth does not derive from the blastopore and whose embryo has indeterminate cleavage (692)

diabetes mellitus a serious disorder in which cells are unable to obtain glucose from the blood; caused by a deficiency of insulin or lack of response to insulin (987)

diaphragm a dome-shaped muscle that is attached to the lower ribs and that functions as the main muscle in respiration (886)

diatom a unicellular alga that has a double shell that contains silica (466)

dicot a dicotyledonous plant; an angiosperm that has two cotyledons, net venation, and flower parts in groups of four or five (515)

differentiation the process in which the structure and function of the parts of an organism change to enable specialization of those parts (419)

diffusion the movement of particles from regions of higher density to regions of lower density (75)

digestion the breaking down of food into chemical substances that can be used for energy (900)

diploid a cell that contains two haploid sets of chromosomes (121)

directional selection a natural selection process in which one genetic variation is selected and that causes a change in the overall genetic composition of the population (332)

dispersion in optics, the process of separating a wave (such as white light) of different frequencies into its individual component waves (the different colors) (321)

diurnal describes animals that are active during the day and sleep at night (732)

DNA deoxyribonucleic acid, the material that contains the information that determines inherited characteristics (37)

DNA fingerprint the pattern of bands that results when an individual's DNA fragments are separated (237)

DNA helicase an enzyme that unwinds the DNA double helix during DNA replication (198)

DNA polymerase an enzyme that catalyzes the formation of the DNA molecule (199)

DNA replication the process of making a copy of DNA (198)

dormancy a state in which seeds, spores, bulbs, and other reproductive organs stop growth and development and reduce their metabolism, especially respiration (584)

double fertilization the process by which one of the two sperm nuclei fuses with the egg nucleus to produce a diploid zygote and the other fuses with the polar nuclei to produce a triploid endosperm (540)

double helix the spiral-staircase structure characteristic of the DNA molecule (194)

down feather a soft feather that covers the body of young birds and provides insulation to adult birds (785)

ecology the study of the interactions of living organisms with one another and with their environment (9)

ecosystem a community of organisms and their abiotic environment (340)

ectoderm the outermost of the three germ layers of an embryo that develops into the epidermis and epidermal tissues, the nervous system, external sense organs, and the mucous membranes lining the mouth and anus (596)

ectothermic describes the ability of an organism to maintain its body temperature by gaining heat from the environment (724)

electron microscope a microscope that focuses a beam of electrons to magnify objects (51)

electron transport chain a series of molecules, found in the inner membranes of mitochondria and chloroplasts, through which electrons pass in a process that causes protons to build up on one side of the membrane (100)

electrophoresis the process by which electrically charged particles suspended in a liquid move through the liquid because of the influence of an electric field (231)

element a substance that cannot be separated or broken down into simpler substances by chemical means; all atoms of an element have the same atomic number (28)

embryo an organism in an early stage of development of plants and animals; in humans, a developing individual is referred to as an embryo from the second through the eighth week of pregnancy (504)

endocrine gland a ductless gland that secretes hormones into the blood (975)

endocytosis the process by which a cell membrane surrounds a particle and encloses the particle in a vesicle to bring the particle into the cell (83)

endoderm the innermost germ layer of the animal embryo; develops into the epithelium of the pharynx, respiratory tract, digestive tract, bladder, and urethra (596)

endoplasmic reticulum a system of membranes that is found in a cell's cytoplasm and that assists in the production, processing, and transport of proteins and in the production of lipids (63)

endosperm a triploid (3n) tissue that develops in the seeds of angiosperms and that provides food for a developing embryo (514)

endospore a thick-walled protective spore that forms inside a bacterial cell and resists harsh conditions (443)

endosymbiosis a mutually beneficial relationship in which one organism lives within another (259)

endothermic describes the ability of an organism to maintain body temperature by producing heat internally (724)

energy the capacity to do work (38)

energy pyramid a triangular diagram that shows an ecosystem's loss of energy, which results as energy passes through the ecosystem's food chain; each row in the pyramid represents a trophic (feeding) level in an ecosystem, and the area of a row represents food chain the pathway of energy transfer through various stages as a result of the feeding patterns of a series of organisms (346)

envelope a membranelike layer that covers the capsids of some viruses (435)

enzyme a type of protein that speeds up metabolic reactions in plant and animals without being permanently changed or destroyed (40)

epidermis the outer surface layer of cells of a plant or animal (553)

epididymis the long, coiled tube that is on the surface of a testis and in which sperm mature (997)

epinephrine a hormone that is released by the adrenal medulla and that rapidly stimulates the metabolism in emergencies, decreases insulin secretion, and stimulates pulse and blood pressure; also called adrenaline (985)

epithelial tissue a tissue composed of cells that form a barrier between an organism and its external environment (846)

equilibrium in chemistry, the state in which a chemical reaction and the reverse chemical reaction occur at the same rate such that the concentrations of reactants and products do not change (74)

esophagus a long, straight tube that connects the pharynx to the stomach (907)

Eubacteria a classification kingdom that contains all prokaryotes except archaebacteria (258)

eukaryote an organism made up of cells that have a nucleus enclosed by a membrane, multiple chromosomes, and a mitotic cycle; eukaryotes include animals, plants, and fungi but not bacteria or cyanobacteria (58)

evolution a change in the characteristics of a population from one generation to the next; the gradual development of organisms from other organisms since the beginnings of life (9)

excretion the process of eliminating metabolic wastes (912)

exocytosis the process by which a substance is released from the cell through a vesicle that transports the substance to the cell surface and then fuses with the membrane to let the substance out (83)

exon the portion of the DNA sequence in a gene that contains the sequence of amino acids in a chain and the beginning and the end of a coding sequence (218)

exoskeleton a hard, external, supporting structure that develops from the ectoderm (607)

experiment a procedure that is carried out under controlled conditions to discover, demonstrate, or test a fact, theory, or general truth (17)

extensor a muscle that extends a joint (856)

external fertilization the union of gametes outside the bodies of the parents, as in many fishes and amphibians (610)

extinct describes a species of organisms that has died out completely (282)

F

F₁ generation the first generation of offspring obtained from an experimental cross of two organisms (164)

F₂ generation the second generation of offspring, obtained from an experimental cross of two organisms; the offspring of the F₁ generation (164)

facilitated diffusion the transport of substances through a cell membrane along a concentration gradient with the aid of carrier proteins (80)

fallopian tube a tube through which eggs move from the ovary to the uterus (1000)

family the taxonomic category below the order and above the genus (302)

fermentation the breakdown of carbohydrates by enzymes, bacteria, yeasts, or mold in the absence of oxygen (108)

fertilization the union of a male and female gamete to form a zygote (153)

fetus a developing human from seven or eight weeks after fertilization until birth (1006)

fixed action pattern behavior an innate behavior that is characteristic of certain species (826)

flagellum a long, hairlike structure that grows out of a cell and enables the cell to move (57)

flexor a muscle that bends a limb or other body part (856)

flower the reproductive structure of a flowering plant that usually consists of a pistil, stamens, petals, and sepals (505)

fluke a parasitic flatworm of the class Trematoda (632)

follicle a small, narrow cavity or sac in an organ or tissue, such as the ones on the skin that contain hair roots or the ones in the ovaries that contain the developing eggs (1001)

food chain the pathway of energy transfer through various stages as a result of the feeding patterns of a series of organisms (346)

food web a diagram that shows the feeding relationships between organisms in an ecosystem (347)

foot an appendage that some invertebrates use to move; the lower part of a vertebrate's leg (431)

fossil the trace or remains of an organism that lived long ago, preserved in sedimentary rock (258)

frond the leaf of a fern or palm (511)

fruit a mature plant ovary; the plant organ in which the seeds are enclosed (514)

fundamental niche the largest ecological niche where an organism or species can live without competition (365)

G

gamete a haploid reproductive cell that unites with another haploid reproductive cell to form a zygote (118)

gametophyte in alternation of generations, the phase in which gametes are formed; a haploid individual that produces gametes (154)

gastrovascular cavity a cavity that serves both digestive and circulatory purposes in some cnidarians (605)

gemmule an asexual reproductive structure produced by some freshwater sponges (621)

gene a segment of DNA that is located in a chromosome and that codes for a specific hereditary trait (8)

gene cloning the process of isolating a gene sequence in the genome of an organism and inserting the gene sequence into a plasmid vector for production in large numbers (229)

gene expression the manifestation of the genetic material of an organism in the form of specific traits (208)

gene flow the movement of genes into or out of a population due to interbreeding (328)

genetic code the rule that describes how a sequence of nucleotides, read in groups of three consecutive nucleotides (triplets) that correspond to specific amino acids, specifies the amino acid sequence of a protein (211)

genetic drift the random change in allele frequency in a population (328)

genetic engineering a technology in which the genome of a living cell is modified for medical or industrial use (228)

genetics the science of heredity and of the mechanisms by which traits are passed from parents to offspring (162)

genital herpes a sexually transmitted disease that is caused by a herpes simplex virus (1010)

genotype the entire genetic makeup of an organism; also the combination of genes for one or more specific traits (168)

genus the level of classification that comes after family and that contains similar species (301)

germination the beginning of growth or development in a seed, spore, or zygote, especially after a period of inactivity (572)

gestation in mammals, the process of carrying young from fertilization to birth (1005)

gestation period in mammals, the length of time between fertilization and birth (810)

gill in mushrooms, a structure that is located on the underside of the cap and bears the spores; in aquatic animals, a respiratory structure that consists of many blood vessels surrounded by a membrane that allows for gas exchange (605)

gill slit a perforation between two gill arches through which water taken in through the mouth of a fish passes over the gills and out of the fish's body (747)

glucagon a hormone that is produced in the pancreas and that raises the blood glucose level (986)

glycolysis the anaerobic breakdown of glucose pyruvic acid, which makes a small amount of energy available to cells in the form of ATP (105)

glycoprotein a protein to which carbohydrate molecules are attached (435)

Golgi apparatus cell organelle that helps make and package materials to be transported out of the cell (64)

gonorrhea a sexually transmitted disease that is caused by bacteria and that results in inflammation of the mucous membranes in the urinary and reproductive tracts (1008)

gradualism a model of evolution in which gradual change over a long period of time leads to biological diversity (282)

grain the edible seed or seedlike fruit of a cereal grass (518)

greenhouse effect the warming of the surface of Earth and the lower atmosphere as a result of carbon dioxide and water vapor, which absorb and reradiate infrared radiation (388)

ground tissue a type of plant tissue other than vascular tissue that makes up much of the inside of a plant (552)

groundwater the water that is beneath the Earth's surface (351)

guard cell one of a pair of specialized cells that border a stoma and regulate gas exchange (503)

gymnosperm a woody vascular seed plant whose seeds are not enclosed by an ovary or fruit (512)

 H

habitat the place where an organism usually lives (340)

hair in mammals, one of the many long and thin structures that grow out from the skin (800)

hair follicle a depression in the skin that encloses a hair and its root (862)

half-life the time required for half of a sample of a radioactive substance to disintegrate by radioactive decay or by natural processes (252)

haploid describes a cell, nucleus, or organism that has only one set of unpaired chromosomes (121)

Hardy-Weinberg principle the principle that states that the frequency of alleles in a population does not change unless evolutionary forces act on the population (326)

Haversian canal a channel containing blood vessels in compact bone tissue (852)

heart attack the death of heart tissues due to a blockage of their blood supply (884)

heartwood the nonconducting older wood in the center of a tree trunk (557)

helper T cell a white blood cell necessary for B cells to develop normal levels of antibodies (927)

herbaceous plant a plant that is soft and green instead of woody (556)

herbivore an organism that eats only plants (346)

heredity the passing of genetic traits from parent to offspring (8)

hermaphrodite an organism that has both male and female reproductive organs (609)

heterotroph an organism that obtains organic food molecules by eating other organisms or their by products and that cannot synthesize organic compounds from inorganic materials (95)

heterozygous describes an individual that has two different alleles for a trait (167)

histamine a chemical that stimulates the autonomous nervous system, secretion of gastric juices, and dilation of capillaries (925)

HIV human immunodeficiency virus, the virus that causes AIDS (12)

homeostasis the maintenance of a constant internal state in a changing environment; a constant internal state that is maintained in a changing environment by continually making adjustments to the internal and external environment (8)

hominid a member of the family Hominidae of the order Primates; characterized by opposable thumbs, no tail, relatively long lower limbs, and bipedalism; examples include modern humans and their ancestors (733)

homologous chromosomes chromosomes that have the same sequence of genes, that have the same structure, and that pair during meiosis (120)

homologous structures anatomical structures that share a common ancestry (286)

homozygous describes an individual that has identical alleles for a trait on both homologous chromosomes (167)

hormone a substance that is made in one cell or tissue and that causes a change in another cell or tissue located in a different part of the body (580)

Human Genome Project a research effort to sequence and locate the entire collection of genes in human cells (233)

hydrostatic skeleton in many invertebrates, the cavity that is filled with water and that has a support function (607)

hypertonic describes a solution whose solute concentration is higher than the solute concentration inside a cell (77)

hypha a nonreproductive filament of a fungus (421)

hypothalamus the region of the brain that coordinates the activities of the nervous and endocrine systems and that controls many body activities related to homeostasis (952)

hypothesis a theory or explanation that is based on observations and that can be tested (16)

hypotonic describes a solution whose solute concentration is lower than the solute concentration inside a cell (77)

immunity the ability to resist or to recover from an infection or disease (931)

implantation the process by which the newly fertilized egg in the blastocyst stage embeds itself in the lining of the uterus (1004)

imprinting learning that occurs early and quickly in a young animal's life and that cannot be changed once learned (829)

incomplete dominance a condition in which a trait in an individual is intermediate between the phenotype of the individual's two parents because the dominant allele is unable to express itself fully (177)

independent assortment the random distribution of the pairs of genes on different chromosomes to the gametes (146)

independent variable the factor that is deliberately manipulated in an experiment (17)

inflammatory response a protective response of tissues affected by disease or injury, characterized by redness, swelling, and pain (925)

innate behavior an inherited behavior that does not depend on the environment or experience (826)

insulin a hormone that is produced by a group of specialized cells in the pancreas and that lowers blood glucose levels (986)

interferon a protein that is produced by cells infected by a virus and that can protect uninfected cells from reproduction of the virus (926)

internal fertilization fertilization of an egg by sperm that occurs inside the body of a female (610)

interneuron a neuron located between the afferent neuron and the final neuron in a neural chain (953)

interphase a period between two mitotic or meiotic divisions during which the cell grows, copies its DNA, and synthesizes proteins (125)

intron a section of DNA that does not code for an amino acid and that is transcribed into RNA but is removed before it is translated (218)

invertebrate an animal that does not have a backbone (424)

ion an atom, radical, or molecule that has gained or lost one or more electrons and has a negative or positive charge (30)

ion channel a pore in a cell membrane through which ions can pass (78)

isotonic solution a solution whose solute concentration is equal to the solute concentration inside a cell (77)

J

joint a place where two or more bones meet (854)

K

karyotype an array of the chromosomes found in an individual's cells at metaphase of mitosis and arranged in homologous pairs and in order of diminishing size (122)

keratin a hard protein that forms hair, bird feathers, nails, and horns (861)

kingdom the highest taxonomic category, which contains a group of similar phyla (302)

Koch's postulates a four-stage procedure that Robert Koch formulated for identifying specific pathogens and determining the cause of a given disease (930)

Krebs cycle a series of biochemical reactions that convert pyruvic acid into carbon dioxide and water; it is the major pathway of oxidation in animal, bacterial, and plant cells, and it releases energy (106)

krill a small marine crustacean that is the main food source of the baleen whale (681)

K-strategist a species characterized by slow maturation, few young, slow population growth, reproduction late in life, and a population density near the carrying capacity of the environment (325)

L

lac operon a gene system whose operator gene and three structural genes control lactose metabolism in *E. coli* (216)

larynx the area of the throat that contains the vocal cords and produces vocal sounds (886)

lateral line a faint line visible on both sides of a fish's body that runs the length of the body and marks the location of sense organs that detect vibrations in water (753)

law of independent assortment the law that states that genes separate independently of one another in meiosis (169)

law of segregation Mende's law that states that the pairs of homologous chromosomes separate in meiosis so that only one chromosome from each pair is present in each gamete (169)

learning the development of behaviors through experience or practice (827)

lichen a mass of fungal and algal cells that grow together in a symbiotic relationship and that are usually found on rocks or trees (491)

life cycle all of the events in the growth and development of an organism until the organism reaches sexual maturity (152)

ligament a type of tissue that holds together the bones in a joint (854)

light microscope a microscope that uses a beam of visible light passing through one or more lenses to magnify an object (57)

limnetic zone the area in a freshwater habitat that is away from the shore but still close to the surface (376)

lipase an enzyme that breaks down fat molecules into fatty acids and glycerol (909)

lipid a type of biochemical that does not dissolve in water, including fats and steroids; lipids store energy and make up cell membranes (35)

lipid bilayer the basic structure of a biological membrane, composed of two layers of phospholipids (60)

littoral zone a shallow zone in a freshwater habitat where light reaches the bottom and nurtures plants (376)

logistic model a model of population growth that assumes that finite resource levels limit population growth (323)

lung the central organ of the respiratory system in which oxygen from the air is exchanged with carbon dioxide from the blood (758)

lymphatic system a collection of organs whose primary function is to collect extracellular fluid and return it to the blood; the organs in this system include the lymph nodes and the lymphatic vessels (875)

lysosome a cell organelle that contains digestive enzymes (64)

M

macrophage an immune system cell that engulfs pathogens and other materials (926)

magnification the increase of an object's apparent size by using lenses or mirrors (51)

Malpighian tubule an excretory tube that opens into the back part of the intestine of most insects and certain arthropods (669)

mammary gland a gland that is located in the chest of a female mammal and that secretes milk (806)

mandible a type of mouthpart found in some arthropods and used to pierce and suck food; the lower part of the jaw (673)

mantle in biology, a layer of tissue that covers the body of many invertebrates (643)

mass extinction an episode during which large numbers of species become extinct (263)

medusa a free-swimming, jellyfish-like, and often umbrella-shaped sexual stage in the life cycle of a cnidarian; also a jellyfish or a hydra (622)

meiosis a process in cell division during which the number of chromosomes decreases to half the original number by two divisions of the nucleus, which results in the production of sex cells (gametes or spores) (144)

melanin a pigment that helps determine skin color (862)

membrane potential the difference in electric potential between the two sides of a cell membrane (945)

menstrual cycle the female reproductive cycle, characterized by a monthly change of the lining of the uterus and the discharge of blood (1002)

menstruation the discharge of blood and discarded tissue from the uterus during the menstrual cycle (1003)

meristem a region of undifferentiated plant cells that are capable of dividing and developing into specialized plant tissues (507)

mesoderm in an embryo, the middle layer of cells that gives rise to muscles, blood, and various systems (596)

mesophyll in leaves, the tissue between epidermal layers, where photosynthesis occurs (559)

metabolism the sum of all chemical processes that occur in an organism (7)

metamorphosis a phase in the life cycle of many animals during which a rapid change from the immature organism to the adult takes place; an example is the change from larva to adult in insects (675)

mineral a natural, usually inorganic solid that has a characteristic chemical composition, an orderly internal structure, and a characteristic set of physical properties (905)

mitochondrion in eukaryotic cells, the cell organelle that is surrounded by two membranes and that is the site of cellular respiration, which produces ATP (65)

mitosis in eukaryotic cells, a process of cell division that forms two new nuclei, each of which has the same number of chromosomes (125)

molecule the smallest unit of a substance that keeps all of the physical and chemical properties of that substance; it can consist of one atom or two or more atoms bonded together (29)

molting the shedding of an exoskeleton, skin, feathers, or horns to be replaced by new parts (668)

monocot a monocotyledonous plant; a plant that produces seeds that have only one cotyledon (515)

monohybrid cross a cross between individuals that involves one pair of contrasting traits (164)

monosaccharide a simple sugar that is the basic subunit of a carbohydrate (34)

motor neuron a nerve cell that conducts nerve impulses from the central nervous system to the muscles and glands (950)

mRNA messenger RNA, a single-stranded RNA molecule that encodes the information to make a protein (211)

mucous membrane the layer of epithelial tissue that covers internal surfaces of the body and that secretes mucus (924)

multiple alleles more than two alleles (versions of the gene) for a genetic trait (178)

muscle tissue the tissue made of cells that can contract and relax to produce movement (847)

mutation a change in the nucleotide-base sequence of a gene or DNA molecule (8)

mutualism a relationship between two species in which both species benefit (265)

mycelium the mass of fungal filaments, or hyphae, that forms the body of a fungus (483)

mycorrhiza a symbiotic association between fungi and plant roots (265)

myofibril a fiber that is found in striated muscle cells and that is responsible for muscle contraction (857)

myosin the most abundant protein in muscle tissue and the main constituent of the thick filaments of muscle fibers (857)

natural killer cell a type of white blood cell that is present in individuals who have not been immunized and that kills a variety of cells (926)

natural selection the process by which individuals that have favorable variations and are better adapted to their environment survive and reproduce more successfully than less well adapted individuals do (9)

nauplius the free-swimming larva of most crustaceans (680)

nematocyst in cnidarians, a stinging cell that is used to inject a toxin into prey (623)

nephridium a tubule through which some invertebrates eliminate wastes (644)

nephron the functional unit of the kidney (913)

nerve a collection of nerve fibers through which impulses travel between the central nervous system and other parts of the body (944)

nervous tissue the tissue of the nervous system, which consists of neurons, their supporting cells, and connective tissue (847)

neuron a nerve cell that is specialized to receive and conduct electrical impulses (944)

neurotransmitter a chemical substance that transmits nerve impulses across a synapse (948)

neutrophil a large leukocyte that contains a lobed nucleus and many cytoplasmic granules (926)

niche the position (way of life) of a species in an ecosystem in terms of the physical characteristics (such as size, location, temperature, and pH) of the area where the species lives and the function of the species in the biological community (365)

nitrogen fixation the process by which gaseous nitrogen is converted into ammonia, a compound that organisms can use to make amino acids and other nitrogen-containing organic molecules (363)

nonvascular plant the three groups of plants (liverworts, hornworts, and mosses) that lack specialized conducting tissues and true roots, stems, and leaves (504)

norepinephrine a chemical that is both a neurotransmitter produced by the sympathetic nerve endings in the autonomic nervous system and a hormone secreted by the adrenal medulla to stimulate the functions of the circulatory and respiratory systems (abbreviation, NE) (985)

normal distribution a distribution of numerical data whose graph forms a bell-shaped curve that is symmetrical about the mean (331)

notochord the rod-shaped supporting axis found in the dorsal part of the embryos of all chordates, including vertebrates (700)

nucleic acid an organic compound, either RNA or DNA, whose molecules are made up of one or two chains of nucleotides and carry genetic information (37)

nucleotide in a nucleic-acid chain, a subunit that consists of a sugar, a phosphate, and a nitrogenous base (37)

nucleus in a eukaryotic cell, biology, a membrane-bound organelle that contains the cell's DNA and that has a role in processes such as growth, metabolism, and reproduction (58)

nutrient a substance or compound that provides nourishment (or food) or raw materials needed for life processes (900)

nymph an immature stage of some insects that is similar in function and structure to the adult (675)

observation the process of obtaining information by using the senses; the information obtained by using the senses (14)

omnivore an organism that eats both plants and animals (346)

oogenesis the production, growth, and maturation of an egg, or ovum (148)

open circulatory system a type of circulatory system in which the circulatory fluid is not contained entirely within vessels; a heart pumps fluid through vessels that empty into spaces called sinuses (606)

operator a short sequence of viral or bacterial DNA to which a repressor binds to prevent transcription (mRNA synthesis) of the adjacent gene in an operon (216)

operculum in fish, a hard plate that is attached to each side of the head, that covers gills, and that is open at the rear (756)

operon a unit of gene regulation and transcription in bacterial DNA that consists of a promoter, an operator, and one or more structural genes (216)

optic nerve the nerve that connects the retina of the eye to the brain and that transmits impulses that contribute to the sense of sight (958)

order the taxonomic category below the class and above the family (302)

organ a collection of tissues that carry out a specialized function of the body (419)

organ system a group of organs that work together to perform body functions (419)

organelle one of the small bodies that are found in the cytoplasm of a cell and that are specialized to perform a specific function (58)

osmosis the diffusion of water or another solvent from a more dilute solution (of a solute) to a more concentrated solution (of the solute) through a membrane that is permeable to the solvent (76)

ossicle one of the small, calcium carbonate plates that make up the endoskeleton of an echinoderm (694)

osteocyte a bone cell (852)

ovarian cycle a series of hormone-induced changes in which the ovaries prepare and release a mature ovum each month (1001)

ovary in the female reproductive system of animals, an organ that produces eggs; in flowering plants, the lower part of a pistil that produces eggs in ovules (538)

oviparous describes organisms that produce eggs that develop and hatch outside the body of the mother (777)

ovoviviparous describes organisms that produce eggs that develop and hatch inside the body of the mother (777)

ovulation the release of an ovum from a follicle of the ovary (1001)

ovule a structure in the ovary of a seed plant that contains an embryo sac and that develops into a seed after fertilization (534)

ovum a mature egg cell (149)

P

P generation parental generation, the first two individuals that mate in a genetic cross (164)

paleontologist a scientist who studies fossils (285)

Pangaea a single landmass that existed for about 40 million years before it began to break apart and form the continents that we know today (722)

parapodium in polychaetes, one of the two appendages that are used for locomotion or gas exchange (652)

parasitism a relationship between two species in which one species, the parasite, benefits from the other species, the host, and usually harms the host (362)

passive transport the movement of substances across a cell membrane without the use of energy by the cell (74)

pathogen a virus, microorganism, or other substance that causes disease; an infectious agent (454)

pedigree a diagram that shows the occurrence of a genetic trait in several generations of a family (175)

pedipalp one of the second pair of appendages that are beside the mouth of an arachnid and that are used for chewing and handling prey (670)

pelvic inflammatory disease a pelvic infection of the upper female reproductive system, including the uterus, ovaries, fallopian tubes, and other structures; it is a sexually transmitted disease (1009)

penis the male organ that transfers sperm to a female and that carries urine out of the body (998)

pepsin an enzyme that is found in gastric juices and that helps break down proteins into smaller molecules (908)

perennial a plant whose underground vegetative parts live for more than two years and whose upper parts die and regrow seasonally or annually (573)

periosteum the fibrous tissue that covers bones (851)

peripheral nervous system all of the parts of the nervous system except for the brain and the spinal cord (the central nervous system); includes the cranial nerves and nerves of the neck, chest, lower back, and pelvis (950)

petal one of the ring or rings of the usually brightly colored, leaf-shaped parts of a flower (538)

petiole the stalk that attaches a leaf to the stem of a plant (558)

pH a value used to express the acidity or alkalinity of a solution; it is defined as the logarithm of the reciprocal of the concentration of hydronium ions; a pH of 7 is neutral, a pH of less than 7 is acidic, and a pH of greater than 7 is basic (16)

pharynx in flatworms, the muscular tube that leads from the mouth to the gastrovascular cavity; in animals with a digestive tract, the passage from the mouth to the larynx and esophagus (886)

phenotype an organism's appearance or other detectable characteristic that results from the organism's genotype and the environment (166)

phloem the tissue that conducts food (sugars, amino acids, and mineral nutrients) in vascular plants (507)

phospholipid a lipid that contains phosphorus and that is a structural component in cell membranes (60)

photoperiodism the response of plants to seasonal changes in the relative length of nights and days (583)

photosynthesis the process by which plants, algae, and some bacteria use sunlight, carbon dioxide, and water to produce carbohydrates and oxygen (94)

phylogenetic tree a branching diagram that shows how organisms are related through evolution (602)

phylogeny the evolutionary history of a species or taxonomic group (307)

phylum the taxonomic group below kingdom and above class (302)

pigment a substance that gives another substance or a mixture its color (98)

pilus a short, thick appendage that allows a bacterium to attach to another bacterium (442)

pioneer species a species that colonizes an uninhabited area and that starts an ecological cycle in which many other species become established (343)

pistil the female reproductive part of a flower that produces seeds and consists of an ovary, style, and stigma (538)

pith the tissue that is located in the center of the stem of most vascular plants and that is used for storage (556)

pituitary gland an endocrine gland that is located at the base of the brain, stores and releases hormones produced by the hypothalamus, and secretes hormones under the control of the hypothalamus (983)

placenta the structure that attaches a developing fetus to the uterus and that enables the exchange of nutrients, wastes, and gases between the mother and the fetus (1005)

plankton the mass of mostly microscopic organisms that float or drift freely in the waters of aquatic (freshwater and marine) environments (378)

plant propagation the practice of reproducing plants from seeds or from vegetative parts (544)

planula the free-swimming, ciliated larva of a cnidarian (625)

plasma in biology, the liquid component of blood (876)

plasma cell a type of white blood cell that produces antibodies (929)

plasmid a circular DNA molecule that is usually found in bacteria and that can replicate independent of the main chromosome (229)

plasmodium the multinucleate cytoplasm of a slime mold that is surrounded by a membrane and that moves as a mass (470)

plastron the bottom, or ventral, portion of a turtle's shell (782)

platelet a fragment of a cell that is needed to form blood clots (877)

point mutation a mutation in which only one nucleotide or nitrogenous base in a gene is changed (219)

pollen grain the structure that contains the male gametophyte of seed plants (534)

pollen tube a tubular structure that grows from a pollen grain, enters the embryo sac, and allows the male reproductive cells to move to the ovule (534)

pollination the transfer of pollen from the male reproductive structures (the anthers) to the tip of a female reproductive structure (the pistil) of a flower in angiosperms or to the ovule in gymnosperms (534)

polygenic trait a characteristic of an organism that is determined by many genes (177)

polyp a form of a cnidarian that has a cylindrical, hollow body and that is usually attached to a rock or to another object (622)

population a group of organisms of the same species that live in a specific geographical area and interbreed (278)

population density the number of individuals of the same species that live in a given unit of area (321)

predation an interaction between two species in which one species, the predator, feeds on the other species, the prey (362)

prediction a statement made in advance that expresses the results that will be obtained from testing a hypothesis if the hypothesis is supported; the expected outcome if a hypothesis is accurate (16)

preen gland in birds, a special gland that secretes oil that a bird spreads over its feathers to clean and waterproof them (785)

pregnancy the period of time between conception and birth (1005)

primary growth the growth that occurs as a result of cell division at the tips of stems and roots and that gives rise to primary tissue (574)

primary productivity the total amount of organic material that the autotrophic organisms of an ecosystem produce (345)

primary succession succession that begins in an area that previously did not support life (343)

primate a member of the order primates, the group of mammals that includes humans, apes, and monkeys; typically distinguished by highly developed brains, forward-directed eyes, use of the hands, and varied locomotion (731)

prion an infectious particle that consists only of a protein and that does not contain DNA or RNA (441)

probability the likelihood that a possible future event will occur in any given instance of the event; the mathematical ratio of the number of times one outcome of any event is likely to occur to the number of possible outcomes of the event (173)

probe a strand of RNA or single-stranded DNA that has been labeled with a radioactive element or fluorescent dye and that is used to bind with and identify a specific gene in genetic engineering (231)

producer an organism that can make organic molecules from inorganic molecules; a photosynthetic or chemosynthetic autotroph that serves as the basic food source in an ecosystem (345)

profundal zone the zone in a freshwater habitat to which little sunlight penetrates (376)

proglottid one of the many body sections of a tapeworm; contains reproductive organs (631)

prokaryote an organism that consists of a single cell that does not have a nucleus or cell organelles; an example is a bacterium (57)

prosimian a member of a suborder of primates that are primarily arboreal and nocturnal, such as a lemur, loris, or tarsier (731)

prostate gland a gland in males that contributes to the seminal fluid (998)

protein an organic compound that is made of one or more chains of amino acids and that is a principal component of all cells (36)

protist an organism that belongs to the kingdom Protista (261)

protostome an organism whose embryonic blastopore develops into the mouth, whose coelom arises by schizocoely, and whose embryo has determinate cleavage (692)

protozoan a single-celled protist that can be aquatic or parasitic, that has organelles enclosed by a membrane, and that can move independently; examples include amebas and paramecia (461)

provirus viral DNA that has attached to a host cell's chromosome and that is replicated with the chromosome's DNA (436)

pseudocoelomate an animal that has a pseudocoelom, or false body cavity (600)

pseudopodium a retractable, temporary cytoplasmic extension that functions in food ingestion and movement in certain ameboid cells (464)

psychoactive drug a substance that has a significant effect on the mind or on behavior (961)

pulmonary vein the vein that carries oxygenated blood from the lungs to the heart (759)

pulse the rhythmic pressure of the blood against the walls of a vessel, particularly an artery (883)

punctuated equilibrium a model of evolution in which short periods of drastic change in species, including mass extinctions and rapid speciation, are separated by long periods of little or no change (282)

Punnett square a graphic used to predict the results of a genetic cross (170)

pupa the immobile, nonfeeding stage between the larva and the adult of insects that have complete metamorphosis; as a pupa, the organism is usually enclosed in a cocoon or chrysalis and undergoes important anatomical changes (675)

R

radial symmetry a body plan in which the parts of an animal's body are organized in a circle around a central axis (598)

radiometric dating a method of determining the age of an object by estimating the relative percentages of a radioactive (parent) isotope and a stable (daughter) isotope (252)

radula a rasping, tonguelike organ that is covered with chitinous teeth and that is used for feeding by many mollusks (643)

realized niche the range of resources that a species uses, the conditions that the species can tolerate, and the functional roles that the species plays as a result of competition in the species' fundamental niche (367)

reasoning the act of drawing a conclusion from facts or assumption (828)

receptor protein a protein that binds specific signal molecules, which causes the cell to respond (84)

recessive describes a trait or an allele that is expressed only when two recessive alleles for the same characteristic are inherited (167)

recombinant DNA DNA molecules that are artificially created by combining DNA from different sources (228)

red blood cell a disc-shaped cell that has no nucleus, that contains hemoglobin, and that transports oxygen in the circulatory system (876)

reflex an involuntary and almost immediate movement in response to a stimulus (952)

replication fork a Y-shaped point that results when the two strands of a DNA double helix separate so that the DNA molecule can be replicated (199)

repressor a regulatory protein that binds to an operator and blocks transcription of the genes of an operon (216)

reproduction the process of producing offspring (7)

reproductive isolation the inability of members of a population to successfully interbreed with members of another population of the same or a related species (281)

resolution in microscopes, the ability to form images with fine detail (51)

respiration the exchange of oxygen and carbon dioxide between living cells and their environment; includes breathing and cellular respiration (605)

resting potential the electric potential across the cell membrane of a nerve cell or muscle cell when the cell is not active (946)

restriction enzyme an enzyme that destroys foreign DNA molecules by cutting them at specific sites (229)

retina the light-sensitive inner layer of the eye, which receives images formed by the lens and transmits them through the optic nerve to the brain (958)

Rh factor one of several blood-group antigens carried on the surface of red blood cells (879)

rhizoid a rootlike structure in nonvascular plants, such as mosses or liverworts, that holds the plants in place and aids in absorption (486)

rhizome a horizontal, underground stem that provides a mechanism for asexual reproduction (510)

ribosome a cell organelle composed of RNA and protein; the site of protein synthesis (56)

RNA ribonucleic acid, a natural polymer that is present in all living cells and that plays a role in protein synthesis (37)

RNA polymerase an enzyme that starts (catalyzes) the formation of RNA by using a strand of a DNA molecule as a template (209)

rod one of the two types of light-detecting cells in the eye; rods can detect dim light and play a major role in noncolor and night vision (958)

root the mainly underground organ of vascular plants that holds plants in place and absorbs and stores water and minerals from the soil (507)

root cap the protective layer of cells that covers the tip of a root (555)

root hair an extension of the epidermis of a root that increases the root's surface area for absorption (555)

rRNA ribosomal RNA, an organelle that contains most of the RNA in the cell and that is responsible for ribosome function (212)

r-strategist a species that is adapted for living in an environment where changes are rapid and unpredictable; characterized by rapid growth, high fertility, short life span, small body size, and exponential population growth (324)

 S

sapwood the tissue of the secondary xylem that is distributed around the outside of a tree trunk and is active in transporting sap (557)

sarcomere the basic unit of contraction in skeletal and cardiac muscle (857)

scanning electron microscope a microscope that produces an enlarged, three-dimensional image of an object by using a beam of electrons rather than light (54)

sebum the oily secretion of the sebaceous glands (864)

second messenger a molecule that is generated when a specific substance attaches to a receptor on the outside of a cell membrane, which produces a change in cellular function (85)

secondary growth plant growth that results from cell division in the cambia, or lateral meristems, and that causes the stems and roots to thicken (574)

secondary succession the process by which one community replaces another community that has been partially or totally destroyed (343)

seed a plant embryo that is enclosed in a protective coat (504)

seed coat the protective, outer covering of a seed (535)

seed plant a plant that produces seeds (504)

semen the fluid that contains sperm and various secretions produced by the male reproductive organs (998)

semicircular canal the fluid-filled canal in the inner ear that helps maintain balance and coordinate movements (959)

seminal vesicle one of two glandular structures in male vertebrates that hold and secrete seminal fluid (998)

seminiferous tubule one of the many tubules in the testis where sperm are produced (996)

sensory neuron a neuron that carries stimuli from a sense organ to the central nervous system (950)

sensory receptor a specialized structure that contains the ends of sensory neurons and that responds to specific types of stimuli (956)

sepal in a flower, one of the outermost rings of modified leaves that protect the flower bud (538)

septum a dividing wall, or partition, such as the wall between adjacent cells in a fungal hypha, the internal wall between adjacent segments of an annelid, and the thick wall between the right and left chambers of the heart (760)

sessile describes an organism that remains attached to a surface for its entire life and does not move (618)

seta one of the external bristles or spines that project from the body of an annelid (652)

sex chromosome one of the pair of chromosomes that determine the sex of an individual (122)

sex-linked trait a trait that is determined by a gene found on one of the sex chromosomes, such as the X chromosome or the Y chromosome in humans (175)

sexual reproduction reproduction in which gametes from two parents unite (150)

sexual selection an evolutionary mechanism by which traits that increase the ability of individuals to attract or acquire mates appear with increasing frequency in a population; selection in which a mate is chosen on the basis of a particular trait (836)

shoot the portion of a plant that grows mostly above the ground; includes the stems and leaves (507)

sieve tube in the phloem of a flowering plant, a conducting tube that is made up of a series of sieve-tube members stacked end to end (554)

sink any place where a plant stores or uses organic nutrients, such as sugar or starches (564)

sinoatrial node a mass of cardiac muscle cells that lies at the junction of the superior vena cava with the right atrium and that initiates and regulates contraction of the heart (abbreviation, SA node) (882)

siphon a hollow tube of bivalves used for sucking in and expelling sea water (647)

skin gill a transparent structure that projects from the surface of a sea star and that enables respiration (695)

sodium-potassium pump a carrier protein that uses ATP to actively transport sodium ions out of a cell and potassium ions into the cell (81)

solution a homogeneous mixture of two or more substances uniformly dispersed throughout a single phase (32)

sorus a cluster of spores or sporangia (532)

source a part of a plant that makes sugars and other organic compounds and from which these compounds are transported to other parts of the plant (564)

speciation the formation of new species as a result of evolution by natural selection (291)

species a group of organisms that are closely related and naturally mate to produce fertile offspring; also the level of classification below genus and above subspecies (9)

sperm the male gamete (sex cell) (148)

spermatogenesis the process by which male gametes form (148)

spicule a needle of silica or calcium carbonate in the skeleton of some sponges (620)

spinal cord a column of nerve tissue running from the base of the brain through the vertebral column (952)

spindle a network of microtubules that forms during mitosis and moves chromatids to the poles (128)

spinneret an organ that spiders and certain insect larvae use to produce silky threads for webs and cocoons (670)

spiracle an external opening in an insect or arthropod, used in respiration (668)

spirillum a spiral-shaped bacterium (443)

spongin a fibrous protein that contains sulfur and composes the fibers of the skeleton of some sponges (620)

sporangium a specialized sac, case, capsule, or other structure that produces spores (463)

spore a reproductive cell or multicellular structure that is resistant to environmental conditions and that can develop into an adult without fusion with another cell (154)

sporophyte in plants and algae that have alternation of generations, the diploid individual or generation that produces haploid spores (154)

sporozoite a sporozoan that has been released from the oocyst and is ready to penetrate a new host cell (473)

stabilizing selection a type of natural selection in which the average form of a trait is favored and becomes more common (332)

stamen the male reproductive structure of a flower that produces pollen and consists of an anther at the tip of a filament (538)

steroid a type of lipid that consists of four carbon rings to which various functional groups are attached and that usually has a physiological action (977)

stimulant a drug that increases the activity of the body or the activity of some part of the body (962)

stolon in plants, a creeping stem that can develop roots and shoots at its nodes or at its tip to form new individuals; the creeping hypha of some fungi that gives rise to new individuals (486)

stoma one of many openings in a leaf or a stem of a plant that enable gas exchange to occur (plural, stomata) (502)

stroke a sudden loss of consciousness or paralysis that occurs when the blood flow to the brain is interrupted (884)

subcutaneous tissue the layer of cells that lies beneath the skin (863)

substrate a part, substance, or element that lies beneath and supports another part, substance, or element; the reactant in reactions catalyzed by enzymes (41)

succession the replacement of one type of community by another at a single location over a period of time (343)

swim bladder in bony fishes, a gas-filled sac that is used to control buoyancy (756)

symbiosis a relationship in which two different organisms live in close association with each other (364)

synapse the junction at which the end of the axon of a neuron meets the end of a dendrite or the cell body of another neuron or meets another cell (948)

syphilis a sexually transmitted disease caused by the bacterium Treponema pallidum (1009)

T

target cell a specific cell to which a hormone is directed to produce a specific effect (977)

taxonomy the science of describing, naming, and classifying organisms (300)

teleost a group of ray-finned fishes that have a caudal fin, scales, and a swim bladder; the largest group of bony fishes (757)

tendon a tough connective tissue that attaches a muscle to a bone or to another body part (856)

terrestrial describes an organism that lives on land (721)

test cross the crossing of an individual of unknown genotype with a homozygous recessive individual to determine the unknown genotype (172)

testes the primary male reproductive organs, which produce sperm cells and testosterone (singular, testis) (996)

thalamus the part of the brain that directs incoming sensory and motor signals to the proper region (952)

thecodont the extinct reptile from which dinosaurs evolved (722)

theory an explanation for some phenomenon that is based on observation, experimentation, and reasoning (19)

therapsid the extinct order of mammal-like reptiles that likely gave rise to mammals (728)

thorax in higher vertebrates, the part of the body between the neck and the abdomen; in other animals, the body region behind the head; in arthropods, the mid-body region (666)

thylakoid a membrane system found within chloroplasts that contains the components for photosynthesis (99)

tissue a group of similar cells that perform a common function (419)

tissue culture the technique for growing living cells in an artificial medium (544)

tolerance the condition of drug addiction in which greater amounts of a drug are needed to achieve the desired effect (962)

toxin a substance that is produced by one organism and that is poisonous to other organisms (449)

trachea in insects, myriapods, and spiders, one of a network of air tubes; in vertebrates, the tube that connects the pharynx to the lungs (668)

transcription the process of forming a nucleic acid by using another molecule as a template; particularly the process of synthesizing RNA by using one strand of a DNA molecule as a template (208)

transfer RNA an RNA molecule that transfers amino acids to the growing end of a polypeptide chain during translation (212)

transformation the transfer of genetic material in the form of DNA fragments from one cell to another or from one organism to another (191)

transgenic animal an animal into which cloned genetic material has been transferred (241)

translation the portion of protein synthesis that takes place at ribosomes and that uses the codons in mRNA molecules to specify the sequence of amino acids in polypeptide chains (208)

translocation the movement of a segment of DNA from one chromosome to another, which results in a change in the position of the segment; also the movement of soluble nutrients from one part of a plant to another (564)

transpiration the process by which plants release water vapor into the air through stomata; also the release of water vapor into the air by other organisms (351)

trochophore a free-swimming, ciliated larva of many worms and some mollusks (642)

trophic level one of the steps in a food chain or food pyramid; examples include producers and primary, secondary, and tertiary consumers (345)

tropism the movement of all or part of an organism in response to an external stimulus, such as light or heat; movement is either toward or away from the stimulus (582)

true-breeding describes organisms or genotypes that are homozygous for a specific trait and thus always produce offspring that have the same phenotype for that trait (164)

 U

ungulate a hoofed mammal (814)

uracil one of the four bases that combine with sugar and phosphate to form a nucleotide subunit of RNA; uracil pairs with adenine (208)

urea the principal nitrogenous product of the metabolism of proteins that forms in the liver from amino acids and from compounds of ammonia and that is found in urine and other body fluids (912)

ureter one of the two narrow tubes that carry urine from the kidneys to the urinary bladder (914)

urethra the tube that carries urine from the urinary bladder to the outside of the body (914)

urinary bladder a hollow, muscular organ that stores urine (914)

urine the liquid excreted by the kidneys, stored in the bladder, and passed through the urethra to the outside of the body (914)

uterus in female mammals, the hollow, muscular organ in which a fertilized egg is embedded and in which the embryo and fetus develop (1000)

 V

vaccination the administration of treated microorganisms into humans or animals to induce an immune response (931)

vaccine a substance prepared from killed or weakened pathogens and introduced into a body to produce immunity (190)

vagina the canal in the female that extends from the vulva to the cervix and that receives the penis during sexual intercourse (1000)

valve a fold of membranes that controls the flow of a fluid (874)

vas deferens a duct through which sperm move from the epididymis to the ejaculatory duct at the base of the penis (997)

vascular bundle in a plant, a strand of conducting tissue that contains both xylem and phloem (556)

vascular cambium in a plant, the lateral meristem that produces secondary xylem and phloem (575)

vascular plant a plant that has a vascular system composed of xylem and phloem, specialized tissues that conduct materials from one part of the plant to another (504)

vascular system a conducting system of tissues that transport water and other materials in plants or in animals (504)

vascular tissue the specialized conducting tissue that is found in higher plants and that is made up mostly of xylem and phloem (422)

vector in biology, any agent, such as a plasmid or a virus, that can incorporate foreign DNA and transfer that DNA from one organism to another; an intermediate host that transfers a pathogen or a parasite to another organism (229)

vegetative part any nonreproductive part of a plant (516)

vegetative reproduction a type of asexual reproduction in which new plants grow from nonreproductive plant parts (541)

vein in biology, a vessel that carries blood to the heart (873)

vena cava one of the two large veins that carry blood from the body tissues to the heart (881)

ventricle one of the two large muscular chambers that pump blood out of the heart (881)

vertebra one of the 33 bones in the spinal column (backbone) (712)

vertebrate an animal that has a backbone; includes mammals, birds, reptiles, amphibians, and fish (267)

vesicle a small cavity or sac that contains materials in a eukaryotic cell; forms when part of the cell membrane surrounds the materials to be taken into the cell or transported within the cell (63)

vessel in plants, a tubelike structure in the xylem that is composed of connected cells that conduct water and mineral nutrients; in animals, a tube or duct that carries blood or another bodily fluid (554)

vestigial structure a structure in an organism that is reduced in size and function and that may have been complete and functional in the organism's ancestors (286)

villus one of the many tiny projections from the cells in the lining of the small intestine; increases the surface area of the lining for absorption (909)

viroid an infectious agent that consists of a small strand of RNA and that causes disease in plants (441)

virulent describes a microorganism or virus that causes disease and that is highly infectious (190)

virus a nonliving, infectious particle composed of a nucleic acid and a protein coat; it can invade and destroy a cell (434)

visceral mass the central section of a mollusk's body that contains the mollusk's organs (643)

vitamin an organic compound that participates in biochemical reactions and that builds various molecules in the body; some vitamins are called coenzymes and activate specific enzymes (904)

water vascular system a system of canals filled with a watery fluid that circulates throughout the body of an echinoderm (695)

weaning the time when an animal's dependence on its mother for food (milk) and protection comes to an end (806)

white blood cell a type of cell in the blood that destroys bacteria, viruses, and toxic proteins and helps the body develop immunities (877)

withdrawal the set of symptoms associated with the removal of an addictive drug from the body (962)

yeast a very small, unicellular fungus that ferments carbohydrates into alcohol and carbon dioxide; used to ferment beer and to leaven bread and used as a source of vitamins and proteins (487)

zygosporangium in members of the phylum Zygomycota, a sexual structure that is formed by the fusion of two gametangia and that contains one or more zygotes that resulted from the fusion of gametes produced by the gametangia (486)

zygospore in some algae, a thick-walled protective structure that contains a zygote that resulted from the fusion of two gametes (462)

zygote the cell that results from the fusion of gametes; a fertilized egg (121)

Spanish Glossary

A

abiotic factor / factor abiótic un factor ambiental que no está asociado con las actividades de los seres vivos (340)

ABO blood group system / sistema de grupo sanguíneo ABO un sistema que se usa para clasificar la sangre humana en función de los antígenos que se encuentran en la superficie de los glóbulos rojos (878)

acanthodian / acantodio un pez antiguo; el primer vertebrado con mandíbulas que se conoce (715)

acid / ácido cualquier compuesto que aumenta el número de iones de hidrógeno cuando se disuelve en agua; los ácidos cambian el color del papel tornasol a rojo y forman sales al reaccionar con bases y con algunos metales (33)

acid rain / lluvia ácida precipitación con un pH inferior al normal, que tiene una concentración inusualmente alta de ácido sulfúrico y ácido nítrico como resultado de la contaminación química del aire por fuentes tales como los escapes de los automóviles y la quema de combustibles fósiles (386)

acoelomate / acelomado un animal que no tiene celoma, o cavidad en el cuerpo (600)

actin / actina una proteína responsable de la contracción y relajación de los músculos (857)

action potential / potencial de acción un cambio súbito en la polaridad de la membrana de una neurona, célula glandular o fibra muscular, el cual facilita la transmisión de impulsos eléctricos (946)

activation energy / energía de activación la cantidad mínima de energía que se requiere para iniciar una reacción química (39)

active site / sitio activo el sitio en una enzima que se une al sustrato (41)

active transport / transporte activo el movimiento de substancias químicas, normalmente a través de la membrana celular, en contra de un gradiente de concentración; requiere que la célula gaste energía (81)

adaptation / adaptación el proceso de adaptarse a un ambiente; un cambio anatómico, fisiológico o en la conducta que mejora la capacidad de supervivencia de una población (279)

addiction / adicción una dependencia fisiológica o psicológica de una substancia, tal como el alcohol o las drogas (962)

adductor muscle / músculo aductor el músculo grueso que une las dos válvulas en los moluscos y hace que la concha se abra (647)

adhesion / adhesión la fuerza de atracción entre dos cuerpos de diferentes substancias que están en contacto (31)

adrenal gland / glándula suprarrenal una de las dos glándulas endocrinas ubicadas arriba de cada riñón (985)

aerobic / aeróbico término que describe un proceso que requiere oxígeno (104)

aggregation / conglomerado un grupo de células u otros organismos (418)

agnathan / agnato un miembro de una clase de peces primitivos, sin mandíbulas (714)

AIDS / SIDA síndrome de inmunodeficiencia adquirida, enfermedad causada por una infección de VIH, la cual resulta en un sistema inmunológico ineficiente (934)

allele / alelo una de las formas alternativas de un gene que rige un carácter, como por ejemplo, el color del cabello (167)

allergy / alergia una reacción física a un antígeno, el cual puede ser una substancia común que produce una reacción ligera o que no produce ninguna reacción en la población general (936)

alternation of generations / alternacia de generaciones la alternación entre reproducción sexual y asexual que se da en ciertas plantas y animales (463)

alveoli / alveolo cualquiera de las diminutas células de aire de los pulmones, en donde ocurre el intercambio de oxígeno y dióxido de carbono (886)

amino acid / aminoácido cualquiera de las 20 distintas moléculas orgánicas que contienen un grupo carboxilo y un grupo amino y que se combinan para formar proteínas (36)

amino acid–based hormone / hormona derivada de aminoácidos una hormona que está formada por aminoácidos simples, péptidos o proteínas (977)

amniotic egg / huevo amniótico un tipo de huevo que es producido por los reptiles, las aves y los mamíferos que ponen huevos y que contiene una gran cantidad de yema; normalmente está rodeado por una cáscara áspera y dura, dentro de la cual se desarrollan el embrión y sus membranes embrionarias (775)

amylase / amilasa enzima que descompone los almidones en azúcares (907)

anaerobic / anaeróbico término que describe un proceso que no requiere oxígeno (104)

anemia / anemia condición en la que se reduce la capacidad de los glóbulos rojos de transportar oxígeno y la producción de glóbulos rojos disminuye (877)

angiosperm / angiosperma una planta que da flores y que produce semillas dentro de la fruta (514)

annual / anual planta que completa su ciclo de vida, se reproduce y muere en una estación de cultivo (573)

annual ring / anillo anual en el xilema secundario (madera), el anillo de crecimiento que se forma en una estación (575)

anther / antera en las plantas que dan flores, la punta del estambre, que contiene los sacos de polen donde se forman los granos (538)

antheridium / anteridio una estructura reproductiva que produce células sexuales masculinas en las plantas que no dan flores ni producen semillas (530)

antibiotic / antibiótico una substancia que inhibe el crecimiento de algunos microorganismos o los mata (443)

antibody / anticuerpo una proteína que reacciona ante un antígeno específico o que inactiva o destruye toxinas (929)

anticodon / anticodón una región del ARNt formada por tres bases que complementan el codón del ARNm (212)

antigen / antígeno una substancia que estimula una respuesta inmunológica (927)

antigen shifting / cambio antigénico la producción de antígenos nuevos por un virus cuando éste muta con el paso del tiempo (932)

aorta / aorta la arteria principal del cuerpo; transporta sangre del ventrículo izquierdo a la circulación sistémica (882)

apical dominance / dominancia apical inhibición del crecimiento lateral de un brote en el tallo de una planta debido a la producción de auxina en el brote terminal (581)

apical meristem / meristemo apical la región de crecimiento en la punta de los tallos y raíces de las plantas (574)

appendage / apéndice una estructura que se extiende del cuerpo principal, como por ejemplo, una extremidad, un tentáculo, una aleta o un ala (664)

appendicular skeleton / esqueleto apendicular los huesos de los brazos y piernas (850)

aquifer / acuífero un cuerpo rocoso o sedimento que almacena agua subterránea y permite que fluya (393)

archaebacteria / arqueobacteria un organismo procariótico que se diferencia de otros procariotes por la composición de su membrana y pared celular (258)

archegonium / arquegonio una estructura reproductiva femenina de ciertas plantas pequeñas y no vasculares, que produce un solo óvulo y en el cual ocurren la fertilización y el desarrollo (530)

artery / arteria un vaso sanguíneo que transporta sangre del corazón a los órganos del cuerpo (873)

arthropod / artrópodo miembro del phylum Arthropoda, el cual incluye a animales invertebrados tales como insectos, crustáceos y arácnidos, caracterizados por tener un cuerpo segmentado y un par de apéndices (266)

ascus / asca el saco de esporas donde los ascomicetos producen acosporas (487)

asexual reproduction / reproducción asexual reproducción que no involucra la unión de gametos, en la que un solo progenitor produce descendencia que es genéticamente igual al progenitor (150)

asymmetrical / asimétrico de forma irregular; sin simetría (598)

atom / átomo la unidad más pequeña de un elemento que conserva las propiedades de ese elemento (28)

ATP / ATP adenosín trifosfato; molécula orgánica que funciona como la fuente principal de energía para los procesos celulares; formada por una base nitrogenada, un azúcar y tres grupos fosfato (37)

atrium / aurícula una cámara que recibe la sangre que regresa al corazón (881)

autoimmune disease / enfermedad autoinmune una enfermedad en la que el sistema inmunológico ataca las células del propio organismo (933)

autosome / autosoma cualquier cromosoma que no es un cromosoma sexual (122)

autotroph / autótrofo un organismo que produce sus propios nutrientes a partir de substancias inorgánicas o del ambiente, en lugar de consumir otros organismos (94)

auxin / auxina una hormona vegetal que regula el alargamiento de las células (580)

axial skeleton / esqueleto axial los huesos del cráneo y la columna vertebral (850)

axon / axón una extensión alargada de una neurona que transporta impulsos hacia fuera del cuerpo de la célula (944)

B

B cell / célula B un glóbulo blanco de la sangre que madura en los huesos y fabrica anticuerpos (927)

bacillus / bacilo una bacteria que tiene forma de bastón (443)

bacteriophage / bacteriófago un virus que infecta a las bacterias (192)

basal disk / disco basal un área de los celenterados, como las hidras, medusas y corales, que les permite adherirse a las superficies (624)

base / base cualquier compuesto que aumenta el número de iones de hidróxido cuando se disuelve en agua; las bases cambian el color del papel tornasol a azul y forman sales al reaccionar con ácidos (33)

base pairing rules / regla de apareamiento de las bases las reglas que establecen que en el ADN, la citosina se une con la guanina y la adenina se une a la timina, y que en el ARN, la adenina se une con el uracilo (197)

basidium / basidio una estructura que produce esporas asexuales en los basidiomicetos (488)

behavior / conducta una acción que un individuo realiza en respuesta a un estímulo o a su ambiente (824)

biennial / bienal una planta que tiene un ciclo de vida de dos años (573)

bilateral symmetry / simetría bilateral una condición en la que dos mitades iguales de un cuerpo son imágenes de espejo una de otra (598)

binary fission / fisión binaria una forma de reproducción asexual de los organismos unicelulares, por medio de la cual la célula se divide en dos células del mismo tamaño (119)

binomial nomenclature / nomenclatura binomial un sistema para darle a cada organismo un nombre científico de dos palabras, el cual está formando por el género seguido de la especie (300)

biodiversity / biodiversidad la variedad de organismos que se encuentran en un área determinada, la variación genetica dentro de una población, la variedad de especies en una comunidad o la variedad de comunidades en un ecosistema (341)

biogeochemical cycle / ciclo biogeoquímico la circulación de substancias del ambiente a los seres vivos y de los seres vivos al ambiente (350)

biological magnification / magnificación biológica la acumulación de cantidades cada vez mayores de substancias tóxicas en cada eslabón sucesivo de la cadena alimenticia (391)

biological species / especie biológica un grupo de organismos que sólo se reproducen entre ellos mismos y que normalmente están limitados a una región geográfica (305)

biology / biología el estudio científico de los seres vivos y sus interacciones con el medio ambiente (6)

biomass / biomasa materia orgánica que puede ser una fuente de energía; la masa total de los organismos en un área determinada (349)

biome / bioma una región extensa caracterizada por un tipo de clima específico y ciertos tipos de comunidades de plantas y animales (372)

biotic factor / factor biótico un factor ambiental que está asociado con las actividades de los seres vivos o que resulta de ellas (340)

blastocyst / blastocisto la etapa de bástula modificada de los embriones de los mamíferos (1004)

blastopore / blastoporo una abertura que se desarrolla en la blástula (692)

blastula / blástula la etapa en la que se encuentra un embrión antes de la gastrulación (596)

blood pressure / presión sanguínea la fuerza que la sangre ejerce en las paredes de las arterias (882)

body cavity / cavidad corporal cualquier cavidad que aloja órganos, tales como la cavidad torácica, abdominal o pélvica (849)

body plan / plan corporal la forma, simetría y organización interna de un animal (598)

bone marrow / médula ósea tejido blando que se encuentra en el interior de los huesos, donde se producen los glóbulos blancos y los glóbulos rojos (851)

brain / encéfalo la masa de tejido nervioso que es el centro principal de control del sistema nervioso (950)

brain stem / tronco encefálico la porción del cerebro que tiene forma de tronco, la cual conecta los hemisferios cerebrales con la médula espinal y mantiene las funciones necesarias del cuerpo, tales como la respiración y la circulación (951)

bronchus / bronquio uno de los dos tubos que conectan los pulmones con la tráquea (886)

budding / gemación reproducción asexual en la que una parte del organismo progenitor se separa y forma un nuevo organismo (487)

bulbourethral gland / glándula bulbouretral una de las dos glándulas del aparato reproductor masculino que añaden líquido al semen durante la eyaculación (998)

calorie / caloría la cantidad de energía que se requiere para aumentar la temperatura de 1 g de agua en 1°C; la Caloría que se usa para indicar el contenido energético de los alimentos es la kilocaloría (900)

Calvin cycle / ciclo de Calvin una vía bioquímica de la fotosíntesis en la que el dióxido de carbono se convierte en glucosa usando ATP (102)

cancer / cáncer un tumor en el cual las células comienzan a dividirse a una tasa incontrolable y se vuelven invasivas (12)

capillary / capilar diminuto vaso sanguíneo que permite el intercambio entre la sangre y las células de los tejidos (873)

capsid / cápside una cubierta de proteína que rodea el centro de ácido nucleico de un virus (435)

capsule / cápsula en los musgos, la parte que contiene las esporas; en las bacterias, una capa protectora de polisacáridos que se encuentra alrededor de la pared celular (443)

carapace / caparazón una placa parecida a un escudo que cubre el cefalotórax de algunos crustáceos (782)

carbohydrate / carbohidrato cualquier compuesto orgánico que está hecho de carbono, hidrógeno y oxígeno y que proporciona nutrientes a las células de los seres vivos (34)

carbon fixation / fijación del carbono la síntesis de compuestos orgánicos a partir del dióxido de carbono, como ocurre durante la fotosíntesis (102)

carnivore / carnívoro un animal que se alimenta de otros animales (346)

carotenoid / carotenoide una clase de pigmentos que se encuentran presentes principalmente en las plantas y que ayudan en la fotosíntesis (98)

carrier protein / proteína transportadora una proteína que transporta substancias a través de la membrana celular (80)

carrying capacity / capacidad de carga la población más grande que un ambiente puede sostener en cualquier momento dado (322)

cartilage / cartílago un tejido conectivo flexible y fuerte (715)

caste / casta un grupo de insectos en una colonia que tienen una función específica (679)

cell / célula en biología, la unidad más pequeña que puede realizar todos los procesos vitales; las células están cubiertas por una membrana y tienen un núcleo y citoplasma (7)

cell cycle / ciclo celular el ciclo de vida de una célula; en los eucariotes, consiste de un período de crecimiento celular en el que el ADN se sintetiza, y un período de división celular en el que ocurre la mitosis (125)

cell membrane / membrana celular una capa de fosfolípidos que cubre la superficie de la célula y funciona como una barrera entre el interior de la célula y el ambiente de la célula (56)

cell theory / teoría celular la teoría que establece que todos los seres vivos están formados por células, que las células son las unidades fundamentales de los organismos y que las células provienen únicamente de células existentes (55)

cell wall / pared celular una estructura rígida que rodea la membrana celular y le brinda soporte a la célula (57)

cellular respiration / respiración celular el proceso por medio del cual las células producen energía a partir de los carbohidratos; el oxígeno atmosférico se combina con la glucosa para formar agua y dióxido de carbono (95)

central nervous system / sistema nervioso central el cerebro y la médula espinal; su principal función es controlar el flujo de información en el cuerpo (950)

central vacuole / vacuola central una cavidad o bolsa grande que se encuentra en las células vegetales o en los protozoarios y que contiene aire o alimentos parcialmente digeridos (66)

centromere / centrómero la región de un cromosoma que mantiene unidas las dos cromátidas hermanas durante la mitosis (119)

cephalization / cefalización la concentración de tejido nervioso y órganos sensoriales en la parte anterior de un organismo (599)

cephalothorax / cefalotórax en los arácnidos y algunos crustáceos, la parte del cuerpo constituida por la cabeza y el tórax (666)

cereal / cereal cualquier hierba que produce granos que pueden ser usados como alimento, tales como el arroz, trigo, maíz, avena o centeno (518)

cerebellum / cerebelo una porción posterior del cerebro que coordina el movimiento de los músculos y controla las actividades subconscientes y algunas funciones de equilibrio (951)

cerebral ganglion / ganglio cerebral uno de un par de conjuntos de células nerviosas que funcionan como si fueran un cerebro primitivo en la parte anterior de algunos invertebrados, tales como los anélidos (651)

cerebrum / cerebro la parte superior del encéfalo que recibe las sensaciones y controla el movimiento (951)

chelicera / quelíceros en los arácnidos, uno de los dos apéndices usados para atacar a las presas (670)

chitin / quitina un carbohidrato que forma parte del exoesqueleto de los artrópodos y de otros organismos, como por ejemplo, insectos, crustáceos, hongos y algunas algas (482)

chlamydia / clamidia una enfermedad bacteriana transmitida sexualmente caracterizada por dolor al orinar y descargas vaginales (1009)

chlorofluorocarbons / clorofluorocarbonos hidrocarburos en los que algunos o todos los átomos de hidrógeno son reemplazados por cloro y flúor; se usan en líquidos refrigerantes para refrigeradores y aires acondicionados y en solventes para limpieza; su uso está restringido porque destruyen las moléculas de ozono de la estratosfera (abreviatura: CFC) (387)

chlorophyll / clorofila un pigmento verde presente en la mayoría de las células vegetales que les da a las plantas su color verde característico y que reacciona con la luz del Sol, el dióxido de carbono y el agua para formar carbohidratos (98)

chloroplast / cloroplasto un organelo que se encuentra en las células vegetales y en las células de las algas, en el cual se lleva a cabo la fotosíntesis (66)

choanocyte / coanocito cualquiera de las células flageladas que cubren las cavidades de una esponja (619)

chordate / cordado un animal que, en alguna etapa de su ciclo de vida, tiene un nervio dorsal, un notocrodio y bolsas faríngeas; entre los ejemplos se encuentran los mamíferos, aves, reptiles, anfibios, peces y algunas formas marinas inferiores (700)

chromatid / cromátida una de las dos hebras de un cromosoma que se vuelve visible durante la meiosis o mitosis (119)

chromosome / cromosoma en una célula eucariótica, una de las estructuras del núcleo que está hecha de ADN y proteína; en una célula procariótica, el anillo principal de ADN (119)

chrysalis / crisálida la pupa de cubierta dura de ciertos insectos, como las mariposas (675)

cilium / cilio una estructura parecida a un pelo ordenada en hileras muy comprimidas, que se proyecta a partir de la superficie de algunas células (58)

cladistics / cladística un sistema de clasificación filogénica en el que los únicos criterios de agrupación de los taxa son los caracteres comunes derivados y la ascendencia (327)

cladogram / cladograma un diagrama basado en modelos de caracteres comunes derivados, que muestra las relaciones evolutivas entre grupos de organismos (309)

class / clase una categoría taxonómica que contiene órdenes con características comunes (302)

cleavage / segmentación en el desarrollo biológico, una serie de divisiones celulares que ocurren inmediatamente después de que un óvulo es fecundado (1004)

climate / clima las condiciones promedio del tiempo en un área durante un largo período de tiempo (371)

clone / clon un organismo producido por reproducción asexual que es genéticamente idéntico a su progenitor; clonar significa hacer un duplicado genético (150)

closed circulatory system / aparato circulatorio cerrado un aparato circulatorio en el que el corazón hace que la sangre circule a través de una red de vasos que forman un circuito cerrado; la sangre no sale de los vasos sanguíneos y los materiales pasan a través de las paredes de los vasos por difusión (606)

cnidocyte / cnidocito una célula urticante de los cnidarios (623)

coccus / coco una bacteria que tiene forma de esfera (443)

cochlea / cóclea un tubo enrollado que se encuentra en el oído interno y es esencial para poder oír (959)

codominance / codominancia una condición en la que los dos alelos de un gene están totalmente expresados (178)

codon / codón en el ADN, una secuencia de tres nucleótidos que codifica un aminoácido o indica una señal de inicio o una señal de terminación (211)

coelom / celoma una cavidad del cuerpo que contiene los órganos internos (600)

coelomate / celomado un animal que tiene una cavidad en el cuerpo donde se encuentran los órganos internos (600)

coevolution / coevolución la evolución de dos o más especies que se debe a su influencia mutua, a menudo de un modo que hace que la relación sea más mutuamente beneficiosa (362)

cohesion / cohesión la fuerza que mantiene unidas a las moléculas de un solo material (31)

colon / colon una sección del intestino grueso (910)

colonial organism / organismo colonial un conjunto de células genéticamente idénticas que están asociadas permanentemente, pero en el que no se da una gran integración de las actividades celulares (418)

commensalism / comensalismo una relación entre dos organismos en la que uno se beneficia y el otro no es afectado (364)

community / comunidad un grupo de varias especies que viven en el mismo hábitat e interactúan unas con otras (340)

competition / competencia la relación entre dos especies (o individuos) en la que ambas especies (o individuos) intentan usar el mismo recurso limitado, de modo que ambas resultan afectadas negativamente por la relación (365)

competitive exclusion / exclusión competitiva la exclusión de una especie por otra debido a la competencia (369)

complement system / sistema complementario un sistema de proteínas que circulan en el torrente sanguíneo y que se combinan con anticuerpos para proteger en contra de los antígenos (926)

complementary base pairing / complementariedad de las bases nitrogenadas una característica de los ácidos nucleicos en que la secuencia de bases de una hebra está acoplada con la secuencia de bases de la otra (197)

compound / compuesto una substancia formada por átomos de dos o más elementos diferentes unidos por enlaces químicos (29)

compound eye / ojo compuesto un ojo compuesto por muchos detectores de luz separados por células de pigmentos (666)

concentration gradient/gradiente de concentración una diferencia en la concentración de una substancia a través de una distancia (74)

conditioning / condicionamiento aprendizaje por asociación (827)

cone / cono en los animales, un fotorreceptor de la retina que distingue colores y es muy sensible a la luz brillante (958)

cone / cono en las plantas, una estructura portadora de semillas (511)

connective tissue / tejido conectivo un tejido que tiene mucha substancia intracelular, y que conecta y sostiene otros tejidos (847)

consumer / consumidor un organismo que se alimenta de otros organismos o de materia orgánica, en lugar de producir sus propios nutrientes o de obtenerlos de fuentes inorgánicas (345)

continental drift / deriva continental la hipótesis que establece que alguna vez los continentes formaron una sola masa de tierra, se dividieron y se fueron a la deriva hasta terminar en sus ubicaciones actuales (268)

contour feather / pluma de contorno una las plumas más externas que cubren a un ave y que sirven para determinar su forma (784)

control group / grupo de control en un experimento, un grupo que sirve como estándar de comparación con otro grupo, al cual el grupo de control es idéntico excepto por un factor (17)

convergent evolution / evolución convergente el proceso por medio del cual especies no relacionadas se vuelven más parecidas a medida que se adaptan al mismo tipo de ambiente (307)

cork / corcho la capa externa de corteza de cualquier planta leñosa (553)

cork cambium / cámbium de corcho una capa de tejido que se encuentra debajo de la capa de corcho, en la cual se producen las células de corcho (575)

coronary artery / arteria coronaria una de las dos arterias que suministran sangre directamente al corazón (882)

corpus luteum / cuerpo lúteo la estructura que se forma a partir de los folículos rotos del ovario después de la ovulación; libera hormonas (1002)

cortex / corteza en las plantas, el tejido primario ubicado en la epidermis; en los animales, la porción externa de un órgano (555)

cotyledon / cotiledón la hoja embrionaria de una semilla (535)

countercurrent flow / flujo a contracorriente en las branquias de los peces, un arreglo por el cual el agua fluye alejándose de la cabeza y la sangre fluye hacia la cabeza (747)

crossing-over / entrecruzamien el intercambio de material genético entre cromosomas homólogos durante la meiosis; puede resultar en la recombinación genética (144)

cud / bolo alimenticio de los rumiante comida parcialmente digerida que es regurgitada, masticada y tragada nuevamente con el fin de digerirla más por los mamíferos que tienen rumen (814)

cuticle / cutícula una capa cerosa o grasosa e impermeable ubicada en la pared externa de las células de la epidermis (502)

cyanobacteria / cianobacteria una bacteria que efectúa la fotosíntesis, como por ejemplo, el alga verdiazul (258)

cystic fibrosis / fibrosis quística un trastorno genético mortal en el que se secretan cantidades excesivas de moco, lo cual bloquea los conductos intestinales y bronquiales y causa dificultad al respirar (12)

cytokinesis / citocinesis la división del citoplasma de una célula; la citocinesis ocurre después de que el núcleo de la célula se divide por mitosis o meiosis (125)

cytoplasm / citoplasma la región de la célula dentro de la membrana, que incluye el líquido, el citoesqueleto y los organelos, pero no el núcleo (56)

cytoskeleton / citoesqueleto red citoplásmica de filamentos de proteínas que juega un papel esencial en el movimiento, forma y división de la célula (56)

cytotoxic T cell / célula T citotóxica un tipo de célula T que reconoce y destruye las células infectadas por un virus (927)

decomposer / descomponedor un organismo que desintegra la materia orgánica de organismos muertos y se alimenta de ella; entre los ejemplos se encuentran las bacterias y los hongos (347)

dendrite / dendrita la extensión citoplásmica de una neurona que recibe estímulos (944)

density-dependent factor / factor dependiente de la densidad una variable afectada por el número de organismos presentes en un área determinada (322)

density-independent factor / factor independiente de la densidad una variable que afecta a una población independientemente de la densidad de la población, por ejemplo, el clima (324)

deoxyribose / desoxirribosa azúcar de cinco carbonos que es un componente de los nucleótidos de ADN (194)

dependent variable / variable dependiente en un experimento, la variable que se cambia o que se determina al manipular dos o más factores (las variables independientes) (17)

depressant / depresor un medicamento que reduce la actividad funcional y produce relajación muscular (964)

derived character / carácter derivado una característica especial de un grupo particular de organismos (307)

dermal tissue / tejido dérmico la cubierta exterior de una planta (552)

dermis / dermis la capa de piel que está debajo de la epidermis (862)

detritivore / detritívoro un consumidor que se alimenta de plantas y animales muertos (346)

deuterostome / deuteróstomo un animal cuya boca no se deriva del blastoporo y cuyo embrión presenta segmentación indeterminada (692)

diabetes mellitus / diabetes mellitus un serio trastorno en el que las células no pueden obtener glucosa de la sangre; es causado por una deficiencia de insulina o por una falta de respuesta a la insulina (987)

diaphragm / diafragma un músculo en forma de cúpula que está unido a las costillas inferiores y que es el músculo principal de la respiración (886)

diatom / diatomea un alga unicelular que tiene una concha doble la cual contiene sílice (466)

dicot / dicotiledónea una angiosperma con dos cotiledones, venación en forma de red y partes florales en grupos de cuatro o cinco (515)

differentiation / diferenciación el proceso por medio del cual la estructura y función de las partes de un organismo cambian para permitir la especialización de dichas partes (419)

diffusion / difusión el movimiento de partículas de regiones de mayor densidad a regiones de menor densidad (75)

digestion / digestión la descomposición de la comida en substancias químicas que se usan para generar energía (900)

diploid / diploide una célula que contiene dos juegos de cromosomas haploides (121)

directional selection / selección direccional un proceso de selección natural en el cual se selecciona una variación genética que origina un cambio en la composición genética global de la población (332)

dispersion / dispersión en óptica, el proceso de separar una onda que tiene diferentes frecuencias (por ejemplo, la luz blanca) de las ondas individuales que la componen (los distintos colores) (321)

diurnal / diurno término que describe animales que son activos durante el día y duermen en la noche (732)

DNA / ADN ácido desoxirribonucleico, el material que contiene la información que determina las características que se heredan (37)

DNA fingerprint / huella de ADN el patrón de bandas que se obtiene cuando los fragmentos de ADN de un individuo se separan (237)

DNA helicase / ADN helicasa una enzima que separa las hebras de la doble hélice del ADN durante la replicación del ADN (198)

DNA polymerase / ADN polimerasa una enzima que actúa como catalizadora en la formación de la molécula de ADN (199)

DNA replication / replicación del ADN el proceso de hacer una copia del ADN (198)

dormancy / letargo un estado en el que las semillas, esporas, bulbos y otros órganos reproductores dejan de crecer y desarrollarse y reducen su metabolismo, sobre todo la respiración (584)

double fertilization / fecundación doble el proceso por medio del cual uno de los dos núcleos de los espermatozoides se une con el núcleo del óvulo para producir un cigoto diploide, y el otro núcleo se une con el núcleo polar para producir un endosperma triploide (540)

double helix / doble hélice la estructura en forma de escalera en espiral característica de la molécula del ADN (194)

down feather / plumón una pluma suave que cubre el cuerpo de las crías de las aves y sirve como aislante en las aves adultas (785)

ecology / ecología el estudio de las interacciones de los seres vivos entre sí mismos y entre sí mismos y su ambiente (9)

ecosystem / ecosistema una comunidad de organismos y su ambiente abiótico (340)

ectoderm / ectodermo la capa más externa de las tres capas germinales de un embrión, que al desarrollarse se convierte en la epidermis y en los tejidos epidérmicos, el sistema nervioso, los órganos externos de los sentidos y las membranas mucosas que cubren la boca y el ano (596)

ectothermic / ectotérmico término que describe la capacidad de un organismo de mantener su temperatura corporal al obtener calor del ambiente (724)

electron microscope / microscopio electrónico microscopio que enfoca un haz de electrones para aumentar la imagen de los objetos (51)

electron transport chain / cadena de transporte de electrones una serie de moléculas que se encuentran en las membranas internas de las mitocondrias y cloroplastos y a través de las cuales pasan los electrones en un proceso que hace que los protones se acumulen en un lado de la membrana (100)

electrophoresis / electroforesis el proceso por medio del cual las partículas con carga eléctrica que están suspendidas en un líquido se mueven por todo el líquido debido a la influencia de un campo eléctrico (231)

element / elemento una substancia que no se puede separar o descomponer en substancias más simples por medio de métodos químicos; todos los átomos de un elemento tienen el mismo número atómico (28)

embryo / embrión un organismo en una de las primeras etapas del desarrollo de las plantas y animales; en los humanos, un individuo en desarrollo se denomina embrión de la segunda a la octava semana de embarazo (504)

endocrine gland / glándula endocrina una glándula sin conductos que secreta hormonas a la sangre (975)

endocytosis / endocitosis el proceso por medio del cual la membrana celular rodea una partícula y la encierra en una vesícula para llevarla al interior de la célula (83)

endoderm / endodermo la capa germinal interna del embrión de un animal; al desarrollarse se convierte en el epitelio de la faringe, tracto respiratorio, tracto digestivo, vejiga y uretra (596)

endoplasmic reticulum / retículo endoplásmico un sistema de membranas que se encuentra en el citoplasma de la célula y que tiene una función en la producción, procesamiento y transporte de proteínas y en la producción de lípidos (63)

endosperm / endosperma un tejido triploide (3n) que se desarrolla en las semillas de las angiospermas y que provee alimento para el embrión en desarrollo (514)

endospore / endospora una espora protectiva que tiene una pared gruesa, se forma dentro de una célula bacteriana y resiste condiciones adversas (443)

endosymbiosis / endosimbiosis una relación mutuamente beneficiosa en la que un organismo vive dentro de otro (259)

endothermic / endotérmico término que describe la capacidad de un organismo de mantener su temperatura corporal al producir calor internamente (724)

energy / energía la capacidad de realizar un trabajo (38)

energy pyramid / pirámide de energía un diagrama con forma de triángulo que muestra la pérdida de energía que ocurre en un ecosistema a medida que la energía pasa a través de la cadena alimenticia del ecosistema; cada hilera de la pirámide representa un nivel trófico (de alimentación) en el ecosistema, y el área de la hilera representa la energía almacenada en ese nivel trófico (348)

envelope / envoltura una capa similar a una membrana que cubre las cápsides de algunos virus (435)

enzyme / enzima un tipo de proteína que acelera las reacciones metabólicas en las plantas y animales, sin ser modificada permanentemente ni ser destruida (40)

epidermis / epidermis la superficie externa de las células de una planta o animal (553)

epididymis / epidídimo el conducto largo y enrollado que se encuentra en la superficie de los testículos, en el que los espermatozoides maduran (997)

epinephrine / epinefrina una hormona liberada por la médula suprarrenal que estimula el metabolismo rápidamente en casos de emergencia, disminuye la secreción de insulina y estimula el pulso y la presión sanguínea; también se llama adrenalina (985)

epithelial tissue / tejido epitelial un tejido compuesto por células que forman una barrera entre un organismo y su ambiente externo (846)

equilibrium / equilibrio en química, el estado en el que un proceso químico y el proceso químico inverso ocurren a la misma tasa, de modo que las concentraciones de los reactivos y los productos no cambian (74)

esophagus / esófago un conducto largo y recto que conecta la faringe con el estómago (907)

Eubacteria / Eubacteria un reino de clasificación que agrupa a todos los procariotes, excepto a las arqueobacterias (258)

eukaryote / eucariote un organismo cuyas células tienen un núcleo contenido en una membrana, múltiples cromosomas y un ciclo mitótico; los eucariotes incluyen animales, plantas y hongos, pero no bacterias ni algas (58)

evolution / evolución un cambio en las características de una población de una generación a la siguiente; el desarrollo gradual de organismos a partir de otros organismos desde los inicios de la vida (9)

excretion / excreción el proceso de eliminar desechos metabólicos (912)

exocytosis / exocitosis el proceso por medio del cual una substancia se libera de la célula a través de una vesícula que la transporta a la superficie de la célula en donde se fusiona con la membrana para dejar salir a la substancia (83)

exon / exón la porción de la secuencia del ADN de un gene que contiene la secuencia de aminoácidos en una cadena, y el inicio y el fin de una secuencia de codificación (218)

exoskeleton / exoesqueleto una estructura de soporte, dura y externa, que se desarrolla a partir del ectodermo (607)

experiment / experimento un procedimiento que se lleva a cabo bajo condiciones controladas para descubrir, demostrar o probar un hecho, teoría o verdad general (17)

extensor / extensor un músculo que extiende una articulación (856)

external fertilization / fecundación externa la unión de gametos afuera del cuerpo de los padres, como en el caso de muchos peces y anfibios (610)

extinct / extinto término que describe a una especie de un organismo que ha desaparecido completamente (282)

F₁ generation / generación F₁ la primera generación de descendencia que se obtiene de la cruza experimental de dos organismos (164)

F₂ generation / generación F₂ la segunda generación de descendencia que se obtiene de la cruza experimental de dos organismos de una generación F₁ (164)

facilitated diffusion / difusión facilitada el transporte de substancias a través de la membrana celular de una región de mayor concentración a una de menor concentración con la ayuda de proteínas transportadoras (80)

fallopian tube / trompa de Falopio un conducto a través del cual se mueven los óvulos del ovario al útero (1000)

family / familia la categoría taxonómica debajo del orden y arriba del genero (302)

fermentation / fermentación la descomposición de carbohidratos por enzimas, bacterias, levaduras o mohos, en ausencia de oxígeno (108)

fertilization / fecundación la unión de un gameto masculino y femenino para formar un cigoto (153)

fetus / feto un ser humano en desarrollo de las semanas siete a ocho después de la fecundación hasta el nacimiento (1006)

fixed action pattern behavior / conducta de patrón fijo de acción una conducta innata que es característica de ciertas especies (826)

flagellum / flagelo una estructura larga parecida a una cola, que crece hacia el exterior de una célula y le permite moverse (57)

flexor / flexor un músculo que dobla una extremidad u otra parte del cuerpo (856)

flower / flor la estructura reproductiva de una planta que da flores, que normalmente consiste en un pistilo, estambres, pétalos y sépalos (505)

fluke / trematodo un gusano plano parasítico de la clase Trematoda (632)

follicle / folículo una bolsa o cavidad angosta y pequeña en un órgano o tejido, como las que se encuentran en la piel y contienen las raíces de los pelos, o las que se encuentran en los ovarios y contienen los óvulos en desarrollo (1001)

food chain / cadena alimenticia la vía de transferencia de energía través de varias etapas, que ocurre como resultado de los patrones de alimentación de una serie de organismos (346)

food web / red alimenticia un diagrama que muestra las relaciones de alimentación entre los organismos de un ecosistema (347)

foot / pie un apéndice que algunos invertebrados usan para moverse; la parte inferior de la pierna de un vertebrado (431)

fossil / fósil los rastros o restos de un organismo que vivió hace mucho tiempo, conservados en rocas sedimentarias (258)

frond / fronda la hoja de un helecho o palma (511)

fruit / fruto un ovario maduro de planta; el órgano de una planta donde se encuentran contenidas las semillas (514)

fundamental niche / nicho fundamental el nicho ecológico más grande en el que un organismo o especie vive sin experimentar competencia (365)

G

gamete / gameto una célula reproductiva haploide que se une con otra célula reproductiva haploide para formar un cigoto (118)

gametophyte / gametofito en generaciones alternadas, la fase en la que los gametos se forman; un individuo haploide que produce gametos (154)

gastrovascular cavity / cavidad gastrovascular una cavidad que tiene funciones digestivas y circulatorias en algunos cnidarios (605)

gemmule / gémula una estructura asexual reproductiva producida por algunas esponjas de agua dulce (621)

gene / gene un segmento de ADN ubicado en un cromosoma, que codifica para un carácter hereditario específico (8)

gene cloning / clonación de genes el proceso por medio del cual se aísla la secuencia de un gene del genoma de un organismo y esta secuencia se inserta en un vector plásmido para producir el gene en grandes cantidades (229)

gene expression / expresión de los genes la manifestación del material genético de un organismo en forma de caracteres específicos (208)

gene flow / flujo de genes el movimiento de genes a una población o fuera de ella debido al entrecruzamiento (328)

genetic code / código genético la regla que describe la forma en que una secuencia de nucleótidos, leídos en grupos de tres nucleótidos consecutivos (triplete) que corresponden a aminoácidos específicos, especifica la secuencia de aminoácidos de una proteín (211)

genetic drift / deriva genética el cambio aleatorio en la frecuencia de los alelos de una población (328)

genetic engineering / ingeniería genética una tecnología en la que el genoma de una célula viva se modifica con fines médicos o industriales (228)

genetics / genética la ciencia de la herencia y de los mecanismos por los cuales los caracteres son transmitidos de padres a hijos (162)

genital herpes / herpes genital una enfermedad transmitida sexualmente causada por el virus herpes simplex (1010)

genotype / genotipo la constitución genética completa de un organismo; también, la combinación de genes para uno o más caracteres específicos (168)

genus / género el nivel de clasificación que viene después de la familia y que contiene especies similares (301)

germination / germinación el comienzo del crecimiento o desarrollo de una semilla, espora o cigoto, sobre todo después de un período de inactividad (572)

gestation / gestación en los mamíferos, el proceso de llevar a las crías de la fecundación al nacimiento (1005)

gestation period / período de gestación en los mamíferos, el tiempo que transcurre entre la fecundación y el nacimiento (810)

gill / branquia en los animales acuáticos, una estructura respiratoria que está formada por muchos vasos sanguíneos rodeados por una membrana que permite el intercambio gaseoso (605)

gill / laminilla en los hongos, una estructura que se ubica en la parte inferior de sombrerete y que contiene las esporas (605)

gill slit / apertura branquial una perforación entre dos arcos branquiales a través de la cual el agua que un pez toma por la boca pasa sobre las branquias y hacia el exterior del cuerpo del pez (747)

glucagon / glucagón una hormona producida en el páncreas que aumenta el nivel de glucosa en la sangre (986)

glycolysis / glicólisis la descomposición anaeróbica de ácido pirúvico glucosa, la cual hace que una pequeña cantidad de energía en forma de ATP esté disponible para las células (105)

glycoprotein / glicoproteína una proteína que tienen unidas moléculas de carbohidratos (435)

Golgi apparatus / aparato de Golgi un organelo celular que ayuda a hacer y a empacar los materiales que serán transportados al exterior de la célula (64)

gonorrhea / gonorrea una enfermedad transmitida sexualmente producida por bacterias, que resulta en la inflamación de las membranas mucosas de los tractos urinario y reproductor (1008)

gradualism / gradualismo un modelo de evolución en el que un cambio gradual a través de un largo período de tiempo conlleva a la diversidad biológica (282)

grain / grano la semilla comestible, o fruta similar a una semilla, de un cereal (518)

greenhouse effect / efecto de invernadero el calentamiento de la superficie terrestre y de la parte más baja de la atmósfera, el cual se produce cuando el dióxido de carbono, el vapor de agua y otros gases del aire absorben radiación infrarroja y la vuelven a irradiar (388)

ground tissue / tejido basal un tipo de tejido vegetal diferente del tejido vascular y que constituye gran parte del interior de una planta (552)

groundwater / agua subterránea el agua que está debajo de la superficie de la Tierra (351)

guard cell / célula oclusiva una de las dos células especializadas que se encuentran al borde de un estoma y regulan el intercambio gaseoso (503)

gymnosperm / gimnosperma una planta leñosa y vascular, la cual produce semillas que no están contenidas en un ovario o fruto (512)

H

habitat / hábitat el lugar donde un organismo vive normalmente (340)

hair / pelo en los mamíferos, una de las muchas estructuras largas y delgadas que crecen a partir de la piel (800)

hair follicle / folículo piloso una depresión en la piel que contiene un pelo y su raíz (862)

half-life / vida media el tiempo que tarda la mitad de una muestra de una substancia radiactiva en desintegrarse por desintegración radiactiva o por procesos naturales (252)

haploid / haploide término que describe a una célula, núcleo u organismo que tiene sólo un juego de cromosomas que no están asociados en pares (121)

Hardy-Weinberg principle / principio de Hardy-Weinberg el principio que establece que la frecuencia de alelos en una población no cambia a menos que fuerzas evolutivas actúen en la población (326)

Haversian canal/canal de Havers un canal que contiene vasos sanguíneos en los huesos compactos (852)

heart attack / ataque cardíaco la muerte de los tejidos del corazón debido a una obstrucción de su suministro sanguíneo (884)

heartwood / duramen la madera más vieja y que no conduce la electricidad, que se encuentra en el centro de un tronco de árbol (557)

helper T cell / célula T auxiliar un glóbulo blanco de la sangre necesario para que las células B desarrollen niveles normales de un anticuerpo (927)

herbaceous plant / planta herbácea una planta que es suave y verde, en vez de leñosa (556)

herbivore / herbívoro un organismo que sólo come plantas (346)

heredity / herencia la transmisión de caracteres genéticos de padres a hijos (8)

hermaphrodite / hermafrodita un organismo que tiene órganos reproductores tanto masculinos como femeninos (609)

heterotroph / heterótrofo un organismo que obtiene moléculas de alimento al comer otros organismos o sus productos secundarios y que no puede sintetizar compuestos orgánicos a partir de materiales inorgánicos (95)

heterozygous / heterocigoto término que describe a un individuo que tiene dos alelos distintos para un mismo carácter (167)

histamine / histamina una substancia química que estimula el sistema nervioso autónomo, la secreción de jugos gástricos y la dilatación de capilares (925)

HIV / VIH virus de inmunodeficiencia humana; el virus que causa el SIDA (12)

homeostasis / homeostasis la capacidad de mantener un estado interno constante en un ambiente en cambio; un estado interno constante que se mantiene en un ambiente en cambio al hacer ajustes continuos al ambiente interno y externo (8)

hominid / homínido miembro de la familia Hominidae del orden de los primates; caracterizado por tener pulgares oponibles y extremidades inferiores relativamente largas, ser bípedo y no tener cola; incluye a los seres humanos modernos y a sus ancestros (733)

homologous chromosomes / cromosomas homólogos cromosomas con la misma secuencia de genes, que tienen la misma estructura y que se acoplan durante la meiosis (120)

homologous structures / estructuras homólogas estructuras que comparten un ancestro común (286)

homozygous / homocigoto término que describe a un individuo que tiene alelos idénticos para un carácter en los dos cromosomas homólogos (167)

hormone / hormona una substancia que es producida en una célula o tejido, la cual causa un cambio en otra célula o tejido ubicado en una parte diferente del cuerpo (580)

Human Genome Project / Proyecto del Genoma Humano un esfuerzo de investigación para determinar la secuencia y ubicación de todo el conjunto de genes de las células humanas (233)

hydrostatic skeleton / esqueleto hidrostático la cavidad llena de agua de muchos invertebrados que tiene una función de sostén (607)

hypertonic / hipertónico término que describe una solución cuya concentración de soluto es más alta que la concentración del soluto en el interior de la célula (77)

hypha / hifa un filamento no-reproductor de un hongo (421)

hypothalamus / hipotálamo la región del cerebro que coordina las actividades de los sistemas nervioso y endocrino y que controla muchas actividades del cuerpo relacionadas con la homeostasis (952)

hypothesis / hipótesis una teoría o explicación basada en observaciones y que se puede probar (16)

hypotonic / hipotónico término que describe una solución cuya concentración de soluto es más baja que la concentración del soluto en el interior de la célula (77)

I

immunity / inmunidad la capacidad de resistir una infección o enfermedad, o de recuperarse de ella (931)

implantation / implantación el proceso por medio del cual el óvulo fecundado en la etapa de blastocisto se adhiere a la cubierta interior del útero (1004)

imprinting / impresión aprendizaje que ocurre rápidamente al inicio de la vida de un animal joven y que una vez que se aprende no se puede cambiar (829)

incomplete dominance / dominancia incompleta una condición en la que un carácter de un individuo es intermedio entre el fenotipo de los dos padres del individuo porque el alelo dominante no puede expresarse por completo (177)

independent assortment / distribución independiente la distribución al azar de pares de genes de diferentes cromosomas a los gametos (146)

independent variable / variable independiente el factor que se manipula deliberadamente en un experimento (17)

inflammatory response / reacción inflamatoria una reacción de protección de los tejidos afectados por una enfermedad o lesión, caracterizada por enrojecimiento, inflamación y dolor (925)

innate behavior / conducta innata una conducta heredada que no depende del ambiente ni de la experiencia (826)

insulin / insulina una hormona que es producida por un grupo de células especializadas en el páncreas y que reduce los niveles de glucosa en la sangre (986)

interferon / interferón una proteína que producen las células infectadas por un virus y que puede proteger a las células que no han sido infectadas contra la reproducción del virus (926)

internal fertilization / fecundación interna fecundación de un óvulo por un espermatozoide, la cual ocurre dentro del cuerpo de la hembra (610)

interneuron / interneurona una neurona ubicada entre la neurona aferente y la neurona final en una cadena neural (953)

interphase / interfase un período entre dos divisiones mitóticas o meióticas durante las cuales la célula crece, copia su ADN y sintetiza proteínas (125)

intron / intrón una sección del ADN que no codifica para ningún aminoácido y que se transcribe al ARN pero se elimina antes de ser traducida (218)

invertebrate / invertebrado un animal que no tiene columna vertebral (424)

ion / ion un átomo, radical o molécula que ha ganado o perdido uno o más electrones y que tiene una carga negativa o positiva (30)

ion channel / canal iónico un poro en la membrana celular a través del cual pueden pasar los iones (78)

isotonic solution / solución isotónica una solución cuya concentración de soluto es igual a la concentración de soluto en el interior de la célula (77)

J

joint / articulación un lugar donde se unen dos o más huesos (854)

K

karyotype / cariotipo una distribución de cromosomas que se encuentra en las células de un individuo en la metafase o en la mitosis, los cuales están ordenados en pares homólogos y en orden de mayor a menor (122)

keratin / queratina una proteína dura que forma el cabello, las plumas de las aves, las uñas y los cuernos (861)

kingdom / reino la categoría taxonómica más alta, que contiene un grupo de phyla similares (302)

Koch's postulates / postulados de Koch un procedimiento de cuatro etapas que formuló Robert Koch para identificar patógenos específicos y para determinar la causa de una determinada enfermedad (930)

Krebs cycle / ciclo de Krebs una serie de reacciones bioquímicas que convierten el ácido pirúvico en dióxido de carbono y agua; es la vía principal de oxidación en las células animales, bacterianas y vegetales, y libera energía (106)

krill / krill un crustáceo marino pequeño que es la fuente principal de alimentos de la ballena barbada (681)

K-strategist / estratega K una especie caracterizada por maduración lenta, pocas crías, crecimiento lento de la población, reproducción en etapas tardías de la vida y una densidad de población cercana a la capacidad de carga del ambiente (325)

lac operon / operón lac un sistema de genes cuyo gene operador y sus tres genes estructurales controlan el metabolismo de la lactosa en E. Coli (216)

larynx / laringe el área de la garganta que contiene las cuerdas vocales y que produce sonidos vocales (886)

lateral line / línea lateral una línea apenas visible que se encuentra a ambos lados del cuerpo de un pez y que recorre la longitud del cuerpo, marcando la ubicación de los órganos de los sentidos que detectan vibraciones en el agua (753)

law of independent assortment / ley de la distribución independiente la ley que establece que los genes se separan de manera independiente durante la meiosis (169)

law of segregation / ley de la segregación la ley de Mendel que establece que pares de cromosomas homólogos se separan en la meiosis de modo que sólo un cromosoma de cada par esté presente en cada gameto (169)

learning / aprendizaje el desarrollo de conductas por medio de la experiencia o práctica (827)

lichen / liquen una masa de células de hongos y de algas que crecen juntas en una relación simbiótica y que normalmente se encuentran en rocas o árboles (491)

life cycle / ciclo de vida todos los sucesos en el crecimiento y desarrollo de un organismo hasta que el organismo llega a su madurez sexual (152)

ligament / ligamento un tipo de tejido que mantiene unidos los huesos en una articulación (854)

light microscope / microscopio óptico un microscopio que usa un rayo de luz visible que pasa a través de uno o más lentes para magnificar la imagen de un objeto (57)

limnetic zone / zona limnética el área de un hábitat de agua dulce que está lejos de la costa, pero que se encuentra cerca de la superficie (376)

lipase / lipasa una enzima que descompone moléculas de grasa en ácidos grasos y glicerol (909)

lipid / lípido un tipo de substancia bioquímica que no se disuelve en agua, como por ejemplo, las grasas y esteroides; los lípidos almacenan energía y forman las membranas celulares (35)

lipid bilayer / bicapa lipídica la estructura básica de la membrana biológica, formada por dos capas de fosfolípidos (60)

littoral zone / zona litoral una zona poco profunda del hábitat de agua dulce donde la luz llega al fondo y nutre a las plantas (376)

logistic model / modelo logístico un modelo del crecimiento de la población que supone que los niveles finitos de recursos limitan el crecimiento de la población (323)

lung / pulmón el órgano central del aparato respiratorio en el que el oxígeno del aire se intercambia con el dióxido de carbono de la sangre (758)

lymphatic system / sistema linfático un conjunto de órganos cuya función principal es recolectar el fluido extracelular y regresarlo a la sangre; los órganos de este sistema incluyen los nodos linfáticos y los vasos linfáticos (875)

lysosome / lisosoma un organelo celular que contiene enzimas digestivas (64)

macrophage / macrófago una célula del sistema inmunológico que envuelve a los patógenos y otros materiales (926)

magnification / magnificación el aumento del tamaño aparente de un objeto mediante el uso de lentes o espejos (51)

Malpighian tubule / tubo de Malpighi un tubo excretorio que se abre hacia la parte trasera del intestino de la mayoría de los insectos y ciertos artrópodos (669)

mammary gland / glándula mamaria una glándula que se encuentra en el pecho de los mamíferos hembra y que secreta leche (806)

mandible / mandíbula un tipo de parte de la boca que se encuentra en algunos artrópodos y que se usa para perforar y chupar la comida; la parte inferior de la quijada (673)

mantle / manto en biología, una capa de tejido que cubre el cuerpo de muchos invertebrados (643)

mass extinction / extinción masiva un episodio durante el cual grandes cantidades de especies se extinguen (263)

medusa / medusa una etapa sexual del ciclo de vida de un cnidario, que nada libremente, tiene la apariencia de un aguamala y la forma de un paraguas; también, un aguamala o hidra (622)

meiosis / meiosis un proceso de división celular durante el cual el número de cromosomas disminuye a la mitad del número original por medio de dos divisiones del núcleo, lo cual resulta en la producción de células sexuales (gametos o esporas) (144)

melanin / melanina un pigmento que ayuda a determinar el color de la piel (862)

membrane potential / potencial de membrana la diferencia en potencial eléctrico entre los dos lados de una membrana celular (945)

menstrual cycle / ciclo menstrual el ciclo reproductor femenino, caracterizado por un cambio mensual en el revestimiento del útero y una descarga de sangre (1002)

menstruation / menstruación la descarga de sangre y tejido de desecho del útero durante el ciclo menstrual (1003)

meristem / meristemo una región de células vegetales no diferenciadas que son capaces de dividirse y desarrollarse en tejidos vegetales especializados (507)

mesoderm / mesodermo en un embrión, la capa de células intermedia que da origen a los músculos, sangre y varios sistemas (596)

mesophyll / mesófilo en las hojas, el tejido que se encuentra entre capas de epidermis, donde ocurre la fotosíntesis (559)

metabolism / metabolismo la suma de todos los procesos químicos que ocurren en un organismo (7)

metamorphosis / metamorfosis una fase del ciclo de vida de muchos animales durante la cual ocurre un cambio rápido del organismo inmaduro al adulto; un ejemplo es el cambio de larva a adulto en los insectos (675)

mineral / mineral un sólido natural, normalmente inorgánico, que tiene una composición química característica, una estructura interna ordenada y propiedades físicas y químicas características (905)

mitochondrion / mitocondria en las células eucarióticas, el organelo celular rodeado por dos membranas que es el lugar donde se lleva a cabo la respiración celular, la cual produce ATP (65)

mitosis / mitosis en las células eucarióticas, un proceso de división celular que forma dos núcleos nuevos, cada uno de los cuales posee el mismo número de cromosomas (125)

molecule / molécula la unidad más pequeña de una substancia que conserva todas las propiedades físicas y químicas de esa substancia; puede estar formada por un átomo o por dos o más átomos enlazados uno con el otro (29)

molting / mudar la muda de un exoesqueleto, piel, plumas o cuernos, los cuales son reemplazados por partes nuevas (668)

monocot / monocotiledónea una planta que produce semillas que sólo tienen un cotiledón (515)

monohybrid cross / cruza monohíbrida una cruza entre individuos que involucra un par de caracteres contrastantes (164)

monosaccharide/monosacárido un azúcar simple que es una subunidad fundamental de los carbohidratos (34)

motor neuron / neurona motora una célula nerviosa que transmite impulsos nerviosos del sistema nervioso central a los músculos y a las glándulas (950)

mRNA / ARNm ARNm mensajero; una molécula de ARN de una sola hebra que codifica la información para hacer una proteína (211)

mucous membrane / membrana mucosa la capa de tejido epitelial que cubre las superficies internas del cuerpo y que secreta moco (924)

multiple alleles / alelos múltiples más de dos alelos (versiones del gene) para un carácter genético (178)

muscle tissue / tejido muscular el tejido formado por células que se contraen y relajan para producir movimiento (847)

mutation / mutación un cambio en la secuencia de la base de nucleótidos de un gene o de una molécula de ADN (8)

mutualism / mutualismo una relación entre dos especies en la que ambas se benefician (265)

mycelium / micelio una masa de filamentos de hongos, o hifas, que forma el cuerpo de un hongo (483)

mycorrhiza / micorriza una asociación simbiótica entre los hongos y las raíces de las plantas (265)

myofibril / miofibrilla una fibra que se encuentra en las células de los músculos estriados, la cual es responsable de la contracción muscular (857)

myosin / miosina la proteína más abundante en los tejidos musculares, la cual es el elemento constitutivo principal de los filamentos gruesos de las fibras musculares (857)

natural killer cell / célula asesina natural un glóbulo blanco presente en individuos que no han sido inmunizados, el cual destruye una variedad de células (926)

natural selection / selección natural el proceso por medio del cual los individuos que tienen condiciones favorables y que están mejor adaptados a su ambiente sobreviven y se reproducen con más éxito que los individuos que no están tan bien adaptados (9)

nauplius / naupilia la larva de la mayoría de los crustáceos, la cual nada libremente (680)

nematocyst / nematocisto en los cnidarios, una célula urticante que se usa para inyectar una toxina en una presa (623)

nephridium / nefridio túbulo a través del cual algunos invertebrados eliminan desechos (644)

nephron / nefrona la unidad funcional del riñón (913)

nerve / nervio un conjunto de fibras nerviosas a través de las cuales se desplazan los impulsos entre el sistema nervioso central y otras partes del cuerpo (944)

nervous tissue / tejido nervioso el tejido del sistema nervioso, formado por neuronas, sus células de apoyo y el tejido conectivo (847)

neuron / neurona una célula nerviosa que está especializada en recibir y transmitir impulsos eléctricos (944)

neurotransmitter / neurotransmisor una substancia química que transmite impulsos nerviosos por una sinapsis (948)

neutrophil / neutrófilo un leucocito grande que contiene un núcleo lobulado y muchos gránulos citoplásmicos (926)

niche / nicho la posición única que ocupa una especie, tanto en lo que se refiere al uso de su hábitat como en cuanto a su función dentro de una comunidad ecológica (365)

nitrogen fixation / fijación de nitrógeno el proceso por medio del cual el nitrógeno gaseoso se transforma en amoniaco, un compuesto que los organismos utilizan para elaborar aminoácidos y otras moléculas orgánicas que contienen nitrógeno (363)

nonvascular plant / planta no vascular los tres tipos de plantas (hepáticas, milhojas y musgos) que carecen de tejidos transportadores y de raíces, tallos y hojas verdaderas (504)

norepinephrine / norepinefrina una substancia química que es un neurotransmisor producido por las terminaciones nerviosas simpáticas en el sistema nervioso, y también una hormona secretada por la médula suprarrenal para estimular las funciones de los aparatos circulatorio y respiratorio (abreviatura: NE) (985)

normal distribution / distribución normal una distribución de datos numéricos cuya gráfica forma una curva en forma de campana que es simétrica respecto a la media (331)

notochord / notocordio el eje de soporte que tiene forma de bastoncillo y está ubicado en la parte dorsal de los embriones de todos los cordados, incluyendo los vertebrados (700)

nucleic acid / ácido nucleico un compuesto orgánico, ya sea ARN o ADN, cuyas moléculas están formadas por una o más cadenas de nucleótidos y que contiene información genética (37)

nucleotide / nucleótido en una cadena de ácidos nucleicos, una subunidad formada por un azúcar, un fosfato y una base nitrogenada (37)

nucleus / núcleo en una célula eucariótica, biología, un organelo cubierto por una membrana, el cual contiene el ADN de la célula y participa en procesos tales como el crecimiento, metabolismo y reproducción (58)

nutrient / nutriente una substancia o compuesto que proporciona nutrición (o alimento) o materias primas que se necesitan para llevar a cabo procesos vitales (900)

nymph / ninfa una etapa inmadura de algunos insectos que es similar en función y estructura al adulto (675)

observation / observación el proceso de obtener información por medio de los sentidos; la información que se obtiene al usar los sentidos (14)

omnivore / omnívoro un organismo que come tanto plantas como animales (346)

oogenesis / oogénesis la producción, crecimiento y maduración de un óvulo (148)

open circulatory system / aparato circulatorio abierto un tipo de aparato circulatorio en el que el fluido circulatorio no está totalmente contenido en los vasos sanguíneos; un corazón bombea fluido por los vasos sanguíneos, los cuales se vacían en espacios llamados senos (606)

operator / operador una secuencia corta de ADN viral o bacteriano a la que se une un represor para impedir la transcripción (síntesis de ARNm) del gene adyacente en un operón (216)

operculum / opérculo en los peces, una placa dura que se encuentra adherida a cada lado de la cabeza, cubre las branquias y está abierta en la parte trasera (756)

operon / operón una unidad de regulación y transcripción de los genes en el ADN bacteriano, formada por un promotor, un operador y uno o más genes estructurales (216)

optic nerve / nervio óptico el nervio que conecta la retina del ojo con el cerebro y que transmite impulsos que contribuyen al sentido de la vista (958)

order / orden la categoría taxonómica que se encuentra debajo de la clase y arriba de la familia (302)

organ / órgano un conjunto de tejidos que desempeñan una función especializada en el cuerpo (419)

organ system / aparato (o sistema) de órganos un grupo de órganos que trabajan en conjunto para desempeñar funciones corporales (419)

organelle / organelo uno de los cuerpos pequeños que se encuentran en el citoplasma de una célula y que están especializados para llevar a cabo una función específica (58)

osmosis / ósmosis la difusión de agua u otro solvente de una solución más diluida (de un soluto) a una solución más concentrada (del soluto) a través de una membrana que es permeable al solvente (76)

ossicle / osículo una de las pequeñas placas de carbonato de calcio que forman el endoesqueleto de un equinodermo (694)

osteocyte / osteocito una célula ósea (852)

ovarian cycle / ciclo ovárico una serie de cambios inducidos por hormonas en los cuales los ovarios preparan y liberan un óvulo maduro todos los meses (1001)

ovary / ovario en el aparato reproductor femenino de los animales, un órgano que produce óvulos; en las plantas con flores, la parte inferior del pistilo que produce óvulos (538)

oviparous / ovíparo término que describe organismos que producen huevos que se desarrollan fuera del cuerpo de la madre, y cuyas crías también salen del cascarón fuera del cuerpo de la madre (777)

ovoviviparous / ovovivíparo término que describe a organismos que producen huevos que se desarrollan dentro del cuerpo de la madre, y cuyas crías también salen del cascarón dentro del cuerpo de la madre (777)

ovulation / ovulación la liberación de un óvulo de un folículo del ovario (1001)

ovule / óvulo una estructura del ovario de una planta con semillas que contiene un saco embrionario y se desarrolla para convertirse en una semilla después de la fecundación (534)

ovum / óvulo una célula sexual madura (149)

 P

P generation / generación P generación parental; los primeros dos individuos que se aparean en una cruza genética (164)

paleontologist / paleontólogo un científico que estudia los fósiles (285)

Pangaea / Pangea una sola masa de tierra que existió durante aproximadamente 40 millones de años y luego comenzó a separarse para formar los continentes, tal como los conocemos en la actualidad (722)

parapodium / parapodio en los poliquetos, uno de los dos apéndices que se usan para locomoción o para el intercambio de gases (652)

parasitism / parasitismo una relación entre dos especies en la que una, el parásito, se beneficia de la otra, el huésped, y normalmente lo daña (362)

passive transport / transporte pasivo el movimiento de substancias a través de una membrana celular sin que la célula tenga que usar energía (74)

pathogen / patógeno un virus, microorganismo u otra substancia que causa enfermedades; un agente infeccioso (454)

pedigree / pedigrí un diagrama que muestra la incidencia de un carácter genético en varias generaciones de una familia (175)

pedipalp / pedipalpo uno de los dos pares de apéndices que se encuentran junto a la boca de un arácnido y que sirven para masticar y manipular a las presas (670)

pelvic inflammatory disease / enfermedad pélvica inflamatoria una infección pélvica del aparato reproductor femenino superior, incluyendo al útero, ovarios, trompas de Falopio y otras estructuras; es una enfermedad que se transmite sexualmente (1009)

penis / pene el órgano masculino que transfiere espermatozoides a una hembra y que lleva la orina hacia el exterior del cuerpo (998)

pepsin / pepsina una enzima que se encuentra en los jugos gástricos y que sirve para descomponer proteínas en moléculas más pequeñas (908)

perennial / perenne una planta cuyas partes vegetativas subterráneas viven más de dos años y cuyas partes superiores mueren y vuelven a crecer estacional o anualmente (573)

periosteum / periostio el tejido fibroso que cubre los huesos (851)

peripheral nervous system / sistema nervioso periférico todas las partes del sistema nervioso, excepto el encéfalo y la médula espinal (el sistema nervioso central); incluye los nervios craneales y los nervios del cuello, pecho, espalda baja y pelvis (950)

petal / pétalo una de las partes de una flor que normalmente tienen colores brillantes y forma de hoja, las cuales forman uno de los anillos de una flor (538)

petiole / pecíolo el pedúnculo que une una hoja al tallo de una planta (558)

pH / pH un valor que expresa la acidez o la alcalinidad (basicidad) de un sistema; cada número entero de la escala indica un cambio de 10 veces en la acidez; un pH de 7 es neutro, un pH de menos de 7 es ácido y un pH de más de 7 es básico (16)

pharynx / faringe en los gusanos planos, el tubo muscular que va de la boca a la cavidad gastrovascular; en los animales que tienen tracto digestivo, el conducto que va de la boca a la laringe y al esófago (886)

phenotype / fenotipo la apariencia de un organismo u otra característica perceptible que resulta debido al genotipo del organismo y a su ambiente (166)

phloem / floema el tejido que transporta alimento (azúcares, aminoácidos y nutrientes minerales) en las plantas vasculares (507)

phospholipid / fosfolípido un lípido que contiene fósforo y que es un componente estructural de la membrana celular (60)

photoperiodism / fotoperiodicidad la respuesta de las plantas a los cambios de estación durante la duración relativa de las noches y de los días (583)

photosynthesis / fotosíntesis el proceso por medio del cual las plantas, algas y algunas bacterias utilizan la luz solar, dióxido de carbono y agua para producir carbohidratos y oxígeno (94)

phylogenetic tree / árbol filogenético un diagrama ramificado que muestra cómo se relacionan los organismos a través de la evolución (602)

phylogeny / filogenia la historia evolutiva de una especie o grupo taxonómico (307)

phylum / phylum el grupo taxonómico que se ubica debajo del reino y arriba de la clase (302)

pigment / pigmento una substancia que le da color a otra substancia o mezcla (98)

pilus / pilus un apéndice corto y grueso que le permite a una bacteria unirse a otra (442)

pioneer species / especie pionera una especie que coloniza un área deshabitada y empieza un ciclo ecológico en el cual se establecen muchas otras especies (343)

pistil / pistilo la parte reproductora femenina de una flor, la cual produce semillas y está formada por el ovario, estilo y estigma (538)

pith / médula el tejido que se ubica en el centro del tallo de la mayoría de las plantas vasculares y que se utiliza para almacenamiento (556)

pituitary gland / glándula pituitaria una glándula endocrina que se ubica en la base del encéfalo, almacena y libera hormonas producidas por el hipotálamo y secreta hormonas bajo el control del hipotálamo (983)

placenta / placenta la estructura que une al feto en desarrollo con el útero y que permite el intercambio de nutrientes, desechos y gases entre la madre y el feto (1005)

plankton / plancton la masa de organismos casi microscópicos que flotan o se encuentran a la deriva en aguas (dulces y marinas) de ambientes acuáticos (378)

plant propagation / propagación vegetal la práctica de reproducir plantas a partir de semillas o partes vegetativas (544)

planula / plánula la larva ciliada de un cnidario, la cual nada libremente (625)

plasma / plasma en biología, el componente líquido de la sangre (876)

plasma cell / célula plasmática un tipo de glóbulo blanco que produce anticuerpos (929)

plasmid / plásmido una molécula de ADN circular que se encuentra comúnmente en las bacterias y que puede duplicarse independientemente del cromosoma principal (229)

plasmodium / plasmodio el citoplasma plurinucleado de un moho de fango, el cual está rodeado por una membrana y se mueve como si fuera una masa (470)

plastron / plastrón la porción inferior, o ventral, del caparazón de una tortuga (782)

platelet / plaqueta el fragmento de una célula que se necesita para formar coágulos sanguíneos (877)

point mutation / mutación puntual una mutación en la que sólo cambia un nucleótido o una base nitrogenada en un gene (219)

pollen grain / grano de polen la estructura que contiene el gametofito masculino en las plantas con semilla (534)

pollen tube / tubo de polen una estructura tubular que crece a partir de un grano de polen, entra al saco embrionario y permite que las células reproductoras masculinas se muevan al óvulo (534)

pollination / polinización la transferencia de polen de las estructuras reproductoras masculinas (las anteras) de una flor a la punta de la estructura reproductora femenina (el pistilo) en las angiospermas o al óvulo en las gimnospermas (534)

polygenic trait / carácter poligénico una característica de un organismo que está determinada por muchos genes (177)

polyp / pólipo una forma de un cnidario que tiene un cuerpo hueco y cilíndrico y que normalmente está unido a una roca o a otro objeto (622)

population / población un grupo de organismos de la misma especie que viven en un área geográfica específica y se reproducen entre sí (278)

population density / densidad de población el número de individuos de la misma especie que viven en una unidad superficial determinada (321)

predation / depredación la interacción entre dos especies en la que una especie, el depredador, se alimenta de la otra especie, la presa (362)

prediction / predicción una afirmación que se hace por anticipado, la cual expresa los resultados que se obtendrán al poner a prueba una hipótesis si ésta es corroborada; el resultado esperado si la hipótesis es correcta (16)

preen gland / glándula uropígea en las aves, una glándula especial que secreta grasa que el ave esparce en sus plumas para limpiarlas e impermeabilizarlas (785)

pregnancy / embarazo el período de tiempo que transcurre entre la concepción y el nacimiento (1005)

primary growth / crecimiento primario el crecimiento que ocurre como resultado de la división celular en las puntas de los tallos y raíces y que da lugar al tejido primario (574)

primary productivity / productividad primaria la cantidad total de material orgánico que producen los organismos autótrofos de un ecosistema (345)

primary succession / sucesión primaria sucesión que comienza en un área donde previamente no podía existir la vida (343)

primate / primate un miembro del orden de los primates, el grupo de mamíferos entre los que se encuentran los seres humanos, simios y monos; normalmente se distinguen por tener cerebros muy desarrollados, ojos que miran hacia delante, uso de las manos y locomoción variada (731)

prion / prión una partícula infecciosa formada únicamente por una proteína y que no contiene ni ADN ni ARN (441)

probability / probabilidad término que describe qué tan probable es que ocurra un posible evento futuro en un caso dado del evento; la proporción matemática del número de veces que es posible que ocurra un resultado de cualquier evento respecto al número de resultados posibles del evento (173)

probe / sonda una hebra de ARN o una sola hebra de ADN que se ha marcado con un elemento radiactivo o con un color fluorescente, y que se usa en ingeniería genética para enlazarse con un gene específico e identificarlo (231)

producer / productor un organismo que elabora moléculas orgánicas a partir de moléculas inorgánicas; un autótrofo fotosintético o quimiosintético que funciona como la fuente fundamental de alimento en un ecosistema (345)

profundal zone / zona profunda la zona de un hábitat de agua dulce en la que entra poca luz solar (376)

proglottid / proglótido una de las muchas secciones corporales de una tenia; contiene los órganos reproductores (631)

prokaryote / procariote un organismo que está formado por una sola célula y que no tiene núcleo ni organelos celulares; un ejemplo es una bacteria (57)

prosimian / prosimio un miembro de un suborden de los primates que es principalmente arbóreo y nocturno, tal como el lémur, lorí o tarsius (731)

prostate gland / glándula próstata una glándula que contribuye al fluido seminal en los machos (998)

protein / proteína un compuesto orgánico que está hecho de una o más cadenas de aminoácidos y que es el principal componente de todas las células (36)

protist / protista un organismo que pertenece al reino Protista (261)

protostome / protóstomo un organismo cuyo blastoporo embriónico se desarrolla para convertirse en la boca, cuyo celoma surge por esquizocelia y cuyo embrión tiene segmentación determinada (692)

protozoan / protozoario un protista unicelular que puede ser acuático o parasítico, tiene organelos cubiertos por una membrana y se puede mover independientemente; entre los ejemplos se encuentran las amebas y los paramecios (461)

provirus / provirus ADN viral que se ha unido al cromosoma de una célula huésped y se ha duplicado con el ADN del cromosoma (436)

pseudocoelomate / pseudo-celomado un animal que tiene un pseudoceloma, o cavidad falsa del cuerpo (600)

pseudopodium / pesudópodo una extensión citoplásmica retráctil y temporal que tiene una función en la ingestión de alimentos y en el movimiento de algunas células ameboides (464)

psychoactive drug / droga psi-coactiva una substancia que tiene un efecto considerable en la mente o en el comportamiento (961)

pulmonary vein / vena pul-monar la vena que lleva sangre oxigenada de los pulmones al corazón (759)

pulse / pulso la presión rítmica de la sangre contra las paredes de un vaso sanguíneo, particularmente de una arteria (883)

punctuated equilibrium / equi-librio puntuado un modelo de evolución en el que períodos cortos en los que ocurren cambios drásticos en una especie (incluyendo extinciones masivas y especiación rápida) están separados por períodos largos en los que ocurren muy pocos cambios o en los que no ocurre ningún cambio (282)

Punnett square / cuadro de Punnett una gráfica que se usa para predecir los resultados de una cruza genética (170)

pupa / pupa la etapa inmóvil y que no se alimenta, entre la larva y el adulto de insectos que experimentan una metamorfosis completa; en la etapa de pupa, el organismo normalmente está encerrado en un capullo o crisálida y sufre importantes cambios anatómicos (675)

R

radial symmetry / simetría radial un plan corporal en el que las partes del cuerpo del animal están organizadas en un círculo alrededor de un eje central (598)

radiometric dating / datación radiométrica un método para determinar la edad de un objeto estimando los porcentajes relativos de un isótopo radiactivo (precursor) y un isótopo estable (hijo) (252)

radula / rádula un órgano áspero similar a una lengua, que está cubierto de dientes quitinosos y que muchos moluscos utilizan para alimentarse (643)

realized niche / nicho realizado la gama de recursos que una especie usa, las condiciones que la especie tolera y los papeles funcionales que la especie juega como resultado de la competencia en su nicho fundamental (367)

reasoning / razonamiento el acto de sacar una conclusión a partir de hechos o suposiciones (828)

receptor protein / proteína receptora una proteína que liga moléculas señal específicas, lo cual hace que la célula responda (84)

recessive / recesivo término que describe un carácter o un alelo que se expresa sólo cuando se heredan dos alelos recesivos de la misma característica (167)

recombinant DNA / ADN recombinante moléculas de ADN que son creadas artificialmente al combinar ADN de diferentes fuentes (228)

red blood cell / glóbulo rojo una célula que tiene forma de disco y que no tiene núcleo, contiene hemoglobina y transporta oxígeno en el aparato circulatorio (876)

reflex / reflejo un movimiento involuntario y prácticamente inmediato en respuesta a un estímulo (952)

replication fork / horquilla de replicación un punto que tiene forma de Y, el cual se produce cuando las dos hebras de una doble hélice de ADN se separan de modo que la molécula de ADN pueda duplicarse (199)

repressor / represor una proteína reguladora que se une a un operador y bloquea la transcripción de los genes de un operón (216)

reproduction / reproducción el proceso de producir descendencia (7)

reproductive isolation / aislamiento reproductivo la incapacidad de los miembros de una población de reproducirse exitosamente con miembros de otra población de la misma especie o de una especie relacionada (281)

resolution / resolución en los microscopios, la capacidad de formar imágenes con detalles precisos (51)

respiration / respiración en biología, el intercambio de oxígeno y dióxido de carbono entre células vivas y su ambiente; incluye la respiración y la respiración celular (605)

resting potential / potencial de reposo el potencial eléctrico que existe en la membrana celular de una célula nerviosa o de una célula muscular cuando la célula no está activa (946)

restriction enzyme / enzima de restricción una enzima que destruye moléculas de ADN extraño cortándolas en sitios específicos (229)

retina / retina la capa interna del ojo, sensible a la luz, que recibe imágenes formadas por el lente ocular y las transmite al cerebro por medio del nervio óptico (958)

Rh factor / factor Rh uno de varios antígenos de los grupos sanguíneos que son transportados en la superficie de los glóbulos rojos (879)

rhizoid / rizoide una estructura parecida a una raíz que se encuentra en las plantas no vasculares, tales como los musgos o las hepáticas, la cual mantiene a las plantas en su lugar y tienen una función en la absorción (486)

rhizome / rizoma un tallo horizontal subterráneo que proporciona un mecanismo de reproducción asexual (510)

ribosome / ribosoma un organelo celular compuesto de ARN y proteína; el sitio donde ocurre la síntesis de proteínas (56)

RNA / ARN ácido ribonucleico; un polímero natural que se encuentra en todas las células vivas y que juega un papel en la síntesis de proteínas (37)

RNA polymerase / ARN polimerasa una enzima que comienza (cataliza) la formación de ARN usando una hebra de una molécula de ADN como plantilla (209)

rod / bastoncillo uno de los dos tipos de células detectoras de luz que hay en el ojo; los bastoncillos pueden detectar luz tenue y juegan un papel importante en la visión nocturna y sin color (958)

root / raíz el órgano principalmente subterráneo de las plantas vasculares, el cual mantiene a las plantas en su lugar y absorbe y almacena agua y minerales del suelo (507)

root cap / cofia la capa protectora de células que cubre la punta de una raíz (555)

root hair / pelo radicular una extensión de la epidermis de una raíz, la cual aumenta el área superficial de la raíz para la absorción (555)

rRNA / ARNr ARN ribosomal; un organelo que contiene la mayor parte del ARN en la célula y que es responsable del funcionamiento de los ribosomas (212)

r-strategist / estratega r una especie que está adaptada para vivir en un ambiente donde los cambios son rápidos e impredecibles; se caracteriza por presentar crecimiento rápido, alta fertilidad, período de vida corto, cuerpo pequeño y crecimiento exponencial de la población (324)

S

sapwood / albura el tejido del xilema secundario que se distribuye en el exterior del tronco de un árbol y que tiene una función en el transporte de la savia (557)

sarcomere / sarcómero la unidad fundamental de contracción del músculo esquelético y cardíaco (857)

scanning electron microscope / microscopio electrónico de barrido un microscopio que produce una imagen agrandada tridimensional de un objeto usando un haz de electrones en lugar de luz (54)

sebum / sebo la secreción grasosa de las glándulas sebáceas (864)

second messenger / mensajero secundario una molécula que se genera cuando una substancia específica se une a un receptor en el exterior de la membrana celular, lo cual produce un cambio en la función celular (85)

secondary growth / crecimiento secundario crecimiento de las plantas que ocurre como resultado de la división celular en los cámbiums, o meristemos laterales, y que hace que se engruesen los tallos y las raíces (574)

secondary succession / sucesión secundaria el proceso por medio del cual una comunidad reemplaza a otra, la cual ha sido parcial o totalmente destruida (343)

seed / semilla el embrión de una planta que está encerrado en una cubierta protectora (504)

seed coat / cubierta seminal la cubierta exterior y protectora de una semilla (535)

seed plant / planta con semillas una planta que produce semillas (504)

semen / semen el fluido que contiene espermatozoides y varias secreciones producidas por los órganos reproductores masculinos (998)

semicircular canal / canal semicircular el canal lleno de fluido ubicado en el oído interno, el cual ayuda a mantener el equilibrio y a coordinar los movimientos (959)

seminal vesicle / vesícula seminal una de las dos estructuras glandulares en los vertebrados macho, las cuales acumulan y secretan fluido seminal (998)

seminiferous tubule / túbulo seminífero uno de los muchos túbulos que hay en los testículos, en donde se producen los espermatozoides (996)

sensory neuron / neurona sensorial una neurona que lleva estímulos de un órgano sensorial al sistema nervioso central (950)

sensory receptor / receptor sensorial una estructura especializada que contiene los extremos de las neuronas sensoriales y que responde a tipos específicos de estímulos (956)

sepal / sépalo en una flor, uno de los anillos más externos de hojas modificadas que protegen el capullo de la flor (538)

septum / septo una pared divisoria, o partición, tal como la pared que se encuentra entre células adyacentes en las hifas de los hongos, la pared interna que se encuentra entre segmentos adyacentes de un anélido y la pared gruesa que se encuentra entre las cámaras derecha e izquierda del corazón (760)

sessile / sésil término que describe a un organismo que permanece unido a una superficie durante toda su vida y no se mueve (618)

seta / seda una de las cerdas o espinas externas que se proyectan del cuerpo de un anélido (652)

sex chromosome / cromosoma sexual uno de los dos cromosomas que determinan el sexo de un individuo (122)

sex-linked trait / carácter ligado al sexo un carácter que es determinado por un gene que se encuentra en uno de los cromosomas sexuales, tal como el cromosoma X o el cromosoma Y en los seres humanos (175)

sexual reproduction / reproducción sexual reproducción en la que se unen los gametos de los dos padres (150)

sexual selection / selección sexual un mecanismo evolutivo por medio del cual los caracteres que aumentan la capacidad de los individuos de atraer o adquirir una pareja aparecen con más frecuencia en una población; selección en la que se elige una pareja con base en un carácter o caracteres particulares (6)

shoot / brote la porción de una planta que crece principalmente sobre el suelo; incluye los tallos y las hojas (507)

sieve tube / tubo criboso en el floema de una planta que da flores, un tubo de transporte, que está hecho de una serie de miembros de tubos cribosos apilados de un extremo al otro (554)

sink / sumidero cualquier lugar donde una planta almacena o usa nutrientes orgánicos, tales como azúcares o almidones (564)

sinoatrial node / nodo sinoauricular una masa de células de músculo cardíaco que se encuentra en la unión entre la vena cava superior y el atrio derecho, y que inicia y regula la contracción del corazón (abreviatura: nodo SA) (882)

siphon / sifón un conducto hueco que los bivalvos utilizan para sorber y expulsar el agua de mar (647)

skin gill / branquia dérmicas una estructura transparente que protege la superficie de las estrellas de mar y les permite respirar (695)

sodium-potassium pump / bomba de sodio-potasio una proteína transportadora que utiliza el ATP para efectuar el transporte activo de iones de sodio hacia el exterior de la célula y de iones de potasio hacia el interior de la célula (81)

solution / solución una mezcla homogénea de dos o más sustancias dispersas de manera uniforme en una sola fase (32)

sorus / soro un grupo de esporas o esporangios (532)

source / fuente la parte de una planta que elabora azúcares y otros compuestos orgánicos y a partir de la cual estos compuestos se transportan a otras partes de la planta (564)

speciation / especiación la formación de especies nuevas como resultado de la evolución por selección natural (291)

species / especie un grupo de organismos que tienen un parentesco cercano y que pueden aparearse de modo natural para producir descendencia fértil; tambien, el nivel de clasificación debajo de género y arriba de subespecie (9)

sperm / espermatozoide el gameto masculino (célula sexual) (148)

spermatogenesis / espermatogénesis el proceso por medio del cual se forman los gametos masculinos (148)

spicule / espícula una aguja de sílice o carbonato de calcio que se encuentra en el esqueleto de algunas esponjas (620)

spinal cord / médula espinal una columna de tejido nervioso que se origina en la base del cerebro y corre a lo largo de la columna vertebral (952)

spindle / huso mitótico una red de microtúbulos que se forma durante la mitosis y que mueve cromátidas a los polos (128)

spinneret / hilera un órgano que utilizan las arañas y algunas larvas de insectos para producir hilos sedosos con los que hacen redes y capullos (670)

spiracle / espiráculo una abertura externa de un insecto o artrópodo, que se usa en la respiración (668)

spirillum / espirilo una bacteria que tiene forma en espiral (443)

spongin / espongina una proteína fibrosa que contiene azufre y que forma las fibras del esqueleto de algunas esponjas (620)

sporangium / esporangio una bolsa, cubierta, cápsula u otra estructura especializada que produce esporas (463)

spore / espora una célula reproductora o estructura multicelular que resiste las condiciones ambientales y que se puede desarrollar para convertirse en un adulto sin necesidad de fusionarse con otra célula (154)

sporophyte / esporofito en las plantas y algas que tienen generaciones alternas, el individuo o generación diploide que produce esporas haploides (154)

sporozoite / esporozoito un esporozoario que ha sido liberado de un ooquiste y está listo para entrar a una nueva célula huésped (473)

stabilizing selection / selección de estabilización un tipo de selección natural en la que se favorece la forma promedio de un carácter, el cual se vuelve más común (332)

stamen / estambre la estructura reproductora masculina de una flor, que produce polen y está formada por una antera ubicada en la punta del filamento (538)

steroid / esteroide un tipo de lípido que está formado por cuatro anillos de carbono a los cuales se encuentran unidos diversos grupos funcionales, y que normalmente tiene acción fisiológica (977)

stimulant / estimulante una droga que aumenta la actividad del cuerpo o la actividad de alguna parte del cuerpo (962)

stolon / estolón en las plantas, un tallo que crece y que al desarrollarse da lugar a raíces y brotes en los nodos o en la punta para formar individuos nuevos; la hifa de algunos hongos, la cual crece y da lugar a individuos nuevos (486)

stoma / estoma una de las muchas aberturas de una hoja o de un tallo de una planta, la cual permite que se lleve a cabo el intercambio de gases (502)

stroke / ataque de apoplejía una pérdida súbita de la conciencia o parálisis que ocurre cuando se interrumpe el flujo sanguíneo al cerebro (884)

subcutaneous tissue / tejido subcutáneo la capa de células que se encuentra debajo de la piel (863)

substrate / sustrato una parte, sustancia o elemento que se encuentra debajo de otra parte, sustancia o elemento y lo sostiene; el reactivo en reacciones que son catalizadas por enzimas (41)

succession / sucesión el reemplazo de un tipo de comunidad por otro en un mismo lugar a lo largo de un período de tiempo (343)

swim bladder / vejiga natatoria en los peces óseos, una bolsa llena de gas que se usa para controlar la flotabilidad (756)

symbiosis / simbiosis una relación en la que dos organismos diferentes viven estrechamente asociados uno con el otro (364)

synapse / sinapsis el punto en el cual el extremo del axón de una neurona se une con el extremo de una dendrita o con el cuerpo de la célula de otra neurona, o bien, se encuentra con otra célula (948)

syphilis / sífilis una enfermedad transmitida sexualmente, producida por la bacteria Treponema pallidum (1009)

target cell / célula blanco una célula específica a la que se dirige una hormona para producir un efecto específico (977)

taxonomy / taxonomía la ciencia de describir, nombrar y clasificar organismos (300)

teleost / teleósteo un grupo de peces con aletas rayadas que tienen una aleta caudal, escamas y una vejiga natatoria; el grupo más grande de peces óseos (757)

tendon / tendón un tejido conectivo duro que une un músculo con un hueso o con otra parte del cuerpo (856)

terrestrial / terrestre término que describe a un organismo que vive en la tierra (721)

test cross / cruza de prueba el cruzamiento de un individuo cuyo genotipo se desconoce con un individuo homocigoto recesivo para determinar el genotipo desconocido (172)

testes / testículos los principales órganos reproductores masculinos, los cuales producen espermatozoides y testosterona (996)

thalamus / tálamo la parte del cerebro que dirige a la región apropiada las señales sensoriales y motoras que se reciben (952)

thecodont / tecodonte el reptil extinto a partir del cual evolucionaron los dinosaurios (722)

theory / teoría una explicación sobre algún fenómeno que está basada en la observación, experimentación y razonamiento (19)

therapsid / terápsido el orden extinto de reptiles parecidos a mamíferos que posiblemente dio origen a los mamíferos (728)

thorax / tórax en los vertebrados superiores, la parte del cuerpo que se encuentra entre el cuello y el abdomen; en otros animales, la región del cuerpo que se encuentra detrás de la cabeza; en los artrópodos, la región media del cuerpo (666)

thylakoid / tilacoide un sistema de membranas que se encuentra dentro de los cloroplastos y que contiene los componentes para que se lleve a cabo la fotosíntesis (99)

tissue / tejido un grupo de células similares que llevan a cabo una función común (419)

tissue culture / cultivo de tejidos la técnica de cultivar células vivas en un medio artificial (544)

tolerance / tolerancia el estado de adicción a una droga en el que se necesitan mayores cantidades de la droga para obtener el efecto deseado (962)

toxin / toxina una sustancia que un organismo produce y que es venenosa para otros organismos (449)

trachea / tráquea en los insectos, miriápodos y arañas, uno de una red de conductos de aire; en los vertebrados, el conducto que une la faringe a los pulmones (668)

transcription / transcripción el proceso de formar un ácido nucleico usando otra molécula como plantilla; en particular, el proceso de sintetizar ARN usando una de las hebras de la molécula de ADN como plantilla (208)

transfer RNA / ARN de transferencia una molécula de ARN que transfiere aminoácidos al extremo en crecimiento de una cadena de polipéptidos durante la traducción (212)

transformation / transformación la transferencia de material genético en forma de fragmentos de ADN de una célula a otra o de un organismo a otro (191)

transgenic animal / animal transgénico un animal al que se le ha transferido material genético clonado (241)

translation / traducción la porción de la síntesis de proteínas que tiene lugar en los ribosomas y que usa los codones de las moléculas de ARNm para especificar la secuencia de aminoácidos en las cadenas de polipéptidos (208)

translocation / translocación el movimiento de un segmento de ADN de un cromosoma a otro, lo cual resulta en un cambio en la posición del segmento; también, el movimiento de nutrientes solubles de una parte a otra de una planta (564)

transpiration / transpiración el proceso por medio del cual las plantas liberan vapor de agua al aire por medio de los estomas; también, la liberación de vapor de agua al aire por otros organismos (351)

trochophore / trocófora una larva ciliada de muchos gusanos y algunos moluscos, la cual nada libremente (642)

trophic level / nivel trófico uno de los pasos de la cadena alimenticia o de la pirámide alimenticia; entre los ejemplos se encuentran los productores y los consumidores primarios, secundarios y terciarios (345)

tropism / tropismo el movimiento de un organismo o de una parte de él en respuesta a un estímulo externo, como por ejemplo, la luz o el calor; el movimiento puede ser hacia el estímulo o en sentido opuesto a él (582)

true-breeding / variedad pura término que describe organismos o genotipos que son homocigotos para un carácter específico y, por lo tanto, producen descendencia que tiene el mismo fenotipo para ese carácter (164)

ungulate / ungulado un animal que tiene pezuñas (814)

uracil / uracilo una de las cuatro bases que se combinan con un azúcar y un fosfato para formar una subunidad de nucleótido de ADN; el uracilo se une a la adenina (208)

urea / urea el principal producto nitrogenado que se obtiene del metabolismo de las proteínas, se forma en el hígado a partir de aminoácidos y compuestos de amoníaco y se encuentra en la orina y otros fluidos del cuerpo (912)

ureter / uréter uno de los dos tubos angostos que llevan orina de los riñones a la vejiga urinaria (914)

urethra / uretra el tubo que lleva orina de la vejiga urinaria al exterior del cuerpo (914)

urinary bladder / vejiga urinaria un órgano hueco y muscular que almacena orina (914)

urine / orina el líquido que excretan los riñones, se almacena en la vejiga y pasa a través de la uretra hacia el exterior del cuerpo (914)

uterus / útero en los mamíferos hembras, el órgano hueco y muscular en el que se incrusta el óvulo fecundado y en el que se desarrollan el embrión y el feto (1000)

vaccination / vacunación la administración a seres humanos o animales de organismos que han sido tratados para inducir una respuesta inmunológica (931)

vaccine / vacuna una sustancia que se prepara a partir de organismos patógenos muertos o debilitados y se introduce al cuerpo para producir inmunidad (190)

vagina / vagina el canal de las hembras que se extiende de la vulva al cuello del útero y que recibe al pene durante el coito (1000)

valve / válvula un pliegue de membranas que controla el flujo de un fluido (874)

vas deferens / conducto deferente un conducto a través del cual los espermatozoides se mueven del epidídimo al conducto eyaculatorio que está en la base del pene (997)

vascular bundle / haz vascular en una planta, una hebra de tejido de transporte que contiene tanto xilema como floema (556)

vascular cambium / cámbium vascular en una planta, el meristemo lateral que produce xilema y floema secundarios (575)

vascular plant / planta vascular una planta que tiene un sistema vascular formado por xilema y floema, tejidos especializados que transportan materiales de una parte de la planta a otra (504)

vascular system / sistema vascular un sistema de transporte de los tejidos que lleva agua y otros materiales en las plantas o en los animales (504)

vascular tissue / tejido vascular el tejido especializado de transporte que se encuentra en las plantas superiores y que está formado principalmente por xilema y floema (422)

vector / vector en biología, cualquier agente, como por ejemplo un plásmido o un virus, que tiene la capacidad de incorporar ADN extraño y de transferir ese ADN de un organismo a otro; un huésped intermediario que transfiere un organismo patógeno o un parásito a otro organismo (229)

vegetative part / parte vegetativa cualquier parte no reproductiva de una planta (516)

vegetative reproduction / reproducción vegetativa un tipo de reproducción asexual en el que crecen plantas nuevas a partir de partes plantas que no se reproducen (541)

vein / vena en biología, un vaso que lleva sangre al corazón (873)

vena cava / vena cava una de las dos venas grandes que llevan sangre de los tejidos del cuerpo al corazón (881)

ventricle / ventrículo una de las dos cámaras musculares grandes que bombean sangre hacia el exterior del corazón (881)

vertebra / vértebra uno de los 33 huesos de la columna vertebral (espina dorsal) (712)

vertebrate / vertebrado un animal que tiene columna vertebral; incluye a los mamíferos, aves, reptiles, anfibios y peces (267)

vesicle / vesícula una cavidad o bolsa pequeña que contiene materiales en una célula eucariótica; se forma cuando parte de la membrana celular rodea los materiales que van a ser llevados al interior la célula o transportados dentro de ella (63)

vessel / vaso en las plantas, una estructura tubular que se encuentra en el xilema y que está formada por células conectadas que transportan agua y nutrientes minerales; en los animales, un tubo o conducto que lleva sangre y otros fluidos del cuerpo (554)

vestigial structure / estructura vestigial una estructura de un organismo, cuyo tamaño y función están reducidos, pero que es posible que haya estado completa y que haya sido funcional en los ancestros del organismo (286)

villus / vellosidad intestina una de las muchas proyecciones diminutas de las células que se encuentran en la pared interior del intestino delgado; aumenta el área superficial de la pared para absorción (909)

viroid / viroide un agente infeccioso que está constituido por una hebra pequeña de ARN y que produce enfermedades en las plantas (441)

virulent / virulento término que describe a un microorganismo o virus que causa enfermedades y que es altamente infeccioso (190)

virus / virus una partícula infecciosa sin vida formada por un ácido nucleico y una cubierta de proteína; puede invadir una célula y destruirla (434)

visceral mass / masa visceral la sección central del cuerpo de un molusco, la cual contiene sus órganos (643)

vitamin / vitamina un compuesto orgánico que participa en las reacciones bioquímicas y que forma varias moléculas en el cuerpo; algunas vitaminas se llaman coenzimas y activan enzimas específicas (904)

water vascular system / sistema vascular acuoso un sistema de canales que están llenos de un fluido acuoso que circula por todo el cuerpo de los equinodermos (695)

weaning / destete el momento en el que termina la dependencia que tiene un animal en su madre para obtener alimento (leche) y protección (806)

white blood cell / glóbulo blanco un tipo de célula de la sangre que destruye bacterias, virus y proteínas tóxicas, y que ayuda al cuerpo a desarrollar inmunidad (877)

withdrawal / abstinencia la serie de síntomas asociados con la remoción del cuerpo de una droga adictiva (962)

yeast / levadura un hongo unicelular muy pequeño que fermenta a los carbohidratos y los convierte en alcohol y dióxido de carbono; se usa para fermentar cerveza, para hacer pan y como fuente de vitaminas y proteínas (487)

zygosporangium / en los miembros del phylum Zygomycota, una estructura sexual que se forma debido a la fusión de dos gametangios y que contiene uno o más cigotos que se formaron a partir de la fusión de gametos producidos por los gametangios (486)

zygospore / zigospora en algunas algas, una estructura protectora que tiene una pared gruesa y que contiene un cigoto que se formó a partir de la fusión de dos gametos (462)

zygote / cigoto la célula que resulta debido a la fusión de los gametos; el óvulo fecundado (121)

Index

Note: Boldface page numbers refer to primary discussions. Page numbers followed by f refer to figures. Page numbers followed by t refer to tables.

Index

Index

cells *(continued)*
vacuoles of, 66, 66*f*, 77, 468, 469*f*, 543*f*
white blood cells, 877, 877*f*, 922*f*, 926–929, 926*f*, 928*f*
cell theory, 55
cellular respiration, 95, 95*f*, **104–110,** 104*f*. *See also* **metabolism**
aerobic ATP production, 110, 110*f*
aerobic electron transport chain, 107, 107*f*
alcoholic fermentation, 108–109, 108*f*
carbon cycle, 352, 352*f*
glycolysis, 104–105, 104*f*, 105*f*, 108, 108*f*
Krebs cycle, 106–107, 106*f*, 110*f*
lactic acid fermentation, 108, 108*f*
cellulose, 34, 346, 522, 901
cell walls
absence in animals, 596
in archaebacteria, 415
in eubacteria, 258, 414, 443
in fungi, 421, 482
kingdom classification and, 412
in plant cells, 66, 66*f*, 132, 132*f*
in prokaryotes, 57, 57*f*
centipedes, 362, 679, 679*f*
central nervous system (CNS), **950–953,** 950*f*, 951*f*, 952*f*. *See also* **human nervous system**
central vacuole, 66, 66*f*, 77
centrioles, 128–129, 128*f*, 130*f*
centromeres, 119, 119*f*, 121*f*
meiosis, 144–145, 144*f*, 145*f*
mitosis, 129, 130
cephalization, 599
cephalopods, 649–650, 649*f*
cephalothorax, 666, 670–671, 671*t*
cereals, 518–519, 518*f*, 519*f*, 576–577
cerebellum, 755*f*, 805*f*, 951, 951*f*
cerebral cortex, 957, 957*f*
cerebral ganglion, 651
cerebrum, 755*f*, 805*f*, 951, 951*f*
cervical cancer, 440, 1010
CFCs (chlorofluorocarbons), 387, 396
cf **gene,** 182
characters, derived and ancestral, 307–309

Chargaff, Erwin, 196
Chase, Martha, 192–193, 192*f*
cheetahs, 328–329, 328*f*, 831*f*
chelicerae, 670, 670*f*, 671*f*
chemical bonding, 29–30
chemical reactions, 38–39, 39*f*, 95, 95*f*, 97
chemoautotrophs, 444, 450
chemoreceptors, 956*t*, 960, 966
chimpanzees, 121, 806, 806*f*, 813, 833–834. *See also* **primates**
chitin, 421, 482, 678
chlamydia, 1008*t*, 1009
Chlamydomonas, 152, 152*f*, 462, 462*f*
chlorinated hydrocarbons, 391, 391*f*
chlorofluorocarbons (CFCs), 387, 396
chlorophyll, 35, 98–99, 98*f*
chloroplasts, 66, 66*f*
in euglenas, 467, 467*f*
evolution of, 259–260
in leaves, 559, 559*f*
photosynthesis in, 97, 99, 99*f*
thylakoid membranes, 99–101, 99*f*, 100*f*
choanocytes, 619, 619*f*
cholesterol, 35, 181*t*, 884, 884*f*
chordates, 693*f*, **700–702,** 700*f*, 701*f*, 706–707. *See also* **vertebrates**
chorion, 775, 775*f*, 1005, 1005*f*
chorionic villi sampling, 123, 123*f*
chromatids, 119, 119*f*, 121*f*
meiosis, 144–145, 144*f*, 145*f*
mitosis, 128–130, 128*f*, 131*f*
chromatography, paper, 568–569
chromosomes, 118–124
during cell division, 118–119, 128–132, 128*f*, 130*f*
in cell nucleus, 62
chromosome number, 120–123, 121*t*, 122*f*
crossing-over, 144, 144*f*, 146–147, 147*f*
homologous, 120
independent assortment, 146, 146*f*
during meiosis, 144–149, 144*f*, 145*f*, 146*f*
mutations, 124, 124*f*, 219–220, 219*f*
sex chromosomes, 122, 175, 181, 737

structure, 37, 37*f*, 119*f*
viruses in, 436–437, 436*f*
chrysalis, 675, 675*f*
Chthamalus stellatus, 368, 368*f*
chytrid fungus, 18
cilia
in ciliates, 420, 461*t*, 468, 468*f*, 469*f*
in human respiratory system, 58, 58*f*
in mollusks, 642, 642*f*, 644–645, 644*f*
ciliates, 420, 461*t*, 468, 468*f*, 469*f*. *See also Paramecium*
circulatory systems, 871–884. *See also* **hearts**
amphibians, 759–760, 759*f*, 760*f*
birds, 786–787, 786*f*, 787*f*
echinoderms, 695, 695*f*
fishes, 746, 748, 748*f*
human, 848*t*, 871–884
insects, 677*f*
mammals, 803
mollusks, 644, 644*f*
open vs. closed, 606, 606*f*, 652
reptiles, 776, 776*f*
vertebrates, 713, 713*f*
cirrhosis, 911, 964
cladistics, 307–310, 308*f*, 310*f*
cladograms, 308–309, 308*f*
clams, 647, 647*f*, 650*f*, 660–661, 661*f*
classes, 302, 302*f*
classical conditioning, 827
classification of organisms, 299–315
algae, 420, 465
animals, 412, 412*f*, 416, 416*f*, 417, 417*f*, 424–425, 602, 602*f*
arachnids, 303–304, 665, 665*t*
bacteria, 57, 57*f*, 413–414, 413*f*
binomial nomenclature, 300
biological species concept, 305–306
birds, 301, 417*t*
cladistics, 307–310, 308*f*, 310*f*
dichotomous keys, 304, 314–315
domains, 413, 413*f*, 417*t*
earthworms, 471*t*, 651*f*
echinoderms, 602*f*, 693, 693*f*
evolutionary systematics, 310, 310*f*
field guides, 304
fishes, 717

flatworms, 629
fungi, 412, 412*f*, 417*t*, 421, 482
green algae, 461*t*
insects, 425, 673, 673*f*
kingdoms, 412, 412*f*
mushrooms, 417*t*, 421, 485*t*,
scientific names, 301, 301*t*, 303
trees, 301, 301*f*, 314–315
vascular plants, 417*t*, 423, 423*f*
cleavage (of eggs), 597, 597*f*, **1004,** 1004*f*
climate, 372–374
biomes, 372–374, 372*f*, 373*f*
global warming, 388–389, 388*f*, 389*f*
temperature and moisture, 371, 371*f*
cloaca, 763*f*, 781*f*, 789*f*
cloning, 241–242
by asexual reproduction, 150
confirmation of, 231, 231*f*
gene cloning, 229, 229*f*
genetically engineered animals, 11, 240–242, 240*f*, 241*f*
genomic imprinting, 242, 242*f*
by parthenogenesis, 153
mammary cell cloning, 240*f*, 241
problems with, 242
screening cells, 230, 230*f*
twins, 153
closed circulatory systems, 606, 606*f*
closed ecosystems, 358
Clostridium **spp.,** 448, 448*f*, 450
clotting, 876–878, 878*f*
club mosses, 511, 511*f*
cnidarians, 425, 602*f*, **622–628,** 638–639
cnidocytes, 623, 623*f*
coacervates, 256–257, 257*f*
cocaine, 961*t*, 962–963, 962*f*, 963*f*
coccus, 443, 443*f*
cochlea, 959, 959*f*
codeine, 521*t*, 961*t*
codominance, 178
codons, 211–213, 211*f*
coelacanths, 757, 757*f*
coelom, 600, 607, 642, 644*f*, 652, 713
coelomates, 600, 600*f*
coenzyme A, 106, 106*f*
coevolution, 362, 362*f*
Cohen, Stanley, 228
cohesion, 31, 31*f*

Index

extinctions, 262–263,
267f, 277
formation of, 284–285
hominids, 733–738, 733f,
735f, 736f, 738f
intermediaries, 283
mammoth, 710f
punctuated equilibrium
and gradualism in,
282, 282f
frameshift mutations,
220
Franklin, Rosalind, 196,
196f
**frequency distribution
curves,** 336
FROGLOG, 769, 769f
frogs. *See also* **amphibians**
disappearance of, 14
glass, 158f
life cycle, 761, 761f
mating activity, 291f,
763f, 833, 835
saltwater, 81
structures, 121t, 762f, 763f
fronds, 511, 511f, 532,
532f, 533, 533f
fruit flies, 121, 149, 326f,
332
fruits, definition, 516, 516f
fundamental niches,
366–367, 366f, 369
fungi (singular, *fungus*),
481–492
absorption of nutrients
by, 483, 483f
acid rain and, 386
ascomycetes, 421, 487,
487f, 490, 491
basidiomycetes, 421,
485t, 488, 488f, 489f,
490
cells walls, 421, 482
chitin in, 482
chromosome number,
121, 121t
as decomposers, 347
diseases caused by, 483,
483f
as eukaryotes, 416–417,
417t
evolution of, 265, 265f
fermentation reactions
in, 108–109, 108f
forest, 342, 342f
hyphae, 421, 421f,
486–489
kingdom classification,
412, 412f, 417t, 421,
482
lichens, 342, 491–492,
491f
mitosis in, 482

molds, 420–421, 470,
470f, 484, 485t
mycelium, 483, 483f, 489f
mycorrhizae, 265, 265f,
490, 490f, 502
nuclear mitosis, 482
reproduction, 484–488,
484f, 485t, 486f, 487f,
488f, 489, 489f
soil, 18
toxins from, 483, 488
zygomycetes, 421, 485t,
486, 486f, 490
fur, 800–801, 801f, 804f

 G

Galápagos Islands, 278,
278f, 290
gallbladders, 908–909, 908f
gametes, 118, 121, 146–147,
146f, 173. *See also* **egg
cells; sperm cells
gametophytes**
alternation of generations,
154, 154f, 506, 506f
angiosperms, 538, 540,
540f
gymnosperms, 512, 537,
537f
nonvascular plants,
508–509, 530–531,
530f, 531f
seedless vascular plants,
510–511, 511f, 532,
532f, 533, 533f
seed plants, 534, 534f
ganglia, 606, 607f
gastric juices, 908
gastrin, 908, 986
gastropods, 646, 646f
**gastrovascular cavities,
604–605,** 604f, 607,
623, 622f, 623f
gastrulation, 597, 597f
**gated sodium ion
channels,** 78–79, 78f
Gause, G. F., 369, 369f
geckos, 778–779, 778f
gel electrophoresis,
231–232, 231f
gemmules, 621
gene alterations, 219, 219f
gene cloning, 229, 229f.
See also **cloning
gene expression.** *See*
**protein synthesis
gene flow,** 328
gene rearrangements,
219, 219f
gene regulation, 215–220,
215f, 217f, 218f, 219f
genes
alleles, 167–168, 326–330
in archaebacteria, 415

in bacteria, 414
cf gene, 182
codominance, 178
definition, 8, 119
dominant, 167, 168, 168f,
170–171, 176–178
dominant and recessive,
167, 167f, 168, 168f,
170–171, 175–178
genetic variation, 146–147,
146f, 147f, 151, 288
heterozygous and
homozygous, 167–168,
167f, 171–172, 171f,
176
human, 233
introns and exons, 218,
218f
jumping, 216
karyotype analysis, 122,
122f, 123, 123f, 577f
law of independent
assortment, 169, 169f
multiple alleles, 178, 178f
mutations, 219–220, 219f
nitrogen-fixing, 354
nomenclature, 168
oncogenes, 127
persistence, 330
protein synthesis,
208–209, 208f
recessive, 167, 167f, 168,
170–171, 170f, 175,
176
transposons, 216,
219–220
tumor supressor, 127
in viruses, 437
gene sequencing, 220
gene therapy, 13, 182
**genetically modified
crops,** 238–239. *See also*
**genetic engineering
genetic code,** 211, 214
genetic disorders, 180–182
albinism, 175–176, 175f
chromosome mutations,
124, 124f
cystic fibrosis, 13, 13f,
180, 181t, 327, 330
Down syndrome,
122–123, 122f
gene therapy for, 182
hemophilia, 181, 181t,
234, 330, 878
Huntington's disease,
181, 181t
hypercholesterolemia,
181t
phenylketonuria (PKU),
182
sickle cell anemia, 8,
180–181, 180f, 181t,
224, 329
Tay-Sachs disease, 181t
treating, 182

genetic diversity
in cheetahs, 328–329
conjugation, 443, 463,
469f
disease resistance and,
329
from sexual reproduc-
tion, 151
genetic drift, 328–329
**genetic engineering,
227–242**
in agriculture, 238–242,
238f, 240f, 241f, 354
in animals, 240–242,
240f, 241f
bacteria in, 228–229,
229f, 230f, 232, 450,
450f
basic steps of, 228–229,
228f, 230f
cloning and screening
cells, 230, 230f
cloning problems, 242
in crops, 238–239, 238f,
239f, 578
cutting and recombining
DNA, 230, 230f
of drugs, 228–229, 229f,
230f, 234, 234f
food supply and, 11
nitrogen-fixing genes, 354
polymerase chain reac-
tion (PCR), 236, 236f
Southern blot confirma-
tion, 231–232, 231f
of vaccines, 235–236, 235f
**genetic recombination,
146–147,** 147f. *See also*
**genetic engineering
genetics, 161–176.** *See also*
heredity
Mendel's breeding experi-
ments, 162–165, 163f,
164f, 165t
Mendel's hypotheses,
166–168, 166f, 167f
Mendel's ratios, 165, 165f
probability, 173–174,
173f, 174f
genetic variation, 146–147,
146f, 147f, 151, 288
genital herpes, 235–236,
235f, 1008t, **1010,** 1010f
genital warts, 1008t
genome, human, 11, 233
genomic imprinting, 242,
242f
genotypes, 168, 170–171
genus, definition, 301
geometric progression,
278, 278f
germ cell cysts, 149
germination, 572–573,
572f, 584

Index

gestation periods,
810–811, 811*f*, 1005
Giardia lamblia, 472*t*
gibberellins, 581
gill filaments, 747, 747*f*
gills
amphibians, 605, 605*f*,
761, 764
echinoderms, 695, 695*f*
fishes, 746–747, 747*f*
mollusks, 644–645, 644*f*,
647–648
opercula, 754*f*, 756, 756*f*
skin, 695, 695*f*
surface area of, 719
gill slits, 747, 747*f*
Gilman, Alfred, 978
ginkgo, 513, 513*f*
gizzards, 654–655, 677*f*
Glacier Bay, succession in,
343–344, 343*f*
glands
endocrine, 975, 975*f*,
982–988
exocrine, 975
glass sponges, 620, 620*f*
global change, 386, 386*f*,
387*f*, **388–389**, 389*f*,
global warming, 388–389,
388*f*, 389*f*
glomerulus, 913, 913*f*,
glucagon, 978, 978*f*, 986,
986*f*
glucose
active transport of, 82
in foods, 901
glycolysis, 104–105, 104*f*,
105*f*, 108, 108*f*, 860
insulin and, 986–987, 986*f*
structure of, 34, 34*f*
glutamate, 948
glycogen, 34, 109, 860, 901
glycolysis
in cellular respiration,
104–105, 104*f*, 105*f*,
108, 108*f*
enzymes, 105
during exercise, 109–110,
110*f*, 860
glycoproteins, 435, 437–439,
437*f*, 439*f*
glyphosate, 238–239
gnetophytes, 513, 513*f*, 554
goiter, 984, 984*f*
Golgi apparatus, 58*f*, 64,
64*f*, 83, 132
gonorrhea, 1008–1009,
1008*t*, 1009*f*
gorillas, 732*f*, 733*f*, 836,
836*f*
Gould, Stephen Jay, 282
gradualism, 282, 282*f*
grains, **518–519**, 518*f*,
519*f*, 576–577, 576*f*, 577*f*

Gram staining, 443,
454–455, 445*f*
grasses, 518–519, 518*f*,
519*f*, 576–577
grasshoppers, **676–677**
external structures, 674*f*,
676, 676*f*, 678
internal structures, 607*f*,
677*f*
life cycle, 675, 675*f*
sex chromosomes in, 122
Graves' disease, 933*t*
gravitropism, 582, 582*f*
green algae
characteristics, 419, 419*f*,
465, 465*f*
classification, 461*t*
reproduction, 462–463,
462*f*, 463*f*
greenhouse effect,
388–389, 388*f*, 389*f*
Griffith, Frederick,
190–191, 191*f*
grizzly bears, 804*f*, 805*f*
ground tissue, 552–553,
552*f*, 553*f*
ground water, 351, 351*f*,
393, 393*f*
growth factors, 234*f*
growth hormones, 234*f*,
240, 983, 983*t*
Guadalupe Mountains
National Park, 375*f*
guanine, 195–197, 195*f*, 208*f*
guard cells, 503, 503*f*, 561,
561*f*
gymnosperms, **512–513**
characteristics, 423, 423*f*,
512
fertilization, 536
kinds of, 513, 513*f*
life cycle, 534–536, 534*f*,
537, 537*f*
meiosis in, 537, 537*f*
mitosis in, 537, 537*f*

habitat, 340, 377. *See also*
ecosystems
Haeckel, Ernst, 340
hagfish, 751, 751*f*
hair, **800–801**, 801*f*
cells, 959
facial, 985
follicles, 861*f*, 862–863
heredity, 177, 177*f*
Haldane, J. B. S., 254
half-life, 252, 252*f*
hantavirus, 441
haploid cells, 121, **144–145**,
144*f*, 152, 152*f*, 153*f*
hardened arteries, 884
884*f*

Hardy-Weinberg
principle, 326–329, 327*f*
Harte, John, 14–19
Haversian canals, 852, 852*f*
hawk (Swainson's), 832,
832*f*
head lice, 668
heart attacks, 884
heart rate, 883, 883*f*
hearts, **880–884**
in amphibians, 719,
759–760, 759*f*, 760*f*
in birds, 786, 786*f*
cardiac muscle tissue,
846*f*, 847
in earthworms, 655*f*
in fishes, 746, 748, 748*f*
in humans, 880–884, 880*f*,
881*f*, 882*f*, 883*f*, 888*f*
in insects, 677*f*
in mammals, 803, 805*f*
in reptiles, 776, 776*f*, 781*f*
heartwood, 557, 557*f*
heat, 31, 348. *See also*
temperature
height, 331, 331*f*
helicases, 198–199, 198*f*
Helix, 646
helper T cells, **927–929**,
928*f*, 934, 934*f*, 934*f*
hemocytometers, 405*f*
hemodialysis, 915–916, 916*f*
hemoglobin
amino acid differences
in, 287, 287*f*
multiple copies of, 218
oxygen transport, 36,
877, 888, 888*f*
in sickle cell anemia, 8,
180, 180*f*, 224
hemophilia, **181**, 181*t*,
234, 330, 878
hepatitis viruses, 236,
440, 440*t*, 910, 1008*t*
herbaceous plants, 556,
556*f*, 573, 573*f*
herbicide resistance,
238–239
herbivores, 346, 363
heredity, **161–185**. *See also*
genetics
allele frequencies, 326
codominance, 178
definition, 8
determining genotypes,
172, 172*f*
environmental influ-
ences, 179, 179*f*
intermediate traits, 177
law of independent
assortment, 169, 169*f*
law of segregation, 169
mitochondria from the
mother, 737
multiple alleles, 178, 178*f*

origin of, 257
polygenic traits, 177, 177*f*
probability, 173–174,
173*f*, 174*f*
Punnett squares,
170–172, 170*f*
sex-linked traits, 175
hermaphrodites, 148,
609–610
barnacles, 682
earthworms, 655
flatworms, 630*f*
mollusks, 645, 647
sea cucumbers, 698
sponges, 621, 621*f*
tunicates, 701
herpes virus, 235–236,
235*f*, 1008*t*, 1010, 1010*f*
Hershey, Alfred, 192–193,
192*f*
Hershey-Chase experi-
ment, 192–193, 192*f*
heterotrophs, 95, 445, **594**
bacteria, 414, 445
fungi, 421, 482
protists, 461*t*
heterozygous individuals,
167–168, 167*f*
crosses, 171, 171*f*
determining genotypes,
172
Hardy-Weinberg
principle, 326–329,
327*f*
phenotypes, 168, 176
high blood pressure, 883
hinge joints, 854*t*
histamine, 925, 936
HIV (human immuno-
deficiency virus),
934–935, 934*f*, 1010.
See also **AIDS**
macrophages in HIV
infection, 86, 438
structure of, 435, 437,
437*f*, 438*f*
testing for, 934
transmission, 438–439,
440*t*, 935
vaccines, 12
Hodgkin's disease, 521
homeostasis, 8, 31, 40,
561, 849, 889
hominids, **733–738**
australopithecines, 733,
733*f*, 734*f*
brains of, 733
definition, 733
DNA dating, 737
fossils, 733–738, 733*f*,
735*f*, 736*f*, 738*f*
Homo spp., 735–738,
735*f*, 736*f*
migrations, 736*f*, 737
movement, 733, 733*f*

other early, 734, 734*f*
skulls, 743
homologous chromo-somes, 120
homologous structures, 286, 286*f*
Homo spp., 735–738, 735*f*, 736*f*
homozygous individuals, 167, 167*f*, 170, 170*f*, 172, 326–329
honeybees, 153, 300, 300*f*, 303*f*, **303–304,** 539, 539*f*, 609
Hooke, Robert, 50, 55
hookworms, 633–634
hooves, 810, 811*f*
hormones, 974–993. *See also* **endocrine system**
  adrenocorticotropic hor-mone (ACTH), 983*t*
  aldosterone, 985
  amino-acid-based vs. steroid, 977–979, 978*f*, 979*f*
  anabolic steroids, 981
  activation of enzymes, 978, 978*f*
  body fat and, 987
  calcitonin, 984
  cortisol, 979, 979*f*, 985
  daily rhythms, 988, 988*f*
  endorphins, 976
  enkephalins, 976
  epinephrine, 975, 985, 992–993
  estrogens, 987, 1001–1003, 1002*f*
  and flight-or-flight response, 955, 975, 985
  follicle stimulating hor-mone (FSH), 983*t*, 996, 1001–1002, 1002*f*
  functions of, 974
  gastrin, 908, 986
  glucagon, 978, 978*f*, 986, 986*f*
  growth hormones, 234*f*, 240, 983, 983*t*
  insulin, 228–229, 229*f*, 234, 234*f*, 903, 986–987, 986*f*
  leptin, 987
  luteinizing hormone (LH), 983*t*, 996, 1001–1002, 1002*f*
  melatonin, 988, 988*f*
  negative feedback, 980, 980*f*
  neuropeptides, 976
  neurotransmitters and, 975–976
  oxytocin, 980, 983*t*
  parathyroid, 984

parathyroid hormone (PTH), 984
from the pituitary gland, 983, 983*t*
plant growth, 580–581, 580*f*
progesterone, 1002, 1002*f*
prostaglandins, 976
receptors, 978–979, 978*f*, 979*f*
reproduction and, 988
resistin, 903
as signal molecules, 84
target cells for, 977–979, 977*f*, 978*f*, 979*f*
testosterone, 981, 987
thyroid, 979, 979*f*
horns, 810
hornworts, 509, 509*f*
horsetails, 309, 511, 511*f*
host cell specificity, 437
human body structure, 845–864. *See also* **human skeletal system**
  body cavities, 849, 849*f*
  endothermy, 849
  kinds of tissues, 846–847, 846*f*
  organ systems, 848–849, 848*t*
  skin, 861–864, 861*f*, 864*f*, 924
human circulatory system, 871–884
  arteries, 873, 873*f*, 881–882, 881*f*, 884, 884*f*, 888, 888*f*, 913*f*, 1005*f*
  capillaries, 873–874, 873*f*, 885*f*
  heart, 880–884, 880*f*, 881*f*, 882*f*, 883*f*, 888*f*
  kidneys, 913*f*
  lymphatic system, 874–875, 875*f*
  pulmonary and systemic circuits, 880, 880*f*, 888*f*
  transport and distribution, 872, 872*f*
  veins, 873–874, 873*f*, 874*f*
human digestive system, 906–911, 906*f*
  esophagus, 906*f*, 907, 907*f*
  large intestine, 910, 910*f*
  liver, 755*f*, 909, 911, 964
  mouth and throat, 907, 907*f*
  small intestine, 909, 909*f*
  stomach, 906*f*, 907–908, 908*f*
  water absorption, 910
human excretory system, 910, **912–916,** 912*f*, 913*f*, 914*f*, 916*f*
human genome, 11, 233

human growth hormone, 234*f*
human heart, 880–884, 880*f*, 881*f*, 882*f*, 883*f*, 888*f*
human immune system, 848*t*, **924–936**
  allergic reactions, 239, 936
  antigen shifting, 932
  autoimmune disease, 933, 933*t*, 987
  cells involved in, 926–929
  HIV infection, 934–935, 934*f*
  inflammatory response, 925, 925*f*
  long-term protection, 931, 931*f*
  lymphatic system, 874–875, 875*f*
  macrophages, 596, 926, 926*f*, 931
  nonspecific defenses, 924–926, 925*f*, 926*f*
  organ transplants and, 915–916
  recognizing invaders, 927, 927*f*
  vaccination, 931–932
  white blood cells, 877, 877*f*, 922*f*, 926–929, 926*f*, 927*f*, 928*f*
human immuno-deficiency virus (HIV). *See* **HIV.**
human nervous system, 848*t*, **943–971**
  autonomic, 955, 955*t*
  brain, 950–952, 951*f*, 957, 957*f*
  drug addiction and, 962–963, 962*f*, 963*f*
  drug effects on, 961, 961*t*, 964–966, 964*t*, 965*f*
  membrane potential, 945–948, 947*f*
  nerve impulses, 945–947, 947*f*
  neuron structure, 944–945, 944*f*, 945*f*
  neurotransmitters, 948–949, 949*f*, 962–963, 966, 975–976
  optic nerve, 958, 958*f*
  sensory system, 956–960
  somatic, 954, 954*f*
  spinal cord, 952–953, 952*f*
  synaptic transmission, 948–949, 948*f*, 949*f*
human papilloma virus, 440
human reproduction, 848*t*, **995–1007.** *See also* **pregnancy**
  childbirth, 1007, 1007*f*

female reproductive system, 148–149, 999–1003, 999*f*, 1000*f*, 1001*f*, 1002*f*
fertilization, 596*f*, 997–998, 1002–1004, 1004*f*
fetal development, 1006, 1006*f*
fetus, 994*f*
male reproductive system, 148, 996–998, 996*f*, 997*f*, 998*f*
human respiratory system, 885–890
  amount of air respired, 886*f*
  circulation to lungs, 880, 880*f*
  diseases of, 890, 890*f*
  fetal lungs, 886*f*
  path of air, 885-886, 885*f*
  regulation of breathing rate, 887, 887*f*
  role of muscles in breathing, 887, 887*f*
  structure of, 885*f*
  transport of carbon dioxide, 889, 889*f*
  transport of oxygen, 888, 888*f*
humans
  chromosome number of, 62, 121*t*
  language development of, 834
  number of genes in, 233
  ovarian follicles in, 149*f*, 1001, 1001*f*
  trophic levels of, 349, 349*f*
  vestigial and homologous structures of, 286, 286*f*
human skeletal system, 850–855
  appendicular skeleton, 850–851, 850*f*
  axial skeleton, 850, 850*f*
  bone growth, 852, 852*f*
  bone marrow, 851, 851*f*, 875, 875*f*
  bone structure, 851, 851*f*
  joints, 854–855, 854*f*, 854*t*, 855*f*
  movement of, 856, 856*f*
  osteoporosis, 853, 853*f*, 987
  skeletal muscle tissue, 846*f*, 847, 857, 857*f*
hummingbirds, 44, 362*f*, 539, 790*t*
Huntington's disease, 181, 181*t*
Hybodus, 716*f*

Index

Index

Index

ozone, 254, 264, 264*f*, **387,** 387*f*
ozone hole, 387, 387*f*

Paine, Robert, 370
pain receptors, 956–957, 956*t*
palea, 576*f*
paleontologists, 285
palindromes, 230
palisade layer, 559, 559*f*
pancreas, 755*f*, 906*f*, **908–909,** 908*f*, 986, 986*f*
Pangaea, 722, 722*f*
pangolins, 816*t*
paper chromatography, 568–569
paralysis, 953
Paramecium
  characteristics, 469*f*
  competition among, 369, 369*f*, 468, 468*f*
  micrographs of, 7*f*, 48*f*, 56*f*, 261*f*
Paramecium aurelia, 369*f*, 468*f*
Paramecium bursaria, 261*f*, 369, 369*f*
Paramecium caudatum, 369, 369*f*, 468*f*, 469*f*
Paramecium syngens, 305
parapodia, 652–653
parasites, **362**
  amphibian decline and, 18
  of birds, 634
  within communities, 362
  fungi, 421, 483, 483*f*
  identifying, 634
  leeches, 656, 656*f*
  mollusks, 647
  of plants, 502, 502*f*
  protists, 464, 471, 471*f*, 471*t*, 473–474
  as symbiotic species, 364
  worms, 631–634, 631*f*, 632*f*, 633*f*
parasympathetic division, 955, 955*t*
parathyroid gland, 984, 984*f*
parathyroid hormone (PTH), 984
parental care, 806, 806*f*, 831*f*
parthenogenesis, 153, 609
passive transport, **74–80**
  diffusion, 74–75, 75*f*
  electrical charge and, 78, 79, 79*t*
  facilitated diffusion, 80, 80*f*
  ion channels, 78–79, 78*f*, 79*t*
  osmosis, 76–77, 76*f*, 77*t*

pathogens, 924, 928*f*, **929–931,** 931*f*
Pavlov, Ivan, 827
peas, 121*t*, 162–163, 163*t*
pedicellaria, 697–698
pedigrees, 175–176, 175*f*, 176*f*
pedipalps, 670, 670*f*, 671*f*
pellicles, 467, 467*f*, 469*f*
pelvic inflammatory disease (PID), 1009, 1009*f*
penicillin, 449
Penicillium, 121, 121*t*, 449, 483, 483*f*, 485
penis, 997*f*, 998, 998*f*
pepsin, 42, 42*f*, 908
peptidoglycan, 258, 414, 443, 445, 446*f*
Perca flavescens, 754*f*, 755*f*
perennials, 573, 573*f*
periosteum, 851, 851*f*
peripheral nervous system (PNS), 954–955, 955*t*. *See also* **human nervous system**
peristaltic contractions, 907, 907*f*, 909
pesticides, 238–239, 332, **390–391,** 391*f*, 678
petals, 538, 538*f*
P generation, 164, 164*f*
pH, **33**
  acid rain and, 386
  amphibians and, 16, 16*f*
  enzyme activity, 42, 42*f*
  pH scale, 33, 33*f*
phages, 192, 192*f*, **435–436,** 436*f*
pharyngeal slits, 700, 700*f*, 701*f*
pharynx, 885*f*, 886, 906*f*, 907
phenotypes, 168, 170–171, 175–176, 330
phenylketonuria (PKU), 182
pheromones, 679
phloem, 507, 510, 554, 554*f*, 556, 557, 557*f*
phospholipids, 35, 60, 60*f*
phosphorus, 353, 579*t*
phosphorus cycle, 353
photoperiodism, 583, 583*f*
photosynthesis, **97–103**
  absorption of light energy, 98, 98*f*
  by bacteria, 444, 444*f*
  Calvin cycle, 102–103, 102*f*
  CAM, 543*f*
  in carbon cycle, 352*f*
  C_3 vs. C_4, 559, 559*f*
  electron transport chains, 100–101, 100*f*
  in energy flow, 94

enzymes in, 99–101, 99*f*, 100*f*
evolution of, 260, 265, 265*f*
factors affecting, 103
oxygen production, 99, 99*f*, 101, 258, 258*f*
by plankton, 378
primary productivity, 345
by protists, 461*t*, 467, 472
summary of, 97, 97*f*
phototropism, 582, 582*f*
phyla (singular, *phylum*), 302, 302*f*
Phyllium pulchrifolium, 274*f*
phylogenetic trees, **602,** 602*f*, 665*f*, 714*f*, 732*f*
Physalia, 624, 624*f*
Phytophthora infestans, 470
phytoremediation, 699
pigments
  bile, 910
  definition, 568
  jumping genes, 216
  melanin, 175, 862–863
  in photosynthesis, 35, 98–101, 98*f*
  polygenic traits, 331
  sepia, 650
pili (singular, *pilus*), 442–443, 442*f*, 446*f*
pill bugs, 680, 686–687
Pinctada, 648, 648*f*
pineal gland, 988
pine trees, 417*t*, 505*f*, 513, 535*f*, 536, 536*f*
pioneer species, 343
Pisaster, 370, 370*f*
pistils, 163, 163*f*, 538, 538*f*, 540*f*
Pisum sativum, 121*t*, 162, 163, 163*t*
pith, 556–557, 556*f*, 557*f*
pit organ, 780*f*
pituitary glands, 982*f*, **983,** 983*f*, 1002
placenta, 810, **1005,** 1005*f*, 1007*f*
placental mammals, 730, 730*f*, **810–816,** 813*f*, 814*f*, 815*f*, 816*t*
placoderms, 714*f*, 715
planarians, 629–631, 630*f*
plankton, 378, 420, 465, 467
Plantae (kingdom), 412, 412*f*, 417*f*, **422–423,** 423*f*
plant growth, **571–589**
  development, 578
  germination, 572–573, 572*f*, 584
  hormonal control, 580–581, 580*f*
  life span, 573

meristems, 574–575, 574*f*, 577*f*
mineral nutrient requirements, 548, 579, 579*t*
photoperiodism, 583, 583*f*
response to temperature, 584, 584*f*
tissue culture, 544, 544*t*, 578
tropisms, 582, 582*f*
plantlets, 542*f*
plant propagation, **544,** 544*t*, 578. *See also* **genetic engineering**
plants, **501–527.** *See also* **photosynthesis**
  angiosperms, 423, 423*f*, 514–515, 515*f*, 515*t*, 538–540
  cells, 66, 66*f*, 70*f*, 132, 132*f*
  characteristics, 422–423
  day-neutral, 583, 583*f*
  deciduous, 573
  evolution on land, 265–266, 265*f*, 502
  food, 516–519
  gymnosperms, 423, 423*f*, 512–513, 513*f*
  herbaceous, 556, 556*f*, 573, 573*f*
  life cycles, 506–507, 506*f*
  Mendel experiments, 162–165, 163*f*, 163*t*, 164*f*, 165*t*
  mineral nutrients, 548, 579, 579*t*
  movement of organic compounds, 563, 563*f*, 564, 564*f*
  mycorrhizae, 490, 490*f*, 502
  nonfood uses, 520–522
  nonvascular, 423, 504, 508–509, 509*f*, 530–531, 531*f*
  predation effects, 363, 363*f*
  propagation of, 544, 544*t*, 578
  reproduction on land, 503, 506, 506*f*, 508
  roots, 507, 507*f*, 542*f*, 555, 555*f*
  seedless vascular, 423, 510–511, 510*f*, 511*f*, 532–533, 533*f*
  seed plant reproduction, 534–540
  seeds, 504–505, 505*f*
  self-fertilizing, 328
  tissue types, 552–554, 552*f*, 553*f*, 554*f*
  trophic levels, 345–347, 345*f*, 346*f*, 348

Acknowledgments *continued*

Academic Reviewers *(continued from page iv)*

Chris Nice, Ph.D.
Assistant Professor
Department of Biology
Texas State University
San Marcos, Texas

Martin Nickels, Ph.D.
Professor
Department of Anthropology
Illinois State University
Normal, Illinois

Eva Oberdörster, Ph.D.
Lecturer
Department of Biological
   Sciences
Southern Methodist University
Dallas, Texas

Brian L. Pagenkopf, Ph.D.
Assistant Professor
Department of Chemistry
University of Texas
Austin, Texas

Barron Rector, Ph.D.
*Assistant Professor and Extension
   Range Specialist*
Texas Agricultural Extension
   Service
Texas A&M University
College Station, Texas

Michael Ryan, Ph.D.
*Clark Hubbs Regents Professor
   in Zoology*
The University of Texas
Austin, Texas

Dork Sahagian
*Research Professor, Stratigraphy and
   Basin Analysis, Geodynamics*
Global Analysis, Interpretation,
   and Modeling Program
University of New Hampshire
Durham, New Hampshire

Miles Silman, Ph.D.
Assistant Professor of Biology
Department of Biology
Wake Forest University
Winston-Salem, North Carolina

Richard Storey, Ph.D.
*Dean of the Faculty and Professor
   of Biology*
Colorado College
Colorado Springs, Colorado

Mary Wicksten, Ph.D.
Professor of Biology
Department of Biology
Texas A&M University
College Station, Texas

Teacher Reviewers

Robert Akeson
Science Teacher
Boston Latin School
Boston, Massachusetts

Robert Baronak
Biology Teacher
Donegal High School
Mount Joy, Pennsylvania

Betty Bates
Biology Teacher
DeKalb High School
DeKalb, Texas

David Blinn
Biology Teacher
Wrenshall High School
Wrenshall, Minnesota

Cindy Copolo, Ph.D.
Science Education Consultant
Summit Solutions
Bahama, North Carolina

Linda Culp
Science Chair and Science Teacher
Thorndale High School
Thorndale, Texas

Katherine Cummings
Science Teacher
Currituck County High School
Currituck, North Carolina

Alonda Droege
Biology Teacher
Highline High School
Burien, Washington

Benjamin Ebersole
Science Teacher
Donegal High School
Mount Joy, Pennsylvania

Richard Filson
Science Teacher
Edison High School
Stockton, California

Randa Flinn
Science Teacher
Northeast High School
Fort Lauderdale, Florida

Arthur Goldsmith
Biology and Earth Sciences Teacher
Hallandale High School
Hallandale, Florida

Marguerite A. Graham
Science Teacher
Gulliver Preparatory School
Miami, Florida

Teacher Reviewers *(continued from page 1136)*

Stacey Jeffress
Environmental Science Teacher
El Dorado High School
El Dorado, Arkansas

Bruce Katt
Biology Teacher
Georgetown High School
Georgetown, Texas

Debbie Keel
Science Teacher
Cypress Creek High School
Houston, Texas

Mike Lubich
Science Teacher
Mapletown High School
Greensboro, Pennsylvania

Thomas Manerchia
Environmental Science Teacher,
  Retired
Archmere Academy
Claymont, Delaware

Jason Marsh
Biology Teacher
Moose Lake Community School
Moose Lake, Minnesota

Ray McLarty
Science Teacher
Hononegah High School
Rockdon, IL

Betsy McGrew
Science Teacher
Star Charter School
Austin, Texas

Tammie Niffenegger
Science Chair and Science Teacher
Port Washington High School
Port Washington, Wisconsin

Donna Norwood
Science Teacher
Monroe High School
Charlotte, North Carolina

Denice Lee Sandefur
Science Chairperson
Nucla High School
Nucla, Colorado

Bert Sherwood
Secondary Science Instructional
  Specialist
Socorro Independent School
  District
El Paso, Texas

Joe Stanaland
Biology Teacher
Lake Travis High School
Austin, Texas

Tyson Yager
Science Teacher
Wichita High School East
Wichita, Kansas

Credits

Staff Credits

Executive Editor, High School Sciences and Health
Mark Grayson

Managing Editor
Debbie Starr

Editorial Development Team
Karen Ross, Senior Editor
Roxanne McCauley
Dyanne Semerjibashian

Copyeditors
Dawn Marie Spinozza,
   Copyediting Manager
Anne-Marie De Witt
Simon Key
Jane A. Kirschman
Kira J. Watkins

Editorial Support Staff
Mary Anderson
Soojinn Choi
Jeanne Graham
Shannon Oehler
Stephanie Sanchez
Tanu'e White

Editorial Interns
Kristina Bigelow
Erica Garza
Sarah Ray
Kenneth G. Raymond
Kyle Stock
Audra Teinert

Online Products
Bob Tucek, Executive Editor
Wesley M. Bain
Catherine Gallagher
Douglas P. Rutley

Book Design
Kay Selke, Director of Book
   Design

Media Design
Richard Metzger, Design Director
Chris Smith, Developmental
   Designer

Cover Design
Kay Selke, Director of Book
   Design

Image Acquisitions
Curtis Riker, Director
Jeanie Taylor, Photo Research
   Manager
Diana Goetting, Senior Photo
   Researcher

Publishing Services
Carol Martin, Director
Jeff Robinson, Manager, Ancillary
   Design

Production
Eddie Dawson, Senior
   Production Manager
Beth Sample, Project Manager
Teresa McQueen, Senior
   Production Coordinator

Technology Services
Laura Likon, Director
Juan Baquera, Manager
JoAnn Stringer, Manager

Senior Technology Services Analysts
Katrina Gnader
Lana Kaupp

Production Artists
Sara Buller
Margaret Sanchez
Patty Zepeda

eMedia
Melanie Baccus, eMedia
   Coordinator
Ed Blake, Design Director
Kimberly Cammerata, Design
   Manager
Lydia Doty, Senior Project
   Manager
Marsh Flournoy, Technology
   Project Manager
Dakota Smith, Quality Assurance
   Analyst
Cathy Kuhles, Technical Assistant
Tara F. Ross, Senior Project
   Manager
Ken Whiteside, Manager,
   Application Development

Photography Credits

Abbreviation Code
AA=Animals, Animals; BC=Bruce Coleman, Inc.; DRK=DRK Photo; ES=Earth Scenes; GH=Grant Heilman Photography; GI=Getty Images; HRW/SD=Holt, Rinehart and Winston/Sam Dudgeon; MP=Minden Pictures; PA=Peter Arnold, Inc.; PR=Photo Researchers; PT=Phototake; RLM=Robert & Linda Mitchell; SP/FOCA= Sergio Purtell/Foca Co., NY, NY; TF=Tim Fuller; VU=Visuals Unlimited

Abbreviations used: (t) top, (b) bottom, (c) center, (l) left, (r) right, (bkgd) background

COVER AND TITLE Biotic/Photonica; **BACK COVER** John Shaw/Bruce Coleman, Inc.; **ii** Don Riepe/PA; **iii** UNEP (Brunner)/PA; **v** Art Wolfe/GI; **vii** (t) Norbert Wu/PA, (b) Arthur Tilley/GI; **viii** (t) E.R. Degginger/AA/ES, (b) Jerry L. Ferrara/PR; **ix** R.H. Virdee/GH; **xi** (t) William H. Mullins/NASC/PR., (b) D. Cavagnaro/VU; **xii** PhotoDisc; **xiii** (t) William H. Mullins/NASC/PR., (b) Dan Guravich/PR; **xiv** SP/FOCA.; **xv** Tom McHugh/PR; **xvi** A. Witte & C. Mahaney/GI; **xvii** (t) Marty Cordano/DRK, (c) Jeff Rotman/PA, (b) Scott Bauer/Agricultural Research Service, USDA; **xviii** HRW/SD; **xix** Patti Murray/AA/ES; **xx** HRW/SD; **xxi** C. Milkins/OSF/AA/ES; **CHAPTER 1: 2-3** (bkgd) Dr. Dennis Kunkel/PT; **2** PhotoDisc; **3** (b) C. Meckes/Ottawa/PR, (c) Wood Ransaville Harlin, Inc., (t) Corbis; **4** TF; **6** UNEP/Brunner/PA; **7** (b) TF, (c) Zig Leszczynski/AA/ES, (t) M. Abbey/VU; **8** (b) Walter Hodges/GI, (tl) Dan Guravich/NASC/PR; **9** (b) Gerard Lacz/PA, (t) Jamie Harron-Papilio/Corbis; **11** Mauro Fermariello/SPL/PR; **12** Scott Camazine/PR; **13** Simon Fraser/RVI, Newcastle-Upon-Tyne/SPL/PR; **14** Zig Leszczynski/AA/ES; **15** (l) John Harte, (r) Jeff Smith/Fotosmith; **16** (b) RLM, (t) HRW/SD; **17** HRW/SD; **18** (b) J. Reid/Custom Medical Stock Photo, (t) Dan Nedrelo; **19** Will & Deni/PR; **20** Sinclair Stammers/SPL/PR;

21 TF; **22** TF; **24** HRW/SD; **CHAPTER 2: 26** Georg Gerster/PR; **29** HRW/SD; **30** (bc) Dennis Kunkel/PT, (bl) SP/FOCA, (tl) Paul Hermansen/GI; **31** Marc Epstein/DRK; **33** HRW/SD; **34** SP/FOCA; **35** HRW/SD; **36** TF; **38** (l) HRW/SD, (r) G.I. Bernard/AA/ES; **43** Georg Gerster/PR; **44** Georg Gerster/PR; **46** HRW/SD; **CHAPTER 3: 48** Robert Brons/BPS/GI; **51** (b, c) HRW/SD, (cl) Dr. Dennis Kunkel/PT, (cr) Victoria Smith/HRW, (tr) M. Abbey/PR; **52** (l) SP/FOCA/HRW, (r) Manfred Kage/PA; **53** (bl) Sinclair Stammers/SPL/PR, (br) Dr. Tony Brain/SPL/PR, (t) Victoria Smith/HRW; **54** (l) Philippe/Plailly/Eurelios/SPL/PR, (r) David M. Philips/VU; **56** Michael Abbey/PR; **57** (b) Chris Bjornberg/PR, (t) John Cardmore/BPS/GI; **58** Michael Gabridge/VU; **59** Volker Steger/Christian Bardele/SPL/PR; **60** Dr. Dennis Kunkel/PT; **62** Don Fawcett/VU; **63** R. Boldender/D. Fawcett/VU; **65** Don Fawcett/VU; **67** Robert Brons/BPS/GI; **68** (b) Michael Gabridge/VU, (t) Robert Brons/BPS/GI; **70** E.R. Degginger/Color-Pic, Inc.; **71** Barry Bomzer/GI; **CHAPTER 4: 72** Dr. Dennis Kunkel/PT; **75** HRW/SD; **76** HRW/SD; **78** Victor Scocozza/GI; **79** Dr. David Scott/PT; **81** Doug Wechsler; **84** TF; **85** HRW/SD; **86** Nibsc/PR; **87** Dr. Dennis Kunkel/PT; **88** Dr. Dennis Kunkel/PT; **90** Ward's Natural Science; **CHAPTER 5: 92** Steve Gettle/ENP Images; **99** Dr. E. R. Degginger/Color-Pic; **101** C. Milkins/OSF/AA/ES; **103** Alan & Linda Detrick/PR; **108** John Cowell/GH; **109** AFP/Corbis; **111** Steve Gettle/ENP Images; **112** (b) BPA/SS/PR, (t) Steve Gettle/ENP Images; **114** HRW/SD, (br) TF, (tl) CNRI/SPL/PR; **115** HRW/SD; **CHAPTER 6: 116** Professors P.M. Motta & J. van Blerkom/SPL/PR; **118** (t) HRW/SD, (b) Stephen J. Krasemann/DRK; **119** Institut Pasteur/CNRI/PT; **120** (b) David M. Phillips/VU, (l, r) PhotoDisc; **121** (bc) Dennis Kunkel/PT, (bl) Evelyn Gallaroo/PA, HRW/SD, (t) David M. Philips/VU; **122** (bl) Custom Medical Stock Photo, (br) TF, (tl) CNRI/SPL/PR; **123** ISM/PT; **127** Nancy Kedersha/PR; **129** Ariel Skelley/The Stock Market; **130** John D. Cunningham/VU; **131** (bl, c, br) John D. Cunningham/VU, (tr) David M. Philips/VU; **132** (cr) HRW/SD, (tl) R. Calentine/VU; **133** Professors P.M. Motta & J. van Blerkom/SPL/PR; **134** Professors P.M. Motta & J. van Blerkom/SPL/PR; **136** (l) HRW/SD, John D. Cunningham/VU; **137** HRW/SD; **138** (b) Dr. Tony Brian/SPL/PR, (t) Moredun